Common Formulas

Distance

$d = rt$

d = distance traveled
t = time
r = rate

Temperature

$F = \frac{9}{5}C + 32$

F = degrees Fahrenheit
C = degrees Celsius

Simple Interest

$I = Prt$

I = interest
P = principal
r = annual interest rate
t = time in years

Compound Interest

$A = P\left(1 + \frac{r}{n}\right)^{nt}$

A = balance
P = principal
r = annual interest rate
n = compoundings per year
t = time in years

Coordinate Plane: Midpoint Formula

Midpoint of line segment joining (x_1, y_1) and (x_2, y_2)

$\left(\frac{x_1 + x_2}{2}, \frac{y_1 + y_2}{2}\right)$

Coordinate Plane: Distance Formula

d = distance between points (x_1, y_1) and (x_2, y_2)

$d = \sqrt{(x_2 - x_1)^2 + (y_2 - y_1)^2}$

Quadratic Formula

Solutions of $ax^2 + bx + c = 0$

$$x = \frac{-b \pm \sqrt{b^2 - 4ac}}{2a}$$

Rules of Exponents

$a^0 = 1$

$a^m \cdot a^n = a^{m+n}$

$(ab)^m = a^m \cdot b^m$

$(a^m)^n = a^{mn}$

$\frac{a^m}{a^n} = a^{m-n}, \quad a \neq 0$

$\left(\frac{a}{b}\right)^m = \frac{a^m}{b^m}, \quad b \neq 0$

$a^{-n} = \frac{1}{a^n}, \quad a \neq 0$

$\left(\frac{a}{b}\right)^{-n} = \frac{b^n}{a^n}, \quad a \neq 0, \ b \neq 0$

Basic Rules of Algebra

Commutative Property of Addition

$a + b = b + a$

Commutative Property of Multiplication

$ab = ba$

Associative Property of Addition

$(a + b) + c = a + (b + c)$

Associative Property of Multiplication

$(ab)c = a(bc)$

Left Distributive Property

$a(b + c) = ab + ac$

Right Distributive Property

$(a + b)c = ac + bc$

Additive Identity Property

$a + 0 = a$

Multiplicative Identity Property

$a \cdot 1 = 1 \cdot a = a$

Additive Inverse Property

$a + (-a) = 0$

Multiplicative Inverse Property

$a \cdot \frac{1}{a} = 1, \quad a \neq 0$

Properties of Equality

Addition Property of Equality

If $a = b$, then $a + c = b + c$.

Multiplication Property of Equality

If $a = b$, then $ac = bc$.

Cancellation Property of Addition

If $a + c = b + c$, then $a = b$.

Cancellation Property of Multiplication

If $ac = bc$, and $c \neq 0$, then $a = b$.

Zero Factor Property

If $ab = 0$, then $a = 0$ or $b = 0$.

Elementary and Intermediate Algebra

A Combined Course

Third Edition

Ron Larson
The Pennsylvania State University
The Behrend College

Robert P. Hostetler
The Pennsylvania State University
The Behrend College

With the assistance of
David E. Heyd
The Pennsylvania State University
The Behrend College

Property of
FAMILY OF FAITH
LIBRARY

Houghton Mifflin Company
Boston New York

Sponsoring Editor: Jack Shira
Managing Editor: Cathy Cantin
Senior Associate Editor: Maureen Ross
Associate Editor: Laura Wheel
Assistant Editor: Carolyn Johnson
Supervising Editor: Karen Carter
Project Editor: Patty Bergin
Editorial Assistant: Christine E. Lee
Art Supervisor: Gary Crespo
Marketing Manager: Ros Kane
Senior Manufacturing Coordinator: Sally Culler
Composition and Art: Meridian Creative Group

We have included examples and exercises that use real-life data as well as technology output from a variety of software. This would not have been possible without the help of many people and organizations. Our wholehearted thanks go to all for their time and effort.

Printed in the U.S.A.

Library of Congress Catalog Card Number: 99-71985

ISBN: 0-395-97632-4

23456789–DOW–03 02 01 00

Contents

A Word from the Authors

Welcome to *Elementary and Intermediate Algebra: A Combined Course*, Third Edition. In this revision, we have continued to focus on developing students' proficiency and conceptual understanding of algebra. We hope you enjoy the Third Edition.

In response to suggestions from elementary and intermediate algebra instructors, we have revised and reorganized the coverage of topics for the Third Edition. We combined the content of the first two chapters of the previous edition and streamlined them into Chapter 1 "The Real Number System" for the Third Edition. To improve the flow of the material, the business applications have been incorporated into Section 3.4 "Ratios and Proportions." "Geometric and Scientific Applications" is now Section 3.5. Compound inequalities and set notation are now introduced in Section 3.6 "Linear Inequalities." And a new section, "Absolute Value Equations and Inequalities," has been added to Chapter 3. In order to be more efficient and to improve the flow of the text, Chapter 4, which previously introduced the coordinate plane and graphs of equations, now includes Section 4.3 "Relations, Functions, and Graphs," Section 4.4 "Slope and Graphs of Linear Equations," Section 4.5 "Equations of Lines," and Section 4.6 "Graphs of Linear Inequalities." "Systems of Equations" has been moved forward to Chapter 7. "Variation" is now covered in Chapter 8 "Rational Expressions, Equations and Functions." And finally, "Graphs of Quadratic Functions" now appears in Chapter 10 "Quadratic Equations and Inequalities."

In order to address the diverse needs and abilities of students, we offer a straightforward approach to the presentation of difficult concepts. In the Third Edition, the emphasis is on helping students learn a variety of techniques—symbolic, numeric, and visual—for solving problems. We are committed to providing students with a successful and meaningful course of study.

Our approach begins with Motivating the Chapter, a new feature that introduces each chapter. These multipart problems are designed to show students the relevance of algebra to the world around them. Each Motivating the Chapter feature is a real-life application that requires students to apply the concepts of the chapter in order to solve each part of the problem. Problem-solving and critical thinking skills are emphasized here and throughout the text in applications that appear in the examples and exercise sets.

To improve the usefulness of the text as a study tool, we added Objectives, which highlight the main concepts that students will learn throughout the section. Each objective is restated in the margin at the point where the concept is introduced, to help keep students focused as they read the section. The Chapter Summary was revised for the Third Edition to make it a more comprehensive and effective study tool. It now highlights the important mathematical vocabulary (Key Terms) and primary concepts (Key Concepts) of the chapter. For easy reference, the Key Terms are correlated to the chapter by page number and the Key Concepts by section number.

As students proceed through each chapter they have many opportunities to assess their understanding. They can check their progress after each section with the Exercise sets (which are correlated to Examples in the section), midway through the chapter with the Mid-Chapter Quiz, and at the end of the chapter with the Review Exercises (which are correlated to the sections) and the Chapter Test. The exercises and test items were carefully chosen and graded in difficulty to allow students to gain confidence as they progress. In addition, students can assess their understanding of previously learned concepts through the Integrated Review exercises that precede the section exercise sets and the Cumulative Tests that follow Chapters 3, 6, 9, and 12.

In the Third Edition, we combined the Technology and Discovery features of the Second Edition. Technology Tips provide point-of-use instructions for using a graphing utility. Technology Discovery features encourage students to explore mathematical concepts with graphing utilities and scientific calculators. Both are highlighted and can easily be omitted without loss of continuity in coverage of material.

To show students the practical uses of algebra, we highlight the connections between the mathematical concepts and the real world in the multitude of applications found throughout the text. We believe that students can overcome their difficulties in mathematics if they are encouraged and supported throughout the learning process. Too often, students become frustrated and lose interest in the material when they cannot follow the text. With this in mind, every effort has been made to write a readable text that can be understood by every student. We hope that your students find our approach engaging and effective.

Ron Larson

Ron Larson

Robert P. Hostetler

Robert P. Hostetler

Features

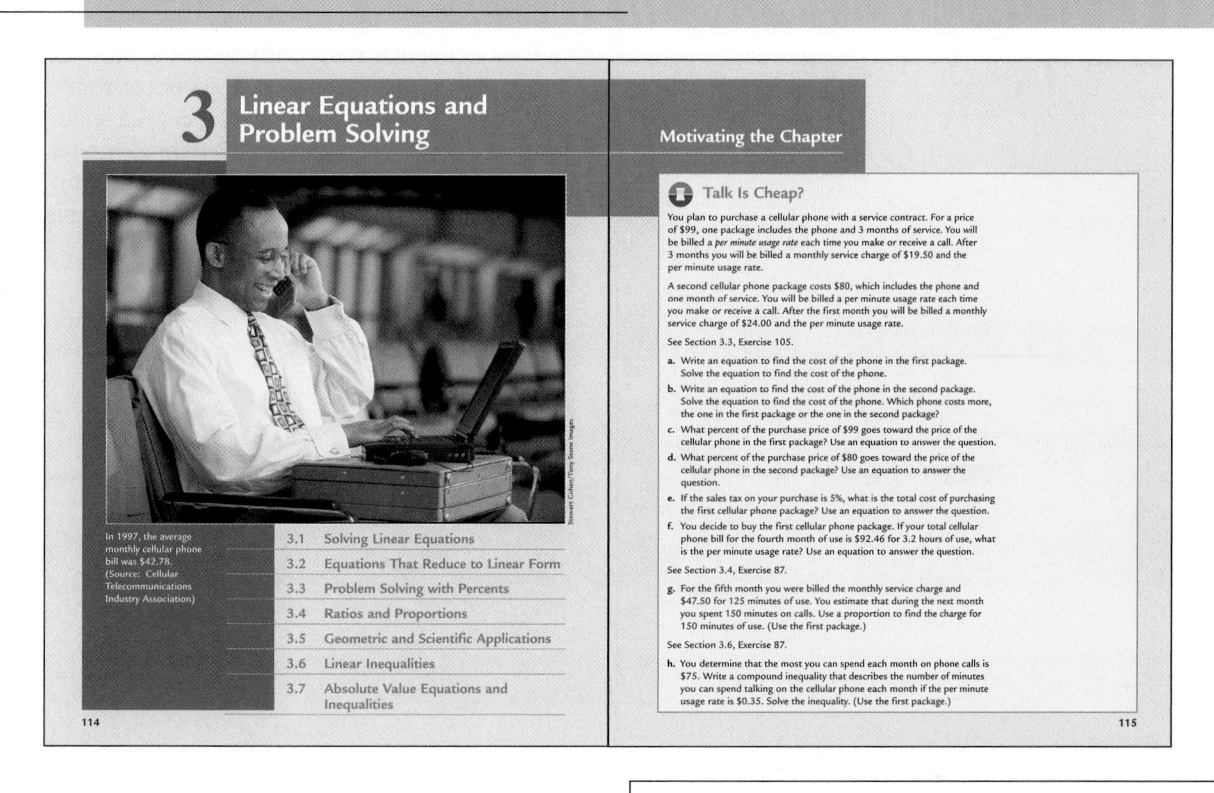

3 Linear Equations and Problem Solving

In 1997, the average monthly cellular phone bill was $42.78. (Source: Cellular Telecommunications Industry Association)

Stewart Cohen/Tony Stone Images

114

Motivating the Chapter

Talk Is Cheap?

You plan to purchase a cellular phone with a service contract. For a price of $99, one package includes the phone and 3 months of service. You will be billed a *per minute usage rate* each time you make or receive a call. After 3 months you will be billed a monthly service charge of $19.50 and the per minute usage rate.

A second cellular phone package costs $80, which includes the phone and one month of service. You will be billed a per minute usage rate each time you make or receive a call. After the first month you will be billed a monthly service charge of $24.00 and the per minute usage rate.

See Section 3.3, Exercise 105.

a. Write an equation to find the cost of the phone in the first package. Solve the equation to find the cost of the phone.

b. Write an equation to find the cost of the phone in the second package. Solve the equation to find the cost of the phone. Which phone costs more, the one in the first package or the one in the second package?

c. What percent of the purchase price of $99 goes toward the price of the cellular phone in the first package? Use an equation to answer the question.

d. What percent of the purchase price of $80 goes toward the price of the cellular phone in the second package? Use an equation to answer the question.

e. If the sales tax on your purchase is 5%, what is the total cost of purchasing the first cellular phone package? Use an equation to answer the question.

f. You decide to buy the first cellular phone package. If your total cellular phone bill for the fourth month of use is $92.46 for 3.2 hours of use, what is the per minute usage rate? Use an equation to answer the question.

See Section 3.4, Exercise 87.

g. For the fifth month you were billed the monthly service charge and $47.50 for 125 minutes of use. You estimate that during the next month you spent 150 minutes on calls. Use a proportion to find the charge for 150 minutes of use. (Use the first package.)

See Section 3.6, Exercise 87.

h. You determine that the most you can spend each month on phone calls is $75. Write a compound inequality that describes the number of minutes you can spend talking on the cellular phone each month if the per minute usage rate is $0.35. Solve the inequality. (Use the first package.)

115

Chapter Opener *New*

Every chapter opens with *Motivating the Chapter*. Each of these multipart problems incorporates the concepts presented in the chapter in the context of a single real-world application. *Motivating the Chapter* problems are correlated to sections and exercises and can be assigned as students work through the chapter or can be assigned as individual or group projects. The icon identifies an exercise that relates back to *Motivating the Chapter*.

Section Opener *New*

Every section begins with a list of learning objectives. Each objective is restated in the margin at the point where it is covered.

Historical Note

Historical notes featuring mathematicians or mathematical artifacts are included throughout the text.

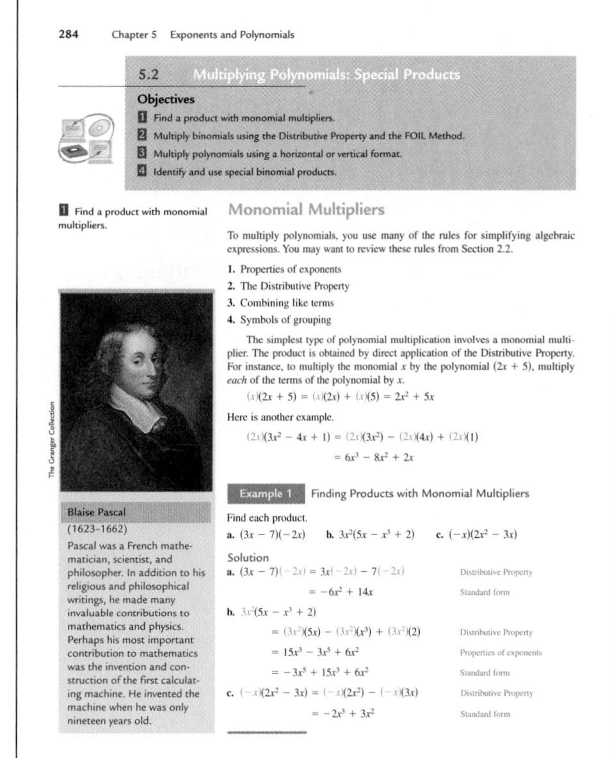

284 Chapter 5 Exponents and Polynomials

5.2 Multiplying Polynomials: Special Products

Objectives

1. Find a product with monomial multipliers.
2. Multiply binomials using the Distributive Property and the FOIL Method.
3. Multiply polynomials using a horizontal or vertical format.
4. Identify and use special binomial products.

1 Find a product with monomial multipliers.

Monomial Multipliers

To multiply polynomials, you use many of the rules for simplifying algebraic expressions. You may want to review these rules from Section 2.2.

1. Properties of exponents
2. The Distributive Property
3. Combining like terms
4. Symbols of grouping

The simplest type of polynomial multiplication involves a monomial multiplier. The product is obtained by direct application of the Distributive Property. For instance, to multiply the monomial x by the polynomial $(2x + 5)$, multiply *each* of the terms of the polynomial by x.

$$(x)(2x + 5) = (x)(2x) + (x)(5) = 2x^2 + 5x$$

Here is another example.

$$(2x)(3x^2 - 4x + 1) = (2x)(3x^2) - (2x)(4x) + (2x)(1) = 6x^3 - 8x^2 + 2x$$

The Granger Collection

Blaise Pascal (1623–1662)

Pascal was a French mathematician, scientist, and philosopher. In addition to his religious and philosophical writings, he made many invaluable contributions to mathematics and physics. Perhaps his most important contribution to mathematics was the invention and construction of the first calculating machine. He invented the machine when he was only nineteen years old.

Example 1 Finding Products with Monomial Multipliers

Find each product.

a. $(3x - 7)(-2x)$ **b.** $3x^2(5x - x^3 + 2)$ **c.** $(-x)(2x^2 - 3x)$

Solution

a. $(3x - 7)(-2x) = 3x(-2x) - 7(-2x)$ Distributive Property

$= -6x^2 + 14x$ Standard form

b. $3x^2(5x - x^3 + 2)$

$= (3x^2)(5x) - (3x^2)(x^3) + (3x^2)(2)$ Distributive Property

$= 15x^3 - 3x^5 + 6x^2$ Properties of exponents

$= -3x^5 + 15x^3 + 6x^2$ Standard form

c. $(-x)(2x^2 - 3x) = (-x)(2x^2) - (-x)(3x)$ Distributive Property

$= -2x^3 + 3x^2$ Standard form

Example 2 Solving a Linear Equation Involving Parentheses

Solve $3(2x - 1) + x = 11$.

Solution

$3(2x - 1) + x = 11$	Original equation
$3 \cdot 2x - 3 \cdot 1 + x = 11$	Distributive Property
$6x - 3 + x = 11$	Simplify.
$6x + x - 3 = 11$	Collect like terms.
$7x - 3 = 11$	Combine like terms.
$7x - 3 + 3 = 11 + 3$	Add 3 to both sides.
$7x = 14$	Combine like terms.
$\frac{7x}{7} = \frac{14}{7}$	Divide both sides by 7.
$x = 2$	Simplify.

Check

$3(2x - 1) + x = 11$	Original equation
$3[2(2) - 1] + 2 \stackrel{?}{=} 11$	Substitute 2 for x.
$3(4 - 1) + 2 \stackrel{?}{=} 11$	Simplify.
$3(3) + 2 \stackrel{?}{=} 11$	Simplify.
$9 + 2 \stackrel{?}{=} 11$	Simplify.
$11 = 11$	Solution checks. ✓

The solution is 2.

Example 3 Solving a Linear Equation Involving Parentheses

Solve $5(x + 2) = 2(x - 1)$.

Solution

$5(x + 2) = 2(x - 1)$	Original equation
$5x + 10 = 2x - 2$	Distributive Property
$5x - 2x + 10 = -2$	Subtract $2x$ from both sides.
$3x + 10 = -2$	Combine like terms.
$3x = -2 - 10$	Subtract 10 from both sides.
$3x = -12$	Combine like terms.
$x = -4$	Divide both sides by 3.

The solution is -4. Check this in the original equation.

Examples

Each example was carefully chosen to illustrate a particular mathematical concept or problem-solving technique. The examples cover a wide variety of problems and are titled for easy reference. Many examples include detailed, step-by-step solutions with side comments, which explain the key steps of the solution process.

Applications

A wide variety of real-life applications are integrated throughout the text in examples and exercises. These applications demonstrate the relevance of algebra in the real world. Many of the applications use current, real data. The icon indicates an example involving a real-life application.

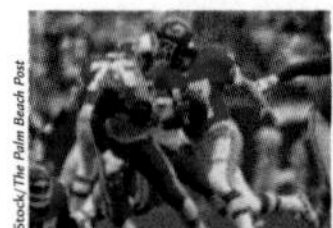

Post Stock/The Palm Beach Post

Each year since 1967, the winners of the American Football Conference and the National Football Conference have played in the Super Bowl. The first Super Bowl was played between the Green Bay Packers and the Kansas City Chiefs.

Example 3 Super Bowl Scores

The scores of the winning and losing football teams for the Super Bowl games from 1981 through 1999 are given in the table below. Plot these points on a rectangular coordinate system. (Source: National Football League)

Year	1981	1982	1983	1984	1985	1986	1987
Winning score	27	26	27	38	38	46	39
Losing score	10	21	17	9	16	10	20

Year	1988	1989	1990	1991	1992	1993	1994
Winning score	42	20	55	20	37	52	30
Losing score	10	16	10	19	24	17	13

Year	1995	1996	1997	1998	1999
Winning score	49	27	35	31	34
Losing score	26	17	21	24	19

Solution

Plot the years on the x-axis and the winning and losing scores on the y-axis. In Figure 4.5, the winning scores are shown as black dots, and the losing scores are shown as blue dots. Note that the break in the x-axis indicates that the numbers between 0 and 1981 have been omitted.

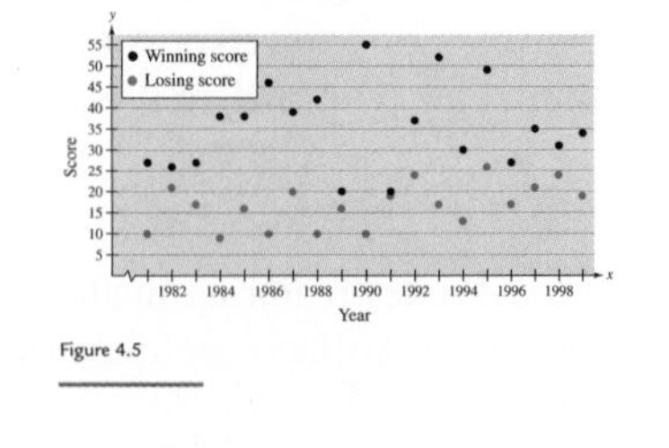

Figure 4.5

140 Chapter 3 Linear Equations and Problem Solving

From Example 3, you can see that there are three basic types of percent problems. Each can be solved by substituting the two given quantities into the percent equation and solving for the third quantity.

Question	*Given*	*Percent Equation*
a is what percent of b?	a and b	Solve for p.
What number is p percent of b?	p and b	Solve for a.
a is p percent of what number?	a and p	Solve for b.

For instance, part (b) of Example 3 fits the form "a is p percent of what number?"

In most real-life applications, the base number b and the number a are much more disguised than they are in Example 3. It sometimes helps to think of a as a "new" amount and b as the "original" amount.

Example 4 Real Estate Commission

A real estate agency receives a commission of $5167.50 for the sale of a $79,500 house. What percent commission is this?

Solution

Verbal Model: Commission = Percent (in decimal form) · Sale price

Labels: Commission = 5167.50 (dollars)
Percent = p (in decimal form)
Sale price = 79,500 (dollars)

Equation: $5167.50 = p \cdot (79{,}500)$

$$\frac{5167.50}{79{,}500} = p$$

$$0.065 = p$$

So, the real estate agency receives a commission of 6.5%.

Example 5 Cost-of-Living Raise

A union negotiates for a cost-of-living raise of 7%. What is the raise for a union member whose salary is $17,240? What is this person's new salary?

Solution

Verbal Model: Raise = Percent (in decimal form) · Salary

Labels: Raise = a (dollars)
Percent = 7% = 0.07 (in decimal form)
Salary = 17,240 (dollars)

Equation: $a = 0.07(17{,}240) = 1206.80$

So, the raise is $1206.80 and the new salary is 17,240.00 + 1206.80 or $18,446.80.

Problem Solving

This text provides many opportunities for students to sharpen their problem-solving skills. In both the examples and the exercises, students are asked to apply verbal, numerical, analytical, and graphical approaches to problem-solving. In the spirit of the AMATYC and NCTM standards, students are taught a five-step strategy for solving applied problems, which begins with constructing a verbal model and ends with checking the answer.

FEATURES

Geometry

Coverage and integration of geometry in examples and exercises have been enhanced throughout the Third Edition.

162 Chapter 3 Linear Equations and Problem Solving

Example 1 Using a Geometric Formula

A sailboat has a triangular sail with an area of 96 square feet and a base that is 16 feet long, as shown in Figure 3.3. What is the height of the sail?

Solution

Because the sail is triangular, and you are given its area, you should begin with the formula for the area of a triangle.

$A = \frac{1}{2}bh$ — Area of a triangle

$96 = \frac{1}{2}(16)h$ — Substitute 96 for A and 16 for b.

$96 = 8h$ — Simplify.

$12 = h$ — Divide both sides by 8.

The height of the sail is 12 feet.

h

$b = 16$ ft

Figure 3.3

In Example 1, notice that b and h are measured in feet. When they are multiplied in the formula $\frac{1}{2}bh$, the resulting area is measured in *square* feet.

$$A = \frac{1}{2}(16 \text{ feet})(12 \text{ feet}) = 96 \text{ feet}^2$$

Note that square feet can be written as feet^2.

Example 2 Using a Geometric Formula

The local municipality is planning to develop the street along which you own a rectangular lot that is 500 feet deep and has an area of 100,000 square feet. To help pay for the new sewer system, each lot owner will be assessed $5.50 per foot of lot frontage.

a. Find the length of the frontage of your lot.

b. How much will you be assessed for the new sewer system?

Solution

a. To solve this problem, it helps to begin by drawing a diagram such as the one shown in Figure 3.4. In the diagram, label the depth of the property as $l = 500$ feet and the unknown frontage as w.

$A = lw$ — Area of a rectangle

$100{,}000 = 500(w)$ — Substitute 100,000 for A and 500 for l.

$200 = w$ — Divide both sides by 500 and simplify.

The frontage of the rectangular plot is 200 feet.

b. If each foot of frontage costs $5.50, then your total assessment will be $200(5.50) = \$1100$.

500 ft

w

Figure 3.4

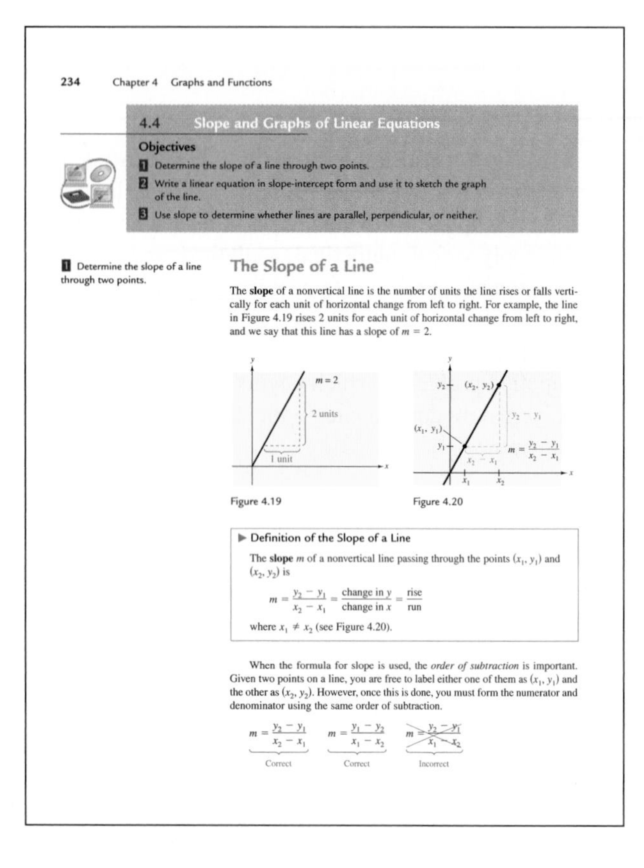

234 Chapter 4 Graphs and Functions

4.4 Slope and Graphs of Linear Equations

Objectives

1. Determine the slope of a line through two points.
2. Write a linear equation in slope-intercept form and use it to sketch the graph of the line.
3. Use slope to determine whether lines are parallel, perpendicular, or neither.

1 Determine the slope of a line through two points.

The Slope of a Line

The **slope** of a nonvertical line is the number of units the line rises or falls vertically for each unit of horizontal change from left to right. For example, the line in Figure 4.19 rises 2 units for each unit of horizontal change from left to right, and we say that this line has a slope of $m = 2$.

Figure 4.19

Figure 4.20

Definition of the Slope of a Line

The **slope** m of a nonvertical line passing through the points (x_1, y_1) and (x_2, y_2) is

$$m = \frac{y_2 - y_1}{x_2 - x_1} = \frac{\text{change in } y}{\text{change in } x} = \frac{\text{rise}}{\text{run}}$$

where $x_1 \neq x_2$ (see Figure 4.20).

When the formula for slope is used, the *order of subtraction* is important. Given two points on a line, you are free to label either one of them as (x_1, y_1) and the other as (x_2, y_2). However, once this is done, you must form the numerator and denominator using the same order of subtraction.

$m = \frac{y_2 - y_1}{x_2 - x_1}$ Correct $\quad m = \frac{y_1 - y_2}{x_1 - x_2}$ Correct $\quad m = \frac{y_2 - y_1}{x_1 - x_2}$ Incorrect

Definitions and Rules

All important definitions, rules, formulas, properties, and summaries of solution methods are highlighted for emphasis. Each of these features is also titled for easy reference.

Graphics

Visualization is a critical problem-solving skill. To encourage the development of this skill, students are shown how to use graphs to reinforce algebraic and numeric solutions and to interpret data. The numerous figures in examples and exercises throughout the text were computer generated for accuracy.

Technology Tips

Point-of-use instructions for using graphing utlities appear in the margins. They provide convenient reference for students using graphing technology. In addition, they encourage the use of graphing technology as a tool for visualization of mathematical concepts, for verification of other solution methods, and for facilitation of computations. The *Technology Tips* can easily be omitted without loss of continuity in coverage.

Section 3.6 Linear Inequalities 175

2 Become familiar with the properties of inequalities.

Properties of Inequalities

The procedures for solving linear inequalities in one variable are much like those for solving linear equations. To isolate the variable, you can use the **properties of inequalities.** These properties are similar to the properties of equality, but there are two important exceptions. *When both sides of an inequality are multiplied or divided by a negative number, the direction of the inequality symbol must be reversed.* Here is an example.

$-2 < 5$	Original inequality
$(-3)(-2) > (-3)(5)$	Multiply both sides by -3 and reverse inequality.
$6 > -15$	Simplify.

Technology: Tip

Linear inequalities can be graphed using a graphing utility. The inequality $x > -2$ is shown in the graph below. Notice that the graph appears above the x-axis. Consult the user's manual of your graphing utility for directions.

Two inequalities that have the same solution set are called **equivalent.** The following list describes operations that can be used to create equivalent inequalities.

Properties of Inequalities

Let a, b, and c be real numbers, variables, or algebraic expressions.

Property	*Verbal and Algebraic Descriptions*
Addition:	Add the same quantity to both sides. If $a < b$, then $a + c < b + c$.
Subtraction:	Subtract the same quantity from both sides. If $a < b$, then $a - c < b - c$.
Multiplication:	Multiply both sides by a *positive* quantity. If $a < b$ and c is positive, then $ac < bc$. Multiply both sides by a *negative* quantity and reverse the inequality symbol. If $a < b$ and c is negative, then $ac > bc$.
Division:	Divide both sides by a *positive* quantity. If $a < b$ and c is positive, then $\frac{a}{c} < \frac{b}{c}$. Divide both sides by a *negative* quantity and reverse the inequality symbol. If $a < b$ and c is negative, then $\frac{a}{c} > \frac{b}{c}$.
Transitive:	If $a < b$ and $b < c$, then $a < c$.

Each of the properties above is true if the symbol $<$ is replaced by $\leq$ and the symbol $>$ is replaced by $\geq$. Moreover, the letters a, b, and c can be real numbers, variables, or algebraic expressions. Note that you cannot multiply or divide both sides of an inequality by zero.

In Examples 1 and 2, the original equations each involved a second-degree (quadratic) polynomial and each had *two different* solutions. You will sometimes encounter second-degree polynomial equations that have only one (repeated) solution. This occurs when the left side of the equation is a perfect square trinomial, as shown in Example 3.

Technology: Discovery

Write the function in Example 3 in general form. Graph this function on your graphing utility.

$$y = x^2 - 8x + 16$$

What are the x-intercepts of the function?

Write the function in Example 4 in general form. Graph this function on your graphing utility.

$$y = x^2 + 9x + 14$$

What are the x-intercepts of the function?

How do the x-intercepts relate to the solutions of the equations? What can you conclude about the solutions to the equations and the x-intercepts?

Example 3 A Quadratic Equation with a Repeated Solution

Solve $x^2 - 8x + 20 = 4$.

Solution

$x^2 - 8x + 20 = 4$ Original equation

$x^2 - 8x + 16 = 0$ Write in general form.

$(x - 4)^2 = 0$ Factor.

$x - 4 = 0$ Set factor equal to 0.

$x = 4$ Solve for x.

Note that even though the left side of this equation has two factors, the factors are the same. Thus, the only solution of the equation is 4.

$x^2 - 8x + 20 = 4$ Original equation

$(4)^2 - 8(4) + 20 \stackrel{?}{=} 4$ Substitute 4 for x.

$16 - 32 + 20 \stackrel{?}{=} 4$ Simplify.

$4 = 4$ Solution checks. ✓

Example 4 Solving a Polynomial Equation

Solve $(x + 3)(x + 6) = 4$.

Solution

Begin by multiplying the factors on the left side.

$(x + 3)(x + 6) = 4$ Original equation

$x^2 + 9x + 18 = 4$ Multiply factors.

$x^2 + 9x + 14 = 0$ General form

$(x + 2)(x + 7) = 0$ Factor left side of equation.

$x + 2 = 0$ Set 1st factor equal to 0.

$x = -2$ Solve for x.

$x + 7 = 0$ Set 2nd factor equal to 0.

$x = -7$ Solve for x.

The equation has two solutions: -2 and -7. Check these in the original equation.

Technology Discovery

Utilizing the power of technology (scientific calculator and graphing utility), *Technology Discovery* invites students to engage in active exploration of mathematical concepts and discovery of mathematical relationships. These activities encourage students to use their critical thinking skills and help them develop an intuitive understanding of theoretical concepts. *Technology Discovery* features can easily be omitted without loss of continuity in coverage.

Study Tips

Study Tips offer students specific point-of-use suggestions for studying algebra, as well as pointing out common errors and discussing alternative solution methods. They appear in the margins.

Discussing the Concept

Each section concludes with a *Discussing the Concept* feature. Designed as a section wrap-up activity to give students an opportunity to think, talk, and write about mathematics, each of these activities encourages students to synthesize the mathematical concepts presented in the section. *Discussing the Concept* can be assigned as an independent or collaborative activity or can be used as a basis for a class discussion.

3 Solve a linear equation involving decimals.

Many real-life applications of linear equations involve decimal coefficients. To solve such an equation, you can clear it of decimals in much the same way you clear an equation of fractions. Multiply both sides by a power of 10 that converts all decimal coefficients to integers, as shown in the next example.

Study Tip

There are other ways to solve the decimal equation in Example 10. You could first clear the equation of decimals by multiplying both sides by 100. Or, you could keep the decimals and use a graphing utility to do the arithmetic operations. The method you choose is a matter of personal preference.

Example 10 Solving a Linear Equation Involving Decimals

Solve $0.3x + 0.2(10 - x) = 0.15(30)$.

Solution

$0.3x + 0.2(10 - x) = 0.15(30)$ Original equation

$0.3x + 2 - 0.2x = 4.5$ Distributive Property

$0.1x + 2 = 4.5$ Combine like terms.

$10(0.1x + 2) = 10(4.5)$ Multiply both sides by 10.

$x + 20 = 45$ Clear decimals.

$x = 25$ Subtract 20 from both sides.

Check

$0.3x + 0.2(10 - x) = 0.15(30)$ Original equation

$0.3(25) + 0.2(10 - 25) \stackrel{?}{=} 0.15(30)$ Substitute 25 for x.

$0.3(25) + 0.2(-15) \stackrel{?}{=} 0.15(30)$ Perform subtraction within parentheses.

$7.5 - 3.0 \stackrel{?}{=} 4.5$ Multiply.

$4.5 = 4.5$ Solution checks. ✓

The solution is 25.

Discussing the Concept **Error Analysis**

Suppose you are teaching an algebra class and one of your students hands in the following problem. Find the error in the solution. Write an explanation for the student.

$4(x + 2) - 8 = 3x$ Given equation

$4x + 8 - 8 = 3x$ Distributive Property

$4x = 3x$ Additive inverse

$4 = 3$ Divide both sides by x.

No solution because 4 is not equal to 3.

Explain what happens when you divide both sides of an equation by a variable factor.

68 Chapter 2 Fundamentals of Algebra

2.1 Exercises

Integrated Review *Concepts, Skills, and Problem Solving*

Keep mathematically in shape by doing these exercises *before* the problems of this section.

Properties and Definitions

In Exercises 1–4, identify the property illustrated by the equation.

1. $x(5) = 5x$

2. $10 - 10 = 0$

3. $3(t + 2) = 3t + 3 \cdot 2$

4. $7 + (8 + z) = (7 + 8) + z$

Simplifying Expressions

In Exercises 5–10, evaluate the expression.

5. $10 - |-7|$ **6.** $6 - (10 - 12)$

7. $\dfrac{3 - (5 - 20)}{4}$ **8.** $\dfrac{6}{7} - \dfrac{4}{7}$

9. $-\frac{3}{4}\left(\frac{28}{33}\right)$ **10.** $\frac{5}{8} \div \frac{3}{16}$

Problem Solving

11. You plan to save \$50 per month for 10 years. How much money will you set aside during the 10 years?

12. It is necessary to cut a 120-foot rope into eight pieces of equal length. What is the length of each piece?

Developing Skills

In Exercises 1–4, write an algebraic expression for the given statement. See Example 1.

1. The distance traveled in t hours if the average speed is 60 miles per hour

2. The cost of an amusement park ride for a family of n people if the cost per person is \$1.25

3. The cost of m pounds of meat if the cost per pound is \$2.19

4. The total weight of x 50-pound bags of fertilizer

In Exercises 5–8, identify the variables and constants in the expression.

5. $x + 3$ **6.** $y + 1$

7. $x + z$ **8.** $3^2 + z$

In Exercises 9–22, identify the terms of the expression. See Example 2.

9. $4x + 3$ **10.** $6x - 1$

11. $3x^2 + 5$ **12.** $5 - 3t^2$

13. $\frac{5}{3} - 3y^3$ **14.** $6x - \frac{2}{3}$

15. $2x - 3y + 1$ **16.** $x^2 + 18xy + y^2$

17. $3(x + 5) + 10$ **18.** $16 - (x + 1)$

19. $\dfrac{x}{4} + \dfrac{5}{x}$ **20.** $10 - \dfrac{t}{6}$

21. $\dfrac{3}{x + 2} - 3x + 4$ **22.** $x^2 + \dfrac{3x + 1}{x - 1} + 4$

In Exercises 23–32, identify the coefficient of the term. See Example 3.

23. $-6x$ **24.** $25y$

25. $-\frac{1}{3}y$ **26.** $\frac{1}{8}n$

27. $-\dfrac{3x}{2}$ **28.** $\dfrac{3x}{4}$

29. $2\pi x^2$ **30.** πt^4

31. $4.7u$ **32.** $-5.32b$

In Exercises 33–50, expand the expression as a product of factors. See Example 4.

33. y^5 **34.** x^6

35. 2^2x^4 **36.** 5^3x^2

37. $4y^2z^3$ **38.** $3uv^4$

39. $(a^2)^3$ **40.** $(z^3)^3$

41. $4x^3 \cdot x^4$ **42.** $a^2y^2 \cdot y^3$

43. $(ab)^3$ **44.** $2(xz)^4$

Integrated Review

Each exercise set (except in Chapter 1) is preceded by *Integrated Review* exercises. These exercises are designed to help students keep up with concepts and skills learned in previous chapters. Answers to all *Integrated Review* problems are given in the back of the book.

Exercises

The exercise sets have been reorganized in the Third Edition. Each exercise set is grouped into three categories: *Developing Skills*, *Solving Problems*, and *Explaining Concepts*. The exercise sets offer a diverse variety of computational, conceptual, and applied problems to accommodate many teaching and learning styles. Designed to build competence, skill, and understanding, each exercise set is graded in difficulty to allow students to gain confidence as they progress. Detailed solutions to all odd-numbered exercises are given in the *Student Solutions Guide*, and answers to all odd-numbered exercises are given in the back of the book.

Section 4.1 Ordered Pairs and Graphs 211

61. *Organizing Data* With an initial cost of \$5000, a company will produce x units at \$35 per unit. Write an equation that relates the total cost of producing x units to the number of units produced. Plot the cost for producing 100, 150, 200, 250, and 300 units.

62. *Organizing Data* An employee earns \$10 plus \$0.50 for every x units produced per hour. Write an equation that relates the employee's total hourly wage to the number of units produced. Plot the hourly wage for producing 2, 5, 8, 10, and 20 units per hour.

63. *Organizing Data* The table gives the normal temperature y (in degrees Fahrenheit) for Anchorage, Alaska for each month x of the year. The months are numbered 1 through 12, with $x = 1$ corresponding to January. (Source: National Oceanic and Atmospheric Administration)

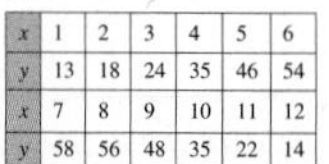

x	1	2	3	4	5	6
y	13	18	24	35	46	54
x	7	8	9	10	11	12
y	58	56	48	35	22	14

(a) Plot the data given in the table.

(b) Did you use the same scale on both axes? Explain.

(c) Using the graph, find the three consecutive months when the normal temperature changes the least.

64. *Organizing Data* The table gives the speed of a car x (in kilometers per hour) and the approximate stopping distance y (in meters).

x	50	70	90	110	130
y	20	35	60	95	148

(a) Plot the data given in the table.

(b) The x-coordinates increase at equal increments of 20 kilometers per hour. Describe the pattern for the y-coordinates. What are the implications for the driver?

65. *Graphical Interpretation* The table gives the numbers of hours x that a student studied for five different algebra exams and the resulting scores y.

x	3.5	1	8	4.5	0.5
y	72	67	95	81	53

(a) Plot the data given in the table.

(b) Use the graph to describe the relationship between the number of hours studied and the resulting exam score.

66. *Graphical Interpretation* The table gives the net income y per share of common stock of the H. J. Heinz Company for the years 1988 through 1997. The year is represented by x. (Source: H. J. Heinz Company 1997 Annual Report)

x	1988	1989	1990	1991	1992
y	\$0.97	\$1.11	\$1.26	\$1.42	\$1.60
x	1993	1994	1995	1996	1997
y	\$1.02	\$1.57	\$1.59	\$1.75	\$0.81

(a) Plot the data given in the table.

(b) Use the graph to find the year that had the greatest increase and the year that had the greatest decrease in the income per share.

Graphical Estimation In Exercises 67–70, use the scatter plot showing new privately-owned housing unit starts (in thousands) in the United States from 1985 through 1997. (Source: U.S. Bureau of the Census)

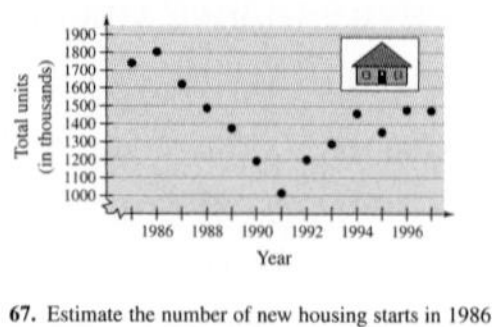

67. Estimate the number of new housing starts in 1986.

68. Estimate the number of new housing starts in 1991.

69. Estimate the increase and the percent increase in housing starts from 1993 to 1994.

70. Estimate the decrease and the percent decrease in housing starts from 1994 to 1995.

CHAPTER SUMMARY

Key Terms

algebraic expression, *p. 62*
variables, *p. 62*
constants, *p. 62*
terms, *p. 62*
coefficient, *p. 62*
evaluate an algebraic expression, *p. 65*
like terms, *p. 75*
simplify an algebraic expression, *p. 77*
verbal mathematical model, *p. 86*
equation, *p. 99*
solution, *p. 99*
solution set, *p. 99*
identity, *p. 99*
conditional equation, *p. 99*
equivalent equations, *p. 101*

Key Concepts

2.1 Exponential form

Repeated multiplication can be expressed in exponential form using a base a and an exponent n, where a is a real number, variable, or algebraic expression and n is a positive integer.

$$a^n = a \cdot a \cdot \cdot \cdot a$$

2.1 Evaluating algebraic expressions

To evaluate an algebraic expression, replace every occurrence of the variable in the expression with the appropriate real number and perform the operations.

2.2 Rules of exponents

Let m and n be positive integers, and let a and b be real numbers, variables, or algebraic expressions.

1. $a^m \cdot a^n = a^{m+n}$ 2. $(a^m)^n = a^{m \cdot n}$
3. $(ab)^m = a^m \cdot b^m$

2.2 Basic rules of algebra

Commutative Property:
Addition $a + b = b + a$
Multiplication $ab = ba$

Associative Property:
Addition $(a + b) + c = a + (b + c)$
Multiplication $(ab)c = a(bc)$

Distributive Property:
$a(b + c) = ab + ac$ $a(b - c) = ab - ac$
$(a + b)c = ac + bc$ $(a - b)c = ac - bc$

Identities:
Additive $a + 0 = a$
Multiplicative $a \cdot 1 = a$

Inverses:
Additive $a + (-a) = 0$
Multiplicative $a \cdot \frac{1}{a} = 1,\ a \neq 0$

2.2 Combining like terms

To combine like terms, add their respective coefficients and attach the common variable factor.

2.2 Simplifying an algebraic expression

To simplify an algebraic expression, remove symbols of grouping and combine like terms.

2.3 Translating phrases: verbal to algebraic

From the verbal description, write a verbal mathematical model. Assign labels to the known and unknown quantities, and write an algebraic model.

2.3 Other problem-solving strategies

Other problem-solving strategies are (1) guess, check, and revise, (2) make a table/look for a pattern, (3) draw a diagram, and (4) solve a simpler problem.

2.4 Checking solutions of equations

To check a solution, substitute the given solution for each occurrence of the variable in the original equation. Evaluate each side of the equation. If both sides are equivalent, the solution checks.

2.4 Properties of equality

Addition: Add (or subtract) the same quantity to (from) both sides of the equation.

Multiplication: Multiply (or divide) both sides of the equation by the same nonzero quantity.

109

Chapter Summary

The *Chapter Summary* has been completely revised in the Third Edition. Designed to be an effective study tool for students preparing for exams, it highlights the *Key Terms* (referenced by page) and the *Key Concepts* (referenced by section) presented in the chapter.

Review Exercises

The *Review Exercises* at the end of each chapter have been reorganized in the Third Edition. They are grouped into two categories: *Reviewing Skills* and *Solving Problems*. Exercises in *Reviewing Skills* are correlated to sections in the chapter. The *Review Exercises* offer students additional practice in preparation for exams. Answers to all odd-numbered exercises are given in the back of the book.

368 Chapter 6 Factoring and Solving Equations

REVIEW EXERCISES

Reviewing Skills

6.1 In Exercises 1–4, find the greatest common factor of the expressions.

1. 20, 60, 150 **2.** $3x^4, 21x^2$
3. $18ab^2, 27a^2b$ **4.** $14z^2, 1, 21z$

In Exercises 5–20, factor the polynomial.

5. $3x - 6$ **6.** $7 + 21x$
7. $3t - t^2$ **8.** $u^2 - 6u$
9. $5x^2 + 10x^3$ **10.** $7y - 21y^4$
11. $8a - 12a^3$ **12.** $6u - 9u^2 + 15u^3$
13. $x(x + 1) - 3(x + 1)$
14. $2u(u - 2) + 5(u - 2)$
15. $y^3 + 3y^2 + 2y + 6$ **16.** $z^3 - 5z^2 + z - 5$
17. $x^3 + 2x^2 + x + 2$ **18.** $x^3 - 5x^2 + 5x - 25$
19. $x^2 - 4x + 3x - 12$ **20.** $2x^2 + 6x - 5x - 15$

6.2 In Exercises 21–30, factor the trinomial.

21. $x^2 - 3x - 28$ **22.** $x^2 - 3x - 40$
23. $u^2 + 5u - 36$ **24.** $y^2 + 15y + 56$
25. $x^2 + 9xy - 10y^2$ **26.** $u^2 + uv - 5v^2$
27. $y^2 - 6xy - 27x^2$ **28.** $v^2 + 18uv + 32u^2$
29. $4x^2 - 24x + 32$ **30.** $x^3 + 9x^2 + 18x$

In Exercises 31–34, find all values of b such that the trinomial is factorable.

31. $x^2 + bx + 9$ **32.** $y^2 + by + 25$
33. $z^2 + bz + 11$ **34.** $x^2 + bx + 14$

6.3 In Exercises 35–44, factor the trinomial.

35. $5 - 2x - 3x^2$ **36.** $8x^2 - 18x + 9$
37. $50 - 5x - x^2$ **38.** $7 + 5x - 2x^2$
39. $6x^2 + 7x + 2$ **40.** $16x^2 + 13x - 3$
41. $6u^3 + 3u^2 - 30u$ **42.** $8x^3 - 8x^2 + 30x$
43. $2x^2 - 3x + 1$ **44.** $3x^2 + 8x + 4$

In Exercises 45–48, find all values of b such that the trinomial is factorable.

45. $x^2 + bx - 24$ **46.** $2x^2 + bx - 16$
47. $3x^2 + bx - 20$ **48.** $3x^2 + bx + 1$

In Exercises 49 and 50, find two values of c such that the trinomial is factorable.

49. $2x^2 - 4x + c$ **50.** $5x^2 + 6x + c$

6.4 In Exercises 51 and 52, insert the missing factors.

51. $x^3 - x = x(\quad)(\quad)$
52. $u^4 - v^4 = (u^2 + v^2)(\quad)(\quad)$

In Exercises 53–72, factor the polynomial completely.

53. $a^2 - 100$ **54.** $36 - b^2$
55. $25 - 4y^2$ **56.** $16b^2 - 1$
57. $(u + 1)^2 - 4$ **58.** $(y - 2)^2 - 9$
59. $x^2 - 8x + 16$ **60.** $y^2 + 24y + 144$
61. $x^2 + 6x + 9$ **62.** $v^2 - 10v + 25$
63. $9s^2 + 12s + 4$ **64.** $u^2 - 2uv + v^2$
65. $s^3t - st^3$ **66.** $y^3z + 4y^2z^2 + 4yz^3$
67. $a^3 + 1$ **68.** $z^3 + 8$
69. $27 - 8t^3$ **70.** $z^3 - 125$
71. $-16a^3 - 16a^2 - 4a$ **72.** $5t - 125t^3$

6.5 In Exercises 73–86, solve the polynomial equation.

73. $x^2 - 81 = 0$ **74.** $121 - y^2 = 0$
75. $x^2 - 12x + 36 = 0$ **76.** $2t^2 - 3t - 2 = 0$
77. $4s^2 + s - 3 = 0$ **78.** $y^3 - y^2 - 6y = 0$
79. $x(2x - 3) = 0$ **80.** $3x(5x + 1) = 0$
81. $(z - 2)^2 - 4 = 0$ **82.** $(x + 1)^2 - 16 = 0$
83. $x(7 - x) = 12$ **84.** $x(x + 5) = 24$
85. $u^3 + 5u^2 - u = 5$ **86.** $a^3 - 3a^2 - a = -3$

FEATURES

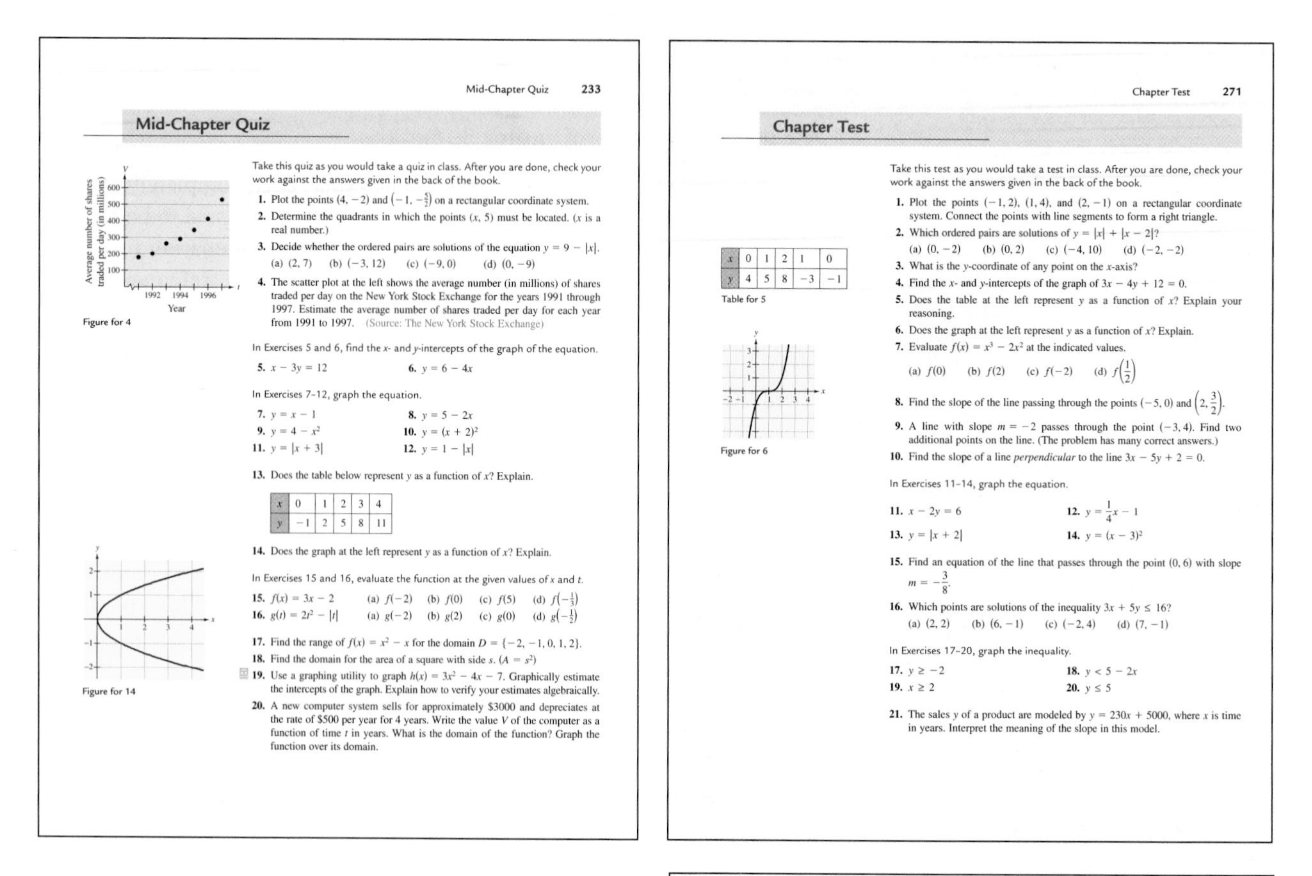

Mid-Chapter Quiz

Take this quiz as you would take a quiz in class. After you are done, check your work against the answers given in the back of the book.

1. Plot the points $(4, -2)$ and $\left(-1, -\frac{5}{2}\right)$ on a rectangular coordinate system.
2. Determine the quadrants in which the points $(x, 5)$ must be located. (x is a real number.)
3. Decide whether the ordered pairs are solutions of the equation $y = 9 - |x|$.
   (a) $(2, 7)$ (b) $(-3, 12)$ (c) $(-9, 0)$ (d) $(0, -9)$
4. The scatter plot at the left shows the average number (in millions) of shares traded per day on the New York Stock Exchange for the years 1991 through 1997. Estimate the average number of shares traded per day for each year from 1991 to 1997. (Source: The New York Stock Exchange)

Figure for 4

In Exercises 5 and 6, find the x- and y-intercepts of the graph of the equation.

5. $x - 3y = 12$
6. $y = 6 - 4x$

In Exercises 7–12, graph the equation.

7. $y = x - 1$
8. $y = 5 - 2x$
9. $y = 4 - x^2$
10. $y = (x + 2)^2$
11. $y = |x + 3|$
12. $y = 1 - |x|$

13. Does the table below represent y as a function of x? Explain.

x	0	1	2	3	4
y	−1	2	5	8	11

14. Does the graph at the left represent y as a function of x? Explain.

Figure for 14

In Exercises 15 and 16, evaluate the function at the given values of x and t.

15. $f(x) = 3x - 2$ (a) $f(-2)$ (b) $f(0)$ (c) $f(5)$ (d) $f\left(-\frac{1}{3}\right)$
16. $g(t) = 2t^2 - |t|$ (a) $g(-2)$ (b) $g(2)$ (c) $g(0)$ (d) $g\left(-\frac{1}{2}\right)$
17. Find the range of $f(x) = x^2 - x$ for the domain $D = \{-2, -1, 0, 1, 2\}$.
18. Find the domain for the area of a square with side s. $(A = s^2)$
19. Use a graphing utility to graph $h(x) = 3x^2 - 4x - 7$. Graphically estimate the intercepts of the graph. Explain how to verify your estimates algebraically.
20. A new computer system sells for approximately \$3000 and depreciates at the rate of \$500 per year for 4 years. Write the value V of the computer as a function of time t in years. What is the domain of the function? Graph the function over its domain.

Chapter Test

Take this test as you would take a test in class. After you are done, check your work against the answers given in the back of the book.

1. Plot the points $(-1, 2)$, $(1, 4)$, and $(2, -1)$ on a rectangular coordinate system. Connect the points with line segments to form a right triangle.
2. Which ordered pairs are solutions of $y = |x| + |x - 2|$?
   (a) $(0, -2)$ (b) $(0, 2)$ (c) $(-4, 10)$ (d) $(-2, -2)$
3. What is the y-coordinate of any point on the x-axis?
4. Find the x- and y-intercepts of the graph of $3x - 4y + 12 = 0$.
5. Does the table at the left represent y as a function of x? Explain your reasoning.

x	0	1	2	1	0
y	4	5	8	−3	−1

Table for 5

6. Does the graph at the left represent y as a function of x? Explain.

Figure for 6

7. Evaluate $f(x) = x^3 - 2x^2$ at the indicated values.
   (a) $f(0)$ (b) $f(2)$ (c) $f(-2)$ (d) $f\left(\frac{1}{2}\right)$
8. Find the slope of the line passing through the points $(-5, 0)$ and $\left(2, \frac{3}{2}\right)$.
9. A line with slope $m = -2$ passes through the point $(-3, 4)$. Find two additional points on the line. (The problem has many correct answers.)
10. Find the slope of a line *perpendicular* to the line $3x - 5y + 2 = 0$.

In Exercises 11–14, graph the equation.

11. $x - 2y = 6$
12. $y = \frac{1}{4}x - 1$
13. $y = |x + 2|$
14. $y = (x - 3)^2$

15. Find an equation of the line that passes through the point $(0, 6)$ with slope $m = -\frac{3}{8}$.
16. Which points are solutions of the inequality $3x + 5y \le 16$?
   (a) $(2, 2)$ (b) $(6, -1)$ (c) $(-2, 4)$ (d) $(7, -1)$

In Exercises 17–20, graph the inequality.

17. $y \ge -2$
18. $y < 5 - 2x$
19. $x \ge 2$
20. $y \le 5$

21. The sales y of a product are modeled by $y = 230x + 5000$, where x is time in years. Interpret the meaning of the slope in this model.

Mid-Chapter Quiz

Each chapter contains a *Mid-Chapter Quiz*. This feature allows students to perform a self-assessment midway through the chapter. Answers to all questions in the *Mid-Chapter Quiz* are given in the back of the book.

Chapter Test

Each chapter ends with a *Chapter Test*. This feature allows students to perform a self-assessment at the end of the chapter. Answers to all questions in the *Chapter Test* are given in the back of the book.

Cumulative Test

The *Cumulative Tests* that follow Chapters 3, 6, 9, and 12 provide a comprehensive self-assessment tool that helps students check their mastery of previously covered material. Answers to all questions in the *Cumulative Tests* are given in the back of the book.

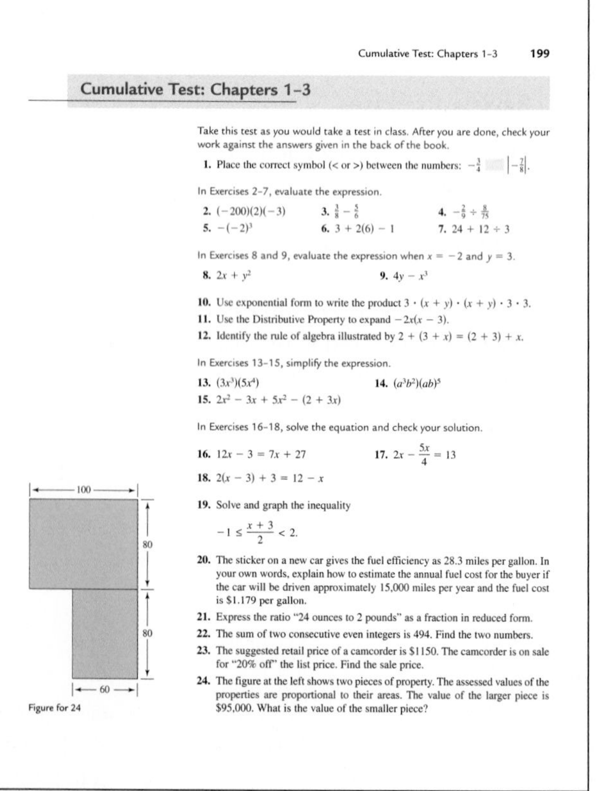

Cumulative Test: Chapters 1–3

Take this test as you would take a test in class. After you are done, check your work against the answers given in the back of the book.

1. Place the correct symbol (< or >) between the numbers: $-\frac{3}{4}$ ☐ $\left|-\frac{7}{8}\right|$.

In Exercises 2–7, evaluate the expression.

2. $(-200)(2)(-3)$
3. $\frac{3}{8} - \frac{5}{6}$
4. $-\frac{2}{9} \div \frac{8}{75}$
5. $-(-2)^3$
6. $3 + 2(6) - 1$
7. $24 + 12 \div 3$

In Exercises 8 and 9, evaluate the expression when $x = -2$ and $y = 3$.

8. $2x + y^2$
9. $4y - x^3$

10. Use exponential form to write the product $3 \cdot (x + y) \cdot (x + y) \cdot 3 \cdot 3$.
11. Use the Distributive Property to expand $-2x(x - 3)$.
12. Identify the rule of algebra illustrated by $2 + (3 + x) = (2 + 3) + x$.

In Exercises 13–15, simplify the expression.

13. $(3x^3)(5x^4)$
14. $(a^3b^2)(ab)^5$
15. $2x^2 - 3x + 5x^2 - (2 + 3x)$

In Exercises 16–18, solve the equation and check your solution.

16. $12x - 3 = 7x + 27$
17. $2x - \frac{5x}{4} = 13$
18. $2(x - 3) + 3 = 12 - x$

19. Solve and graph the inequality
$$-1 \le \frac{x + 3}{2} < 2.$$
20. The sticker on a new car gives the fuel efficiency as 28.3 miles per gallon. In your own words, explain how to estimate the annual fuel cost for the buyer if the car will be driven approximately 15,000 miles per year and the fuel cost is \$1.179 per gallon.
21. Express the ratio "24 ounces to 2 pounds" as a fraction in reduced form.
22. The sum of two consecutive even integers is 494. Find the two numbers.
23. The suggested retail price of a camcorder is \$1150. The camcorder is on sale for "20% off" the list price. Find the sale price.
24. The figure at the left shows two pieces of property. The assessed values of the properties are proportional to their areas. The value of the larger piece is \$95,000. What is the value of the smaller piece?

Figure for 24

Supplements

Elementary and Intermediate Algebra: A Combined Course, Third Edition, by Larson and Hostetler is accompanied by a comprehensive supplements package, which includes resources for both students and instructors. All items are keyed to the text.

Printed Resources

For the Student

Study and Solutions Guide by Carolyn Neptune, Johnson County Community College, and Gerry C. Fitch, Louisiana State University
(0-395-97646-4)

- Detailed, step-by-step solutions to all Integrated Review exercises and to all odd-numbered exercises in the section exercise sets and in the review exercises
- Detailed, step-by-step solutions to all Mid-Chapter Quiz, Chapter Test, and Cumulative Test questions

Graphing Calculator Keystroke Guide by Benjamin N. Levy and Laurel Technical Services
(0-395-87777-6)

- Keystroke instructions for the following graphing calculators: (Texas Instruments) *TI-80*, *TI-81*, *TI-82*, *TI-83*, *TI-85*, and *TI-92*; (Casio) *fx-7700GE*, *fx-9700GE*, and *CFX-9800G*; (Hewlett Packard) *HP-38G*; and (Sharp) *EL-9200/9300*
- Examples with step-by-step solutions
- Extensive graphics screen output
- Technology tips

For the Instructor

Instructor's Annotated Edition
(0-395-97644-8)

- Includes entire student edition
- Instructor's answer section, which includes answers to all even-numbered exercises, Technology Discovery boxes, Technology Tip boxes, and Discussing the Concept activities
- Annotations at point of use that offer strategies and suggestions for teaching the course and point out common student errors

Test Item File and Instructor's Resource Guide by Cheryl A. Leech, The Pennsylvania State University, The Behrend College, and Ann R. Kraus, The Pennsylvania State University, The Behrend College
(0-395-97643-X)

- Printed test bank with approximately 3300 test items, coded by level of difficulty
- Technology required test items, coded for easy reference
- Chapter test forms with answer key
- Two final exams
- Transparency masters

- Notes to the instructor, which include information on standardized tests such as the Texas Academic Skills Program (TASP), the Florida College Level Academic Skills Test (CLAST), and the California State University Entry Level Mathematics (ELM) Exam. A list of skills covered by the test and the corresponding sections in the text where the topics are covered are also provided.
- Alternative assessment strategies

Media Resources

For Students and Instructors

Web Site (***www.hmco.com***)
Contains, but is not limited to, the following student and instructor resources:
- Study guide (for students), which includes section summaries, additional examples with solutions, and starter exercises with answers
- Chapter projects and additional real-life applications
- Geometry review
- ACE Algebra Tutor
- Graphing calculator programs
- Math Matters and Career Interviews

HM^3 Tutor
(Instructor's version Windows: 0-618-04208-3)
This networkable, interactive tutorial software offers the following features:
- Algorithmically generated practice and quiz problems
- A variety of multiple-choice and free-response questions, varying in degree of difficulty
- Animated examples and interactivity within lessons
- Hints and full solutions available for every problem
- Integrated classroom management system (for instructors), which includes a syllabus builder and the capability to track and report student performance
- Non-networkable student version (Windows: 0-395-97656-1)

For the Student

Videotape Series by Dana Mosely
(0-395-97659-6)
- Comprehensive section-by-section coverage
- Detailed explanations of important concepts
- Numerous examples and applications, often illustrated via computer-generated animations
- Discussion of study skills

For the Instructor

Computerized Test Bank
(Windows: 0-395-97654-5; Macintosh: 0-395-97655-3)
- Test-generating software for IBM and Macintosh computers
- Approximately 3300 test items
- Also available as a printed test bank

Acknowledgments

We would like to thank the many people who have helped us prepare the Third Edition of this text. Their encouragement, criticisms, and suggestions have been invaluable to us.

Third Edition Reviewers

Mary Kay Best, Coastal Bend College; Beverly Broomell, Suffolk County Community College; Connie L. Buller, Metropolitan Community College; David L. Byrd, Enterprise State Junior College; E. Judith Cantey, Jefferson State Community College; Kelly E. Champagne, Nicholls State University; Sally Copeland, Johnson County Community College; Maggie W. Flint, Northeast State Technical Community College; Fletcher Gross, University of Utah; William Hoard, Front Range Community College; Judith Kasabian, El Camino College; Harvey W. Lambert, University of Nevada–Reno; Jennifer L. Laveglia, Bellevue Community College; Carol McVey, Florence–Darlington Technical College; Aaron Montgomery, Purdue University North Central; William Naegele, South Suburban College; Jeanette O'Rourke, Middlesex County College; Judith Pranger, Binghamton University; Scott Reed, College of Lake County; Eveline Robbins, Gainesville College; Kent Sandefer, Mohave Community College; Robert L. Sartain, Howard Payne University; Jon W. Scott, Montgomery College; John Seims, Mesa Community College; Ralph Selensky, Eastern Arizona College; Charles I. Sherrill, Community College of Aurora; Joel Siegel, Sierra College; Allan Struck, California State University–Chico; Katherine R. Struve, Columbus State Community College; Marilyn Treder, Rochester Community and Technical College; Bettie Truitt, Black Hawk College; Christine Walker, Utah Valley State College; Maureen Watson, Nicholls State University; Matrid Whiddon, Edison Community College; Betsey S. Whitman, Framingham State College; George J. Witt, Glendale Community College.

We would also like to thank the staff of Larson Texts, Inc. and the staff of Meridian Creative Group, who assisted in proofreading the manuscript, preparing and proofreading the art package, and checking and typesetting the supplements.

On a personal level, we are grateful to our wives, Deanna Gilbert Larson and Eloise Hostetler, for their love, patience, and support. Also, a special thanks goes to R. Scott O'Neil.

If you have suggestions for improving this text, please feel free to write to us. Over the past two decades we have received many useful comments from both instructors and students, and we value these comments very much.

Ron Larson
Robert P. Hostetler

How to Study Algebra

Your success in algebra depends on your active participation both in class and outside of class. Because the material you learn each day builds on the material you learned previously, it is important that you keep up with the course work every day and develop a clear plan of study. To help you learn how to study algebra, we have prepared a set of guidelines that highlight key study strategies.

Preparing for Class

The syllabus your instructor provides is an invaluable resource that outlines the major topics to be covered in the course. Use it to help you prepare. As a general rule, you should set aside two to four hours of study time for each hour spent in class. Being prepared is the first step toward success in algebra. Before class,

- ❒ Review your notes from the previous class.
- ❒ Read the portion of the text that will be covered in class.
- ❒ Use the objectives listed at the beginning of each section to keep you focused on the main ideas of the section.
- ❒ Pay special attention to the definitions, rules, and concepts highlighted in boxes. Also, be sure you understand the meanings of mathematical symbols and terms written in boldface type. Keep a vocabulary journal for easy reference.
- ❒ Read through the solved examples. Use the side comments given in the solution steps to help you follow the solution process. Also, read the *Study Tips* given in the margins.
- ❒ Make notes of anything you do not understand as you read through the text. If you still do not understand after your instructor covers the topic in question, ask questions before your instructor moves on to a new topic.
- ❒ If you are using technology in this course, read the *Technology Tips* and try the *Technology Discovery* exercises.

Keeping Up

Another important step toward success in algebra involves your ability to keep up with the work. It is very easy to fall behind, especially if you miss a class. To keep up with the course work, be sure to

- ❒ Attend every class. Bring your text, a notebook, and a pen or pencil. If you miss a class, get the notes from a classmate as soon as possible and review them carefully.
- ❒ Take notes in class. After class, read through your notes and add explanations so that your notes make sense to *you*.
- ❒ Reread the portion of your text that was covered in class. This time, work each example *before* reading through the solution.

- ❐ Do your homework as soon as possible, while concepts are still fresh in your mind.
- ❐ Use your notes from class, the text discussion, the examples, and the *Study Tips* as you do your homework. Many exercises are keyed to specific examples in the text for easy reference.

Getting Extra Help

It can be very frustrating when you do not understand concepts and are unable to complete homework assignments. However, there are many resources available to help you with your study of algebra.

- ❐ Your instructor may have office hours. If you are feeling overwhelmed and need help, make an appointment to discuss your difficulties with your instructor.
- ❐ Find a study partner or a study group. Sometimes it helps to work through problems with another person.
- ❐ Arrange to get regular assistance from a tutor. Many colleges have a math resource center available on campus, as well.
- ❐ Consult one of the many ancillaries available with this text: the *Student Solutions Guide*, tutorial software, videotapes, and additional study resources available at our website at *www.hmco.com*.

Preparing for an Exam

The last step toward success in algebra lies in how you prepare for and complete exams. If you have followed the suggestions given above, then you are almost ready for exams. Do not assume that you can cram for the exam the night before—this seldom works. As a final preparation for the exam,

- ❐ Read the *Chapter Summary*, which is keyed to each section, and review the concepts and terms.
- ❐ Work through the *Review Exercises* if you need extra practice on material from a particular section.
- ❐ Take the *Mid-Chapter Quiz* and the *Chapter Test* as if you were in class. You should set aside at least one hour per test. Check your answers against the answers given in the back of the book.
- ❐ Review your notes and the portion of the text that will be covered on the exam.
- ❐ Avoid studying up until the last minute. This will only make you anxious.
- ❐ Once the exam begins, read through the directions and the entire exam before beginning. Work the problems that you know how to do first to avoid spending too much time of the exam on any one problem. Time management is extremely important when taking an exam.
- ❐ If you finish early, use the remaining exam time to go over your work.
- ❐ When you get an exam back, review it carefully and go over your errors. Rework the problems you answered incorrectly. Discovering the mistakes you made will help you improve your test-taking ability.

1 The Real Number System

Jerry Driendl

The average daily temperature in Pittsburgh, Pennsylvania, in December of 1998 was 37.8°F. This was 6.3°F above the normal average daily temperature of 31.5°F. (Source: The National Weather Service)

Motivating the Chapter

December in Pennsylvania

A city in Pennsylvania has an average daily high temperature of 0° Celsius for the month of December. The temperature records for the first 14 days of December 1997 are given in the table.

Day	1	2	3	4	5	6	7
Low temperature (°C)	-3°	-5°	-12°	-20°	-6°	$-\frac{4}{3}^\circ$	0°
High temperature (°C)	2.5°	1°	-4°	-15°	0°	5°	8°

Day	8	9	10	11	12	13	14
Low temperature (°C)	0°	2°	-1°	-2°	-9°	-10°	-8°
High temperature (°C)	2°	7.2°	3°	2°	-4°	$-4\frac{1}{2}^\circ$	-4°

Here are some of the types of questions you will be able to answer as you study this chapter. You will be asked to answer Questions (a) to (f) in Section 1.1, Exercise 71.

a. Write the set A of *integer* high temperatures.

b. Write the set B of *rational* low temperatures.

c. Write the set C of *nonnegative* low temperatures.

d. Write the high temperatures in *increasing* order.

e. Write the low temperatures in *decreasing* order.

f. What day(s) had high and low temperatures that were opposite numbers?

You will be asked to answer Questions (g) to (l) in Section 1.3, Exercise 153.

g. What day had a high temperature of greatest departure from the monthly average high temperature of 0°?

h. What successive days had the greatest change in high temperature?

i. What successive days had the greatest change in low temperature?

j. Find the average high temperature for the 14 days.

k. Find the average low temperature for the 14 days.

l. In which of the preceding problems is the concept of absolute value used?

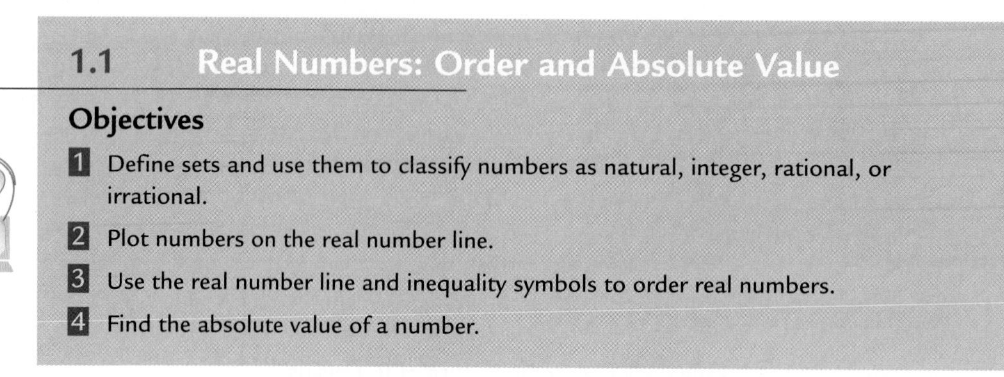

1.1 Real Numbers: Order and Absolute Value

Objectives

1. Define sets and use them to classify numbers as natural, integer, rational, or irrational.
2. Plot numbers on the real number line.
3. Use the real number line and inequality symbols to order real numbers.
4. Find the absolute value of a number.

1 Define sets and use them to classify numbers as natural, integer, rational, or irrational.

Sets and Real Numbers

The ability to communicate precisely is an essential part of a modern society, and it is the primary goal of this text. Specifically, this section introduces the language used to communicate numerical concepts.

The formal term that is used in mathematics to talk about a collection of objects is the word **set.** For instance, the set $\{1, 2, 3\}$ contains the three numbers 1, 2, and 3. Note that a pair of braces $\{\ \ \}$ is used to list the members of the set. Parentheses ( ) and brackets [ ] are used to represent other ideas.

The set of numbers that is used in arithmetic is called the set of **real numbers.** The term *real* distinguishes real numbers from *imaginary* numbers—a type of number that is used in some mathematics courses. You will not study imaginary numbers in Elementary Algebra.

Study Tip

Whenever we formally introduce a mathematical term in this text, the word will occur in boldface type. Be sure you understand the meaning of each new word; it is important that each word become part of your mathematical vocabulary.

If each member of a set A is also a member of a set B, then A is called a **subset** of B. The set of real numbers has many important subsets, each with a special name. For instance, the set

$\{1, 2, 3, 4, . . .\}$ A subset of the set of real numbers

is the set of **natural numbers** or **positive integers.** Note that the three dots indicate that the pattern continues. For instance, the set also contains the numbers 5, 6, 7, and so on. Every positive integer is a real number, but there are many real numbers that are not positive integers. For example, the numbers -2, 0, and $\frac{1}{2}$ are real numbers, but they are not positive integers.

Positive integers can be used to describe many things that you encounter in everyday life. For instance, you might be taking four classes this term, or you might be paying \$180 a month for rent. But even in everyday life, positive integers cannot describe some concepts accurately. For instance, you could have a zero balance in your checking account, or the temperature could be $-10°$ (ten degrees below zero). To describe such quantities you need to expand the set of positive integers to include **zero** and the **negative integers.** The expanded set is called the set of **integers.**

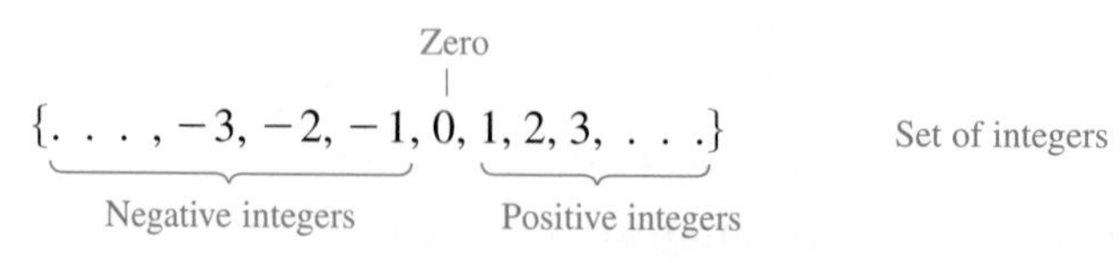

The set of integers is also a subset of the set of real numbers.

Even with the set of integers, there are still many quantities in everyday life that you cannot describe accurately. The costs of many items are not in whole-dollar amounts, but in parts of dollars, such as \$1.19 or \$39.98. You might work $8\frac{1}{2}$ hours, or you might miss the first half of a movie. To describe such quantities, you can expand the set of integers to include **fractions.** The expanded set is called the set of **rational numbers.** In the formal language of mathematics, a real number is **rational** if it can be written as a ratio of two integers. So, $\frac{3}{4}$ is a rational number; so is 0.5 $\left(\text{it can be written as } \frac{1}{2}\right)$; and so is every integer. A real number that is not rational is called **irrational** and cannot be written as the ratio of two integers. One example of an irrational number is $\sqrt{2}$, which is read as the positive square root of 2. Another example is π (the Greek letter pi), which represents the ratio of the circumference of a circle to its diameter. Each of the sets of numbers mentioned—natural numbers, integers, rational numbers, and irrational numbers—is a subset of the set of real numbers, as shown in Figure 1.1.

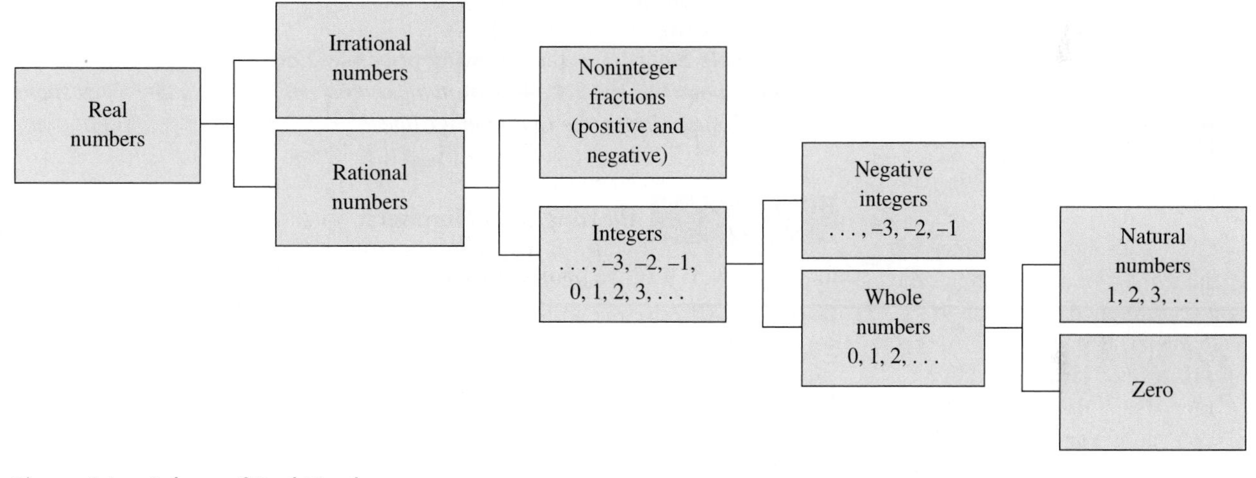

Figure 1.1 *Subsets of Real Numbers*

Study Tip

In *decimal form,* you can recognize rational numbers as decimals that terminate

$\frac{1}{2} = 0.5$ or $\frac{3}{8} = 0.375$

or repeat

$\frac{4}{3} = 1.3\overline{3}$ or $\frac{2}{11} = 0.18\overline{18}$.

Irrational numbers are represented by decimals that neither terminate nor repeat, as in

$\sqrt{2} = 1.414256237\ldots$

or

$\pi = 3.14159265359\ldots.$

Example 1 Classifying Real Numbers

Determine which numbers in the following set are (a) natural numbers, (b) integers, (c) rational numbers, and (d) irrational numbers.

$$\left\{\frac{1}{2}, -1, 0, 4, -\frac{5}{8}, \frac{4}{2}, -\frac{3}{1}, 0.86, \sqrt{2}, \sqrt{9}\right\}$$

Solution

a. Natural numbers: $\left\{4, \frac{4}{2} = 2, \sqrt{9} = 3\right\}$

b. Integers: $\left\{-1, 0, 4, \frac{4}{2} = 2, -\frac{3}{1} = -3, \sqrt{9} = 3\right\}$

c. Rational numbers: $\left\{\frac{1}{2}, -1, 0, 4, -\frac{5}{8}, \frac{4}{2}, -\frac{3}{1}, 0.86, \sqrt{9} = 3\right\}$

d. Irrational numbers: $\left\{\sqrt{2}\right\}$

e. Real numbers: $\left\{\frac{1}{2}, -1, 0, 4, -\frac{5}{8}, \frac{4}{2}, -\frac{3}{1}, 0.86, \sqrt{2}, \sqrt{9}\right\}$

2 Plot numbers on the real number line.

The Real Number Line

The diagram used to represent the real numbers is called the **real number line.** It consists of a horizontal line with a point (the **origin**) labeled 0. Numbers to the left of 0 are **negative** and numbers to the right of 0 are **positive,** as shown in Figure 1.2. The real number zero is neither positive nor negative. Thus, the term **nonnegative** implies that a number may be positive *or* zero.

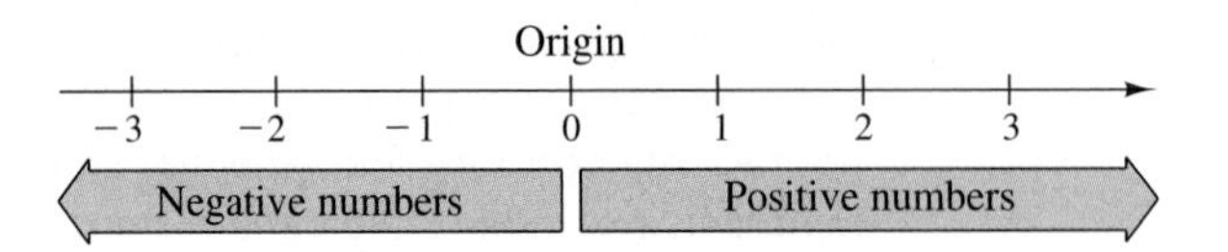

Figure 1.2 *The Real Number Line*

Drawing the point on the real number line that corresponds to a real number is called **plotting** the real number.

Example 2 illustrates the following principle. *Each point on the real number line corresponds to exactly one real number, and each real number corresponds to exactly one point on the real number line.*

Technology: Tip

The Greek letter pi, denoted by the symbol π, is the ratio of the circumference of a circle to its diameter. Because π cannot be written as a ratio of two integers, it is an irrational number. You can get an approximation of π on a scientific or graphing calculator by using the following keystrokes.

Keystroke	*Display*
π	3.141592654

Between which two integers would you plot π on the real number line?

Example 2 Plotting Real Numbers

a. In Figure 1.3(a), the point corresponds to the real number $-\frac{1}{2}$.

b. In Figure 1.3(b), the point corresponds to the real number 2.

c. In Figure 1.3(c), the point corresponds to the real number $-\frac{3}{2}$.

d. In Figure 1.3(d), the point corresponds to the real number 1.

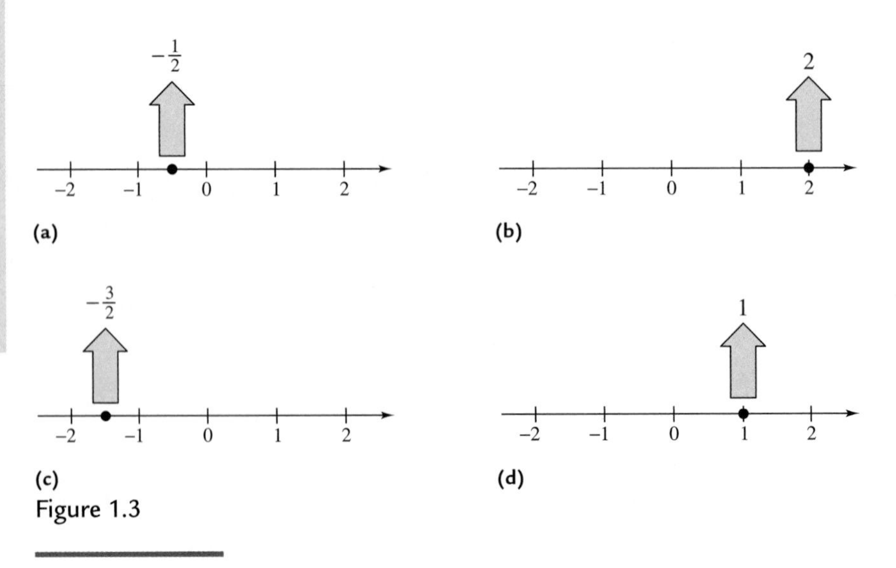

Figure 1.3

3 Use the real number line and inequality symbols to order real numbers.

Ordering Real Numbers

If you choose any two numbers on the real number line, one of the numbers must be to the left of the other number. The number to the left is **less than** the number to the right, and the number to the right is **greater than** the number to the left. For example, from Figure 1.4 you can see that -3 is less than 2 because -3 lies to the left of 2 on the number line. A "less than" comparison is denoted by the **inequality symbol** <. For instance, "-3 is less than 2" is denoted by $-3 < 2$.

Similarly, the inequality symbol > is used to denote a "greater than" comparison. For instance, "2 is greater than -3" is denoted by $2 > -3$. The inequality symbol $\leq$ means **less than or equal to,** and the inequality symbol $\geq$ means **greater than or equal to.**

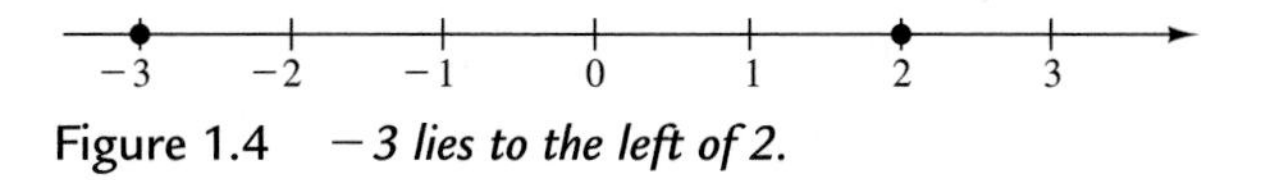

Figure 1.4 *−3 lies to the left of 2.*

When you are asked to **order** two numbers, you are simply being asked to say which of the two numbers is greater.

Example 3 Ordering Integers

Place the correct inequality symbol (< or >) between the two numbers.

a. 3 ▒ 5 **b.** -3 ▒ -5 **c.** 4 ▒ 0

d. -2 ▒ 2 **e.** 1 ▒ -4

Solution

See Figure 1.5.

a. $3 < 5$, because 3 lies to the *left* of 5.

b. $-3 > -5$, because -3 lies to the *right* of -5.

c. $4 > 0$, because 4 lies to the *right* of 0.

d. $-2 < 2$, because -2 lies to the *left* of 2.

e. $1 > -4$, because 1 lies to the *right* of -4.

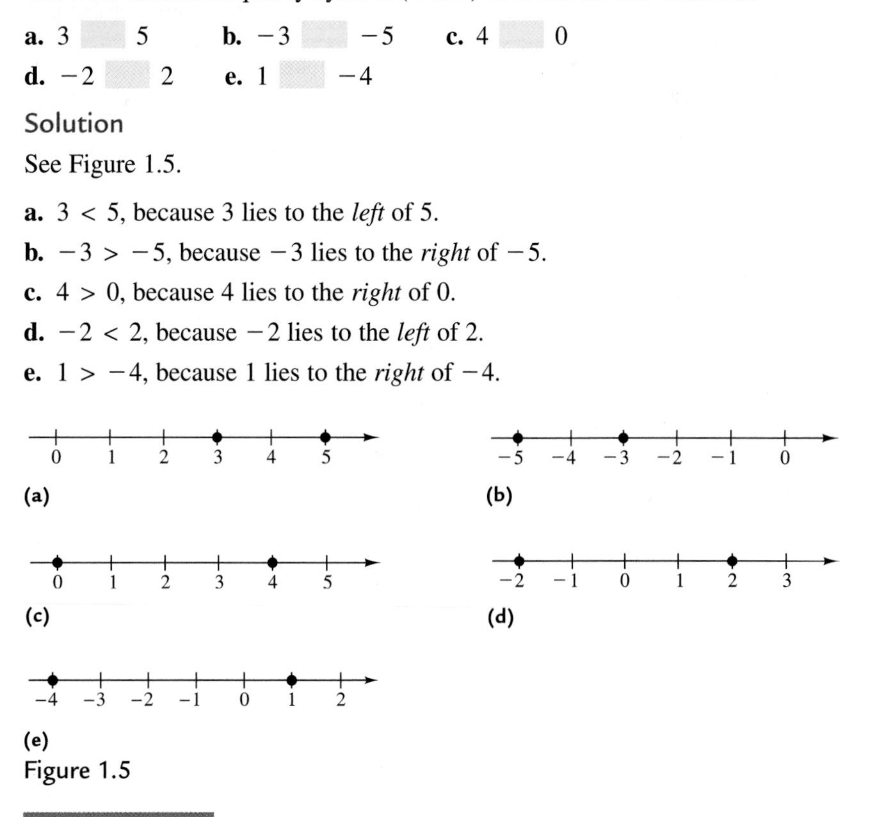

Figure 1.5

There are two ways to order fractions: you can write both fractions with the same denominator, or you can rewrite both fractions in decimal form. Here are two examples.

$$\frac{1}{3} = \frac{4}{12} \text{ and } \frac{1}{4} = \frac{3}{12} \quad \Rightarrow \quad \frac{1}{3} > \frac{1}{4}$$

$$\frac{11}{131} \approx 0.084 \text{ and } \frac{19}{209} \approx 0.091 \quad \Rightarrow \quad \frac{11}{131} < \frac{19}{209}$$

The symbol $\approx$ means "is approximately equal to."

Example 4 Ordering Fractions and Decimals

Place the correct inequality symbol (< or >) between the two numbers.

a. $\frac{1}{3} \quad \square \quad \frac{1}{5}$

b. $-\frac{3}{2} \quad \square \quad \frac{1}{2}$

c. $-3.1 \quad \square \quad 2.8$

d. $-1.09 \quad \square \quad -1.90$

Solution

See Figure 1.6.

a. $\frac{1}{3} > \frac{1}{5}$, because $\frac{1}{3} = \frac{5}{15}$ lies to the *right* of $\frac{1}{5} = \frac{3}{15}$.

b. $-\frac{3}{2} < \frac{1}{2}$, because $-\frac{3}{2}$ lies to the *left* of $\frac{1}{2}$.

c. $-3.1 < 2.8$, because -3.1 lies to the *left* of 2.8.

d. $-1.09 > -1.90$, because -1.09 lies to the *right* of -1.90.

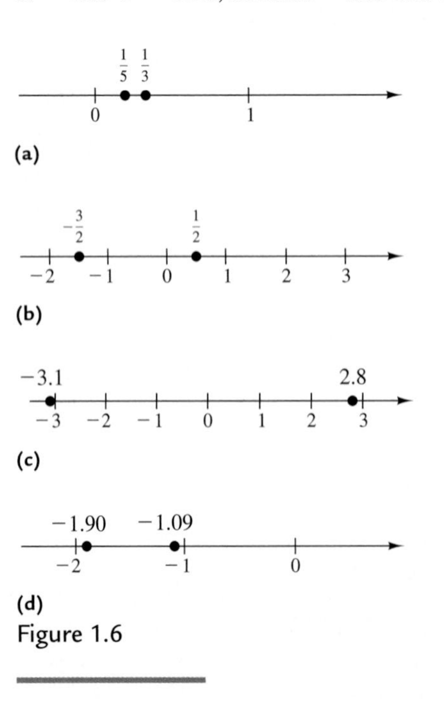

Figure 1.6

4 Find the absolute value of a number.

Absolute Value

Two real numbers are **opposites** of each other if they lie the same distance from, but on opposite sides of, zero. For example, -2 is the opposite of 2, and 4 is the opposite of -4, as shown in Figure 1.7.

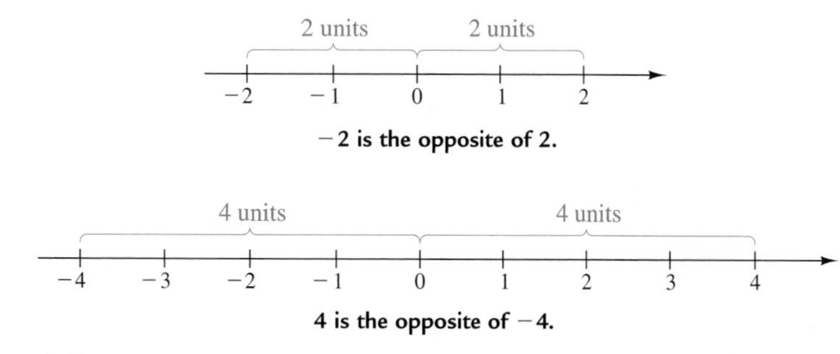

Figure 1.7

Parentheses are useful for denoting the opposite of a negative number. For example, $-(-3)$ means the opposite of -3, which you know to be 3. That is,

$-(-3) = 3.$ The opposite of -3 is 3.

For any real number, its distance from zero (on the real number line) is its **absolute value.** A pair of vertical bars, $|\ \ |$, is used to denote absolute value. Here are two examples.

$|5| =$ "distance between 5 and 0" $= 5$

$|-8| =$ "distance between -8 and 0" $= 8$ See Figure 1.8.

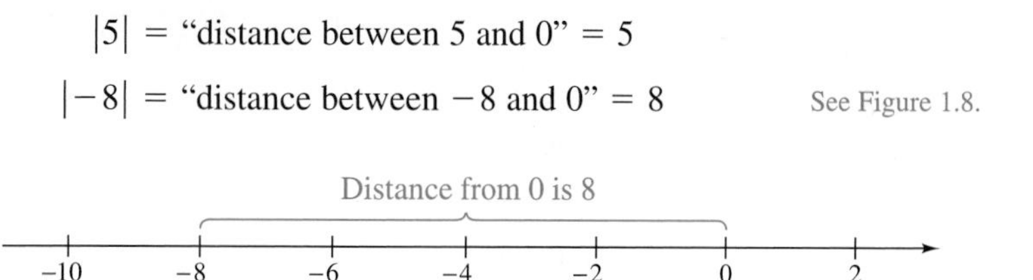

Figure 1.8

Because opposite numbers lie the same distance from 0 on the real number line, they have the same absolute value. Thus,

$|5| = 5$ and $|-5| = 5.$ See Figure 1.9.

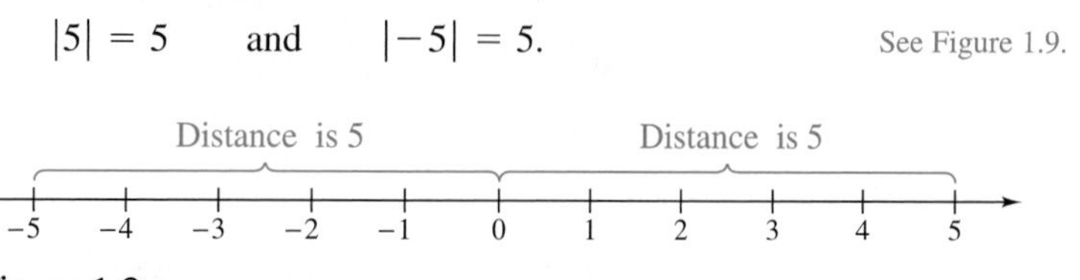

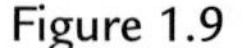

Figure 1.9

You can write this more simply as $|5| = |-5| = 5$.

The absolute value of a real number is either positive or zero (never negative). Moreover, zero is the only real number whose absolute value is 0. That is, $|0| = 0$.

The word **expression** means a collection of numbers and symbols such as $3 + 5$ or $|-4|$. When asked to **evaluate** an expression, you are to find the *number* that is equal to the expression.

Example 5 Evaluating Absolute Value

Evaluate the following expressions.

a. $|-10|$

b. $\left|\frac{3}{4}\right|$

c. $|-3.2|$

d. $-|-6|$

Solution

a. $|-10| = 10$, because the distance between -10 and 0 is 10.

b. $\left|\frac{3}{4}\right| = \frac{3}{4}$, because the distance between $\frac{3}{4}$ and 0 is $\frac{3}{4}$.

c. $|-3.2| = 3.2$, because the distance between -3.2 and 0 is 3.2.

d. $-|-6| = -(6) = -6$.

Note in Example 6(d) that $-|-6| = -6$ does not contradict the fact that the absolute value of a real number cannot be negative.

Example 6 Comparing Absolute Values

Place the correct symbol (<, >, or =) between the two numbers.

a. $|-9| \quad \square \quad |9|$

b. $0 \quad \square \quad |-5|$

c. $-4 \quad \square \quad -|-4|$

d. $|12| \quad \square \quad |-15|$

Solution

a. $|-9| = |9|$, because both are equal to 9.

b. $0 < |-5|$, because $|-5| = 5$ and 0 is less than 5.

c. $-4 = -|-4|$, because both numbers are equal to -4.

d. $|12| < |-15|$, because $|12| = 12$ and $|-15| = 15$, and 12 is less than 15.

Discussing the Concept **Interpreting Inequalities**

Is the statement "$5 \geq 5$" true? Compare your answer with those of the other students in your class. Write an explanation to support your answer.

1.1 Exercises

Developing Skills

In Exercises 1–4, determine which numbers in the set are (a) natural numbers, (b) integers, and (c) rational numbers. Plot the numbers on the real number line. See Examples 1 and 2.

1. $\left\{-3, 2, -\frac{3}{2}, \frac{9}{3}, 4.5\right\}$

2. $\left\{100, -82, -\frac{24}{3}, -8.2\right\}$

3. $\left\{-\frac{5}{2}, 6.5, -4.5, \frac{8}{4}, \frac{3}{4}\right\}$

4. $\left\{8, -1, \frac{4}{3}, -3.25, -\frac{10}{2}\right\}$

In Exercises 5–12, write the real numbers shown by the points on the real number line and place the correct inequality symbol (< or >) between the two numbers. See Examples 3 and 4.

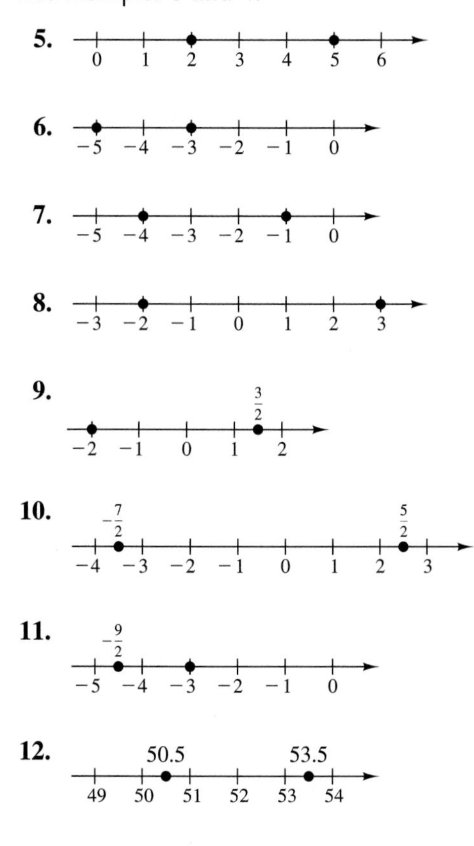

In Exercises 13–24, show each real number as a point on the real number line and place the correct inequality symbol (< or >) between the real numbers. See Examples 3 and 4.

13. $3 \quad \square \quad -4$

14. $6 \quad \square \quad -2$

15. $4 \quad \square \quad -\frac{7}{2}$

16. $2 \quad \square \quad \frac{3}{2}$

17. $0 \quad \square \quad -\frac{7}{16}$

18. $-\frac{7}{3} \quad \square \quad -\frac{7}{2}$

19. $-4.6 \quad \square \quad 1.5$

20. $28.60 \quad \square \quad -3.75$

21. $\frac{7}{16} \quad \square \quad \frac{5}{8}$

22. $-\frac{3}{8} \quad \square \quad -\frac{5}{8}$

23. $-2\pi \quad \square \quad -10$

24. $2 \quad \square \quad \pi$

In Exercises 25–28, on the real number line, what is the distance between *a* and zero?

25. $a = 2$

26. $a = 5$

27. $a = -4$

28. $a = -10$

In Exercises 29–34, find the opposite of the number. Plot the number and its opposite on the real number line. What is the distance of each from 0?

29. 5

30. 2

31. -3.8

32. -7.5

33. $-\frac{5}{2}$

34. $\frac{3}{4}$

In Exercises 35–38, find the absolute value of the real number and its distance from 0.

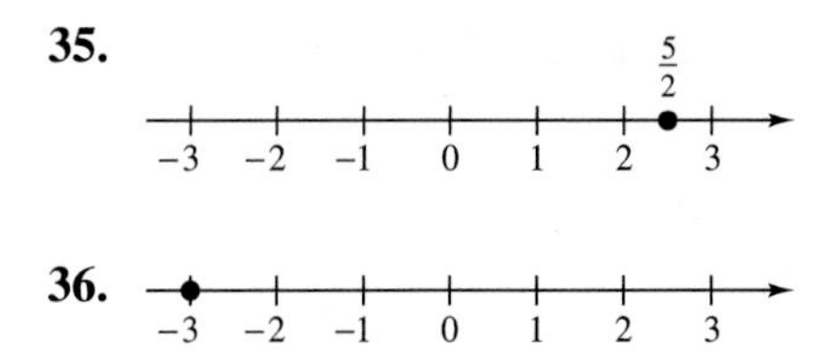

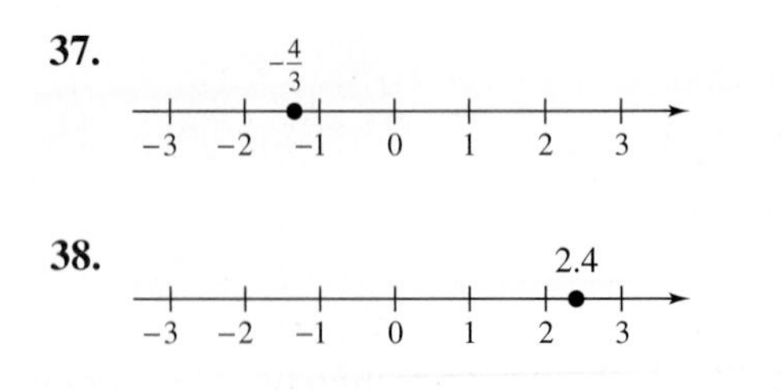

In Exercises 39–50, evaluate the expression. See Example 5.

39. $|7|$

40. $|-6|$

41. $|-3.4|$

42. $|-16.2|$

43. $\left|-\frac{7}{2}\right|$

44. $\left|-\frac{9}{16}\right|$

45. $-|4.09|$

46. $-|-43.8|$

47. $-|-23.6|$

48. $-|91.3|$

49. $|-3.2|$

50. $|0|$

In Exercises 51–62, place the correct symbol (<, >, or =) between the two real numbers. See Example 6.

51. $|-15| \quad \square \quad |15|$

52. $|525| \quad \square \quad |-525|$

53. $|-4| \quad \square \quad |3|$

54. $|16| \quad \square \quad |-25|$

55. $|32| \quad \square \quad |-50|$

56. $|1026| \quad \square \quad |800|$

57. $\left|\frac{3}{16}\right| \quad \square \quad \left|\frac{3}{2}\right|$

58. $\left|-\frac{7}{8}\right| \quad \square \quad \left|\frac{4}{3}\right|$

59. $-|-48.5| \quad \square \quad |-48.5|$

60. $-|-64| \quad \square \quad |-50|$

61. $|-\pi| \quad \square \quad -|-2\pi|$

62. $|-4.9| \quad \square \quad |-10.2|$

In Exercises 63–66, show the numbers on the real number line.

63. $\frac{5}{2}, \pi, -2, -|-3|$

64. $3.7, \frac{16}{3}, |-1.9|, -\frac{1}{2}$

65. $-4, \frac{7}{3}, |-3|, 0$

66. $|-2.3|, 3.2, -2.3, -|3.2|$

In Exercises 67–70, find all real numbers whose distance from a is given by d.

67. $a = 8,\ d = 12.5$

68. $a = 21.3,\ d = 6$

69. $a = -2,\ d = 3.5$

70. $a = 42.5,\ d = 7$

Explaining Concepts

71. Answer parts (a) to (f) of Motivating the Chapter on page 1.

72. Explain why $\frac{8}{4}$ is a natural number, but $\frac{7}{4}$ is not.

73. How many numbers are 3 units from 0 on the real number line? Explain your answer.

74. Which real number lies farther from 0?

(a) -25 (b) 10

Explain your answer.

75. Which real number lies farther from -7?

(a) 3 (b) -10

Explain your answer.

76. Explain how to determine the smaller of two distinct real numbers.

77. Select the smaller real number and explain your answer.

(a) $\frac{3}{8}$ (b) 0.35

In Exercises 78–83, determine if the statement is true or false. Explain your reasoning.

78. The absolute value of any real number is always positive.

79. The absolute value of a number is equal to the absolute value of its opposite.

80. The absolute value of a rational number is a rational number.

81. A given real number corresponds to exactly one point on the real number line.

82. The opposite of a positive number is a negative number.

83. Every rational number is an integer.

The symbol indicates an exercise in which you are asked to answer parts of the Motivating the Chapter problem found on the Chapter Opener pages.

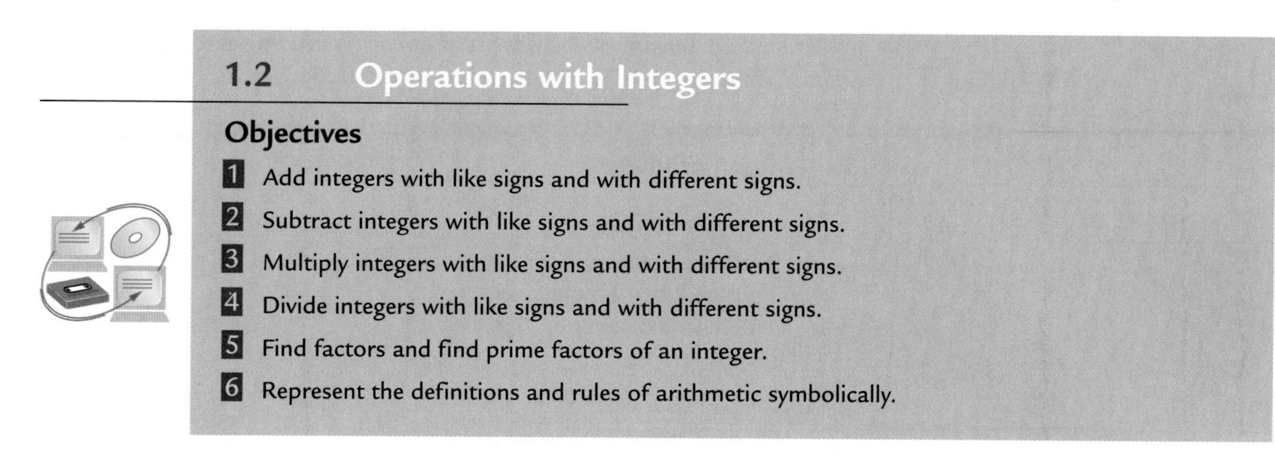

1.2 Operations with Integers

Objectives

1. Add integers with like signs and with different signs.
2. Subtract integers with like signs and with different signs.
3. Multiply integers with like signs and with different signs.
4. Divide integers with like signs and with different signs.
5. Find factors and find prime factors of an integer.
6. Represent the definitions and rules of arithmetic symbolically.

1 Add integers with like signs and with different signs.

Adding Integers

In this section, you will study the four operations of arithmetic (addition, subtraction, multiplication, and division) on the set of integers. There are many examples of these operations in real life. For instance, suppose that your business had a gain of \$550 during one week and a loss of \$600 the next week. Over the 2-week period, your business would have had a combined profit of

$$550 + (-600) = -50,$$

which means you had a loss of \$50.

The number line is a good visual model for demonstrating addition of integers. To add a positive integer, move right, for a negative integer, move left.

Study Tip

As you continue through this chapter, try to capture the overall picture of a *mathematical system,* and note the particular features discussed in each section.

Example 1 Adding Integers Using a Number Line

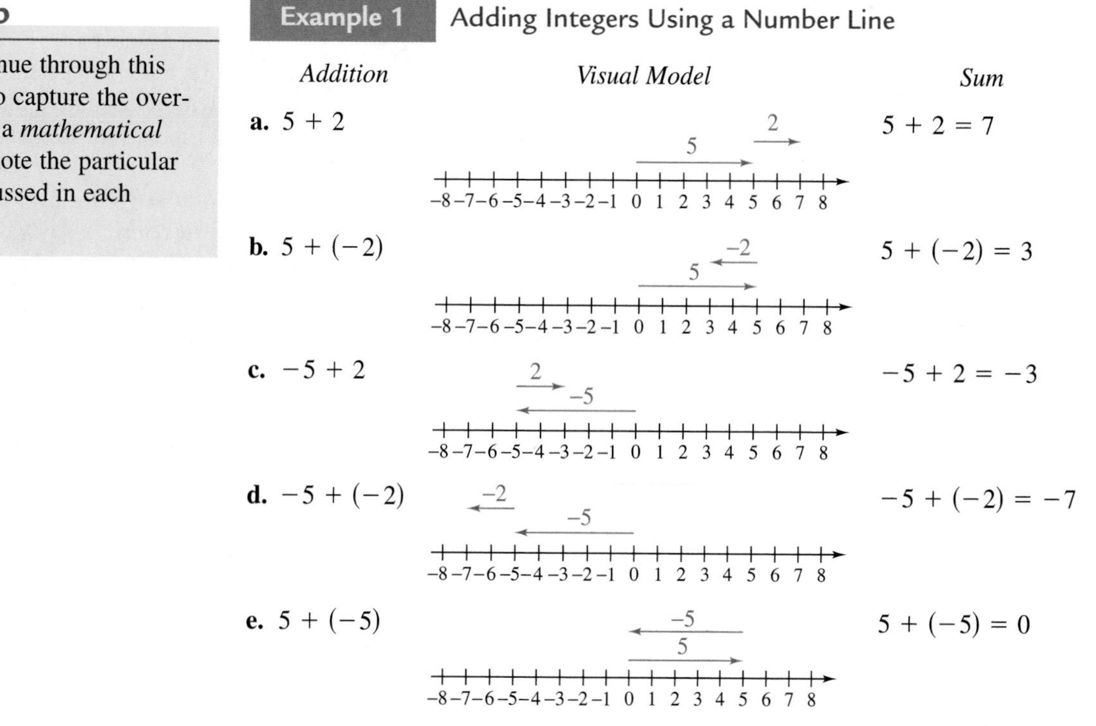

Addition	*Visual Model*	*Sum*
a. $5 + 2$		$5 + 2 = 7$
b. $5 + (-2)$		$5 + (-2) = 3$
c. $-5 + 2$		$-5 + 2 = -3$
d. $-5 + (-2)$		$-5 + (-2) = -7$
e. $5 + (-5)$		$5 + (-5) = 0$

Example 1 illustrates a *graphical approach* to adding integers. It is more common to use an *analytic approach*, as summarized by the following rules.

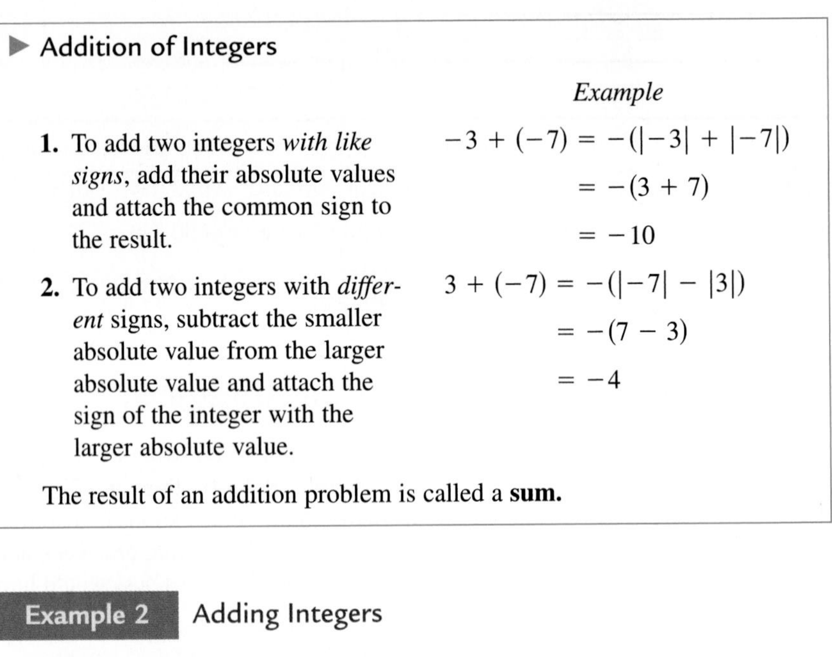

▶ Addition of Integers

	Example
1. To add two integers *with like signs*, add their absolute values and attach the common sign to the result.	$-3 + (-7) = -(\lvert -3 \rvert + \lvert -7 \rvert)$ $= -(3 + 7)$ $= -10$
2. To add two integers with *different* signs, subtract the smaller absolute value from the larger absolute value and attach the sign of the integer with the larger absolute value.	$3 + (-7) = -(\lvert -7 \rvert - \lvert 3 \rvert)$ $= -(7 - 3)$ $= -4$

The result of an addition problem is called a **sum.**

Example 2 Adding Integers

a. Different signs: $22 + (-17) = 22 - 17 = 5$

b. Different signs: $-84 + 14 = -(84 - 14) = -70$

c. Like signs: $-138 + (-62) = -(138 + 62) = -200$

There are different ways to add three or more integers. You can use the **carrying algorithm** with a vertical format with nonnegative integers, as shown in Figure 1.10, or you can add them two at a time, as illustrated in Example 3.

Example 3 Balance in a Checking Account 

Find the balance in the checking account after each of the following transactions.

a. The balance is \$28. A deposit of \$60 is made and a check is written for \$40.

b. The balance is \$120. A check is written for \$132 and a deposit of \$10 is made.

Solution

a. $\$28 + \$60 + (-\$40) = (\$28 + \$60) + (-\$40)$

$= \$88 + (-\$40)$

$= \$48$ Balance

b. $\$120 + (-\$132) + \$10 = -\$12 + \$10$

$= -\$2$ Account is overdrawn.

$$\begin{array}{r} {\scriptstyle 1\,1} \\ 148 \\ 62 \\ +\,536 \\ \hline 746 \end{array}$$

Figure 1.10 *Carrying Algorithm*

2 Subtract integers with like signs and with different signs.

Subtracting Integers

Subtraction can be thought of as "taking away." For instance, $8 - 5$ can be thought of as "8 take away 5," which leaves 3. On the number line, you first move 8 units to the right, then 5 units to the left, as shown in Figure 1.11(a). This same result is accomplished by "adding the opposite of 5 to 8," as shown in Figure 1.11(b).

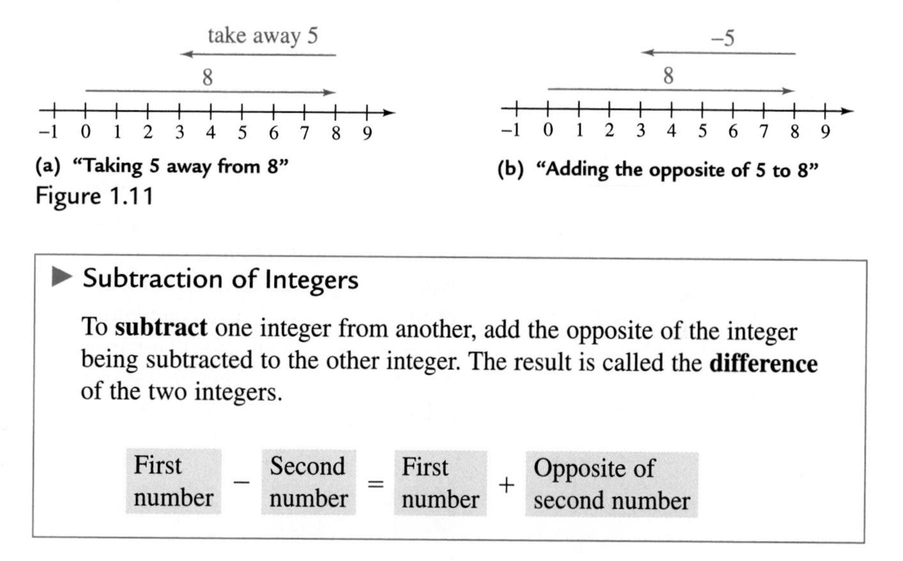

(a) "Taking 5 away from 8" **(b) "Adding the opposite of 5 to 8"**

Figure 1.11

▶ Subtraction of Integers

To **subtract** one integer from another, add the opposite of the integer being subtracted to the other integer. The result is called the **difference** of the two integers.

First number − Second number = First number + Opposite of second number

The **opposite** of an integer is also called its **additive inverse.** For instance, the additive inverse of 5 is -5. The name *additive inverse* comes from the fact that the sum of an integer and its additive inverse is 0. For instance, $5 + (-5) = 0$.

Example 4 Subtracting Integers

a. $3 - 8 = 3 + (-8) = -5$ — Add opposite of 8.

b. $10 - (-13) = 10 + 13 = 23$ — Add opposite of -13.

c. $-5 - 12 = -5 + (-12) = -17$ — Add opposite of 12.

d. $-4 - (-17) - 23 = -4 + 17 + (-23) = -10$ — Add opposite of -17 and opposite of 23.

Be sure you understand that the terminology involving subtraction is not the same as that used for negative numbers. For instance, -5 is read as "negative 5," but $8 - 5$ is read as "8 subtract 5." It is important to distinguish between the operation and the signs of the numbers involved. For instance, in

$$-3 - 5$$

the operation is subtraction and the numbers are -3 and 5.

For subtraction problems involving only two nonnegative integers, you can use the **borrowing algorithm** shown in Figure 1.12.

```
   3 10 15
   4  1  5
 - 2  7  6
 ---------
   1  3  9
```

Figure 1.12 *Borrowing Algorithm*

Example 5 Temperature Change

The temperature at 4 P.M. was 15°. By midnight, the temperature had decreased by 18°. What was the temperature at midnight?

Solution

To find the temperature at midnight, subtract 18 from 15.

$$15 - 18 = 15 + (-18) = -3$$

The temperature at midnight was $-3°$.

This text includes several examples and exercises that use a calculator. As each new calculator application is encountered, you will be given general instructions for using a calculator. These instructions, however, may not agree precisely with the steps required by *your* calculator, so be sure you are familiar with the use of the keys on your own calculator.

For each of the calculator examples in the text, we will give two possible keystroke sequences: one for a standard *scientific* calculator and one for a *graphing* calculator.

Example 6 Evaluating Expressions with a Calculator

Evaluate the following with a calculator.

a. $-4 - 5$ **b.** $2 - (3 - 9)$

	Keystrokes	*Display*	
a.	4 [+/−] [−] 5 [=]	−9	Scientific
	[(−)] 4 [−] 5 [ENTER]	−9	Graphing

	Keystrokes	*Display*	
b.	2 [−] [(] 3 [−] 9 [)] [=]	8	Scientific
	2 [−] [(] 3 [−] 9 [)] [ENTER]	8	Graphing

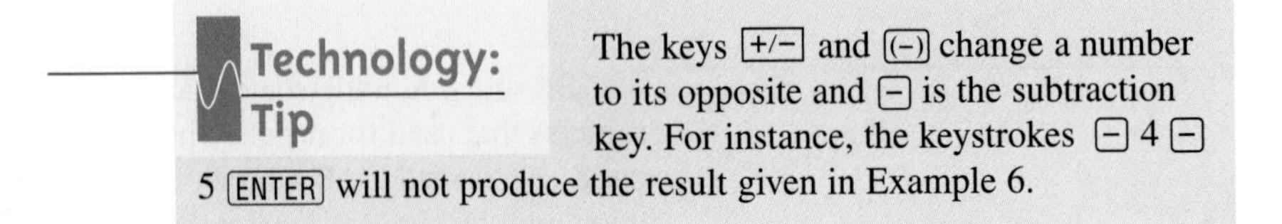

Technology: Tip

The keys [+/−] and [(−)] change a number to its opposite and [−] is the subtraction key. For instance, the keystrokes [−] 4 [−] 5 [ENTER] will not produce the result given in Example 6.

3 Multiply integers with like signs and with different signs.

Multiplying Integers

Multiplication of two integers can be described as repeated addition or subtraction. The result of multiplying one number by another is called a **product.** Here are two examples.

Multiplication	*Repeated Addition*
$3 \times 5 = 15$	$\underbrace{5 + 5 + 5}_{\text{Add 5 three times.}} = 15$
$4 \times (-2) = -8$	$\underbrace{(-2) + (-2) + (-2) + (-2)}_{\text{Add } -2 \text{ four times.}} = -8$

Multiplication is denoted in a variety of ways. For instance,

$$7 \times 3, \quad 7 \cdot 3, \quad 7(3), \quad (7)3, \quad \text{and} \quad (7)(3)$$

all denote the product of "7 times 3," which is 21.

▶ Rules for Multiplying Integers

1. The product of an integer and zero is 0.
2. The product of two integers with *like* signs is *positive.*
3. The product of two integers with *different* signs is *negative.*

To find the product of more than two numbers, first find the product of their absolute values. If there is an *even* number of negative factors, then the product is positive. If there is an *odd* number of negative factors, then the product is negative. For instance,

$5(-3)(-4)(7) = 420.$ Even number of negative factors

Example 7 Multiplying Integers

a. $-6 \cdot 9 = -54$ (Negative) · (positive) = (negative)

b. $(-5)(-7) = 35$ (Negative) · (negative) = (positive)

c. $3(-12) = -36$ (Positive) · (negative) = (negative)

d. $-12 \cdot 0 = 0$ (Negative) · (zero) = (zero)

e. $(-2)(8)(-3)(-1) = -(2 \cdot 8 \cdot 3 \cdot 1)$ Odd number of negative factors

$= -48$ Answer is negative.

Be careful to distinguish properly between expressions such as $3(-5)$ and $3 - 5$ or $-3(-5)$ and $-3 - 5$. The first of each pair is a multiplication problem, whereas the second is a subtraction problem.

Multiplication	*Subtraction*
$3(-5) = -15$	$3 - 5 = -2$
$-3(-5) = 15$	$-3 - 5 = -8$

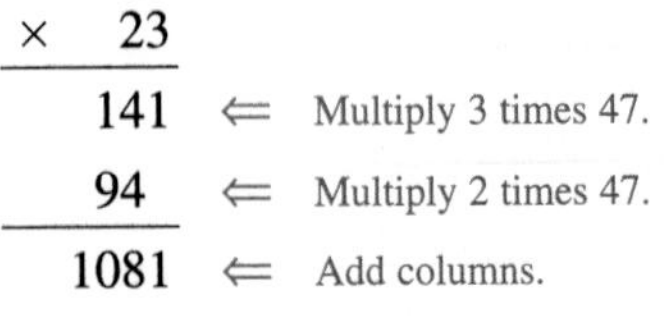

Figure 1.13 *Vertical Multiplication Algorithm*

To multiply two integers having two or more digits, we suggest the **vertical multiplication algorithm** demonstrated in Figure 1.13. The sign of the product is determined by the usual multiplication rule.

Example 8 Geometry: Volume of a Box

Find the volume of the rectangular box shown in Figure 1.14.

Study Tip

Formulas for geometric figures can be found on the inside front cover of this text.

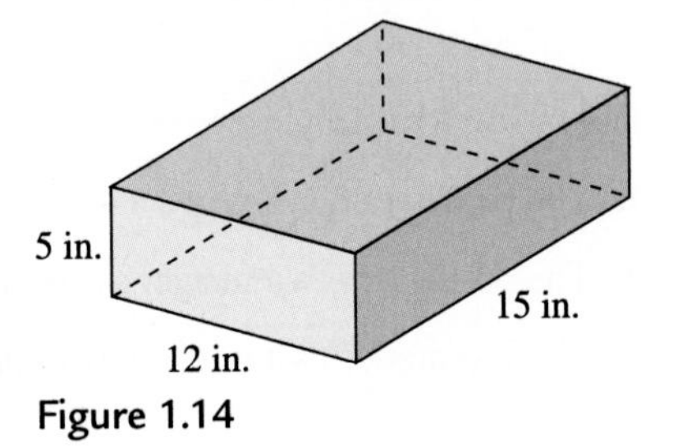

Figure 1.14

Solution

To find the volume, multiply the length, width, and height of the box.

$$\begin{aligned}\text{Volume} &= (\text{Length}) \cdot (\text{Width}) \cdot (\text{Height}) \\ &= (15 \text{ in.}) \cdot (12 \text{ in.}) \cdot (5 \text{ in.}) \\ &= 900 \text{ cubic inches}\end{aligned}$$

So, the box has a volume of 900 cubic inches.

4 Divide integers with like signs and with different signs.

Dividing Integers

Just as subtraction can be expressed in terms of addition, you can express division in terms of multiplication. Here are some examples.

Division		*Related Multiplication*
$12 \div 4 = 3$	because	$12 = 3 \cdot 4$
$15 \div 3 = 5$	because	$15 = 5 \cdot 3$
$15 \div (-3) = -5$	because	$15 = (-5) \cdot (-3)$
$-15 \div (-3) = 5$	because	$-15 = 5 \cdot (-3)$

The result of dividing one integer by another is called the **quotient** of the integers. Division is denoted by the symbol ÷, or by /, or by a horizontal line. For example,

$$30 \div 6, \quad 30/6, \quad \text{and} \quad \frac{30}{6}$$

all denote the quotient of 30 and 6, which is 5. Using the form $30 \div 6$, 30 is called the **dividend** and 6 is the **divisor.** In the forms 30/6 and $\frac{30}{6}$, 30 is the **numerator** and 6 is the **denominator.**

It is important to know how to use 0 in a division problem. Zero divided by a nonzero integer is always 0. For instance,

$$\frac{0}{13} = 0 \quad \text{because} \quad 0 = 0 \cdot 13.$$

On the other hand, division by zero is *undefined.*

Because division can be described in terms of multiplication, the rules for dividing two integers with like or unlike signs are the same as those for multiplying such integers.

Technology: Discovery

Does $\frac{1}{0} = 0$? Does $\frac{2}{0} = 0$? Write each division above in terms of multiplication. What does this tell you about division by zero? What does your calculator display when you perform the division?

Rules for Dividing Integers

1. Zero divided by a nonzero integer is 0, whereas a nonzero integer divided by zero is *undefined.*
2. The quotient of two nonzero integers with *like* signs is *positive.*
3. The quotient of two nonzero integers with *different* signs is *negative.*

Example 9 Dividing Integers

a. $\frac{-42}{-6} = 7$ because $-42 = 7(-6)$.

b. $36 \div (-9) = -4$ because $(-4)(-9) = 36$.

c. $\frac{0}{-13} = 0$ because $(0)(-13) = 0$.

d. $-105 \div 7 = -15$ because $(-15)(7) = -105$.

When dividing large numbers, the **long division algorithm** can be used. For instance, the long division algorithm shown in Figure 1.15 shows that

$$\frac{351}{13} = 27.$$

```
    27
13 ) 351
     26
      91
      91
```

Figure 1.15 *Long Division Algorithm*

Remember that division can be checked by multiplying the answer by the divisor. So it is true that

$$\frac{351}{13} = 27 \quad \text{because} \quad 27(13) = 351.$$

All four operations on integers (addition, subtraction, multiplication, and division) are used in the following real-life example.

Example 10 Stock Purchase

On Monday you bought \$500 worth of stock in a company. During the rest of the week, you recorded the following gains and losses in your stock's value.

Tuesday	Wednesday	Thursday	Friday
Gained \$15	Lost \$18	Lost \$23	Gained \$10

a. What was the value of the stock at the close of Tuesday?

b. What was the value of the stock at the close of Wednesday?

c. What was the value of the stock at the end of the week?

d. What would the total loss have been if Thursday's loss had occurred each of the four days?

e. What was the average daily gain (or loss) for the four days recorded?

Solution

a. Because the original value of the stock was \$500, and the stock gained \$15 by the close of Tuesday, its value at the close of Tuesday was

$$500 + 15 = \$515.$$

b. Using the result of part (a), the value at the close of Wednesday was

$$515 - 18 = \$497.$$

c. The value of the stock at the end of the week was

$$500 + 15 - 18 - 23 + 10 = \$484.$$

d. The loss on Thursday was \$23. If this loss had occurred each day, the total loss would have been

$$4(23) = \$92.$$

e. To find the average of the four gains and losses, we add and divide by 4. Thus, the average is

$$\text{Average} = \frac{15 + (-18) + (-23) + 10}{4} = \frac{-16}{4} = -4.$$

This means that during the four days, the stock had an average loss of \$4 per day.

Study Tip

To find the **average** of n numbers, add the numbers and divide the result by n.

5 Find factors and find prime factors of an integer.

Factors and Prime Numbers

The set of positive integers

$$\{1, 2, 3, \ldots\}$$

is one subset of the real numbers that has intrigued mathematicians for many centuries.

Historically, an important number concept has been *factors* of positive integers. From experience, you know that in a multiplication problem such as $3 \cdot 7 = 21$, the numbers 3 and 7 are called *factors* of 21.

$$\underbrace{3 \cdot 7}_{\text{Factors}} = \underbrace{21}_{\text{Product}}$$

It is also correct to call the numbers 3 and 7 *divisors* of 21, because 3 and 7 each divide evenly into 21.

Corbis-Bettmann

Pythagoras
(580–500 B.C.)

In the 5th century B.C., the followers of the Greek mathematician Pythagoras believed that numbers revealed the basic structure of the universe. They devoted their lives to the study, discovery, and proof of number patterns. Their works laid the foundation for the field of mathematics now called *number theory*.

▶ Factor (or Divisor)

If a and b are positive integers, then a is a **factor** (or **divisor**) of b if and only if there is a positive integer c such that $a \cdot c = b$.

The concept of factors allows you to classify positive integers into three groups: *prime* numbers, *composite* numbers, and the number 1.

▶ Prime and Composite Numbers

1. A positive integer greater than 1 with no factors other than itself and 1 is called a **prime number,** or simply a **prime.**
2. A positive integer greater than 1 with more than two factors is called a **composite number,** or simply a **composite.**

The numbers 2, 3, 5, 7, and 11 are primes because they have only themselves and 1 as factors. The numbers 4, 6, 8, 9, and 10 are composites because each has more than two factors. The number 1 is neither prime nor composite because 1 is its only factor.

Every composite number can be expressed as a *unique* product of prime factors. Here are some examples.

$$6 = 2 \cdot 3, \quad 15 = 3 \cdot 5, \quad 18 = 2 \cdot 3 \cdot 3, \quad 42 = 2 \cdot 3 \cdot 7, \quad 124 = 2 \cdot 2 \cdot 31$$

According to the definition of a prime number, is it possible for any negative number to be prime? Consider the number -2. Is it prime? Are its only factors one and itself? No, because

$$-2 = 1(-2),$$

$$-2 = (-1)(2),$$

$$\text{or} \quad -2 = (-1)(1)(2).$$

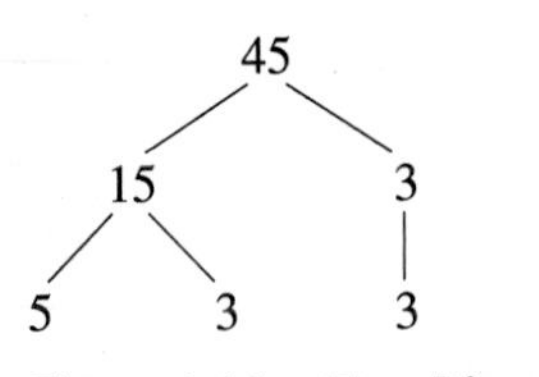

Figure 1.16 *Tree Diagram*

One strategy for factoring a composite number into prime numbers is to begin by finding the smallest prime number that is a factor of the composite number. Dividing this factor into the number yields a *companion* factor. For instance, 3 is the smallest prime number that is a factor of 45 and its companion factor is $15 = 45 \div 3$. Because 15 is also a composite number, continue hunting for factors and companion factors until each factor is prime. As shown in Figure 1.16, a *tree diagram* is a nice way to record your work. From the tree diagram, you can see that the prime factorization of 45 is $45 = 3 \cdot 3 \cdot 5$.

Example 11 Prime Factorization

Find the prime factorization of each of the following.

a. 84 **b.** 78 **c.** 133 **d.** 43

Solution

a. 4 is a recognized divisor of 84. So, $84 = 4 \cdot 21 = 2 \cdot 2 \cdot 3 \cdot 7$.

b. 2 is a recognized divisor of 78. So, $78 = 2 \cdot 39 = 2 \cdot 3 \cdot 13$.

c. If you do not recognize a divisor of 133, you can get started by dividing any of the prime numbers 2, 3, 5, 7, 11, 13, etc., into 133. You will find 7 to be the first prime to divide 133. So, $133 = 7 \cdot 19$ (19 is prime).

d. In this case, none of the primes less than 43 divides 43. So, 43 is prime.

Other aids to finding prime factors of a number n include the following divisibility tests.

Divisibility Tests

	Example
1. A number is divisible by 2 if it is *even.*	364 is divisible by 2 because it is even.
2. A number is divisible by 3 if the sum of its digits is divisible by 3.	261 is divisible by 3 because $2 + 6 + 1 = 9$.
3. A number is divisible by 9 if the sum of its digits is divisible by 9.	738 is divisible by 9 because $7 + 3 + 8 = 18$.
4. A number is divisible by 5 if its units digit is 0 or 5.	325 is divisible by 5 because its units digit is 5.
5. A number is divisible by 10 if its units digit is 0.	120 is divisible by 10 because its units digit is 0.

When a number is **divisible** by 2, it means that 2 divides into the number without leaving a remainder.

6 Represent the definitions and rules of arithmetic symbolically.

Summary of Definitions and Rules

So far in this chapter, we have described rules and procedures more with words than with symbols. For instance, subtraction is verbally defined as "adding the opposite of the number being subtracted." As you move to higher and higher levels of mathematics, it becomes more and more convenient to use symbols to describe rules and procedures. For instance, subtraction is symbolically defined as $a - b = a + (-b)$.

At its simplest level, algebra is a symbolic form of arithmetic. This arithmetic–algebra connection can be illustrated in the following way.

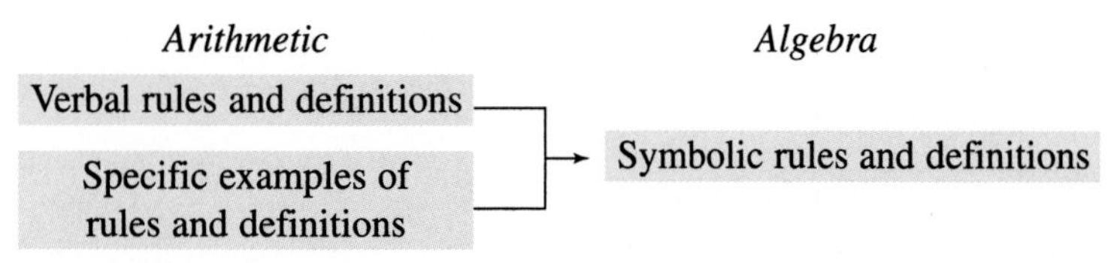

An illustration of this connection is shown in Example 12.

Example 12 Writing a Rule of Arithmetic in Symbolic Form

Write an example and an algebraic description of the arithmetic rule: *The product of two integers with unlike signs is negative.*

Solution

Example

For the integers -3 and 7,

$$\begin{aligned}(-3) \cdot 7 &= 3 \cdot (-7)\\ &= -(3 \cdot 7)\\ &= -21.\end{aligned}$$

Algebraic Description

If a and b are positive integers, then

$$\underbrace{(-a) \cdot b}_{\text{Unlike signs}} = \underbrace{a \cdot (-b)}_{\text{Unlike signs}} = \underbrace{-(a \cdot b)}_{\text{Negative product}}.$$

The following list summarizes the algebraic versions of important definitions and rules of arithmetic. In each case a specific example is included for clarification.

▶ Arithmetic Summary

Definitions: Let a, b, and c be integers.

	Definition	*Example*
1.	Subtraction: $a - b = a + (-b)$	$5 - 7 = 5 + (-7)$
2.	Multiplication: (a is a positive integer) $a \cdot b = \underbrace{b + b + \cdots + b}_{a \text{ terms}}$	$3 \cdot 5 = 5 + 5 + 5$
3.	Division: $(b \neq 0)$ $a \div b = c$, if and only if $a = c \cdot b$.	$12 \div 4 = 3$ because $12 = 3 \cdot 4$
4.	Less than: $a < b$ if there is a positive real number c such that $a + c = b$.	$-2 < 1$ because $-2 + 3 = 1$
5.	Absolute value: $\lvert a \rvert = \begin{cases} a, & \text{if } a \geq 0 \\ -a, & \text{if } a < 0 \end{cases}$	$\lvert -3 \rvert = -(-3) = 3$
6.	Divisor: a is a divisor of b if and only if there is an integer c such that $a \cdot c = b$.	7 is a divisor of 21 because $7 \cdot 3 = 21$

Rules: Let a and b be integers.

	Rule	*Example*
1.	Addition:	
	(a) If a and b have *like* signs, evaluate $\lvert a \rvert + \lvert b \rvert$ and attach the common sign to the result.	$3 + 7 = \lvert 3 \rvert + \lvert 7 \rvert = 10$
	(b) If a and b have *different* signs, evaluate which difference, $\lvert a \rvert - \lvert b \rvert$ or $\lvert b \rvert - \lvert a \rvert$, is positive and attach the sign of the integer with the larger absolute value.	$-5 + 8 = \lvert 8 \rvert - \lvert -5 \rvert = 8 - 5 = 3$
2.	Multiplication:	
	(a) $a \cdot 0 = 0 = 0 \cdot a$	$3 \cdot 0 = 0 = 0 \cdot 3$
	(b) Like signs: $a \cdot b > 0$	$(-2)(-5) = 10$
	(c) Different signs: $a \cdot b < 0$	$(2)(-5) = -10$
3.	Division:	
	(a) $\frac{0}{a} = 0$	$\frac{0}{4} = 0$
	(b) $\frac{a}{0}$ is undefined.	$\frac{6}{0}$ is undefined.
	(c) Like signs: $\frac{a}{b} > 0$	$\frac{-2}{-3} = \frac{2}{3}$
	(d) Different signs: $\frac{a}{b} < 0$	$\frac{-5}{7} = -\frac{5}{7}$

Example 13 Using Definitions and Rules

a. Use the definition of subtraction to complete the following.

$4 - 9 = \square$

b. Use the definition of multiplication to complete the following.

$6 + 6 + 6 + 6 = \square$

c. Use the definition of absolute value to complete the following.

$|-9| = \square$

d. Use the rule for adding integers with unlike signs to complete the following.

$-7 + 3 = \square$

e. Use the rule for multiplying integers with unlike signs to complete the following.

$-9 \times 2 = \square$

Solution

a. $4 - 9 = 4 + (-9) = -5$

b. $6 + 6 + 6 + 6 = 4 \cdot 6 = 24$

c. $|-9| = -(-9) = 9$

d. $-7 + 3 = -(|-7| - |3|) = -4$

e. $-9 \times 2 = -18$

Discussing the Concept **Finding a Pattern**

Complete the patterns below. Decide which rules the patterns demonstrate. Use a calculator to confirm your answers.

$3 \cdot (3) = 9$	$-3 \cdot (3) = -9$
$3 \cdot (2) = 6$	$-3 \cdot (2) = -6$
$3 \cdot (1) = 3$	$-3 \cdot (1) = -3$
$3 \cdot (0) = 0$	$-3 \cdot (0) = 0$
$3 \cdot (-1) = \square$	$-3 \cdot (-1) = \square$
$3 \cdot (-2) = \square$	$-3 \cdot (-2) = \square$
$3 \cdot (-3) = \square$	$-3 \cdot (-3) = \square$

1.2 Exercises

Developing Skills

In Exercises 1–4, find the required sum and demonstrate the addition on the real number line. See Example 1.

1. $2 + 7$
2. $10 + (-3)$
3. $-6 + 4$
4. $(-8) + (-3)$

In Exercises 5–32, find the sum. See Example 2.

5. $-1 + 0$
6. $-3 + 0$
7. $14 + (-14)$
8. $-45 + 45$
9. $(-14) + 13$
10. $(-20) + 19$
11. $-23 + 4$
12. $10 + (-10)$
13. $-18 + (-12)$
14. $-34 + (-16)$
15. $-32 + 16$
16. $-75 + 100$
17. $5 + |-3|$
18. $49 + (-|-17|)$
19. $-|-12| + |-16|$
20. $|-10| + |35|$
21. $-10 + 6 + 34$
22. $-15 + (-3) + 8$
23. $-82 + (-36) + 82$
24. $15 + (-75) + (-75)$
25. $32 + (-32) + (-16)$
26. $-312 + (-564) + (-100)$
27. $1200 + 1300 + (-275)$
28. $104 + 203 + 613 + (-214)$
29. $1875 + (-3143) + 5826$
30. $4365 + (-2145) + (-1873) + 40{,}084$
31. $|-890| + (-|-82|) + 90$
32. $-770 + |492| + (-|-383|)$

In Exercises 33–54, find the difference. See Example 4.

33. $12 - 9$
34. $4 - (-1)$
35. $-4 - (-4)$
36. $9 - (-6)$
37. $55 - 20$
38. $39 - 13$
39. $45 - 35$
40. $27 - 57$
41. $-71 - 32$
42. $-84 - 106$
43. $-10 - (-4)$
44. $2500 - (-600)$
45. $-210 - 400$
46. $-110 - (-30)$
47. $-942 - (-942)$
48. $-12 - (-7)$
49. $|15| - |-7|$
50. $|-100| - |25|$
51. $23 - |15|$
52. $-125 - |165|$
53. $-32 - (-18)$
54. $|515 - 160 - 480|$

55. Find the sum of 250 and -300.
56. Find the sum of -40 and -60.
57. Subtract -120 from 380.
58. Find the absolute value of the sum of -35 and 15.
59. What number must be added to 10 to obtain -5?
60. What number must be subtracted from -12 to obtain 24?

In Exercises 61–64, write each multiplication as repeated addition and find the product.

61. $3 \cdot 2$
62. 4×5
63. $5 \times (-3)$
64. $6(-2)$

In Exercises 65–80, find the product. See Example 7.

65. 7×3
66. $0 \cdot 2$
67. $4(-8)$
68. $10(-5)$
69. $(-6)(-12)$
70. $(-20)(-8)$
71. $(310)(-3)$
72. $(-500)(-6)$
73. $5(-3)(-6)$
74. $-7(3)(-1)$
75. $(-2)(-3)(-5)$
76. $(-10)(-4)(-2)$
77. $|3(-5)(6)|$
78. $|6(20)(4)|$
79. $|(-3)4|$
80. $|8(-9)|$

In Exercises 81–86, use the vertical multiplication algorithm to find the product.

81. 26×13

82. $(-14) \times 24$

83. $75(-63)$

84. $(-13)(-20)$

85. $(-72)(866)$

86. $(-14)(-585)$

In Exercises 87–100, perform the division, if possible. If not possible, state the reason. See Example 9.

87. $27 \div 9$

88. $-35 \div (-5)$

89. $72 \div (-12)$

90. $(-28) \div 4$

91. $\frac{8}{0}$

92. $\frac{0}{8}$

93. $\frac{-81}{-3}$

94. $\frac{-125}{-25}$

95. $\frac{6}{-1}$

96. $\frac{-12}{1}$

97. $\frac{0}{81}$

98. $\frac{32}{0}$

99. $-180 \div (-45)$

100. $(-27) \div (-27)$

In Exercises 101–106, use the long division algorithm to find the quotient.

101. $1440 \div 45$

102. $-1312 \div (-16)$

103. $1440 \div (-45)$

104. $-1312 \div 16$

105. $2750 \div 25$

106. $22{,}010 \div 71$

In Exercises 107–112, use a calculator to perform the specified operation(s).

107. $5(1650) - 3710$

108. $516 - (-125)$

109. $\frac{44{,}290}{515}$

110. $\frac{33{,}511}{47}$

111. $\frac{169{,}290}{162}$

112. $\frac{1{,}027{,}500}{250}$

In Exercises 113 and 114, find the product mentally. Explain your strategy.

113. $(-2)(532)(500)$

114. $72(8)(25)$

In Exercises 115–124, is the number prime or composite?

115. 240

116. 257

117. 643

118. 533

119. 3911

120. 1281

121. 8324

122. 3555

123. 1321

124. 1323

In Exercises 125–134, write the prime factorization. See Example 11.

125. 12

126. 52

127. 210

128. 561

129. 192

130. 245

131. 525

132. 264

133. 2535

134. 1521

Solving Problems

135. *Temperature Change* The temperature at 6 A.M. was -10°F. By noon, the temperature had increased by 22°F. What was the temperature at noon?

136. *Balance in an Account* At the beginning of a month, your balance was \$2750. During the month you withdrew \$350 and \$500, deposited \$450, and earned interest of \$6.42. What was your balance at the end of the month?

137. *Profit* Your company lost \$650,000 during the first 6 months of the year. By the end of the year, you had an overall profit of \$362,000. What was your profit during the second 6 months of the year?

138. *Flying Altitude* An airliner flying at an altitude of 31,000 feet is instructed to descend to an altitude of 24,000 feet (see figure). How many feet must the aircraft descend?

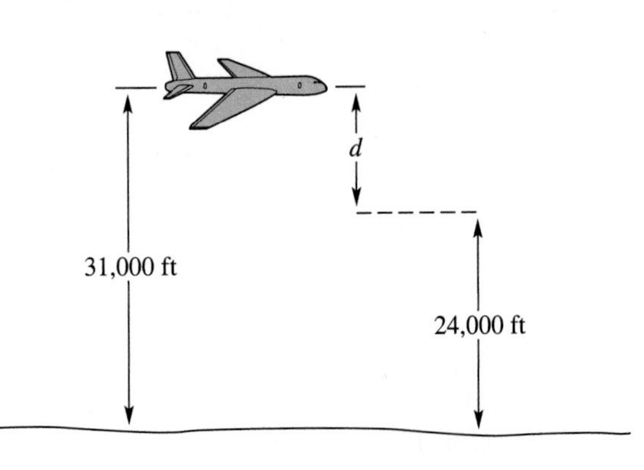

139. *Endowments* The bar graph gives the market values of Penn State's endowment and similar funds on June 30 in the years 1986 to 1997. (Source: Penn State Intercom, October 9, 1997)

(a) Estimate the increase in the market value from 1990 to 1995.

(b) How much greater was the increase from 1996 to 1997 than the increase from 1995 to 1996?

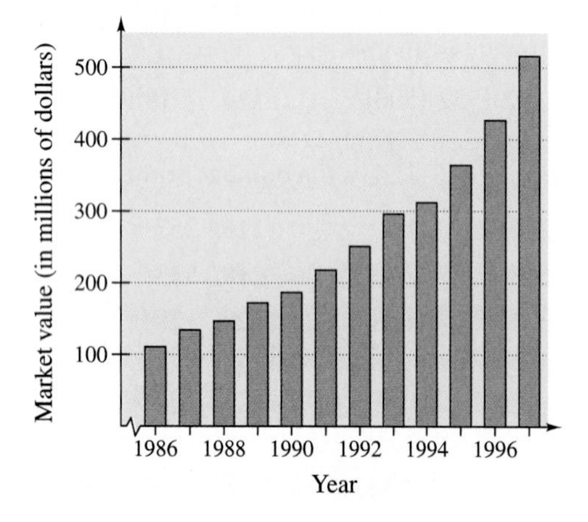

140. *Charitable Giving* The bar graph gives the sources of charitable giving for the year 1996. (Source: *USA Today*)

(a) How much greater is the giving by individuals than the giving by foundations?

(b) The giving by individuals is approximately (to the nearest integer) how many times greater than the giving by corporations?

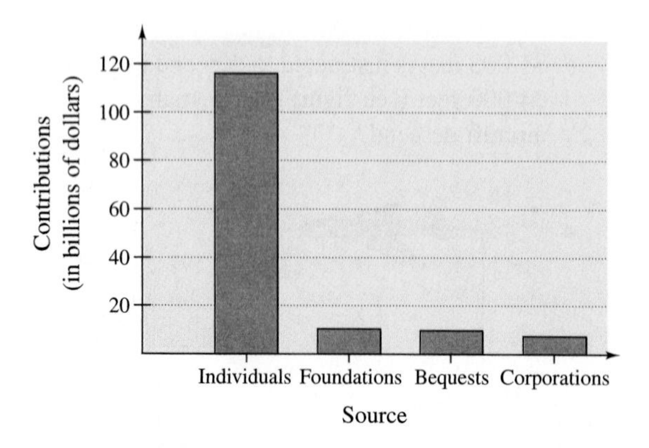

141. *Temperature Change* The temperature measured by a weather balloon is decreasing approximately 3° for each 1000-foot increase in altitude. The balloon rises 8000 feet. Describe its total temperature change.

142. *Stock Prices* The Dow Jones average loses 11 points on each of 4 consecutive days. What is the cumulative loss during the 4 days?

143. *Savings Plan* After you save $50 per month for 10 years, what is the total amount you have saved?

144. *Loss Leaders* To attract customers, a grocery store runs a sale on bananas. The bananas are *loss leaders*, which means the store loses money on the bananas but hopes to make it up on other items. The store sells 800 pounds at a loss of 26 cents per pound. What is the total loss?

145. *Geometry* Find the area of the football field.

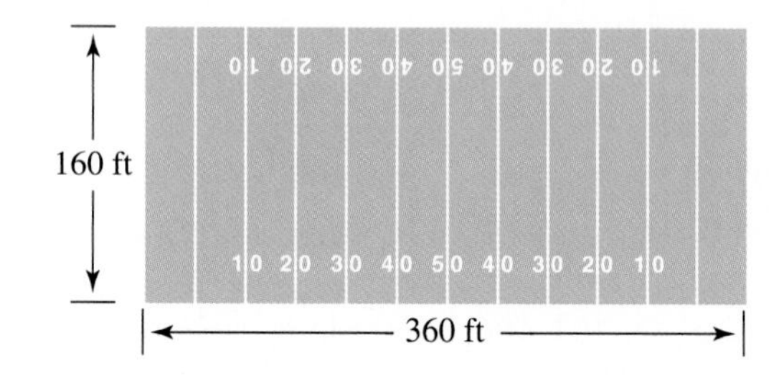

146. *Geometry* Find the area of the garden.

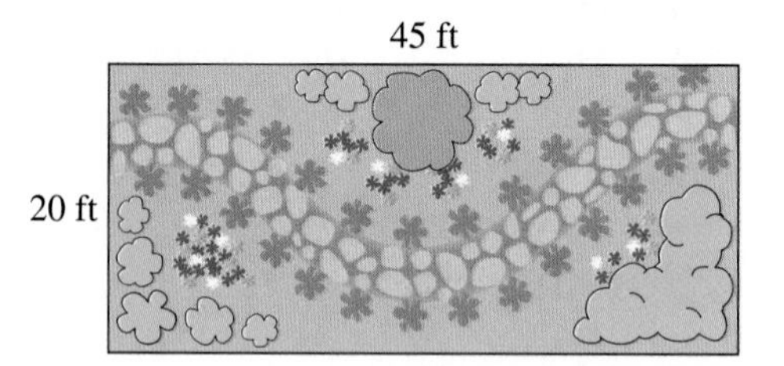

147. *Average Speed* A commuter train travels a distance of 195 miles between two cities in 3 hours. What is the average speed of the train in miles per hour?

148. *Exam Scores* A student has a total of 328 points after four 100-point exams.

(a) What is the average number of points scored per exam?

(b) The scores on the four exams are 87, 73, 77, and 91. Plot each of the scores and the average score on the real number line.

(c) Find the difference between each score and the average score. Find the sum of these distances and give a possible explanation of the result.

Geometry In Exercises 149 and 150, find the volume of the rectangular solid. The volume is found by multiplying the length, width, and height of the solid. See Example 8.

149.

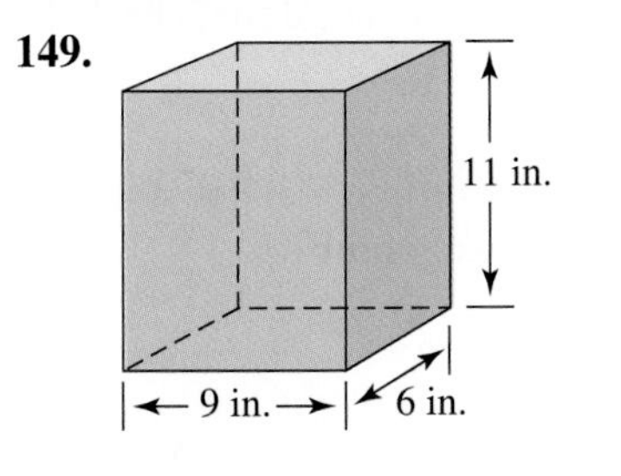

150.

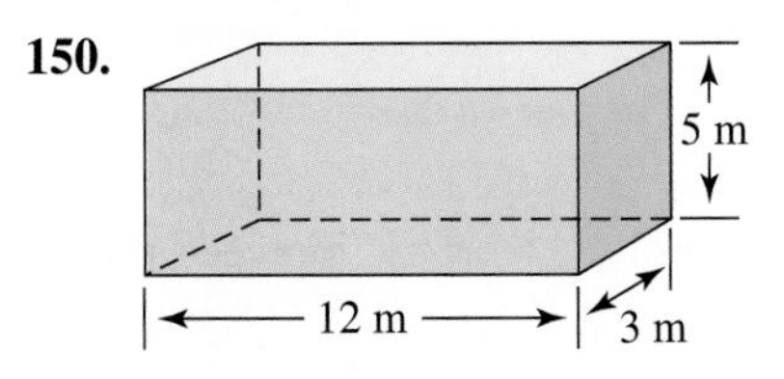

In Exercises 151 and 152, an addition problem is shown visually on the real number line. (a) Write the addition problem and find the sum. (b) State the rule for the addition of integers demonstrated. (c) Suppose the numbers represent the yards gained in two consecutive downs of a football game. How would the sportscasters announce the plays?

151.

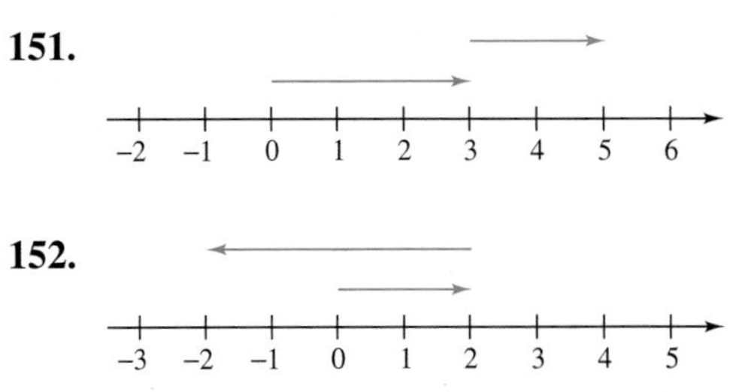

152.

−3 −2 −1 0 1 2 3 4 5

Explaining Concepts

153. *Writing* What is the only even prime number? Explain why there are no other even prime numbers.

154. *Investigation* Twin primes are prime numbers that differ by 2. For instance, 3 and 5 are twin primes. How many other twin primes are less than 100?

155. *Think About It* The number 1997 is not divisible by a prime number that is less than 45. Explain why this implies that 1997 is a prime number.

156. *The Sieve of Eratosthenes* Write the integers from 1 through 100 in 10 lines of 10 numbers each.

(a) Cross out the number 1. Cross out all multiples of 2 other than 2 itself. Do the same for 3, 5, and 7.

(b) Of what type are the remaining numbers? Explain why this is the only type of number left.

157. Explain why the sum of two negative numbers is a negative number.

158. Write the rule for adding two numbers of opposite sign. How do you determine the sign of the sum?

159. If a negative number is used as a factor 25 times, what is the sign of the product?

160. If a negative number is used as a factor 16 times, what is the sign of the product?

161. Write a verbal description of what is meant by $3(-5)$.

162. Write the rules for determining the sign of the product or quotient of real numbers.

163. Explain why the product of an even integer and any other integer is even. What can you conclude about the product of two odd integers?

164. Explain how to check the result of a division problem.

165. An integer n is divided by 2 and the quotient is an even integer. What does this tell you about n? Give an example.

166. Which of the following is (are) undefined: $\frac{1}{1}, \frac{0}{1}, \frac{1}{0}, \frac{0}{0}$?

167. The **proper factors** of a number are all its factors less than the number itself. A number is **perfect** if the sum of its proper factors is equal to the number. A number is **abundant** if the sum of its proper factors is greater than the number. Which numbers less than 25 are perfect? Which are abundant? Compare your answers with those in your group and resolve any differences. Try to find the first perfect number greater than 25.

Mid-Chapter Quiz

Take this test as you would take a test in class. After you are done, check your work against the answers given in the back of the book.

In Exercises 1–4, show each real number as a point on the real line and place the correct inequality symbol (< or >) between the real numbers.

1. $-2.5 \quad \square \quad -4$

2. $\frac{3}{16} \quad \square \quad \frac{3}{8}$

3. $-3.1 \quad \square \quad 2.7$

4. $2\pi \quad \square \quad 6$

In Exercises 5 and 6, evaluate the expression.

5. $-|-0.75|$

6. $|25.2|$

In Exercises 7 and 8, place the correct symbol (<, > or =) between the real numbers.

7. $\left|\frac{7}{2}\right| \quad \square \quad |-3.5|$

8. $\left|\frac{3}{4}\right| \quad \square \quad -|0.75|$

In Exercises 9 and 10, copy the number line, write the opposites of a and b, and plot the opposites on the number line.

9.

10.

In Exercises 11–16, evaluate the expression.

11. $-15 - 12$

12. $-15 - (-12)$

13. $25 + |-75|$

14. $-6(10)$

15. $\dfrac{-45}{-3}$

16. $\dfrac{-24}{6}$

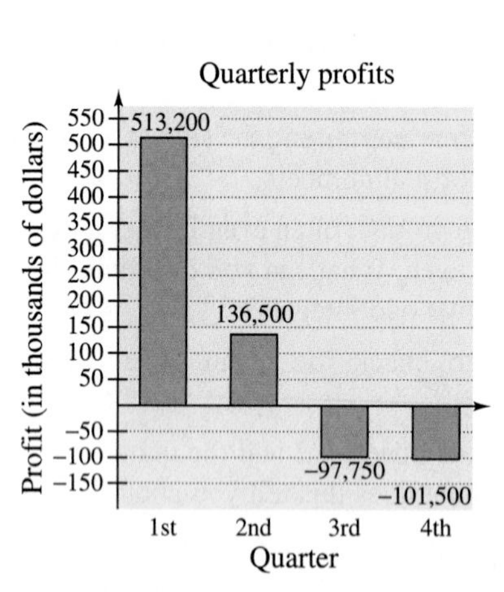

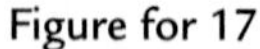
Figure for 17

17. A company's quarterly profits are shown in the bar graph at the left. What is the company's total profit for the year?

18. A cord of wood is a pile 8 feet long, 4 feet wide, and 4 feet high. The volume of a rectangular solid is its length times its width times its height. Find the number of cubic feet in a cord of wood.

19. It is necessary to cut a 90-foot rope into six pieces of equal length. What is the length of each piece?

20. Consider the statement, "The sum of two negative integers is positive." Is the statement true or false? If it is false, suggest any change that would make it true.

1.3 Operations with Rational Numbers

Objectives

1. Rewrite fractions as equivalent fractions.
2. Add and subtract fractions.
3. Multiply and divide fractions.
4. Add, subtract, multiply, and divide decimals.

1 Rewrite fractions as equivalent fractions.

Rewriting Fractions

A **fraction** is a number that is written as a quotient, with a *numerator* and a *denominator*. The terms *fraction* and *rational number* are related, but are not exactly the same. The term *fraction* refers to a number's form, whereas the term *rational number* refers to its classification. For instance, the number 2 is a fraction when it is written as $\frac{2}{1}$, but it is a rational number regardless of how it is written.

Rules of Signs for Fractions

1. If the numerator and denominator of a fraction have *like* signs, the value of the fraction is *positive*.
2. If the numerator and denominator of a fraction have *unlike* signs, the value of the fraction is *negative*.

Study Tip

To find the **greatest common factor** of two natural numbers, list their factors, identify the factors that are common, and then identify the greatest of the common factors. For instance, the greatest common factor of 16 and 24 is 8.

Example 1 Positive and Negative Fractions

a. All of the following fractions are positive and are equivalent to $\frac{2}{3}$.

$$\frac{2}{3}, \frac{-2}{-3}, -\frac{-2}{3}, -\frac{2}{-3}$$

b. All of the following fractions are negative and are equivalent to $-\frac{2}{3}$.

$$-\frac{2}{3}, \frac{-2}{3}, \frac{2}{-3}, -\frac{-2}{-3}$$

In both arithmetic and algebra, it is often beneficial to write a fraction in **simplest form** or **reduced form,** which means that the numerator and denominator have no common factors (other than 1). By finding the prime factors of the numerator and the denominator, you can determine what common factor(s) to divide out.

Writing a Fraction in Simplest Form

To write a fraction in simplest form, divide both the numerator and denominator by their greatest common factor (GCF).

You can obtain an **equivalent fraction** by multiplying the numerator and denominator by the same nonzero number or by dividing the numerator and denominator by the same nonzero number. Here are some examples.

Fraction	*Equivalent Fraction*	*Operation*
$\frac{9}{12} = \frac{\cancel{3}^{1} \cdot 3}{\cancel{3}_{1} \cdot 4}$	$\frac{3}{4}$	Divide numerator and denominator by 3. (See Figure 1.17.)
$\frac{6}{5} = \frac{6 \cdot 2}{5 \cdot 2}$	$\frac{12}{10}$	Multiply numerator and denominator by 2.
$\frac{-8}{12} = -\frac{\cancel{2}^{1} \cdot \cancel{2}^{1} \cdot 2}{\cancel{2}_{1} \cdot \cancel{2}_{1} \cdot 3}$	$-\frac{2}{3}$	Divide numerator and denominator by GCF of 4.

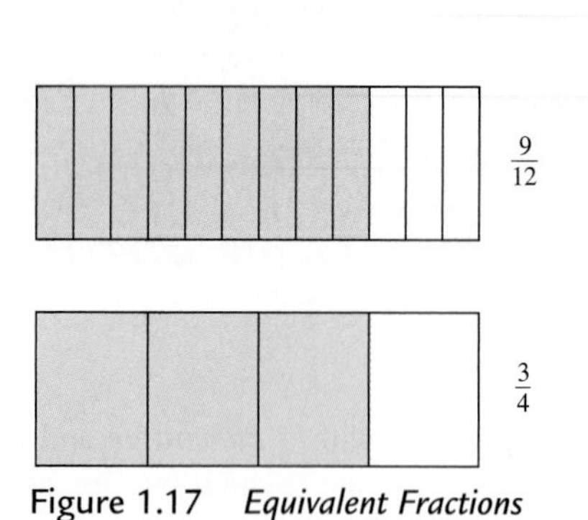

Figure 1.17 *Equivalent Fractions*

Example 2 Writing Fractions in Simplest Form

Write each fraction in simplest form.

a. $\frac{18}{24}$ **b.** $\frac{35}{21}$ **c.** $\frac{24}{72}$

Solution

a. $\frac{18}{24} = \frac{\cancel{2}^{1} \cdot \cancel{3}^{1} \cdot 3}{2 \cdot 2 \cdot \cancel{2}_{1} \cdot \cancel{3}_{1}} = \frac{3}{4}$ Divide out GCF of 6.

b. $\frac{35}{21} = \frac{5 \cdot \cancel{7}^{1}}{3 \cdot \cancel{7}_{1}} = \frac{5}{3}$ Divide out GCF of 7.

c. $\frac{24}{72} = \frac{\cancel{2}^{1} \cdot \cancel{2}^{1} \cdot \cancel{2}^{1} \cdot \cancel{3}^{1}}{\cancel{2}_{1} \cdot \cancel{2}_{1} \cdot \cancel{2}_{1} \cdot \cancel{3}_{1} \cdot 3} = \frac{1}{3}$ Divide out GCF of 24.

Example 3 Writing Equivalent Fractions

Write an equivalent fraction with the indicated denominator.

a. $\frac{2}{3} = \frac{\square}{15}$ **b.** $\frac{4}{7} = \frac{\square}{42}$ **c.** $\frac{9}{15} = \frac{\square}{35}$

Solution

a. $\frac{2}{3} = \frac{2 \cdot 5}{3 \cdot 5} = \frac{10}{15}$ Multiply numerator and denominator by 5.

b. $\frac{4}{7} = \frac{4 \cdot 6}{7 \cdot 6} = \frac{24}{42}$ Multiply numerator and denominator by 6.

c. $\frac{9}{15} = \frac{\cancel{3} \cdot 3}{\cancel{3} \cdot 5} = \frac{3 \cdot 7}{5 \cdot 7} = \frac{21}{35}$ Reduce first, then multiply by $\frac{7}{7}$.

2 Add and subtract fractions.

Adding and Subtracting Fractions

To add fractions with *like* denominators such as $\frac{3}{12}$ and $\frac{4}{12}$, add the numerators and write the sum over the like denominator.

$$\frac{3}{12}+\frac{4}{12}=\frac{3+4}{12}=\frac{7}{12}$$ Add the numbers in the numerator.

To add fractions with *unlike* denominators such as $\frac{1}{4}$ and $\frac{1}{3}$, rewrite the fractions as equivalent fractions with a common denominator.

$$\frac{1}{4}+\frac{1}{3}=\frac{1\cdot 3}{4\cdot 3}+\frac{1\cdot 4}{3\cdot 4}$$ Rewrite fractions in equivalent form.

$$=\frac{3}{12}+\frac{4}{12}$$ Rewrite with like denominators.

$$=\frac{7}{12}$$ Add numerators.

To find a common denominator for two or more fractions, find the **least common multiple** (LCM) of their denominators. For instance, the least common multiple of 8 and 12 is 24. To see this, consider all multiples of 8 (8, 16, 24, 32, 40, 48, . . .) and all multiples of 12 (12, 24, 36, 48, . . .). The numbers 24 and 48 are common multiples and the number 24 is the smallest of the common multiples.

$$\frac{3}{8}+\frac{-5}{12}=\frac{3(3)}{8(3)}+\frac{(-5)(2)}{12(2)}$$ LCM of 8 and 12 is 24.

$$=\frac{9}{24}+\frac{-10}{24}$$ Rewrite with like denominators.

$$=\frac{9-10}{24}$$ Add numerators.

$$=\frac{-1}{24}$$ Simplify.

$$=-\frac{1}{24}$$

Study Tip

Adding fractions with unlike denominators is an example of a basic problem-solving strategy that is used in mathematics—rewriting a given problem in a simpler or more familiar form.

▶ Addition and Subtraction of Fractions

1. To add two fractions *with like denominators,* add their numerators and write the sum over the like (or common) denominator.
2. To add two fractions *with unlike denominators,* rewrite both fractions so that they have like denominators. Then use the rule for adding fractions with like denominators.
3. To subtract two fractions, add the opposite fraction and proceed as in addition.

Example 4 Adding and Subtracting Fractions

a. $1\frac{4}{5} + \frac{11}{15}$ **b.** $\frac{7}{9} - \frac{11}{12}$

Solution

a. To begin, rewrite the **mixed number** $1\frac{4}{5}$ as a fraction.

$$1\frac{4}{5} = 1 + \frac{4}{5} = \frac{5}{5} + \frac{4}{5} = \frac{9}{5}$$

Then add the two fractions as follows.

$$1\frac{4}{5} + \frac{11}{15} = \frac{9}{5} + \frac{11}{15} \qquad \text{Rewrite } 1\tfrac{4}{5} \text{ as } \tfrac{9}{5}.$$

$$= \frac{9(3)}{5(3)} + \frac{11}{15} \qquad \text{LCM of 5 and 15 is 15.}$$

$$= \frac{27}{15} + \frac{11}{15} \qquad \text{Rewrite with like denominators.}$$

$$= \frac{38}{15} \qquad \text{Add numerators.}$$

b. $$\frac{7}{9} - \frac{11}{12} = \frac{7(4)}{9(4)} + \frac{-11(3)}{12(3)} \qquad \text{LCM of 9 and 12 is 36.}$$

$$= \frac{28}{36} + \frac{-33}{36} \qquad \text{Rewrite with like denominators.}$$

$$= \frac{28 + (-33)}{36} \qquad \text{Add numerators.}$$

$$= \frac{-5}{36} \qquad \text{Simplify.}$$

$$= -\frac{5}{36}$$

Study Tip

In Example 4(a), a common shortcut for writing $1\frac{4}{5}$ as $\frac{9}{5}$ is to multiply 1 by 5, add the result to 4, and then divide by 5, as follows.

$$1\frac{4}{5} = \frac{1(5) + 4}{5} = \frac{9}{5}$$

You can add or subtract *two* fractions, without first finding a common denominator, by using the following rule.

▶ Alternative Rule for Adding or Subtracting Two Fractions

If a, b, c, and d are integers with $b \neq 0$ and $d \neq 0$, then

$$\frac{a}{b} + \frac{c}{d} = \frac{ad + bc}{bd} \quad \text{or} \quad \frac{a}{b} - \frac{c}{d} = \frac{ad - bc}{bd}.$$

On page 31, the sum of $\frac{3}{8}$ and $\frac{-5}{12}$ was found using the least common multiple of 8 and 12. Compare those solution steps with the following steps, which use the alternative rule for adding or subtracting two fractions.

$$\frac{3}{8} + \frac{-5}{12} = \frac{3(12) + 8(-5)}{8(12)}$$ Apply alternative rule.

$$= \frac{36 - 40}{96}$$ Simplify.

$$= \frac{-4}{96}$$ Simplify.

$$= -\frac{1}{24}$$ Write in simplest form.

Technology: Tip

When you use a scientific or graphing calculator to add or subtract fractions, your answer may appear in decimal form. An answer such as 0.583333333 is not as exact as $\frac{7}{12}$ and may introduce roundoff error. Refer to the user's manual for your calculator for instructions on adding and subtracting fractions and displaying answers in fraction form.

Example 5 Subtracting Fractions

$$\frac{5}{16} - \left(-\frac{7}{30}\right) = \frac{5}{16} + \frac{7}{30}$$ Add the opposite.

$$= \frac{5(30) + 16(7)}{16(30)} = \frac{150 + 112}{480}$$ Apply alternative rule.

$$= \frac{262}{480}$$ Simplify.

$$= \frac{131}{240}$$ Write in simplest form.

Example 6 Combining Three or More Fractions

Evaluate the following.

$$\frac{5}{6} - \frac{7}{15} + \frac{3}{10} - 1$$

Solution

The least common denominator of 6, 15, and 10 is 30. So, you can rewrite the given expression as follows.

$$\frac{5}{6} - \frac{7}{15} + \frac{3}{10} - 1 = \frac{5(5)}{6(5)} + \frac{(-7)(2)}{15(2)} + \frac{3(3)}{10(3)} + \frac{(-1)(30)}{30}$$

$$= \frac{25}{30} + \frac{-14}{30} + \frac{9}{30} + \frac{-30}{30}$$ Rewrite with like denominators.

$$= \frac{25 - 14 + 9 - 30}{30}$$ Add numerators.

$$= \frac{-10}{30} = -\frac{1}{3}$$ Simplify.

3 Multiply and divide fractions.

Multiplying and Dividing Fractions

The procedure for multiplying fractions is simpler than those for adding and subtracting fractions. Regardless of whether the fractions have like or unlike denominators, you can find the product of two fractions by multiplying the numerators and multiplying the denominators.

> ### ▶ Multiplication of Fractions
> To multiply two fractions, multiply the two numerators to form the numerator of the product, and multiply the two denominators to form the denominator of the product.

Example 7 Multiplying Fractions

a. $\frac{5}{8} \cdot \frac{3}{2} = \frac{5(3)}{8(2)}$ Multiply numerators and denominators.

$= \frac{15}{16}$ Simplify.

b. $\left(-\frac{7}{9}\right)\left(-\frac{5}{21}\right) = \frac{7}{9} \cdot \frac{5}{21}$ Product of two negatives is positive.

$= \frac{7(5)}{9(21)}$ Multiply numerators and denominators.

$= \frac{\cancel{7}(5)}{9(3)(\cancel{7})}$ Factor and simplify fraction.

$= \frac{5}{27}$ Write in simplest form.

c. $\left(3\frac{1}{5}\right)\left(-\frac{7}{6}\right)\left(\frac{5}{3}\right) = \left(\frac{16}{5}\right)\left(-\frac{7}{6}\right)\left(\frac{5}{3}\right)$ Rewrite mixed number as a fraction.

$= -\frac{(8)(\cancel{2})(7)(\cancel{5})}{(\cancel{5})(3)(\cancel{2})(3)}$ Factor and simplify fraction.

$= -\frac{56}{9}$ Write in simplest form.

Technology: Tip

Try verifying some of the products shown in Example 7 with your calculator. Using a *TI-83*, you can verify that $\left(3\frac{1}{5}\right)\left(-\frac{7}{6}\right)\left(\frac{5}{3}\right) = -\frac{56}{9}$ as follows.

(3 + 1 ÷ 5) × ((–) 7 ÷ 6) ×
(5 ÷ 3) MATH ▶FRAC ENTER

The **reciprocal** or **multiplicative inverse** of a number is the number by which it must be multiplied to obtain 1. For instance, the reciprocal of 3 is $\frac{1}{3}$ because $3(\frac{1}{3}) = 1$. Similarly, the reciprocal of $-\frac{2}{3}$ is $-\frac{3}{2}$ because

$$\left(-\frac{2}{3}\right)\left(-\frac{3}{2}\right) = 1.$$

To divide two fractions, multiply the first fraction by the reciprocal of the second fraction. Another way of saying this is "invert the divisor and multiply."

▶ Division of Fractions

The quotient of $\frac{a}{b}$ and $\frac{c}{d}$ is

$$\frac{a}{b} \div \frac{c}{d} = \frac{a}{b} \cdot \frac{d}{c}.$$

Example 8 Dividing Fractions

Perform the following divisions and write the answers in simplest form.

a. $\frac{5}{8} \div \frac{20}{12}$ **b.** $\frac{6}{13} \div \left(-\frac{9}{26}\right)$

Solution

a. $\frac{5}{8} \div \frac{20}{12} = \frac{5}{8} \cdot \frac{12}{20}$ Invert divisor and multiply.

$= \frac{(5)(12)}{(8)(20)}$ Multiply numerators and denominators.

$= \frac{\cancel{(5)}(3)\cancel{(4)}}{(8)\cancel{(4)}\cancel{(5)}}$ Factor and simplify fraction.

$= \frac{3}{8}$ Write in simplest form.

b. $\frac{6}{13} \div \left(-\frac{9}{26}\right) = \frac{6}{13} \cdot \left(-\frac{26}{9}\right)$ Invert divisor and multiply.

$= -\frac{(6)(26)}{(13)(9)}$ Multiply numerators and denominators.

$= -\frac{(2)\cancel{(3)}(2)\cancel{(13)}}{\cancel{(13)}\cancel{(3)}(3)}$ Factor and simplify fraction.

$= -\frac{4}{3}$ Write in simplest form.

Study Tip

The Division of Fractions Rule works for **complex fractions** (compare to Example 8a).

$$\frac{5}{8} \div \frac{20}{12} = \frac{5/8}{20/12}$$

Invert bottom fraction and multiply. $= \frac{5}{8} \cdot \frac{12}{20}$

4 Add, subtract, multiply, and divide decimals.

Operations with Decimals

Rational numbers can be represented as **terminating** or **repeating decimals.** Here are some examples.

Terminating Decimals	*Repeating Decimals*
$\frac{1}{4} = 0.25$	$\frac{1}{6} = 0.1666\ldots$ or $0.1\overline{6}$
$\frac{3}{8} = 0.375$	$\frac{1}{3} = 0.3333\ldots$ or $0.\overline{3}$
$\frac{2}{10} = 0.2$	$\frac{1}{12} = 0.0833\ldots$ or $0.08\overline{3}$
$\frac{5}{16} = 0.3125$	$\frac{8}{33} = 0.2424\ldots$ or $0.\overline{24}$

Note that the *bar* notation is used to indicate the *repeated* digit (or digits) in the decimal notation. You can obtain the decimal representation of any fraction by long division. For instance, the decimal representation of $\frac{5}{12}$ is $0.41\overline{6}$, as can be seen from the following long division algorithm.

$$
\begin{array}{r}
0.4166\ldots = 0.41\overline{6} \\
12\,\overline{)\,5.000} \\
\underline{4\,8} \\
20 \\
\underline{12} \\
80 \\
\underline{72} \\
80
\end{array}
$$

Technology: Tip

You can use a calculator to round decimals. For instance, to round 0.9375 to two decimal places on a scientific calculator, enter

[FIX] [2] .9375 [=].

On a *TI-83* graphing calculator, enter

round (.9375, 2) [ENTER].

Without using a calculator, round −0.88247 to three decimal places. Verify your answer with a calculator. Name the rounding and decision digits.

For calculations involving decimals such as 0.41666 . . . , you must **round the decimal.** For instance, rounded to two decimal places, the number 0.41666 . . . is 0.42. Similarly, rounded to three decimal places, the number 0.41666 . . . is 0.417.

▶ Rounding a Decimal

1. Determine the number of digits of accuracy you wish to keep. The digit in the last position you keep is called the **rounding digit,** and the digit in the first position you discard is called the **decision digit.**
2. If the decision digit is 5 or greater, round up by adding 1 to the rounding digit.
3. If the decision digit is 4 or less, round down by leaving the rounding digit unchanged.

Given decimal	*Rounded to three places*
0.9763	0.976
0.9768	0.977
0.9765	0.977

Example 9 Operations with Decimals

a. Add 0.583, 1.06, and 2.9104.

b. Multiply -3.57 and 0.032.

Solution

a. To add decimals, align the decimal points and proceed as in integer addition.

$$\begin{array}{r} ^{1\,1} \\ 0.583 \\ 1.06 \\ +\ 2.9104 \\ \hline 4.5534 \end{array}$$

b. To multiply decimals, use integer multiplication and then place the decimal point (in the product) so that the number of decimal places equals the sum of the decimal places in the two factors.

$$\begin{array}{rl} -3.57 & \text{Two decimal places} \\ \times \quad 0.032 & \text{Three decimal places} \\ \hline 714 & \\ 1071 & \\ \hline -0.11424 & \text{Five decimal places} \end{array}$$

Example 10 Dividing Decimal Fractions

Divide 1.483 by 0.56.

Solution

To divide 1.483 by 0.56, convert the divisor to an integer by moving its decimal point to the right. Move the decimal point in the dividend an equal number of places to the right. Place the decimal point in the quotient directly above the new decimal point in the dividend and then divide as with integers.

$$\begin{array}{r} 2.648 \\ 56\,\overline{)\,148.3} \\ \underline{112} \\ 36\;3 \\ \underline{33\;6} \\ 2\;70 \\ \underline{2\;24} \\ 460 \\ \underline{448} \end{array}$$

Rounded to two decimal places, the answer is 2.65. This answer can be written as

$$\frac{1.483}{0.56} \approx 2.65$$

where the symbol $\approx$ means **is approximately equal to.**

The following example is similar to the stock investment example on page 18. The difference is that this time the gains and losses are given in fractional form.

Example 11 Stock Purchase

On Monday you bought 50 shares of stock at $\$48\frac{1}{2}$ per share. During the week the stock rose and fell, as shown in the table.

Tuesday	Wednesday	Thursday	Friday
Up $\frac{3}{8}$	Up $1\frac{3}{4}$	Down $\frac{1}{2}$	Up $2\frac{7}{8}$

a. What was the value of the stock at the close on Tuesday?

b. What was the value of the stock at the close on Wednesday?

c. What was the value of the stock at the end of the week?

d. What would the total gain have been if Friday's gain had occurred each of the 4 days?

e. What was the average daily gain (loss) for the 4 days recorded?

Alan Schein/The Stock Market

There are three major stock exchanges in the United States: the New York Stock Exchange (NYSE), the American Stock Exchange (AMEX), and the National Association of Securities Dealers Automated Quotations (NASDAQ).

Solution

a. Because the original value of the stock was $50\left(48\frac{1}{2}\right) = \$2425$, and each of the 50 shares gained $\$\frac{3}{8}$ by the close of Tuesday, the total value of your stock at the close of Tuesday was

$$2425 + 50\left(\frac{3}{8}\right) = 2425 + 18.75 = \$2443.75.$$

b. Using the result of part(a), the value at the close of Wednesday was

$$2443.75 + 50\left(1\frac{3}{4}\right) = 2443.75 + 87.5 = \$2531.25.$$

c. The value of the stock at the end of the week was

$$2425 + 50\left(\frac{3}{8}\right) + 50\left(1\frac{3}{4}\right) + 50\left(-\frac{1}{2}\right) + 50\left(2\frac{7}{8}\right) = \$2650.00.$$

d. The gain on Friday was $2\frac{7}{8}$ per share. If this gain had occurred each day, the total gain would have been

$$4(50)\left(2\frac{7}{8}\right) = \$575.00.$$

e. The total gain for the 4 days was $2650 - 2425 = \$225$. Thus, the average daily gain for the 4 days was $\frac{225}{4} = \$56.25$.

▶ Summary of Rules for Fractions

Let a, b, c, and d be real numbers.

Rule	*Example*
1. Addition of fractions:	
$\frac{a}{b} + \frac{c}{d} = \frac{ad + bc}{bd}, \quad b \neq 0, \quad d \neq 0$	$\frac{1}{3} + \frac{2}{7} = \frac{1 \cdot 7 + 3 \cdot 2}{3 \cdot 7} = \frac{13}{21}$
2. Subtraction of fractions:	
$\frac{a}{b} - \frac{c}{d} = \frac{ad - bc}{bd}, \quad b \neq 0, \quad d \neq 0$	$\frac{1}{3} - \frac{2}{7} = \frac{1 \cdot 7 - 3 \cdot 2}{3 \cdot 7} = \frac{1}{21}$
3. Multiplication of fractions:	
$\frac{a}{b} \cdot \frac{c}{d} = \frac{a \cdot c}{b \cdot d}, \quad b \neq 0, \quad d \neq 0$	$\frac{1}{3} \cdot \frac{2}{7} = \frac{1(2)}{3(7)} = \frac{2}{21}$
4. Division of fractions:	
$\frac{a}{b} \div \frac{c}{d} = \frac{a}{b} \cdot \frac{d}{c}, \quad b \neq 0, \quad d \neq 0, \quad c \neq 0$	$\frac{1}{3} \div \frac{2}{7} = \frac{1}{3} \cdot \frac{7}{2} = \frac{7}{6}$
5. Rule of signs for fractions:	
$\frac{a}{b} = \frac{-a}{-b}$	$\frac{12}{4} = \frac{-12}{-4}$
$\frac{-a}{b} = \frac{a}{-b} = -\frac{a}{b}$	$\frac{-12}{4} = \frac{12}{-4} = -\frac{12}{4}$
6. Equivalent fractions:	
$\frac{a}{b} = \frac{c}{d}$, if and only if $ad = bc$; $\quad b \neq 0, \quad d \neq 0$	$\frac{1}{4} = \frac{3}{12}$ because $1 \cdot 12 = 4 \cdot 3$

Discussing the Concept — To Round or Not to Round?

When using a calculator to perform operations with decimals, you should try to get in the habit of rounding your answers *only* after all the calculations are done. If you round the answer at a preliminary stage, you can introduce unnecessary roundoff error. Suppose $l = 5.24$, $w = 3.03$, and $h = 2.749$ are the dimensions of a box. Find the volume, $l \cdot w \cdot h$, by multiplying the given numbers and then rounding the answer to one decimal place. Now use a second method, first rounding each dimension to one decimal place and then multiplying the numbers. Compare your answers, and explain which of these techniques produces the more accurate answer.

1.3 Exercises

Developing Skills

In Exercises 1–10, find the greatest common factor.

1. 20, 45
2. 45, 90
3. 28, 52
4. 48, 64
5. 18, 84, 90
6. 84, 98, 192
7. 240, 300, 360
8. 117, 195, 507
9. 134, 225, 315, 945
10. 80, 144, 214, 504

In Exercises 11–18, write the fraction in simplest form. See Example 2.

11. $\frac{2}{8}$
12. $\frac{21}{28}$
13. $\frac{12}{18}$
14. $\frac{16}{56}$
15. $\frac{60}{192}$
16. $\frac{45}{225}$
17. $\frac{28}{350}$
18. $\frac{88}{154}$

In Exercises 19–22, each figure is divided into regions of equal area. Find the sum of the two fractions indicated by the shaded regions of the same color.

19.

20.

21.

22.

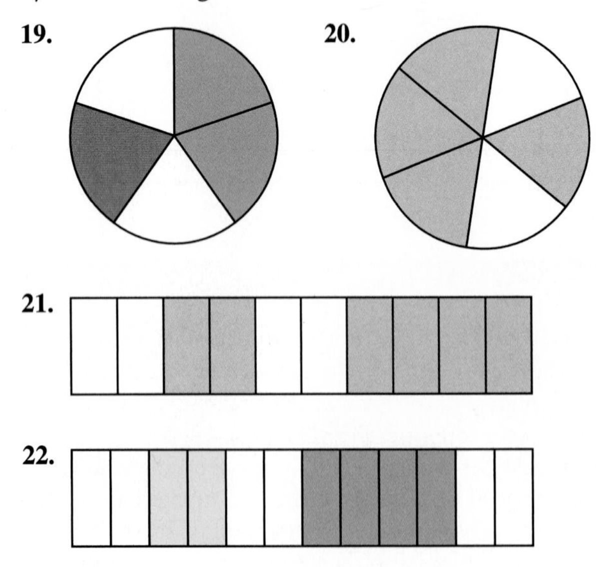

In Exercises 23–34, add or subtract. Write the result in simplest form. See Example 4.

23. $\frac{7}{15} + \frac{2}{15}$
24. $\frac{13}{35} + \frac{5}{35}$
25. $\frac{9}{11} + \frac{5}{11}$
26. $\frac{5}{6} + \frac{13}{6}$
27. $\frac{9}{16} - \frac{3}{16}$
28. $\frac{15}{32} - \frac{7}{32}$
29. $-\frac{23}{11} + \frac{12}{11}$
30. $\frac{46}{13} - \frac{20}{13}$
31. $\frac{3}{4} - \frac{5}{4}$
32. $\frac{3}{8} - \frac{5}{8}$
33. $\frac{13}{15} + \left|-\frac{11}{15}\right| - \frac{4}{15}$
34. $\frac{5}{8} - \left(-\frac{13}{8}\right) + \frac{3}{8}$

In Exercises 35–38, write an equivalent fraction with the indicated denominator. See Example 3.

35. $\frac{3}{8} = \frac{\square}{16}$
36. $\frac{4}{5} = \frac{\square}{15}$
37. $\frac{6}{15} = \frac{\square}{25}$
38. $\frac{21}{49} = \frac{\square}{28}$

In Exercises 39–58, add or subtract. Write the result in simplest form. See Examples 4, 5, and 6.

39. $\frac{1}{2} + \frac{1}{3}$
40. $\frac{3}{5} + \frac{1}{2}$
41. $\frac{1}{4} - \frac{1}{3}$
42. $\frac{2}{3} - \frac{1}{6}$
43. $\frac{3}{16} + \frac{3}{8}$
44. $\frac{2}{3} + \frac{4}{9}$
45. $-\frac{1}{8} - \frac{1}{6}$
46. $\frac{13}{8} - \frac{3}{4}$
47. $4 - \frac{8}{3}$
48. $\frac{17}{25} + 2$
49. $-\frac{7}{8} - \frac{5}{6}$
50. $-\frac{5}{12} - \frac{1}{9}$
51. $-\frac{5}{6} - \left(-\frac{3}{4}\right)$
52. $\frac{3}{4} - \frac{2}{5}$
53. $\frac{5}{12} - \frac{3}{8} + \frac{5}{16}$
54. $-\frac{3}{7} + \frac{5}{14} + \frac{3}{4}$
55. $2 - \frac{25}{6} + \frac{3}{4}$
56. $3 + \frac{12}{3} + \frac{1}{9}$
57. $1 + \frac{2}{3} - \frac{5}{6}$
58. $2 - \frac{15}{16} - \frac{7}{8}$

In Exercises 59–66, write the mixed number as a fraction. See Example 4.

59. $4\frac{3}{5}$ **60.** $7\frac{2}{3}$

61. $3\frac{7}{10}$ **62.** $-1\frac{3}{4}$

63. $8\frac{2}{3}$ **64.** $2\frac{5}{8}$

65. $-10\frac{5}{11}$ **66.** $3\frac{1}{100}$

In Exercises 67–74, add or subtract. Write the result in simplest form. See Example 4(a).

67. $3\frac{1}{2} + 5\frac{2}{3}$ **68.** $5\frac{3}{4} + 8\frac{1}{10}$

69. $1\frac{3}{16} - 2\frac{1}{4}$ **70.** $5\frac{7}{8} - 2\frac{1}{2}$

71. $15\frac{5}{6} - 20\frac{1}{4}$ **72.** $6 - 3\frac{5}{8}$

73. $-5\frac{2}{3} - 4\frac{5}{12}$ **74.** $-2\frac{3}{4} - 3\frac{1}{5}$

In Exercises 75 and 76, determine the unknown fractional part of the pie graph.

75.

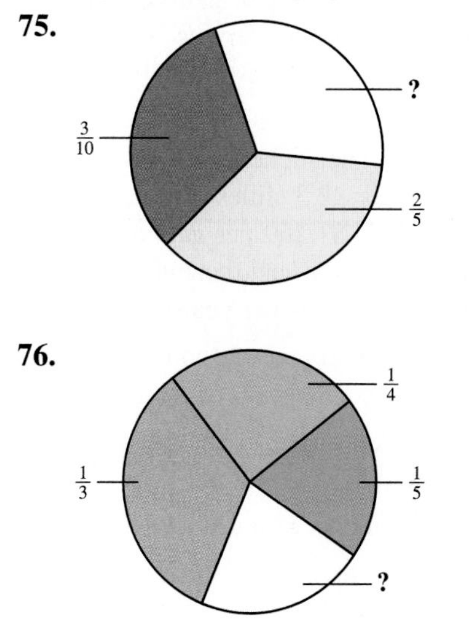

76.

In Exercises 77–94, evaluate the expression. Write the result in simplest form. See Example 7.

77. $\frac{1}{2} \times \frac{3}{4}$ **78.** $-\frac{2}{3} \times \frac{5}{7}$

79. $\frac{2}{3}\left(-\frac{9}{16}\right)$ **80.** $\left(-\frac{3}{4}\right)\left(-\frac{4}{9}\right)$

81. $\left(-\frac{7}{16}\right)\left(-\frac{12}{5}\right)$ **82.** $\left(\frac{5}{3}\right)\left(-\frac{3}{5}\right)$

83. $\left(-\frac{3}{2}\right)\left(-\frac{15}{16}\right)\left(\frac{12}{25}\right)$ **84.** $\left(\frac{1}{2}\right)\left(-\frac{4}{15}\right)\left(-\frac{5}{24}\right)$

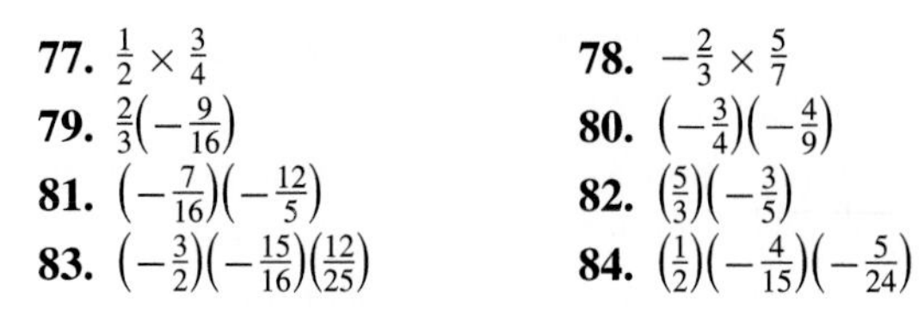

85. $\left(\frac{11}{12}\right)\left(-\frac{9}{44}\right)$ **86.** $\left(\frac{5}{18}\right)\left(\frac{3}{4}\right)$

87. $9\left(\frac{4}{15}\right)$ **88.** $24\left(-\frac{7}{18}\right)$

89. $\left(-\frac{3}{11}\right)\left(-\frac{11}{3}\right)$ **90.** $35\left(\frac{3}{5}\right)\left(\frac{5}{3}\right)$

91. $2\frac{3}{4} \times 3\frac{2}{3}$ **92.** $-5\frac{2}{3} \times 4\frac{1}{2}$

93. $2\frac{4}{5} \times 6\frac{2}{3}$ **94.** $-8\frac{1}{2} \times 3\frac{2}{5}$

In Exercises 95–98, find the reciprocal of the number. Show that the product of the number and its reciprocal is 1.

95. 7 **96.** 14

97. $\frac{4}{7}$ **98.** $-\frac{5}{9}$

In Exercises 99–114, evaluate the expression and write the result in simplest form. If it is not possible, explain why. See Example 8.

99. $\frac{3}{8} \div \frac{3}{4}$ **100.** $\frac{5}{16} \div \frac{25}{8}$

101. $-\frac{5}{12} \div \frac{45}{32}$ **102.** $\left(-\frac{16}{21}\right) \div \left(-\frac{12}{27}\right)$

103. $\frac{3}{5} \div 0$ **104.** $\frac{3}{5} \div \frac{7}{5}$

105. $-10 \div \frac{1}{9}$ **106.** $0 \div (-33)$

107. $\dfrac{-\frac{7}{15}}{-\frac{14}{25}}$ **108.** $\dfrac{-\frac{5}{9}}{0}$

109. $\dfrac{-5}{\frac{15}{16}}$ **110.** $\dfrac{-\frac{35}{12}}{-14}$

111. $3\frac{3}{4} \div 1\frac{1}{2}$ **112.** $2\frac{4}{9} \div 5\frac{1}{3}$

113. $3\frac{3}{4} \div 2\frac{5}{8}$ **114.** $1\frac{5}{6} \div 2\frac{1}{3}$

In Exercises 115–124, write the fraction in decimal form. (Use the bar notation for repeating digits.)

115. $\frac{3}{4}$ **116.** $\frac{5}{8}$

117. $\frac{9}{16}$ **118.** $\frac{7}{20}$

119. $\frac{2}{3}$ **120.** $\frac{5}{6}$

121. $\frac{7}{12}$ **122.** $\frac{8}{15}$

123. $\frac{5}{11}$ **124.** $\frac{5}{21}$

In Exercises 125–134, evaluate the expression. Round the answer to two decimal places. See Examples 9 and 10.

125. $1.21 + 4.06 - 3.00$

126. $-3.4 + 1.062 - 5.13$

127. $-0.0005 - 2.01 + 0.111$

128. $132.1 + (-25.45)$

129. $(-6.3)(9.05)$ **130.** $(-0.05)(-85.95)$

131. $(-0.09)(-0.45)$ **132.** $3.7(-14.8)$

133. $4.69 \div 0.12$ **134.** $1.062 \div (-2.1)$

Estimation In Exercises 135 and 136, estimate the sum to the nearest integer.

135. $\frac{3}{11} + \frac{7}{10}$ **136.** $\frac{5}{8} + \frac{9}{7}$

Solving Problems

137. *Stock Price* On Monday, a stock closed at $\$52\frac{5}{8}$ per share. On Tuesday, it closed at $\$54\frac{1}{4}$ per share. Determine the increase in the price.

138. *Sewing* A pattern requires $3\frac{1}{6}$ yards of material to make a skirt and an additional $2\frac{3}{4}$ yards to make a matching jacket. Find the total amount of material required.

139. *Livestock Feed* During the months of January, February, and March, an animal shelter bought $8\frac{3}{4}$ tons, $7\frac{1}{5}$ tons, and $9\frac{3}{8}$ tons of feed, respectively. Find the total amount of feed purchased during the first quarter of the year.

140. *Recipe* You are making a batch of cookies. You have placed 2 cups of flour, $\frac{1}{3}$ cup shortening, $\frac{1}{3}$ cup butter, $\frac{1}{2}$ cup brown sugar, and $\frac{1}{3}$ cup granulated sugar in a mixing bowl. How many cups of ingredients are in the mixing bowl?

141. *Volume* The fuel gauge on a gasoline tank indicates that the tank is $\frac{3}{8}$ full. What fraction of the tank is empty?

142. *Work Progress* The highway workers have a sign beside a construction project indicating what fraction of the work has been completed. At the beginnings of May and June the fractions of work completed were $\frac{5}{16}$ and $\frac{2}{3}$, respectively. What fraction of the work was completed during the month of May?

143. *Grocery Purchase* At a convenience store you buy two gallons of milk at $2.23 per gallon and three loaves of bread at $1.23 per loaf. You give the clerk a 20-dollar bill. How much change will you receive? (Assume there is no sales tax.)

144. *Telephone Charge* A telephone company charges $1.16 for the first minute and $0.85 for each additional minute. Find the cost of a 7-minute phone call.

145. *Making Breadsticks* You make 60 ounces of dough for breadsticks. If each breadstick requires $\frac{5}{4}$ ounces of dough, how many breadsticks can you make?

146. *Gasoline Price* The prices per gallon of regular unleaded gasoline at three service stations are $1.259, $1.369, and $1.279, respectively. Find the average price per gallon.

147. *Annual Fuel Cost* The sticker on a new car gives the fuel efficiency as 22.3 miles per gallon. The average cost of fuel is $1.259 per gallon. Estimate the annual fuel cost for a car that will be driven approximately 12,000 miles per year.

148. *Unit Price* A $2\frac{1}{2}$-pound can of food costs $4.95. What is the cost per pound?

149. *Stock Purchase* You buy 200 shares of stock at $\$23\frac{5}{8}$ per share and 300 shares at $\$86\frac{1}{4}$ per share.

(a) Estimate the total cost of the stock.

(b) Use a calculator to find the total cost of the stock.

150. *Walking Time* Your apartment is $\frac{3}{4}$ mile from the subway. If you walk at the rate of $3\frac{1}{4}$ miles per hour, how long does it take you to walk to the subway?

151. *Estimation* Each day for a week, you practiced the saxophone for $\frac{2}{3}$ hour.

(a) Explain how to use mental math to estimate the number of hours of practice in a week.

(b) Determine the actual number of hours you practiced during the week. Write the result in decimal form, rounding to one decimal place.

152. *Estimation* Use mental math to determine whether $\left(5\frac{3}{4}\right) \times \left(4\frac{1}{8}\right)$ is less than 20. Explain your reasoning.

Explaining Concepts

153. Answer parts (g) to (l) of Motivating the Chapter on page 1.

154. Is it true that the sum of two fractions of like signs is positive? If not, give an example that shows the statement is false.

155. Is it true that $\frac{2}{3} + \frac{3}{2} = (2 + 3)/(3 + 2) = 1$? Explain your answer.

156. In your own words, describe the rule for determining the sign of the product of two fractions.

157. Two-thirds of a pizza was eaten at dinner.

(a) How much of the pizza was left?

(b) For a midnight snack you ate $\frac{1}{2}$ of the pizza that was left. What fraction of the whole pizza did you eat as a midnight snack?

(c) Make a sketch of the pizza and show how it could be cut to obtain the results of parts (a) and (b).

158. Is it true that $\frac{2}{3} = 0.67$? Explain your answer.

159. Use the figure to determine how many one-fourths are in 3. Explain how to obtain the same result by division.

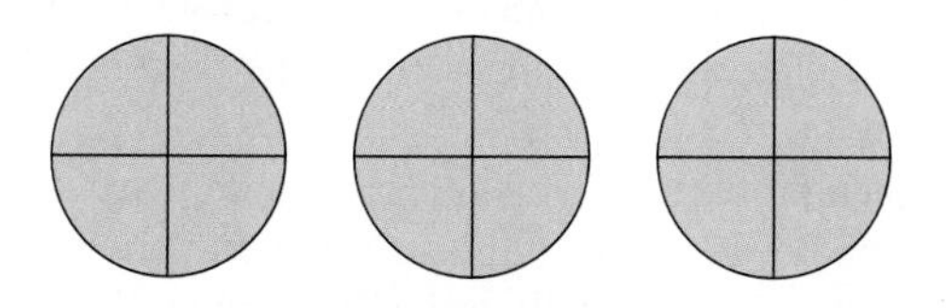

160. Use the figure to determine how many one-sixths are in $\frac{2}{3}$. Explain how to obtain the same result by division.

In Exercises 161–166, determine whether the statement is true or false.

161. The reciprocal of every nonzero integer is an integer.

162. The reciprocal of every nonzero rational number is a rational number.

163. The product of two nonzero rational numbers is a rational number.

164. The product of two positive rational numbers is greater than either factor.

165. If $u > v$, then $u - v > 0$.

166. If $u > 0$ and $v > 0$, then $u - v > 0$.

167. *Think About It* Determine the placement of the digits 3, 4, 5, and 6 in the following addition problem so that you obtain the specified sum. Use each number only once.

$$\frac{\square}{\square} + \frac{\square}{\square} = \frac{13}{10}$$

168. If the fractions represented by the points R and P are multiplied, what point on the number line best represents their product: M, S, N, P, or T? (Source: National Council of Teachers of Mathematics)

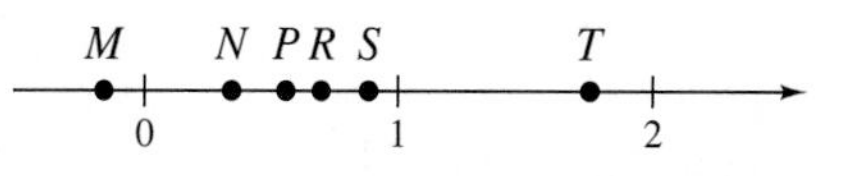

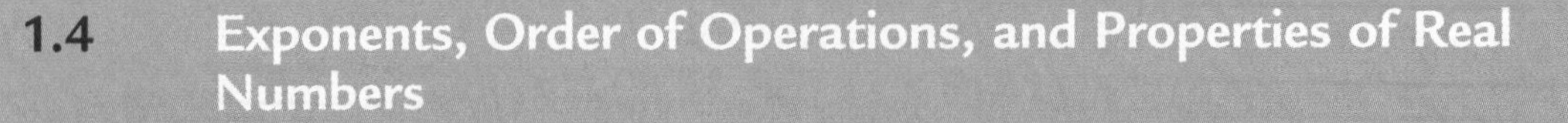

1.4 Exponents, Order of Operations, and Properties of Real Numbers

Objectives

1. Rewrite repeated multiplication in exponential form and evaluate exponential expressions.
2. Evaluate expressions using order of operations.
3. Identify and use the properties of real numbers.

1 Rewrite repeated multiplication in exponential form and evaluate exponential expressions.

Exponents

In Section 1.2, you learned that multiplication by a positive integer can be described as repeated addition.

Repeated Addition	*Multiplication*
$\underbrace{7 + 7 + 7 + 7}_{\text{4 terms of 7}}$	4×7

In a similar way, repeated multiplication can be described in **exponential form.**

Repeated Multiplication	*Exponential Form*
$\underbrace{7 \cdot 7 \cdot 7 \cdot 7}_{\text{4 factors of 7}}$	7^4

In the exponential form 7^4, 7 is the **base** and it specifies the repeated factor. The number 4 is the **exponent** and it indicates how many times the base occurs as a factor.

When you write the exponential form 7^4, you can say that you are raising 7 to the fourth **power.** When a number is raised to the first power, you usually do not write the exponent 1. For instance, we usually write 5^1 simply as 5. Here are some examples of how exponential expressions are read.

Exponential Expression	*Verbal Statement*
7^2	"seven to the second power" or "seven squared"
4^3	"four to the third power" or "four cubed"
$(-2)^4$	"negative two to the fourth power"
-2^4	"the opposite of two to the fourth power"

It is important to recognize how exponential forms such as $(-2)^4$ and -2^4 differ.

$$(-2)^4 = (-2)(-2)(-2)(-2)$$ The negative sign is part of the base.

$$= 16$$ The value of the expression is positive.

$$-2^4 = -(2 \cdot 2 \cdot 2 \cdot 2)$$ The negative sign is not part of the base.

$$= -16$$ The value of the expression is negative.

Keep in mind that an exponent applies only to the factor (number) directly preceding it. Parentheses are needed to include a negative sign or other factors as part of the base.

Technology: Discovery

When a negative number is raised to a power, the use of parentheses is very important. To discover why, use a calculator to evaluate $(-5)^4$ and -5^4. Write a statement explaining the results. Then use a calculator to evaluate $(-5)^3$ and -5^3. If necessary, write a new statement explaining your discoveries.

Example 1 Evaluating Exponential Expressions

a. $2^5 = 2 \cdot 2 \cdot 2 \cdot 2 \cdot 2$ — Rewrite expression as a product.

$= 32$ — Simplify.

b. $\left(\frac{2}{3}\right)^4 = \frac{2}{3} \cdot \frac{2}{3} \cdot \frac{2}{3} \cdot \frac{2}{3}$ — Rewrite expression as a product.

$= \frac{2 \cdot 2 \cdot 2 \cdot 2}{3 \cdot 3 \cdot 3 \cdot 3}$ — Multiply fractions.

$= \frac{16}{81}$ — Simplify.

c. $(-3)^3 = (-3)(-3)(-3)$ — Rewrite expression as a product.

$= -27$ — Simplify.

d. $(-3)^4 = (-3)(-3)(-3)(-3)$ — Rewrite expression as a product.

$= 81$ — Simplify.

e. $-3^4 = -(3 \cdot 3 \cdot 3 \cdot 3)$ — Rewrite expression as a product.

$= -81$ — Simplify.

In parts (c) and (d) of Example 1, note that when a negative number is raised to an *odd* power, the result is *negative*, and when a negative number is raised to an *even* power, the result is *positive.*

Example 2 Transporting Capacity 

A truck can transport a load of motor oil that is 6 cases high, 6 cases wide, and 6 cases long. Each case contains 6 quarts of motor oil. How many quarts can the truck transport?

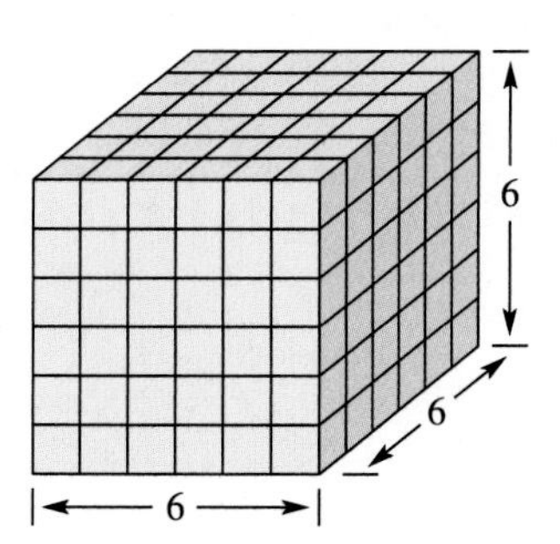

Figure 1.18

Solution

A sketch can help you solve this problem. From Figure 1.18, you can see that 6 occurs as a factor four times. That is, there are $6 \cdot 6 \cdot 6$ cases of motor oil and each case contains 6 quarts, which implies that the total number of quarts is

$$(6 \cdot 6 \cdot 6) \cdot 6 = 6^4 = 1296.$$

So, the truck can transport 1296 quarts of oil.

2 Evaluate expressions using order of operations.

Order of Operations

Up to this point in the text, you have studied five operations of arithmetic—addition, subtraction, multiplication, division, and exponentiation (repeated multiplication). When you use more than one operation in a given problem, you face the question of which operation to do first. For example, without further guidelines, you could evaluate $4 + 3 \cdot 5$ in two ways.

Technology: Discovery

To discover if your calculator performs the established order of operations, evaluate $7 + 5 \cdot 3 - 2^4 \div 4$ exactly as it appears. Does your calculator display 5 or 18? If your calculator performs the established order of operations, it will display 18.

Corbis-Bettmann

Gottfried Wilhelm von Leibniz (1646–1716)

The symbols used to represent operations were introduced over time. The minus and plus signs were first used in Germany in 1489. The plus sign evolved from the Latin word *et*, meaning "and." The equal sign, the square root notation, and variables were introduced in the 1500s in Europe. In the 1630s, × was introduced to indicate multiplication. Gottfried Wilhelm von Leibniz thought it was easily confused with the variable *x* and proposed the dot symbol (·) in 1698.

Add First	*Multiply First*
$4 + 3 \cdot 5 \stackrel{?}{=} (4 + 3) \cdot 5$	$4 + 3 \cdot 5 \stackrel{?}{=} 4 + (3 \cdot 5)$
$= 7 \cdot 5$	$= 4 + 15$
$= 35$	$= 19$

According to the established **order of operations,** the second evaluation is correct. The reason for this is that multiplication has a higher priority than addition. The accepted priorities for order of operations are summarized below.

▶ Order of Operations

1. Perform operations inside *symbols of grouping*—( ) or [ ]—or *absolute value symbols*, starting with the innermost symbol.
2. Evaluate all *exponential* expressions.
3. Perform all *multiplications* and *divisions* from left to right.
4. Perform all *additions* and *subtractions* from left to right.

In the priorities for order of operations, note that the highest priority is given to **symbols of grouping** such as parentheses or brackets. This means that when you want to be sure that you are communicating an expression correctly, you can insert symbols of grouping to specify which operations you intend to be performed first. For instance, if you want to make sure that $4 + 3 \cdot 5$ will be evaluated correctly, you can write it as $4 + (3 \cdot 5)$.

Example 3 Order of Operations

a. $7 - [(5 \cdot 3) + 2^3] = 7 - [15 + 2^3]$ Multiply inside the parentheses.

$= 7 - [15 + 8]$ Evaluate exponential expression.

$= 7 - 23$ Add inside the brackets.

$= -16$ Subtract.

b. $36 \div (3^2 \cdot 2) - 6 = 36 \div (9 \cdot 2) - 6$ Evaluate exponential expression.

$= 36 \div 18 - 6$ Multiply inside the parentheses.

$= 2 - 6$ Divide.

$= -4$ Subtract.

When you use symbols of grouping in an expression, we suggest that you alternate between parentheses and brackets. For instance, the expression

$$10 - (3 - [4 - (5 + 7)])$$

is easier to understand than $10 - (3 - (4 - (5 + 7)))$.

Study Tip

Often in mathematics, there is no "best way" to solve a problem. For instance, here is a different way to evaluate the expression in Example 4(b) using the Distributive Property. Which way do you prefer?

$$\frac{8}{3}\left(\frac{1}{6}+\frac{1}{4}\right)$$
$$=\frac{8}{3}\cdot\frac{1}{6}+\frac{8}{3}\cdot\frac{1}{4}$$
$$=\frac{8}{18}+\frac{8}{12}$$
$$=\frac{16}{36}+\frac{24}{36}$$
$$=\frac{40}{36}$$
$$=\frac{10}{9}$$

Example 4 Order of Operations

a. $\frac{3}{7}\div\frac{8}{7}+\left(-\frac{3}{5}\right)\left(\frac{1}{3}\right)=\frac{3}{7}\cdot\frac{7}{8}+\left(-\frac{3}{5}\right)\left(\frac{1}{3}\right)$ Invert divisor and multiply.

$=\frac{3}{8}+\left(-\frac{1}{5}\right)$ Multiply fractions.

$=\frac{15}{40}+\frac{-8}{40}$ Find common denominator.

$=\frac{7}{40}$ Add fractions.

b. $\frac{8}{3}\left(\frac{1}{6}+\frac{1}{4}\right)=\frac{8}{3}\left(\frac{2}{12}+\frac{3}{12}\right)$ Find common denominator.

$=\frac{8}{3}\left(\frac{5}{12}\right)$ Add inside the parentheses.

$=\frac{40}{36}$ Multiply fractions.

$=\frac{10}{9}$ Simplify.

Example 5 Order of Operations

Evaluate the expression $6+\frac{8+7}{3^2-4}-(-5)$.

Solution

Using the established order of operations, you can evaluate the expression as follows.

$6+\frac{8+7}{3^2-4}-(-5)=6+\frac{8+7}{9-4}-(-5)$ Evaluate exponential expression.

$=6+\frac{15}{9-4}-(-5)$ Add in numerator.

$=6+\frac{15}{5}-(-5)$ Subtract in denominator.

$=6+3-(-5)$ Divide.

$=9+5$ Add.

$=14$ Add.

In Example 5, note that a fraction bar acts as a symbol of grouping. For instance,

$$\frac{8+7}{3^2-4} \quad \text{means} \quad (8+7)\div(3^2-4), \quad \text{not} \quad 8+7\div 3^2-4.$$

3 Identify and use the properties of real numbers.

Properties of Real Numbers

You are now ready for the symbolic versions of the properties that we know are true about operations with real numbers. These properties are referred to as **properties of real numbers.** The table gives a verbal description and an illustrative example for each property. Bear in mind that the letters a, b, c, etc., represent real numbers, even though we have used only rational numbers to this point.

▶ **Properties of Real Numbers:** Let a, b, and c be real numbers.

Property	*Example*
1. Commutative Property of Addition: Two real numbers can be added in either order.	
$a + b = b + a$	$3 + 5 = 5 + 3$
2. Commutative Property of Multiplication: Two real numbers can be multiplied in either order.	
$ab = ba$	$4 \cdot (-7) = -7 \cdot 4$
3. Associative Property of Addition: When three real numbers are added, it makes no difference which two are added first.	
$(a + b) + c = a + (b + c)$	$(2 + 6) + 5 = 2 + (6 + 5)$
4. Associative Property of Multiplication: When three real numbers are multiplied, it makes no difference which two are multiplied first.	
$(ab)c = a(bc)$	$(3 \cdot 5) \cdot 2 = 3 \cdot (5 \cdot 2)$
5. Distributive Property: Multiplication distributes over addition.	
$a(b + c) = ab + ac$	$3(8 + 5) = 3 \cdot 8 + 3 \cdot 5$
$(a + b)c = ac + bc$	$(3 + 8)5 = 3 \cdot 5 + 8 \cdot 5$
6. Additive Identity Property: The sum of zero and a real number equals the number itself.	
$a + 0 = 0 + a = a$	$3 + 0 = 0 + 3 = 3$
7. Multiplicative Identity Property: The product of 1 and a real number equals the number itself.	
$a \cdot 1 = 1 \cdot a = a$	$4 \cdot 1 = 1 \cdot 4 = 4$
8. Additive Inverse Property: The sum of a real number and its opposite is zero.	
$a + (-a) = 0$	$3 + (-3) = 0$
9. Multiplicative Inverse Property: The product of a nonzero real number and its reciprocal is 1.	
$a \cdot \frac{1}{a} = 1,\ a \neq 0$	$8 \cdot \frac{1}{8} = 1$

Technology: Tip

The Multiplicative Inverse Property is one rule listed on page 48: It states that *The product of a nonzero real number and its reciprocal is 1.*

$$a \cdot \frac{1}{a} = 1, \ a \neq 0$$

This property can be illustrated on your calculator by using the reciprocal key 1/x or x^{-1}. Try doing this with $a = \frac{2}{3}$.

Example 6 Identifying Properties of Real Numbers

Name the property of real numbers that justifies the given statement.

a. $3(a + 2) = 3 \cdot a + 3 \cdot 2$

b. $5 \cdot \frac{1}{5} = 1$

c. $7 + (5 + b) = (7 + 5) + b$

d. $(b + 3) + 0 = b + 3$

e. $5(c - 3) = 5 \cdot c - 5 \cdot 3$

Solution

a. This statement is justified by the Distributive Property.

b. This statement is justified by the Multiplicative Inverse Property.

c. This statement is justified by the Associative Property of Addition.

d. This statement is justified by the Additive Identity Property.

e. This statement is justified by the Distributive Property, which works for subtraction also.

Example 7 Using the Properties of Real Numbers

Complete each statement using the specified property of real numbers.

a. Multiplicative Identity Property:

$(3b)1 = \square$

b. Associative Property of Addition:

$(c + 2) + 7 = \square$

c. Additive Inverse Property:

$0 = 3a + \square$

d. Distributive Property:

$3 \cdot a + 3 \cdot 4 = \square$

Solution

a. By the Multiplicative Identity Property, you can write

$$(3b)1 = 3b.$$

b. By the Associative Property of Addition, you can write

$$(c + 2) + 7 = c + (2 + 7).$$

c. By the Additive Inverse Property, you can write

$$0 = 3a + (-3a).$$

d. By the Distributive Property, you can write

$$3 \cdot a + 3 \cdot 4 = 3(a + 4).$$

One of the distinctive things about algebra is that its rules make sense. You don't have to accept them on "blind faith"—instead, you can learn the reasons that the rules work. For instance, the next example looks at some basic differences among the operations of addition, multiplication, subtraction, and division.

Example 8 Properties of Real Numbers

In the summary of properties of real numbers on page 48, why are all the properties listed in terms of addition and multiplication and not subtraction and division?

Solution

The reason for this is that subtraction and division lack many of the properties listed in the summary. For instance, subtraction and division are not commutative. To see this, consider the following.

$$7 - 5 \neq 5 - 7 \quad \text{and} \quad 12 \div 4 \neq 4 \div 12$$

Similarly, subtraction and division are not associative.

$$9 - (5 - 3) \neq (9 - 5) - 3 \quad \text{and} \quad 12 \div (4 \div 2) \neq (12 \div 4) \div 2$$

Example 9 Geometry: Area

You measure the width of a billboard and find that it is 60 feet. You are told that its height is 22 feet less than its width. Write an expression for the area of the billboard. Use the Distributive Property to rewrite the expression.

Solution

Figure 1.19

Begin by drawing and labeling a diagram, as shown in Figure 1.19. To find the area of the billboard, multiply the width by the height.

$$\text{Width} \times \text{height} = 60(60 - 22)$$

To rewrite the expression $60(60 - 22)$ using the Distributive Property, distribute 60 over the subtraction.

$$60(60 - 22) = 60(60) - 60(22)$$

Discussing the Concept — Order of Operations

Using the established order of operations, the value of $7 \cdot 8 + 12$ is

$$7 \cdot 8 + 12 = 56 + 12 = 68.$$

By inserting parentheses into the expression, you can obtain a value of

$$7 \cdot (8 + 12) = 7(20) = 140.$$

Using the established order of operations, which of the following expressions has a value of 72? For those that don't, decide whether you can insert parentheses into the expression so that its value is 72.

a. $4 + 2^3 - 7$ **b.** $4 + 8 \cdot 6$

c. $93 - 25 - 4$ **d.** $70 + 10 \div 5$

e. $60 + 20 \div 2 + 32$ **f.** $35 \cdot 2 + 2$

1.4 Exercises

Developing Skills

In Exercises 1–4, rewrite in exponential form.

1. $2 \cdot 2 \cdot 2 \cdot 2 \cdot 2$
2. $(-5) \cdot (-5) \cdot (-5) \cdot (-5)$
3. $\left(-\frac{1}{4}\right) \cdot \left(-\frac{1}{4}\right) \cdot \left(-\frac{1}{4}\right)$
4. $(1.6) \cdot (1.6) \cdot (1.6) \cdot (1.6) \cdot (1.6)$

In Exercises 5–10, rewrite as a product.

5. $(-3)^6$
6. $\left(\frac{3}{8}\right)^5$
7. $(9.8)^3$
8. $(0.01)^8$
9. $\left(-\frac{1}{2}\right)^5$
10. $\left(\frac{3}{11}\right)^4$

In Exercises 11–14, is the value positive or negative?

11. -2^2
12. $(-2)^4$
13. -5^3
14. $-(-5)^3$

In Exercises 15–24, evaluate the expression. See Example 1.

15. 3^2
16. 4^3
17. 2^6
18. 5^3
19. $(-5)^3$
20. $-(-3)^2$
21. $\left(\frac{1}{4}\right)^3$
22. $\left(\frac{4}{5}\right)^3$
23. $(-1.2)^3$
24. $(1.5)^4$

In Exercises 25–64, evaluate the expression. If it is not possible, state the reason. Write fractional answers in simplest form. See Examples 3, 4, and 5.

25. $4 - 6 + 10$
26. $5 - (8 - 15)$
27. $-|2 - (6 + 5)|$
28. $125 - |10 - (25 - 3)|$
29. $15 + 3 \cdot 4$
30. $25 - 32 \div 4$
31. $(16 - 5) \div (3 - 5)$
32. $(10 - 16) \cdot (20 - 26)$
33. $(45 \div 10) \cdot 2$
34. $[360 - (8 + 12)] \div 10$
35. $5 + (2^2 \cdot 3)$
36. $181 - (13 \cdot 3^2)$
37. $(-6)^2 - (5^2 \cdot 4)$
38. $(-3)^3 + (12 \div 2^2)$
39. $\left(3 \cdot \frac{5}{9}\right) + 1 - \frac{1}{3}$
40. $\frac{2}{3}\left(\frac{3}{4}\right) + 2 - \frac{1}{2}$
41. $4\left(-\frac{2}{3} + \frac{4}{3}\right)$
42. $18\left(\frac{1}{2} + \frac{2}{3}\right)$
43. $\frac{3}{2}\left(\frac{2}{3} + \frac{1}{6}\right)$
44. $\frac{7}{25}\left(\frac{7}{16} - \frac{1}{8}\right)$
45. $\dfrac{3 \cdot 6 - 4 \cdot 6}{5 + 1}$
46. $\dfrac{3 + [15 \div (-3)]}{16}$
47. $\frac{7}{3}\left(\frac{2}{3}\right) \div \frac{28}{15}$
48. $\frac{3}{8}\left(\frac{1}{5}\right) \div \frac{25}{32}$
49. $\dfrac{1 - 3^2}{-2}$
50. $\dfrac{3^2 + 4^2}{5}$
51. $\dfrac{3^2 - 4^2}{0}$
52. $\dfrac{0}{3^2 - 4^2}$
53. $\dfrac{5^2 + 12^2}{13}$
54. $\dfrac{4^2 - 2^3}{4}$
55. 2.1×10^2
56. 4.85×10^4
57. 5.84×10^3
58. 3.28×10^5
59. $\dfrac{8.4}{10^3}$
60. $\dfrac{6.23}{10^2}$
61. $\dfrac{732}{10^2}$
62. $\dfrac{8235}{10^4}$
63. $\dfrac{0}{5^2 + 1}$
64. $\dfrac{3^2 + 1}{0}$

In Exercises 65–68, use a calculator to evaluate the expression. Round your result to two decimal places.

65. $3.4^2 - 6(1.2)^3$
66. $300\left(1 + \dfrac{0.1}{12}\right)^{24}$
67. $1000 \div \left(1 + \dfrac{0.09}{4}\right)^8$
68. $\dfrac{1.32 + 4(3.68)}{1.5}$

In Exercises 69–72, explain why the statement is true. (The symbol $\neq$ means *is not equal to.*)

69. $4 \cdot 6^2 \neq 24^2$
70. $4 - (6 - 2) \neq 4 - 6 - 2$
71. $-3^2 \neq (-3)(-3)$
72. $\dfrac{8 - 6}{2} \neq 4 - 6$

In Exercises 73–92, identify the property of real numbers that justifies the statement. See Example 6.

73. $6(-3) = -3(6)$
74. $16 + 10 = 10 + 16$
75. $x + 10 = 10 + x$
76. $8x = x(8)$
77. $0 + 15 = 15$
78. $1 \cdot 4 = 4$

79. $-16 + 16 = 0$

80. $(2 \cdot 3)4 = 2(3 \cdot 4)$

81. $(10 + 3) + 2 = 10 + (3 + 2)$

82. $25 + (-25) = 0$

83. $4(3 \cdot 10) = (4 \cdot 3)10$

84. $(32 + 8) + 5 = 32 + (8 + 5)$

85. $7(\frac{1}{7}) = 1$

86. $14 + (-14) = 0$

87. $6(3 + x) = 6 \cdot 3 + 6x$

88. $(14 + 2)3 = 14 \cdot 3 + 2 \cdot 3$

89. $(4 + x)(2 - x) = 4(2 - x) + x(2 - x)$

90. $\frac{1}{a}(3 + y) = \frac{1}{a}(3) + \frac{1}{a}(y)$

91. $x + (y + 3) = (x + y) + 3$

92. $[(x + y)u]v = (x + y)(uv)$

In Exercises 93–96, use the Commutative Property of Addition or Multiplication to rewrite the expression. See Example 7.

93. $5(u + v) =$ ____ **94.** $y + 5 =$ ____

95. $3 + x =$ ____ **96.** $10(-3) =$ ____

In Exercises 97–100, use the Distributive Property to rewrite the expression. See Example 7.

97. $6(x + 2) =$ ____

98. $5(u + v) =$ ____

99. $(4 + y)25 =$ ____

100. $x(4 - y) =$ ____

In Exercises 101–104, use the Associative Property of Addition or Multiplication to rewrite the expression. See Example 7.

101. $3x + (2y + 5) =$ ____

102. $12(3 \cdot 4) =$ ____

103. $(6x)y =$ ____

104. $10 + (x + 2y) =$ ____

In Exercises 105–112, find (a) the additive inverse and (b) the multiplicative inverse of the quantity.

105. 50 **106.** 12

107. -1 **108.** $-\frac{1}{2}$

109. $2x$ **110.** $5y$

111. ab **112.** uv

In Exercises 113–116, simplify the expression.

113. $3(6 + 10)$ **114.** $4(8 - 3)$

115. $\frac{2}{3}(9 + 24)$ **116.** $\frac{1}{2}(4 - 2)$

In Exercises 117–120, explain why the statement is true.

117. $5(x + 3) \neq 5x + 3$

118. $7(x - 2) \neq 7x - 2$

119. $\frac{8}{0} \neq 0$

120. $5(\frac{1}{5}) \neq 0$

In Exercises 121–124, identify the property of real numbers used to justify each rewritten step.

121.
$$\begin{aligned} 4(2 + x) &= 4(x + 2) \\ &= 4x + 8 \end{aligned}$$

122.
$$\begin{aligned} 3 + 10(x + 1) &= 3 + 10x + 10 \\ &= 3 + 10 + 10x \\ &= (3 + 10) + 10x \\ &= 13 + 10x \end{aligned}$$

123.
$$\begin{aligned} 7x + 9 + 2x &= 7x + 2x + 9 \\ &= (7x + 2x) + 9 \\ &= (7 + 2)x + 9 \\ &= 9x + 9 \\ &= 9(x + 1) \end{aligned}$$

124.
$$\begin{aligned} 2(x + 3) + x &= 2x + 2 \cdot 3 + x \\ &= 2x + x + 6 \\ &= (2 + 1)x + 6 \\ &= 3x + 6 \\ &= 3(x + 2) \end{aligned}$$

Solving Problems

Geometry In Exercises 125 and 126, find the area.

125.

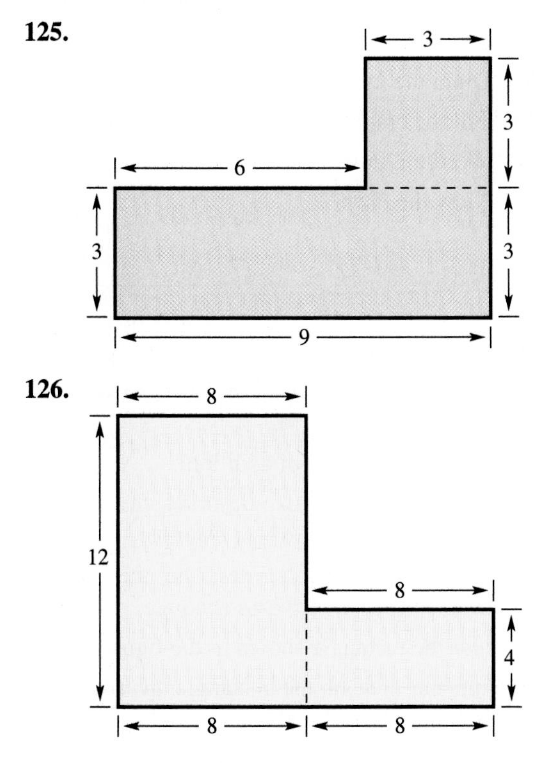

126.

127. *Interpreting a Pie Graph* The portions of the total expenses for a company are shown in the pie graph. What portion of the total expenses is spent on utilities? If the total expenses are $450,000, how much is spent on utilities?

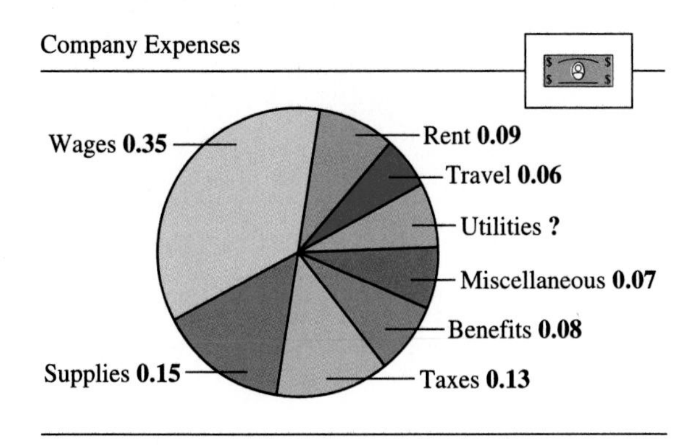

128. *Forecasting* The projected number of elementary and secondary school teachers for the year 2006 is 3.43×10^6. Evaluate this quantity. (Source: U.S. National Center for Education Statistics)

129. *Total Cost* A car is purchased for $750 down and 48 monthly payments of $215 each. What is the total amount paid for the car?

130. *Think About It* A child suggests the following plan for an allowance during a month with 30 days. The first day of the month she will receive 1 cent, the second day 2 cents, the third day 4 cents, and so on. If the amount continues to double each day, what will her allowance be on day 30?

131. *Sales Tax* You purchase an item for x dollars. There is a 6% sales tax, which implies that the total amount you must pay is $x + 0.06x$.

(a) Use the Distributive Property to rewrite the expression.

(b) How much must you pay if the item costs $25.95?

132. *Geometry* The width of a movie screen is 30 feet and its height is 8 feet less than the width. Write an expression for the area of the movie screen. Use the Distributive Property to rewrite the expression.

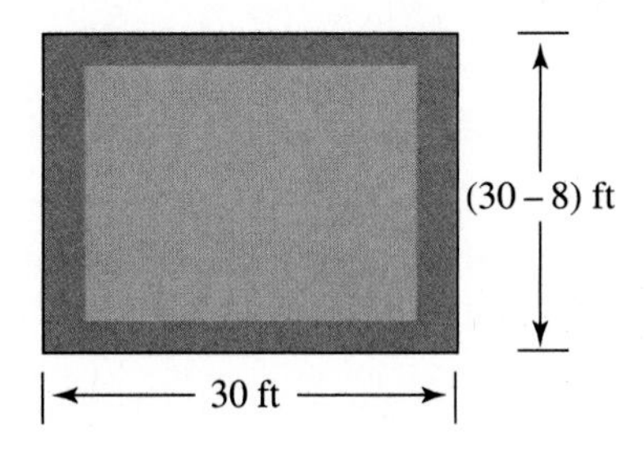

133. *Geometry* Write an expression for the perimeter of the triangle shown in the figure. Use the properties of real numbers to simplify the expression.

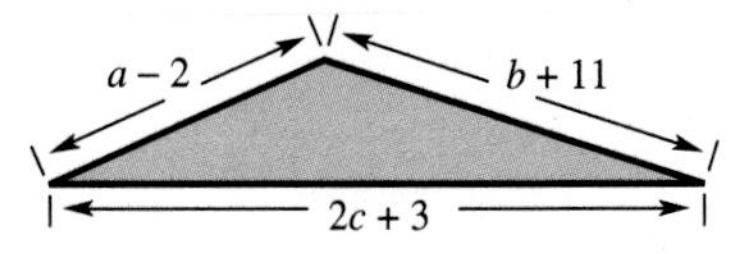

134. *Geometry* Find the area of the yellow rectangle in two ways. Explain how the results are related to the Distributive Property.

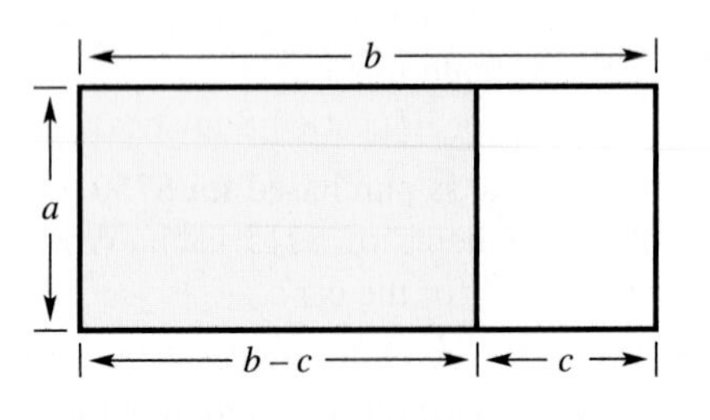

Think About It In Exercises 135 and 136, determine whether the order in which the two activities are performed is "commutative." That is, do you obtain the same result regardless of which activity is performed first?

135. (a) "Drain the used oil from the engine."

(b) "Fill the crankcase with 5 quarts of new oil."

136. (a) "Weed the flower beds."

(b) "Mow the lawn."

Explaining Concepts

137. Consider the expression 3^5.

(a) What is the number 3 called?

(b) What is the number 5 called?

138. Are -6^2 and $(-6)^2$ equal? Explain.

139. Are $2 \cdot 5^2$ and 10^2 equal? Explain.

140. In your own words, describe the priorities for the established order of operations.

141. In the expression $12 + 48 \div 6 - 5$, where would you insert symbols of grouping to help someone understand that the value of the expression is 15?

142. *Error Analysis* Find the error.

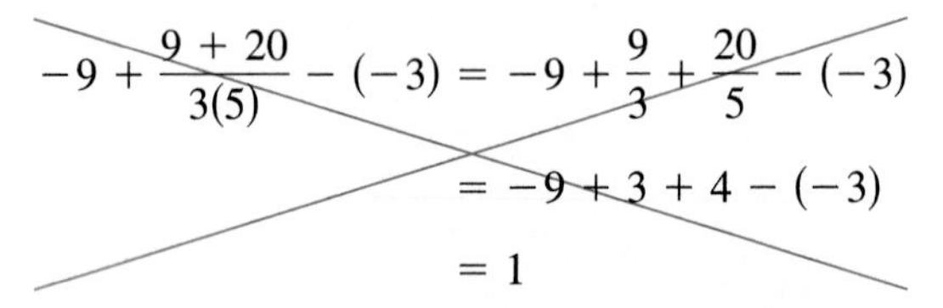

$$-9 + \frac{9 + 20}{3(5)} - (-3) = -9 + \frac{9}{3} + \frac{20}{5} - (-3)$$
$$= -9 + 3 + 4 - (-3)$$
$$= 1$$

143. In your own words, state the Commutative Properties of Addition and Multiplication. Give an example of each.

144. In your own words, state the Associative Properties of Addition and Multiplication. Give an example of each.

145. Consider the operation of addition.

(a) In your own words, describe the Additive Identity Property. Give an example.

(b) In your own words, describe the Additive Inverse Property. Give an example.

146. Consider the rectangle shown in the figure.

(a) Find the area of the rectangle by adding the areas of regions I and II.

(b) Find the area of the rectangle by multiplying its length by its width.

(c) Explain how the results of parts (a) and (b) relate to the Distributive Property.

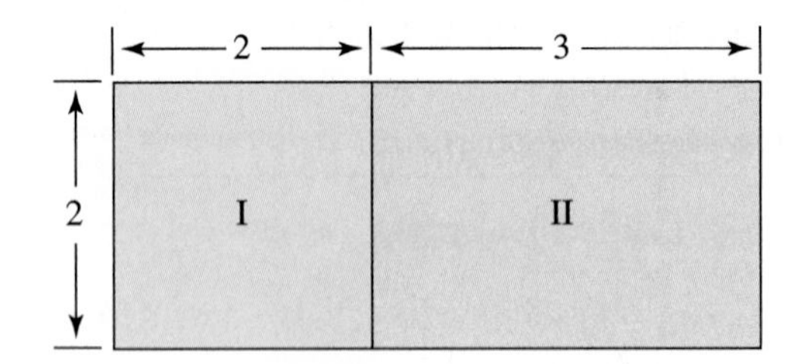

CHAPTER SUMMARY

Key Terms

real numbers, *p. 2*
natural numbers, *p. 2*
integers, *p. 2*
rational numbers, *p. 3*
irrational numbers, *p. 3*
real number line, *p. 4*
inequality symbol, *p. 5*
opposites, *p. 7*
absolute value, *p. 7*
expression, *p. 7*
evaluate, *p. 7*
additive inverse, *p. 13*
factor, *p. 19*
prime number, *p. 19*
greatest common factor, *p. 29*
reciprocal, *p. 35*
exponent, *p. 44*

Key Concepts

1.1 Ordering of real numbers

Use the real number line and an inequality symbol (<, >, ≤, or ≥) to order real numbers.

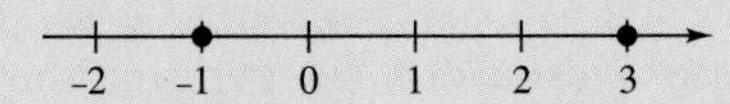

$-1 < 3$

1.1 Absolute value

The absolute value of a number is its distance from zero on the real number line. The absolute value is either positive or zero.

1.2 Addition and subtraction of integers

To add integers with like signs, add their absolute values and attach the common sign to the result.

To add integers with different signs, subtract the smaller absolute value from the larger absolute value and attach the sign of the integer with the larger absolute value.

To subtract one integer from another, add the opposite of the integer being subtracted to the other integer.

1.2 Rules for multiplying and dividing integers

1. The product of an integer and zero is 0.
2. Zero divided by a nonzero integer is 0, whereas a nonzero integer divided by zero is undefined.
3. The product or quotient of two nonzero integers with like signs is positive.
4. The product or quotient of two nonzero integers with different signs is negative.

1.3 Addition and subtraction of fractions

1. To add two fractions with like denominators, add their numerators and write the sum over the like (or common) denominator.
2. To add two fractions with unlike denominators, rewrite both fractions so that they have like denominators. Then use the rule for adding fractions with like denominators.
3. To subtract two fractions, add the opposite fraction and proceed as in addition.

1.3 Multiplication of fractions

To multiply two fractions, multiply the two numerators to form the numerator of the product, and multiply the two denominators to form the denominator of the product.

1.3 Division of fractions

To divide two fractions, invert the divisor and multiply.

1.4 Order of operations

1. Perform operations inside symbols of grouping—( ) or [ ]—or absolute value symbols, starting with the innermost symbol.
2. Evaluate all exponential expressions.
3. Perform all multiplications and divisions from left to right.
4. Perform all additions and subtractions from left to right.

1.4 Properties of real numbers

Commutative Property of Addition $a + b = b + a$

Commutative Property of Multiplication $ab = ba$

Associative Property of Addition

$(a + b) + c = a + (b + c)$

Associative Property of Multiplication $(ab)c = a(bc)$

Distributive Property

$a(b + c) = ab + ac$ $\quad a(b - c) = ab - ac$

$(a + b)c = ac + bc$ $\quad (a - b)c = ac - bc$

Additive Identity Property $a + 0 = a$

Multiplicative Identity Property $a \cdot 1 = a$

Additive Inverse Property $a + (-a) = 0$

Multiplicative Inverse Property $a \cdot \frac{1}{a} = 1, \quad a \neq 0$

REVIEW EXERCISES

Reviewing Skills

1.1 In Exercises 1–4, plot each real number as a point on the real number line and place the correct inequality symbol (< or >) between the real numbers.

1. $-\frac{1}{10} \;\square\; 4$ **2.** $\frac{25}{3} \;\square\; \frac{5}{3}$

3. $-3 \;\square\; -7$ **4.** $10.6 \;\square\; -3.5$

In Exercises 5–8, find the opposite of the number, and determine the distance of the number and its opposite from 0.

5. 152 **6.** -10.4

7. $-\frac{7}{3}$ **8.** $\frac{2}{3}$

In Exercises 9–12, evaluate the expression.

9. $|-8.5|$ **10.** $|3.4|$

11. $-|-8.5|$ **12.** $|-9.6|$

In Exercises 13–16, place the correct symbol (<, >, or =) between the real numbers.

13. $|-84| \;\square\; |84|$ **14.** $|-10| \;\square\; |4|$

15. $\left|\frac{3}{10}\right| \;\square\; -\left|\frac{4}{5}\right|$ **16.** $|2.3| \;\square\; -|2.3|$

1.2 In Exercises 17–36, perform the indicated operations with integers, if possible. If it is not possible, state the reason.

17. $32 + 68$ **18.** $14 + 54$

19. $16 + (-5)$ **20.** $-125 + 30$

21. $350 - 125 + 15$ **22.** $35 - 25 - 10$

23. $-114 + 76 - 230$ **24.** $-448 - 322 + 100$

25. $|-86| - |124|$ **26.** $67 + |-53|$

27. 15×3 **28.** -22×4

29. $-300(-5)$ **30.** $8(320)$

31. $31(-6)(3)$ **32.** $(-46)(-5)(-2)$

33. $\frac{-162}{9}$ **34.** $\frac{-52}{-4}$

35. $815 \div 0$ **36.** $-48 \div 6$

37. Subtract -549 from 613.

38. Find the absolute value of the sum of 693 and -420.

39. What must you add to 75 to obtain -27?

40. What must you subtract from -83 to obtain 43?

In Exercises 41 and 42, use the long division algorithm to find the quotient.

41. $33{,}768 \div -72$ **42.** $-144{,}512 \div -32$

In Exercises 43–46, use a calculator to perform the operations.

43. $7(5207) - 52{,}318$ **44.** $783(1995) + 75(-832)$

45. $\frac{345{,}582}{438}$ **46.** $\frac{1{,}111{,}521}{89}$

In Exercises 47–50, decide whether the number is prime or composite.

47. 839 **48.** 909

49. 1764 **50.** 1847

In Exercises 51–54, write the prime factorization of the number.

51. 378 **52.** 858

53. 1612 **54.** 1787

1.3 In Exercises 55–58, find the greatest common factor.

55. 54, 90 **56.** 154, 220

57. 63, 84, 441 **58.** 99, 132, 253

In Exercises 59–62, write an equivalent fraction with the indicated denominator.

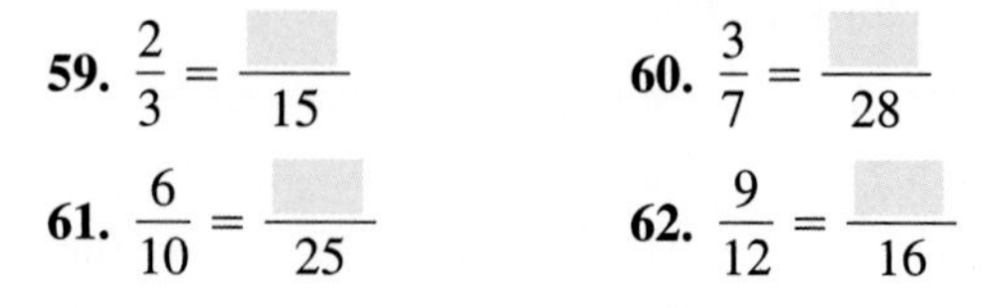

59. $\frac{2}{3} = \frac{\square}{15}$ **60.** $\frac{3}{7} = \frac{\square}{28}$

61. $\frac{6}{10} = \frac{\square}{25}$ **62.** $\frac{9}{12} = \frac{\square}{16}$

In Exercises 63–74, evaluate the expression. Write the result in simplest form.

63. $\frac{3}{25} + \frac{7}{25}$

64. $\frac{9}{64} + \frac{7}{64}$

65. $\frac{27}{16} - \frac{15}{16}$

66. $-\frac{5}{12} + \frac{1}{12}$

67. $-\frac{5}{9} + \frac{2}{3}$

68. $\frac{7}{15} - \frac{2}{25}$

69. $\frac{25}{32} + \frac{7}{24}$

70. $-\frac{7}{8} - \frac{11}{12}$

71. $5 - \frac{15}{4}$

72. $\frac{12}{5} - 3$

73. $5\frac{3}{4} - 3\frac{5}{8}$

74. $-3\frac{7}{10} + 1\frac{1}{20}$

In Exercises 75–86, evaluate the expression. If it is not possible, explain why.

75. $\frac{5}{8} \cdot \frac{-2}{15}$

76. $\frac{3}{32} \cdot \frac{32}{3}$

77. $35\left(\frac{1}{35}\right)$

78. $-\frac{5}{12}\left(-\frac{4}{25}\right)$

79. $\frac{5}{14} \div \frac{15}{28}$

80. $-\frac{7}{10} \div \frac{4}{15}$

81. $\frac{-\frac{3}{4}}{-\frac{7}{8}}$

82. $\frac{\frac{15}{32}}{-5}$

83. $\frac{\frac{5}{9}}{0}$

84. $\frac{0}{12}$

85. $\frac{5.25}{0.25}$

86. $(5.2)(16.8)$

In Exercises 87–90, use a calculator to evaluate the expression. Round your answer to two decimal places.

87. $(5.8)^4 - (3.2)^5$

88. $\frac{(15.8)^3}{(2.3)^8}$

89. $\frac{3000}{(1.05)^{10}}$

90. $500\left(1 + \frac{0.07}{4}\right)^{40}$

1.4 In Exercises 91–94, evaluate the exponential expression.

91. 7^3

92. $(-5)^2$

93. $(-7)^3$

94. $-(-2)^4$

In Exercises 95–98, insert the correct symbol (<, >, or =) between the numbers.

95. $2^2 \quad \square \quad 2^4$

96. $(-3)^2 \quad \square \quad (-3)^3$

97. $\frac{3}{4} \quad \square \quad \left(\frac{3}{4}\right)^2$

98. $\left(\frac{2}{3}\right)^3 \quad \square \quad \left(\frac{2}{3}\right)^2$

In Exercises 99–114, evaluate the expression using the order of operations.

99. $\left(\frac{3}{5}\right)^4$

100. $\frac{2}{6^3}$

101. $240 - (4^2 \cdot 5)$

102. $5^2 - (625 \cdot 5^2)$

103. $3^2(10 - 2^2)$

104. $-5(16 - 5^2)$

105. $\left(\frac{3}{4}\right)\left(\frac{5}{6}\right) + 4$

106. $75 - 24 \div 2^3$

107. $122 - [45 - (32 + 8) - 23]$

108. $-58 - (48 - 12) - (-30 - 4)$

109. $\frac{6 \cdot 4 - 36}{4}$

110. $\frac{144}{2 \cdot 3 \cdot 3}$

111. $\frac{54 - 4 \cdot 3}{6}$

112. $\frac{3 \cdot 5 + 125}{10}$

113. $\frac{78 - |-78|}{5}$

114. $\frac{300}{15 - |-15|}$

In Exercises 115–122, identify the property of real numbers that justifies the statement.

115. $123 - 123 = 0$

116. $9 \cdot \frac{1}{9} = 1$

117. $14(3) = 3(14)$

118. $5(3x) = (5 \cdot 3)x$

119. $17 \cdot 1 = 17$

120. $10 + 6 = 6 + 10$

121. $-2(7 + x) = -2 \cdot 7 + (-2)x$

122. $2 + (3 + x) = (2 + 3) + x$

Solving Problems

123. *Think About It* Which is smaller: $\frac{2}{3}$ or 0.6?

124. *Think About It* An integer n is divisible by 3 and the quotient is also divisible by 3. What does this tell you about n? Give some examples.

125. *True or False?* The sum of two integers, one negative and one positive, is negative. Explain.

126. *True or False?* The product of two integers, one negative and one positive, is negative. Explain.

127. *Think About It* You rotate the tires on your truck, including the spare, so that all five tires are used equally. After 40,000 miles, how many miles has each tire been driven?

128. *Total Cost* You have purchased a television set. In addition to a down payment of \$75 you must make nine monthly payments of \$25 each. What is the total amount you will pay for the product?

129. *Reading a Table* The costs of adult and student tickets for a concert are \$25 and \$10, respectively. The following table gives the numbers of tickets sold the first 4 days of sales.

Day	1	2	3	4
Adult	162	98	148	186
Student	98	64	81	105

(a) Find the revenue from ticket sales each day.

(b) Find the revenue from ticket sales for each type of ticket.

(c) Find the total revenue from ticket sales using part (a). Find the total revenue from ticket sales using part (b). Do your answers agree? Does this provide a sufficient check for your work?

130. *Reading a Graph* The bar graph shows the popular votes (in millions) cast for president in the presidential elections from 1972 through 1996. (Source: U.S. Bureau of the Census)

(a) Estimate the total popular votes cast for the candidates in 1980.

(b) Estimate the difference between the votes cast for the winning candidate and for the independent candidate in 1992.

(c) Estimate the total popular votes cast for each of the elections. Did the number of votes cast increase with time?

(d) Describe how you can use the graph to determine whether the winning candidate received more than one-half of the popular vote. Did this always occur for the elections shown on the graph?

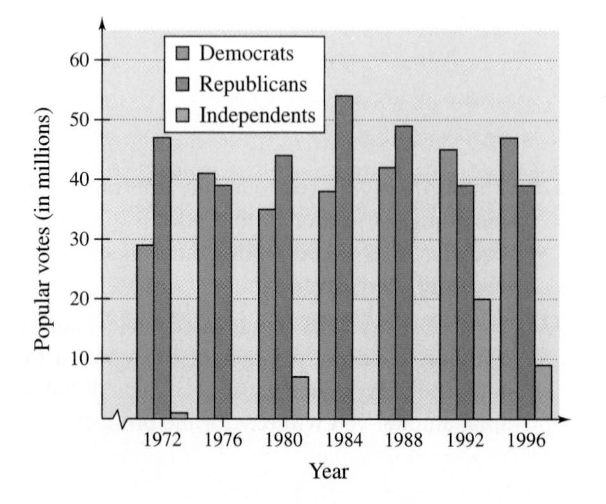

131. *Reading a Table* Initially, a share of stock cost $\$35\frac{1}{4}$. The daily changes in closing values during the week are shown in the table. Determine the closing price of a share on Friday.

Day	Mon	Tue	Wed	Thu	Fri
Change	$-\frac{3}{8}$	$-\frac{1}{2}$	$-\frac{1}{8}$	$+1\frac{1}{4}$	$+\frac{1}{2}$

132. *Fuel Consumption* The morning and evening readings of the fuel gauge on a car were $\frac{7}{8}$ and $\frac{1}{3}$. What fraction of the tank of fuel was used that day?

133. *Telephone Charge* A telephone call costs \$0.64 for the first minute plus \$0.72 for each additional minute. Find the cost of a 5-minute call.

134. *Snowfall Rate* During an 8-hour period, $6\frac{3}{4}$ inches of snow fell. What was the average rate of snowfall per hour?

135. *Depreciation* After 3 years, the value of a \$16,000 car is given by $16{,}000\left(\frac{3}{4}\right)^3$.

(a) What is the value of the car after 3 years?

(b) How much has the car depreciated during the 3 years?

136. *Geometry* The volume of water in a hot tub is given by $V = 6^2 \cdot 3$. How many cubic feet of water will the hot tub hold? Find the total weight of the water in the tub. (Use the fact that 1 cubic foot of water weighs 62.4 pounds.)

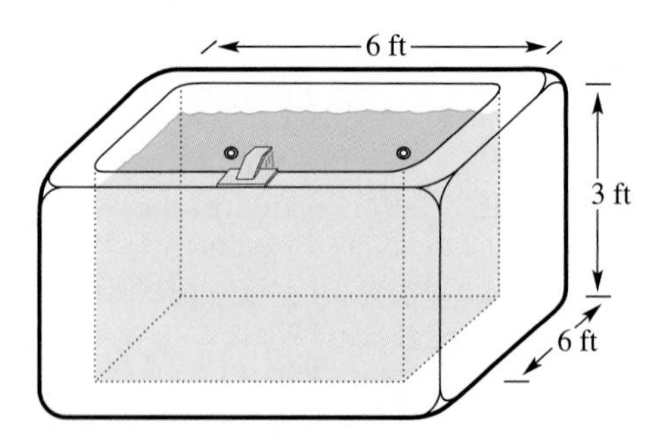

Chapter Test

Take this test as you would take a test in class. After you are done, check your work against the answers given in the back of the book.

1. Which of the following are (a) natural numbers? (b) integers? (c) rational numbers?

$-10, 8, \frac{3}{4}, \frac{12}{4}, 6.5$

2. Place the correct inequality symbol (< or >) between the real numbers.

$-\frac{3}{5} \quad \square \quad -|-2|$

In Exercises 3–16, evaluate the expression.

3. $16 + (-20)$

4. $-50 - (-60)$

5. $7 + |-3|$

6. $64 - (25 - 8)$

7. $-5(32)$

8. $\dfrac{-72}{-9}$

9. $\dfrac{12 + 9}{7}$

10. $-\dfrac{(-2)(5)}{10}$

11. $\frac{5}{6} - \frac{1}{8}$

12. $-27\left(\frac{5}{6}\right)$

13. $\dfrac{7}{16} \div \dfrac{21}{28}$

14. $\dfrac{-8.1}{0.3}$

15. $-\left(\frac{2}{3}\right)^2$

16. $35 - (50 \div 5^2)$

In Exercises 17–20, state the property of real numbers that justifies the statement.

17. $3(4 + 6) = 3 \cdot 4 + 3 \cdot 6$

18. $5 \cdot \frac{1}{5} = 1$

19. $3 + (4 + 8) = (3 + 4) + 8$

20. $3(x + 2) = (x + 2)3$

21. Write the fraction $\frac{30}{72}$ in simplest form.

22. Explain why -3^4 is not equal to $(-3)^4$.

23. State the order of operations for the expression $32 - 3 \cdot 2^3$.

24. Copy the figure shown below. Then shade two-thirds of the figure. Write two different fractions that are represented by the shaded region.

2 Fundamentals of Algebra

Joseph E. Ramir

Party and special event rentals represent one facet of the equipment rental industry. Today consumers can rent just about any type of equipment from home and garden tools to fine china.

Motivating the Chapter

Beachwood Rental

Beachwood Rental is a rental company specializing in equipment for parties and special events. A wedding ceremony is to be held under a canopy that contains 15 rows of 12 chairs.

See Section 2.1, Exercise 89.

a. Let c represent the rental cost of a chair. Write an expression that represents the cost of renting all of the chairs under the canopy. The table at the right lists the rental prices for two types of chairs. Use the expression you wrote to find the cost of renting the plastic chairs and the cost of renting the wood chairs.

Chair rental	
Plastic	\$1.95
Wood	\$2.95

b. The table at the right lists the available canopy sizes. The rental rate for a canopy is $115 + 0.25t$ dollars, where t represents the size of the canopy in square feet. Find the cost of each canopy. (*Hint*: The total area under a 20 by 20 foot canopy is $20 \cdot 20 = 400$ square feet.)

Canopy sizes	
Canopy 1	20 by 20 feet
Canopy 2	20 by 30 feet
Canopy 3	30 by 40 feet
Canopy 4	30 by 60 feet
Canopy 5	40 by 60 feet

The figure at the right shows the arrangement of the chairs under the canopy. Beachwood Rental recommends the following.

Width of center aisle—Three times the space between rows
Width of side aisle—Two times the space between rows
Depth of rear aisle—Two times the space between rows
Depth of front region—Seven feet more than three times the space between rows

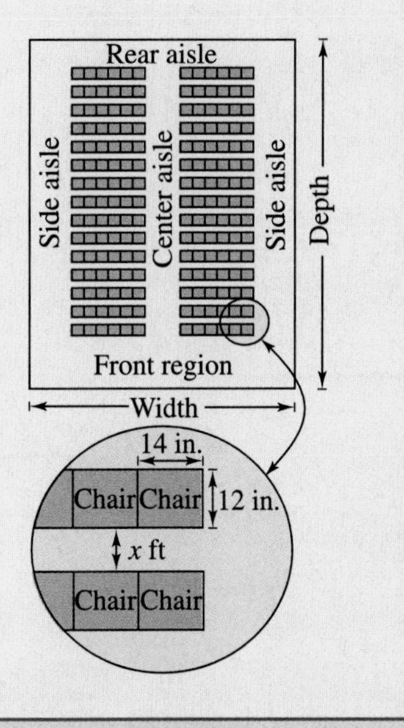

See Section 2.3, Exercise 80.

c. Let x represent the space between rows of chairs. Write an expression for the width of the center aisle. Write an expression for the width of a side aisle.

d. Each chair is 14 inches wide. Convert the width of a chair to feet. Write an expression for the width of the canopy.

e. Write an expression for the depth of the rear aisle. Write an expression for the depth of the front region.

f. Each chair is 12 inches deep. Convert the depth of a chair to feet. Write an expression for the depth of the canopy.

g. If $x = 2$ feet, what is the width of the center aisle? What are the width and depth of the canopy? What size canopy do you need? What is the total rental cost of the canopy and chairs if the wood chairs are used?

h. What could be done to save on the rental cost?

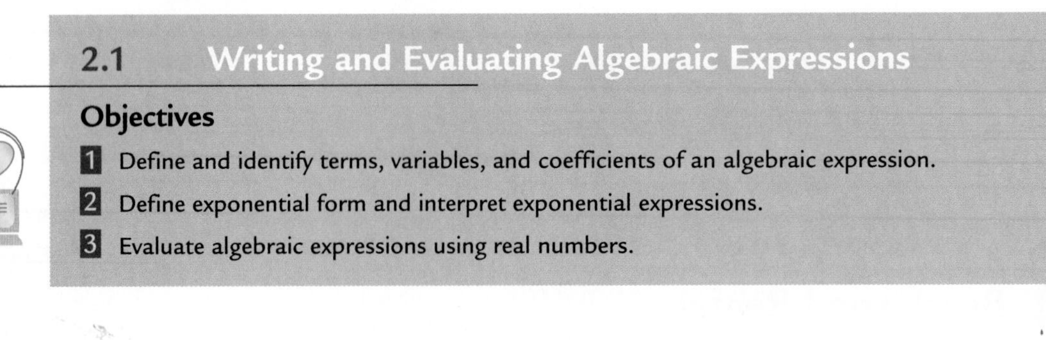

2.1 Writing and Evaluating Algebraic Expressions

Objectives

1. Define and identify terms, variables, and coefficients of an algebraic expression.
2. Define exponential form and interpret exponential expressions.
3. Evaluate algebraic expressions using real numbers.

1 Define and identify terms, variables, and coefficients of an algebraic expression.

Variables and Algebraic Expressions

One of the distinguishing characteristics of algebra is its use of symbols to represent quantities whose numerical values are unknown. Here is a simple example.

Example 1 Writing an Algebraic Expression

You accept a part-time job for \$6 per hour. The job offer states that you will be expected to work between 15 and 30 hours a week. Because you don't know how many hours you will work during a week, your total income for a week is unknown. Moreover, your income will probably *vary* from week to week. By representing the variable quantity (the number of hours worked) by the letter x, you can represent the weekly income by the following *algebraic expression.*

\$6 per hour — Number of hours worked

$$6x$$

In the product $6x$, the number 6 is a *constant* and the letter x is a *variable.*

Algebraic Expression

A collection of letters (**variables**) and real numbers (**constants**) combined by using addition, subtraction, multiplication, or division is an **algebraic expression.**

Some examples of algebraic expressions are

$$3x + y, \quad -5a^3, \quad 2W - 7, \quad \frac{x}{y+3}, \quad \text{and} \quad x^2 - 4x + 5.$$

The **terms** of an algebraic expression are those parts that are separated by *addition.* For example, the expression $x^2 - 4x + 5$ has three terms: x^2, $-4x$, and 5. Note that $-4x$, rather than $4x$, is a term of $x^2 - 4x + 5$ because

$$x^2 - 4x + 5 = x^2 + (-4x) + 5. \quad \text{To subtract, add the opposite.}$$

For variable terms such as x^2 and $-4x$, the numerical factor is the **coefficient** of the term. Here, the coefficient of x^2 is 1 and the coefficient of $-4x$ is -4.

Example 2 Identifying the Terms of an Algebraic Expression

Identify the terms of each algebraic expression.

a. $x + 2$ **b.** $3x + \frac{1}{2}$ **c.** $2y - 5x - 7$

d. $5(x - 3) + 3x - 4$ **e.** $4 - 6x + \frac{x + 9}{3}$

Solution

	Algebraic Expression	*Terms*
a.	$x + 2$	$x, 2$
b.	$3x + \frac{1}{2}$	$3x, \frac{1}{2}$
c.	$2y - 5x - 7$	$2y, -5x, -7$
d.	$5(x - 3) + 3x - 4$	$5(x - 3), 3x, -4$
e.	$4 - 6x + \frac{x + 9}{3}$	$4, -6x, \frac{x + 9}{3}$

The terms of an algebraic expression depend on the way the expression is written. Rewriting the expression can (and, in fact, usually does) change its terms. For instance, the expression $2 + 4 - x$ has three terms, but the equivalent expression $6 - x$ has only two terms.

Example 3 Identifying Coefficients

Identify the coefficient of each of the following terms.

a. $-5x^2$ **b.** x^3 **c.** $\frac{2x}{3}$ **d.** $-\frac{x}{4}$ **e.** $-x^3$

Solution

	Term	*Coefficient*	*Comment*
a.	$-5x^2$	-5	Note that $-5x^2 = (-5)x^2$.
b.	x^3	1	Note that $x^3 = 1 \cdot x^3$.
c.	$\frac{2x}{3}$	$\frac{2}{3}$	Note that $\frac{2x}{3} = \frac{2}{3}(x)$.
d.	$-\frac{x}{4}$	$-\frac{1}{4}$	Note that $-\frac{x}{4} = -\frac{1}{4}(x)$.
e.	$-x^3$	-1	Note that $-x^3 = (-1)x^3$.

2 Define exponential form and interpret exponential expressions.

Exponential Form

You know from Section 1.4 that a number raised to a power can be evaluated by repeated multiplication. For example, 7^4 represents the product obtained by multiplying 7 by itself four times.

$$\underset{\text{Base}}{7}^{\overset{\text{Exponent}}{4}} = \underbrace{7 \cdot 7 \cdot 7 \cdot 7}_{4 \text{ factors}}$$

In general, for any positive integer n and any real number a, you have

$$a^n = \underbrace{a \cdot a \cdot a \cdot \cdot \cdot a}_{n \text{ factors}}.$$

This rule applies to factors that are *variables* as well as to factors that are *algebraic expressions.*

▶ Definition of Exponential Form

Let n be a positive integer and let a be a real number, a variable, or an algebraic expression.

$$a^n = \underbrace{a \cdot a \cdot a \cdot \cdot \cdot a}_{n \text{ factors}}.$$

In this definition remember that the letter a can be a number, a variable, or an algebraic expression. It may be helpful to think of a as a box into which you can place any algebraic expression.

$$\square^n = \square \cdot \square \cdot \cdot \cdot \square$$

The box may contain a number, a variable, or an algebraic expression.

Example 4 Interpreting Exponential Expressions

a. $3^4 = 3 \cdot 3 \cdot 3 \cdot 3$

b. $3x^4 = 3 \cdot x \cdot x \cdot x \cdot x$

c. $(-3x)^4 = (-3x)(-3x)(-3x)(-3x) = (-3)(-3)(-3)(-3) \cdot x \cdot x \cdot x \cdot x$

d. $(y + 2)^3 = (y + 2)(y + 2)(y + 2)$

e. $(5x)^2y^3 = (5x)(5x)y \cdot y \cdot y = 5 \cdot 5 \cdot x \cdot x \cdot y \cdot y \cdot y$

Be sure you understand the priorities for order of operations involving exponents. Here are two examples that tend to cause problems.

Expression	*Correct Evaluation*	*Incorrect Evaluation*
-3^2	$-(3 \cdot 3) = -9$	~~$(-3)(-3) = 9$~~
$3x^2$	$3 \cdot x \cdot x$	~~$(3x)(3x)$~~

3 Evaluate algebraic expressions using real numbers.

Evaluating Algebraic Expressions

In applications of algebra, you are often required to **evaluate** an algebraic expression. This means you are to find the *value* of an expression when its variables are replaced by real numbers. For instance, when $x = 2$, the value of the expression $2x + 3$ is as follows.

Expression	*Replace x by 2.*	*Value of Expression*
$2x + 3$	$2(2) + 3$	7

When finding the value of an algebraic expression, be sure to replace every occurrence of the specified variable with the appropriate real number. For instance, when $x = -2$, the value of $x^2 - x + 3$ is

$$(-2)^2 - (-2) + 3 = 4 + 2 + 3 = 9.$$

Example 5 Evaluating Algebraic Expressions

Evaluate each expression when $x = -3$ and $y = 5$.

a. $-x$ **b.** $x - y$ **c.** $3x + 2y$

d. $y - 2(x + y)$ **e.** $y^2 - 3y$

Solution

a. When $x = -3$, the value of $-x$ is

$-x = -(-3)$ Substitute -3 for x.

$= 3.$ Simplify.

b. When $x = -3$ and $y = 5$, the value of $x - y$ is

$x - y = -3 - 5$ Substitute -3 for x and 5 for y.

$= -8.$ Simplify.

c. When $x = -3$ and $y = 5$, the value of $3x + 2y$ is

$3x + 2y = 3(-3) + 2(5)$ Substitute -3 for x and 5 for y.

$= -9 + 10$ Simplify.

$= 1.$ Simplify.

d. When $x = -3$ and $y = 5$, the value of $y - 2(x + y)$ is

$y - 2(x + y) = 5 - 2[(-3) + 5]$ Substitute -3 for x and 5 for y.

$= 5 - 2(2)$ Simplify.

$= 1.$ Simplify.

e. When $y = 5$, the value of $y^2 - 3y$ is

$y^2 - 3y = (5)^2 - 3(5)$ Substitute 5 for y.

$= 25 - 15$ Simplify.

$= 10.$ Simplify.

Study Tip

As shown in parts (a) and (d) of Example 5, it is a good idea to use parentheses when substituting a negative number for a variable.

Technology: Tip

Absolute value expressions can be evaluated on a graphing calculator using the key [ABS]. To evaluate $|-3|$, you can use the following keystrokes.

[ABS] [(–)] 3 [ENTER]

When evaluating an expression such as $|3 - 6|$, parentheses should surround the entire expression, as shown in the following keystrokes.

[ABS] [(] 3 [–] 6 [)] [ENTER]

Display: 3

Example 6 Evaluating Algebraic Expressions

Evaluate each expression when $x = 4$ and $y = -6$.

a. y^2 **b.** $-y^2$ **c.** $y - x$ **d.** $|y - x|$ **e.** $|x - y|$

Solution

a. When $y = -6$, the value of the expression y^2 is

$$y^2 = (-6)^2 = 36.$$

b. When $y = -6$, the value of the expression $-y^2$ is

$$-y^2 = -(y^2) = -(-6)^2 = -36.$$

c. When $x = 4$ and $y = -6$, the value of the expression $y - x$ is

$$y - x = (-6) - 4 = -6 - 4 = -10.$$

d. When $x = 4$ and $y = -6$, the value of the expression $|y - x|$ is

$$|y - x| = |-6 - 4| = |-10| = 10.$$

e. When $x = 4$ and $y = -6$, the value of the expression $|x - y|$ is

$$|x - y| = |4 - (-6)| = |4 + 6| = |10| = 10.$$

Example 7 Evaluating Algebraic Expressions

Evaluate each expression when $x = -5$, $y = -2$, and $z = 3$.

a. $\dfrac{y + 2z}{5y - xz}$ **b.** $(y + 2z)(z - 3y)$

Solution

a. When $x = -5$, $y = -2$, and $z = 3$, the value of the expression is

$$\frac{y + 2z}{5y - xz} = \frac{-2 + 2(3)}{5(-2) - (-5)(3)} \qquad \text{Substitute for } x, y, \text{ and } z.$$

$$= \frac{-2 + 6}{-10 + 15} \qquad \text{Simplify.}$$

$$= \frac{4}{5}. \qquad \text{Simplify.}$$

b. When $y = -2$ and $z = 3$, the value of the expression is

$$(y + 2z)(z - 3y) = [(-2) + 2(3)][3 - 3(-2)] \qquad \text{Substitute for } y \text{ and } z.$$

$$= (-2 + 6)(3 + 6) \qquad \text{Simplify.}$$

$$= 4(9) \qquad \text{Simplify.}$$

$$= 36. \qquad \text{Simplify.}$$

On occasion you may need to evaluate an algebraic expression for *several* values of x. In such cases, a table format is a useful way to organize the values of the expression.

Technology: Tip

If you have a graphing calculator, try using it to store and evaluate the expression given in Example 8. For instance, here are the steps that will evaluate $-9x + 6$ when $x = 2$ on a *TI-83*.

- Use the [Y=] key to store the expression as Y_1.
- [2nd] [QUIT]
- Store 2 in X.
  2 [STO▶] [X,T,Θ,n] [ENTER]
- Display Y_1.
  [VARS] [Y-VARS] [ENTER]
  [ENTER]
  and then press [ENTER] again.

Example 8 Repeated Evaluation of an Expression

Complete the following table by evaluating the expression $5x + 2$ for each value of x given in the table.

x	-1	0	1	2
$5x + 2$				

Solution

Begin by substituting each value of x into the expression.

When $x = -1$: $5x + 2 = 5(-1) + 2 = -5 + 2 = -3$
When $x = 0$: $5x + 2 = 5(0) + 2 = 0 + 2 = 2$
When $x = 1$: $5x + 2 = 5(1) + 2 = 5 + 2 = 7$
When $x = 2$: $5x + 2 = 5(2) + 2 = 10 + 2 = 12$

Once you have evaluated the expression for each value of x, fill in the table with the values.

x	-1	0	1	2
$5x + 2$	-3	2	7	12

Discussing the Concept **Error Analysis**

Suppose you are teaching an algebra class and one of your students hands in the following problem. What is the error in this work?

Evaluate $y - 2(x - y)$ when $x = 2$ and $y = -4$.

$$\begin{aligned} y - 2(x - y) &= -4 - 2(2 - 4) \\ &= -4 - 2(-2) \\ &= -4 + 4 \\ &= 0 \end{aligned}$$

What are some possible related errors? Discuss ways of helping students avoid these types of errors.

2.1 Exercises

Integrated Review *Concepts, Skills, and Problem Solving*

Keep mathematically in shape by doing these exercises *before* the problems of this section.

Properties and Definitions

In Exercises 1–4, identify the property illustrated by the equation.

1. $x(5) = 5x$

2. $10 - 10 = 0$

3. $3(t + 2) = 3t + 3 \cdot 2$

4. $7 + (8 + z) = (7 + 8) + z$

Simplifying Expressions

In Exercises 5–10, evaluate the expression.

5. $10 - |-7|$ **6.** $6 - (10 - 12)$

7. $\dfrac{3 - (5 - 20)}{4}$ **8.** $\dfrac{6}{7} - \dfrac{4}{7}$

9. $-\frac{3}{4}\left(\frac{28}{33}\right)$ **10.** $\frac{5}{8} \div \frac{3}{16}$

Problem Solving

11. You plan to save \$50 per month for 10 years. How much money will you set aside during the 10 years?

12. It is necessary to cut a 120-foot rope into eight pieces of equal length. What is the length of each piece?

Developing Skills

In Exercises 1–4, write an algebraic expression for the given statement. See Example 1.

1. The distance traveled in t hours if the average speed is 60 miles per hour

2. The cost of an amusement park ride for a family of n people if the cost per person is \$1.25

3. The cost of m pounds of meat if the cost per pound is \$2.19

4. The total weight of x 50-pound bags of fertilizer

In Exercises 5–8, identify the variables and constants in the expression.

5. $x + 3$ **6.** $y + 1$

7. $x + z$ **8.** $3^2 + z$

In Exercises 9–22, identify the terms of the expression. See Example 2.

9. $4x + 3$ **10.** $6x - 1$

11. $3x^2 + 5$ **12.** $5 - 3t^2$

13. $\frac{5}{3} - 3y^3$ **14.** $6x - \frac{2}{3}$

15. $2x - 3y + 1$ **16.** $x^2 + 18xy + y^2$

17. $3(x + 5) + 10$ **18.** $16 - (x + 1)$

19. $\dfrac{x}{4} + \dfrac{5}{x}$ **20.** $10 - \dfrac{t}{6}$

21. $\dfrac{3}{x + 2} - 3x + 4$ **22.** $x^2 + \dfrac{3x + 1}{x - 1} + 4$

In Exercises 23–32, identify the coefficient of the term. See Example 3.

23. $-6x$ **24.** $25y$

25. $-\frac{1}{3}y$ **26.** $\frac{1}{8}n$

27. $-\dfrac{3x}{2}$ **28.** $\dfrac{3x}{4}$

29. $2\pi x^2$ **30.** πt^4

31. $4.7u$ **32.** $-5.32b$

In Exercises 33–50, expand the expression as a product of factors. See Example 4.

33. y^5 **34.** x^6

35. 2^2x^4 **36.** 5^3x^2

37. $4y^2z^3$ **38.** $3uv^4$

39. $(a^2)^3$ **40.** $(z^3)^3$

41. $4x^3 \cdot x^4$ **42.** $a^2y^2 \cdot y^3$

43. $(ab)^3$ **44.** $2(xz)^4$

45. $(x + y)^2$

46. $(s - t)^5$

47. $\left(\frac{a}{3s}\right)^4$

48. $\left(\frac{2}{x + 1}\right)^3$

49. $[3(r + s)^2][3(r + s)]^2$

50. $[2(a - b)^3][2(a - b)](a - b)^2$

In Exercises 51–60, rewrite the product in exponential form.

51. $2 \cdot u \cdot u \cdot u \cdot u$

52. $\frac{1}{3} \cdot x \cdot x \cdot x \cdot x \cdot x$

53. $(2u) \cdot (2u) \cdot (2u) \cdot (2u)$

54. $\frac{1}{3}x \cdot \frac{1}{3}x \cdot \frac{1}{3}x \cdot \frac{1}{3}x \cdot \frac{1}{3}x$

55. $a \cdot a \cdot a \cdot b \cdot b$

56. $y \cdot y \cdot z \cdot z \cdot z \cdot z$

57. $3 \cdot (x - y) \cdot (x - y) \cdot 3 \cdot 3$

58. $(u - v) \cdot (u - v) \cdot 8 \cdot 8 \cdot 8 \cdot (u - v)$

59. $\left(\frac{x^2}{2}\right)\left(\frac{x^2}{2}\right)\left(\frac{x^2}{2}\right)$

60. $\frac{r - s}{5} \cdot \frac{r - s}{5} \cdot \frac{r - s}{5} \cdot \frac{r - s}{5}$

In Exercises 61–78, evaluate the algebraic expression for the given values of the variables. If it is not possible, state the reason. See Examples 5, 6, and 7.

Expression	*Values*
61. $2x - 1$	(a) $x = \frac{1}{2}$ (b) $x = 4$
62. $3x - 2$	(a) $x = \frac{4}{3}$ (b) $x = -1$
63. $2x^2 - 5$	(a) $x = -2$ (b) $x = 3$
64. $64 - 16t^2$	(a) $t = 2$ (b) $t = 3$
65. $3x - 2y$	(a) $x = 4, y = 3$ (b) $x = \frac{2}{3}, y = 1$
66. $10u - 3v$	(a) $u = 3, v = 10$ (b) $u = -2, v = -7$
67. $x - 3(x - y)$	(a) $x = 3, y = 3$ (b) $x = 4, y = -4$
68. $-3x + 2(x + y)$	(a) $x = -2, y = 2$ (b) $x = 0, y = 5$
69. $b^2 - 4ac$	(a) $a = 2, b = -3, c = -1$ (b) $a = -4, b = 6, c = -2$
70. $a^2 + 2ab$	(a) $a = -2, b = 3$ (b) $a = -2, b = 4$
71. $\frac{x - 2y}{x + 2y}$	(a) $x = 4, y = 2$ (b) $x = 4, y = -2$
72. $\frac{-y}{x^2 + y^2}$	(a) $x = 0, y = 5$ (b) $x = 1, y = -3$
73. $\frac{5x}{y - 3}$	(a) $x = 2, y = 4$ (b) $x = 2, y = 3$
74. $\frac{2x - y}{y^2 + 1}$	(a) $x = 1, y = 2$ (b) $x = 1, y = 3$

75. *Area of a Triangle*

$\frac{1}{2}bh$ (a) $b = 3, h = 5$ (b) $b = 2, h = 10$

76. *Volume of a Rectangular Prism*

lwh (a) $l = 4, w = 2, h = 9$ (b) $l = 10, w = 5, h = 20$

77. *Distance traveled*

rt (a) $r = 50, t = 3.5$ (b) $r = 35, t = 4$

78. *Simple interest*

Prt (a) $P = 1000, r = 0.08, t = 3$ (b) $P = 500, r = 0.07, t = 5$

79. *Finding a Pattern*

(a) Complete the following table by evaluating the expression $3x - 2$. See Example 8.

x	-1	0	1	2	3	4
$3x - 2$						

(b) Use the table to find the increase in the value of the expression for each 1-unit increase in x.

(c) From the pattern of parts (a) and (b), predict the increase in the algebraic expression $\frac{2}{3}x + 4$ for each 1-unit increase in x. Then verify your prediction.

80. *Finding a Pattern*

(a) Complete the table by evaluating the expression $3 - 2x$. See Example 8.

x	-1	0	1	2	3	4
$3 - 2x$						

(b) Use the table to find the change in the value of the expression for each 1-unit increase in x.

(c) From the pattern of parts (a) and (b), predict the change in the algebraic expression $4 - \frac{3}{2}x$ for each 1-unit increase in x. Then verify your prediction.

Solving Problems

Geometry In Exercises 81–84, find an expression for the area of the figure. Then evaluate the expression for the given value(s) of the variable(s).

81. $n = 8$ **82.** $x = 10, y = 3$

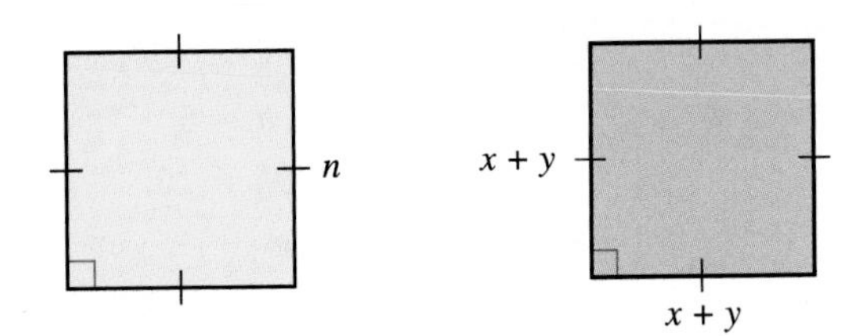

83. $a = 5, b = 4$ **84.** $x = 9$

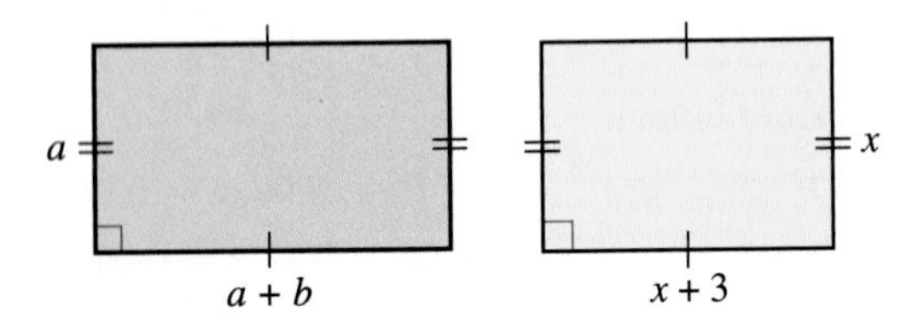

85. *Exploration* A convex polygon with n sides has

$$\frac{n(n-3)}{2}, \quad n \geq 4$$

diagonals. Verify the formula for (a) a square (2 diagonals), (b) a pentagon (5 diagonals), and (c) a hexagon (9 diagonals).

86. *Think About It* Explain why the formula in Exercise 85 will always yield a natural number for the number of diagonals.

87. *Iteration and Exploration* Once an expression has been evaluated for a specified value, the expression can be repeatedly evaluated by using the result of the previous evaluation as the input for the next evaluation.

(a) The procedure for repeated evaluation of the algebraic expression $\frac{1}{2}x + 3$ can be accomplished on a graphing utility in the following way.

- Clear the display.
- Enter 2 in the display and press ENTER.
- Enter $\frac{1}{2}$ * ANS + 3 and press ENTER.
- Each time ENTER is pressed the utility will evaluate the expression at the value of x obtained in the preceding computation. Continue the process six more times. What value does the expression appear to be approaching?

(b) Repeat part (a) starting with $x = 12$.

88. *Exploration* Repeat Exercise 87 using the expression $\frac{3}{4}x + 2$.

Explaining Concepts

89. Answer parts (a) and (b) of Motivating the Chapter on page 61.

90. Discuss the difference between terms and factors.

91. Is $3x$ a term of $4 - 3x$? Explain.

92. In the expression $(10x)^3$, what is $10x$ called? What is 3 called?

93. Is it possible to evaluate the expression

$$\frac{x+2}{y-3}$$

when $x = 5$ and $y = 3$? Explain.

The symbol [graphing utility icon] indicates an exercise in which you are instructed to use a graphing utility.

2.2 Simplifying Algebraic Expressions

Objectives

1. Apply the rules of exponents to rewrite exponential expressions.
2. Use the basic rules of algebra to combine like terms of an algebraic expression.
3. Simplify an algebraic expression by rewriting the terms.
4. Use the Distributive Property to remove symbols of grouping.

1 Apply the rules of exponents to rewrite exponential expressions.

Rules of Exponents

To simplify algebraic expressions, you often need to use some rules for operating with exponential expressions. Consider the following illustrations.

1. Multiplying exponential forms with like bases: *Rule*

$$a^3 \cdot a^2 = \underbrace{(a \cdot a \cdot a)}_{\text{3 factors}} \cdot \underbrace{(a \cdot a)}_{\text{2 factors}}$$

$$= \underbrace{a \cdot a \cdot a \cdot a \cdot a}_{\text{5 factors}} = a^5 = a^{2+3}$$

Add exponents.

2. Raising an exponential form to a power: *Rule*

$$(a^3)^2 = \underbrace{a^3 \cdot a^3}_{\text{2 factors of } a^3}$$

$$= \underbrace{(a \cdot a \cdot a)}_{\text{3 factors}} \cdot \underbrace{(a \cdot a \cdot a)}_{\text{3 factors}} = a^6 = a^{2 \cdot 3}$$

Multiply exponents.

3. Raising a product to a power: *Rule*

$$(a \cdot b)^3 = \underbrace{(a \cdot b) \cdot (a \cdot b) \cdot (a \cdot b)}_{\text{3 factors of } (a \cdot b)}$$

$$= \underbrace{(a \cdot a \cdot a)}_{\text{3 factors}} \cdot \underbrace{(b \cdot b \cdot b)}_{\text{3 factors}} = a^3 \cdot b^3$$

Apply exponent to each factor.

These illustrations suggest the following rules for exponential forms.

> **Rules of Exponents**
>
> Let m and n be positive integers, and let a and b be real numbers, variables, or variable expressions. Then, the following are true.
>
> **1.** $a^m \cdot a^n = a^{m+n}$ **2.** $(a^m)^n = a^{m \cdot n}$ **3.** $(ab)^m = a^m \cdot b^m$

Rules 1 and 3 can be extended to three or more factors such as $a^m \cdot a^n \cdot a^k = a^{m+n+k}$ and $(abc)^m = a^m \cdot b^m \cdot c^m$.

Example 1 Simplifying Products Involving Exponential Forms

Simplify each expression.

a. $5^2 \cdot 5^6 \cdot 5$ **b.** b^4b^2b **c.** $3^2x^3 \cdot x$

d. $(-9x^2)(-3x^5)$ **e.** $(2x^2y)(-xy^4)$

Solution

a. $5^2 \cdot 5^6 \cdot 5 = 5^{2+6+1} = 5^9$

b. $b^4b^2b = b^{4+2+1} = b^7$

c. $3^2x^3 \cdot x = (3^2)(x^{3+1}) = 9x^4$

d. $(-9x^2)(-3x^5) = (-9)(-3)(x^2 \cdot x^5) = 27(x^{2+5}) = 27x^7$

e. $(2x^2y)(-xy^4) = (2)(-1)(x^2 \cdot x)(y \cdot y^4) = -2x^{2+1}y^{1+4} = -2x^3y^5$

Be sure you see the difference between the expressions

$x^3 \cdot x^4$ and $x^3 + x^4$.

The first is a *product* of exponential forms, whereas the second is a *sum* of exponential forms. The rule for multiplying exponential forms having the same base can be applied to the first expression, but *not* to the second expression.

Example 2 Applying the Rules of Exponents

Use the rules of exponents to simplify each of the following.

a. $(2^3)^4$ **b.** $(y^2)^3$ **c.** $[(x + 2)^3]^3$ **d.** $(3x)^3$

e. $(-x)^4$ **f.** $(2x^2)^3$ **g.** $x(x^3y^2)^3$

Solution

a. $(2^3)^4 = 2^{3 \cdot 4} = 2^{12} = 4096$

b. $(y^2)^3 = y^{2 \cdot 3} = y^6$

c. $[(x + 2)^3]^3 = (x + 2)^{3 \cdot 3} = (x + 2)^9$

d. $(3x)^3 = 3^3 \cdot x^3 = 27x^3$

e. $(-x)^4 = (-1)^4x^4 = x^4$

f. $(2x^2)^3 = 2^3(x^2)^3 = 2^3x^{2 \cdot 3} = 8x^6$

g. $x(x^3y^2)^3 = x(x^{3 \cdot 3}y^{2 \cdot 3}) = x(x^9y^6) = x^{1+9}y^6 = x^{10}y^6$

It is important to recognize that the Rules of Exponents apply to products and not to sums or differences. Note the following illustrations.

Product	*Example*
$x^5 \cdot x^4 = x^{5+4}$	$2^5 \cdot 2^4 \stackrel{?}{=} 2^{5+4}$
	$512 = 512$
Sum	
$x^5 + x^4 \neq x^{5+4}$	$2^5 + 2^4 \neq 2^{5+4}$
	$48 \neq 512$

2 Use the basic rules of algebra to combine like terms of an algebraic expression.

Basic Rules of Algebra

Knowing the rules of exponents, you are now ready to combine algebraic expressions using the basic rules of algebra. You'll discover as you review the following table of rules that they are the same as the properties of real numbers given on page 48. The only difference is that the *input* for algebra rules can be real numbers, variables, or algebraic expressions.

▶ Basic Rules of Algebra

Let a, b, and c represent real numbers, variables, or algebraic expressions.

Property	*Example*
Commutative Property of Addition:	
$a + b = b + a$	$3x + x^2 = x^2 + 3x$
Commutative Property of Multiplication:	
$ab = ba$	$(5 + x)x^3 = x^3(5 + x)$
Associative Property of Addition:	
$(a + b) + c = a + (b + c)$	$(2x + 7) + x^2 = 2x + (7 + x^2)$
Associative Property of Multiplication:	
$(ab)c = a(bc)$	$(2x \cdot 5y) \cdot 7 = 2x \cdot (5y \cdot 7)$
Distributive Property:	
$a(b + c) = ab + ac$	$4x(7 + 3x) = 4x \cdot 7 + 4x \cdot 3x$
$(a + b)c = ac + bc$	$(2y + 5)y = 2y \cdot y + 5 \cdot y$
Additive Identity Property:	
$a + 0 = 0 + a = a$	$3y^2 + 0 = 0 + 3y^2 = 3y^2$
Multiplicative Identity Property:	
$a \cdot 1 = 1 \cdot a = a$	$(-2x^3) \cdot 1 = 1 \cdot (-2x^3) = -2x^3$
Additive Inverse Property:	
$a + (-a) = 0$	$3y^2 + (-3y^2) = 0$
Multiplicative Inverse Property:	
$a \cdot \dfrac{1}{a} = 1, \quad a \neq 0$	$(x^2 + 2) \cdot \dfrac{1}{x^2 + 2} = 1$

Because subtraction is defined as "adding the opposite," the Distributive Property is also true for subtraction. That is,

$$a(b - c) = ab - ac \quad \text{and} \quad (a - b)c = ac - bc.$$

Example 3 Applying the Basic Rules of Algebra

Use the indicated rule to complete the statement.

a. Additive Identity Property: $(x - 2) + \square = x - 2$

b. Commutative Property of Multiplication: $5(y + 6) = \square$

c. Commutative Property of Addition: $5(y + 6) = \square$

d. Distributive Property: $5(y + 6) = \square$

e. Associative Property of Addition: $(x^2 + 3) + 7 = \square$

f. Additive Inverse Property: $\square + 3x^2 = 0$

Solution

a. $(x - 2) + 0 = x - 2$

b. $5(y + 6) = (y + 6)5$

c. $5(y + 6) = 5(6 + y)$

d. $5(y + 6) = 5y + 5(6)$

e. $(x^2 + 3) + 7 = x^2 + (3 + 7)$

f. $-3x^2 + 3x^2 = 0$

Example 4 illustrates some common uses of the Distributive Property. Study this example carefully. Such uses of the Distributive Property are very important in algebra. Applying the Distributive Property as illustrated in Example 4 is called **expanding** an algebraic expression.

Example 4 Using the Distributive Property

Use the Distributive Property to expand each expression.

a. $2(7 - x)$ **b.** $(10 - 2y)3$ **c.** $2x(x + 4y)$ **d.** $-(1 - 2y + x)$

Solution

a. $2(7 - x) = 2 \cdot 7 - 2 \cdot x$
$= 14 - 2x$

b. $(10 - 2y)3 = 10(3) - 2y(3)$
$= 30 - 6y$

c. $2x(x + 4y) = 2x(x) + 2x(4y)$
$= 2x^2 + 8xy$

d. $-(1 - 2y + x) = (-1)(1 - 2y + x)$
$= (-1)(1) - (-1)(2y) + (-1)(x)$
$= -1 + 2y - x$

Study Tip

In Example 4(d) the negative sign is distributed over each term in the parentheses by multiplying each term by -1.

In the next example, note how area can be used to demonstrate the Distributive Property.

Example 5 The Distributive Property and Area

Write the area of each component part of the figure. Then demonstrate the Distributive Property by writing the total area of each figure in two ways.

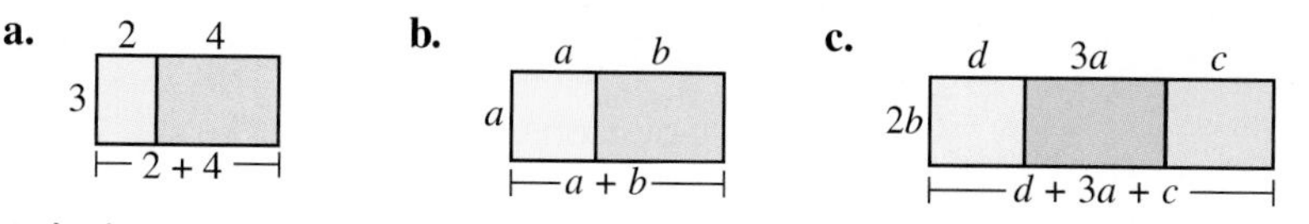

Solution

a.

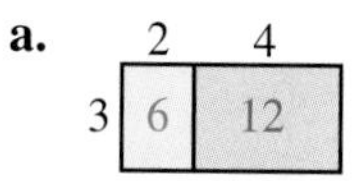

The total area is $3(2 + 4) = 3 \cdot 2 + 3 \cdot 4$.

b.

a b

a | a^2 | $a \cdot b$

The total area is $a(a + b) = a \cdot a + a \cdot b$.

c.

d $3a$ c

$2b$ | $2bd$ | $6ab$ | $2bc$

The total area is $2b(d + 3a + c) = 2bd + 6ab + 2bc$.

Two or more terms of an algebraic expression can be combined only if they are *like terms.*

Definition of Like Terms

In an algebraic expression, two terms are said to be **like terms** if they are both constant terms or if they have the same variable factor(s). Factors such as x in $5x$ and ab in $6ab$ are called **variable factors.**

The terms $5x$ and $-3x$ are like terms because they have the same variable factor, x. Similarly, $3x^2y$, $-x^2y$, and $\frac{1}{3}(x^2y)$ are like terms because they have the same variable factor, x^2y.

Study Tip

Notice in Example 6(b) that x^2 and $3x$ are *not* like terms because the variable x is not raised to the same power in both terms.

Example 6 Identifying Like Terms in Expressions

Expression	*Like Terms*
a. $5xy + 1 - xy$	$5xy$ and $-xy$
b. $12 - x^2 + 3x - 5$	12 and -5
c. $7x - 3 - 2x + 5$	$7x$ and $-2x$, -3 and 5

To combine like terms in an algebraic expression, you can simply add their respective coefficients and attach the common variable factor. This is actually an application of the Distributive Property, as shown in Example 7.

Example 7 Using the Distributive Property

Simplify each expression by combining like terms.

a. $5x + 2x - 4$ **b.** $-5 + 8 + 7y - 5y$ **c.** $2y - 3x - 4x$

Solution

a. $5x + 2x - 4 = (5 + 2)x - 4$ Distributive Property

$= 7x - 4$ Simplest form

b. $-5 + 8 + 7y - 5y = (-5 + 8) + (7 - 5)y$ Distributive Property

$= 3 + 2y$ Simplest form

c. $2y - 3x - 4x = 2y - x(3 + 4)$ Distributive Property

$= 2y - x(7)$ Simplify.

$= 2y - 7x$ Simplest form

Often, you need to use other rules of algebra before you can apply the Distributive Property to combine like terms. This is illustrated in the next example.

Example 8 Using Rules of Algebra to Combine Like Terms

Simplify each expression by combining like terms.

a. $7x + 3y - 4x$ **b.** $12a - 5 - 3a + 7$ **c.** $y - 4x - 7y + 9y$

Solution

a. $7x + 3y - 4x = 3y + 7x - 4x$ Commutative Property

$= 3y + (7x - 4x)$ Associative Property

$= 3y + (7 - 4)x$ Distributive Property

$= 3y + 3x$ Simplest form

b. $12a - 5 - 3a + 7 = 12a - 3a - 5 + 7$ Commutative Property

$= (12a - 3a) + (-5 + 7)$ Associative Property

$= (12 - 3)a + (-5 + 7)$ Distributive Property

$= 9a + 2$ Simplest form

c. $y - 4x - 7y + 9y = -4x + (y - 7y + 9y)$ Collect like terms.

$= -4x + (1 - 7 + 9)y$ Distributive Property

$= -4x + 3y$ Simplest form

Study Tip

As you gain experience with the rules of algebra, you may want to combine some of the steps in your work. For instance, you might feel comfortable listing only the following steps to solve part (b) of Example 8.

$12a - 5 - 3a + 7$

$= (12a - 3a) + (-5 + 7)$

$= 9a + 2$

3 Simplify an algebraic expression by rewriting the terms.

Simplifying Algebraic Expressions

Simplifying an algebraic expression by rewriting it in a more usable form is one of the three most frequently used skills in algebra. You will study the other two—solving an equation and sketching the graph of an equation—later in this text.

To "simplify an algebraic expression" generally means to remove symbols of grouping and combine like terms. For instance, the expression $x + (3 + x)$ can be simplified as $2x + 3$.

Example 9 Simplifying Algebraic Expressions

Simplify each expression.

a. $-3(-5x)$

b. $7(-x)$

Solution

a. $-3(-5x) = (-3)(-5)x$	Associative Property
$= 15x$	Simplest form
b. $7(-x) = 7(-1)(x)$	Coefficient of $-x$ is -1.
$= -7x$	Simplest form

Example 10 Simplifying Algebraic Expressions

Simplify each expression.

a. $\frac{5x}{3} \cdot \frac{3}{5} = \left(\frac{5}{3} \cdot x\right) \cdot \frac{3}{5}$	Coefficient of $\frac{5x}{3}$ is $\frac{5}{3}$.
$= \left(\frac{5}{3} \cdot \frac{3}{5}\right) \cdot x$	Commutative and Associative Properties
$= 1 \cdot x$	Multiplicative Inverse
$= x$	Multiplicative Identity
b. $x^2(-2x^3) = (-2)(x^2 \cdot x^3)$	Commutative and Associative Properties
$= -2x^{2+3}$	Rule of exponents
$= -2x^5$	Simplest form
c. $(-2x)(4x) = (-2 \cdot 4)(x \cdot x)$	Commutative and Associative Properties
$= -8(x^{1+1})$	Rule of exponents
$= -8x^2$	Simplest form
d. $(2rs)(r^2s) = 2(r \cdot r^2)(s \cdot s)$	Commutative and Associative Properties
$= 2(r^{1+2})(s^{1+1})$	Rule of exponents
$= 2r^3s^2$	Simplest form

4 Use the Distributive Property to remove symbols of grouping.

Symbols of Grouping

The main tool for removing symbols of grouping is the Distributive Property, as illustrated in Example 11. You may want to review order of operations in Section 1.4.

Study Tip

When a parenthetical expression is preceded by a *plus* sign, you can remove the parentheses without changing the signs of the terms inside.

$$3y + (-2y + 7)$$
$$= 3y - 2y + 7$$

When a parenthetical expression is preceded by a *minus* sign, however, you must change the sign of each term to remove the parentheses.

$$3y - (2y - 7)$$
$$= 3y - 2y + 7$$

Remember that $-(2y - 7)$ is equal to $(-1)(2y - 7)$, and the Distributive Property can be used to "distribute the minus sign."

Example 11 Removing Symbols of Grouping

Simplify each expression.

a. $-(2y - 7)$ **b.** $5x + (x - 7)2$

c. $-2(4x - 1) + 3x$ **d.** $3(y - 5) - (2y - 7)$

Solution

a. $-(2y - 7) = -2y + 7$ Distributive Property

b. $5x + (x - 7)2 = 5x + 2x - 14$ Distributive Property

$= 7x - 14$ Combine like terms.

c. $-2(4x - 1) + 3x = -8x + 2 + 3x$ Distributive Property

$= -8x + 3x + 2$ Commutative Property

$= -5x + 2$ Combine like terms.

d. $3(y - 5) - (2y - 7) = 3y - 15 - 2y + 7$ Distributive Property

$= (3y - 2y) + (-15 + 7)$ Group like terms.

$= y - 8$ Combine like terms.

Example 12 Removing Nested Symbols of Grouping

Simplify each expression.

a. $2[-2(1 - 3x)] = 2[-2 + 6x]$ Distributive Property

$= -4 + 12x$ Distributive Property

b. $5x - 2[4x + 3(x - 1)]$

$= 5x - 2[4x + 3x - 3]$ Distributive Property

$= 5x - 2[7x - 3]$ Combine like terms.

$= 5x - 14x + 6$ Distributive Property

$= -9x + 6$ Combine like terms.

c. $-7y + 3[2y - (3 - 2y)] - 5y + 4$

$= -7y + 3[2y - 3 + 2y] - 5y + 4$ Distributive Property

$= -7y + 3[4y - 3] - 5y + 4$ Combine like terms.

$= -7y + 12y - 9 - 5y + 4$ Distributive Property

$= (-7y + 12y - 5y) + (-9 + 4)$ Group like terms.

$= -5$ Combine like terms.

Example 13 Simplifying Algebraic Expressions

Simplify each expression.

a. $(-3x)(5x^4) + 7x^5 = (-3)(5)x \cdot x^4 + 7x^5$ — Commutative and Associative Properties

$= -15x^5 + 7x^5$ — Rule of exponents

$= -8x^5$ — Combine like terms.

b. $2x(x + 3y) + 4(5 - xy) = 2x^2 + 6xy + 20 - 4xy$ — Distributive Property

$= 2x^2 + 6xy - 4xy + 20$ — Commutative Property

$= 2x^2 + 2xy + 20$ — Combine like terms.

The next example illustrates the use of the Distributive Property with fractional expressions.

Example 14 Simplifying Fractional Expressions

Simplify each expression.

a. $\frac{3x}{5} - \frac{x}{5} = \frac{3}{5}x - \frac{1}{5}x$ — Write with fractional coefficients.

$= \left(\frac{3}{5} - \frac{1}{5}\right)x$ — Distributive Property

$= \frac{2}{5}x$ — Subtract fractions.

b. $\frac{x}{4} + \frac{2x}{7} = \frac{1}{4}x + \frac{2}{7}x$ — Write with fractional coefficients.

$= \left(\frac{1}{4} + \frac{2}{7}\right)x$ — Distributive Property

$= \left[\frac{1(7)}{4(7)} + \frac{2(4)}{7(4)}\right]x$ — Common denominator

$= \frac{15}{28}x$ — Simplest form

Discussing the Concept A Mathematical Riddle

What is the largest number that can be written using the three digits 2, 3, and 4? The number 432 seems to be the obvious answer. However, if you allow the digits to be exponents, then you can obtain numbers that are much larger than 432. For instance, consider the numbers

$$(32)^4 = 1{,}048{,}576 \text{ and } 3^{24} \approx 282{,}430{,}000{,}000.$$

Create the largest number you can using the three digits 2, 3, and 4. Compare your number with those of other students.

2.2 Exercises

Integrated Review *Concepts, Skills, and Problem Solving*

Keep mathematically in shape by doing these exercises *before* the problems of this section.

Properties and Definitions

1. Complete the following property of exponents.

$a^m \cdot a^n =$ ▭

2. Name the property demonstrated by the statement: $\frac{1}{2}(4x + 10) = 2x + 5$.

Simplifying Expressions

In Exercises 3–10, perform the operation.

3. $0 - (-12)$ **4.** $60 - (-60)$

5. $-12 - 2 + |-3|$

6. $-730 + 1820 + 3150 + (-10{,}000)$

7. Find the sum of 72 and -37.

8. Subtract 600 from 250.

9. $\frac{5}{16} - \frac{3}{10}$ **10.** $\frac{9}{16} + 2\frac{3}{12}$

Problem Solving

11. *Profit* A company showed a loss of \$1,530,000 during the first 6 months of a given year. If the company ended the year with an overall profit of \$832,000, what was the profit during the last two quarters of the year?

12. *Average Speed* A family on vacation traveled 676 miles in 13 hours. Determine their average speed in miles per hour.

Developing Skills

In Exercises 1–26, simplify the expression. See Examples 1 and 2.

1. $u^2 \cdot u^4$ **2.** $z^3 \cdot z$

3. $3x^3 \cdot x^4$ **4.** $4y^3 \cdot y$

5. $5x(x^6)$ **6.** $(-6x^2)x^4$

7. $(-5z^3)(3z^2)$ **8.** $(-2x^2)(-4x)$

9. $(-xz)(-2y^2z)$ **10.** $(6u^2v)(3uv^2)$

11. $2b^4(-ab)(3b^2)$ **12.** $4xy(-3x^2)(-2y^3)$

13. $(t^2)^4$ **14.** $(v^3)^2$

15. $5(uv)^5$ **16.** $3(pq)^4$

17. $(-2s)^3$ **18.** $(-3z)^2$

19. $(a^2b)^3(ab^2)^4$ **20.** $(st)^5(s^2t)^4$

21. $(u^2v^3)(-2uv^2)^4$ **22.** $(-3y^2z)^2(2yz^2)^3$

23. $[(x - 3)^4]^2$ **24.** $[(t + 1)^2]^5$

25. $(x - 2y)^3(x - 2y)^3$ **26.** $(x - 3)^2(x - 3)^5$

Think About It In Exercises 27–30, decide whether the expressions are equal. Explain your reasoning.

27. $x^5 \cdot x^3 \stackrel{?}{=} x^{15}$ **28.** $(-2x)^4 \stackrel{?}{=} -2x^4$

29. $-3x^3 \stackrel{?}{=} -27x^3$ **30.** $(xy)^2 \stackrel{?}{=} xy^2$

In Exercises 31–44, identify the basic rule (or rules) of algebra illustrated by the equation. See Example 3.

31. $x + 2y = 2y + x$

32. $-10(xy^2) = (-10x)y^2$

33. $(9x)y = 9(xy)$

34. $rt + 0 = rt$

35. $(x^2 + y^2) \cdot 1 = x^2 + y^2$

36. $(3x + 2y) + z = 3x + (2y + z)$

37. $2zy = 2yz$

38. $x(y + z) = xy + xz$

39. $(5m + 3) - (5m + 3) = 0$

40. $16xy \cdot \dfrac{1}{16xy} = 1, \quad xy \neq 0$

41. $(x + y) \cdot \dfrac{1}{(x + y)} = 1, \quad x + y \neq 0$

42. $(x + 2)(x + y) = x(x + y) + 2(x + y)$

43. $x^2 + (y^2 - y^2) = x^2$

44. $3y + (z^3 - z^3) = 3y$

In Exercises 45–54, complete the statement. State the rule of algebra that you used. See Example 3.

45. $(x + 10) - \_\_\_\_ = 0$

46. $(-5r)s = -5(\_\_\_\_)$

47. $v(2) = \_\_\_\_$

48. $(4x - 3y) + \_\_\_\_ = 4x - 3y$

49. $5(t - 2) = 5(\_\_\_\_) + 5(\_\_\_\_)$

50. $(2z - 3) + \_\_\_\_ = 0$

51. $5x(\_\_\_\_) = 1, \quad x \neq 0$

52. $(s - 5)(\_\_\_\_) = s - 5$

53. $12 + (8 - x) = \_\_\_\_ - x$

54. $(2x - y)(-3) = -3\_\_\_\_$

In Exercises 55–68, use the Distributive Property to expand the expression. See Example 4.

55. $-5(2x - y)$

56. $2(16 + 8z)$

57. $(x + 2)(3)$

58. $(4 - t)(-6)$

59. $4(x + xy + y^2)$

60. $6(r - t + s)$

61. $3(x^2 + x)$

62. $4(2y^2 - y)$

63. $-4y(3y - 4)$

64. $-z(5 - 2z)$

65. $-(u - v)$

66. $-(x + y)$

67. $x(3x - 4y)$

68. $r(2r^2 - t)$

In Exercises 69–72, write the area of each component part of the figure. Then demonstrate the Distributive Property by writing the total area of each figure in two ways. See Example 5.

69.

70.

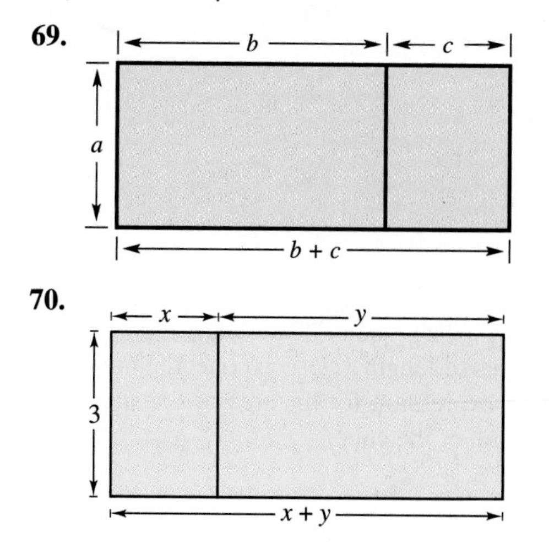

71.

72.

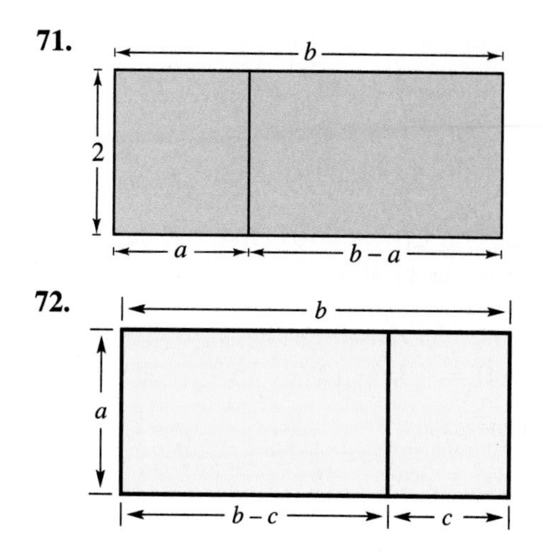

In Exercises 73 and 74, identify the terms of the expression and the coefficient of each term.

73. $6x^2 - 3xy + y^2$

74. $-4xy + 2xz - yz$

In Exercises 75–78, identify the like terms. See Example 6.

75. $16t^3 + 4 - 5 + 3t^3$

76. $a^2 + 5ab^2 - 3b^2 + 7a^2b - ab^2 + a^2$

77. $6x^2y + 2xy - 4x^2y$

78. $-\frac{1}{4}x^2 - 3x + \frac{3}{4}x^2 + x$

Think About It In Exercises 79 and 80, state why the two expressions are not like terms.

79. $\frac{1}{2}x^2y, \frac{5}{2}xy^2$

80. $-16x^2y^3, 7x^2y$

In Exercises 81–100, simplify the expression by combining like terms. See Examples 7 and 8.

81. $3y - 5y$

82. $-16x + 25x$

83. $x + 5 - 3x$

84. $7s + 3 - 3s$

85. $2x + 9x + 4$

86. $10x - 4 - 5x$

87. $5r + 6 - 2r + 1$

88. $2t - 4 + 8t + 9$

89. $x^2 - 2xy + 4 + xy$

90. $r^2 + 3rs - 6 - rs$

91. $5z - 5 + 10z + 2z + 16$

92. $7x - 4x + 8 + 3x - 6$

93. $z^3 + 2z^2 + z + z^2 + 2z + 1$

94. $3x^2 - x^2 + 4x + 3x^2 - x + x^2$

95. $2x^2y + 5xy^2 - 3x^2y + 4xy + 7xy^2$

96. $6rt - 3r^2t + 2rt^2 - 4rt - 2r^2t$

97. $3\left(\frac{1}{x}\right) - \frac{1}{x} + 8$

98. $1.2\left(\frac{1}{x}\right) + 3.8\left(\frac{1}{x}\right) - 4x$

99. $5\left(\frac{1}{t}\right) + 6\left(\frac{1}{t}\right) - 2t$

100. $16\left(\frac{a}{b}\right) - 6\left(\frac{a}{b}\right) + \frac{3}{2} - \frac{1}{2}$

True or False? In Exercises 101–104, decide whether the statement is true or false.

101. $3(x - 4) \stackrel{?}{=} 3x - 4$

102. $-3(x - 4) \stackrel{?}{=} -3x - 12$

103. $6x - 4x \stackrel{?}{=} 2x$

104. $12y^2 + 3y^2 \stackrel{?}{=} 36y^2$

Mental Math In Exercises 105–108, use the Distributive Property to perform the required arithmetic *mentally*. For example, suppose you work in an industry where the wage is \$14 per hour and time-and-one-half for overtime. Thus, your hourly wage for overtime is

$14(1.5) = 14\left(1 + \frac{1}{2}\right) = 14 + 7 = \$21.$

105. $8(52) = 8(50 + 2)$ **106.** $6(29) = 6(30 - 1)$

107. $5(7.98) = 5(8 - 0.02)$

108. $12(11.95) = 12(12 - 0.05)$

In Exercises 109–122, simplify the expression. See Examples 9 and 10.

109. $2(6x)$ **110.** $7(5a)$

111. $-(-4x)$ **112.** $-(5t)$

113. $(-2x)(-3x)$ **114.** $-4(-3y)$

115. $(-5z)(2z^2)$ **116.** $(10t)(-4t^2)$

117. $\frac{18a}{5} \cdot \frac{15}{6}$ **118.** $\frac{5x}{8} \cdot \frac{16}{5}$

119. $\left(-\frac{3x^2}{2}\right)(4x^3)$ **120.** $\left(\frac{4x}{3}\right)\left(\frac{3x}{2}\right)$

121. $(12xy^2)(-2x^3y^2)$ **122.** $(7r^2s^3)(3rs)$

In Exercises 123–142, simplify the expression by removing symbols of grouping and combining like terms. See Examples 11, 12, and 13.

123. $2(x - 2) + 4$ **124.** $-3(x + 1) - 2$

125. $6(2s - 1) + s + 4$ **126.** $(2x - 1)(2) + x$

127. $m - 3(m - 5)$ **128.** $5l - 6(3l - 5)$

129. $-6(1 - 2x) + 10(5 - x)$

130. $3(r - 2s) - 5(3r - 5s)$

131. $\frac{2}{3}(12x + 15) + 16$ **132.** $\frac{3}{8}(4 - y) - \frac{5}{2} + 10$

133. $3 - 2[6 + (4 - x)]$

134. $10x + 5[6 - (2x + 3)]$

135. $7x(2 - x) - 4x$ **136.** $-6x(x - 1) + x^2$

137. $4x^2 + x(5 - x)$ **138.** $-z(z - 2) + 3z^2 + 5$

139. $-3t(4 - t) + t(t + 1)$

140. $-2x(x - 1) + x(3x - 2)$

141. $3t[4 - (t - 3)] + t(t + 5)$

142. $4y[5 - (y + 1)] + 3y(y + 1)$

In Exercises 143–150, use the Distributive Property to simplify the expression. See Example 14.

143. $\frac{2x}{3} - \frac{x}{3}$ **144.** $\frac{7y}{8} - \frac{3y}{8}$

145. $\frac{4z}{5} + \frac{3z}{5}$ **146.** $\frac{5t}{12} + \frac{7t}{12}$

147. $\frac{x}{3} - \frac{5x}{4}$ **148.** $\frac{5x}{7} + \frac{2x}{3}$

149. $\frac{3x}{10} - \frac{x}{10} + \frac{4x}{5}$ **150.** $\frac{3z}{4} - \frac{z}{2} - \frac{z}{3}$

Solving Problems

Balance in an Account In Exercises 151 and 152, the balance in an account with an initial deposit of P dollars, at an annual interest rate of r for t years, is $P(1 + r)^t$. Find the balance for the given values of P, r, and t.

151. $P = 10{,}000,\quad r = 0.08,\quad t = 10$

152. $P = 5000,\quad r = 0.06,\quad t = 30$

153. *Geometry* The square and cube shown below have edges of length x. Use exponential notation to write an expression for the area of the square and the volume of the cube.

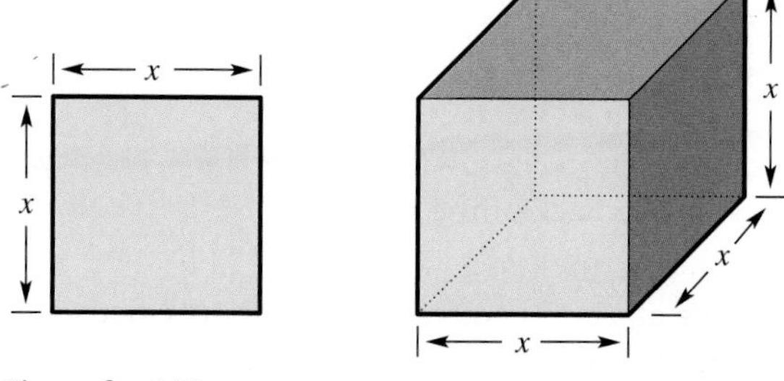

Figure for 153

154. The formulas give the moments of inertia of two solids. Simplify each expression.

(a) $\frac{1}{2}m(2a)^2(2L)$ (b) $k\pi a^2L\left(\frac{a^2}{2}\right)$

155. *Geometry* Write an expression for the perimeter of the triangle shown in the figure. Use the rules of algebra to simplify the expression.

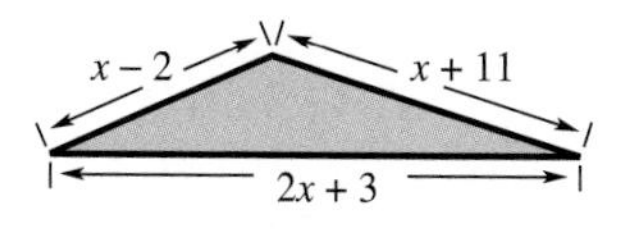

156. *Area of a Trapezoid* The area of a trapezoid with parallel bases of lengths b_1 and b_2 and height h (see figure) is $\frac{1}{2}h(b_1 + b_2)$.

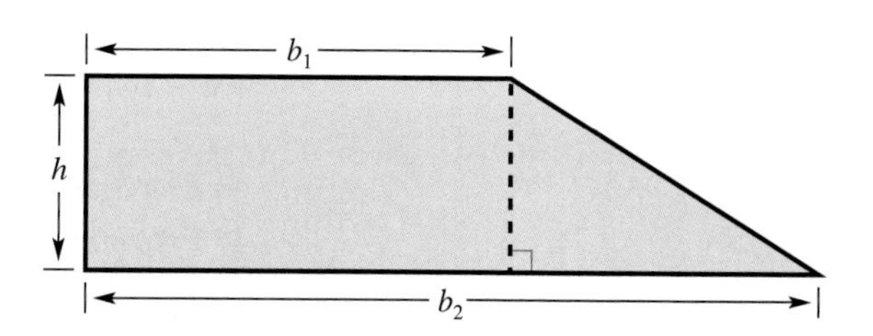

(a) Show that the area can also be expressed as $b_1h + \frac{1}{2}(b_2 - b_1)h$, and give a geometric explanation for the area represented by each term in this expression.

(b) Find the area of a trapezoid with $b_1 = 7$, $b_2 = 12$, and $h = 3$.

Area of a Trapezoid In Exercises 157 and 158, use the formula for the area of a trapezoid, $\frac{1}{2}h(b_1 + b_2)$, to find the area of the trapezoidal house lot and tile.

157. **158.**

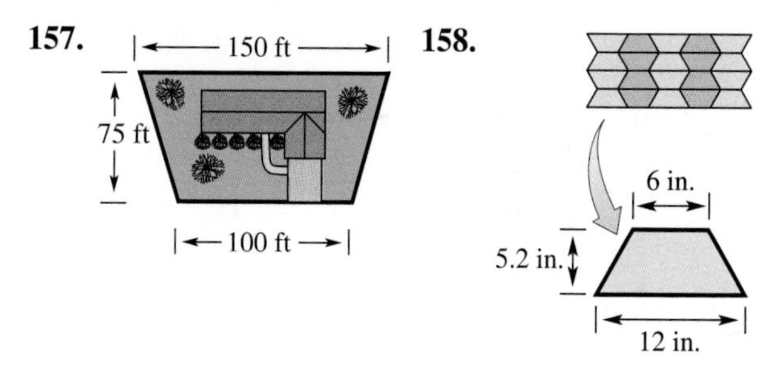

Explaining Concepts

159. Discuss the difference between $(6x)^4$ and $6x^4$.

160. The expressions $4x$ and x^4 each represent repeated operations. What are the operations? Write the expressions showing the repeated operations.

161. Which of the following are equivalent? Explain.

(a) $12x^8$ (b) $12(x^3)^5$ (c) $12x^3x^5$

(d) $3 \cdot 2^2(x^2)^4$ (e) $3 \cdot 5x^8$

162. In your own words, state the definition of like terms. Give an example of like terms and an example of unlike terms.

163. Describe how to combine like terms. What operations are used? Give an example of an expression that can be simplified by combining like terms.

164. Explain why $3(x + 9) \neq 3x + 9$.

165. In your own words, describe the procedure for removing nested symbols of grouping.

166. In your own words, describe the priorities for order of operations.

167. Does the expression $[x - (3 \cdot 4)] \div 5$ change if the parentheses are removed? Does it change if the brackets are removed? Explain.

168. Explain the error in the equation.

$$\frac{x}{3} + \frac{4x}{3} = \frac{5x}{6}$$

Mid-Chapter Quiz

Take this quiz as you would take a quiz in class. After you are done, check your work against the answers given in the back of the book.

In Exercises 1 and 2, evaluate the algebraic expression for the specified values of the variables. If it is not possible, state the reason.

1. $x^2 - 3x$ (a) $x = 3$ (b) $x = -2$ (c) $x = 0$

2. $\dfrac{x}{y - 3}$ (a) $x = 2, y = 4$ (b) $x = 0, y = -1$ (c) $x = 5, y = 3$

3. Identify the coefficients of the terms (a) $-5xy^2$ and (b) $\dfrac{5z}{16}$.

4. Rewrite the expression in exponential form.

(a) $3y \cdot 3y \cdot 3y \cdot 3y$ (b) $2 \cdot (x - 3) \cdot (x - 3) \cdot 2 \cdot 2$

In Exercises 5–10, simplify the expression.

5. $x^4 \cdot x^3$ **6.** $(v^2)^5$ **7.** $(-3y)^2y^3$

8. $8(x - 4)^2(x - 4)^4$ **9.** $\dfrac{2z^2}{3y} \cdot \dfrac{5z}{7y^3}$ **10.** $\left(\dfrac{x}{y}\right)^2\left(\dfrac{x}{y}\right)^5$

In Exercises 11–14, identify the rule of algebra illustrated by the equation.

11. $-3(2y) = (-3 \cdot 2)y$ **12.** $(x + 2)y = xy + 2y$

13. $3y \cdot \dfrac{1}{3y} = 1, \quad y \neq 0$ **14.** $x - x^2 + 2 = -x^2 + x + 2$

In Exercises 15 and 16, use the Distributive Property to expand the expression.

15. $2(3x - 1)$ **16.** $-4(2y - 3)$

In Exercises 17 and 18, simplify the expression by combining like terms.

17. $y^2 - 3xy + y + 7xy$ **18.** $10\left(\dfrac{1}{u}\right) - 7\left(\dfrac{1}{u}\right) + 3u$

In Exercises 19 and 20, simplify the expression by removing symbols of grouping and combining like terms.

19. $5(a - 2b) + 3(a + b)$ **20.** $4x + 3[2 - 4(x + 6)]$

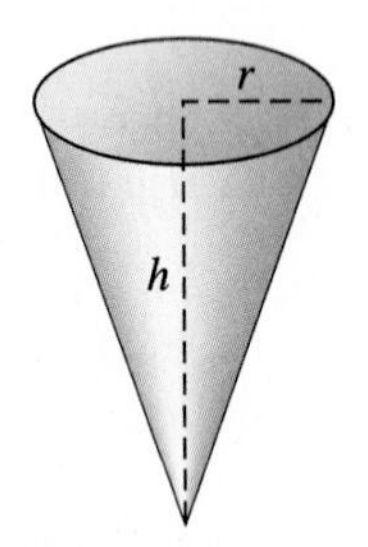

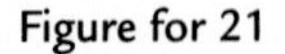
Figure for 21

21. Simplify the following expression for the moment of inertia of a cone of height h and radius r (see figure).

$$\left(\frac{1}{3}\pi r^2h\right)\left(\frac{3}{10}r^2\right)$$

22. Evaluate the expression $4 \cdot 10^4 + 5 \cdot 10^3 + 7 \cdot 10^2$.

2.3 Algebra and Problem Solving

Objectives

1. Define algebra as a problem-solving language.
2. Construct verbal mathematical models from written statements.
3. Translate verbal phrases into algebraic expressions.
4. Identify hidden operations when constructing algebraic expressions.
5. Use problem-solving strategies to solve an application problem.

1 Define algebra as a problem-solving language.

What Is Algebra?

Algebra is a problem-solving language that is used to solve real-life problems. It has four basic components, which tend to nest within each other, as indicated in Figure 2.1.

1. Symbolic representations and applications of the rules of arithmetic
2. Rewriting (reducing, simplifying, factoring) algebraic expressions into equivalent forms
3. Creating and solving equations
4. Studying relationships among variables by the use of functions and graphs

Rules of arithmetic

Algebraic expressions: rewriting into equivalent forms

Algebraic equations: creating and solving

Functions and graphs: relationships among variables

Figure 2.1

Notice that one of the components deals with expressions and another deals with equations. As you study algebra, it is important to understand the difference between simplifying or rewriting an algebraic *expression*, and solving an algebraic *equation*. In general, remember that a mathematical expression *has no equal sign*, whereas a mathematical equation *must have an equal sign*.

When you use an equal sign to *rewrite* an expression, you are merely indicating the *equivalence* of the new expression and the previous one.

Original Expression	*equals*	*Equivalent Expression*
$(a + b)c$	$=$	$ac + bc$

2 Construct verbal mathematical models from written statements.

Constructing Verbal Models

In the first two sections of this chapter, you studied techniques for rewriting and simplifying algebraic expressions. In this section you will study ways to *construct* algebraic expressions from written statements by first constructing a **verbal mathematical model.**

Let's take another look at Example 1 in Section 2.1 (page 62). In that example you are paid \$6 per hour and your weekly pay can be represented by the verbal model

[Pay per hour] · [Number of hours] = [6 dollars] · [x hours] = $6x$.

Note the hidden operation of multiplication in this expression. Nowhere in the verbal problem does it say you are to multiply 6 times x. It is *implied* in the problem. This is often the case when algebra is used to solve real-life problems.

Telegraph Colour Library/ FPG International

In 1995, 1 million tons of aluminum containers were recycled. This accounted for more than 35% of all aluminum containers produced. (Source: Franklin Associates, Ltd.)

Example 1 Constructing an Algebraic Expression

You are paid 5¢ for each aluminum soda can and 3¢ for each glass soda bottle you collect. Write an algebraic expression that represents the total weekly income for this recycling activity.

Solution

Before writing an algebraic expression for the weekly income, it is helpful to construct an informal verbal model. For instance, the following verbal model could be used.

[Pay per can] · [Number of cans] + [Pay per bottle] · [Number of bottles]

Note that the word *and* in the problem indicates addition. Because both the number of cans and the number of bottles can vary from week to week, you can use the two variables c and b, respectively, to write the following algebraic expression.

[5 cents] · [c cans] + [3 cents] · [b bottles] = $5c + 3b$

In Example 1, notice that c is used to represent the number of *cans* and b is used to represent the number of *bottles*. When writing algebraic expressions, choose variables that can be identified with the unknown quantities.

The number of one kind of item can be expressed in terms of the number of another kind of item. Suppose the number of cans in Example 1 was said to be "three times the number of bottles." In this case, only one variable is needed and the model could be written as

[3 cents] · [b bottles] + [5 cents] · [$3 \cdot b$ cans] $= 3b + 5(3b)$

$$= 3b + 15b$$

$$= 18b.$$

3 Translate verbal phrases into algebraic expressions.

Translating Phrases

When translating verbal sentences and phrases into algebraic expressions, it is helpful to watch for key words and phrases that indicate the four different operations of arithmetic. The following list gives several examples.

▶ Translating Phrases into Algebraic Expressions

Key Words and Phrases	*Verbal Description*	*Algebraic Expression*
Addition:		
Sum, plus, greater, increased by, more than,	The sum of 6 and x	$6 + x$
exceeds, total of	Eight more than y	$y + 8$
Subtraction:		
Difference, minus, less, decreased by,	Five decreased by a	$5 - a$
subtracted from, reduced by, the remainder	Four less than z	$z - 4$
Multiplication:		
Product, multiplied by, twice, times, percent of	Five times x	$5x$
Division:		
Quotient, divided by, ratio, per	The ratio of x to 3	$\frac{x}{3}$

Example 2 Translating Phrases Having Specified Variables

Translate each of the following into an algebraic expression.

a. Three less than m

b. y decreased by 10

c. The product of 5 and x

d. The quotient of n and 7

Solution

a. Three less than m

$m - 3$ — Think: 3 subtracted from what?

b. y decreased by 10

$y - 10$ — Think: What is subtracted from y?

c. The product of 5 and x

$5x$ — Think: 5 times what?

d. The quotient of n and 7

$\frac{n}{7}$ — Think: n is divided by what?

Example 3 Translating Phrases Having Specified Variables

Translate each of the following into an algebraic expression.

a. Six times the sum of x and 7

b. The product of 4 and x, divided by 3

c. k decreased by the product of 8 and m

Solution

a. Six times the sum of x and 7

$6(x + 7)$ Think: 6 multiplied by what?

b. The product of 4 and x, divided by 3

$\frac{4x}{3}$ Think: What is divided by 3?

c. k decreased by the product of 8 and m

$k - 8m$ Think: What is subtracted from k?

In most applications of algebra, the variables are not specified and it is your task to assign variables to the *appropriate* quantities. Although similar to the translations in Examples 2 and 3, the translations in the next example may seem more difficult because variables have not been assigned to the unknown quantities.

Study Tip

Any variable, such as b, k, n, r, or x, can be chosen to represent an unspecified number. The choice is a matter of preference. In Example 4, x was chosen as the variable.

Example 4 Translating Phrases Having No Specified Variable

Translate each of the following into a variable expression.

a. The sum of 3 and a number

b. Five decreased by the product of 3 and a number

c. The difference of a number and 3, divided by 12

Solution

In each case, let x be the unspecified number.

a. The sum of 3 and a number

$3 + x$ Think: 3 added to what?

b. Five decreased by the product of 3 and a number

$5 - 3x$ Think: What is subtracted from 5?

c. The difference of a number and 3, divided by 12

$\frac{x - 3}{12}$ Think: What is divided by 12?

A good way to learn algebra is to do it *forward* and *backward*. In the next example, algebraic expressions are translated into verbal form. Keep in mind that other key words could be used to describe the operations in each expression. Your goal is to use key words or phrases that keep the verbal expressions clear and concise.

Example 5 Translating Algebraic Expressions into Verbal Form

Without using a variable, write a verbal description for each of the following.

a. $7x - 12$ **b.** $7(x - 12)$ **c.** $5 + \frac{x}{2}$ **d.** $\frac{5 + x}{2}$ **e.** $(3x)^2$

Solution

a. *Algebraic expression:* $7x - 12$

Primary operation: Subtraction

Terms: $7x$ and 12

Verbal description: Twelve less than the product of 7 and a number

b. *Algebraic expression:* $7(x - 12)$

Primary operation: Multiplication

Factors: 7 and $(x - 12)$

Verbal description: Twelve is subtracted from a number and the result is multiplied by 7.

c. *Algebraic expression:* $5 + \frac{x}{2}$

Primary operation: Addition

Terms: 5 and $\frac{x}{2}$

Verbal description: Five added to the quotient of a number and 2

d. *Algebraic expression:* $\frac{5 + x}{2}$

Primary operation: Division

Numerator, denominator: Numerator is $5 + x$; denominator is 2

Verbal description: The sum of 5 and a number, divided by 2

e. *Algebraic expression:* $(3x)^2$

Primary operation: Raise to a power

Base, power: $3x$ is the base, 2 is the power

Verbal description: The product of 3 and x, squared

Translating algebraic expressions into verbal phrases is more difficult than it may appear. It is easy to write a phrase that is ambiguous. For instance, what does the phrase "the sum of 5 and a number times 2" mean? Without further information, this phrase could mean

$$5 + 2x \quad \text{or} \quad 2(5 + x).$$

4 Identify hidden operations when constructing algebraic expressions.

Verbal Models with Hidden Operations

Most real-life problems do not contain verbal expressions that clearly identify all the arithmetic operations involved. You need to rely on past experience and the physical nature of the problem in order to identify the operations hidden in the problem statement. Multiplication is the operation most commonly hidden in real life applications. Watch for *hidden operations* in the next two examples.

Example 6 Discovering Hidden Operations

a. A cash register contains n nickels and d dimes. Write an expression for this amount of money in cents.

b. Write an expression showing how far a person can ride a bicycle in t hours if the person travels at a constant rate of 15 miles per hour.

c. A person paid x dollars plus 6% sales tax for an automobile. Write an expression for the total cost of the automobile.

Solution

a. The amount of money is a sum of products.

Verbal Model: [Value of nickel] · [Number of nickels] + [Value of dime] · [Number of dimes]

Labels: Value of nickel = 5 (cents)
Number of nickels = n
Value of dime = 10 (cents)
Number of dimes = d

Expression: $5n + 10d$ (cents)

b. The distance traveled is a product.

Verbal Model: [Rate of travel] · [Time traveled]

Labels: Rate of travel = 15 (miles per hour)
Time traveled = t (hours)

Expression: $15t$ (miles)

c. The total cost is a sum.

Verbal Model: [Cost of automobile] + [Percent of sales tax] · [Cost of automobile]

Labels: Percent of sales tax = 0.06 (decimal form)
Cost of automobile = x (dollars)

Expression: $x + 0.06x = (1 + 0.06)x$

$= 1.06x$

Study Tip

In Example 6(b), the final answer is listed in terms of miles. This makes sense in the following way.

$$15\frac{\text{miles}}{\text{hours}} \cdot t \text{ hours}$$

Note that the hours "cancel," leaving the answer in terms of miles. This technique, called *unit analysis*, can be very helpful in determining the final unit of measure.

Notice in part (c) of Example 6 that the equal sign is used to denote the equivalence of the three expressions. It is not an equation to be solved.

5 Use problem-solving strategies to solve an application problem.

Additional Problem-Solving Strategies

In addition to constructing verbal models, there are other problem-solving strategies that can help you succeed in this course.

Summary of Additional Problem-Solving Strategies

1. **Guess, Check, and Revise** Guess a reasonable solution based on the given data. Check the guess, and revise it, if necessary. Continue guessing, checking, and revising until a correct solution is found.
2. **Make a Table/Look for a Pattern** Make a table using the data in the problem. Look for a number pattern. Then use the pattern to complete the table or find a solution.
3. **Draw a Diagram** Draw a diagram that shows the facts from the problem. Use the diagram to visualize the action of the problem. Use algebra to find a solution. Then check the solution against the facts.
4. **Solve a Simpler Problem** Construct a simpler problem that is similar to the given problem. Solve the simpler problem. Then use the same procedure to solve the given problem.

Example 7 Guess, Check, and Revise

You deposit \$500 in an account that earns 6% simple interest. The balance in the account after t years is

$$A = 500(1 + 0.06)^t.$$

How long will it take for your investment to double?

Solution

You can solve this problem using a guess, check, and revise strategy. For instance, you might guess that it takes 10 years for your investment to double. The balance in 10 years is

$$A = 500(1 + 0.06)^{10} \approx \$895.42.$$

Because the amount has not yet doubled, you increase your guess to 15 years.

$$A = 500(1 + 0.06)^{15} \approx \$1198.28$$

Because this amount is more than double the investment, your next guess should be a number between 10 and 15. After trying several more numbers, you can determine that your balance doubles in about 11.9 years.

Another strategy that works well for a problem like Example 7 is to make up a table of data values. Your calculator or graphing utility would work well to create the following table.

t	2	4	6	8	10	12
A	561.80	631.24	709.26	796.92	895.42	1006.10

Example 8 Make a Table/Look for a Pattern

Find the following products. Then describe the pattern and use your description to find the product of 14 and 16.

$$1 \cdot 3,\ 2 \cdot 4,\ 3 \cdot 5,\ 4 \cdot 6,\ 5 \cdot 7,\ 6 \cdot 8,\ 7 \cdot 9$$

Solution

One way to help find a pattern is to organize the results in a table.

Numbers	$1 \cdot 3$	$2 \cdot 4$	$3 \cdot 5$	$4 \cdot 6$	$5 \cdot 7$	$6 \cdot 8$	$7 \cdot 9$
Product	3	8	15	24	35	48	63

From the table, you can see that each of the products is 1 less than a perfect square. For instance, 3 is 1 less than 2^2 or 4, 8 is 1 less than 3^2 or 9, 15 is 1 less than 4^2 or 16, and so on.

If this pattern continues for other numbers, you can hypothesize that the product of 14 and 16 is 1 less than 15^2 or 225. That is,

$$14 \cdot 16 = 15^2 - 1 = 224.$$

You can confirm this result by actually multiplying 14 and 16.

Example 9 Draw a Diagram

The outer dimensions of a rectangular apartment are 25 feet by 40 feet. The combination living-room, dining-room, and kitchen areas occupy two-fifths of the apartment's area. Find the area of the remaining rooms.

Solution

For this problem, it helps to draw a diagram, as shown in Figure 2.2. From the figure, you can see that the total area in the apartment is

$$\begin{aligned} \text{Area} &= (\text{length})(\text{width}) \\ &= (40)(25) \\ &= 1000 \text{ square feet.} \end{aligned}$$

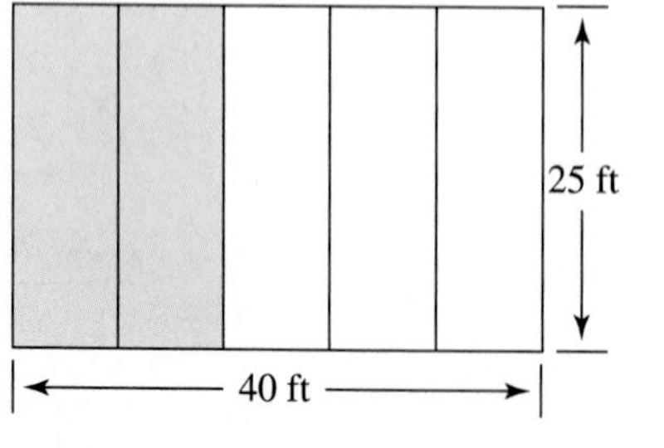

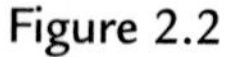
Figure 2.2

The area occupied by the living room, dining room, and kitchen is

$$\frac{2}{5}(1000) = 400 \text{ square feet.}$$

This implies that the remaining rooms must have a total area of 600 square feet.

Example 10 Solve a Simpler Problem

You are driving on an interstate highway and are traveling at an average speed of 60 miles per hour. How far will you travel in $12\frac{1}{2}$ hours?

Solution

One way to solve the problem is to use the formula that relates distance, rate, and time. Suppose, however, that you have forgotten the formula. To help you remember, you could solve some simpler problems.

- If you travel 60 miles per hour for 1 hour, you will travel 60 miles.
- If you travel 60 miles per hour for 2 hours, you will travel 120 miles.
- If you travel 60 miles per hour for 3 hours, you will travel 180 miles.

From these examples, it appears that you can find the total miles traveled by multiplying the rate times the time. So, if you travel 60 miles per hour for $12\frac{1}{2}$ hours, you will travel a distance of

$$(60)(12.5) = 750 \text{ miles.}$$

Hidden operations are often involved when variable names (labels) are assigned to two unknown quantities. A good strategy is to use a *specific* case to help you write a model for the *general* case. For instance, a specific case of finding three consecutive integers

$$3, 3 + 1, \text{ and } 3 + 2$$

may help you write a general case for finding three consecutive integers n, $n + 1$, and $n + 2$. This strategy is illustrated in Examples 11 and 12.

Example 11 Using a Specific Case to Find a General Case

In each of the following, use the given variable to label the unknown quantity.

a. A person's weekly salary is d dollars. What is the annual salary?

b. A person's annual salary is y dollars. What is the monthly salary?

Solution

a. There are 52 weeks in a year.

Specific case: If the weekly salary is \$200, then the annual salary (in dollars) is $52 \cdot 200$.

General case: If the weekly salary is d dollars, then the annual salary (in dollars) is $52 \cdot d$ or $52d$.

b. There are 12 months in a year.

Specific case: If the annual salary is \$24,000, then the monthly salary (in dollars) is $24{,}000 \div 12$.

General case: If the annual salary is y dollars, then the monthly salary (in dollars) is $y \div 12$ or $y/12$.

Example 12 Using a Specific Case to Find a General Case

In each of the following, use the given variable to label the unknown quantity.

a. One person is k inches shorter than another person. The first person is 60 inches tall. How tall is the second person?

b. A consumer buys g gallons of gasoline for a total of d dollars. What is the price per gallon?

c. A person drives on the highway at an average speed of 60 miles per hour for t hours. How far has the person traveled?

Solution

a. The first person is k inches shorter than the second person.

Specific case: If the first person is 10 inches shorter than the second person, then the second person is $60 + 10$ inches tall.

General case: If the first person is k inches shorter than the second person, then the second person is $60 + k$ inches tall.

b. To obtain the price per gallon, divide the price by the number of gallons.

Specific case: If the total price is $11.50 and the total number of gallons is 10, then the price per gallon is $11.50 \div 10$ dollars per gallon.

General case: If the total price is d dollars and the total number of gallons is g, then the price per gallon is $d \div g$ or d/g dollars per gallon.

c. To obtain the distance driven, multiply the speed by the number of hours.

Specific case: If the person has driven for 2 hours at a speed of 60 miles per hour, then the person has traveled $60 \cdot 2$ miles.

General case: If the person has driven for t hours at a speed of 60 miles per hour, then the person has traveled $60t$ miles.

Discussing the Concept **Enough Information?**

Most of the verbal problems you encounter in a mathematics text have precisely the right amount of information necessary to solve the problem. In real life, however, you may need to collect additional information. Decide what additional information would be needed to solve the following problem.

During a given week, a person worked 48 hours for the same employer. The hourly rate for overtime is $12. Write an expression for the person's gross pay for the week, including any pay received for overtime.

2.3 Exercises

Integrated Review *Concepts, Skills, and Problem Solving*

Keep mathematically in shape by doing these exercises *before* the problems of this section.

Properties and Definitions

1. The product of two real numbers is -35 and one of the factors is 5. What is the sign of the other factor?

2. Determine the sum of the digits of 744. Since this sum is divisible by 3, the number 744 is divisible by what number?

3. *True or False?* -4^2 is positive.

4. *True or False?* $(-4)^2$ is positive.

Simplifying Expressions

In Exercises 5–10, perform the operation.

5. $(-6)(-13)$ **6.** $|4(-6)(5)|$

7. $\left(-\frac{4}{3}\right)\left(-\frac{9}{16}\right)$ **8.** $\frac{7}{8} \div \frac{3}{16}$

9. $\left|-\frac{5}{9}\right| + 2$ **10.** $-7\frac{3}{5} - 3\frac{1}{2}$

Problem Solving

11. *Buying a Coat* A coat costs $133.50, including tax. If you can save $30 a week, how many weeks must you save in order to buy the coat? How much money will you have left?

12. *Perimeter* The length of a rectangle is $1\frac{1}{2}$ times its width. If its width is 8 meters, find its perimeter.

Developing Skills

In Exercises 1–6, match the verbal phrase with the correct algebraic expression.

(a) $11 + \frac{1}{3}x$ (b) $3x - 12$

(c) $3(x - 12)$ (d) $12 - 3x$

(e) $11x + \frac{1}{3}$ (f) $12x + 3$

1. Twelve decreased by 3 times a number

2. Eleven more than $\frac{1}{3}$ of a number

3. Eleven times a number plus $\frac{1}{3}$

4. Three increased by 12 times a number

5. The difference between 3 times a number and 12

6. Three times the difference of a number and 12

In Exercises 7–30, translate the phrase into an algebraic expression. (Let x represent the real number.) See Examples 1, 2, 3, and 4.

7. A number increased by 5

8. 25 more than a number

9. A number decreased by 25

10. A number decreased by 7

11. Six less than a number

12. Ten more than a number

13. Twice a number

14. The product of 30 and a number

15. A number divided by 3

16. A number divided by 100

17. The ratio of a number to 50

18. One-fourth of a number

19. Three-tenths of a number

20. Twenty-five hundredths of a number

21. A number is tripled and the product is increased by 5

22. A number is increased by 5 and the sum is tripled

23. Eight more than 5 times a number

24. The quotient of a number divided by 5 is decreased by 15

25. Ten times the sum of a number and 4

26. Seven more than 5 times a number

27. The absolute value of the sum of a number and 4

28. The absolute value of 4 less than twice a number

29. The square of a number, increased by 1

30. Twice the square of a number, increased by 4

In Exercises 31–42, write a verbal description of the algebraic expression. Use words only—do not use the variable. (There is more than one correct answer.) See Example 5.

31. $x - 10$
32. $x + 9$
33. $3x + 2$
34. $4 - 7x$
35. $7x + 4$
36. $9 - \frac{1}{4}x$
37. $3(2 - x)$
38. $-10(t - 6)$
39. $\dfrac{t + 1}{2}$
40. $\dfrac{1}{2} - \dfrac{t}{5}$
41. $x^2 + 5$
42. $x^3 - 1$

In Exercises 43–50, translate the phrase into a mathematical expression. Simplify the expression.

43. The sum of x and 3 is multiplied by x.

44. The sum of 6 and n is multiplied by 5.

45. The sum of 25 and x is added to x.

46. The sum of 4 and x is added to the sum of x and -8.

47. Nine is subtracted from x and the result is multiplied by 3.

48. The square of x is added to the product of x and $x + 1$.

49. The product of 8 times the sum of x and 24 is divided by 2.

50. Fifteen is subtracted from x and the difference is multiplied by 4.

Problem Solving

51. *Total Amount of Money* A cash register contains d dimes. Write an algebraic expression that represents the total amount of money (in dollars). See Example 6.

52. *Amount of Money* A cash register contains d dimes and q quarters. Write an algebraic expression that represents the total amount of money (in dollars).

53. *Sales Tax* The sales tax on a purchase of L dollars is 6%. Write an algebraic expression that represents the total amount of sales tax. (To find 6% of a quantity, multiply the quantity by 0.06.)

54. *Income Tax* The state income tax on a gross income of I dollars is 2.2%. Write an algebraic expression that represents the total amount of income tax. (To find 2.2% of a quantity, multiply the quantity by 0.022.)

55. *Travel Time* A truck travels 100 miles at an average speed of r miles per hour (see figure). Write an algebraic expression that represents the total travel time.

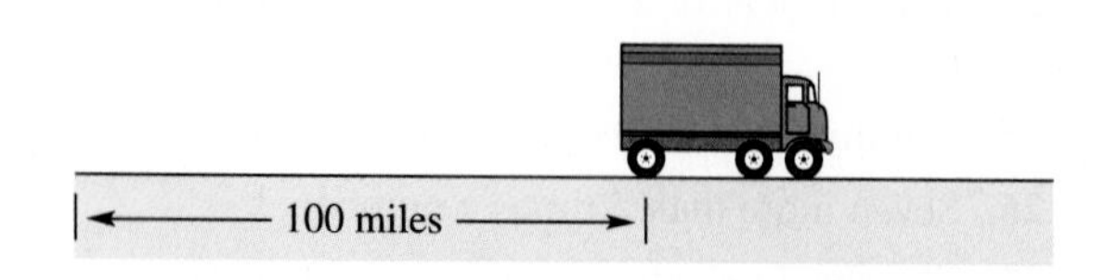

56. *Distance Traveled* A plane travels at the rate of r miles per hour for 3 hours. Write an algebraic expression that represents the total distance traveled by the plane.

57. *Camping Fee* A campground charges \$15 for adults and \$2 for children. Write an algebraic expression that represents the total camping fee for m adults and n children.

58. *Hourly Wage* The hourly wage for an employee is \$12.50 per hour plus 75 cents for each of the q units produced during the hour. Write an algebraic expression that represents the total hourly earnings for the employee.

Guess, Check, and Revise In Exercises 59–62, an expression for the balance in an account is given. Guess, check, and revise to determine the time (in years) necessary for the investment of \$1000 to double. See Example 7.

59. Interest rate: 7%
$1000(1 + 0.07)^t$

60. Interest rate: 5%
$1000(1 + 0.05)^t$

61. Interest rate: 6%
$1000(1 + 0.06)^t$

62. Interest rate: 8%
$1000(1 + 0.08)^t$

Finding a Pattern In Exercises 63 and 64, complete the table. The third row in the table is the difference between consecutive entries of the second row. Describe the pattern of the third row. See Example 8.

63.

n	0	1	2	3	4	5
$2n - 1$						
Differences						

64.

n	0	1	2	3	4	5
$7n + 5$						
Differences						

65. *Finding a Pattern* What would the pattern be in the third row of the table in Exercise 63 if the algebraic expression were $3n + 5$ rather than $2n - 1$?

66. *Finding a Pattern* What would the pattern be in the third row of the table in Exercise 64 if the algebraic expression were $an + b$ rather than $7n + 5$?

Exploration In Exercises 67 and 68, find a and b such that the expression $an + b$ yields the table values.

67.

n	0	1	2	3	4	5
$an + b$	4	9	14	19	24	29

68.

n	0	1	2	3	4	5
$an + b$	1	5	9	13	17	21

Geometry In Exercises 69–72, write an algebraic expression that represents the area of the region. Use the rules of algebra to simplify the expression.

69.

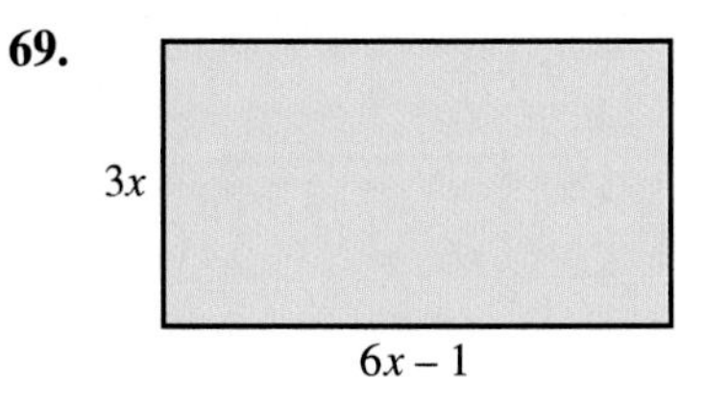

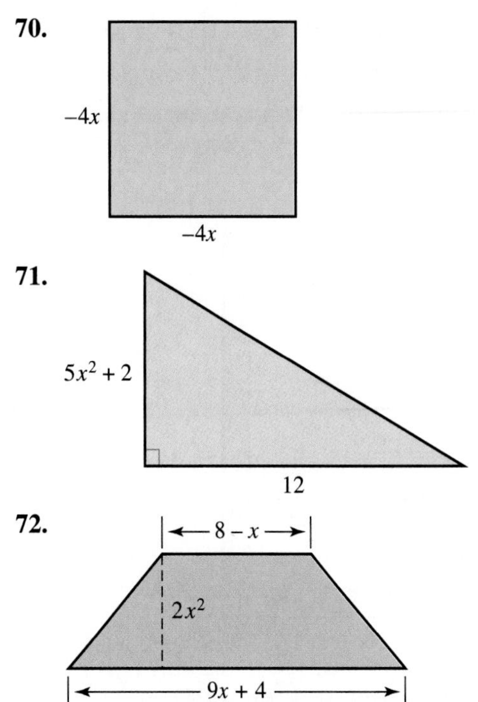

Drawing a Diagram In Exercises 73 and 74, draw figures satisfying the specified conditions. See Example 9.

73. The sides of a square have length a centimeters. Draw the square. Draw the rectangle obtained by extending two of the parallel sides of the square 6 centimeters. Find expressions for the perimeter and area of each figure.

74. The dimensions of a rectangular lawn are 150 feet by 250 feet. The property owner has the option of buying a rectangular strip x feet wide along one 250-foot side of the lawn. Draw diagrams representing the lawn before and after the purchase. Write an expression for the area of each.

75. *Geometry* A computer screen has sides of length s inches (see figure). Write an algebraic expression that represents the area of the screen. Express the area in the correct unit of measure.

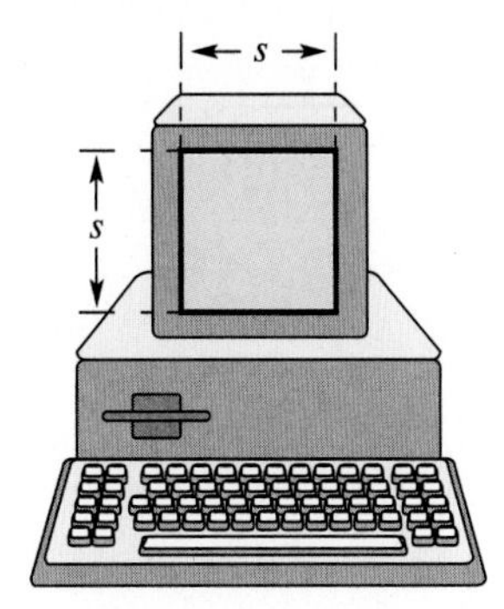

76. *Geometry* A rectangle has sides of length $3w$ and w. Write an algebraic expression that represents the perimeter of the rectangle.

77. *Geometry* Write an algebraic expression that represents the perimeter of the picture frame in the figure.

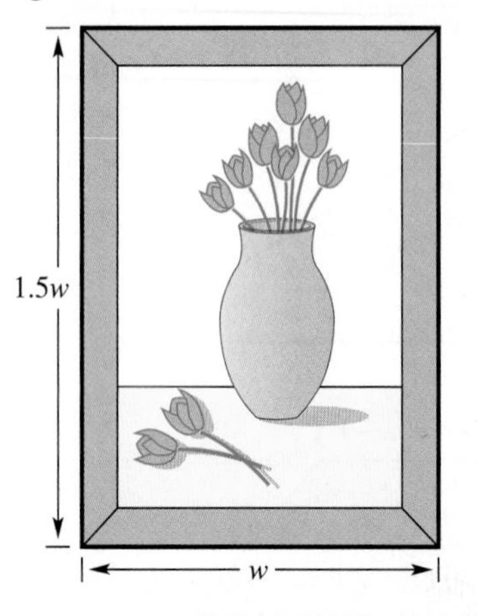

78. *Geometry* A square has sides of length s. Write an algebraic expression that represents the perimeter of the square.

79. *Fibonacci Sequence* Each term in the Fibonacci Sequence is the sum of the previous two terms.

(a) Let m and n represent two consecutive terms in the sequence. Write an algebraic expression for the next term.

(b) The first three terms in the sequence are 1, 1, and 2. Write the next five terms.

Explaining Concepts

80. Answer parts (c) to (h) of Motivating the Chapter on page 61.

81. The word *difference* indicates what operation?

82. The word *quotient* indicates what operation?

83. Determine which of the following are equivalent to the expression $n + 4$.

(a) 4 more than n

(b) the sum of n and 4

(c) n less than 4

(d) the ratio of n to 4

(e) the total of 4 and n

84. Determine whether order is important when translating each of the following phrases into an algebraic expression. Explain.

(a) x is increased by 10

(b) 10 is decreased by x

(c) the product of x and 10

(d) the quotient of x and 10

85. Give two interpretations of "the quotient of 5 and a number times 3."

2.4 Introduction to Equations

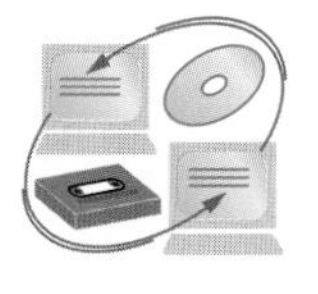

Objectives

1. Distinguish between an algebraic expression and an algebraic equation.
2. Check whether a given value is a solution of an equation.
3. Use properties of equality to solve an equation.
4. Use a verbal model to construct an algebraic equation.

1 Distinguish between an algebraic expression and an algebraic equation.

Equations

An **equation** is a statement in which two mathematical expressions are equal. Here are some examples:

$$x = 3, \quad 5x - 2 = 8, \quad 3x - 12 = 3(x - 4), \quad \text{and} \quad x^2 - 9 = 0.$$

To **solve** an equation involving x means to find all values of x for which the equation is true. Such values are called **solutions,** and solutions **satisfy** the equation. For instance, 3 is a solution of $x = 3$ because $3 = 3$ is a true statement.

The **solution set** of an equation is the set of all solutions of the equation. Sometimes an equation will have the set of all real numbers as its solution set. Such an equation is called an **identity.** For instance, the equation

$$3x - 12 = 3(x - 4) \qquad \text{Identity}$$

is an identity because the equation is true for all real values of x. Try values such as 0, 1, -2, and 5 in this equation to see that each one is a solution.

An equation whose solution set is not the entire set of real numbers is called a **conditional equation.** For instance, the equation

$$x^2 - 9 = 0 \qquad \text{Conditional equation}$$

is a conditional equation because it has only two solutions, 3 and -3.

Be sure that you understand the distinction between an algebraic expression and an algebraic equation. The differences are summarized in the following table.

Algebraic Expression	Algebraic Equations	
	Conditional Equation	Identity
• Example: $4(x - 1)$ • Contains *no* equal sign • Can sometimes be *simplified* to an equivalent form: $4(x - 1)$ simplifies to $4x - 4$ • Can be evaluated for any real number for which the expression is defined	• Example: $4(x - 1) = 12$ • Contains an equal sign and is true for only certain values of the variable • Solution is found by isolating the variable x: $4(x - 1) = 12$ $4x - 4 = 12$ $4x = 16$ $x = 4$	• Example: $3(x - 2) = 3x - 6$ • Contains an equal sign and is true for all real values of the variable • Rewriting one side to be identical to the other shows that every real number is a solution: $3(x - 2) = 3x - 6$ $3x - 6 = 3x - 6$

2 Check whether a given value is a solution of an equation.

Examples 1 and 2 show how to **check** whether a given value of x is a solution of an equation.

Example 1 Checking a Solution of an Equation

Determine whether -2 is a solution of $x^2 - 5 = 4x + 7$.

Solution

$$x^2 - 5 = 4x + 7 \qquad \text{Original equation}$$
$$(-2)^2 - 5 \stackrel{?}{=} 4(-2) + 7 \qquad \text{Substitute } -2 \text{ for } x.$$
$$4 - 5 \stackrel{?}{=} -8 + 7 \qquad \text{Simplify.}$$
$$-1 = -1 \qquad \text{Solution checks. ✓}$$

Because both sides of the equation turn out to be the same number, you can conclude that -2 is a solution of the original equation.

Study Tip

When checking a solution, we suggest that you write a question mark over the equal sign to indicate that you are not sure of the validity of the equation.

Just because you have found one solution of an equation, you should not conclude that you have found all of the solutions. For instance, you can check that 6 is also a solution of the equation in Example 1 as follows.

$$x^2 - 5 = 4x + 7 \qquad \text{Original equation}$$
$$(6)^2 - 5 \stackrel{?}{=} 4(6) + 7 \qquad \text{Substitute 6 for } x.$$
$$36 - 5 \stackrel{?}{=} 24 + 7 \qquad \text{Simplify.}$$
$$31 = 31 \qquad \text{Solution checks. ✓}$$

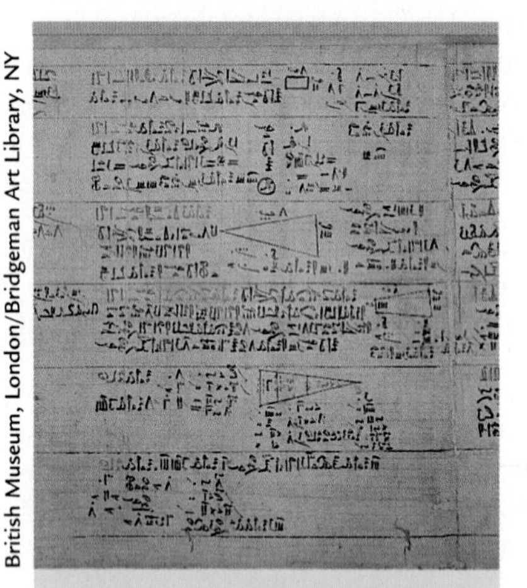

British Museum, London/Bridgeman Art Library, NY

Ahmes Papyrus

An ancient Egyptian papyrus, discovered in 1858, contains one of the earliest examples of mathematical writing in existence. The papyrus itself dates back to about 1650 B.C., but it is actually a copy of writings from two centuries earlier. The algebraic equations on the papyrus were written in words.

Example 2 A Trial Solution That Does Not Check

Determine whether 2 is a solution of $x^2 - 5 = 4x + 7$.

Solution

$$x^2 - 5 = 4x + 7 \qquad \text{Original equation}$$
$$(2)^2 - 5 \stackrel{?}{=} 4(2) + 7 \qquad \text{Substitute 2 for } x.$$
$$4 - 5 \stackrel{?}{=} 8 + 7 \qquad \text{Simplify.}$$
$$-1 \neq 15 \qquad \text{2 is not a solution. ✗}$$

Because the two sides of the equation turn out to be different, you can conclude that 2 is not a solution of the original equation.

3 Use properties of equality to solve an equation.

Forming Equivalent Equations

It is helpful to think of an equation as having two sides that are in balance. Consequently, when you try to solve an equation, you must be careful to maintain that balance by performing the same operation on both sides.

Two equations that have the same set of solutions are called **equivalent.** For instance, the equations $x = 3$ and $x - 3 = 0$ are equivalent because both have only one solution—the number 3. When any one of the operations in the following list is applied to an equation, the resulting equation is equivalent to the original equation.

▶ Forming Equivalent Equations: Properties of Equality

An equation can be transformed into an *equivalent equation* using one or more of the following procedures.

	Original Equation	*Equivalent Equation(s)*
1. *Simplify either side:* Remove symbols of grouping, combine like terms, or reduce fractions on one or both sides of the equation.	$3x - x = 8$	$2x = 8$
2. *Apply the Addition Property of Equality:* Add (or subtract) the same quantity to (from) *both* sides of the equation.	$x - 2 = 5$	$x - 2 + 2 = 5 + 2$ $x = 7$
3. *Apply the Multiplication Property of Equality:* Multiply (or divide) *both* sides of the equation by the same *nonzero* quantity.	$3x = 9$	$\frac{3x}{3} = \frac{9}{3}$ $x = 3$
4. *Interchange the sides of the equation.*	$7 = x$	$x = 7$

The second and third operations in this list can be used to eliminate terms or factors in an equation. For example, to solve the equation $x - 5 = 1$, we need to eliminate the term -5 on the left side. This is accomplished by adding its opposite, 5, to both sides.

$$x - 5 = 1 \qquad \text{Original equation}$$

$$x - 5 + 5 = 1 + 5 \qquad \text{Add 5 to both sides.}$$

$$x + 0 = 6 \qquad \text{Combine like terms.}$$

$$x = 6 \qquad \text{Solution}$$

All four of the equations listed above are equivalent, and we call them the **steps** of the solution.

The next example shows how the properties of equality can be used to solve equations. You will get many more opportunities to practice these skills in the next chapter. For now, your goal should be to understand why each step in the solution is valid. For instance, the second step in part (a) is valid because the Addition Property of Equality states that you can add the same quantity to both sides of an equation.

Example 3 Operations Used to Solve Equations

Identify the property of equality used to solve each equation.

a. $x - 2 = 3$ Original equation

$x - 2 + 2 = 3 + 2$ Add 2 to both sides.

$x = 5$ Solution

b. $\frac{x}{5} = -2$ Original equation

$\frac{x}{5}(5) = -2(5)$ Multiply both sides by 5.

$x = -10$ Solution

c. $4x = 9$ Original equation

$\frac{4x}{4} = \frac{9}{4}$ Divide both sides by 4.

$x = \frac{9}{4}$ Solution

d. $\frac{5}{3}x = 7$ Original equation

$\frac{3}{5} \cdot \frac{5}{3}x = \frac{3}{5} \cdot 7$ Multiply both sides by $\frac{3}{5}$.

$x = \frac{21}{5}$ Solution

Study Tip

In Example 3(c), both sides of the equation are divided by 4 to eliminate the coefficient 4 on the left side. You could just as easily *multiply* both sides by $\frac{1}{4}$. Both techniques are legitimate—which one you decide to use is a matter of personal preference.

Solution

a. The Addition Property of Equality is used to add 2 to both sides of the equation in the second step. Adding 2 eliminates the term -2 from the left side of the equation.

b. The Multiplication Property of Equality is used to multiply both sides of the equation by 5 in the second step. Multiplying by 5 eliminates the denominator from the left side of the equation.

c. The Multiplication Property of Equality is used to divide both sides of the equation by 4 $\left(\text{or multiply both sides by } \frac{1}{4}\right)$ in the second step. Dividing by 4 eliminates the coefficient from the left side of the equation.

d. The Multiplication Property of Equality is used to multiply both sides of the equation by $\frac{3}{5}$ in the second step. Multiplying by the reciprocal of the fraction $\frac{5}{3}$ eliminates the fraction from the left side of the equation.

4 Use a verbal model to construct an algebraic equation.

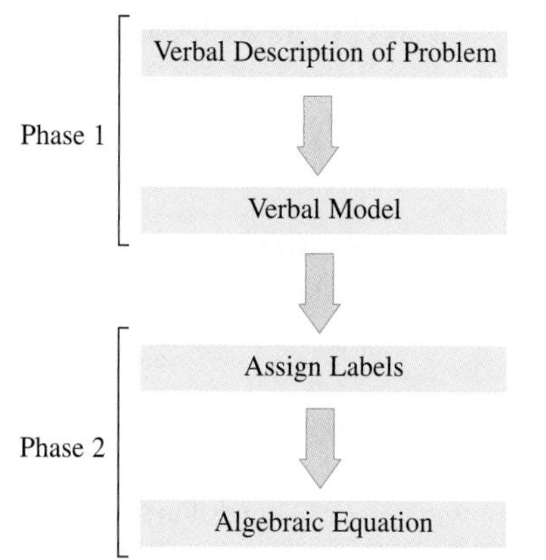

Constructing Equations

It is helpful to use two phases in constructing equations that model real life. In the first phase, you translate the verbal description into a *verbal model.* In the second phase, you assign labels and translate the verbal model into a *mathematical model* or *algebraic equation.* Here are two examples of verbal models.

1. The sale price of a basketball is \$28. The sale price is \$7 less than the original price. What is the original price?

Verbal Model: [Sale price] = [Original price] − [Discount]

[\$28] = [Original price] − [\$7]

2. The original price of a basketball is \$35. The original price is discounted by \$7. What is the sale price?

Verbal Model: [Sale price] = [Original price] − [Discount]

[Sale price] = [\$35] − [\$7]

Example 4 Using Verbal Models to Construct Equations

Write an algebraic equation for the problem.

The total income that an employee received in 1998 was \$31,550. How much was the employee paid each week? Assume that each weekly paycheck contained the same amount, and that the year consisted of 52 weeks.

Solution

Verbal Model: [Income for year] = [52] · [Weekly pay]

Labels: Income for year = 31,550 (dollars)
Weekly pay = x (dollars)

Algebraic Model: $31{,}550 = 52x$

When you construct an equation, be sure to check that both sides of the equation represent the *same* unit of measure. For instance, in Example 4, both sides of the equation $31{,}550 = 52x$ represent dollar amounts.

Example 5 Using Verbal Models to Construct Equations

Write an algebraic equation for the following problem.

Tickets for a concert cost $15 for each floor seat and $10 for each stadium seat. There were 800 seats on the main floor, and these were sold out. If the total revenue from ticket sales was $52,000, how many stadium seats were sold?

Solution

Verbal Model: Total revenue = Revenue from floor seats + Revenue from stadium seats

Labels:

Total revenue $= 52{,}000$	(dollars)
Price per floor seat $= 15$	(dollars per seat)
Number of floor seats $= 800$	(seats)
Price per stadium seat $= 10$	(dollars per seat)
Number of stadium seats $= x$	(seats)

Algebraic Model: $52{,}000 = 15(800) + 10x$

In Example 5, you can use the following *unit analysis* to check that both sides of the equation are measured in dollars.

$$52{,}000 \text{ dollars} = \left(\frac{15 \text{ dollars}}{\text{seat}}\right)(800 \text{ seats}) + \left(\frac{10 \text{ dollars}}{\text{seat}}\right)(x \text{ seats})$$

In the next chapter, you will study techniques for solving the equations constructed in Examples 4 and 5.

Discussing the Concept **Red Herring**

When constructing an equation to represent a word problem, you are occasionally given too much information. The unnecessary information in a word problem is sometimes called a "red herring." Find the red herring in the following problem.

Returning to college after spring break, a student travels 3 hours and stops for lunch. If it takes 45 minutes to complete the last 36 miles of the 180-mile trip, find the average speed during the first 3 hours of the trip.

Decide what question to ask that uses all the given information.

2.4 Exercises

Integrated Review *Concepts, Skills, and Problem Solving*

Keep mathematically in shape by doing these exercises *before* the problems of this section.

Properties and Definitions

1. If the numerator and denominator of a fraction have *unlike* signs, the sign of the fraction is ___.
2. If a negative number is used as a factor eight times, what is the sign of the product? Explain.
3. Complete the Commutative Property: $6 + 10 =$ ___.
4. Name the property illustrated by $6\left(\frac{1}{6}\right) = 1$.

Simplifying Expressions

In Exercises 5–10, simplify the expression.

5. $t^2 \cdot t^5$
6. $(-3y^3)y^2$
7. $(u^3)^2$
8. $2(ab)^5$
9. $(3a^2)(4ab)$
10. $2(x + 3)^2(x + 3)^3$

Graphs and Models

Geometry In Exercises 11 and 12, write expressions for the perimeter and area of the figure.

11. 12.

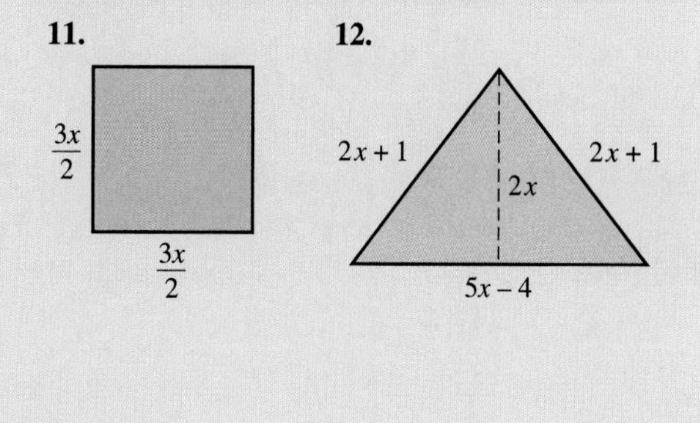

Developing Skills

In Exercises 1–16, determine whether the value of x is a solution of the equation. See Examples 1 and 2.

	Equation	*Values*	
1.	$2x - 6 = 0$	(a) $x = 3$	(b) $x = 1$
2.	$5x - 25 = 0$	(a) $x = 10$	(b) $x = 5$
3.	$2x + 4 = 2$	(a) $x = 0$	(b) $x = -1$
4.	$3x + 10 = 4$	(a) $x = -2$	(b) $x = 2$
5.	$x + 5 = 2x$	(a) $x = -1$	(b) $x = 5$
6.	$2x - 3 = 5x$	(a) $x = 0$	(b) $x = -1$
7.	$x + 3 = 2(x - 4)$	(a) $x = 11$	(b) $x = -5$
8.	$5x - 1 = 3(x + 5)$	(a) $x = 8$	(b) $x = -2$
9.	$2x + 10 = 7(x + 1)$	(a) $x = \frac{3}{5}$	(b) $x = \frac{2}{3}$
10.	$3(3x + 2) = 9 - x$	(a) $x = -\frac{3}{4}$	(b) $x = \frac{3}{10}$
11.	$x^2 - 4 = x + 2$	(a) $x = 3$	(b) $x = -2$
12.	$x^2 = 8 - 2x$	(a) $x = 2$	(b) $x = -4$
13.	$\frac{2}{x} - \frac{1}{x} = 1$	(a) $x = 3$	(b) $x = \frac{1}{3}$
14.	$\frac{4}{x} + \frac{2}{x} = 1$	(a) $x = 0$	(b) $x = 6$
15.	$\frac{5}{x - 1} + \frac{1}{x} = 5$	(a) $x = 3$	(b) $x = \frac{1}{6}$
16.	$\frac{3}{x - 2} = x$	(a) $x = -1$	(b) $x = 3$

In Exercises 17–26, use a calculator to determine whether the value of x is a solution of the equation.

	Equation	*Values*	
17.	$x + 3 = 3.5$	(a) $x = 1.2$	(b) $x = 4.8$
18.	$x - 6 = 1.4$	(a) $x = -4.6$	(b) $x = 7.4$
19.	$40x - 490 = 0$	(a) $x = 12.25$	(b) $x = -12.25$
20.	$20x - 560 = 0$	(a) $x = 27.5$	(b) $x = -27.5$
21.	$2x^2 - x - 10 = 0$	(a) $x = \frac{5}{2}$	(b) $x = -1.09$
22.	$22x - 5x^2 = 17$	(a) $x = 1$	(b) $x = 3.4$
23.	$\frac{1}{x} - \frac{9}{x - 4} = 1$	(a) $x = 0$	(b) $x = -2$

	Equation	*Values*	
24.	$x = \dfrac{3}{4x + 1}$	(a) $x = -0.25$	(b) $x = 0.75$
25.	$x^3 - 1.728 = 0$	(a) $x = \frac{6}{5}$	(b) $x = -\frac{6}{5}$
26.	$4x^2 - 10.24 = 0$	(a) $x = \frac{8}{5}$	(b) $x = -\frac{8}{5}$

In Exercises 27–34, justify each step of the solution. See Example 3.

27.
$$\begin{aligned} 5x + 12 &= 22 \\ 5x + 12 - 12 &= 22 - 12 \\ 5x &= 10 \\ \frac{5x}{5} &= \frac{10}{5} \\ x &= 2 \end{aligned}$$

28.
$$\begin{aligned} 14 - 3x &= 5 \\ 14 - 3x - 14 &= 5 - 14 \\ 14 - 14 - 3x &= -9 \\ -3x &= -9 \\ \frac{-3x}{-3} &= \frac{-9}{-3} \\ x &= 3 \end{aligned}$$

29.
$$\begin{aligned} \frac{2}{3}x &= 12 \\ \frac{3}{2}\left(\frac{2}{3}x\right) &= \frac{3}{2}(12) \\ x &= 18 \end{aligned}$$

30.
$$\begin{aligned} \frac{4}{5}x &= -28 \\ \frac{5}{4}\left(\frac{4}{5}x\right) &= \frac{5}{4}(-28) \\ x &= -35 \end{aligned}$$

31.
$$\begin{aligned} 2(x - 1) &= x + 3 \\ 2x - 2 &= x + 3 \\ -x + 2x - 2 &= -x + x + 3 \\ x - 2 &= 3 \\ x - 2 + 2 &= 3 + 2 \\ x &= 5 \end{aligned}$$

32.
$$\begin{aligned} x + 6 &= -6(4 - x) \\ x + 6 &= -24 + 6x \\ -x + x + 6 &= -x - 24 + 6x \\ 6 &= 5x - 24 \\ 6 + 24 &= 5x - 24 + 24 \\ 30 &= 5x \\ \frac{30}{5} &= \frac{5x}{5} \\ 6 &= x \end{aligned}$$

33.
$$\begin{aligned} x &= -2(x + 3) \\ x &= -2x - 6 \\ 2x + x &= 2x - 2x - 6 \\ 3x &= 0 - 6 \\ 3x &= -6 \\ \frac{3x}{3} &= \frac{-6}{3} \\ x &= -2 \end{aligned}$$

34.
$$\begin{aligned} \frac{x}{3} &= x + 1 \\ 3\left(\frac{x}{3}\right) &= 3(x + 1) \\ x &= 3x + 3 \\ -3x + x &= -3x + 3x + 3 \\ -2x &= 0 + 3 \\ -2x &= 3 \\ \frac{-2x}{-2} &= \frac{3}{-2} \\ x &= -\frac{3}{2} \end{aligned}$$

In Exercises 35–38, use a property of equality to solve the equation. Check your solution. See Examples 1, 2, and 3.

35. $x + 4 = 6$

36. $x - 10 = 5$

37. $3x = 30$

38. $\dfrac{x}{4} = 12$

Solving Problems

In Exercises 39–44, write a verbal description of the algebraic equation. Use words only; do not use the variable. (There is more than one correct answer.)

39. $x + 8 = 25$ **40.** $x - 9 = 52$

41. $10(x - 3) = 8x$ **42.** $2(x - 5) = 12$

43. $\dfrac{x + 1}{3} = 8$ **44.** $\dfrac{x - 2}{10} = 6$

In Exercises 45–68, construct an equation for the word problem. Do *not* solve the equation. See Examples 4 and 5.

45. *Test Score* After your instructor added 6 points to each student's test score, your score is 94. What was your original score?

46. *Rainfall* With the 1.2-inch rainfall today, the total for the month is 4.5 inches. How much had been recorded for the month before today's rainfall?

47. *Computer Purchase* You have $3650 saved for the purchase of a new computer that will cost $4532. How much more must you save?

48. *List Price* The sale price of a coat is $225.98. If the discount is $64, what is the list (original) price?

49. The sum of a number and 12 is 45. What is the number?

50. The sum of 3 times a number and 4 is 16. What is the number?

51. Four times the sum of a number and 6 is 100. What is the number?

52. Find a number such that 6 times the number subtracted from 120 is 96.

53. Find a number such that 2 times the number decreased by 14 equals the number divided by 3.

54. The sum of a number and 8, divided by 4, is 32. What is the number?

55. *Travel Costs* A company pays its sales representatives 32 cents per mile if they use their personal cars. A sales representative submitted a bill to be reimbursed for $135.36 for driving. How many miles did the sales representative drive?

56. *Pocket Change* A student has n quarters and seven $1 bills totaling $8.75. How many quarters does the student have?

57. *Dimensions of a Mirror* The width of a rectangular mirror is one-third its length, as shown in the figure. The perimeter of the mirror is 96 inches. What are the dimensions of the mirror?

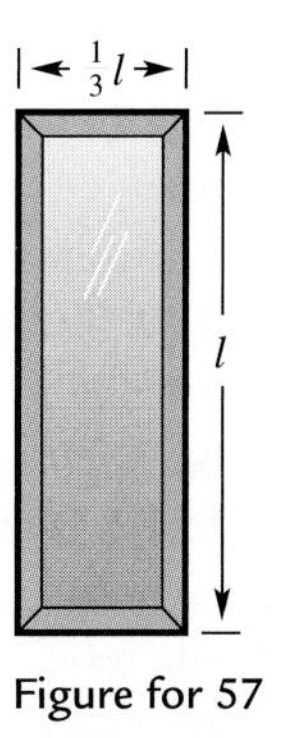

Figure for 57

58. *Height of a Box* Find the height of a rectangular box if its base is 4 feet by 6 feet and its volume is 72 cubic feet (see figure).

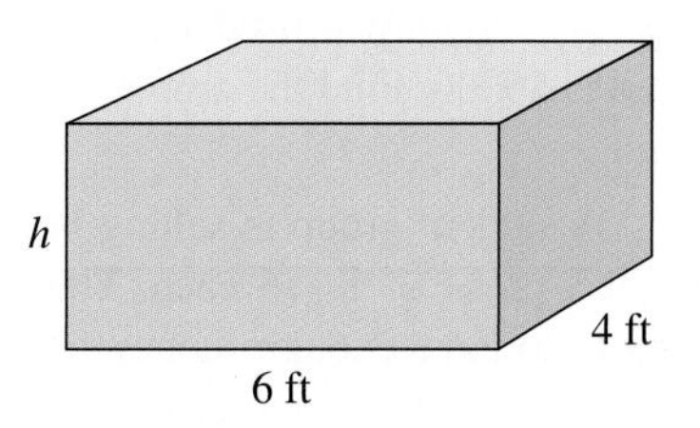

59. *Average Speed* After traveling for 3 hours, your family is still 25 miles from completing a 160-mile trip (see figure). What was the average speed during the first 3 hours?

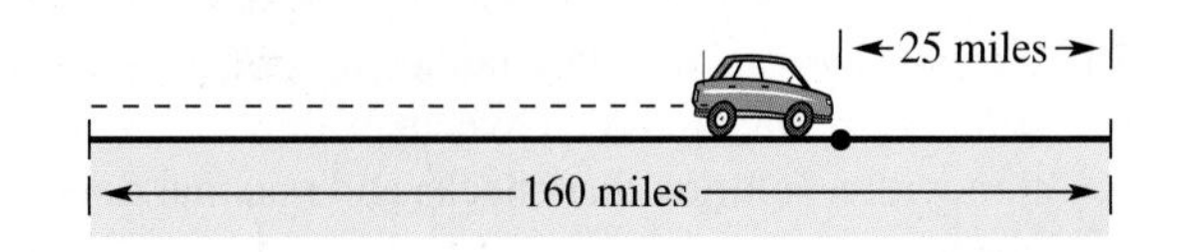

60. *Average Speed* After traveling for 4 hours, you are still 24 miles from completing a 200-mile trip. If it requires one-half hour to travel the last 24 miles, find the average speed during the first 4 hours of the trip.

61. *Average Speed* A group of students plans to take two cars to a soccer game. The first car leaves on time, travels at an average speed of 45 miles per hour, and arrives at the destination in 3 hours. Determine the average speed of the students in the second car if they leave one-half hour after the first car and arrive at the game at the same time as the students in the first car.

62. *Dow Jones Average* The Dow Jones average fell 58 points during a week and was 8695 at the close of the market on Friday. What was the average at the close of the market on the previous Friday?

63. *Price of a Product* The price of a product has increased by \$45 over the past year. It is now selling for \$375. What was the price 1 year ago?

64. *Thunderstorm* You hear thunder 3 seconds after seeing the lightning. How far away is the lightning, if the speed of sound is 1100 feet per second?

65. *Annual Depreciation* A corporation buys equipment with an initial purchase price of \$750,000. It is estimated that its useful life will be 3 years and at that time its value will be \$75,000. The total depreciation is divided equally among the three years. (Depreciation is the difference between the initial price of an item and its current value.) Determine the amount of depreciation declared each year.

66. *Car Payments* Suppose you make 48 monthly payments of \$158 each to buy a used car. The total amount financed is \$6000. Find the amount of interest that you paid.

67. *Fund Raising* A student group is selling boxes of greeting cards at a profit of \$1.75 each. The group needs \$2000 more to have enough money for a trip to Washington, D.C. How many boxes does the group need to sell to earn \$2000?

68. *Price of a Product* The price of a product increased \$1432 during the past year. The price of the product was \$9850 two years ago and \$10,120 one year ago. What is its current price?

Unit Analysis In Exercises 69–74, simplify the expression. State the units of the simplified value.

69. $\dfrac{3 \text{ dollars}}{\text{unit}} \cdot (5 \text{ units})$

70. $\dfrac{25 \text{ miles}}{\text{gallon}} \cdot (15 \text{ gallons})$

71. $\dfrac{3 \text{ dollars}}{\text{pound}} \cdot (5 \text{ pounds})$

72. $\dfrac{12 \text{ dollars}}{\text{hour}} \cdot \dfrac{1 \text{ hour}}{60 \text{ minutes}} \cdot (45 \text{ minutes})$

73. $\dfrac{5 \text{ feet}}{\text{second}} \cdot \dfrac{60 \text{ seconds}}{\text{minute}} \cdot (20 \text{ minutes})$

74. $\dfrac{100 \text{ centimeters}}{\text{meter}} \cdot (2.4 \text{ meters})$

Explaining Concepts

75. In your own words, explain the difference between a conditional equation and an identity.

76. Explain how to decide whether a real number is a solution of an equation. Give an example of an equation with a solution that checks and one that does not check.

77. Explain the difference between simplifying an expression and solving an equation. Give an example of each.

78. In your own words, explain what is meant by the term *equivalent equations.*

79. Describe, from memory, the steps that are used to transform an equation into an equivalent equation.

80. Describe a real-life problem that uses the following verbal model.

Revenue of \$840	=	\$35 per case	·	Number of cases

CHAPTER SUMMARY

Key Terms

algebraic expression, *p. 62*
variables, *p. 62*
constants, *p. 62*
terms, *p. 62*
coefficient, *p. 62*
evaluate an algebraic expression, *p. 65*
like terms, *p. 75*
simplify an algebraic expression, *p. 77*
verbal mathematical model, *p. 86*
equation, *p. 99*
solution, *p. 99*
solution set, *p. 99*
identity, *p. 99*
conditional equation, *p. 99*
equivalent equations, *p. 101*

Key Concepts

2.1 Exponential form

Repeated multiplication can be expressed in exponential form using a base a and an exponent n, where a is a real number, variable, or algebraic expression and n is a positive integer.

$$a^n = a \cdot a \cdot \cdot \cdot a$$

2.1 Evaluating algebraic expressions

To evaluate an algebraic expression, replace every occurrence of the variable in the expression with the appropriate real number and perform the operations.

2.2 Rules of exponents

Let m and n be positive integers, and let a and b be real numbers, variables, or algebraic expressions.

1. $a^m \cdot a^n = a^{m+n}$ 2. $(a^m)^n = a^{m \cdot n}$
3. $(ab)^m = a^m \cdot b^m$

2.2 Basic rules of algebra

Commutative Property:

Addition $a + b = b + a$

Multiplication $ab = ba$

Associative Property:

Addition $(a + b) + c = a + (b + c)$

Multiplication $(ab)c = a(bc)$

Distributive Property:

$a(b + c) = ab + ac$ $a(b - c) = ab - ac$

$(a + b)c = ac + bc$ $(a - b)c = ac - bc$

Identities:

Additive $a + 0 = a$

Multiplicative $a \cdot 1 = a$

Inverses:

Additive $a + (-a) = 0$

Multiplicative $a \cdot \dfrac{1}{a} = 1, \ a \neq 0$

2.2 Combining like terms

To combine like terms, add their respective coefficients and attach the common variable factor.

2.2 Simplifying an algebraic expression

To simplify an algebraic expression, remove symbols of grouping and combine like terms.

2.3 Translating phrases: verbal to algebraic

From the verbal description, write a verbal mathematical model. Assign labels to the known and unknown quantities, and write an algebraic model.

2.3 Other problem-solving strategies

Other problem-solving strategies are (1) guess, check, and revise, (2) make a table/look for a pattern, (3) draw a diagram, and (4) solve a simpler problem.

2.4 Checking solutions of equations

To check a solution, substitute the given solution for each occurrence of the variable in the original equation. Evaluate each side of the equation. If both sides are equivalent, the solution checks.

2.4 Properties of equality

Addition: Add (or subtract) the same quantity to (from) both sides of the equation.

Multiplication: Multiply (or divide) both sides of the equation by the same nonzero quantity.

REVIEW EXERCISES

Reviewing Skills

2.1 In Exercises 1–4, identify the terms and the coefficients of the algebraic expression.

1. $4 - \frac{1}{2}x^3$

2. $5x^2 - 3x + 10$

3. $y^2 - 10yz + \frac{2}{3}z^2$

4. $\dfrac{x + 2y}{3} - \dfrac{4x}{y}$

In Exercises 5–8, rewrite the product in exponential form.

5. $5z \cdot 5z \cdot 5z$

6. $\frac{3}{8}y \cdot \frac{3}{8}y \cdot \frac{3}{8}y \cdot \frac{3}{8}y$

7. $a(b - c) \cdot a(b - c)$

8. $3 \cdot (y - x) \cdot (y - x) \cdot 3 \cdot 3$

In Exercises 9–12, evaluate the algebraic expression for the specified value(s) of the variable(s).

	Expression	*Values*	
9.	$x^2 - 2x + 5$	(a) $x = 0$	(b) $x = 2$
10.	$x^3 - 8$	(a) $x = 2$	(b) $x = 4$
11.	$x^2 - x(y + 1)$	(a) $x = 2, y = -1$	
		(b) $x = 1, y = 2$	
12.	$\dfrac{x + 5}{y}$	(a) $x = -5, y = 3$	
		(b) $x = 2, y = -1$	

2.2 In Exercises 13–22, simplify the exponential expression.

13. $x^2 \cdot x \cdot x^4$

14. $y^2 \cdot y^3 \cdot y$

15. $(x^3)^2$

16. $(t^4)^3$

17. $t^4(-2t^2)$

18. $u^2(3u^2)$

19. $(xy)(-5x^2y^3)$

20. $(3uv)(-2uv^2)$

21. $(-2y^2)^3(8y)$

22. $(-3x)^2(5x^2)$

In Exercises 23–28, identify the rule of algebra illustrated by the equation.

23. $xy \cdot \dfrac{1}{xy} = 1$

24. $u(vw) = (uv)w$

25. $(x - y)(2) = 2(x - y)$

26. $(a + b) + 0 = a + b$

27. $2x + (3y - z) = (2x + 3y) - z$

28. $x(y + z) = xy + xz$

In Exercises 29–36, use the Distributive Property to expand the expression.

29. $4(x + 3y)$

30. $3(8s - 12t)$

31. $-5(2u - 3v)$

32. $-3(-2x - 8y)$

33. $x(8x + 5y)$

34. $-u(3u - 10v)$

35. $-(-a + 3b)$

36. $(7 - 2j)(-6)$

In Exercises 37–48, simplify the expression by combining like terms.

37. $3a - 5a$

38. $6c - 2c$

39. $3p - 4q + q + 8p$

40. $10x - 4y - 25x + 6y$

41. $\frac{1}{4}s - 6t + \frac{7}{2}s + t$

42. $\frac{2}{3}a + \frac{3}{5}a - \frac{1}{2}b + \frac{2}{3}b$

43. $x^2 + 3xy - xy + 4$

44. $uv^2 + 10 - 2uv^2 + 2$

45. $5x - 5y + 3xy - 2x + 2y$

46. $y^3 + 2y^2 + 2y^3 - 3y^2 + 1$

47. $5\left(1 + \dfrac{r}{n}\right)^2 - 2\left(1 + \dfrac{r}{n}\right)^2$

48. $-7\left(\dfrac{1}{u}\right) + 4\left(\dfrac{1}{u^2}\right) + 3\left(\dfrac{1}{u}\right)$

In Exercises 49–60, simplify the expression by removing symbols of grouping and combining like terms.

49. $5(u - 4) + 10$

50. $16 - 3(v + 2)$

51. $3s - (r - 2s)$

52. $50x - (30x + 100)$

53. $-3(1 - 10z) + 2(1 - 10z)$

54. $8(15 - 3y) - 5(15 - 3y)$

55. $\frac{1}{3}(42 - 18z) - 2(8 - 4z)$

56. $\frac{1}{4}(100 + 36s) - (15 - 4s)$

57. $10 - [8(5 - x) + 2]$

58. $3[2(4x - 5) + 4] - 3$

59. $2[x + 2(y - x)]$

60. $2t[4 - (3 - t)] + 5t$

2.3 In Exercises 61–70, translate the phrase into an algebraic expression. Let x represent the number.

61. Two-thirds of a real number, plus 5

62. One hundred, decreased by 5 times a number

63. Ten less than twice a number

64. The ratio of a number to 10

65. Fifty, increased by the product of 7 and a number

66. Ten decreased by the quotient of a number and 2

67. The sum of a number and 10 divided by 8

68. The product of 15 and a number, decreased by 2

69. The sum of the square of a real number and 64

70. The absolute value of the sum of a number and -10

In Exercises 71–74, write a verbal description of the expression without using the variable. (There is more than one correct answer.)

71. $x + 3$

72. $3x - 2$

73. $\dfrac{y-2}{3}$

74. $4(x + 5)$

2.4 In Exercises 75–84, check whether each value of x is a solution of the equation.

	Equation	*Values*	
75.	$5x + 6 = 36$	(a) $x = 3$	(b) $x = 6$
76.	$17 - 3x = 8$	(a) $x = 3$	(b) $x = -3$
77.	$3x - 12 = x$	(a) $x = -1$	(b) $x = 6$
78.	$8x + 24 = 2x$	(a) $x = 0$	(b) $x = -4$
79.	$4(2 - x) = 3(2 + x)$	(a) $x = \frac{2}{7}$	(b) $x = -\frac{2}{3}$
80.	$5x + 2 = 3(x + 10)$	(a) $x = 14$	(b) $x = -10$
81.	$\dfrac{4}{x} - \dfrac{2}{x} = 5$	(a) $x = -1$	(b) $x = \frac{2}{5}$
82.	$\dfrac{x}{3} + \dfrac{x}{6} = 1$	(a) $x = \frac{2}{9}$	(b) $x = -\frac{2}{9}$
83.	$x(x - 7) = -12$	(a) $x = 3$	(b) $x = 4$
84.	$x(x + 1) = 2$	(a) $x = 1$	(b) $x = -2$

Solving Problems

85. *Depreciation* You pay P dollars for new equipment. Its value after 5 years is given by

$$P\left(\tfrac{9}{10}\right)\left(\tfrac{9}{10}\right)\left(\tfrac{9}{10}\right)\left(\tfrac{9}{10}\right)\left(\tfrac{9}{10}\right).$$

Simplify the expression.

86. *Area* The height of a triangle is $1\frac{1}{2}$ times its base. Its area is given by $\frac{1}{2}b\left(\frac{3}{2}b\right)$. Simplify the expression.

87. *Income Tax* The income tax rate on a taxable income of I dollars is 28%. Write an algebraic expression that represents the total amount of income tax. (To find 28% of a quantity, multiply the quantity by 0.28.)

88. *Total Amount of Money* A person has n nickels and q quarters. Write an algebraic expression that represents the total amount of money in dollars.

89. *Geometry* The face of a tape deck has the dimensions shown in the figure. Find an algebraic expression that represents the area of the face of the tape deck excluding the compartment holding the cassette.

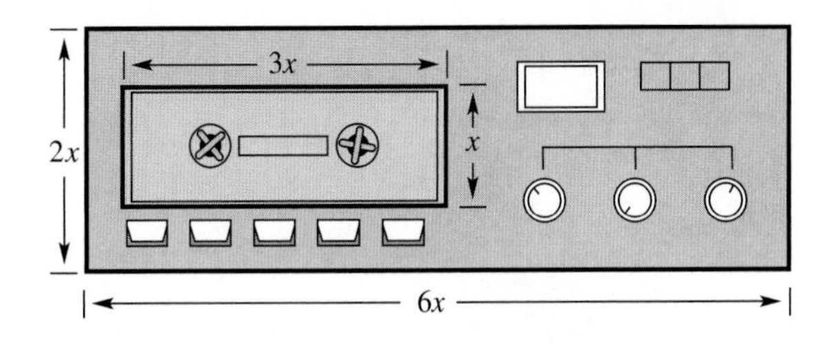

90. *Geometry* Find the perimeter of the figure.

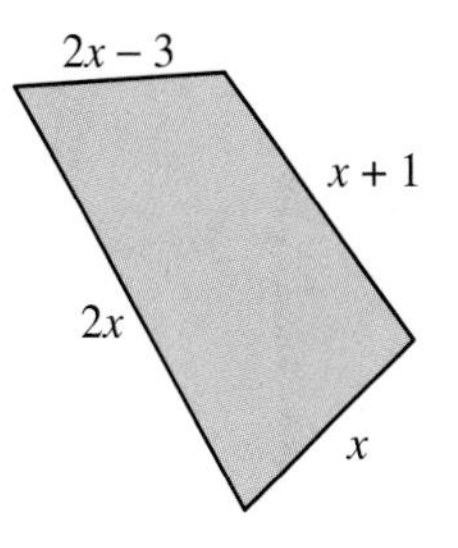

91. *Distance Traveled* A car travels for 10 hours at an average speed of s miles per hour. Write an algebraic expression that represents the total distance traveled.

92. *Sum* Simplify the algebraic expression that represents the sum of three consecutive odd integers, $2n - 1$, $2n + 1$, and $2n + 3$.

93. *Rental Income* Write an expression that represents the rent for n months if the monthly rent is \$625.

94. Perform the indicated operations and simplify.

$7 \cdot 10^4 + 2 \cdot 10^3 + 8 \cdot 10^1$

95. *Finding a Pattern*

(a) Complete the table. The third row is the difference of consecutive entries of the second row. The fourth row is the difference of consecutive entries of the third row.

n	0	1	2	3	4	5
$n^2 + 3n + 2$						
Differences						
Differences						

(b) Describe the patterns for the third and fourth rows.

96. *Finding a Pattern* Find values for a and b such that the expression $an + b$ agrees with the values given in the table.

n	0	1	2	3	4	5
$an + b$	4	9	14	19	24	29

In Exercises 97–100, write an equation that represents the statement. (Identify the letters you choose as labels.)

97. *Sum* The sum of a number and its reciprocal is $\frac{37}{6}$.

98. *Distance* An automobile travels 135 miles in t hours with an average speed of 45 miles per hour (see figure).

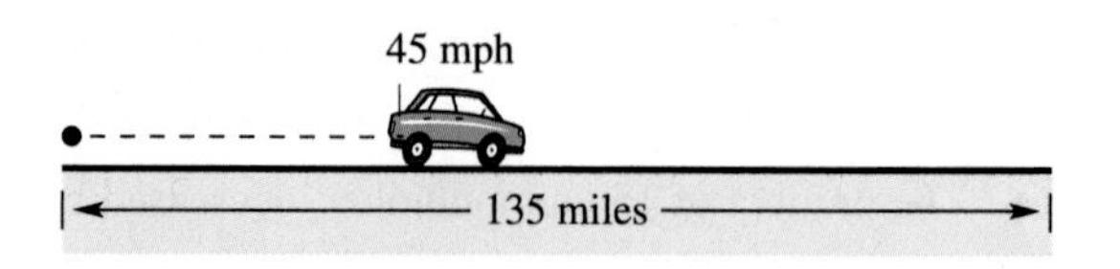

99. *Geometry* The area of the shaded region in the figure is 24 square inches.

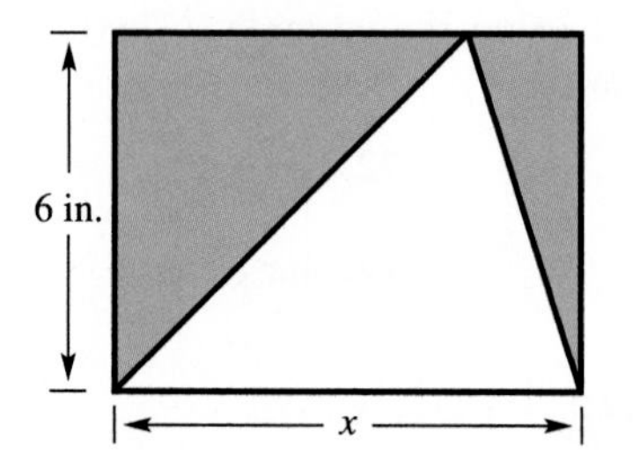

100. *Geometry* The perimeter of the face of the rectangular traffic light is 72 inches (see figure).

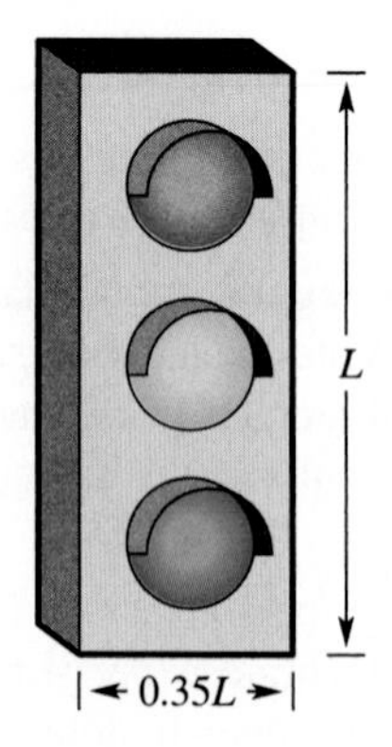

Chapter Test

Take this test as you would take a test in class. After you are done, check your work against the answers given in the back of the book.

1. Identify the terms and coefficients of the expression.

$2x^2 - 7xy + 3y^3$

2. Rewrite the following product in exponential form.

$x \cdot (x + y) \cdot x \cdot (x + y) \cdot x$

In Exercises 3–6, identify the rule of algebra demonstrated.

3. $(5x)y = 5(xy)$

4. $2 + (x - y) = (x - y) + 2$

5. $7xy - 7xy = 0$

6. $1 \cdot (x + 5) = (x + 5)$

In Exercises 7 and 8, use the Distributive Property to expand the expression.

7. $3(x + 8)$

8. $-y(3 - 2y)$

In Exercises 9–14, simplify the expression.

9. $(c^2)^4$

10. $-5uv(2u^3)$

11. $3b - 2a + a - 10b$

12. $15(u - v) - 7(u - v)$

13. $3z - (4 - z)$

14. $2[10 - (t + 1)]$

15. Evaluate the expression when $x = 3$ and $y = -12$.

(a) $x^3 - 2$ (b) $x^2 + 4(y + 2)$

16. Explain why it is not possible to evaluate $\frac{a + 2b}{3a - b}$ when $a = 2$ and $b = 6$.

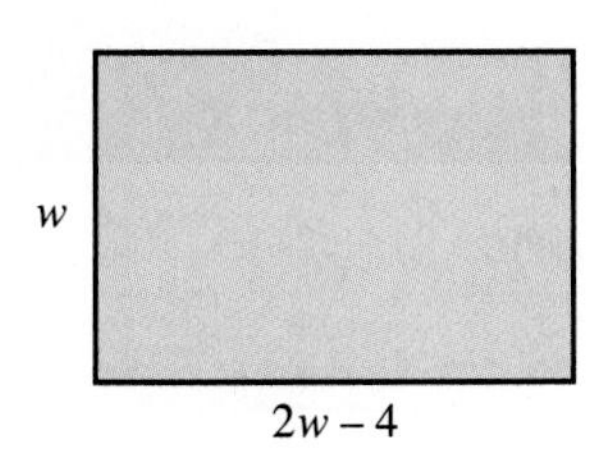

Figure for 18

17. Translate the phrase, "one-fifth of a number, increased by two," into an algebraic expression. Let n represent the number.

18. (a) Write expressions for the perimeter and area of the rectangle at the left.

(b) Simplify the expressions.

(c) Identify the unit of measure for each expression.

(d) Evaluate each expression when $w = 12$ feet.

19. Write an algebraic expression for the income from a concert if the prices of the tickets for adults and children are \$3 and \$2, respectively. Let n represent the number of adults in attendance and let m represent the number of children.

20. Determine whether the values of x are solutions of $6(3 - x) - 5(2x - 1) = 7$.

(a) $x = -2$ (b) $x = 1$

3 Linear Equations and Problem Solving

Stewart Cohen/Tony Stone Images

In 1997, the average monthly cellular phone bill was $42.78. (Source: Cellular Telecommunications Industry Association)

Motivating the Chapter

Talk Is Cheap?

You plan to purchase a cellular phone with a service contract. For a price of $99, one package includes the phone and 3 months of service. You will be billed a *per minute usage rate* each time you make or receive a call. After 3 months you will be billed a monthly service charge of $19.50 and the per minute usage rate.

A second cellular phone package costs $80, which includes the phone and one month of service. You will be billed a per minute usage rate each time you make or receive a call. After the first month you will be billed a monthly service charge of $24.00 and the per minute usage rate.

See Section 3.3, Exercise 105.

a. Write an equation to find the cost of the phone in the first package. Solve the equation to find the cost of the phone.

b. Write an equation to find the cost of the phone in the second package. Solve the equation to find the cost of the phone. Which phone costs more, the one in the first package or the one in the second package?

c. What percent of the purchase price of $99 goes toward the price of the cellular phone in the first package? Use an equation to answer the question.

d. What percent of the purchase price of $80 goes toward the price of the cellular phone in the second package? Use an equation to answer the question.

e. If the sales tax on your purchase is 5%, what is the total cost of purchasing the first cellular phone package? Use an equation to answer the question.

f. You decide to buy the first cellular phone package. If your total cellular phone bill for the fourth month of use is $92.46 for 3.2 hours of use, what is the per minute usage rate? Use an equation to answer the question.

See Section 3.4, Exercise 87.

g. For the fifth month you were billed the monthly service charge and $47.50 for 125 minutes of use. You estimate that during the next month you spent 150 minutes on calls. Use a proportion to find the charge for 150 minutes of use. (Use the first package.)

See Section 3.6, Exercise 87.

h. You determine that the most you can spend each month on phone calls is $75. Write a compound inequality that describes the number of minutes you can spend talking on the cellular phone each month if the per minute usage rate is $0.35. Solve the inequality. (Use the first package.)

3.1 Solving Linear Equations

Objectives

1 Solve a linear equation in standard form.

2 Solve a linear equation in nonstandard form.

3 Use a linear equation to solve an application problem.

1 Solve a linear equation in standard form.

Linear Equations in the Standard Form $ax + b = c$

This is an important step in your study of algebra. In the first two chapters, you were introduced to the rules of algebra, and you learned to use these rules to rewrite and simplify algebraic expressions. In Sections 2.3 and 2.4, you gained experience in translating verbal expressions and problems into algebraic forms. You are now ready to use these skills and experiences to *solve equations.*

In this section, you will learn how the rules of algebra and the properties of equality can be used to solve the most common type of equation—a linear equation in one variable.

Diophantus

(250 A.D.)

Diophantus, a Greek of Alexandria who lived around 250 A.D., is often called the "Father of Algebra." He was the first to use abbreviated word forms in equations. Diophantus introduced this symbolism in the *Arithmetica,* a collection of problems comprising 13 books.

Definition of Linear Equation

A **linear equation** in one variable x is an equation that can be written in the standard form

$$ax + b = c$$

where a, b, and c are real numbers with $a \neq 0$.

A linear equation in one variable is also called a **first-degree equation** because its variable has an (implied) exponent of 1. Some examples of linear equations in standard form are

$$2x = 3, \quad x - 7 = 5, \quad 4x + 6 = 0, \quad \text{and} \quad \frac{x}{2} - 1 = \frac{5}{3}.$$

Remember that to *solve* an equation involving x means that you are to find all values of x that satisfy the equation. For the linear equation $ax + b = c$, the goal is to *isolate* x by rewriting the equation in the form

$x =$ a number . Isolate the variable x.

To obtain this form, you are to use the techniques discussed in Section 2.4. That is, beginning with the original equation, you write a sequence of equivalent equations, each having the same solution as the original equation. For instance, to solve the linear equation $x - 2 = 0$, you can add 2 to both sides of the equation to obtain $x = 2$. As mentioned in Section 2.4, each equivalent equation is called a **step** of the solution.

Example 1 Solving a Linear Equation

Solve $3x - 5 = 10$.

Solution

$3x - 5 = 10$	Original equation
$3x - 5 + 5 = 10 + 5$	Add 5 to both sides.
$3x = 15$	Combine like terms.
$\frac{3x}{3} = \frac{15}{3}$	Divide both sides by 3.
$x = 5$	Simplify.

It appears that the solution is 5. Here is the check.

Check

$3x - 5 = 10$	Original equation
$3(5) - 5 \stackrel{?}{=} 10$	Substitute 5 for x.
$15 - 5 \stackrel{?}{=} 10$	Simplify.
$10 = 10$	Solution checks. ✓

In Example 1, be sure you see that solving an equation has two basic stages. The first stage is to *find* the solution (or solutions). The second stage is to *check* that each solution you find actually satisfies the original equation. You can improve your accuracy in algebra by developing the habit of checking each solution.

A common question in algebra is

"How do I know which step to do *first* to isolate x?"

The answer is that you need practice. By solving many linear equations, you will find that your skill will improve. The key thing to remember is that you can "get rid of" terms and factors by using *inverse* operations. Here are some guidelines and examples.

Guideline	*Equation*	*Inverse Operation*
1. Subtract to remove a sum.	$x + 3 = 4$	Subtract 3 from both sides.
2. Add to remove a difference.	$x - 5 = 7$	Add 5 to both sides.
3. Divide to remove a product.	$4x = 20$	Divide both sides by 4.
4. Multiply to remove a quotient.	$\frac{x}{8} = 2$	Multiply both sides by 8.

For additional examples, review Example 3 on page 102. In each case of that example, note how inverse operations are used to isolate the variable.

Example 2 Solving a Linear Equation in Standard Form

Solve $2x + 7 = 4$.

Solution

$$2x + 7 = 4 \qquad \text{Original equation}$$

$$2x + 7 - 7 = 4 - 7 \qquad \text{Subtract 7 from both sides.}$$

$$2x = -3 \qquad \text{Combine like terms.}$$

$$\frac{2x}{2} = -\frac{3}{2} \qquad \text{Divide both sides by 2.}$$

$$x = -\frac{3}{2} \qquad \text{Simplify.}$$

Check

$$2x + 7 = 4 \qquad \text{Original equation}$$

$$2\left(-\frac{3}{2}\right) + 7 \stackrel{?}{=} 4 \qquad \text{Substitute } -\tfrac{3}{2} \text{ for } x.$$

$$-3 + 7 \stackrel{?}{=} 4 \qquad \text{Simplify.}$$

$$4 = 4 \qquad \text{Solution checks. } \checkmark$$

So, the solution is $-\frac{3}{2}$.

Example 3 Solving a Linear Equation in Standard Form

Solve $5x - 3 = 9$.

Solution

$$5x - 3 = 9 \qquad \text{Original equation}$$

$$5x - 3 + 3 = 9 + 3 \qquad \text{Add 3 to both sides.}$$

$$5x = 12 \qquad \text{Combine like terms.}$$

$$\frac{5x}{5} = \frac{12}{5} \qquad \text{Divide both sides by 5.}$$

$$x = \frac{12}{5} \qquad \text{Simplify.}$$

Check

$$5x - 3 = 9 \qquad \text{Original equation}$$

$$5\left(\frac{12}{5}\right) - 3 \stackrel{?}{=} 9 \qquad \text{Substitute } \tfrac{12}{5} \text{ for } x.$$

$$12 - 3 \stackrel{?}{=} 9 \qquad \text{Simplify.}$$

$$9 = 9 \qquad \text{Solution checks. } \checkmark$$

So, the solution is $\frac{12}{5}$.

Study Tip

To eliminate a fractional coefficient, it may be easier to multiply both sides by the *reciprocal* of the fraction than to divide by the fraction itself. Here is an example.

$$-\frac{2}{3}x = 4$$

$$\left(-\frac{3}{2}\right)\left(-\frac{2}{3}\right)x = \left(-\frac{3}{2}\right)4$$

$$x = -\frac{12}{2}$$

$$x = -6$$

Example 4 Solving a Linear Equation in Standard Form

Solve $\frac{x}{3} - 1 = -4$.

Solution

$\frac{x}{3} - 1 = -4$ Original equation

$\frac{x}{3} - 1 + 1 = -4 + 1$ Add 1 to both sides.

$\frac{x}{3} = -3$ Combine like terms.

$3\left(\frac{x}{3}\right) = 3(-3)$ Multiply both sides by 3.

$x = -9$ Simplify.

Check

$\frac{x}{3} - 1 = -4$ Original equation

$\frac{-9}{3} - 1 \stackrel{?}{=} -4$ Substitute -9 for x.

$-3 - 1 \stackrel{?}{=} -4$ Simplify.

$-4 = -4$ Solution checks. ✓

So, the solution is -9.

Technology: Tip

Remember to check your solution in the original equation. This can be done efficiently with a graphing utility.

As you gain experience in solving linear equations, you will probably find that you can perform some of the solution steps in your head. For instance, you might solve the equation given in Example 4 by writing only the following steps.

$\frac{x}{3} - 1 = -4$ Original equation

$\frac{x}{3} = -3$ Add 1 to both sides.

$x = -9$ Multiply both sides by 3.

2 Solve a linear equation in nonstandard form.

Solving a Linear Equation in Nonstandard Form

The definition of a linear equation contains the phrase "that can be written in the standard form $ax + b = c$." This suggests that some linear equations may come in nonstandard or disguised form.

A common form of linear equations is one in which the variable terms are not combined into one term. In such cases, you can begin the solution by rewriting the equation in standard form. Note how this is done in the next two examples.

Example 5 Solving a Linear Equation in Nonstandard Form

Solve $3y + 8 - 5y = 4$.

Solution

$$3y + 8 - 5y = 4 \quad \text{Original equation}$$
$$3y - 5y + 8 = 4 \quad \text{Collect like terms.}$$
$$-2y + 8 = 4 \quad \text{Combine like terms.}$$
$$-2y + 8 - 8 = 4 - 8 \quad \text{Subtract 8 from both sides.}$$
$$-2y = -4 \quad \text{Combine like terms.}$$
$$\frac{-2y}{-2} = \frac{-4}{-2} \quad \text{Divide both sides by } -2.$$
$$y = 2 \quad \text{Simplify.}$$

Check

$$3y + 8 - 5y = 4 \quad \text{Original equation}$$
$$3(2) + 8 - 5(2) \stackrel{?}{=} 4 \quad \text{Substitute 2 for } y.$$
$$6 + 8 - 10 \stackrel{?}{=} 4 \quad \text{Simplify.}$$
$$4 = 4 \quad \text{Solution checks. } \checkmark$$

So, the solution is 2.

Study Tip

In Example 5, note that the variable in the equation doesn't always have to be x. Any letter can be used.

The solution for Example 5 began by collecting like terms. You can use any of the rules of algebra to attain your goal of "isolating the variable." The next example shows how to solve a linear equation using the Distributive Property.

Example 6 Using the Distributive Property

Solve $x + 6 = 2(x - 3)$.

Solution

$$x + 6 = 2(x - 3) \quad \text{Original equation}$$
$$x + 6 = 2x - 6 \quad \text{Apply Distributive Property.}$$
$$x - 2x + 6 = 2x - 2x - 6 \quad \text{Subtract } 2x \text{ from both sides.}$$
$$-x + 6 = -6 \quad \text{Combine like terms.}$$
$$-x + 6 - 6 = -6 - 6 \quad \text{Subtract 6 from both sides.}$$
$$-x = -12 \quad \text{Combine like terms.}$$
$$(-1)(-x) = (-1)(-12) \quad \text{Multiply both sides by } -1.$$
$$x = 12 \quad \text{Simplify.}$$

The solution is 12. Check this in the original equation.

Study Tip

You can isolate the variable term on either side of the equal sign. For instance, Example 6 could have been solved this way.

$$x + 6 = 2(x - 3)$$
$$x + 6 = 2x - 6$$
$$x - x + 6 = 2x - x - 6$$
$$6 = x - 6$$
$$6 + 6 = x - 6 + 6$$
$$12 = x$$

The examples in this section would indicate that an equation that can be written in the standard form $ax + b = c$, where $a \neq 0$, has exactly one solution. Note the following steps in finding the solution.

$$ax + b = c \qquad \text{Original equation}$$

$$ax = c - b \qquad \text{Subtract } b \text{ from both sides.}$$

$$x = \frac{c - b}{a} \qquad \text{Divide both sides by } a.$$

So, the *linear* equation has exactly one solution: $x = (c - b)/a$.

It may not be possible to rewrite some nonstandard forms of linear equations in the form $ax + b = c$, where $a \neq 0$. These types of equations have either no solution or infinitely many solutions.

Study Tip

In the *No Solution* equation the result is not true because $3 \neq 8$. This means that there is no value of x that will make the equation true.

In the *Infinitely Many Solutions* equation the result is true. This means that *any* real number is a solution to the equation.

No Solution	*Infinitely Many Solutions*
$2x + 3 \stackrel{?}{=} 2(x + 4)$	$2(x + 3) = 2x + 6$
$2x + 3 \stackrel{?}{=} 2x + 8$	$2x + 6 = 2x + 6$ Identity equation
$2x - 2x + 3 \stackrel{?}{=} 2x - 2x + 8$	$2x - 2x + 6 - 6 = 2x - 2x + 6 - 6$
$3 \neq 8$	$0 = 0$

Watch out for these types of equations in the exercise set.

3 Use a linear equation to solve an application problem.

Applications

Example 7 Geometry: Dimensions of a Dog Pen

You have 96 feet of fencing to enclose a rectangular pen for your dog. To provide sufficient running space for the dog to exercise, the pen is to be three times as long as it is wide. Find the dimensions of the pen.

Solution

Begin by drawing and labeling a diagram as shown in Figure 3.1. The perimeter of a rectangle is the sum of the widths and lengths.

Verbal Model: Perimeter = 2(width) + 2(length)

Algebraic Model: $96 = 2x + 2(3x)$

You can solve this equation as follows.

$$96 = 2x + 6x \qquad \text{Multiply.}$$

$$96 = 8x \qquad \text{Combine like terms.}$$

$$\frac{96}{8} = \frac{8x}{8} \qquad \text{Divide both sides by 8.}$$

$$12 = x \qquad \text{Simplify.}$$

So, the width of the pen is 12 feet and its length is 36 feet.

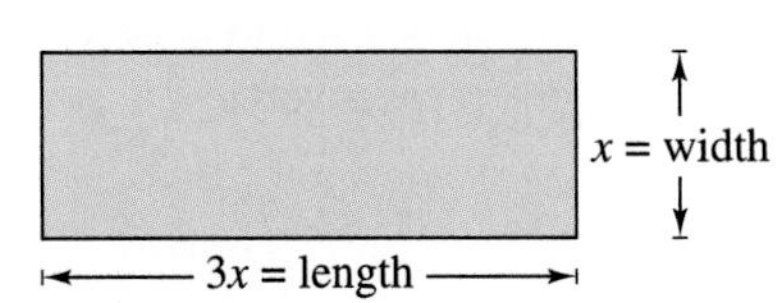

Figure 3.1

Example 8 Ticket Sales

Tickets for a concert were \$40 for each floor seat and \$20 for each stadium seat. There were 800 seats on the main floor, and these were sold out. The total revenue from ticket sales was \$92,000. How many stadium seats were sold?

Solution

Verbal Model: Total revenue = Revenue from floor seats + Revenue from stadium seats

Labels:
Total revenue = 92,000 (dollars)
Price per floor seat = 40 (dollars per seat)
Number of floor seats = 800 (seats)
Price per stadium seat = 20 (dollars per seat)
Number of stadium seats = x (seats)

Algebraic Model: $92{,}000 = 40(800) + 20x$

Now that you have written an algebraic equation to represent the problem, you can solve the equation as follows.

$$92{,}000 = 40(800) + 20x$$ Original equation

$$92{,}000 = 32{,}000 + 20x$$ Simplify.

$$92{,}000 - 32{,}000 = 32{,}000 - 32{,}000 + 20x$$ Subtract 32,000 from both sides.

$$60{,}000 = 20x$$ Combine like terms.

$$\frac{60{,}000}{20} = \frac{20x}{20}$$ Divide both sides by 20.

$$3000 = x$$ Simplify.

There were 3000 stadium seats sold. To check this solution, you should go back to the original statement of the problem and substitute 3000 stadium seats and 800 floor seats into the equation. You will find that the total revenue is \$92,000.

Two integers are called **consecutive integers** if they differ by 1. Hence, for any integer n, its next two larger consecutive integers are $n + 1$ and $(n + 1) + 1$ or $n + 2$. Thus, you can denote three consecutive integers by n, $n + 1$, and $n + 2$.

▶ Expressions for Special Types of Integers

Let n be an integer. Then the following expressions can be used to denote even integers, odd integers, and consecutive integers, respectively.

1. $2n$ denotes an *even* integer.
2. $2n - 1$ and $2n + 1$ denote *odd* integers.
3. The set $\{n, n + 1, n + 2\}$ denotes three *consecutive* integers.

Example 9 Consecutive Integers

Find three consecutive integers whose sum is 48.

Solution

Verbal Model: $\boxed{\text{First integer}} + \boxed{\text{Second integer}} + \boxed{\text{Third integer}} = 48$

Labels: First integer $= n$
Second integer $= n + 1$
Third integer $= n + 2$

Algebraic Model: $n + (n + 1) + (n + 2) = 48$

You can solve this equation as follows.

$$n + (n + 1) + (n + 2) = 48 \quad \text{Original equation}$$

$$3n + 3 = 48 \quad \text{Combine like terms.}$$

$$3n + 3 - 3 = 48 - 3 \quad \text{Subtract 3 from both sides.}$$

$$3n = 45 \quad \text{Combine like terms.}$$

$$\frac{3n}{3} = \frac{45}{3} \quad \text{Divide both sides by 3.}$$

$$n = 15 \quad \text{Simplify.}$$

The solution is $n = 15$. This implies that the three consecutive integers are 15, 16, and 17. Check this in the original statement of the problem.

Discussing the Concept — Solutions That Don't Make Sense

When solving a word problem, be sure to ask yourself whether your solution makes sense. Decide why the following answers don't make sense.

a. A problem asks you to find the volume of an oil drum. The answer you obtain is 20 square feet.

b. A problem asks you to find the price per bar of a packet of candy bars. The answer you obtain is 0.42¢.

c. A problem asks you to find the net weight of a carton of oranges. The answer you obtain is 12.5 liters.

d. A problem asks you to find the height of the ceiling of a room. The answer you obtain is 3 square meters.

3.1 Exercises

Integrated Review *Concepts, Skills, and Problem Solving*

Keep mathematically in shape by doing these exercises *before* the problems of this section.

Properties and Definitions

1. Complete the following properties of exponents.

(a) $(ab)^n =$ (b) $(a^m)^n =$

2. Identify the property illustrated by $(2x + 5) + 8 = 2x + (5 + 8)$.

Simplifying Expressions

In Exercises 3–10, simplify the expression.

3. $(u^2)^4$ **4.** $(-3a^3)^2$

5. $-3(x - 5)^2(x - 5)^3$ **6.** $(4rs)(-5r^2)(2s^3)$

7. $\dfrac{2m^2}{3n} \cdot \dfrac{3m}{5n^3}$ **8.** $\dfrac{5(x + 3)^2}{10(x + 8)}$

9. $-3(3x - 2y) + 5y$ **10.** $3v - (4 - 5v)$

Problem Solving

11. The length of a relay race is $\frac{3}{4}$ mile. The last change of runners occurs at the $\frac{2}{3}$ mile marker. How far does the last person run?

12. During the months of January, February, and March, a farmer bought $10\frac{1}{3}$ tons, $7\frac{3}{5}$ tons, and $12\frac{5}{6}$ tons of soybeans, respectively. Find the total amount of soybeans purchased during the first quarter of the year.

Developing Skills

In Exercises 1–8, solve the equation mentally.

1. $x + 6 = 14$ **2.** $u - 3 = 8$

3. $x - 9 = 4$ **4.** $a + 5 = 11$

5. $7y = 28$ **6.** $4z = -36$

7. $4s = 12$ **8.** $6z = 18$

In Exercises 9–12, justify each step of the solution. See Examples 1–6.

9.
$$5x + 15 = 0$$
$$5x + 15 - 15 = 0 - 15$$
$$5x = -15$$
$$\frac{5x}{5} = \frac{-15}{5}$$
$$x = -3$$

10.
$$7x - 14 = 0$$
$$7x - 14 + 14 = 0 + 14$$
$$7x = 14$$
$$\frac{7x}{7} = \frac{14}{7}$$
$$x = 2$$

11.
$$-2x + 5 = 13$$
$$-2x + 5 - 5 = 13 - 5$$
$$-2x = 8$$
$$\frac{-2x}{-2} = \frac{8}{-2}$$
$$x = -4$$

12.
$$22 - 3x = 10$$
$$22 - 3x + 3x = 10 + 3x$$
$$22 = 10 + 3x$$
$$22 - 10 = 10 + 3x - 10$$
$$12 = 3x$$
$$\frac{12}{3} = \frac{3x}{3}$$
$$4 = x$$

In Exercises 13–56, solve the equation and check your solution. (Some equations have no solution.) See Examples 1–6.

13. $5x = 30$ **14.** $-14x = 42$

15. $9x = -21$ **16.** $12x = 18$

17. $8x - 4 = 20$ **18.** $-7x + 24 = 3$

19. $25x - 4 = 46$

20. $15x - 18 = 12$

21. $10 - 4x = -6$

22. $6x + 1 = -11$

23. $6x - 4 = 0$

24. $8z + 10 = 0$

25. $3y - 2 = 2y$

26. $24 - 5x = x$

27. $4 - 7x = 5x$

28. $2s - 13 = 28s$

29. $4 - 5t = 16 + t$

30. $3x + 4 = x + 10$

31. $-3t + 5 = -3t$

32. $4z + 2 = 4z$

33. $15x - 3 = 15 - 3x$

34. $2x - 5 = 7x + 10$

35. $4z = 10$

36. $-6t = 0$

37. $8t - 4 = -6$

38. $4z - 8 = 2$

39. $4x - 6 = 4x - 6$

40. $5 - 3x = 5 - 3x$

41. $2x + 4 = -3x + 6$

42. $4y + 4 = -y + 5$

43. $2x = -3x$

44. $2x = 3x - 3$

45. $2x - 5 + 10x = 3$

46. $-4x + 10 + 10x = 4$

47. $\frac{x}{3} = 10$

48. $-\frac{x}{2} = 3$

49. $x - \frac{1}{3} = \frac{4}{3}$

50. $x + \frac{5}{2} = \frac{9}{2}$

51. $t - \frac{1}{3} = \frac{1}{2}$

52. $z + \frac{2}{5} = -\frac{3}{10}$

53. $3t + 1 - 2t = t + 1$

54. $7z - 5z - 8 = 2z - 8$

55. $2y - 18 = -5y - 4$

56. $6 - 21x = 12 - 21x$

Solving Problems

57. *Geometry* The length of a tennis court is 6 feet more than twice the width (see figure). Find the width of the court if the length is 78 feet.

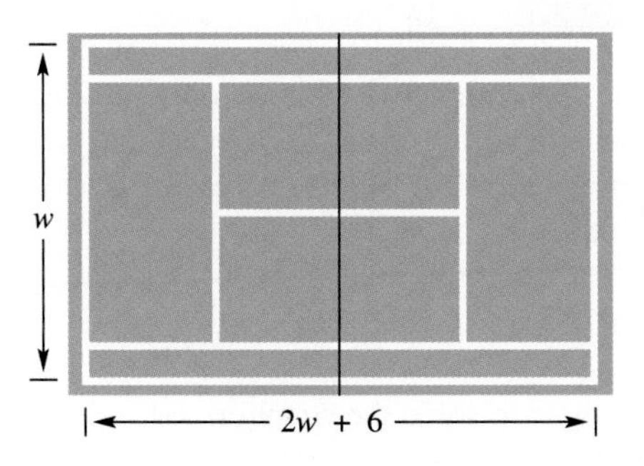

58. *Geometry* The perimeter of a rectangle is 240 inches. Find the dimensions of the rectangle if the length is twice the width.

59. *Geometry* You are asked to cut a 12-foot board into three pieces. Two pieces are to have the same length and the third is to be twice as long as the others. How long are the pieces?

60. *Geometry* The sign below has the shape of an equilateral triangle. The perimeter of the sign is 225 centimeters. Find the length of the sides of the sign. (An equilateral triangle is one whose sides have the same length.)

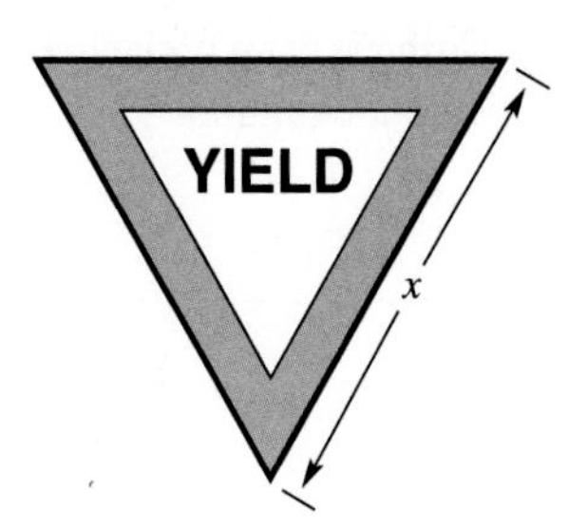

61. *Car Repair* The bill (including parts and labor) for the repair of your car is shown below. Some of the bill is unreadable. From what is given, can you determine how many hours were spent on labor? Explain.

Parts	$285.00
Labor ($32 per hour)	$
Total	**$357.00**

62. *Car Repair* The bill for the repair of your car was $439. The cost for parts was $265. The cost for labor was $29 per hour. How many hours did the repair work take?

63. *Ticket Sales* Tickets for a community theater are $10 for main floor seats and $8 for balcony seats. There are 400 seats on the main floor, and these were sold out for the evening performance. The total revenue from ticket sales was $5200. How many balcony seats were sold?

64. *Ticket Sales* Tickets for a marching band competition are $5 for 50-yard-line seats and $3 for bleacher seats. Eight hundred 50-yard-line seats were sold. The total revenue from ticket sales was $5500. How many bleacher seats were sold?

65. *Summer Jobs* You have two summer jobs. In the first job, you work 40 hours a week and earn \$9.25 an hour. In the second job, you earn \$7.50 an hour and can work as many hours as you want. If you want to earn a combined total of \$425 a week, how many hours must you work at the second job?

66. *Summer Jobs* You have two summer jobs. In the first job, you work 30 hours a week and earn \$8.75 an hour. In the second job, you earn \$11.00 an hour and can work as many hours as you want. If you want to earn a combined total of \$400 a week, how many hours must you work at the second job?

67. Find a number such that the sum of that number and 45 is 75.

68. Five times the sum of a number and 16 is 100. Find the number.

69. The sum of two consecutive odd integers is 72. Find the two integers.

70. The sum of three consecutive even integers is 192. Find the three integers.

71. *Finding a Pattern* The length of a rectangle is t times its width (see figure). The rectangle has a perimeter of 1200 meters, which implies that

$$2w + 2(tw) = 1200$$

where w is the width of the rectangle.

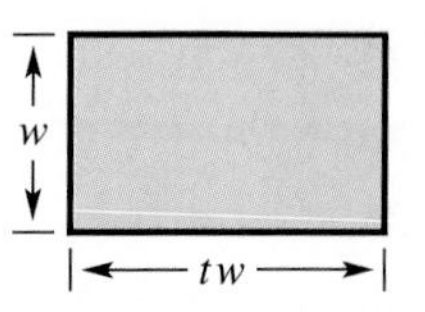

(a) Complete the table.

t	1	1.5	2	3	4	5
Width						
Length						
Area						

(b) Use the completed table to draw a conclusion concerning the area of a rectangle of given perimeter as the length increases relative to its width.

Explaining Concepts

72. Give two examples of linear equations and two examples of nonlinear equations.

73. The scale below is balanced. Each blue box weighs 1 ounce. How much does the red box weigh? If you removed three blue boxes from each side, would the scale still balance? What property of equality does this illustrate?

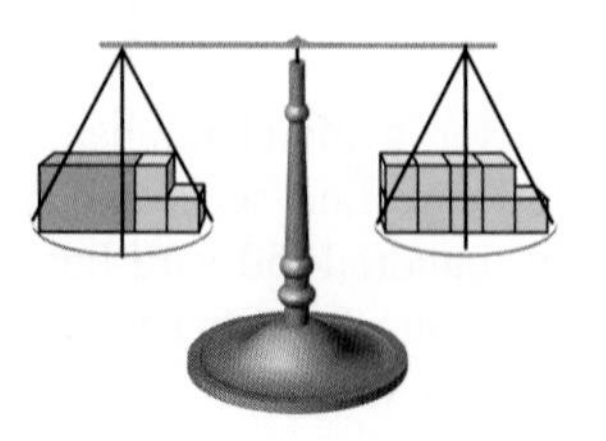

74. In your own words, describe the steps that can be used to transform an equation into an equivalent equation.

75. Explain how to solve the equation

$x + 5 = 32.$

What property of equality are you using?

76. Explain how to solve the equation

$3x = 5.$

What property of equality are you using?

77. When solving a word problem to determine the average speed of a moving van on a trip from Pittsburgh to Chicago, you obtained an answer of 134.5 kilometers per hour. Can this answer be correct? Explain.

78. *True or False?* Multiplying both sides of an equation by 0 yields an equivalent equation.

79. *True or False?* Subtracting 0 from both sides of an equation yields an equivalent equation.

3.2 Equations That Reduce to Linear Form

Objectives

1. Solve a linear equation containing symbols of grouping.
2. Solve a linear equation involving fractions.
3. Solve a linear equation involving decimals.

1 Solve a linear equation containing symbols of grouping.

Equations Containing Symbols of Grouping

In this section you will continue your study of linear equations by looking at more complicated types of linear equations. To solve a linear equation that contains symbols of grouping, *first remove the symbols of grouping from each side* by the Distributive Property. Then combine like terms and proceed to solve the resulting linear equation in the usual way using properties of equality.

Example 1 Solving a Linear Equation Involving Parentheses

Solve the linear equation (and then display the equation).

Solution

$$4(x - 3) = 8 \quad \text{Original equation}$$
$$4 \cdot x - 4 \cdot 3 = 8 \quad \text{Distributive Property}$$
$$4x - 12 = 8 \quad \text{Simplify.}$$
$$4x - 12 + 12 = 8 + 12 \quad \text{Add 12 to both sides.}$$
$$4x = 20 \quad \text{Combine like terms.}$$
$$\frac{4x}{4} = \frac{20}{4} \quad \text{Divide both sides by 4.}$$
$$x = 5 \quad \text{Simplify.}$$

Check

$$4(x - 3) = 8 \quad \text{Original equation}$$
$$4(5 - 3) \stackrel{?}{=} 8 \quad \text{Substitute 5 for } x.$$
$$4(2) \stackrel{?}{=} 8 \quad \text{Simplify.}$$
$$8 = 8 \quad \text{Solution checks. ✓}$$

The solution is 5.

Notice in the check of Example 1 that you do not need to use the Distributive Property to remove the parentheses. Simply evaluate the expression within the parentheses and then multiply.

Example 2 Solving a Linear Equation Involving Parentheses

Solve $3(2x - 1) + x = 11$.

Solution

$3(2x - 1) + x = 11$	Original equation
$3 \cdot 2x - 3 \cdot 1 + x = 11$	Distributive Property
$6x - 3 + x = 11$	Simplify.
$6x + x - 3 = 11$	Collect like terms.
$7x - 3 = 11$	Combine like terms.
$7x - 3 + 3 = 11 + 3$	Add 3 to both sides.
$7x = 14$	Combine like terms.
$\frac{7x}{7} = \frac{14}{7}$	Divide both sides by 7.
$x = 2$	Simplify.

Check

$3(2x - 1) + x = 11$	Original equation
$3[2(2) - 1] + 2 \stackrel{?}{=} 11$	Substitute 2 for x.
$3(4 - 1) + 2 \stackrel{?}{=} 11$	Simplify.
$3(3) + 2 \stackrel{?}{=} 11$	Simplify.
$9 + 2 \stackrel{?}{=} 11$	Simplify.
$11 = 11$	Solution checks. ✓

The solution is 2.

Example 3 Solving a Linear Equation Involving Parentheses

Solve $5(x + 2) = 2(x - 1)$.

Solution

$5(x + 2) = 2(x - 1)$	Original equation
$5x + 10 = 2x - 2$	Distributive Property
$5x - 2x + 10 = -2$	Subtract $2x$ from both sides.
$3x + 10 = -2$	Combine like terms.
$3x = -2 - 10$	Subtract 10 from both sides.
$3x = -12$	Combine like terms.
$x = -4$	Divide both sides by 3.

The solution is -4. Check this in the original equation.

Example 4 Solving a Linear Equation Involving Parentheses

Solve $2(x - 7) - 3(x + 4) = 4 - (5x - 2)$.

Solution

$2(x - 7) - 3(x + 4) = 4 - (5x - 2)$	Original equation
$2x - 14 - 3x - 12 = 4 - 5x + 2$	Distributive Property
$-x - 26 = -5x + 6$	Combine like terms.
$-x + 5x - 26 = 6$	Add $5x$ to both sides.
$4x - 26 = 6$	Combine like terms.
$4x = 6 + 26$	Add 26 to both sides.
$4x = 32$	Combine like terms.
$x = 8$	Divide both sides by 4.

The solution is 8. Check this in the original equation.

The linear equation in the next example involves both brackets and parentheses. Watch out for nested symbols of grouping such as these. The innermost symbols of grouping should be removed first.

Technology: Tip

Try using your graphing utility to check the solution found in Example 5. You will need to *nest* some parentheses inside other parentheses. This will give you practice working with nested parentheses on a graphing utility.

Left side

$$5\left(-\frac{1}{3}\right) - 2\left(4\left(-\frac{1}{3}\right) + 3\left(\left(-\frac{1}{3}\right) - 1\right)\right)$$

Right side

$$8 - 3\left(-\frac{1}{3}\right)$$

Example 5 An Equation Involving Nested Symbols of Grouping

Solve $5x - 2[4x + 3(x - 1)] = 8 - 3x$.

Solution

$5x - 2[4x + 3(x - 1)] = 8 - 3x$	Original equation
$5x - 2[4x + 3x - 3] = 8 - 3x$	Distributive Property
$5x - 2[7x - 3] = 8 - 3x$	Combine like terms inside brackets.
$5x - 14x + 6 = 8 - 3x$	Distributive Property
$-9x + 6 = 8 - 3x$	Combine like terms.
$-9x + 3x + 6 = 8$	Add $3x$ to both sides.
$-6x + 6 = 8$	Combine like terms.
$-6x = 8 - 6$	Subtract 6 from both sides.
$-6x = 2$	Combine like terms.
$x = \frac{2}{-6}$	Divide both sides by -6.
$x = -\frac{1}{3}$	Simplify.

The solution is $-\frac{1}{3}$. Check this in the original equation.

2 Solve a linear equation involving fractions.

Equations Involving Fractions or Decimals

To solve a linear equation that contains one or more fractions, it is usually best to first *clear the equation of fractions.*

▶ Clearing an Equation of Fractions

An equation such as

$$\frac{x}{a} + \frac{b}{c} = d$$

that contains one or more fractions can be cleared of fractions by multiplying both sides by the least common multiple (LCM) of a and c.

For example, the equation

$$\frac{3x}{2} - \frac{1}{3} = 2$$

can be cleared of fractions by multiplying both sides by 6, the LCM of 2 and 3. Notice how this is done in the next example.

Study Tip

For an equation that contains a *single numerical* fraction such as $2x - \frac{3}{4} = 1$, you can simply add $\frac{3}{4}$ to both sides and then solve for x. You do not need to clear the fraction.

$$2x - \frac{3}{4} + \frac{3}{4} = 1 + \frac{3}{4} \quad \text{Add } \tfrac{3}{4}.$$

$$2x = \frac{7}{4} \quad \text{Combine terms.}$$

$$x = \frac{7}{8} \quad \text{Multiply by } \tfrac{1}{2}.$$

Example 6 Solving a Linear Equation Involving Fractions

Solve $\frac{3x}{2} - \frac{1}{3} = 2$.

Solution

$$\frac{3x}{2} - \frac{1}{3} = 2 \quad \text{Original equation}$$

$$6\left(\frac{3x}{2} - \frac{1}{3}\right) = 6 \cdot 2 \quad \text{Multiply both sides by LCM 6.}$$

$$6 \cdot \frac{3x}{2} - 6 \cdot \frac{1}{3} = 12 \quad \text{Distributive Property}$$

$$9x - 2 = 12 \quad \text{Clear fractions.}$$

$$9x = 14 \quad \text{Add 2 to both sides.}$$

$$x = \frac{14}{9} \quad \text{Divide both sides by 9.}$$

The solution is $\frac{14}{9}$. Check this in the original equation.

To check a fraction solution like $\frac{14}{9}$ in Example 1, it is helpful to rewrite the variable term as a product.

$$\frac{3}{2} \cdot x - \frac{1}{3} = 2 \quad \text{Write fraction as a product.}$$

In this form the substitution of $\frac{14}{9}$ for x is easier to calculate.

Example 7 Solving a Linear Equation Involving Fractions

Solve $\frac{x}{5} + \frac{3x}{4} = 19$.

Solution

$$\frac{x}{5} + \frac{3x}{4} = 19 \qquad \text{Original equation}$$

$$20\left(\frac{x}{5}\right) + 20\left(\frac{3x}{4}\right) = 20(19) \qquad \text{Multiply both sides by LCM 20.}$$

$$4x + 15x = 380 \qquad \text{Simplify.}$$

$$19x = 380 \qquad \text{Combine like terms.}$$

$$x = 20 \qquad \text{Divide both sides by 19.}$$

Check

$$\frac{x}{5} + \frac{3x}{4} = 19 \qquad \text{Original equation}$$

$$\frac{20}{5} + \frac{3(20)}{4} \stackrel{?}{=} 19 \qquad \text{Substitute 20 for } x.$$

$$4 + 15 \stackrel{?}{=} 19 \qquad \text{Simplify.}$$

$$19 = 19 \qquad \text{Solution checks. ✓}$$

The solution is 20.

Study Tip

Notice in Example 8 that to clear all fractions in the equation, you multiply by 12 which is the LCM of 3, 4, and 2.

Example 8 Solving a Linear Equation Involving Fractions

Solve $\frac{2}{3}\left(x + \frac{1}{4}\right) = \frac{1}{2}$.

Solution

$$\frac{2}{3}\left(x + \frac{1}{4}\right) = \frac{1}{2} \qquad \text{Original equation}$$

$$\frac{2}{3}x + \frac{2}{12} = \frac{1}{2} \qquad \text{Distributive Property}$$

$$12 \cdot \frac{2}{3}x + 12 \cdot \frac{2}{12} = 12 \cdot \frac{1}{2} \qquad \text{Multiply both sides by LCM 12.}$$

$$8x + 2 = 6 \qquad \text{Simplify.}$$

$$8x = 4 \qquad \text{Subtract 2 from both sides.}$$

$$x = \frac{4}{8} \qquad \text{Divide both sides by 8.}$$

$$x = \frac{1}{2} \qquad \text{Simplify.}$$

The solution is $\frac{1}{2}$. Check this in the original equation.

A common type of linear equation is one that equates two fractions. To solve such equations, consider the fractions to be **equivalent** and use **cross-multiplication.** That is, if

$$\frac{a}{b} = \frac{c}{d}, \quad \text{then} \quad a \cdot d = b \cdot c.$$

Note how cross-multiplication is used in the next example.

Example 9 Using Cross-Multiplication

Use cross-multiplication to solve $\frac{x+2}{3} = \frac{8}{5}$.

Solution

$\frac{x+2}{3} = \frac{8}{5}$ Original equation

$5(x+2) = 3(8)$ Cross-multiply.

$5x + 10 = 24$ Distributive Property

$5x = 14$ Subtract 10 from both sides.

$x = \frac{14}{5}$ Divide both sides by 5.

Check

$\frac{x+2}{3} = \frac{8}{5}$ Original equation

$\frac{\left(\frac{14}{5} + 2\right)}{3} \stackrel{?}{=} \frac{8}{5}$ Substitute $\frac{14}{5}$ for x.

$\frac{\left(\frac{14}{5} + \frac{10}{5}\right)}{3} \stackrel{?}{=} \frac{8}{5}$ Write 2 as $\frac{10}{5}$.

$\frac{\frac{24}{5}}{3} \stackrel{?}{=} \frac{8}{5}$ Simplify.

$\frac{24}{5}\left(\frac{1}{3}\right) \stackrel{?}{=} \frac{8}{5}$ Invert and multiply.

$\frac{8}{5} = \frac{8}{5}$ Solution checks. ✓

The solution is $\frac{14}{5}$.

Bear in mind that cross-multiplication can only be used with equations written in a form that equates two fractions. Try rewriting the equation in Example 6 in this form and then use cross-multiplication to solve for x.

More extensive applications of cross-multiplication will be discussed when you study ratios and proportions later in this chapter.

3 Solve a linear equation involving decimals.

Many real-life applications of linear equations involve decimal coefficients. To solve such an equation, you can clear it of decimals in much the same way you clear an equation of fractions. Multiply both sides by a power of 10 that converts all decimal coefficients to integers, as shown in the next example.

Study Tip

There are other ways to solve the decimal equation in Example 10. You could first clear the equation of decimals by multiplying both sides by 100. Or, you could keep the decimals and use a graphing utility to do the arithmetic operations. The method you choose is a matter of personal preference.

Example 10 Solving a Linear Equation Involving Decimals

Solve $0.3x + 0.2(10 - x) = 0.15(30)$.

Solution

$0.3x + 0.2(10 - x) = 0.15(30)$ Original equation

$0.3x + 2 - 0.2x = 4.5$ Distributive Property

$0.1x + 2 = 4.5$ Combine like terms.

$10(0.1x + 2) = 10(4.5)$ Multiply both sides by 10.

$x + 20 = 45$ Clear decimals.

$x = 25$ Subtract 20 from both sides.

Check

$0.3x + 0.2(10 - x) = 0.15(30)$ Original equation

$0.3(25) + 0.2(10 - 25) \stackrel{?}{=} 0.15(30)$ Substitute 25 for x.

$0.3(25) + 0.2(-15) \stackrel{?}{=} 0.15(30)$ Perform subtraction within parentheses.

$7.5 - 3.0 \stackrel{?}{=} 4.5$ Multiply.

$4.5 = 4.5$ Solution checks. ✓

The solution is 25.

Discussing the Concept **Error Analysis**

Suppose you are teaching an algebra class and one of your students hands in the following problem. Find the error in the solution. Write an explanation for the student.

$4(x + 2) - 8 = 3x$ Given equation

$4x + 8 - 8 = 3x$ Distributive Property

$4x = 3x$ Additive inverse

$4 = 3$ Divide both sides by x.

No solution because 4 is not equal to 3.

Explain what happens when you divide both sides of an equation by a variable factor.

3.2 Exercises

Integrated Review *Concepts, Skills, and Problem Solving*

Keep mathematically in shape by doing these exercises *before* the problems of this section.

Properties and Definitions

1. In your own words, describe how you add the following fractions.

(a) $\frac{1}{5} + \frac{7}{5}$ (b) $\frac{1}{5} + \frac{7}{3}$

2. Make up two examples of algebraic expressions.

Simplifying Expressions

In Exercises 3–10, simplify the expression.

3. $(-2x)^2x^4$

4. $-y^2(-2y)^3$

5. $5z^3(z^2)^2$

6. $(a + 3)^2(a + 3)^5$

7. $\frac{5x}{3} - \frac{2x}{3} - 4$

8. $2x^2 - 4 + 5 - 3x^2$

9. $-y^2(y^2 + 4) + 6y^2$

10. $5t(2 - t) + t^2$

Problem Solving

11. At the beginning of the day, a gasoline tank was full. The tank holds 20 gallons. At the end of the day the fuel gauge indicates that the tank is $\frac{5}{8}$ full. How many gallons of gasoline were used?

12. You buy a pickup truck for \$1800 down and 36 monthly payments of \$625 each.

(a) What is the total amount you will pay?

(b) The final cost of the pickup is \$19,999. How much extra did you pay in finance charges and other fees?

Developing Skills

In Exercises 1–52, solve the equation and check your solution. (Some of the equations have no solution.) See Examples 1–8.

1. $-5(t + 3) = 0$

2. $9(y - 7) = 0$

3. $2(y - 4) = 12$

4. $-3(x + 1) = 18$

5. $2(x - 3) = 4$

6. $4(x + 1) = 24$

7. $7(x + 5) = 49$

8. $25(z - 2) = 60$

9. $4 - (z + 6) = 8$

10. $25 - (y + 3) = 15$

11. $3 - (2x - 4) = 3$

12. $16 - (3x - 10) = 5$

13. $-3(t + 5) = 0$

14. $4(z - 2) = 0$

15. $-4(t + 5) = -2(2t + 10)$

16. $4(z - 2) = 2(2z - 4)$

17. $3(x + 4) = 10(x + 4)$

18. $-8(x - 6) = 3(x - 6)$

19. $7 = 3(x + 2) - 3(x - 5)$

20. $24 = 12(z + 1) - 3(4z - 2)$

21. $7x - 2(x - 2) = 12$

22. $15(x + 1) - 8x = 29$

23. $6 = 3(y + 1) - 4(1 - y)$

24. $100 = 4(y - 6) - (y - 1)$

25. $7(2x - 1) = 4(1 - 5x) + 6$

26. $-3(5x + 2) + 5(1 + 3x) = 0$

27. $2[(3x + 5) - 7] = 3(5x - 2)$

28. $6[x - (2x + 3)] = 8 - 5x$

29. $4x + 3[x - 2(2x - 1)] = 4 - 3x$

30. $16 + 4[5x - 4(x + 2)] = 7 - 2x$

31. $\frac{x}{2} = \frac{3}{2}$

32. $\frac{t}{4} = \frac{3}{8}$

33. $\frac{y}{5} = \frac{3}{5}$

34. $\frac{z}{3} = -\frac{5}{3}$

35. $\frac{y}{5} = -\frac{3}{10}$

36. $\frac{v}{4} = \frac{4}{3}$

37. $\frac{6x}{25} = \frac{3}{5}$

38. $-\frac{8x}{9} = \frac{2}{3}$

39. $\frac{5x}{4} + \frac{1}{2} = 0$

40. $\frac{y}{4} - \frac{5}{8} = 2$

41. $\frac{x}{5} - \frac{x}{2} = 1$

42. $\frac{x}{3} + \frac{x}{4} = 1$

43. $2s + \frac{3}{2} = 2s + 2$

44. $\frac{3}{4} + 5s = -2 + 5s$

45. $3x + \frac{1}{4} = \frac{3}{4}$

46. $2x - \frac{3}{8} = \frac{5}{8}$

47. $\frac{1}{5}x + 1 = \frac{3}{10}x - 4$

48. $\frac{1}{8}x + 3 = \frac{1}{4}x + 5$

49. $\frac{2}{3}(z + 5) - \frac{1}{4}(z + 24) = 0$

50. $\frac{3x}{2} + \frac{1}{4}(x - 2) = 10$

51. $\frac{100 - 4u}{3} = \frac{5u + 6}{4} + 6$

52. $\frac{8 - 3x}{2} - 4 = \frac{x}{6}$

In Exercises 53–62, solve the equation by first cross-multiplying. See Example 9.

53. $\frac{t + 4}{6} = \frac{2}{3}$

54. $\frac{x - 6}{10} = \frac{3}{5}$

55. $\frac{x - 2}{5} = \frac{2}{3}$

56. $\frac{2x + 1}{3} = \frac{5}{2}$

57. $\frac{5x - 4}{4} = \frac{2}{3}$

58. $\frac{10x + 3}{6} = \frac{1}{2}$

59. $\frac{x}{4} = \frac{1 - 2x}{3}$

60. $\frac{x + 1}{6} = \frac{3x}{10}$

61. $\frac{10 - x}{2} = \frac{x + 4}{5}$

62. $\frac{2x + 3}{5} = \frac{3 - 4x}{8}$

In Exercises 63–72, solve the equation. Round the solution to two decimal places. See Example 10.

63. $0.2x + 5 = 6$

64. $4 - 0.3x = 1$

65. $0.234x + 1 = 2.805$

66. $275x - 3130 = 512$

67. $0.02x - 0.96 = 1.50$

68. $1.35x + 14.50 = 6.34$

69. $\frac{x}{3.25} + 1 = 2.08$

70. $\frac{3x}{4.5} = \frac{1}{8}$

71. $\frac{x}{3.155} = 2.850$

72. $2x + \frac{1}{3.7} = \frac{3}{4}$

Solving Problems

73. *Time to Complete a Task* Two people can complete 80% of a task in t hours, where t must satisfy the equation

$$\frac{t}{10} + \frac{t}{15} = 0.8.$$

Solve this equation for t.

74. *Time to Complete a Task* The time to complete a task is given by the solution of the equation

$$\frac{t}{10} + \frac{t}{15} = 1.$$

Find the required time t.

75. *Course Grade* To get an A in a course you must have an average of at least 90 points for four tests of 100 points each.

(a) For the first three tests, your scores are 87, 92, and 84. What must you score on the fourth exam to earn a 90% average for the course?

(b) Is it possible for you to get an A if your scores on the first three tests are 87, 69, and 89? Explain.

76. *Course Grade* Repeat Exercise 75 if the fourth test is weighted so that it counts for twice as much as each of the first three tests.

In Exercises 77–80, use the following equation and solve for x.

$$p_1x + p_2(a - x) = p_3a$$

77. *Mixture Problem* Determine the number of quarts of a 10% solution that must be mixed with a 30% solution to obtain 100 quarts of a 25% solution. ($p_1 = 0.1, p_2 = 0.3, p_3 = 0.25$, and $a = 100$.)

78. *Mixture Problem* Determine the number of gallons of a 25% solution that must be mixed with a 50% solution to obtain 5 gallons of a 30% solution. ($p_1 = 0.25$, $p_2 = 0.5$, $p_3 = 0.3$, and $a = 5$.)

79. *Mixture Problem* An 8-quart automobile cooling system is filled with coolant that is 40% antifreeze. Determine the amount that must be withdrawn and replaced with pure antifreeze so that the 8 quarts of coolant will be 50% antifreeze. ($p_1 = 1$, $p_2 = 0.4$, $p_3 = 0.5$, and $a = 8$.)

80. *Mixture Problem* A grocer mixes two kinds of nuts costing \$2.49 per pound and \$3.89 per pound to make 100 pounds of a mixture costing \$3.19 per pound. How many pounds of the nuts costing \$2.49 per pound must be put into the mixture? ($p_1 = 2.49$, $p_2 = 3.89$, $p_3 = 3.19$, and $a = 100$.)

In Exercises 81 and 82, use $W_1x = W_2(a - x)$.

81. *Balancing a Seesaw* Find the position of the fulcrum so that the seesaw shown in the figure will balance. ($W_1 = 90$, $W_2 = 60$, and $a = 10$.)

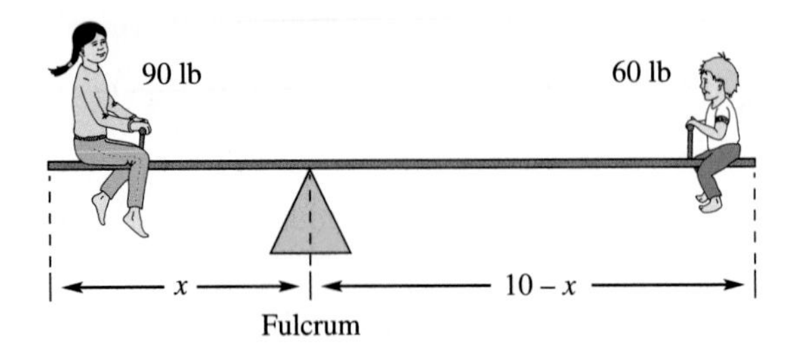

82. *Raising a Weight* The fulcrum of a 6-foot-long lever is 6 inches from a weight, as shown in the figure. Find the maximum weight that a 190-pound person can lift using this lever. ($W_1 = 190$, $x = 5\frac{1}{2}$, and $a = 6$.)

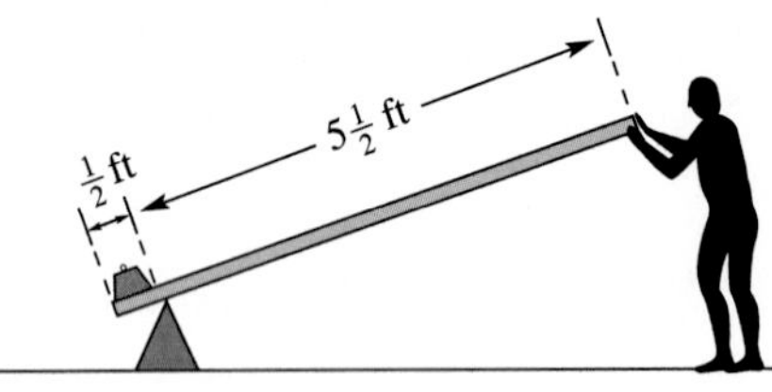

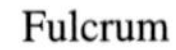

83. *Fireplace Construction* A fireplace is 93 inches wide. Each brick in the fireplace has a length of 8 inches and there is $\frac{1}{2}$ inch of mortar between adjoining bricks. Let n be the number of bricks per row.

(a) Explain why the number of bricks per row is the solution of the equation $8n + \frac{1}{2}(n - 1) = 93$.

(b) Find the number of bricks per row in the fireplace.

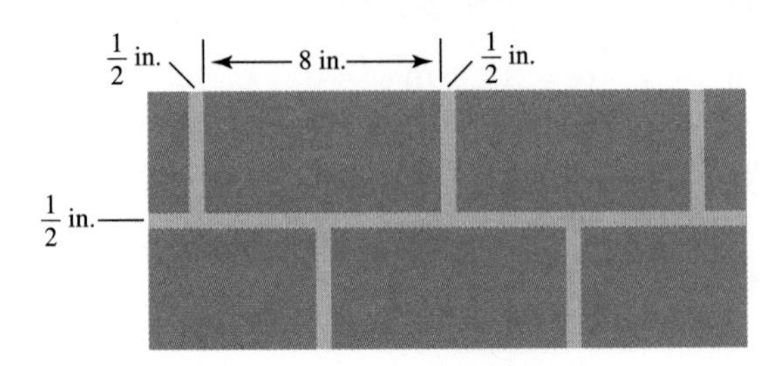

84. *Data Analysis* The table gives the projected number N (in millions) of persons 65 years of age or older in the United States. (Source: U.S. Bureau of the Census)

Year	2000	2010	2020	2030
N	34.7	39.4	53.2	69.4

A model for the data is

$$N = 1.2t + 31.5$$

where t represents time in years, with $t = 0$ corresponding to the year 2000. According to the model, in what year will the population of those 65 or older exceed 75 million?

Explaining Concepts

85. In your own words, describe the procedure for removing symbols of grouping. Give some examples.

86. Describe the error in the following.

$$-2(x - 5) = 8$$

$$-2x - 5 = 8$$

87. You could solve $3(x - 7) = 15$ by applying the Distributive Property as the first step. However, there is another way to begin. What is it?

88. What is meant by the least common multiple of the denominators of two or more fractions? Discuss the method for finding the least common multiple of the denominators of fractions.

89. When solving an equation that contains fractions, what is accomplished by multiplying both sides of the equation by the least common multiple of the denominators of the fractions?

90. When simplifying an algebraic *expression* involving fractions, why can't you simplify the expression by multiplying by the least common multiple of the denominators?

3.3 Problem Solving with Percents

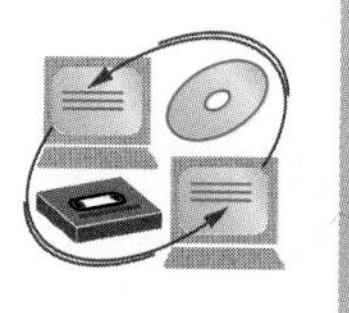

Objectives

1. Convert percents to decimals and fractions and convert decimals and fractions to percents.
2. Solve linear equations involving percents.
3. Solve application problems involving markups and discounts.

1 Convert percents to decimals and fractions and convert decimals and fractions to percents.

Percents

In applications involving percents, you usually must convert the percents to decimal (or fraction) form before performing any arithmetic operations. Consequently, you need to be able to convert from percents to decimals (or fractions), and vice versa. The following verbal model can be used to perform the conversions.

Decimal or fraction · 100% = Percent

For example, the decimal 0.38 corresponds to 38 percent. That is,

$$0.38(100\%) = 38\%.$$

Example 1 Converting Decimals and Fractions to Percents

Convert each number to a percent.

a. $\frac{3}{5}$ **b.** 1.20

Solution

a. *Verbal Model:* Fraction · 100% = Percent

Equation: $\frac{3}{5}(100\%) = \frac{300}{5}\%$

$$= 60\%$$

So, the fraction $\frac{3}{5}$ corresponds to 60%.

b. *Verbal Model:* Decimal · 100% = Percent

Equation: $(1.20)(100\%) = 120\%$

So, the decimal 1.20 corresponds to 120%.

Note in Example 1(b) that it is possible to have percents that are larger than 100%. It is also possible to have percents that are less than 1% such as $\frac{1}{2}\%$ or 0.78%.

Study Tip

In Examples 1 and 2, there is a quick way to convert between percent form and decimal form.

- To convert from percent form to decimal form, move the decimal point 2 places to the left. For instance,

  $3.5\% = 0.035.$

- To convert from decimal form to percent form, move the decimal point 2 places to the right. For instance,

  $1.20 = 120\%.$

- Decimal-to-fraction or fraction-to-decimal conversions can be done on a calculator. Consult your user's guide.

Example 2 Converting Percents to Decimals and Fractions

a. Convert 3.5% to a decimal.

b. Convert 55% to a fraction.

Solution

a. *Verbal Model:* Decimal · 100% = Percent

Label: $x =$ decimal

Equation: $x(100\%) = 3.5\%$

$$x = \frac{3.5\%}{100\%}$$

$$x = 0.035$$

So, 3.5% corresponds to the decimal 0.035.

b. *Verbal Model:* Fraction · 100% = Percent

Label: $x =$ fraction

Equation: $x(100\%) = 55\%$

$$x = \frac{55\%}{100\%}$$

$$x = \frac{11}{20}$$

So, 55% corresponds to the fraction $\frac{11}{20}$.

Some percents occur so commonly that it is helpful to memorize their conversions. For instance, 100% corresponds to 1 and 200% corresponds to 2. The table below shows the decimal and fraction conversions for several percents.

Percent	10%	$12\frac{1}{2}\%$	20%	25%	$33\frac{1}{3}\%$	50%	$66\frac{2}{3}\%$	75%
Decimal	0.1	0.125	0.2	0.25	$0.\overline{3}$	0.5	$0.\overline{6}$	0.75
Fraction	$\frac{1}{10}$	$\frac{1}{8}$	$\frac{1}{5}$	$\frac{1}{4}$	$\frac{1}{3}$	$\frac{1}{2}$	$\frac{2}{3}$	$\frac{3}{4}$

Percent means *per hundred* or *parts of 100.* (The Latin word for 100 is *centum.*) For example, 20% means 20 parts of 100, which is equivalent to the fraction $20/100$ or $\frac{1}{5}$. In applications involving percent, many people like to state percent in terms of a portion. For instance, the statement "20% of the population lives in apartments" is often stated as "1 out of every 5 people lives in an apartment."

2 Solve linear equations involving percents.

The Percent Equation

The primary use of percents is to compare two numbers. For example, 2 is 50% of 4, and 5 is 25% of 20. The following model is helpful.

Verbal Model: a = p percent of b

Labels: b = base number
p = percent (in decimal form)
a = number being compared to b

Equation: $a = p \cdot b$

Example 3 Solving Percent Equations

a. What number is 30% of 70?

b. Fourteen is 25% of what number?

c. One hundred thirty-five is what percent of 27?

Solution

a. *Verbal Model:* What number = 30% of 70

Label: a = unknown number

Equation: $a = (0.3)(70) = 21$

So, 21 is 30% of 70.

b. *Verbal Model:* 14 = 25% of what number

Label: b = unknown number

Equation:

$$14 = 0.25b$$

$$\frac{14}{0.25} = b$$

$$56 = b$$

So, 14 is 25% of 56.

c. *Verbal Model:* 135 = What percent of 27

Label: p = unknown percent (in decimal form)

Equation:

$$135 = p(27)$$

$$\frac{135}{27} = p$$

$$5 = p$$

So, 135 is 500% of 27.

From Example 3, you can see that there are three basic types of percent problems. Each can be solved by substituting the two given quantities into the percent equation and solving for the third quantity.

Question	*Given*	*Percent Equation*
a is what percent of b?	a and b	Solve for p.
What number is p percent of b?	p and b	Solve for a.
a is p percent of what number?	a and p	Solve for b.

For instance, part (b) of Example 3 fits the form "a is p percent of what number?"

In most real-life applications, the base number b and the number a are much more disguised than they are in Example 3. It sometimes helps to think of a as a "new" amount and b as the "original" amount.

Example 4 Real Estate Commission

A real estate agency receives a commission of \$5167.50 for the sale of a \$79,500 house. What percent commission is this?

Solution

Verbal Model: Commission = Percent (in decimal form) · Sale price

Labels:
Commission = 5167.50 (dollars)
Percent = p (in decimal form)
Sale price = 79,500 (dollars)

Equation:

$$5167.50 = p \cdot (79{,}500)$$

$$\frac{5167.50}{79{,}500} = p$$

$$0.065 = p$$

So, the real estate agency receives a commission of 6.5%.

Example 5 Cost-of-Living Raise

A union negotiates for a cost-of-living raise of 7%. What is the raise for a union member whose salary is \$17,240? What is this person's new salary?

Solution

Verbal Model: Raise = Percent (in decimal form) · Salary

Labels:
Raise = a (dollars)
Percent = 7% = 0.07 (in decimal form)
Salary = 17,240 (dollars)

Equation: $a = 0.07(17{,}240) = 1206.80$

So, the raise is \$1206.80 and the new salary is 17,240.00 + 1206.80 or \$18,446.80.

Example 6 Course Grade

You missed an A in your chemistry course by only three points. Your point total for the course is 402. How many points were possible in the course? (Assume that you needed 90% of the course total for an A.)

Solution

Verbal Model: Your points + 3 points = Percent (in decimal form) · Total points

Labels:
Your points = 402 (points)
Percent = 90% = 0.9 (in decimal form)
Total points for course = b (points)

Equation:
$$402 + 3 = 0.9b$$
$$405 = 0.9b$$
$$\frac{405}{0.9} = b$$
$$450 = b$$

Check

$402 + 3 = 0.9b$ Original equation

$402 + 3 \stackrel{?}{=} 0.9(450)$ Substitute 450 for b.

$405 = 405$ Solution checks. ✓

So, there were 450 total points for the course.

3 Solve application problems involving markups and discounts.

Markups and Discounts

You may have had the experience of buying an item at one store and later finding that you could have paid less for the same item at another store. The basic reason for this price difference is **markup,** which is the difference between the **cost** (the amount a retailer pays for the item) and the **price** (the amount at which the retailer sells the item to the consumer). A verbal model for this problem is as follows.

Selling price = Cost + Markup

In such a problem, the markup may be known or it may be expressed as a percent of the cost. This percent is called the **markup rate.**

Markup = Markup rate · Cost

Markup is one of those "hidden products" referred to in Section 2.3.

In business and economics, the terms *cost* and *price* do not mean the same thing. The cost of an item is the amount a business pays for the item. The price of an item is the amount for which the business sells the item.

Example 7 Finding the Selling Price

A sporting goods store uses a markup rate of 55% on all items. The cost of a golf bag is \$45. What is the selling price of the bag?

Solution

Verbal Model: Selling price = Cost + Markup

Labels:

Selling price $= x$	(dollars)
Cost $= 45$	(dollars)
Markup rate $= 0.55$	(rate in decimal form)
Markup $= (0.55)(45)$	(dollars)

Equation:

$$x = 45 + (0.55)(45)$$
$$= 45 + 24.75$$
$$= \$69.75$$

The selling price is \$69.75. Check this in the original statement of the problem.

In Example 7, you are given the cost and are asked to find the selling price. Example 8 illustrates the reverse problem. That is, in Example 8 you are given the selling price and asked to find the cost.

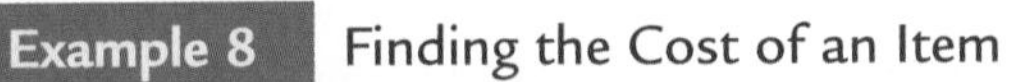

Example 8 Finding the Cost of an Item

The selling price of a pair of ski boots is \$98. The markup rate is 60%. What is the cost of the boots?

Solution

Verbal Model: Selling price = Cost + Markup

Labels:

Selling price $= 98$	(dollars)
Cost $= x$	(dollars)
Markup rate $= 0.60$	(rate in decimal form)
Markup $= 0.60x$	(dollars)

Equation:

$$98 = x + 0.60x$$
$$98 = 1.60x$$
$$\frac{98}{1.60} = x$$
$$\$61.25 = x$$

The cost is \$61.25. Check this in the original statement of the problem.

Example 9 Finding the Markup Rate 

A pair of shoes sells for \$60. The cost of the shoes is \$24. What is the markup rate?

Solution

Verbal Model: Selling price = Cost + Markup

Labels:
Selling price = 60 (dollars)
Cost = 24 (dollars)
Markup rate = p (rate in decimal form)
Markup = $p(24)$ (dollars)

Equation:

$$60 = 24 + p(24)$$

$$36 = 24p$$

$$\frac{36}{24} = p$$

$$1.5 = p$$

Because $p = 1.5$, it follows that the markup rate is 150%.

The mathematics of a discount is similar to that of a markup. The model for this situation is

Sale price = List price − Discount

where the **discount** is given in dollars, and the **discount rate** is given as a percent of the list price. Notice the "hidden product" in the discount.

Discount = Discount rate · List price

Example 10 Finding the Discount Rate

During a midsummer sale, a lawn mower listed at \$199.95 is on sale for \$139.95. What is the discount rate?

Solution

Verbal Model: Discount = Discount rate · List price

Labels:
Discount = 199.95 − 139.95 = 60 (dollars)
List price = 199.95 (dollars)
Discount rate = p (rate in decimal form)

Equation:

$$60 = p(199.95)$$

$$0.30 \approx p$$

Because $p \approx 0.30$, it follows that the discount rate is 30%.

Example 11 Finding the Sale Price

A drug store advertises 40% off the prices of all summer tanning products. A bottle of suntan oil lists for \$3.49. What is the sale price?

Solution

Verbal Model: Sale price = List price − Discount

Labels:
List price $= 3.49$ (dollars)
Discount rate $= 0.4$ (rate in decimal form)
Discount $= 0.4(3.49)$ (dollars)
Sale price $= x$ (dollars)

Equation: $x = 3.49 - (0.4)(3.49)$

$\approx \$2.09$

The sale price is \$2.09. Check this in the original statement of the problem.

The following guidelines summarize the problem-solving strategy that we recommend for word problems.

Guidelines for Solving Word Problems

1. Write a *verbal model* that describes the problem.
2. Assign *labels* to fixed quantities and variable quantities.
3. Rewrite the verbal model as an *algebraic equation* using the assigned labels.
4. *Solve* the algebraic equation.
5. *Check* to see that your solution satisfies the word problem as stated.

Discussing the Concept — Comparing Growth Patterns

In the year 2000, your starting annual salary is \$28,000. You are given two options for an 8-year contract. In the first option, you will be given a \$1500 raise each year. In the second option, you will be given a 5% raise each year.

a. Make a table showing your salaries and raises for both options for each of the 8 years of the contract. Which option would you choose for an 8-year contract?

b. Which option would you choose if it were a 3-year contract? a 4-year contract? Explain.

c. In which year would the raise for the second option surpass the raise for the first option? Why do you think this happens?

3.3 Exercises

Integrated Review *Concepts, Skills, and Problem Solving*

Keep mathematically in shape by doing these exercises *before* the problems of this section.

Properties and Definitions

1. Explain how to put the two numbers 63 and -28 in order.

2. For any real number, its distance from ______ on the real number line is its absolute value.

Simplifying Expressions

In Exercises 3–6, evaluate the expression.

3. $8 - |-7 + 11| + (-4)$

4. $34 - [54 - (-16 + 4) + 6]$

5. Subtract 230 from -300.

6. Find the absolute value of the difference of 17 and -12.

In Exercises 7 and 8, use the Distributive Property to expand the expression.

7. $4(2x - 5)$

8. $-z(xz - 2y^2)$

In Exercises 9 and 10, evaluate the algebraic expression for the specified values of the variables. (If not possible, state the reason.)

9. $x^2 - y^2$

(a) $x = 4, y = 3$

(b) $x = -5, y = 3$

10. $\dfrac{z^2 + 2}{x^2 - 1}$

(a) $x = 1, z = 1$

(b) $x = 2, z = 2$

Problem Solving

11. A telephone company charges \$1.37 for the first minute and \$0.95 for each additional minute. Find the cost of a 15-minute phone call.

12. A train travels at the rate of r miles per hour for 5 hours. Write an algebraic expression that represents the total distance traveled by the train.

Developing Skills

In Exercises 1–12, complete the table showing the equivalent forms of a percent. See Examples 1 and 2.

	Percent	*Parts out of 100*	*Decimal*	*Fraction*
1.	40%			
2.	15%			
3.	7.5%			
4.	75%			
5.		63		
6.		10.5		
7.			0.155	
8.			0.80	
9.				$\frac{3}{5}$
10.				$\frac{3}{20}$
11.	150%			
12.			1.25	

In Exercises 13–20, change the decimal to a percent. See Example 1.

13. 0.62 **14.** 0.57

15. 0.20 **16.** 0.38

17. 0.075 **18.** 0.005

19. 2.5 **20.** 1.75

In Exercises 21–28, change the percent to a decimal. See Example 2.

21. 12.5% **22.** 95%

23. 125% **24.** 8.5%

25. 250% **26.** 0.3%

27. $\frac{3}{4}\%$ **28.** $33\frac{1}{3}\%$

In Exercises 29–36, change the fraction to a percent. See Example 1.

29. $\frac{4}{5}$ **30.** $\frac{1}{4}$

31. $\frac{5}{4}$ **32.** $\frac{6}{5}$

33. $\frac{5}{6}$ **34.** $\frac{2}{3}$

35. $\frac{7}{20}$ **36.** $\frac{3}{2}$

In Exercises 37–40, what percent of the figure is shaded? (There are a total of 360° in a circle.)

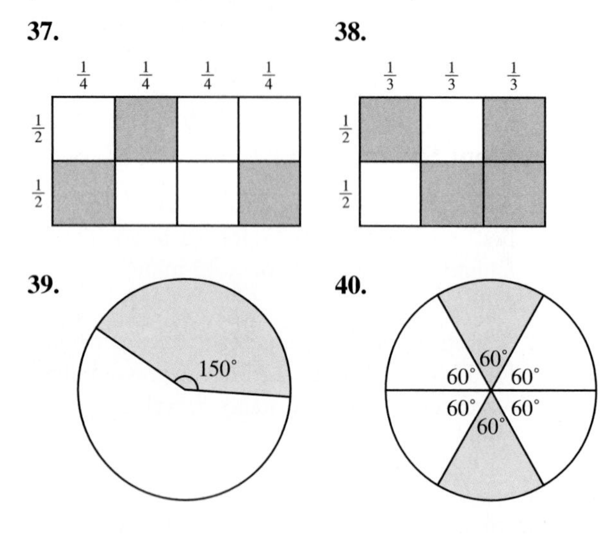

In Exercises 41–64, solve the percent equation. See Example 3.

41. What number is 30% of 150?

42. What number is 62% of 1200?

43. What number is 9.5% of 816?

44. What number is $33\frac{1}{3}\%$ of 516?

45. What number is $\frac{3}{4}\%$ of 56?

46. What number is 0.2% of 100,000?

47. What number is 200% of 88?

48. What number is 325% of 450?

49. 903 is 43% of what number?

50. 425 is 85% of what number?

51. 275 is $12\frac{1}{2}\%$ of what number?

52. 210 is 250% of what number?

53. 594 is 450% of what number?

54. 814 is $66\frac{2}{3}\%$ of what number?

55. 2.16 is 0.6% of what number?

56. 51.2 is 0.08% of what number?

57. 576 is what percent of 800?

58. 1950 is what percent of 5000?

59. 45 is what percent of 360?

60. 38 is what percent of 5700?

61. 22 is what percent of 800?

62. 110 is what percent of 110?

63. 1000 is what percent of 200?

64. 148.8 is what percent of 960?

In Exercises 65–74, find the missing quantities. See Examples 7, 8, and 9.

	Cost	*Selling Price*	*Markup*	*Markup Rate*
65.	$26.97	$49.95		
66.		$224.87	$75.08	
67.		$74.38		81.5%
68.	$680.00			$33\frac{1}{3}\%$
69.		$125.98	$56.69	
70.	$71.97	$119.95		
71.		$15,900.00	$2650.00	
72.		$350.00	$80.77	
73.	$107.97			85.2%
74.		$69.99		55.5%

In Exercises 75–84, find the missing quantities. See Examples 10 and 11.

	List Price	*Sale Price*	*Discount*	*Discount Rate*
75.	$39.95	$29.95		
76.	$18.95		$8.00	
77.		$18.95		20%
78.		$259.97	$135.00	
79.	$189.99		$30.00	
80.	$50.99	$45.99		
81.	$119.96			50%
82.	$84.95			65%
83.		$695.00	$300.00	
84.		$189.00		40%

Solving Problems

85. *Rent Payment* You spend 17% of your monthly income of $3200 for rent. What is your monthly payment?

86. *Cost of Housing* You budget 30% of your annual after-tax income for housing. If your after-tax income is $38,500, what amount can you spend on housing?

87. *Retirement Plan* You budget $7\frac{1}{2}$% of your gross income for an individual retirement plan. Your annual gross income is $45,800. How much will you put in your retirement plan each year?

88. *Enrollment* Thirty-five percent of the students enrolled in a college are freshmen. The enrollment of the college is 2800. Find the number of freshmen.

89. *Snowfall* During the winter, there was 120 inches of snow. Of that amount, 86 inches fell in December. What percent of the snow fell in December?

90. *Layoff* Because of slumping sales, a small company laid off 30 of its 153 employees.

(a) What percent of the work force was laid off?

(b) Complete the statement: "About 1 out of every ____ workers was laid off."

91. *Unemployment Rate* During a recession, 72 out of 1000 workers in the population were unemployed. Find the unemployment rate (as a percent).

92. *Inflation Rate* You purchase a lawn tractor for $3750 and 1 year later you note that the cost has increased to $3900. Determine the inflation rate (as a percent) for the tractor.

93. *Original Price* A coat sells for $250 during a 20% off storewide clearance sale. What was the original price of the coat?

94. *Membership Drive* Because of a membership drive for a public television station, the current membership is 125% of what it was a year ago. The current number of members is 7815. How many members did the station have last year?

95. *Price* The price of a new van is approximately 110% of what it was 3 years ago. The current price is $26,850. What was the approximate price 3 years ago?

96. *Decision Making* A new car you want to buy costs $17,800. If you wait another month to buy the car, the price will increase by 6%. However, to buy it now you will have to pay an interest penalty of $450 for the early withdrawal of a certificate of deposit. Should you buy the car now or wait another month? Explain.

97. *Eligible Voters* The news media reported that 6432 votes were cast in the last election and that this represented 63% of the eligible voters of a district. How many eligible voters are in the district?

98. *Defective Parts* A quality control engineer tested several parts and found two to be defective. The engineer reported that 2.5% were defective. How many were tested?

99. *Course Grade* You were six points shy of a B in your mathematics course. Your point total for the course is 394. How many points were possible in the course? (Assume that you needed 80% of the course total for a B.)

100. *Target Size* A circular target is attached to a rectangular board, as shown in the figure. The radius of the circle is $4\frac{1}{2}$ inches, and the measurements of the board are 12 inches by 15 inches. What percentage of the board is covered by the target? (The area of a circle is $A = \pi r^2$, where r is the radius of the circle.)

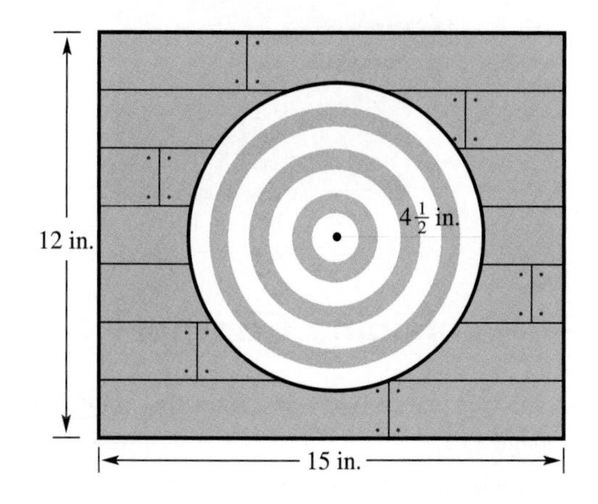

101. *Analyzing Data* In 1995 there were 697.1 million visits to office-based physicians. The figure classifies the age groups of those making the visits. Approximate the number of Americans in each of the classifications. (Source: U.S. National Center for Health Statistics)

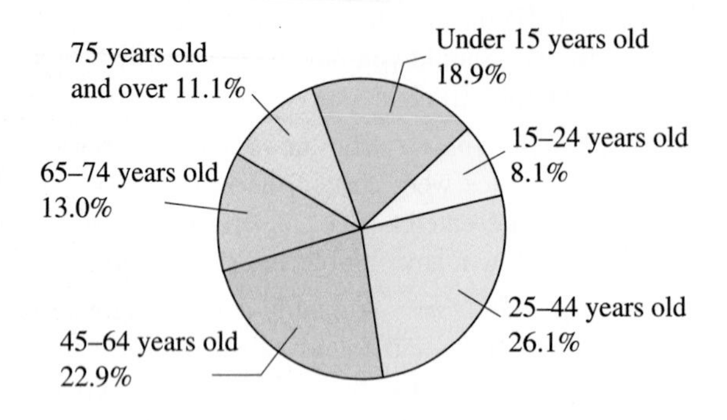

102. *Graphical Estimation* The graph shows the number (in thousands) of criminal cases commenced in the United States District Courts from 1990 through 1996. (Source: Administrative Office of the U.S. Courts)

(a) Determine the percent increase in cases from 1991 to 1992.

(b) Determine the percent decrease in cases from 1992 to 1995.

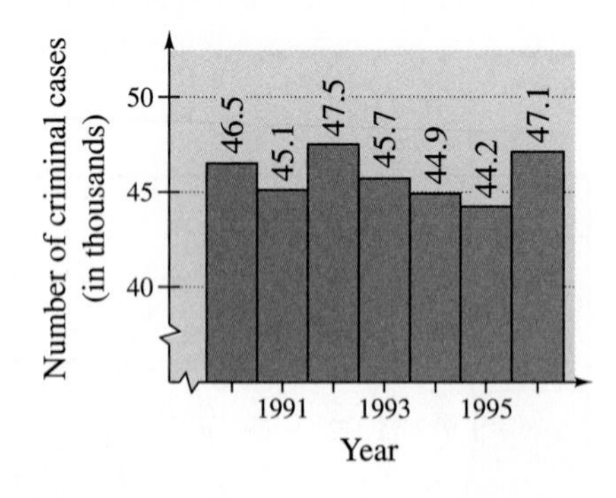

103. *Interpreting a Table* The table shows the numbers of women scientists and the percents of women scientists in the United States in three fields for the years 1983 and 1996. (Source: U.S. Bureau of Labor Statistics)

(a) Find the total number of mathematicians and computer scientists (men and women) in 1996.

(b) Find the total number of chemists (men and women) in 1983.

(c) Explain how the number of women in biology can increase while the percent of women in biology decreases.

Field	1983 Number	1983 %	1996 Number	1996 %
Math/Computer	137,000	29.6%	411,600	30.6%
Chemistry	22,800	23.3%	42,600	30.6%
Biology	22,400	40.8%	45,200	39.0%

104. *Population Analysis* The table gives the approximate population (in millions) of Bangladesh for each decade from 1950 through 1990. Approximate the percent growth rate for each decade. If the growth rate of the 1980s continued until the year 2010, approximate the population in 2010. (Source: U.S. Bureau of the Census, International Data Base)

Year	1950	1960	1970	1980	1990
Population	45.6	54.6	67.4	88.1	110.1

Explaining Concepts

105. Answer parts (a)–(f) of Motivating the Chapter on page 115.

106. Explain the meaning of the word "percent."

107. Explain the concept of "rate."

108. In your own words, explain how to change a percent to a fraction. Give an example.

109. In your own words, explain how to change a decimal to a percent. Give an example.

110. In your own words, explain how to change a fraction to a percent. Give an example.

111. Can any positive decimal be written as a percent? Explain.

112. Is it true that $\frac{1}{2}\% = 50\%$? Explain.

3.4 Ratios and Proportions

Objectives

1. Compare relative sizes using ratios.
2. Find the unit price of a consumer item.
3. Solve a proportion that equates two ratios.
4. Solve application problems using the Consumer Price Index.

1 Compare relative sizes using ratios.

Setting Up Ratios

A **ratio** is a comparison of one number to another by division. For example, in a class of 29 students made up of 16 women and 13 men, the ratio of women to men is 16 to 13 or $\frac{16}{13}$. Some other ratios for this class are as follows.

Men to women: $\dfrac{13}{16}$ Men to students: $\dfrac{13}{29}$ Students to women: $\dfrac{29}{16}$

Note the order implied by a ratio. The ratio of a to b means a/b, whereas the ratio of b to a means b/a.

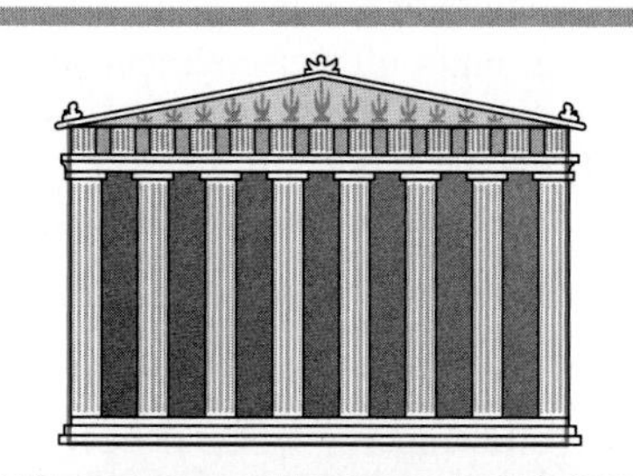

The Golden Ratio

In ancient Greek art and architecture, rectangles can be found in which the ratio of the longer side to the shorter side is

$$\frac{\sqrt{5}+1}{2} \approx 1.618.$$

This ratio is known as the Golden Ratio. The ancient Greeks were not the only ones to use this ratio. Leonardo DaVinci used the Golden Ratio in his paintings. And the Golden Ratio has also been discovered in nature in the growth patterns of living things such as plants and seashells.

▶ **Definition of Ratio**

The **ratio** of the real number a to the real number b is given by

$$\frac{a}{b}.$$

The ratio of a to b is sometimes written as $a : b$.

Example 1 Writing Ratios in Fractional Form

a. The ratio of 7 to 5 is given by $\frac{7}{5}$.

b. The ratio of 12 to 8 is given by $\frac{12}{8} = \frac{3}{2}$.

Note that the fraction $\frac{12}{8}$ can be written in reduced form as $\frac{3}{2}$.

c. The ratio of $3\frac{1}{2}$ to $5\frac{1}{4}$ is given by

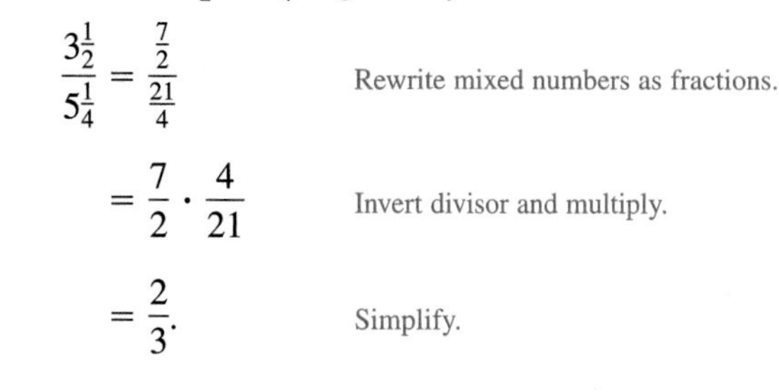

$$\frac{3\frac{1}{2}}{5\frac{1}{4}} = \frac{\frac{7}{2}}{\frac{21}{4}} \qquad \text{Rewrite mixed numbers as fractions.}$$

$$= \frac{7}{2} \cdot \frac{4}{21} \qquad \text{Invert divisor and multiply.}$$

$$= \frac{2}{3}. \qquad \text{Simplify.}$$

There are many real-life applications of ratios. For instance, ratios are used to describe opinion surveys (for/against), populations (male/female, unemployed/employed), and mixtures (oil/gasoline, water/alcohol).

When comparing two *measurements* by a ratio, you should use the same unit of measurement in both the numerator and the denominator. For example, to find the ratio of 4 feet to 8 inches, you could convert 4 feet to 48 inches (by multiplying by 12) to obtain

$$\frac{4 \text{ feet}}{8 \text{ inches}} = \frac{48 \text{ inches}}{8 \text{ inches}} = \frac{48}{8} = \frac{6}{1}.$$

Or you could convert 8 inches to $\frac{8}{12}$ feet (by dividing by 12) to obtain

$$\frac{4 \text{ feet}}{8 \text{ inches}} = \frac{4 \text{ feet}}{\frac{8}{12} \text{ feet}} = 4 \cdot \frac{12}{8} = \frac{6}{1}.$$

If you use different units of measurement in the numerator and denominator, then you *must* include the units. If you use the same units of measurement in the numerator and denominator, then it is not necessary to write the units. A list of common conversion factors is given on the inside back cover.

Example 2 Comparing Measurements

Find a ratio to compare the relative sizes of the following.

a. 5 gallons to 7 gallons **b.** 3 meters to 40 centimeters

c. 200 cents to 3 dollars **d.** 30 months to $1\frac{1}{2}$ years

Solution

a. Because the units of measurement are the same, the ratio is $\frac{5}{7}$.

b. Because the units of measurement are different, begin by converting meters to centimeters *or* centimeters to meters. Here, it is easier to convert meters to centimeters by multiplying by 100.

$$\frac{3 \text{ meters}}{40 \text{ centimeters}} = \frac{3(100) \text{ centimeters}}{40 \text{ centimeters}}$$ Convert meters to centimeters.

$$= \frac{300}{40}$$ Multiply numerator.

$$= \frac{15}{2}$$ Simplify.

c. Because 200 cents is the same as 2 dollars, the ratio is

$$\frac{200 \text{ cents}}{3 \text{ dollars}} = \frac{2 \text{ dollars}}{3 \text{ dollars}} = \frac{2}{3}.$$

d. Because $1\frac{1}{2}$ years $= 18$ months, the ratio is

$$\frac{30 \text{ months}}{1\frac{1}{2} \text{ years}} = \frac{30 \text{ months}}{18 \text{ months}} = \frac{30}{18} = \frac{5}{3}.$$

2 Find the unit price of a consumer item.

Unit Prices

As a consumer, you must be able to determine the unit prices of items you buy in order to make the best use of your money. The **unit price** of an item is given by the ratio of the total price to the total units.

$$\text{Unit price} = \frac{\text{Total price}}{\text{Total units}}$$

To state unit prices, we usually use the word *per.* For instance, the unit price for a particular brand of coffee might be 4.69 dollars *per* pound, or \$4.69 per pound.

Example 3 Finding a Unit Price

Find the unit price (in dollars per ounce) for a 5-pound, 4-ounce box of detergent that sells for \$4.62.

Solution

Begin by writing the weight in ounces. That is,

$$\begin{aligned} 5 \text{ pounds} + 4 \text{ ounces} &= 5 \cancel{\text{pounds}}\left(\frac{16 \text{ ounces}}{1 \cancel{\text{pound}}}\right) + 4 \text{ ounces} \\ &= 80 \text{ ounces} + 4 \text{ ounces} \\ &= 84 \text{ ounces.} \end{aligned}$$

Next, determine the unit price as follows.

Verbal Model: $\text{Unit price} = \frac{\text{Total price}}{\text{Total units}}$

Unit Price: $\frac{\$4.62}{84 \text{ ounces}} = \$0.055 \text{ per ounce}$

Example 4 Comparing Unit Prices

Which has the lower unit price: a 12-ounce box of breakfast cereal for \$2.69 or a 16-ounce box of the same cereal for \$3.49?

Solution

The unit price for the smaller box is

$$\text{Unit price} = \frac{\text{total price}}{\text{total units}} = \frac{\$2.69}{12 \text{ ounces}} \approx \$0.224 \text{ per ounce.}$$

The unit price for the larger box is

$$\text{Unit price} = \frac{\text{total price}}{\text{total units}} = \frac{\$3.49}{16 \text{ ounces}} \approx \$0.218 \text{ per ounce.}$$

So, the larger box has a slightly lower unit price.

3 Solve a proportion that equates two ratios.

Solving Proportions

A **proportion** is a statement that equates two ratios. For example, if the ratio of a to b is the same as the ratio of c to d, we can write the proportion as

$$\frac{a}{b} = \frac{c}{d}.$$

In typical applications, you know the values for three of the letters (quantities) and are required to find the value of the fourth. To solve such a fractional equation, you can use the *cross-multiplication* procedure introduced in Section 3.2.

> **Solving a Proportion**
>
> If $\frac{a}{b} = \frac{c}{d}$, then $ad = bc$. The quantities a and d are called the **extremes** of the proportion, whereas b and c are called the **means** of the proportion.

Example 5 Solving Proportions

Solve the following proportions for x.

a. $\frac{50}{x} = \frac{2}{28}$

b. $\frac{x}{3} = \frac{10}{6}$

Solution

a.

$\frac{50}{x} = \frac{2}{28}$ Proportion

$50(28) = 2x$ Cross-multiply.

$\frac{1400}{2} = x$ Divide both sides by 2.

$700 = x$ Simplify.

So, the ratio of 50 to 700 is the same as the ratio of 2 to 28.

b.

$\frac{x}{3} = \frac{10}{6}$ Proportion

$x = \frac{30}{6}$ Multiply both sides by 3.

$x = 5$ Simplify.

So, the ratio of 5 to 3 is the same as the ratio of 10 to 6.

To solve an equation, you want to isolate the variable. In Example 5(b) this was done by multiplying both sides by 3 instead of cross-multiplying. In this case, multiplying both sides by 3 was the only step needed to isolate the x-variable. However, either method is valid for solving the equation.

Example 6 Geometry: Similar Triangles

A triangular lot has perpendicular sides of lengths 100 feet and 210 feet. You are to make a proportional sketch of this lot using 8 inches as the length of the shorter side. How long should you make the other side?

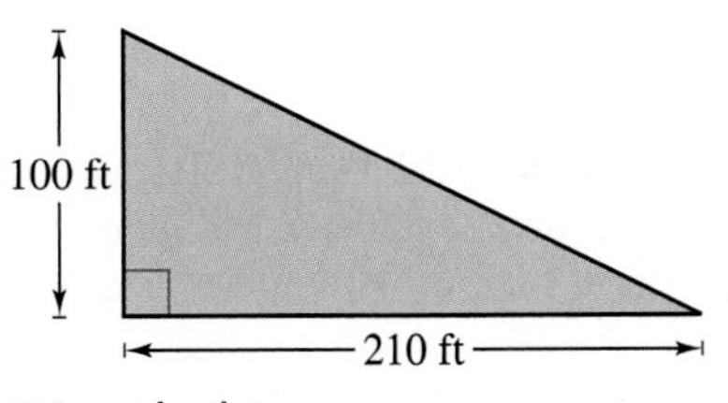

Triangular lot

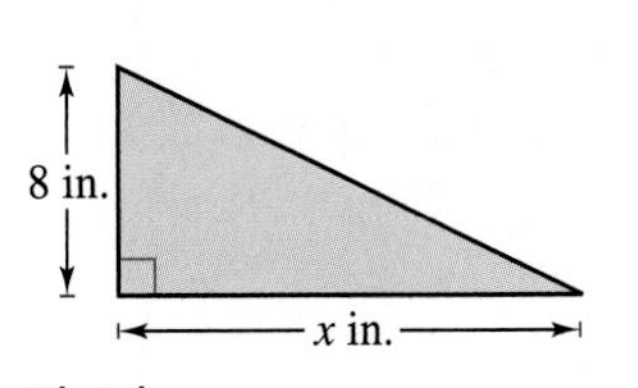

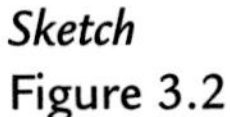

Sketch
Figure 3.2

Solution

This is a case of similar triangles in which the ratios of the corresponding sides are equal. The triangles are shown in Figure 3.2.

$$\frac{\text{Shorter side of lot}}{\text{Longer side of lot}} = \frac{\text{shorter side of sketch}}{\text{longer side of sketch}}$$ Proportion for similar triangles

$$\frac{100}{210} = \frac{8}{x}$$ Substitute.

$$x \cdot 100 = 210 \cdot 8$$ Cross-multiply.

$$x = \frac{1680}{100} = 16.8$$ Divide both sides by 100.

So, the length of the longer side of the sketch should be 16.8 inches.

Example 7 Resizing a Picture

You have a 7-by-8-inch picture of a graph that you want to paste into a term paper, but you have only a 6-by-6-inch space in which to put it. You go to the copier that has five options for resizing your graph: 64%, 78%, 100%, 121%, and 129%.

a. Which option should you choose?

b. What are the measurements of the resized picture?

Solution

a. Because the longest side must be *reduced* from 8 inches to no more than 6 inches, consider the proportion

$$\frac{\text{New length}}{\text{Old length}} = \frac{\text{new percent}}{\text{old percent}}$$ Proportion

$$\frac{6}{8} = \frac{x}{100}$$ Substitute.

$$\frac{6}{8} \cdot 100 = x$$ Multiply both sides by 100.

$$75 = x.$$

To guarantee a fit, you should choose the 64% option, because 78% is greater than the required 75%.

b. To find the measurements of the resized picture, multiply by 64% or 0.64.

$$\text{Length} = 0.64(8) = 5.12 \text{ inches}$$

$$\text{Width} = 0.64(7) = 4.48 \text{ inches}$$

The size of the reduced picture is 5.12 inches by 4.48 inches.

4 Solve application problems using the Consumer Price Index.

The Consumer Price Index

The rate of inflation is important to all of us. Simply stated, *inflation* is an economic condition in which the price of a fixed amount of goods or services increases. So, a fixed amount of money buys less in a given year than in previous years.

The most widely used measurement of inflation in the United States is the *Consumer Price Index* (CPI), often called the *Cost-of-Living Index.* The table below shows the "All Items" or general index for the years 1950 to 1997. (Source: U.S. Bureau of Labor Statistics)

Year	CPI	Year	CPI	Year	CPI	Year	CPI
1950	24.1	1962	30.2	1974	49.3	1986	109.6
1951	26.0	1963	30.6	1975	53.8	1987	113.6
1952	26.5	1964	31.0	1976	56.9	1988	118.3
1953	26.7	1965	31.5	1977	60.6	1989	124.0
1954	26.9	1966	32.4	1978	65.2	1990	130.7
1955	26.8	1967	33.4	1979	72.6	1991	136.2
1956	27.2	1968	34.8	1980	82.4	1992	140.3
1957	28.1	1969	36.7	1981	90.9	1993	144.5
1958	28.9	1970	38.8	1982	96.5	1994	148.2
1959	29.1	1971	40.5	1983	99.6	1995	152.4
1960	29.6	1972	41.8	1984	103.9	1996	156.9
1961	29.9	1973	44.4	1985	107.6	1997	160.5

To determine (from the CPI) the change in the buying power of a dollar from one year to another, use the following proportion.

$$\frac{\text{Price in year } n}{\text{Price in year } m} = \frac{\text{index in year } n}{\text{index in year } m}$$

For instance, if you paid $15,000 for a house in 1950, then the amount you could expect to pay for the same house in 1990 is given by the following proportion.

$$\frac{\text{Price in 1990}}{\text{Price in 1950}} = \frac{\text{index in 1990}}{\text{index in 1950}}$$

$$\frac{x}{15,000} = \frac{130.7}{24.1} \quad \Rightarrow \quad x \approx \$81,350$$

$$x = \frac{130.7}{24.1} \cdot 15,000$$

$$x \approx \$81,350$$

Example 8 Using the Consumer Price Index

You purchased a piece of jewelry for $750 in 1990. What would you expect the replacement value of the jewelry to have been in 1996?

Solution

To answer this question, you can use the Consumer Price Index, as follows.

Verbal Model: $\dfrac{\text{Price in 1996}}{\text{Price in 1990}} = \dfrac{\text{index in 1996}}{\text{index in 1990}}$

Labels:
Price in 1996 $= x$ (dollars)
Price in 1990 $= 750$ (dollars)
Index in 1996 $= 156.9$
Index in 1990 $= 130.7$

Proportion: $\dfrac{x}{750} = \dfrac{156.9}{130.7}$

$$x = 750 \cdot \frac{156.9}{130.7}$$

$$x \approx 900$$

You should expect the replacement value of the jewelry to have been approximately $900 in 1996. Check this solution in the original statement of the problem.

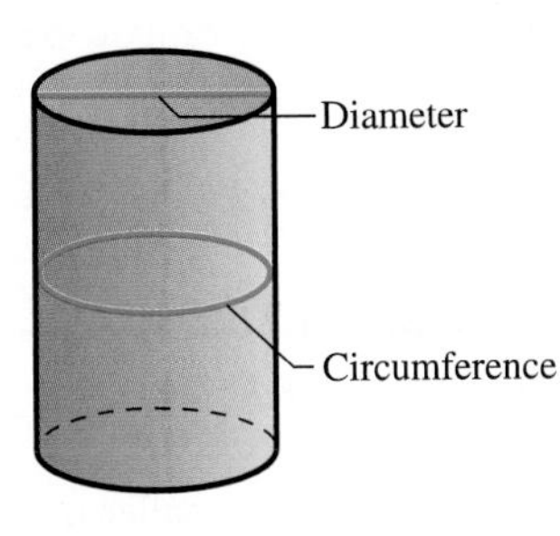

Discussing the Concept The Value of Pi

One of the best known ratios in mathematics is denoted by the Greek letter π, pronounced "pie." This number represents the ratio of the circumference of *any* circle to its diameter. To estimate the value of π, try the following experiment.

Measure the diameter of a circular cylinder (such as a soda can), and then measure the circumference of the cylinder, as shown in the figure at the left. Then use your graphing utility to approximate the value of π as follows.

$$\pi = \frac{\text{Circumference}}{\text{Diameter}}$$

Try to find this ratio for several different sizes of circular objects. Does the ratio depend on the size of the circumference of the circle?

3.4 Exercises

Integrated Review *Concepts, Skills, and Problem Solving*

Keep mathematically in shape by doing these exercises *before* the problems of this section.

Properties and Definitions

1. Explain how to write $\frac{15}{12}$ in reduced form.

2. Explain how to divide $\frac{3}{5}$ by $\frac{x}{2}$.

3. Complete the Associative Property: $(3x)y =$ ______.

4. Name the property illustrated by $x^2 + 0 = x^2$.

Simplifying Expressions

In Exercises 5–10, evaluate the expression.

5. $3^2 - (-4)$

6. $(-5)^3 + 3$

7. 9.3×10^6

8. $\frac{-|7 + 3^2|}{4}$

9. $(-4)^2 - (30 \div 50)$

10. $(8 \cdot 9) + (-4)^3$

Writing Models

In Exercises 11 and 12, translate the sentence into an algebraic expression.

11. A number is decreased by 10 and the difference is doubled.

12. The area of a triangle with base b and height $\frac{1}{2}(b + 6)$

Developing Skills

In Exercises 1–8, write the ratio as a fraction in simplest form. See Example 1.

1. 36 to 9

2. 24 to 32

3. 27 to 54

4. 50 to 15

5. 14 : 21

6. 60 : 45

7. 144 : 16

8. 12 : 30

In Exercises 9–26, find a ratio to compare the relative sizes. (Use the same units of measurement for both quantities.) See Example 2.

9. Thirty-six inches to 24 inches

10. Fifteen feet to 12 feet

11. Forty dollars to $60

12. Twenty-four pounds to 30 pounds

13. One quart to 1 gallon

14. Three inches to 2 feet

15. Seven nickels to 3 quarters

16. Twenty-four ounces to 3 pounds

17. Three hours to 90 minutes

18. Twenty-one feet to 35 yards

19. Seventy-five centimeters to 2 meters

20. Two meters to 75 centimeters

21. Sixty milliliters to 1 liter

22. Fifty cubic centimeters to 1 liter

23. Ninety minutes to 2 hours

24. Five and one-half pints to 2 quarts

25. Three thousand pounds to 5 tons

26. Twelve thousand pounds to 2 tons

In Exercises 27–30, find the unit price (in dollars per ounce). See Example 3.

27. A 20-ounce can of pineapple for 79¢

28. An 18-ounce box of cereal for $3.19

29. A 1-pound, 4-ounce loaf of bread for $1.29

30. A 1-pound package of cheese for $2.89

In Exercises 31–36, which product has the smaller unit price? See Example 4.

31. (a) A $27\frac{3}{4}$-ounce can of spaghetti sauce for $1.19

(b) A 32-ounce jar of spaghetti sauce for $1.45

32. (a) A 16-ounce package of margarine quarters for $1.29

(b) A 3-pound tub of margarine for $3.29

33. (a) A 10-ounce package of frozen green beans for 59¢
    (b) A 16-ounce package of frozen green beans for 89¢
34. (a) An 18-ounce jar of peanut butter for $1.39
    (b) A 28-ounce jar of peanut butter for $2.19
35. (a) A 2-liter bottle (67.6 ounces) of soft drink for $1.09
    (b) Six 12-ounce cans of soft drink for $1.69
36. (a) A 1-quart container of oil for $1.29
    (b) A 2.5-gallon container of oil for $11.20

In Exercises 37–52, solve the proportion. See Example 5.

37. $\frac{5}{3} = \frac{20}{y}$ **38.** $\frac{9}{x} = \frac{18}{5}$

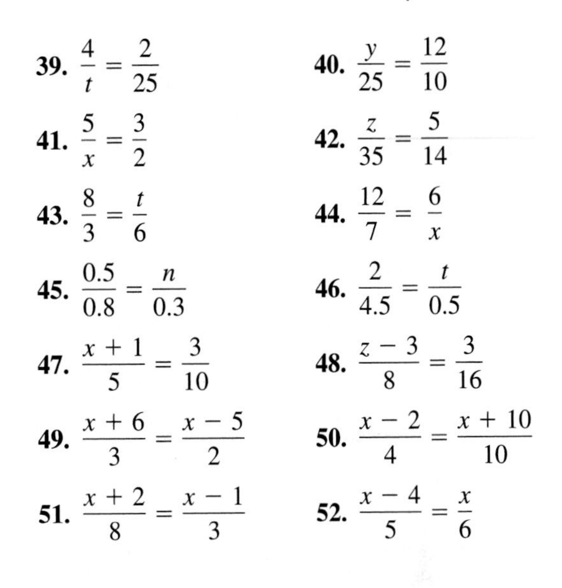

39. $\frac{4}{t} = \frac{2}{25}$ **40.** $\frac{y}{25} = \frac{12}{10}$

41. $\frac{5}{x} = \frac{3}{2}$ **42.** $\frac{z}{35} = \frac{5}{14}$

43. $\frac{8}{3} = \frac{t}{6}$ **44.** $\frac{12}{7} = \frac{6}{x}$

45. $\frac{0.5}{0.8} = \frac{n}{0.3}$ **46.** $\frac{2}{4.5} = \frac{t}{0.5}$

47. $\frac{x+1}{5} = \frac{3}{10}$ **48.** $\frac{z-3}{8} = \frac{3}{16}$

49. $\frac{x+6}{3} = \frac{x-5}{2}$ **50.** $\frac{x-2}{4} = \frac{x+10}{10}$

51. $\frac{x+2}{8} = \frac{x-1}{3}$ **52.** $\frac{x-4}{5} = \frac{x}{6}$

Solving Problems

In Exercises 53–62, express the statement as a ratio in simplest form. (Use the same units of measurement for both quantities.)

53. *Study Hours* You study 6 hours per day and are in class 3 hours per day. Find the ratio of the number of study hours to class hours.

54. *Income Tax* You have $16.50 of state tax withheld from your paycheck per week when your gross pay is $750. Find the ratio of tax to gross pay.

55. *Price-Earnings Ratio* The ratio of the price of a stock to its earnings is called the *price-earnings ratio.* A certain stock sells for $78 per share and earns $6.50 per share. What is the price-earnings ratio of this stock?

56. *Student-Teacher Ratio* There are 2921 students and 127 faculty members at your school. Find the ratio of the number of students to the number of faculty members.

57. *Compression Ratio* The *compression ratio* of an engine is the ratio of the expanded volume of gas in one of its cylinders to the compressed volume of gas in the cylinder (see figure). A cylinder in a certain diesel engine has an expanded volume of 345 cubic centimeters and a compressed volume of 17.25 cubic centimeters. What is the compression ratio of this engine?

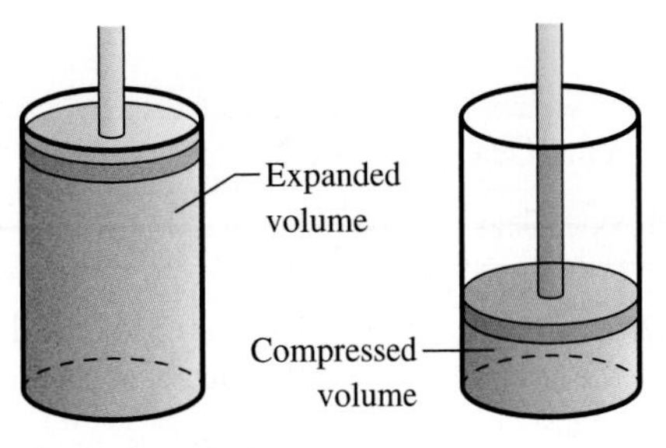

Figure for 57

58. *Turn Ratio* The *turn ratio* of a transformer is the ratio of the number of turns on the secondary winding to the number of turns on the primary winding (see figure). A transformer has a primary winding with 250 turns and a secondary winding with 750 turns. What is its turn ratio?

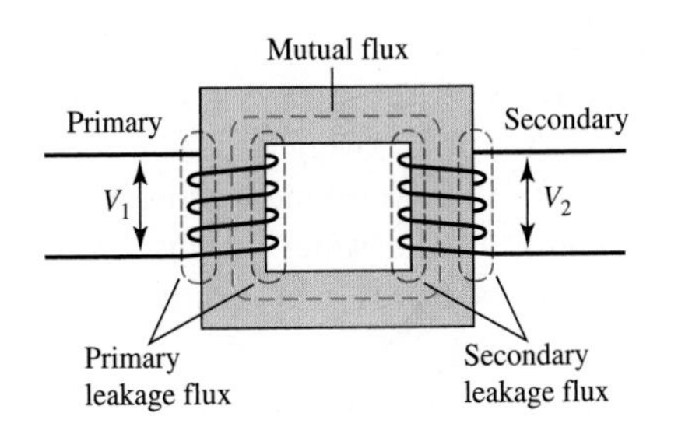

59. *Gear Ratio* The *gear ratio* of two gears is the ratio of the number of teeth on one gear to the number of teeth on the other gear. Find the gear ratio of the larger gear to the smaller gear for the gears in the figure.

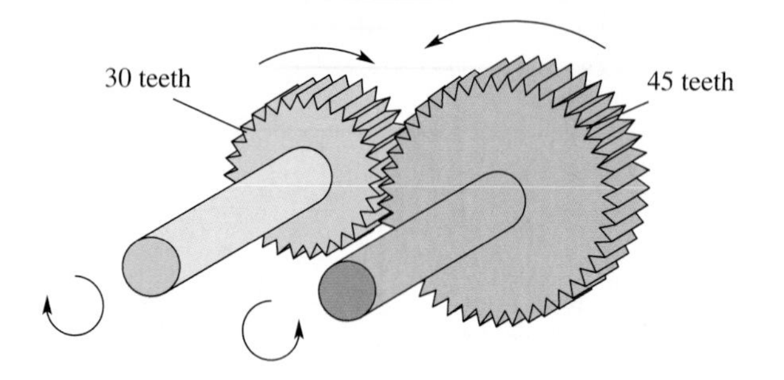

60. *Gear Ratio* On a five-speed bicycle, the ratio of the pedal gear to the axle gear depends on which axle gear is engaged. Use the table to find the gear ratios for the five different gears. For which gear is it easiest to pedal? Why?

Gear	1st	2nd	3rd	4th	5th
Teeth on Pedal Gear	52	52	52	52	52
Teeth on Axle Gear	28	24	20	17	14

61. *Geometry* Find the ratio of the area of the larger pizza to the area of the smaller pizza in the figure. (*Note:* The area of a circle is $A = \pi r^2$.)

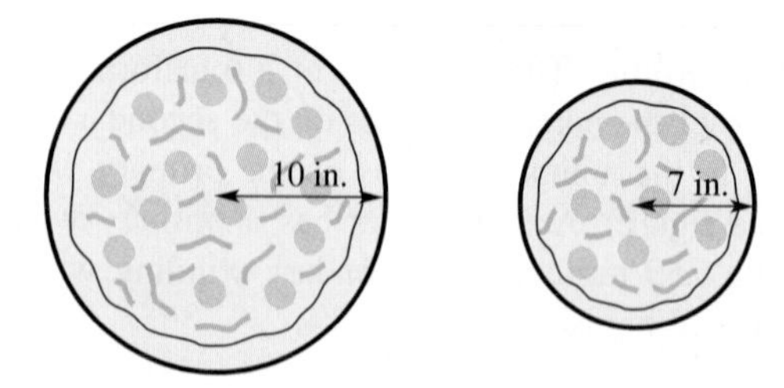

62. *Specific Gravity* The *specific gravity* of a substance is the ratio of its weight to the weight of an equal volume of water. Kerosene weighs 0.82 grams per cubic centimeter and water weighs 1 gram per cubic centimeter. What is the specific gravity of kerosene?

63. *Gasoline Cost* A car uses 20 gallons of gasoline for a trip of 500 miles. How many gallons would be used on a trip of 400 miles?

64. *Amount of Fuel* A tractor requires 4 gallons of diesel fuel to plow for 90 minutes. How many gallons of fuel would be required to plow for 8 hours?

65. *Building Material* One hundred cement blocks are required to build a 16-foot wall. How many blocks are needed to build a 40-foot wall?

66. *Force on a Spring* A force of 50 pounds stretches a spring 4 inches. How much force is required to stretch the spring 6 inches?

67. *Real Estate Taxes* The tax on a property with an assessed value of \$65,000 is \$825. Find the tax on a property with an assessed value of \$90,000.

68. *Real Estate Taxes* The tax on a property with an assessed value of \$65,000 is \$1100. Find the tax on a property with an assessed value of \$90,000.

69. *Polling Results* In a poll, 624 people from a sample of 1100 indicated they would vote for a certain candidate. How many votes can the candidate expect to receive from 40,000 votes cast?

70. *Quality Control* A quality control engineer found two defective units in a sample of 50. At this rate, what is the expected number of defective units in a shipment of 10,000 units?

71. *Pumping Time* A pump can fill a 750-gallon tank in 35 minutes. How long will it take to fill a 1000-gallon tank with this pump?

72. *Increasing a Recipe* Two cups of flour are required to make one batch of cookies. How many cups are required for $2\frac{1}{2}$ batches?

73. *Amount of Gasoline* The gasoline-to-oil ratio for a two-cycle engine is 40 to 1. How much gasoline is required to produce a mixture that contains one-half pint of oil?

74. *Pounds of Sand* The ratio of cement to sand in an 80-pound bag of dry mix is 1 to 4. Find the number of pounds of sand in the bag. (*Note:* Dry mix is composed of only cement and sand.)

75. *Map Scale* Use the map to approximate the distance between Philadelphia and Pittsburgh.

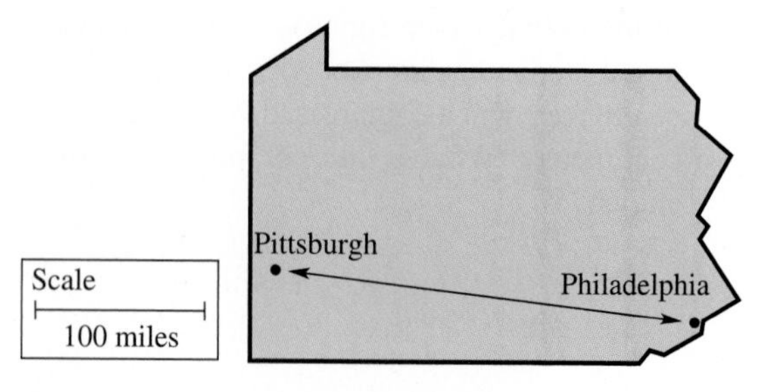

76. *Map Scale* On a map, $1\frac{1}{2}$ inches represents 40 miles. Estimate the distance between two cities that are 4 inches apart on the map.

Similar Triangles In Exercises 77 and 78, find the length x of the side of the larger triangle. (Assume that the two triangles are similar, and use the fact that corresponding sides of similar triangles are proportional.)

77.

78.

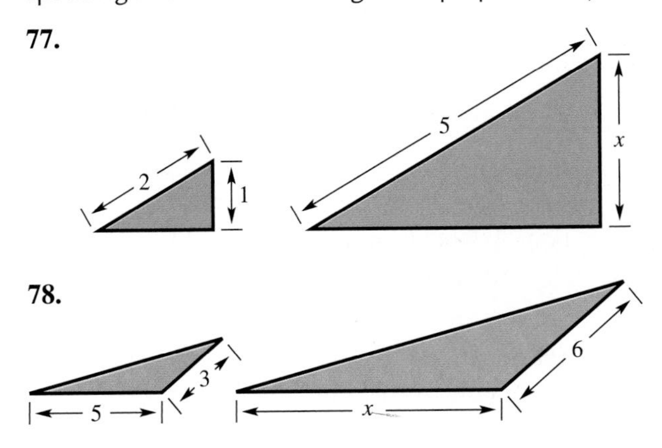

79. *Shadow Length* In the figure, how long is the man's shadow? (*Hint:* Use similar triangles to create a proportion.)

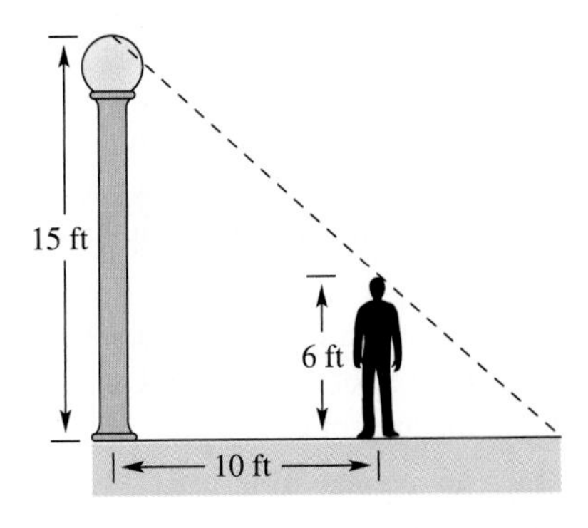

80. *Shadow Length* A man who is 6 feet tall walks directly toward the tip of the shadow of a tree (see figure). When the man is 100 feet from the tree, he starts forming his own shadow beyond the shadow of the tree. The length of the shadow of the tree beyond this point is 8 feet. Find the height of the tree. (*Hint:* Use similar triangles to create a proportion.)

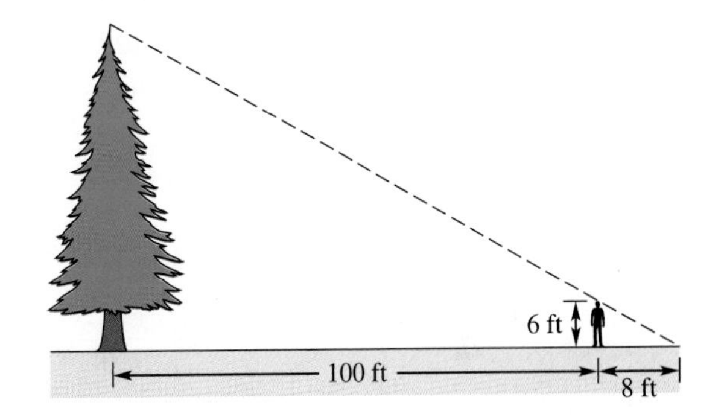

81. *Resizing a Picture* You have an 8-by-10-inch photo that must be reduced to a size of 1.6 by 2 inches for the school yearbook. What percent does the photo need to be reduced by in order to fit the allotted space?

82. *Resizing a Picture* You have a 7-by-5-inch photo of the math club that must be reduced to a size of 5.6 by 4 inches for the school yearbook. What percent does the photo need to be reduced by in order to fit the allotted space?

In Exercises 83–86, use the Consumer Price Index table on page 154 to estimate the price of the item in the indicated year.

83. The 1988 price of a lawn tractor that cost $2875 in 1978

84. The 1993 price of a watch that cost $58 in 1960

85. The 1960 price of a gallon of milk that cost $2.75 in 1996

86. The 1970 price of a coat that cost $225 in 1992

Explaining Concepts

87. Answer part (g) of Motivating the Chapter on page 115.

88. In your own words, describe the term *ratio*.

89. You are told that the ratio of men to women in a class is 2 to 1. Does this information tell you the total number of people in the class? Explain.

90. Explain the following statement. "When setting up a ratio, be sure you are comparing apples to apples and not apples to oranges."

91. In your own words, describe the term *proportion*.

92. Create a proportion problem. Exchange problems with another student and solve the problem you receive.

Mid-Chapter Quiz

Take this quiz as you would take a quiz in class. After you are done, check your work against the answers given in the back of the book.

In Exercises 1–10, solve the equation.

1. $120 - 3y = 0$

2. $10(y - 8) = 0$

3. $3x + 1 = x + 20$

4. $6x + 8 = 8 - 2x$

5. $-10x + \frac{2}{3} = \frac{7}{3} - 5x$

6. $\frac{x}{5} + \frac{x}{8} = 1$

7. $\frac{9 + x}{3} = 15$

8. $4 - 0.3(1 - x) = 7$

9. $\frac{x + 3}{6} = \frac{4}{3}$

10. $\frac{x + 7}{5} = \frac{x + 9}{7}$

In Exercises 11 and 12, solve the equation. Round the solution to two decimal places. In your own words, explain how to check the solution.

11. $32.86 - 10.5x = 11.25$

12. $\frac{x}{5.45} + 3.2 = 12.6$

13. What number is 62% of 25?

14. What number is $\frac{1}{2}\%$ of 8400?

15. 300 is what percent of 150?

16. 145.6 is 32% of what number?

17. The perimeter of a rectangle is 60 meters. Find the measurements of the rectangle if the length is $1\frac{1}{2}$ times the width.

18. You have two jobs. In the first job, you work 40 hours a week and earn \$7.50 per hour. In the second job, you earn \$6.00 per hour and can work as many hours as you want. If you want to earn \$360 a week, how many hours must you work at the second job?

19. A region has an area of 42 square meters. It must be divided into three subregions so that the second has twice the area of the first, and the third has twice the area of the second. Determine the area of each subregion.

20. To get an A in a course, you must have an average of at least 90 points for three tests of 100 points each. For the first two tests, your scores are 84 and 93. What must you score on the third test to earn a 90% average for the course?

21. The price of a television set is approximately 108% of what it was 2 years ago. The current price is \$535. What was the approximate price 2 years ago?

22. The figure at the left shows where charitable giving went for the year 1996. What percent of the total giving went to religious organizations? (Source: *USA Today*)

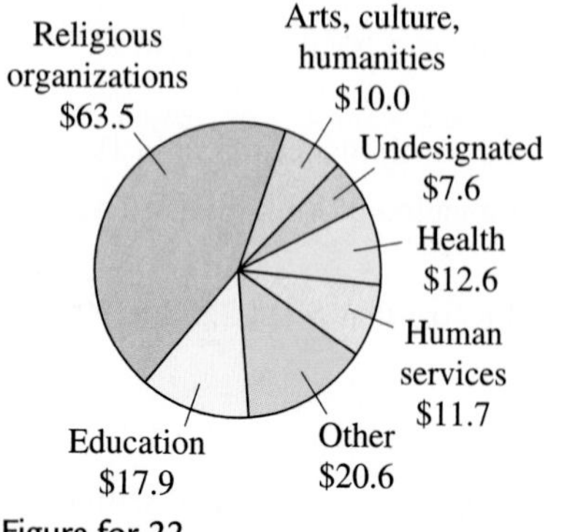

Figure for 22

23. A large round pizza has a radius of $r = 15$ inches and a small round pizza has a radius of $r = 8$ inches. Find the ratio of the area of the large pizza to the area of the small pizza. (*Hint:* The area of a circle is $A = \pi r^2$.)

3.5 Geometric and Scientific Applications

Objectives

1. Use a common formula to solve an application problem.
2. Solve a mixture problem involving hidden products.
3. Solve a work-rate problem.

1 Use a common formula to solve an application problem.

Using Formulas

Some formulas occur so frequently in problem solving that it is to your benefit to memorize them. For instance, the following formulas for area, perimeter, and volume are often used to create verbal models for word problems. In the geometry formulas below, A represents area, P represents perimeter, C represents circumference, and V represents volume.

Study Tip

When solving problems involving perimeter, area, or volume, be sure you list the units of measurement for your answers.

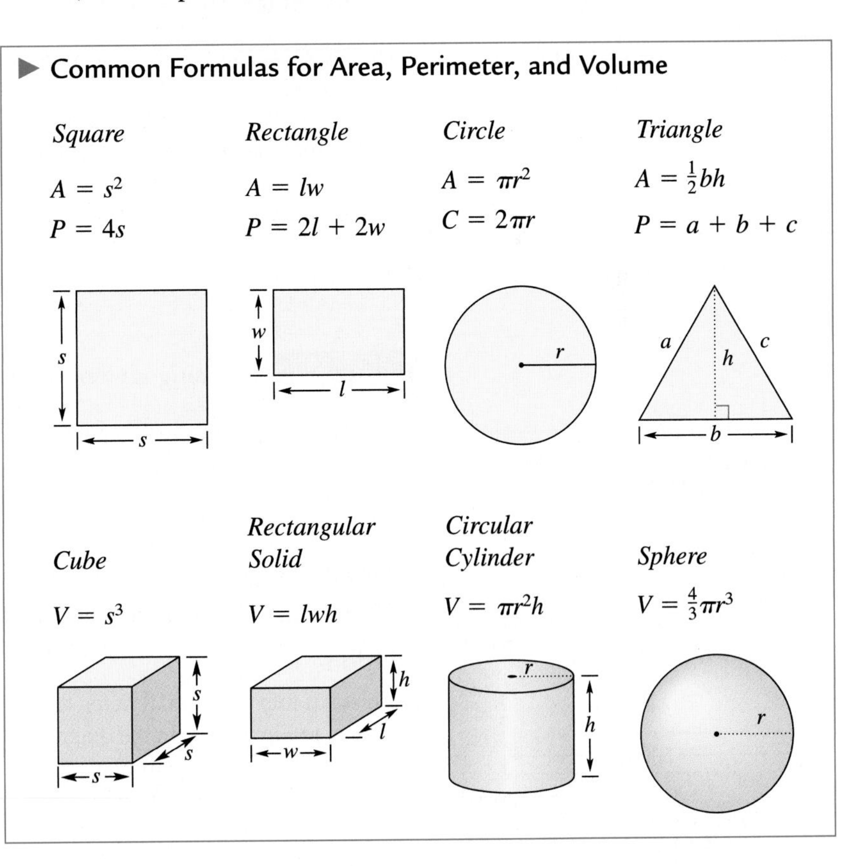

Common Formulas for Area, Perimeter, and Volume
Square
$A = s^2$
$P = 4s$
Rectangle
$A = lw$
$P = 2l + 2w$
Circle
$A = \pi r^2$
$C = 2\pi r$
Triangle
$A = \frac{1}{2}bh$
$P = a + b + c$
s
w
l
r
a
c
h
b
Cube
$V = s^3$
Rectangular Solid
$V = lwh$
Circular Cylinder
$V = \pi r^2 h$
Sphere
$V = \frac{4}{3}\pi r^3$

- *Perimeter* is always measured in linear units, such as inches, feet, miles, centimeters, meters, and kilometers.
- *Area* is always measured in square units, such as square inches, square feet, square centimeters, and square meters.
- *Volume* is always measured in cubic units, such as cubic inches, cubic feet, cubic centimeters, and cubic meters.

Example 1 Using a Geometric Formula 

A sailboat has a triangular sail with an area of 96 square feet and a base that is 16 feet long, as shown in Figure 3.3. What is the height of the sail?

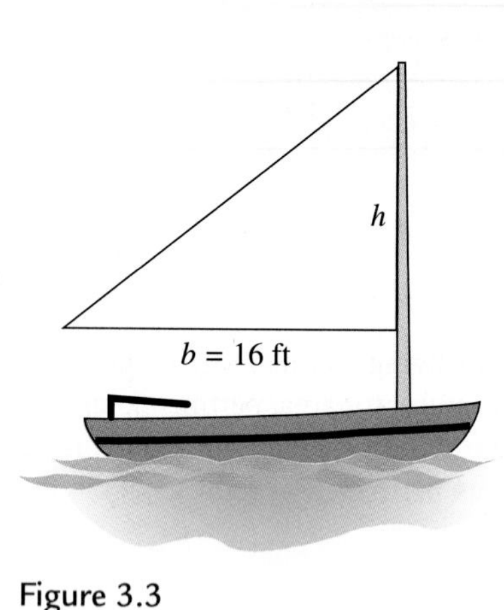

Figure 3.3

Solution

Because the sail is triangular, and you are given its area, you should begin with the formula for the area of a triangle.

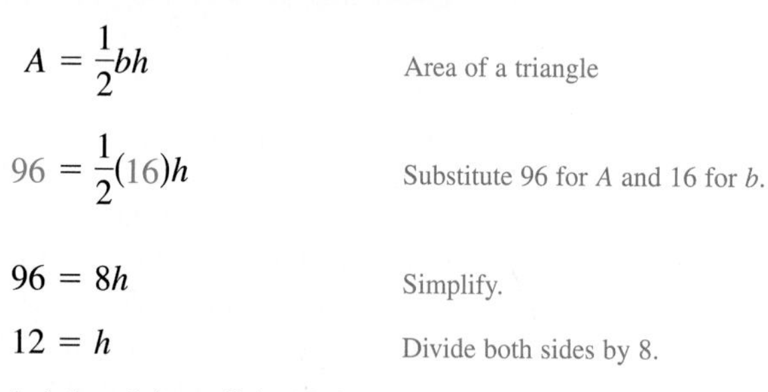

$A = \frac{1}{2}bh$ Area of a triangle

$96 = \frac{1}{2}(16)h$ Substitute 96 for A and 16 for b.

$96 = 8h$ Simplify.

$12 = h$ Divide both sides by 8.

The height of the sail is 12 feet.

In Example 1, notice that b and h are measured in feet. When they are multiplied in the formula $\frac{1}{2}bh$, the resulting area is measured in *square* feet.

$$A = \frac{1}{2}(16 \text{ feet})(12 \text{ feet}) = 96 \text{ feet}^2$$

Note that square feet can be written as feet^2.

Example 2 Using a Geometric Formula 

The local municipality is planning to develop the street along which you own a rectangular lot that is 500 feet deep and has an area of 100,000 square feet. To help pay for the new sewer system, each lot owner will be assessed \$5.50 per foot of lot frontage.

a. Find the length of the frontage of your lot.

b. How much will you be assessed for the new sewer system?

Solution

a. To solve this problem, it helps to begin by drawing a diagram such as the one shown in Figure 3.4. In the diagram, label the depth of the property as $l = 500$ feet and the unknown frontage as w.

$A = lw$ Area of a rectangle

$100{,}000 = 500(w)$ Substitute 100,000 for A and 500 for l.

$200 = w$ Divide both sides by 500 and simplify.

The frontage of the rectangular plot is 200 feet.

b. If each foot of frontage costs \$5.50, then your total assessment will be $200(5.50) = \$1100$.

Figure 3.4

▶ Miscellaneous Common Formulas

Temperature: F = degrees Fahrenheit, C = degrees Celsius

$$F = \frac{9}{5}C + 32$$

Simple Interest: I = interest, P = principal, r = interest rate, t = time

$$I = Prt$$

Distance: d = distance traveled, r = rate, t = time

$$d = rt$$

In some applications, it helps to rewrite a common formula by solving for a different variable. For instance, you can obtain a formula for C (degrees Celsius) in terms of F (degrees Fahrenheit) as follows.

$$F = \frac{9}{5}C + 32 \qquad \text{Temperature formula}$$

$$F - 32 = \frac{9}{5}C \qquad \text{Subtract 32 from both sides.}$$

$$\frac{5}{9}(F - 32) = C \qquad \text{Multiply both sides by } \tfrac{5}{9}.$$

$$C = \frac{5}{9}(F - 32) \qquad \text{Formula}$$

Technology: Tip

You can use a graphing utility to solve simple interest problems by using the program found at our website *www.hmco.com.* Use the program to check the results of Example 3. Then use the program and the guess, check, and revise method to find P when $I = \$5269$, $r = 11\%$, and $t = 5$ years.

An amount of \$5000 is deposited in an account paying simple interest. After 6 months, the account has earned \$162.50 in interest. What is the annual interest rate for this account?

Solution

$$I = Prt \qquad \text{Simple interest formula}$$

$$162.50 = 5000(r)\left(\tfrac{1}{2}\right) \qquad \text{Substitute for } I, P, \text{ and } t.$$

$$162.50 = 2500r \qquad \text{Simplify.}$$

$$\frac{162.50}{2500} = r \qquad \text{Divide both sides by 2500.}$$

$$0.065 = r \qquad \text{Simplify.}$$

The annual interest rate is $r = 0.065$ (or 6.5%). Check this solution in the original statement of the problem.

One of the most familiar rate problems and most often used formulas in real life is the one that relates distance, rate (or speed), and time: $d = rt$. For instance, if you are traveling at a constant (or average) rate of 50 miles per hour for 45 minutes, the total distance traveled is given by

$$\left(50\,\frac{\text{miles}}{\text{hour}}\right)\cdot\left(\frac{45}{60}\text{ hour}\right) = 37.5 \text{ miles.}$$

As with all problems involving applications, be sure to check that the units in the model make sense. For instance, in this problem the rate is given in *miles per hour*. Therefore, in order for the solution to be given in *miles*, we must convert the time (from minutes) to *hours*. In the model, you can think of canceling the two "hours," as follows.

$$\left(50\,\frac{\text{miles}}{\cancel{\text{hour}}}\right)\cdot\left(\frac{45}{60}\,\cancel{\text{hour}}\right) = 37.5 \text{ miles}$$

Example 4 A Distance-Rate-Time Problem

You can jog at an average rate of 8 kilometers per hour. How long will it take you to jog 14 kilometers?

Solution

Verbal Model: Distance = Rate · Time

Labels:
Distance = 14 (kilometers)
Rate = 8 (kilometers per hour)
Time = t (hours)

Equation:

$$14 = 8(t)$$

$$\frac{14}{8} = t$$

$$1.75 = t$$

It will take you 1.75 hours (or 1 hour and 45 minutes). Check this in the original statement of the problem.

If you are having trouble solving a distance-rate-time problem, consider making a table such as that shown below for Example 4.

Distance = Rate · Time

Rate (km/hr)	8	8	8	8	8	8	8	8
Time (hours)	0.25	0.50	0.75	1.00	1.25	1.50	1.75	2.00
Distance (kilometers)	2	4	6	8	10	12	14	16

2 Solve a mixture problem involving hidden products.

Solving Mixture Problems

Many real-world problems involve combinations of two or more quantities that make up a new or different quantity. Such problems are called **mixture problems.** They are usually composed of the sum of two or more "hidden products" that involve *rate factors*. Here is the generic form of the verbal model for mixture problems.

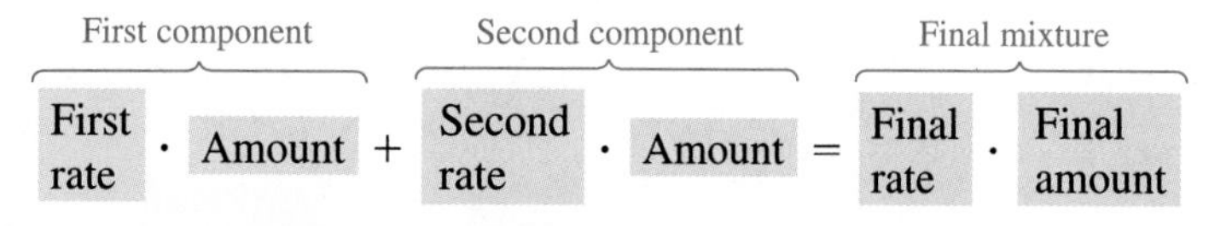

The rate factors are usually expressed as *percents* or *percent of measure* such as dollars per pound, jobs per hour, or gallons per minute.

Example 5 A Nut Mixture Problem

A grocer wants to mix cashew nuts worth \$7 per pound with 15 pounds of peanuts worth \$2.50 per pound. To obtain a nut mixture worth \$4 per pound, how many pounds of cashews are needed? How many pounds of mixed nuts will be produced for the grocer to sell?

Solution

In this problem, the rates are the *unit prices* for the nuts.

Verbal Model: Total cost of cashews + Total cost of peanuts = Total cost of mixed nuts

Labels:
Unit price of cashews = 7 (dollars per pound)
Unit price of peanuts = 2.5 (dollars per pound)
Unit price of mixed nuts = 4 (dollars per pound)
Amount of cashews = x (pounds)
Amount of peanuts = 15 (pounds)
Amount of mixed nuts = $x + 15$ (pounds)

Equation:
$$7(x) + 2.5(15) = 4(x + 15)$$
$$7x + 37.5 = 4x + 60$$
$$3x = 22.5$$
$$x = \frac{22.5}{3}$$
$$x = 7.5$$

The grocer needs 7.5 pounds of cashews. This will result in 7.5 + 15 or 22.5 pounds of mixed nuts. You can check these results as follows.

$$\underbrace{(\$7.00/\text{lb})(7.5 \text{ lb})}_{\text{Cashews}} + \underbrace{(\$2.50/\text{lb})(15 \text{ lb})}_{\text{Peanuts}} = \underbrace{(\$4.00/\text{lb})(22.5 \text{ lb})}_{\text{Mixed Nuts}}$$
$$\$52.50 + \$37.50 = \$90.00$$
$$\$90.00 = \$90.00$$

Solution checks. ✓

Example 6 A Solution Mixture Problem

A pharmacist needs to strengthen a 15% alcohol solution so that it contains 32% alcohol. How much pure alcohol should be added to 100 milliliters of the 15% solution? (See Figure 3.5.)

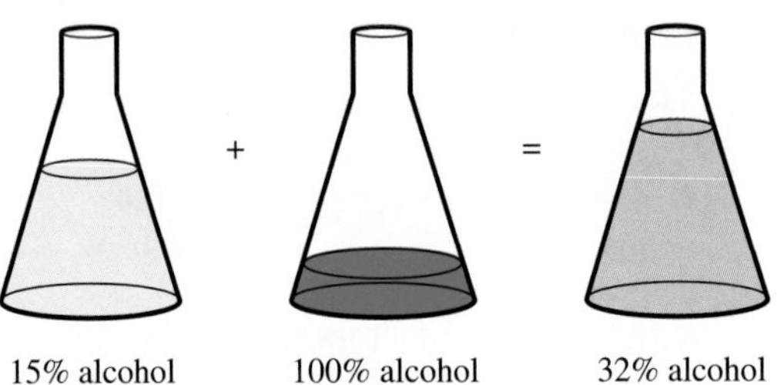

Figure 3.5

Solution

In this problem, the rates are the alcohol *percents* of the solutions.

Verbal Model: Amount of alcohol in original solution + Amount of alcohol in pure solution = Amount of alcohol in final solution

Labels:
Alcohol percent of original solution = 0.15 (% in decimal form)
Alcohol amount of original solution = 100 (milliliters)
Alcohol percent of pure solution = 1.00 (% in decimal form)
Alcohol amount of pure solution = x (milliliters)
Alcohol percent of final solution = 0.32 (% in decimal form)
Alcohol amount of final solution = $x + 100$ (milliliters)

Equation: $0.15(100) + 1.00(x) = 0.32(100 + x)$

$$15 + x = 32 + 0.32x$$

$$0.68x = 17$$

$$x = \frac{17}{0.68}$$

$$x = 25 \text{ ml}$$

The pharmacist should add 25 milliliters of pure alcohol to the original solution. You can check this in the original statement of the problem as follows.

$$\underbrace{0.15(100)}_{\text{Original}} + \underbrace{1.00(25)}_{\text{Pure}} = \underbrace{0.32(125)}_{\text{Final}}$$

$$15 + 25 = 40$$

$$40 = 40 \quad \text{Solution checks. } \checkmark$$

Remember that mixture problems are sums of two or more hidden products that involve different rates. Watch for such problems in the exercises.

3 Solve a work-rate problem.

Solving Work-Rate Problems

Although not generally referred to as such, most **work-rate problems** are actually *mixture* problems because they involve two or more rates. In work-rate problems, the work rate is the *reciprocal* of the time needed to do the entire job. For instance, if it takes 7 hours to complete a job, the per-hour work rate is

$$\frac{1}{7} \text{ job per hour.}$$

Similarly, if it takes $4\frac{1}{2}$ minutes to complete a job, the per-minute rate is

$$\frac{1}{4\frac{1}{2}} = \frac{1}{\frac{9}{2}} = \frac{2}{9} \text{ job per minute.}$$

Example 7 A Work-Rate Problem

Consider two machines in a paper manufacturing plant. Machine 1 can produce 2000 pounds of paper in 3 hours. Machine 2 is newer and can produce 2000 pounds of paper in $2\frac{1}{2}$ hours. How long will it take the two machines working together to produce 2000 pounds of paper?

Solution

Verbal Model: Work done = Portion done by machine 1 + Portion done by machine 2

Labels:

Work done = 1	(job)
Rate (machine 1) = $\frac{1}{3}$	(job per hour)
Time (machine 1) = t	(hours)
Rate (machine 2) = $\frac{2}{5}$	(job per hour)
Time (machine 2) = t	(hours)

Equation:

$$1 = \left(\frac{1}{3}\right)(t) + \left(\frac{2}{5}\right)(t)$$

$$1 = \left(\frac{1}{3} + \frac{2}{5}\right)(t)$$

$$1 = \left(\frac{11}{15}\right)(t)$$

$$\frac{15}{11} = t$$

It would take $\frac{15}{11}$ hours (or about 1.36 hours) for the machines to complete the job working together. Check this solution in the original statement of the problem.

Note in Example 7 that the "2000 pounds" of paper was unnecessary information. We simply represented the 2000 pounds as "one complete job." This unnecessary information was a red herring.

Example 8 A Fluid-Rate Problem

An above-ground swimming pool has a capacity of 15,600 gallons, as shown in Figure 3.6. A drain pipe can empty the pool in $6\frac{1}{2}$ hours. At what rate (in gallons per minute) does the water flow through the drain pipe?

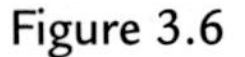

Figure 3.6

Solution

To begin, change the time from hours to minutes by multiplying by 60. That is, $6\frac{1}{2}$ hours is equal to $(6.5)(60)$ or 390 minutes.

Verbal Model: Volume of pool = Rate · Time

Labels:
Volume = 15,600 (gallons)
Rate = r (gallons per minute)
Time = 390 (minutes)

Equation:

$$15{,}600 = r(390)$$

$$\frac{15{,}600}{390} = r$$

$$40 = r$$

The water is flowing through the drain pipe at the rate of 40 gallons per minute. You can check this as follows.

$$\left(\frac{40 \text{ gallons}}{\cancel{\text{minute}}}\right)(390 \ \cancel{\text{minutes}}) = 15{,}600 \text{ gallons}$$

Discussing the Concept Creating a Formula

You are to design a package from a rectangular piece of material measuring 9 inches by 12 inches by cutting out a square from each corner and folding up the sides, as shown in the figure. Calculate the volumes of the different boxes formed by cutting out 1-inch, 2-inch, and 3-inch squares. Create a formula for the volume of a box made in this manner.

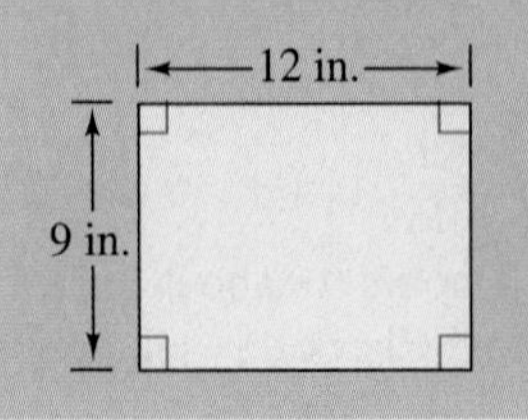

3.5 Exercises

Integrated Review — *Concepts, Skills, and Problem Solving*

Keep mathematically in shape by doing these exercises *before* the problems of this section.

Properties and Definitions

1. If n is an integer, distinguish between $2n$ and $2n + 1$.

2. Demonstrate the Addition Property of Equality for the equation

$2x - 3 = 10$.

Simplifying Expressions

In Exercises 3–10, simplify the expression.

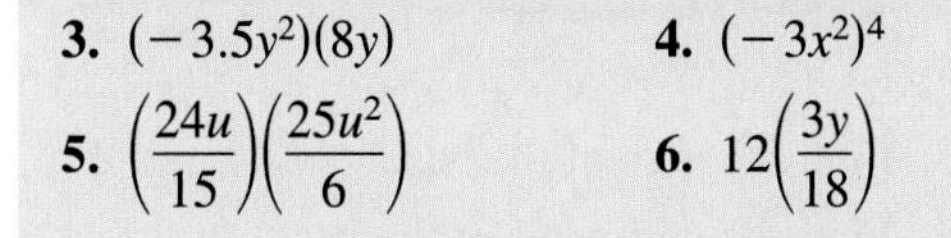

3. $(-3.5y^2)(8y)$

4. $(-3x^2)^4$

5. $\left(\frac{24u}{15}\right)\left(\frac{25u^2}{6}\right)$

6. $12\left(\frac{3y}{18}\right)$

7. $5x(2 - x) + 3x$

8. $3t - 4(2t - 8)$

9. $3(v - 4) + 7(v - 4)$

10. $5[6 - 2(x - 3)]$

Problem Solving

11. *Sales Tax* You buy a computer for $2750 and your total bill is $2915. Find the sales tax rate.

12. *Comparing Prices* A mail-order catalog lists an area rug for $109.95, plus a shipping charge of $14.25. A local store has a sale on the same rug with 20% off a list price of $139.99. Which is the better bargain?

Developing Skills

In Exercises 1–16, solve for the specified variable.

1. Solve for h: $A = \frac{1}{2}bh$

2. Solve for L: $P = 2L + 2W$

3. Solve for R: $E = IR$

4. Solve for r: $C = 2\pi r$

5. Solve for l: $V = lwh$

6. Solve for h: $V = \pi r^2 h$

7. Solve for r: $A = P + Prt$

8. Solve for L: $S = L - RL$

9. Solve for C: $S = C + RC$

10. Solve for P: $A = P\left(1 + \frac{r}{n}\right)^{nt}$

11. Solve for b: $A = \frac{1}{2}(a + b)h$

12. Solve for m_2: $F = \alpha\frac{m_1 m_2}{r^2}$

13. Solve for r: $V = \frac{1}{3}\pi h^2(3r - h)$

14. Solve for b: $V = \frac{4}{3}\pi a^2 b$

15. Solve for a: $h = v_0 t + \frac{1}{2}at^2$

16. Solve for a: $S = \frac{n}{2}[2a + (n - 1)d]$

In Exercises 17 and 18, evaluate the formula for the specified values of the variables. (List the *units* of the answer.)

17. *Volume of a Right Circular Cylinder:* $V = \pi r^2 h$
$r = 5$ meters, $h = 4$ meters

18. *Electric Power:* $I = \frac{P}{V}$

$P = 1500$ watts, $V = 110$ volts

In Exercises 19–24, find the missing distance, rate, or time. See Example 4.

	Distance, d	*Rate, r*	*Time, t*
19.		55 mi/hr	3 hr
20.		32 ft/sec	10 sec
21.	500 km	90 km/hr	
22.	128 ft	16 ft/sec	
23.	5280 ft		$\frac{5}{2}$ sec
24.	432 mi		9 hr

Solving Problems

In Exercises 25–32, use a common geometric formula to solve the problem. See Examples 1 and 2.

25. *Geometry* Each room in the floor plan of a house is square (see figure). The perimeter of the bathroom is 32 feet. The perimeter of the kitchen is 80 feet. Find the area of the living room.

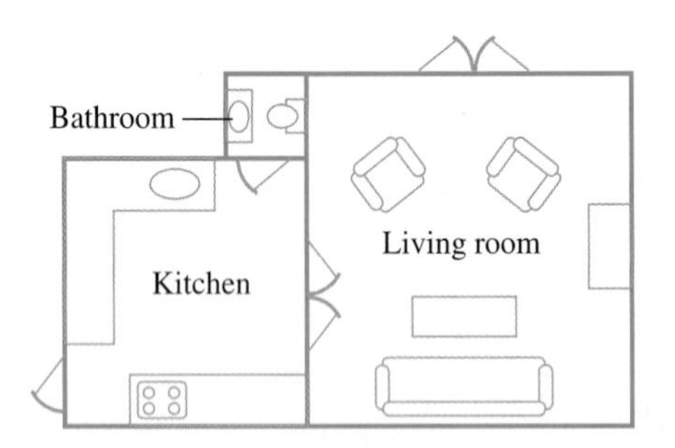

26. *Geometry* A rectangle has a perimeter of 10 feet and a width of 2 feet. Find the length of the rectangle.

27. *Geometry* A triangle has an area of 48 square meters and a height of 12 meters. Find the length of the base.

28. *Geometry* The perimeter of a square is 48 feet. Find its area.

29. *Geometry* The circumference of the wheel in the figure is 30π inches. Find the diameter of the wheel.

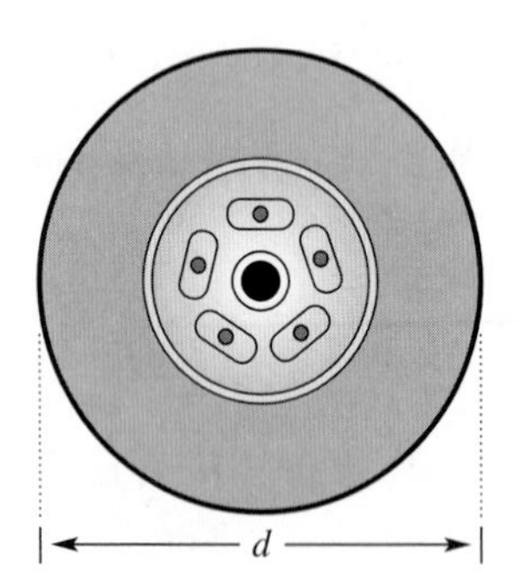

30. *Geometry* A circle has a circumference of 15 meters. What is the radius of the circle? Round your result to two decimal places.

31. *Geometry* A circle has a circumference of 25 meters. Find the radius and area of the circle. Round your results to two decimal places.

32. *Geometry* The volume of a right circular cylinder is $V = \pi r^2 h$. Find the volume of a right circular cylinder that has a radius of 2 meters and a height of 3 meters. List the units of measurement for your result.

Geometry In Exercises 33–36, use the closed rectangular box shown in the figure to answer the question.

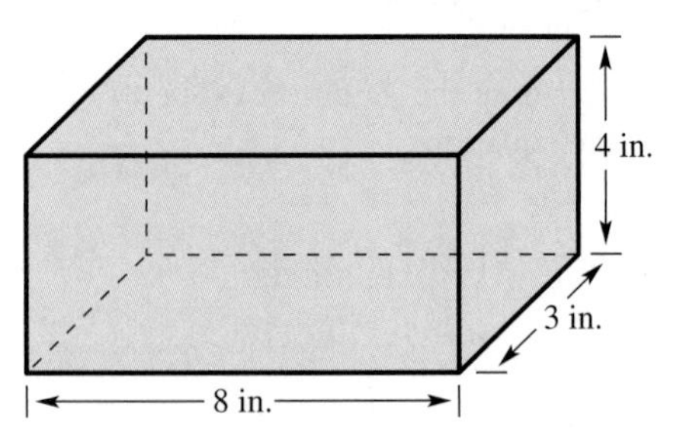

33. Find the area of the base.

34. Find the perimeter of the base.

35. Find the volume of the box.

36. Find the surface area of the box. (*Note*: This is the combined area of the six surfaces.)

Simple Interest In Exercises 37–46, use the formula for simple interest. See Example 3.

37. Find the interest on a \$1000 bond paying an annual rate of 9% for 6 years.

38. A \$1000 corporate bond pays an annual rate of $7\frac{1}{2}\%$. Find the interest on the bond if it matures in $3\frac{1}{2}$ years.

39. You borrow \$15,000 for $\frac{1}{2}$ year. You promise to pay back the principal and the interest in one lump sum. The annual interest rate is 13%. What is your payment?

40. You have a balance of \$650 on your credit card that you cannot pay this month. The annual interest rate on an unpaid balance is 19%. Find the lump sum of principal and interest due in 1 month.

41. Find the annual rate on a savings account that earns \$110 interest in 1 year on a principal of \$1000.

42. Find the annual interest rate on a certificate of deposit that earned $128.98 interest in 1 year on a principal of $1500.

43. Find the principal required to earn $408 interest in 4 years, if the annual interest rate is $8\frac{1}{2}\%$.

44. How long must $1000 be invested at an annual interest rate of $7\frac{1}{2}\%$ to earn $225 interest?

45. *Mixture Problem* Six thousand dollars is divided between two investments earning 7% and 9% simple interest. (There is more risk in the 9% fund.) Your goal is to have a total annual interest income of $500. What is the smallest amount you can invest at 9% in order to meet your objective?

46. *Mixture Problem* An inheritance of $30,000 is divided into two investments earning 8.5% and 10% simple interest, respectively. (There is more risk in the 10% fund.) Your goal is to have a total annual interest income of $2700. What is the smallest amount you can invest at 10% in order to meet your objective?

In Exercises 47–56, use the distance formula to solve the problem. See Example 4.

47. *Space Shuttle Time* The speed of the space shuttle is 17,000 miles per hour (see figure). How long will it take the shuttle to travel a distance of 3000 miles?

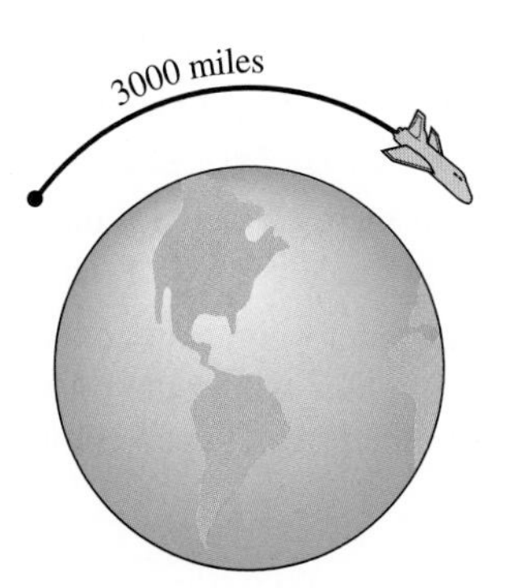

48. *Speed of Light* The speed of light is 670,616,625.6 miles per hour, and the distance between the earth and the sun is 93,000,000 miles. How long does it take light from the sun to reach the earth?

49. *Distance* Two cars start at a given point and travel in the same direction at average speeds of 45 miles per hour and 52 miles per hour (see figure). How far apart will they be in 4 hours?

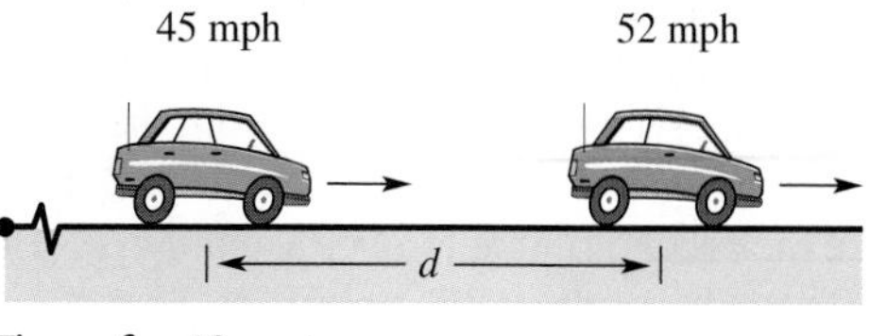

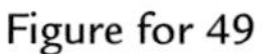

Figure for 49

50. *Distance* Two planes leave an airport at approximately the same time and fly in opposite directions (see figure). Their speeds are 510 miles per hour and 600 miles per hour. How far apart will the planes be after $1\frac{1}{2}$ hours?

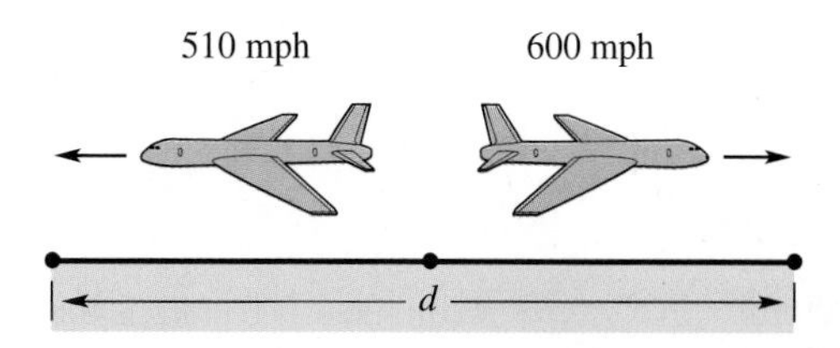

51. *Speed* Determine the average speed of an experimental plane that can travel 3000 miles in 2.6 hours.

52. *Speed* Determine the average speed of an Olympic runner who completes the 10,000-meter race in 27 minutes and 45 seconds.

53. *Time* Two cars start at the same point and travel in the same direction at average speeds of 40 miles per hour and 55 miles per hour. How much time must elapse before the two cars are 5 miles apart?

54. *Time* Suppose that on the first part of a 225-mile automobile trip you averaged 55 miles per hour. On the last part of the trip you averaged 48 miles per hour because of increased traffic congestion. The total trip took 4 hours and 15 minutes. Find the amount of time at each speed.

55. *Think About It* A truck traveled at an average speed of 60 miles per hour on a 200-mile trip to pick up a load of freight. On the return trip, with the truck fully loaded, the average speed was 40 miles per hour.

(a) Guess the average speed for the round trip.

(b) Calculate the average speed for the round trip. Is the result the same as in part (a)? Explain.

56. *Time* A jogger leaves a given point on a fitness trail running at a rate of 4 miles per hour. Ten minutes later a second jogger leaves from the same location running at 5 miles per hour. How long will it take the second runner to overtake the first? How far will each have run at that point?

In Exercises 57–60, determine the numbers of units of solutions 1 and 2 required to obtain the desired amount and percent concentration of the final solution. See Example 6.

	Concentration Solution 1	*Concentration Solution 2*	*Concentration Final Solution*	*Amount of Final Solution*
57.	10%	30%	25%	100 gal
58.	25%	50%	30%	5 L
59.	15%	45%	30%	10 qt
60.	70%	90%	75%	25 gal

61. *Number of Stamps* You have 100 stamps that have a total value of $27.80. Some of the stamps are worth 20¢ each and the others are worth 33¢ each. How many stamps of each type do you have?

62. *Number of Stamps* You have 20 stamps that have a total value of $6.08. Some of the stamps are worth 20¢ each and others are worth 33¢ each. How many stamps of each type do you have?

63. *Number of Coins* A person has 20 coins in nickels and dimes with a combined value of $1.60. Determine the number of coins of each type.

64. *Number of Coins* A person has 50 coins in dimes and quarters with a combined value of $7.70. Determine the number of coins of each type.

65. *Nut Mixture* A grocer mixes two kinds of nuts that cost $2.49 and $3.89 per pound to make 100 pounds of a mixture that costs $3.47 per pound. How many pounds of each kind of nut are put into the mixture? See Example 5.

66. *Flower Order* A floral shop receives an order for flowers that totals $384. The prices per dozen for the roses and carnations are $18 and $12, respectively. The order contains twice as many roses as carnations. How many of each type of flower are in the order?

67. *Antifreeze* The cooling system in a truck contains 4 gallons of coolant that is 30% antifreeze. How much must be withdrawn and replaced with 100% antifreeze to bring the coolant in the system to 50% antifreeze?

68. *Ticket Sales* Ticket sales for a play total $1700. The number of tickets sold to adults is three times the number sold to children. The prices of the tickets for adults and children are $5 and $2, respectively. How many of each type were sold?

69. *Interpreting a Table* An agricultural corporation must purchase 100 tons of cattle feed. The feed is to be a mixture of soybeans, which cost $200 per ton, and corn, which costs $125 per ton. Complete the following table, where x is the number of tons of corn in the mixture.

Corn, x	Soybeans, $100 - x$	Price per ton of the mixture
0		
20		
40		
60		
80		
100		

(a) How does an increase in the number of tons of corn affect the number of tons of soybeans in the mixture?

(b) How does an increase in the number of tons of corn affect the price per ton of the mixture?

(c) If there were equal weights of corn and soybeans in the mixture, how would the price of the mixture relate to the price of each component?

70. *Interpreting a Table* A metallurgist is making 5 ounces of an alloy of metal A, which costs \$52 per ounce, and metal B, which costs \$16 per ounce. Complete the following table, where x is the number of ounces of metal A in the alloy.

Metal A, x	Metal B, $5 - x$	Price per ounce of the alloy
0		
1		
2		
3		
4		
5		

(a) How does an increase in the number of ounces of metal A in the alloy affect the number of ounces of metal B in the alloy?

(b) How does an increase in the number of ounces of metal A in the alloy affect the price of the alloy?

(c) If there were equal amounts of metal A and metal B in the alloy, how would the price of the alloy relate to the price of each of the components?

71. *Work Rate* You can mow a lawn in 2 hours using a riding mower, and in 3 hours using a push mower. Using both machines together, how long will it take you and a friend to mow the lawn? See Example 7.

72. *Work Rate* One person can complete a typing project in 6 hours, and another can complete the same project in 8 hours. If they both work on the project, in how many hours can it be completed?

73. *Work Rate* One worker can complete a task in h hours while a second can complete the task in $3h$ hours. Show that by working together they can complete the task in $t = \frac{3}{4}h$ hours.

74. *Age Problem* Your age is three times that of one of your cousins. What is the age of your cousin if your combined ages total 32?

75. *Age Problem* A mother was 30 years old when her son was born. How old will the son be when his age is $\frac{1}{3}$ his mother's age?

76. *Age Problem* The difference in age between a father and daughter is 32 years. Determine the age of the father when his age is twice that of his daughter.

77. *Poll Results* One thousand people were surveyed in an opinion poll. Candidates A and B received approximately the same number of votes. Candidate C received twice as many votes as each of the other two candidates. How many votes did each candidate receive?

78. *Poll Results* One thousand people were surveyed in an opinion poll. The numbers of votes for candidates A, B, and C had ratios 5 to 3 to 2, respectively. How many people voted for each candidate?

Explaining Concepts

79. In your own words, describe the units of measure used for perimeter, area, and volume. Give some examples of each.

80. If the height of a triangle is doubled, does the area of the triangle double? Explain.

81. If the radius of a circle is doubled, does its circumference double? Does its area double? Explain.

82. It takes you 4 hours to drive 180 miles. Explain how to use mental math to find your average speed. Then explain how your method is related to the formula $d = rt$.

83. It takes you 5 hours to complete a job. What portion do you complete each hour?

84. Create a mixture problem. Exchange problems with another student and solve the problem you receive.

3.6 Linear Inequalities

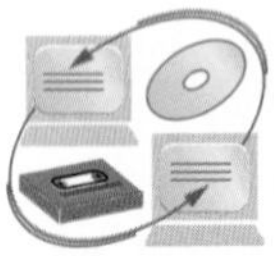

Objectives

1. Graph the solution set of an inequality on the real number line.
2. Become familiar with the properties of inequalities.
3. Solve a linear inequality and graph its solution set.
4. Solve a compound inequality and graph its solution set.
5. Translate a verbal statement into a linear inequality and solve the application problem.

1 Graph the solution set of an inequality on the real number line.

Inequalities and Their Graphs

In this section you will study **algebraic inequalities,** which are inequalities that contain one or more variable terms. Here are some examples.

$$x \le 3, \quad x \ge -2, \quad x - 5 < 2, \quad \text{and} \quad 5x - 7 < 3x + 9$$

Each of these inequalities is a **linear inequality** in the variable x because the (implied) exponent of x is 1.

As with an equation, you can **solve an inequality** in the variable x by finding all values of x for which the inequality is true. Such values are **solutions** and are said to **satisfy** the inequality. The **solution set** of an inequality is the set of all real numbers that are solutions of the inequality.

Often, the solution set of an inequality will consist of infinitely many real numbers. To get a visual image of the solution set, it is helpful to sketch its **graph** on the real number line. For instance, the graph of the solution set of $x < 2$ consists of all points on the real number line that are to the left of 2. A parenthesis is used to *exclude* an endpoint from the solution interval. A square bracket is used to *include* an endpoint in the solution interval. This is illustrated in the next example.

Example 1 Graphs of Inequalities

Inequality / *Graph of Solution Set* / *Verbal Description*

a. $x < 2$

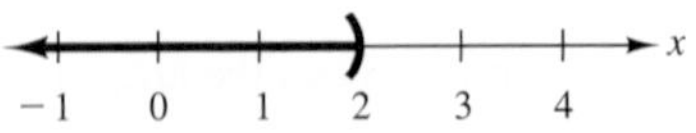

x is less than 2.

b. $x \ge -2$

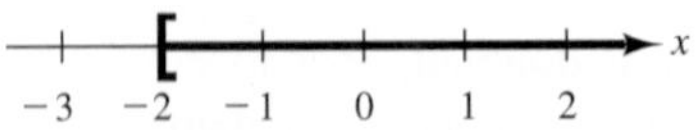

x is greater than or equal to -2.

c. $-1 \le x \le 2$

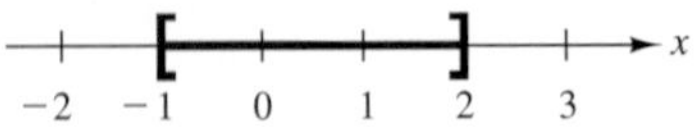

x is greater than or equal to -1 *and* less than or equal to 2.

d. $2 \le x < 5$

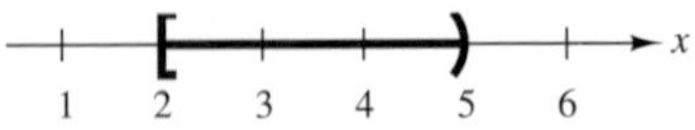

x is greater than or equal to 2 *and* less than 5.

e. $-3 < x \le -1$

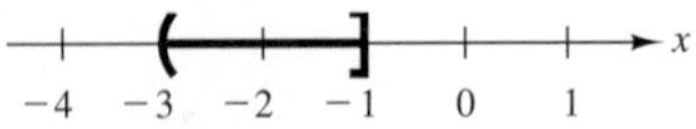

x is greater than -3 *and* less than or equal to -1.

2 Become familiar with the properties of inequalities.

Properties of Inequalities

The procedures for solving linear inequalities in one variable are much like those for solving linear equations. To isolate the variable, you can use the **properties of inequalities.** These properties are similar to the properties of equality, but there are two important exceptions. *When both sides of an inequality are multiplied or divided by a negative number, the direction of the inequality symbol must be reversed.* Here is an example.

$-2 < 5$ Original inequality

$(-3)(-2) > (-3)(5)$ Multiply both sides by -3 and reverse inequality.

$6 > -15$ Simplify.

Two inequalities that have the same solution set are called **equivalent.** The following list describes operations that can be used to create equivalent inequalities.

Linear inequalities can be graphed using a graphing utility. The inequality $x > -2$ is shown in the graph below. Notice that the graph appears above the x-axis. Consult the user's manual of your graphing utility for directions.

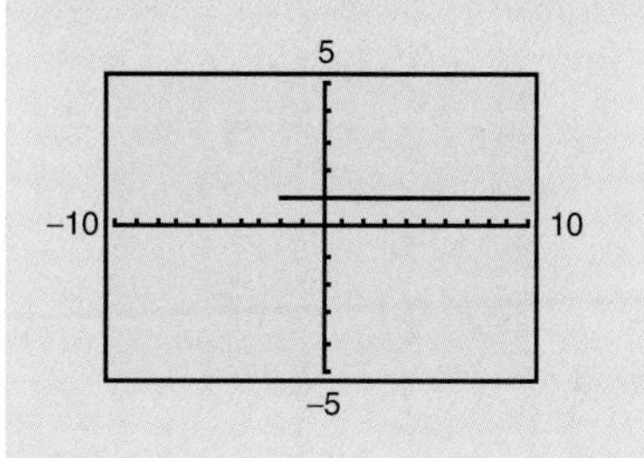

▶ Properties of Inequalities

Let a, b, and c be real numbers, variables, or algebraic expressions.

Property	*Verbal and Algebraic Descriptions*
Addition:	Add the same quantity to both sides.
	If $a < b$, then $a + c < b + c$.
Subtraction:	Subtract the same quantity from both sides.
	If $a < b$, then $a - c < b - c$.
Multiplication:	Multiply both sides by a *positive* quantity.
	If $a < b$ and c is positive, then $ac < bc$.
	Multiply both sides by a *negative* quantity and reverse the inequality symbol.
	If $a < b$ and c is negative, then $ac > bc$.
Division:	Divide both sides by a *positive* quantity.
	If $a < b$ and c is positive, then $\frac{a}{c} < \frac{b}{c}$.
	Divide both sides by a *negative* quantity and reverse the inequality symbol.
	If $a < b$ and c is negative, then $\frac{a}{c} > \frac{b}{c}$.
Transitive:	If $a < b$ and $b < c$, then $a < c$.

Each of the properties above is true if the symbol $<$ is replaced by $\leq$ and the symbol $>$ is replaced by $\geq$. Moreover, the letters a, b, and c can be real numbers, variables, or algebraic expressions. Note that you cannot multiply or divide both sides of an inequality by zero.

3 Solve a linear inequality and graph its solution set.

Solving Inequalities

The solution set of a linear inequality can be written in set notation. For the solution $x > 1$, the set notation is $\{x|x > 1\}$ and is read "the set of all x such that x is greater than 1."

In Examples 4 and 5, pay special attention to the steps in which the inequality symbol is reversed. Remember that when you multiply or divide an inequality by a negative number, you must reverse the inequality symbol.

Study Tip

Checking the solution set of an inequality is not as simple as checking the solution set of an equation. (There are usually too many x-values to substitute back into the original inequality.) You can, however, get an indication of the validity of a solution set by substituting a few convenient values of x. For instance, in Example 2, the solution of $x + 5 < 8$ was found to be $x < 3$. Try checking that $x = 0$ satisfies the original inequality, whereas $x = 4$ does not.

Example 2 Solving a Linear Inequality

Solve and graph the inequality $x + 5 < 8$.

Solution

$x + 5 < 8$ Original inequality

$x + 5 - 5 < 8 - 5$ Subtract 5 from both sides.

$x < 3$ Solution set

The solution set is $x < 3$ or, in set notation, $\{x|x < 3\}$. The graph of the solution set is shown in Figure 3.7.

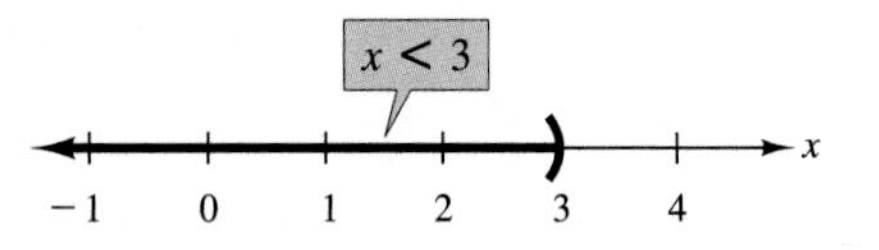

Figure 3.7 *All real numbers that are less than 3*

Example 3 Solving a Linear Inequality

Solve and graph the inequality $3y - 1 \le -7$.

Solution

$3y - 1 \le -7$ Original inequality

$3y - 1 + 1 \le -7 + 1$ Add 1 to both sides.

$3y \le -6$ Combine like terms.

$\frac{3y}{3} \le \frac{-6}{3}$ Divide both sides by (positive) 3.

$y \le -2$ Solution set

The solution set is $y \le -2$ or, in set notation, $\{y|y \le -2\}$. The graph of the solution set is shown in Figure 3.8.

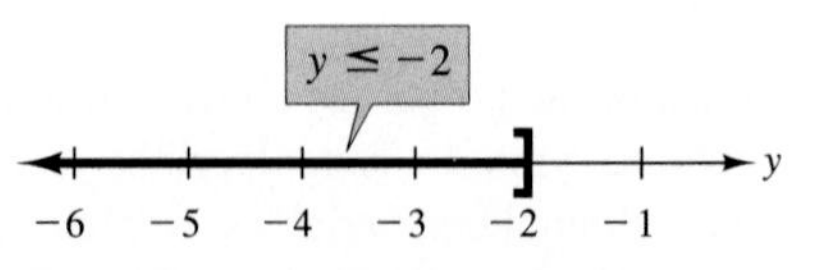

Figure 3.8 *All real numbers that are less than or equal to −2*

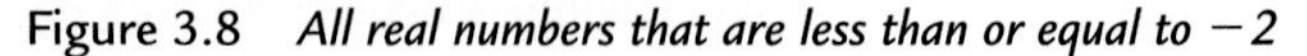

Example 4 Solving a Linear Inequality

Solve and graph the inequality $12 - 2x > 10$.

Solution

$$\begin{aligned} 12 - 2x &> 10 && \text{Original inequality} \\ 12 - 12 - 2x &> 10 - 12 && \text{Subtract 12 from both sides.} \\ -2x &> -2 && \text{Combine like terms.} \\ \frac{-2x}{-2} &< \frac{-2}{-2} && \text{Divide both sides by } -2 \text{ and reverse inequality.} \\ x &< 1 && \text{Solution set} \end{aligned}$$

The solution set is $x < 1$ or, in set notation, $\{x | x < 1\}$. The graph of the solution set is shown in Figure 3.9.

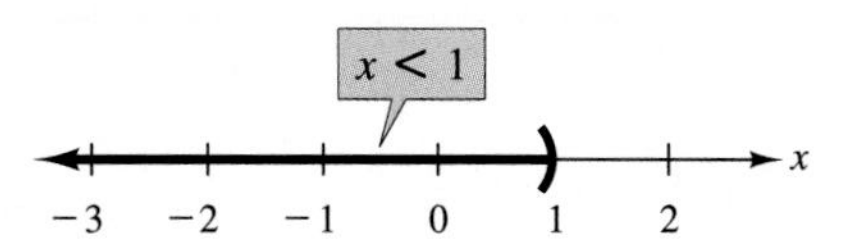

Figure 3.9 *All real numbers that are less than 1*

Example 5 Solving a Linear Inequality

Solve and graph the inequality $1 - \frac{3x}{2} \geq x - 4$.

Solution

$$\begin{aligned} 1 - \frac{3x}{2} &\geq x - 4 && \text{Original inequality} \\ 2 - 3x &\geq 2x - 8 && \text{Multiply both sides by 2.} \\ -3x &\geq 2x - 10 && \text{Subtract 2 from both sides.} \\ -5x &\geq -10 && \text{Subtract } 2x \text{ from both sides.} \\ x &\leq 2 && \text{Divide both sides by } -5 \text{ and reverse inequality.} \end{aligned}$$

The solution set is $x \leq 2$ or, in set notation, $\{x | x \leq 2\}$. The graph of the solution set is shown in Figure 3.10.

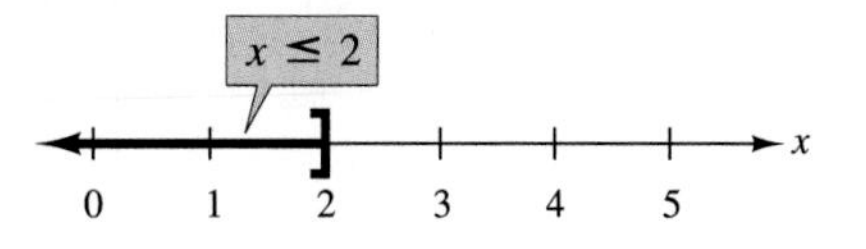

Figure 3.10 *All real numbers that are less than or equal to 2*

4 Solve a compound inequality and graph its solution set.

Solving a Compound Inequality

Two inequalities joined by the word *and* or *or* constitute a **compound inequality.** When the two inequalities are joined by the word *and*, the solution set consists of all real numbers that satisfy *both* inequalities. The solution set for the compound inequality

$x \geq -2$ and $x \leq 3$ Compound inequality

is all real numbers greater than or equal to -2 *and* less than or equal to 3. This compound inequality can be written more simply as the **double inequality**

$-2 \leq x \leq 3.$ Double inequality

The graph of the solution set is shown in Figure 3.11.

Study Tip

Compound inequalities formed by the word *and* are called **conjunctive** and they are the only kind that have the potential to form a double inequality. Compound inequalities joined by the word *or* are called **disjunctive** and they cannot be reformed into double inequalities.

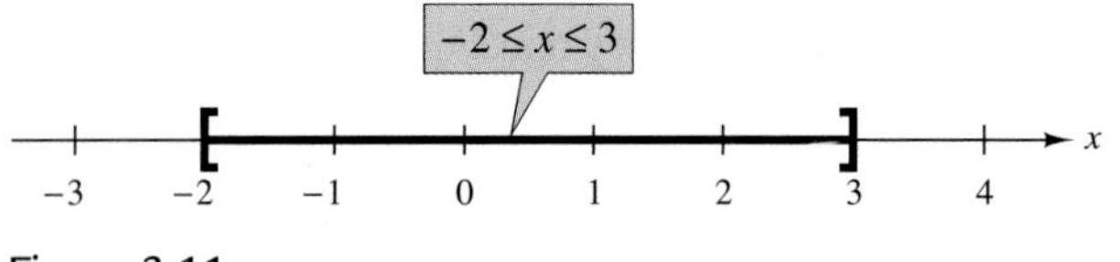

Figure 3.11

When the two inequalities are joined by the word *or*, the solution set consists of all real numbers that satisfy *either* inequality. The solution set for the compound inequality

$x < -1$ or $x \geq 4$ Compound inequality

is all real numbers less than -1 *or* greater than or equal to 4. The graph of the solution set is shown in Figure 3.12.

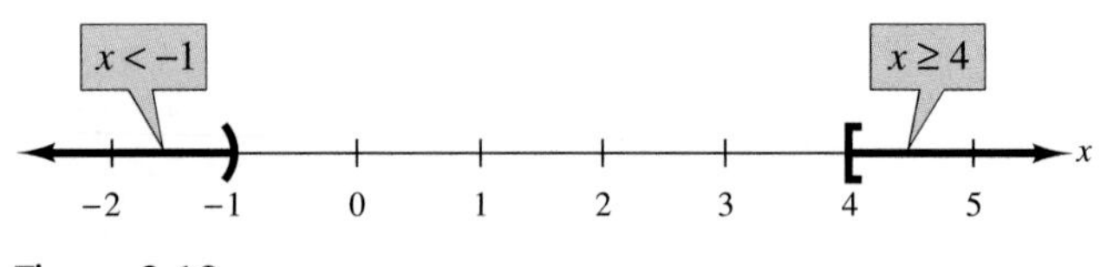

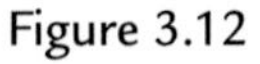
Figure 3.12

Example 6 Solving a Double Inequality

Solve the compound inequality $5 < 2x \leq 8$.

Solution

$5 < 2x \leq 8$ Original inequality

$\frac{5}{2} < \frac{2}{2}x \leq \frac{8}{2}$ Divide all parts by 2.

$\frac{5}{2} < x \leq 4$ Solution set

The solution set is $\frac{5}{2} < x \leq 4$ or, in set notation, $\{x \mid \frac{5}{2} < x \leq 4\}$. The graph of the solution set is shown in Figure 3.13.

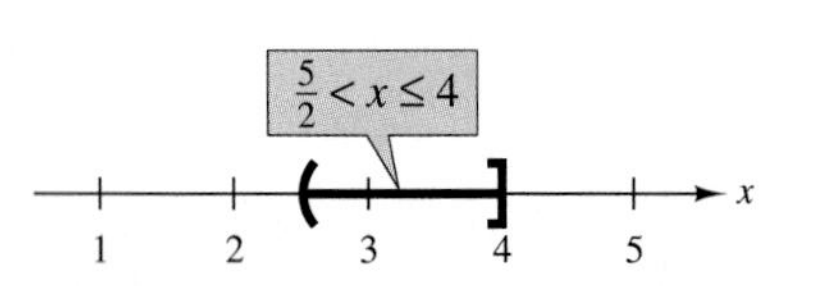

Figure 3.13 *All real numbers that are greater than $\frac{5}{2}$ and less than or equal to 4*

Compound inequalities can be written using *set notation.* In set notation, the word *and* is represented by the symbol $\cap$, which is read as **intersection.** The word *or* is represented by the symbol $\cup$, which is read as **union.** A graphical representation is shown in Figure 3.14.

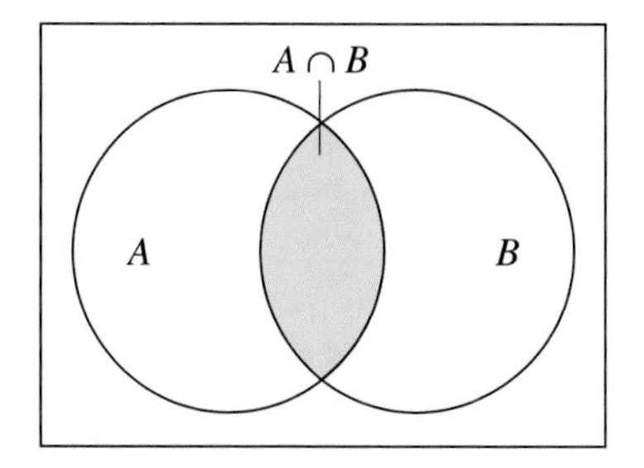

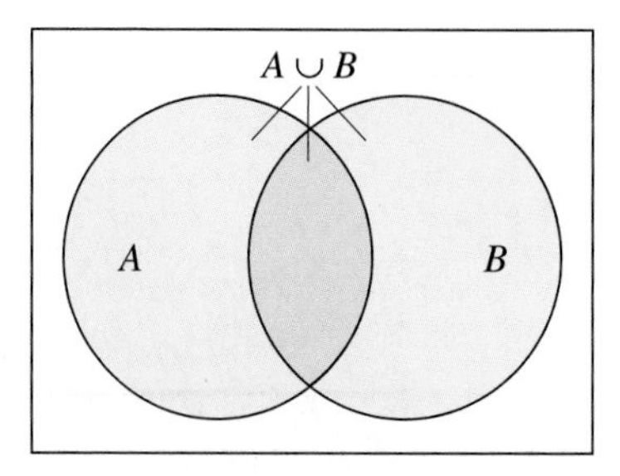

Intersection of two sets *Union of two sets*

Figure 3.14

If A and B are sets, then x is in $A \cap B$ if it is in both A *and* B. Similarly, x is in $A \cup B$ if it is in A *or* B, or possibly in both.

Example 7 Writing a Compound Inequality Using Union

A solution set is shown on the number line in Figure 3.15.

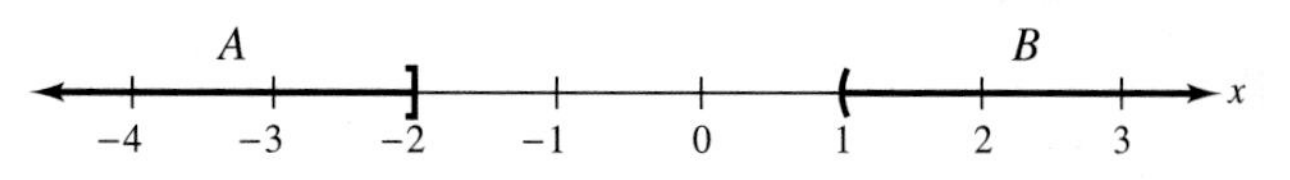

Figure 3.15

a. Write the solution set as a compound inequality.

b. Write the solution set using set notation and union.

Solution

a. As a compound inequality, you can write the solution set as $x \le -2$ *or* $x > 1$.

b. Using set notation, you can write the left interval as $A = \{x \mid x \le -2\}$ and the right interval as $B = \{x \mid x > 1\}$. So, using the union symbol, the entire solution set can be written as $A \cup B$.

Example 8 Writing a Compound Inequality Using Intersection

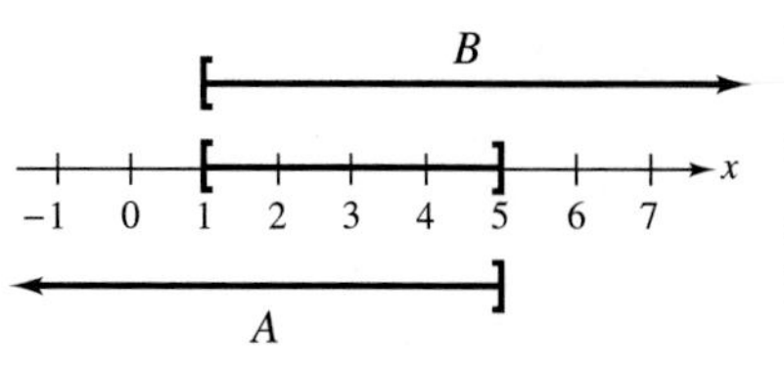

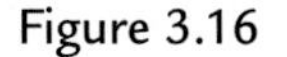

Figure 3.16

Write the compound inequality $1 \le x \le 5$ using set notation and intersection.

Solution

Consider the two sets $A = \{x \mid x \le 5\}$ and $B = \{x \mid x \ge 1\}$. These two sets overlap, as shown on the number line in Figure 3.16. The compound inequality $1 \le x \le 5$ consists of all numbers that are in $x \le 5$ *and* $x \ge 1$, which means that it can be written as $A \cap B$.

Example 9 Solving a Conjunctive Inequality

Solve the compound inequality $-3 \le 6x - 1$ and $6x - 1 < 3$.

Solution

Begin by rewriting the compound inequality as $-3 \le 6x - 1 < 3$.

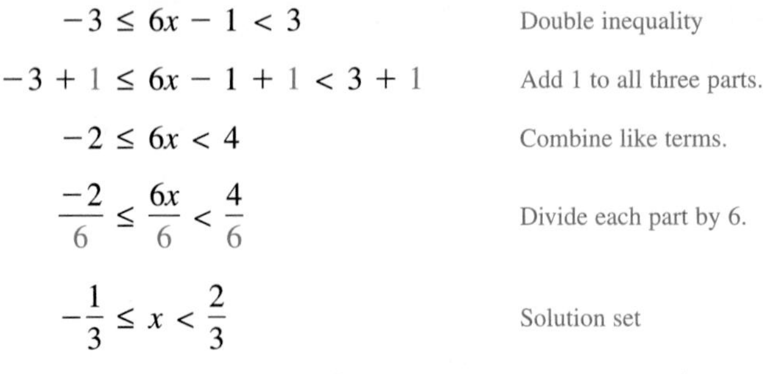

$$-3 \le 6x - 1 < 3 \qquad \text{Double inequality}$$

$$-3 + 1 \le 6x - 1 + 1 < 3 + 1 \qquad \text{Add 1 to all three parts.}$$

$$-2 \le 6x < 4 \qquad \text{Combine like terms.}$$

$$\frac{-2}{6} \le \frac{6x}{6} < \frac{4}{6} \qquad \text{Divide each part by 6.}$$

$$-\frac{1}{3} \le x < \frac{2}{3} \qquad \text{Solution set}$$

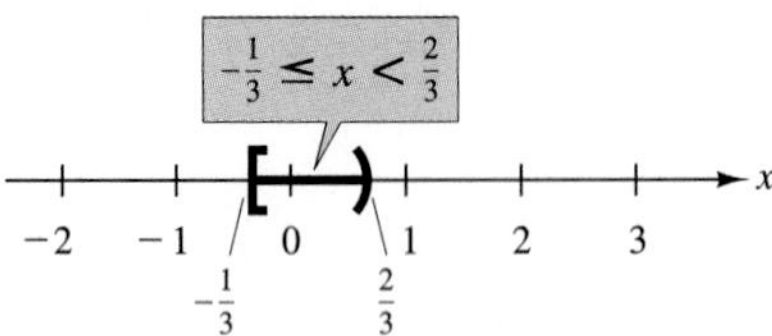

Figure 3.17 *All real numbers that are greater than or equal to $-\frac{1}{3}$ and less than $\frac{2}{3}$*

The solution set is $-\frac{1}{3} \le x < \frac{2}{3}$ or, in set notation, $\{x \mid -\frac{1}{3} \le x < \frac{2}{3}\}$. The graph of the solution set is shown in Figure 3.17.

The double inequality in Example 9 could have been solved in two parts as follows.

$$-3 \le 6x - 1 \quad \text{and} \quad 6x - 1 < 3$$

$$-2 \le 6x \qquad\qquad 6x < 4$$

$$-\frac{1}{3} \le x \qquad\qquad x < \frac{2}{3}$$

The solution set consists of all real numbers that satisfy *both* inequalities. In other words, the solution set is the set of all values of x for which $-\frac{1}{3} \le x < \frac{2}{3}$.

Example 10 Solving a Disjunctive Inequality

Solve the compound inequality

$$-2x + 3 < -5 \quad \text{or} \quad -2x + 3 > 5.$$

Solution

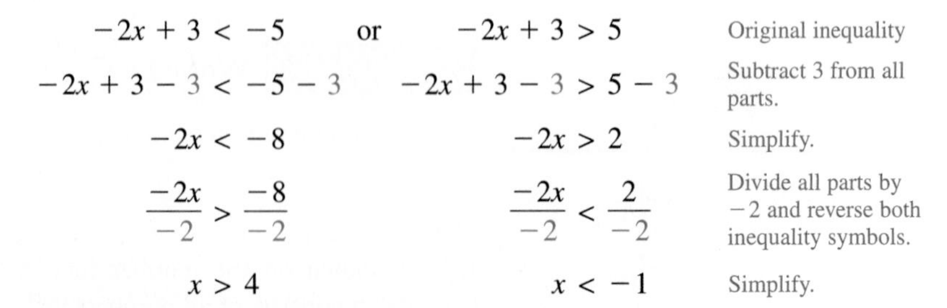

$$-2x + 3 < -5 \quad \text{or} \quad -2x + 3 > 5 \qquad \text{Original inequality}$$

$$-2x + 3 - 3 < -5 - 3 \qquad -2x + 3 - 3 > 5 - 3 \qquad \text{Subtract 3 from all parts.}$$

$$-2x < -8 \qquad -2x > 2 \qquad \text{Simplify.}$$

$$\frac{-2x}{-2} > \frac{-8}{-2} \qquad \frac{-2x}{-2} < \frac{2}{-2} \qquad \text{Divide all parts by } -2 \text{ and reverse both inequality symbols.}$$

$$x > 4 \qquad x < -1 \qquad \text{Simplify.}$$

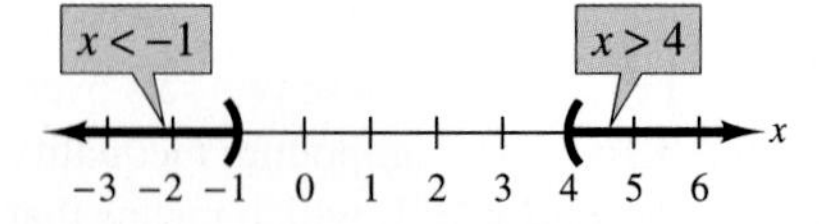

Figure 3.18 *All real numbers that are less than −1 or greater than 4*

The solution set is $x < -1$ *or* $x > 4$ or, in set notation, $\{x \mid x < -1 \text{ or } x > 4\}$. The graph of the solution set is shown in Figure 3.18.

5 Translate a verbal statement into a linear inequality and solve the application problem.

Applications

Before looking at applications, we give some examples of the translations of verbal statements into inequalities. Study the meanings of the key phrases in the next example.

Example 11 Translating Verbal Statements

Verbal Statement	*Inequality*
a. x is at most 2.	$x \le 2$
b. x is no more than 2.	$x \le 2$
c. x is at least 2.	$x \ge 2$
d. x is more than 2.	$x > 2$
e. x is less than 2.	$x < 2$

When translating inequalities, remember that "at most" means "less than or equal to," and "at least" means "greater than or equal to." Also, be sure to distinguish between the *sum* "2 more than a number" $(x + 2)$ and the *inequality* "2 is more than a number" $(2 > x)$. It is generally preferable to read an inequality from left to right.

Example 12 Course Grade 

Suppose you are taking a college course in which your grade is based on six 100-point exams. To earn an A in the course, you must have a total of at least 90% of the points. On the first five exams, your scores were 85, 92, 88, 96, and 87. How many points do you have to obtain on the sixth test in order to earn an A in the course?

Solution

Verbal Model: Total points $\ge$ 90% of 600

Labels: Score for sixth exam $= x$ (points)
Total points $= (85 + 92 + 88 + 96 + 87) + x$ (points)

Inequality:
$$(85 + 92 + 88 + 96 + 87) + x \ge 0.9(600)$$
$$448 + x \ge 540$$
$$x \ge 540 - 448$$
$$x \ge 92$$

You must get at least 92 points on the sixth exam to earn an A in the course. Check this solution in the original statement of the problem.

Example 13 Car Rental

Michael Newman/PhotoEdit

In 1996, $14.5 billion was spent on passenger car rentals in the United States. (Source: U.S. Bureau of the Census)

A subcompact car can be rented from Company A for \$190 per week with no extra charge for mileage. A similar car can be rented from Company B for \$100 per week, plus 20¢ for each mile driven. How many miles must you drive in a week to make the rental fee for Company A less than that for Company B?

Solution

Verbal Model: [Weekly cost for A] < [Weekly cost for B]

Labels:
Number of miles driven in 1 week $= m$ (miles)
Weekly cost for A $= 190$ (dollars)
Weekly cost for B $= 100 + 0.2m$ (dollars)

Inequality:
$$190 < 100 + 0.2m$$
$$90 < 0.2m$$
$$450 < m$$

Note that the inequality $450 < m$ is equivalent to writing $m > 450$. So, the car from Company A is cheaper if you plan to drive more than 450 miles in a week. The table confirms this conclusion.

Miles driven	447	448	449	450	451	452	453
Company A	\$190.00	\$190.00	\$190.00	\$190.00	\$190.00	\$190.00	\$190.00
Company B	\$189.40	\$189.60	\$189.80	\$190.00	\$190.20	\$190.40	\$190.60

Discussing the Concept — Misuse of a Compound Inequality

Suppose you are to find the solution set of an inequality that says: "$x - 2$ is greater than 3 and less than -3." Determine if it is appropriate to write this inequality in the form

$$3 < x - 2 < -3.$$

If not, explain why. Then find the solution set using appropriate steps.

3.6 Exercises

Integrated Review *Concepts, Skills, and Problem Solving*

Keep mathematically in shape by doing these exercises *before* the problems of this section.

Properties and Definitions

1. Name the property illustrated by $3x(x + 1) = 3x^2 + 3x$.

2. Complete the Associative Property: $(x + 2) - 4 =$ ______.

3. If $a < 0$, then $|a| =$ ______.

4. If $a < 0$ and $b > 0$, then $a \cdot b$ ______ 0.

In Exercises 5–8, place the correct inequality symbol between the two real numbers.

5. $-\frac{1}{2}$ ______ -7 **6.** $-\frac{1}{3}$ ______ $-\frac{1}{6}$

7. $-\pi$ ______ -3 **8.** -6 ______ $-\frac{13}{2}$

Graphs and Models

In Exercises 9 and 10, write expressions for the perimeter and area of the triangle. Then simplify the expressions.

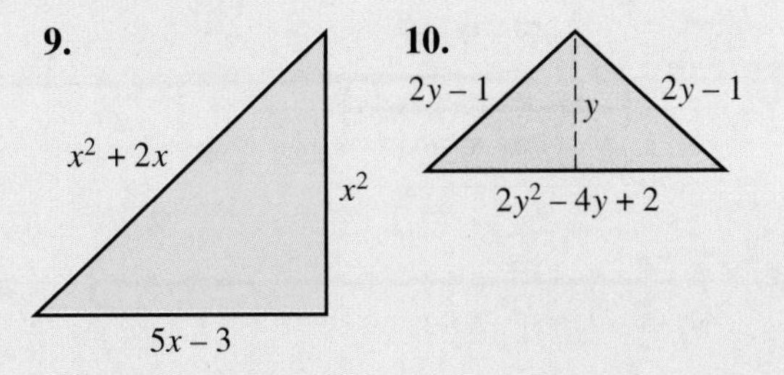

Problem Solving

11. A company had a first-quarter loss of \$312,500, a second-quarter profit of \$275,500, a third-quarter profit of \$297,750, and a fourth-quarter profit of \$71,300. What was the profit for the year?

12. A family on vacation traveled 371 miles in 7 hours. Determine their average speed.

Developing Skills

In Exercises 1–6, describe the inequality verbally and sketch its graph. See Example 1.

1. $x \geq 3$ **2.** $z > 8$

3. $x \leq 10$ **4.** $-3 < x < 4$

5. $-\frac{3}{2} < y \leq 5$ **6.** $-3 \geq t > -3.8$

In Exercises 7–14, determine whether the value of x is a solution of the inequality.

Inequality	Values	
7. $5x - 12 > 0$	(a) $x = 3$	(b) $x = -3$
	(c) $x = \frac{5}{2}$	(d) $x = \frac{3}{2}$
8. $2x + 1 < 3$	(a) $x = 0$	(b) $x = 4$
	(c) $x = -4$	(d) $x = -3$
9. $3 - \frac{1}{2}x > 0$	(a) $x = 10$	(b) $x = 6$
	(c) $x = -\frac{3}{4}$	(d) $x = 0$
10. $\frac{2}{3}x + 4 < 6$	(a) $x = 7$	(b) $x = 0$
	(c) $x = -\frac{1}{2}$	(d) $x = 3$
11. $0 < \frac{x - 2}{4} < 2$	(a) $x = 4$	(b) $x = 10$
	(c) $x = 0$	(d) $x = \frac{7}{2}$
12. $-1 < \frac{3 - x}{2} \leq 1$	(a) $x = 0$	(b) $x = 3$
	(c) $x = 1$	(d) $x = 5$
13. $-12 \leq 3(x + 4) \leq 6$	(a) $x = -5$	(b) $x = -1$
	(c) $x = -7$	(d) $x = \frac{2}{3}$
14. $0 \leq 2(x - 4) \leq 12$	(a) $x = 0$	(b) $x = 15$
	(c) $x = 5$	(d) $x = 10$

In Exercises 15–20, match the inequality with its graph. [The graphs are labeled (a), (b), (c), (d), (e), and (f).]

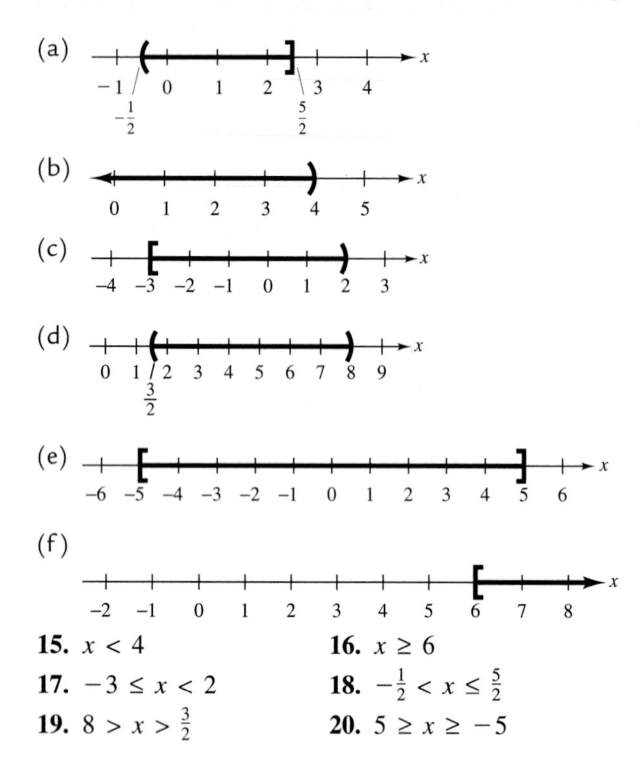

15. $x < 4$

16. $x \geq 6$

17. $-3 \leq x < 2$

18. $-\frac{1}{2} < x \leq \frac{5}{2}$

19. $8 > x > \frac{3}{2}$

20. $5 \geq x \geq -5$

In Exercises 21–58, solve and graph the inequality. See Examples 2–6, 9, and 10.

21. $t - 3 \geq 2$

22. $t + 1 < 6$

23. $x + 4 \leq 6$

24. $z - 2 > 0$

25. $4x < 12$

26. $2x > 3$

27. $-10x < 40$

28. $-6x > 18$

29. $\frac{2}{3}x \leq 12$

30. $-\frac{5}{8}x \geq 10$

31. $2x - 5 > 7$

32. $3x + 2 \leq 14$

33. $4 - 2x < 3$

34. $14 - 3x > 5$

35. $2x - 5 > -x + 6$

36. $25x + 4 \leq 10x + 19$

37. $6 < 3(y + 1) - 4(1 - y)$

38. $8 > 2(3x - 1) - (7x - 3)$

39. $-2(z + 1) \geq 3(z + 1)$

40. $8(t - 3) < 4(t - 3)$

41. $10(1 - y) < -4(y - 2)$

42. $6(3 - z) \geq 5(3 + z)$

43. $\frac{x}{4} + \frac{1}{2} > 0$

44. $\frac{y}{4} - \frac{5}{8} < 2$

45. $\frac{x}{5} - \frac{x}{2} \leq 1$

46. $\frac{x}{3} + \frac{x}{4} \geq 1$

47. $1 < 2x + 3 < 9$

48. $-9 \leq -3x + 6 < 12$

49. $-4 < 2x - 3 < 4$

50. $0 \leq 4x + 3 < 5$

51. $6 > \frac{x - 2}{-3} > -2$

52. $-2 < \frac{x - 4}{-2} \leq 3$

53. $\frac{3}{4} > x + 1 > \frac{1}{4}$

54. $-\frac{1}{3} < x - 2 < \frac{1}{4}$

55. $-5 < 2x + 3$ and $2x + 3 \leq 9$

56. $-1 \leq 5x - 11$ and $5x - 11 < 4$

57. $4 - 3x < -8$ or $4 - 3x > 7$

58. $2x - 7 \leq -12$ or $2x - 7 > -1$

In Exercises 59–66, write the compound inequality using set notation and intersection or union. See Examples 7 and 8.

59. $x < -5$ or $x > 3$

60. $x \leq 1$ or $x \geq 2$

61. $x < 6$ and $x > 0$

62. $x > 4$ and $x < 10$

63. $-3 \leq x \leq 7$

64. $12 > x > 6$

65. $x < 7$ or $x > 8$

66. $x \geq -4$ or $x \leq -10$

In Exercises 67–76, translate the verbal statement into a linear inequality. See Example 11.

67. x is nonnegative.

68. P is no more than 2.

69. y is more than -6.

70. z is at least 3.

71. x is at least 4.

72. t is less than 8.

73. y is no more than 25.

74. x is greater than or equal to -2 and less than 5.

75. x is greater than 0 and less than or equal to 6.

76. x is greater than -4 and less than or equal to 3.

Solving Problems

77. *Planet Distances* Mars is farther from the sun than Venus, and Venus is farther from the sun than Mercury. What can be said about the relationship between the distances of Mars and Mercury from the sun? Identify the property of inequalities that is demonstrated.

78. *Budgets* Department A's budget is less than Department B's budget, and Department B's budget is less than Department C's budget. What can you say about the relationship between the budgets of Departments A and C? Identify the property of inequalities that is demonstrated.

79. *Cellular Phone Cost* The cost of a cellular phone call is \$0.46 for the first minute and \$0.31 for each additional minute. The total cost of the call cannot exceed \$4. Find the interval of time that is available for the call.

80. *Budget for a Trip* You have \$2500 budgeted for a trip. The transportation for the trip will cost \$900. To stay within your budget, all other costs must be no more than what amount?

81. *Annual Operating Budget* A utility company has a fleet of vans. The annual operating cost C (in dollars) per van is $C = 0.45m + 3200$ where m is the number of miles traveled by a van in a year. What number of miles will yield an annual operating cost per van that is less than \$12,000?

82. *Profit* The revenue for selling x units of a product is $R = 115.95x$. The cost of producing x units is $C = 95x + 750$. In order to obtain a profit, the revenue must be greater than the cost. For what values of x will this product produce a profit?

83. *Geometry* The lengths of the sides of the triangle in the figure are a, b, and c. Find the inequality that relates $a + b$ and c.

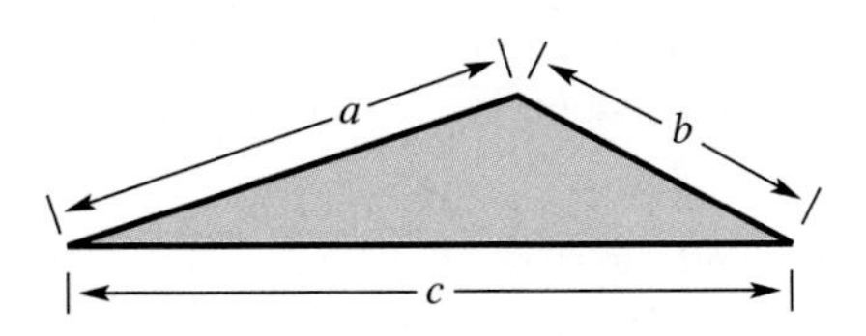

84. *Cargo Weight* The weight of a truck is 4350 pounds. The legal gross weight of the loaded truck is 6000 pounds. Find an interval for the number of bushels of grain that the truck can haul if each bushel weighs 48 pounds.

85. *Distance* The minimum and maximum speeds on an interstate highway are 45 miles per hour and 65 miles per hour. You travel nonstop for 4 hours on this highway. Assuming that you stay within the speed limits, give an interval for the distance you traveled.

86. *Comparing Distances* You live 3 miles from college and 2 miles from the business where you work (see figure). Let d represent the distance between your work and the college. Write an inequality involving d.

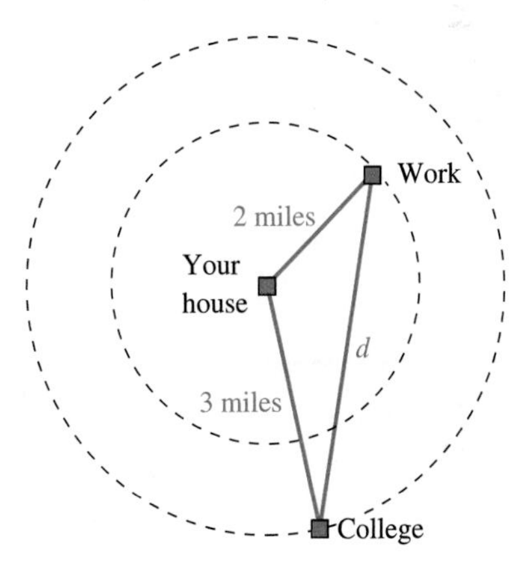

Explaining Concepts

87. Answer part (h) of Motivating the Chapter on page 115.

88. Give a verbal description of each of the symbols $<, \leq, >, \geq,$ and $=$.

89. Is adding -5 to both sides of an inequality the same as subtracting 5 from both sides? Explain.

90. Is dividing both sides of an inequality by 5 the same as multiplying both sides by $\frac{1}{5}$? Explain.

91. How many numbers are in the solution set of a linear inequality? Give an example.

92. Explain the effect on an inequality when both sides are multiplied or divided by a negative number. Give examples demonstrating your explanation.

93. Compare solving linear equations to solving linear inequalities.

94. Write an inequality symbol equivalent to $\not<$.

95. Write an inequality symbol equivalent to $\not\geq$.

In Exercises 96–99, determine whether the statement is true or false. Explain your answer.

96. The inequality $x + 6 > 0$ is equivalent to $x > -6$.

97. The inequality $-\frac{1}{2}x + 6 > 0$ is equivalent to $x > 12$.

98. The statement that z is nonnegative is equivalent to the inequality $z > 0$.

99. The statement that u is at least 10 is equivalent to the inequality $u \geq 10$.

3.7 Absolute Value Equations and Inequalities

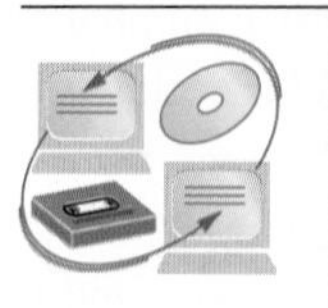

Objectives

1 Solve an equation involving absolute value.

2 Solve an inequality involving absolute value.

1 Solve an equation involving absolute value.

Solving Equations Involving Absolute Value

Consider the **absolute value equation**

$$|x| = 4.$$

The only solutions of this equation are -4 and 4, because these are the only two real numbers whose distance from zero is 4. (See Figure 3.19.) In other words, the absolute value equation $|x| = 4$ has exactly two solutions: $x = -4$ and $x = 4$.

Figure 3.19

Study Tip

Recall from Section 1.1 that the absolute value of a difference, $|a - b|$, denotes the distance between the real numbers a and b. So, $|x|$ can be written as $|x - 0|$ and means the distance between x and 0. Similarly, $|x - 2|$ denotes the distance between x and 2.

▶ Solving an Absolute Value Equation

Let x be a variable or a variable expression and let a be a real number such that $a \geq 0$. The solutions of the equation $|x| = a$ are given by $x = -a$ and $x = a$. That is,

$$|x| = a \quad \Rightarrow \quad x = -a \quad \text{and} \quad x = a.$$

Example 1 Solving Absolute Value Equations

Solve each absolute value equation.

a. $|x| = 8$ **b.** $|x| = 0$ **c.** $|y| = -2$

Solution

a. This equation is equivalent to the two linear equations

$$x = -8 \quad \text{and} \quad x = 8. \qquad \text{Equivalent linear equations}$$

So, the absolute value equation has two solutions: -8 and 8.

b. This equation is equivalent to the two linear equations

$$x = 0 \quad \text{and} \quad x = 0. \qquad \text{Equivalent linear equations}$$

Because both equations are the same, you can conclude that the absolute value equation has only one solution: 0.

c. This absolute value equation has *no solution* because it is not possible for the absolute value of a real number to be negative.

Study Tip

The strategy for solving absolute value equations is to *rewrite* the equation in *equivalent forms* that can be solved by previously learned methods. This is a common strategy in mathematics. That is, when you encounter a new type of problem, you try to rewrite the problem so that it can be solved by techniques you already know.

Example 2 Solving an Absolute Value Equation

Solve $|4x - 5| = 15$.

Solution

$|4x - 5| = 15$		Original equation
$4x - 5 = -15$ or	$4x - 5 = 15$	Equivalent equations
$4x = -10$	$4x = 20$	Add 5 to both sides.
$x = -\frac{5}{2}$	$x = 5$	Divide both sides by 4.

Check

$|4x - 5| = 15$	$|4x - 5| = 15$	Original equation
$|4(-\frac{5}{2}) - 5| = 15$	$|4(5) - 5| = 15$	Substitute $-\frac{5}{2}$ and 5 for x.
$|-15| = 15$	$|15| = 15$	Simplify.
$15 = 15$	$15 = 15$	Solutions check.

The solutions are $x = -\frac{5}{2}$ and $x = 5$.

When solving absolute value equations, remember that it is possible that they have no solution. For instance, the equation $|4x - 5| = -15$ has no solution because the absolute value of a real number cannot be negative. Do not make the mistake of trying to solve such an equation by writing the "equivalent" linear equations as $4x - 5 = -15$ and $4x - 5 = 15$. These equations have solutions, but they are both extraneous.

The equation in the next example is not given in the **standard form**

$$|ax + b| = c, \quad c \geq 0.$$

Notice that the first step in solving such an equation is to write it in standard form.

Example 3 An Absolute Value Equation in Nonstandard Form

Solve $|2x + 4| + 2 = 6$.

Solution

$|2x + 4| + 2 = 6$		Original equation
$|2x + 4| = 4$		Standard form
$2x + 4 = -4$ or	$2x + 4 = 4$	Equivalent equations
$2x = -8$	$2x = 0$	Subtract 4 from both sides.
$x = -4$	$x = 0$	Divide both sides by 2.

The solutions are $x = -4$ and $x = 0$. Check these in the original equation.

2 Solve an inequality involving absolute value.

Solving Inequalities Involving Absolute Value

To see how to solve inequalities involving absolute value, consider the following comparisons.

$\lvert x \rvert = 1$	$\lvert x \rvert < 1$	$\lvert x \rvert > 1$
$x = -1$ and $x = 1$	$-1 < x < 1$	$x < -1$ or $x > 1$

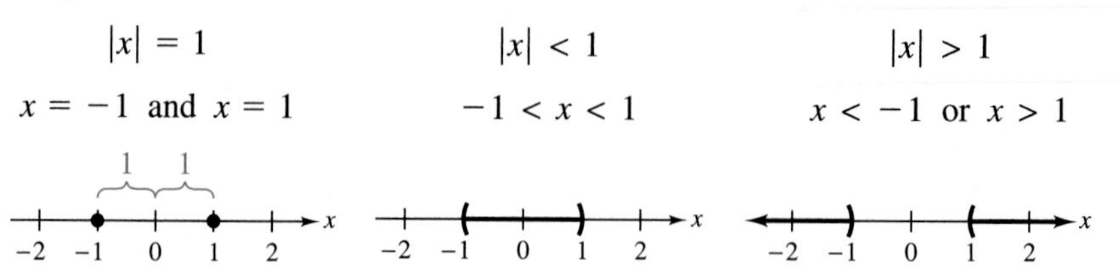

Notice that an **absolute value inequality** can be rewritten as a compound inequality. These comparisons suggest the following rules for solving inequalities involving absolute value.

Study Tip

- A "less than" inequality $\lvert ax + b \rvert < c$ is a *conjunction* and can be solved as the double inequality

$$-c < ax + b < c.$$

- A "greater than" inequality $\lvert ax + b \rvert > c$ is a *disjunction* and must be solved as a compound inequality.

▶ Solving an Absolute Value Inequality

Let x be a variable or an algebraic expression and let a be a real number such that $a > 0$.

1. The solutions of $\lvert x \rvert < a$ are all values of x that lie *between* $-a$ and a. That is,

$$\lvert x \rvert < a \quad \text{if and only if} \quad -a < x < a.$$

2. The solutions of $\lvert x \rvert > a$ are all values of x that are *less than* $-a$ or *greater than* a. That is,

$$\lvert x \rvert > a \quad \text{if and only if} \quad x < -a \text{ or } x > a.$$

These rules are also valid if $<$ is replaced by $\leq$ and $>$ is replaced by $\geq$.

Example 4 Solving a Conjunctive Absolute Value Inequality

Solve $\lvert x + 3 \rvert \leq 6$.

Solution

$\lvert x + 3 \rvert \leq 6$ Original inequality

$-6 \leq x + 3 \leq 6$ Equivalent double inequality

$-6 - 3 \leq x + 3 - 3 \leq 6 - 3$ Subtract 3 from all three parts.

$-9 \leq x \leq 3$ Combine like terms.

The solution set consists of all real numbers greater than or equal to -9 and less than or equal to 3. The set notation for this solution set is $\{x \mid -9 \leq x \leq 3\}$. The graph of this solution set is shown in Figure 3.20.

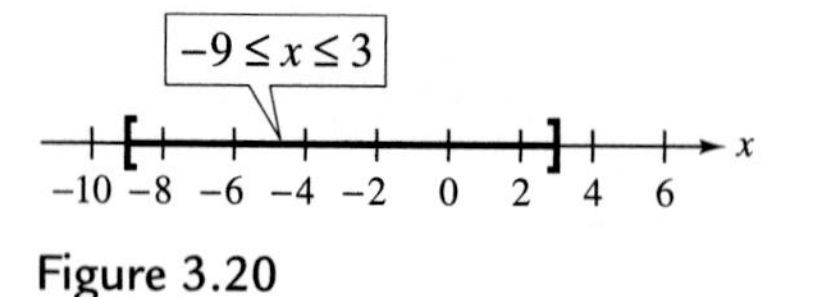

Figure 3.20

Keep in mind that a conjunctive ($\leq$) absolute value inequality can be solved as a double inequality where a disjunctive ($\geq$) one must be solved as a compound inequality.

Example 5 Solving a Disjunctive Absolute Value Inequality

Solve $|4x - 5| > 13$.

Solution

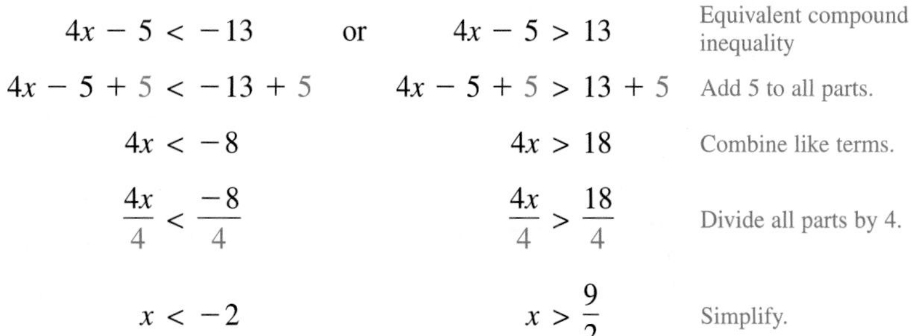
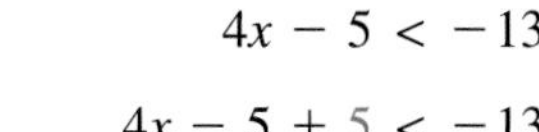

$|4x - 5| > 13$			Original inequality
$4x - 5 < -13$	or	$4x - 5 > 13$	Equivalent compound inequality
$4x - 5 + 5 < -13 + 5$		$4x - 5 + 5 > 13 + 5$	Add 5 to all parts.
$4x < -8$		$4x > 18$	Combine like terms.
$\frac{4x}{4} < \frac{-8}{4}$		$\frac{4x}{4} > \frac{18}{4}$	Divide all parts by 4.
$x < -2$		$x > \frac{9}{2}$	Simplify.

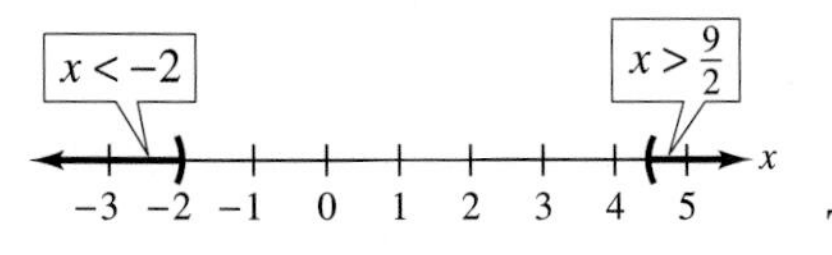

Figure 3.21

The solution set consists of all real numbers less than -2 or greater than $\frac{9}{2}$ or in set notation $\{x \mid x < -2 \text{ or } x > \frac{9}{2}\}$. (See Figure 3.21.)

Example 6 Solving a Conjunctive Absolute Value Inequality

Solve $\left|3.6 - \frac{x}{2}\right| \le 0.5$.

Solution

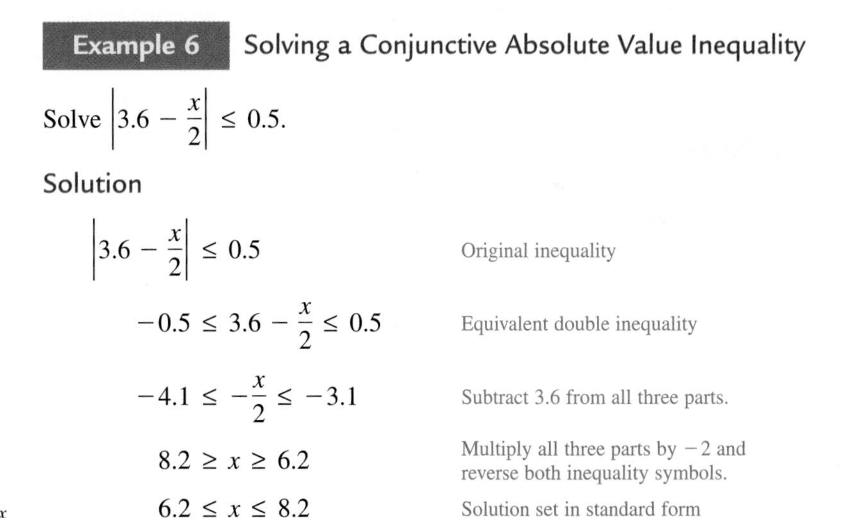

$\left|3.6 - \frac{x}{2}\right| \le 0.5$	Original inequality
$-0.5 \le 3.6 - \frac{x}{2} \le 0.5$	Equivalent double inequality
$-4.1 \le -\frac{x}{2} \le -3.1$	Subtract 3.6 from all three parts.
$8.2 \ge x \ge 6.2$	Multiply all three parts by -2 and reverse both inequality symbols.
$6.2 \le x \le 8.2$	Solution set in standard form

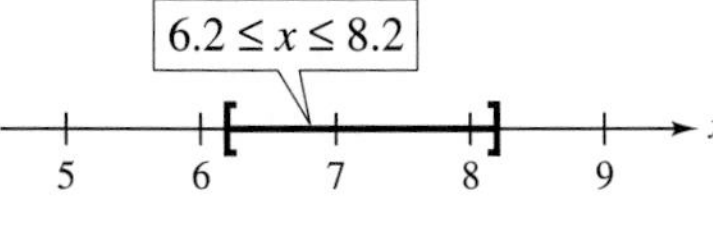

Figure 3.22

The solution set consists of all real numbers greater than or equal to 6.2 and less than or equal to 8.2 or in set notation $\{x \mid 6.2 \le x \le 8.2\}$. (See Figure 3.22.)

Discussing the Concept Solution Sets

Without doing any calculations, which of the following absolute value equations and inequalities do not have a solution? Explain. Find the solution set for the remaining equations and inequalities.

a. $|3x + 7| = 5$ **b.** $|3 - x| = -3$ **c.** $\left|\frac{x}{4} + 2\right| = 0$

d. $|-x + 1| \le 4$ **e.** $|5x - 9| > -1$ **f.** $\left|\frac{x}{3} - \frac{2}{3}\right| < -5$

3.7 Exercises

Integrated Review *Concepts, Skills, and Problem Solving*

Keep mathematically in shape by doing these exercises *before* the problems of this section.

Properties and Definitions

1. If n is an integer, how do the numbers $2n$ and $2n - 1$ differ? Explain.

2. Are $-3x^2$ and $(-3x)^2$ equal? Explain.

3. Explain how to write $\frac{27}{12}$ in reduced form.

4. Explain how to divide $\frac{2}{3}$ by $\frac{5}{3}$.

Order of Real Numbers

In Exercises 5–10, place the correct inequality symbol (< or >) between the two real numbers.

5. $3 \quad -2$ **6.** $-3 \quad -2$

7. $-\frac{1}{2} \quad -3$ **8.** $-\frac{1}{3} \quad -\frac{2}{3}$

9. $\frac{1}{2} \quad \frac{5}{16}$ **10.** $4 \quad \frac{45}{11}$

Problem Solving

In Exercises 11 and 12, determine whether there is more than a $500 difference between the budgeted amount and the actual expense.

11. Wages
Budgeted: $76,300
Actual: $75,926

12. Taxes
Budgeted: $37,800
Actual: $39,632

Developing Skills

In Exercises 1–4, determine whether the value is a solution of the equation.

	Equation	*Value*
1.	$\lvert x + 3\rvert = 10$	$x = -13$
2.	$\lvert x - 12\rvert = 10$	$x = 18$
3.	$\lvert 3 - 2y\rvert = 2$	$y = 1$
4.	$\left\lvert \frac{1}{3}z + 11\right\rvert = 14$	$z = 9$

In Exercises 5–8, transform the absolute value equation into two linear equations.

5. $\lvert u - 3\rvert = 7$ **6.** $\lvert m + 4\rvert = 3$

7. $\left\lvert \frac{1}{2}x + 7\right\rvert = \frac{3}{2}$ **8.** $\lvert 3k - 5\rvert = 7$

In Exercises 9–34, solve the equation. (Some of the equations have no solution.) See Examples 1–3.

9. $\lvert x\rvert = 7$ **10.** $\lvert y\rvert = 6$

11. $\lvert v\rvert = 15$ **12.** $\lvert z\rvert = 3$

13. $\lvert m\rvert = -3$ **14.** $\lvert a\rvert = -8$

15. $\lvert 3x\rvert = 18$ **16.** $\left\lvert \frac{3}{2}x\right\rvert = 12$

17. $\lvert x - 12\rvert = 4$ **18.** $\lvert y - 10\rvert = 25$

19. $\lvert s + 3\rvert = 11$ **20.** $\lvert a + 6\rvert = 2$

21. $\lvert 16 - y\rvert = 3$ **22.** $\lvert 3 - x\rvert = 2$

23. $\lvert 2x + 4\rvert = -8$ **24.** $\lvert 20 - 5t\rvert = 0$

25. $\lvert 2x - 3\rvert = 21$ **26.** $\lvert 3x + 5\rvert = 17$

27. $\lvert 3x + 2\rvert = 5$ **28.** $\lvert 4 - 3x\rvert = 16$

29. $\lvert 5x - 9\rvert - 4 = 0$ **30.** $\lvert 4x + 3\rvert + 1 = 12$

31. $\left\lvert 5 - \frac{2}{3}x\right\rvert = 3$ **32.** $\left\lvert \frac{1}{2}x + 3\right\rvert = 9$

33. $\lvert 0.25x - 2\rvert = 4$ **34.** $\lvert 3.2 - 1.5x\rvert = 2$

Think About It In Exercises 35 and 36, write a single equation that is equivalent to the two equations.

35. $x + 3 = 8,\ x + 3 = -8$

36. $3t - 5 = 7,\ 3t - 5 = -7$

Think About It In Exercises 37 and 38, write a single equation that is equivalent to the statement.

37. The distance between x and 4 is 2.

38. The distance between t and 10 is 4.

In Exercises 39–46, determine whether the *x*-values are solutions of the inequality.

Inequality	*Values*	
39. $\lvert x\rvert < 2$	(a) $x = 1$	(b) $x = -3$
	(c) $x = 5$	(d) $x = -\frac{1}{2}$
40. $\lvert x\rvert \le 10$	(a) $x = -12$	(b) $x = -6$
	(c) $x = 15$	(d) $x = 4$
41. $\lvert x\rvert \ge 5$	(a) $x = 2$	(b) $x = -7$
	(c) $x = 25$	(d) $x = -3$
42. $\lvert x\rvert > 8$	(a) $x = 12$	(b) $x = -6$
	(c) $x = 7.9$	(d) $x = 8.1$
43. $\lvert x - 4\rvert < 2$	(a) $x = 2$	(b) $x = 1.5$
	(c) $x = 0$	(d) $x = 5$
44. $\lvert x + 1\rvert \le 4$	(a) $x = 10$	(b) $x = -10$
	(c) $x = -2$	(d) $x = 3$
45. $\lvert x + 5\rvert \ge 3$	(a) $x = -8$	(b) $x = -5$
	(c) $x = -2$	(d) $x = -4$
46. $\lvert x - 4\rvert > 6$	(a) $x = 9$	(b) $x = 10$
	(c) $x = 11$	(d) $x = -1$

In Exercises 47–50, transform the absolute value inequality into a double inequality or two separate inequalities.

47. $\lvert z + 2\rvert < 1$ **48.** $\lvert x - 7\rvert \le 3$

49. $\lvert 5 - h\rvert \ge 2$ **50.** $\lvert 8 - x\rvert > 10$

In Exercises 51–54, sketch a graph that shows the real numbers that satisfy the statement.

51. All real numbers greater than -3 *and* less than 3

52. All real numbers greater than or equal to 2 *and* less than 8

53. All real numbers less than or equal to 5 *or* greater than 10

54. All real numbers less than -2 *or* greater than or equal to 4

In Exercises 55–70, solve the inequality and sketch the solution on the real number line. See Examples 4 and 5.

55. $\lvert y\rvert < 3$ **56.** $\lvert x\rvert < 5$

57. $\lvert y\rvert \ge 2$ **58.** $\lvert x\rvert \ge 5$

59. $\lvert t\rvert > 5$ **60.** $\lvert y\rvert > 7$

61. $\lvert y - 2\rvert \le 2$ **62.** $\lvert u - 3\rvert \le 4$

63. $\lvert x + 1\rvert < 4$ **64.** $\lvert z + 3\rvert < 5$

65. $\lvert 2x\rvert < 12$ **66.** $\lvert 4z\rvert \le 16$

67. $\lvert y - 5\rvert > 2$ **68.** $\lvert v - 4\rvert > 4$

69. $\lvert m + 2\rvert \ge 3$ **70.** $\lvert s + 3\rvert \ge 5$

In Exercises 71–74, match the inequality with its graph. [The graphs are labeled (a), (b), (c), and (d).]

71. $\lvert x - 3\rvert \le 2$ **72.** $\lvert x - 3\rvert < 4$

73. $\lvert x - 3\rvert > 2$ **74.** $\lvert x - 3\rvert \ge 3$

(a) 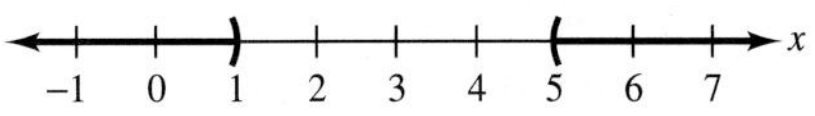

(b) 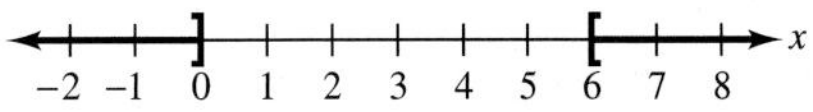

(c) 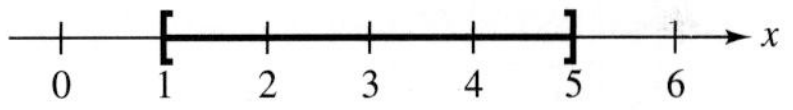

(d) 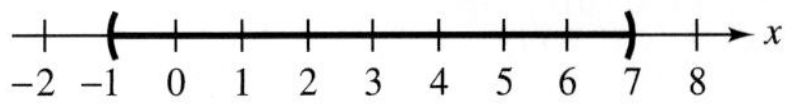

In Exercises 75–78, write an absolute value inequality that represents the interval.

75. 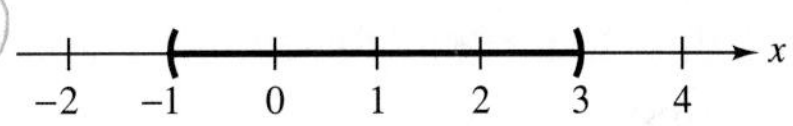

76. 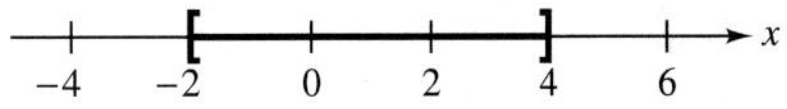

77. 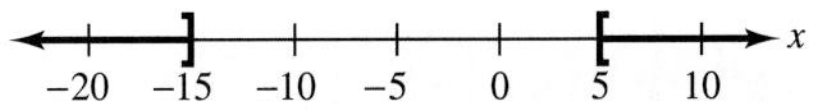

78. 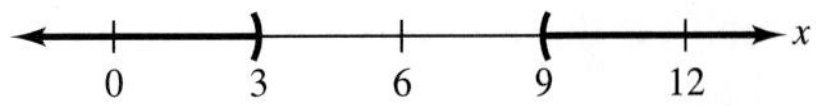

In Exercises 79–84, write an absolute value inequality that represents the verbal statement.

79. The set of all real numbers x whose distance from 0 is less than 2.

80. The set of all real numbers x whose distance from 0 is more than 2.

81. The set of all real numbers x whose distance from 0 is more than 5.

82. The set of all real numbers x whose distance from 0 is at least 5.

83. The set of all real numbers x whose distance from 4 is more than 2.

84. The set of all real numbers x whose distance from 6 is no more than 2.

Solving Problems

85. *Temperature* The temperature of a room satisfies the inequality

$$|t - 72| < 1.5$$

where t is in degrees Fahrenheit. Sketch the graph of the solution set of the inequality.

86. *Time Study* A time study was conducted to determine the length of the useful life of a car battery. Approximately two-thirds of the batteries had lifetimes satisfying the inequality

$$|L - 52| < 4$$

where L is time in months. Sketch the graph of the solution set of the inequality.

87. *Accuracy of Measurements* In a machine shop, the diameter of a machined part must be within 0.005 centimeter of specifications. Let $(s - x)$ represent the difference between the specification s and the measured diameter x of the machined part.

(a) Write an absolute value inequality that describes the values of x that are within specifications.

(b) The diameter of a part is specified to be $s = 3.5$ centimeters. Describe the acceptable diameters for this piece.

88. *Think About It* When you buy a 16-ounce bag of chips, you expect to get *precisely* 16 ounces. Suppose the actual weight w (in ounces) of a "16-ounce" bag of chips is given by $|w - 16| \le \frac{1}{2}$. If you buy four 16-ounce bags, what is the greatest amount you can expect to get? What is the least? Explain.

Explaining Concepts

89. Give a graphical description of the absolute value of a real number.

90. Give an example of an absolute value equation that has only one solution.

91. In your own words, explain how to solve an absolute value equation. Illustrate your explanation with an example.

92. Give a verbal description of the solution of the inequality $|x| > 3$.

CHAPTER SUMMARY

Key Terms

Key Concepts

3.1 Solving a linear equation

Solve a linear equation using inverse operations to isolate the variable.

3.1 Expressions for special types of integers

1. $2n$ denotes an *even* integer.
2. $2n - 1$ and $2n + 1$ denote *odd* integers.
3. The set $\{n, n + 1, n + 2\}$ denotes three *consecutive* integers.

3.2 Equations containing symbols of grouping

Solve a linear equation by first removing the symbols of grouping using the Distributive Property.

3.2 Equations involving fractions or decimals

1. Clear an equation of fractions by multiplying both sides by the least common multiple (LCM) of the denominators.
2. Use cross-multiplication to solve a linear equation that equates two fractions. That is, if $\frac{a}{b} = \frac{c}{d}$, then $a \cdot d = b \cdot c$.
3. To solve a linear equation with decimal coefficients, multiply both sides by a power of 10 that converts all decimal coefficients to integers.

3.3 The percent equation

The percent equation $a = p \cdot b$ compares two numbers.

b is the base number.

p is the percent in decimal form.

a is the number being compared to b.

3.3 Markups and discounts

1. A markup is the difference between the cost (what the retailer pays) and the price (what the consumer pays).
2. A discount is the amount off the list price (what the consumer pays).

3.3 Guidelines for solving word problems

1. Write a *verbal model* that describes the problem.
2. Assign *labels* to fixed quantities and variable quantities.
3. Rewrite the verbal model as an *algebraic equation* using the assigned labels.
4. *Solve* the algebraic equation.
5. *Check* to see that your solution satisfies the word problem as stated.

3.4 Solving a proportion

A proportion equates two ratios.

If $\frac{a}{b} = \frac{c}{d}$, then $ad = bc$.

3.6 Properties of inequalities

Let a, b, and c be real numbers, variables, or algebraic expressions.

Addition: If $a < b$, then $a + c < b + c$.

Subtraction: If $a < b$, then $a - c < b - c$.

Multiplication: If $a < b$ and $c > 0$, then $ac < bc$.
If $a < b$ and $c < 0$, then $ac > bc$.

Division: If $a < b$ and $c > 0$, then $\frac{a}{c} < \frac{b}{c}$.
If $a < b$ and $c < 0$, then $\frac{a}{c} > \frac{b}{c}$.

Transitive: If $a < b$ and $b < c$, then $a < c$.

3.6 Solving a linear inequality or a compound inequality

Solve a linear inequality or a compound inequality by performing inverse operations on all parts of the inequality.

3.7 Solving an absolute value equation

Solve an absolute value equation by rewriting as two linear equations.

3.7 Solving an absolute value inequality

Solve an absolute value inequality by rewriting as a compound inequality.

REVIEW EXERCISES

Reviewing Skills

3.1 In Exercises 1–4, solve the equation mentally.

1. $y - 25 = 10$

2. $z + 5 = 12$

3. $\frac{x}{4} = 7$

4. $6u = 30$

In Exercises 5 and 6, justify each step of the solution.

5.
$$10x - 12 = 18$$
$$10x - 12 + 12 = 18 + 12$$
$$10x = 30$$
$$\frac{10x}{10} = \frac{30}{10}$$
$$x = 3$$

6.
$$\frac{t}{4} + \frac{t}{3} = 1$$
$$3t + 4t = 12$$
$$7t = 12$$
$$\frac{7t}{7} = \frac{12}{7}$$
$$t = \frac{12}{7}$$

In Exercises 7–20, solve the linear equation and check your solution.

7. $x + 10 = 13$

8. $x - 3 = 8$

9. $5 - x = 2$

10. $3 = 8 - x$

11. $10x = 50$

12. $-3x = 21$

13. $8x + 7 = 39$

14. $12x - 5 = 43$

15. $24 - 7x = 3$

16. $13 + 6x = 61$

17. $15x - 4 = 16$

18. $3x - 8 = 2$

19. $\frac{x}{5} = 4$

20. $-\frac{x}{14} = \frac{1}{2}$

3.2 In Exercises 21–30, solve the linear equation and check your solution.

21. $3x - 2(x + 5) = 10$

22. $4x + 2(7 - x) = 5$

23. $2x + 3 = 5x - 2$

24. $8(x - 2) = 3(x + 2)$

25. $\frac{2}{3}x - \frac{1}{6} = \frac{9}{2}$

26. $\frac{1}{8}x + \frac{3}{4} = \frac{5}{2}$

27. $\frac{x}{3} - \frac{1}{9} = 2$

28. $\frac{1}{2} - \frac{x}{8} = 7$

29. $\frac{u}{10} + \frac{u}{5} = 6$

30. $\frac{x}{3} + \frac{x}{5} = 1$

In Exercises 31–34, solve the equation. Round your result to two decimal places.

31. $516x - 875 = 3250$

32. $2.825x + 3.125 = 12.5$

33. $\frac{x}{4.625} = 48.5$

34. $5x + \frac{1}{4.5} = 18.125$

3.3 In Exercises 35 and 36, complete the table.

35.

Percent	Parts out of 100	Decimal	Fraction
35%			

36.

Percent	Parts out of 100	Decimal	Fraction
			$\frac{4}{5}$

37. What number is 125% of 16?

38. What number is 0.8% of 3250?

39. 150 is $37\frac{1}{2}$% of what number?

40. 323 is 95% of what number?

41. 150 is what percent of 250?

42. 130.6 is what percent of 3265?

3.4 In Exercises 43–46, find a ratio that compares the relative sizes of the quantities. (Use the same units of measurement for both quantities.)

43. Eighteen inches to 4 yards

44. One pint to 2 gallons

45. Two hours to 90 minutes

46. Four meters to 150 centimeters

In Exercises 47–52, solve the proportion.

47. $\frac{7}{16} = \frac{z}{8}$

48. $\frac{x}{12} = \frac{5}{4}$

49. $\dfrac{x+2}{4} = -\dfrac{1}{3}$

50. $\dfrac{x-4}{1} = \dfrac{9}{4}$

51. $\dfrac{x-3}{2} = \dfrac{x+6}{5}$

52. $\dfrac{x+1}{3} = \dfrac{x+2}{4}$

3.5 In Exercises 53 and 54, solve the formula for the specified variable.

53. Solve for θ: $A = \dfrac{r^2\theta}{2}$

54. Solve for n: $S = a + (n-1)d$

In Exercises 55–60, find the missing distance, rate, or time.

	Distance, d	*Rate, r*	*Time, t*
55.		65 mi/hr	8 hr
56.		4.7 m/sec	3 sec
57.	400 mi	50 mi/hr	
58.	855 m	5 m/min	
59.	3000 mi		50 hr
60.	1000 km		25 hr

3.6 In Exercises 61–64, write the linear inequality symbolically.

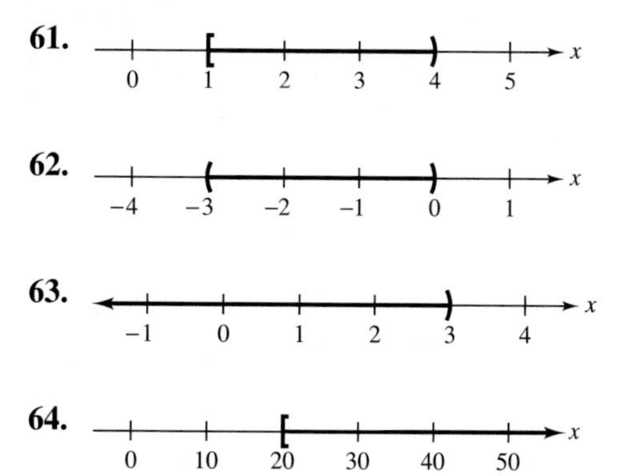

In Exercises 65–80, solve and graph the linear inequality.

65. $x + 5 \geq 7$

66. $x - 2 \leq 1$

67. $3x - 8 < 1$

68. $4x + 3 > 15$

69. $-11x \leq -22$

70. $-7x \geq 21$

71. $\frac{4}{5}x > 8$

72. $\frac{2}{3}n < -4$

73. $14 - \frac{1}{2}t < 12$

74. $32 + \frac{7}{8}k > 11$

75. $3 - 3y \geq 2(4 + y)$

76. $4 - 3y \leq 8(10 - y)$

77. $-2 < 2x + 6 \leq 2$

78. $-5 \leq 3 - 4x < 5$

79. $3 > \dfrac{x+1}{-2} > 0$

80. $5 \geq \dfrac{x-3}{3} > 2$

In Exercises 81–88, write the compound inequality using set notation and intersection or union.

81. $x < -3$ or $x > 7$

82. $x \leq -2$ or $x \geq -1$

83. $x < 0$ and $x > -6$

84. $x > -3$ and $x < 4$

85. $-8 \leq x \leq -5$

86. $5 > x > -1$

87. $x < 2$ or $x > 3$

88. $x \geq 6$ or $x \leq -1$

In Exercises 89–94, write a linear inequality that represents the statement.

89. z is at least 10.

90. x is nonnegative.

91. y is more than 8 but less than 12.

92. The area A is no more than 100 square feet.

93. The volume V is less than 12 cubic feet.

94. The perimeter P is at least 24 inches.

3.7 In Exercises 95–100, solve the absolute value equation.

95. $|x - 25| = 5$

96. $|x - 150| = 100$

97. $|7t| = 42$

98. $\left|\frac{3}{2}z\right| = 72$

99. $|3u + 24| = 0$

100. $|4x - 3| - 13 = 0$

In Exercises 101–106, solve the absolute value inequality and sketch the solution on the real number line.

101. $\left|\frac{1}{2}v\right| \leq 3$

102. $|3u| \geq 12$

103. $|y - 4| > 3$

104. $|k + 4| < 6$

105. $|2n - 3| < 5$

106. $|1 - 3n| \geq 5$

Solving Problems

107. *Driving Distances* On a 1200-mile trip, you drive about $1\frac{1}{2}$ times as much as your friend. Approximate the number of miles each of you drive.

108. *Hourly Wage* Your hourly wage is \$8.30 per hour plus 60 cents for each unit you produce. How many units must you produce in an hour so that your hourly wage is \$15.50?

109. *Geometry* The length of a rectangle is 30 meters greater than its width. Find the measurements of the rectangle if its perimeter is 260 meters.

110. *Geometry* A 10-foot board is cut so that one piece is 4 times as long as the other. Find the length of each piece.

111. *Revenue* The revenues for a corporation (in millions of dollars) in the years 1997 and 1998 were \$4521.4 and \$4679.0, respectively. Determine the percent increase in revenue from 1997 to 1998.

112. *Price Increase* The manufacturer's suggested retail price for a car is \$18,459. Estimate the price of a comparably equipped car for the next model year if the price will increase by $4\frac{1}{2}\%$.

113. *Analyzing Data* The figure gives the living arrangements for women 65 years old or older for the year 1995. Find the number of women in each category if there were approximately 19,844,000 women who were at least 65 in 1995. (Source: U.S. Bureau of the Census)

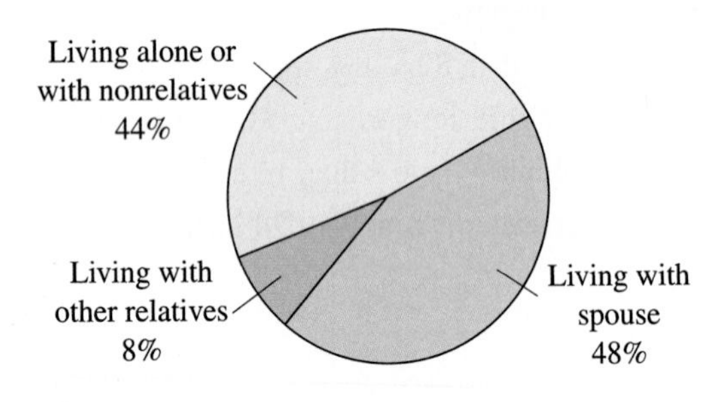

114. *Analyzing Data* The figure gives the living arrangements for men 65 years old or older for the year 1995. Find the number of men in each category if there were approximately 13,689,000 men who were at least 65 in 1995. (Source: U.S. Bureau of the Census)

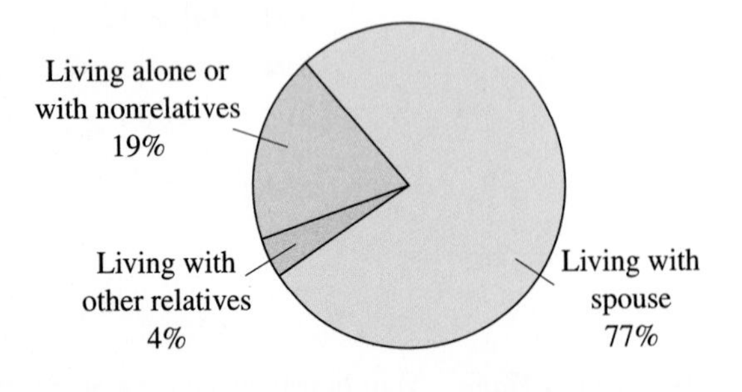

In Exercises 115–118, use a proportion to solve the problem.

115. *Real Estate Tax* The tax on a property with an assessed value of \$75,000 is \$1150. Find the tax on a property with an assessed value of \$110,000.

116. *Recipe Proportions* One and one-half cups of milk are needed to make one batch of pudding. How much is required to make three batches?

117. *Map Distance* The scale on the map in the figure represents 100 miles on the map. Use the map to approximate the distance between St. Petersburg and Tallahassee.

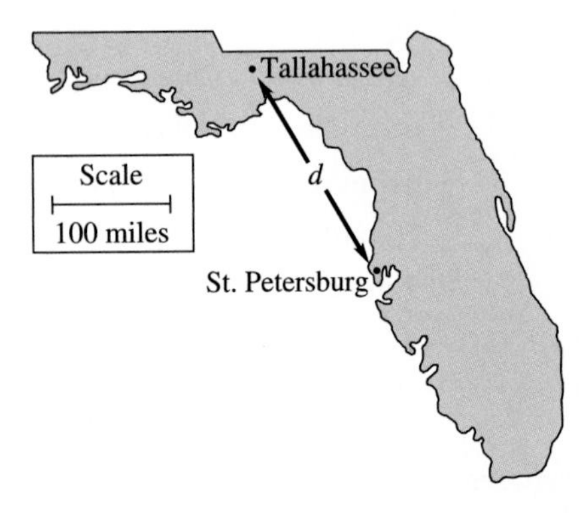

118. *Geometry* Solve for the length x in the figure. Assume that the two triangles are similar, and use the fact that corresponding sides of similar triangles are proportional.

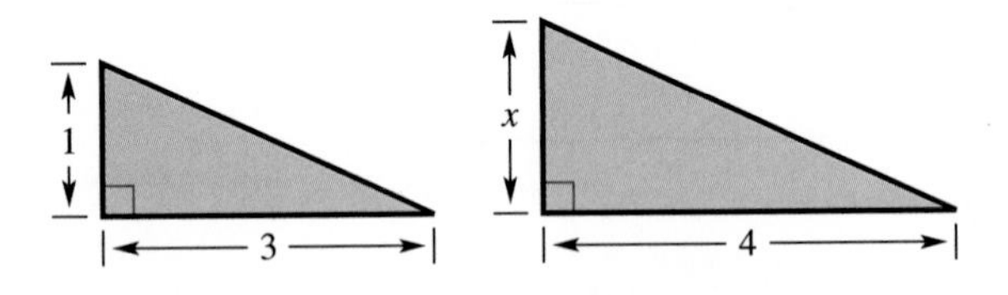

119. *Time* A train's average speed is 60 miles per hour. How long will it take the train to travel 562 miles?

120. *Distance* An airplane has an average speed of 475 miles per hour. How far will it travel in $2\frac{1}{3}$ hours?

121. *Speed* You can walk 20 kilometers in 3 hours and 47 minutes. What is your average speed?

122. *Speed* For the first hour of a 350-mile trip, your average speed is 40 miles per hour. Determine the average speed that must be maintained for the remainder of the trip if you want the average speed for the entire trip to be 50 miles per hour.

123. *Number of Coins* You have 30 coins in dimes and quarters with a combined value of $5.55. Determine the number of coins of each type.

124. *Poll Results* Thirteen hundred people were surveyed in an opinion poll. Candidates A and B received the same number of votes. Candidate C received $1\frac{1}{4}$ times as many votes as each of the other two candidates. How many votes did each candidate receive?

125. *Measurements of a Swimming Pool* The width of a rectangular swimming pool is 4 feet less than its length. The perimeter of the pool is 112 feet. Find the measurements of the pool.

126. *Measurements of a Triangle* The perimeter of an isosceles triangle is 65 centimeters. Find the length of the two equal sides if each is 10 centimeters longer than the third side. (An isosceles triangle has two sides of equal length.)

Simple Interest In Exercises 127–130, use the simple interest formula.

127. Find the total interest you will earn on a $1000 corporate bond that matures in 5 years and has an annual interest rate of 9.5%.

128. Find the annual interest rate on a certificate of deposit that pays $60 per year in interest on a principal of $750.

129. Find the principal required to have an annual interest income of $25,000 if the annual interest rate is 8.75%.

130. You invest $2500 in a certificate of deposit that has an annual interest rate of 7%. After 6 months, the interest is computed and added to the principal. During the second 6 months the interest is computed using the original investment plus the interest earned during the first 6 months. What is the total interest earned during the first year of the investment?

131. *Work Rate* Find the time for two people working together to complete a task that, if they worked individually, would take them 5 hours and 6 hours, respectively.

132. *Work Rate* Suppose the person in Exercise 131 who can complete the task in 5 hours has already worked 1 hour when the second person starts. How long will they work together to complete the task?

Chapter Test

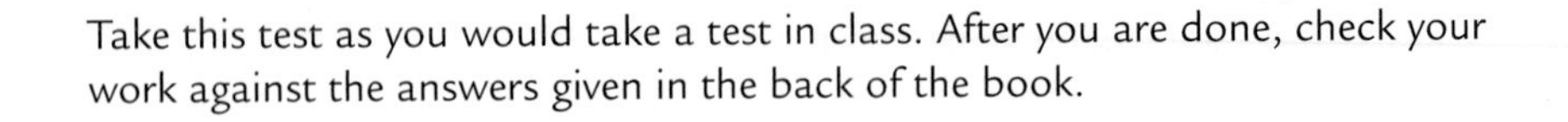

Take this test as you would take a test in class. After you are done, check your work against the answers given in the back of the book.

In Exercises 1–6, solve the equation and check your solution.

1. $4x - 3 = 18$

2. $10 - (2 - x) = 2x + 1$

3. $\frac{3x}{4} = \frac{5}{2} + x$

4. $\frac{t+2}{3} = \frac{2t}{5}$

5. $|x - 5| = 2$

6. $|3x - 4| = 5$

7. Solve $4.08(x + 10) = 9.50(x - 2)$. Round the result to two decimal places.

8. The bill (including parts and labor) for the repair of a home appliance is \$142. The cost for parts is \$62. How many hours were spent repairing the appliance if the cost of labor is \$32 per hour?

9. Express the fraction $\frac{3}{8}$ as a percent and as a decimal.

10. 324 is 27% of what number?

11. 90 is what percent of 250?

12. Express the ratio of 40 inches to 2 yards as a fraction in simplest form. Use the same units for both quantities, and explain how you made this conversion.

13. Solve the proportion $\frac{2x}{3} = \frac{x+4}{5}$.

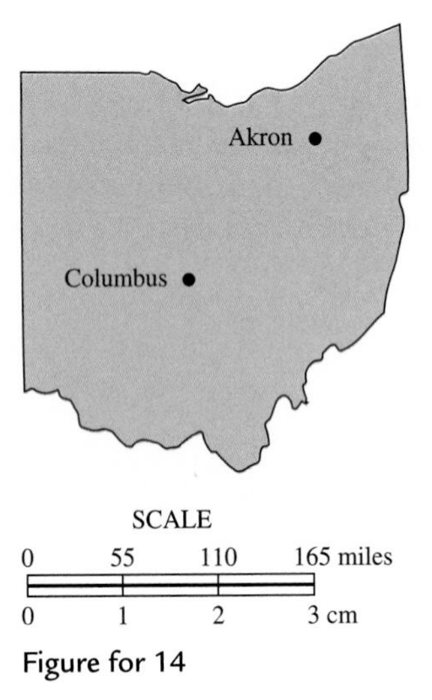

Figure for 14

14. On the map at the left, 1 centimeter represents 55 miles. Approximate the distance between Akron and Columbus.

15. You traveled 264 miles in $5\frac{1}{2}$ hours. What was your average speed?

16. You can paint a building in 9 hours. Your friend would require 12 hours. Working together, how long will it take the two of you to paint the building?

17. Solve for R in the formula $S = C + RC$.

18. How much must you deposit to earn \$500 per year at 8% simple interest?

In Exercises 19–24, solve and graph the inequality.

19. $x + 3 \leq 7$

20. $-\frac{2x}{3} > 4$

21. $-3 < 2x - 1 \leq 3$

22. $2 \geq \frac{3 - x}{2} > -1$

23. $|x + 4| \leq 3$

24. $|2x - 1| > 3$

Cumulative Test: Chapters 1–3

Take this test as you would take a test in class. After you are done, check your work against the answers given in the back of the book.

1. Place the correct symbol (< or >) between the numbers: $-\frac{3}{4}$ ▢ $\left|-\frac{7}{8}\right|$.

In Exercises 2–7, evaluate the expression.

2. $(-200)(2)(-3)$ **3.** $\frac{3}{8} - \frac{5}{6}$ **4.** $-\frac{2}{9} \div \frac{8}{75}$

5. $-(-2)^3$ **6.** $3 + 2(6) - 1$ **7.** $24 + 12 \div 3$

In Exercises 8 and 9, evaluate the expression when $x = -2$ and $y = 3$.

8. $2x + y^2$ **9.** $4y - x^3$

10. Use exponential form to write the product $3 \cdot (x + y) \cdot (x + y) \cdot 3 \cdot 3$.

11. Use the Distributive Property to expand $-2x(x - 3)$.

12. Identify the rule of algebra illustrated by $2 + (3 + x) = (2 + 3) + x$.

In Exercises 13–15, simplify the expression.

13. $(3x^3)(5x^4)$ **14.** $(a^3b^2)(ab)^5$

15. $2x^2 - 3x + 5x^2 - (2 + 3x)$

In Exercises 16–18, solve the equation and check your solution.

16. $12x - 3 = 7x + 27$ **17.** $2x - \dfrac{5x}{4} = 13$

18. $2(x - 3) + 3 = 12 - x$

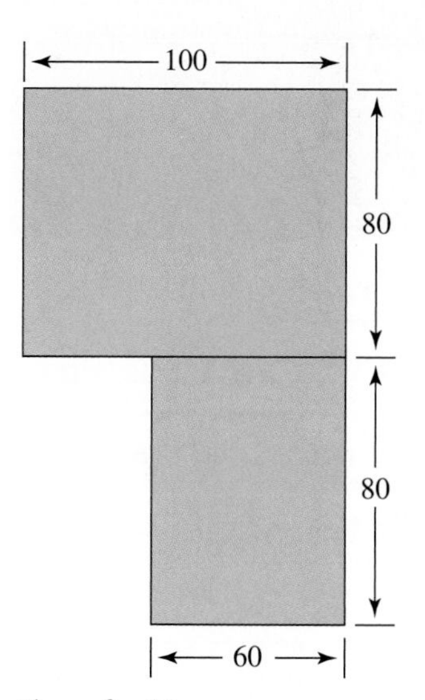

Figure for 24

19. Solve and graph the inequality

$$-1 \le \frac{x + 3}{2} < 2.$$

20. The sticker on a new car gives the fuel efficiency as 28.3 miles per gallon. In your own words, explain how to estimate the annual fuel cost for the buyer if the car will be driven approximately 15,000 miles per year and the fuel cost is \$1.179 per gallon.

21. Express the ratio "24 ounces to 2 pounds" as a fraction in reduced form.

22. The sum of two consecutive even integers is 494. Find the two numbers.

23. The suggested retail price of a camcorder is \$1150. The camcorder is on sale for "20% off" the list price. Find the sale price.

24. The figure at the left shows two pieces of property. The assessed values of the properties are proportional to their areas. The value of the larger piece is \$95,000. What is the value of the smaller piece?

4 Graphs and Functions

Rudi Von Briel/PhotoEdit

In 1996, $30.7 billion was spent on advertising. Of that, advertising agencies received $21.4 billion. (Source: U.S. Bureau of the Census)

Motivating the Chapter

Salary Plus Commission

You work as a sales representative for an advertising agency. You are paid a weekly salary, plus a commission on all ads placed by your accounts. The table shows your sales and your total weekly earnings.

	Week 1	Week 2	Week 3	Week 4
Weekly sales	\$24,000	\$7000	\$0	\$36,000
Weekly earnings	\$980	\$640	\$500	\$1220

See Section 4.3, Exercise 72

a. Rewrite the data as a set of ordered pairs.

b. Does the table represent a function? If so, identify the dependent and independent variables.

c. Describe what you consider to be appropriate domain and range values.

See Section 4.5, Exercise 108

d. Explain how to determine whether the function is linear or not.

e. Determine the slope of this function. What is the *rate* at which the weekly pay increases for each unit increase in ad sales? What is the rate called in the context of the problem?

f. Write an equation that describes the linear relationship between weekly sales and weekly earnings.

g. Sketch a graph of the equation. Identify the y-intercept and explain its meaning in the context of the problem. Identify the x-intercept. Does the x-intercept have any meaning in the context of the problem? If so, what is it?

See Section 4.6, Exercise 70

h. What amount of ad sales is needed to guarantee a weekly pay of at least \$840?

4.1 Ordered Pairs and Graphs

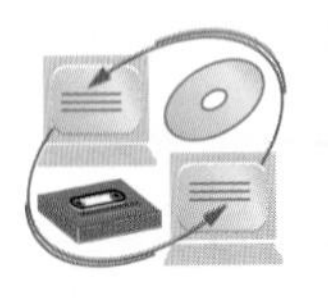

Objectives

1. Plot and find the coordinates of a point on a rectangular coordinate system.
2. Construct a table of values for an equation and determine whether an ordered pair is a solution point of the equation.
3. Use the verbal problem-solving method to plot points on a rectangular coordinate system.

1 Plot and find the coordinates of a point on a rectangular coordinate system.

The Rectangular Coordinate System

Just as you can represent real numbers by points on the real number line, you can represent ordered pairs of real numbers by points in a plane. This plane is called a **rectangular coordinate system** or the **Cartesian plane,** after the French mathematician René Descartes (1596–1650).

A rectangular coordinate system is formed by two real lines intersecting at right angles, as shown in Figure 4.1. The horizontal number line is usually called the ***x*-axis** and the vertical number line is usually called the ***y*-axis.** (The plural of axis is *axes*.) The point of intersection of the two axes is the **origin,** and the axes separate the plane into four regions called **quadrants.**

Corbis-Bettmann

René Descartes
(1596–1650)
Descartes made many contributions to philosophy, science, and mathematics. The idea of representing points in the plane by pairs of real numbers and representing curves in the plane by equations was described by Descartes in his book *La Géométrie,* published in 1637.

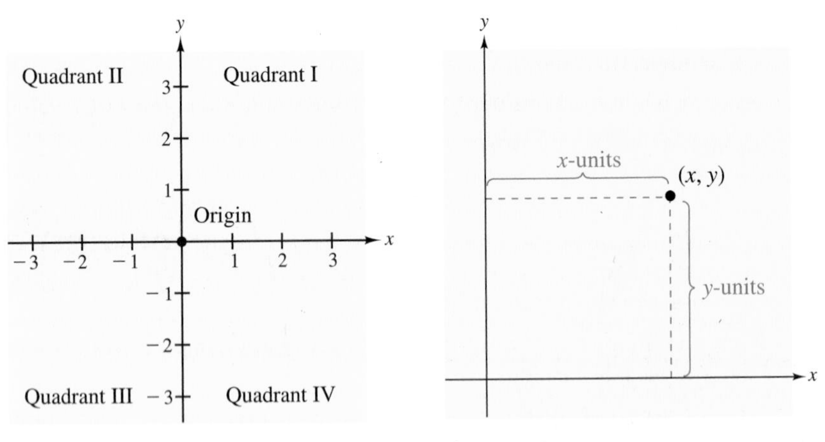

Figure 4.1

Figure 4.2

Each point in the plane corresponds to an **ordered pair** (x, y) of real numbers x and y, called the **coordinates** of the point. The first number (or ***x*-coordinate**) tells how far to the left or right the point is from the vertical axis, and the second number (or ***y*-coordinate**) tells how far up or down the point is from the horizontal axis, as shown in Figure 4.2.

A positive x-coordinate implies that the point lies to the *right* of the vertical axis; a negative x-coordinate implies that the point lies to the *left* of the vertical axis; and an x-coordinate of zero implies that the point lies *on* the vertical axis. Similar statements can be made about y-coordinates. A positive y-coordinate implies that the point lies *above* the horizontal axis, and a negative y-coordinate implies that the point lies *below* the horizontal axis.

Locating a point in a plane is called **plotting** the point. This procedure is demonstrated in Example 1.

Example 1 Plotting Points on a Rectangular Coordinate System

Plot the points $(-1, 2)$, $(3, 0)$, $(2, -1)$, $(3, 4)$, $(0, 0)$, and $(-2, -3)$ on a rectangular coordinate system.

Solution

The point $(-1, 2)$ is 1 unit to the *left* of the vertical axis and 2 units *above* the horizontal axis.

x-coordinate ↘ $(-1, 2)$ ↙ *y*-coordinate

Similarly, the point $(3, 0)$ is 3 units to the *right* of the vertical axis and *on* the horizontal axis. (It is on the horizontal axis because the *y*-coordinate is zero.) The other four points can be plotted in a similar way, as shown in Figure 4.3.

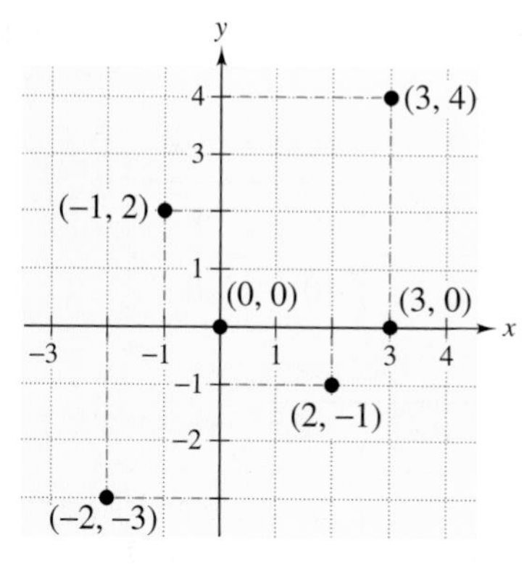

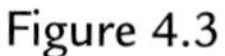
Figure 4.3

In Example 1 you were given the coordinates of several points and asked to plot the points on a rectangular coordinate system. Example 2 looks at the reverse problem. That is, you are given points on a rectangular coordinate system and asked to determine their coordinates.

Example 2 Finding Coordinates of Points

Determine the coordinates for each of the points shown in Figure 4.4.

Solution

Point *A* lies 3 units to the *left* of the vertical axis and 2 units *above* the horizontal axis. So, point *A* must be given by the ordered pair $(-3, 2)$. The coordinates of the other four points can be determined in a similar way, and we summarize the results as follows.

Point	*Position*	*Coordinates*
A	3 units *left*, 2 units *up*	$(-3, 2)$
B	3 units *right*, 1 unit *up*	$(3, 1)$
C	0 units *left* (or *right*), 4 units *down*	$(0, -4)$
D	2 units *right*, 3 units *up*	$(2, 3)$
E	2 units *left*, 2 units *down*	$(-2, -2)$
F	2 units *right*, 3 units *down*	$(2, -3)$

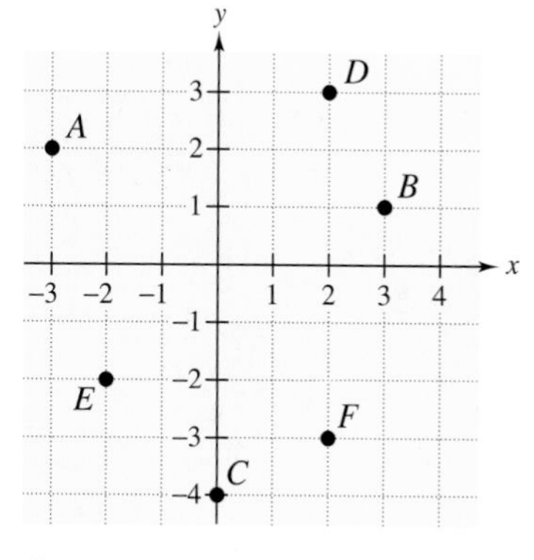

Figure 4.4

In Example 2, note that point *A* $(-3, 2)$ and point *F* $(2, -3)$ are different points. The order in which the numbers appear in an ordered pair is important.

Each year since 1967, the winners of the American Football Conference and the National Football Conference have played in the Super Bowl. The first Super Bowl was played between the Green Bay Packers and the Kansas City Chiefs.

Example 3 Super Bowl Scores

The scores of the winning and losing football teams for the Super Bowl games from 1981 through 1999 are given in the table below. Plot these points on a rectangular coordinate system. (Source: National Football League)

Year	1981	1982	1983	1984	1985	1986	1987
Winning score	27	26	27	38	38	46	39
Losing score	10	21	17	9	16	10	20

Year	1988	1989	1990	1991	1992	1993	1994
Winning score	42	20	55	20	37	52	30
Losing score	10	16	10	19	24	17	13

Year	1995	1996	1997	1998	1999
Winning score	49	27	35	31	34
Losing score	26	17	21	24	19

Solution

Plot the years on the x-axis and the winning and losing scores on the y-axis. In Figure 4.5, the winning scores are shown as black dots, and the losing scores are shown as blue dots. Note that the break in the x-axis indicates that the numbers between 0 and 1981 have been omitted.

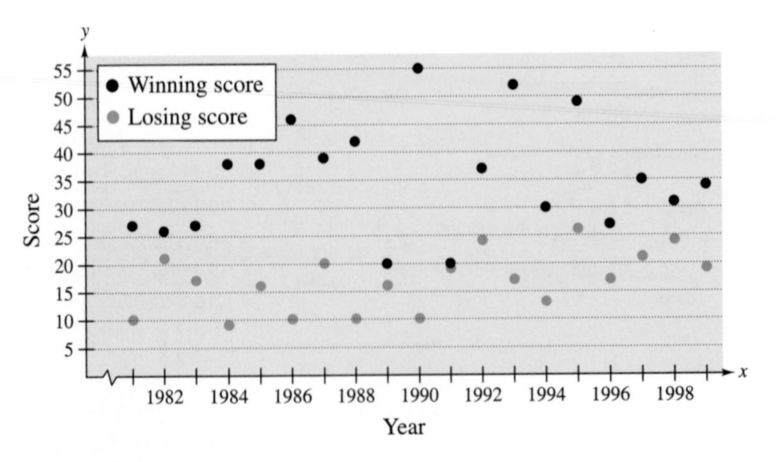

Figure 4.5

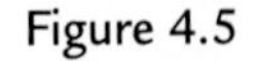

2 Construct a table of values for an equation and determine whether an ordered pair is a solution point of the equation.

Ordered Pairs as Solutions of Equations

In Example 3, the relationship between the year and the Super Bowl scores was given by a **table of values.** In mathematics, the relationship between the variables x and y is often given by an equation. From the equation, you must then construct your own table of values. For instance, consider the equation

$$y = 2x + 1.$$

To construct a table of values for this equation, choose several x-values and then calculate the corresponding y-values. For example, if you choose $x = 1$, the corresponding y-value is

$$y = 2(1) + 1 \qquad \text{Substitute 1 for } x.$$

$$y = 3. \qquad \text{Simplify.}$$

The corresponding ordered pair $(x, y) = (1, 3)$ is a **solution point** (or simply a **solution**) of the equation. The table below is a table of values (and the corresponding solution points) using x-values of -3, -2, -1, 0, 1, 2, and 3. These x-values are arbitrary. You should try to use x-values that are convenient and simple to use.

Choose x	Calculate y	Solution Points
$x = -3$	$y = 2(-3) + 1 = -5$	$(-3, -5)$
$x = -2$	$y = 2(-2) + 1 = -3$	$(-2, -3)$
$x = -1$	$y = 2(-1) + 1 = -1$	$(-1, -1)$
$x = 0$	$y = 2(0) + 1 = 1$	$(0, 1)$
$x = 1$	$y = 2(1) + 1 = 3$	$(1, 3)$
$x = 2$	$y = 2(2) + 1 = 5$	$(2, 5)$
$x = 3$	$y = 2(3) + 1 = 7$	$(3, 7)$

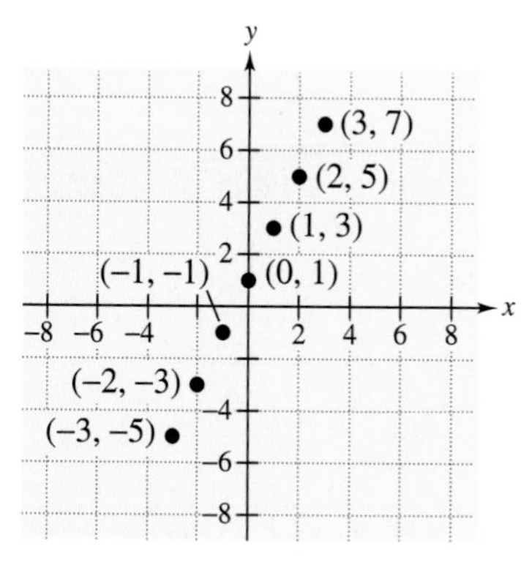

Figure 4.6

Once you have constructed a table of values, you can get a visual idea of the relationship between the variables x and y by plotting the solution points on a rectangular coordinate system. For instance, the solution points shown in the table are plotted in Figure 4.6.

In many places throughout this course, you will see that approaching a problem in different ways can help you understand the problem better. For instance, the discussion above looks at solutions of an equation in three ways.

▶ **Three Approaches to Problem Solving**

1. **Algebraic Approach** Use algebra to find several solutions.
2. **Numerical Approach** Construct a table that shows several solutions.
3. **Graphical Approach** Draw a graph that shows several solutions.

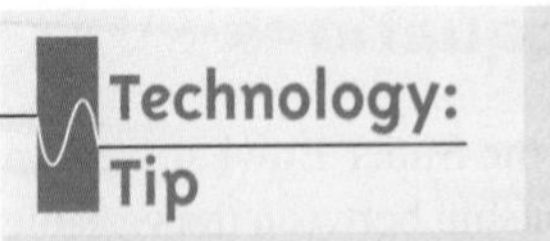

Consult the user's guide for your graphing utility to see if your graphing utility has a table feature. By using the table feature in the *ask* mode, you can create a table of values for an equation.

When making up a table of values for an equation, it is helpful first to solve the equation for y. For instance, the equation $4x + 2y = -8$ can be solved for y as follows.

$$4x + 2y = -8 \qquad \text{Original equation}$$

$$4x - 4x + 2y = -8 - 4x \qquad \text{Subtract } 4x \text{ from both sides.}$$

$$2y = -8 - 4x \qquad \text{Combine like terms.}$$

$$\frac{2y}{2} = \frac{-8 - 4x}{2} \qquad \text{Divide both sides by 2.}$$

$$y = -4 - 2x \qquad \text{Simplify.}$$

This procedure is further demonstrated in Example 4.

Example 4 Making Up a Table of Values

Make up a table of values showing five solution points for the equation

$$6x - 2y = 4.$$

Then plot the solution points on a rectangular coordinate system. (Choose x-values of -2, -1, 0, 1, and 2.)

Solution

$$6x - 2y = 4 \qquad \text{Original equation}$$

$$6x - 6x - 2y = 4 - 6x \qquad \text{Subtract } 6x \text{ from both sides.}$$

$$-2y = -6x + 4 \qquad \text{Combine like terms.}$$

$$\frac{-2y}{-2} = \frac{-6x + 4}{-2} \qquad \text{Divide both sides by } -2.$$

$$y = 3x - 2 \qquad \text{Simplify.}$$

Now, using the equation $y = 3x - 2$, you can construct a table of values, as shown below.

x	-2	-1	0	1	2
$y = 3x - 2$	-8	-5	-2	1	4
Solution points	$(-2, -8)$	$(-1, -5)$	$(0, -2)$	$(1, 1)$	$(2, 4)$

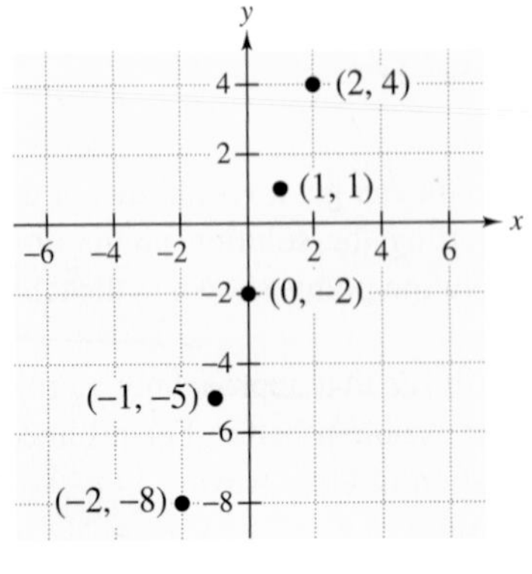

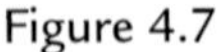
Figure 4.7

Finally, from the table you can plot the five solution points on a rectangular coordinate system, as shown in Figure 4.7.

▶ Guidelines for Verifying Solutions

To verify that an ordered pair (x, y) is a solution to an equation with variables x and y, use the following steps.

1. Substitute the values of x and y into the equation.
2. Simplify both sides of the equation.
3. If both sides simplify to the same number, the ordered pair is a solution. If the two sides yield different numbers, the ordered pair is not a solution.

Example 5 Verifying Solutions of an Equation

Determine which of the following ordered pairs is a solution of $x + 3y = 6$.

a. $(1, 2)$ **b.** $\left(-2, \frac{8}{3}\right)$ **c.** $(-6, 0)$ **d.** $(0, 2)$ **e.** $\left(\frac{8}{3}, \frac{10}{9}\right)$

Solution

a. For the ordered pair $(x, y) = (1, 2)$, substitute $x = 1$ and $y = 2$ into the original equation.

$$x + 3y = 6 \qquad \text{Original equation}$$

$$1 + 3(2) \stackrel{?}{=} 6 \qquad \text{Substitute } x = 1 \text{ and } y = 2.$$

$$7 \neq 6 \qquad \text{Simplify.}$$

Because the substitution does not satisfy the original equation, the ordered pair $(1, 2)$ *is not* a solution of the equation.

b. For the ordered pair $(x, y) = \left(-2, \frac{8}{3}\right)$, substitute $x = -2$ and $y = \frac{8}{3}$ into the original equation.

$$x + 3y = 6 \qquad \text{Original equation}$$

$$(-2) + 3\left(\frac{8}{3}\right) \stackrel{?}{=} 6 \qquad \text{Substitute } x = -2 \text{ and } y = \tfrac{8}{3}.$$

$$-2 + 8 = 6 \qquad \text{Simplify.}$$

$$6 = 6 \qquad \text{Simplify.}$$

Because the substitution satisfies the original equation, the ordered pair $\left(-2, \frac{8}{3}\right)$ *is* a solution of the equation.

c. The ordered pair $(-6, 0)$ *is not* a solution of the original equation because

$$-6 + 3(0) = -6 \neq 6.$$

d. The ordered pair $(0, 2)$ *is* a solution of the original equation because

$$0 + 3(2) = 6.$$

e. The ordered pair $\left(\frac{8}{3}, \frac{10}{9}\right)$ *is* a solution of the original equation because

$$\frac{8}{3} + 3\left(\frac{10}{9}\right) = \frac{8}{3} + \frac{10}{3} = \frac{18}{3} = 6.$$

3 Use the verbal problem-solving method to plot points on a rectangular coordinate system.

Application

Example 6 A Business Application

You are setting up a small business to assemble computer keyboards. Your initial cost is \$120,000, and your unit cost to assemble each keyboard is \$40. Write an equation that relates your total cost to the number of keyboards produced. Then plot the total costs of producing 1000, 2000, 3000, 4000, and 5000 keyboards.

Solution

The total cost equation must represent both the unit cost and the initial cost. A verbal model for this problem is as follows.

Verbal Model: Cost = Unit cost · Number of keyboards + Initial cost

Labels:
Cost $= C$ (dollars)
Unit cost $= 40$ (dollars per keyboard)
Number of keyboards $= x$ (keyboards)
Initial cost $= 120{,}000$ (dollars)

Algebraic Model: $C = 40x + 120{,}000$

Using this equation, you can construct the following table of values.

x	1000	2000	3000	4000	5000
$C = 40x + 120{,}000$	160,000	200,000	240,000	280,000	320,000

From the table you can plot the ordered pairs, as shown in Figure 4.8.

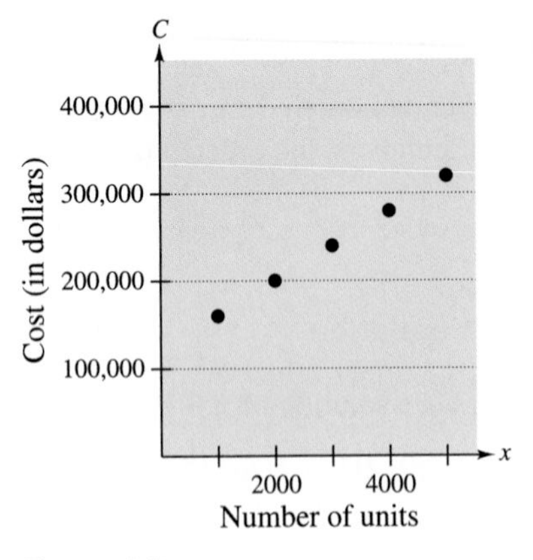

Figure 4.8

Discussing the Concept **Misleading Graphs**

Although graphs can help us visualize relationships between two variables, they can also be misleading. The graphs shown below represent the same data points. Which graph is misleading? Why?

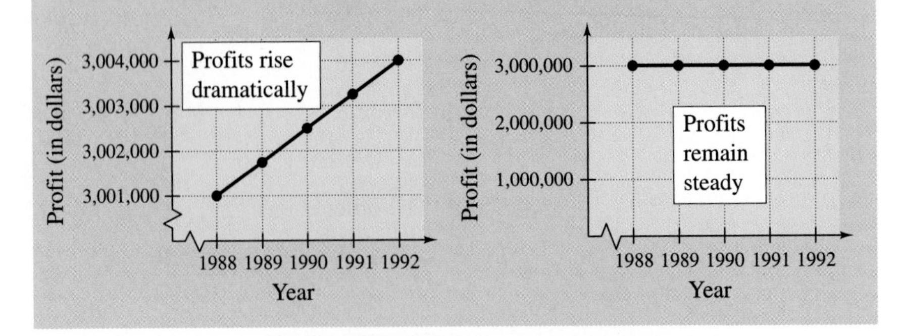

4.1 Exercises

Integrated Review *Concepts, Skills, and Problem Solving*

Keep mathematically in shape by doing these exercises *before* the problems of this section.

Properties and Definitions

1. Is $3x = 7$ a linear equation? Explain. Is $x^2 + 3x = 2$ a linear equation? Explain.
2. Explain how to check whether $x = 3$ is a solution to the equation $5x - 4 = 11$.

Solving Equations

In Exercises 3–10, solve the equation.

3. $-y = 10$
4. $10 - t = 6$
5. $3x - 42 = 0$
6. $64 - 16x = 0$
7. $125(r - 1) = 625$
8. $2(3 - y) = 7y + 5$
9. $20 - \frac{1}{9}x = 4$
10. $0.35x = 70$

Problem Solving

11. The total cost of a lot and house is $154,000. The cost of constructing the house is 7 times the cost of the lot. What is the cost of the lot?
12. You have two summer jobs. In the first job, you work 40 hours a week and earn $9.50 an hour. In the second job, you work as many hours as you want and earn $8 an hour. If you plan to earn $450 a week, how many hours a week should you work at the second job?

Developing Skills

In Exercises 1–10, plot the points on a rectangular coordinate system. See Example 1.

1. $(3, 2), (-4, 2), (2, -4)$
2. $(-1, 6), (-1, -6), (4, 6)$
3. $(-10, -4), (4, -4), (4, 3)$
4. $(-6, 4), (0, 0), (3, -2)$
5. $(-3, 4), (0, -1), (2, -2), (5, 0)$
6. $(-1, 3), (0, 2), (-4, -4), (-1, 0)$
7. $\left(\frac{3}{2}, -1\right), \left(-3, \frac{3}{4}\right), \left(\frac{1}{2}, -\frac{1}{2}\right)$
8. $\left(-\frac{2}{3}, 4\right), \left(\frac{1}{2}, -\frac{5}{2}\right), \left(-4, -\frac{5}{4}\right)$
9. $(3, -4), \left(\frac{5}{2}, 0\right), (0, 3)$
10. $\left(\frac{5}{2}, 2\right), \left(-3, \frac{4}{3}\right), \left(\frac{3}{4}, \frac{9}{4}\right)$

In Exercises 11–14, determine the coordinates of the points. See Example 2.

11. 12.

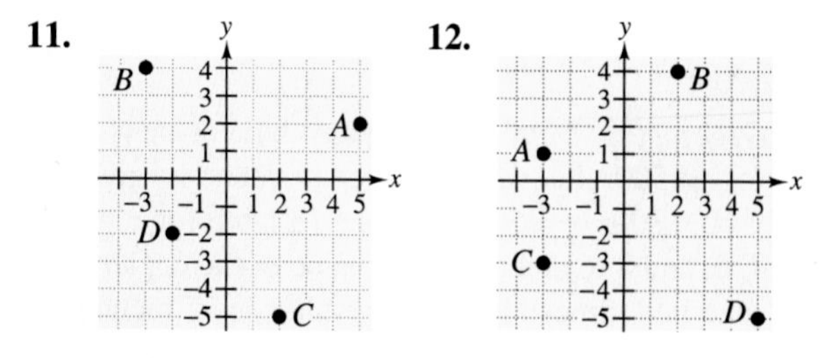

13. 14.

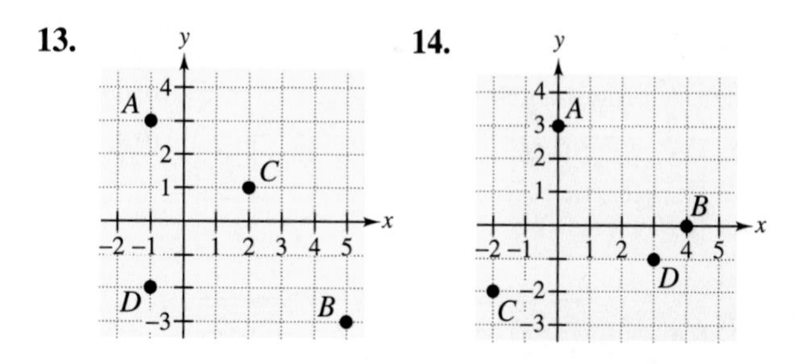

In Exercises 15–20, determine the quadrant in which the point is located.

15. $(-3, 1)$
16. $(4, -3)$
17. $\left(-\frac{1}{8}, -\frac{2}{7}\right)$
18. $\left(\frac{3}{11}, \frac{7}{8}\right)$
19. $(-100, -365.6)$
20. $(-157.4, 305.6)$

In Exercises 21–26, determine the quadrant or quadrants in which the point must be located.

21. $(-5, y)$, y is a real number.
22. $(6, y)$, y is a real number.
23. $(x, -2)$, x is a real number.
24. $(x, 3)$, x is a real number.
25. (x, y), $xy < 0$
26. (x, y), $xy > 0$

In Exercises 27–30, find the coordinates of the point.

27. The point is on the y-axis and 3 units above the x-axis.

28. The point is on the x-axis and 2 units to the left of the y-axis.

29. The point is 5 units to the left of the y-axis and 10 units below the x-axis.

30. The point is 12 units to the right of the y-axis and 4 units below the x-axis.

In Exercises 31–38, plot the points and connect them with line segments to form the figure.

31. Triangle: $(-1, 1), (2, -1), (3, 4)$

32. Triangle: $(0, 3), (-1, -2), (4, 8)$

33. Square: $(2, 4), (5, 1), (2, -2), (-1, 1)$

34. Rectangle: $(2, 1), (4, 2), (-1, 7), (1, 8)$

35. Parallelogram: $(5, 2), (7, 0), (1, -2), (-1, 0)$

36. Parallelogram: $(-1, 1), (0, 4), (4, -2), (5, 1)$

37. Rhombus: $(0, 0), (3, 2), (2, 3), (5, 5)$

38. Rhombus: $(0, 0), (1, 2), (2, 1), (3, 3)$

In Exercises 39–44, complete the table. Plot the results on a rectangular coordinate system. See Example 4.

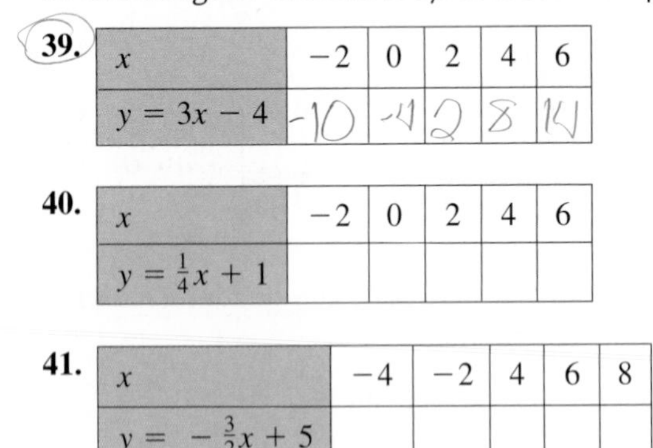

39.

x	-2	0	2	4	6
$y = 3x - 4$	-10	-4	2	8	14

40.

x	-2	0	2	4	6
$y = \frac{1}{4}x + 1$					

41.

x	-4	-2	4	6	8
$y = -\frac{3}{2}x + 5$					

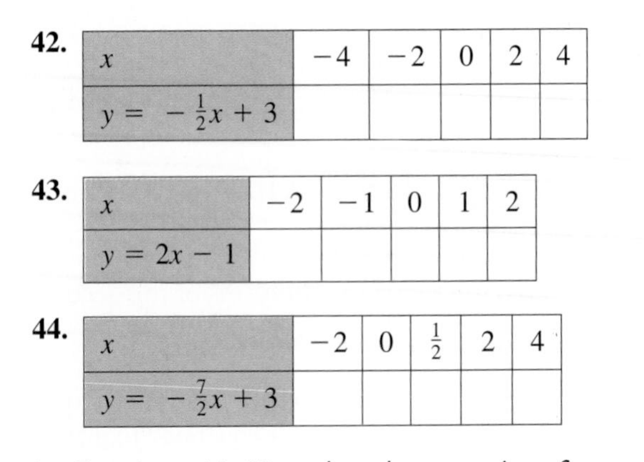

42.

x	-4	-2	0	2	4
$y = -\frac{1}{2}x + 3$					

43.

x	-2	-1	0	1	2
$y = 2x - 1$					

44.

x	-2	0	$\frac{1}{2}$	2	4
$y = -\frac{7}{2}x + 3$					

In Exercises 45–50, solve the equation for y. See Example 4.

45. $6x - 3y = 3$ **46.** $2x + y = 1$ **47.** $x + 4y = 8$

48. $x - 2y = -6$ **49.** $4x - 5y = 3$ **50.** $4y - 3x = 7$

In Exercises 51–58, determine whether each ordered pair is a solution of the equation. See Example 5.

51. $y = 2x + 4$ (a) $(3, 10)$ (b) $(-1, 3)$ (c) $(0, 0)$ (d) $(-2, 0)$

52. $y = 5x - 2$ (a) $(2, 0)$ (b) $(-2, -12)$ (c) $(6, 28)$ (d) $(1, 1)$

53. $2y - 3x + 1 = 0$ (a) $(1, 1)$ (b) $(5, 7)$ (c) $(-3, -1)$ (d) $(-3, -5)$

54. $x - 8y + 10 = 0$ (a) $(-2, 1)$ (b) $(6, 2)$ (c) $(0, -1)$ (d) $(2, -4)$

55. $y = \frac{2}{3}x$ (a) $(6, 6)$ (b) $(-9, -6)$ (c) $(0, 0)$ (d) $\left(-1, \frac{2}{3}\right)$

56. $y = 5x - 2$ (a) $\left(-\frac{4}{5}, 1\right)$ (b) $(0, 0)$ (c) $\left(-\frac{2}{5}, -4\right)$ (d) $\left(\frac{3}{5}, 1\right)$

57. $y = 3 - 4x$ (a) $\left(-\frac{1}{2}, 5\right)$ (b) $(1, 7)$ (c) $(0, 0)$ (d) $\left(-\frac{3}{4}, 0\right)$

58. $y = \frac{3}{2}x + 1$ (a) $\left(0, \frac{3}{2}\right)$ (b) $(4, 7)$ (c) $\left(\frac{2}{3}, 2\right)$ (d) $(-2, -2)$

Solving Problems

59. *Organizing Data* The distance y (in centimeters) a spring is compressed by a force x (in kilograms) is given by $y = 0.066x$. Complete a table for $x = 20$, 40, 60, 80, and 100 to determine the distance the spring is compressed for each of the specified forces. Plot the results on a rectangular coordinate system.

60. *Organizing Data* A company buys a new copier for \$9500. Its value y after x years is given by $y = -800x + 9500$. Complete a table for $x = 0$, 2, 4, 6, and 8 to determine the value of the copier at the specified times. Plot the results on a rectangular coordinate system.

61. *Organizing Data* With an initial cost of \$5000, a company will produce x units at \$35 per unit. Write an equation that relates the total cost of producing x units to the number of units produced. Plot the cost for producing 100, 150, 200, 250, and 300 units.

62. *Organizing Data* An employee earns \$10 plus \$0.50 for every x units produced per hour. Write an equation that relates the employee's total hourly wage to the number of units produced. Plot the hourly wage for producing 2, 5, 8, 10, and 20 units per hour.

63. *Organizing Data* The table gives the normal temperature y (in degrees Fahrenheit) for Anchorage, Alaska for each month x of the year. The months are numbered 1 through 12, with $x = 1$ corresponding to January. (Source: National Oceanic and Atmospheric Administration)

x	1	2	3	4	5	6
y	13	18	24	35	46	54
x	7	8	9	10	11	12
y	58	56	48	35	22	14

(a) Plot the data given in the table.

(b) Did you use the same scale on both axes? Explain.

(c) Using the graph, find the three consecutive months when the normal temperature changes the least.

64. *Organizing Data* The table gives the speed of a car x (in kilometers per hour) and the approximate stopping distance y (in meters).

x	50	70	90	110	130
y	20	35	60	95	148

(a) Plot the data given in the table.

(b) The x-coordinates increase at equal increments of 20 kilometers per hour. Describe the pattern for the y-coordinates. What are the implications for the driver?

65. *Graphical Interpretation* The table gives the numbers of hours x that a student studied for five different algebra exams and the resulting scores y.

x	3.5	1	8	4.5	0.5
y	72	67	95	81	53

(a) Plot the data given in the table.

(b) Use the graph to describe the relationship between the number of hours studied and the resulting exam score.

66. *Graphical Interpretation* The table gives the net income y per share of common stock of the H. J. Heinz Company for the years 1988 through 1997. The year is represented by x. (Source: H. J. Heinz Company 1997 Annual Report)

x	1988	1989	1990	1991	1992
y	\$0.97	\$1.11	\$1.26	\$1.42	\$1.60
x	1993	1994	1995	1996	1997
y	\$1.02	\$1.57	\$1.59	\$1.75	\$0.81

(a) Plot the data given in the table.

(b) Use the graph to find the year that had the greatest increase and the year that had the greatest decrease in the income per share.

Graphical Estimation In Exercises 67–70, use the scatter plot showing new privately-owned housing unit starts (in thousands) in the United States from 1985 through 1997. (Source: U.S. Bureau of the Census)

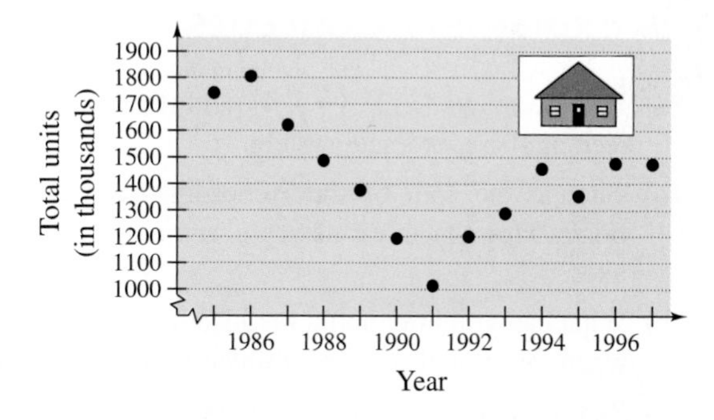

67. Estimate the number of new housing starts in 1986.

68. Estimate the number of new housing starts in 1991.

69. Estimate the increase and the percent increase in housing starts from 1993 to 1994.

70. Estimate the decrease and the percent decrease in housing starts from 1994 to 1995.

Graphical Estimation In Exercises 71–74, use the scatter plot showing the per capita personal income in the United States from 1990 through 1997. (Source: U.S. Bureau of Economic Analysis)

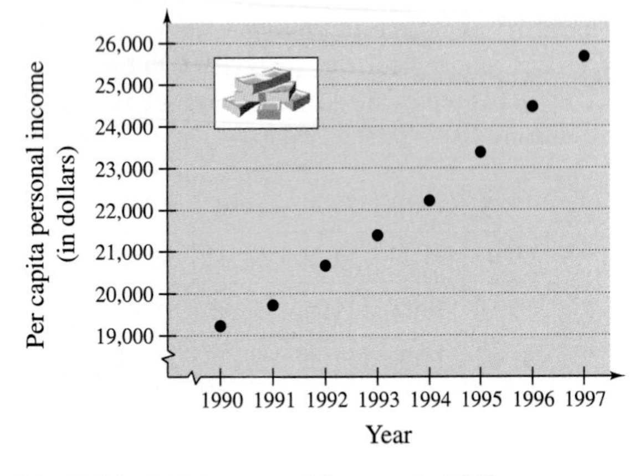

71. Estimate the personal income in 1992.

72. Estimate the personal income in 1995.

73. Estimate the percent increase in personal income from 1996 to 1997.

74. Estimate the percent increase in personal income from 1980 to 1990 if the per capita income in 1980 was $11,892.

Graphical Estimation In Exercises 75 and 76, use the bar graph, which compares the percents of gross domestic product spent on health care in several countries in 1995. (Source: Organization for Economic Cooperation and Development)

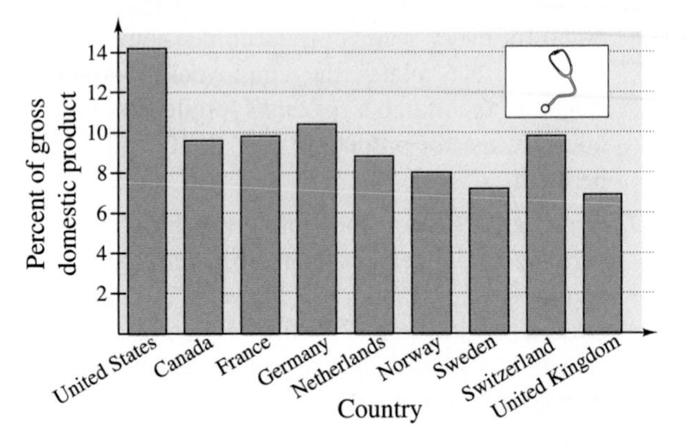

75. Estimate the percent of gross domestic product spent on health care in Sweden.

76. Estimate the percent of gross domestic product spent on health care in the United States.

Explaining Concepts

77. (a) Plot the points $(3, 2)$, $(-5, 4)$, and $(6, -4)$ on a rectangular coordinate system.

(b) Change the sign of the x-coordinate of each point. Plot the three new points on the same axes.

(c) What can you infer about the location of a point when the sign of the x-coordinate is changed?

78. (a) Plot the points $(3, 2)$, $(-5, 4)$, and $(6, -4)$ on a rectangular coordinate system.

(b) Change the sign of the y-coordinate of each point. Plot the three new points on the same axes.

(c) What can you infer about the location of a point when the sign of the y-coordinate is changed?

79. Discuss the significance of the word "ordered" when referring to an ordered pair (x, y).

80. When the point (x, y) is plotted, what does the x-coordinate measure? What does the y-coordinate measure?

81. What is the x-coordinate of any point on the y-axis? What is the y-coordinate of any point on the x-axis?

82. Describe the signs of the x- and y-coordinates of points that lie in the first and second quadrants.

83. Describe the signs of the x- and y-coordinates of points that lie in the third and fourth quadrants.

84. In a rectangular coordinate system, must the scales on the x-axis and y-axis be the same? If not, give an example in which the scales differ.

85. Review the tables in Exercises 39–44 and observe that in some cases the y-coordinates of the solution points increase and in others the y-coordinates decrease. What factor in the equation causes this? Explain.

4.2 Graphs of Equations in Two Variables

Objectives

1. Sketch the graph of an equation using the point-plotting method.
2. Find and use x- and y-intercepts as aids to sketching graphs.
3. Use the verbal problem-solving method to write an equation and sketch its graph.

1 Sketch the graph of an equation using the point-plotting method.

The Graph of an Equation

You have already seen that the solutions of an equation involving two variables can be represented by points on a rectangular coordinate system. The set of all such points is called the **graph** of the equation.

To see how to sketch a graph, let's begin with an example. For instance, consider the equation

$$y = 2x - 1.$$

To begin sketching the graph of this equation, construct a table of values, as shown at the left. Next, plot the solution points on a rectangular coordinate system, as shown in Figure 4.9(a). Finally, find a pattern for the plotted points and use the pattern to connect the points with a smooth curve or line, as shown in Figure 4.9(b).

x	$y = 2x - 1$
-3	-7
-2	-5
-1	-3
0	-1
1	1
2	3
3	5

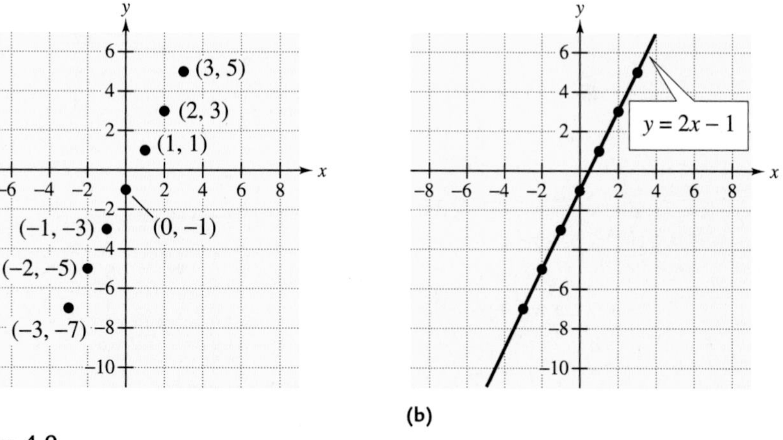

(a) (b)

Figure 4.9

Technology: Tip

To graph an equation using a graphing utility, do the following steps.

(1) Select a viewing rectangle.

(2) Solve the given equation for y in terms of x.

(3) Enter the equation in the equation editor.

(4) Display the graph.

Consult the user's guide of your graphing utility for specific instructions.

▶ The Point-Plotting Method of Sketching a Graph

1. If possible, rewrite the equation by isolating one of the variables.
2. Make up a table of values showing several solution points.
3. Plot these points on a rectangular coordinate system.
4. Connect the points with a smooth curve or line.

Example 1 Sketching the Graph of an Equation

Sketch the graph of $3x + y = 5$.

Solution

To begin, rewrite the equation so that y is isolated on the left.

$$3x + y = 5 \qquad \text{Original equation}$$

$$3x - 3x + y = -3x + 5 \qquad \text{Subtract } 3x \text{ from both sides.}$$

$$y = -3x + 5 \qquad \text{Simplify.}$$

Next, create a table of values, as shown below.

x	-2	-1	0	1	2	3
$y = -3x + 5$	11	8	5	2	-1	-4
Solution	$(-2, 11)$	$(-1, 8)$	$(0, 5)$	$(1, 2)$	$(2, -1)$	$(3, -4)$

Plot the six solution points, as shown in Figure 4.10(a). It appears that all six points lie on a line, so you can complete the sketch by drawing a line through the six points, as shown in Figure 4.10(b).

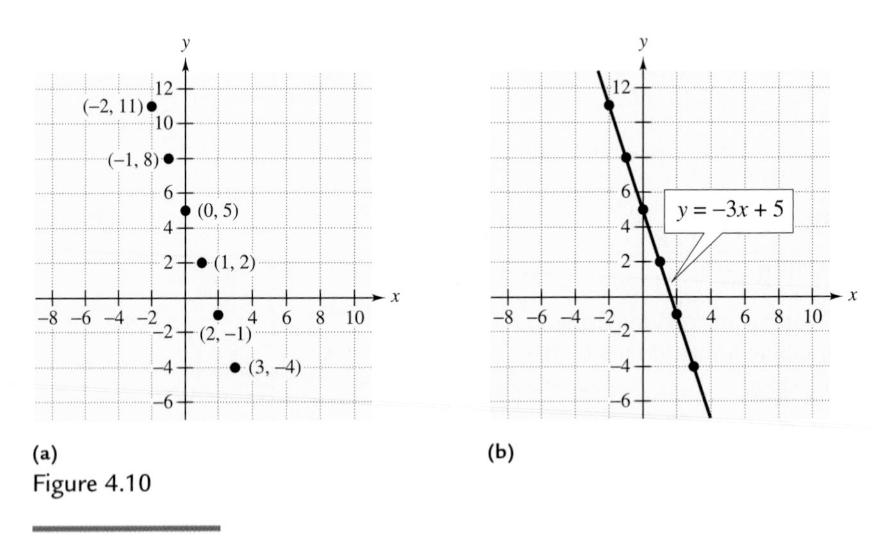

Figure 4.10

When creating a table of values, you are generally free to choose any x-values. When doing this, however, remember that the more x-values you choose, the easier it will be to recognize a pattern.

The equation in Example 1 is an example of a **linear equation** in two variables—the variables are raised to the first power and the graph of the equation is a straight line. As shown in the next two examples, graphs of nonlinear equations are not straight lines.

Technology: Discovery

Most graphing utilities have the following standard viewing window.

Xmin = -10
Xmax = 10
Xscl = 1
Ymin = -10
Ymax = 10
Yscl = 1

What happens when the equation $x + y = 12$ is graphed in a standard viewing window?

To see where the equation crosses the x- and y-axes, you need to change the viewing window. What changes would you make to the viewing window to see where the line intersects the axes?

Graph each of the following equations on a graphing utility and describe the viewing window used.

a. $y = \left|\frac{1}{2}x + 6\right|$

b. $y = 2x^2 + 5x + 10$

c. $y = 10 - x$

d. $y = -3x^3 + 5x + 8$

Example 2 Sketching the Graph of a Nonlinear Equation

Sketch the graph of $x^2 + y = 4$.

Solution

To begin, rewrite the equation so that y is isolated on the left.

$$x^2 + y = 4 \quad \text{Original equation}$$

$$x^2 - x^2 + y = -x^2 + 4 \quad \text{Subtract } x^2 \text{ from both sides.}$$

$$y = -x^2 + 4 \quad \text{Simplify.}$$

Next, create a table of values, as shown below. Be careful with the signs of the numbers when creating a table. For instance, when $x = -3$, the value of y is

$$y = -(-3)^2 + 4$$
$$= -9 + 4$$
$$= -5.$$

x	-3	-2	-1	0	1	2	3
$y = -x^2 + 4$	-5	0	3	4	3	0	-5
Solution	$(-3, -5)$	$(-2, 0)$	$(-1, 3)$	$(0, 4)$	$(1, 3)$	$(2, 0)$	$(3, -5)$

Plot the seven solution points, as shown in Figure 4.11(a). Finally, connect the points with a smooth curve, as shown in Figure 4.11(b).

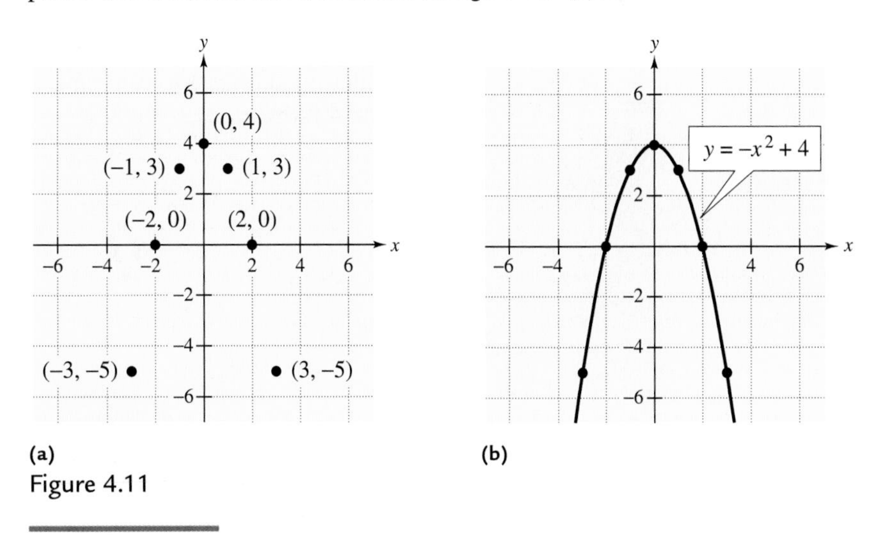

Figure 4.11

The graph of the equation in Example 2 is called a **parabola.** You will study this type of graph in a later chapter.

Example 3 examines the graph of an equation that involves an absolute value. Remember that to find the absolute value of a number, you disregard the sign of the number. For instance, $|-5| = 5$, $|2| = 2$, and $|0| = 0$.

Example 3 The Graph of an Equation Having an Absolute Value

Sketch the graph of $y = |x - 1|$.

Solution

This equation is already written in a form with y isolated on the left. You can begin by creating a table of values, as shown below. Be sure to check the values in this table to make sure that you understand how the absolute value is working. For instance, when $x = -2$, the value of y is

$$
\begin{aligned}
y &= |-2 - 1| \\
&= |-3| \\
&= 3.
\end{aligned}
$$

Similarly, when $x = 2$, the value of y is $|2 - 1|$ or 1.

x	-2	-1	0	1	2	3	4
$y = \|x - 1\|$	3	2	1	0	1	2	3
Solution	$(-2, 3)$	$(-1, 2)$	$(0, 1)$	$(1, 0)$	$(2, 1)$	$(3, 2)$	$(4, 3)$

Plot the seven solution points, as shown in Figure 4.12(a). It appears that the points lie in a "V-shaped" pattern, with the point $(1, 0)$ lying at the bottom of the "V." Following this pattern, you can connect the points to form the graph shown in Figure 4.12(b).

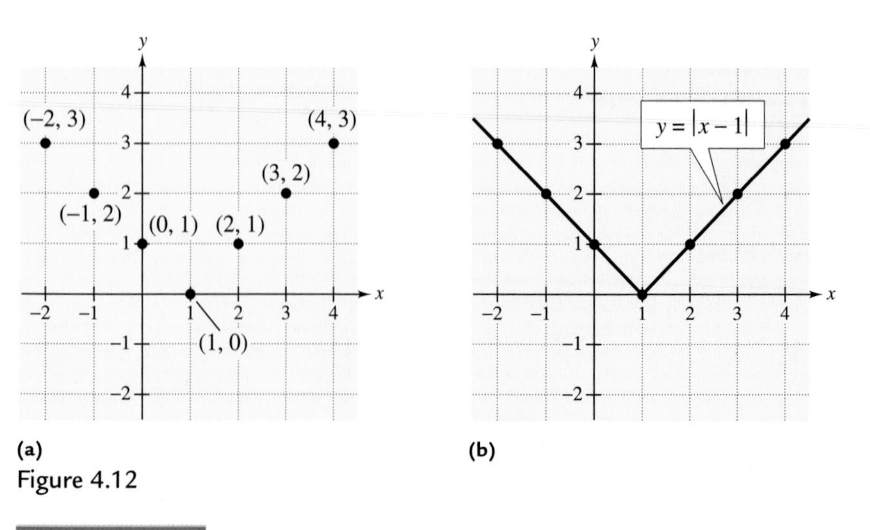

Figure 4.12

2 Find and use x- and y-intercepts as aids to sketching graphs.

Intercepts: An Aid to Sketching Graphs

Two types of solution points that are especially useful are those having zero as either the x- or y-coordinate. Such points are called **intercepts** because they are the points at which the graph intersects the x- or y-axis.

> **Definition of Intercepts**
>
> The point $(a, 0)$ is called an ***x*-intercept** of the graph of an equation if it is a solution point of the equation. To find the x-intercept(s), let y be zero and solve the equation for x.
>
> The point $(0, b)$ is called a ***y*-intercept** of the graph of an equation if it is a solution point of the equation. To find the y-intercept(s), let x be zero and solve the equation for y.

Example 4 Finding the Intercepts of a Graph

Find the intercepts and sketch the graph of

$$y = 2x - 5.$$

Solution

To find any x-intercepts, let $y = 0$ and solve the resulting equation for x.

$y = 2x - 5$ Original equation

$0 = 2x - 5$ Let $y = 0$.

$\frac{5}{2} = x$ Solve equation for x.

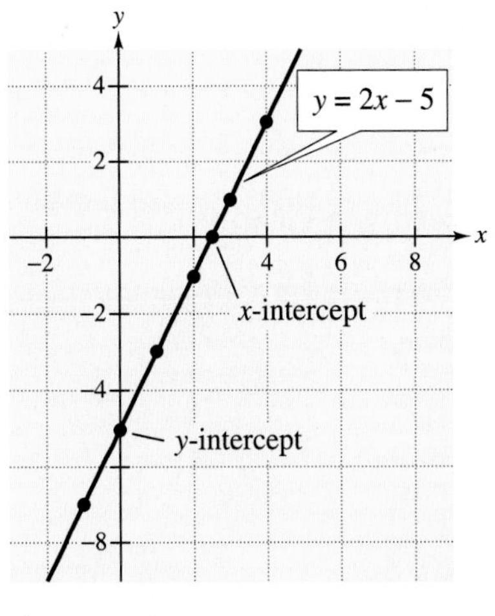

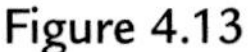
Figure 4.13

So, the graph has one x-intercept, which occurs at the point $\left(\frac{5}{2}, 0\right)$. To find any y-intercepts, let $x = 0$ and solve the resulting equation for y.

$y = 2x - 5$ Original equation

$y = 2(0) - 5$ Let $x = 0$.

$y = -5$ Solve equation for y.

So, the graph has one y-intercept, which occurs at the point $(0, -5)$. To sketch the graph of the equation, create a table of values (as follows). Then plot the points and connect the points with a line, as shown in Figure 4.13.

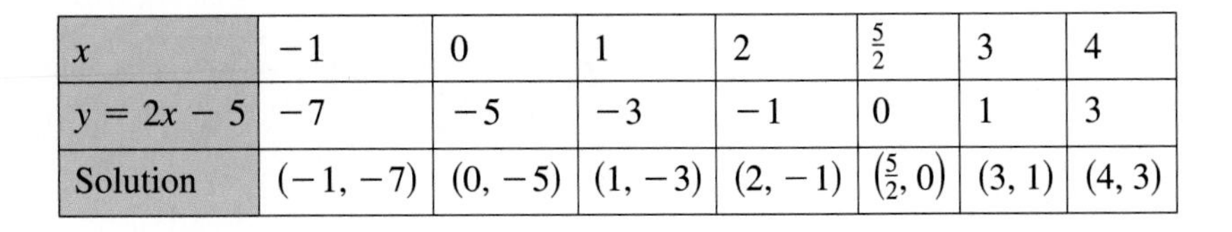

x	-1	0	1	2	$\frac{5}{2}$	3	4
$y = 2x - 5$	-7	-5	-3	-1	0	1	3
Solution	$(-1, -7)$	$(0, -5)$	$(1, -3)$	$(2, -1)$	$\left(\frac{5}{2}, 0\right)$	$(3, 1)$	$(4, 3)$

When you create a table of values, include any intercepts you have found. You should also include points to the left and to the right of the intercepts.

3 Use the verbal problem-solving method to write an equation and sketch its graph.

Application

Example 5 Depreciation

The value of a \$25,500 van depreciates over 10 years. At the end of the 10 years, the salvage value is expected to be \$1500. Find an equation that relates the value of the van to the number of years. Then sketch the graph of the equation. (The depreciation is the same each year.)

Solution

The total depreciation over the 10 years is $25{,}500 - 1500 = \$24{,}000$. Because the same amount is depreciated each year, it follows that the annual depreciation is $24{,}000/10 = \$2400$.

Verbal Model: [Value after t years] = [Original value] − [Annual depreciation] · [Number of years]

Labels:

Value after t years $= y$ (dollars)
Original value $= 25{,}500$ (dollars)
Annual depreciation $= 2400$ (dollars per year)
Number of years $= t$ (years)

Algebraic Model: $y = 25{,}500 - 2400t$

A sketch of the graph of this equation is shown in Figure 4.14. Note that the y-intercept $(0, 25{,}500)$ corresponds to the original value of the van.

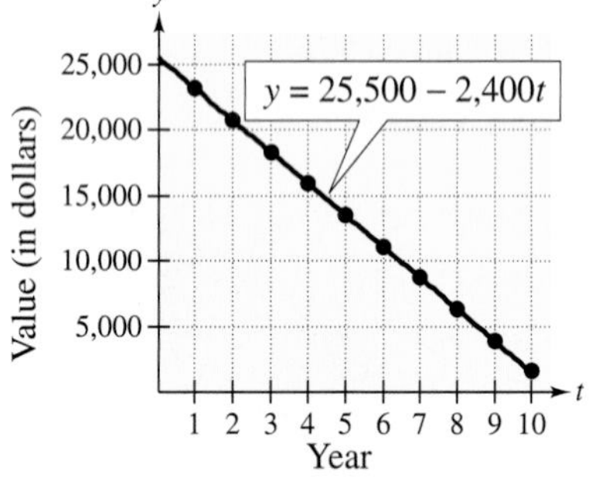

Figure 4.14

Discussing the Concept — Interpreting a Graph

The graph below shows the speed of a delivery van on a 22-minute trip. Write a short paragraph describing the trip.

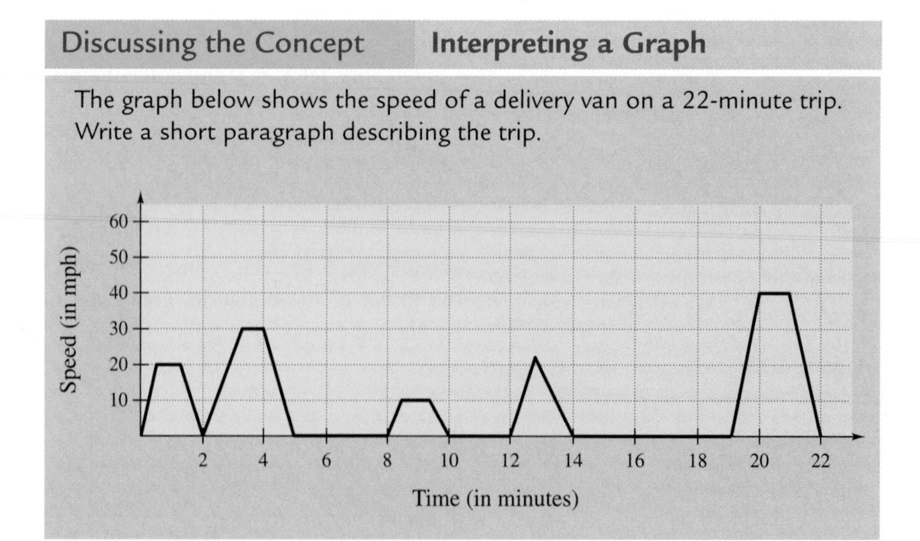

4.2 Exercises

Integrated Review *Concepts, Skills, and Problem Solving*

Keep mathematically in shape by doing these exercises *before* the problems of this section.

Properties and Definitions

1. If $x - 2 > 5$ and c is an algebraic expression, then what is the relationship between $x - 2 + c$ and $5 + c$?

2. If $x - 2 < 5$ and $c < 0$, then what is the relationship between $(x - 2)c$ and $5c$?

3. Complete the Multiplicative Inverse Property: $x(1/x) =$ ____ .

4. Name the property illustrated by $x + y = y + x$.

Simplifying Expressions

In Exercises 5–10, simplify the expression.

5. $-3(3x - 2y) + 5y$ **6.** $3z - (4 - 5z)$

7. $-y^2(y^2 + 4) + 6y^2$ **8.** $5t(2 - t) + t^2$

9. $3[6x - 5(x - 2)]$ **10.** $5(t - 2) - 5(t - 2)$

Problem Solving

11. A company pays its sales representatives \$30 per day plus 32 cents per mile for the use of their personal cars. A sales representative submits a bill for \$50.80 for driving her own car.

(a) How many miles did she drive?

(b) How many days did she drive? Explain.

(c) Suppose the bill had been submitted for \$96.80. Could you determine how many days and miles were claimed? Explain.

12. The width of a rectangular mirror is $\frac{3}{5}$ its length. The perimeter of the mirror is 80 inches. What are the measurements of the mirror?

Developing Skills

In Exercises 1–8, match the equation with its graph. [The graphs are labeled (a), (b), (c), (d), (e), (f), (g), and (h).]

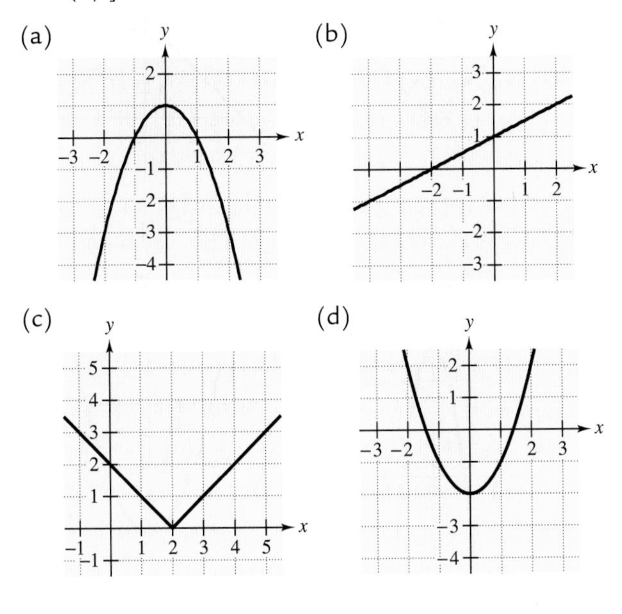

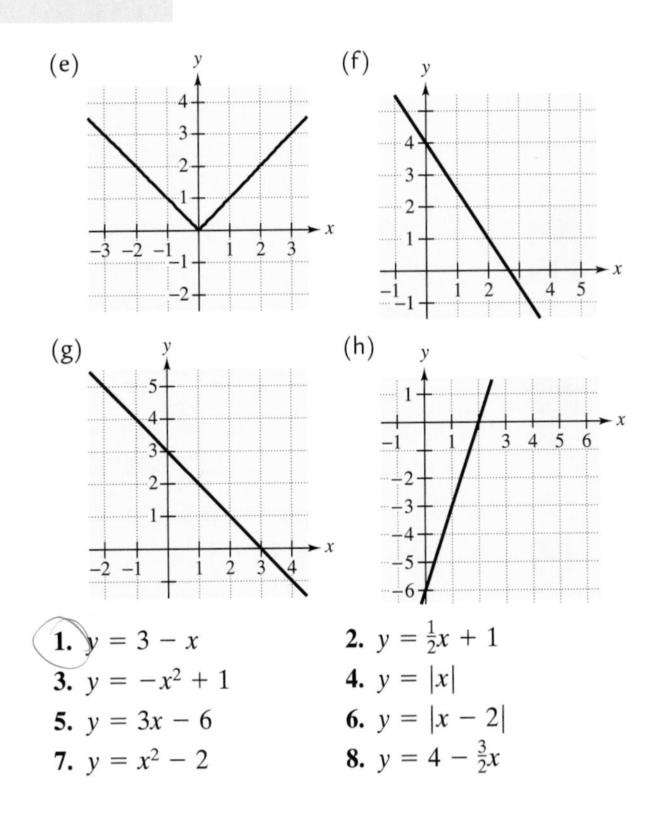

1. $y = 3 - x$ **2.** $y = \frac{1}{2}x + 1$

3. $y = -x^2 + 1$ **4.** $y = |x|$

5. $y = 3x - 6$ **6.** $y = |x - 2|$

7. $y = x^2 - 2$ **8.** $y = 4 - \frac{3}{2}x$

In Exercises 9–16, complete the table and use the results to sketch a graph of the equation. See Examples 1–3.

9. $y = 9 - x$

x	-2	-1	0	1	2
y					

10. $y = x - 1$

x	-2	-1	0	1	2
y					

11. $y = 4x - 2$

x	-2	-1	0	1	2
y					

12. $y = 7 - \frac{3}{2}x$

x	-2	0	2	4	6
y					

13. $x + 2y = 4$

x	-2	0	2	4	6
y					

14. $3x - 2y = 6$

x	-2	0	2	4	6
y					

15. $y = |x + 1|$

x	-3	-2	-1	0	1
y					

16. $y = (x - 1)^2$

x	-1	0	1	2	3
y					

In Exercises 17–24, estimate the x- and y-intercepts from the graph. Check your results algebraically.

17. $4x - 2y = -8$ **18.** $5y - 2x = 10$

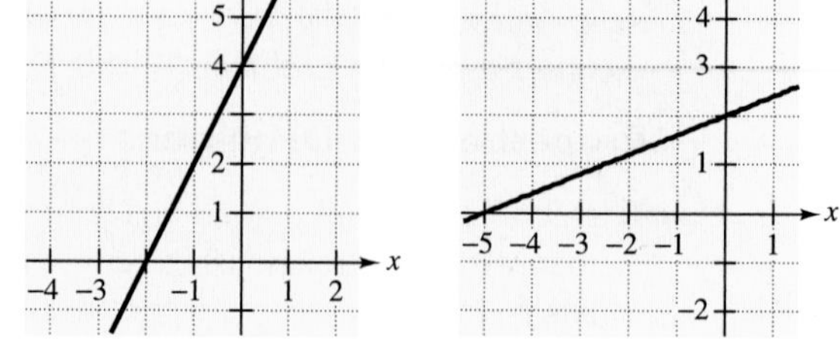

19. $x + 3y = 6$ **20.** $4x + 3y = 12$

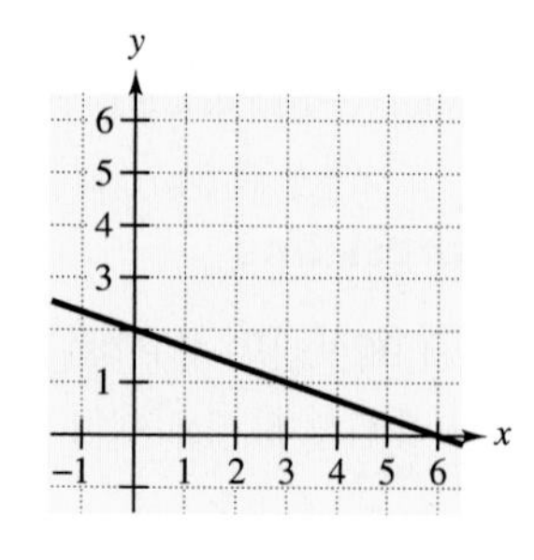

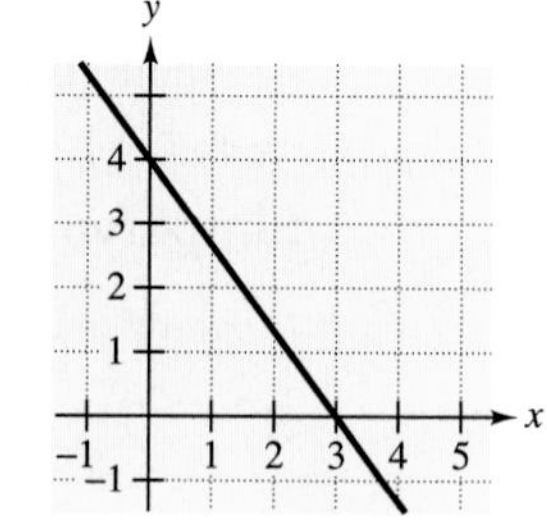

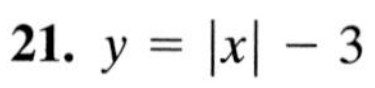

21. $y = |x| - 3$

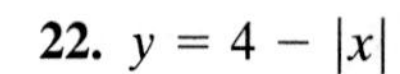

22. $y = 4 - |x|$

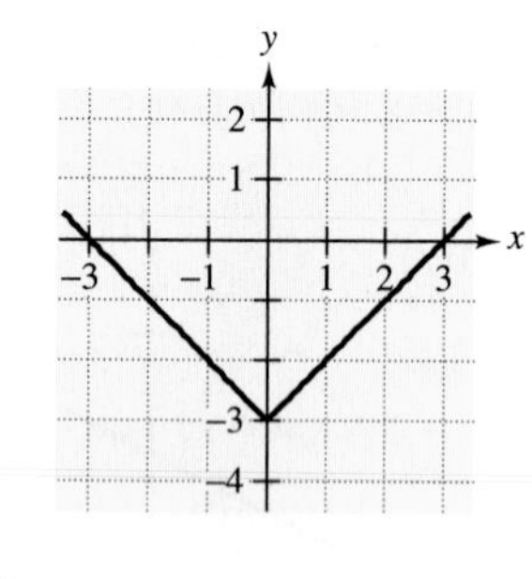

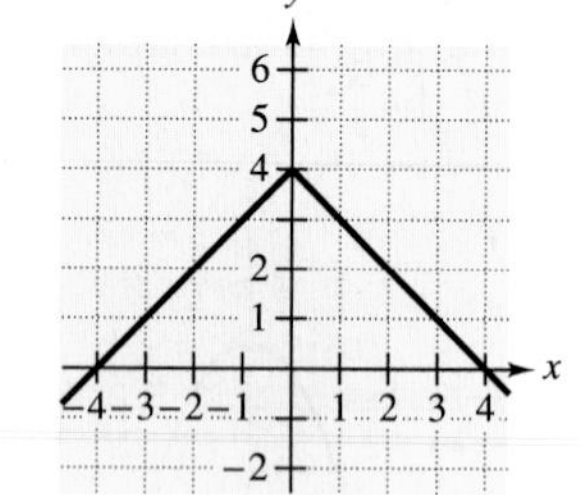

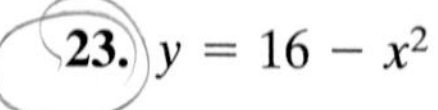

23. $y = 16 - x^2$ **24.** $y = x^2 - 4$

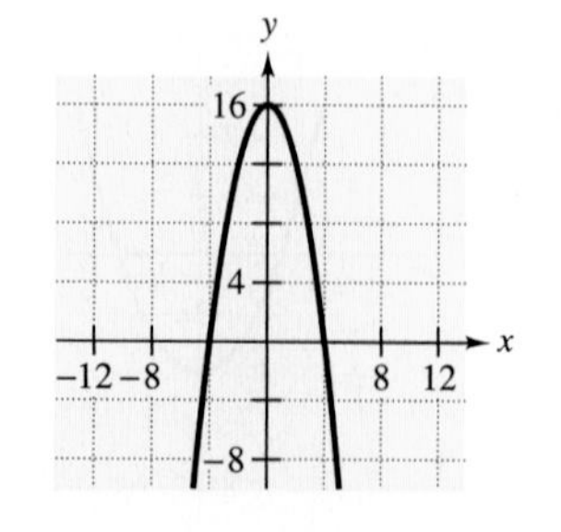

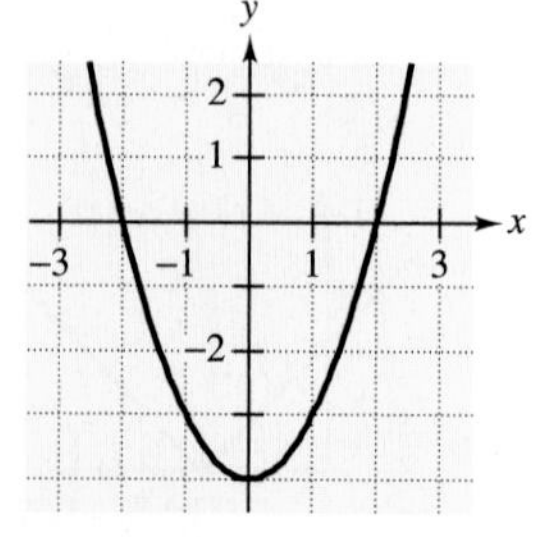

In Exercises 25–36, find the x- and y-intercepts (if any) of the graph of the equation. See Example 4.

25. $y = 6x + 2$ **26.** $y = -3x + 5$

27. $y = \frac{1}{2}x - 1$ **28.** $y = -\frac{1}{2}x + 3$

29. $x - y = 1$ **30.** $x + y = 10$

31. $2x + y = 4$ **32.** $3x - 2y = 1$

33. $2x + 6y - 9 = 0$ **34.** $2x - 5y + 50 = 0$

35. $\frac{3}{4}x - \frac{1}{2}y = 3$ **36.** $\frac{1}{2}x + \frac{2}{3}y = 1$

In Exercises 37–62, sketch the graph of the equation and label the coordinates of at least three solution points.

37. $y = 2 - x$ **38.** $y = x + 3$

39. $y = x - 1$ **40.** $y = 5 - x$

41. $y = 3x$ **42.** $y = -2x$

43. $2x - y = 4$ **44.** $2x + y = -2$

45. $10x + 5y = 20$ **46.** $7x - 7y = 14$

47. $4x + y = 2$ **48.** $y - 2x = 3$

49. $y = \frac{3}{8}x + 15$ **50.** $y = 14 - \frac{2}{3}x$

51. $y = \frac{2}{3}x - 5$ **52.** $y = \frac{3}{2}x + 3$

53. $y = x^2$ **54.** $y = -x^2$

55. $y = -x^2 + 9$ **56.** $y = x^2 - 1$

57. $y = (x - 3)^2$ **58.** $y = -(x + 2)^2$

59. $y = |x - 5|$ **60.** $y = |x + 3|$

61. $y = 5 - |x|$ **62.** $y = |x| - 3$

In Exercises 63–66, use a graphing utility to graph both equations on the same screen. Are the graphs identical? If so, what rule of algebra is being illustrated?

63. $y_1 = \frac{1}{3}x - 1$, $y_2 = -1 + \frac{1}{3}x$ **64.** $y_1 = 3(\frac{1}{4}x)$, $y_2 = (3 \cdot \frac{1}{4})x$

65. $y_1 = 2(x - 2)$, $y_2 = 2x - 4$ **66.** $y_1 = 2 + (x + 4)$, $y_2 = (2 + x) + 4$

In Exercises 67–74, use a graphing utility to graph the equation. (Use a standard setting.)

67. $y = 4x$ **68.** $y = -2x$

69. $y = -\frac{1}{3}x$ **70.** $y = \frac{1}{2}x$

71. $y = -2x^2 + 5$ **72.** $y = x^2 - 7$

73. $y = |x + 1| - 2$ **74.** $y = 4 - |x - 2|$

In Exercises 75–78, use a graphing utility to graph the equation. Use the viewing window given.

75. $y = 25 - 5x$

Xmin = -5
Xmax = 7
Xscl = 1
Ymin = -5
Ymax = 30
Yscl = 5

76. $y = 8x + 20$

Xmin = -5
Xmax = 5
Xscl = 1
Ymin = -10
Ymax = 30
Yscl = 5

77. $y = 2.3x - 4.1$

Xmin = -5
Xmax = 5
Xscl = 1
Ymin = -10
Ymax = 5
Yscl = 1

78. $y = 1.7 - 0.1x$

Xmin = -10
Xmax = 25
Xscl = 5
Ymin = -5
Ymax = 5
Yscl = .5

In Exercises 79–82, use a graphing utility to graph the equation and find a viewing window that yields a graph that matches the one shown.

79. $y = \frac{1}{2}x + 2$ **80.** $y = 2x - 1$

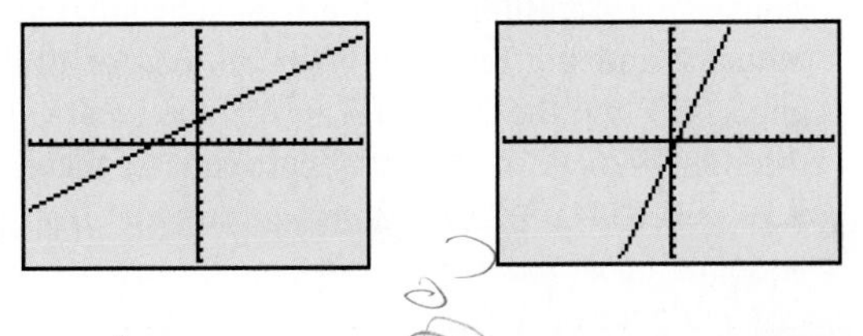

81. $y = \frac{1}{4}x^2 - 4x + 12$ **82.** $y = 16 - 4x - x^2$

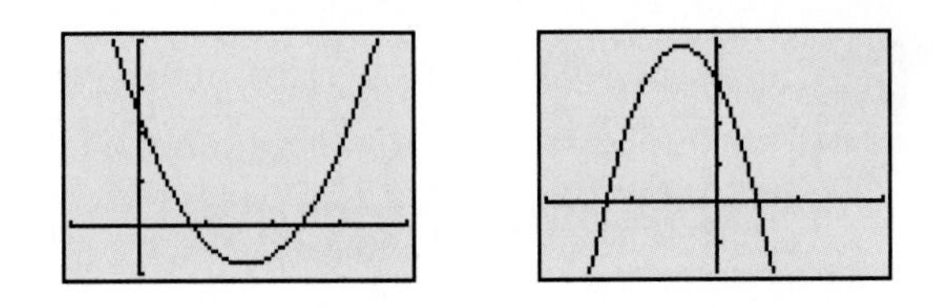

Solving Problems

83. *Creating a Model* Let y represent the distance traveled by a car that is moving at a constant speed of 35 miles per hour. Let t represent the number of hours the car has traveled. Write an equation that relates y to t and sketch its graph.

84. *Creating a Model* The cost of printing a book is \$500, plus \$5 per book. Let C represent the total cost and let x represent the number of books. Write an equation that relates C and x and sketch its graph.

85. *Modeling Data* The table gives the life expectancy (in years) in the United States for a child at birth for various years. A model for the life expectancy during this period is $y = 0.2t + 66.7$, with $t = 0$ corresponding to 1950. (Source: U.S. Bureau of the Census)

t	-10	0	10	20	30	40	45
y	62.9	68.2	69.7	70.8	73.7	75.4	75.8

(a) Graph the data and the model.

(b) Predict the life expectancy for a child born in 2010.

86. *Modeling Data* The table gives the number of passengers x (in millions) in the U.S. scheduled airline industry and the revenue y (in billions of dollars) generated by the passengers for the years 1990 through 1996. A model that approximates these data is $y = 0.1344x - 3.7122$. (Source: Air Transport Association of America)

Year	1990	1991	1992	1993	1994	1995	1996
x	465.6	452.3	475.1	488.5	528.8	547.8	581.2
y	58.5	57.1	59.8	63.9	65.4	69.6	75.3

(a) Plot the points that represent the actual data.

(b) On the same axes, graph the model.

(c) Predict passenger-generated revenue for a year when there are 600 million passengers.

87. *Graphical Comparisons* The graphs of two types of depreciation are shown. In one type, called *straight-line depreciation,* the value depreciates by the same amount each year. In the other type, called *declining balances,* the value depreciates by the same percent each year. Which is which?

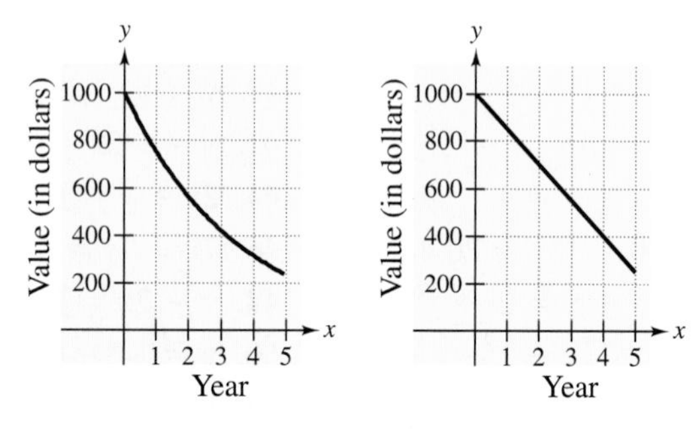

Figures for 87

88. *Graphical Interpretation* In Exercise 87, what is the original cost of the equipment that is being depreciated?

89. *Writing* Compare the benefits and disadvantages of the two types of depreciation shown in Exercise 87.

90. *Interpreting Intercepts* The model $5F - 9C = 160$ relates the temperature in degrees Celsius C and degrees Fahrenheit F.

(a) Graph the equation where F is measured on the horizontal axis.

(b) Explain what the intercepts represent.

Explaining Concepts

91. In your own words, define what is meant by the *graph* of an equation.

92. How many solution points can an equation in two variables have? How many points do you need to plot the general shape of the graph?

93. In your own words, describe the point-plotting method of sketching the graph of an equation.

94. In your own words, describe how you can check that an ordered pair (x, y) is a solution of an equation.

95. Explain how to find the x- and y-intercepts of a graph.

96. You are walking toward an object. Let x represent the time (in seconds) and let y represent the distance (in feet) between you and the object. Sketch a possible graph that shows how x and y are related.

97. *Research Project* Use a newspaper or a weekly news magazine to find examples of misleading graphs and explain why they are misleading.

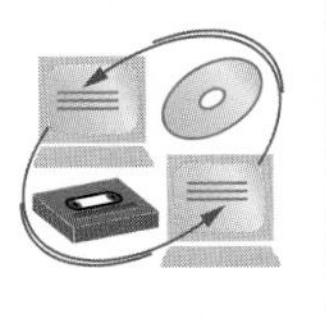

4.3 Relations, Functions, and Graphs

Objectives

1. Identify the domain and range of a relation.
2. Determine if a relation is a function by inspection or by using the Vertical Line Test.
3. Use function notation and evaluate a function.

1 Identify the domain and range of a relation.

Relations

Many everyday occurrences involve pairs of quantities that are matched with each other by some rule of correspondence. For instance, each person is matched with a birth month (person, month); the number of hours worked is matched with a paycheck (hours, pay); an instructor is matched with a course (instructor, course); and the time of day is matched with the outside temperature (time, temperature). In each instance, sets of ordered pairs can be formed. Such sets of ordered pairs are called **relations.**

> **Definition of a Relation**
>
> A **relation** is any set of ordered pairs. The set of first components in the ordered pairs is the **domain** of the relation. The set of second components is the **range** of the relation.

In mathematics, relations are commonly described by ordered pairs of *numbers*. The set of x-coordinates is the domain and the set of y-coordinates is the range. In the relation

$$\{(3, 5), (1, 2), (4, 4), (0, 3)\}$$

the domain D and range R are the sets

$$D = \{3, 1, 4, 0\} \quad \text{and} \quad R = \{5, 2, 4, 3\}.$$

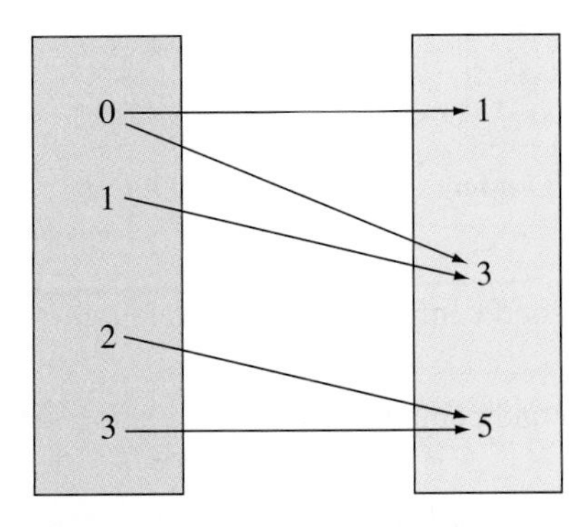

Figure 4.15

Example 1 Analyzing a Relation

Find the domain and range of the relation

$$\{(0, 1), (1, 3), (2, 5), (3, 5), (0, 3)\}.$$

Then sketch a graphic representation of the relation.

Solution

The domain and range are

$$D = \{0, 1, 2, 3\} \quad \text{and} \quad R = \{1, 3, 5\}.$$

A graphic representation is shown in Figure 4.15.

2 Determine if a relation is a function by inspection or by using the Vertical Line Test.

Functions

In the study of mathematics and its applications, the focus is mainly on a special type of relation, called a **function.**

> **Definition of a Function**
>
> A **function** is a relation in which no two ordered pairs have the same first component and a different second component.

This definition means that a given first component cannot be paired with two different second components. For instance, the pairs $(1, 3)$ and $(1, -1)$ cannot be part of a function.

Consider the relations described at the beginning of this section.

Relation	*Ordered Pairs*	*Sample Relation*
1	(person, month)	{(A, May), (B, Dec), (C, Oct), . . .}
2	(hours, pay)	{(12, 84), (4, 28), (6, 42), (15, 105), . . .}
3	(instructor, course)	{(A, MATH001), (A, MATH002), . . .}
4	(time, temperature)	{(8, 70°), (10, 78°), (12, 78°), . . .}

The first relation *is* a function because each person has only one birth month. The second relation *is* a function because the given number of hours worked at a particular job can yield only *one* paycheck amount. The third relation *is not* a function because an instructor can teach more than one course. The fourth relation *is* a function. Note that the ordered pairs (10, 78°) and (12, 78°) do not violate the definition of a function.

Study Tip

The ordered pairs of a relation can be thought of in the form (input, output). For a *function*, a given input cannot yield two different outputs. For instance, if the input is a person's name and the output is that person's month of birth, then your name as the input can yield only your month of birth as the output.

Example 2 Testing Whether a Relation Is a Function

Which of the relations are functions?

a. Input: a, b, c
Output: 2, 3, 4
$\{(a, 2), (b, 3), (c, 4)\}$

b.

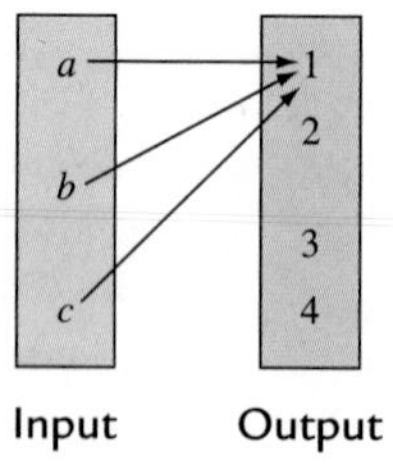

c.

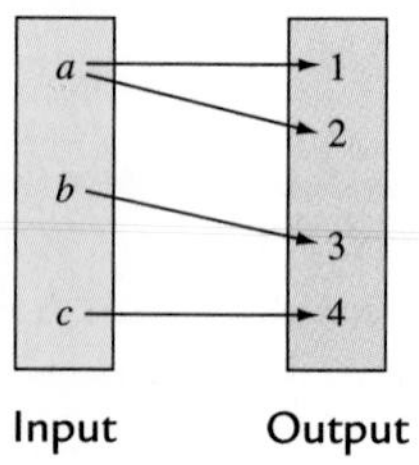

Solution

a. No first component has two different second components, so the relation *is* a function.

b. No first component has two different second components, so the relation *is* a function.

c. Because the first component a is paired with two different second components, this relation *is not* a function.

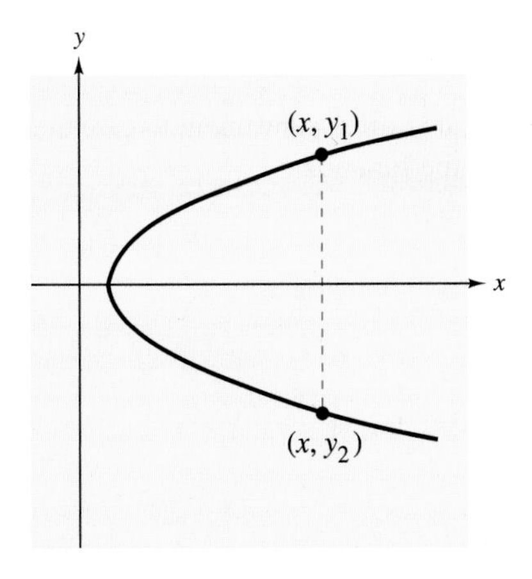

Figure 4.16

In algebra, it is common to represent functions by equations in two variables rather than by ordered pairs. For instance, the equation $y = x^2$ represents the variable y as a function of x. The variable x is the **independent variable** (the input) and y is the **dependent variable** (the output). In this context, the domain of the function is the set of all *allowable* values for x, and the range is the *resulting* set of all values taken on by the dependent variable y.

From the graph of an equation, it is easy to determine whether the equation represents y as a function of x. For instance, the graph in Figure 4.16 *does not* represent a function of x because the indicated value of x is paired with two y-values. Graphically, this means that a vertical line intersects the graph more than once. See Figure 4.16.

> **Vertical Line Test**
>
> A graph is not the graph of a function if a vertical line can be drawn that intersects the graph at more than one point.

Example 3 Using the Vertical Line Test for Functions

Determine if the relation is a function using the Vertical Line Test.

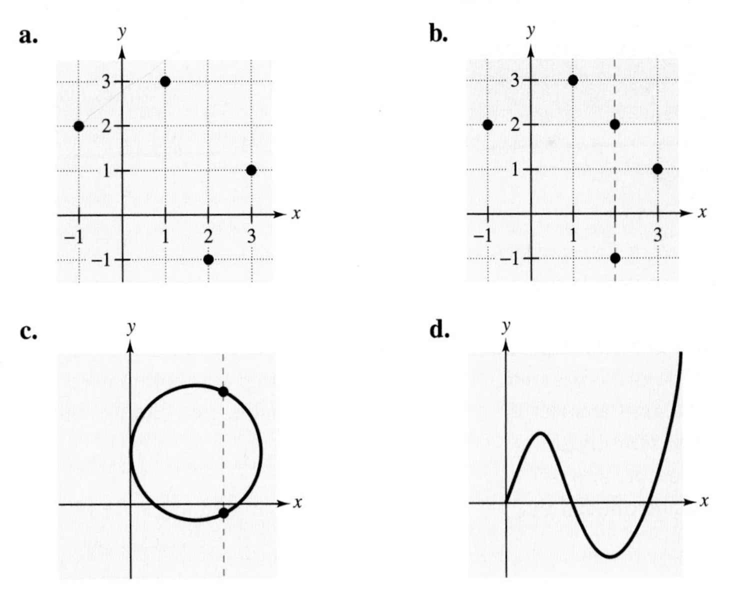

Solution

a. This relation *is* a function. No vertical line intersects more than one point on the graph.

b. This relation *is not* a function. A vertical line intersects more than one point on the graph.

c. This relation *is not* a function. A vertical line intersects more than one point on the graph.

d. This relation *is* a function. No vertical line intersects more than one point on the graph.

3 Use function notation and evaluate a function.

Function Notation

To discuss functions represented by equations, it is often convenient to give them names using **function notation.** For instance, the function

$$y = 2x - 6$$

can be given the name "f" and written in function notation as

$$f(x) = 2x - 6.$$

The Granger Collection

Leonhard Euler
(1707–1783)
Leonhard Euler, a Swiss mathematician, is considered to have been the most prolific and productive mathematician in history. One of his greatest influences on mathematics was his use of symbols, or notation. The notation $y = f(x)$ was introduced by Euler.

▶ $f(x)$ Notation

In the notation $f(x)$:

f is the **name** of the function.
x is a **domain** value.
$f(x)$ is a **range** value y for a given x.

The symbol $f(x)$ is read as **the value of f at x** or simply **f of x.**

The process of finding $f(x)$ for a given value of x is called **evaluating the function.** This is accomplished by substituting the given x-value (input) into the equation and obtaining the value of $f(x)$ (output). Here's an illustration.

Function	*x-Values (input)*	*f(x)-Values (output)*
$f(x) = 4 - 3x$	$x = -2$	$f(-2) = 4 - 3(-2) = 4 + 6 = 10$
	$x = -1$	$f(-1) = 4 - 3(-1) = 4 + 3 = 7$
	$x = 0$	$f(0) = 4 - 3(0) = 4 - 0 = 4$
	$x = 2$	$f(2) = 4 - 3(2) = 4 - 6 = -2$
	$x = 3$	$f(3) = 4 - 3(3) = 4 - 9 = -5$

Although f and x are often used as a convenient function name and independent (input) variable, you can use other letters. For instance, the equations

$$f(x) = x^2 - 3x + 5, \quad f(t) = t^2 - 3t + 5, \quad \text{and} \quad g(s) = s^2 - 3s + 5$$

all describe the same function. In fact, the letters used are just "place holders" and the same function is well described by the form

$$f(\quad) = (\quad)^2 - 3(\quad) + 5.$$

You can evaluate $f(-2)$ as

$$\begin{aligned} f(-2) &= (-2)^2 - 3(-2) + 5 \\ &= 4 + 6 + 5 \\ &= 15. \end{aligned}$$

It is important to put parentheses around the x-value (input) and then simplify the result.

Example 4 Evaluating a Function

Given $f(x) = x^2 + 1$ and $g(x) = 3x - x^2$, find the following.

a. $f(-2)$ **b.** $f(0)$ **c.** $g(2)$ **d.** $g(0)$

Solution

a. $f(x) = x^2 + 1$ Given function

$f(-2) = (-2)^2 + 1$ Substitute -2 for x.

$= 4 + 1 = 5$ Simplify.

b. $f(x) = x^2 + 1$ Given function

$f(0) = (0)^2 + 1$ Substitute 0 for x.

$= 0 + 1 = 1$ Simplify.

c. $g(x) = 3x - x^2$ Given function

$g(2) = 3(2) - (2)^2$ Substitute 2 for x.

$= 6 - 4 = 2$ Simplify.

d. $g(x) = 3x - x^2$ Given function

$g(0) = 3(0) - (0)^2$ Substitute 0 for x.

$= 0 - 0 = 0$ Simplify.

The domain of a function may be explicitly described along with the function, or it may be *implied* by the context in which the function is used. For instance, if weekly pay is a function of hours worked, the implied domain is typically the interval $0 \le x \le 40$. Certainly x cannot be negative in this context.

Example 5 Finding the Range of a Function

Determine the range R for the specified domain D of the function. Graph the function over the given domain.

$$f(x) = 2 - x^2, \qquad D = \{-2, -1, 0, 1\}$$

Solution

To find the range, substitute the values of the domain into the function.

$$f(-2) = 2 - (-2)^2 = 2 - 4 = -2$$

$$f(-1) = 2 - (-1)^2 = 2 - 1 = 1$$

$$f(0) = 2 - (0)^2 = 2 - 0 = 2$$

$$f(1) = 2 - (1)^2 = 2 - 1 = 1$$

The range of the function over the given domain is $R = \{-2, 1, 2\}$. The graph of the function is shown in Figure 4.17.

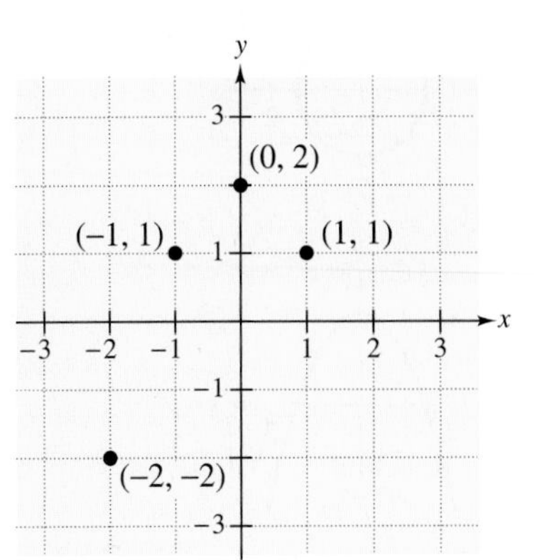

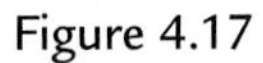
Figure 4.17

Example 6 Finding an Equation to Represent a Function

Is the area of a square a *function* of the length of one of its sides? If so, find an equation that represents this function.

Solution

Figure 4.18 shows a square.

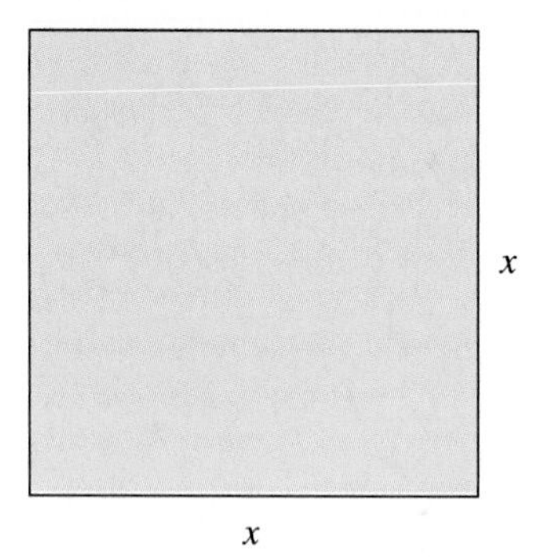

Figure 4.18

For this square, let the variable A represent the area of the square, and let the variable x represent the length of any one of its sides. (Remember that, by definition, all sides of a square have the same length.) Because the area of a square is completely determined by the lengths of its sides, you can see that A *is* a function of x. The equation that represents the function is

$$A(x) = x^2.$$

Discussing the Concept Matching Equations with Graphs

Match the equations with their graphs. Discuss which equations are easier to graph using a graphing utility. Which equations represent y as a function of x? What can you conclude?

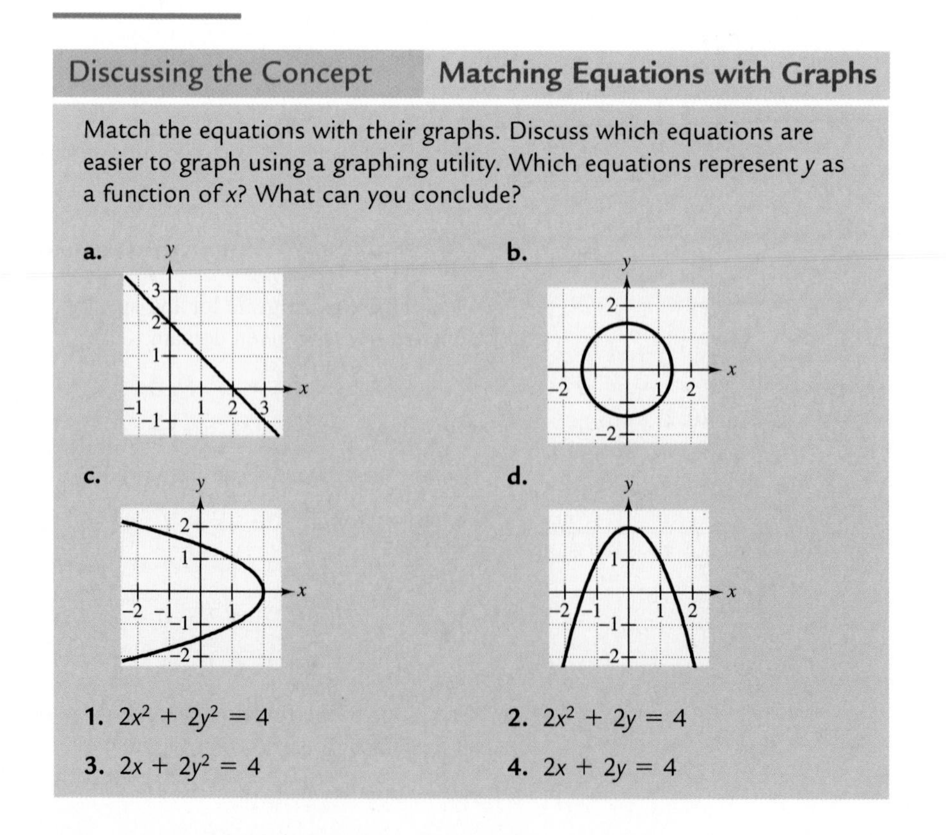

1. $2x^2 + 2y^2 = 4$
2. $2x^2 + 2y = 4$
3. $2x + 2y^2 = 4$
4. $2x + 2y = 4$

4.3 Exercises

Integrated Review *Concepts, Skills, and Problem Solving*

Keep mathematically in shape by doing these exercises *before* the problems of this section.

Properties and Definitions

1. If $a < b$ and $b < c$, then what is the relationship between a and c? Name this property.

2. Demonstrate the Multiplicative Property of Equality for the equation $7x = 21$.

Simplifying Expressions

In Exercises 3–6, simplify the expression.

3. $4s - 6t + 7s + t$ **4.** $2x^2 - 4 + 5 - 3x^2$

5. $\frac{5}{3}x - \frac{2}{3}x - 4$

6. $3x^2y + xy - xy^2 - 6xy$

Solving Equations

In Exercises 7–10, solve the equation.

7. $3x + 9 = 0$ **8.** $\dfrac{x}{4} + \dfrac{x}{3} = \dfrac{1}{3}$

9. $\dfrac{2x - 3}{4} = \dfrac{3}{2}$ **10.** $-(4 - 3x) = 2(x - 1)$

Problem Solving

11. An inheritance of $7500 is invested in a mutual fund and at the end of 1 year the value of the investment is $8190. What simple interest rate would yield the same growth?

12. Determine the average speed of an aircraft that can travel 2500 miles in 3 hours.

Developing Skills

In Exercises 1–6, find the domain and range of the relation. See Example 1.

1. $\{(-4, 3), (2, 5), (1, 2), (4, -3)\}$

2. $\{(-1, 5), (8, 3), (4, 6), (-5, -2)\}$

3. $\{(2, 16), (-9, -10), (\frac{1}{2}, 0)\}$

4. $\{(\frac{2}{3}, -4), (-6, \frac{1}{4}), (0, 0)\}$

5. $\{(-1, 3), (5, -7), (-1, 4), (8, -2), (1, -7)\}$

6. $\{(1, 1), (2, 4), (3, 9), (-2, 4), (-1, 1)\}$

In Exercises 7–24, is the relation a function? See Example 2.

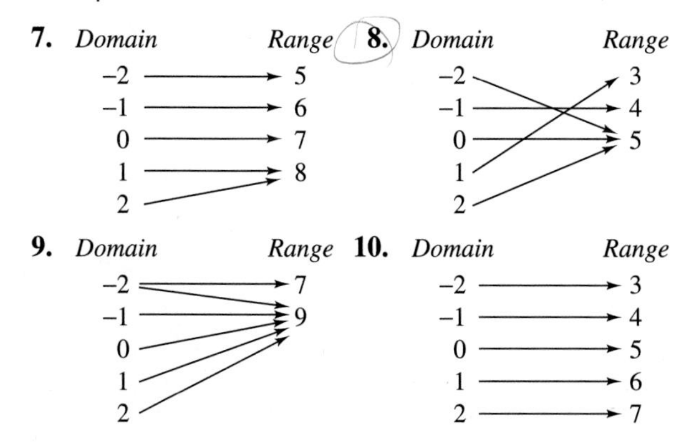

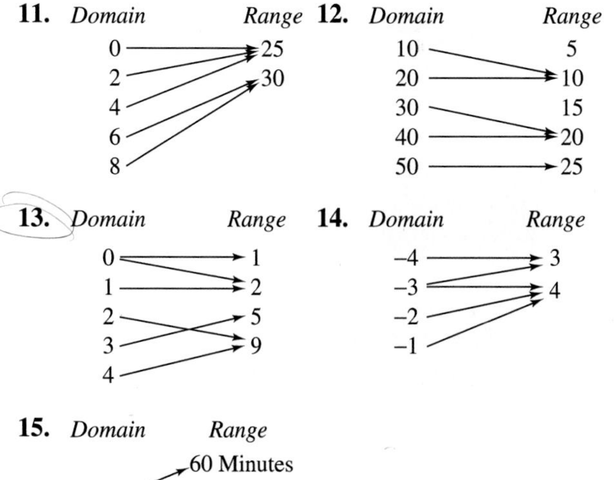

15. *Domain* *Range*

CBS → 60 Minutes, Chicago Hope, Dan Rather

ABC → 20/20, The Practice, Peter Jennings

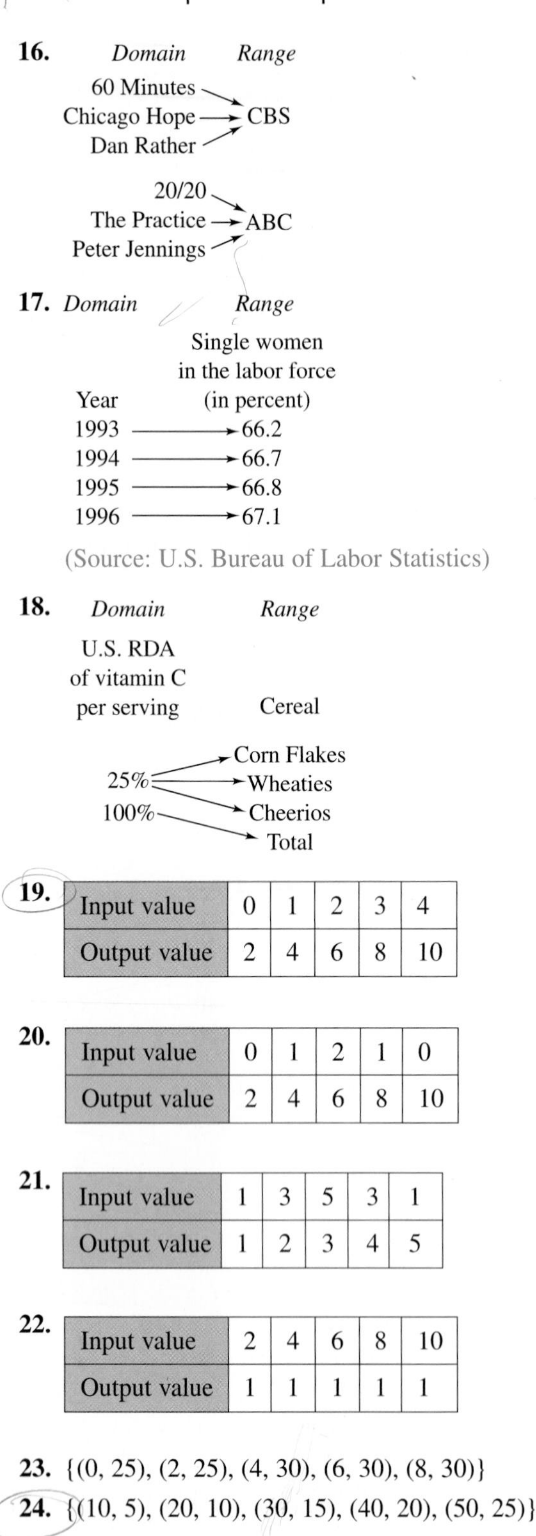

19.

Input value	0	1	2	3	4
Output value	2	4	6	8	10

20.

Input value	0	1	2	1	0
Output value	2	4	6	8	10

21.

Input value	1	3	5	3	1
Output value	1	2	3	4	5

22.

Input value	2	4	6	8	10
Output value	1	1	1	1	1

23. $\{(0, 25), (2, 25), (4, 30), (6, 30), (8, 30)\}$

24. $\{(10, 5), (20, 10), (30, 15), (40, 20), (50, 25)\}$

In Exercises 25–36, use the Vertical Line Test to determine whether y is a function of x. See Example 3.

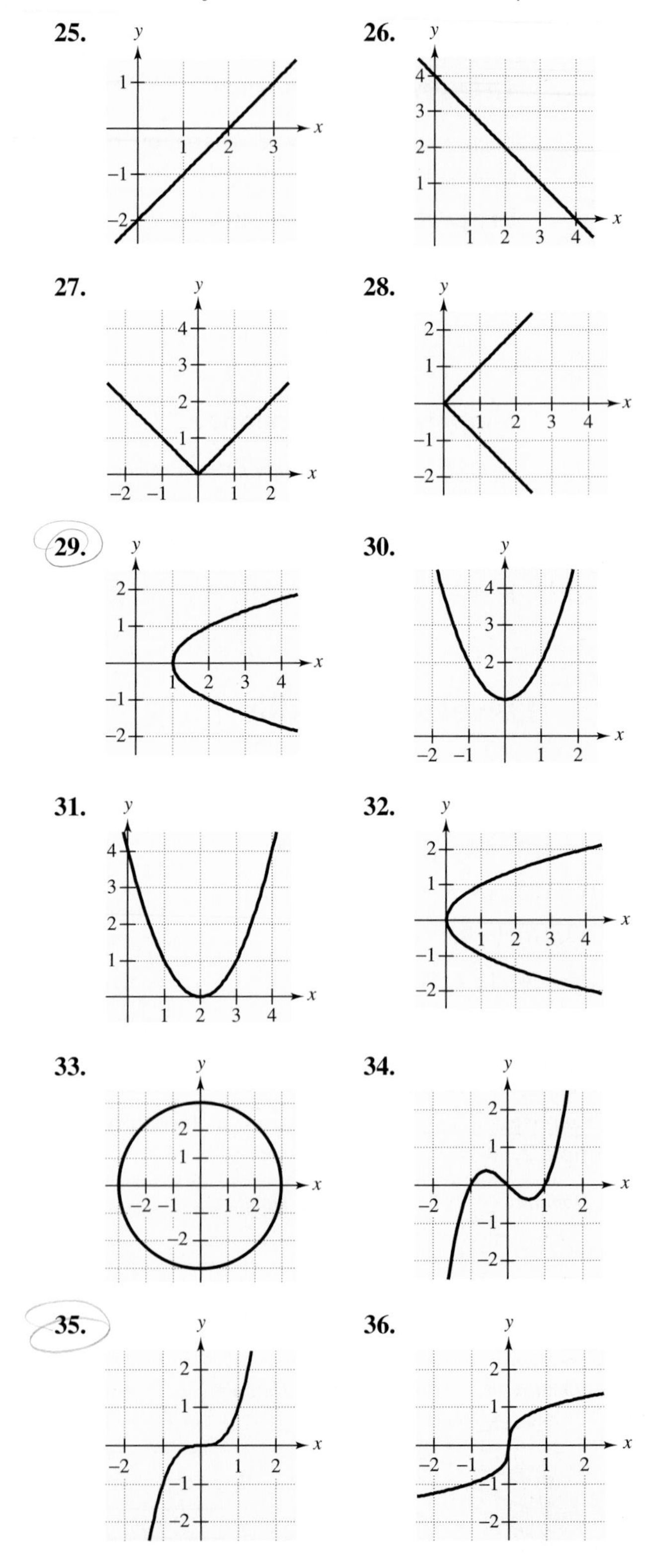

In Exercises 37–52, evaluate the function. See Example 4.

37. $f(x) = \frac{1}{2}x$ (a) $f(2)$ (b) $f(5)$ (c) $f(-4)$ (d) $f\left(-\frac{2}{3}\right)$

38. $g(x) = -\frac{4}{5}x$ (a) $g(5)$ (b) $g(0)$ (c) $g(-3)$ (d) $g\left(-\frac{5}{4}\right)$

39. $f(x) = 2x - 1$ (a) $f(0)$ (b) $f(3)$ (c) $f(-3)$ (d) $f\left(-\frac{1}{2}\right)$

40. $f(t) = 3 - 4t$ (a) $f(0)$ (b) $f(1)$ (c) $f(-2)$ (d) $f\left(\frac{3}{4}\right)$

41. $f(x) = 4x + 1$ (a) $f(1)$ (b) $f(-1)$ (c) $f(-4)$ (d) $f\left(-\frac{4}{3}\right)$

42. $g(t) = 5 - 2t$ (a) $g\left(\frac{5}{2}\right)$ (b) $g(-10)$ (c) $g(0)$ (d) $g\left(\frac{3}{4}\right)$

43. $h(t) = \frac{1}{4}t - 1$ (a) $h(200)$ (b) $h(-12)$ (c) $h(8)$ (d) $h\left(-\frac{5}{2}\right)$

44. $f(s) = 4 - \frac{2}{3}s$ (a) $f(60)$ (b) $f(-15)$ (c) $f(-18)$ (d) $f\left(\frac{1}{2}\right)$

45. $f(v) = \frac{1}{2}v^2$ (a) $f(-4)$ (b) $f(4)$ (c) $f(0)$ (d) $f(2)$

46. $g(u) = -2u^2$ (a) $g(0)$ (b) $g(2)$ (c) $g(3)$ (d) $g(-4)$

47. $g(x) = 2x^2 - 3x + 1$ (a) $g(0)$ (b) $g(-2)$ (c) $g(1)$ (d) $g\left(\frac{1}{2}\right)$

48. $h(x) = x^2 + 4x - 1$ (a) $h(0)$ (b) $h(-4)$ (c) $h(10)$ (d) $h\left(\frac{3}{2}\right)$

49. $g(u) = |u + 2|$ (a) $g(2)$ (b) $g(-2)$ (c) $g(10)$ (d) $g\left(-\frac{5}{2}\right)$

50. $h(s) = |s| + 2$ (a) $h(4)$ (b) $h(-10)$ (c) $h(-2)$ (d) $h\left(\frac{3}{2}\right)$

51. $h(x) = x^3 - 1$ (a) $h(0)$ (b) $h(1)$ (c) $h(3)$ (d) $h\left(\frac{1}{2}\right)$

52. $f(x) = 16 - x^4$ (a) $f(-2)$ (b) $f(2)$ (c) $f(1)$ (d) $f(3)$

In Exercises 53–60, determine the range R of the function for the specified domain D. Graph the function over the given domain. See Example 5.

53. $g(x) = 4 - x$
$D = \{0, 1, 2, 3, 4\}$

54. $g(x) = x + 1$
$D = \{-2, -1, 0, 1, 2\}$

55. $h(t) = 100$
$D = \{-1, 0, 1, 2, 3\}$

56. $f(x) = x^2$
$D = \{-2, -1, 0, 1, 2\}$

57. $f(x) = x^3$
$D = \{-2, -1, 0, 1, 2\}$

58. $h(t) = t^3 - 3t^2 + 3t - 1$
$D = \{-1, 0, 1, 2, 3\}$

59. $g(s) = |s|$
$D = \{-2, -1, 0, 1, 2\}$

60. $g(u) = |u + 2| - |u|$
$D = \{-3, -1, 1, 3, 5\}$

Solving Problems

61. *Demand Function* The demand for a product is a function of its price. Consider the demand function $f(p) = 20 - 0.5p$, where p is the price in dollars.

(a) Find $f(10)$ and $f(15)$.

(b) Describe the effect a price increase has on demand.

62. *Maximum Load* The maximum safe load L (in pounds) for a wooden beam 2 inches wide and d inches high is $L(d) = 100d^2$.

d	2	4	6	8
$L(d)$				

(a) Complete the table.

(b) Describe the effect of an increase in height on the maximum safe load.

63. *Distance* The function $d(t) = 50t$ gives the distance (in miles) that a car will travel in t hours at an average speed of 50 miles per hour. Find the distance traveled for (a) $t = 2$, (b) $t = 4$, and (c) $t = 10$.

64. *Speed of Sound* The function $S(h) = 1116 - 4.04h$ approximates the speed of sound (in feet per second) at altitude h (in thousands of feet). Use the function to approximate the speed of sound for (a) $h = 0$, (b) $h = 10$, and (c) $h = 30$.

Interpreting a Graph In Exercises 65–68, use the information in the graph. (Source: National Center for Education Statistics)

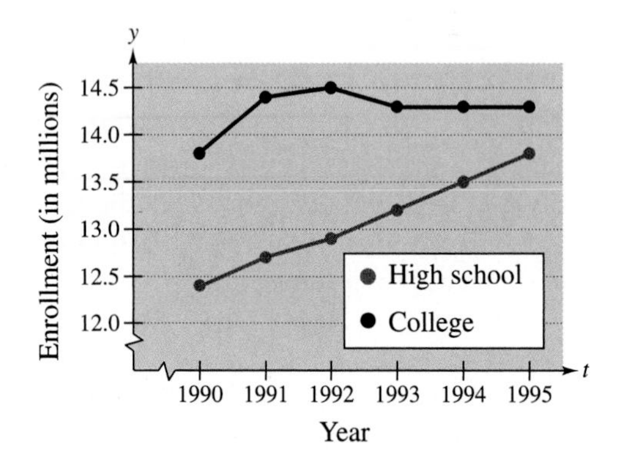

65. Is the high school enrollment a function of the year?

66. Is the college enrollment a function of the year?

67. Let $f(t)$ represent the number of high school students in year t. Find $f(1992)$.

68. Let $g(t)$ represent the number of college students in year t. Find $g(1990)$.

69. *Geometry* Write the formula for the perimeter P of a square with sides of length s. Is P a function of s? Explain.

70. *Geometry* Write the formula for the volume V of a cube with sides of length t. Is V a function of t? Explain.

71. *Time Between Sunrise and Sunset* The graph approximates the length of time L (in hours) between sunrise and sunset in Erie, Pennsylvania over a period of 1 year. The variable t represents the day of the year.

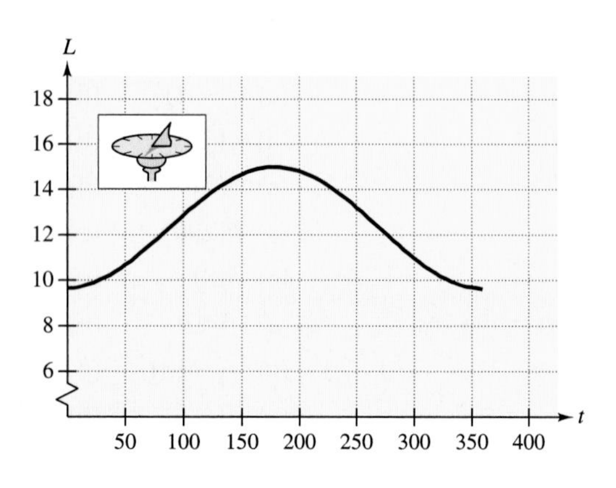

(a) Is L a function of t?

(b) Estimate the range for this relation.

Explaining Concepts

72. Answer parts (a)–(c) of Motivating the Chapter on page 201.

73. Explain the difference between a relation and a function. Give an example of a relation that is not a function.

74. Is it possible to find a function that is not a relation? If it is, find one.

75. Explain the meaning of the terms *domain* and *range* in the context of a function.

76. Give an example of a function defined by an equation in two variables. Give an example of a function that is not defined by an equation in two variables.

77. State the Vertical Line Test. Explain how this test can be used to determine if a relation is a function.

78. Describe some advantages of using function notation.

79. Is it possible for the number of elements in the domain to be greater than the number of elements in the range? Explain.

80. *Terminology* Do the statements use the word *function* in a way that is mathematically correct? Explain your reasoning.

(a) The amount of money in your savings account is a function of your salary.

(b) The speed at which a free-falling baseball strikes the ground is a function of the height from which it is dropped.

Mid-Chapter Quiz

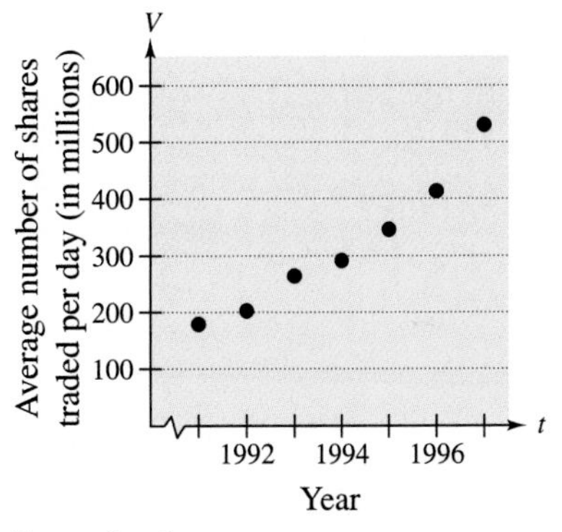

Figure for 4

Take this quiz as you would take a quiz in class. After you are done, check your work against the answers given in the back of the book.

1. Plot the points $(4, -2)$ and $\left(-1, -\frac{5}{2}\right)$ on a rectangular coordinate system.

2. Determine the quadrants in which the points $(x, 5)$ must be located. (x is a real number.)

3. Decide whether the ordered pairs are solutions of the equation $y = 9 - |x|$.

(a) $(2, 7)$ (b) $(-3, 12)$ (c) $(-9, 0)$ (d) $(0, -9)$

4. The scatter plot at the left shows the average number (in millions) of shares traded per day on the New York Stock Exchange for the years 1991 through 1997. Estimate the average number of shares traded per day for each year from 1991 to 1997. (Source: The New York Stock Exchange)

In Exercises 5 and 6, find the x- and y-intercepts of the graph of the equation.

5. $x - 3y = 12$ **6.** $y = 6 - 4x$

In Exercises 7–12, graph the equation.

7. $y = x - 1$ **8.** $y = 5 - 2x$

9. $y = 4 - x^2$ **10.** $y = (x + 2)^2$

11. $y = |x + 3|$ **12.** $y = 1 - |x|$

13. Does the table below represent y as a function of x? Explain.

x	0	1	2	3	4
y	-1	2	5	8	11

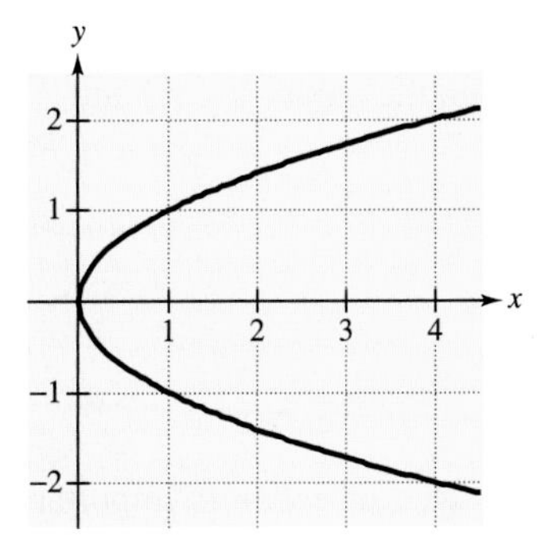

Figure for 14

14. Does the graph at the left represent y as a function of x? Explain.

In Exercises 15 and 16, evaluate the function at the given values of x and t.

15. $f(x) = 3x - 2$ (a) $f(-2)$ (b) $f(0)$ (c) $f(5)$ (d) $f\left(-\frac{1}{3}\right)$

16. $g(t) = 2t^2 - |t|$ (a) $g(-2)$ (b) $g(2)$ (c) $g(0)$ (d) $g\left(-\frac{1}{2}\right)$

17. Find the range of $f(x) = x^2 - x$ for the domain $D = \{-2, -1, 0, 1, 2\}$.

18. Find the domain for the area of a square with side s. ($A = s^2$)

19. Use a graphing utility to graph $h(x) = 3x^2 - 4x - 7$. Graphically estimate the intercepts of the graph. Explain how to verify your estimates algebraically.

20. A new computer system sells for approximately \$3000 and depreciates at the rate of \$500 per year for 4 years. Write the value V of the computer as a function of time t in years. What is the domain of the function? Graph the function over its domain.

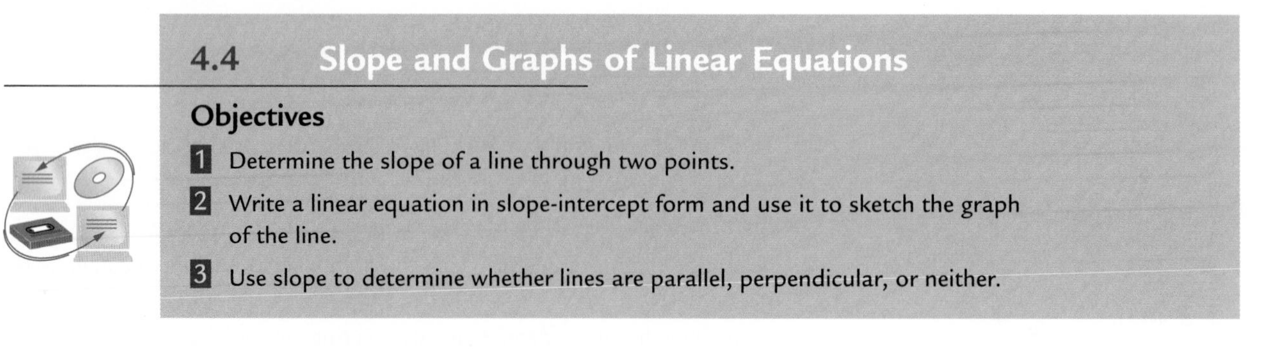

4.4 Slope and Graphs of Linear Equations

Objectives

1. Determine the slope of a line through two points.
2. Write a linear equation in slope-intercept form and use it to sketch the graph of the line.
3. Use slope to determine whether lines are parallel, perpendicular, or neither.

1 Determine the slope of a line through two points.

The Slope of a Line

The **slope** of a nonvertical line is the number of units the line rises or falls vertically for each unit of horizontal change from left to right. For example, the line in Figure 4.19 rises 2 units for each unit of horizontal change from left to right, and we say that this line has a slope of $m = 2$.

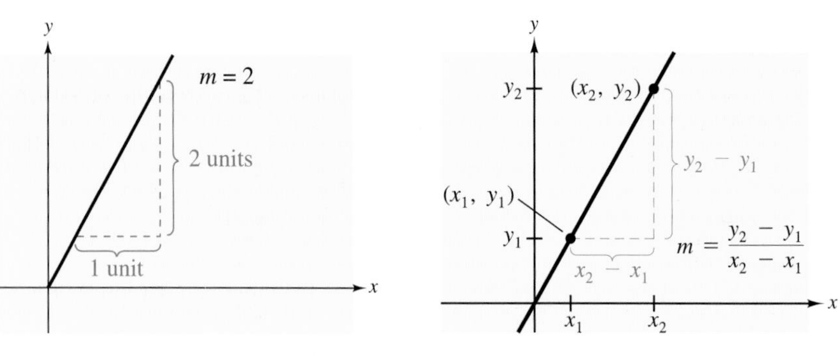

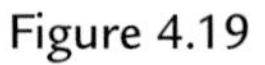

Figure 4.19

Figure 4.20

> **Definition of the Slope of a Line**
>
> The **slope** m of a nonvertical line passing through the points (x_1, y_1) and (x_2, y_2) is
>
> $$m = \frac{y_2 - y_1}{x_2 - x_1} = \frac{\text{change in } y}{\text{change in } x} = \frac{\text{rise}}{\text{run}}$$
>
> where $x_1 \neq x_2$ (see Figure 4.20).

When the formula for slope is used, the *order of subtraction* is important. Given two points on a line, you are free to label either one of them as (x_1, y_1) and the other as (x_2, y_2). However, once this is done, you must form the numerator and denominator using the same order of subtraction.

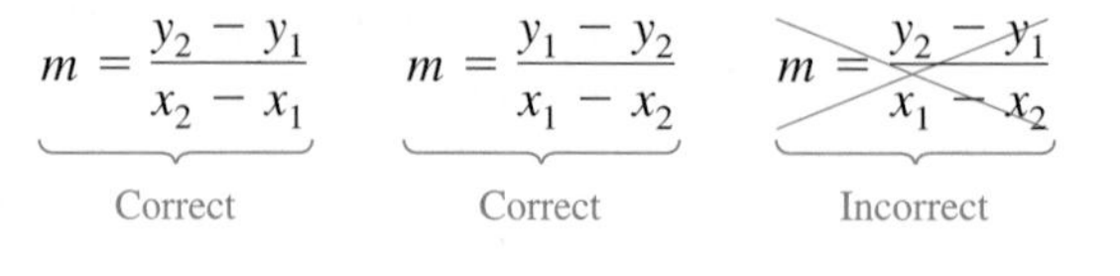

Example 1 Finding the Slope of a Line Through Two Points

Find the slope of the line passing through each pair of points.

a. $(-2, 0)$ and $(3, 1)$ **b.** $(-1, 2)$ and $(2, 2)$ **c.** $(0, 0)$ and $(1, -1)$

Solution

a. Let $(x_1, y_1) = (-2, 0)$ and $(x_2, y_2) = (3, 1)$.

$$m = \frac{y_2 - y_1}{x_2 - x_1}$$

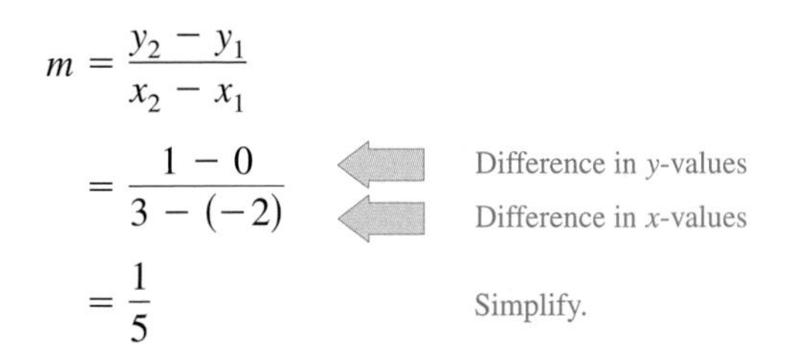

$$= \frac{1 - 0}{3 - (-2)}$$ Difference in y-values / Difference in x-values

$$= \frac{1}{5}$$ Simplify.

b. The slope of the line through $(-1, 2)$ and $(2, 2)$ is

$$m = \frac{2 - 2}{2 - (-1)}$$ Difference in y-values / Difference in x-values

$$= \frac{0}{3} = 0.$$ Simplify.

c. The slope of the line through $(0, 0)$ and $(1, -1)$ is

$$m = \frac{-1 - 0}{1 - 0}$$ Difference in y-values / Difference in x-values

$$= \frac{-1}{1} = -1.$$ Simplify.

The graphs of the three lines are shown in Figure 4.21.

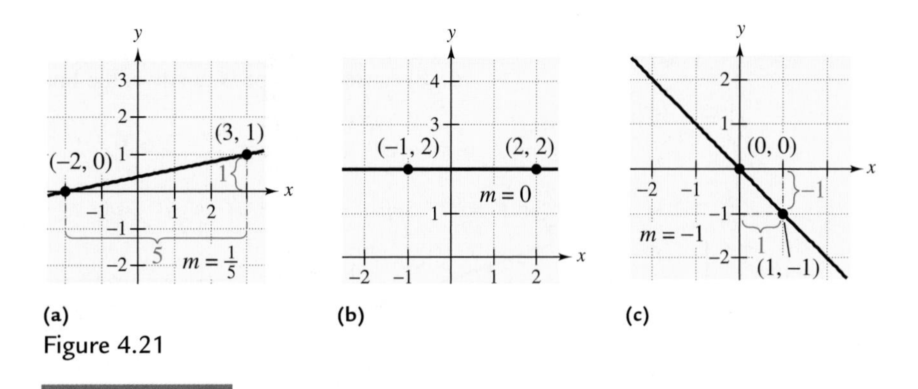

Figure 4.21

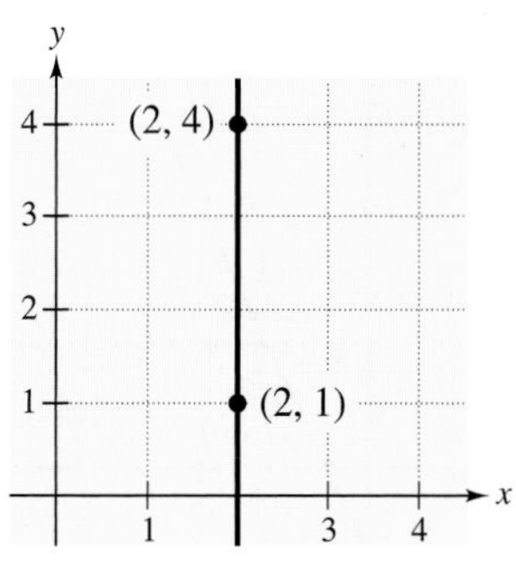

Figure 4.22 *Slope is undefined.*

The definition of slope does not apply to vertical lines. For instance, consider the points $(2, 4)$ and $(2, 1)$ on the vertical line shown in Figure 4.22. Applying the formula for slope, you have

$$\frac{4 - 1}{2 - 2} = \frac{3}{0}.$$ Undefined division by zero

Because division by zero is not defined, the slope of a vertical line is not defined.

From the slopes of the lines shown in Figures 4.21 and 4.22, you can make several generalizations about the slope of a line.

Slope of a Line

1. A line with positive slope $(m > 0)$ *rises* from left to right.
2. A line with negative slope $(m < 0)$ *falls* from left to right.
3. A line with zero slope $(m = 0)$ is *horizontal.*
4. A line with undefined slope is *vertical.*

Example 2 Using Slope to Describe Lines

Describe the lines through the pairs of points.

a. $(3, -2), (3, 3)$ **b.** $(-2, 5), (1, 4)$ **c.** $(-4, -3), (0, -3)$ **d.** $(1, 0), (4, 6)$

Solution

a. Because the slope is undefined, the line is vertical.

$$m = \frac{3 - (-2)}{3 - 3} = \frac{5}{0}$$ Undefined slope (See Figure 4.23a.)

b. Because the slope is negative, the line falls from left to right.

$$m = \frac{4 - 5}{1 - (-2)} = -\frac{1}{3} < 0$$ Negative slope (See Figure 4.23b.)

c. Because the slope is zero, the line is horizontal.

$$m = \frac{-3 - (-3)}{0 - (-4)} = \frac{0}{4} = 0$$ Zero slope (See Figure 4.23c.)

d. Because the slope is positive, the line rises from left to right.

$$m = \frac{6 - 0}{4 - 1} = \frac{6}{3} = 2 > 0$$ Positive slope (See Figure 4.23d.)

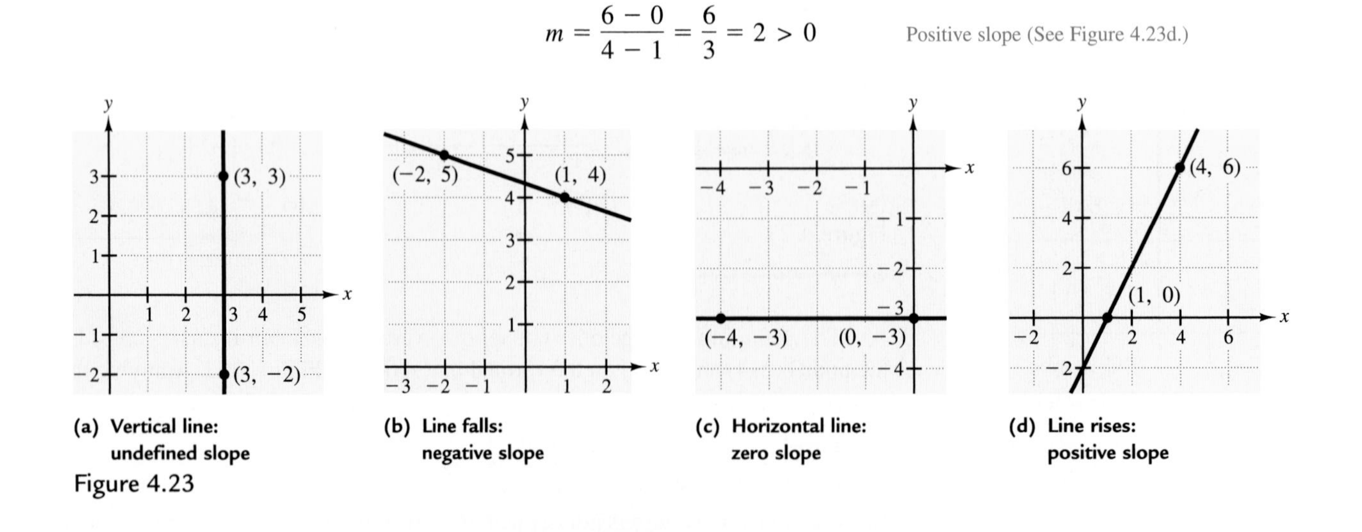

(a) Vertical line: undefined slope

(b) Line falls: negative slope

(c) Horizontal line: zero slope

(d) Line rises: positive slope

Figure 4.23

Any two points on a nonvertical line can be used to calculate its slope. This is demonstrated in the next two examples.

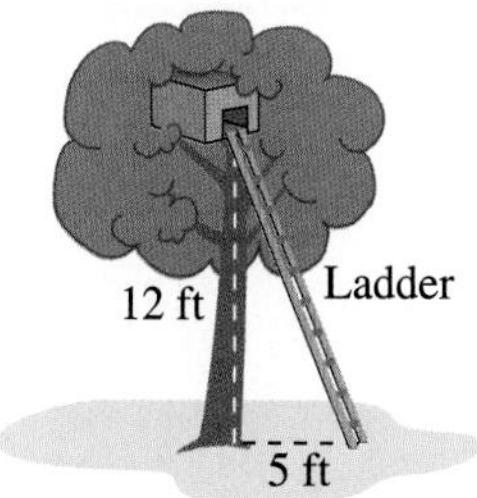
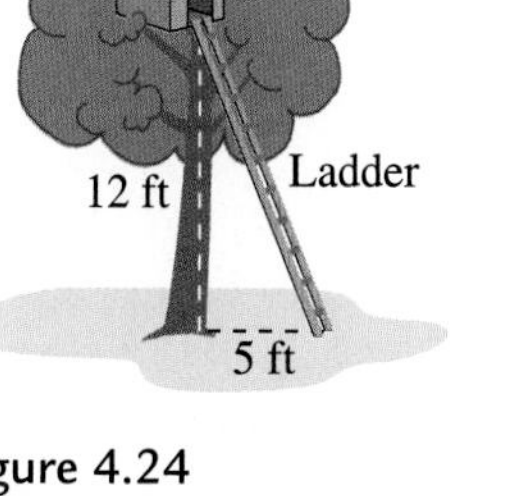

Figure 4.24

Example 3 Finding the Slope of a Ladder

Find the slope of the ladder leading up to the tree house in Figure 4.24.

Solution

Consider the tree trunk as the y-axis and the level ground as the x-axis. The endpoints of the ladder are $(0, 12)$ and $(5, 0)$. So, the slope of the ladder is

$$m = \frac{y_2 - y_1}{x_2 - x_1} = \frac{0 - 12}{5 - 0} = -\frac{12}{5}.$$

Example 4 Finding the Slope of a Line

Sketch the graph of the line $3x - 2y = 4$. Then find the slope of the line. (Choose two different pairs of points on the line and show that the same slope is obtained using either pair.)

Solution

Begin by solving the equation for y.

$$y = \frac{3}{2}x - 2 \qquad y \text{ is a function of } x.$$

Then, construct a table of values as shown below.

x	-2	0	2	4
$y = \frac{3}{2}x - 2$	-5	-2	1	4
Solution points	$(-2, -5)$	$(0, -2)$	$(2, 1)$	$(4, 4)$

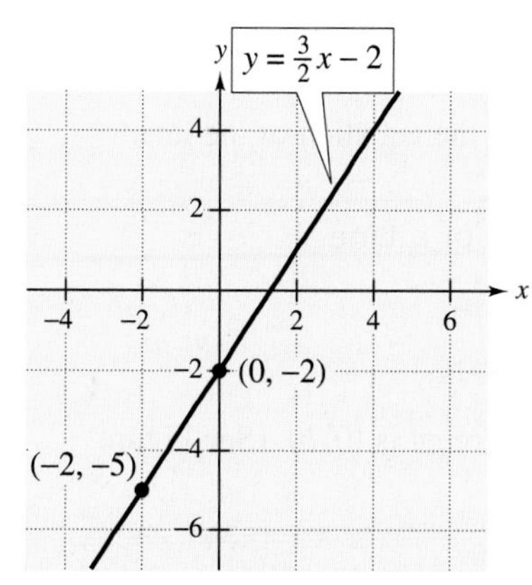

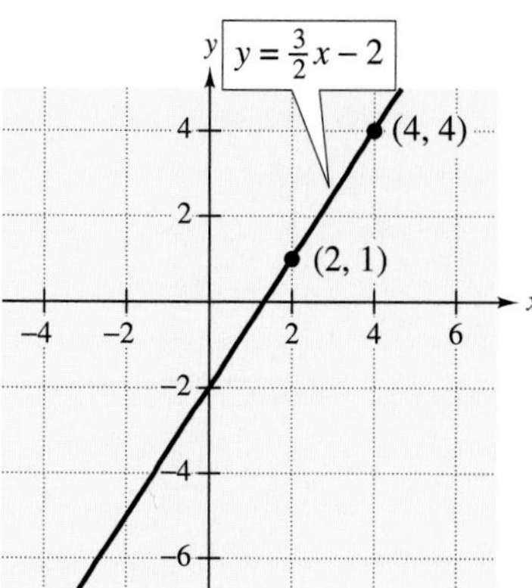

Figure 4.25

From the solution points shown in the table, sketch the graph of the line, as shown in Figure 4.25. To calculate the slope of the line using two different sets of points, first use the points $(-2, -5)$ and $(0, -2)$ to obtain a slope of

$$m = \frac{-2 - (-5)}{0 - (-2)} = \frac{3}{2}.$$

Next, use the points $(2, 1)$ and $(4, 4)$ to obtain a slope of

$$m = \frac{4 - 1}{4 - 2} = \frac{3}{2}.$$

Try some other pairs of points on the line to see that you obtain a slope of $m = \frac{3}{2}$ regardless of which two points you use.

2 Write a linear equation in slope-intercept form and use it to sketch the graph of the line.

Slope as a Graphing Aid

You have seen in Section 4.1 that before creating a table of values for an equation, it is helpful first to solve the equation for y. When doing this for a linear equation, you obtain some very useful information. Consider the results of Example 4.

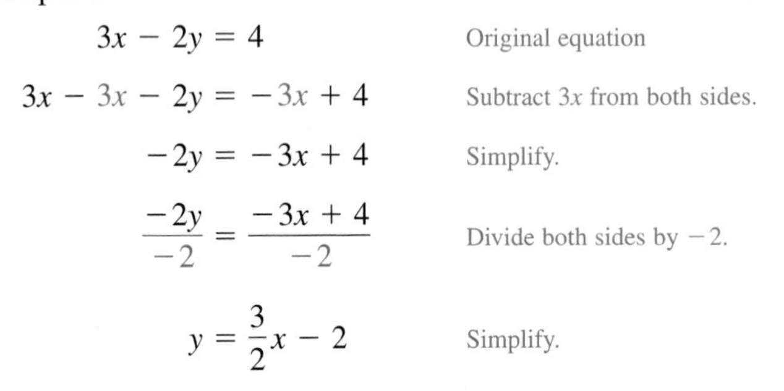

$$3x - 2y = 4 \quad \text{Original equation}$$

$$3x - 3x - 2y = -3x + 4 \quad \text{Subtract } 3x \text{ from both sides.}$$

$$-2y = -3x + 4 \quad \text{Simplify.}$$

$$\frac{-2y}{-2} = \frac{-3x + 4}{-2} \quad \text{Divide both sides by } -2.$$

$$y = \frac{3}{2}x - 2 \quad \text{Simplify.}$$

Technology: Tip

Setting the viewing window on a graphing utility affects the appearance of a line's slope. When you are using a graphing utility, remember that you cannot judge whether a slope is steep or shallow unless you use a *square* setting—a setting that shows equal spacing of the units on both axes. For many graphing utilities, a square setting is obtained by using the ratio of 10 vertical units to 15 horizontal units.

Observe that the coefficient of x is the slope of the graph for this equation (see Example 4). Moreover, the constant term, -2, gives the y-intercept of the graph.

$$y = \frac{3}{2}x + -2$$

slope $\quad$ y-intercept $(0, -2)$

This form is called the **slope-intercept form** of the equation of the line.

> **▶ Slope-Intercept Form of the Equation of a Line**
>
> The graph of the equation
>
> $y = mx + b$ $\quad$ y is a linear function of x.
>
> is a line whose slope is m and whose y-intercept is $(0, b)$. (See Figure 4.26.)

Study Tip

Remember that slope is a *rate of change.* In the slope-intercept equation

$$y = mx + b$$

the slope m is the rate of change of y with respect to x.

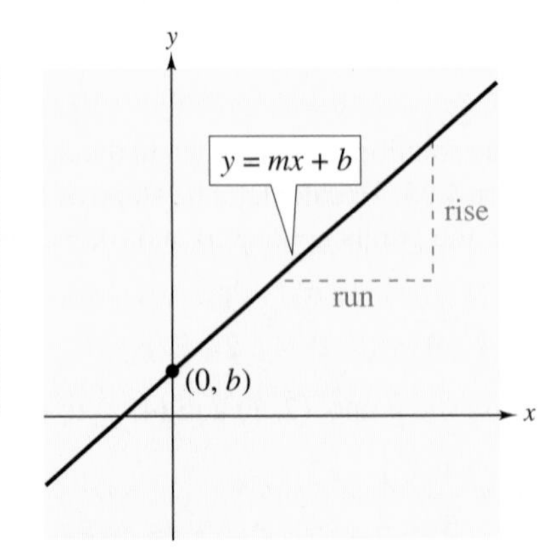

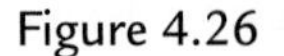

Figure 4.26

The slope-intercept form of the equation of a line identifies y as a function of x. So, the term **linear function** is often used as an alternative description of the slope-intercept form of the equation of a line.

So far, you have been plotting several points to sketch the equation of a line. However, now that you can recognize equations of lines (linear functions), you don't have to plot as many points—two points are enough. (You might remember from geometry that *two points are all that are necessary to determine a line.*) The next example shows how to use the slope to help sketch a line.

Example 5 Using the Slope and y-Intercept to Sketch a Line

Use the slope and y-intercept to sketch the graph of

$$x - 3y = -6.$$

Solution

First, write the equation in slope-intercept form.

$$x - 3y = -6 \quad \text{Original equation}$$

$$-3y = -x - 6 \quad \text{Subtract } x \text{ from both sides.}$$

$$y = \frac{-x - 6}{-3} \quad \text{Divide both sides by } -3.$$

$$y = \frac{1}{3}x + 2 \quad \text{Simplify to slope-intercept form.}$$

So, the slope of the line is $m = \frac{1}{3}$ and the y-intercept is $(0, b) = (0, 2)$. Now you can sketch the graph of the equation. First, plot the y-intercept, as shown in Figure 4.27(a). Then, using a slope of $\frac{1}{3}$,

$$m = \frac{1}{3} = \frac{\text{change in } y}{\text{change in } x}$$

locate a second point on the line by moving 3 units to the right and 1 unit up (or 1 unit up and 3 units to the right), also shown in Figure 4.27(a). Finally, obtain the graph by drawing a line through the two points [Figure 4.27(b)].

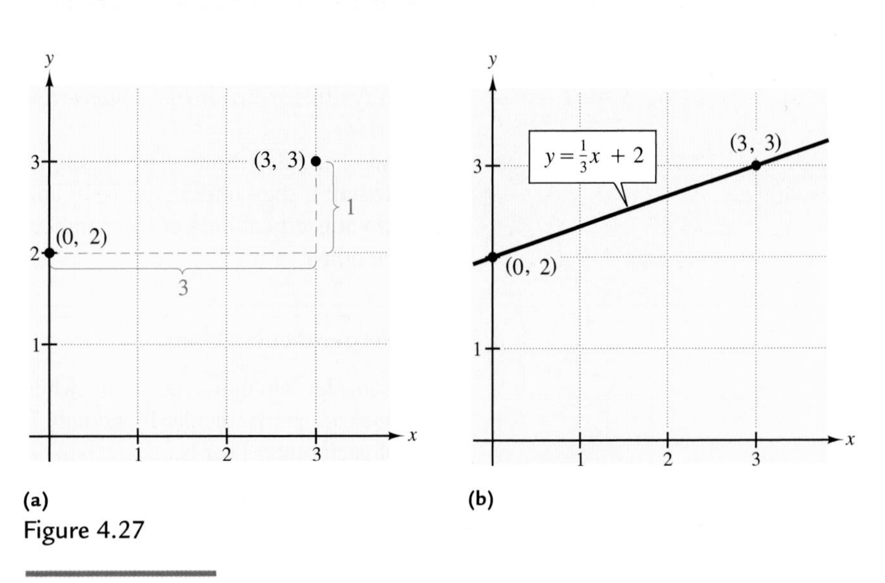

Figure 4.27

3 Use slope to determine whether lines are parallel, perpendicular, or neither.

Parallel and Perpendicular Lines

You know from geometry that two lines in a plane are **parallel** if they do not intersect. What this means in terms of their slopes is shown in Example 6.

Example 6 Lines That Have the Same Slope

On the same set of axes, sketch the lines $y = 3x$ and $y = 3x - 4$.

Solution

For the line

$$y = 3x$$

the slope is $m = 3$ and the y-intercept is $(0, 0)$. For the line

$$y = 3x - 4$$

the slope is also $m = 3$ and the y-intercept is $(0, -4)$. The graphs of these two lines are shown in Figure 4.28.

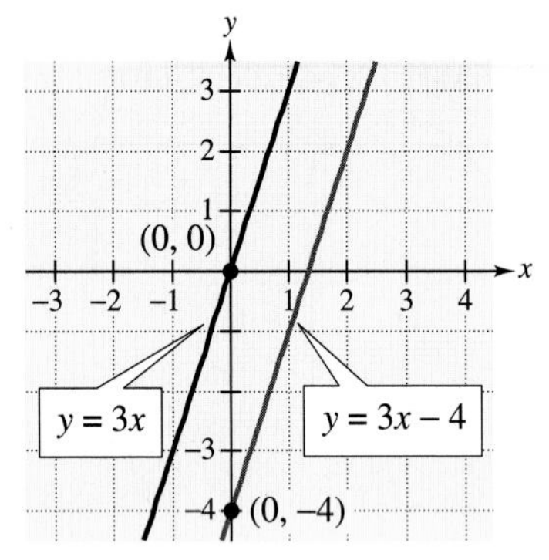

Figure 4.28

In Example 6, notice that the two lines have the same slope *and* that the two lines appear to be parallel. The following rule states that this is always the case. That is, two (nonvertical) lines are parallel *if and only if* they have the same slope.

> **▶ Parallel Lines**
>
> Two distinct nonvertical lines are parallel if and only if they have the same slope.

The phrase "if and only if" in this rule is used in mathematics as a way to write two statements in one. The first statement says that *if two distinct nonvertical lines have the same slope, they must be parallel.* The second (or reverse) statement says that *if two distinct nonvertical lines are parallel, they must have the same slope.*

Another rule resulting from geometry is that two lines in a plane are **perpendicular** if they intersect at right angles. In terms of their slopes, this means that two nonvertical lines are perpendicular if their slopes are negative reciprocals of each other.

> **▶ Perpendicular Lines**
>
> Consider two nonvertical lines whose slopes are m_1 and m_2. The two lines are perpendicular if and only if their slopes are *negative reciprocals* of each other. That is,
>
> $$m_1 = -\frac{1}{m_2}, \text{ or equivalently, } m_1 \cdot m_2 = -1.$$

Example 7 Parallel or Perpendicular?

Determine whether the pairs of lines are parallel, perpendicular, or neither.

a. $y = -3x - 2,\ y = \frac{1}{3}x + 1$

b. $y = \frac{1}{2}x + 1,\ y = \frac{1}{2}x - 1$

Solution

a. The first line has a slope of $m_1 = -3$ and the second line has a slope of $m_2 = \frac{1}{3}$. Because these slopes are negative reciprocals of each other, the two lines must be perpendicular, as shown in Figure 4.29.

b. Both lines have a slope of $m = \frac{1}{2}$. So, the two lines must be parallel, as shown in Figure 4.30.

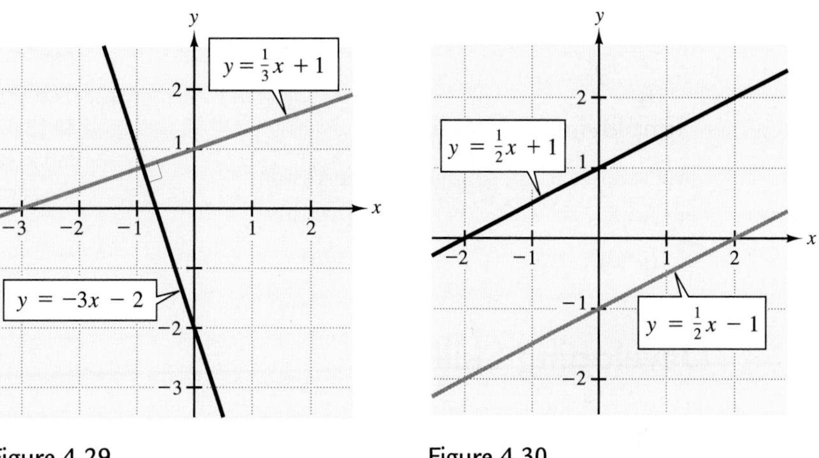

Figure 4.29

Figure 4.30

Discussing the Concept — Creating a Linear Equation

Plot the following data points for a telephone call.

Minutes	1	3	8	15
Cost of call	\$1.22	\$1.66	\$2.76	\$4.30

Decide if these data points fit a straight line. If so, what are the slope and *y*-intercept? If not, why not?

4.4 Exercises

Integrated Review *Concepts, Skills, and Problem Solving*

Keep mathematically in shape by doing these exercises *before* the problems of this section.

Properties and Definitions

1. Two equations that have the same set of solutions are called ______.

2. Use the Addition Property of Equality to fill in the blank.

$$5x - 2 = 6$$

$$5x = 6 + \_\_\_$$

Simplifying Expressions

In Exercises 3–10, simplify the expression.

3. $(x^2)^3 \cdot x^3$ **4.** $(y^2z^3)(z^2)$

5. $(u^4v^2)^2$ **6.** $(ab)^4$

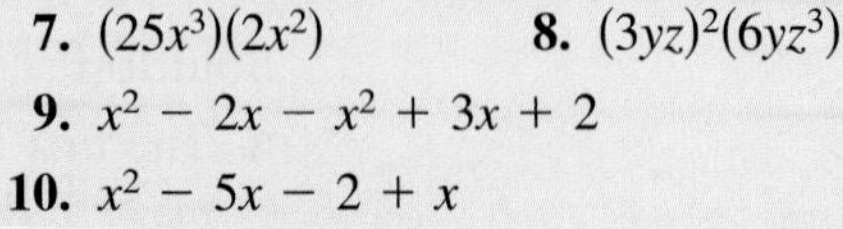

7. $(25x^3)(2x^2)$ **8.** $(3yz)^2(6yz^3)$

9. $x^2 - 2x - x^2 + 3x + 2$

10. $x^2 - 5x - 2 + x$

Problem Solving

11. A builder must cut a 10-foot board into three pieces. Two are to have the same length and the third is to be three times as long as the two of equal length. Find the lengths of the three pieces.

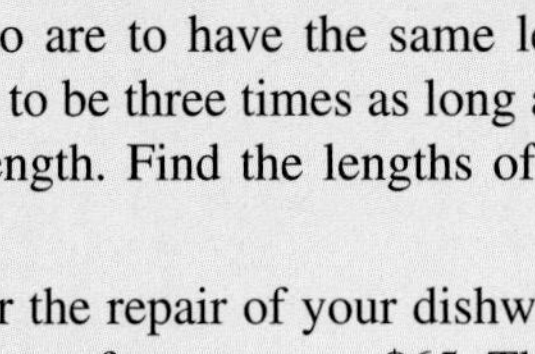

12. The bill for the repair of your dishwasher was \$113. The cost for parts was \$65. The cost for labor was \$32 per hour. How many hours did the repair work take?

Developing Skills

In Exercises 1–10, estimate the slope (if it exists) of the line from its graph.

1.

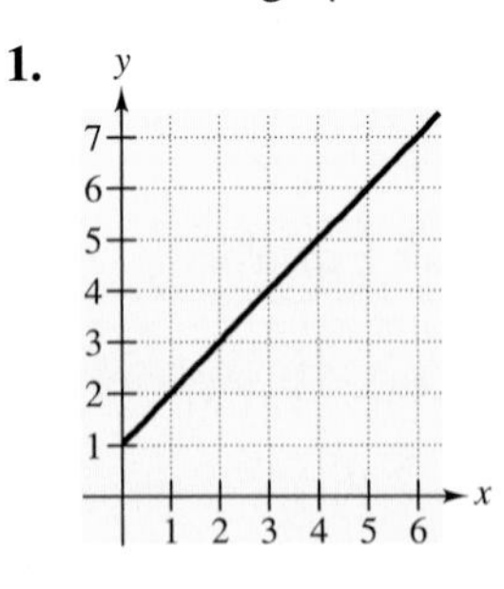

2.

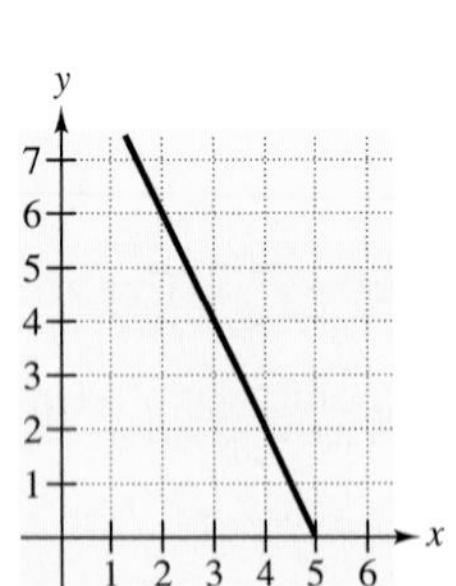

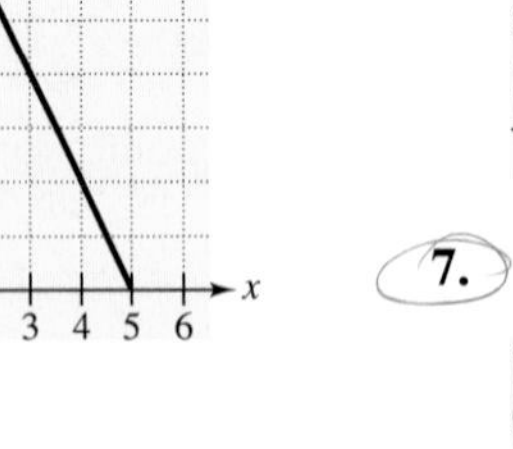

3.

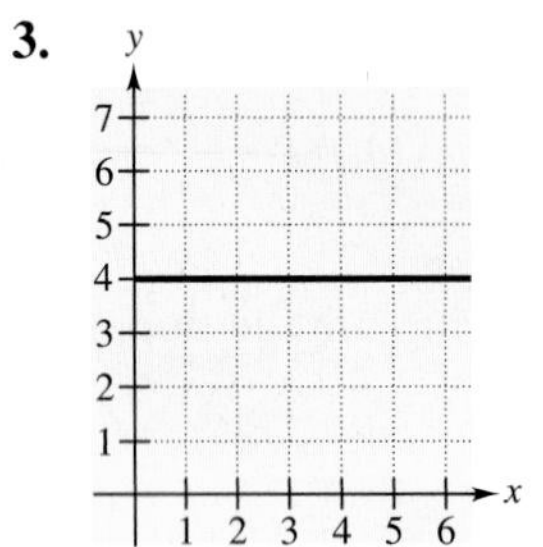

4.

5.

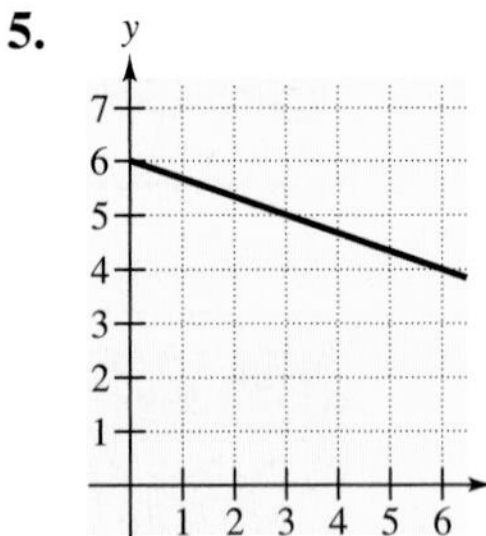

6.

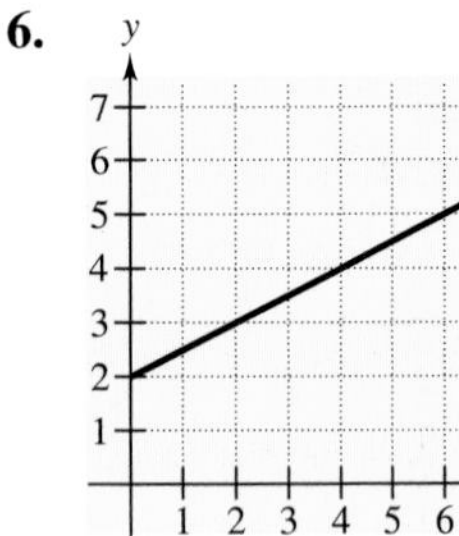

7.

8.

9. **10.**

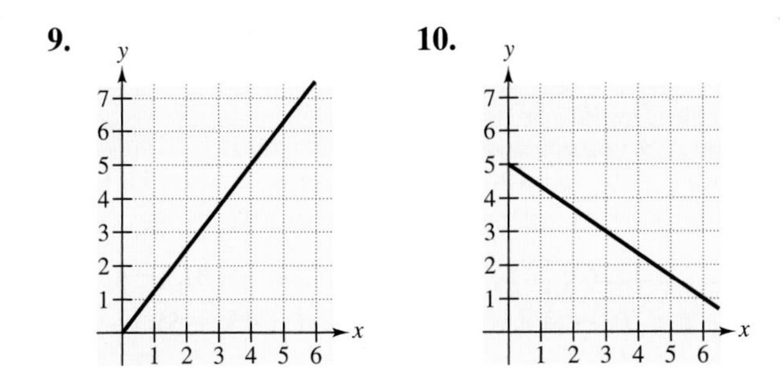

In Exercises 11 and 12, match the line in the figure with its slope.

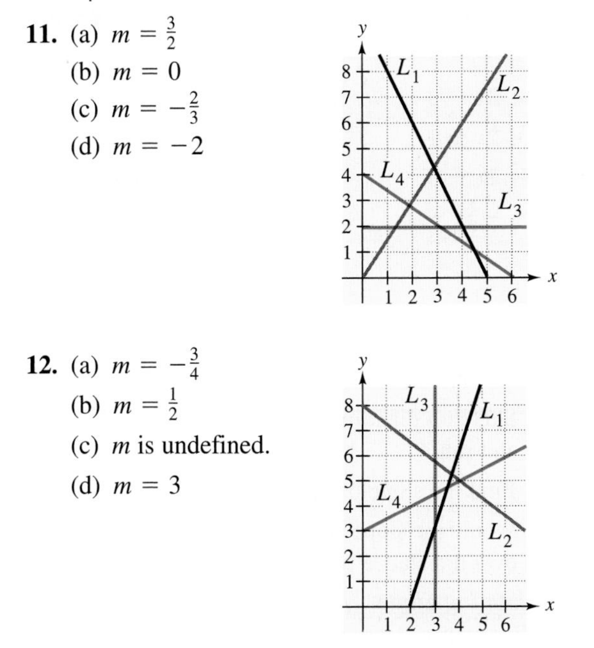

11. (a) $m = \frac{3}{2}$
(b) $m = 0$
(c) $m = -\frac{2}{3}$
(d) $m = -2$

12. (a) $m = -\frac{3}{4}$
(b) $m = \frac{1}{2}$
(c) m is undefined.
(d) $m = 3$

In Exercises 13–32, plot the points and find the slope (if possible) of the line passing through the pair of points. State whether the line rises, falls, is horizontal, or is vertical. See Examples 1 and 2.

13. $(0, 0), (4, 5)$
14. $(0, 0), (-3, 6)$
15. $(0, 0), (8, -4)$
16. $(0, 0), (-1, -3)$
17. $(0, 6), (8, 0)$
18. $(0, -6), (8, 0)$
19. $(-3, -2), (1, 6)$
20. $(2, 4), (4, -4)$
21. $(-6, -1), (-6, 4)$
22. $(-4, -10), (-4, 0)$
23. $(3, -4), (8, -4)$
24. $(1, 2), (-2, -2)$
25. $\left(\frac{1}{4}, \frac{3}{2}\right), \left(\frac{9}{2}, -3\right)$
26. $\left(-\frac{5}{4}, -\frac{1}{4}\right), \left(\frac{7}{8}, \frac{3}{4}\right)$
27. $(3.2, -1), (-3.2, 4)$
28. $(1.4, 0), (1.4, 3)$
29. $(3.5, -1), (5.75, 4.25)$
30. $(0, 6.4), (5, 6.4)$
31. $(a, 3), (4, 3),\ a \neq 4$
32. $(4, a), (4, 2),\ a \neq 2$

In Exercises 33 and 34, complete the table. Use two different pairs of solution points to show that the same slope is obtained using either pair.

x	-2	0	2	4
y				
Solution points				

33. $y = -2x - 2$
34. $y = 3x + 4$

In Exercises 35–38, use the slope formula to find the value of y such that the line through the two points will have the given slope.

35. Points: $(3, -2), (0, y)$
Slope: $m = -8$

36. Points: $(-3, y), (8, 2)$
Slope: $m = 2$

37. Points: $(-4, y), (7, 6)$
Slope: $m = \frac{5}{2}$

38. Points: $(0, 10), (6, y)$
Slope: $m = -\frac{1}{3}$

In Exercises 39–50, a point on a line and the slope of the line are given. Plot the point and use the slope to find two additional points on the line. (There are many correct answers.)

39. $(2, 1)$
$m = 0$

40. $(-3, 4)$
m is undefined.

41. $(1, -6)$
$m = 2$

42. $(-2, -4)$
$m = 1$

43. $(0, 1)$
$m = -2$

44. $(-2, 4)$
$m = -3$

45. $(-4, 0)$
$m = \frac{2}{3}$

46. $(-1, -1)$
$m = -\frac{1}{4}$

47. $(3, 5)$
$m = -\frac{1}{2}$

48. $(1, 3)$
$m = \frac{4}{3}$

49. $(-8, 1)$
m is undefined.

50. $(-3, -1)$
$m = 0$

In Exercises 51–56, sketch the graph of a line through the point $(0, 2)$ having the given slope.

51. $m = 0$
52. m is undefined.
53. $m = 3$
54. $m = -1$
55. $m = -\frac{2}{3}$
56. $m = \frac{3}{4}$

In Exercises 57–62, plot the x- and y-intercepts and sketch the graph of the line.

57. $2x - 3y + 6 = 0$
58. $3x + 4y + 12 = 0$
59. $-5x + 2y - 10 = 0$
60. $3x - 7y - 21 = 0$
61. $6x - 4y + 12 = 0$
62. $5y - 2x - 20 = 0$

In Exercises 63–76, write the equation in slope-intercept form. Use the slope and y-intercept to graph the line. See Example 5.

63. $x + y = 0$
64. $x - y = 0$
65. $\frac{1}{2}x + y = 0$
66. $3x - y = 0$
67. $2x - y - 3 = 0$
68. $x - y + 2 = 0$
69. $x - 3y + 6 = 0$
70. $3x - 2y - 2 = 0$
71. $x + 2y - 2 = 0$
72. $10x + 6y - 3 = 0$
73. $3x - 4y + 2 = 0$
74. $2x + 3y = 0$
75. $y + 5 = 0$
76. $y - 3 = 0$

In Exercises 77–80, determine if the lines L_1 and L_2 passing through the given pairs of points are parallel, perpendicular, or neither.

77. L_1: $(0, -1), (5, 9)$
L_2: $(0, 3), (4, 1)$

78. L_1: $(-2, -1), (1, 5)$
L_2: $(1, 3), (5, -5)$

79. L_1: $(3, 6), (-6, 0)$
L_2: $(0, -1), \left(5, \frac{7}{3}\right)$

80. L_1: $(4, 8), (-4, 2)$
L_2: $(3, -5), \left(-1, \frac{1}{3}\right)$

In Exercises 81–84, sketch the graphs of the two lines on the same rectangular coordinate system. Determine whether the lines are parallel, perpendicular, or neither. Use a graphing utility to verify your result. (Use a square setting.) See Examples 6 and 7.

81. $y_1 = 2x - 3$
$y_2 = 2x + 1$

82. $y_1 = -\frac{1}{3}x - 3$
$y_2 = -\frac{1}{3}x + 1$

83. $y_1 = 2x - 3$
$y_2 = -\frac{1}{2}x + 1$

84. $y_1 = -\frac{1}{3}x - 3$
$y_2 = 3x + 1$

Solving Problems

85. *Roof Pitch* Determine the slope (pitch) of the roof of the house in the figure.

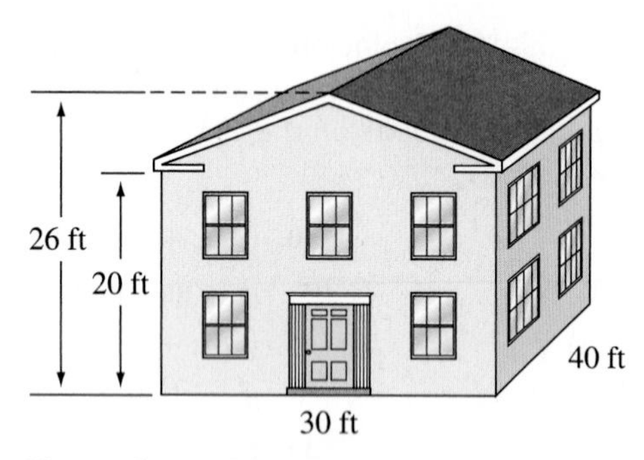

86. *Slope of a Ladder* Find the slope of the ladder in the figure.

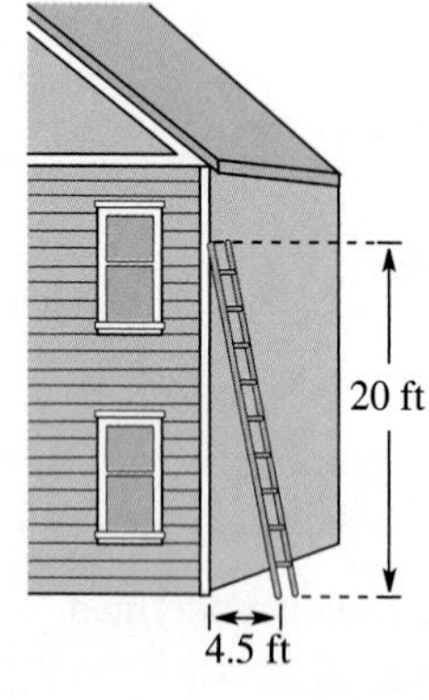

87. *Sketching a Diagram* A subway track rises 3 feet over a 200-foot horizontal distance.

(a) Sketch a diagram of the track and label the rise and run.

(b) Find the slope of the track.

(c) Would the slope be steeper if the track rose 3 feet over a distance of 100 feet? Explain.

88. *Estimating Slope* An airplane leaves an airport. As it flies over a town, its altitude is 4 miles. The town is about 20 miles from the airport. Approximate the slope of the linear path followed during takeoff.

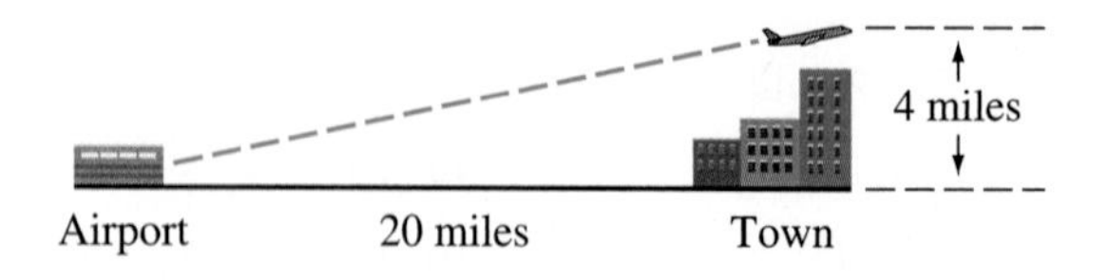

89. *Graphical Interpretation* The graph gives the net sales (in billions of dollars) for Wal-Mart for 1993 through 1997. (Source: 1997 Wal-Mart Annual Report)

(a) Find the slopes of the four line segments.

(b) Find the slope of the line segment connecting the first and last points of the line graph. Explain the meaning of this slope.

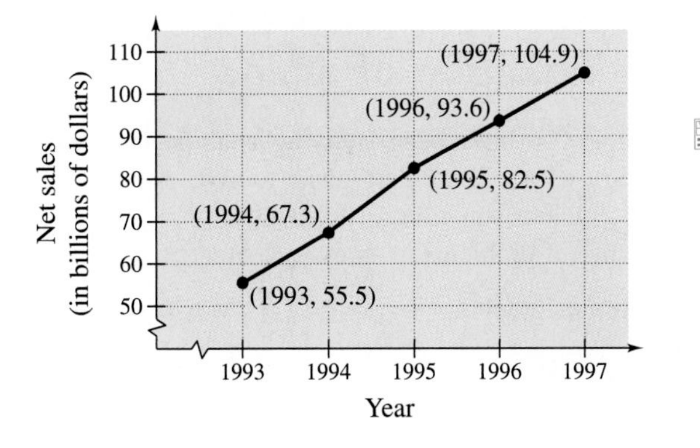

90. *Comparing Models* Based on different assumptions, the marketing department of a company develops two models to predict the annual profit of the company over the next 10 years. The models are

$P_1 = 0.2t + 2.4$ and $P_2 = 0.3t + 2.4$

where P_1 and P_2 represent profit in millions of dollars and t is time in years $(0 \leq t \leq 10)$.

(a) Interpret the slopes of the two linear models.

(b) Which model predicts a faster increase in profits?

(c) Use each model to predict profits when $t = 10$.

(d) Use a graphing utility to graph the models on the same screen. Use the following viewing window.

Xmin = 0
Xmax = 10
Xscl = 1
Ymin = 0
Ymax = 7
Yscl = 1

91. *Misleading Graphs* Use a graphing utility to graph the line $y = 0.75x - 2$ for each viewing window.

Xmin = -10	Xmin = 0
Xmax = 10	Xmax = 1
Xscl = 2	Xscl = 0.5
Ymin = -100	Ymin = -2
Ymax = 100	Ymax = -1.5
Yscl = 10	Yscl = 0.1

(a) Do the lines appear to have the same slope?

(b) Does either of the lines appear to have a slope of 0.75? If not, find a setting that will make the line appear to have a slope of 0.75.

(c) Describe real-life situations in which it would be to your advantage to use the two given settings.

92. *Rate of Change* The following are the slopes of lines representing annual sales y in terms of time t in years. Use the slopes to determine any change in annual sales for a 1-year increase in time t.

(a) $m = 76$ (b) $m = 0$ (c) $m = -14$

Explaining Concepts

93. Is the slope of a line a ratio? Explain.

94. Explain how you can visually determine the sign of the slope of a line by observing the graph of the line.

95. *True or False?* If both the x- and y-intercepts of a line are positive, then the slope of the line is positive.

96. Which slope is steeper: -5 or 2? Explain.

97. Is it possible to have two perpendicular lines with positive slopes? Explain.

98. The slope of a line is $\frac{3}{2}$. If x is increased by 8 units, how much will y change? Explain.

99. When a quantity y is increasing or decreasing at a constant rate over time t, the graph of y versus t is a line. What is another name for the rate of change?

100. Is it possible to use a graphing utility in function mode to graph the equation $x - 5 = 0$? Explain.

101. Explain how to use slopes to determine if the points $(-2, -3)$, $(1, 1)$, and $(3, 4)$ lie on the same line.

102. When determining the slope of the line through two points, does the order of subtracting coordinates of the points matter? Explain.

4.5 Equations of Lines

Objectives

1. Write an equation of a line using the point-slope form.
2. Write the equations of horizontal and vertical lines.
3. Use a linear model to solve an application problem.

1 Write an equation of a line using the point-slope form.

The Point-Slope Equation of a Line

In Sections 4.1 through 4.4, you have been studying analytic (or coordinate) geometry. Analytic geometry uses a coordinate plane to give visual representations of algebraic concepts, such as equations or functions.

There are two basic types of problems in analytic geometry.

1. Given an equation, sketch its graph.

   Algebra ⇨ Geometry

2. Given a graph, write its equation.

   Geometry ⇨ Algebra

In Section 4.4, you worked primarily with the first type of problem. In this section, you will study the second type. Specifically, you will learn how to write the equation of a line when you are given its slope and a point on the line. Before we give a general formula for doing this, consider the following example.

Example 1 Writing an Equation of a Line

A line has a slope of $\frac{5}{3}$ and passes through the point $(2, 1)$. Find its equation.

Solution

Begin by sketching the line, as shown in Figure 4.31. The slope of a line is the same through any two points on the line. So, to find an equation of the line, let (x, y) represent *any* point on the line. Now, using the representative point (x, y) and the given point $(2, 1)$, it follows that the slope of the line is

$$m = \frac{y - 1}{x - 2}.$$ Difference in y-coordinates / Difference in x-coordinates

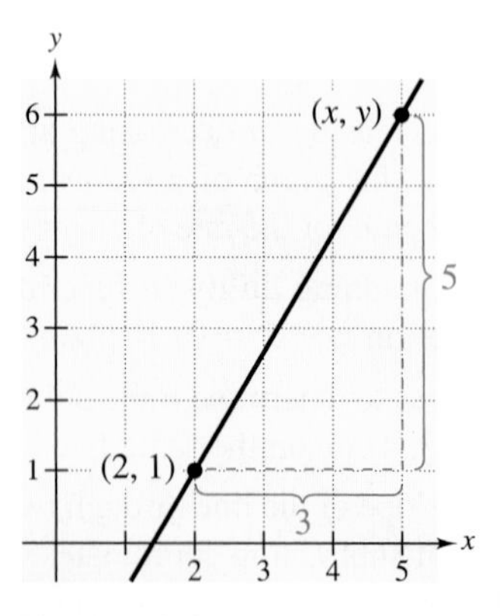

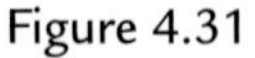

Figure 4.31

By substituting $\frac{5}{3}$ for m, you obtain the equation of the line.

$\frac{5}{3} = \frac{y - 1}{x - 2}$ Slope formula

$5(x - 2) = 3(y - 1)$ Cross-multiply.

$5x - 10 = 3y - 3$ Distributive Property

$5x - 3y = 7$ Equation of line

So, an equation for the line is $5x - 3y = 7$.

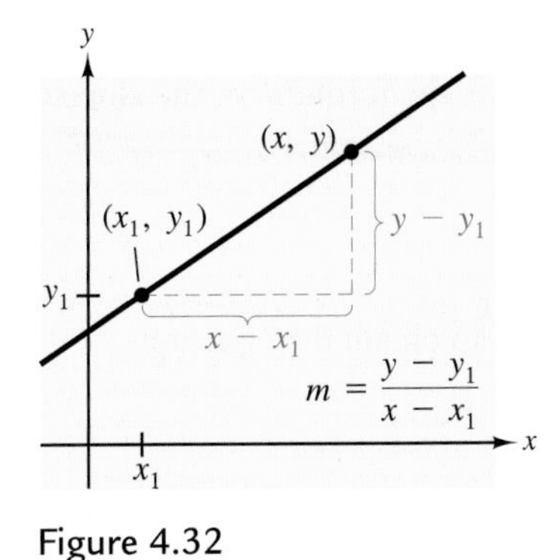

Figure 4.32

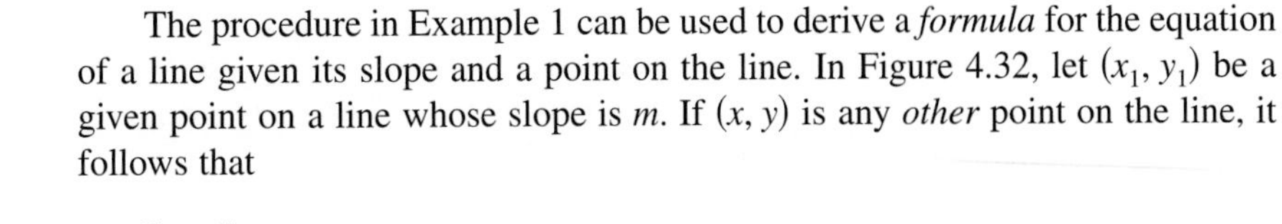

The procedure in Example 1 can be used to derive a *formula* for the equation of a line given its slope and a point on the line. In Figure 4.32, let (x_1, y_1) be a given point on a line whose slope is m. If (x, y) is any *other* point on the line, it follows that

$$\frac{y - y_1}{x - x_1} = m.$$

This equation in variables x and y can be rewritten in the form

$$y - y_1 = m(x - x_1)$$

which is called the **point-slope form** of the equation of a line.

▶ Point-Slope Form of the Equation of a Line

The **point-slope form** of the equation of a line with slope m and passing through the point (x_1, y_1) is

$$y - y_1 = m(x - x_1).$$

Example 2 The Point-Slope Form of the Equation of a Line

Find an equation of the line with slope 3 and passing through the point $(1, -2)$.

Solution

Use the point-slope form with $(x_1, y_1) = (1, -2)$ and $m = 3$.

$$y - y_1 = m(x - x_1) \qquad \text{Point-slope form}$$

$$y - (-2) = 3(x - 1) \qquad \text{Substitute } -2 \text{ for } y_1\text{, 1 for } x_1\text{, and 3 for } m.$$

$$y + 2 = 3x - 3 \qquad \text{Simplify.}$$

$$y = 3x - 5 \qquad \text{Equation of line}$$

So, an equation of the line is $y = 3x - 5$. Note that this is the slope-intercept form of the equation. The graph of this line is shown in Figure 4.33.

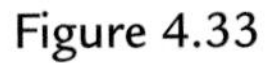

Figure 4.33

In Example 2, note that we concluded that $y = 3x - 5$ is "an" equation of the line rather than saying it is "the" equation of the line. The reason for this is that every equation can be written in many equivalent forms. For instance,

$$y = 3x - 5, \quad 3x - y = 5, \quad \text{and} \quad 3x - y - 5 = 0$$

are all equations of the line in Example 2. The first of these equations $(y = 3x - 5)$ is in the slope-intercept form

$$y = mx + b \qquad \text{Slope-intercept form}$$

and it provides the most information about the line. The last of these equations $(3x - y - 5 = 0)$ is in the general form of the equation of a line.

$$ax + by = 0 \qquad \text{General form}$$

A program that uses the two-point form to find the equation of a line is available at our website *www.hmco.com.* Programs for several models of calculators are available.

The program prompts for the coordinates of the two points and then outputs the slope and the y-intercept of the line that passes through the two points. Verify Example 3 using this program.

The point-slope form can be used to find an equation of a line passing through any two points (x_1, y_1) and (x_2, y_2). First, use the formula for the slope of a line passing through these two points.

$$m = \frac{y_2 - y_1}{x_2 - x_1}$$

Then, knowing the slope, use the point-slope form to obtain the equation

$$y - y_1 = \frac{y_2 - y_1}{x_2 - x_1}(x - x_1). \qquad \text{Two-point form}$$

This is sometimes called the **two-point form** of an equation of a line.

Example 3 A Line Passing Through Two Points

Find an equation of the line that passes through the points $(3, 1)$ and $(-3, 4)$.

Solution

Let $(x_1, y_1) = (3, 1)$ and $(x_2, y_2) = (-3, 4)$. The slope of a line passing through these points is

$$m = \frac{y_2 - y_1}{x_2 - x_1} \qquad \text{Formula for slope}$$

$$= \frac{4 - 1}{-3 - 3} \qquad \text{Substitute for } x_1, y_1, x_2, \text{ and } y_2.$$

$$= \frac{3}{-6} \qquad \text{Simplify.}$$

$$= -\frac{1}{2}. \qquad \text{Simplify.}$$

Now, use the point-slope form to find an equation of the line.

$$y - y_1 = m(x - x_1) \qquad \text{Point-slope form}$$

$$y - 1 = -\frac{1}{2}(x - 3) \qquad \text{Substitute 1 for } y_1, \text{ 3 for } x_1, \text{ and } -\tfrac{1}{2} \text{ for } m.$$

$$y - 1 = -\frac{1}{2}x + \frac{3}{2} \qquad \text{Simplify.}$$

$$y = -\frac{1}{2}x + \frac{5}{2} \qquad \text{Equation of line}$$

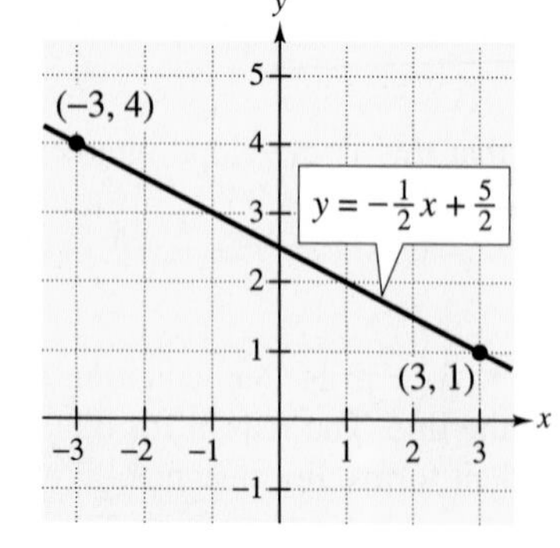

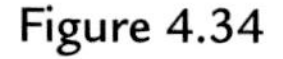
Figure 4.34

The graph of this line is shown in Figure 4.34.

In Example 3, it does not matter which of the two points is labeled (x_1, y_1) and which is labeled (x_2, y_2). Try switching these labels to $(x_1, y_1) = (-3, 4)$ and $(x_2, y_2) = (3, 1)$ and reworking the problem to see that you obtain the same equation.

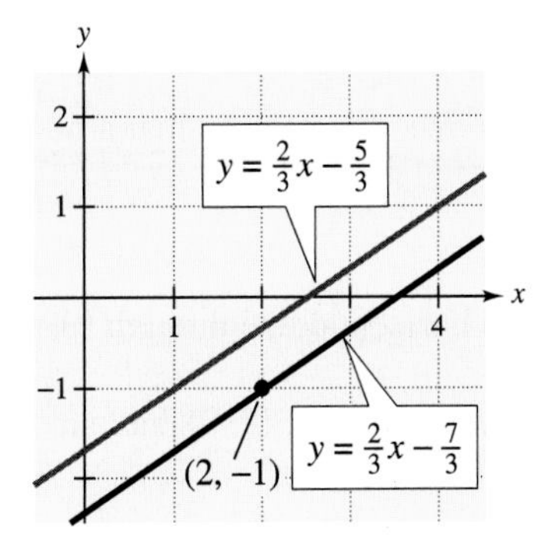

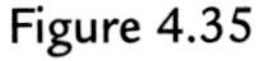

Figure 4.35

Example 4 Equations of Parallel Lines

Find an equation of the line that passes through the point $(2, -1)$ and is parallel to the line

$$2x - 3y = 5,$$

as shown in Figure 4.35.

Solution

To begin, write the given equation in slope-intercept form.

$$2x - 3y = 5 \qquad \text{Given equation}$$

$$-3y = -2x + 5 \qquad \text{Subtract } 2x \text{ from both sides.}$$

$$y = \frac{2}{3}x - \frac{5}{3} \qquad \text{Divide both sides by } -3.$$

Because the line has a slope of $m = \frac{2}{3}$, it follows that any parallel line must have the same slope. So, an equation of the line through $(2, -1)$, parallel to the given line is

$$y - y_1 = m(x - x_1) \qquad \text{Point-slope form}$$

$$y - (-1) = \frac{2}{3}(x - 2) \qquad \text{Substitute } -1 \text{ for } y_1,\ 2 \text{ for } x_1, \text{ and } \tfrac{2}{3} \text{ for } m.$$

$$y + 1 = \frac{2}{3}x - \frac{4}{3} \qquad \text{Distributive Property}$$

$$y = \frac{2}{3}x - \frac{7}{3}. \qquad \text{Equation of line}$$

Technology: Tip

With a graphing utility, parallel lines appear to be parallel in both *square* and *nonsquare* window settings. Verify this by graphing $y = 2x - 3$ and $y = 2x + 1$ in both a square and a nonsquare window.

Such is not the case with perpendicular lines, as you can see by graphing $y = 2x - 3$ and $y = -\frac{1}{2}x + 1$ in a square and a nonsquare setting.

Example 5 Equations of Perpendicular Lines

Find an equation of the line that passes through the point $(2, -1)$ and is perpendicular to the line

$$2x - 3y = 5,$$

as shown in Figure 4.36.

Solution

From Example 4, the given line has a slope of $\frac{2}{3}$. Hence, any line perpendicular to this line must have a slope of $-\frac{3}{2}$. So, the equation of the required line through $(2, -1)$ has the following form.

$$y - y_1 = m(x - x_1) \qquad \text{Point-slope form}$$

$$y - (-1) = -\frac{3}{2}(x - 2) \qquad \text{Substitute } -1 \text{ for } y_1,\ 2 \text{ for } x_1, \text{ and } -\tfrac{3}{2} \text{ for } m.$$

$$y + 1 = -\frac{3}{2}x + 3 \qquad \text{Distributive Property}$$

$$y = -\frac{3}{2}x + 2 \qquad \text{Equation of line}$$

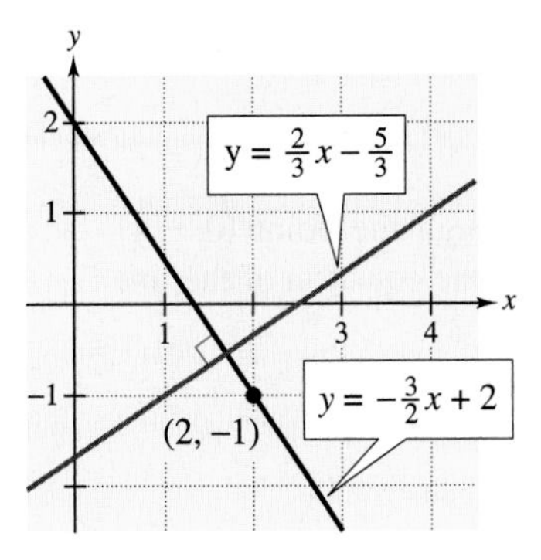

Figure 4.36

2 Write the equations of horizontal and vertical lines.

Equations of Horizontal and Vertical Lines

From the slope-intercept form of the equation of a line, you can see that a horizontal line ($m = 0$) has an equation of the form

$$y = (0)x + b \quad \text{or} \quad y = b. \qquad \text{Horizontal line}$$

This is consistent with the fact that each point on a horizontal line through $(0, b)$ has a y-coordinate of b.

In a similar way, each point on a vertical line through $(a, 0)$ has an x-coordinate of a. So, a vertical line has an equation of the form

$$x = a. \qquad \text{Vertical line}$$

The equation of a vertical line cannot be written in slope-intercept form because the slope of a vertical line is undefined. However, *every* line has an equation that can be written in the **general form**

$$ax + by + c = 0 \qquad \text{General form}$$

where a and b are not *both* zero.

Example 6 Writing Equations of Horizontal and Vertical Lines

Write an equation for each of the following lines.

a. Vertical line through $(-3, 2)$

b. Line passing through $(-1, 2)$ and $(4, 2)$

c. Line passing through $(0, 2)$ and $(0, -2)$

d. Horizontal line through $(0, -4)$

Solution

a. Because the line is vertical and passes through the point $(-3, 2)$, every point on the line has an x-coordinate of -3. So, the equation of the line is

$$x = -3. \qquad \text{Vertical line}$$

b. Because both points have the same y-coordinate, the line through $(-1, 2)$ and $(4, 2)$ is horizontal. So, its equation is

$$y = 2. \qquad \text{Horizontal line}$$

c. Because both points have the same x-coordinate, the line through $(0, 2)$ and $(0, -2)$ is vertical. So, its equation is

$$x = 0. \qquad \text{Vertical line (}y\text{-axis)}$$

d. Because the line is horizontal and passes through the point $(0, -4)$, every point on the line has a y-coordinate of -4. So, the equation of the line is

$$y = -4. \qquad \text{Horizontal line}$$

The graphs of the lines are shown in Figure 4.37.

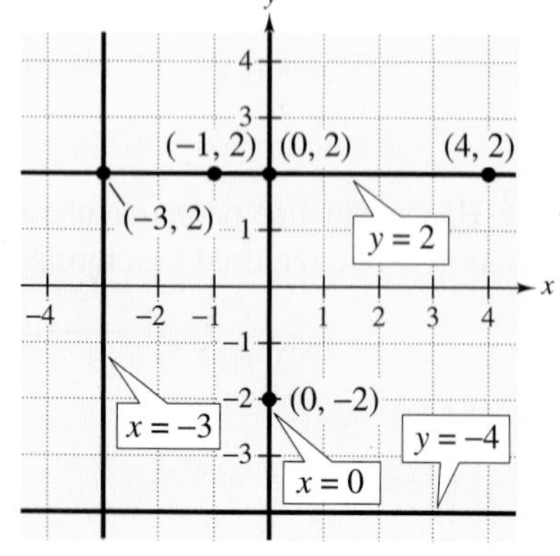

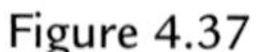
Figure 4.37

In Example 6(c), note that the equation $x = 0$ represents the y-axis. In a similar way, you can show that the equation $y = 0$ represents the x-axis.

3 Use a linear model to solve an application problem.

Application

Example 7 Total Sales

During the first year of operation, a company had sales of \$146 million. During the second year, the company had sales of \$154 million. Using this information only and assuming this trend continues, what would you estimate the sales to be during the third year? the fifth year?

Solution

To solve this problem, use a *linear model,* with y representing the total sales and t representing the year. That is, in Figure 4.38, let (1, 146) and (2, 154) be two points on the line representing the sales for the company. The slope of this line is

$$m = \frac{154 - 146}{2 - 1} = 8.$$

With this slope, you can use the point-slope form to find an equation of the line.

$$y - y_1 = m(t - t_1) \qquad \text{Point-slope form}$$

$$y - 146 = 8(t - 1) \qquad \text{Substitute 146 for } y_1\text{, 1 for } t_1\text{, and 8 for } m.$$

$$y - 146 = 8t - 8 \qquad \text{Distributive Property}$$

$$y = 8t + 138 \qquad \text{Equation of line}$$

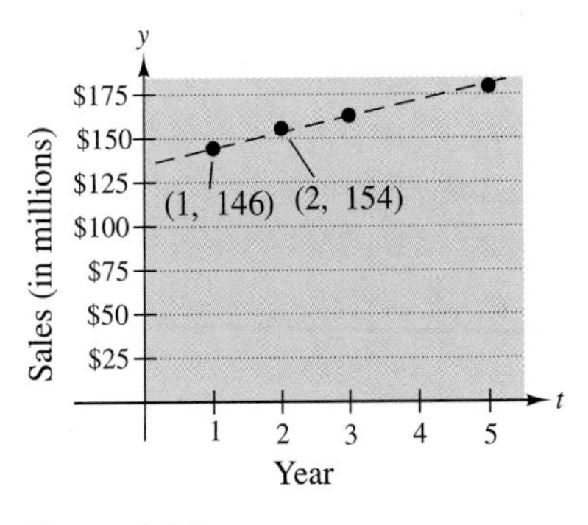

Figure 4.38

Using this model, an estimate of the sales during the third year ($t = 3$) is

$$y = 8(3) + 138 = \$162 \text{ million}.$$

An estimate of the sales during the fifth year ($t = 5$) is found similarly.

$$y = 8(5) + 138 = \$178 \text{ million}.$$

The estimation method illustrated in Example 7 is called **linear extrapolation.** Note in Figure 4.39 that for linear extrapolation, the estimated point lies *to the right* of the given points. When the estimated point lies *between* two given points, the method is called **linear interpolation.**

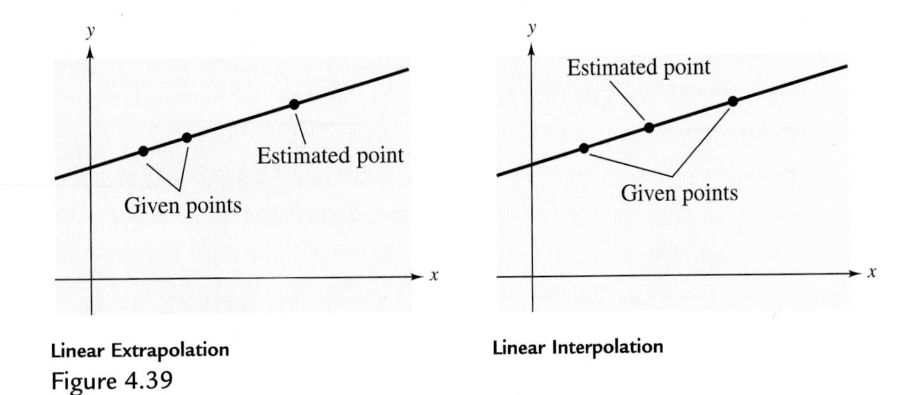

Linear Extrapolation **Linear Interpolation**

Figure 4.39

You have now studied several formulas that relate to equations of lines. In the summary below, remember that the formulas that deal with slope cannot be applied to vertical lines. For instance, the lines $x = 2$ and $y = 3$ are perpendicular, but they do not follow the "negative reciprocal property" of perpendicular lines because the line $x = 2$ is vertical (and has no slope).

Summary of Equations of Lines

1. Slope of the line through (x_1, y_1) and (x_2, y_2):

$$m = \frac{y_2 - y_1}{x_2 - x_1}$$

2. General form of an equation of a line:

$$ax + by + c = 0$$

3. Equation of a vertical line:

$$x = a$$

4. Equation of a horizontal line:

$$y = b$$

5. Slope-intercept form of an equation of a line:

$$y = mx + b$$

6. Point-slope form of an equation of a line:

$$y - y_1 = m(x - x_1)$$

7. Parallel lines have *equal* slopes:

$$m_1 = m_2$$

8. Perpendicular lines have *negative reciprocal* slopes:

$$m_1 = -\frac{1}{m_2}$$

Discussing the Concept — Versatility of $y = mx + b$

In this chapter, it has been shown that, of the forms of the equation of a line, $y = mx + b$ is better suited for *sketching the graph* of a given equation. On the other hand, $y - y_1 = m(x - x_1)$ is better suited for *creating the equation* of a line, given its slope and a point on the line. Show how $y = mx + b$ can be used to determine the equation of a line that passes through $(-3, 2)$ and has slope of -2. Compare this procedure with the use of the point-slope form. Which procedure do you prefer? Explain why.

4.5 Exercises

Integrated Review *Concepts, Skills, and Problem Solving*

Keep mathematically in shape by doing these exercises *before* the problems of this section.

Properties and Definitions

1. Find the greatest common factor of 180 and 300 and explain how you arrived at your answer.
2. Find the least common multiple of 180 and 300 and explain how you arrived at your answer.

Simplifying Expressions

In Exercises 3-6, simplify the expression.

3. $4(3 - 2x)$ **4.** $x^2(xy^3)$

5. $3x - 2(x - 5)$ **6.** $u - [3 + (u - 4)]$

Solving Equations

In Exercises 7–10, solve for y in terms of x.

7. $3x + y = 4$ **8.** $4 - y + x = 0$

9. $4x - 5y = -2$ **10.** $3x + 4y - 5 = 0$

Developing Skills

In Exercises 1–14, find an equation for the line that passes through the point and has the specified slope. Sketch the line. See Example 1.

1. $(0, 0), m = -2$ **2.** $(0, -2), m = 3$

3. $(6, 0), m = \frac{1}{2}$ **4.** $(0, 10), m = -\frac{1}{4}$

5. $(-2, 1), m = 2$ **6.** $(3, -5), m = -1$

7. $(-8, -1), m = -\frac{1}{4}$ **8.** $(12, 4), m = -\frac{2}{3}$

9. $\left(\frac{1}{2}, -3\right), m = 0$ **10.** $\left(-\frac{5}{4}, 6\right), m = 0$

11. $\left(0, \frac{3}{2}\right), m = \frac{2}{3}$ **12.** $\left(0, -\frac{5}{2}\right), m = \frac{3}{4}$

13. $(2, 4), m = -0.8$ **14.** $(6, -3), m = 0.67$

In Exercises 15–26, use the point-slope form to write an equation of the line passing through the point and having the specified slope. (Write your answer in slope-intercept form.) See Example 2.

15. $(0, -4), m = 3$ **16.** $(0, 7), m = -1$

17. $(-3, 6), m = -2$ **18.** $(-1, 4), m = 4$

19. $(9, 0), m = -\frac{1}{3}$ **20.** $(0, -2), m = \frac{4}{3}$

21. $(-10, 4), m = 0$ **22.** $(-2, -5), m = 0$

23. $(8, 1), m = -\frac{3}{4}$ **24.** $(-3, 2), m = \frac{1}{3}$

25. $(-2, 1), m = \frac{2}{3}$ **26.** $(1, 3), m = -\frac{1}{2}$

In Exercises 27–38, find the slope of the line. If it is not possible, explain why.

27. $y = \frac{3}{8}x - 4$ **28.** $y = -3x + 10$

29. $y - 2 = 5(x + 3)$ **30.** $y + 3 = -2(x - 6)$

31. $y + \frac{5}{6} = \frac{2}{3}(x + 4)$ **32.** $y - \frac{1}{4} = \frac{5}{8}\left(x - \frac{13}{5}\right)$

33. $3x + y = 0$ **34.** $y - 6 = 0$

35. $2x - y = 0$ **36.** $x + 5 = 0$

37. $3x - 2y + 10 = 0$ **38.** $5x + 4y - 8 = 0$

In Exercises 39–42, write the slope-intercept form of the line.

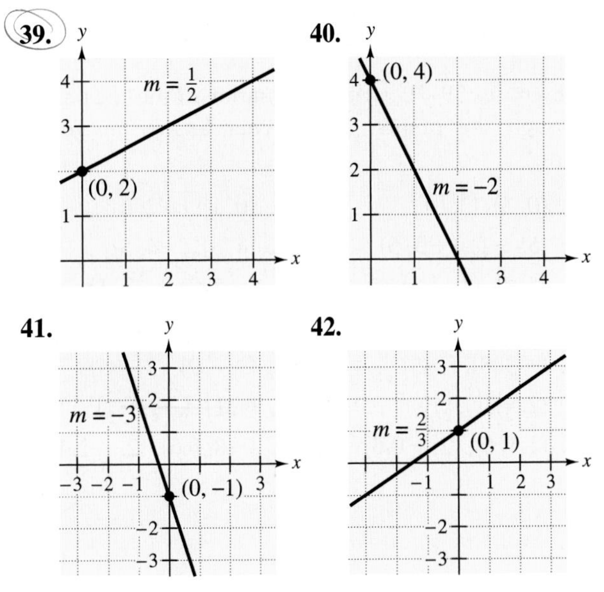

In Exercises 43–46, write the point-slope form of the equation of the line.

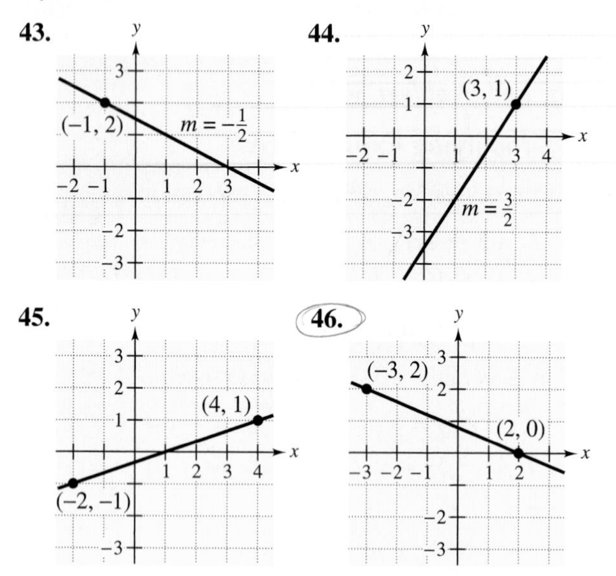

In Exercises 47–58, write an equation of the line through the points. Sketch a graph of the line. See Example 3.

47. (0, 0), (4, 4)

48. (0, 0), (−2, 4)

49. (0, 0), (2, −4)

50. (6, −1), (3, 3)

51. (2, 3), (6, 5)

52. (−4, 6), (−2, 3)

53. (−6, 2), (3, 5)

54. (−9, 7), (−4, 4)

55. (5, −1), (3, 2)

56. (0, 3), (5, 3)

57. $\left(\frac{5}{2}, -1\right), \left(\frac{9}{2}, 7\right)$

58. $\left(4, \frac{5}{3}\right), \left(-1, \frac{2}{3}\right)$

In Exercises 59–72, write an equation of the line passing through the points. (Write your answer in general form.)

59. (0, 3), (3, 0)

60. (0, 1), (−2, 0)

61. (5, −1), (−5, 5)

62. (4, 3), (−4, 5)

63. (5, 4), (1, −4)

64. (−5, 7), (−2, 1)

65. (5, −1), (7, −4)

66. (3, 5), (1, 6)

67. (−3, 8), (2, 5)

68. (9, −9), (7, −5)

69. $\left(2, \frac{1}{2}\right), \left(\frac{1}{2}, \frac{5}{2}\right)$

70. $\left(\frac{1}{4}, 1\right), \left(-\frac{3}{4}, -\frac{2}{3}\right)$

71. (1, 0.6), (2, −0.6)

72. (−8, 0.6), (2, −2.4)

In Exercises 73–82, write an equation of the line through the indicated point (a) parallel to the given line and (b) perpendicular to the given line. See Examples 4 and 5.

73. (2, 1)
$x - y = 3$

74. (−3, 2)
$x + y = 7$

75. (−12, 4)
$3x + 4y = 7$

76. (15, −2)
$5x + 3y = 0$

77. (1, 3)
$2x + y = 0$

78. (5, −2)
$x + 5y = 3$

79. (−1, 0)
$y + 3 = 0$

80. (2, 5)
$x - 4 = 0$

81. (4, −1)
$3y - 2x = 7$

82. (−6, 5)
$4x - 5y = 2$

In Exercises 83–90, write an equation for each line. See Example 6.

83. Vertical line through (−2, 4)

84. Horizontal line through (7, 3)

85. Horizontal line through $\left(\frac{1}{2}, \frac{2}{3}\right)$

86. Vertical line through $\left(\frac{1}{4}, 0\right)$

87. Line passing through (4, 1) and (4, 8)

88. Line passing through (−1, 5) and (6, 5)

89. Line passing through (1, −8) and (7, −8)

90. Line passing through (3, 0) and (3, 5)

Graphical Exploration In Exercises 91–94, use a graphing utility to graph the lines. Use the square setting. Are the lines parallel, perpendicular, or neither?

91. $y = -0.4x + 3$
$y = \dfrac{5}{2}x - 1$

92. $y = \dfrac{2x - 3}{3}$
$y = \dfrac{4x + 3}{6}$

93. $y = 0.4x + 1$
$y = x + 2.5$

94. $y = \frac{3}{4}x - 5$
$y = -\frac{3}{4}x + 2$

95. *Graphical Exploration* Use a graphing utility to graph the following equations on the same screen. Use the square setting. What can you conclude?

(a) $y = \frac{1}{3}x + 2$ (b) $y = 4x + 2$

(c) $y = -3x + 2$ (d) $y = -\frac{1}{4}x + 2$

Solving Problems

96. *Writing a Linear Model* A sales representative receives a salary of \$2000 per month plus a commission of 2% of the total monthly sales. Write the wages W as a linear function of sales S.

97. *Writing a Linear Model* A sales representative is reimbursed \$225 per day for lodging and meals plus \$0.28 per mile driven. Write the daily cost C to the company as a function of x, the number of miles driven.

98. *Writing a Linear Model* A sales representative is reimbursed \$250 per day for lodging and meals plus \$0.43 per mile driven. Write the daily cost C to the company as a function of x, the number of miles driven.

99. *Writing a Linear Model* A sales representative receives a salary of \$2300 per month plus a commission of 3% of the total monthly sales. Write the wages W as a linear function of sales S.

100. *Writing and Graphing a Linear Model* A car travels for t hours at an average speed of 50 miles per hour. Write the distance d as a linear function of t. Graph the function for $0 \leq t \leq 5$.

101. *Writing and Using a Linear Model* A store is offering a 20% discount on all items in its inventory.

(a) Write the sale price S for an item as a linear function of its list price L.

(b) Use a graphing utility to graph the model.

(c) Use the graph to estimate the sale price of an item whose list price is \$49.98. Confirm your estimate algebraically.

102. *Writing and Using a Linear Model* A small business purchased a plain paper copier for \$5400. After 1 year, its depreciated value is \$4300. The depreciation is linear. See Example 7.

(a) Write the value V of the copier as a linear function of time t in years.

(b) Use the model to estimate the value after 3 years.

103. *Writing and Using a Linear Model* A business purchased a new machine for \$200,000. After 1 year, its depreciated value is \$170,000. The depreciation is linear.

(a) Write the value V of the machine as a linear function of time t in years.

(b) Use the model to estimate the value after 5 years.

104. *Writing and Using a Linear Model* A real estate office handles an apartment complex with 50 units. When the rent per unit is \$480 per month, all 50 units are occupied. However, when the rent is \$525 per month, the average number of occupied units drops to 47. Assume that the relationship between the monthly rent p and the demand x is linear.

(a) Represent the given information as two ordered pairs of the form (x, p). Plot these ordered pairs.

(b) Write the rent p as a linear function of the demand x. Graph the line and describe the relationship between the rent and the demand.

(c) (*Linear Extrapolation*) Predict the number of units occupied if the rent is raised to \$555.

(d) (*Linear Interpolation*) Predict the number of units occupied if the rent is lowered to \$495.

105. *Writing and Using a Linear Model* A small liberal arts college had an enrollment of 1200 students in 1990. During the next 10 years the enrollment increased by approximately 50 students per year.

(a) Write the enrollment N as a function of the year t. (Let $t = 0$ represent 1990.)

(b) (*Linear Interpolation*) Use the model to predict the enrollment in the year 2004.

(c) (*Linear Interpolation*) Use the model to estimate the enrollment in 1998.

106. *Think About It* Find the slope of the line for the equation $5x + 7y - 21 = 0$. Use the same process to find a formula for the slope of the line $ax + by + c = 0$ where $b \neq 0$.

107. *Graphical Interpretation* Match each situation labeled (a), (b), (c), and (d) with one of the graphs labeled (e), (f), (g), and (h). Then find the slope of the line and write a verbal description of the slope in the context of the real-life situation.

(a) A friend is paying you \$10 per week to repay a \$100 loan.

(b) An employee is paid \$12.50 per hour plus \$1.50 for each unit produced per hour.

(c) A sales representative receives \$40 per day for food plus \$0.32 for each mile traveled.

(d) A typewriter purchased for \$600 depreciates \$100 per year.

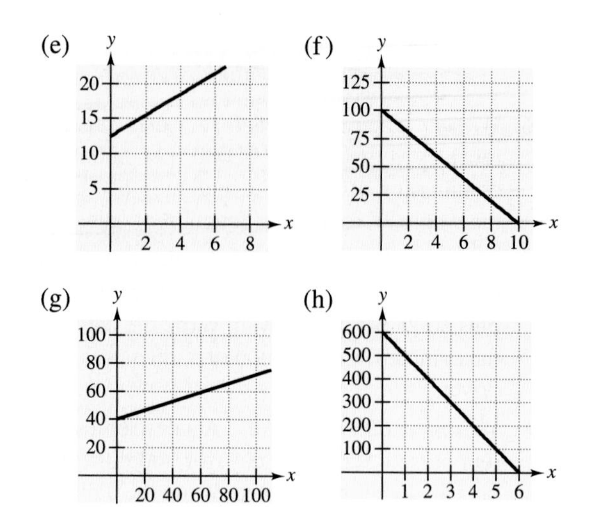

Explaining Concepts

108. Answer parts (d)–(g) of Motivating the Chapter on page 201.

109. Can any pair of points on a line be used to calculate the slope of the line? Explain.

110. Can the equation of a vertical line be written in slope-intercept form? Explain.

111. In the equation $y = mx + b$, what do m and b represent?

112. In the equation $y - y_1 = m(x - x_1)$, what do x_1 and y_1 represent?

113. Explain how to find analytically the x-intercept of the line given by $y = mx + b$.

114. What is implied about the graphs of the lines $a_1x + b_1y + c_1 = 0$ and $a_2x + b_2y + c_2 = 0$ if $a_1/b_1 = a_2/b_2$?

115. *Research Project* In a news magazine or newspaper, find an example of data that are *increasing* linearly with time. Write a linear function that models the data. Repeat the project for data that are *decreasing*.

4.6 Graphs of Linear Inequalities

Objectives

1 Determine whether an ordered pair is a solution of a linear inequality in two variables.

2 Sketch the graph of a linear inequality in two variables.

1 Determine whether an ordered pair is a solution of a linear inequality in two variables.

Linear Inequalities in Two Variables

A **linear inequality** in two variables x and y is an inequality that can be written in one of the following forms.

$$ax + by < c, \quad ax + by > c, \quad ax + by \leq c, \quad \text{and} \quad ax + by \geq c$$

Here are some examples.

$$x - y > 2, \quad 3x - 2y \leq 6, \quad x \geq 5, \quad \text{and} \quad y < -1$$

An ordered pair (x_1, y_1) is a **solution** of a linear inequality in x and y if the inequality is true when x_1 and y_1 are substituted for x and y, respectively. For instance, the ordered pair $(3, 2)$ is a solution of the inequality $x - y > 0$ because $3 - 2 > 0$ is a true statement.

Example 1 Verifying Solutions of Linear Inequalities

Decide whether the points are solutions of $3x - y \geq -1$.

a. $(0, 0)$ **b.** $(1, 4)$ **c.** $(-1, 2)$

Solution

a. $3x - y \geq -1$ Original inequality

$3(0) - 0 \overset{?}{\geq} -1$ Substitute 0 for x and 0 for y.

$0 \geq -1$ Inequality is satisfied. ✓

Because the inequality is satisfied, the point $(0, 0)$ *is* a solution.

b. $3x - y \geq -1$ Original inequality

$3(1) - 4 \overset{?}{\geq} -1$ Substitute 1 for x and 4 for y.

$-1 \geq -1$ Inequality is satisfied. ✓

Because the inequality is satisfied, the point $(1, 4)$ *is* a solution.

c. $3x - y \geq -1$ Original inequality

$3(-1) - 2 \overset{?}{\geq} -1$ Substitute -1 for x and 2 for y.

$-5 \not\geq -1$ Inequality is not satisfied. ✗

Because the inequality is not satisfied, the point $(-1, 2)$ *is not* a solution.

2 Sketch the graph of a linear inequality in two variables.

The Graph of a Linear Inequality

The **graph** of an inequality is the collection of all solution points of the inequality. To sketch the graph of a linear inequality such as

$3x - 2y < 6$ Original inequality

begin by sketching the graph of the *corresponding linear equation*

$3x - 2y = 6.$ Corresponding equation

Use a *dashed* line for the inequalities $<$ and $>$ and a *solid* line for the inequalities $\leq$ and $\geq$. The graph of the equation separates the plane into two **half-planes.** In each half-plane, one of the following must be true.

1. All points in the half-plane are solutions of the inequality.
2. No point in the half-plane is a solution of the inequality.

So, you can determine whether the points in an entire half-plane satisfy the inequality by simply testing *one* point in the region.

Example 2 Sketching the Graph of a Linear Inequality

Sketch the graphs of the linear inequalities.

a. $x > -2$ **b.** $y \leq 3$

Solution

a. The graph of the corresponding equation $x = -2$ is a vertical line. The points (x, y) that satisfy the inequality $x > -2$ are those lying to the right of this line, as shown in Figure 4.40.

b. The graph of the corresponding equation $y = 3$ is a horizontal line. The points (x, y) that satisfy the inequality $y \leq 3$ are those lying below (or on) this line, as shown in Figure 4.41.

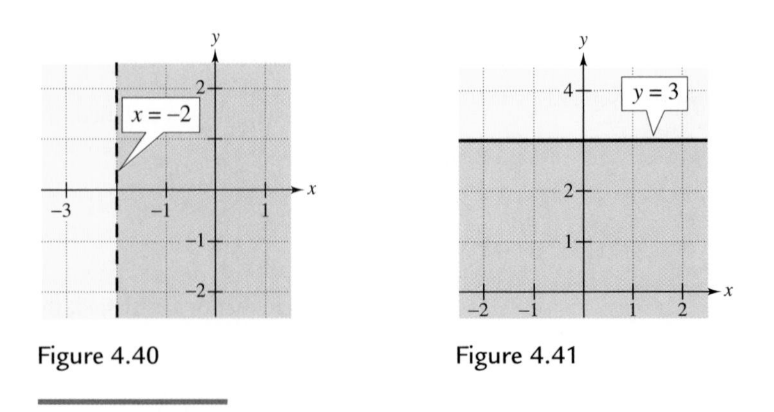

Figure 4.40 Figure 4.41

Notice that a dashed line is used for the graph of $x > -2$ and a solid line is used for the graph of $y \leq 3$.

Some guidelines for sketching the graph of a linear inequality in two variables are listed below.

▶ Guidelines for Graphing a Linear Inequality

1. Replace the inequality sign by an equal sign and sketch the graph of the resulting equation. (Use a dashed line for $<$ and $>$ and a solid line for $\leq$ and $\geq$.)

2. Test one point in each of the half-planes formed by the graph in Step 1. If the point satisfies the inequality, then shade the entire half-plane to denote that every point in the region satisfies the inequality.

Study Tip

You can use any point that is not on the line as a test point. However, the origin is often the most convenient because it is easy to evaluate expressions in which 0 is substituted for each variable.

Example 3 Sketching the Graph of a Linear Inequality

Sketch the graph of the linear inequality

$x - y < 2.$ Original inequality

Solution

The graph of the corresponding equation

$x - y = 2$ Corresponding equation

is a line, as shown in Figure 4.42. Because the origin $(0, 0)$ does not lie on the line, use it as the test point.

$x - y < 2$ Original inequality

$0 - 0 \stackrel{?}{<} 2$ Substitute 0 for x and 0 for y.

$0 < 2$ Inequality is satisfied. ✓

Because $(0, 0)$ satisfies the inequality, the graph consists of the half-plane lying above the line. Try checking a point below the line. Regardless of the point you choose, you will see that it is not a solution.

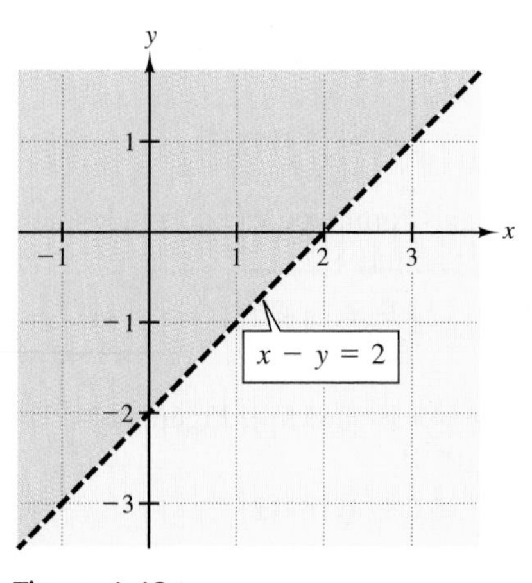

Figure 4.42

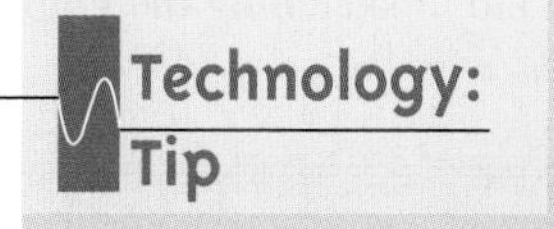

Many graphing utilities are capable of graphing linear inequalities. Consult the user's guide of your graphing utility for specific instructions.

The graph of $y \leq -x + 2$ is shown below.

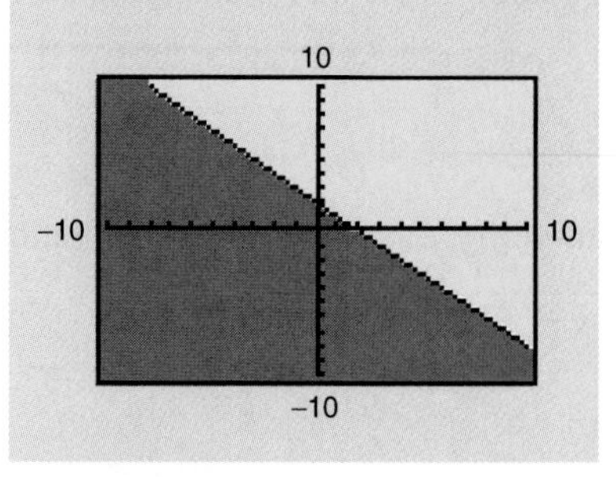

For a linear inequality in two variables, you can sometimes simplify the graphing procedure by writing the inequality in *slope-intercept* form. For instance, by writing $x - y < 2$ in the form $y > x - 2$, you can see that the solution points lie *above* the line $y = x - 2$, as shown in Figure 4.42. Similarly, by writing the inequality $3x - 2y > 5$ in the form

$$y < \frac{3}{2}x - \frac{5}{2}$$

you can see that the solutions lie *below* the line $y = \frac{3}{2}x - \frac{5}{2}$, as shown in Figure 4.43.

Study Tip

The solution of the inequality

$$y < \tfrac{3}{2}x - \tfrac{5}{2}$$

is a half-plane with the line

$$y = \tfrac{3}{2}x - \tfrac{5}{2}$$

as its boundary. The y-values that are less than this equation make the inequality true. So, you want to shade the half-plane with the smaller y-values, as shown in Figure 4.43.

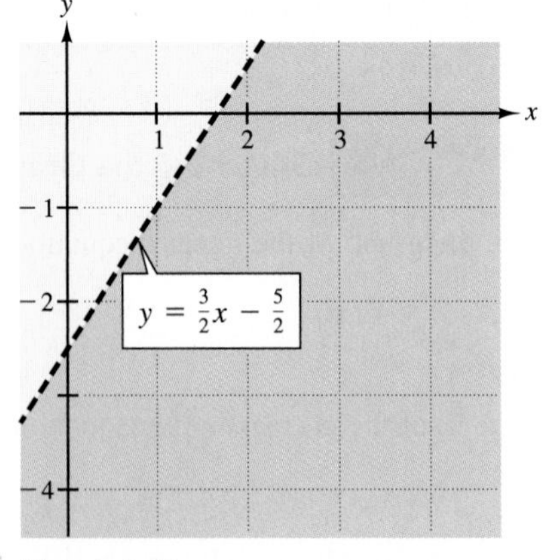

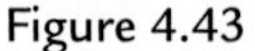
Figure 4.43

Example 4 Sketching the Graph of a Linear Inequality

Use the slope-intercept form of a linear equation as an aid in sketching the graph of the inequality $5x + 4y \le 12$.

Solution

To begin, rewrite the inequality in slope-intercept form.

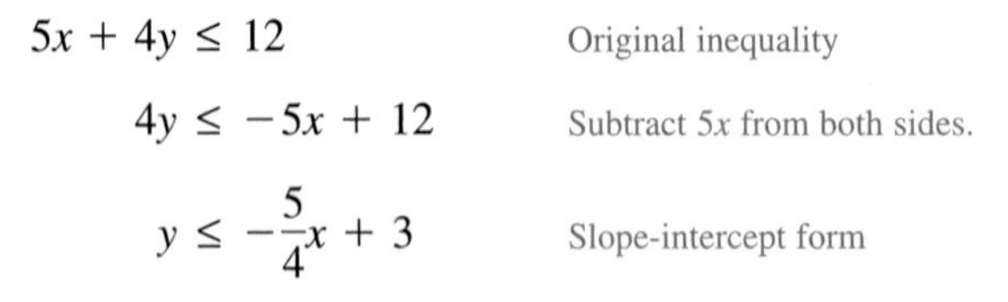

$5x + 4y \le 12$ Original inequality

$4y \le -5x + 12$ Subtract $5x$ from both sides.

$y \le -\frac{5}{4}x + 3$ Slope-intercept form

From this form, you can conclude that the solution is the half-plane lying *on* or *below* the line

$$y = -\frac{5}{4}x + 3.$$

The graph is shown in Figure 4.44. You can verify this by testing the solution point (0, 0).

$5x + 4y \le 12$ Original inequality

$5(0) + 4(0) \overset{?}{\le} 12$ Substitute 0 for x and 0 for y.

$0 \le 12$ Inequality is satisfied. ✓

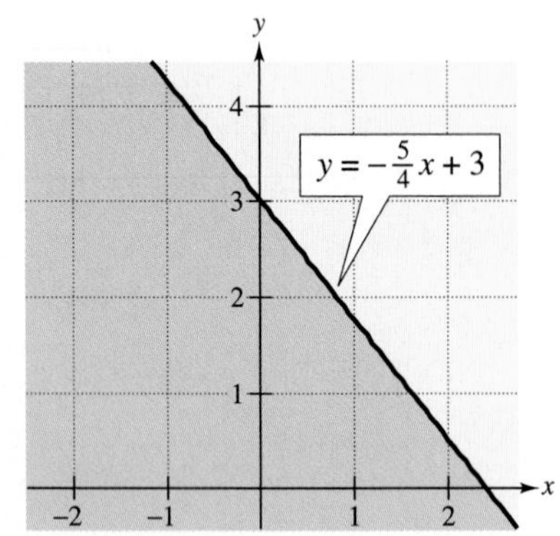

Figure 4.44

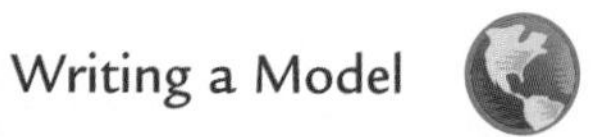

Example 5 Writing a Model

You have two part-time jobs. One is at a fast-food restaurant, which pays \$6 per hour, and the other is babysitting for \$4 per hour. Between the two jobs, you want to earn at least \$100 a week. Write an inequality that shows the number of hours that you need to work. Then sketch a graph of the inequality.

Solution

To write the inequality, use the problem-solving method.

Verbal Model: Hourly pay job 1 · Number of hours job 1 + Hourly pay job 2 · Number of hours job 2 ≥ Earnings in a week

Labels:

Hourly pay job 1 $= 6$	(dollars per hour)
Number of hours job 1 $= x$	(hours)
Hourly pay job 2 $= 4$	(dollars per hour)
Number of hours job 2 $= y$	(hours)
Earnings in a week $= 100$	(dollars)

Algebraic Inequality: $6x + 4y \geq 100$

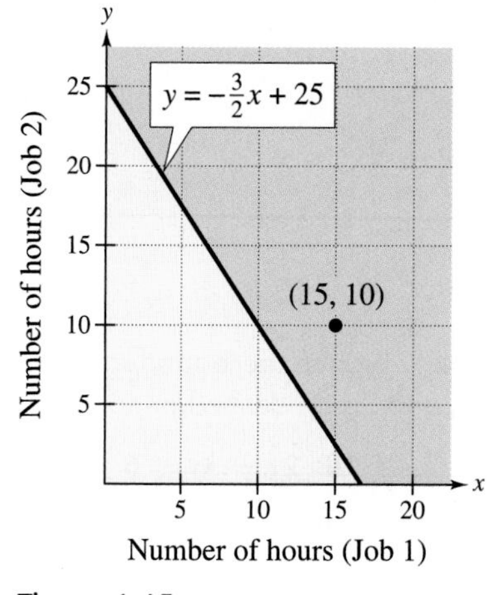

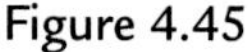
Figure 4.45

To sketch the graph, rewrite the inequality in slope-intercept form.

$$6x + 4y \geq 100 \quad \text{Original inequality}$$

$$4y \geq -6x + 100 \quad \text{Subtract } 6x \text{ from both sides.}$$

$$y \geq \frac{-6x + 100}{4} \quad \text{Divide both sides by 4.}$$

$$y \geq -\frac{3}{2}x + 25 \quad \text{Slope-intercept form}$$

From the graph of the inequality, shown in Figure 4.45, you can see that the point (15, 10) is one solution point. This means that if you work 15 hours at the restaurant and babysit for 10 hours, you will earn at least \$100.

Discussing the Concept — A Double Inequality

Determine how to use a graphing utility to find the solution set to the double inequality

$$2x \leq y \leq x + 2.$$

Choose a point in the region shaded by the utility, and show that it satisfies the double inequality.

4.6 Exercises

Integrated Review *Concepts, Skills, and Problem Solving*

Keep mathematically in shape by doing these exercises *before* the problems of this section.

Properties and Definitions

In Exercises 1–4, complete the property of inequalities by inserting the correct inequality symbol. (Consider a, b, and c to be real numbers, variables, or algebraic expressions.)

1. If $a < b$, then $a + 5$ ☐ $b + 5$.

2. If $a < b$, then $2a$ ☐ $2b$.

3. If $a < b$, then $-3a$ ☐ $-3b$.

4. If $a < b$ and $b < c$, then a ☐ c.

Solving Inequalities

In Exercises 5–10, solve the inequality and graph it on the real number line.

5. $x + 3 > 0$ **6.** $2 - x \geq 0$

7. $2t - 11 \leq 5$ **8.** $\frac{3}{2}y + 8 < 20$

9. $5 < 2x + 3 < 15$ **10.** $-2 < -\frac{x}{4} < 1$

Problem Solving

11. Assume the sales commission rate is 4.5%. Determine the sales of an employee who earned $544.50 as a sales commission.

12. One person can complete a typing project in 3 hours, and another can complete the same project in 4 hours. If they both work on the project, in how many hours can it be completed?

Developing Skills

In Exercises 1–8, which points are solutions? See Example 1.

Inequality	*Points*	
1. $x + y > 5$	(a) $(0, 0)$	(b) $(3, 6)$
	(c) $(-6, 20)$	(d) $(3, 2)$
2. $2x - y > 3$	(a) $(3, 0)$	(b) $(2, 6)$
	(c) $(-6, -20)$	(d) $(3, 3)$
3. $-3x + 5y \leq 12$	(a) $(1, 2)$	(b) $(2, -3)$
	(c) $(1, 3)$	(d) $(2, 8)$
4. $5x + 3y < 100$	(a) $(25, 10)$	(b) $(6, 10)$
	(c) $(0, -12)$	(d) $(4, 5)$
5. $3x - 2y < 2$	(a) $(1, 3)$	(b) $(2, 0)$
	(c) $(0, 0)$	(d) $(3, -5)$
6. $y - 2x > 5$	(a) $(4, 13)$	(b) $(8, 1)$
	(c) $(0, 7)$	(d) $(1, -3)$
7. $5x + 4y \geq 6$	(a) $(-2, 4)$	(b) $(5, 5)$
	(c) $(7, 0)$	(d) $(-2, 5)$
8. $5y + 8x \leq 14$	(a) $(-3, 8)$	(b) $(7, -6)$
	(c) $(1, 1)$	(d) $(3, 0)$

In Exercises 9–12, state whether the boundary of the graph of the inequality should be dashed or solid.

9. $2x + 3y < 6$ **10.** $2x + 3y \leq 6$

11. $2x + 3y \geq 6$ **12.** $2x + 3y > 6$

In Exercises 13–16, match the inequality with its graph. [The graphs are labeled (a), (b), (c), and (d).]

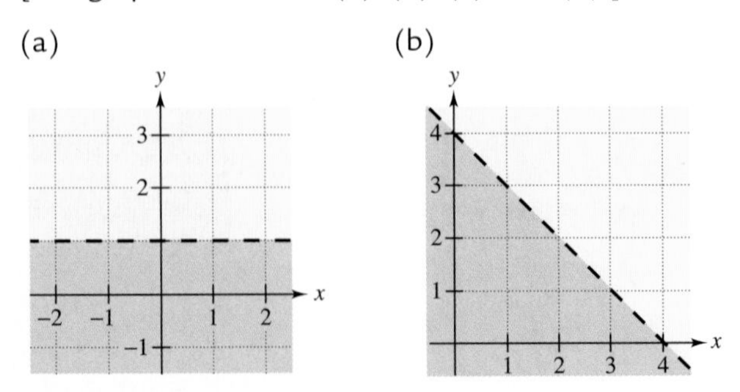

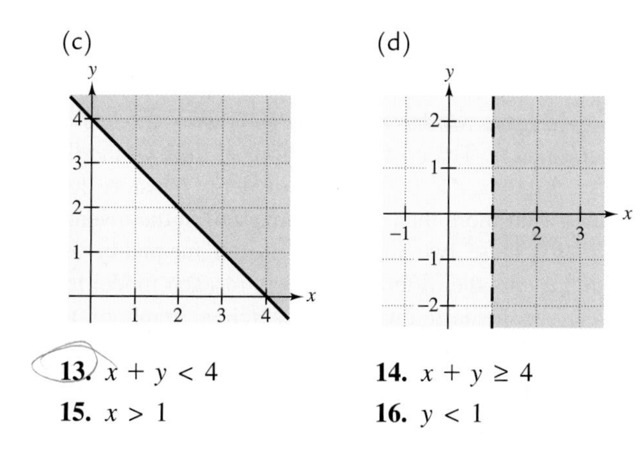

13. $x + y < 4$

14. $x + y \geq 4$

15. $x > 1$

16. $y < 1$

In Exercises 17–20, match the inequality with its graph. [The graphs are labeled (a), (b), (c), and (d).]

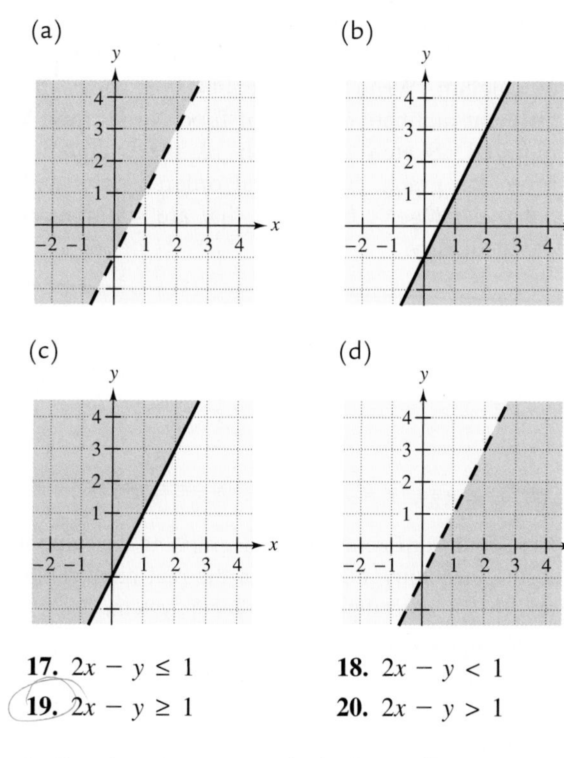

17. $2x - y \leq 1$

18. $2x - y < 1$

19. $2x - y \geq 1$

20. $2x - y > 1$

In Exercises 21–50, graph the inequality. See Examples 2–4.

21. $y \geq 3$

22. $x \leq 0$

23. $x > \frac{3}{2}$

24. $y < -2$

25. $y < \frac{1}{2}x$

26. $y > -\frac{2}{3}x$

27. $x - y < 0$

28. $x + y > 0$

29. $y \leq 2x - 1$

30. $y \geq -x + 3$

31. $y \leq x - 2$

32. $y \geq 0.6x + 1$

33. $y > x - 2$

34. $y < -x + 3$

35. $y > -2x + 10$

36. $y < 3x + 1$

37. $y \geq \frac{2}{3}x + \frac{1}{3}$

38. $y \leq -\frac{3}{4}x + 2$

39. $2x + y - 3 \geq 3$

40. $x - 2y + 6 \leq 0$

41. $-3x + 2y - 6 < 0$

42. $x + 4y + 2 \geq 2$

43. $5x + 2y < 5$

44. $5x + 2y > 5$

45. $x \geq 3y - 5$

46. $x > -2y + 10$

47. $y - 3 < \frac{1}{2}(x - 4)$

48. $y + 1 < -2(x - 3)$

49. $\dfrac{x}{3} + \dfrac{y}{4} < 1$

50. $\dfrac{x}{-2} + \dfrac{y}{2} > 1$

In Exercises 51–58, use a graphing utility to graph the inequality.

51. $y \geq 2x - 1$

52. $y \leq 4 - 0.5x$

53. $y \leq -2x + 4$

54. $y \geq x - 3$

55. $y \geq \frac{1}{2}x + 2$

56. $y \leq -\frac{2}{3}x + 6$

57. $6x + 10y - 15 \leq 0$

58. $3x - 2y + 4 \geq 0$

In Exercises 59–64, write an inequality that represents the graph.

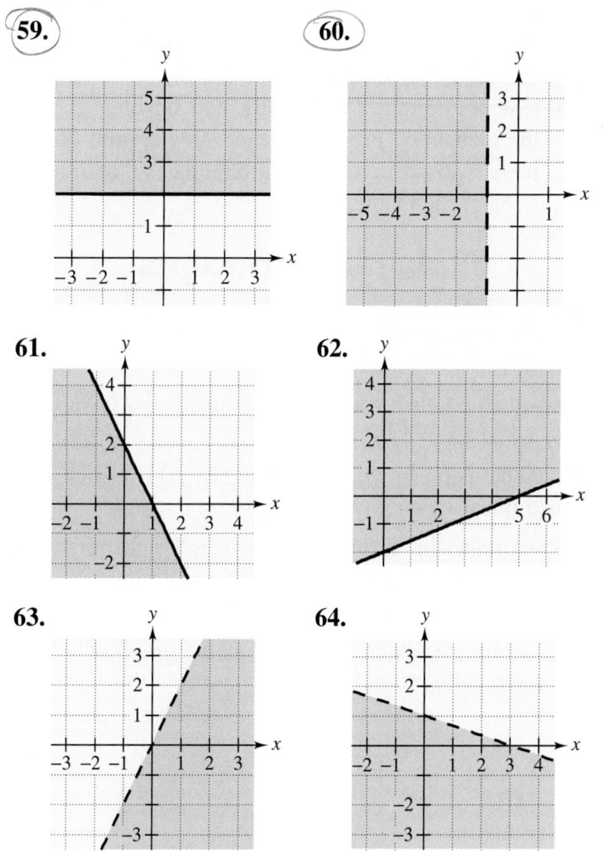

Solving Problems

65. *Writing a Model* You have two part-time jobs. One is at a grocery store, which pays $7 per hour, and the other is mowing lawns, which pays $5 per hour. Between the two jobs, you want to earn at least $140 a week. Write an inequality that shows the different numbers of hours you can work at each job, and sketch the graph of the inequality. From the graph, find several ordered pairs with positive integer coordinates that are solutions of the inequality.

66. *Writing a Model* A cash register must have at least $25 in change consisting of d dimes and q quarters. Write an inequality that shows the different numbers of coins that can be in the cash register, and sketch the graph of the inequality. From the graph, find several ordered pairs with positive integer coordinates that are solutions of the inequality.

67. *Writing a Model* Each table produced by a furniture company requires 1 hour in the assembly center. The matching chair requires $1\frac{1}{2}$ hours in the assembly center. A total of 12 hours per day is available in the assembly center. Write an inequality that shows the different numbers of hours that can be spent assembling tables and chairs, and sketch a graph of the inequality. From the graph, find several ordered pairs with positive integer coordinates that are solutions of the inequality.

68. *Writing a Model* A store sells two models of computers. The costs to the store of the two models are $2000 and $3000, and the owner of the store does not want more than $30,000 invested in the inventory for these two models. Write an inequality that represents the different numbers of each model that can be held in inventory. Sketch a graph of the inequality. From the graph, find several ordered pairs with positive integer coordinates that are solutions of the inequality.

69. *Dietetics* A dietitian is asked to design a special diet supplement using two foods. Each ounce of food X contains 20 units of calcium and each ounce of food Y contains 10 units of calcium. The minimum daily requirement in the diet is 300 units of calcium. Write an inequality that shows the different numbers of units of food X and food Y required. Sketch the graph of the inequality. From the graph, find several ordered pairs with positive integer coordinates that are solutions of the inequality.

Explaining Concepts

70. Answer part (h) of Motivating the Chapter on page 201.

71. List the four forms of a linear inequality in variables x and y.

72. What is meant by saying that (x_1, y_1) is a solution of a linear inequality in x and y?

73. Explain the difference between graphs that have dashed lines and those that have solid lines.

74. After graphing the boundary, explain how you determine which half-plane is the graph of a linear inequality.

75. Explain the difference between graphing the solution to the inequality $x \geq 1$ (a) on the real number line and (b) on a rectangular coordinate system.

76. Write the inequality whose graph consists of all points above the x-axis.

77. Does $2x < 2y$ have the same graph as $y > x$? Explain.

78. Write an inequality whose graph has no points in the first quadrant.

CHAPTER SUMMARY

Key Terms

rectangular coordinate system, *p. 202*
ordered pair, *p. 202*
x-coordinate, *p.202*
y-coordinate, *p. 202*
solution point, *p. 205*
x-intercept, *p. 217*
y-intercept, *p. 217*
relation, *p. 223*
domain, *p. 223*
range, *p. 223*
function, *p. 224*
independent variable, *p. 225*
dependent variable, *p. 225*
slope, *p. 234*
slope-intercept form, *p. 238*
parallel lines, *p. 240*
perpendicular lines, *p. 240*
point-slope form, *p. 247*
half-plane, *p. 258*

Key Concepts

4.1 Rectangular coordinate system

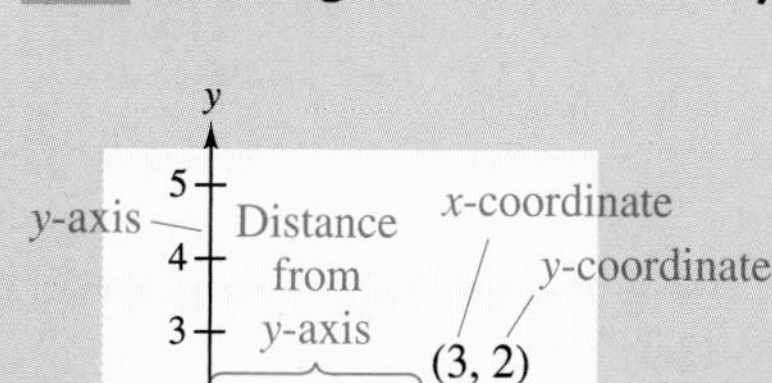

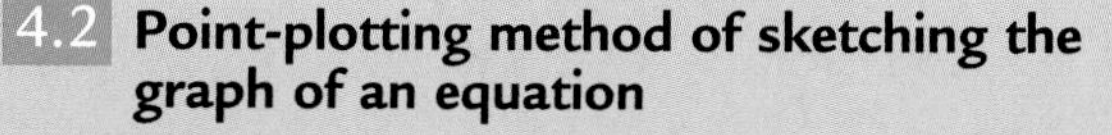

4.2 Point-plotting method of sketching the graph of an equation

1. If possible, rewrite the equation by isolating one of the variables.
2. Make up a table of values showing several solution points.
3. Plot these points on a rectangular coordinate system.
4. Connect the points with a smooth curve or line.

4.2 Finding x- and y-intercepts

1. To find the x-intercept(s), let $y = 0$ and solve the resulting equation for x.
2. To find the y-intercept(s), let $x = 0$ and solve the resulting equation for y.

4.3 Vertical Line Test

A graph is not the graph of a function if a vertical line can be drawn that intersects the graph at more than one point.

4.3 Function notation

In the notation $f(x)$:

f is the name of the function.

x is a domain value.

$f(x)$ is a range value y for a given x.

The symbol $f(x)$ is read as "the value of f at x" or simply "f of x."

4.4 Slope of a line

The slope of a nonvertical line passing through points (x_1, y_1) and (x_2, y_2) is

$$m = \frac{y_2 - y_1}{x_2 - x_1} = \frac{\text{change in } y}{\text{change in } x} = \frac{\text{rise}}{\text{run}}$$

where $x_1 \neq x_2$.

1. If $m > 0$, the line rises from left to right.
2. If $m < 0$, the line falls from left to right.
3. If $m = 0$, the line is horizontal.
4. If m is undefined ($x_1 = x_2$), the line is vertical.

4.4 Slope-intercept form

The slope-intercept form of the equation of a line is

$$y = mx + b,$$

where the slope of the line is m and the y-intercept is $(0, b)$.

4.5 Point-slope form

The point-slope form of the equation of a line with slope m and passing through the point (x_1, y_1) is

$$y - y_1 = m(x - x_1).$$

4.6 Graphing a linear inequality

1. Replace the inequality sign by an equal sign and sketch the graph of the corresponding equation. (Use a dashed line for < and > and a solid line for ≤ and ≥.)
2. Test one point in each of the half-planes formed by the graph in Step 1. If the point satisfies the inequality, then shade the entire half-plane to denote that every point in the region satisfies the inequality.

REVIEW EXERCISES

Reviewing Skills

4.1 In Exercises 1–4, plot the points on a rectangular coordinate system.

1. $(-1, 6), (4, -3), (-2, 2), (3, 5)$

2. $(0, -1), (-4, 2), (5, 1), (3, -4)$

3. $(-2, 0), \left(\frac{3}{2}, 4\right), (-1, -3)$

4. $\left(3, -\frac{5}{2}\right), \left(-5, 2\frac{3}{4}\right), (4, 6)$

In Exercises 5–12, determine the quadrant(s) in which the point must be located or the axis on which the point is located.

5. $(-5, 3)$

6. $(4, -6)$

7. $(4, 0)$

8. $(0, -3)$

9. $(x, 5),\ x < 0$

10. $(-3, y),\ y > 0$

11. $(-6, y)$, y is a real number.

12. $(x, -1)$, x is a real number.

In Exercises 13–16, solve the equation for y.

13. $3x + 4y = 12$

14. $2x + 3y = 6$

15. $x - 2y = 8$

16. $-x - 3y = 9$

In Exercises 17 and 18, construct a table of values for the equation that shows four solution points. Use x-values of -1, 0, 1, and 2.

17. $2x - y = 1$

18. $6x + 3y = -3$

In Exercises 19–22, determine whether the ordered pairs are solution points of the equation.

19. $x - 3y = 4$ (a) $(1, -1)$ (b) $(0, 0)$ (c) $(2, 1)$ (d) $(5, -2)$

20. $y - 2x = -1$ (a) $(3, 7)$ (b) $(0, -1)$ (c) $(-2, -5)$ (d) $(-1, 0)$

21. $y = \frac{2}{3}x + 3$ (a) $(3, 5)$ (b) $(-3, 1)$ (c) $(-6, 0)$ (d) $(0, 3)$

22. $y = \frac{1}{4}x + 2$ (a) $(-4, 1)$ (b) $(-8, 0)$ (c) $(12, 5)$ (d) $(0, 2)$

4.2 In Exercises 23–34, sketch the graph of the equation using the point-plotting method, and label any x- and y-intercepts of the graph.

23. $y = 7$

24. $x = -2$

25. $y = 3x$

26. $y = -2x$

27. $y = 4 - \frac{1}{2}x$

28. $y = \frac{3}{2}x - 3$

29. $y - 2x - 4 = 0$

30. $3x + 2y + 6 = 0$

31. $y = 2x - 1$

32. $y = 5 - 4x$

33. $y = \frac{1}{4}x + 2$

34. $y = -\frac{2}{3}x - 2$

In Exercises 35–38, use a graphing utility to graph the equation. (Use a standard setting.)

35. $y = \frac{7}{8}x + 1$

36. $y = 5 - 2x$

37. $y = -\frac{1}{4}x^2 + x$

38. $y = x(x^2 - 4)$

In Exercises 39–42, graph the equation using a graphing utility.

39. $y = 250 - 50x$

Xmin = -5
Xmax = 5
Xscl = 1
Ymin = 0
Ymax = 500
Yscl = 25

40. $y = 800 + 9x$

Xmin = 0
Xmax = 100
Xscl = 10
Ymin = 500
Ymax = 2000
Yscl = 200

41. $y = -2x^2 + 112x + 50$

Xmin = -10
Xmax = 75
Xscl = 5
Ymin = -200
Ymax = 2000
Yscl = 200

42. $y = |x - 3| + |x - 6|$

Xmin = -2
Xmax = 11
Xscl = 1
Ymin = 0
Ymax = 12
Yscl = 1

4.3 In Exercises 43–46, identify the domain and range of the relation.

43. $\{(8, 3), (-2, 7), (5, 1), (3, 8)\}$

44. $\{(0, 1), (-1, 3), (4, 6), (-7, 5)\}$

45. $\{(2, -3), (-2, 3), (7, 0), (-4, -2)\}$

46. $\{(1, 7), (-3, 4), (6, 5), (-2, -9)\}$

In Exercises 47–50, determine if the relation is a function.

47.

Input value	0	2	4	6	2
Output value	0	1	1	2	3

48.

Input value	−6	−3	0	3	6
Output value	1	0	1	4	2

49.

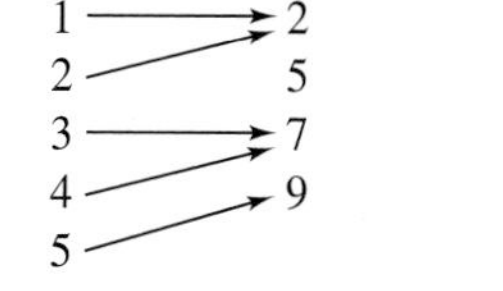

50. *Domain* 5, 7, 9, 11, 13 *Range* 5, 9, 13, 17, 19

In Exercises 51–56, determine whether the relation represents y as a function of x.

51. $x = y^2 - 4$

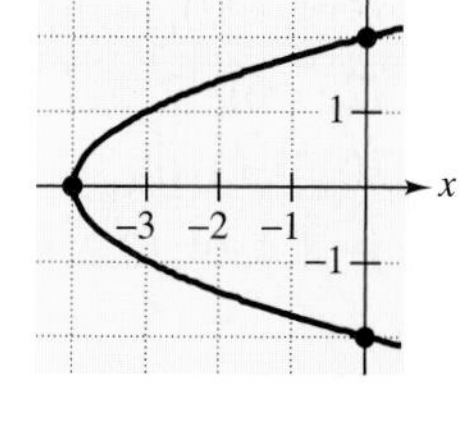

52. $y = x^3 - 3x + 2$

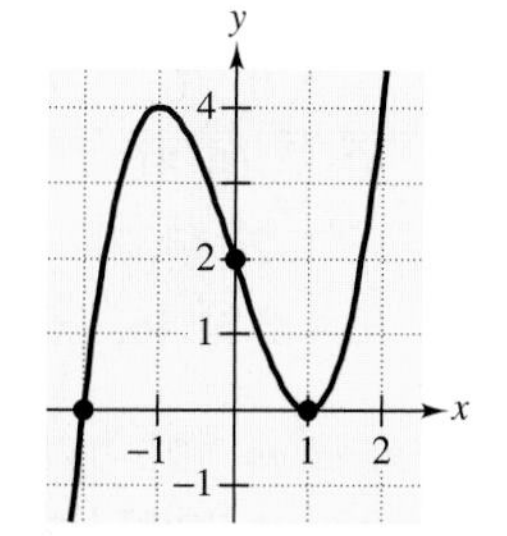

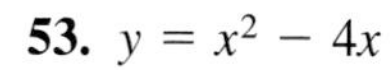

53. $y = x^2 - 4x$

54. $x = y^3 - y$

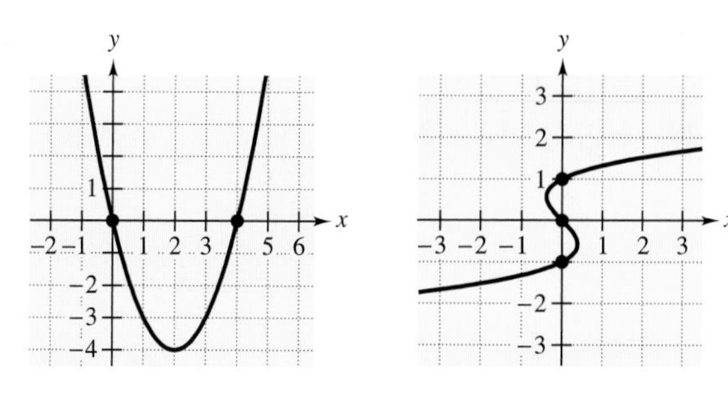

55. **56.**

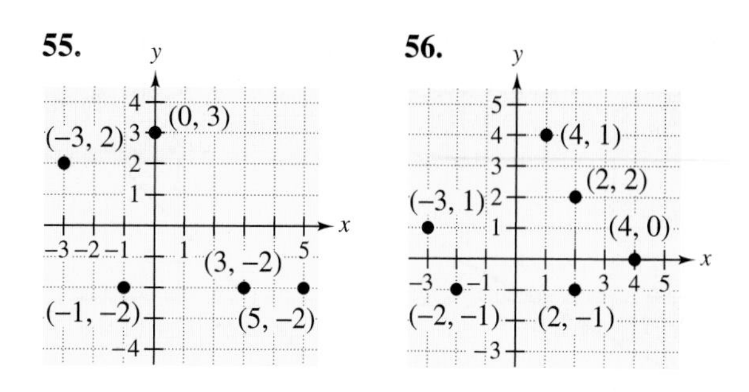

In Exercises 57–62, evaluate the function.

57. $f(x) = |2x + 3|$ (a) $f(0)$ (b) $f(5)$ (c) $f(-4)$ (d) $f\left(-\frac{3}{2}\right)$

58. $g(t) = -16t^2 + 64$ (a) $g(0)$ (b) $g\left(\frac{1}{4}\right)$ (c) $g(1)$ (d) $g(2)$

59. $h(u) = u(u - 3)^2$ (a) $h(0)$ (b) $h(3)$ (c) $h(-1)$ (d) $h\left(\frac{3}{2}\right)$

60. $f(x) = 25$ (a) $f(-1)$ (b) $f(7)$ (c) $f(10)$ (d) $f\left(-\frac{4}{3}\right)$

61. $f(x) = 2x - 7$ (a) $f(-1)$ (b) $f(3)$ (c) $f\left(\frac{1}{2}\right)$ (d) $f(-4)$

62. $f(x) = |x| - 4$ (a) $f(-1)$ (b) $f(1)$ (c) $f(-4)$ (d) $f(2)$

In Exercises 63–70, use a graphing utility to graph the function. Identify any intercepts of the graph.

63. $f(x) = \frac{3}{2}x + 2$

64. $f(x) = 5 - \frac{1}{2}x$

65. $g(x) = \frac{1}{5}(25 - x^2)$

66. $f(x) = x^2 + 2x - 15$

67. $g(x) = x^2(x - 4)$

68. $h(x) = x(x^2 - 4)$

69. $f(x) = 4 - \frac{1}{2}|x|$

70. $h(x) = |x^2 - 9|$

4.4 In Exercises 71 and 72, estimate the slope of the line from its graph.

71. **72.**

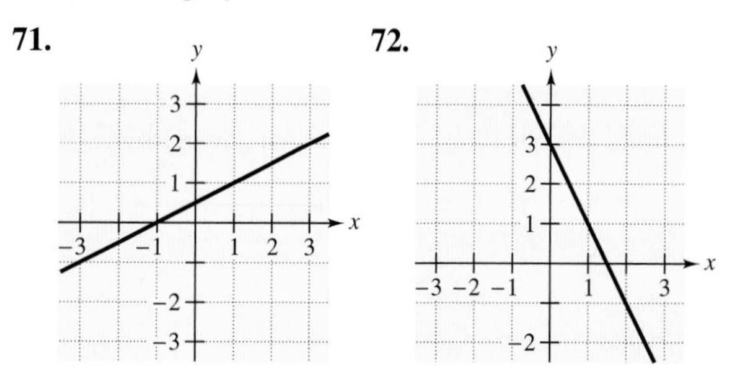

In Exercises 73–76, match the equation with its graph. [The graphs are labeled (a), (b), (c), and (d).]

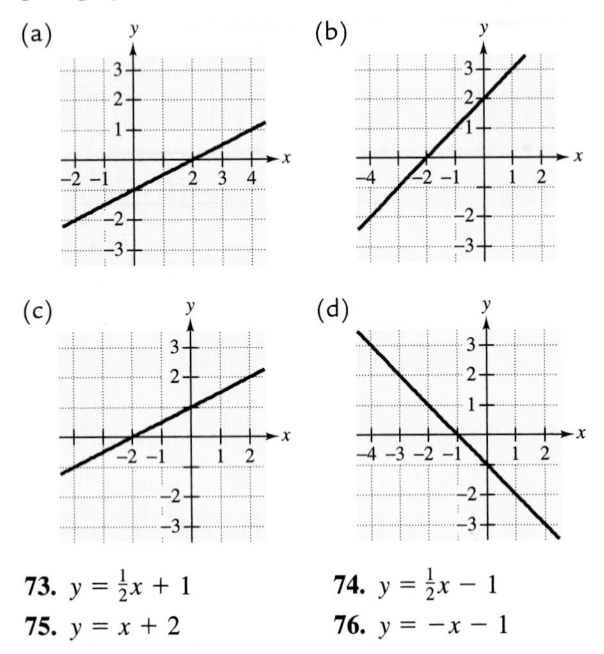

73. $y = \frac{1}{2}x + 1$ **74.** $y = \frac{1}{2}x - 1$

75. $y = x + 2$ **76.** $y = -x - 1$

In Exercises 77–88, determine the slope of the line through the points.

77. $(2, 1), (14, 6)$ **78.** $(-2, 2), (3, -10)$

79. $(-1, 0), (6, 2)$ **80.** $(1, 6), (4, 2)$

81. $(4, 0), (4, 6)$ **82.** $(1, 3), (4, 3)$

83. $(-2, 5), (1, 1)$ **84.** $(-6, 1), (10, 5)$

85. $(1, -4), (5, 10)$ **86.** $(-3, 3), (8, 6)$

87. $\left(0, \frac{5}{2}\right), \left(\frac{5}{6}, 0\right)$ **88.** $(0, 0), \left(3, \frac{4}{5}\right)$

In Exercises 89–92, write the linear equation in slope-intercept form and graph the line.

89. $3x + 6y = 12$ **90.** $2x - y = -1$

91. $5y - 2x = 5$ **92.** $7x + 21y = -14$

In Exercises 93–96, use the slope to determine whether the lines are parallel, perpendicular, or neither.

93. $y = 4x - 8,\ y = -\frac{1}{4}x + 2$

94. $y = x + 3,\ y = 5 - x$

95. $4x - y = 7,\ 8x - 2y = 3$

96. $2x - 3y = 7,\ 3y - 2x = 5$

In Exercises 97–102, a point on a line and the slope of the line are given. Find two additional points on the line. (There are many correct answers.)

97. $(3, -1),\ m = -2$ **98.** $(-2, 5),\ m = 3$

99. $(2, 3),\ m = \frac{3}{4}$ **100.** $(-3, 1),\ m = -\frac{2}{3}$

101. $\left(\frac{4}{3}, 4\right),\ m = 0$ **102.** $\left(6, \frac{7}{2}\right),\ m$ is undefined.

4.5 In Exercises 103–112, write an equation of the line passing through the point and with the specified slope using the point-slope form if possible. (Write your answer in general form.)

103. $(4, -1),\ m = 2$ **104.** $(-5, 2),\ m = 3$

105. $(1, 2),\ m = -4$ **106.** $(7, -3),\ m = -1$

107. $(-5, -2),\ m = \frac{4}{5}$ **108.** $(12, -4),\ m = -\frac{1}{6}$

109. $(-1, 3),\ m = -\frac{8}{3}$ **110.** $(4, -2),\ m = \frac{8}{5}$

111. $(3, 8),\ m$ is undefined.

112. $(-4, 6),\ m = 0$

In Exercises 113–116, determine the slope of the line. If it is not possible, explain why.

113. $y = 5x + 3$ **114.** $y = 6$

115. $y + 10 = \frac{4}{3}(x - 4)$ **116.** $x + 5 = 0$

In Exercises 117–124, find an equation of the line passing through the points. (Write your answer in general form.)

117. $(-4, 0), (0, -2)$ **118.** $(-4, -2), (4, 6)$

119. $(0, 8), (6, 8)$ **120.** $(2, -6), (2, 5)$

121. $(-1, 2), (4, 7)$ **122.** $\left(0, \frac{4}{3}\right), (3, 0)$

123. $(2.4, 3.3), (6, 7.8)$ **124.** $(-1.4, 0), (3.2, 9.2)$

In Exercises 125–128, find an equation of the line through the point (a) parallel to the given line and (b) perpendicular to the given line.

125. $(-6, 3),\ 2x + 3y = 1$

126. $\left(\frac{1}{5}, -\frac{4}{5}\right),\ 5x + y = 2$

127. $\left(\frac{3}{8}, 4\right),\ 4x + 3y = 16$

128. $(-2, 1),\ 5x = 2$

Graphical Exploration In Exercises 129–132, use a graphing utility to graph the lines. Use the square setting. Are the lines parallel, perpendicular, or neither?

129. $y = -0.5x + 4$
$y = 2x - 3$

130. $y = \frac{1}{4}(x - 4)$
$y = \frac{1}{8}(2x + 15)$

131. $y = 2.5x + 1$
$y = \frac{1}{2}(5x - 4)$

132. $y = 3x - 7$
$y = -2.5x + 6$

4.6 In Exercises 133 and 134, determine whether the ordered pair is a solution of the inequality.

133. $x - y > 4$ (a) $(-1, -5)$ (b) $(0, 0)$ (c) $(3, -2)$ (d) $(8, 1)$

134. $y - 2x \le -1$ (a) $(0, 0)$ (b) $(-2, 1)$ (c) $(-3, 4)$ (d) $(-1, -6)$

In Exercises 135–140, sketch the graph of the linear inequality.

135. $x - 2 \ge 0$

136. $y + 3 < 0$

137. $2x + y < 1$

138. $3x - 4y > 2$

139. $x \le 4y - 2$

140. $x \ge 3 - 2y$

In Exercises 141–144, write an inequality that represents the graph.

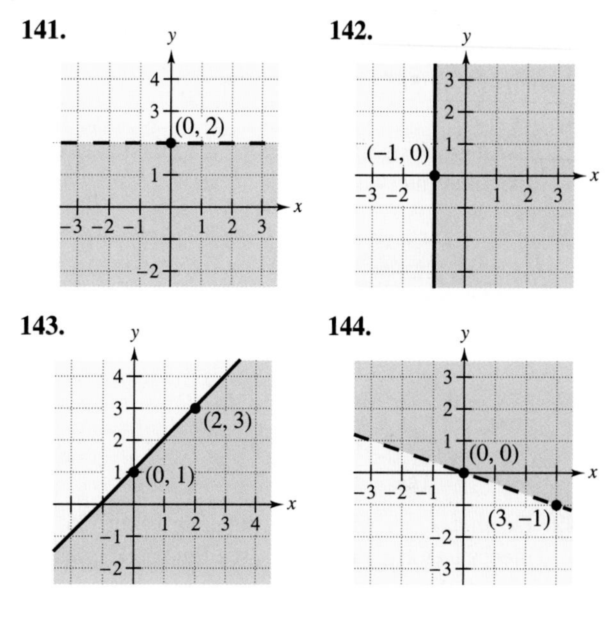

Solving Problems

145. *Organizing Data* The data from a study measuring the relationship between the wattage x of a standard light bulb and the energy rate y (in lumens) are given in the table. (Source: Standard Handbook for Mechanical Engineers)

x	25	40	60	100	150	200
y	266	470	840	1750	2700	4000

(a) Plot the data given in the table.

(b) Describe the relationship between x and y.

(c) Estimate the value of y when $x = 125$. Is it easier to use the table or the graph to estimate this value? Explain your reasoning.

146. *Graphical Interpretation* The line graph shows the average salaries (in thousands of dollars) for professional players in baseball, basketball, and football in the United States for the years 1990 through 1995. (Source: Major League Baseball Players Association, National Basketball Association, National Football League Players Association)

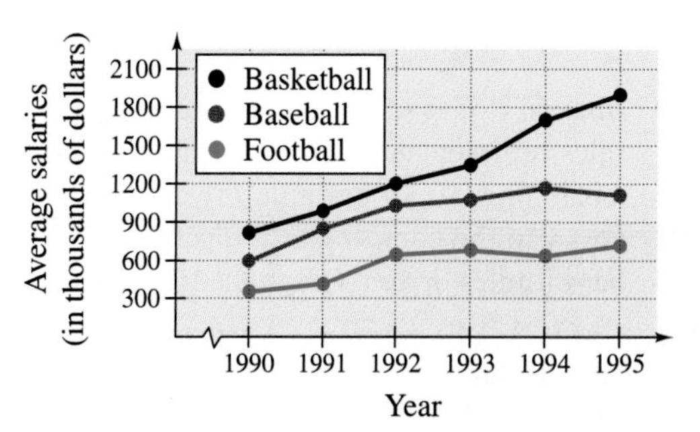

Figure for 146

(a) Approximate the average salary for professional basketball players in 1991.

(b) In what year did the average salary of professional baseball players first exceed 1 million dollars?

(c) Approximate the percentage increase in average salaries for football players from 1990 to 1991.

147. *Writing a Model* The cost of sending a package is \$2.25 plus \$0.75 per pound. Write the total cost C as a function of the weight x. Use a graphing utility to graph the function for $0 < x \le 20$.

148. *Geometry* The volume V of the segment of a sphere of radius 5 is a function of the height h of the segment (see figure). The formula for the function is

$$V(h) = \pi\left(5h^2 - \frac{h^3}{3}\right), \quad 0 < h \le 5.$$

Use a graphing utility to graph the function over the specified interval.

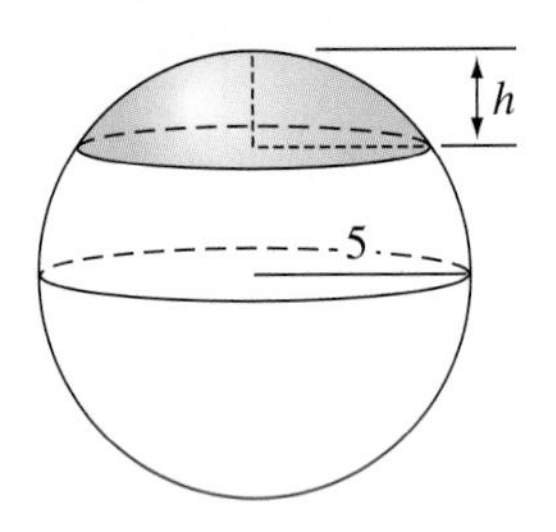

149. *Geometry* A wire 24 inches long is cut into four pieces to form a rectangle (see figure). Express the area A of the rectangle as a function of x.

150. *Business Expense* A company reimburses its sales representatives $150 per day for lodging and meals plus $0.30 per mile driven. Write the daily cost y to the company as a function of x, the number of miles driven. Graph the function.

151. *Graphical Estimation* A person who weighs 180 pounds begins a diet of 1500 calories per day. The person's weight y after dieting for x weeks is approximated by

$$y = 0.014x^2 - 1.218x + 180, \quad 0 \le x \le 26.$$

(a) Use a graphing utility to graph the function.

(b) Approximate the person's weight after 26 weeks.

(c) Approximate the time required for the person to lose 5 pounds.

152. *Slope of a Ramp* The floor of a truck is 4 feet above ground level. The end of the ramp used in loading the truck rests on the ground 6 feet behind the truck. Determine the slope of the ramp.

153. *Slope of a Path* An aircraft is on its approach to an airport. Radar shows its altitude to be 15,000 feet when it is 10 miles from touchdown. Approximate the slope of the linear path followed during landing.

154. *Graphical Interpretation* The velocity (in feet per second) of a ball thrown vertically upward from ground level is modeled by $v = -32t + 48$, where t is time in seconds.

(a) Graph the expression for the velocity.

(b) Interpret the slope of the line in the context of this real-life setting.

(c) Find the velocity when $t = 0$ and $t = 1$.

(d) Find the time when the ball reaches its maximum height. (*Hint:* Find the time when $v = 0$.)

155. *Writing a Model* A company produces a product for which the variable cost is $5.35 per unit and the fixed cost is $16,000. The product is sold for $8.20 per unit. (Let x represent the number of units produced and sold.)

(a) Write an equation that represents the total cost C as a linear function of x.

(b) Write an equation that represents the profit P as a linear function of x.

156. *Writing a Model* Each week a company produces x VCRs and y camcorders. The assembly times for the two types of units are 2 and 3 hours, respectively. The time available in a week is 120 hours. Write an inequality that shows the different numbers of VCRs and camcorders that can be produced. Sketch a graph of the inequality. From the graph, find several ordered pairs that are solutions of the inequality.

Chapter Test

Take this test as you would take a test in class. After you are done, check your work against the answers given in the back of the book.

1. Plot the points $(-1, 2)$, $(1, 4)$, and $(2, -1)$ on a rectangular coordinate system. Connect the points with line segments to form a right triangle.
2. Which ordered pairs are solutions of $y = |x| + |x - 2|$?
   (a) $(0, -2)$ (b) $(0, 2)$ (c) $(-4, 10)$ (d) $(-2, -2)$
3. What is the y-coordinate of any point on the x-axis?
4. Find the x- and y-intercepts of the graph of $3x - 4y + 12 = 0$.
5. Does the table at the left represent y as a function of x? Explain your reasoning.

x	0	1	2	1	0
y	4	5	8	-3	-1

Table for 5

6. Does the graph at the left represent y as a function of x? Explain.

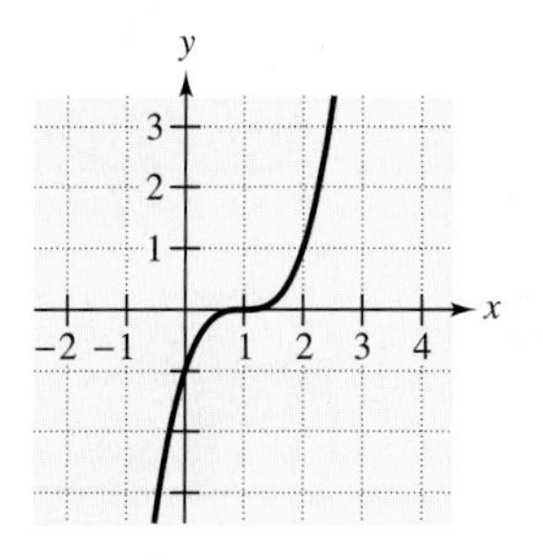

Figure for 6

7. Evaluate $f(x) = x^3 - 2x^2$ at the indicated values.
   (a) $f(0)$ (b) $f(2)$ (c) $f(-2)$ (d) $f\left(\frac{1}{2}\right)$
8. Find the slope of the line passing through the points $(-5, 0)$ and $\left(2, \frac{3}{2}\right)$.
9. A line with slope $m = -2$ passes through the point $(-3, 4)$. Find two additional points on the line. (The problem has many correct answers.)
10. Find the slope of a line *perpendicular* to the line $3x - 5y + 2 = 0$.

In Exercises 11–14, graph the equation.

11. $x - 2y = 6$
12. $y = \frac{1}{4}x - 1$
13. $y = |x + 2|$
14. $y = (x - 3)^2$

15. Find an equation of the line that passes through the point $(0, 6)$ with slope $m = -\frac{3}{8}$.
16. Which points are solutions of the inequality $3x + 5y \leq 16$?
    (a) $(2, 2)$ (b) $(6, -1)$ (c) $(-2, 4)$ (d) $(7, -1)$

In Exercises 17–20, graph the inequality.

17. $y \geq -2$
18. $y < 5 - 2x$
19. $x \geq 2$
20. $y \leq 5$

21. The sales y of a product are modeled by $y = 230x + 5000$, where x is time in years. Interpret the meaning of the slope in this model.

5 Exponents and Polynomials

John Madere/The Stock Market

FedEx receives an average of more than 3.2 million packages daily worldwide. (Source: FedEx)

Motivating the Chapter

Packaging Restrictions

A shipping company has the following restrictions on the dimensions and weight of packages.

1. The maximum weight is 150 pounds.
2. The maximum length is 108 inches.
3. The sum of the length and girth can be at most 130 inches.

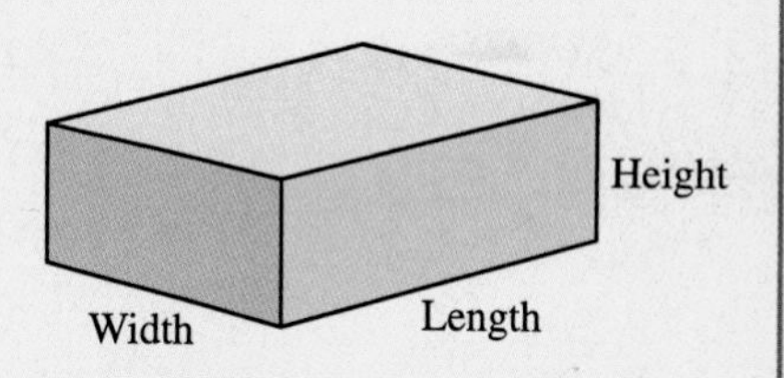

The girth of a package is the distance around the package, as shown in the figure.

Girth = 2(Height + Width)

You are shipping a package that has a height of x inches. The length of the package is twice the square of the height and the width is 5 inches more than 3 times the height.

See Section 5.1, Exercise 111

a. Write an expression for the length of the package in terms of the height x. Write an expression for the width of the package in terms of the height x.

b. Write an expression for the *perimeter* of the base of the package. Simplify the expression.

c. Write an expression for the *girth* of the package. Simplify the expression. Write an expression for the sum of the length and the girth. If the height of the package is 5 inches, does the package meet the second and third restrictions? Explain.

See Section 5.2, Exercise 135

d. Write an expression for the *surface area* of the package. Simplify the expression. (The surface area is the sum of the areas of the six sides of the package.)

e. The length of the package is changed to match its width (5 inches more than 3 times its height). Write an expression for the area of the base. Simplify the expression.

f. Write an expression for the *volume* of the package. Simplify the expression.

See Section 5.4, Exercise 94

g. Suppose the width of the package is the same as in part (a) and the area of the base is known to be $6x^2 + 7x - 5$. What is the new length of the package?

5.1 Adding and Subtracting Polynomials

Objectives

1. Determine the degree and leading coefficient of a polynomial.
2. Add polynomials using a horizontal or vertical format.
3. Subtract polynomials using a horizontal or vertical format.

1 Determine the degree and leading coefficient of a polynomial.

Basic Definitions

To work with polynomials, you need to know the following rules for exponents, which were discussed in Section 2.2.

1. $a^m \cdot a^n = a^{m+n}$ — Multiply factors having the same base.

2. $(a^m)^n = a^{m \cdot n}$ — Raise a power to a power.

3. $(ab)^m = a^m b^m$ — Raise a product to a power.

Additional rules for exponents will be introduced later in this chapter.

Remember that the *terms* of an algebraic expression are those parts separated by addition. An algebraic expression whose terms are all of the form ax^k, where a is any real number and k is a nonnegative integer, is called a **polynomial in one variable,** or simply a **polynomial.** Here are some examples of polynomials in one variable.

$$2x + 5, \quad x^2 - 3x + 7, \quad \text{and} \quad x^3 + 8$$

In the term ax^k, a is the **coefficient** of the term and k is the **degree** of the term. Because a polynomial is an algebraic sum, the coefficients take on the signs between the terms. For instance,

$$x^4 + 2x^3 - 5x^2 + 7 = (1)x^4 + 2x^3 + (-5)x^2 + (0)x + 7$$

has coefficients 1, 2, -5, 0, and 7. For this polynomial, the last term, 7, is the **constant term.** Polynomials are usually written in the order of descending powers of the variable. This is called **standard form.** Here are two examples.

Nonstandard Form	*Standard Form*
$4 + x$	$x + 4$
$3x^2 - 5 - x^3 + 2x$	$-x^3 + 3x^2 + 2x - 5$

The **degree of a polynomial** is the degree of the term with the highest power, and the coefficient of this term is the **leading coefficient** of the polynomial. For instance, the polynomial

Leading coefficient

$$-3x^4 + 4x^2 + x + 7$$

is of fourth degree, and its leading coefficient is -3. The reasons why the degree of a polynomial is important will become clear as you study factoring and problem solving in Chapter 6.

▶ Definition of a Polynomial in x

Let $a_n, a_{n-1}, \ldots, a_2, a_1, a_0$ be real numbers and let n be a *nonnegative integer.* A **polynomial in x** is an expression of the form

$$a_nx^n + a_{n-1}x^{n-1} + \cdots + a_2x^2 + a_1x + a_0$$

where $a_n \neq 0$. The polynomial is of **degree** n, and the number a_n is the **leading coefficient.** The number a_0 is the **constant term.**

Example 1 Identifying Polynomials

Identify which of the following are polynomials, and for any that are not polynomials, state why.

a. $3x^4 - 8x + x^{-1}$ **b.** $x^2 - 3x + 1$

c. $x^3 + 3x^{1/2}$ **d.** $-\frac{1}{3}x + \frac{x^3}{4}$

Solution

a. $3x^4 - 8x + x^{-1}$ is *not* a polynomial because the third term, x^{-1}, has a negative exponent.

b. $x^2 - 3x + 1$ is a polynomial of degree 2 with integer coefficients.

c. $x^3 + 3x^{1/2}$ is *not* a polynomial because the exponent in the second term, $3x^{1/2}$, is not an integer.

d. $-\frac{1}{3}x + \frac{x^3}{4}$ is a polynomial of degree 3 with rational coefficients.

Example 2 Determining Degrees and Leading Coefficients

Write the polynomial in standard form and determine the degree and leading coefficient.

	Polynomial	*Standard Form*	*Degree*	*Leading Coefficient*
a.	$4x^2 - 5x^7 - 2 + 3x$	$-5x^7 + 4x^2 + 3x - 2$	7	-5
b.	$4 - 9x^2$	$-9x^2 + 4$	2	-9
c.	8	8	0	8
d.	$2 + x^3 - 5x^2$	$x^3 - 5x^2 + 2$	3	1

In part (c), note that a polynomial with *only* a constant term has a degree of zero.

A polynomial with only one term is called a **monomial.** Polynomials with two *unlike* terms are called **binomials,** and those with three *unlike* terms are called **trinomials.** For example, $3x^2$ is a *monomial*, $-3x + 1$ is a *binomial*, and $4x^3 - 5x + 6$ is a *trinomial.*

2 Add polynomials using a horizontal or vertical format.

Adding Polynomials

As with algebraic expressions, the key to adding two polynomials is to recognize *like* terms—those having the *same degree.* By the Distributive Property, you can then combine the like terms using either a horizontal or a vertical arrangement of terms. For instance, the polynomials $2x^2 + 3x + 1$ and $x^2 - 2x + 2$ can be added horizontally to obtain

$$\begin{aligned}(2x^2 + 3x + 1) + (x^2 - 2x + 2) &= (2x^2 + x^2) + (3x - 2x) + (1 + 2)\\ &= (2 + 1)x^2 + (3 - 2)x + (1 + 2)\\ &= 3x^2 + x + 3\end{aligned}$$

or they can be added vertically to obtain the same result.

$$\begin{array}{r} 2x^2 + 3x + 1 \\ \underline{x^2 - 2x + 2} \\ 3x^2 + x + 3 \end{array} \qquad \text{Vertical arrangement}$$

Study Tip

When you use the vertical arrangement to add polynomials, be sure that you line up the *like terms.*

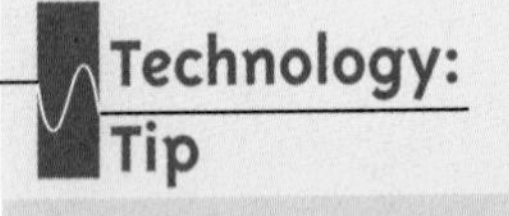

Technology: Tip

You can use a graphing utility to check the results of adding or subtracting polynomials. For instance, try graphing

$y = (2x + 1) + (-3x - 4)$

and

$y = -x - 3$

on the same screen, as shown below. Because both graphs are the same, you can reason that

$(2x + 1) + (-3x - 4) = -x - 3.$

This graphing technique is called "graph the left side and graph the right side."

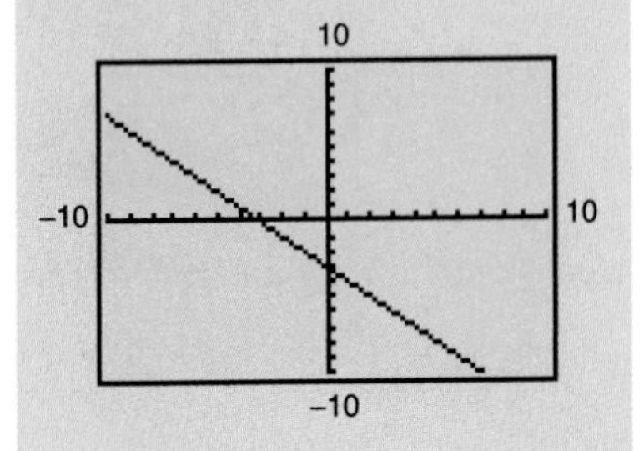

Example 3 Adding Polynomials Horizontally

Use a horizontal arrangement to find the sum.

a. $(2x^2 + 4x - 1) + (x^2 - 3)$ — Original polynomials

$= (2x^2 + x^2) + (4x) + (-1 - 3)$ — Group like terms.

$= 3x^2 + 4x - 4$ — Combine like terms.

b. $(x^3 + 2x^2 + 4) + (3x^2 - x + 5)$ — Original polynomials

$= (x^3) + (2x^2 + 3x^2) + (-x) + (4 + 5)$ — Group like terms.

$= x^3 + 5x^2 - x + 9$ — Combine like terms.

c. $(2x^2 - x + 3) + (4x^2 - 7x + 2) + (-x^2 + x - 2)$ — Original polynomials

$= (2x^2 + 4x^2 - x^2) + (-x - 7x + x) + (3 + 2 - 2)$ — Group like terms.

$= 5x^2 - 7x + 3$ — Combine like terms.

Example 4 Adding Polynomials Vertically

Use a vertical arrangement to find the sum.

a. $(-4x^3 - 2x^2 + x - 5) + (2x^3 + 3x + 4)$

b. $(5x^3 + 2x^2 - x + 7) + (3x^2 - 4x + 7) + (-x^3 + 4x^2 - 2x - 8)$

Solution

a.
$$\begin{array}{rrrr} -4x^3 & -\ 2x^2 & +\ x & -\ 5 \\ 2x^3 & & +\ 3x & +\ 4 \\ \hline -2x^3 & -\ 2x^2 & +\ 4x & -\ 1 \end{array}$$

b.
$$\begin{array}{rrrr} 5x^3 & +\ 2x^2 & -\ x & +\ 7 \\ & 3x^2 & -\ 4x & +\ 7 \\ -x^3 & +\ 4x^2 & -\ 2x & -\ 8 \\ \hline 4x^3 & +\ 9x^2 & -\ 7x & +\ 6 \end{array}$$

3 Subtract polynomials using a horizontal or vertical format.

Subtracting Polynomials

To subtract one polynomial from another, you *add the opposite* by changing the sign of each term of the polynomial that is being subtracted and then adding the resulting like terms. Note how $(x^2 - 1)$ is subtracted from $(2x^2 - 4)$.

$$(2x^2 - 4) - (x^2 - 1) = 2x^2 - 4 - x^2 + 1 \qquad \text{Distributive Property}$$
$$= (2x^2 - x^2) + (-4 + 1) \qquad \text{Group like terms.}$$
$$= x^2 - 3 \qquad \text{Combine like terms.}$$

Recall from the Distributive Property that

$$-(x^2 - 1) = (-1)(x^2 - 1) = -x^2 + 1.$$

Example 5 Subtracting Polynomials Horizontally

Perform the following operations.

a. $(2x^2 + 3) - (3x^2 - 4)$
b. $(4x^4 - x^2 + 1) - (x^4 - 2x^3 - x^2)$
c. $(3x^3 - 4x^2 + 3) - (x^3 + 3x^2 - x - 4)$
d. $(x^2 - 2x + 1) - [(x^2 + x - 3) + (-2x^2 - 4x)]$

Solution

a. $(2x^2 + 3) - (3x^2 - 4) = 2x^2 + 3 - 3x^2 + 4$ Distributive Property
$= (2x^2 - 3x^2) + (3 + 4)$ Group like terms.
$= -x^2 + 7$ Combine like terms.

b. $(4x^4 - x^2 + 1) - (x^4 - 2x^3 - x^2)$ Original polynomials
$= 4x^4 - x^2 + 1 - x^4 + 2x^3 + x^2$ Distributive Property
$= (4x^4 - x^4) + (2x^3) + (-x^2 + x^2) + (1)$ Group like terms.
$= 3x^4 + 2x^3 + 1$ Combine like terms.

c. $(3x^3 - 4x^2 + 3) - (x^3 + 3x^2 - x - 4)$ Original polynomials
$= 3x^3 - 4x^2 + 3 - x^3 - 3x^2 + x + 4$ Distributive Property
$= (3x^3 - x^3) + (-4x^2 - 3x^2) + (x) + (3 + 4)$ Group like terms.
$= 2x^3 - 7x^2 + x + 7$ Combine like terms.

d. $(x^2 - 2x + 1) - [(x^2 + x - 3) + (-2x^2 - 4x)]$ Original polynomials
$= (x^2 - 2x + 1) - [(x^2 - 2x^2) + (x - 4x) + (-3)]$ Group like terms.
$= (x^2 - 2x + 1) - [-x^2 - 3x - 3]$ Combine like terms.
$= x^2 - 2x + 1 + x^2 + 3x + 3$ Distributive Property
$= (x^2 + x^2) + (-2x + 3x) + (1 + 3)$ Group like terms.
$= 2x^2 + x + 4$ Combine like terms.

Be especially careful to use the correct signs when subtracting one polynomial from another. One of the most common mistakes in algebra is to forget to change signs correctly when subtracting one expression from another. Here is an example.

$$(x^2 + 3) - (x^2 + 2x - 2) \neq x^2 + 3 - x^2 \underset{\uparrow \text{ Wrong sign}}{\overset{\text{Wrong sign} \downarrow}{+}} 2x \;\underset{\uparrow \text{ Wrong sign}}{-}\; 2 \qquad \text{Common error}$$

Note that the error is forgetting to change all of the signs in the polynomial that is being subtracted. Here is the correct way to perform the subtraction.

$$(x^2 + 3) - (x^2 + 2x - 2) = x^2 + 3 - x^2 \overset{\text{Correct sign} \downarrow}{-} 2x \underset{\uparrow \text{ Correct sign}}{+} 2 \qquad \text{Correct}$$

Just as you did for addition, you can use a vertical arrangement to subtract one polynomial from another. (The vertical arrangement doesn't work well with subtractions involving three or more polynomials.) When using a vertical arrangement, write the polynomial being subtracted underneath the one it is being subtracted from. Be sure to align like terms in vertical columns.

Example 6 Subtracting Polynomials Vertically

Use a vertical arrangement to perform the following operations.

a. $(3x^2 + 7x - 6) - (3x^2 + 7x)$

b. $(5x^3 - 2x^2 + x) - (4x^2 - 3x + 2)$

c. $(4x^4 - 2x^3 + 5x^2 - x + 8) - (3x^4 - 2x^3 + 3x - 4)$

Solution

a.

$$\begin{array}{rr} (3x^2 + 7x - 6) \\ \underline{-(3x^2 + 7x \quad\;\;)} \end{array} \Rightarrow \begin{array}{r} 3x^2 + 7x - 6 \\ \underline{-3x^2 - 7x \qquad} \\ -6 \end{array} \quad \begin{array}{l} \\ \text{Change signs and add.} \\ \text{Combine like terms.} \end{array}$$

b.

$$\begin{array}{r} (5x^3 - 2x^2 + x \qquad) \\ \underline{-(\qquad 4x^2 - 3x + 2)} \end{array} \Rightarrow \begin{array}{r} 5x^3 - 2x^2 + x \qquad \\ \underline{\qquad - 4x^2 + 3x - 2} \\ 5x^3 - 6x^2 + 4x - 2 \end{array} \quad \begin{array}{l} \\ \text{Change signs and add.} \\ \text{Combine like terms.} \end{array}$$

c.

$$\begin{array}{r} (4x^4 - 2x^3 + 5x^2 - x + 8) \\ \underline{-(3x^2 - 2x^3 \qquad + 3x - 4)} \end{array} \Rightarrow \begin{array}{r} 4x^4 - 2x^3 + 5x^2 - x + 8 \\ \underline{-3x^4 + 2x^3 \qquad - 3x + 4} \\ x^4 \qquad + 5x^2 - 4x + 12 \end{array}$$

In Example 6, try using a horizontal arrangement to perform the subtractions.

Example 7 Combining Polynomials

Perform the indicated operations.

a. $(3x^2 - 7x + 2) - (4x^2 + 6x - 1) + (-x^2 + 4x + 5)$

b. $(-2x^2 + 4x - 3) - [(4x^2 - 5x + 8) - (-x^2 + x + 3)]$

c. $3(x^2 - 2x + 1) - 2(x^2 + x - 3)$

Solution

a. $(3x^2 - 7x + 2) - (4x^2 + 6x - 1) + (-x^2 + 4x + 5)$

$$= 3x^2 - 7x + 2 - 4x^2 - 6x + 1 - x^2 + 4x + 5$$

$$= (3x^2 - 4x^2 - x^2) + (-7x - 6x + 4x) + (2 + 1 + 5)$$

$$= -2x^2 - 9x + 8$$

b. $(-2x^2 + 4x - 3) - [(4x^2 - 5x + 8) - (-x^2 + x + 3)]$

$$= (-2x^2 + 4x - 3) - [4x^2 - 5x + 8 + x^2 - x - 3]$$

$$= (-2x^2 + 4x - 3) - [(4x^2 + x^2) + (-5x - x) + (8 - 3)]$$

$$= (-2x^2 + 4x - 3) - [5x^2 - 6x + 5]$$

$$= -2x^2 + 4x - 3 - 5x^2 + 6x - 5$$

$$= (-2x^2 - 5x^2) + (4x + 6x) + (-3 - 5)$$

$$= -7x^2 + 10x - 8$$

c. $3(x^2 - 2x + 1) - 2(x^2 + x - 3) = 3x^2 - 6x + 3 - 2x^2 - 2x + 6$

$$= (3x^2 - 2x^2) + (-6x - 2x) + (3 + 6)$$

$$= x^2 - 8x + 9$$

Example 8 Geometry: Area of a Region

Find a polynomial that represents the area of the shaded region in Figure 5.1.

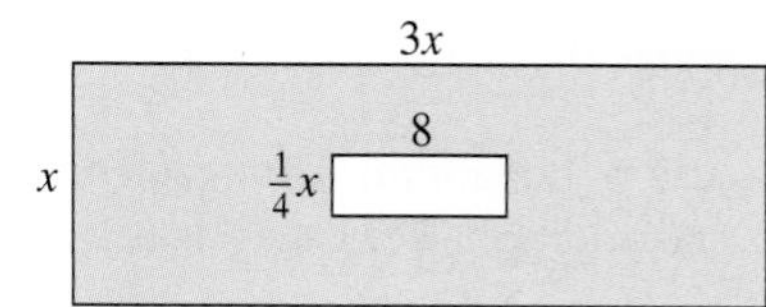

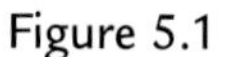

Figure 5.1

Solution

Area of shaded region	=	Area of outer rectangle	−	Area of inner rectangle

$$\text{Area} = 3x(x) - 8\left(\frac{1}{4}x\right) = 3x^2 - 2x$$

Discussing the Concept **Adding Polynomials**

Write a paragraph that explains how the adage "You can't add apples and oranges" might relate to adding two polynomials. Include several examples to illustrate the applicability of this statement. Share your paragraph and examples with fellow students to see if they make sense to others.

5.1 Exercises

Integrated Review *Concepts, Skills, and Problem Solving*

Keep mathematically in shape by doing these exercises *before* the problems of this section.

Properties and Definitions

1. In your own words, state the definition of an algebraic expression.

2. State the definition of a term of an algebraic expression.

Simplifying Expressions

In Exercises 3–6, use the Distributive Property to expand the expression.

3. $10(x - 1)$ **4.** $4(3 - 2z)$

5. $-\frac{1}{2}(4 - 6x)$ **6.** $-25(2x - 3)$

In Exercises 7–10, simplify the expression.

7. $8y - 2x + 7x - 10y$

8. $\frac{5}{6}x - \frac{2}{3}x + 8$

9. $10(x - 1) - 3(x + 2)$

10. $-3[x + (2 + 3x)]$

Graphs

In Exercises 11 and 12, graph the function. Use a graphing utility to verify your graph.

11. $g(x) = 2 + \frac{3}{2}x$ **12.** $h(t) = (t + 1)(t - 3)$

Developing Skills

In Exercises 1–8, determine whether the expression is a polynomial. If it is not, explain why. See Example 1.

1. $9 - z$ **2.** $t^2 - 4$

3. $x^{2/3} + 8$ **4.** $9 - z^{1/2}$

5. $6x^{-1}$ **6.** $1 - 4x^{-2}$

7. $z^{-1} + z^2 - 2$ **8.** $t^3 - 3t + 4$

In Exercises 9–18, write the polynomial in standard form. Then determine its degree and leading coefficient. See Example 2.

9. $5 - 32x$ **10.** $2x - 3$

11. $x^3 - 4x^2 + 9$ **12.** $9 - 2y^4$

13. $8x + 2x^5 - x^2 - 1$ **14.** $5x^3 - 3x^2 + 10$

15. 10 **16.** -32

17. $v_0t - 16t^2$ (v_0 is a constant.) **18.** $64 - \frac{1}{2}at^2$ (a is a constant.)

In Exercises 19–24, determine whether the polynomial is a monomial, a binomial, or a trinomial.

19. $x^2 - 2x + 3$ **20.** $-6y$

21. $x^3 - 4$ **22.** $u^2 - 3u + 5$

23. 5 **24.** $16 - z^2$

In Exercises 25–30, give an example of a polynomial that fits the description. (*Note:* There are many correct answers.)

25. A binomial of degree 3

26. A trinomial of degree 4

27. A monomial of degree 2

28. A binomial of degree 5

29. A trinomial of degree 6

30. A monomial of degree 0

In Exercises 31–44, use a horizontal arrangement to perform the polynomial addition. See Example 3.

31. $(11x - 2) + (3x + 8)$

32. $(-2x + 4) + (x - 6)$

33. $(3z^2 - z + 2) + (z^2 - 4)$

34. $(6x^4 + 8x) + (4x - 6)$

35. $b^2 + (b^3 - 2b^2 + 3) + (b^3 - 3)$

36. $(3x^2 - x) + 5x^3 + (-4x^3 + x^2 - 8)$

37. $(12 - 3t - 7t^2) + (1 + 3t - t^2)$

38. $(3 + 6x + 8x^2 + 9x^3) + (3 - 2x + 4x^2 - 5x^3)$

39. $(2ab - 3) + (a^2 - 2ab) + (4b^2 - a^2)$

40. $(uv - 3) + (4uv + 1)$

41. $\left(\frac{2}{3}y^2 - \frac{3}{4}\right) + \left(\frac{5}{6}y^2 + 2\right)$

42. $\left(\frac{3}{4}x^3 - \frac{1}{2}\right) + \left(\frac{1}{8}x^3 + 3\right)$

43. $(0.1t^3 - 3.4t^2) + (1.5t^3 - 7.3)$

44. $(0.7x^2 - 0.2x + 2.5) + (7.4x - 3.9)$

In Exercises 45–60, use a vertical arrangement to perform the polynomial addition. See Example 4.

45. $\begin{array}{r} 2x + 5 \\ \underline{3x + 8} \end{array}$

46. $\begin{array}{r} 10x - 7 \\ \underline{6x + 4} \end{array}$

47. $\begin{array}{r} -2x + 10 \\ \underline{x - 38} \end{array}$

48. $\begin{array}{r} 4x^2 + 13 \\ \underline{3x^2 - 11} \end{array}$

49. $\begin{array}{rrr} -x^3 & & +\ 3 \\ \underline{3x^3} & \underline{+\ 2x^2} & \underline{+\ 5} \end{array}$

50. $\begin{array}{rrr} 2z^3 & +\ 3z & -\ 2 \\ \underline{} & \underline{z^2 - 2z} & \underline{} \end{array}$

51. $\begin{array}{r} 3x^4 - 2x^3 - 4x^2 + 2x - 5 \\ \underline{x^2 - 7x + 5} \end{array}$

52. $\begin{array}{rrrr} x^5 & & -\ 4x^3 + x + 9 \\ \underline{} & \underline{2x^4 + 3x^3} & \underline{\ -\ 3} \end{array}$

53. $(x^2 - 4) + (2x^2 + 6)$

54. $(x^3 + 2x - 3) + (4x + 5)$

55. $(2 - 3y) + (y^4 + 3y + 2)$

56. $(a^2 + 3a - 2) + (5a - a^2 - 6a) + (a^2 + 2)$

57. $(x^2 - 2x + 2) + (x^2 + 4x) + 2x^2$

58. $(5y + 10) + (y^2 - 3y - 2) + (2y^2 + 4y - 3)$

59. Add $8y^3 + 7$ to $5 - 3y^3$.

60. Add $2z - 8z^2 - 3$ to $z^2 + 5z$.

Comparing Two Formats In Exercises 61 and 62, add the two polynomials using the horizontal arrangement and the vertical arrangement. Which format do you prefer? Explain.

61. $(6x^2 + 5) + (3 - 2x^2)$

62. $(0.5x^4 - 6.2x^2 + 7.1) +$
$(3.2x^4 + 8x^3 - 16x + 10.5)$

In Exercises 63–72, use a horizontal arrangement to perform the polynomial subtraction. See Example 5.

63. $(11x - 8) - (2x + 3)$

64. $(9x + 2) - (15x - 4)$

65. $(x^2 - x) - (x - 2)$

66. $(x^2 - 4) - (x^2 - 4)$

67. $(4 - 2x - x^3) - (3 - 2x + 2x^3)$

68. $(t^4 - 2t^2) - (3t^2 - t^4 - 5)$

69. $10 - (u^2 + 5)$

70. $(z^3 + z^2 + 1) - z^2$

71. $(x^5 - 3x^4 + x^3 - 5x + 1) - (4x^5 - x^3 + x - 5)$

72. $(t^4 + 5t^3 - t^2 + 8t - 10) -$
$(t^4 + t^3 + 2t^2 + 4t - 7)$

In Exercises 73–86, use a vertical arrangement to perform the polynomial subtraction. See Example 6.

73. $\begin{array}{r} 2x - 2 \\ \underline{-\ (x - 1)} \end{array}$

74. $\begin{array}{r} 9x + 7 \\ \underline{-\ (3x + 9)} \end{array}$

75. $\begin{array}{r} 2x^2 - x + 2 \\ \underline{-(3x^2 + x - 1)} \end{array}$

76. $\begin{array}{r} y^4 - 2 \\ \underline{-\ (y^4 + 2)} \end{array}$

77. $\begin{array}{r} -\ 3x^3 - 4x^2 + 2x - 5 \\ \underline{-\ (2x^4 + 2x^3 - 4x + 5)} \end{array}$

78. $\begin{array}{r} 12x^3 + 25x^2 - 15 \\ \underline{-\ (-2x^3 + 18x^2 - 3x)} \end{array}$

79. $(2 - x^3) - (2 + x^3)$

80. $(4z^3 - 6) - (-z^3 + z - 2)$

81. $(4t^3 - 3t + 5) - (3t^2 - 3t - 10)$

82. $(-s^2 - 3) - (2s^2 + 10s)$

83. $(6x^3 - 3x^2 + x) - [(x^3 + 3x^2 + 3) + (x - 3)]$

84. $(y^2 - y) - [(2y^2 + y) - (4y^2 - y + 2)]$

85. Subtract $7x^3 - 4x + 5$ from $10x^3 + 15$.

86. Subtract $y^5 - y^4$ from $y^2 + 3y^4$.

In Exercises 87–100, perform the operations. See Example 7.

87. $(6x - 5) - (8x + 15)$

88. $(2x^2 + 1) + (x^2 - 2x + 1)$

89. $-(x^3 - 2) + (4x^3 - 2x)$

90. $-(5x^2 - 1) - (-3x^2 + 5)$

91. $2(x^4 + 2x) + (5x + 2)$

92. $(z^4 - 2z^2) + 3(z^4 + 4)$

93. $(15x^2 - 6) - (-8x^3 - 14x^2 - 17)$

94. $(15x^4 - 18x - 19) - (-13x^4 - 5x + 15)$

95. $5z - [3z - (10z + 8)]$

96. $(y^3 + 1) - [(y^2 + 1) + (3y - 7)]$

97. $2(t^2 + 5) - 3(t^2 + 5) + 5(t^2 + 5)$

98. $-10(u + 1) + 8(u - 1) - 3(u + 6)$

99. $8v - 6(3v - v^2) + 10(10v + 3)$

100. $3(x^2 - 2x + 3) - 4(4x + 1) - (3x^2 - 2x)$

Solving Problems

Geometry In Exercises 101 and 102, find the perimeter of the figure.

101. **102.**

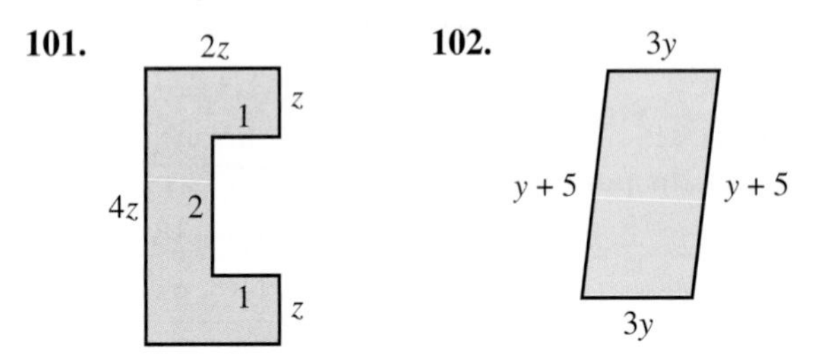

In Exercises 103–108, write a polynomial that represents the area of the shaded portion of the figure. See Example 8.

103. **104.** **105.** **106.**

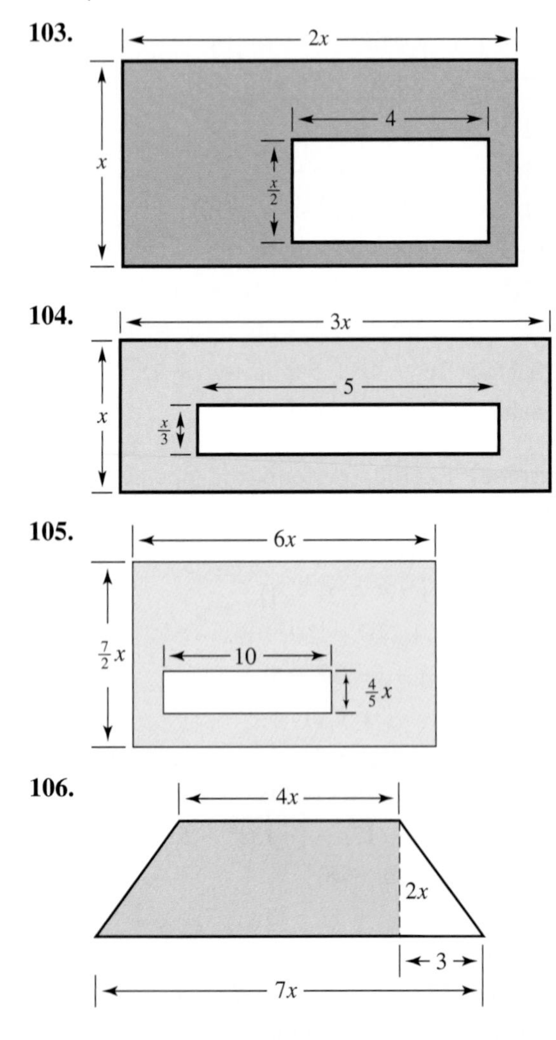

107. **108.**

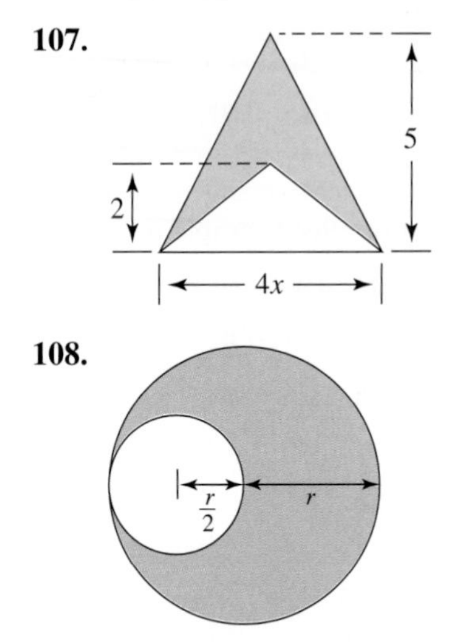

109. *Comparing Models* From 1990 through 1995, the per capita consumption (in pounds) of beef B and chicken C in the United States can be modeled by

$$B = 0.29t^2 - 1.43t + 64.11, \quad 0 \le t \le 5$$

and

$$C = -0.32t^2 + 2.98t + 42.17, \quad 0 \le t \le 5$$

where $t = 0$ represents 1990. (Source: U.S. Department of Agriculture)

(a) Add the polynomials to find a model for the total per capita consumption T of beef and chicken.

(b) Use a graphing utility to graph the models B, C, and T.

(c) Use the graphs in part (b) to determine whether the consumption represented by the model T is increasing or decreasing.

110. *Comparing Business Models* The cost of producing x units of a product is $C = 100 + 30x$. The revenue for selling x units is $R = 90x - x^2$, where $0 \le x \le 40$. The profit is given by the revenue minus the cost.

(a) Perform the subtraction required to find the polynomial representing profit.

(b) Use a graphing utility to graph the polynomial representing profit.

(c) Determine the profit when $x = 30$ units are produced and sold. Use the graph in part (b) to predict the change in profit if x is some value other than 30.

Explaining Concepts

111. Answer parts (a)–(c) of Motivating the Chapter on page 273.

112. Explain the difference between the degree of a term of a polynomial and the degree of a polynomial.

113. Determine which of the two statements is always true. Is the statement not selected always false? Explain.

(a) "A polynomial is a trinomial."

(b) "A trinomial is a polynomial."

114. In your own words, define "like terms." What is the only factor of like terms that can differ?

115. Describe how to combine like terms. What operations are used?

116. Is a polynomial an algebraic expression? Explain.

117. In your own words, explain how to subtract polynomials. Give an example.

118. Is the sum of two binomials always a binomial? Explain.

5.2 Multiplying Polynomials: Special Products

Objectives

1. Find a product with monomial multipliers.
2. Multiply binomials using the Distributive Property and the FOIL Method.
3. Multiply polynomials using a horizontal or vertical format.
4. Identify and use special binomial products.

1 Find a product with monomial multipliers.

Monomial Multipliers

To multiply polynomials, you use many of the rules for simplifying algebraic expressions. You may want to review these rules from Section 2.2.

1. Properties of exponents
2. The Distributive Property
3. Combining like terms
4. Symbols of grouping

The Granger Collection

Blaise Pascal

(1623–1662)

Pascal was a French mathematician, scientist, and philosopher. In addition to his religious and philosophical writings, he made many invaluable contributions to mathematics and physics. Perhaps his most important contribution to mathematics was the invention and construction of the first calculating machine. He invented the machine when he was only nineteen years old.

The simplest type of polynomial multiplication involves a monomial multiplier. The product is obtained by direct application of the Distributive Property. For instance, to multiply the monomial x by the polynomial $(2x + 5)$, multiply *each* of the terms of the polynomial by x.

$$(x)(2x + 5) = (x)(2x) + (x)(5) = 2x^2 + 5x$$

Here is another example.

$$(2x)(3x^2 - 4x + 1) = (2x)(3x^2) - (2x)(4x) + (2x)(1)$$
$$= 6x^3 - 8x^2 + 2x$$

Example 1 Finding Products with Monomial Multipliers

Find each product.

a. $(3x - 7)(-2x)$ **b.** $3x^2(5x - x^3 + 2)$ **c.** $(-x)(2x^2 - 3x)$

Solution

a. $(3x - 7)(-2x) = 3x(-2x) - 7(-2x)$ Distributive Property

$= -6x^2 + 14x$ Standard form

b. $3x^2(5x - x^3 + 2)$

$= (3x^2)(5x) - (3x^2)(x^3) + (3x^2)(2)$ Distributive Property

$= 15x^3 - 3x^5 + 6x^2$ Properties of exponents

$= -3x^5 + 15x^3 + 6x^2$ Standard form

c. $(-x)(2x^2 - 3x) = (-x)(2x^2) - (-x)(3x)$ Distributive Property

$= -2x^3 + 3x^2$ Standard form

2 Multiply binomials using the Distributive Property and the FOIL Method.

Multiplying Binomials

To multiply two binomials, you can use both (left and right) forms of the Distributive Property. For example, if you treat the binomial $(5x + 7)$ as a single quantity, you can multiply $(3x - 2)$ by $(5x + 7)$ as follows.

$$\begin{aligned}(3x - 2)(5x + 7) &= 3x(5x + 7) - 2(5x + 7)\\ &= (3x)(5x) + (3x)(7) - (2)(5x) - 2(7)\\ &= 15x^2 + 21x - 10x - 14\end{aligned}$$

Product of First terms	Product of Outer terms	Product of Inner terms	Product of Last terms

$$= 15x^2 + 11x - 14$$

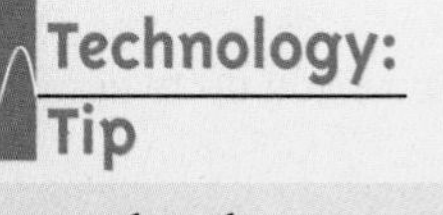

Remember that you can use a graphing utility to check whether you have performed a polynomial operation correctly. For instance, to check if

$$(x - 1)(x + 5) = x^2 + 4x - 5$$

you can "graph the left side and graph the right side" on the same screen, as shown below. Because both graphs are the same, you can reason that the multiplication was performed correctly.

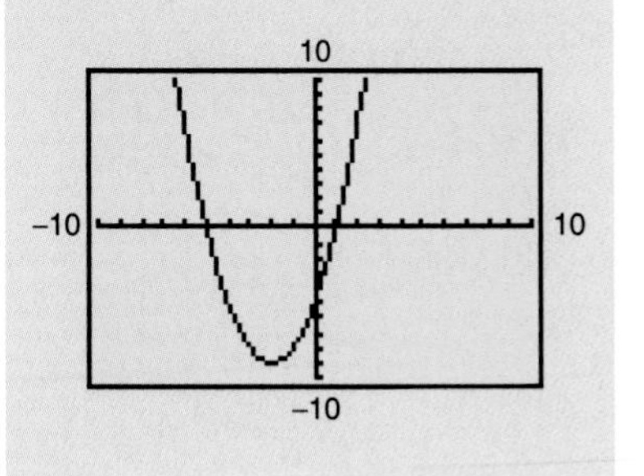

With practice you should be able to multiply two binomials without writing out all of the steps above. In fact, the four products in the boxes above suggest that you can write the product of two binomials in just one step. This is referred to as the **FOIL Method.** Note that the words *first*, *outer*, *inner*, and *last* refer to the positions of the terms in the original product.

First
Outer
$(3x - 2)(5x + 7)$
Inner
Last

Example 2 Multiplying with the Distributive Property

Use the Distributive Property to find the product.

a. $(x - 1)(x + 5)$ **b.** $(2x + 3)(x - 2)$

Solution

a.
$$\begin{aligned}(x - 1)(x + 5) &= x(x + 5) - (1)(x + 5) && \text{Right Distributive Property}\\ &= x^2 + 5x - x - 5 && \text{Left Distributive Property}\\ &= x^2 + (5x - x) - 5 && \text{Group like terms.}\\ &= x^2 + 4x - 5 && \text{Combine like terms.}\end{aligned}$$

b.
$$\begin{aligned}(2x + 3)(x - 2) &= 2x(x - 2) + 3(x - 2) && \text{Right Distributive Property}\\ &= 2x^2 - 4x + 3x - 6 && \text{Left Distributive Property}\\ &= 2x^2 + (-4x + 3x) - 6 && \text{Group like terms.}\\ &= 2x^2 - x - 6 && \text{Combine like terms.}\end{aligned}$$

Example 3 Multiplying Binomials Using the FOIL Method

Use the FOIL Method to find the product.

a. $(x - 4)(x + 4)$ **b.** $(3x + 5)(2x + 1)$

Solution

$$\begin{aligned} &\quad\quad\quad\quad\quad\ \ \text{F}\quad \text{O}\quad \text{I}\quad \text{L} \\ \textbf{a. } (x - 4)(x + 4) &= x^2 + 4x - 4x - 16 \\ &= x^2 - 16 \end{aligned}$$

Combine like terms.

$$\begin{aligned} &\quad\quad\quad\quad\quad\ \ \text{F}\quad \text{O}\quad \text{I}\quad \text{L} \\ \textbf{b. } (3x + 5)(2x + 1) &= 6x^2 + 3x + 10x + 5 \\ &= 6x^2 + 13x + 5 \end{aligned}$$

Combine like terms.

In Example 3(a), note that the outer and inner products add up to zero.

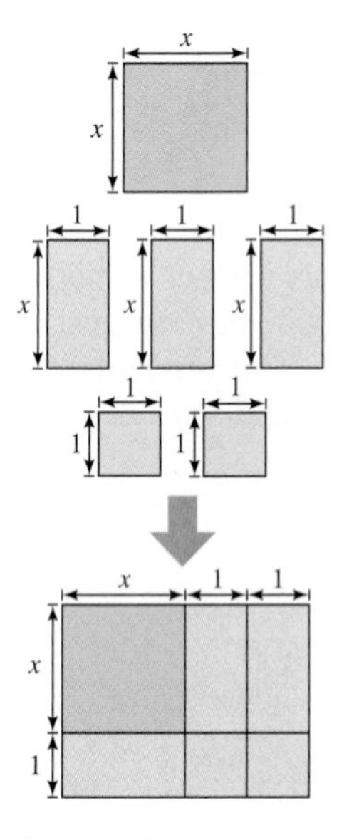

Figure 5.2

Example 4 A Geometric Model of a Polynomial Product

What polynomial product is represented in Figure 5.2?

Solution

On the top you have the sum $x^2 + (x + x + x) + (1 + 1)$, or $x^2 + 3x + 2$. On the bottom you have the product $(x + 1)(x + 2)$, which represents the area of the rectangular figure. So, the display shows that

$$x^2 + 3x + 2 = (x + 1)(x + 2).$$

Example 5 Simplifying Polynomial Expressions

Simplify each expression and write the result in standard form.

a. $(4x + 5)^2$ **b.** $(3x^2 - 2)(4x + 7) - (4x)^2$

Solution

a. $(4x + 5)^2 = (4x + 5)(4x + 5)$ Repeated multiplication

$= 16x^2 + 20x + 20x + 25$ Multiply binomials.

$= 16x^2 + 40x + 25$ Combine like terms.

b. $(3x^2 - 2)(4x + 7) - (4x)^2$

$= 12x^3 + 21x^2 - 8x - 14 - (4x)^2$ Multiply binomials.

$= 12x^3 + 21x^2 - 8x - 14 - 16x^2$ Square monomial.

$= 12x^3 + 5x^2 - 8x - 14$ Combine like terms.

3 Multiply polynomials using a horizontal or vertical format.

Multiplying Polynomials

The FOIL Method for multiplying two binomials is simply a device for guaranteeing that *each term of one binomial is multiplied by each term of the other binomial.*

$$(ax + b)(cx + d) = \underset{F}{ax(cx)} + \underset{O}{ax(d)} + \underset{I}{b(cx)} + \underset{L}{b(d)}$$

This same rule applies to the product of two polynomials: *each term of one polynomial must be multiplied by each term of the other polynomial.* This can be accomplished using either a horizontal or vertical format.

Example 6 Multiplying Polynomials (Horizontal Format)

Use a horizontal format to find the product.

a. $(x - 4)(x^2 - 4x + 2)$ **b.** $(2x^2 - 7x + 1)(4x + 3)$

Solution

a. $(x - 4)(x^2 - 4x + 2)$

$= x(x^2 - 4x + 2) - 4(x^2 - 4x + 2)$ Distributive Property

$= x^3 - 4x^2 + 2x - 4x^2 + 16x - 8$ Distributive Property

$= x^3 - 8x^2 + 18x - 8$ Combine like terms.

b. $(2x^2 - 7x + 1)(4x + 3)$

$= (2x^2 - 7x + 1)(4x) + (2x^2 - 7x + 1)(3)$ Distributive Property

$= 8x^3 - 28x^2 + 4x + 6x^2 - 21x + 3$ Distributive Property

$= 8x^3 - 22x^2 - 17x + 3$ Combine like terms.

Example 7 Multiplying Polynomials (Vertical Format)

Use a vertical format to find the product: $(3x^2 + x - 5)(2x - 1)$.

Solution

With a vertical format, line up like terms in the same vertical columns, just as you align digits in whole number multiplication.

$$\begin{array}{rrrrl} & 3x^2 & + x & - 5 & \text{Place polynomial with most terms on top.} \\ \times & & 2x & - 1 & \\ \hline & -3x^2 & - x & + 5 & \Leftarrow -1(3x^2 + x - 5) \\ 6x^3 & + 2x^2 & - 10x & & \Leftarrow 2x(3x^2 + x - 5) \\ \hline 6x^3 & - x^2 & - 11x & + 5 & \text{Combine like terms in columns.} \end{array}$$

When multiplying two polynomials, it is best to write each in standard form before using either the horizontal or vertical format. This is illustrated in the next example.

Example 8 Multiplying Polynomials

Multiply the polynomials.

$(x + 3x^2 - 4)(5 + 3x - x^2)$

Solution

$$\begin{array}{rrrrrl}
 & & 3x^2 & + x & - 4 & \text{Standard form} \\
\times & & -x^2 & + 3x & + 5 & \text{Standard form} \\
\hline
 & & 15x^2 & + 5x & - 20 & 5(3x^2 + x - 4) \\
 & 9x^3 & + 3x^2 & - 12x & & 3x(3x^2 + x - 4) \\
-3x^4 & - x^3 & + 4x^2 & & & -x^2(3x^2 + x - 4) \\
\hline
-3x^4 & + 8x^3 & + 22x^2 & - 7x & - 20 & \text{Combine like terms.}
\end{array}$$

Example 9 Multiplying Polynomials

Find the product.

$(x - 3)^3$

Solution

To raise $(x - 3)$ to the third power, you can use two steps. First, because $(x - 3)^3 = (x - 3)^2(x - 3)$, find the product $(x - 3)^2$.

$$\begin{aligned}
(x - 3)^2 &= (x - 3)(x - 3) && \text{Repeated multiplication} \\
&= x^2 - 3x - 3x + 9 && \text{FOIL} \\
&= x^2 - 6x + 9 && \text{Combine like terms.}
\end{aligned}$$

Now, using a vertical arrangement, find $(x - 3)^3$.

$$\begin{array}{rrrrl}
 & x^2 & - 6x & + 9 & (x - 3)^2 \\
\times & & x & - 3 & \\
\hline
 & -3x^2 & + 18x & - 27 & -3(x^2 - 6x + 9) \\
x^3 & - 6x^2 & + 9x & & x(x^2 - 6x + 9) \\
\hline
x^3 & - 9x^2 & + 27x & - 27 & \text{Combine like terms.}
\end{array}$$

So, $(x - 3)^3 = x^3 - 9x^2 + 27x - 27$.

Use a graphing utility to graph $y = (x - 3)^3$ and $y = x^3 - 9x^2 + 27x - 27$ and verify that these two expressions are equal.

4 Identify and use special binomial products.

Special Products

Some binomial products such as those in Examples 3(a) and 5(a) have special forms that occur frequently in algebra. Let's look at those products again. The product $(x + 4)(x - 4)$ is called a **product of the sum and difference of two terms.** With such products, the two middle terms cancel, as follows.

$$(x + 4)(x - 4) = x^2 - 4x + 4x - 16$$ Sum and difference of two terms

$$= x^2 - 16$$ Product has no middle term.

Another common type of product is the **square of a binomial.** With this type of product, the middle term is always twice the product of the terms in the binomial.

$$(4x + 5)^2 = (4x + 5)(4x + 5)$$ Square of a binomial

$$= 16x^2 + 20x + 20x + 25$$

$$= 16x^2 + 40x + 25$$ Middle term is twice the product of the terms of the binomial.

You should learn to recognize the patterns of these two special products. We give the general form of these special products in the following statements. The FOIL Method can be used to verify each rule.

▶ Special Products

Let a and b be real numbers, variables, or algebraic expressions.

Special Product	*Example*
Sum and Difference of Two Terms:	
$(a + b)(a - b) = a^2 - b^2$	$(2x - 5)(2x + 5) = 4x^2 - 25$
Square of a Binomial:	
$(a + b)^2 = a^2 + 2ab + b^2$	$(3x + 4)^2 = 9x^2 + 2(3x)(4) + 16$
	$= 9x^2 + 24x + 16$
$(a - b)^2 = a^2 - 2ab + b^2$	$(x - 7)^2 = x^2 - 2(x)(7) + 49$
	$= x^2 - 14x + 49$

When a binomial is squared, the resulting middle term is always *twice* the product of the two terms.

$$(a + b)^2 = a^2 + 2(ab) + b^2$$

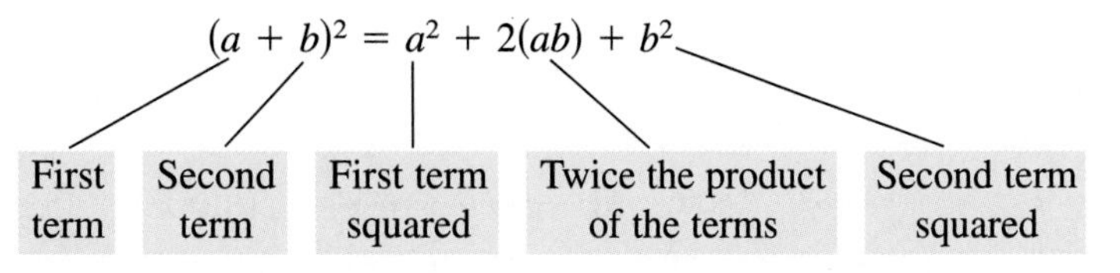

Be sure to include the middle term. For instance, $(a + b)^2$ is *not* equal to $a^2 + b^2$.

Example 10 Finding Sum and Difference Products

Find each product.

a. $(x + 2)(x - 2)$ **b.** $(5x - 6)(5x + 6)$ **c.** $(2 + 3x)(2 - 3x)$

Solution

Sum Difference (1st term)² (2nd term)²

$$\textbf{a. } (x + 2)(x - 2) = (x)^2 - (2)^2$$
$$= x^2 - 4$$

Difference Sum (1st term)² (2nd term)²

$$\textbf{b. } (5x - 6)(5x + 6) = (5x)^2 - (6)^2$$
$$= 25x^2 - 36$$

Sum Difference (1st term)² (2nd term)²

$$\textbf{c. } (2 + 3x)(2 - 3x) = (2)^2 - (3x)^2$$
$$= 4 - 9x^2$$

Example 11 Squaring a Binomial

Find each product.

a. $(4x - 9)^2$ **b.** $(3x + 7)^2$ **c.** $(6 - 5x^2)^2$

Solution

1st term 2nd term (1st term)² Twice the product of the terms (2nd term)²

$$\textbf{a. } (4x - 9)^2 = (4x)^2 - 2(4x)(9) + (9)^2$$
$$= 16x^2 - 72x + 81$$

1st term 2nd term (1st term)² Twice the product of the terms (2nd term)²

$$\textbf{b. } (3x + 7)^2 = (3x)^2 + 2(3x)(7) + (7)^2$$
$$= 9x^2 + 42x + 49$$

1st term 2nd term (1st term)² Twice the product of the terms (2nd term)²

$$\text{c. } (6 - 5x^2)^2 = (6)^2 - 2(6)(5x^2) + (5x^2)^2$$
$$= 36 - 60x^2 + (5)^2(x^2)^2$$
$$= 36 - 60x^2 + 25x^4$$

Example 12 Finding the Measurements of a Golf Tee

A landscaper wants to reshape a square tee area for the 9th hole of a golf course. The new tee area is to have one side 2 feet longer and the adjacent side 6 feet longer than the original tee. (See Figure 5.3.) If the new tee has 204 square feet more area than the original tee, what are the measurements of the original 9th hole tee?

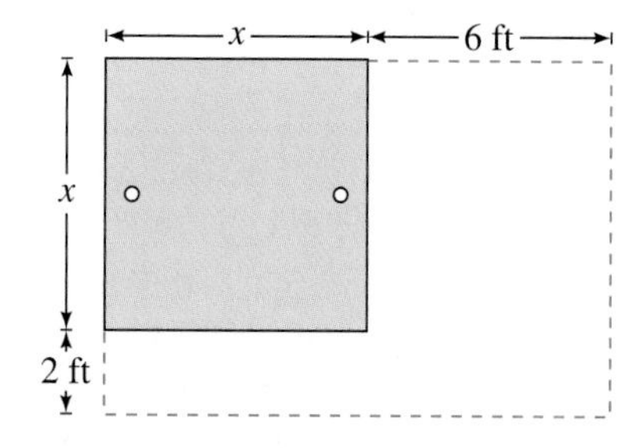

Figure 5.3

Solution

Verbal Model: New area = Old area + 204

Labels: Original length = original width = x (feet)
New length = $x + 6$ (feet)
New width = $x + 2$ (feet)

Equation: $(x + 6)(x + 2) = x^2 + 204$ — x^2 is original area.

$x^2 + 8x + 12 = x^2 + 204$ — Multiply factors.

$8x + 12 = 204$ — Subtract x^2 from both sides.

$8x = 192$ — Subtract 12 from both sides.

$x = 24$ — Simplify.

The original tee measured 24 feet by 24 feet.

Discussing the Concept — Pascal's Triangle

The following triangular pattern of numbers on the left shows the first seven rows of **Pascal's Triangle,** named after the French mathematician Blaise Pascal (1623–1662).Try to discover the pattern formed by the numbers in the triangle. Then use the pattern on the right to write out the expansion of $(x + 1)^7$.

1
1 1
1 2 1
1 3 3 1
1 4 6 4 1
1 5 10 10 5 1
1 6 15 20 15 6 1

$$(x + 1)^0 = 1$$
$$(x + 1)^1 = x + 1$$
$$(x + 1)^2 = x^2 + 2x + 1$$
$$(x + 1)^3 = x^3 + 3x^2 + 3x + 1$$
$$(x + 1)^4 = x^4 + 4x^3 + 6x^2 + 4x + 1$$
$$(x + 1)^5 = x^5 + 5x^4 + 10x^3 + 10x^2 + 5x + 1$$
$$(x + 1)^6 = x^6 + 6x^5 + 15x^4 + 20x^3 + 15x^2 + 6x + 1$$

Discuss how to determine the number of terms in the expansion of $(x + 1)^{14}$.

5.2 Exercises

Integrated Review *Concepts, Skills, and Problem Solving*

Keep mathematically in shape by doing these exercises *before* the problems of this section.

Properties and Definitions

1. Relative to the x- and y-axes, explain the meaning of each coordinate of the point $(3, -2)$.

2. A point lies 4 units from the x-axis and 3 units from the y-axis. Give the ordered pair for such a point in each quadrant.

Simplifying Expressions

In Exercises 3–8, simplify the expression.

3. $\frac{3}{4}x - \frac{5}{2} + \frac{3}{2}x$ **4.** $4 - 2(3 - x)$

5. $2(x - 4) + 5x$ **6.** $4(3 - y) + 2(y + 1)$

7. $-3(z - 2) - (z - 6)$

8. $(u - 2) - 3(2u + 1)$

Problem Solving

9. Your sales commission rate is 5.5%. Your commission is \$1600. How much did you sell?

10. A jogger leaves a location on a fitness trail running at a rate of 4 miles per hour. Fifteen minutes later, a second jogger leaves from the same location running at 5 miles per hour. How long will it take the second runner to overtake the first and how far will each have run at that point? Use a diagram to help answer the question.

Graphs and Models

In Exercises 11 and 12, use a graphing utility to graph the function. Identify any intercepts.

11. $g(x) = 4 - \frac{1}{2}x$ **12.** $f(x) = x(x - 4)$

Developing Skills

In Exercises 1–50, multiply and simplify. See Examples 1–3 and 5.

1. $x(-2x)$ **2.** $y(-3y)$ **3.** $t^2(4t)$ **4.** $3u(u^4)$

5. $\left(\frac{x}{4}\right)(10x)$ **6.** $9x\left(\frac{x}{12}\right)$

7. $(-2b^2)(-3b)$ **8.** $(-4m)(3m^2)$

9. $y(3 - y)$ **10.** $z(z - 3)$

11. $-x(x^2 - 4)$ **12.** $-t(10 - 3t)$

13. $3t(2t - 5)$ **14.** $-5u(u^2 + 4)$

15. $-4x(3 + 3x^2 - 6x^3)$ **16.** $5v(5 - 4v + 5v^2)$

17. $3x(x^2 - 2x + 1)$ **18.** $y(4y^2 + 2y - 3)$

19. $2x(x^2 - 2x + 8)$ **20.** $-3x(x - 3)$

21. $4t^3(t - 3)$ **22.** $-2t^4(t + 6)$

23. $x^2(4x^2 - 3x + 1)$ **24.** $y^2(2y^2 + y - 5)$

25. $-3x^3(4x^2 - 6x + 2)$ **26.** $5u^4(2u^3 - 3u + 3)$

27. $-2x(-3x)(5x + 2)$ **28.** $4x(-2x)(x^2 - 1)$

29. $2x(6x^4) - 3x^2(2x^2)$ **30.** $-8y(-5y^4) - 2y^2(5y^3)$

31. $(x + 3)(x + 4)$ **32.** $(x - 5)(x + 10)$

33. $(3x - 5)(2x + 1)$ **34.** $(7x - 2)(4x - 3)$

35. $(2x - y)(x - 2y)$ **36.** $(x + y)(x + 2y)$

37. $(2x + 4)(x + 1)$ **38.** $(4x + 3)(2x - 1)$

39. $(6 - 2x)(4x + 3)$ **40.** $(8x - 6)(5 - 4x)$

41. $(3x - 2y)(x - y)$ **42.** $(7x + 5y)(x + y)$

43. $(3x^2 - 4)(x + 2)$ **44.** $(5x^3 - 2)(x - 1)$

45. $(2x^3 + 4)(x^2 + 6)$ **46.** $(7x^2 - 3)(2x^2 - 4)$

47. $(3s + 1)(3s + 4) - (3s)^2$

48. $(2t + 5)(4t - 2) - (2t)^2$

49. $(4x^2 - 1)(2x + 8) + (-x)^3$

50. $(3 - 3x^2)(4 - 5x^2) - (-x^2)^2$

In Exercises 51–64, multiply using a horizontal format. See Example 6.

51. $(x + 10)(x + 2)$ **52.** $(x - 1)(x + 3)$

53. $(2x - 5)(x + 2)$ **54.** $(3x - 2)(2x - 3)$

55. $(x + 1)(x^2 + 2x - 1)$ **56.** $(x - 3)(x^2 - 3x + 4)$

57. $(x^3 - 2x + 1)(x - 5)$ **58.** $(x + 1)(x^2 - x + 1)$

59. $(x - 2)(x^2 + 2x + 4)$ **60.** $(x + 9)(x^2 - x - 4)$

61. $(x^2 + 3)(x^2 - 6x + 2)$ **62.** $(x^2 + 3)(x^2 - 2x + 3)$

63. $(3x^2 + 1)(x^2 - 4x - 2)$

64. $(x^2 + 2x + 5)(4x^3 - 2)$

In Exercises 65–80, multiply using a vertical format. See Examples 7–9.

65. $\begin{array}{r} x + 3 \\ \times\ x - 2 \\ \hline \end{array}$ **66.** $\begin{array}{r} 2x - 1 \\ \times\ 5x + 1 \\ \hline \end{array}$

67. $\begin{array}{r} x^2 - 3x + 9 \\ \times \qquad x + 3 \\ \hline \end{array}$ **68.** $\begin{array}{r} 4x^4 - 6x^2 + 9 \\ \times \qquad 2x + 3 \\ \hline \end{array}$

69. $(x^2 - x + 2)(x^2 + x - 2)$

70. $(x^2 + 2x + 5)(2x^2 - x - 1)$

71. $(x^3 + x + 3)(x^2 + 5x - 4)$

72. $(x^2 + x + 1)(x^2 - x - 1)$

73. $(x - 2)^3$ **74.** $(x + 3)^3$

75. $(x - 1)^2(x - 1)^2$ **76.** $(x + 4)^2(x + 4)^2$

77. $(x + 2)^2(x - 4)$ **78.** $(x - 4)^2(x - 1)$

79. $(u - 1)(2u + 3)(2u + 1)$

80. $(2x + 5)(x - 2)(5x - 3)$

In Exercises 81–110, use a special product pattern to find the product. See Examples 10 and 11.

81. $(x + 3)(x - 3)$ **82.** $(x - 5)(x + 5)$

83. $(x + 4)(x - 4)$ **84.** $(y + 9)(y - 9)$

85. $(2u + 3)(2u - 3)$ **86.** $(3z + 4)(3z - 4)$

87. $(4t - 6)(4t + 6)$ **88.** $(3u + 7)(3u - 7)$

89. $(2x + 3y)(2x - 3y)$ **90.** $(5u + 12v)(5u - 12v)$

91. $(4u - 3v)(4u + 3v)$ **92.** $(8a - 5b)(8a + 5b)$

93. $(2x^2 + 5)(2x^2 - 5)$ **94.** $(4t^2 + 6)(4t^2 - 6)$

95. $(x + 6)^2$ **96.** $(a - 2)^2$

97. $(t - 3)^2$ **98.** $(x + 10)^2$

99. $(3x + 2)^2$ **100.** $(2x - 8)^2$

101. $(8 - 3z)^2$ **102.** $(1 - 5t)^2$

103. $(2x - 5y)^2$ **104.** $(4s + 3t)^2$

105. $(6t + 5s)^2$ **106.** $(3u - 8v)^2$

107. $[(x + 1) + y]^2$ **108.** $[(x - 3) - y]^2$

109. $[u - (v - 3)]^2$ **110.** $[2u + (v + 1)]^2$

In Exercises 111 and 112, perform the multiplication and simplify.

111. $(x + 2)^2 - (x - 2)^2$ **112.** $(u + 5)^2 + (u - 5)^2$

Think About It In Exercises 113 and 114, is the equation an identity? Explain.

113. $(x + y)^3 = x^3 + 3x^2y + 3xy^2 + y^3$

114. $(x - y)^3 = x^3 - 3x^2y + 3xy^2 - y^3$

In Exercises 115 and 116, use the results of Exercises 113 and 114 to find the product.

115. $(x + 2)^3$ **116.** $(x + 1)^3$

Solving Problems

117. *Finding a Pattern* Perform each multiplication.

(a) $(x - 1)(x + 1)$

(b) $(x - 1)(x^2 + x + 1)$

(c) $(x - 1)(x^3 + x^2 + x + 1)$

(d) Use the pattern formed in the first three products to guess the product

$(x - 1)(x^4 + x^3 + x^2 + x + 1)$.

Verify your guess by multiplying.

118. *Geometry* The base of a triangular sail is $2x$ feet and its height is $x + 10$ feet (see figure). Find the area A of the sail.

119. *Geometry* The height of a rectangular sign is twice its width w (see figure). Find (a) the perimeter and (b) the area of the rectangle.

Figure for 118

Figure for 119

Geometry In Exercises 120–123, what polynomial product is represented? Explain. See Example 4.

120.

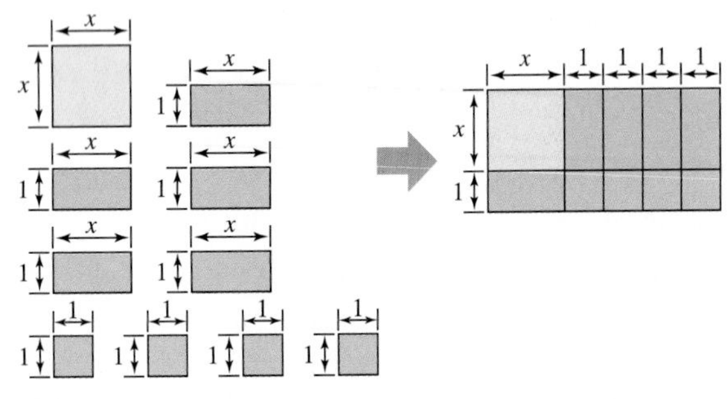

121.

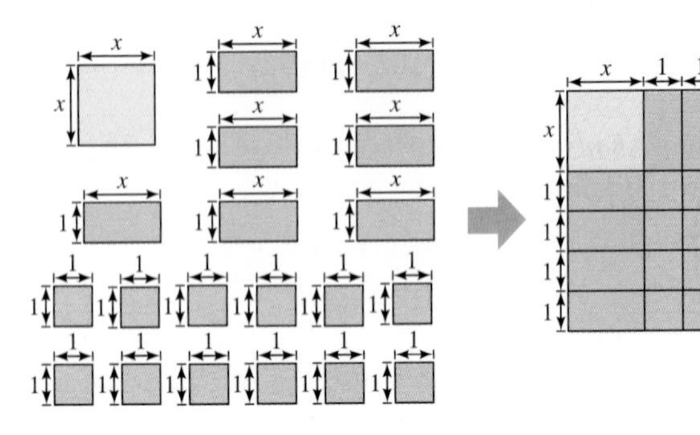

122.

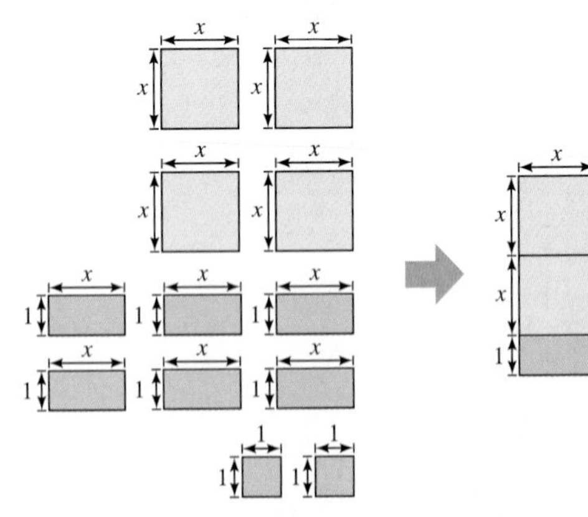

123.

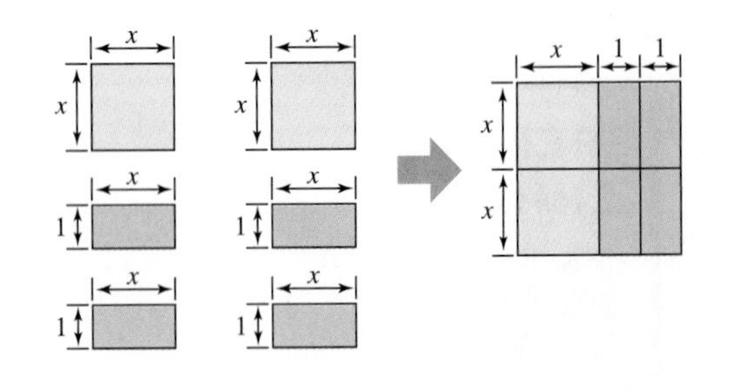

124. *Geometry* Add the areas of the four rectangular regions shown in the figure. What special product does the geometric model represent?

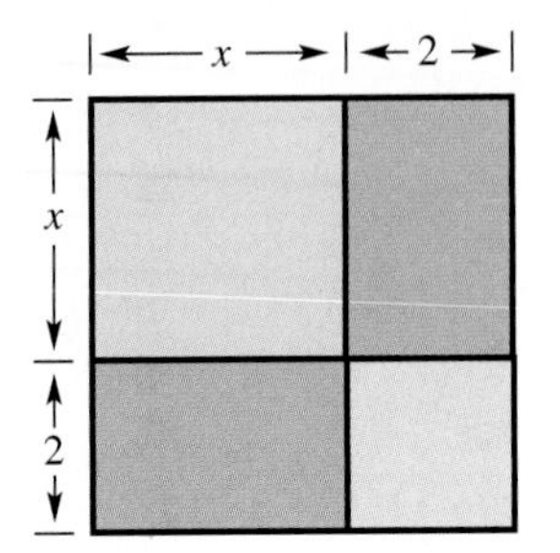

125. *Geometry* Add the areas of the four rectangular regions shown in the figure. Notice how this demonstrates the FOIL Method for finding the product $(x + a)(x + b)$.

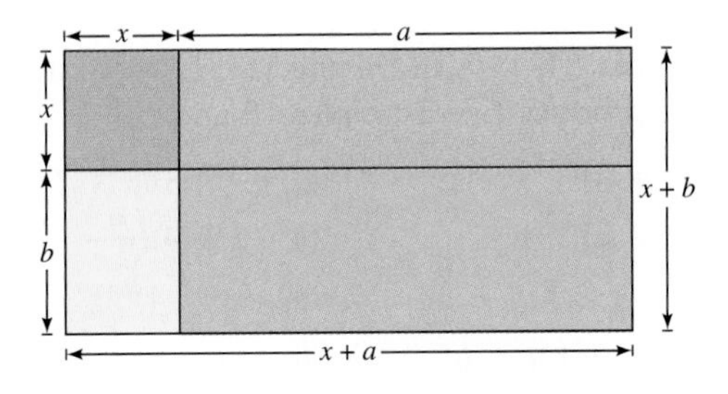

Geometry In Exercises 126 and 127, find a polynomial product that represents the area of the region. Then simplify the product.

126.

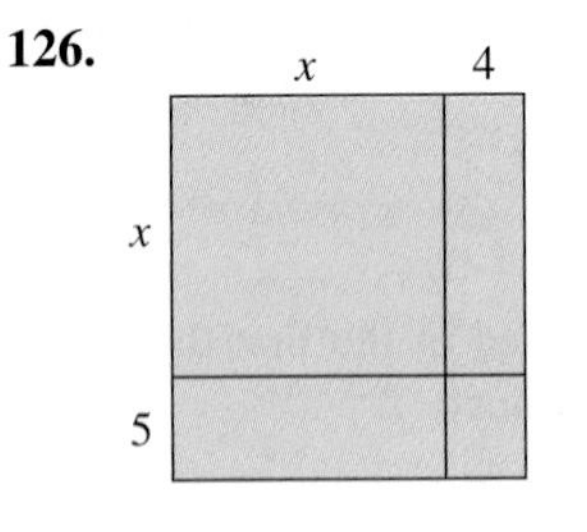

127.

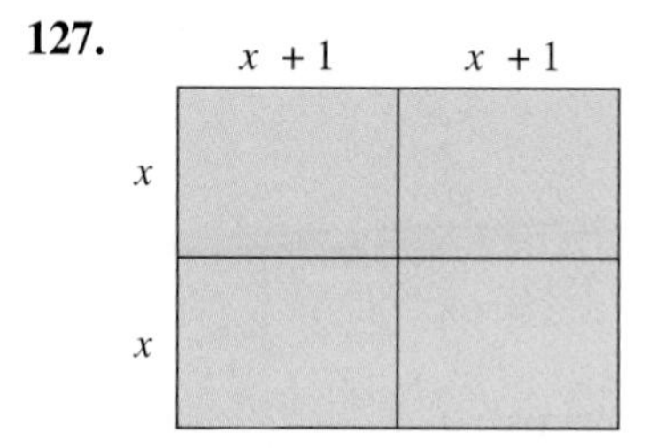

Geometry In Exercises 128 and 129, find two different expressions that represent the area of the shaded portion of the figure.

128. **129.**

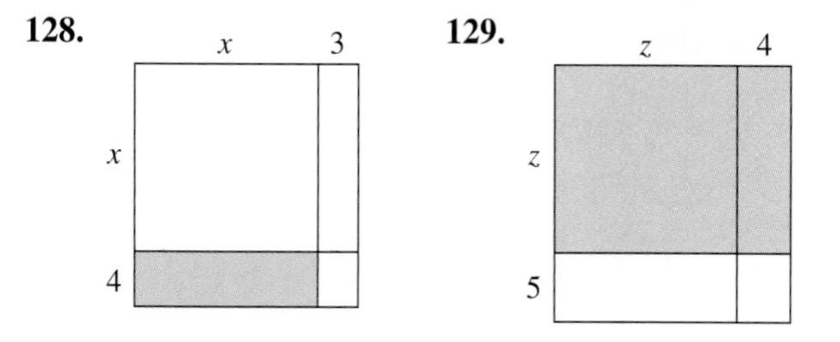

130. *Measurements of a Living Room* A contractor plans to enlarge a living room by enclosing the wrap-around porch and knocking out the walls between the porch and the living room (see figure). If the enlarged living room is 112 square feet larger than the original living room, what are the measurements of the enlarged living room?

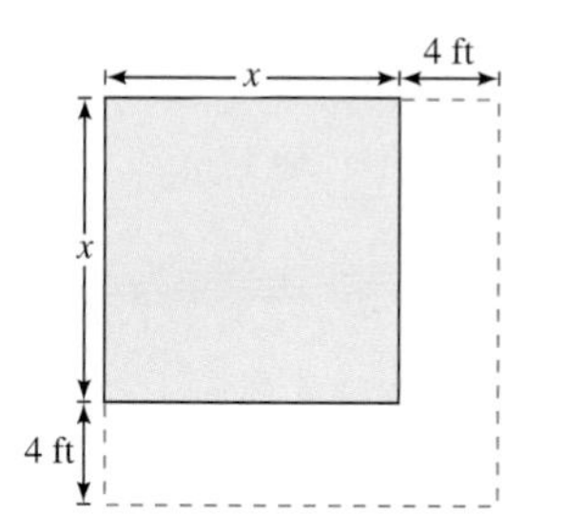

131. *Using Mathematical Models* For the years 1980 through 1996, each American's share S of the debt of the federal government is modeled by

$$S = 16.99t^2 + 767.47t + 3525.39, \quad 0 \leq t \leq 16$$

where $t = 0$ represents 1980. The population P (in millions) during the same period can be modeled by

$$P = 2.37t + 226.94, \quad 0 \leq t \leq 16.$$

(Source: U.S. Bureau of the Census)

(a) Use a graphing utility to graph the model of the per capita debt S.

(b) Multiply the polynomials representing the population P and the per capita debt S.

(c) Use the product in part (b) to estimate the total federal debt for 1990. (*Note*: The answer will be in millions of dollars.)

132. *Interpreting Graphs* When x units of a product are sold, the revenue R is given by $R = x(900 - 0.5x)$.

(a) Use a graphing utility to graph the expression.

(b) Multiply the factors in the expression for revenue and use a graphing utility to graph the product. Verify that the graph is the same as in part (a).

(c) Find the revenue if 500 units are sold. Use the graph to determine if revenue would increase or decrease if more units were sold.

133. *Compound Interest* After 2 years, a \$500 investment compounded annually at interest rate r, will yield an amount $500(1 + r)^2$. Find this product.

134. *Compound Interest* Repeat Exercise 133 if \$1200 is invested.

Explaining Concepts

135. Answer parts (d)–(f) of Motivating the Chapter on page 273.

136. Explain why an understanding of the Distributive Property is essential in multiplying polynomials. Illustrate your explanation with an example.

137. Describe the properties of exponents that are used to multiply polynomials. Give examples.

138. Discuss any differences between the expressions $(3x)^2$ and $3x^2$.

139. Explain the meaning of each letter of "FOIL" as it relates to multiplying two binomials.

140. What is the degree of the product of two polynomials of degrees m and n? Explain.

141. A polynomial with m terms is multiplied by a polynomial with n terms. How many *monomial-by-monomial* products must be found? Explain.

142. *True or False?* Because the product of two monomials is a monomial, it follows that the product of two binomials is a binomial.

143. *True or False?* $(x + 2)^2 = x^2 + 4$

Mid-Chapter Quiz

Take this quiz as you would take a quiz in class. After you are done, check your work against the answers given in the back of the book.

1. Explain why $x^2 + 2x - 3x^{-1}$ is not a polynomial.
2. Determine the degree and the leading coefficient of the polynomial $-3x^4 + 2x^2 - x$.
3. Give an example of a trinomial in one variable of degree 5.
4. *True or False?* The product of two binomials is a binomial. If false, give an example to show it is false.

In Exercises 5–14, perform the indicated operation and simplify.

5. $(y^2 + 3y - 1) + (4 + 3y)$
6. $(3v^2 - 5) - (v^3 + 2v^2 - 6v)$
7. $9s - [6 - (s - 5) + 7s]$
8. $-3(4 - x) + 4(x^2 + 2) - (x^2 - 2x)$
9. $2r^2(5r)$
10. $m^3(-2m)$
11. $(2y - 3)(y + 5)$
12. $(x + 4)(2x^2 - 3x - 2)$
13. $(4 - 3x)^2$
14. $(2u - 3)(2u + 3)$

In Exercises 15–18, perform the indicated operation using a vertical format.

15. $\begin{array}{rrrrr} & 5x^4 & + 2x^2 & + x & - 3 \\ + & 3x^3 & - 2x^2 & - 3x & + 5 \\ \hline \end{array}$

16. $\begin{array}{rrrr} 2x^3 + & x^2 & & - 8 \\ - & (5x^2 & - 3x & - 9) \\ \hline \end{array}$

17. $\begin{array}{rrrr} & 3x^2 + 7x & + & 1 \\ \times & 2x & - & 5 \\ \hline \end{array}$

18. $\begin{array}{rr} & 5x^3 - 6x^2 + 3 \\ \times & x^2 - 3x \\ \hline \end{array}$

19. Find the perimeter of the figure.

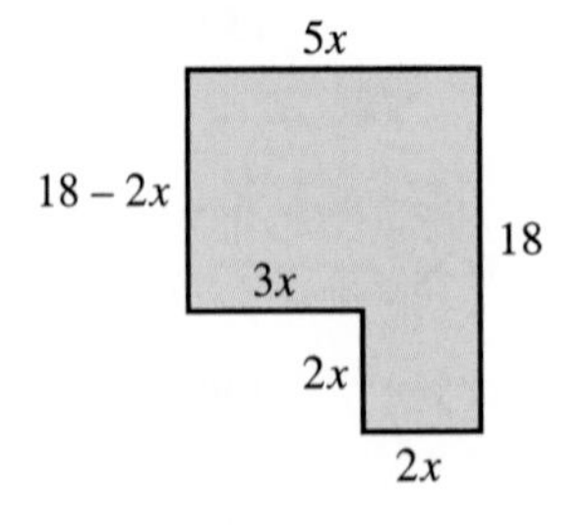

20. Find the area of the figure.

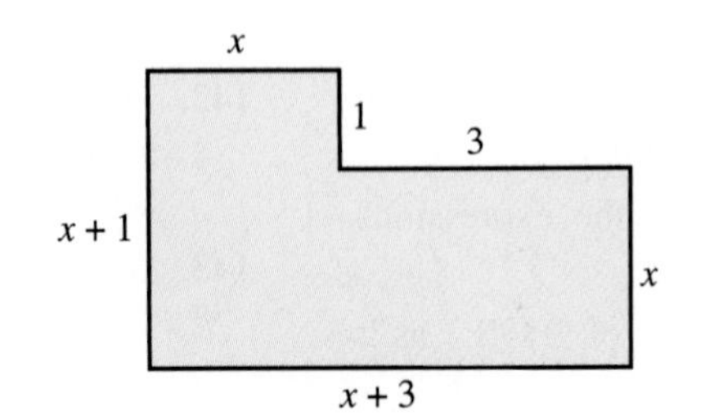

5.3 Negative Exponents and Scientific Notation

Objectives

1 Use the negative exponent rule to rewrite exponential expressions.

2 Use rules of exponents to rewrite expressions without negative exponents.

3 Write numbers in scientific notation.

1 Use the negative exponent rule to rewrite exponential expressions.

Negative Exponents

This section extends the properties of exponents to include **negative exponents.** Consider the property

$$a^m \cdot a^n = a^{m+n}, \quad a \neq 0.$$

If this property is to hold for negative exponents, then the statement

$$a^2 \cdot a^{-2} = a^{2+(-2)} = a^0 = 1$$

implies that a^{-2} is the *reciprocal* of a^2. In other words, it must be true that

$$a^{-2} = \frac{1}{a^2}.$$

Informally, you can think of this property as allowing you to "move" powers from the numerator to the denominator (or vice versa) by changing the sign of the exponent.

Negative Exponent Rule

Let n be an integer and let a be a real number, variable, or algebraic expression such that $a \neq 0$.

$$a^{-n} = \frac{1}{a^n}$$

Study Tip

Be sure you see that the negative exponent rule allows you to move only *factors* in a numerator (or denominator), *not terms.* For example,

$$\frac{x^{-2} \cdot y}{4} = \frac{y}{4x^2}$$

whereas the following is not true.

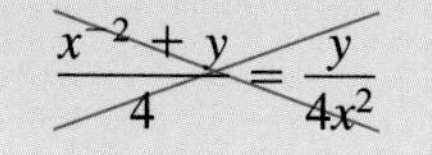

Example 1 Monomials Involving Negative Exponents

a. $6^{-2} = \frac{1}{6^2} = \frac{1}{36}$ Move 6^{-2} to denominator and change the sign of the exponent.

b. $x^{-7} = \frac{1}{x^7}$ Move x^{-7} to denominator and change the sign of the exponent.

c. $5x^{-4} = \frac{5}{x^4}$ Move x^{-4} to denominator and change the sign of the exponent.

d. $\frac{1}{2x^{-3}} = \frac{x^3}{2}$ Move x^{-3} to numerator and change the sign of the exponent.

e. $x^{-2}y^3 = \frac{y^3}{x^2}$ Move x^{-2} to denominator and change the sign of the exponent.

2 Use rules of exponents to rewrite expressions without negative exponents.

Rules of Exponents

All of the rules of exponents apply to negative exponents. For convenience, these rules are summarized below. Remember that these rules apply to real numbers, variables, or algebraic expressions.

Technology: Discovery

Use a calculator to evaluate the expressions below.

$$\frac{3.4^{5.8}}{3.4^{1.6}} \quad \text{and} \quad 3.4^{4.2}$$

How are these two expressions related? Use your calculator to verify other rules of exponents.

▶ Rules of Exponents

Let m and n be integers, and let a and b be real numbers, variables, or algebraic expressions such that $a \neq 0$ and $b \neq 0$.

Property	*Example*
1. $a^m a^n = a^{m+n}$	$y^2 \cdot y^4 = y^{2+4} = y^6$
2. $\dfrac{a^m}{a^n} = a^{m-n}$	$\dfrac{x^7}{x^4} = x^{7-4} = x^3$
3. $(ab)^m = a^m b^m$	$(5x)^4 = 5^4 x^4$
4. $\left(\dfrac{a}{b}\right)^m = \dfrac{a^m}{b^m}$	$\left(\dfrac{2}{x}\right)^3 = \dfrac{2^3}{x^3}$
5. $(a^m)^n = a^{mn}$	$(y^3)^{-4} = y^{3(-4)} = y^{-12}$
6. $a^{-n} = \dfrac{1}{a^n}$	$y^{-4} = \dfrac{1}{y^4}$
7. $a^0 = 1$	$(x^2 + 1)^0 = 1$

Study Tip

There is more than one way to solve problems such as those in Example 2. For instance, you might prefer to write Example 2(c) as

$$\frac{y^{-2}}{3y^{-5}} = \frac{y^5}{3y^2} = \frac{y^{5-2}}{3} = \frac{y^3}{3}.$$

Example 2 Using Rules of Exponents

a. $x^3(2x^{-4}) = 2(x^3)(x^{-4})$ Regroup factors.

$= 2x^{3+(-4)}$ Apply rules of exponents.

$= 2x^{-1}$ Simplify.

$= \dfrac{2}{x}$ Simplify.

b. $(-3ab^4)(4ab^{-3}) = (-3)(4)(a)(a)(b^4)(b^{-3})$ Regroup factors.

$= (-12)(a^{1+1})(b^{4-3})$ Apply rules of exponents.

$= -12a^2b$ Simplify.

c. $\dfrac{y^{-2}}{3y^{-5}} = \dfrac{1}{3}y^{-2-(-5)}$ Apply rules of exponents.

$= \dfrac{1}{3}y^3$ Simplify.

$= \dfrac{y^3}{3}$ Simplify.

Example 3 Using Rules of Exponents

Use rules of exponents to rewrite each expression without negative exponents. (Assume that no variable is equal to zero.)

a. $3x^{-1}(-4x^2y)^0$ **b.** $\left(\dfrac{5x^3}{y^{-1}}\right)^2$ **c.** $\left(\dfrac{a^2}{3}\right)^{-2}$ **d.** $\left(\dfrac{x^{-2}y^3}{2}\right)^{-3}$

Solution

a. This problem is a little tricky. Note that the factor $(-4x^2y)$ is raised to the *zero* power. Because any nonzero number raised to the zero power is 1, you can write

$$3x^{-1}(-4x^2y)^0 = 3x^{-1}(1) = \frac{3}{x}.$$

b. This problem can also be tricky. The important thing to realize is that the *entire fraction* $(5x^3/y^{-1})$ is raised to the second power. This means that you must apply the exponent 2 to each factor of the numerator and denominator, as follows.

$$\left(\frac{5x^3}{y^{-1}}\right)^2 = \frac{(5x^3)^2}{(y^{-1})^2} = \frac{5^2(x^3)^2}{y^{-2}} = \frac{25x^6}{y^{-2}} = 25x^6y^2$$

c. In this problem, note that the *entire expression* $(a^2/3)$ is raised to the -2 power.

$$\left(\frac{a^2}{3}\right)^{-2} = \frac{a^{-4}}{3^{-2}} = \frac{3^2}{a^4} = \frac{9}{a^4}$$

d. In this problem, note that the *entire expression* $(x^{-2}y^3/2)$ is raised to the -3 power.

$$\left(\frac{x^{-2}y^3}{2}\right)^{-3} = \frac{x^6y^{-9}}{2^{-3}}$$ Property of exponents

$$= \frac{2^3x^6}{y^9} = \frac{8x^6}{y^9}$$ Convert to positive exponent.

Technology: Tip Calculators and Negative Exponents

The keystrokes used to evaluate expressions with negative exponents vary. For instance, to evaluate 13^{-2} on a calculator, you can try one of the following keystroke sequences.

Keystrokes

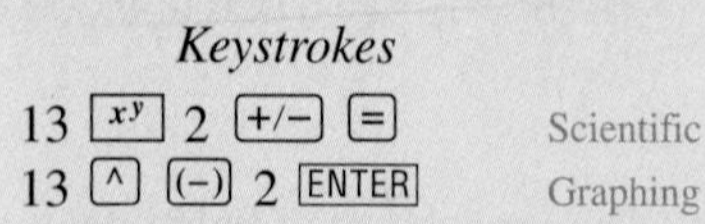

Scientific
Graphing

With either of these sequences, your calculator should display .00591716. If it doesn't, consult the user's guide for your calculator to find the correct keystrokes.

3 Write numbers in scientific notation.

Scientific Notation

Exponents provide an efficient way of writing and computing with the very large (or very small) numbers used in science. For instance, a drop of water contains more than 33 billion billion molecules. That is 33 followed by 18 zeros. It is convenient to write such numbers in **scientific notation.** This notation has the form $c \times 10^n$, where $1 \leq c < 10$ and n is an integer. So, the number of molecules in a drop of water can be written in scientific notation as follows.

33,000,000,000,000,000,000 = 3.3 × 10^19
19 places

The *positive* exponent 19 indicates that the number is large (10 or more) and that the decimal point has been moved 19 places.

A *negative* exponent in scientific notation indicates that the number is *small* (less than 1). For instance, the mass (in grams) of one electron is approximately as follows.

$$9.0 \times 10^{-28} = 0.0000000000000000000000000009$$

28 places

Example 4 Converting from Decimal to Scientific Notation

Write the decimal number in scientific notation.

a. 1,260,000 **b.** 0.0000782 **c.** 836,100,000.0

Solution

a. $1{,}260{,}000. = 1.26 \times 10^6$ Large number yields positive exponent.

Six places

b. $0.0000782 = 7.82 \times 10^{-5}$ Small number yields negative exponent.

Five places

c. $836{,}100{,}000.0 = 8.361 \times 10^8$ Large number yields positive exponent.

Eight places

Example 5 Converting from Scientific to Decimal Notation

The storage capacity of a 6-gigabyte computer is

$$6 \times 10^9 = 6{,}000{,}000{,}000 \text{ bytes.}$$ Positive exponent yields large number.

Nine places

The probability of being dealt a royal flush in poker is

$$1.54 \times 10^{-6} = 0.00000154.$$ Negative exponent yields small number.

Six places

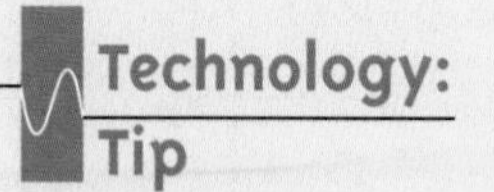

Calculators and Scientific Notation

Most scientific calculators automatically switch to scientific notation when they are displaying large (or small) numbers that exceed the display range. Try multiplying 98,900,000 × 5000. If your calculator follows standard conventions, its display should show

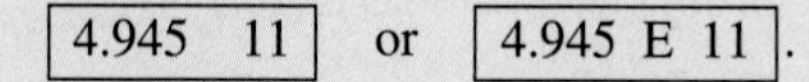

This means that $c = 4.945$ and the exponent of 10 is $n = 11$, which implies that the number is 4.945×10^{11}. For *entering* numbers in scientific notation, your calculator should have an exponential entry key labeled EE or EXP.

Example 6 Using Scientific Notation

Use a calculator to evaluate 78,000 × 2,400,000,000.

Solution

Because $78{,}000 = 7.8 \times 10^4$ and $2{,}400{,}000{,}000 = 2.4 \times 10^9$, you can evaluate the product as follows.

7.8 EXP 4 × 2.4 EXP 9 = Scientific

7.8 EE 4 × 2.4 EE 9 ENTER Graphing

After these keystrokes have been entered, the calculator display should show 1.872 14. So, the product of the two numbers is

$$(7.8 \times 10^4)(2.4 \times 10^9) = 1.872 \times 10^{14} = 187{,}200{,}000{,}000{,}000.$$

Use a calculator to evaluate $(7.8)(2.4) \times (10^4)(10^9)$. Your answer should be the same as in Example 6. This illustrates the commutative property of multiplication.

Discussing the Concept Exponential Expressions

Find as many equivalent pairs as possible among the following exponential expressions.

$$\frac{2}{x^{-3}},\ \frac{1}{2x^3},\ 2x^{-3},\ \frac{1}{(2x)^{-3}},\ 2x^3,\ \frac{x^3}{8},\ \frac{1}{8x^3},\ \frac{x^{-3}}{2},\ 8x^3,\ (2x)^{-3}$$

Next, use a calculator, with $x = 3$, to illustrate the equivalence of each pair. Organize your work into a table with four columns—an expression, the equivalent expression, and each expression evaluated at $x = 3$. Compare your results with those of other students in your class.

5.3 Exercises

Integrated Review *Concepts, Skills, and Problem Solving*

Keep mathematically in shape by doing these exercises *before* the problems of this section.

Properties and Definitions

1. In your own words, define the graph of the function $y = f(x)$.
2. Describe the point-plotting method of sketching a graph for $y = f(x)$.
3. Find the coordinates of two points on the graph of $g(x) = \sqrt{x}$.
4. Describe the procedure for finding the x- and y-intercepts of the graph of $f(x) = 3(x - 2)$.

Simplifying Expressions

In Exercises 5–8, simplify the expression. (Assume that no denominator is zero.)

5. $x^2 \cdot x^3$
6. $(y^2z^3)(z^2)^4$
7. $\left(\frac{x^2}{y}\right)^3$
8. $\frac{a^2b^3}{c} \cdot \frac{2a}{3}$

Graphing Equations

In Exercises 9–12, use a graphing utility to graph the function. Identify any intercepts.

9. $f(x) = 4 - 3x$
10. $g(x) = |2x + 1|$
11. $g(x) = x^2 - 2x + 1$
12. $h(x) = \sqrt{x + 4}$

Developing Skills

In Exercises 1–12, rewrite with positive exponents. See Example 1.

1. 3^{-3}
2. 4^{-2}
3. y^{-5}
4. z^{-2}
5. $8x^{-7}$
6. $6x^{-2}y^{-3}$
7. $7x^{-4}y^{-1}$
8. $9u^{-5}v^{-2}$
9. $\frac{1}{2z^{-4}}$
10. $\frac{7x^2}{y^{-3}}$
11. $\frac{2x}{3y^{-2}}$
12. $\frac{5u^2}{6v^{-4}}$

In Exercises 13–22, rewrite with negative exponents.

13. $\frac{1}{4}$
14. $\frac{1}{3^2}$
15. $\frac{1}{x^2}$
16. $\frac{7}{y^3}$
17. $\frac{10}{t^5}$
18. $\frac{3}{z^n}$
19. $\frac{5}{x^n}$
20. $\frac{9}{y^n}$
21. $\frac{2x^2}{y^4}$
22. $\frac{5x^3}{y^6}$

In Exercises 23–36, rewrite with positive exponents. Then evaluate the expression.

23. 3^{-2}
24. 5^{-3}
25. $(-4)^{-3}$
26. $(-6)^{-2}$
27. $\frac{1}{4^{-2}}$
28. $\frac{1}{16^{-1}}$
29. $\frac{2}{3^{-4}}$
30. $\frac{4}{3^{-2}}$
31. $\frac{2^{-4}}{3^{-2}}$
32. $\frac{4^{-3}}{2}$
33. $\frac{4^{-2}}{3^{-4}}$
34. $\left(\frac{3}{4}\right)^{-3}$
35. $\left(\frac{2}{3}\right)^{-2}$
36. $\left(\frac{5}{4}\right)^{-3}$

In Exercises 37–40, use a calculator to evaluate the expression.

37. 3.8^{-4}
38. 6.2^{-3}
39. $100(1.06)^{-15}$
40. $500(1.08)^{-20}$

In Exercises 41–82, simplify using rules of exponents. Write your answer with positive exponents. (Assume that no variable is zero.) See Examples 2 and 3.

41. $4^{-2} \cdot 4^3$ **42.** $5^{-3} \cdot 5^2$

43. $x^{-4} \cdot x^6$ **44.** $a^{-5} \cdot a^2$

45. $u^{-6} \cdot u^3$ **46.** $t^{-2} \cdot t^2$

47. $xy^{-3} \cdot y^2$ **48.** $u^{-2}v \cdot u^2$

49. $\dfrac{x^2}{x^{-3}}$ **50.** $\dfrac{z^4}{z^{-2}}$

51. $\dfrac{y^{-5}}{y}$ **52.** $\dfrac{x^{-3}}{x^2}$

53. $\dfrac{x^{-4}}{x^{-2}}$ **54.** $\dfrac{t^{-5}}{t^{-1}}$

55. $(y^{-3})^2$ **56.** $(z^{-2})^3$

57. $(s^2)^{-1}$ **58.** $(a^3)^{-3}$

59. $(2x^{-2})^0$ **60.** $(2x^{-5})^0$

61. $\dfrac{b^2 \cdot b^{-3}}{b^4}$ **62.** $\dfrac{c^{-3} \cdot c^4}{c^{-1}}$

63. $(3x^2y)^{-2}$ **64.** $(4x^{-3}y^2)^{-3}$

65. $(4a^{-2}b^3)^{-3}$ **66.** $(-2s^{-1}t^{-2})^{-1}$

67. $(-2x^2)(4x^{-3})$ **68.** $(4y^{-2})(3y^4)$

69. $\left(\dfrac{x}{10}\right)^{-1}$ **70.** $\left(\dfrac{4}{z}\right)^{-2}$

71. $\left(\dfrac{3z^2}{x}\right)^{-2}$ **72.** $\left(\dfrac{x^{-3}y^4}{5}\right)^{-3}$

73. $\dfrac{(2y)^{-4}}{(2y)^{-4}}$ **74.** $\dfrac{(3z)^{-2}}{(3z)^{-2}}$

75. $\dfrac{3}{2} \cdot \left(\dfrac{-2}{3}\right)^{-3}$ **76.** $\dfrac{3}{8} \cdot \left(\dfrac{-5}{2}\right)^{-3}$

77. $\dfrac{(-2x)^{-3}}{-4x^{-2}}$ **78.** $\dfrac{2x^{-3}}{(5x)^{-1}}$

79. $(5x^2y^4z^6)^3(5x^2y^4z^6)^{-3}$ **80.** $(8x^3y^2z^5)^6(8x^3y^2z^5)^{-6}$

81. $(x + y)^{-8}(x + y)^8$ **82.** $(u^2 - v)^4(u^2 - v)^{-4}$

In Exercises 83–92, write in scientific notation. See Example 4.

83. 93,000,000 **84.** 900,000,000

85. 1,637,000,000 **86.** 67.8

87. 0.000435 **88.** 0.008367

89. 0.004392 **90.** 0.00000045

91. 16,000,000 **92.** 0.00082

In Exercises 93–102, write in decimal form. See Example 5.

93. 1.09×10^6 **94.** 2.345×10^8

95. 8.67×10^{-2} **96.** 9.4675×10^4

97. 8.52×10^{-3} **98.** 7.021×10^{-5}

99. 6.21×10^0 **100.** 4.73×10^0

101. $(8 \times 10^3) + (3 \times 10^0) + (5 \times 10^{-2})$

102. $(6 \times 10^4) + (9 \times 10^3) + (4 \times 10^{-1})$

In Exercises 103–114, use a calculator to evaluate the expression. See Example 6.

103. $8{,}000{,}000 \times 623{,}000$

104. $93{,}200{,}000 \times 1{,}657{,}000$

105. $0.000345 \times 8{,}980{,}000{,}000$

106. $345{,}000 \times 0.000086$

107. $3{,}200{,}000^5$ **108.** $75{,}000{,}000^6$

109. $(3.28 \times 10^{-6})^4$ **110.** $(4.5 \times 10^{-5})^3$

111. $\dfrac{848{,}000{,}000}{1{,}620{,}000}$ **112.** $\dfrac{67{,}000{,}000}{0.0052}$

113. $(4.85 \times 10^5)(2.04 \times 10^8)$

114. $\dfrac{8.6 \times 10^4}{3.9 \times 10^7}$

Solving Problems

115. *Time for Light to Travel* Light travels from the sun to the earth in approximately

$$\frac{9.3 \times 10^7}{1.1 \times 10^7}$$

minutes. Write this time in decimal form.

116. *Distance to a Star* The star Beta Andromeda is approximately 76 light-years from the earth (see figure). (A light-year is the distance light can travel in 1 year.) Estimate the distance to this star if a light-year is approximately 5.8746×10^{12} miles.

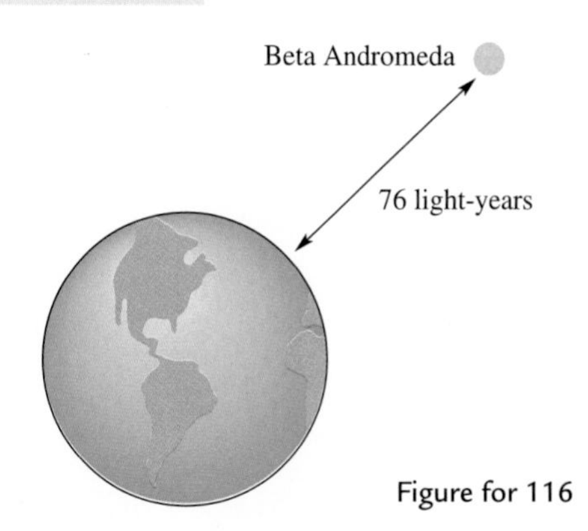

Figure for 116

117. *Solar System* One astronomical unit AU is the mean distance between the sun and the earth (approximately 149,503,000 kilometers). The table gives the mean distances between selected planets and the sun in astronomical units. Approximate each distance in kilometers and give the answer in scientific notation.

Planet	Mercury	Saturn	Neptune	Pluto
AU	0.39	9.56	30.13	39.47

118. *Numerical and Graphical Analysis* A new car is purchased for $24,000. Its value V after t years is

$V = 24{,}000(1.2)^{-t}$.

(a) Use the model to complete the table.

t	0	2	4	6	8
$24{,}000(1.2)^{-t}$					

(b) Graph the data in the table.

(c) *Guess, Check, and Revise* When will the car be valued at less than $1000?

119. *Numerical and Graphical Analysis*

(a) Complete the table by evaluating the indicated powers of 2.

x	-1	-2	-3	-4	-5
2^x					

(b) Graph the data in the table.

(c) Use the table or the graph to describe the value of 2^{-n} when n is very large. Will the value of 2^{-n} ever be negative?

120. *Hydraulic Compression* A hydraulic cylinder in a large press (see figure) contains 2 gallons of oil. When the cylinder is under full pressure, the actual volume of oil is decreased by

$2(150)(20 \times 10^{-6})$

gallons. Write this volume in decimal form.

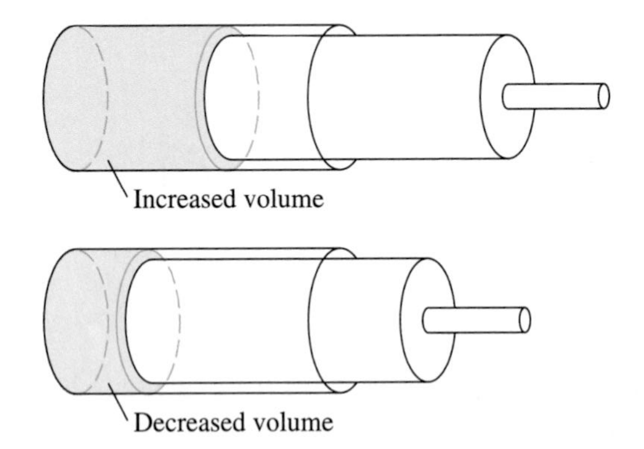

121. *Boltzmann's Constant* The study of the kinetic energy of an ideal gas uses Boltzmann's constant. This constant k is given by

$$k = \frac{8.31 \times 10^7}{6.02 \times 10^{23}}.$$

Perform the division, leaving your result in scientific notation.

Explaining Concepts

True or False? In Exercises 122–127, state whether the equation is true or false. If it is false, find values of x and y that show it to be false.

122. $x^3y^3 = xy^3$

123. $x^{-1}y^{-1} = \dfrac{1}{xy}$

124. $x^{-1} + y^{-1} = \dfrac{1}{x + y}$

125. $\dfrac{x^{-4}}{x^{-3}} = x$

126. $(x \times 10^3)^4 = x^4 \times 10^{12}$

127. $\dfrac{2x \times 10^{-5}}{x \times 10^{-3}} = 2 \times 10^{-2}$

128. Without looking back at page 298, state as many of the seven rules of exponents as you can.

129. Give examples of large and small numbers written in scientific notation.

130. Find the reciprocal of 4×10^{-3}.

131. Justify each step.

$$\begin{aligned}(3 \times 10^5)(4 \times 10^6) &= (3 \times 10^5)(10^6 \times 4)\\ &= 3(10^5 \times 10^6)(4)\\ &= 3(10^{5+6})(4)\\ &= (3 \cdot 4)10^{11}\\ &= 12 \times 10^{11} = 1.2 \times 10^{12}\end{aligned}$$

5.4 Dividing Polynomials

Objectives

1. Use the properties of exponents to divide a monomial expression by a monomial expression.
2. Divide a polynomial expression by a monomial expression.
3. Divide a polynomial expression by a binomial expression.
4. Use synthetic division to divide a polynomial expression by a binomial expression.

1 Use the properties of exponents to divide a monomial expression by a monomial expression.

Dividing a Monomial by a Monomial

In this section, you will learn how to divide a polynomial by a monomial or a binomial. In all cases, assume that the divisor (denominator) is nonzero.

To begin, let's consider division problems in which both the numerator *and* the denominator are monomials. To divide a monomial by a monomial, you make use of the *subtraction* property of exponents illustrated in the following examples. (In each of the following, assume that the variable is *not zero*.)

By Reducing	*By Subtracting Exponents*
$\dfrac{x^4}{x^2} = \dfrac{x \cdot x \cdot \not{x} \cdot \not{x}}{\not{x} \cdot \not{x}} = x^2$	$\dfrac{x^4}{x^2} = x^{4-2} = x^2$
$\dfrac{y^3}{y^3} = \dfrac{\not{y} \cdot \not{y} \cdot \not{y}}{\not{y} \cdot \not{y} \cdot \not{y}} = 1$	$\dfrac{y^3}{y^3} = y^{3-3} = y^0 = 1$
$\dfrac{5y^7}{2y^5} = \dfrac{5 \cdot y \cdot y \cdot \not{y} \cdot \not{y} \cdot \not{y} \cdot \not{y} \cdot \not{y}}{2 \cdot \not{y} \cdot \not{y} \cdot \not{y} \cdot \not{y} \cdot \not{y}} = \dfrac{5y^2}{2}$	$\dfrac{5y^7}{2y^5} = \dfrac{5y^{7-5}}{2} = \dfrac{5y^2}{2}$
$\dfrac{2x^2}{x^5} = \dfrac{2 \cdot \not{x} \cdot \not{x}}{x \cdot x \cdot x \cdot \not{x} \cdot \not{x}} = \dfrac{2}{x^3}$	$\dfrac{2x^2}{x^5} = 2x^{2-5} = 2x^{-3} = \dfrac{2}{x^3}$

The examples above show that you can divide one monomial by another by subtracting exponents.

> **Properties of Exponents**
>
> Let m and n be positive integers and let a represent a real number, a variable, or an algebraic expression.
>
> **1.** $\dfrac{a^m}{a^n} = a^{m-n}$
>
> **2.** $\dfrac{a^n}{a^n} = 1 = a^0$
>
> Note the special definition for raising a *nonzero* quantity to the zero power. That is, if $a \neq 0$, then $a^0 = 1$.

Study Tip

If you remember the first rule at the right, you can use it to derive the second rule. That is,

$$1 = \frac{a^n}{a^n} = a^{n-n} = a^0,$$

$a \neq 0.$

Example 1 Dividing a Monomial by a Monomial

Perform each division. (In each case, assume that $x \neq 0$ and $y \neq 0$.)

a. $2y^8 \div y^5$

b. $16x^4 \div 4x^2$

c. $16x^4 \div 3x$

d. $32y^3 \div 8y^5$

e. $8x^3 \div \frac{1}{2}x^3$

f. $12x^3 \div 4x^4$

g. $3x^3y \div 4x$

Solution

a. $\dfrac{2y^8}{y^5} = 2 \cdot y^{8-5} = 2y^3$

b. $\dfrac{16x^4}{4x^2} = \dfrac{16}{4} \cdot \dfrac{x^4}{x^2} = \dfrac{16}{4} \cdot x^{4-2} = 4x^2$

c. $\dfrac{16x^4}{3x} = \dfrac{16}{3} \cdot \dfrac{x^4}{x} = \dfrac{16}{3} \cdot x^{4-1} = \dfrac{16}{3}x^3$

d. $\dfrac{32y^3}{8y^5} = \dfrac{32}{8} \cdot \dfrac{y^3}{y^5} = \dfrac{32}{8} \cdot y^{3-5} = 4 \cdot y^{-2} = \dfrac{4}{y^2}$

e. $\dfrac{8x^3}{\frac{1}{2}x^3} = \dfrac{8}{\frac{1}{2}} \cdot \dfrac{x^3}{x^3} = 8\left(\dfrac{2}{1}\right)(1) = 16$

f. $\dfrac{12x^3}{4x^4} = \dfrac{12}{4} \cdot \dfrac{x^3}{x^4} = 3 \cdot x^{3-4} = 3 \cdot x^{-1} = \dfrac{3}{x}$

g. $\dfrac{3x^3y}{4x} = \dfrac{3}{4} \cdot x^{3-1} \cdot y = \dfrac{3}{4} \cdot x^2 \cdot y = \dfrac{3x^2y}{4}$

> **Study Tip**
>
> The subtraction property of exponents works only for division of monomials with the *same variable* for a base. For instance, the subtraction property does not apply to x^5/y^3 because no cancellations can occur. So, for the fraction
>
> $$\frac{x^5}{y^3} = \frac{x \cdot x \cdot x \cdot x \cdot x}{y \cdot y \cdot y}$$
>
> no simplifying is possible.

Although the division problems in Example 1 are straightforward, you should study them carefully. Be sure you can justify each step. Also remember that there are often several ways to solve a given problem in algebra. As you gain practice and confidence, you will discover that you like some techniques better than others. For instance, which one of the following techniques seems best to you?

1. $\dfrac{6x^3}{2x} = \dfrac{3 \cdot \cancel{2} \cdot x \cdot x \cdot \cancel{x}}{\cancel{2} \cdot \cancel{x}} = 3x^2, \quad x \neq 0$

2. $\dfrac{6x^3}{2x} = \left(\dfrac{6}{2}\right)\left(\dfrac{x^3}{x}\right) = (3)(x^{3-1}) = 3x^2, \quad x \neq 0$

3. $\dfrac{6x^3}{2x} = \dfrac{3x^3}{x} = 3x^{3-1} = 3x^2, \quad x \neq 0$

When two different people (even math instructors) are writing out the steps of a solution, rarely will the steps be the same, so don't worry if your steps don't look exactly like someone else's. If you feel comfortable with writing more steps, then you should write more steps. Just be sure that each step can be justified by the rules of algebra.

2 Divide a polynomial expression by a monomial expression.

Dividing a Polynomial by a Monomial

The preceding examples show how to divide a *monomial* by a monomial. To divide a *polynomial* by a monomial, use the reverse form of the rule for adding two fractions with a common denominator. In Section 1.3, you added two fractions with like denominators using the rule

$$\frac{a}{c} + \frac{b}{c} = \frac{a+b}{c}.$$ Adding fractions

Here you can use the rule in the *reverse* order and divide a polynomial by a monomial by dividing each term of the polynomial by the monomial. That is,

$$\frac{a+b}{c} = \frac{a}{c} + \frac{b}{c}.$$ Dividing by a monomial

Here is an example.

$$\frac{x^3 - 5x^2}{x^2} = \frac{x^3}{x^2} - \frac{5x^2}{x^2} = x - 5, \quad x \neq 0$$

The essence of this problem is to separate the original division problem into *two* division problems, each involving the division of a monomial by a monomial.

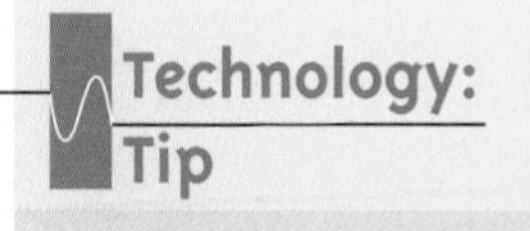

Technology: Tip

As with other types of operations with polynomials, you can use a graphing utility to help check division problems. For instance, graph

$$y = \frac{6x-5}{3} \text{ and } y = 2x - \frac{5}{3}$$

on the same screen, as shown below. Because both graphs are the same, you can reason that

$$\frac{6x-5}{3} = 2x - \frac{5}{3}.$$

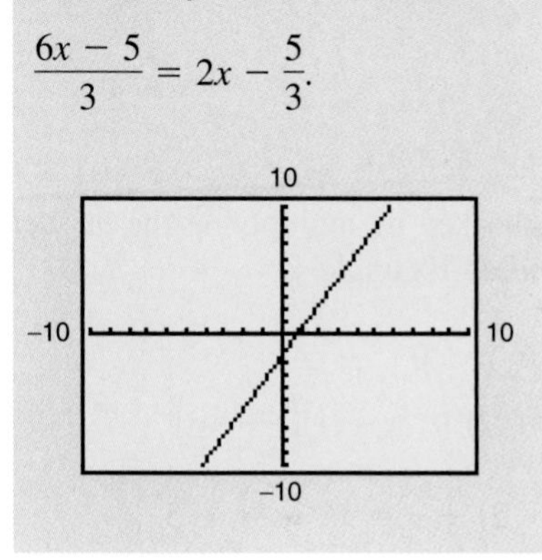

Example 2 Dividing a Polynomial by a Monomial

a. $\dfrac{6x+5}{3} = \dfrac{6x}{3} + \dfrac{5}{3} = 2x + \dfrac{5}{3}$

b. $\dfrac{4x^2 - 3x}{3x} = \dfrac{4x^2}{3x} - \dfrac{3x}{3x} = \dfrac{4x}{3} - 1, \quad x \neq 0$

c. $\dfrac{8x^3 - 6x^2 + 10x}{2x} = \dfrac{8x^3}{2x} - \dfrac{6x^2}{2x} + \dfrac{10x}{2x} = 4x^2 - 3x + 5, \quad x \neq 0$

Example 3 Dividing a Polynomial by a Monomial

Perform the division. (Assume $x \neq 0$.)

a. $(5x^3 - 4x^2 - x + 6) \div 2x$

b. $(8x^4 + 6x^3 + 3x^2 - 2x) \div 3x^2$

Solution

a. $\dfrac{5x^3 - 4x^2 - x + 6}{2x} = \dfrac{5x^3}{2x} - \dfrac{4x^2}{2x} - \dfrac{x}{2x} + \dfrac{6}{2x}$ Divide each term separately.

$= \dfrac{5x^2}{2} - 2x - \dfrac{1}{2} + \dfrac{3}{x}$ Use properties for dividing monomials.

b. $\dfrac{8x^4 + 6x^3 + 3x^2 - 2x}{3x^2} = \dfrac{8x^4}{3x^2} + \dfrac{6x^3}{3x^2} + \dfrac{3x^2}{3x^2} - \dfrac{2x}{3x^2}$ Divide each term separately.

$= \dfrac{8x^2}{3} + 2x + 1 - \dfrac{2}{3x}$ Use properties for dividing monomials.

3 Divide a polynomial expression by a binomial expression.

Dividing a Polynomial by a Binomial

To divide a polynomial by a *binomial*, follow the *long division* pattern used for dividing whole numbers. Recall that you divide 6982 by 27 as follows.

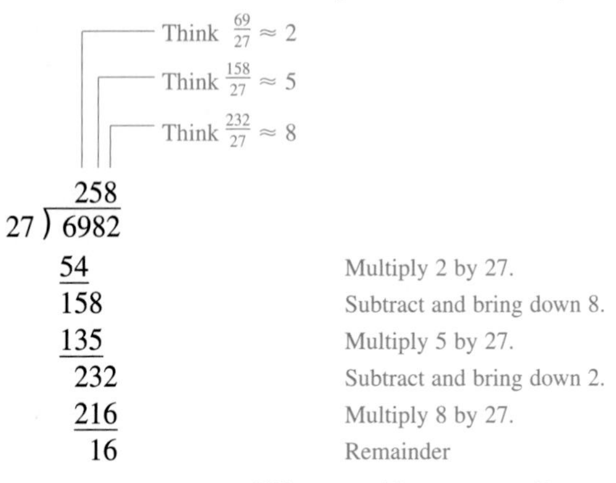

54	Multiply 2 by 27.
158	Subtract and bring down 8.
135	Multiply 5 by 27.
232	Subtract and bring down 2.
216	Multiply 8 by 27.
16	Remainder

You can express the result as $\frac{6982}{27} = 258\frac{16}{27}$ or $258 + \frac{16}{27}$.

Example 4 Dividing a Polynomial by a Binomial

Divide $(x^2 + 3x + 5)$ by $(x + 1)$.

Solution

Think $\frac{x^2}{x} = x$

Think $\frac{2x}{x} = 2$

$$\begin{array}{r|l} x + 2 & \\ x + 1 \,\overline{)\, x^2 + 3x + 5} & \\ \underline{x^2 + x } & \text{Multiply } x(x + 1). \\ 2x + 5 & \text{Subtract and bring down 5.} \\ \underline{2x + 2} & \text{Multiply } 2(x + 1). \\ 3 & \text{Remainder.} \end{array}$$

Considering the remainder as a fractional part of the divisor, you can write

$$\overbrace{\frac{x^2 + 3x + 5}{\underbrace{x + 1}_{\text{Divisor}}}}^{\text{Dividend}} = \overbrace{x + 2}^{\text{Quotient}} + \frac{\overset{\text{Remainder}}{3}}{\underbrace{x + 1}_{\text{Divisor}}}.$$

Remember that a division problem can be checked by multiplying the *quotient* (answer) by the *divisor* to obtain the *dividend*. In Example 4,

$$\overbrace{(x + 1)}^{\text{Divisor}}\overbrace{\left(x + 2 + \frac{3}{x + 1}\right)}^{\text{Quotient}} = (x + 1)(x + 2) + (x + 1)\left(\frac{3}{x + 1}\right)$$

$$= (x^2 + 3x + 2) + 3 = \underbrace{x^2 + 3x + 5}_{\text{Dividend}}.$$

Example 5 A Binomial Divisor

Divide $(6x^3 - 19x^2 + 16x - 4)$ by $(x - 2)$.

Solution

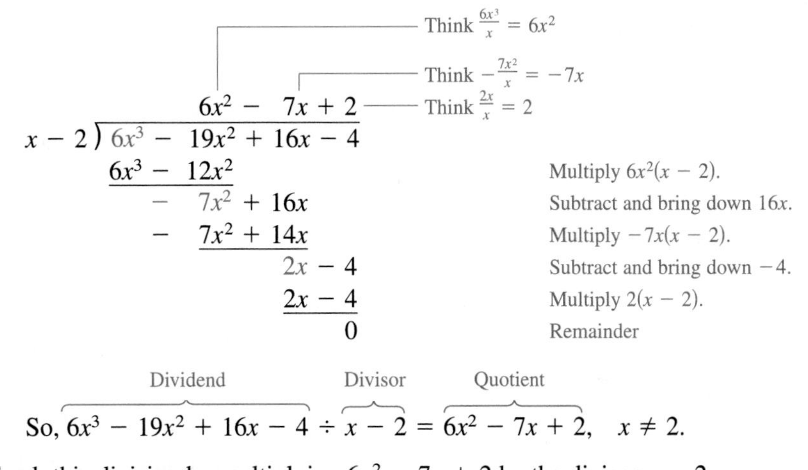

Think $\frac{6x^3}{x} = 6x^2$
Think $-\frac{7x^2}{x} = -7x$
Think $\frac{2x}{x} = 2$
Multiply $6x^2(x - 2)$.
Subtract and bring down $16x$.
Multiply $-7x(x - 2)$.
Subtract and bring down -4.
Multiply $2(x - 2)$.
Remainder

$$
\begin{array}{r}
6x^2 - 7x + 2 \\
x - 2 \overline{) 6x^3 - 19x^2 + 16x - 4} \\
\underline{6x^3 - 12x^2} \qquad\qquad \\
-7x^2 + 16x \qquad \\
\underline{-7x^2 + 14x} \qquad \\
2x - 4 \\
\underline{2x - 4} \\
0
\end{array}
$$

So, $\overbrace{6x^3 - 19x^2 + 16x - 4}^{\text{Dividend}} \div \overbrace{x - 2}^{\text{Divisor}} = \overbrace{6x^2 - 7x + 2}^{\text{Quotient}}, \quad x \neq 2.$

Check this division by multiplying $6x^2 - 7x + 2$ by the divisor $x - 2$.

In Example 5, the remainder is zero. In such cases, the denominator (or divisor) is said to **divide evenly** into the numerator (or dividend).

Example 6 A Binomial Divisor

Divide $(-13x^3 + 10x^4 + 8x - 7x^2 + 4)$ by $(3 - 2x)$.

Solution

First write the divisor and dividend in standard polynomial form.

$$
\begin{array}{r}
-5x^3 - x^2 + 2x - 1 \\
-2x + 3 \overline{) 10x^4 - 13x^3 - 7x^2 + 8x + 4} \\
\underline{10x^4 - 15x^3} \qquad\qquad\qquad\quad \\
2x^3 - 7x^2 \qquad\qquad \\
\underline{2x^3 - 3x^2} \qquad\qquad \\
-4x^2 + 8x \qquad \\
\underline{-4x^2 + 6x} \qquad \\
2x + 4 \\
\underline{2x - 3} \\
7
\end{array}
$$

Multiply $-5x^3(-2x + 3)$.
Subtract and bring down $-7x^2$.
Multiply $-x^2(-2x + 3)$.
Subtract and bring down $8x$.
Multiply $2x(-2x + 3)$.
Subtract and bring down 4.
Multiply $-1(-2x + 3)$.
Remainder

Using the fractional form of the remainder, you can write

$$\frac{\overbrace{10x^4 - 13x^3 - 7x^2 + 8x + 4}^{\text{Dividend}}}{\underbrace{-2x + 3}_{\text{Divisor}}} = \overbrace{-5x^3 - x^2 + 2x - 1}^{\text{Quotient}} + \overbrace{\frac{7}{-2x + 3}}^{\text{Remainder}}.$$

Study Tip

You should always check two things when you begin a long division problem.

1. The divisor and the dividend should be written in standard form—that is, in decreasing powers of the variable.
2. Zero coefficients or spaces should be inserted for any "missing" terms in the dividend.

Example 7 A Binomial Divisor

Use the long division algorithm to simplify

$$\frac{x^3 - 1}{x - 1}.$$

Solution

Because there are no x^2- or x-terms in the dividend, you can line up the subtractions by using *zero* coefficients (or by leaving spaces) for the missing terms.

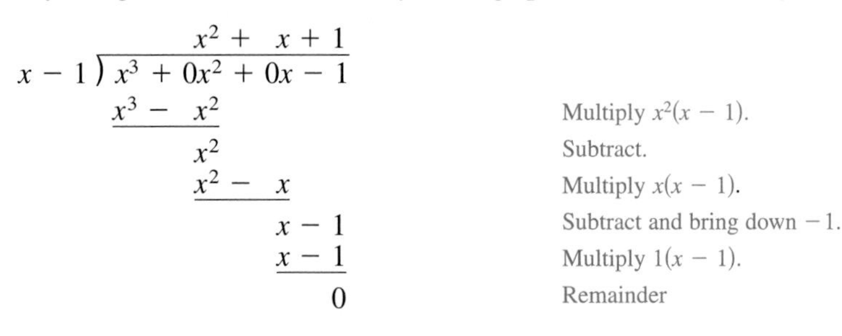

$x^2 + x + 1$
$x - 1 \overline{) x^3 + 0x^2 + 0x - 1}$
$x^3 - x^2$
x^2
$x^2 - x$
$x - 1$
$x - 1$
0
Multiply $x^2(x - 1)$.
Subtract.
Multiply $x(x - 1)$.
Subtract and bring down -1.
Multiply $1(x - 1)$.
Remainder

So, $x - 1$ divides evenly into $x^3 - 1$ and you can write

$$\frac{x^3 - 1}{x - 1} = x^2 + x + 1, \quad x \neq 1.$$

4 Use synthetic division to divide a polynomial expression by a binomial expression.

Synthetic Division

There is a nice shortcut for long division of polynomials by divisors of the form $x - k$. The shortcut is called **synthetic division.** We summarize the pattern for synthetic division of a cubic polynomial as follows. (The pattern for higher-degree polynomials is similar.)

▶ Synthetic Division (for a Cubic Polynomial)

To divide $ax^3 + bx^2 + cx + d$ by $x - k$, use the following pattern.

Divisor → k | ⓐ b c d ← Coefficients of dividend

↓ (ka) ◯ ◯

ⓐ ◯ ◯ ⓡ ← Remainder

Coefficients of quotient

Vertical pattern: Add terms.
Diagonal pattern: Multiply by k.

Synthetic division works *only* for divisors of the form $x - k$. [Remember that $x + k = x - (-k)$.] You cannot use synthetic division to divide a polynomial by a quadratic such as $x^2 - 3$.

Example 8 Using Synthetic Division

Use synthetic division to divide $x^3 + 5x^2 + 4x - 2$ by $x + 2$.

Solution

You should set up the division as follows.

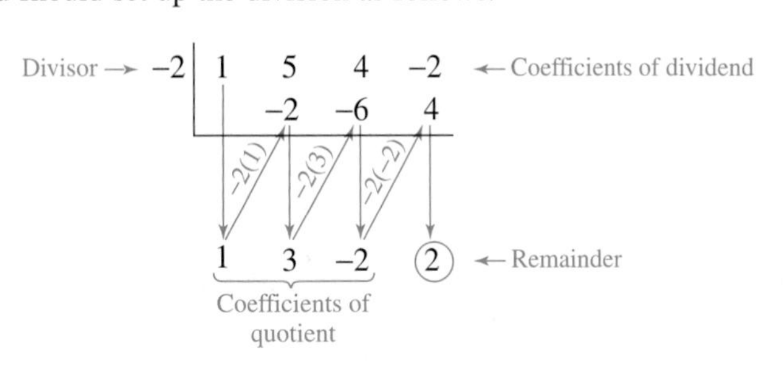

The bottom row of the table shows the coefficients of the quotient and the remainder. So the quotient is

$$(1)x^2 + (3)x + (-2)$$

and the remainder is 2. The result of the division problem is

$$\frac{x^3 + 5x^2 + 4x - 2}{x + 2} = x^2 + 3x - 2 + \frac{2}{x + 2}.$$

You can check the result by multiplying.

Try using synthetic division to perform the divisions in Examples 4, 5, and 7.

Discussing the Concept Creating Practice Problems

You are tutoring a friend in algebra and you want to create some division problems for practice. You want to find several division problems that involve third-degree polynomials that are evenly divisible by first-degree polynomials. For instance, in Example 7, $x^3 - 1$ is evenly divisible by $x - 1$ because the remainder is 0.

$$\frac{x^3 - 1}{x - 1} = x^2 + x + 1$$

Develop a method for finding a third-degree polynomial that is evenly divisible by a first-degree polynomial. Demonstrate your method by finding a third-degree polynomial that is evenly divisible by $2x - 1$.

5.4 Exercises

Integrated Review *Concepts, Skills, and Problem Solving*

Keep mathematically in shape by doing these exercises *before* the problems of this section.

Properties and Definitions

1. Explain how to write the fraction $\frac{24x}{18}$ in reduced form.

2. The point $(-1, 4)$ lies in what quadrant? Explain.

Simplifying Expressions

In Exercises 3–6, simplify the fraction.

3. $\frac{8}{12}$ **4.** $\frac{18}{144}$

5. $\frac{60}{150}$ **6.** $\frac{175}{42}$

In Exercises 7–10, find the product and simplify.

7. $-2x^2(5x^3)$ **8.** $(2z + 1)(2z - 1)$

9. $(x + 7)^2$ **10.** $(x + 4)(2x - 5)$

Creating a Model and Problem Solving

11. Write an algebraic expression that represents the product of two consecutive odd integers, the first of which is $2n + 1$.

12. After traveling for 3 hours, you are still 24 miles from completing a 180-mile trip. It takes you one-half hour to travel the last 24 miles. Find your average speed during the trip.

Developing Skills

In Exercises 1–14, perform the division by cancellation *and* by subtracting exponents. (Assume that no denominator is zero.)

1. $\frac{x^5}{x^2}$ **2.** $\frac{y^7}{y^3}$

3. $\frac{x^2}{x^5}$ **4.** $\frac{y^3}{y^7}$

5. $\frac{z^4}{z^7}$ **6.** $\frac{y^8}{y^3}$

7. $\frac{3u^4}{u^3}$ **8.** $\frac{z^6}{5z^4}$

9. $\frac{2^3y^4}{2^2y^2}$ **10.** $\frac{3^5x^7}{3^3x^4}$

11. $\frac{4^5x^3}{4x^5}$ **12.** $\frac{6z^5}{6z^5}$

13. $\frac{3^4(ab)^2}{3(ab)^3}$ **14.** $\frac{8^2u^4v^5}{8^3u^4v^2}$

In Exercises 15–32, simplify the expression. (Assume that no denominator is zero.) See Example 1.

15. $\frac{-3x^2}{x}$ **16.** $\frac{-4a^6}{-a^2}$

17. $\frac{4}{x^3}$ **18.** $\frac{-16}{v^2}$

19. $\frac{-12z^3}{-3z}$ **20.** $\frac{16y^5}{8y^3}$

21. $\frac{32b^4}{12b^3}$ **22.** $\frac{-7c^2}{8c^5}$

23. $\frac{-22y^2}{4y}$ **24.** $\frac{54x^2}{-24x^4}$

25. $\frac{-18s^4}{-12r^2s}$ **26.** $\frac{-21v^3}{12u^2v}$

27. $\frac{(-3z)^2}{18z^3}$ **28.** $\frac{4a^3}{(-8a)^2}$

29. $\frac{(2x^2y)^3}{(4y^2)^2x^4}$ **30.** $\frac{15(uv^4)^2}{(-3u^3)^3v^5}$

31. $\frac{24u^2v^4}{18u^2v^6}$ **32.** $\frac{15x^3y^0}{27x^3}$

In Exercises 33–72, perform the division and simplify. (Assume that no denominator is zero.) See Examples 2–7.

33. $\frac{3z + 3}{3}$ **34.** $\frac{7x + 7}{7}$

35. $\dfrac{4z - 12}{4}$

36. $\dfrac{8u - 24}{8}$

37. $\dfrac{9x - 5}{3}$

38. $\dfrac{3 - 10x}{5}$

39. $\dfrac{b^2 - 2b}{b}$

40. $\dfrac{3x + 2x^3}{x}$

41. $(5x^2 - 2x) \div x$

42. $(16a^2 + 5a) \div a$

43. $\dfrac{25z^3 + 10z^2}{-5z}$

44. $\dfrac{12c^4 - 36c}{-6c}$

45. $\dfrac{8z^3 + 3z^2 - 2z}{2z}$

46. $\dfrac{3x^3 + 5x^2 - 4x}{3x}$

47. $\dfrac{m^3 + 3m - 4}{m}$

48. $\dfrac{l^2 - 4l + 8}{l}$

49. $\dfrac{4x^2 - 12x}{4x^2}$

50. $\dfrac{14y^4 + 21y^3}{-7y^3}$

51. $\dfrac{6x^4 - 2x^3 + 3x^2 - x + 4}{2x^3}$

52. $\dfrac{9x^5 - 12x^3 + 3x^2 - 5x}{-3x^2}$

53. $\dfrac{x^2 - x - 2}{x + 1}$

54. $\dfrac{x^2 - 5x + 6}{x - 2}$

55. $\dfrac{x^2 + 9x + 20}{x + 4}$

56. $\dfrac{x^2 - 7x - 30}{x - 10}$

57. $\dfrac{3y^2 + 4y - 4}{3y - 2}$

58. $\dfrac{7t^2 - 10t - 8}{7t + 4}$

59. $(18t^2 - 21t - 4) \div (3t - 4)$

60. $(20t^2 + 32t - 16) \div (2t + 4)$

61. $(x^3 - 4x^2 + 9x - 7) \div (x - 2)$

62. $(2x^3 - 2x^2 + 3x + 9) \div (x + 1)$

63. $(7x + 3) \div (x + 2)$

64. $(8x - 5) \div (2x + 1)$

65. $\dfrac{x^3 - 8}{x - 2}$

66. $\dfrac{x^3 + 27}{x + 3}$

67. $\dfrac{x^2 + 9}{x + 3}$

68. $\dfrac{y^2 + 3}{y + 3}$

69. Divide $9x^2 - 1$ by $3x + 1$.

70. Divide $25y^2 - 4$ by $5y - 2$.

71. Divide $x^4 - 1$ by $x - 1$.

72. Divide x^4 by $x - 1$.

In Exercises 73–78, use synthetic division. See Example 8.

73. Divide $4x^2 + 3x + 1$ by $x + 1$.

74. Divide $7x^2 + 4x + 3$ by $x + 2$.

75. $\dfrac{x^3 - 7x + 6}{x - 2}$

76. $\dfrac{x^3 - 28x - 48}{x + 4}$

77. $\dfrac{3t^3 + 7t^2 + 3t - 2}{t + 2}$

78. $\dfrac{2x^3 + 5x^2 - 2x + 3}{x + 3}$

In Exercises 79–84, simplify the expression. (Assume that no denominator is zero.)

79. $\dfrac{4x^3}{x^2} - \dfrac{8x}{4}$

80. $\dfrac{25x^2}{10x} + \dfrac{3x}{2}$

81. $\dfrac{8u^2v}{2u} + \dfrac{(uv)^2}{uv}$

82. $\dfrac{9x^5y}{3x^4} - \dfrac{(x^2y)^3}{x^5y^2}$

83. $\dfrac{x^2 + 2x + 1}{x + 1} - (3x - 4)$

84. $\dfrac{x^2 - 3x + 2}{x - 1} + (4x - 3)$

In Exercises 85–88, determine whether the cancellation is valid.

85. $\dfrac{3 + 4}{3} = \dfrac{\not{3} + 4}{\not{3}} = 4$

86. $\dfrac{4 + 7}{4 + 11} = \dfrac{\not{4} + 7}{\not{4} + 11} = \dfrac{7}{11}$

87. $\dfrac{7 \cdot 12}{19 \cdot 7} = \dfrac{\not{7} \cdot 12}{19 \cdot \not{7}} = \dfrac{12}{19}$

88. $\dfrac{24}{43} = \dfrac{2\not{4}}{\not{4}3} = \dfrac{2}{3}$

Solving Problems

89. *Exploration* Consider the equation

$$(x + 3)(x^2 + 2x - 1) = x^3 + 5x^2 + 5x - 3.$$

(a) Use a graphing utility to verify that the equation is an identity by graphing both the left side and the right side of the equation. Are the graphs the same?

(b) Verify the identity equation by multiplying the polynomials on the left side of the equation.

(c) Verify the identity equation by performing the long division

$$\frac{x^3 + 5x^2 + 5x - 3}{x + 3}.$$

90. *Exploration* Consider the equation

$$2x^3 - 5x^2 + 2x - 5 = (2x - 5)(x^2 + 1).$$

(a) Using a graphing utility to verify that the equation is an identity by graphing both the left side and the right side of the equation. Are the graphs the same?

(b) Verify the identity equation by multiplying the polynomials on the right side of the equation.

(c) Verify the identity equation by performing the long division

$$\frac{2x^3 - 5x^2 + 2x - 5}{2x - 5}.$$

91. *Comparing Ages* You have two children: one is 18 years old and the other is 8 years old. In t years, their ages will be $t + 18$ and $t + 8$.

(a) Use long division to rewrite the ratio of your older child's age to your younger child's age.

(b) Complete the table.

t	0	10	20	30	40	50	60
$\frac{t+18}{t+8}$							

(c) What happens to the values of the ratio as t increases? Use the result of part (a) to explain your conclusion.

92. *Geometry* The area of a rectangle is $x^2 + 5x - 6$. Find the length of the rectangle if its width is $x - 1$.

93. *Geometry* The area of a rectangle is $x^2 + 2x - 15$. Find the width of the rectangle if its length is $x + 5$.

Explaining Concepts

94. Answer part (g) of Motivating the Chapter on page 273.

95. Match each part of the equation with its name:

$$\frac{x^2 + 2}{x - 3} = x + 3 + \frac{11}{x - 3}.$$

(a) Dividend (b) Divisor

(c) Quotient (d) Remainder

96. Explain how you can check the result of a division problem algebraically *and* graphically.

97. Give an example of using the subtraction property of exponents to divide a monomial by a monomial.

98. Describe the method of dividing a polynomial by a monomial.

99. What does it mean when the divisor divides *evenly* into the dividend?

100. If the degree of the dividend is 5 and the degree of the divisor is 3, what is the degree of the quotient? Generalize this result if the degree of the numerator is m and the degree of the denominator is n, where $m > n$.

CHAPTER SUMMARY

Key Terms

polynomial, *p. 274*
constant term, *p. 274*
standard form of a polynomial, *p. 274*
degree of a polynomial, *p. 274*
leading coefficient, *p. 274*
monomial, *p. 275*
binomial, *p. 275*
trinomial, *p. 275*
FOIL Method, *p. 285*
negative exponents, *p. 297*
scientific notation, *p. 300*
synthetic division, *p. 310*

Key Concepts

5.1 Polynomial in x

Let $a_n, a_{n-1}, \ldots, a_2, a_1$ be real numbers and let n be a nonnegative integer. A polynomial in x is an expression of the form

$$a_nx^n + a_{n-1}x^{n-1} + \cdots + a_2x^2 + a_1x + a_0$$

where $a_n \neq 0$. The polynomial is of degree n, and the number a_n is the leading coefficient. The number a_0 is the constant term.

5.1 Adding polynomials

To add polynomials, you combine like terms (those having the same degree) by using the Distributive Property.

5.1 Subtracting polynomials

To subtract polynomials, you add the opposite by changing the sign of each term of the polynomial being subtracted and then adding the resulting like terms.

5.2 Multiplying polynomials

1. To multiply a polynomial by a monomial, apply the Distributive Property.
2. To multiply two binomials, use the FOIL Method. Find the product of the **F**irst terms, the product of the **O**uter terms, the product of the **I**nner terms, and the product of the **L**ast terms.
3. To multiply two polynomials, use the Distributive Property to multiply each term of one polynomial by each term of the other polynomial.

5.2 Special Products

Sum and Difference of Two Terms:

$$(a + b)(a - b) = a^2 - b^2$$

Square of a Binomial:

$$(a + b)^2 = a^2 + 2ab + b^2$$
$$(a - b)^2 = a^2 - 2ab + b^2$$

5.3 Rules of Exponents

1. $a^ma^n = a^{m+n}$
2. $\dfrac{a^m}{a^n} = a^{m-n}$
3. $(ab)^m = a^mb^m$
4. $\left(\dfrac{a}{b}\right)^m = \dfrac{a^m}{b^m}$
5. $(a^m)^n = a^{mn}$
6. $a^{-n} = \dfrac{1}{a^n}$
7. $a^0 = 1$

5.4 Dividing Polynomials

1. To divide a monomial by a monomial, use the properties of exponents.
2. To divide a polynomial by a monomial, divide each term of the polynomial by the monomial.
3. To divide a polynomial by a binomial, follow the long division pattern used for dividing whole numbers.
4. Use synthetic division to divide a polynomial by a binomial of the form $x - k$. [Remember that $x + k = x - (-k)$.]

REVIEW EXERCISES

Reviewing Skills

5.1 In Exercises 1–8, write the polynomial in standard form. Then determine its degree and leading coefficient.

1. $10x - 4 - 5x^3$
2. $2x^2 + 9$
3. $4x^3 - 2x + 5x^4 - 7x^2$
4. $6 - 3x + 6x^2 - x^3$
5. $7x^4 - 1$
6. $12x^2 + 2x - 8x^5 + 1$
7. -2
8. $\frac{1}{4}t^2$

In Exercises 9–12, give an example of a polynomial that satisfies the given conditions. (*Note:* There are many correct answers.)

9. A trinomial of degree 4
10. A monomial of degree 2
11. A binomial of degree 1
12. A trinomial of degree 5

In Exercises 13–32, perform the operations and simplify.

13. $(2x + 3) + (x - 4)$
14. $\left(\frac{1}{2}x + \frac{2}{3}\right) + \left(4x + \frac{1}{3}\right)$
15. $(t - 5) - (3t - 1)$
16. $(y + 3) - (y - 9)$
17. $(2x^3 - 4x^2 + 3) + (x^3 + 4x^2 - 2x)$
18. $(6x^2 - 9x - 5) - (4x^2 - 6x + 1)$
19. $3(2x^2 - 4) - (2x^2 - 5)$
20. $-4(6 - x + x^2) + (3x^2 + x)$
21. $(5x^4 - 7x^3 + x) - (4x^3 + 2x^2 - 4) + (4x + 8x^3 - 2x^4)$
22. $(6x^3 - 4x^2 + 3) + (x^2 - 2x) - (6x^3 - 4x + 6)$
23. $(4 - x^2) + 2(x - 2)$
24. $(z^2 + 6z) - 3(z^2 + 2z)$
25. $(-x^3 - 3x) - 2(2x^3 + x + 1)$
26. $(3u + 4u^2) + 5(u + 1) + 3u^2$
27. $4y^2 - [y - 3(y^2 + 2)]$
28. $(6a^3 + 3a) - 2[a + (a^3 - 2)]$
29. $\begin{array}{r} -x^4 - 2x^2 + 3 \\ \underline{+ (3x^4 - 5x^2 \quad\quad)} \end{array}$
30. $\begin{array}{r} 5z^3 \quad\quad - 4z - 7 \\ \underline{+ (z^2 - 2z \quad\quad)} \end{array}$
31. $\begin{array}{r} 5x^2 + 2x - 27 \\ \underline{- (2x^2 - 2x - 13)} \end{array}$
32. $\begin{array}{r} 12y^4 - 15 \\ \underline{- (18y^4 - \ \ 9)} \end{array}$

5.2 In Exercises 33–46, multiply the polynomials using the Distributive Property or the FOIL Method.

33. $2x(x + 4)$
34. $3y(y + 1)$
35. $(x - 4)(x + 6)$
36. $(u + 5)(u - 2)$
37. $(x + 3)(2x - 4)$
38. $(y + 2)(4y - 3)$
39. $(4x - 3)(3x + 4)$
40. $(6 - 2x)(7x + 10)$
41. $(x^2 + 5x + 2)(2x + 3)$
42. $(s^3 + 4s - 3)(s - 3)$
43. $(2t - 1)(t^2 - 3t + 3)$
44. $(4x + 2)(x^2 + 6x - 5)$
45. $2u(u - 5) - (u + 1)(u - 5)$
46. $(3v - 2)(-2v) + 2v(3v - 2)$

In Exercises 47–60, use a special binomial product to expand the expression.

47. $(x + 3)^2$
48. $(x - 5)^2$
49. $(4x - 7)^2$
50. $(9 - 2x)^2$
51. $\left(\frac{1}{2}x - 4\right)^2$
52. $(4 + 3b)^2$
53. $(u - 6)(u + 6)$
54. $(r + 3)(r - 3)$
55. $(3t - 1)(3t + 1)$
56. $(3a + 8)(3a - 8)$
57. $(2x - y)^2$
58. $(3a + b)^2$
59. $(2x - 4y)(2x + 4y)$
60. $(4u + 5v)(4u - 5v)$

5.3 In Exercises 61–76, evaluate the expression.

61. 4^{-2}
62. 3^{-4}
63. $6^{-4}6^2$
64. $(2^2 \cdot 3^2)^{-1}$
65. $\dfrac{1}{3^{-2}}$
66. $\dfrac{1}{5^{-3}}$
67. $\dfrac{4}{4^{-2}}$
68. $\dfrac{7}{3^{-3}}$
69. $\left(\dfrac{3}{5}\right)^{-3}$
70. $\left(\dfrac{2^{-2}}{3}\right)^2$
71. $\left(-\dfrac{2}{5}\right)^3\left(\dfrac{5}{2}\right)^2$
72. $\dfrac{2^2 \cdot 3^{-2}}{2^{-2} \cdot 3^{-1}}$
73. $(3 \times 10^3)^2$
74. $(4 \times 10^{-3})(5 \times 10^7)$
75. $\dfrac{1.85 \times 10^9}{5 \times 10^4}$
76. $\dfrac{1}{(4 \times 10^{-2})^3}$

In Exercises 77–100, use the rules of exponents to write the expression without negative exponents. (Assume that no variable is zero.)

77. y^{-4}

78. x^{-5}

79. $6t^{-2}$

80. $-4u^{-3}$

81. $\dfrac{1}{7x^{-6}}$

82. $\dfrac{1}{2y^{-4}}$

83. $2x^{-1}y^{-3}$

84. $5u^{-2}v^{-4}$

85. $t^{-4} \cdot t^2$

86. $x^5 \cdot x^{-8}$

87. $4x^{-6}y^2 \cdot x^6$

88. $-2u^5v^{-4} \cdot v^4$

89. $(-3a^2)^{-2}$

90. $(3u^{-3})(9u^6)$

91. $(x^2y^{-3})^2$

92. $5(x + 3)^0$

93. $\dfrac{t^{-4}}{t^{-1}}$

94. $\dfrac{a^3 \cdot a^{-2}}{a^{-1}}$

95. $\dfrac{u^5 \cdot u^{-8}}{u^{-3}}$

96. $\dfrac{x^9 \cdot x^{-6}}{x^{-3}}$

97. $\left(\dfrac{y}{5}\right)^{-2}$

98. $\left(\dfrac{7}{x^4}\right)^{-1}$

99. $(2u^{-2}v)^3(4u^{-5}v^4)^{-1}$

100. $(3x^2y^4)^3(3x^2y^4)^{-3}$

5.4 In Exercises 101–112, divide the polynomials.

101. $\dfrac{8x^3 - 12x}{4x^2}$

102. $\dfrac{18 - 3x + 9x^2}{12x^2}$

103. $(5x^2 + 15x) \div (5x)$

104. $(8u^3 + 4u^2) \div (2u)$

105. $\dfrac{x^2 - x - 6}{x - 3}$

106. $\dfrac{x^2 + x - 20}{x + 5}$

107. $\dfrac{24x^2 - x - 8}{3x - 2}$

108. $\dfrac{21x^2 + 4x + 7}{3x - 2}$

109. $\dfrac{2x^3 + 2x^2 - x + 2}{x - 1}$

110. $\dfrac{6x^4 - 4x^3 - 27x^2 + 18x}{3x - 2}$

111. $\dfrac{x^4 - 3x^2 + 2}{x^2 - 1}$

112. $\dfrac{3x^4}{x^2 - 1}$

Solving Problems

Geometry In Exercises 113–116, find a polynomial that represents the area of the shaded portion of the figure.

113.

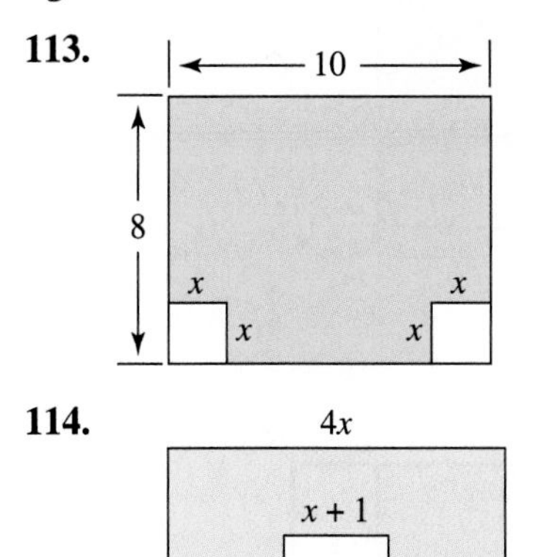

114.

4x
x + 1
3x
x

115. **116.**

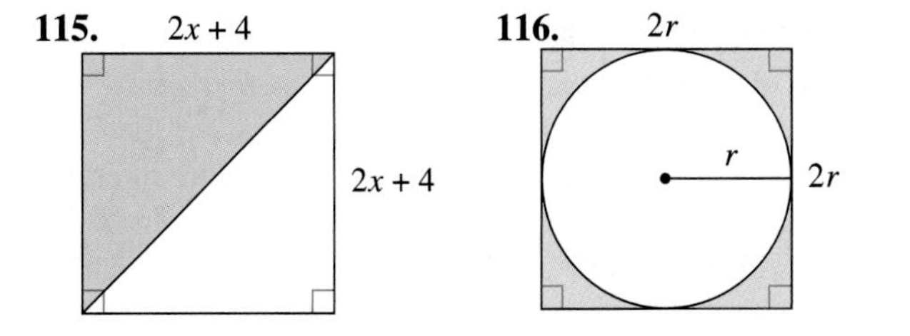

117. *Geometry* The length of a rectangular wall is x units, and its height is $x - 3$ units (see figure). Find (a) the perimeter and (b) the area of the wall.

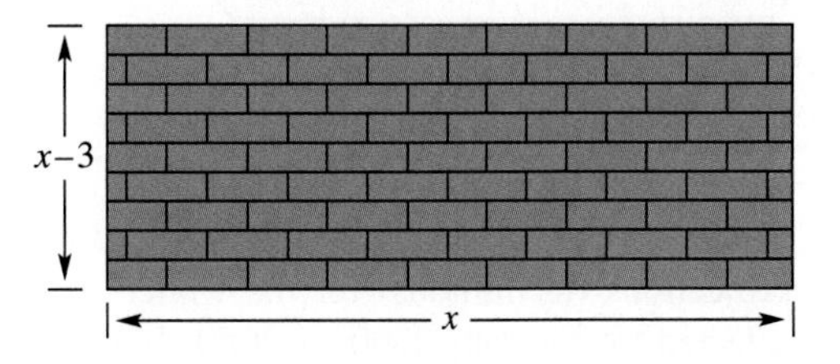

118. *Graphical Interpretation* The cost of producing x units of a product is

$$C = 15 + 26x.$$

The revenue for selling x units is

$$R = 40x - \frac{1}{2}x^2, \quad 0 \le x \le 20.$$

The profit is the difference between revenue and cost.

(a) Perform the subtraction required to find the polynomial representing profit.

(b) Use a graphing utility to graph the polynomial representing profit.

(c) Determine the profit when $x = 14$ units. Use the graph in part (b) to describe the profit when x is less than or greater than 14.

119. *Geometry* The area of a rectangle is

$$2x^2 - 5x - 12.$$

Find the length, if the width is $x - 4$.

120. *Geometry* The area of a rectangle is

$$3x^2 + 5x - 3.$$

Find the width, if the length is $x + 3$.

121. *Metal Expansion* When the temperature of a 150-foot iron steam pipe is increased by 100°C, the length of the pipe, as shown in the figure, increases by $100(150)(10 \times 10^{-6})$ feet. Write the amount of increase in decimal form.

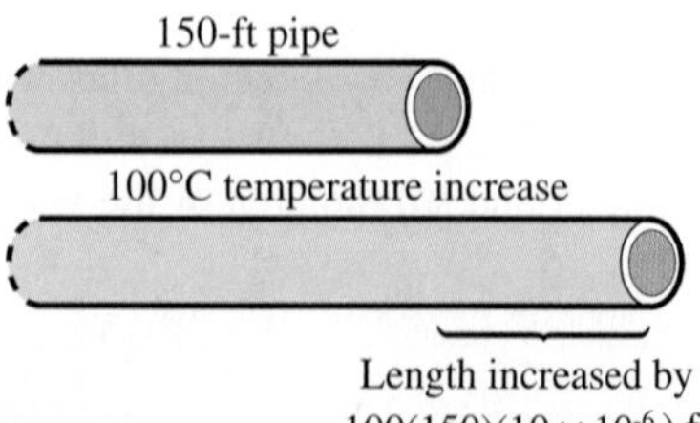

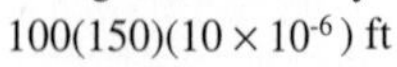

122. *Comparing Models* The table gives population projections (in millions) for the United States for selected years from 2000 to 2050. It gives three series of projections: lowest P_L, middle P_M, and highest P_H. (Source: U.S. Bureau of the Census)

Year	2000	2010	2020	2030	2040	2050
P_L	271.2	281.5	288.8	291.1	287.7	282.5
P_M	274.6	297.7	322.7	347.0	370.0	393.9
P_H	278.1	314.6	357.7	405.1	458.4	518.9

In the following models for the data, $t = 0$ corresponds to the year 2000.

$$P_L = -0.022t^2 + 1.33t + 270.71$$

$$P_M = 2.386t + 274.857$$

$$P_H = 0.028t^2 + 3.40t + 278.18$$

(a) Use a graphing utility to plot the data and graph the models on the same screen.

(b) Find $(P_L + P_H)/2$. Use a graphing utility to graph this polynomial and state which graph from part (a) it most resembles. Does this seem reasonable? Explain.

(c) Find $P_H - P_L$ and sketch its graph. Explain why it is increasing.

123. *Special Product* What special product does the figure illustrate? Explain your reasoning.

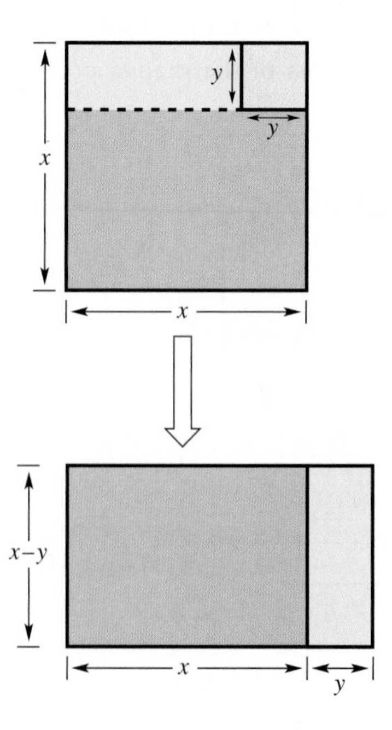

Chapter Test

Take this test as you would take a test in class. After you are done, check your work against the answers given in the back of the book.

1. Explain how to determine the degree and the leading coefficient of $-3x^4 - 5x^2 + 2x - 10$.

2. Give an example of a trinomial in one variable of degree 4.

In Exercises 3–14, perform the indicated operation and simplify. (Assume that no variable or denominator is zero.)

3. $(3z^2 - 3z + 7) + (8 - z^2)$

4. $(8u^3 + 3u^2 - 2u - 1) - (u^3 + 3u^2 - 2u)$

5. $6y - [2y - (3 + 4y - y^2)]$

6. $-5(x^2 - 1) + 3(4x + 7) - (x^2 + 26)$

7. $(5b + 3)(2b - 1)$

8. $4x\left(\dfrac{3x}{2}\right)^2$

9. $(z + 2)(2z^2 - 3z + 5)$

10. $(x - 5)^2$

11. $(2x - 3)(2x + 3)$

12. $\dfrac{15x + 25}{5}$

13. $\dfrac{x^3 - x - 6}{x - 2}$

14. $\dfrac{4x^3 + 10x^2 - 2x - 5}{2x + 1}$

In Exercises 15 and 16, simplify the expression. (Assume that no variable is zero.)

15. $\dfrac{-6a^2b}{-9ab}$

16. $(3x^{-2}y^3)^{-2}$

17. Evaluate the expression *without* using a calculator. Show your work.

(a) 4^{-3} (b) $\dfrac{2^{-3}}{3^{-1}}$ (c) $(1.5 \times 10^5)^2$

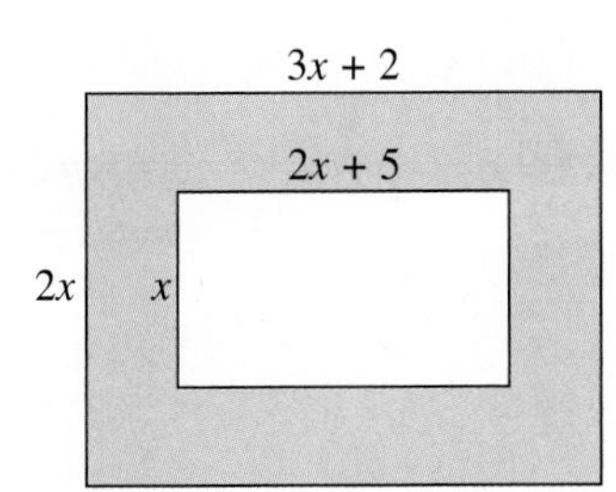

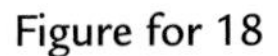

Figure for 18

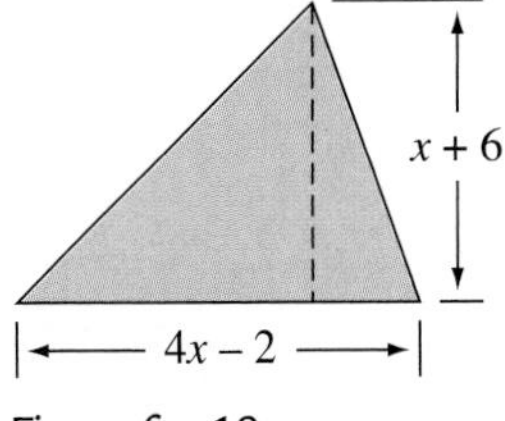

Figure for 19

18. Find the polynomial that represents the area of the shaded region (see figure).

19. Write an expression that represents the area of the triangle (see figure). Explain your reasoning.

20. The mean distance from earth to the moon is 3.84×10^8 meters. Write this distance in decimal form.

21. The standard atmospheric pressure is 101,300 newtons per square meter. Write this pressure in scientific notation.

22. The area of a rectangle is $x^2 - 2x - 3$. Find the width, if the length is $x + 1$.

6 Factoring and Solving Equations

David Frazier/Tony Stone Images

In 1997, approximately 46 billion pounds of potatoes were produced in the United States. (Source: U.S. Department of Agriculture)

Motivating the Chapter

Dimensions of a Potato Storage Bin

A bin used to store potatoes has the form of a rectangular solid with a volume (in cubic feet) given by the polynomial $12x^3 + 64x^2 - 48x$.

See Section 6.3, Exercise 109

a. The height of the bin is $4x$ feet. Write an expression for the area of the base of the bin.

b. Factor the expression for the area of the base of the bin. Use the result to write expressions for the length and width of the bin.

See Section 6.5, Exercise 102

c. The area of the base is 32 square feet. What are the dimensions of the bin?

d. If you were told that the bin has a volume of 256 cubic feet, could you find the dimensions of the bin? Explain your reasoning.

e. A polynomial that represents the volume of the truck bin in cubic feet is $6x^3 + 32x^2 - 24x$. How many truck loads does it take to fill the bin? Explain your reasoning.

6.1 Factoring Polynomials with Common Factors

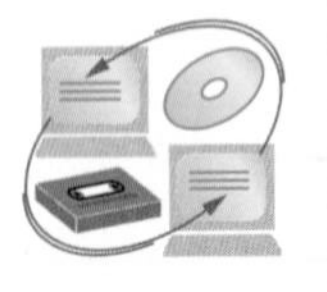

Objectives

1. Find the greatest common factor of two expressions.
2. Factor out the greatest common monomial factor from a polynomial.
3. Factor a polynomial by grouping.

1 Find the greatest common factor of two expressions.

Greatest Common Factor

In Chapter 5, you used the Distributive Property to multiply polynomials. In this chapter, you will study the *reverse* process, which is **factoring.**

Multiplying Polynomials	*Factoring Polynomials*
$\underbrace{2x}_{\text{Factor}}\underbrace{(7-3x)}_{\text{Factor}} \Rightarrow \underbrace{14x-6x^2}_{\text{Product}}$	$\underbrace{14x-6x^2}_{\text{Product}} \Rightarrow \underbrace{2x}_{\text{Factor}}\underbrace{(7-3x)}_{\text{Factor}}$

To factor an expression efficiently, you need to understand the concept of the *greatest common factor* of two (or more) integers or terms. In Section 1.3, you learned that the **greatest common factor** of two or more integers is the greatest integer that is a factor of each integer. For example, the greatest common factor of $12 = 2 \cdot 2 \cdot 3$ and $30 = 2 \cdot 3 \cdot 5$ is $2 \cdot 3 = 6$.

Example 1 Finding the Greatest Common Factor

Find the greatest common factor of $5x^2y^2$ and $30x^3y$.

Solution

From the factorizations

$$5x^2y^2 = 5 \cdot x \cdot x \cdot y \cdot y = (5x^2y)(y)$$

$$30x^3y = 2 \cdot 3 \cdot 5 \cdot x \cdot x \cdot x \cdot y = (5x^2y)(6x)$$

you can conclude that the greatest common factor is $5x^2y$.

Example 2 Finding the Greatest Common Factor

Find the greatest common factor of $8x^5$, $20x^3$, and $16x^4$.

Solution

From the factorizations

$$8x^5 = 2 \cdot 2 \cdot 2 \cdot x \cdot x \cdot x \cdot x \cdot x = (4x^3)(2x^2)$$

$$20x^3 = 2 \cdot 2 \cdot 5 \cdot x \cdot x \cdot x = (4x^3)(5)$$

$$16x^4 = 2 \cdot 2 \cdot 2 \cdot 2 \cdot x \cdot x \cdot x \cdot x = (4x^3)(4x)$$

you can conclude that the greatest common factor is $4x^3$.

2 Factor out the greatest common monomial factor from a polynomial.

Study Tip

To find the greatest common monomial factor of a polynomial, answer these two questions.

1. What is the greatest integer factor common to each coefficient of the polynomial?
2. What is the highest–powered variable factor common to each term of the polynomial?

Common Monomial Factors

Consider the three terms listed in Example 2 as terms of the polynomial

$$8x^5 + 16x^4 + 20x^3.$$

The greatest common factor, $4x^3$, of these terms is the **greatest common monomial factor** of the polynomial. When you use the Distributive Property to remove this factor from each term of the polynomial, you are **factoring out** the common monomial factor.

$$8x^5 + 16x^4 + 20x^3 = 4x^3(2x^2) + 4x^3(4x) + 4x^3(5) \quad \text{Factor each term.}$$

$$= 4x^3(2x^2 + 4x + 5) \quad \text{Factor out common monomial factor.}$$

Example 3 Greatest Common Monomial Factor

Factor out the greatest common monomial factor from $6x - 18$.

Solution

The greatest common integer factor of $6x$ and 18 is 6. There is no common variable factor.

$$6x - 18 = 6(x) - 6(3) \quad \text{Greatest common monomial factor is 6.}$$

$$= 6(x - 3) \quad \text{Factor 6 out of each term.}$$

Example 4 Greatest Common Monomial Factor

Factor out the greatest common monomial factor from

$$10y^3 - 25y^2.$$

Solution

For the terms $10y^3$ and $25y^2$, 5 is the greatest common integer factor and y^2 is the highest–powered common variable factor.

$$10y^3 - 25y^2 = (5y^2)(2y) - (5y^2)(5) \quad \text{Greatest common factor is } 5y^2.$$

$$= 5y^2(2y - 5) \quad \text{Factor } 5y^2 \text{ out of each term.}$$

Example 5 Greatest Common Monomial Factor

Factor out the greatest common monomial factor from

$$45x^3 - 15x^2 - 15.$$

Solution

The greatest common integer factor of $45x^3$, $15x^2$, and 15 is 15. There is no common variable factor.

$$45x^3 - 15x^2 - 15 = 15(3x^3) - 15(x^2) - 15(1)$$

$$= 15(3x^3 - x^2 - 1)$$

Example 6 Common Monomial Factors

Factor each polynomial.

a. $35y^3 - 7y^2 - 14y$

b. $6y^5 + 3y^3 - 2y^2$

c. $3xy^2 - 15x^2y + 12xy$

Solution

a. $35y^3 - 7y^2 - 14y = 7y(5y^2) - 7y(y) - 7y(2)$ — $7y$ is common factor.

$= 7y(5y^2 - y - 2)$ — Factor $7y$ out of each term.

b. $6y^5 + 3y^3 - 2y^2 = y^2(6y^3) + y^2(3y) - y^2(2)$ — y^2 is common factor.

$= y^2(6y^3 + 3y - 2)$ — Factor y^2 out of each term.

c. $3xy^2 - 15x^2y + 12xy = 3xy(y) - 3xy(5x) + 3xy(4)$ — $3xy$ is common factor.

$= 3xy(y - 5x + 4)$ — Factor $3xy$ out of each term.

The greatest common monomial factor of the terms of a polynomial is usually considered to have a positive coefficient. However, sometimes it is convenient to factor a negative number out of a polynomial.

Example 7 A Negative Common Monomial Factor

Factor the polynomial $-2x^2 + 8x - 12$ in two ways.

a. Factor out a common monomial factor of 2.

b. Factor out a common monomial factor of -2.

Solution

a. To factor out the common monomial factor of 2, write the following.

$$-2x^2 + 8x - 12 = 2(-x^2) + 2(4x) - 2(6)$$
$$= 2(-x^2 + 4x - 6)$$

b. To factor -2 out of the polynomial, write the following.

$$-2x^2 + 8x - 12 = -2(x^2) + (-2)(-4x) + (-2)(6)$$
$$= -2(x^2 - 4x + 6)$$

Check this result by multiplying $(x^2 - 4x + 6)$ by -2. When you do, you will obtain the original polynomial.

With experience, you should be able to omit writing the first step shown in Examples 6 and 7. For instance, to factor -2 out of $-2x^2 + 8x - 12$, you could simply write

$$-2x^2 + 8x - 12 = -2(x^2 - 4x + 6).$$

3 Factor a polynomial by grouping.

Factoring by Grouping

There are occasions when the common factor of a polynomial is not simply a monomial. For instance, the polynomial

$$x^2(x - 2) + 3(x - 2)$$

has the common *binomial* factor $(x - 2)$. Factoring out this common factor produces

$$x^2(x - 2) + 3(x - 2) = (x - 2)(x^2 + 3).$$

Example 8 Common Binomial Factors

Factor each polynomial.

a. $5x^2(7x - 1) - 3(7x - 1)$ **b.** $2x(3x - 4) + (3x - 4)$

c. $3y^2(y - 3) + 4(3 - y)$

Solution

a. Each of the terms of this polynomial has a binomial factor of $(7x - 1)$.

$$5x^2(7x - 1) - 3(7x - 1) = (7x - 1)(5x^2 - 3)$$

b. Each of the terms of this polynomial has a binomial factor of $(3x - 4)$.

$$2x(3x - 4) + (3x - 4) = (3x - 4)(2x + 1)$$

Be sure you see that when $(3x - 4)$ is factored out of itself, you are left with the factor 1. This follows from the fact that $(3x - 4)(1) = (3x - 4)$.

c. $3y^2(y - 3) + 4(3 - y) = 3y^2(y - 3) - 4(y - 3)$ Write $4(3 - y)$ as $-4(y - 3)$.

$= (y - 3)(3y^2 - 4)$ Common factor is $(y - 3)$.

In Example 8, the polynomials were already grouped so that it was easy to determine the common binomial factors. In practice, you will have to do the grouping as well as the factoring. To see how this works, consider the expression

$$x^3 + 2x^2 + 3x + 6$$

and try to *factor* it. Note first that there is no common monomial factor to take out of all four terms. But suppose you *group* the first two terms together and the last two terms together.

$x^3 + 2x^2 + 3x + 6 = (x^3 + 2x^2) + (3x + 6)$ Group terms.

$= x^2(x + 2) + 3(x + 2)$ Distributive Property

$= (x + 2)(x^2 + 3)$ Distributive Property

This process is called **factoring by grouping.** Try grouping the polynomial as

$$(x^3 + 3x) + (2x^2 + 6).$$

Show how this grouping is used to factor the polynomial.

Example 9 Factoring by Grouping

Factor $x^3 - 2x^2 + x - 2$.

Solution

$$\begin{aligned} x^3 - 2x^2 + x - 2 &= (x^3 - 2x^2) + (x - 2) && \text{Group terms.} \\ &= x^2(x - 2) + (x - 2) && \text{Distributive Property} \\ &= (x - 2)(x^2 + 1) && \text{Distributive Property} \end{aligned}$$

Example 10 Factoring by Grouping

Factor $3x^2 + 12x - 5x - 20$.

Solution

$$\begin{aligned} 3x^2 + 12x - 5x - 20 &= (3x^2 + 12x) - (5x + 20) && \text{Group terms.} \\ &= 3x(x + 4) - 5(x + 4) && \text{Distributive Property} \\ &= (x + 4)(3x - 5) && \text{Distributive Property} \end{aligned}$$

Study Tip

Notice in Example 11 that the polynomial is not written in standard form. You could have rewritten the polynomial before factoring and still obtained the same result.

$$\begin{aligned} &2x^3 + 4x - x^2 - 2 \\ &= 2x^3 - x^2 + 4x - 2 \\ &= (2x^3 - x^2) + (4x - 2) \\ &= x^2(2x - 1) + 2(2x - 1) \\ &= (2x - 1)(x^2 + 2) \end{aligned}$$

You can always check to see that you have factored an expression correctly by multiplying and comparing the result with the original expression. Try using multiplication to check the results of Examples 9 and 10.

Example 11 Geometry: Area of a Rectangle

The area of the rectangle in Figure 6.1 can be represented by the polynomial $2x^3 + 4x - x^2 - 2$. Factor the polynomial to find the dimensions of the rectangle.

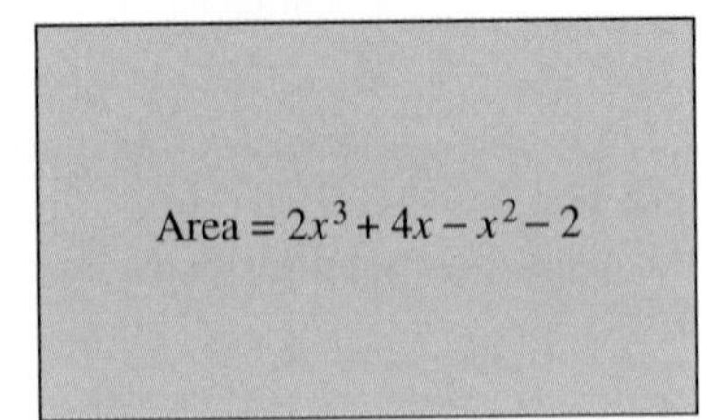

Figure 6.1

Solution

$$\begin{aligned} 2x^3 + 4x - x^2 - 2 &= (2x^3 + 4x) - (x^2 + 2) && \text{Group terms.} \\ &= 2x(x^2 + 2) - (x^2 + 2) && \text{Distributive Property} \\ &= (x^2 + 2)(2x - 1) && \text{Distributive Property} \end{aligned}$$

The dimensions of the rectangle are $(x^2 + 2)$ by $(2x - 1)$.

Discussing the Concept **Factoring by Grouping**

Suppose you are tutoring someone in algebra and you want to create several polynomials for your student to factor. Develop a procedure for creating polynomials that contain a common factor or that can be factored by the grouping method of this section. Create a list of practice problems and have another member of your class factor them.

6.1 Exercises

Integrated Review *Concepts, Skills, and Problem Solving*

Keep mathematically in shape by doing these exercises *before* the problems of this section.

Properties and Definitions

1. In your own words, define a function of x.
2. State the definitions of the domain and range of a function of x.
3. Bearing in mind the Vertical Line Test, sketch a graph for which y is not a function of x.
4. Bearing in mind the Vertical Line Test, sketch a graph for which y is a function of x.

Evaluating Functions

In Exercises 5–8, evaluate the function.

5. $f(x) = \frac{1}{2}x + 1$ (a) $f(0)$ (b) $f(4)$ (c) $f(-3)$ (d) $f\left(-\frac{3}{2}\right)$
6. $g(t) = t(t - 4)$ (a) $g(0)$ (b) $g(4)$ (c) $g(-2)$ (d) $g\left(-\frac{5}{2}\right)$
7. $F(x) = \sqrt{2x + 1}$ (a) $F(0)$ (b) $F(4)$ (c) $F\left(-\frac{1}{2}\right)$ (d) $F(10)$
8. $h(s) = |s - 3|$ (a) $h(0)$ (b) $h(4)$ (c) $h(2)$ (d) $h(-3)$

Problem Solving

9. Determine the commission rate for an employee who earned \$1620 in commissions on sales of \$54,000.
10. One person can complete a typing project in 10 hours, and another can complete the same project in 6 hours. Working together, how long will they take to complete the project?

Graphs

In Exercises 11–14, graph the function and show the coordinates of at least three solution points, including any intercepts.

11. $h(x) = 8 - 4x$
12. $g(x) = 3x - 6$
13. $f(x) = -\frac{1}{2}x^2$
14. $H(x) = |x + 2|$

Developing Skills

In Exercises 1–16, find the greatest common factor. See Examples 1 and 2.

1. $24, 90$
2. $20, 45$
3. $18, 150, 100$
4. $60, 80, 90$
5. $z^2, -z^6$
6. t^4, t^7
7. $2x^2, 12x$
8. $36x^4, 18x^3$
9. u^2v, u^3v^2
10. $r^6s^4, -rs$
11. $9yz^2, -12y^2z^3$
12. $-15x^6y^3, 45xy^3$
13. $14x^2, 1, 7x^4$
14. $5y^4, 10x^2y^2, 15xy$
15. $28a^4b^2, 14a^3b^3, 42a^2b^5$
16. $16x^2y, 12xy^2, 36x^2y^2$

In Exercises 17–60, factor the polynomial. (*Note:* Some of the polynomials have no common factor.) See Examples 3–6.

17. $3x + 3$
18. $5y + 5$
19. $6z - 6$
20. $3x - 3$
21. $8t - 16$
22. $3u + 12$
23. $-25x - 10$
24. $-14y - 7$
25. $24y^2 - 18$
26. $7z^3 + 21$
27. $x^2 + x$
28. $-s^3 - s$
29. $25u^2 - 14u$
30. $36t^4 + 24t^2$
31. $2x^4 + 6x^3$
32. $9z^6 + 27z^4$
33. $7s^2 + 9t^2$
34. $12x^2 - 5y^3$
35. $12x^2 - 2x$
36. $12u + 9u^2$
37. $-10r^3 - 35r$
38. $-144a^2 + 24a$
39. $16a^3b^3 + 24a^4b^3$
40. $6x^4y + 12x^2y$
41. $10ab + 10a^2b$
42. $21x^2z - 35xz$
43. $12x^2 + 16x - 8$
44. $9 - 3y - 15y^2$
45. $100 + 75z - 50z^2$
46. $42t^3 - 21t^2 + 7$
47. $9x^4 + 6x^3 + 18x^2$
48. $32a^5 - 2a^3 + 6a$

49. $5u^2 + 5u^2 + 5u$

50. $11y^3 - 22y^2 + 11y^2$

51. $x(x - 3) + 5(x - 3)$

52. $x(x + 6) + 3(x + 6)$

53. $t(s + 10) - 8(s + 10)$

54. $y(q - 5) - 10(q - 5)$

55. $a^2(b + 2) - b(b + 2)$

56. $x^3(y + 4) + y(y + 4)$

57. $z^3(z + 5) + z^2(z + 5)$

58. $x^3(x - 2) + x(x - 2)$

59. $(a + b)(a - b) + a(a + b)$

60. $(x + y)(x - y) - x(x - y)$

In Exercises 61–68, factor a negative real number from the polynomial and write the polynomial factor with a positive leading coefficient. See Example 7.

61. $5 - 10x$

62. $3 - x$

63. $3000 - 3x$

64. $9 - 2x^2$

65. $4 + 2x - x^2$

66. $18 - 12x - 6x^2$

67. $4 + 12x - 2x^2$

68. $x - 2x^2 - x^4$

In Exercises 69–84, factor the polynomial by grouping. See Examples 8–10.

69. $x^2 + 10x + x + 10$

70. $x^2 - 5x + x - 5$

71. $a^2 - 4a + a - 4$

72. $x^2 + 25x + x + 25$

73. $ky^2 - 4ky + 2y - 8$

74. $ay^2 + 3ay + 3y + 9$

75. $t^3 - 3t^2 + 2t - 6$

76. $3s^3 + 6s^2 + 2s + 4$

77. $x^3 + 2x^2 + x + 2$

78. $x^3 - 5x^2 + x - 5$

79. $6z^3 + 3z^2 - 2z - 1$

80. $4u^3 - 2u^2 - 6u + 3$

81. $x^3 - 3x - x^2 + 3$

82. $x^3 + 7x - 3x^2 - 21$

83. $4x^2 - x^3 - 8 + 2x$

84. $5x^2 + 10x^3 + 4 + 8x$

In Exercises 85–90, complete the factorization.

85. $\frac{1}{4}x + \frac{3}{4} = \frac{1}{4}(\quad)$

86. $\frac{5}{6}x - \frac{1}{6} = \frac{1}{6}(\quad)$

87. $2y - \frac{1}{5} = \frac{1}{5}(\quad)$

88. $3z + \frac{3}{4} = \frac{1}{4}(\quad)$

89. $\frac{7}{8}x + \frac{5}{16}y = \frac{1}{16}(\quad)$

90. $\frac{5}{12}u - \frac{5}{8}v = \frac{1}{24}(\quad)$

In Exercises 91–94, use a graphing utility to graph both functions on the same screen. Use the graphs to verify the factorization.

91. $y_1 = 9 - 3x$
$y_2 = -3(x - 3)$

92. $y_1 = x^2 - 4x$
$y_2 = x(x - 4)$

93. $y_1 = 6x - x^2$
$y_2 = x(6 - x)$

94. $y_1 = x(x + 2) - 3(x + 2)$
$y_2 = (x + 2)(x - 3)$

Solving Problems

In Exercises 95 and 96, factor the polynomial to find the length of the rectangle.

95. Area $= 2x^2 + 2x$

96. Area $= x^2 + 2x + 10x + 20$

$x + 2$

Geometry In Exercises 97–100, write an expression for the area of the shaded region and factor the expression if possible.

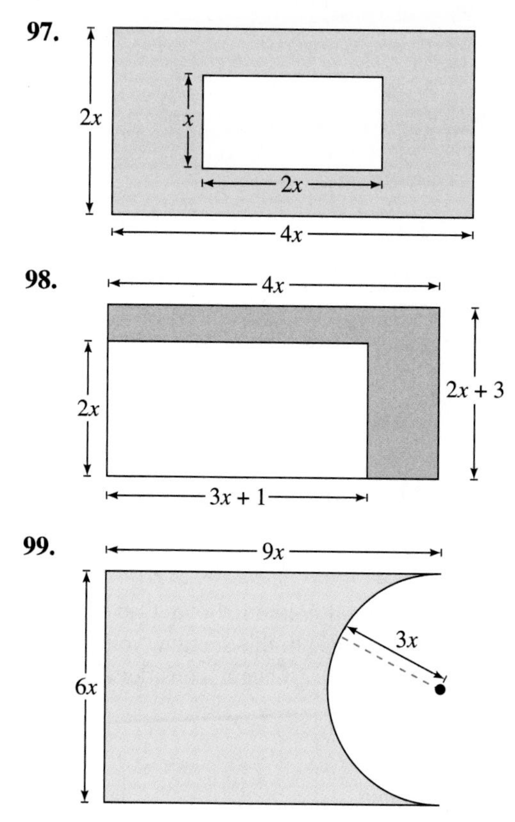

100.

101. *Geometry* The surface area of a right circular cylinder is

$$S = 2\pi r^2 + 2\pi rh.$$

Factor the expression for the surface area.

102. *Simple Interest* The amount after t years when a principal of P dollars is invested at $r\%$ simple interest is given by $P + Prt$. Factor the expression for simple interest.

103. *Chemical Reaction* The rate of change in a chemical reaction is $kQx - kx^2$, where Q is the original amount, x is the new amount, and k is a constant of proportionality. Factor the expression.

104. *Unit Price* The revenue R for selling x units of a product at a price of p dollars per unit is given by $R = xp$. For a particular commodity, the revenue is

$$900x - 0.1x^2.$$

Factor the revenue model and determine an expression that represents the price p in terms of x.

Explaining Concepts

105. Give an example of a polynomial that is written in factored form.

106. Give an example of a trinomial whose greatest common monomial factor is $3x$.

107. In your own words, describe a method for finding the greatest common factor of a polynomial.

108. How do you check your result when factoring a polynomial?

109. Explain how the word *factor* can be used as a noun and as a verb.

110. Give several examples of the use of the Distributive Property in factoring.

111. Give an example of a polynomial with four terms that can be factored by grouping. Explain the steps you used to construct the polynomial.

6.2 Factoring Trinomials

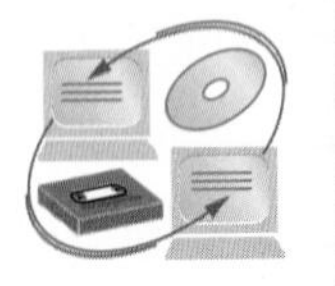

Objectives

1. Factor a trinomial of the form $x^2 + bx + c$.
2. Factor a trinomial in two variables.
3. Factor a trinomial completely.

1 Factor a trinomial of the form $x^2 + bx + c$.

Factoring Trinomials of the Form $x^2 + bx + c$

From Section 5.2, you know that the product of two binomials is often a trinomial. Here are some examples.

Factored Form	F O I L	*Trinomial Form*
$(x - 1)(x + 5)$	$= x^2 + 5x - x - 5$	$= x^2 + 4x - 5$
$(x - 3)(x - 3)$	$= x^2 - 3x - 3x + 9$	$= x^2 - 6x + 9$
$(x + 5)(x + 1)$	$= x^2 + x + 5x + 5$	$= x^2 + 6x + 5$
$(x - 2)(x - 4)$	$= x^2 - 4x - 2x + 8$	$= x^2 - 6x + 8$

Try covering the factored forms in the left-hand column above. Can you determine the factored forms from the trinomial forms? In this section, you will learn how to factor trinomials of the form $x^2 + bx + c$. To begin, consider the following factorization.

$$(x + m)(x + n) = x^2 + nx + mx + mn$$

$$= x^2 + \underbrace{(n + m)}_{\text{Sum of terms}}x + \underbrace{mn}_{\text{Product of terms}}$$

$$= x^2 + \boxed{b}\,x + \boxed{c}$$

So, to *factor* a trinomial $x^2 + bx + c$ into a product of two binomials, you must find two numbers m and n whose product is c and whose sum is b.

There are many different techniques that people use to factor trinomials. The most common is to use *Guess, Check, and Revise* with mental math.

Study Tip

With *any* factoring problem, remember that you can check your result by multiplying. For instance, in Example 1, you can check the result by multiplying $(x + 2)$ by $(x + 3)$ to see that you obtain $x^2 + 5x + 6$.

Example 1 Factoring a Trinomial

Factor $x^2 + 5x + 6$.

Solution

You need to find two numbers whose product is 6 and whose sum is 5. Using mental math, you can determine that the numbers are 2 and 3.

$$x^2 + 5x + 6 = (x + 2)(x + 3)$$

The product of 2 and 3 is 6.

The sum of 2 and 3 is 5.

Example 2 Factoring Trinomials

Factor each trinomial.

a. $x^2 + 5x - 6$ **b.** $x^2 - x - 6$ **c.** $x^2 - 5x + 6$ **d.** $14 + 5x - x^2$

Study Tip

Use a list to help you find the two numbers with the required product and sum. For Example 2(b):

Factors of -6	*Sum*
$1, -6$	-5
$-1, 6$	5
$2, -3$	-1
$-2, 3$	1

Because -1 is the required sum, the correct factorization is

$$x^2 - x - 6 = (x + 2)(x - 3).$$

Solution

a. You need to find two numbers whose product is -6 and whose sum is 5.

The product of -1 and 6 is -6.

$$x^2 + 5x - 6 = (x - 1)(x + 6)$$

The sum of -1 and 6 is 5.

b. You need to find two numbers whose product is -6 and whose sum is -1.

The product of -3 and 2 is -6.

$$x^2 - x - 6 = (x - 3)(x + 2)$$

The sum of -3 and 2 is -1.

c. You need to find two numbers whose product is 6 and whose sum is -5.

The product of -2 and -3 is 6.

$$x^2 - 5x + 6 = (x - 2)(x - 3)$$

The sum of -2 and -3 is -5.

d. It is helpful to first factor out -1. So,

$$14 + 5x - x^2 = -1(x^2 - 5x - 14).$$

Now you need two numbers -7 and 2 whose product is -14 and whose sum is -5. So,

$$14 + 5x - x^2 = -(x^2 - 5x - 14) = -(x - 7)(x + 2).$$

If you have trouble factoring a trinomial, it helps to make a list of all the distinct pairs of factors and then check each sum. For instance, consider the trinomial

$$x^2 - 5x - 24.$$

For this trinomial, you need to find two numbers whose product is -24 and whose sum is -5.

Factors of -24	*Sum*	
$1, -24$	-23	
$-1, 24$	23	
$2, -12$	-10	
$-2, 12$	10	
$3, -8$	-5	Correct choice
$-3, 8$	5	
$4, -6$	-2	
$-4, 6$	2	

So, $x^2 - 5x - 24 = (x + 3)(x - 8)$.

With experience, you will be able to narrow the list of possible factors *mentally* to only two or three possibilities whose sums can then be tested to determine the correct factorization. Here are some suggestions for narrowing the list.

▶ Guidelines for Factoring $x^2 + bx + c$

To factor $x^2 + bx + c$, you need to find two numbers m and n whose product is c and whose sum is b.

$$x^2 + bx + c = (x + m)(x + n)$$

1. If c is *positive*, then m and n have like signs that match the sign of b.
2. If c is *negative*, then m and n have unlike signs.
3. If $|b|$ is small relative to $|c|$, first try those factors of c that are closest to each other in absolute value.

Example 3 Factoring Trinomials

Factor the following trinomials.

a. $x^2 - 2x - 15$

b. $x^2 + 20x + 36$

c. $x^2 + 7x - 30$

Study Tip

Notice that factors may be written in any order. For example, $(x - 5)(x + 3) = (x + 3)(x - 5)$ and $(x + 2)(x + 18) = (x + 18)(x + 2)$ because of the Commutative Property of Multiplication.

Solution

a. You need to find two numbers whose product is -15 and whose sum is -2.

The product of -5 and 3 is -15.

$$x^2 - 2x - 15 = (x - 5)(x + 3)$$

The sum of -5 and 3 is -2.

b. You need to find two numbers whose product is 36 and whose sum is 20.

The product of 2 and 18 is 36.

$$x^2 + 20x + 36 = (x + 2)(x + 18)$$

The sum of 2 and 18 is 20.

c. You need to find two numbers whose product is -30 and whose sum is 7.

The product of -3 and 10 is -30.

$$x^2 + 7x - 30 = (x - 3)(x + 10)$$

The sum of -3 and 10 is 7.

Not all trinomials are factorable using integer factors. For instance,

$$x^2 - 2x - 6$$

is not factorable using integer factors. (List the factors and test them.) Such nonfactorable trinomials are called **prime polynomials.**

2 Factor a trinomial in two variables.

Factoring Trinomials in Two Variables

The first three examples each involved trinomials of the form

$x^2 + bx + c.$ Trinomial in one variable

The next two examples show how to factor trinomials of the form

$x^2 + bxy + cy^2.$ Trinomial in two variables

Note that this trinomial has two variables, x and y. However, from the factorization

$$\begin{aligned} x^2 + bxy + cy^2 &= (x + my)(x + ny) \\ &= x^2 + (m + n)xy + mny^2 \end{aligned}$$

you can see that you still need to find two factors of c whose sum is b.

Example 4 Factoring a Trinomial in Two Variables

Factor the trinomial $x^2 - xy - 12y^2$.

Solution

You need to find two numbers whose product is -12 and whose sum is -1.

The product of -4 and 3 is -12.

$$x^2 - xy - 12y^2 = (x - 4y)(x + 3y)$$

The sum of -4 and 3 is -1.

Check this result by multiplying $(x - 4y)$ by $(x + 3y)$.

Example 5 Factoring a Trinomial in Two Variables

Factor the following trinomials.

a. $y^2 - 6xy + 8x^2$ **b.** $x^2 + 11xy + 10y^2$

Solution

a. You need to find two numbers whose product is 8 and whose sum is -6.

The product of -2 and -4 is 8.

$$y^2 - 6xy + 8x^2 = (y - 2x)(y - 4x)$$

The sum of -2 and -4 is -6.

Check this result by multiplying $(y - 2x)$ by $(y - 4x)$.

b. You need to find two numbers whose product is 10 and whose sum is 11.

The product of 1 and 10 is 10.

$$x^2 + 11xy + 10y^2 = (x + y)(x + 10y)$$

The sum of 1 and 10 is 11.

Check this result by multiplying $(x + y)$ by $(x + 10y)$.

3 Factor a trinomial completely.

Factoring Completely

Some trinomials have a common monomial factor. In such cases you should first factor out the common monomial factor. Then you can try to factor the resulting trinomial by the methods of this section. This "multiple-stage factoring process" is called **factoring completely.** For instance, the trinomial

$$2x^2 - 4x - 6 = 2(x^2 - 2x - 3)$$ Factor out common monomial factor 2.

$$= 2(x - 3)(x + 1)$$ Factor trinomial.

is factored completely.

Example 6 Factoring Completely

Factor each trinomial completely.

a. $2x^2 - 12x + 10$

b. $3x^3 - 27x^2 + 54x$

c. $4y^4 + 32y^3 + 28y^2$

Solution

a. $2x^2 - 12x + 10 = 2(x^2 - 6x + 5)$ Factor out common monomial factor 2.

$= 2(x - 5)(x - 1)$ Factor trinomial.

b. $3x^3 - 27x^2 + 54x = 3x(x^2 - 9x + 18)$ Factor out common monomial factor $3x$.

$= 3x(x - 3)(x - 6)$ Factor trinomial.

c. $4y^4 + 32y^3 + 28y^2 = 4y^2(y^2 + 8y + 7)$ Factor out common monomial factor $4y^2$.

$= 4y^2(y + 1)(y + 7)$ Factor trinomial.

Check these results by multiplying the factors to see that you obtain the original trinomials.

Discussing the Concept Factoring Polynomials

Discuss this question in your class. "Is it possible to factor a polynomial such as

$$x^3 + 5x^2 - 3x - 15$$

by the method used for trinomials in this section?" Try this method on

$$x^3 + 5x^2 - 3x - 15$$

and

$$x^3 - 7x^2 + 2x - 14,$$

and then factor these polynomials by grouping. Which method do you prefer? Explain your preference.

6.2 Exercises

Integrated Review *Concepts, Skills, and Problem Solving*

Keep mathematically in shape by doing these exercises *before* the problems of this section.

Properties and Definitions

1. Explain why a function of x cannot have two y-intercepts.
2. What is the leading coefficient of the polynomial $3x - 7x^2 + 4x^3 - 4$?

Rewriting Algebraic Expressions

In Exercises 3–8, find the product.

3. $y(y + 2)$
4. $-a^2(a - 1)$
5. $(x - 2)(x - 5)$
6. $(v - 4)(v + 7)$
7. $(2x + 5)(2x - 5)$
8. $x^2(x + 1) - 5(x^2 - 2)$

Problem Solving

9. A company showed a loss of $2,500,000 during the first 6 months of a given year. If the company ended the year with an overall profit of $1,475,000, what was the profit during the second 6 months of the year?
10. Computer printer ribbons cost $11.95 per ribbon. If there are 12 ribbons per box and five boxes were ordered, determine the total cost of the order.
11. The revenue from selling x units of a product is $R = 75x$. The cost of producing x units is

    $C = 62.5x + 570.$

    In order to obtain a profit, the revenue must be greater than the cost. For what values of x will this product produce a profit?
12. The minimum and maximum speeds on an interstate highway are 40 miles per hour and 65 miles per hour. You travel nonstop for $3\frac{1}{2}$ hours on this highway. Assuming that you stay within the speed limits, write an inequality for the distance you travel.

Developing Skills

In Exercises 1–8, find the missing factor. Then check your answer by multiplying the factors.

1. $x^2 + 4x + 3 = (x + 3)(\quad)$
2. $x^2 + 5x + 6 = (x + 3)(\quad)$
3. $a^2 + a - 6 = (a + 3)(\quad)$
4. $c^2 + 2c - 3 = (c + 3)(\quad)$
5. $y^2 - 2y - 15 = (y + 3)(\quad)$
6. $y^2 - 4y - 21 = (y + 3)(\quad)$
7. $z^2 - 5z + 6 = (z - 3)(\quad)$
8. $z^2 - 4z + 3 = (z - 3)(\quad)$

In Exercises 9–12, find all possible products of the form $(x + m)(x + n)$ where $m \cdot n$ is the specified product. (Assume that m and n are integers.)

9. $m \cdot n = 11$
10. $m \cdot n = 10$
11. $m \cdot n = 12$
12. $m \cdot n = 18$

In Exercises 13–42, factor the trinomial. (*Note:* Some of the trinomials may be prime.) See Examples 1–5.

13. $x^2 + 6x + 8$
14. $x^2 + 13x + 12$
15. $x^2 - 13x + 40$
16. $x^2 - 9x + 14$
17. $z^2 - 7z + 12$
18. $x^2 + 10x + 24$
19. $y^2 + 5y + 11$
20. $s^2 - 7s - 25$
21. $x^2 - x - 6$
22. $x^2 + x - 6$
23. $x^2 + 2x - 15$
24. $b^2 - 2b - 15$
25. $y^2 - 6y + 10$
26. $c^2 - 6c + 10$
27. $u^2 - 22u - 48$
28. $x^2 - x - 36$
29. $x^2 + 19x + 60$
30. $x^2 + 3x - 70$
31. $x^2 - 17x + 72$
32. $x^2 + 21x + 108$

33. $x^2 - 8x - 240$

34. $r^2 - 30r + 216$

35. $x^2 + xy - 2y^2$

36. $x^2 - 5xy + 6y^2$

37. $x^2 + 8xy + 15y^2$

38. $u^2 - 4uv - 5v^2$

39. $x^2 - 7xz - 18z^2$

40. $x^2 + 15xy + 50y^2$

41. $a^2 + 2ab - 15b^2$

42. $y^2 + 4yz - 60z^2$

In Exercises 43–60, factor the trinomial completely. (*Note:* Some of the trinomials may be prime.) See Example 6.

43. $3x^2 + 21x + 30$

44. $4x^2 - 32x + 60$

45. $4y^2 - 8y - 12$

46. $5x^2 - 20x - 25$

47. $3z^2 + 5z + 6$

48. $7x^2 + 5x + 10$

49. $9x^2 + 18x - 18$

50. $6x^2 - 24x - 6$

51. $x^3 - 13x^2 + 30x$

52. $x^3 + x^2 - 2x$

53. $x^4 - 5x^3 + 6x^2$

54. $x^4 + 3x^3 - 10x^2$

55. $-3y^2x - 9yx + 54x$

56. $-5x^2z + 15xz + 50z$

57. $x^3 + 5x^2y + 6xy^2$

58. $x^2y - 6xy^2 + y^3$

59. $2x^3y + 4x^2y^2 - 6xy^3$

60. $x^4y^2 + 3x^3y^3 + 2x^2y^4$

In Exercises 61–66, find all integer values of b such that the trinomial can be factored.

61. $x^2 + bx + 15$

62. $x^2 + bx + 10$

63. $x^2 + bx - 21$

64. $x^2 + bx - 18$

65. $x^2 + bx + 36$

66. $x^2 + bx - 48$

In Exercises 67–72, find two integer values of c such that the trinomial can be factored. (There are many correct answers.)

67. $x^2 + 3x + c$

68. $x^2 + 5x + c$

69. $x^2 - 6x + c$

70. $x^2 - 15x + c$

71. $x^2 - 9x + c$

72. $x^2 + 12x + c$

Graphical Verification In Exercises 73–76, use a graphing utility to graph the two functions in the same viewing rectangle. What can you conclude?

73. $y_1 = x^2 - x - 6$
$y_2 = (x + 2)(x - 3)$

74. $y_1 = x^2 - 10x + 16$
$y_2 = (x - 2)(x - 8)$

75. $y_1 = x^3 + x^2 - 20x$
$y_2 = x(x - 4)(x + 5)$

76. $y_1 = 2x - x^2 - x^3$
$y_2 = x(1 - x)(2 + x)$

Geometric Model of Factoring In Exercises 77–80, factor the trinomial and draw a geometric model of the result. [The sample shows a geometric model for factoring $x^2 + 3x + 2 = (x + 1)(x + 2)$.]

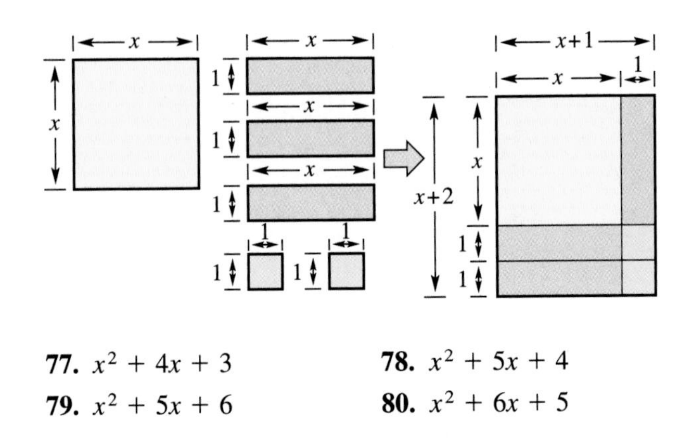

77. $x^2 + 4x + 3$

78. $x^2 + 5x + 4$

79. $x^2 + 5x + 6$

80. $x^2 + 6x + 5$

Solving Problems

81. *Exploration* An open box is to be made from a 4-foot-by-6-foot sheet of metal by cutting equal squares from the corners and turning up the sides (see figure). The volume of the box can be modeled by

$V = 4x^3 - 20x^2 + 24x, \quad 0 < x < 2.$

(a) Factor the trinomial modeling the volume of the box. Use the factored form to explain how the model was found.

(b) Use a graphing utility to graph the trinomial over the specified interval. Use the graph to approximate the size of the squares to be cut from the corners so that the volume of the box is greatest.

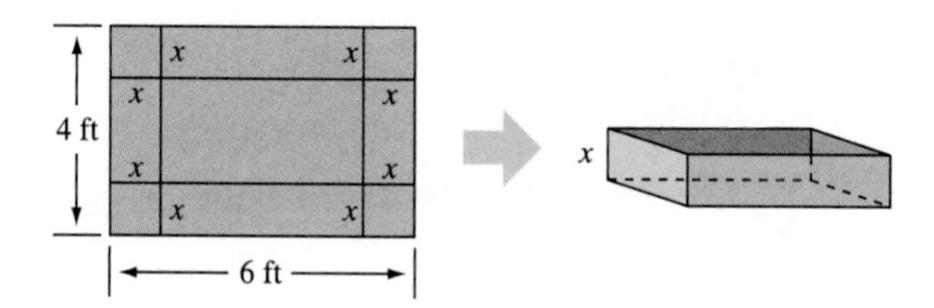

Figure for 81

82. *Geometry* The area of the rectangle in the figure is $x^2 + 30x + 200$. What is the area of the shaded region?

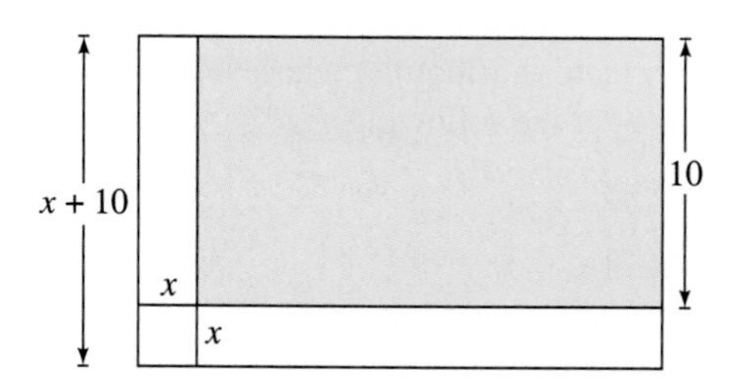

83. *Geometry* The area of the rectangle in the figure is $x^2 + 17x + 70$. What is the area of the shaded region?

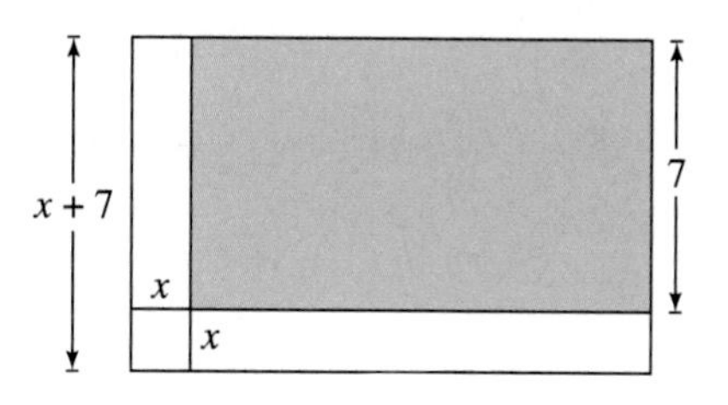

Explaining Concepts

84. State which of the following are factorizations of $2x^2 + 6x - 20$. For each correct factorization, state whether or not it is complete.

(a) $(2x - 4)(x + 5)$ (b) $(2x - 4)(2x + 10)$

(c) $(x - 2)(x + 5)$ (d) $2(x - 2)(x + 5)$

85. In factoring $x^2 - 4x + 3$, why is it unnecessary to test $(x - 1)(x + 3)$ and $(x + 1)(x - 3)$?

86. In your own words, explain how to factor a trinomial of the form $x^2 + bx + c$. Give examples with your explanation.

87. What is meant by a prime trinomial?

88. Can you completely factor a trinomial into two different sets of prime factors? Explain.

89. In factoring the trinomial $x^2 + bx + c$, is the process easier if c is a prime number such as 5 or a composite number such as 120? Explain.

6.3 More About Factoring Trinomials

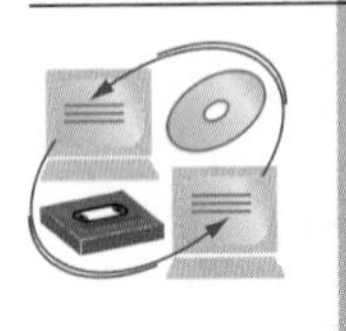

Objectives

1. Factor a trinomial of the form $ax^2 + bx + c$.
2. Factor a trinomial completely.
3. Factor a trinomial by grouping.

1 Factor a trinomial of the form $ax^2 + bx + c$.

Factoring Trinomials of the Form $ax^2 + bx + c$

In this section you will learn how to factor a trinomial whose leading coefficient is *not* 1. To see how this works, consider the following.

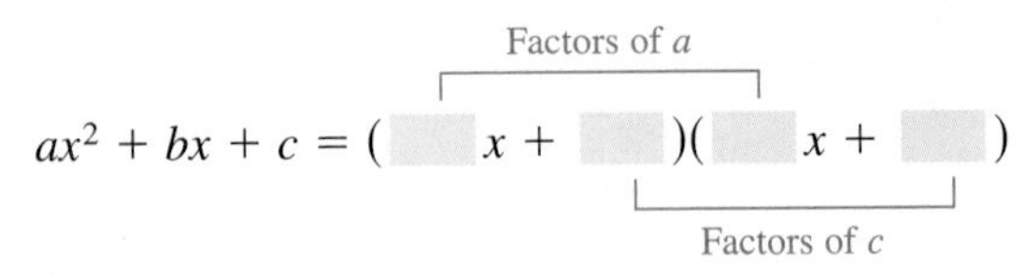

The goal is to find a combination of factors of a and c such that the outer and inner products add up to the middle term bx.

Example 1 Factoring a Trinomial of the Form $ax^2 + bx + c$

Factor $4x^2 - 4x - 3$.

Solution

In this trinomial, $a = 4$ and $c = -3$. You need to find a combination of the factors of 4 and -3 such that the outer and inner products add up to $-4x$. The possible combinations are as follows.

Factors	$O + I$	
Inner product $= 4x$; $(x + 1)(4x - 3)$; Outer product $= -3x$	$-3x + 4x = x$	x does not equal $-4x$.
Inner product $= -4x$; $(x - 1)(4x + 3)$; Outer product $= 3x$	$3x - 4x = -x$	$-x$ does not equal $-4x$.
$(x + 3)(4x - 1)$	$-x + 12x = 11x$	$11x$ does not equal $-4x$.
$(x - 3)(4x + 1)$	$x - 12x = -11x$	$-11x$ does not equal $-4x$.
$(2x + 1)(2x - 3)$	$-6x + 2x = -4x$	$-4x$ equals $-4x$. ✓
$(2x - 1)(2x + 3)$	$6x - 2x = 4x$	$4x$ does not equal $-4x$.

So, the correct factorization is

$$4x^2 - 4x - 3 = (2x + 1)(2x - 3).$$

Example 2 Factoring a Trinomial of the Form $ax^2 + bx + c$

Factor $6x^2 + 5x - 4$.

Solution

In this trinomial, $a = 6$ and $c = -4$. You need to find a combination of the factors of 6 and -4 such that the outer and inner products add up to $5x$.

Factors	$O + I$	
$(x + 1)(6x - 4)$	$-4x + 6x = 2x$	$2x$ does not equal $5x$.
$(x - 1)(6x + 4)$	$4x - 6x = -2x$	$-2x$ does not equal $5x$.
$(x + 4)(6x - 1)$	$-x + 24x = 23x$	$23x$ does not equal $5x$.
$(x - 4)(6x + 1)$	$x - 24x = -23x$	$-23x$ does not equal $5x$.
$(x + 2)(6x - 2)$	$-2x + 12x = 10x$	$10x$ does not equal $5x$.
$(x - 2)(6x + 2)$	$2x - 12x = -10x$	$-10x$ does not equal $5x$.
$(2x + 1)(3x - 4)$	$-8x + 3x = -5x$	$-5x$ does not equal $5x$.
$(2x - 1)(3x + 4)$	$8x - 3x = 5x$	$5x$ equals $5x$. ✓
$(2x + 4)(3x - 1)$	$-2x + 12x = 10x$	$10x$ does not equal $5x$.
$(2x - 4)(3x + 1)$	$2x - 12x = -10x$	$-10x$ does not equal $5x$.
$(2x + 2)(3x - 2)$	$-4x + 6x = 2x$	$2x$ does not equal $5x$.
$(2x - 2)(3x + 2)$	$4x - 6x = -2x$	$-2x$ does not equal $5x$.

So, the correct factorization is $6x^2 + 5x - 4 = (2x - 1)(3x + 4)$.

Study Tip

If the original trinomial has no common monomial factors, then its binomial factors can't have common monomial factors. So, in Example 2, you don't have to test factors such as $(6x - 4)$ that have a common monomial factor of 2.

The following guidelines can help shorten the list of possible factorizations.

▶ Guidelines for Factoring $ax^2 + bx + c$ $(a > 0)$

1. First, factor out any common monomial factor.
2. Because the resulting trinomial has no common monomial factors, you don't have to test any binomial factors that have a common monomial factor.
3. If the middle-term test (O + I) yields the opposite of b, switch the signs of the factors of c.

Using these guidelines, you can shorten the list in Example 2 to the following.

$$(x + 4)(6x - 1) = 6x^2 + 23x - 4$$

$$(x - 4)(6x + 1) = 6x^2 - 23x - 4$$

$$(2x + 1)(3x - 4) = 6x^2 - 5x - 4$$

$$(2x - 1)(3x + 4) = 6x^2 + 5x - 4 \quad \text{Correct factorization}$$

Do you see why you can cut the list from 12 possible factorizations to only four?

As with other types of factoring, you can use a graphing utility to check your results. For instance, graph

$$y = 2x^2 + x - 15 \text{ and}$$
$$y = (2x - 5)(x + 3)$$

on the same screen, as shown below. Because both graphs are the same, you can reason that

$$2x^2 + x - 15$$
$$= (2x - 5)(x + 3).$$

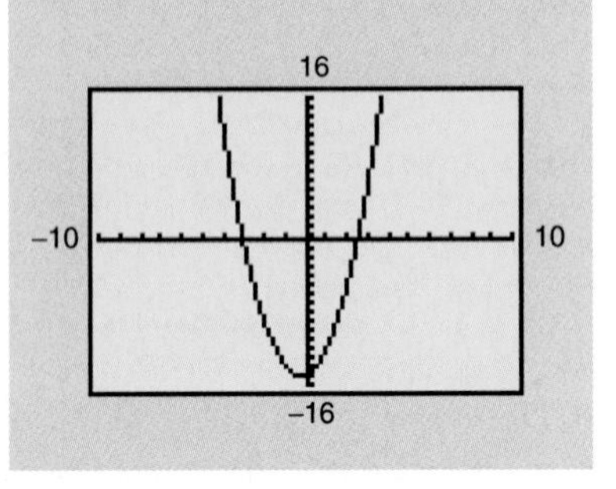

Example 3 Factoring a Trinomial of the Form $ax^2 + bx + c$

Factor $2x^2 + x - 15$.

Solution

In this trinomial, $a = 2$, which factors as $(1)(2)$, and $c = -15$, which factors as $(1)(-15)$, $(-1)(15)$, $(3)(-5)$, and $(-3)(5)$.

$$(2x + 1)(x - 15) = 2x^2 - 29x - 15$$
$$(2x + 15)(x - 1) = 2x^2 + 13x - 15$$
$$(2x + 3)(x - 5) = 2x^2 - 7x - 15$$
$$(2x + 5)(x - 3) = 2x^2 - x - 15 \qquad \text{Middle term has incorrect sign.}$$
$$(2x - 5)(x + 3) = 2x^2 + x - 15 \qquad \text{Correct factorization}$$

So, the correct factorization is

$$2x^2 + x - 15 = (2x - 5)(x + 3).$$

Notice in Example 3 that when the middle term has the incorrect sign, you need only to swap the signs of the second term of each factor.

Factoring Completely

Remember that if a trinomial has a common monomial factor, the common monomial factor should be factored out first. The complete factorization then shows all monomial and binomial factors.

2 Factor a trinomial completely.

Example 4 Factoring Completely

Factor $4x^2 - 30x + 14$.

Solution

Begin by factoring out the common monomial factor.

$$4x^2 - 30x + 14 = 2(2x^2 - 15x + 7)$$

Now, for the trinomial $2x^2 - 15x + 7$, $a = 2$ and $c = 7$. The possible factorizations of this trinomial are listed below.

$$(2x - 7)(x - 1) = 2x^2 - 9x + 7$$
$$(2x - 1)(x - 7) = 2x^2 - 15x + 7 \qquad \text{Correct factorization}$$

So, the complete factorization of the original trinomial is

$$4x^2 - 30x + 14 = 2(2x^2 - 15x + 7)$$
$$= 2(2x - 1)(x - 7).$$

In factoring a trinomial with a negative leading coefficient, we suggest that you first factor -1 out of the trinomial.

Example 5 A Negative Leading Coefficient

Factor $-5x^2 + 7x + 6$.

Solution

Begin by factoring -1 out of the trinomial.

$$-5x^2 + 7x + 6 = (-1)(5x^2 - 7x - 6)$$

Now, for the trinomial $5x^2 - 7x - 6$, $a = 5$ and $c = -6$. After testing the possible factorizations, you can conclude that

$$(x - 2)(5x + 3) = 5x^2 - 7x - 6. \qquad \text{Correct factorization}$$

So, a correct factorization is

$$\begin{aligned} -5x^2 + 7x + 6 &= (-1)(x - 2)(5x + 3) \\ &= (-x + 2)(5x + 3). \end{aligned}$$

Another correct factorization is $(x - 2)(-5x - 3)$.

3 Factor a trinomial by grouping.

Factoring by Grouping

The examples in this and the preceding section have shown how to use *Guess, Check, and Revise* to factor trinomials. An alternative technique that some people like to use is factoring by grouping. Recall from Section 6.1 that the polynomial

$$x^3 + 2x^2 + 3x + 6$$

was factored by first grouping terms and then applying the Distributive Property.

$$\begin{aligned} x^3 + 2x^2 + 3x + 6 &= (x^3 + 2x^2) + (3x + 6) && \text{Group terms.} \\ &= x^2(x + 2) + 3(x + 2) && \text{Distributive Property} \\ &= (x + 2)(x^2 + 3) && \text{Distributive Property} \end{aligned}$$

By rewriting the middle term of the trinomial $2x^2 + x - 15$ as

$$2x^2 + x - 15 = 2x^2 + 6x - 5x - 15$$

you can group the first two terms and the last two terms and factor the trinomial as shown.

$$\begin{aligned} 2x^2 + x - 15 &= 2x^2 + (6x - 5x) - 15 && \text{Rewrite middle term.} \\ &= (2x^2 + 6x) - (5x + 15) && \text{Group terms.} \\ &= 2x(x + 3) - 5(x + 3) && \text{Distributive Property} \\ &= (x + 3)(2x - 5) && \text{Distributive Property} \end{aligned}$$

The key to this method of factoring is knowing how to rewrite the middle term. In general, *to factor a trinomial $ax^2 + bx + c$ by grouping, choose factors of the product ac that add up to b and use these factors to rewrite the middle term.*

Example 6 Factoring a Trinomial by Grouping

Use factoring by grouping to factor the trinomial

$$2x^2 + 5x - 3.$$

Solution

In the trinomial $2x^2 + 5x - 3$, $ac = 2(-3) = -6$, which has factors 6 and -1 that add up to 5. So, rewrite the middle term as $5x = 6x - x$. This produces the following.

$$\begin{aligned} 2x^2 + 5x - 3 &= 2x^2 + 6x - x - 3 && \text{Rewrite middle term.} \\ &= (2x^2 + 6x) - (x + 3) && \text{Group terms.} \\ &= 2x(x + 3) - (x + 3) && \text{Distributive Property} \\ &= (x + 3)(2x - 1) && \text{Distributive Property} \end{aligned}$$

So, the trinomial factors as

$$2x^2 + 5x - 3 = (x + 3)(2x - 1).$$

Example 7 Factoring a Trinomial by Grouping

Use factoring by grouping to factor the trinomial

$$6x^2 - 11x - 10.$$

Solution

In the trinomial $6x^2 - 11x - 10$, $ac = 6(-10) = -60$, which has the factors -15 and 4 that add up to -11. So, rewrite the middle term as $-11x = -15x + 4x$. This produces the following.

$$\begin{aligned} 6x^2 - 11x - 10 &= 6x^2 - 15x + 4x - 10 && \text{Rewrite middle term.} \\ &= (6x^2 - 15x) + (4x - 10) && \text{Group terms.} \\ &= 3x(2x - 5) + 2(2x - 5) && \text{Distributive Property} \\ &= (2x - 5)(3x + 2) && \text{Distributive Property} \end{aligned}$$

So, the trinomial factors as

$$6x^2 - 11x - 10 = (2x - 5)(3x + 2).$$

Discussing the Concept **Factoring Trinomials**

What do you think of the technique of factoring a trinomial by grouping? Many people think it is more efficient than the *Guess, Check, and Revise* strategy, especially when the coefficients a and c have many factors. Try factoring $6x^2 - 13x + 6$, $2x^2 + 5x - 12$, and $3x^2 + 11x - 4$ using both methods. Which method do you prefer? Explain the advantages and disadvantages of each method.

6.3 Exercises

Integrated Review *Concepts, Skills, and Problem Solving*

Keep mathematically in shape by doing these exercises *before* the problems of this section.

Properties and Definitions

1. Is 29 prime or composite?

2. Without dividing 255 by 3, how can you tell whether it is divisible by 3?

Simplifying Expressions

In Exercises 3–6, write the prime factorization.

3. 500 **4.** 315 **5.** 792 **6.** 2275

In Exercises 7 and 8, multiply and simplify.

7. $(2x - 5)(x + 7)$ **8.** $(3x - 2)^2$

Graphs and Models

In Exercises 9 and 10, graph the function and identify any intercepts.

9. $f(x) = (3 + x)(3 - x)$

10. $g(t) = 2t - 1$

11. An equation for the distance y (in inches) a spring is stretched from its equilibrium when a force of x pounds is applied is modeled by $y = 0.066x$.

(a) Graph the model.

(b) Estimate y when a force of 100 pounds is applied.

Developing Skills

In Exercises 1–8, find the missing factor.

1. $5x^2 + 18x + 9 = (x + 3)(\quad)$

2. $5x^2 + 19x + 12 = (x + 3)(\quad)$

3. $5a^2 + 12a - 9 = (a + 3)(\quad)$

4. $5c^2 + 11c - 12 = (c + 3)(\quad)$

5. $2y^2 - 3y - 27 = (y + 3)(\quad)$

6. $3y^2 - y - 30 = (y + 3)(\quad)$

7. $4z^2 - 13z + 3 = (z - 3)(\quad)$

8. $6z^2 - 23z + 15 = (z - 3)(\quad)$

In Exercises 9–12, find all possible products of the form $(5x + m)(x + n)$, where $m \cdot n$ is the specified product. (Assume that m and n are integers.)

9. $m \cdot n = 3$ **10.** $m \cdot n = 21$

11. $m \cdot n = 12$ **12.** $m \cdot n = 36$

In Exercises 13–40, factor the polynomial. (*Note:* Some of the trinomials may be prime.) See Examples 1–3.

13. $2x^2 + 5x + 3$ **14.** $3x^2 + 7x + 2$

15. $4y^2 + 5y + 1$ **16.** $3x^2 + 5x - 2$

17. $2y^2 - 3y + 1$ **18.** $3a^2 - 5a + 2$

19. $2x^2 - x - 3$ **20.** $3z^2 - z - 2$

21. $5x^2 - 2x + 1$ **22.** $4z^2 - 8z + 1$

23. $2x^2 + x + 3$ **24.** $6x^2 - 10x + 5$

25. $5s^2 - 10s + 6$ **26.** $6v^2 + v - 2$

27. $4x^2 + 13x - 12$ **28.** $6y^2 - 7y - 20$

29. $9x^2 - 18x + 8$ **30.** $4a^2 - 16a + 15$

31. $18u^2 - 9u - 2$ **32.** $24s^2 + 37s - 5$

33. $15a^2 + 14a - 8$ **34.** $12x^2 - 8x - 15$

35. $10t^2 - 3t - 18$ **36.** $10t^2 + 43t - 9$

37. $15m^2 + 16m - 15$ **38.** $21b^2 - 40b - 21$

39. $16z^2 - 34z + 15$ **40.** $12x^2 - 41x + 24$

In Exercises 41–50, factor the trinomial. (*Note:* The leading coefficient is negative.) See Example 5.

41. $-2x^2 + x + 3$ **42.** $-5x^2 + x + 4$

43. $4 - 4x - 3x^2$ **44.** $-4x^2 + 17x + 15$

45. $-6x^2 + 7x + 10$ **46.** $2 + x - 6x^2$

47. $1 - 4x - 60x^2$ **48.** $2 + 5x - 12x^2$

49. $16 - 8x - 15x^2$ **50.** $20 + 17x - 10x^2$

In Exercises 51–72, factor the polynomial completely. (*Note:* Some of the polynomials may be prime.) See Examples 4 and 5.

51. $6x^2 - 3x$
52. $3a^4 - 9a^3$
53. $15y^2 + 18y$
54. $24y^3 - 16y$
55. $u(u - 3) + 9(u - 3)$
56. $x(x - 8) - 2(x - 8)$
57. $2v^2 + 8v - 42$
58. $4z^2 - 12z - 40$
59. $-3x^2 - 3x - 60$
60. $5y^2 + 40y + 35$
61. $9z^2 - 24z + 15$
62. $6x^2 + 8x - 8$
63. $4x^2 + 4x + 2$
64. $6x^2 - 6x - 36$
65. $-15x^4 - 2x^3 + 8x^2$
66. $15y^2 - 7y^3 - 2y^4$
67. $3x^3 + 4x^2 + 2x$
68. $5x^3 - 3x^2 - 4x$
69. $6x^3 + 24x^2 - 192x$
70. $35x + 28x^2 - 7x^3$
71. $18u^4 + 18u^3 - 27u^2$
72. $12x^5 - 16x^4 + 8x^3$

In Exercises 73–78, find all integers b such that the trinomial can be factored.

73. $3x^2 + bx + 10$
74. $4x^2 + bx + 3$
75. $2x^2 + bx - 6$
76. $5x^2 + bx - 6$
77. $6x^2 + bx + 20$
78. $8x^2 + bx - 18$

In Exercises 79–84, find two integer values of c such that the trinomial can be factored. (There are many correct answers.)

79. $4x^2 + 3x + c$
80. $2x^2 + 5x + c$
81. $3x^2 - 10x + c$
82. $8x^2 - 3x + c$
83. $6x^2 - 5x + c$
84. $4x^2 - 9x + c$

In Exercises 85–100, factor the trinomial by grouping. See Examples 6 and 7.

85. $3x^2 + 7x + 2$
86. $2x^2 + 5x + 2$
87. $2x^2 + x - 3$
88. $5x^2 - 14x - 3$
89. $6x^2 + 5x - 4$
90. $12y^2 + 11y + 2$
91. $15x^2 - 11x + 2$
92. $12x^2 - 13x + 1$
93. $3a^2 + 11a + 10$
94. $3z^2 - 4z - 15$
95. $16x^2 + 2x - 3$
96. $20c^2 + 19c - 1$
97. $12x^2 - 17x + 6$
98. $10y^2 - 13y - 30$
99. $6u^2 - 5u - 14$
100. $12x^2 + 28x + 15$

Geometric Model of Factoring In Exercises 101 and 102, factor the trinomial and draw a geometric model of the result. [This sample geometric model illustrates the factorization of $2x^2 + 3x + 1$ as $(2x + 1)(x + 1)$.]

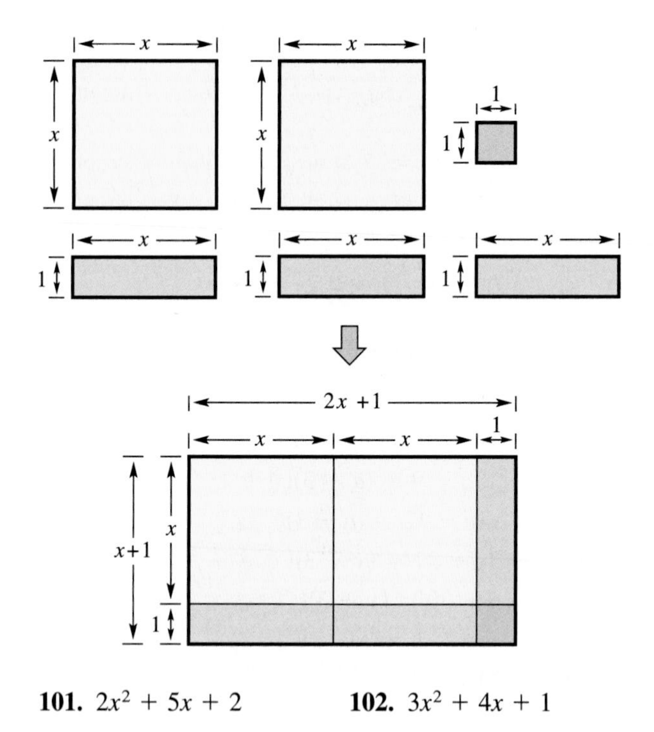

101. $2x^2 + 5x + 2$
102. $3x^2 + 4x + 1$

Solving Problems

103. *Geometry* The sandbox shown in the figure has a height of x and a width of $x + 2$. The volume of the box is $2x^3 + 7x^2 + 6x$. Find the length of the box.

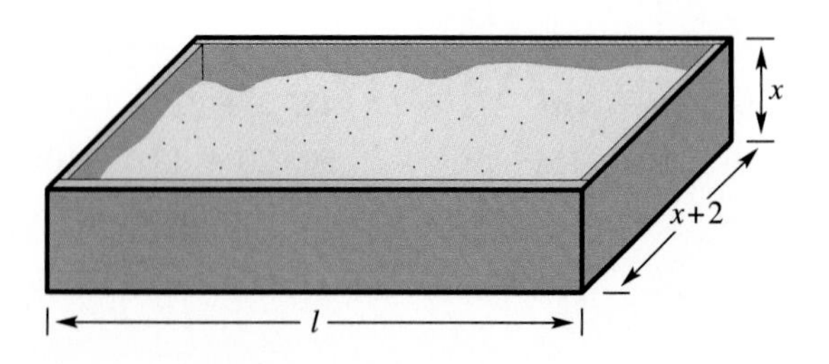

104 *Geometry* The pool shown in the figure has a depth of d and a length of $5d + 2$. The volume of the pool is $15d^3 - 14d^2 - 8d$. Find the width of the pool.

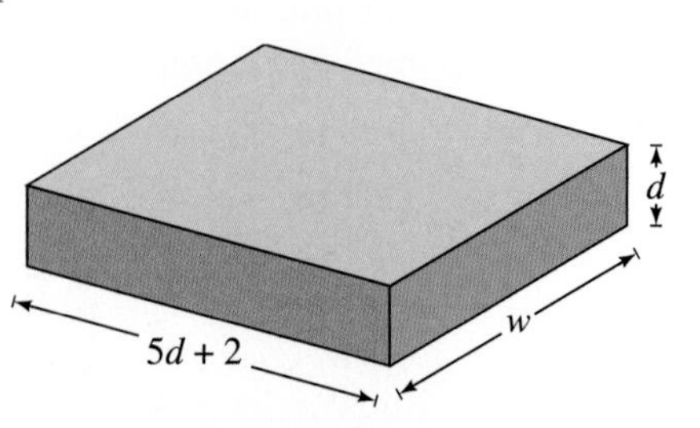

105. *Geometry* The area of the rectangle in the figure is $2x^2 + 9x + 10$. What is the area of the shaded region?

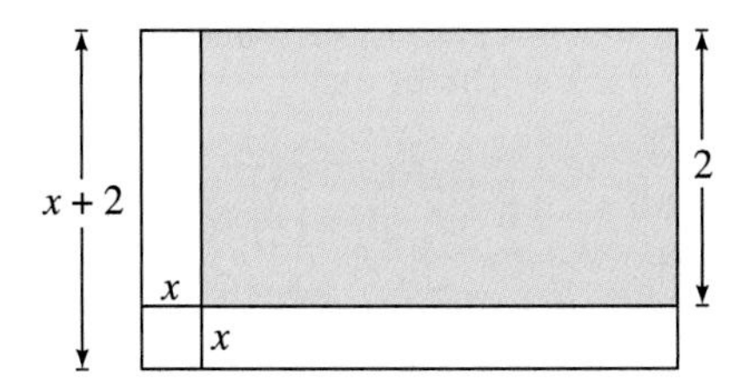

106. *Geometry* The area of the rectangle in the figure is $3x^2 + 10x + 3$. What is the area of the shaded region?

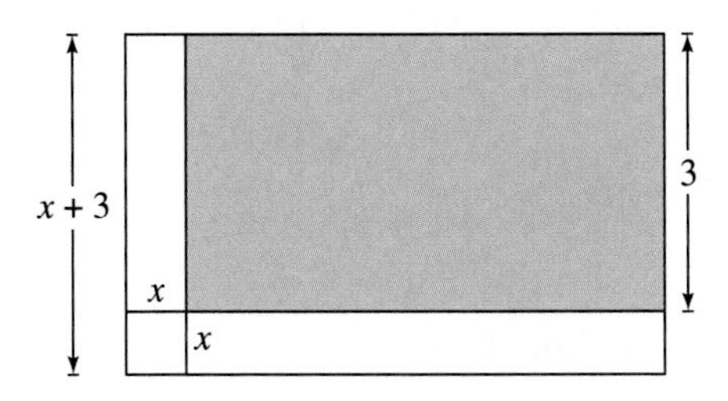

107. *Graphical Exploration* Consider the functions $y_1 = 2x^3 + 3x^2 - 5x$ and $y_2 = x(2x + 5)(x - 1)$.

(a) Factor the trinomial represented by y_1. What is the relationship between y_1 and y_2?

(b) Demonstrate your answer to part (a) graphically by using a graphing utility to graph y_1 and y_2.

(c) Identify the x- and y-intercepts of the graphs of y_1 and y_2.

108. *Beam Deflection* A cantilever beam of length l is fixed at the origin. A load weighing W pounds is attached to the end of the beam (see figure). The deflection y of the beam x units from the origin is given by

$$y = -\frac{1}{10}x^2 - \frac{1}{120}x^3, \quad 0 \le x \le 3.$$

(a) Factor the expression for the deflection. (Write the binomial factor with positive integer coefficients.)

(b) Use a graphing utility to graph the expression for deflection over the specified interval.

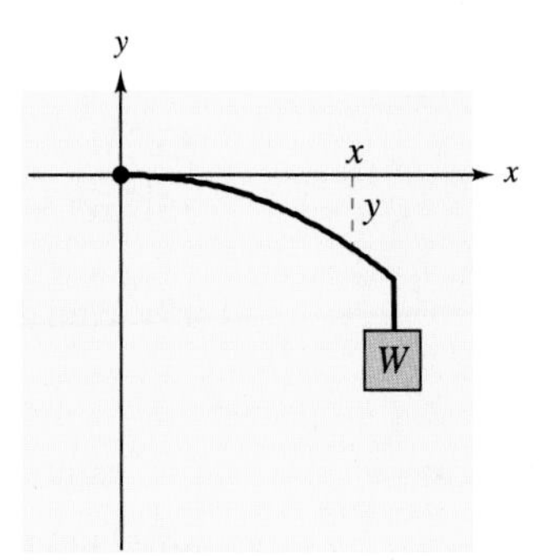

Explaining Concepts

109. Answer parts (a) and (b) of Motivating the Chapter on page 321.

110. Explain the meaning of each letter of FOIL.

111. Without multiplying, why is $(2x + 3)(x + 5)$ not a factorization of $2x^2 + 7x - 15$?

112. Find the error.

$$9x^2 - 9x - 54 = (3x + 6)(3x - 9)$$
$$= 3(x + 2)(x - 3)$$

113. In factoring $ax^2 + bx + c$, how many possible factorizations must be tested if a and c are prime? Explain your reasoning.

114. Give an example of a prime trinomial that is of the form $ax^2 + bx + c$.

115. Give an example of a trinomial of the form $ax^3 + bx^2 + cx$ that has a common monomial factor of $2x$.

116. Can a trinomial with its leading coefficient not equal to 1 have two identical factors? If so, give an example.

Mid-Chapter Quiz

Take this test as you would take a test in class. After you are done, check your work against the answers given in the back of the book.

In Exercises 1–4, find the missing factor.

1. $\frac{2}{3}x - 1 = \frac{1}{3}(\quad)$

2. $x^2y - xy^2 = xy(\quad)$

3. $y^2 + y - 42 = (y + 7)(\quad)$

4. $2x^2 - x - 1 = (x - 1)(\quad)$

In Exercises 5–16, factor the polynomial.

5. $10x^2 + 70$

6. $2a^3b - 4a^2b^2$

7. $x(x + 2) - 3(x + 2)$

8. $t^3 - 3t^2 + t - 3$

9. $y^2 + 11y + 30$

10. $u^2 + u - 30$

11. $x^3 - x^2 - 30x$

12. $2x^2y + 8xy - 64y$

13. $3v^2 - 4v - 2$

14. $6 - 13z - 5z^2$

15. $6x^2 - x - 2$

16. $10s^4 - 14s^3 + 2s^2$

17. Find all integer values of b such that the polynomial

$x^2 + bx + 12$

can be factored. Describe the method you used.

18. Find two values of c such that

$x^2 - 10x + c$

can be factored. Describe the method you used.

19. Find all possible products of the form

$(3x + m)(x + n)$

such that $mn = 6$. Describe the method you used.

20. The area of the rectangle in the figure is $3x^2 + 38x + 80$. What is the area of the shaded region?

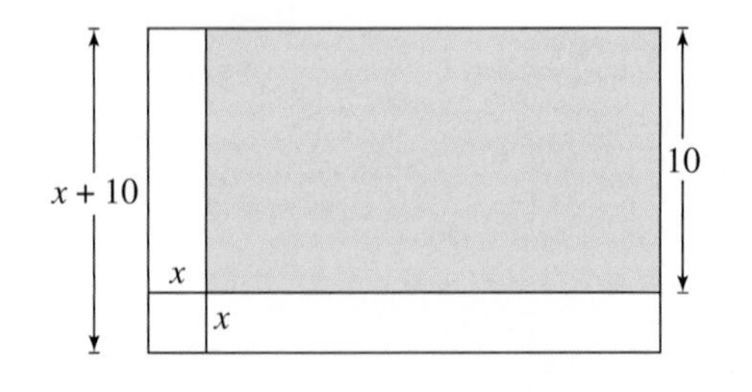

21. Use a graphing utility to graph $y_1 = -2x^2 + 11x - 12$ and $y_2 = (3 - 2x)(x - 4)$ on the same screen. What can you conclude?

6.4 Factoring Polynomials with Special Forms

Objectives

1. Factor the difference of two squares.
2. Recognize repeated factorization.
3. Identify and factor a perfect square trinomial.
4. Factor the sum and difference of two cubes.

1 Factor the difference of two squares.

Difference of Two Squares

One of the easiest special polynomial forms to recognize and to factor is the form $a^2 - b^2$. It is called a **difference of two squares,** and it factors according to the following pattern.

Technology: Discovery

Use your calculator to verify the special polynomial form called the "difference of two squares." To do so, evaluate the equation when $a = 16$ and $b = 9$. Try more values, including negative values. What can you conclude?

> **Difference of Two Squares**
>
> Let a and b be real numbers, variables, or algebraic expressions.
>
> $$a^2 - b^2 = (a + b)(a - b)$$
>
> Difference — Opposite signs

This pattern can be illustrated geometrically, as shown in Figure 6.2. The area of the shaded region on the left is represented by $a^2 - b^2$ (the area of the larger square minus the area of the smaller square). On the right, the *same* area is represented by a rectangle whose width is $a + b$ and whose length is $a - b$.

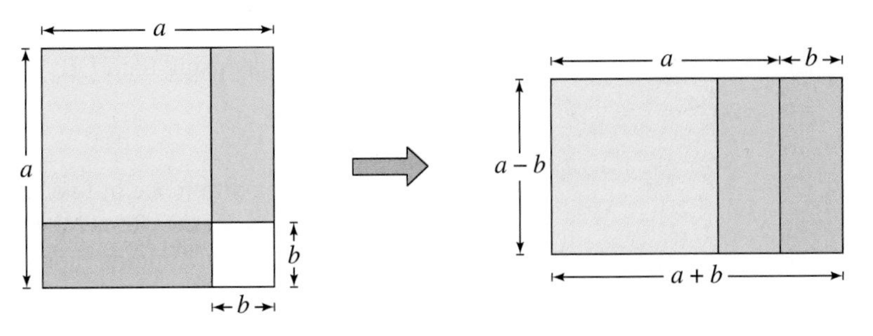

Figure 6.2

To recognize perfect squares, look for coefficients that are squares of integers and for variables raised to *even* powers. Here are some examples.

Original Polynomial		*Difference of Squares*		*Factored Form*
$x^2 - 1$	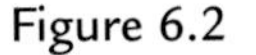	$(x)^2 - (1)^2$	⇨	$(x + 1)(x - 1)$
$4x^2 - 9$	⇨	$(2x)^2 - (3)^2$	⇨	$(2x + 3)(2x - 3)$
$25 - 64x^4$	⇨	$(5)^2 - (8x^2)^2$	⇨	$(5 + 8x^2)(5 - 8x^2)$

Study Tip

When factoring a polynomial, remember that you can check your result by multiplying the factors. For instance, you can check the factorization in Example 1(a) as follows.

$$(x + 6)(x - 6) = x^2 - 36$$

Example 1 Factoring the Difference of Two Squares

Factor each polynomial.

a. $x^2 - 36$ **b.** $x^2 - \frac{4}{25}$ **c.** $81x^2 - 49$

Solution

a. $x^2 - 36 = x^2 - 6^2$ Write as a difference of squares.

$= (x + 6)(x - 6)$ Factored form

b. $x^2 - \frac{4}{25} = x^2 - \left(\frac{2}{5}\right)^2$ Write as a difference of squares.

$= \left(x + \frac{2}{5}\right)\left(x - \frac{2}{5}\right)$ Factored form

c. $81x^2 - 49 = (9x)^2 - 7^2$ Write as a difference of squares.

$= (9x + 7)(9x - 7)$ Factored form

Check your results by using the FOIL Method.

The rule $u^2 - v^2 = (u + v)(u - v)$ applies to polynomials or expressions in which u and v are themselves expressions.

Example 2 Factoring the Difference of Two Squares

Factor $(x + 1)^2 - 4$.

Solution

$(x + 1)^2 - 4 = (x + 1)^2 - 2^2$ Write as a difference of squares.

$= [(x + 1) + 2][(x + 1) - 2]$ Factored form

$= (x + 3)(x - 1)$ Simplify.

Check your result by using the FOIL Method.

Sometimes the difference of two squares can be hidden by the presence of a common monomial factor. Remember that with all factoring techniques, you should first remove any common monomial factors.

Example 3 Removing a Common Monomial Factor First

Factor $20x^3 - 5x$.

Solution

$20x^3 - 5x = 5x(4x^2 - 1)$ Factor out common monomial factor $5x$.

$= 5x[(2x)^2 - 1^2]$ Write as a difference of squares.

$= 5x(2x + 1)(2x - 1)$ Factored form

2 Recognize repeated factorization.

Repeated Factorization

To factor a polynomial completely, you should always check to see whether the factors obtained might themselves be factorable. That is, can any of the factors be factored? For instance, after factoring the polynomial $(x^4 - 1)$ once as the difference of two squares

$$x^4 - 1 = (x^2)^2 - 1^2 \quad \text{Write as a difference of squares.}$$

$$= (x^2 + 1)(x^2 - 1) \quad \text{Factored form}$$

you can see that the second factor is itself the difference of two squares. So, to factor the polynomial *completely*, you must continue the factoring process.

$$x^4 - 1 = (x^2 + 1)(x^2 - 1) \quad \text{Factor as a difference of squares.}$$

$$= (x^2 + 1)(x + 1)(x - 1) \quad \text{Factor completely.}$$

Another example of repeated factoring is shown in the next example.

Study Tip

Note in Example 4 that no attempt was made to factor the *sum of two squares*. A second-degree polynomial that is the sum of two squares cannot be factored as the product of binomials (using integer coefficients). For instance, the second-degree polynomials

$$x^2 + 4$$

and

$$4x^2 + 9$$

cannot be factored using integer coefficients. In general, *the sum of two squares is not factorable.*

Example 4 Factoring Completely

Factor $x^4 - 16$ completely.

Solution

Recognizing $x^4 - 16$ as a difference of two squares, you can write

$$x^4 - 16 = (x^2)^2 - 4^2 \quad \text{Write as a difference of squares.}$$

$$= (x^2 + 4)(x^2 - 4). \quad \text{Factored form}$$

Note that the second factor $(x^2 - 4)$ is itself a difference of two squares and so

$$x^4 - 16 = (x^2 + 4)(x^2 - 4) \quad \text{Factor as a difference of squares.}$$

$$= (x^2 + 4)(x + 2)(x - 2). \quad \text{Factor completely.}$$

Example 5 Factoring Completely

Factor $48x^4 - 3$ completely.

Solution

Start by removing the common monomial factor.

$$48x^4 - 3 = 3(16x^4 - 1) \quad \text{Remove common monomial factor 3.}$$

Recognizing $16x^4 - 1$ as the difference of two squares, you can write

$$48x^4 - 3 = 3(16x^4 - 1) \quad \text{Factor monomial.}$$

$$= 3[(4x^2)^2 - 1^2] \quad \text{Write as a difference of squares.}$$

$$= 3(4x^2 + 1)(4x^2 - 1) \quad \text{Recognize } 4x^2 - 1 \text{ as a difference of squares.}$$

$$= 3(4x^2 + 1)[(2x)^2 - 1^2] \quad \text{Write as a difference of squares.}$$

$$= 3(4x^2 + 1)(2x + 1)(2x - 1). \quad \text{Factor completely.}$$

3 Identify and factor a perfect square trinomial.

Perfect Square Trinomials

A **perfect square trinomial** is the square of a binomial. For instance,

$$x^2 + 4x + 4 = (x + 2)^2$$

is the square of the binomial $(x + 2)$. Perfect square trinomials come in two patterns, one in which the middle term is positive and the other in which the middle term is negative. In both cases, the first and last terms are perfect squares and positive.

▶ Perfect Square Trinomials

Let a and b be real numbers, variables, or algebraic expressions.

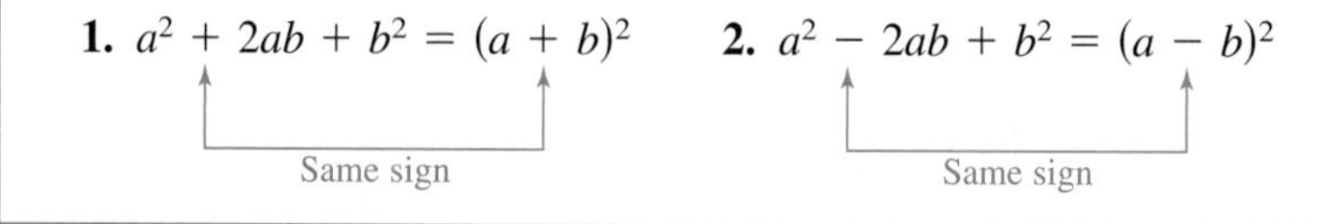

Study Tip

To recognize a perfect square trinomial, remember that the first and last terms must be perfect squares and positive, and the middle term must be twice the product of a and b. (The middle term can be positive or negative.) Watch for squares of fractions.

$$4x^2 - \tfrac{4}{3}x + \tfrac{1}{9}$$

$(2x)^2$, $2(2x)\left(\tfrac{1}{3}\right)$, $\left(\tfrac{1}{3}\right)^2$

Example 6 Identifying Perfect Square Trinomials

Which of the following are perfect square trinomials?

a. $m^2 - 4m + 4$ **b.** $4x^2 - 2x + 1$

c. $y^2 + 6y - 9$ **d.** $x^2 + x + \frac{1}{4}$

Solution

a. This polynomial *is* a perfect square trinomial. It factors as $(m - 2)^2$.

b. This polynomial *is not* a perfect square trinomial because the middle term is not twice the product of $2x$ and 1.

c. This polynomial *is not* a perfect square trinomial because the last term, -9, is not positive.

d. This polynomial *is* a perfect square trinomial. The first and last terms are perfect squares, x^2 and $\left(\frac{1}{2}\right)^2$, and it factors as $\left(x + \frac{1}{2}\right)^2$.

Example 7 Factoring Perfect Square Trinomials

a. $y^2 - 6y + 9 = y^2 - 2(3y) + 3^2$ Recognize the pattern.

$= (y - 3)^2$ Write in factored form.

b. $16x^2 + 40x + 25 = (4x)^2 + 2(4x)(5) + 5^2$ Recognize the pattern.

$= (4x + 5)^2$ Write in factored form.

c. $9x^2 - 24xy + 16y^2 = (3x)^2 - 2(3x)(4y) + (4y)^2$ Recognize the pattern.

$= (3x - 4y)^2$ Write in factored form.

4 Factor the sum and difference of two cubes.

Sum and Difference of Two Cubes

The last type of special factoring that you will study in this section is the sum and difference of two *cubes*. The patterns for these two special forms are summarized below.

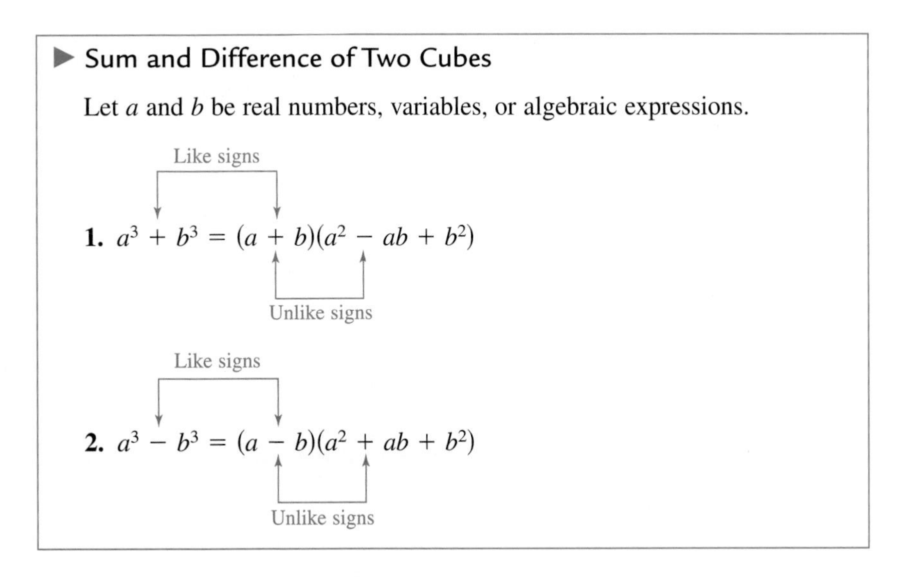

▶ Sum and Difference of Two Cubes

Let a and b be real numbers, variables, or algebraic expressions.

1. $a^3 + b^3 = (a + b)(a^2 - ab + b^2)$

2. $a^3 - b^3 = (a - b)(a^2 + ab + b^2)$

When using either of these factoring patterns, pay special attention to the signs, as indicated above. Remembering the "like" and "unlike" patterns for the signs is helpful.

Study Tip

It is easy to make arithmetic errors when applying the patterns for factoring the sum or difference of two cubes. When you use these patterns, be sure to check your work by multiplying the two factors. For instance, you can check the factorization in Example 8(a) as shown.

$$\begin{array}{r} y^2 - 3y + 9 \\ y + 3 \\ \hline 3y^2 - 9y + 27 \\ y^3 - 3y^2 + 9y \\ \hline y^3 + 27 \end{array}$$

Example 8 Factoring Sums and Differences of Two Cubes

Factor the polynomials.

a. $y^3 + 27$ **b.** $64 - x^3$ **c.** $2x^3 - 16$

Solution

a. $y^3 + 27 = y^3 + 3^3$ Write as sum of two cubes.

$= (y + 3)[y^2 - (y)(3) + 3^2]$ Factored form

$= (y + 3)(y^2 - 3y + 9)$ Simplify.

b. $64 - x^3 = 4^3 - x^3$ Write as difference of two cubes.

$= (4 - x)(4^2 + 4x + x^2)$ Factored form

$= (4 - x)(16 + 4x + x^2)$ Simplify.

c. $2x^3 - 16 = 2(x^3 - 8)$ Factor out common monomial factor 2.

$= 2(x^3 - 2^3)$ Write as a difference of two cubes.

$= 2(x - 2)[x^2 + (x)(2) + 2^2]$ Factored form

$= 2(x - 2)(x^2 + 2x + 4)$ Simplify.

The following guidelines are steps for applying the various procedures involved in factoring polynomials.

▶ Guidelines for Factoring Polynomials

1. Factor out any common factors.
2. Factor according to one of the special polynomial forms: difference of two squares, sum or difference of two cubes, or perfect square trinomials.
3. Factor trinomials, $ax^2 + bx + c$, with $a = 1$ or $a \neq 1$.
4. Factor by grouping—for polynomials with four terms.
5. Check to see whether the factors themselves can be factored.
6. Check the results by multiplying the factors.

Discussing the Concept — A Three-Dimensional View of a Special Product

The figure below shows two cubes: a large cube whose volume is a^3 and a smaller cube whose volume is b^3. If the smaller cube is removed from the larger, the remaining solid has a volume of $a^3 - b^3$ and is composed of three rectangular boxes, labeled Box 1, Box 2, and Box 3. Find the volume of each box and describe how these results are related to the following special product pattern.

$$a^3 - b^3 = (a - b)(a^2 + ab + b^2)$$
$$= (a - b)a^2 + (a - b)ab + (a - b)b^2$$

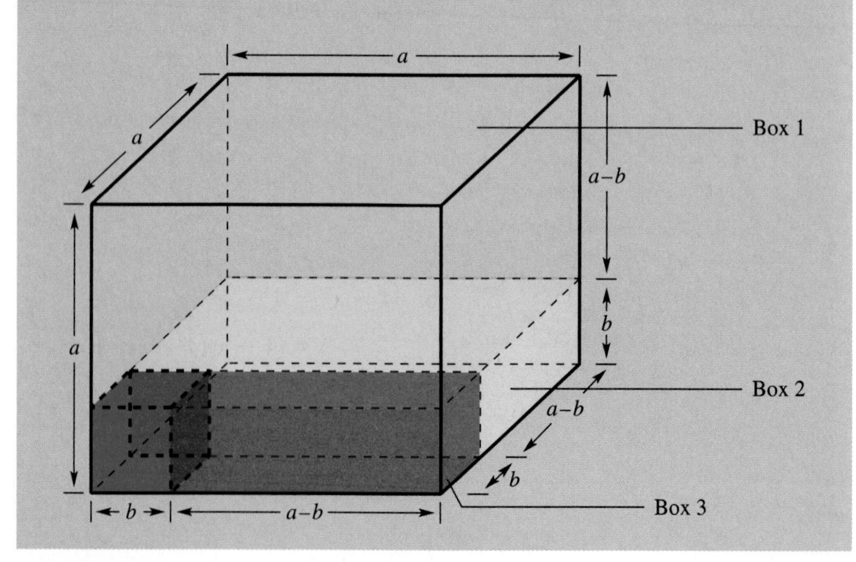

6.4 Exercises

Integrated Review *Concepts, Skills, and Problem Solving*

Keep mathematically in shape by doing these exercises *before* the problems of this section.

Properties and Definitions

In Exercises 1 and 2, determine the quadrant or quadrants in which the point must be located.

1. $(-5, 2)$

2. $(x, 3)$, x is a real number.

3. Find the coordinates of the point on the x-axis and 4 units to the left of the y-axis.

4. Find the coordinates of the point 9 units to the right of the y-axis and 6 units below the x-axis.

Solving Equations

In Exercises 5–10, solve the equation and check your result.

5. $7 + 5x = 7x - 1$

6. $2 - 5(x - 1) = 2[x + 10(x - 1)]$

7. $2(x + 1) = 0$

8. $\frac{3}{4}(12x - 8) = 10$

9. $\frac{x}{5} + \frac{1}{5} = \frac{7}{10}$

10. $\frac{3x}{4} + \frac{1}{2} = 8$

Problem Solving

11. Because of a membership drive for a public television station, the current membership is 120% of what it was a year ago. The current number is 8345. How many members did the station have last year?

12. Suppose you budget 26% of your annual after-tax income for housing. If your after-tax income is $46,750, what amount can you spend on housing?

Developing Skills

In Exercises 1–18, factor the difference of two squares. See Examples 1 and 2.

1. $x^2 - 36$

2. $y^2 - 49$

3. $u^2 - 64$

4. $x^2 - 4$

5. $49 - x^2$

6. $81 - x^2$

7. $u^2 - \frac{1}{4}$

8. $v^2 - \frac{4}{9}$

9. $t^2 - \frac{1}{16}$

10. $u^2 - \frac{25}{81}$

11. $16y^2 - 9$

12. $9z^2 - 25$

13. $100 - 49x^2$

14. $16 - 81x^2$

15. $(x - 1)^2 - 4$

16. $(t + 2)^2 - 9$

17. $25 - (z + 5)^2$

18. $(a - 2)^2 - 16$

In Exercises 19–30, factor completely. See Examples 3–5.

19. $2x^2 - 72$

20. $3x^2 - 27$

21. $8 - 50x^2$

22. $a^3 - 16a$

23. $y^4 - 81$

24. $z^4 - 16$

25. $1 - x^4$

26. $256 - u^4$

27. $3x^4 - 48$

28. $18 - 2x^4$

29. $81x^4 - 16$

30. $81x^4 - 1$

In Exercises 31–48, factor the perfect square trinomial. See Examples 6 and 7.

31. $x^2 - 4x + 4$

32. $x^2 + 10x + 25$

33. $z^2 + 6z + 9$

34. $a^2 - 12a + 36$

35. $4t^2 + 4t + 1$

36. $9x^2 - 12x + 4$

37. $25y^2 - 10y + 1$

38. $16z^2 + 24z + 9$

39. $b^2 + b + \frac{1}{4}$

40. $x^2 + \frac{2}{5}x + \frac{1}{25}$

41. $4x^2 - x + \frac{1}{16}$

42. $4t^2 - \frac{4}{3}t + \frac{1}{9}$

43. $x^2 - 6xy + 9y^2$

44. $16x^2 - 8xy + y^2$

45. $4y^2 + 20yz + 25z^2$

46. $u^2 + 8uv + 16v^2$

47. $9a^2 - 12ab + 4b^2$

48. $49m^2 - 28mn + 4n^2$

Think About It In Exercises 49–54, find two values of b such that the expression is a perfect square trinomial.

49. $x^2 + bx + 1$

50. $x^2 + bx + 100$

51. $x^2 + bx + \frac{16}{25}$

52. $y^2 + by + \frac{1}{9}$

53. $4x^2 + bx + 81$

54. $4x^2 + bx + 9$

Think About It In Exercises 55–58, find a number c such that the expression is a perfect square trinomial.

55. $x^2 + 6x + c$ **56.** $x^2 + 10x + c$

57. $y^2 - 4y + c$ **58.** $z^2 - 14z + c$

In Exercises 59–66, factor the sum or difference of two cubes. See Example 8.

59. $x^3 - 8$ **60.** $x^3 - 27$

61. $y^3 + 64$ **62.** $z^3 + 125$

63. $1 + 8t^3$ **64.** $27s^3 + 1$

65. $27u^3 + 8$ **66.** $64v^3 - 125$

In Exercises 67–108, factor the expression completely. (*Note:* Some of the polynomials may be prime.)

67. $6x - 36$ **68.** $8t + 48$

69. $u^2 + 3u$ **70.** $x^3 - 4x^2$

71. $5y^2 - 25y$ **72.** $12a^2 - 24a$

73. $5y^2 - 125$ **74.** $6x^2 - 54$

75. $y^4 - 25y^2$ **76.** $y^4 - 49y^2$

77. $1 - 4x + 4x^2$ **78.** $9x^2 - 6x + 1$

79. $x^2 - 2x + 1$ **80.** $16 + 6x - x^2$

81. $9x^2 + 10x + 1$ **82.** $4x^3 + 3x^2 + x$

83. $2x^2 + 4x - 2x^3$ **84.** $2y^3 - 7y^2 - 15y$

85. $9t^2 - 16$ **86.** $16t^2 - 144$

87. $36 - (z + 6)^2$ **88.** $(t - 4)^2 - 9$

89. $(t - 1)^2 - 121$ **90.** $(x - 3)^2 - 100$

91. $u^3 + 2u^2 + 3u$ **92.** $u^3 + 2u^2 - 3u$

93. $x^2 + 81$ **94.** $x^2 + 16$

95. $2t^3 - 16$ **96.** $24x^3 - 3$

97. $2 - 16x^3$ **98.** $54 - 2x^3$

99. $x^4 - 81$ **100.** $2x^4 - 32$

101. $1 - x^4$ **102.** $81 - y^4$

103. $x^3 - 4x^2 - x + 4$

104. $y^3 + 3y^2 - 4y - 12$

105. $x^4 + 3x^3 - 16x^2 - 48x$

106. $36x + 18x^2 - 4x^3 - 2x^4$

107. $64 - y^6$ **108.** $1 - y^8$

Graphical Verification In Exercises 109–112, use a graphing utility to graph the two functions on the same screen. What can you conclude?

109. $y_1 = x^2 - 36$
$y_2 = (x + 6)(x - 6)$

110. $y_1 = x^2 - 8x + 16$
$y_2 = (x - 4)^2$

111. $y_1 = x^3 - 6x^2 + 9x$
$y_2 = x(x - 3)^2$

112. $y_1 = x^3 + 27$
$y_2 = (x + 3)(x^2 - 3x + 9)$

Mental Math In Exercises 113–116, evaluate the quantity mentally using the two samples as models.

$$29^2 = (30 - 1)^2$$
$$= 30^2 - 2 \cdot 30 \cdot 1 + 1^2$$
$$= 900 - 60 + 1 = 841$$
$$48 \cdot 52 = (50 - 2)(50 + 2)$$
$$= 50^2 - 2^2 = 2496$$

113. 21^2 **114.** 49^2

115. $59 \cdot 61$ **116.** $28 \cdot 32$

Solving Problems

117. *Geometry* An annulus is the region between two concentric circles. The area of the annulus in the figure is $\pi R^2 - \pi r^2$. Give the complete factorization of the expression for the area.

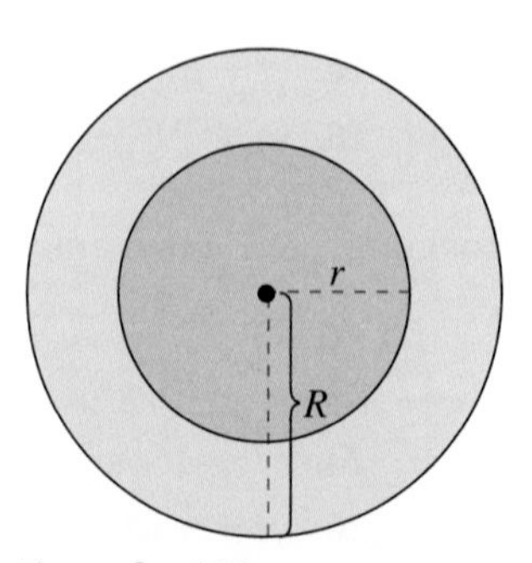

Figure for 117

In Exercises 118 and 119, write the polynomial as the difference of two squares. Use the result to factor the polynomial.

118. $x^2 + 6x + 8 = (x^2 + 6x + 9) - 1$

$= \square^2 - \square^2$

119. $x^2 + 8x + 12 = (x^2 + 8x + 16) - 4$

$= \square^2 - \square^2$

Geometric Factoring Models In Exercises 120 and 121, write the factoring problem represented by the geometric factoring model.

120.

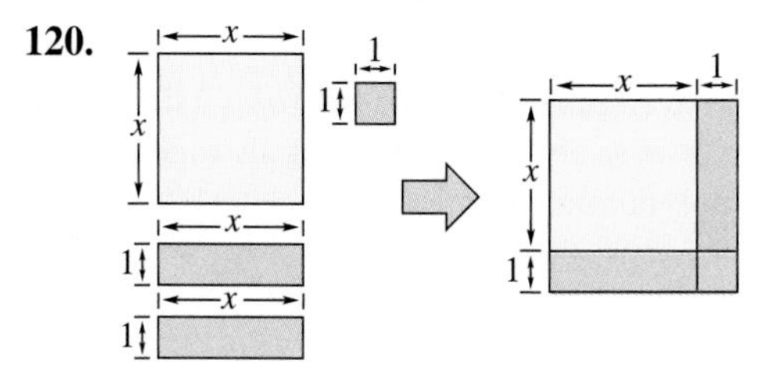

121.

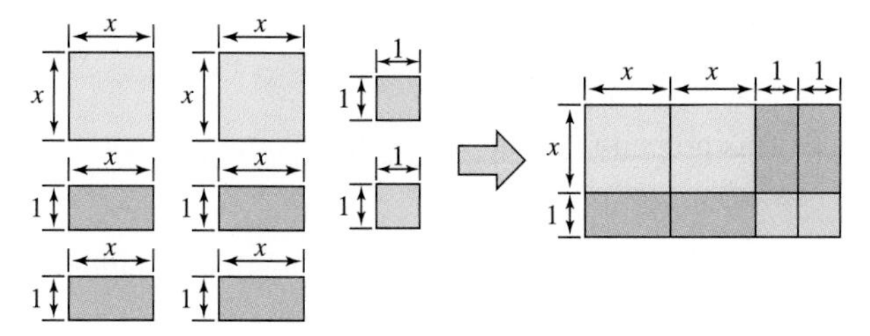

122. *Conjecture*

(a)

$5 \times 5 = \square$ $\quad 8 \times 8 = \square$ $\quad 11 \times 11 = \square$

$6 \times 4 = \square$ $\quad 9 \times 7 = \square$ $\quad 12 \times 10 = \square$

(b) Use the pattern of part (a) to fill in the blanks.

$12 \times 12 = 144$

$\square \times \square = 143$

(c) Use the pattern of parts (a) and (b) to make a conjecture. If possible, prove your conjecture.

123. *Think About It* You design a square flower garden, but later change it to a rectangular shape by increasing one pair of opposite sides by 1 foot and decreasing the other pair of sides by 1 foot. The area of the rectangle is 224 square feet. Use the result of Exercise 122 to determine the dimensions of the square in the original design.

124. *Geometry* From the eight vertices of a cube of dimension x, cubes of dimension y are removed (see figure).

(a) Write an expression for the volume of the solid that remains after the eight cubes at the vertices are removed. (*Hint:* The volume of a rectangular solid is length times width times height.)

(b) Factor the expression for the volume in part (a).

(c) In the context of this problem, y must be less than what multiple of x? Explain your answer geometrically and from the result of part (b).

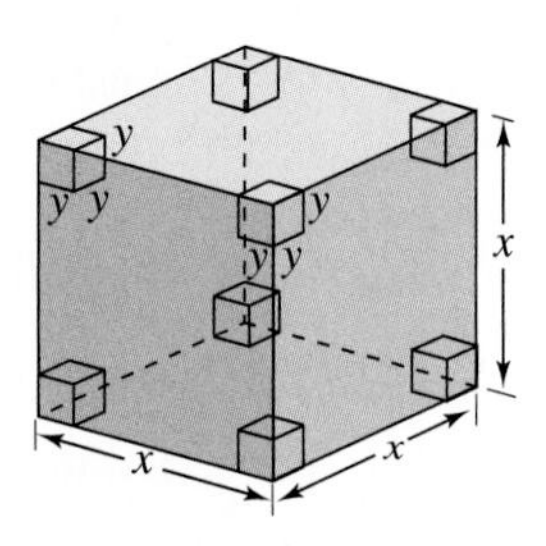

Explaining Concepts

125. Explain how to identify and factor the difference of two squares.

126. Explain how to identify and factor a perfect square trinomial.

127. Is the expression $x(x + 2) - 2(x + 2)$ in factored form? If not, rewrite it in factored form.

128. Is $x^2 + 4$ equal to $(x + 2)^2$? Explain.

129. *True or False?* Because the sum of two squares cannot be factored, it follows that the sum of two cubes cannot be factored. Explain your reasoning.

130. In your own words, state guidelines for factoring polynomials.

6.5 Polynomial Equations and Applications

Objectives

1. Use the Zero-Factor Property to solve an equation.
2. Use factoring to solve a quadratic equation.
3. Solve a polynomial equation by factoring.
4. Solve an application problem by factoring.

1 Use the Zero-Factor Property to solve an equation.

The Zero-Factor Property

You have spent nearly two chapters developing skills for *rewriting* (simplifying and factoring) polynomials. You are now ready to use these skills together with the **Zero-Factor Property** to *solve* polynomial equations.

Study Tip

The Zero-Factor Property is just another way of saying that the only way the product of two or more factors can be zero is if one or more of the factors is zero.

▶ Zero-Factor Property

Let a and b be real numbers, variables, or algebraic expressions. If a and b are factors such that

$$ab = 0$$

then $a = 0$ or $b = 0$. This property also applies to three or more factors.

The Zero-Factor Property is the primary property for solving equations in algebra. For instance, to solve the equation

$(x - 1)(x + 2) = 0$ Original equation

you can use the Zero-Factor Property to conclude that either $(x - 1)$ or $(x + 2)$ must be zero. Setting the first factor equal to zero implies that $x = 1$ is a solution.

$x - 1 = 0 \Rightarrow x = 1$ First solution

Similarly, setting the second factor equal to zero implies that $x = -2$ is a solution.

$x + 2 = 0 \Rightarrow x = -2$ Second solution

So, the equation $(x - 1)(x + 2) = 0$ has exactly two solutions: 1 and -2. You can check these solutions by substituting them into the original equation.

Check

$(x - 1)(x + 2) = 0$ Original equation

$(1 - 1)(1 + 2) \stackrel{?}{=} 0$ Substitute 1 for x.

$(0)(3) = 0$ First solution checks. ✓

$(-2 - 1)(-2 + 2) \stackrel{?}{=} 0$ Substitute -2 for x.

$(-3)(0) = 0$ Second solution checks. ✓

2 Use factoring to solve a quadratic equation.

Solving Quadratic Equations by Factoring

A **quadratic equation** is an equation of the form $ax^2 + bx + c = 0$. Here are some examples.

$$x^2 - 2x - 3 = 0, \quad 2x^2 + x - 1 = 0, \quad \text{and} \quad x^2 - 5x = 0$$

In the next four examples, note how you can combine your factoring skills with the Zero-Factor Property to solve quadratic equations.

Example 1 Using Factoring to Solve a Quadratic Equation

Solve $x^2 - x - 6 = 0$.

Solution

First, check to see that the right side of the equation is zero. Next, factor the left side of the equation. Finally, apply the Zero-Factor Property to find the solutions.

$x^2 - x - 6 = 0$ Original equation

$(x + 2)(x - 3) = 0$ Factor left side of equation.

$x + 2 = 0$ Set 1st factor equal to 0.

$x = -2$ Solve for x.

$x - 3 = 0$ Set 2nd factor equal to 0.

$x = 3$ Solve for x.

Study Tip

In Section 3.1, you learned that the general strategy for solving a linear equation is to *isolate the variable.* Notice in Example 1 that the general strategy for solving a quadratic equation is to factor the equation into linear factors.

Check

$x^2 - x - 6 = 0$ Original equation

$(-2)^2 - (-2) - 6 \stackrel{?}{=} 0$ Substitute -2 for x.

$4 + 2 - 6 \stackrel{?}{=} 0$ Simplify.

$0 = 0$ Solution checks. ✓

$x^2 - x - 6 = 0$ Original equation

$(3)^2 - 3 - 6 \stackrel{?}{=} 0$ Substitute 3 for x.

$9 - 3 - 6 \stackrel{?}{=} 0$ Simplify.

$0 = 0$ Solution checks. ✓

The equation has two solutions: -2 and 3.

Factoring and the Zero-Factor Property allow you to solve a quadratic equation by converting it into two *linear* equations, which you already know how to solve. This is a common strategy of algebra—to break down a given problem into simpler parts, each solved by previously learned methods.

In order for the Zero-Factor Property to be used, a polynomial equation *must* be written in **general form.** That is, the polynomial must be on one side of the equation and zero must be the only term on the other side of the equation. To write $x^2 - 2x = 3$ in general form, subtract 3 from both sides of the equation.

$$x^2 - 2x = 3 \qquad \text{Original equation}$$

$$x^2 - 2x - 3 = 3 - 3 \qquad \text{Subtract 3 from both sides.}$$

$$x^2 - 2x - 3 = 0 \qquad \text{General form}$$

To solve this equation, factor the left side as $(x - 3)(x + 1)$, then form the linear equations $x - 3 = 0$ and $x + 1 = 0$. The solutions of these two linear equations are 3 and -1, respectively. The general strategy for solving a quadratic equation by factoring is summarized in the following guidelines.

▶ Guidelines for Solving Quadratic Equations

1. Write the quadratic equation in general form.
2. Factor the left side of the equation.
3. Set each factor with a variable equal to zero.
4. Solve each linear equation.
5. Check each solution in the original equation.

Example 2 Solving a Quadratic Equation by Factoring

Solve $2x^2 + 5x = 12$.

Solution

$$2x^2 + 5x = 12 \qquad \text{Original equation}$$

$$2x^2 + 5x - 12 = 0 \qquad \text{Write in general form.}$$

$$(2x - 3)(x + 4) = 0 \qquad \text{Factor left side of equation.}$$

$$2x - 3 = 0 \qquad \text{Set 1st factor equal to 0.}$$

$$x = \tfrac{3}{2} \qquad \text{Solve for } x.$$

$$x + 4 = 0 \qquad \text{Set 2nd factor equal to 0.}$$

$$x = -4 \qquad \text{Solve for } x.$$

The solutions are $\frac{3}{2}$ and -4. Check these solutions in the original equation.

Be sure you see that the Zero-Factor Property can be applied only to a product that is equal to *zero*. For instance, you cannot conclude from the equation $x(x - 3) = 10$ that $x = 10$ and $x - 3 = 10$ yield solutions. Instead, you must first write the equation in general form and then factor the left side, as follows.

$$x^2 - 3x - 10 = 0 \quad \Rightarrow \quad (x - 5)(x + 2) = 0$$

Now, from the factored form, you can see that the solutions are 5 and -2.

In Examples 1 and 2, the original equations each involved a second-degree (quadratic) polynomial and each had *two different* solutions. You will sometimes encounter second-degree polynomial equations that have only one (repeated) solution. This occurs when the left side of the equation is a perfect square trinomial, as shown in Example 3.

Technology: Discovery

Write the function in Example 3 in general form. Graph this function on your graphing utility.

$y = x^2 - 8x + 16$

What are the x-intercepts of the function?

Write the function in Example 4 in general form. Graph this function on your graphing utility.

$y = x^2 + 9x + 14$

What are the x-intercepts of the function?

How do the x-intercepts relate to the solutions of the equations? What can you conclude about the solutions to the equations and the x-intercepts?

Example 3 A Quadratic Equation with a Repeated Solution

Solve $x^2 - 8x + 20 = 4$.

Solution

$x^2 - 8x + 20 = 4$	Original equation
$x^2 - 8x + 16 = 0$	Write in general form.
$(x - 4)^2 = 0$	Factor.
$x - 4 = 0$	Set factor equal to 0.
$x = 4$	Solve for x.

Note that even though the left side of this equation has two factors, the factors are the same. Thus, the only solution of the equation is 4.

$x^2 - 8x + 20 = 4$	Original equation
$(4)^2 - 8(4) + 20 \stackrel{?}{=} 4$	Substitute 4 for x.
$16 - 32 + 20 \stackrel{?}{=} 4$	Simplify.
$4 = 4$	Solution checks. ✓

Example 4 Solving a Polynomial Equation

Solve $(x + 3)(x + 6) = 4$.

Solution

Begin by multiplying the factors on the left side.

$(x + 3)(x + 6) = 4$	Original equation
$x^2 + 9x + 18 = 4$	Multiply factors.
$x^2 + 9x + 14 = 0$	General form
$(x + 2)(x + 7) = 0$	Factor left side of equation.
$x + 2 = 0$	Set 1st factor equal to 0.
$x = -2$	Solve for x.
$x + 7 = 0$	Set 2nd factor equal to 0.
$x = -7$	Solve for x.

The equation has two solutions: -2 and -7. Check these in the original equation.

3 Solve a polynomial equation by factoring.

Solving Polynomial Equations by Factoring

The Granger Collection

Niels Henrik Abel
(1802–1829)

In the exploration of algebra, general solutions were found for second-, third-, and fourth-degree polynomial equations. (You will study the Quadratic Formula for second-degree equations in Chapter 10.) Attempts to find an algebraic solution for a fifth-degree polynomial equation met with failure.

In 1824, Niels Henrik Abel published a proof that showed that for any degree greater than four, the general polynomial equation could not be solved algebraically. His work included the concept of a group. Subsequent research in group theory gave mathematicians new ways to explore and describe algebraic structures. This marked the beginning of the modern theory of equations.

Example 5 Solving a Polynomial Equation with Three Factors

Solve $3x^3 = 12x^2 + 15x$.

Solution

$3x^3 = 12x^2 + 15x$		Original equation
$3x^3 - 12x^2 - 15x = 0$		General form
$3x(x^2 - 4x - 5) = 0$		Factor out $3x$.
$3x(x - 5)(x + 1) = 0$		Factor completely.
$3x = 0$	$\Rightarrow x = 0$	Set 1st factor equal to 0.
$x - 5 = 0$	$\Rightarrow x = 5$	Set 2nd factor equal to 0.
$x + 1 = 0$	$\Rightarrow x = -1$	Set 3rd factor equal to 0.

There are three solutions: 0, 5, and -1. Check these in the original equation.

Notice that the equation in Example 5 is a third-degree equation and has three solutions. This is not a coincidence. In general, a polynomial equation can have *at most* as many solutions as its degree. For instance, a second-degree equation can have zero, one, or two solutions, but it cannot have three or more solutions. Notice that the equation in Example 6 is a fourth-degree equation and has four solutions.

Example 6 Solving a Polynomial Equation with Four Factors

Solve $x^4 + x^3 - 4x^2 - 4x = 0$.

Solution

$x^4 + x^3 - 4x^2 - 4x = 0$		Original equation
$x(x^3 + x^2 - 4x - 4) = 0$		Factor out x.
$x[x^2(x + 1) - 4(x + 1)] = 0$		Distributive Property
$x[(x + 1)(x^2 - 4)] = 0$		Factor by grouping.
$x(x + 1)(x + 2)(x - 2) = 0$		Factor completely.
$x = 0$	$\Rightarrow x = 0$	Set 1st factor equal to 0.
$x + 1 = 0$	$\Rightarrow x = -1$	Set 2nd factor equal to 0.
$x + 2 = 0$	$\Rightarrow x = -2$	Set 3rd factor equal to 0.
$x - 2 = 0$	$\Rightarrow x = 2$	Set 4th factor equal to 0.

There are four solutions: 0, -1, -2, and 2. Check these in the original equation.

4 Solve an application problem by factoring.

Applications

Example 7 Consecutive Integers

The product of two consecutive positive integers is 56. What are the integers?

Solution

Verbal Model:

First integer · Second integer = 56

Labels: First integer $= n$
Second integer $= n + 1$

Equation:

$n(n + 1) = 56$ — Original equation

$n^2 + n - 56 = 0$ — General form

$(n + 8)(n - 7) = 0$ — Factor left side of equation.

$n = -8$ or 7 — Solutions

Because the problem states that the integers are positive, discard -8 as a solution and choose $n = 7$. So, the two integers are $n = 7$ and $n + 1 = 8$.

Example 8 A Falling Object Model

A rock is dropped from the top of a 256-foot river gorge, as shown in Figure 6.3. The height (in feet) of the rock is modeled by the equation

$$\text{Height} = -16t^2 + 256$$

where t is the time measured in seconds. How long will it take the rock to hit the bottom of the gorge?

Solution

From Figure 6.3, note that the bottom of the gorge corresponds to a height of 0 feet. So, substitute a height of 0 into the model and solve for t.

$0 = -16t^2 + 256$ — Set height equal to 0.

$16t^2 - 256 = 0$ — General form

$16(t^2 - 16) = 0$ — Factor out 16.

$16(t + 4)(t - 4) = 0$ — Factor left side of equation.

$t + 4 = 0 \Rightarrow t = -4$ — Set 1st factor equal to 0.

$t - 4 = 0 \Rightarrow t = 4$ — Set 2nd factor equal to 0.

Because a time of -4 seconds doesn't make sense in this problem, choose the positive solution and conclude that the rock hits the bottom of the gorge 4 seconds after it is dropped.

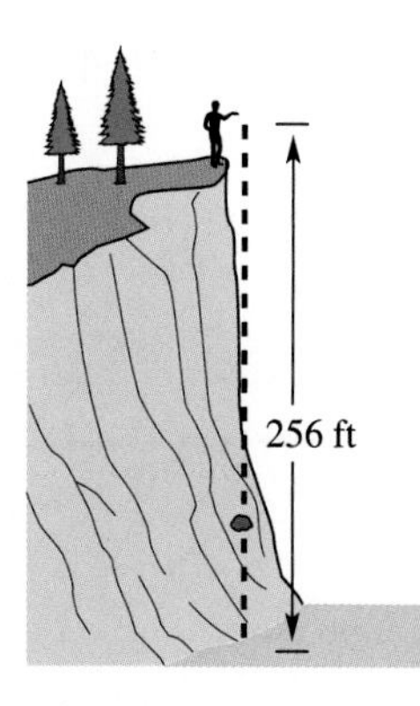

Figure 6.3

In Example 8, the equation is a second-degree equation and, as such, cannot have more than two solutions, 4 and -4. The factor 16 in the equation

$$16(t + 4)(t - 4) = 0$$

does not give us another solution. Setting this factor equal to zero yields the *false* statement $16 = 0$.

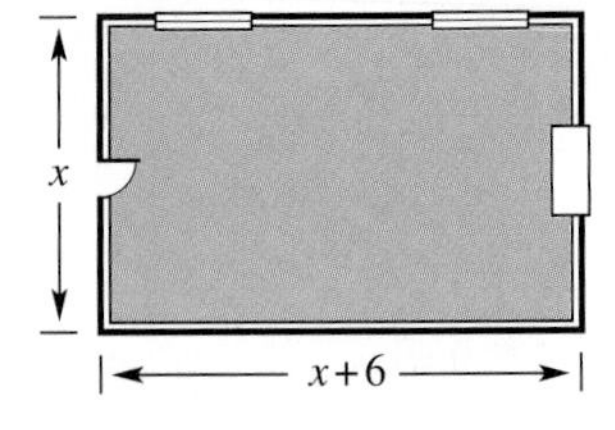

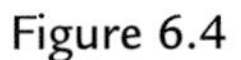
Figure 6.4

Example 9 An Application from Geometry

A rectangular family room has an area of 160 square feet. The length of the room is 6 feet greater than its width. Find the dimensions of the room.

Solution

To begin, make a sketch of the room, as shown in Figure 6.4. Label the width of the room as x and the length of the room as $x + 6$ because the length is 6 feet greater than the width.

Verbal Model: Length · Width = Area

Labels:
Width $= x$ (feet)
Length $= x + 6$ (feet)
Area $= 160$ (square feet)

Equation:

$$x(x + 6) = 160$$

$$x^2 + 6x - 160 = 0$$

$$(x + 16)(x - 10) = 0$$

$$x = -16 \text{ or } 10$$

In this application, the negative solution makes no sense, so discard it and use the positive solution. So, the width of the room is 10 feet and the length of the room is 16 feet.

To check this solution, go back to the original statement of the problem. Note that a length of 16 feet is 6 feet greater than a width of 10 feet. Moreover, a rectangular room with dimensions 16 feet by 10 feet has an area of 160 square feet. So, the solution checks.

Discussing the Concept Misleading Factorization

Suppose a student submits the following steps in solving the equation $x^2 + 3x = 10$:

$x(x + 3) = 10$ Factor.

$x = 10$ and $x + 3 = 10$ Set each factor equal to 10.

$x = 7$

Write an explanation of why the method does not work. Solve the equation correctly and check your answers.

6.5 Exercises

Integrated Review *Concepts, Skills, and Problem Solving*

Keep mathematically in shape by doing these exercises *before* the problems of this section.

Properties and Definitions

In Exercises 1–4, name the property illustrated.

1. $2ab - 2ab = 0$ **2.** $8t \cdot 1 = 8t$

3. $2x(1 - x) = 2x - 2x^2$

4. $3x + (2x + 5) = (3x + 2x) + 5$

Rewriting Expressions

In Exercises 5–10, perform the required operations and/or simplify.

5. (a) $2(-3) + 9$ (b) $(-5)^2 + 3$

6. (a) $4 - \frac{5}{2}$ (b) $\frac{|18 - 25|}{6}$

7. $\left(-\frac{7}{12}\right)\left(\frac{3}{28}\right)$ **8.** $\frac{4}{3} \div \frac{5}{6}$

9. $2t(t - 3) + 4t + 1$ **10.** $2u - 5(2u - 3)$

Problem Solving

11. Find the interest on a \$1000 bond paying an annual percentage rate of 7.5% for 10 years.

12. A car leaves a town 1 hour after a fully-loaded truck. The speed of the truck is approximately 50 miles per hour. If the car overtakes the truck in 2.5 hours, find the speed of the car.

Developing Skills

In Exercises 1–16, use the Zero-Factor Property to solve the equation.

1. $x(x - 5) = 0$ **2.** $z(z - 3) = 0$

3. $(y - 2)(y - 3) = 0$ **4.** $(s - 4)(s - 10) = 0$

5. $(a + 1)(a - 2) = 0$ **6.** $(t - 3)(t + 8) = 0$

7. $(2t - 5)(3t + 1) = 0$ **8.** $(2 - 3x)(5 - 2x) = 0$

9. $\left(\frac{2}{3}x - 4\right)(x + 2) = 0$ **10.** $\left(\frac{3}{4}u - 2\right)\left(\frac{2}{5}u + \frac{1}{2}\right) = 0$

11. $(0.2y - 12)(0.7y + 10) = 0$

12. $(1.5s + 12)(0.75s - 18) = 0$

13. $3x(x + 8)(4x - 5) = 0$

14. $x(x - 3)(x + 25) = 0$

15. $(y - 1)(2y + 3)(y + 12) = 0$

16. $x(5x + 3)(x - 8) = 0$

In Exercises 17–66, solve the equation. See Examples 1–6.

17. $x^2 - 16 = 0$ **18.** $x^2 - 144 = 0$

19. $100 - v^2 = 0$ **20.** $4 - x^2 = 0$

21. $3y^2 - 27 = 0$ **22.** $25z^2 - 100 = 0$

23. $(t - 3)^2 - 25 = 0$ **24.** $1 - (x + 1)^2 = 0$

25. $81 - (u + 4)^2 = 0$ **26.** $(s + 5)^2 - 49 = 0$

27. $2x^2 + 4x = 0$ **28.** $6x^2 + 3x = 0$

29. $4x^2 - x = 0$ **30.** $x - 3x^2 = 0$

31. $y(y - 4) + 3(y - 4) = 0$

32. $u(u + 2) - 3(u + 2) = 0$

33. $x(x - 8) + 2(x - 8) = 0$

34. $x(x + 2) - 3(x + 2) = 0$

35. $m^2 - 2m + 1 = 0$ **36.** $a^2 + 6a + 9 = 0$

37. $x^2 + 14x + 49 = 0$ **38.** $x^2 - 10x + 25 = 0$

39. $4t^2 - 12t + 9 = 0$ **40.** $16x^2 + 56x + 49 = 0$

41. $x^2 - 2x - 8 = 0$ **42.** $x^2 - 8x - 9 = 0$

43. $3 + 5x - 2x^2 = 0$ **44.** $33 + 5y - 2y^2 = 0$

45. $6x^2 + 4x - 10 = 0$ **46.** $12x^2 + 7x + 1 = 0$

47. $z(z + 2) = 15$ **48.** $x(x - 1) = 6$

49. $x(x - 5) = 14$ **50.** $x(x + 4) = -4$

51. $y(2y + 1) = 3$ **52.** $x(5x - 14) = 3$

53. $(x - 3)(x - 6) = 4$ **54.** $(x + 1)(x - 2) = 4$

55. $(x + 1)(x + 4) = 4$ **56.** $(x - 9)(x + 2) = 12$

57. $x^3 + 5x^2 + 6x = 0$ **58.** $x^3 - 3x^2 - 10x = 0$

59. $2t^3 + 5t^2 - 12t = 0$ **60.** $3u^3 - 5u^2 - 2u = 0$

61. $x^2(x - 2) - 9(x - 2) = 0$

62. $y^2(y + 3) - (y + 3) = 0$

63. $x^3 - x^2 - 16x + 16 = 0$

64. $a^3 + 2a^2 - 4a - 8 = 0$

65. $u^4 + 2u^3 - u^2 - 2u = 0$

66. $x^4 - 4x^3 - 9x^2 + 36x = 0$

Graphical Estimation In Exercises 67–74, use the graph to estimate the x-intercepts. Set the polynomial equal to zero and solve. What do you notice?

67. $y = x^2 + 2x - 3$

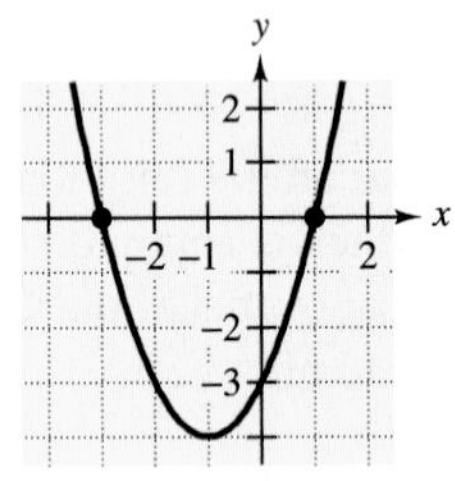

68. $y = 2 + x - x^2$

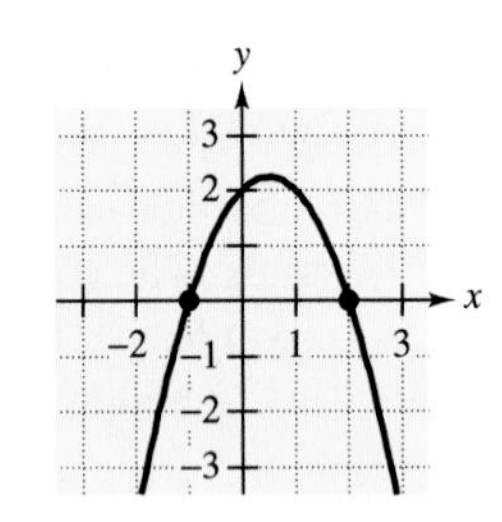

69. $y = 12 + x - x^2$

70. $y = 2x^2 + x - 3$

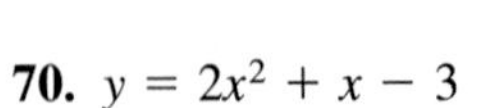

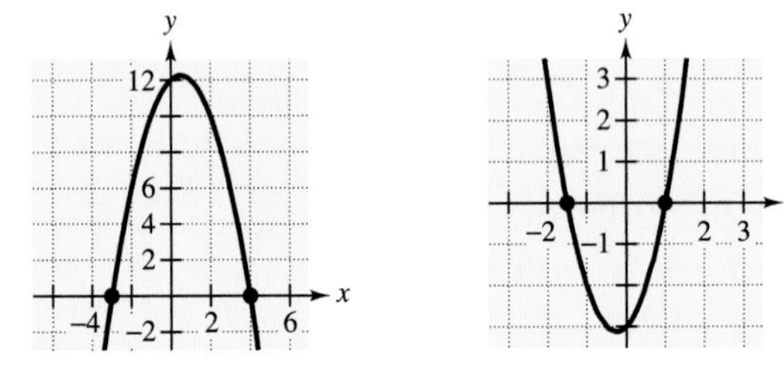

71. $y = x^3 - 6x^2 + 9x$

72. $y = 2x^3 + 3x^2 - 5x$

73. $y = 5x - 3x^2 - 2x^3$

74. $y = x^4 + 2x^3 - 8x^2$

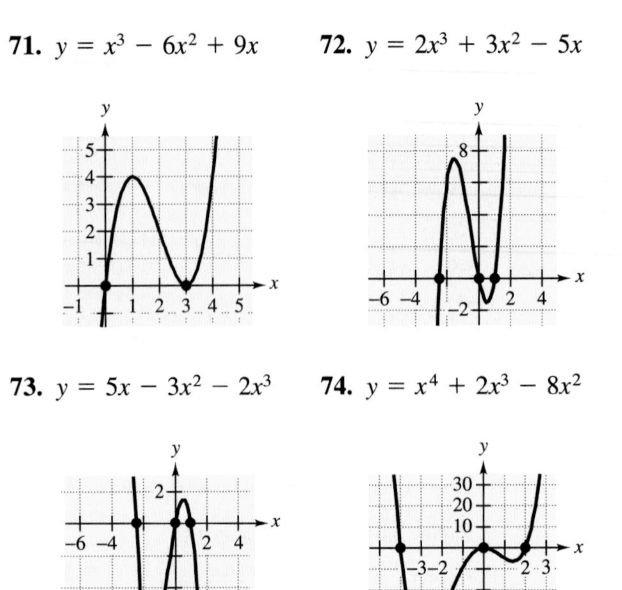

Graphical Estimation In Exercises 75–82, use a graphing utility to graph the function. Use the graph to estimate the x-intercepts. Check your estimates by substituting into the function.

75. $y = x^2 - 4$

76. $y = x^2 - 4x$

77. $y = 2x^2 - 5x - 12$

78. $y = 4x^2 + 3x - 10$

79. $y = x^3 - 4x^2$

80. $y = \frac{1}{4}(x^3 - 2x^2 - x + 2)$

81. $y = x^2(x + 2) - 9(x + 2)$

82. $y = \frac{1}{4}(x^3 + 4x^2 - x - 4)$

Solving Problems

83. Find two consecutive positive integers whose product is 72.

84. Find two consecutive positive integers whose product is 240.

85. Find two consecutive positive even integers whose product is 440.

86. Find two consecutive positive odd integers whose product is 323.

87. *Geometry* The length of a rectangular picture frame is 3 inches greater than its width. The area of the picture frame is 108 square inches. Find the dimensions of the picture frame.

88. *Geometry* The width of a rectangular garden is 4 feet less than its length. The area of the garden is 320 square feet. Find the dimensions of the garden.

89. *Geometry* The length of a rectangle is 2 times its width. The area of the rectangle is 450 square inches. Find the dimensions of the rectangle.

90. *Geometry* The length of a rectangle is $1\frac{1}{2}$ times its width. The area of the rectangle is 600 square inches. Find the dimensions of the rectangle.

91. *Constructing a Box* An open box is to be made from a square piece of material by cutting 2-inch squares from the corners and turning up the sides (see figure).

(a) Show that the volume is given by $V = 2x^2$.

(b) Complete the table.

x	2	4	6	8
V				

(c) Find the dimensions of the original piece of material if $V = 200$ cubic inches.

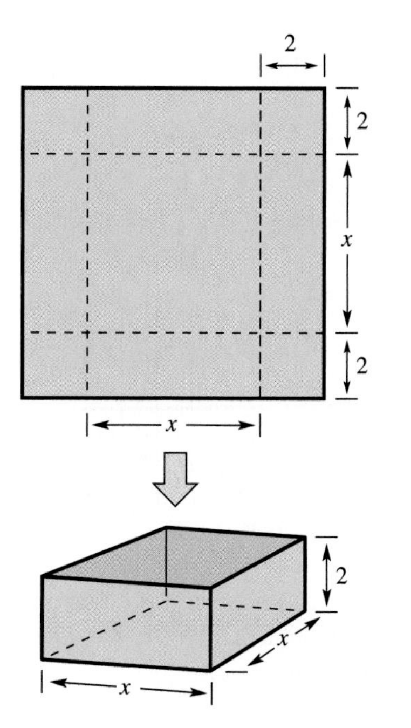

92. *Dimensions of a Box* An open box with a square base is to be constructed from 108 square inches of material. The height of the box is 3 inches. Find the dimensions of the base of the box. (*Hint:* The surface area is given by $S = x^2 + 4xh$.)

93. *Height of an Object* An object is dropped from a weather balloon 1600 feet above the ground (see figure). Find the time t for the object to reach the ground. The height (above ground) of the object is

$$\text{Height} = -16t^2 + 1600$$

where the height is measured in feet and the time t is measured in seconds.

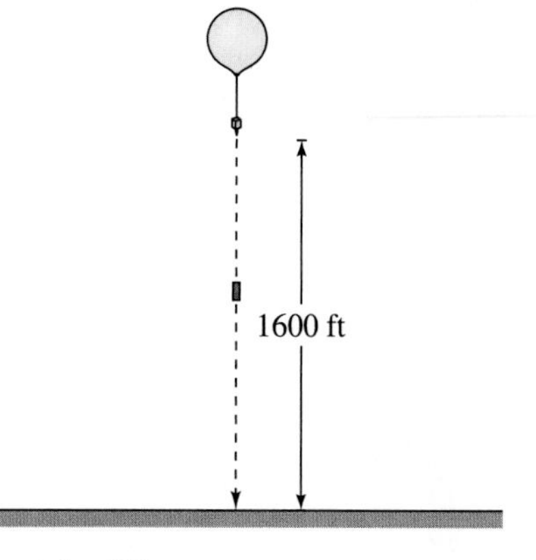

Figure for 93

94. *Height of an Object* An object is dropped from a cliff 400 feet above the ground. Find the time t for the object to reach the ground. The height (above ground) of the object is

$$\text{Height} = -16t^2 + 400$$

where the height is measured in feet and the time t is measured in seconds.

95. *Profit* The profit from selling x units of a product is

$$P = -0.4x^2 + 8x - 10.$$

(a) Use a graphing utility to graph the expression for profit.

(b) Use the graph to estimate any values of x that yield a profit of $P = \$20$.

(c) Use factorization to find any values of x that yield a profit of $P = \$20$.

96. *Revenue* The revenue from selling x units of a product is

$$R = 25x - 0.2x^2.$$

(a) Use a graphing utility to graph the expression for revenue.

(b) Use the graph to estimate the smaller of the two values of x that yield a revenue of $R = \$680$.

(c) Use factorization to find the smaller of the two values of x that yield a revenue of $R = \$680$.

97. *Height of a Diver* A diver jumps from a diving board that is 32 feet above the water (see figure). The height of the diver is modeled by

$$\text{Height} = -16t^2 + 16t + 32$$

where the height is measured in feet and the time t is measured in seconds. How many seconds will it take for the diver to reach the water?

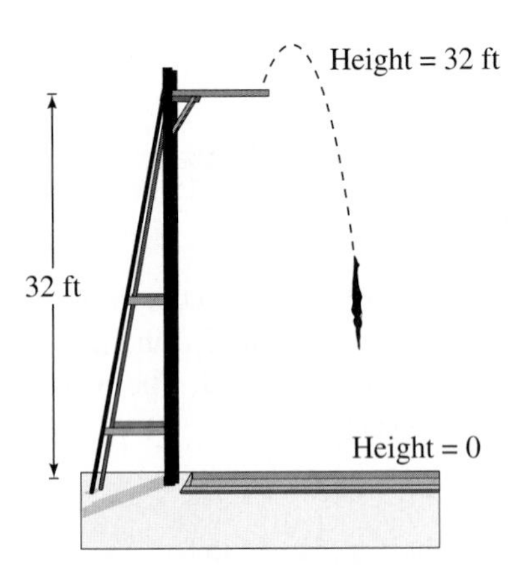

98. *Sum of Natural Numbers*

(a) Find the following sums.

$1 + 2 + 3 + 4 + 5 = \square$

$1 + 2 + 3 + 4 + 5 + 6 + 7 + 8 = \square$

$1 + 2 + 3 + 4 + 5 + 6 + 7 + 8 + 9 + 10 = \square$

(b) Use the following formula for the sum of the first n natural numbers to verify your answers to part (a).

$$1 + 2 + 3 + \cdots + n = \frac{1}{2}n(n + 1)$$

(c) Use the formula in part (b) to find n if the sum of the first n natural numbers is 210.

99. *Modeling Data* The bar graph gives the number of passengers y in millions flying Southwest Airline for the years 1987 through 1996. A model for the data is

$$y = (4.65 + 0.41t)^2, \quad -3 \le t \le 6$$

where t is time in years, with $t = 0$ corresponding to 1990. (Source: *Reader's Digest,* June 1997)

(a) Use a graphing utility to plot the data and graph the model.

(b) Use the graph in part (a) to approximate the solution of the equation

$$100 = (4.65 + 0.41t)^2.$$

Interpret the result in the context of the problem.

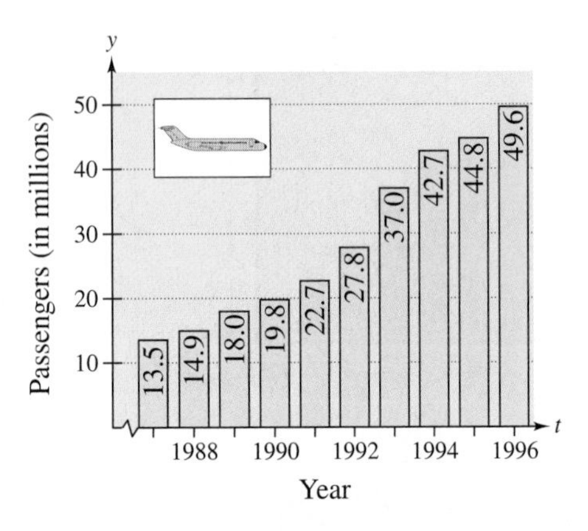

100. *Exploration* If a and b are nonzero real numbers, show that the equation $ax^2 + bx = 0$ must have two different solutions.

101. *Exploration* If a is a nonzero real number, find two solutions of the equation $ax^2 - ax = 0$.

Explaining Concepts

102. Answer parts (c)–(e) of Motivating the Chapter on page 321.

103. Use the Zero-Factor Property to complete the statement. If $ab = 0$, then $\square$.

104. In your own words, describe a strategy for solving a quadratic equation by factoring.

105. Explain the difference between a linear equation and a quadratic equation.

106. Is it possible for a quadratic equation to have one solution? Explain.

107. What is the maximum number of solutions of an nth-degree polynomial equation?

108. *True or False?* The only equation with solutions $x = 2$ and $x = -5$ is $(x - 2)(x + 5) = 0$.

109. *True or False?* If $(5x - 1)(x + 3) = 21$, then $5x - 1 = 21$ or $x + 3 = 21$. Explain.

CHAPTER SUMMARY

Key Terms

factoring, *p. 322*
greatest common factor, *p. 322*
greatest common monomial factor, *p. 323*
factoring out, *p. 323*
prime polynomials, *p. 332*
factoring completely, *p. 334*
quadratic equation, *p. 357*

Key Concepts

6.1 Factoring polynomials with common factors

Use the Distributive Property to remove the greatest common factor from each term of a polynomial.

6.1 Factoring polynomials by grouping

For polynomials with four terms, group the first two terms together and the last two terms together. Factor these two groupings and then look for a common binomial factor.

6.2 Factoring trinomials of the form $x^2 + bx + c$

To factor $x^2 + bx + c$, you need to find two numbers m and n whose product is c and whose sum is b.

$$x^2 + bx + c = (x + m)(x + n)$$

1. If c is positive, then m and n have like signs that match the sign of b.
2. If c is negative, then m and n have unlike signs.
3. If $|b|$ is small relative to $|c|$, first try those factors of c that are closest to each other in absolute value.

6.3 Factoring trinomials of the form $ax^2 + bx + c$ $(a > 0)$

1. First, factor out any common monomial factor.
2. You don't have to test any binomial factors that have a common monomial factor.
3. If the middle-term test (Outer + Inner) yields the opposite of b, switch the signs of the factors of c.

6.4 Factoring a difference of two squares

Let a and b be real numbers, variables, or algebraic expressions. To factor a difference of two squares, use the pattern shown on page 347.

6.4 Factoring perfect square trinomials

Let a and b be real numbers, variables, or algebraic expressions. To factor a perfect square trinomial, use the pattern shown on page 350.

6.4 Sum and difference of two cubes

Let a and b be real numbers, variables, or algebraic expressions. To factor the sum or difference of two cubes, use the patterns shown on page 351.

6.4 Guidelines for factoring polynomials

1. Factor out any common factors.
2. Factor according to one of the special polynomial forms: difference of two squares, sum or difference of two cubes, or perfect square trinomials.
3. Factor trinomials, $ax^2 + bx + c$, with $a = 1$ or $a \neq 1$.
4. Factor by grouping—for polynomials with four terms.
5. Check to see whether the factors themselves can be factored.
6. Check the results by multiplying the factors.

6.5 Zero-Factor Property

Let a and b be real numbers, variables, or algebraic expressions. If a and b are factors such that

$$ab = 0$$

then $a = 0$ or $b = 0$. This property also applies to three or more factors.

6.5 Solving a quadratic equation

To solve a quadratic equation, write the equation in general form. Factor the quadratic into linear factors and apply the Zero-Factor Property.

REVIEW EXERCISES

Reviewing Skills

6.1 In Exercises 1-4, find the greatest common factor of the expressions.

1. 20, 60, 150

2. $3x^4, 21x^2$

3. $18ab^2, 27a^2b$

4. $14z^2, 1, 21z$

In Exercises 5-20, factor the polynomial.

5. $3x - 6$

6. $7 + 21x$

7. $3t - t^2$

8. $u^2 - 6u$

9. $5x^2 + 10x^3$

10. $7y - 21y^4$

11. $8a - 12a^3$

12. $6u - 9u^2 + 15u^3$

13. $x(x + 1) - 3(x + 1)$

14. $2u(u - 2) + 5(u - 2)$

15. $y^3 + 3y^2 + 2y + 6$

16. $z^3 - 5z^2 + z - 5$

17. $x^3 + 2x^2 + x + 2$

18. $x^3 - 5x^2 + 5x - 25$

19. $x^2 - 4x + 3x - 12$

20. $2x^2 + 6x - 5x - 15$

6.2 In Exercises 21-30, factor the trinomial.

21. $x^2 - 3x - 28$

22. $x^2 - 3x - 40$

23. $u^2 + 5u - 36$

24. $y^2 + 15y + 56$

25. $x^2 + 9xy - 10y^2$

26. $u^2 + uv - 5v^2$

27. $y^2 - 6xy - 27x^2$

28. $v^2 + 18uv + 32u^2$

29. $4x^2 - 24x + 32$

30. $x^3 + 9x^2 + 18x$

In Exercises 31-34, find all values of b such that the trinomial is factorable.

31. $x^2 + bx + 9$

32. $y^2 + by + 25$

33. $z^2 + bz + 11$

34. $x^2 + bx + 14$

6.3 In Exercises 35-44, factor the trinomial.

35. $5 - 2x - 3x^2$

36. $8x^2 - 18x + 9$

37. $50 - 5x - x^2$

38. $7 + 5x - 2x^2$

39. $6x^2 + 7x + 2$

40. $16x^2 + 13x - 3$

41. $6u^3 + 3u^2 - 30u$

42. $8x^3 - 8x^2 + 30x$

43. $2x^2 - 3x + 1$

44. $3x^2 + 8x + 4$

In Exercises 45-48, find all values of b such that the trinomial is factorable.

45. $x^2 + bx - 24$

46. $2x^2 + bx - 16$

47. $3x^2 + bx - 20$

48. $3x^2 + bx + 1$

In Exercises 49 and 50, find two values of c such that the trinomial is factorable.

49. $2x^2 - 4x + c$

50. $5x^2 + 6x + c$

6.4 In Exercises 51 and 52, insert the missing factors.

51. $x^3 - x = x(\quad)(\quad)$

52. $u^4 - v^4 = (u^2 + v^2)(\quad)(\quad)$

In Exercises 53-72, factor the polynomial completely.

53. $a^2 - 100$

54. $36 - b^2$

55. $25 - 4y^2$

56. $16b^2 - 1$

57. $(u + 1)^2 - 4$

58. $(y - 2)^2 - 9$

59. $x^2 - 8x + 16$

60. $y^2 + 24y + 144$

61. $x^2 + 6x + 9$

62. $v^2 - 10v + 25$

63. $9s^2 + 12s + 4$

64. $u^2 - 2uv + v^2$

65. $s^3t - st^3$

66. $y^3z + 4y^2z^2 + 4yz^3$

67. $a^3 + 1$

68. $z^3 + 8$

69. $27 - 8t^3$

70. $z^3 - 125$

71. $-16a^3 - 16a^2 - 4a$

72. $5t - 125t^3$

6.5 In Exercises 73-86, solve the polynomial equation.

73. $x^2 - 81 = 0$

74. $121 - y^2 = 0$

75. $x^2 - 12x + 36 = 0$

76. $2t^2 - 3t - 2 = 0$

77. $4s^2 + s - 3 = 0$

78. $y^3 - y^2 - 6y = 0$

79. $x(2x - 3) = 0$

80. $3x(5x + 1) = 0$

81. $(z - 2)^2 - 4 = 0$

82. $(x + 1)^2 - 16 = 0$

83. $x(7 - x) = 12$

84. $x(x + 5) = 24$

85. $u^3 + 5u^2 - u = 5$

86. $a^3 - 3a^2 - a = -3$

Solving Problems

87. *Geometry* The cake box shown in the figure has a height of x and a width of $x + 1$. The volume of the box is $3x^3 + 4x^2 + x$. Find the length of the box.

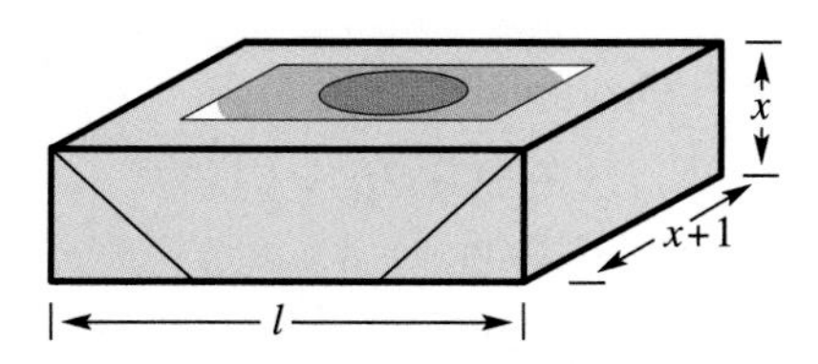

Geometry In Exercises 88 and 89, write an expression for the area of the shaded region and factor the expression.

88.

89.

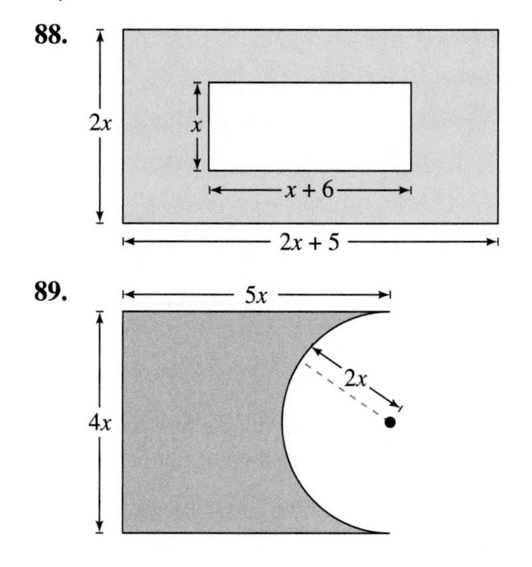

90. *Geometry* A rectangular sheet of metal has dimensions 2 feet by 3 feet. An open box is to be made from the metal by cutting equal squares from the corners and turning up the sides. The volume of the box is

$$V = 4x^3 - 10x^2 + 6x, \quad 0 < x < 1.$$

(a) Sketch the rectangular sheet and the open box. Label the height of the box as x.

(b) Factor the expression for the volume. Show how these factors relate to the dimensions of the box.

(c) Use a graphing utility to graph the volume over the specified interval. Use the graph to approximate the size of the squares to be cut from the corners so that the volume of the box is greatest.

91. *Revenue* The revenue from selling x units of a product is $R = 12x - 0.3x^2$.

(a) Use a graphing utility to graph the revenue function.

(b) Use the graph to estimate the value of x that yields a revenue of $R = \$120$.

(c) Use factorization to find the value of x that yields a revenue of $R = \$120$.

92. *Height of an Object* A rock is thrown vertically upward from a height of 48 feet (see figure) with an initial velocity of 32 feet per second. The height of the rock is given by

$$\text{Height} = -16t^2 + 32t + 48$$

where the height is measured in feet and the time t is measured in seconds. Find the time for the rock to reach the water.

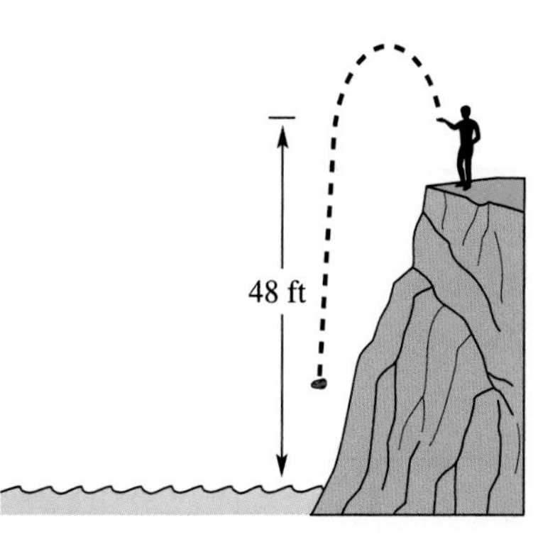

93. *Geometry* The height of a rectangular window is $1\frac{1}{2}$ times its width. The area of the window is 2400 square inches. Find the dimensions of the window.

94. *Geometry* A box with a square base has a surface area of 400 square inches. The height of the box is 5 inches. Find the dimensions of the box. (*Hint:* The surface area is given by $S = 2x^2 + 4xh$.)

95. *Consecutive Integers* The product of two consecutive positive even integers is 168. Find the two integers.

96. *Think About It* In studying for an algebra exam, you and a friend construct trinomials for each other to factor. Explain how you would construct the trinomials and give three examples.

Chapter Test

Take this test as you would take a test in class. After you are done, check your work against the answers given in the back of the book.

In Exercises 1–10, completely factor the polynomial.

1. $7x^2 - 14x^3$

2. $z(z + 7) - 3(z + 7)$

3. $t^2 - 4t - 5$

4. $6x^2 - 11x + 4$

5. $6y^3 + 45y^2 + 75y$

6. $4 - 25v^2$

7. $4x^2 - 20x + 25$

8. $16 - (z + 9)^2$

9. $x^3 + 2x^2 - 9x - 18$

10. $16 - z^4$

11. Find the missing factor: $\frac{2}{5}x - \frac{3}{5} = \frac{1}{5}(\quad)$.

12. Find all values of b such that $x^2 + bx + 5$ can be factored.

13. Find a number c such that $x^2 + 12x + c$ is a perfect square trinomial.

14. Explain why $(x + 1)(3x - 6)$ is not a complete factorization of $3x^2 - 3x - 6$.

In Exercises 15–18, solve the equation.

15. $(x + 4)(2x - 3) = 0$

16. $7x^2 - 14x = 0$

17. $3x^2 + 7x - 6 = 0$

18. $y(2y - 1) = 6$

19. The width of a rectangle is 5 inches less than the length (see figure). The area of the rectangle is 84 square inches. Find the dimensions of the rectangle.

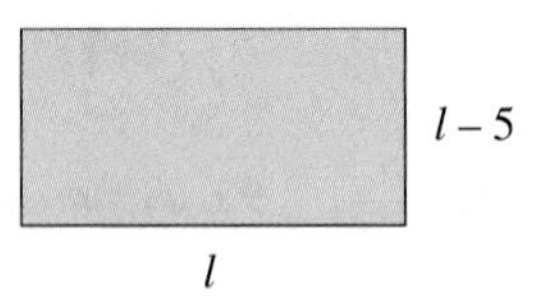

20. An object is dropped from a height of 64 feet. Its height at any time t is modeled by

$$\text{Height} = -16t^2 + 64$$

where the height is measured in feet and the time t is measured in seconds. How long will it take the object to hit the ground? How long will it take the object to fall to a height of 28 feet?

21. The product of two consecutive positive even integers is 624. Find the two integers.

Cumulative Test: Chapters 4–6

Take this test as you would take a test in class. After you are done, check your work against the answers given in the back of the book.

1. Describe how to identify the quadrants in which the points $(-2, y)$ must be located. (y is a real number.)
2. Determine whether the ordered pairs are solution points of the equation $9x - 4y + 36 = 0$.
   (a) $(-1, -1)$ (b) $(8, 27)$ (c) $(-4, 0)$ (d) $(3, -2)$

In Exercises 3 and 4, sketch the graph of the equation and determine any intercepts of the graph.

3. $y = 2 - |x|$
4. $y = \frac{1}{2}x - 2$
5. The slope of a line is $-\frac{1}{4}$ and a point on the line is $(2, 1)$. Find the coordinates of a second point on the line. Explain why there are many correct answers.
6. Find an equation of the line through $\left(0, -\frac{3}{2}\right)$ with slope $m = \frac{5}{6}$.

In Exercises 7 and 8, sketch the lines and determine whether they are parallel, perpendicular, or neither.

7. $y = \frac{2}{3}x - 3, y = -\frac{3}{2}x + 1$
8. $y = 2 - 0.4x, y = -\frac{2}{5}x$
9. Subtract: $(x^3 - 3x^2) - (x^3 + 2x^2 - 5)$
10. Multiply: $(6z)(-7z)(z^2)$
11. Multiply: $(3x + 5)(x - 4)$
12. Multiply: $(5x - 3)(5x + 3)$
13. Expand: $(5x + 6)^2$
14. Divide: $(6x^2 + 72x) \div 6x$
15. Divide: $\dfrac{x^2 - 3x - 2}{x - 4}$
16. Evaluate: $(3^2 \cdot 4^{-1})^2$
17. Factor: $2u^2 - 6u$
18. Factor and simplify: $(x - 2)^2 - 16$
19. Factor completely: $x^3 + 8x^2 + 16x$
20. Factor completely: $x^3 + 2x^2 - 4x - 8$
21. Solve: $u(u - 12) = 0$
22. Solve: $5x^2 - 12x - 9 = 0$
23. Rewrite the expression $\left(\dfrac{x}{2}\right)^{-2}$ using positive exponents.
24. Evaluate the function $g(t) = 2t^2 - |t|$ at the given values of t.
    (a) $g(-2)$ (b) $g(2)$ (c) $g(0)$ (d) $g\left(-\frac{1}{2}\right)$
25. A sales representative is reimbursed \$125 per day for lodging and meals, plus \$0.35 per mile driven. Write a linear equation giving the daily cost C to the company in terms of x, the number of miles driven. Explain the reasoning you used to write the model. Find the cost for a day when the representative drives 70 miles.

7

Systems of Equations

Tony Duffy/Allsport

During the 1996–1997 season, there were 691 women's NCAA soccer teams in the United States. (Source: U.S. Bureau of the Census)

Motivating the Chapter

Soccer Club Fundraiser

A collegiate soccer club has a fundraising dinner. Student tickets sell for $8 and nonstudent tickets sell for $15. There are 115 tickets sold and the total revenue is $1445.

See Section 7.2, Exercise 78

a. Set up a system of linear equations that can be used to determine how many tickets of each type were sold.

b. Solve the system in part (a) by the method of substitution.

c. Solve the system in part (a) by the method of elimination.

The soccer club decides to set goals for the next fundraising dinner. To meet these goals, a "major contributor" category is added. A person donating $100 is considered a major contributor to the soccer club and receives a "free" ticket to the dinner. The club's goals are to have 200 people in attendance, with the number of major contributors being one-fourth the number of students, and to raise $4995.

See Section 7.3, Exercise 65

d. Set up a system of linear equations to determine how many of each kind of ticket would need to be sold for the second fundraising dinner.

e. Solve the system in part (d) by Gaussian elimination.

f. Would it be possible for the soccer club to meet its goals if only 18 people donated $100? Explain.

See Section 7.5, Exercise 100

g. Solve the system in part (d) using matrices.

h. Solve the system in part (d) using determinants.

7.1 Solving Systems of Equations by Graphing and Substitution

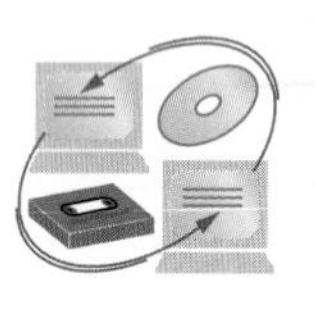

Objectives

1 Determine if an ordered pair is the solution of a system of equations.

2 Use the coordinate plane to solve a system of equations.

3 Solve a system of equations algebraically using the method of substitution.

4 Solve an application problem using a system of equations.

1 Determine if an ordered pair is the solution of a system of equations.

Systems of Equations

Many problems in business and science involve **systems of equations.** These systems consist of two or more equations, each containing two or more variables.

$ax + by = c$ Equation 1

$dx + ey = f$ Equation 2

A **solution** of such a system is an ordered pair (x, y) of real numbers that satisfies *each* equation in the system. When you find the set of all solutions of the system of equations, you are finding the **solution of the system of equations.**

Example 1 Checking Solutions of a System of Equations

Which of the ordered pairs is a solution of the system: (a) (3, 3) or (b) (4, 2)?

$x + y = 6$ Equation 1

$2x - 5y = -2$ Equation 2

Solution

a. To determine whether the ordered pair (3, 3) is a solution of the system of equations, you should substitute $x = 3$ and $y = 3$ into *each* of the equations. Substituting into Equation 1 produces

$3 + 3 = 6.$ ✓ Substitute 3 for x and 3 for y.

Similarly, substituting into Equation 2 produces

$2(3) - 5(3) \neq -2.$ ✗ Substitute 3 for x and 3 for y.

Because the ordered pair (3, 3) fails to check in *both* equations, you can conclude that it *is not* a solution of the original system of equations.

b. By substituting $x = 4$ and $y = 2$ into the original equations, you can determine that the ordered pair (4, 2) is a solution of the first equation

$4 + 2 = 6$ ✓ Substitute 4 for x and 2 for y.

and is also a solution of the second equation

$2(4) - 5(2) = -2.$ ✓ Substitute 4 for x and 2 for y.

So, (4, 2) *is* a solution of the original system of equations.

2 Use the coordinate plane to solve a system of equations.

Solving Systems of Equations by Graphing

A system of two equations in two variables can have exactly one solution, more than one solution, or no solution. In practice, you can gain insight about the location(s) and number of solutions of a system of equations by sketching the graph of each equation in the same coordinate plane. The solutions of the system correspond to the **points of intersection** of the graphs. For instance, Figure 7.1 shows the graphs of three pairs (systems) of equations.

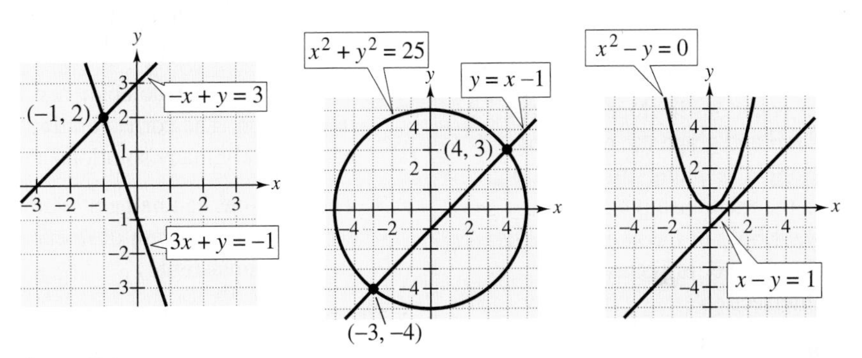

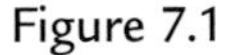
Figure 7.1

Example 2 The Graphical Method of Solving a System

Use the graphical method to solve the system of equations.

$$2x + 3y = 7 \qquad \text{Equation 1}$$

$$2x - 5y = -1 \qquad \text{Equation 2}$$

Solution

Because both equations in the system are linear, you know that they have graphs that are straight lines. To sketch these lines, first write each equation in slope-intercept form, as follows.

$$y = -\frac{2}{3}x + \frac{7}{3} \qquad \text{Slope-intercept form of Equation 1}$$

$$y = \frac{2}{5}x + \frac{1}{5} \qquad \text{Slope-intercept form of Equation 2}$$

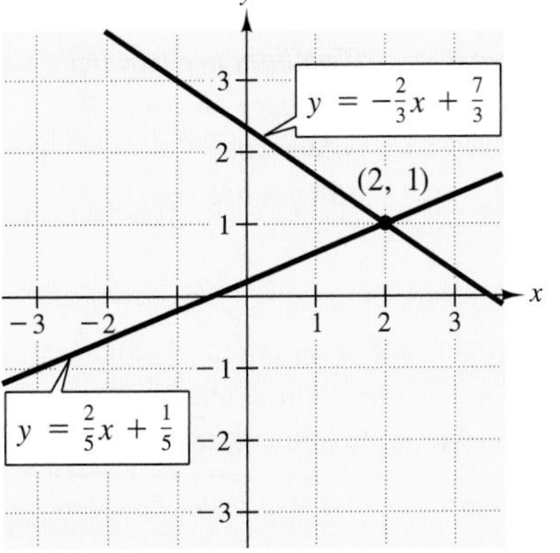

Figure 7.2

The lines corresponding to these two equations are shown in Figure 7.2. From this figure, it appears that the two lines intersect in a single point, (2, 1). You can check these coordinates as follows.

Substitute into 1st Equation	*Substitute into 2nd Equation*
$2x + 3y = 7$	$2x - 5y = -1$
$2(2) + 3(1) \stackrel{?}{=} 7$	$2(2) - 5(1) \stackrel{?}{=} -1$
$7 = 7$ ✓	$-1 = -1$ ✓

Because *both* equations are satisfied, the point (2, 1) *is* a solution of the system.

A good question to ask at this point is, Does every system of two linear equations in two variables have a single solution point? This is the same as asking, Do every two lines in a plane *intersect* in a single point? When you think about it, you can see that the answer to the second question is no. In fact, for two lines in a plane, there are three possible relationships.

Technology: Discovery

Rewrite the following systems of equations in slope-intercept form and graph the equations using a graphing utility. What is the relationship between the slopes of the two lines and the number of points of intersection?

a. $3x + 4y = 12$
$2x - 3y = -9$

b. $-x + 2y = 8$
$2x - 4y = 5$

c. $x + y = 6$
$3x + 3y = 18$

▶ Number of Points of Intersection of Two Lines

1. The two lines can intersect in a single point. The corresponding system of linear equations has a single solution and is called **consistent.**
2. The two lines can coincide and have infinitely many points of intersection. The corresponding consistent system of linear equations has infinitely many solutions and is called **dependent.**
3. The two lines can be parallel and have no point of intersection. The corresponding system of linear equations has no solution and is called **inconsistent.**

These three possibilities are shown in Figure 7.3.

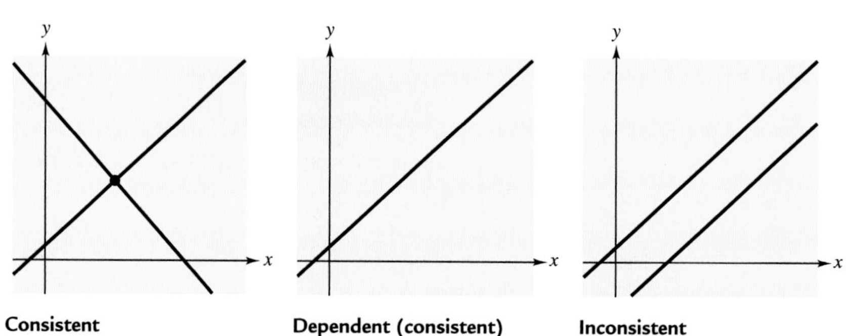

Consistent	**Dependent (consistent)**	**Inconsistent**
Two lines intersect at a single point. (Slopes are not equal.)	*Two lines coincide and have infinitely many points of intersection. (Slopes are equal.)*	*Two lines are parallel and have no point of intersection. (Slopes are equal.)*

Figure 7.3

Note that the word *consistent* is used to mean that the system of linear equations has at least one solution, whereas the word *inconsistent* is used to mean that the system of linear equations has no solution. Here are some examples.

Consistent System	*Consistent System*	*Inconsistent System*
$x + y = 2$	$x + y = 2$	$x + y = 2$
$x + 2y = 4$	$2x + 2y = 4$	$x + y = 4$

You can see from Figure 7.3 that a comparison of the slopes of two lines gives useful information about the number of solutions of the corresponding system of equations. So, to solve a system of equations graphically, it helps to begin by writing the equations in slope-intercept form.

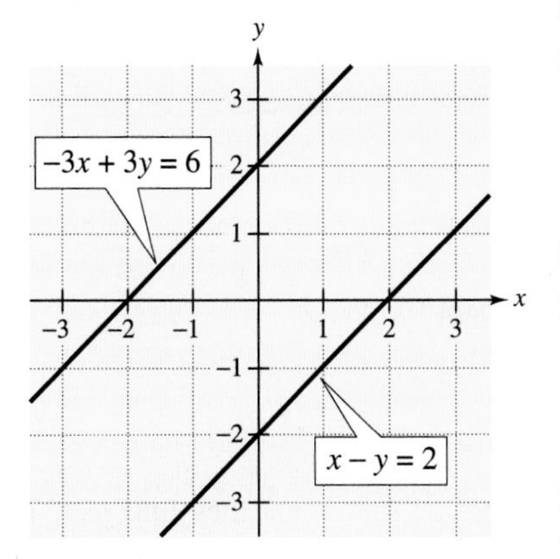

Figure 7.4

Example 3 A System with No Solution

Solve the following system of linear equations.

$$x - y = 2 \qquad \text{Equation 1}$$
$$-3x + 3y = 6 \qquad \text{Equation 2}$$

Solution

Begin by writing each equation in slope-intercept form.

$$y = x - 2 \qquad \text{Slope-intercept form of Equation 1}$$
$$y = x + 2 \qquad \text{Slope-intercept form of Equation 2}$$

From these forms, you can see that the lines representing the two equations are parallel (each has a slope of 1), as shown in Figure 7.4. So, the given system of linear equations has no solution and is an inconsistent system. Try constructing tables of values for the two equations. The tables should help convince you that there is no solution.

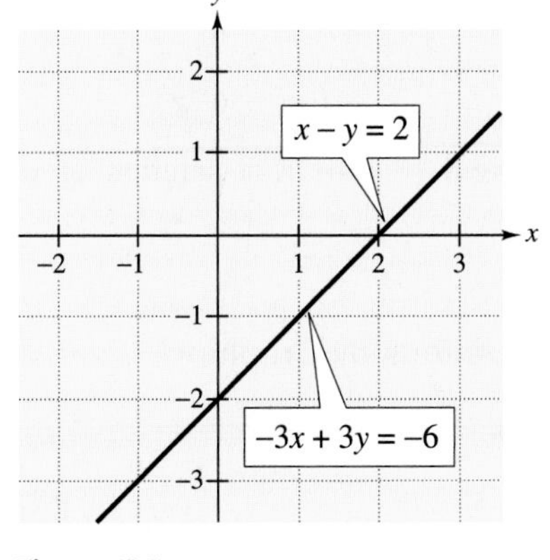

Figure 7.5

Example 4 A System with Infinitely Many Solutions

Solve the following system of linear equations.

$$x - y = 2 \qquad \text{Equation 1}$$
$$-3x + 3y = -6 \qquad \text{Equation 2}$$

Solution

Begin by writing each equation in slope-intercept form.

$$y = x - 2 \qquad \text{Slope-intercept form of Equation 1}$$
$$y = x - 2 \qquad \text{Slope-intercept form of Equation 2}$$

From these forms, you can see that the lines representing the two equations are the same (see Figure 7.5). So, the given system of linear equations is dependent and has infinitely many solutions. You can describe the solution set by saying that each point on the line $y = x - 2$ is a solution of the system of linear equations.

Note in Examples 3 and 4 that if the two lines representing a system of linear equations have the same slope, the system must have either no solution or infinitely many solutions. On the other hand, if the two lines have different slopes, they must intersect in a single point and the corresponding system must have a single solution.

There are two things you should note as you read through Examples 5 and 6. First, your success in applying the graphical method of solving a system of linear equations depends on sketching accurate graphs. Second, once you have made a graph and estimated the point of intersection, it is critical that you check in the original system to see whether the point you have chosen is the correct solution.

Example 5 A System with a Single Solution

Solve the following system of linear equations.

$x + y = 3$ Equation 1

$2x - y = -6$ Equation 2

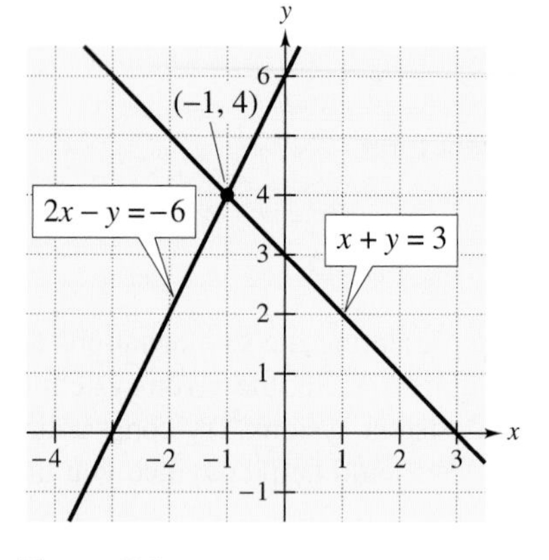

Figure 7.6

Solution

Begin by writing each equation in slope-intercept form.

$y = -x + 3$ Slope-intercept form of Equation 1

$y = 2x + 6$ Slope-intercept form of Equation 2

Because the lines do not have the same slope, you know that they intersect. To find the point of intersection, sketch both lines on the same rectangular coordinate system, as shown in Figure 7.6. From this sketch, it appears that the solution occurs near the point $(-1, 4)$. To check this solution, substitute the coordinates of the point into each of the two given equations.

Substitute into 1st Equation

$$x + y = 3$$
$$-1 + 4 \stackrel{?}{=} 3$$
$$3 = 3 \checkmark$$

Substitute into 2nd Equation

$$2x - y = -6$$
$$2(-1) - 4 \stackrel{?}{=} -6$$
$$-2 - 4 \stackrel{?}{=} -6$$
$$-6 = -6 \checkmark$$

Because *both* equations are satisfied, the point $(-1, 4)$ is a solution of the system.

Example 6 Solving a System of Equations by Graphing

Solve the following system of linear equations.

$2x + y = 4$ Equation 1

$4x + 3y = 9$ Equation 2

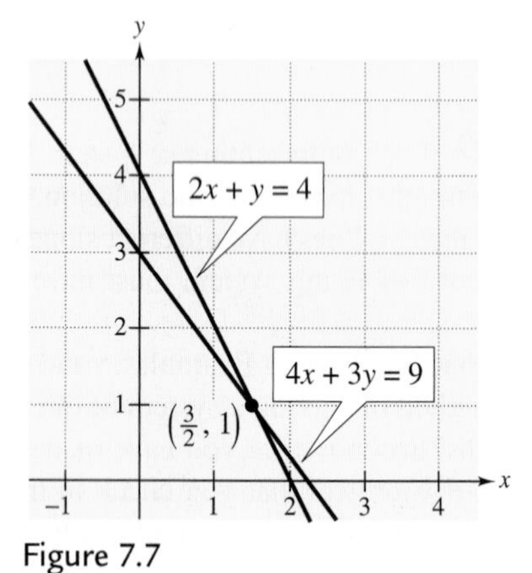

Figure 7.7

Solution

Begin by writing each equation in slope-intercept form.

$y = -2x + 4$ Slope-intercept form of Equation 1

$y = -\frac{4}{3}x + 3$ Slope-intercept form of Equation 2

Because the lines do not have the same slope, you know that they intersect. To find the point of intersect, sketch both lines on the same rectangular coordinate system, as shown in Figure 7.7. From this sketch, it appears that the solution occurs near the point $\left(\frac{3}{2}, 1\right)$. To check this solution, substitute the coordinates of the point into each of the two given equations. This will show that *both* equations are satisfied, so the point $\left(\frac{3}{2}, 1\right)$ is a solution of the system.

3 Solve a system of equations algebraically using the method of substitution.

The Method of Substitution

Solving systems of equations by graphing is useful but less accurate than algebraic methods. In this section, you will study an algebraic method called the **method of substitution.** The goal of the method of substitution is to *reduce a system of two linear equations in two variables to a single equation in one variable.* Examples 7 and 8 illustrate the basic steps of this method.

Example 7 The Method of Substitution

Solve the following system of linear equations.

$-x + y = 1$ Equation 1

$2x + y = -2$ Equation 2

Solution

Begin by solving for y in the first equation.

$-x + y = 1$ Original Equation 1

$y = x + 1$ Revised Equation 1

Next, substitute this expression for y into Equation 2.

$2x + y = -2$ Equation 2

$2x + (x + 1) = -2$ Substitute $x + 1$ for y.

$3x + 1 = -2$ Combine like terms.

$3x = -3$ Subtract 1 from both sides.

$x = -1$ Divide both sides by 3.

At this point, you know that the x-coordinate of the solution is -1. To find the y-coordinate, *back-substitute* the x-value into the revised Equation 1.

$y = x + 1$ Revised Equation 1

$y = -1 + 1$ Substitute -1 for x.

$y = 0$ Simplify.

The solution is $(-1, 0)$. Check this solution by substituting $x = -1$ and $y = 0$ into both of the original equations.

Study Tip

The term **back-substitute** implies that you work backwards. After solving for one of the variables, substitute that value back into one of the equations in the original (or revised) system to find the value of the other variable.

When you use substitution, it does not matter which variable you solve for first, you will obtain the same solution. It's best to choose the variable that is easier to work with. For instance, in the system on the left it is best to solve for x in Equation 2, whereas for the system on the right it is best to solve for y in Equation 1.

$3x - 2y = 1$ Equation 1 $2x + y = 5$

$x + 4y = 3$ Equation 2 $3x - 2y = 11$

Example 8 The Method of Substitution

Solve the following system of linear equations.

$$5x + 7y = 1 \quad \text{Equation 1}$$

$$x + 4y = -5 \quad \text{Equation 2}$$

Solution

For this system, it is convenient to begin by solving for x in the second equation.

$$x + 4y = -5 \quad \text{Original Equation 2}$$

$$x = -4y - 5 \quad \text{Revised Equation 2}$$

Substituting this expression for x into the first equation produces the following.

$$5x + 7y = 1 \quad \text{Equation 1}$$

$$5(-4y - 5) + 7y = 1 \quad \text{Substitute } -4y - 5 \text{ for } x.$$

$$-20y - 25 + 7y = 1 \quad \text{Distributive Property}$$

$$-13y = 26 \quad \text{Simplify.}$$

$$y = -2 \quad \text{Simplify.}$$

Finally, back-substitute this y-value into the revised second equation.

$$x = -4y - 5 \quad \text{Revised Equation 2}$$

$$x = -4(-2) - 5 \quad \text{Substitute } -2 \text{ for } y.$$

$$x = 3 \quad \text{Simplify.}$$

The solution is $(3, -2)$. Check this by substituting $x = 3$ and $y = -2$ into both of the original equations.

The method of substitution demonstrated in Examples 7 and 8 has the following steps.

▶ The Method of Substitution

1. Solve one of the equations for one variable in terms of the other.
2. Substitute the expression obtained in Step 1 into the other equation to obtain an equation in one variable.
3. Solve the equation obtained in Step 2.
4. Back-substitute the solution from Step 3 into the expression obtained in Step 1 to find the value of the other variable.
5. Check your answer to see that it satisfies both of the original equations.

If neither variable has a coefficient of 1 in a system of linear equations, you can still use the method of substitution. It may mean, however, that you have to work with some fractions in the solution steps as in the next example.

Example 9 The Method of Substitution

Solve the following system of linear equations.

$$5x + 3y = 18 \qquad \text{Equation 1}$$

$$2x - 7y = -1 \qquad \text{Equation 2}$$

Solution

Step 1 Because neither variable has a coefficient of 1, you can choose to solve for either variable. For instance, you can begin by solving for x in Equation 2.

$$2x - 7y = -1 \qquad \text{Original Equation 2}$$

$$x = \tfrac{7}{2}y - \tfrac{1}{2} \qquad \text{Revised Equation 2}$$

Step 2 Substitute $\frac{7}{2}y - \frac{1}{2}$ for x in Equation 1 and solve for y.

$$5x + 3y = 18 \qquad \text{Equation 1}$$

$$5\left(\tfrac{7}{2}y - \tfrac{1}{2}\right) + 3y = 18 \qquad \text{Substitute } \tfrac{7}{2}y - \tfrac{1}{2} \text{ for } x.$$

$$\tfrac{35}{2}y - \tfrac{5}{2} + 3y = 18 \qquad \text{Distributive Property}$$

$$35y - 5 + 6y = 36 \qquad \text{Multiply both sides by 2.}$$

$$y = 1 \qquad \text{Simplify.}$$

Step 3 Back-substitute for y in the revised second equation.

$$x = \tfrac{7}{2}y - \tfrac{1}{2} \qquad \text{Revised Equation 2}$$

$$x = \tfrac{7}{2}(1) - \tfrac{1}{2} \qquad \text{Substitute 1 for } y.$$

$$x = 3 \qquad \text{Simplify.}$$

Step 4 The solution is $(3, 1)$. Check this in the original system.

Example 10 The Method of Substitution: No-Solution Case

Solve the following system of linear equations.

$$x - 3y = 2 \qquad \text{Equation 1}$$

$$-2x + 6y = 2 \qquad \text{Equation 2}$$

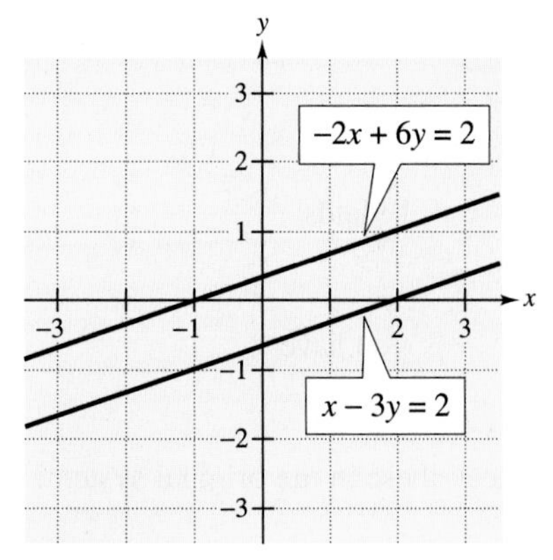

Figure 7.8

Solution

Begin by solving for x in Equation 1 to obtain $x = 3y + 2$. Then, substitute for x in Equation 2.

$$-2x + 6y = 2 \qquad \text{Equation 2}$$

$$-2(3y + 2) + 6y = 2 \qquad \text{Substitute } 3y + 2 \text{ for } x.$$

$$-6y - 4 + 6y = 2 \qquad \text{Simplify.}$$

$$-4 = 2 \qquad \text{False statement.}$$

Because the substitution results in a false statement, the system is inconsistent and has no solution. The graphs in Figure 7.8 confirm this result.

Example 11 The Method of Substitution: Many-Solutions Case

Solve the following system of linear equations.

$$9x + 3y = 15 \quad \text{Equation 1}$$
$$3x + y = 5 \quad \text{Equation 2}$$

Solution

Begin by solving for y in Equation 2 to obtain $y = -3x + 5$. Then, substitute for y in Equation 1.

$$9x + 3y = 15 \quad \text{Equation 1}$$
$$9x + 3(-3x + 5) = 15 \quad \text{Substitute } -3x + 5 \text{ for } y.$$
$$9x - 9x + 15 = 15 \quad \text{Simplify.}$$
$$15 = 15 \quad \text{True statement for any } x$$

Because this last equation is true for any value of x, any solution of Equation 2 is also a solution of Equation 1. So, the given system of linear equations is *dependent* and has infinitely many solutions consisting of all ordered pairs (x, y) such that $y = -3x + 5$. Some sample solutions are $(-1, 8)$, $(0, 5)$, $(1, 2)$, and so on.

Example 12 The Method of Substitution: Nonlinear Case

Solve the system of equations.

$$x^2 + y^2 = 25 \quad \text{Equation 1}$$
$$-x + y = -1 \quad \text{Equation 2}$$

Solution

$$y = x - 1 \quad \text{Solve for } y \text{ in Equation 2.}$$
$$x^2 + y^2 = 25 \quad \text{Equation 1}$$
$$x^2 + (x - 1)^2 = 25 \quad \text{Substitute } x - 1 \text{ for } y.$$
$$2x^2 - 2x - 24 = 0 \quad \text{Simplify.}$$
$$2(x - 4)(x + 3) = 0 \quad \text{Factor.}$$
$$x = 4, -3 \quad \text{Solve for } x.$$

Back-substituting $x = 4$ into the equation $y = x - 1$ yields

$$y = 4 - 1 = 3$$

which implies that $(4, 3)$ is a solution. For $x = -3$, you have

$$y = -3 - 1 = -4$$

which implies that $(-3, -4)$ is a solution. Check these in the original system.

4 Solve an application problem using a system of equations.

Applications

To model a real-life situation with a system of equations, you can use the same basic problem-solving strategy that has been used throughout the text.

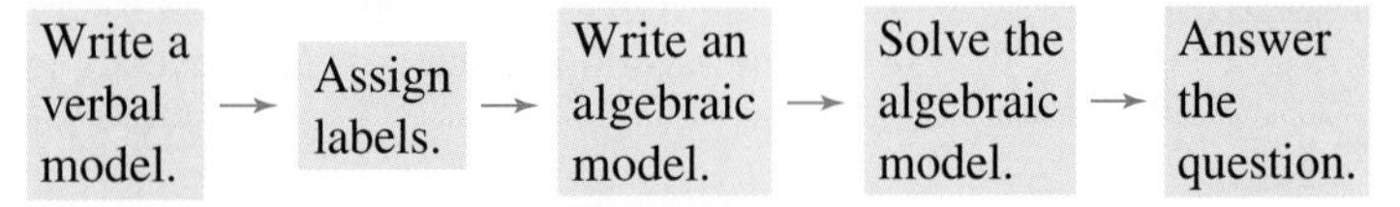
Write a verbal model.
Assign labels.
Write an algebraic model.
Solve the algebraic model.
Answer the question.

After answering the question, remember to check the answer in the original statement of the problem.

Example 13 An Interest Rate Problem

A total of \$12,000 is invested in two funds paying 6% and 8% simple interest. If the interest for 1 year is \$880, how much of the \$12,000 was invested in each fund?

Solution

Verbal Model: [Amount in 6% fund] + [Amount in 8% fund] = [12,000]

[6%] · [Amount in 6% fund] + [8%] · [Amount in 8% fund] = [880]

Labels: Amount in 6% fund $= x$ (dollars)
Amount in 8% fund $= y$ (dollars)

System:

$$x + y = 12{,}000 \quad \text{Equation 1}$$

$$0.06x + 0.08y = 880 \quad \text{Equation 2}$$

Begin by solving for x in the first equation.

$$x + y = 12{,}000 \quad \text{Equation 1}$$

$$x = 12{,}000 - y \quad \text{Revised Equation 1}$$

Substituting this expression for x in the second equation produces the following.

$$0.06x + 0.08y = 880 \quad \text{Equation 2}$$

$$0.06(12{,}000 - y) + 0.08y = 880 \quad \text{Substitute } 12{,}000 - y \text{ for } x.$$

$$720 - 0.06y + 0.08y = 880 \quad \text{Distributive Property}$$

$$0.02y = 160 \quad \text{Simplify.}$$

$$y = 8000 \quad \text{Simplify.}$$

Back-substitute for y in the revised first equation.

$$x = 12{,}000 - y \quad \text{Revised Equation 1}$$

$$x = 12{,}000 - 8000 \quad \text{Substitute 8000 for } y.$$

$$x = 4000 \quad \text{Simplify.}$$

So, \$4000 was invested in the fund paying 6% and \$8000 was invested in the fund paying 8%. Check this in the original statement of the problem.

The total cost C of producing x units of a product usually has two components—the initial cost and the cost per unit. When enough units have been sold so that the total revenue R equals the total cost, the sales are said to have reached the **break-even point.** You can find this break-even point by setting C equal to R and solving for x. In other words, the break-even point corresponds to the point of intersection of the cost and revenue graphs.

Example 14 An Application: Break-Even Analysis

A small business invests \$14,000 in equipment to produce a product. Each unit of the product costs \$0.80 to produce and is sold for \$1.50. How many items must be sold before the business breaks even?

Solution

Verbal Model: Total cost = Cost per unit · Number of units + Initial cost

Total revenue = Price per unit · Number of units

Labels: Total cost $= C$ (dollars)
Cost per unit $= 0.80$ (dollars per unit)
Number of units $= x$ (units)
Initial cost $= 14{,}000$ (dollars)
Total revenue $= R$ (dollars)
Price per unit $= 1.50$ (dollars per unit)

System: $C = 0.80x + 14{,}000$ Equation 1

$R = 1.50x$ Equation 2

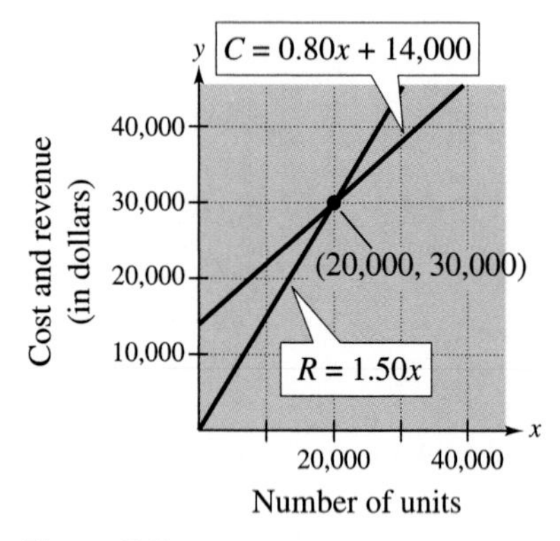

Figure 7.9

Because the break-even point occurs when $R = C$, you have

$$1.5x = 0.8x + 14{,}000$$

$$0.7x = 14{,}000$$

$$x = 20{,}000.$$

So, it follows that the business must sell 20,000 units before it breaks even. Note in Figure 7.9 that sales less than the break-even point correspond to a loss for the business, whereas sales greater than the break-even point correspond to a profit for the business.

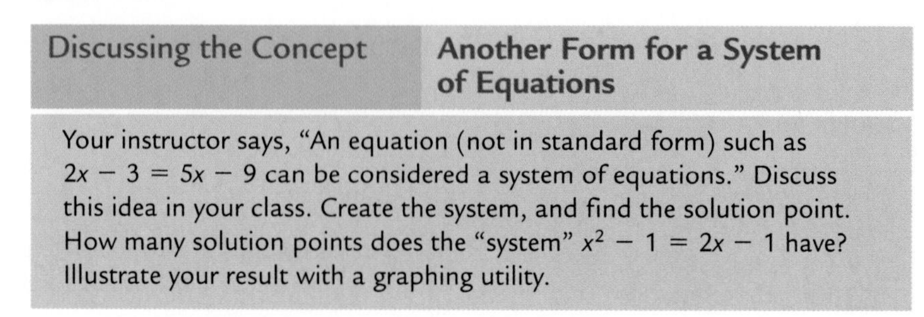

Discussing the Concept **Another Form for a System of Equations**

Your instructor says, "An equation (not in standard form) such as $2x - 3 = 5x - 9$ can be considered a system of equations." Discuss this idea in your class. Create the system, and find the solution point. How many solution points does the "system" $x^2 - 1 = 2x - 1$ have? Illustrate your result with a graphing utility.

7.1 Exercises

Integrated Review *Concepts, Skills, and Problem Solving*

Keep mathematically in shape by doing these exercises *before* the problems of this section.

Properties and Definitions

1. Sketch the graph of a line with positive slope.
2. Sketch the graph of a line with undefined slope.
3. The slope of a line is $-\frac{2}{3}$. What is the slope of a line perpendicular to this line?
4. Two lines have slopes $m = -3$ and $m = \frac{3}{2}$. Which line is steeper? Explain.

Solving Equations

In Exercises 5–8, solve the equation and check your answer.

5. $y - 3(4y - 2) = 1$
6. $x + 6(3 - 2x) = 4$
7. $\frac{1}{2}x - \frac{1}{5}x = 15$
8. $\frac{1}{10}(x - 4) = 6$

In Exercises 9 and 10, solve for y in terms of x.

9. $3x + 4y - 5 = 0$
10. $-2x - 3y + 6 = 0$

Graphing

In Exercises 11–14, sketch the graph of the linear equation.

11. $y = -3x + 2$
12. $4x - 2y = -4$
13. $3x + 2y = 8$
14. $x + 3 = 0$

Developing Skills

In Exercises 1–8, determine whether each ordered pair is a solution of the system of equations. See Example 1.

1. $x + 2y = 9$
   $-2x + 3y = 10$
   (a) $(1, 4)$
   (b) $(3, -1)$

2. $5x - 4y = 34$
   $x - 2y = 8$
   (a) $(0, 3)$
   (b) $(6, -1)$

3. $-2x + 7y = 46$
   $3x + y = 0$
   (a) $(-3, 2)$
   (b) $(-2, 6)$

4. $-5x - 2y = 23$
   $x + 4y = -19$
   (a) $(-3, -4)$
   (b) $(3, 7)$

5. $4x - 5y = 12$
   $3x + 2y = -2.5$
   (a) $(8, 4)$
   (b) $\left(\frac{1}{2}, -2\right)$

6. $2x - y = 1.5$
   $4x - 2y = 3$
   (a) $\left(0, -\frac{3}{2}\right)$
   (b) $\left(2, \frac{5}{2}\right)$

7. $x^2 + y^2 = 169$
   $17x - 7y = 169$
   (a) $(5, -12)$
   (b) $(-7, 10)$

8. $x^2 + y^2 = 17$
   $x + 3y = 7$
   (a) $(4, 1)$
   (b) $(2, 1)$

In Exercises 9–16, determine whether the system is consistent or inconsistent.

9. $x + 2y = 6$
   $x + 2y = 3$

10. $x - 2y = 3$
    $2x - 4y = 7$

11. $2x - 3y = -12$
    $-8x + 12y = -12$

12. $-5x + 8y = 8$
    $7x - 4y = 14$

13. $-x + 4y = 7$
    $3x - 12y = -21$

14. $3x + 8y = 28$
    $-4x + 9y = 1$

15. $5x - 3y = 1$
    $6x - 4y = -3$

16. $9x + 6y = 10$
    $-6x - 4y = 3$

In Exercises 17–20, use a graphing utility to graph the equations in the system. Use the graphs to determine whether the system is consistent or inconsistent. If the system is consistent, determine the number of solutions.

17. $\frac{1}{3}x - \frac{1}{2}y = 1$
    $-2x + 3y = 6$

18. $x + y = 5$
    $x - y = 5$

19. $-2x + 3y = 6$
    $x - y = -1$

20. $2x - 4y = 9$
    $x - 2y = 4.5$

In Exercises 21–28, use the graphs of the equations to determine whether the system has any solutions. Find any solutions that exist.

21. $x + y = 4$
$x + y = -1$

22. $-x + y = 5$
$x + 2y = 4$

23. $5x - 3y = 4$
$2x + 3y = 3$

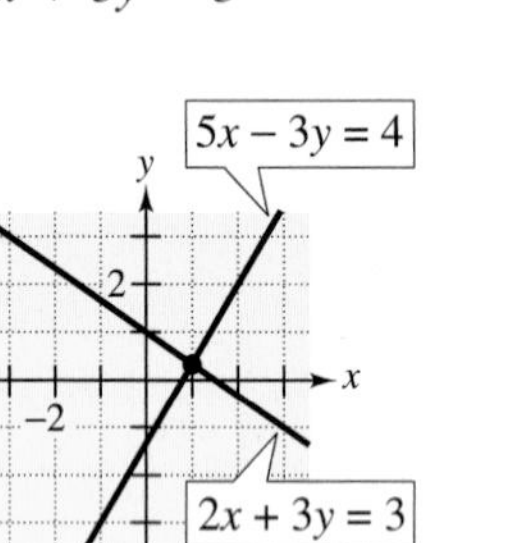

24. $2x - y = 4$
$-4x + 2y = -12$

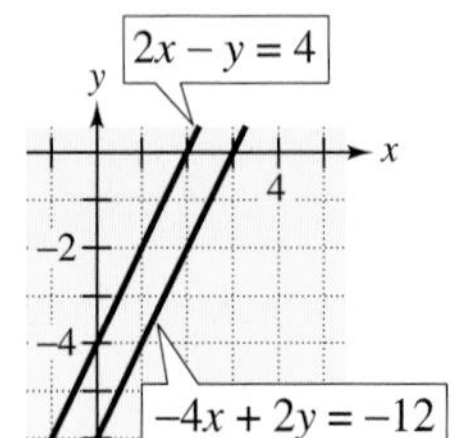

25. $x - 2y = -4$
$-0.5x + y = 2$

26. $2x - 3y = 6$
$4x + 3y = 12$

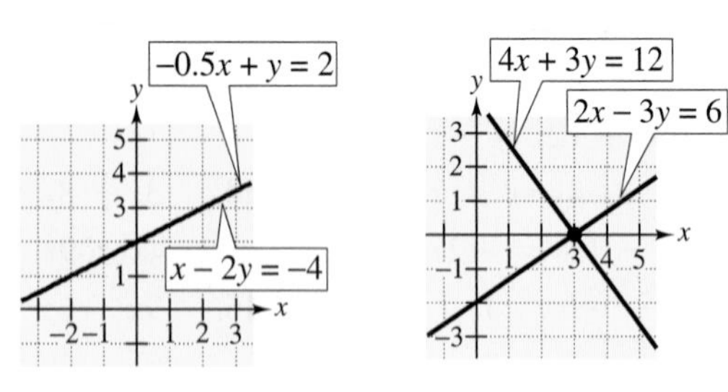

27. $x - 2y = 4$
$x^2 - y = 0$

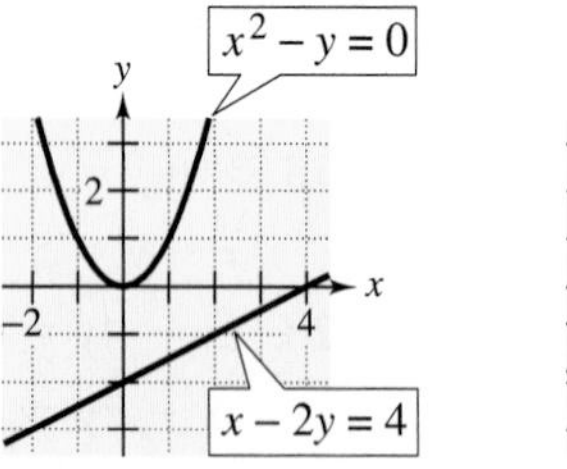

28. $y = \sqrt{x - 2}$
$x - 2y = 1$

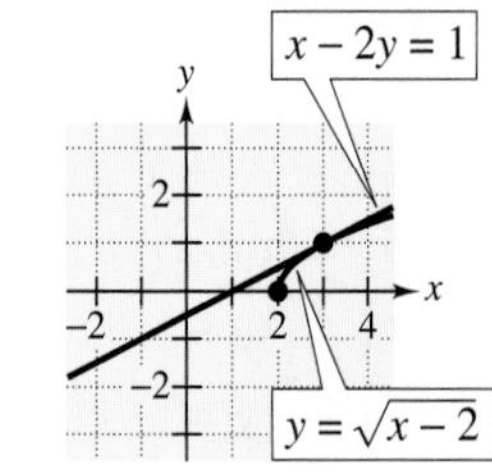

In Exercises 29–42, solve the system of linear equations by graphing. See Examples 2–6.

29. $y = -x + 3$
$y = x + 1$

30. $y = 2x - 4$
$y = -\frac{1}{2}x + 1$

31. $x - y = 2$
$x + y = 2$

32. $x - y = 0$
$x + y = 4$

33. $3x - 4y = 5$
$x = 3$

34. $5x + 2y = 18$
$y = 2$

35. $4x + 5y = 20$
$\frac{4}{5}x + y = 4$

36. $-x + 3y = 7$
$2x - 6y = 6$

37. $2x - 5y = 20$
$4x - 5y = 40$

38. $5x + 3y = 24$
$x - 2y = 10$

39. $x + y = 2$
$3x + 3y = 6$

40. $4x - 3y = -3$
$8x - 6y = -6$

41. $4x + 5y = 7$
$2x - 3y = 9$

42. $7x + 4y = 6$
$5x - 3y = -25$

In Exercises 43–46, use a graphing utility to graph the equations and approximate any solutions of the system of equations.

43. $y = x^2$
$y = 4x - x^2$

44. $y = 8 - x^2$
$y = 6 - x$

45. $y = x^3$
$y = x^3 - 3x^2 + 3x$

46. $y = x^2 - 2x$
$y = x^3 - 4x$

In Exercises 47–76, solve the given system by substitution. See Examples 7–12.

47. $x - 2y = 0$
$3x + 2y = 8$

48. $x - y = 0$
$5x - 2y = 6$

49. $x = 4$
$x - 2y = -2$

50. $y = 2$
$x - 6y = -6$

51. $x + y = 3$
$2x - y = 0$

52. $-x + y = 5$
$x - 4y = 0$

53. $x + y = 2$
$x - 4y = 12$

54. $x - 2y = -1$
$x - 5y = 2$

55. $x + 6y = 19$
$x - 7y = -7$

56. $x - 5y = -6$
$4x - 3y = 10$

57. $8x + 5y = 100$
$9x - 10y = 50$

58. $x + 4y = 300$
$x - 2y = 0$

59. $-13x + 16y = 10$
$5x + 16y = -26$

60. $2x + 5y = 29$
$5x + 2y = 13$

61. $4x - 14y = -15$
$18x - 12y = 9$

62. $5x - 24y = -12$
$17x - 24y = 36$

63. $\frac{1}{5}x + \frac{1}{2}y = 8$
$x + y = 20$

64. $\frac{1}{2}x + \frac{3}{4}y = 10$
$\frac{3}{2}x - y = 4$

65. $y = 2x^2$
$y = -2x + 12$

66. $y = 5x^2$
$y = -15x - 10$

67. $3x + 2y = 30$
$y = 3x^2$

68. $x + 2y = 16$
$x = 5y^2$

69. $x^2 + y = 9$
$x - y = -3$

70. $x - y^2 = 0$
$x - y = 2$

71. $x^2 + y^2 = 100$
$y + x = 2$

72. $x^2 + y^2 = 169$
$x + y = 7$

73. $x^2 - y = 2$
$3x + y = 2$

74. $x^2 + 2y = 6$
$x - y = -4$

75. $x^2 + y^2 = 25$
$2x - y = -5$

76. $x^2 - y^2 = 16$
$3x - y = 12$

In Exercises 77 and 78, use a graphing utility to graph each equation in the system. The graphs appear parallel. Yet, from the slope-intercept forms of the lines, you can find that the slopes are not equal and so the graphs intersect. Find the point of intersection of the two lines.

77. $x - 100y = -200$
$3x - 275y = 198$

78. $35x - 33y = 0$
$12x - 11y = 92$

Think About It In Exercises 79–82, write a system of equations having the given solution.

79. $(4, 5)$

80. $(-2, 6)$

81. $(-1, -2)$

82. $\left(\frac{1}{2}, 3\right)$

Solving Problems

83. *Break-Even Analysis* A small business invests \$8000 in equipment to produce a product. Each unit of the product costs \$1.20 to produce and is sold for \$2.00. How many items must be sold before the business breaks even?

84. *Break-Even Analysis* A business invests \$50,000 in equipment to produce a product. Each unit of the product costs \$19.25 to produce and is sold for \$35.95. How many items must be sold before the business breaks even?

85. *Break-Even Analysis* Suppose you are setting up a small business and have invested \$10,000 to produce an item that will sell for \$3.25. If each unit can be produced for \$1.65, how many units must you sell to break even?

86. *Break-Even Analysis* Suppose you are setting up a small business and have made an initial investment of \$30,000. The unit cost of the product is \$16.40, and the selling price is \$31.40. How many units must you sell to break even?

87. *Simple Interest* A combined total of \$20,000 is invested in two bonds that pay 8% and 9.5% simple interest. The annual interest is \$1675. How much is invested in each bond?

88. *Simple Interest* A combined total of \$12,000 is invested in two bonds that pay 8.5% and 10% simple interest. The annual interest is \$1140. How much is invested in each bond?

89. *Simple Interest* A combined total of \$25,000 is invested in two funds paying 8% and 8.5% simple interest. The annual interest is \$2060. How much is invested in each fund?

Number Problems In Exercises 90–93, find two positive integers that satisfy the given requirements.

90. The sum of two numbers is 80 and their difference is 18.

91. The sum of the larger number and twice the smaller number is 61 and their difference is 7.

92. The sum of two numbers is 52 and the larger number is 8 less than twice the smaller number.

93. The sum of two numbers is 160 and the larger number is 3 times the smaller number.

Geometry In Exercises 94–97, find the dimensions of the rectangle meeting the specified conditions.

	Perimeter	*Condition*
94.	50 feet	The length is 5 feet greater than the width.
95.	320 inches	The width is 20 inches less than the length.
96.	68 yards	The width is $\frac{7}{10}$ of the length.
97.	90 meters	The length is $1\frac{1}{2}$ times the width.

98. *Dimensions of a Corral* You have 250 feet of fencing to enclose two corrals of equal size (see figure). The combined area of the corrals is 2400 square feet. Find the dimensions of each corral.

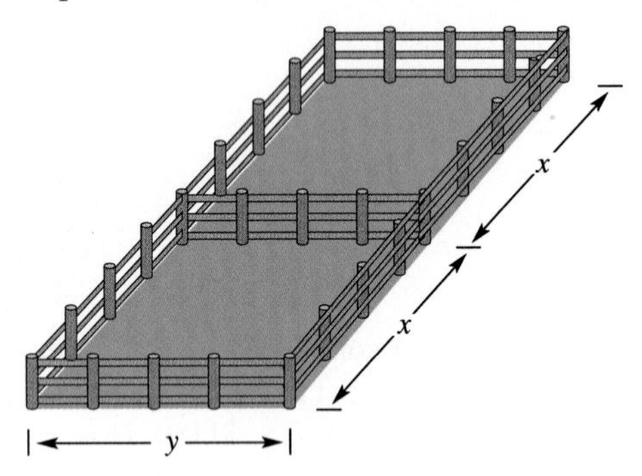

99. *Graphical Analysis* From 1980 through 1996, the northeastern part of the United States grew at a slower rate than the western part. Models that represent the populations of the two regions are

$E = 49{,}088.2 + 194.6t - 2.2t^2$ *Northeast*

$W = 43{,}132.9 + 977.3t$ *West*

where E and W are the populations in thousands and t is the calendar year, with $t = 0$ corresponding to 1980. Use a graphing utility to determine when the population of the West overtook the population of the Northeast. (Source: U.S. Bureau of the Census)

Explaining Concepts

100. What is meant by a solution of a system of equations in two variables?

101. List and explain the basic steps in solving a system of equations by substitution.

102. When solving a system of equations by substitution, how do you recognize that the system has no solution?

103. What does it mean to *back-substitute* when solving a system of equations?

104. Give a geometric description of the solution of a system of equations in two variables.

105. Describe any advantages of the method of substitution over the graphical method of solving a system of equations.

106. Is it possible for a consistent system of linear equations to have exactly two solutions? Explain.

7.2 Solving Systems of Equations by Elimination

Objectives

1 Solve a system of linear equations algebraically using the method of elimination.

2 Use a system of equations to solve an application problem.

1 Solve a system of linear equations algebraically using the method of elimination.

The Method of Elimination

In this section, you will study another way to solve a system of linear equations algebraically—the **method of elimination.** The key step is to obtain, for one of the variables, coefficients that differ only in sign, so that by *adding* the two equations this variable can be eliminated. For instance, by adding the equations in the following system

$$3x + 5y = 7 \quad \text{Equation 1}$$
$$-3x - 2y = -1 \quad \text{Equation 2}$$
$$3y = 6 \quad \text{Add equations.}$$

you eliminate the variable x and obtain a single equation in one variable, y.

Example 1 The Method of Elimination

Solve the following system of equations.

$$4x + 3y = 5 \quad \text{Equation 1}$$
$$2x - 3y = 7 \quad \text{Equation 2}$$

Solution

Begin by noting that the coefficients for y differ only in sign. So, by adding the two equations, you can eliminate y.

$$4x + 3y = 5 \quad \text{Equation 1}$$
$$2x - 3y = 7 \quad \text{Equation 2}$$
$$6x = 12 \quad \text{Add equations.}$$

So, $x = 2$. By back-substituting this value into the first equation, you can solve for y, as follows.

$$4x + 3y = 5 \quad \text{Equation 1}$$
$$4(2) + 3y = 5 \quad \text{Substitute 2 for } x.$$
$$3y = -3 \quad \text{Subtract 8 from both sides.}$$
$$y = -1 \quad \text{Divide both sides by 3.}$$

The solution is $(2, -1)$. Check this in both of the original equations.

Carl Friedrich Gauss
(1777–1855)

Gauss is often ranked with Archimedes and Newton as one of the greatest mathematicians in history. Gauss's doctoral thesis proved the Fundamental Theorem of Algebra. His contributions to mathematics can be found in differential geometry, algebra, complex functions, and potential theory. The process used in the method of elimination was developed by Gauss and is called "Gaussian elimination." Gaussian elimination can be used to solve systems of three or more variables.

Try using substitution to solve the system given in Example 1. Which method do you think is easier: substitution or elimination? Many people find that the method of elimination is more efficient.

▶ The Method of Elimination

1. Obtain coefficients for x (or y) that differ only in sign by multiplying all terms of one or both equations by suitably chosen constants.
2. Add the equations to eliminate one variable and solve the resulting equation.
3. Back-substitute the value obtained in Step 2 into either of the original equations and solve for the other variable.
4. Check your solution in both of the original equations.

Example 2 The Method of Elimination

Solve the system of linear equations.

$4x - 5y = -9$ Equation 1

$3x - y = 7$ Equation 2

Solution

To obtain coefficients of y that differ only in sign, multiply Equation 2 by -5.

$4x - 5y = -9 \Rightarrow 4x - 5y = -9$ Equation 1

$3x - y = 7 \Rightarrow -15x + 5y = -35$ Multiply Equation 2 by -5.

$-11x = -44$ Add equations.

So, $x = 4$. Back-substitute this value into Equation 2 and solve for y.

$3x - y = 7$ Equation 2

$3(4) - y = 7$ Substitute 4 for x.

$y = 5$ Solve for y.

The solution is $(4, 5)$. Check this in the original system of equations.

Substitute into 1st Equation

$4x - 5y = -9$ Equation 1

$4(4) - 5(5) = -9$ Substitute 4 for x and 5 for y.

$16 - 25 = -9$ Solution checks. ✓

Substitute into 2nd Equation

$3x - y = 7$ Equation 2

$3(4) - 5 = 7$ Substitute 4 for x and 5 for y.

$12 - 5 = 7$ Solution checks. ✓

Example 3 Solving a System with Fractional Coefficients

Solve the following system of linear equations.

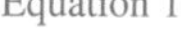

$$\tfrac{3}{2}x - 2y = -2 \qquad \text{Equation 1}$$

$$x - \tfrac{2}{3}y = -1 \qquad \text{Equation 2}$$

Solution

To eliminate the fractional coefficient of x in the first equation, multiply Equation 1 by 2.

$$\tfrac{3}{2}x - 2y = -2 \quad \Rightarrow \quad 3x - 4y = -4 \qquad \text{Revised Equation 1}$$

Now, by multiplying the second equation by -3, you can eliminate the fractional coefficient of y and obtain coefficients of x that differ only in sign.

$$3x - 4y = -4 \quad \Rightarrow \quad 3x - 4y = -4 \qquad \text{Revised Equation 1}$$

$$x - \tfrac{2}{3}y = -1 \quad \Rightarrow \quad -3x + 2y = 3 \qquad \text{Multiply Equation 2 by } -3.$$

$$-2y = -1 \qquad \text{Add equations.}$$

So, $y = \frac{1}{2}$. Back-substitute this value into Equation 2 and solve for y.

$$x - \tfrac{2}{3}y = -1 \qquad \text{Equation 2}$$

$$x - \tfrac{2}{3}\left(\tfrac{1}{2}\right) = -1 \qquad \text{Substitute } \tfrac{1}{2} \text{ for } y.$$

$$x = -\tfrac{2}{3} \qquad \text{Solve for } x.$$

The solution is $\left(-\frac{2}{3}, \frac{1}{2}\right)$. Check this in the original system of equations.

Example 4 The Method of Elimination: Many-Solutions Case

Solve the system of linear equations.

$$-2x + 6y = 3 \qquad \text{Equation 1}$$

$$4x - 12y = -6 \qquad \text{Equation 2}$$

Solution

To obtain coefficients of x that differ only in sign, multiply the first equation by 2.

$$-2x + 6y = 3 \quad \Rightarrow \quad -4x + 12y = 6 \qquad \text{Multiply Equation 1 by 2.}$$

$$4x - 12y = -6 \quad \Rightarrow \quad 4x - 12y = -6 \qquad \text{Equation 2}$$

$$0 = 0 \qquad \text{Add equations.}$$

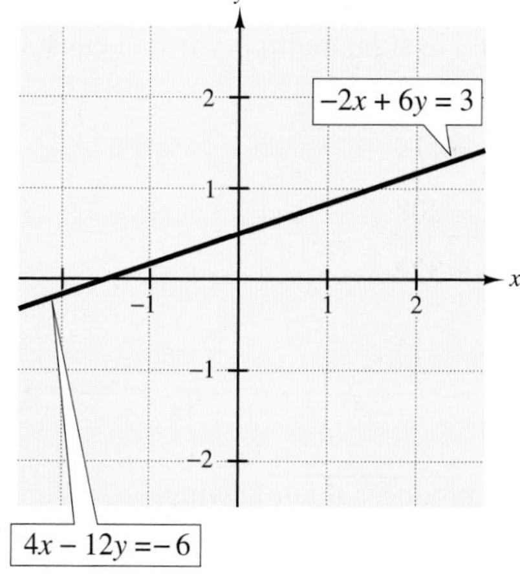

Figure 7.10

Because the two equations turn out to be equivalent, the system has infinitely many solutions. The solution set consists of all points (x, y) lying on the line $-2x + 6y = 3$, as shown in Figure 7.10.

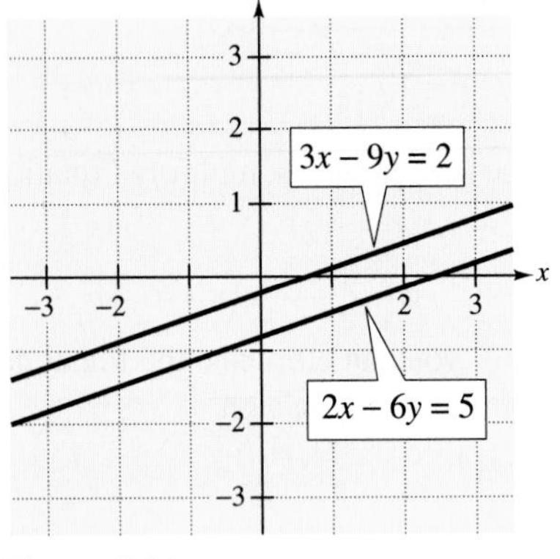

Figure 7.11

Example 5 The Method of Elimination: No-Solution Case

Solve the following system of linear equations.

$2x - 6y = 5$ Equation 1

$3x - 9y = 2$ Equation 2

Solution

To obtain coefficients that differ only in sign, multiply the first equation by 3 and multiply the second equation by -2.

$2x - 6y = 5 \Rightarrow 6x - 18y = 15$ Multiply Equation 1 by 3.

$3x - 9y = 2 \Rightarrow -6x + 18y = -4$ Multiply Equation 2 by -2.

$0 = 11$ False statement

Because there are no values of x and y for which $0 = 11$, you can conclude that the system is inconsistent and has no solution. You can check this result graphically by noting that the equations graph as two parallel lines. (See Figure 7.11.)

Example 6 Solving a System Having Decimal Coefficients

Solve the following system of linear equations.

$0.02x - 0.05y = -0.38$ Equation 1

$0.03x + 0.04y = 1.04$ Equation 2

Solution

To obtain integer coefficients, multiply both equations by 100.

$2x - 5y = -38$ Revised Equation 1

$3x + 4y = 104$ Revised Equation 2

Now, to obtain coefficients of x that differ only in sign, multiply the first equation by 3 and multiply the second equation by -2.

$2x - 5y = -38 \Rightarrow 6x - 15y = -114$ Multiply Equation 1 by 3.

$3x + 4y = 104 \Rightarrow -6x - 8y = -208$ Multiply Equation 2 by -2.

$-23y = -322$ Add equations.

So, the y-coordinate of the solution is

$$y = \frac{-322}{-23} = 14.$$

Back-substituting this value into Equation 2 produces the following.

$3x + 4y = 104$ Revised Equation 2.

$3x + 4(14) = 104$ Substitute 14 for y.

$3x = 48$ Simplify.

$x = 16$ Divide both sides by 3.

The solution is $(16, 14)$. Check this in the original system of equations.

Study Tip

When multiplying an equation by a negative number, be sure to distribute the negative sign to each term of the equation. For instance, in Example 6 the second equation is multiplied by -2.

2 Use a system of equations to solve an application problem.

Applications

To determine whether a real-life problem can be solved using a system of linear equations, consider the following. (1) Does the problem involve more than one unknown quantity? (2) Are there two (or more) equations or conditions to be satisfied? If one or both of these conditions occur, the appropriate mathematical model for the problem may be a system of linear equations.

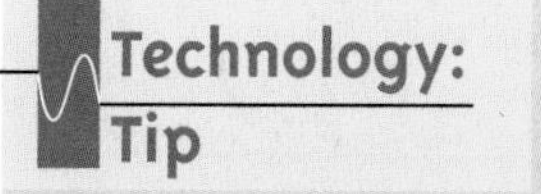

The general solution of the linear system

$ax + by = c$

$dx + ey = f$

is $x = (ce - bf)/(ae - db)$ and $y = (af - cd)/(ae - db)$. If $ae - db = 0$, the system does not have a unique solution. Graphing utility programs for solving such a system can be found at our website *www.hmco.com*. Try using this program to solve the systems in Examples 7 and 8.

Example 7 A Mixture Problem

A company with two stores buys six large delivery vans and five small delivery vans. The first store receives four of the large vans and two of the small vans for a total cost of \$160,000. The second store receives two of the large vans and three of the small vans for a total cost of \$128,000. What is the cost of each type of van?

Solution

The two unknowns in this problem are the costs of the two types of vans.

Verbal Model:

$$4\begin{pmatrix}\text{Cost of}\\ \text{large van}\end{pmatrix} + 2\begin{pmatrix}\text{Cost of}\\ \text{small van}\end{pmatrix} = \$160{,}000$$

$$2\begin{pmatrix}\text{Cost of}\\ \text{large van}\end{pmatrix} + 3\begin{pmatrix}\text{Cost of}\\ \text{small van}\end{pmatrix} = \$128{,}000$$

Labels: Cost of large van $= x$ (dollars)
Cost of small van $= y$ (dollars)

System:
$4x + 2y = 160{,}000$ Equation 1
$2x + 3y = 128{,}000$ Equation 2

To solve this system of linear equations, use the method of elimination. To obtain coefficients of x that differ only in sign, multiply the second equation by -2.

$4x + 2y = 160{,}000 \Rightarrow 4x + 2y = 160{,}000$

$2x + 3y = 128{,}000 \Rightarrow -4x - 6y = -256{,}000$ Multiply Equation 2 by -2.

$-4y = -96{,}000$ Add equations.

$y = 24{,}000$ Divide by -4.

The cost of each small van is $y = \$24{,}000$. Back-substitute this value into Equation 1 to find the cost of each large van.

$4x + 2y = 160{,}000$ Equation 1

$4x + 2(24{,}000) = 160{,}000$ Substitute 24,000 for y.

$4x = 112{,}000$ Simplify.

$x = 28{,}000$ Divide both sides by 4.

The cost of each large van is $x = \$28{,}000$. Check this solution in the original statement of the problem.

Example 8 An Application Involving Two Speeds

You take a motorboat trip on a river (18 miles upstream and 18 miles downstream). You run the motor at the same speed going up and down the river, but because of the river's current, the trip upstream takes $1\frac{1}{2}$ hours and the trip downstream takes only 1 hour. From this information, determine the speed of the current.

Solution

Verbal Model: [Boat speed (still water)] $-$ [Speed of current] $=$ [Upstream speed]

[Boat speed (still water)] $+$ [Speed of current] $=$ [Downstream speed]

Labels: Boat speed in still water $= x$ (miles per hour)
Current speed $= y$ (miles per hour)
Upstream speed $= 18/1.5 = 12$ (miles per hour)
Downstream speed $= 18/1 = 18$ (miles per hour)

System: $x - y = 12$ Equation 1
$x + y = 18$ Equation 2

To solve this system of linear equations, use the method of elimination.

$$x - y = 12 \quad \text{Equation 1}$$
$$x + y = 18 \quad \text{Equation 2}$$
$$2x \quad = 30 \quad \text{Add equations.}$$

So, the speed of the boat in still water is 15 miles per hour. Back-substitute this value into Equation 2 to find the speed of the current.

$$15 + y = 18 \quad \text{Substitute 15 for } x \text{ in Equation 2.}$$
$$y = 3 \quad \text{Subtract 15 from both sides.}$$

The speed of the current is 3 miles per hour. Check this solution in the original statement of the problem.

Discussing the Concept **Discovering the Number of Solutions**

A student claims that by cross-multiplying coefficients, it is easy to tell if a system of linear equations has one, no, or many solutions.

Example 3 [one]	*Example 4 [many]*	*Example 5 [none]*
$\frac{3}{2}x - 2y = -2$	$-2x + 6y = 3$	$2x - 6y = 5$
$1x - \frac{2}{3}y = -1$	$4x - 12y = -6$	$3x - 9y = 2$
$-1 \neq -2$	$24 = 24$	$-18 = -18$

Is the claim valid? Can you further distinguish between the no-solution and many-solution cases? Try the method of elimination on the general system $a_1x + b_1y = c_1$ and $a_2x + b_2y = c_2$ to verify the student's claim.

7.2 Exercises

Integrated Review *Concepts, Skills, and Problem Solving*

Keep mathematically in shape by doing these exercises *before* the problems of this section.

Properties and Definitions

1. Name the property illustrated by $2(x + y) = 2x + 2y$.

2. What property of equality is demonstrated in solving the following equation?

$$x - 4 = 7$$
$$x - 4 + 4 = 7 + 4$$
$$x = 11$$

Solving Inequalities

In Exercises 3–8, solve the inequality.

3. $1 < 2x + 5 < 9$

4. $0 \le \dfrac{x - 4}{2} < 6$

5. $|6x| > 12$

6. $|1 - 2x| < 5$

7. $4x - 12 < 0$

8. $4x + 4 \ge 9$

Problem Solving

9. The annual operating cost of a truck is $C = 0.45m + 6200$, where m is the number of miles traveled by the truck in a year. What number of miles will yield an annual operating cost that is less than $15,000?

10. You must select one of two plans of payment when working for a company. One plan pays $2500 per month. The second pays $1500 per month plus a commission of 4% of your gross sales. Write an inequality whose solution is such that the second option gives the greater monthly wage. Solve the inequality.

Developing Skills

In Exercises 1–12, solve the system of linear equations by elimination. Identify and label each line with its equation and label the point of intersection (if any). See Examples 1–6.

1. $2x + y = 4$
$x - y = 2$

2. $x + 3y = 2$
$-x + 2y = 3$

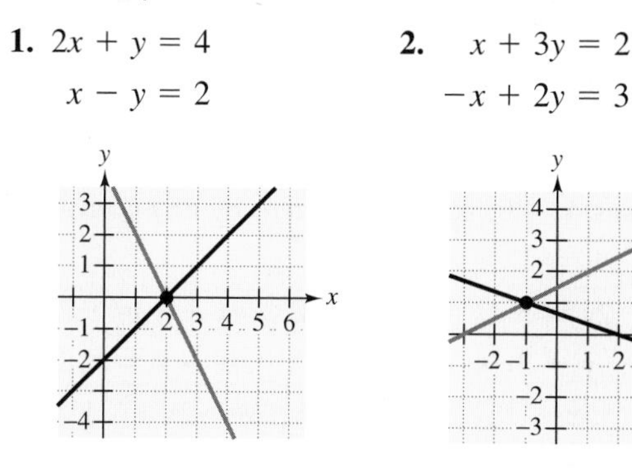

3. $-x + 2y = 1$
$x - y = 2$

4. $x + y = 0$
$3x - 2y = 10$

5. $3x + y = 3$
$2x - y = 7$

6. $-x + 2y = 2$
$3x + y = 15$

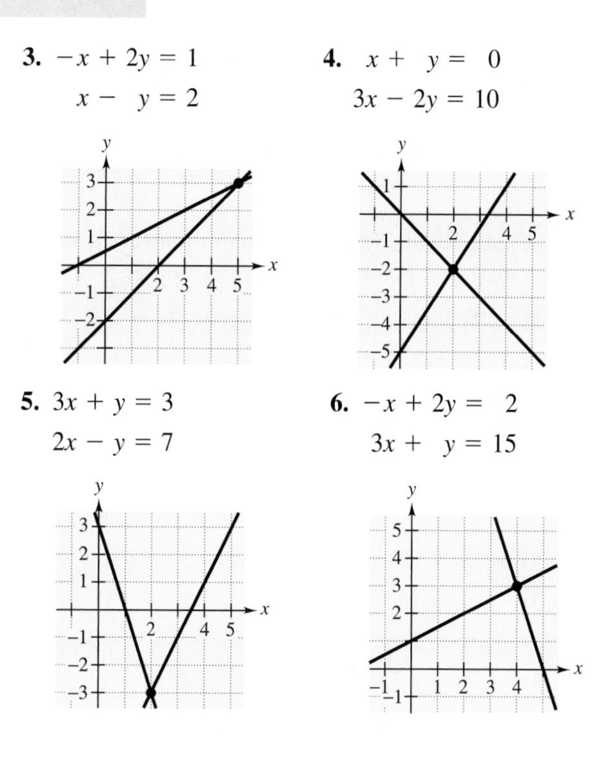

7. $\begin{aligned} x - y &= 1 \\ -3x + 3y &= 8 \end{aligned}$

8. $\begin{aligned} 3x + 4y &= 2 \\ 0.6x + 0.8y &= 1.6 \end{aligned}$

9. $\begin{aligned} x - 3y &= 5 \\ -2x + 6y &= -10 \end{aligned}$

10. $\begin{aligned} x - 4y &= 5 \\ 5x + 4y &= 7 \end{aligned}$

11. $\begin{aligned} 2x - 8y &= -11 \\ 5x + 3y &= 7 \end{aligned}$

12. $\begin{aligned} 3x + 4y &= 0 \\ 9x - 5y &= 17 \end{aligned}$

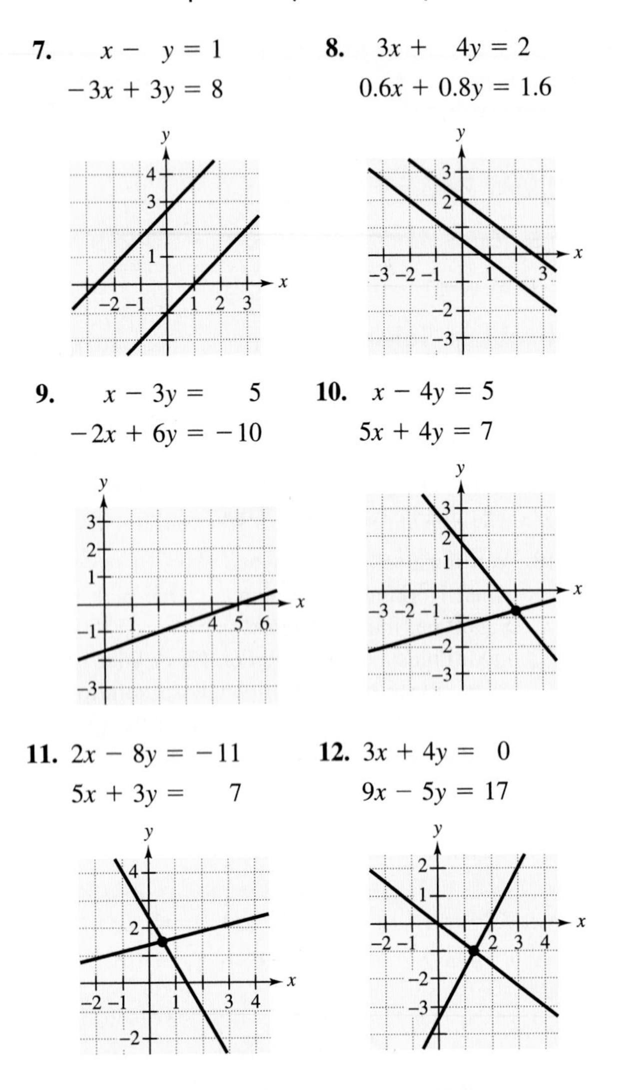

In Exercises 13–40, solve the system of linear equations by the method of elimination. See Examples 1–6.

13. $\begin{aligned} 3x - 2y &= 5 \\ x + 2y &= 7 \end{aligned}$

14. $\begin{aligned} -x + 2y &= 9 \\ x + 3y &= 16 \end{aligned}$

15. $\begin{aligned} 4x + y &= -3 \\ -4x + 3y &= 23 \end{aligned}$

16. $\begin{aligned} -3x + 5y &= -23 \\ 2x - 5y &= 22 \end{aligned}$

17. $\begin{aligned} 3x - 5y &= 1 \\ 2x + 5y &= 9 \end{aligned}$

18. $\begin{aligned} -x + 2y &= 12 \\ x + 6y &= 20 \end{aligned}$

19. $\begin{aligned} 5x + 2y &= 7 \\ 3x - y &= 13 \end{aligned}$

20. $\begin{aligned} 4x + 3y &= 8 \\ x - 2y &= 13 \end{aligned}$

21. $\begin{aligned} x - 3y &= 2 \\ 3x - 7y &= 4 \end{aligned}$

22. $\begin{aligned} 2s - t &= 9 \\ 3s + 4t &= -14 \end{aligned}$

23. $\begin{aligned} 2x + y &= 9 \\ 3x - y &= 16 \end{aligned}$

24. $\begin{aligned} 7r - s &= -25 \\ 2r + 5s &= 14 \end{aligned}$

25. $\begin{aligned} 2u + 3v &= 8 \\ 3u + 4v &= 13 \end{aligned}$

26. $\begin{aligned} 4x - 3y &= 25 \\ -3x + 8y &= 10 \end{aligned}$

27. $\begin{aligned} 12x - 5y &= 2 \\ -24x + 10y &= 6 \end{aligned}$

28. $\begin{aligned} -2x + 3y &= 9 \\ 6x - 9y &= -27 \end{aligned}$

29. $\begin{aligned} \tfrac{2}{3}r - s &= 0 \\ 10r + 4s &= 19 \end{aligned}$

30. $\begin{aligned} x - y &= -\tfrac{1}{2} \\ 4x - 48y &= -35 \end{aligned}$

31. $\begin{aligned} 0.05x - 0.03y &= 0.21 \\ x + y &= 9 \end{aligned}$

32. $\begin{aligned} 0.02x - 0.05y &= -0.19 \\ 0.03x + 0.04y &= 0.52 \end{aligned}$

33. $\begin{aligned} 0.7u - v &= -0.4 \\ 0.3u - 0.8v &= 0.2 \end{aligned}$

34. $\begin{aligned} 0.15x - 0.35y &= -0.5 \\ -0.12x + 0.25y &= 0.1 \end{aligned}$

35. $\begin{aligned} 5x + 7y &= 25 \\ x + 1.4y &= 5 \end{aligned}$

36. $\begin{aligned} 12b - 13m &= 2 \\ -6b + 6.5m &= -2 \end{aligned}$

37. $\begin{aligned} \tfrac{3}{2}x - y &= 4 \\ -x + \tfrac{2}{3}y &= -1 \end{aligned}$

38. $\begin{aligned} 12x - 3y &= 6 \\ 4x - y &= 2 \end{aligned}$

39. $\begin{aligned} 2x &= 25 \\ 4x - 10y &= 0.52 \end{aligned}$

40. $\begin{aligned} 6x - 6y &= 25 \\ 3y &= 11 \end{aligned}$

In Exercises 41–48, solve the system by any convenient method.

41. $\begin{aligned} 3x + 2y &= 5 \\ y &= 2x + 13 \end{aligned}$

42. $\begin{aligned} 4x + y &= -2 \\ -6x + y &= 18 \end{aligned}$

43. $\begin{aligned} y &= 5x - 3 \\ y &= -2x + 11 \end{aligned}$

44. $\begin{aligned} 3y &= 2x + 21 \\ x &= 50 - 4y \end{aligned}$

45. $\begin{aligned} 2x - y &= 20 \\ -x + y &= -5 \end{aligned}$

46. $\begin{aligned} 3x - 2y &= -20 \\ 5x + 6y &= 32 \end{aligned}$

47. $\begin{aligned} \tfrac{3}{2}x + 2y &= 12 \\ \tfrac{1}{4}x + y &= 4 \end{aligned}$

48. $\begin{aligned} x + 2y &= 4 \\ \tfrac{1}{2}x + \tfrac{1}{3}y &= 1 \end{aligned}$

In Exercises 49–54, decide whether the system is consistent or inconsistent.

49. $4x - 5y = 3$
$-8x + 10y = -6$

50. $4x - 5y = 3$
$-8x + 10y = 14$

51. $-2x + 5y = 3$
$5x + 2y = 8$

52. $x + 10y = 12$
$-2x + 5y = 2$

53. $-10x + 15y = 25$
$2x - 3y = -24$

54. $4x - 5y = 28$
$-2x + 2.5y = -14$

In Exercises 55 and 56, determine the value of k such that the system of linear equations is inconsistent.

55. $5x - 10y = 40$
$-2x + ky = 30$

56. $12x - 18y = 5$
$-18x + ky = 10$

In Exercises 57 and 58, find a system of linear equations that has the given solution. (There are many correct answers.)

57. $\left(3, -\frac{3}{2}\right)$

58. $(-8, 12)$

Solving Problems

59. *Break-Even Analysis* You are planning to open a small business. You need an initial investment of $85,000. Each week your costs will be about $7400. If your projected weekly revenue is $8100, how many weeks will it take to break even?

60. *Break-Even Analysis* A small business invests $8000 in equipment to produce a product. Each unit of the product costs $1.20 to produce and is sold for $2.00. How many units must be sold before the business breaks even?

61. *Simple Interest* A combined total of $20,000 is invested in two bonds that pay 8% and 9.5% simple interest. The annual interest is $1675. How much is invested in each bond?

62. *Simple Interest* A total of $4500 is invested in two funds paying 4% and 5% simple interest. The annual interest is $210. How much is invested in each fund?

63. *Average Speed* A van travels for 2 hours at an average speed of 40 miles per hour. How much longer must the van travel at an average speed of 55 miles per hour so that the average speed for the total trip will be 50 miles per hour?

64. *Average Speed* A truck travels for 4 hours at an average speed of 42 miles per hour. How much longer must the truck travel at an average speed of 55 miles per hour so that the average speed for the total trip will be 50 miles per hour?

65. *Air Speed* An airplane flying into a headwind travels 1800 miles in 3 hours and 36 minutes. On the return flight, the same distance is traveled in 3 hours. Find the speed of the plane in still air and the speed of the wind, assuming that both remain constant throughout the round trip.

66. *Air Speed* An airplane flying into a headwind travels the 3000-mile flying distance between two cities in 6 hours and 15 minutes. On the return flight, the distance is traveled in 5 hours. Find the speed of the plane in still air and the speed of the wind, assuming that both remain constant throughout the round trip.

67. *Ticket Sales* Five hundred tickets were sold for a fundraising dinner. The receipts totaled $3312.50. Adult tickets were $7.50 each and children's tickets were $4.00 each. How many tickets of each type were sold?

68. *Ticket Sales* A fundraising dinner was held on two consecutive nights. On the first night, 100 adult tickets and 175 children's tickets were sold, for a total of $937.50. On the second night, 200 adult tickets and 316 children's tickets were sold, for a total of $1790.00. Find the price of each type of ticket.

69. *Gasoline Mixture* Twelve gallons of regular unleaded gasoline plus 8 gallons of premium unleaded gasoline cost $23.08. The price of premium unleaded is 11 cents more per gallon than the price of regular unleaded. Find the price per gallon for each grade of gasoline.

70. *Gasoline Mixture* The total cost of 8 gallons of regular unleaded gasoline and 12 gallons of premium unleaded gasoline is $27.84. Premium unleaded gasoline costs $0.17 more per gallon than regular unleaded. Find the price per gallon for each grade of gasoline.

71. *Alcohol Mixture* How many liters of a 40% alcohol solution must be mixed with how many liters of a 65% solution to obtain 20 liters of a 50% solution?

72. *Acid Mixture* Fifty gallons of 70% acid solution is obtained by mixing an 80% solution with a 50% solution. How many gallons of each solution must be used to obtain the desired mixture?

73. *Nut Mixture* Ten pounds of mixed nuts sell for \$6.95 per pound. The mixture is obtained from two kinds of nuts, with one variety priced at \$5.65 per pound and the other at \$8.95 per pound. How many pounds of each variety of nut were used in the mixture?

74. *Best-Fitting Line* The slope and y-intercept of the line $y = mx + b$ that best fits the three noncollinear points $(0, 0)$, $(1, 1)$, and $(2, 3)$ are given by the solution of the following system of linear equations.

$$5m + 3b = 7$$

$$3m + 3b = 4$$

(a) Solve the system and find the equation of the best-fitting line.

(b) Plot the three points and sketch the graph of the best-fitting line.

75. *Best-Fitting Line* The slope and y-intercept of the line $y = mx + b$ that best fits the three noncollinear points $(0, 4)$, $(1, 2)$, and $(2, 1)$ are given by the solution of the following system of linear equations.

$$3b + 3m = 7$$

$$3b + 5m = 4$$

(a) Solve the system and find the equation of the best-fitting line.

(b) Plot the three points and sketch the graph of the best-fitting line.

76. *U.S. Aircraft Industry* The average hourly wages for those employed in the aircraft industry in the United States from 1994 through 1996 are given in the table. (Source: U.S. Bureau of Labor Statistics)

Year	1994	1995	1996
Wage	\$19.50	\$19.97	\$20.49

(a) Plot the data given in the table, where $x = 0$ corresponds to 1990.

(b) The line $y = mx + b$ that best fits the data is given by the solution of the following system.

$$3b + 15m = 59.96$$

$$15b + 77m = 300.79$$

Solve the system and find the equation of this line. Sketch the graph of the line on the same set of coordinate axes used in part (a).

(c) Explain the meaning of the slope of the line in the context of this problem.

77. *Vietnam Veterans Memorial* "The Wall" in Washington, D.C., designed by Maya Ling Lin when she was a student at Yale University, has two vertical, triangular sections of black granite with a common side (see figure). The top of each section is level with the ground. The bottoms of the two sections can be modeled by the equations $y = \frac{2}{25}x - 10$ and $y = -\frac{5}{61}x - 10$ when the x-axis is superimposed on the top of the wall. Each unit in the coordinate system represents 1 foot. How deep is the memorial at the point where the two sections meet? How long is each section?

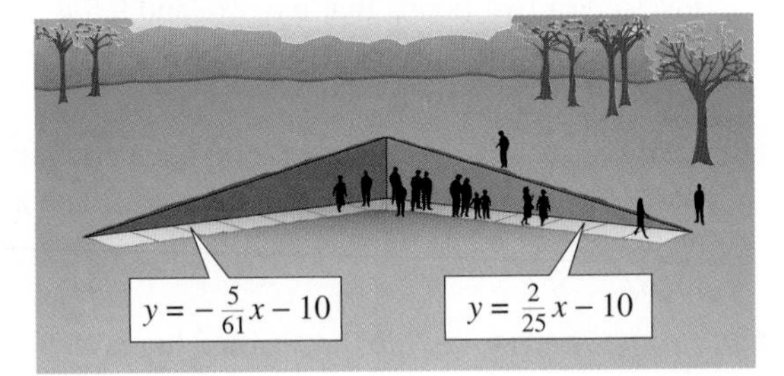

Explaining Concepts

78. Answer parts (a)–(c) of Motivating the Chapter on page 373.

79. When solving a system by elimination, how do you recognize that it has infinitely many solutions?

80. Explain what is meant by an *inconsistent* system of linear equations.

81. In your own words, explain how to solve a system of linear equations by elimination.

82. How can you recognize that a system of linear equations has no solution? Give an example.

83. Under what conditions might substitution be better than elimination for solving a system of linear equations?

7.3 Linear Systems in Three Variables

Objectives

1. Solve a system of equations using row-echelon form with back-substitution.
2. Solve a system of linear equations using elimination with back-substitution.
3. Solve an application problem using elimination with back-substitution.

1 Solve a system of equations using row-echelon form with back-substitution.

Row-Echelon Form

The method of elimination can be applied to a system of linear equations in more than two variables. In fact, this method easily adapts to computer use for solving systems of linear equations with dozens of variables.

When the method of elimination is used to solve a system of linear equations, the goal is to rewrite the system in a form to which back-substitution can be applied. For instance, consider the following two systems of linear equations.

$$\begin{aligned} x - 2y + 2z &= 9 \\ -x + 3y &= -4 \\ 2x - 5y + z &= 10 \end{aligned} \qquad \begin{aligned} x - 2y + 2z &= 9 \\ y + 2z &= 5 \\ z &= 3 \end{aligned}$$

The system on the right is said to be in **row-echelon form,** which means that it has a "stair-step" pattern with leading coefficients of 1. Which of these two systems do you think is easier to solve? After comparing the two systems, it should be clear that it is easier to solve the system on the right because the value of z is already shown and back-substitution will readily yield the values of x and y.

Example 1 Using Back-Substitution

Solve the system of linear equations.

$x - 2y + 2z = 9$ Equation 1

$y + 2z = 5$ Equation 2

$z = 3$ Equation 3

Study Tip

When checking a solution, remember that the solution must satisfy each equation in the original system.

Solution

From Equation 3, you know the value of z. To solve for y, substitute $z = 3$ into Equation 2 to obtain

$y + 2(3) = 5$ Substitute 3 for z.

$y = -1.$ Solve for y.

Finally, substitute $y = -1$ and $z = 3$ into Equation 1 to obtain

$x - 2(-1) + 2(3) = 9$ Substitute -1 for y and 3 for z.

$x = 1.$ Solve for x.

The solution is $x = 1$, $y = -1$, and $z = 3$, which can also be written as the **ordered triple** $(1, -1, 3)$. Check this in the original system of equations.

2 Solve a system of linear equations using elimination with back-substitution.

The Method of Elimination

Two systems of equations are **equivalent** if they have the same solution set. To solve a system that is not in row-echelon form, first convert it to an *equivalent* system that is in row-echelon form. To see how this is done, let's take another look at the method of elimination, as applied to a system of two linear equations.

Example 2 The Method of Elimination

Solve the system of linear equations.

$$3x - 2y = -1 \quad \text{Equation 1}$$
$$x - y = 0 \quad \text{Equation 2}$$

Solution

$$x - y = 0$$
$$3x - 2y = -1$$

You can interchange two equations in the system.

$$-3x + 3y = 0$$

Multiply the first equation by -3.

$$\begin{array}{r} -3x + 3y = 0 \\ 3x - 2y = -1 \\ \hline y = -1 \end{array}$$

You can add the multiple of the first equation to the second equation to obtain a new second equation.

$$x - y = 0$$
$$y = -1$$

New system in row-echelon form

Using back-substitution, you can determine that the solution is $(-1, -1)$. Check the solution in each equation in the original system, as follows.

Equation 1	*Equation 2*
$3x - 2y = -1$	$x - y = 0$
$3(-1) - 2(-1) = -1$	$(-1) - (-1) = 0$

Rewriting a system of linear equations in row-echelon form usually involves a chain of equivalent systems, each of which is obtained by using one of the three basic row operations. This process is called Gaussian elimination.

▶ Operations That Produce Equivalent Systems

Each of the following **row operations** on a system of linear equations produces an *equivalent* system of linear equations.

1. Interchange two equations.
2. Multiply one of the equations by a nonzero constant.
3. Add a multiple of one of the equations to another equation to replace the latter equation.

Example 3 Using Elimination to Solve a System

Solve the system of linear equations.

$$\begin{aligned} x - 2y + 2z &= 9 && \text{Equation 1} \\ -x + 3y \quad &= -4 && \text{Equation 2} \\ 2x - 5y + z &= 10 && \text{Equation 3} \end{aligned}$$

Solution

There are many ways to begin, but we suggest saving the x in the upper left position, because it has a leading coefficient of 1, and eliminating the other x's from the first column.

$$\begin{aligned} x - 2y + 2z &= 9 \\ y + 2z &= 5 \\ 2x - 5y + z &= 10 \end{aligned}$$

Adding the first equation to the second equation produces a new second equation.

$$\begin{aligned} x - 2y + 2z &= 9 \\ y + 2z &= 5 \\ -y - 3z &= -8 \end{aligned}$$

Adding -2 times the first equation to the third equation produces a new third equation.

Now that all but the first x have been eliminated from the first column, go to work on the second column. (You need to eliminate y from the third equation.)

$$\begin{aligned} x - 2y + 2z &= 9 \\ y + 2z &= 5 \\ -z &= -3 \end{aligned}$$

Adding the second equation to the third equation produces a new third equation.

Finally, you need a coefficient of 1 for z in the third equation.

$$\begin{aligned} x - 2y + 2z &= 9 \\ y + 2z &= 5 \\ z &= 3 \end{aligned}$$

Multiplying the third equation by -1 produces a new third equation.

This is the same system that was solved in Example 1, and, as in that example, you can conclude by back-substitution that the solution is

$$x = 1, \quad y = -1, \quad \text{and} \quad z = 3.$$

You can check the solution by substituting $x = 1$, $y = -1$, and $z = 3$ into each equation of the original system, as follows.

Equation 1: $x - 2y + 2z = 9$

$$(1) - 2(-1) + 2(3) = 9 \checkmark$$

Equation 2: $-x + 3y = -4$

$$-(1) + 3(-1) = -4 \checkmark$$

Equation 3: $2x - 5y + z = 10$

$$2(1) - 5(-1) + (3) = 10 \checkmark$$

Example 4 Using Elimination to Solve a System

Solve the following system of linear equations.

$$4x + y - 3z = 11 \quad \text{Equation 1}$$
$$2x - 3y + 2z = 9 \quad \text{Equation 2}$$
$$x + y + z = -3 \quad \text{Equation 3}$$

Solution

$$x + y + z = -3$$
$$2x - 3y + 2z = 9$$
$$4x + y - 3z = 11$$

Interchange the first and third equations.

$$x + y + z = -3$$
$$-5y = 15$$
$$4x + y - 3z = 11$$

Adding -2 times the first equation to the second equation produces a new second equation.

$$x + y + z = -3$$
$$-5y = 15$$
$$-3y - 7z = 23$$

Adding -4 times the first equation to the third equation produces a new third equation.

$$x + y + z = -3$$
$$y = -3$$
$$-3y - 7z = 23$$

Multiplying the second equation by $-\frac{1}{5}$ produces a new second equation.

$$x + y + z = -3$$
$$y = -3$$
$$-7z = 14$$

Adding 3 times the second equation to the third equation produces a new third equation.

$$x + y + z = -3$$
$$y = -3$$
$$z = -2$$

Multiplying the third equation by $-\frac{1}{7}$ produces a new third equation.

Now you can see that $z = -2$ and $y = -3$. Moreover, by back-substituting these values into Equation 1, you can determine that $x = 2$. So, the solution is $x = 2$, $y = -3$, and $z = -2$. You can check this solution as follows.

Equation 1: $4x + y - 3z = 11$

$$4(2) + (-3) - 3(-2) = 11 \checkmark$$

Equation 2: $2x - 3y + 2z = 9$

$$2(2) - 3(-3) + 2(-2) = 9 \checkmark$$

Equation 3: $x + y + z = -3$

$$(2) + (-3) + (-2) = -3 \checkmark$$

The next example involves an inconsistent system—one that has no solution. The key to recognizing an inconsistent system is that at some stage in the elimination process, you obtain a false statement such as $0 = -2$. Watch for such statements when you do the exercises for this section.

Example 5 An Inconsistent System

Solve the system of linear equations.

$$\begin{aligned} x - 3y + z &= 1 && \text{Equation 1} \\ 2x - y - 2z &= 2 && \text{Equation 2} \\ x + 2y - 3z &= -1 && \text{Equation 3} \end{aligned}$$

Solution

$$\begin{aligned} x - 3y + z &= 1 \\ 5y - 4z &= 0 \\ x + 2y - 3z &= -1 \end{aligned}$$

Adding -2 times the first equation to the second equation produces a new second equation.

$$\begin{aligned} x - 3y + z &= 1 \\ 5y - 4z &= 0 \\ 5y - 4z &= -2 \end{aligned}$$

Adding -1 times the first equation to the third equation produces a new third equation.

$$\begin{aligned} x - 3y + z &= 1 \\ 5y - 4z &= 0 \\ 0 &= -2 \end{aligned}$$

Adding -1 times the second equation to the third equation produces a new third equation.

Because the third "equation" is impossible, you can conclude that this system is inconsistent and therefore has no solution. Moreover, because this system is equivalent to the original system, you can conclude that the original system also has no solution.

As with a system of linear equations in two variables, the solution(s) of a system of linear equations in more than two variables must fall into one of three categories.

▶ The Number of Solutions of a Linear System

For a system of linear equations, exactly one of the following is true.

1. There is exactly one solution.
2. There are infinitely many solutions.
3. There is no solution.

Example 6 A System with Infinitely Many Solutions

Solve the system of linear equations.

$$\begin{aligned} x + y - 3z &= -1 \quad &\text{Equation 1}\\ y - z &= 0 \quad &\text{Equation 2}\\ -x + 2y &= 1 \quad &\text{Equation 3} \end{aligned}$$

Solution

Begin by rewriting the system in row-echelon form.

$$\begin{aligned} x + y - 3z &= -1\\ y - z &= 0\\ 3y - 3z &= 0 \end{aligned}$$

Adding the first equation to the third equation produces a new third equation.

$$\begin{aligned} x + y - 3z &= -1\\ y - z &= 0\\ 0 &= 0 \end{aligned}$$

Adding -3 times the second equation to the third equation produces a new third equation.

This means that Equation 3 depends on Equations 1 and 2 in the sense that it gives us no additional information about the variables. So, the original system is equivalent to the system

$$\begin{aligned} x + y - 3z &= -1\\ y - z &= 0. \end{aligned}$$

In this last equation, solve for y in terms of z to obtain $y = z$. Back-substituting for y in the previous equation produces $x = 2z - 1$. Finally, letting $z = a$, where a is any real number, the solutions to the given system are all of the form

$$x = 2a - 1,\ y = a, \text{ and } z = a.$$

So, every ordered triple of the form

$$(2a - 1, a, a), \qquad a \text{ is a real number}$$

is a solution of the system.

In Example 6, there are other ways to write the same infinite set of solutions. For instance, the solutions could have been written as

$$\left(b, \frac{1}{2}(b + 1), \frac{1}{2}(b + 1)\right), \qquad b \text{ is a real number.}$$

Try convincing yourself of this by substituting $a = 0$, $a = 1$, $a = 2$, and $a = 3$ into the solution listed in Example 6. Then substitute $b = -1$, $b = 1$, $b = 3$, and $b = 5$ into the solution listed above. In both cases, you should obtain the same ordered triples. So, when comparing descriptions of an infinite solution set, keep in mind that there is more than one way to describe the set.

3 Solve an application problem using elimination with back-substitution.

Applications

Example 7 Position Equation

The height at time t of an object that is moving in a (vertical) line with constant acceleration a is given by the **position equation**

$$s = \frac{1}{2}at^2 + v_0t + s_0.$$

The height s is measured in feet, the acceleration a is measured in feet per second squared, the time t is measured in seconds, v_0 is the initial velocity (at time $t = 0$), and s_0 is the initial height. Find the values of a, v_0, and s_0, if $s = 164$ feet at 1 second, $s = 180$ feet at 2 seconds, and $s = 164$ feet at 3 seconds.

Solution

By substituting the three values of t and s into the position equation, you obtain three linear equations in a, v_0, and s_0.

When $t = 1$, $s = 164$: $\frac{1}{2}a(1)^2 + v_0(1) + s_0 = 164$

When $t = 2$, $s = 180$: $\frac{1}{2}a(2)^2 + v_0(2) + s_0 = 180$

When $t = 3$, $s = 164$: $\frac{1}{2}a(3)^2 + v_0(3) + s_0 = 164$

By multiplying the first and third equations by 2, this system can be rewritten

$$\begin{aligned} a + 2v_0 + 2s_0 &= 328 \quad \text{Equation 1} \\ 2a + 2v_0 + s_0 &= 180 \quad \text{Equation 2} \\ 9a + 6v_0 + 2s_0 &= 328 \quad \text{Equation 3} \end{aligned}$$

and you can apply elimination to obtain

$$\begin{aligned} a + 2v_0 + 2s_0 &= 328 \\ -2v_0 - 3s_0 &= -476 \\ 2s_0 &= 232. \end{aligned}$$

From the third equation, $s_0 = 116$, so that back-substitution into the second equation yields

$$\begin{aligned} -2v_0 - 3(116) &= -476 \\ -2v_0 &= -128 \\ v_0 &= 64. \end{aligned}$$

Finally, back-substituting $s_0 = 116$ and $v_0 = 64$ into the first equation yields

$$\begin{aligned} a + 2(64) + 2(116) &= 328 \\ a &= -32. \end{aligned}$$

So, the position equation for this object is $s = -16t^2 + 64t + 116$.

Example 8 Data Analysis: Curve-Fitting

Find a quadratic equation

$$y = ax^2 + bx + c$$

whose graph passes through the points $(-1, 3)$, $(1, 1)$, and $(2, 6)$.

Solution

Because the graph of $y = ax^2 + bx + c$ passes through the points $(-1, 3)$, $(1, 1)$, and $(2, 6)$, you can write the following.

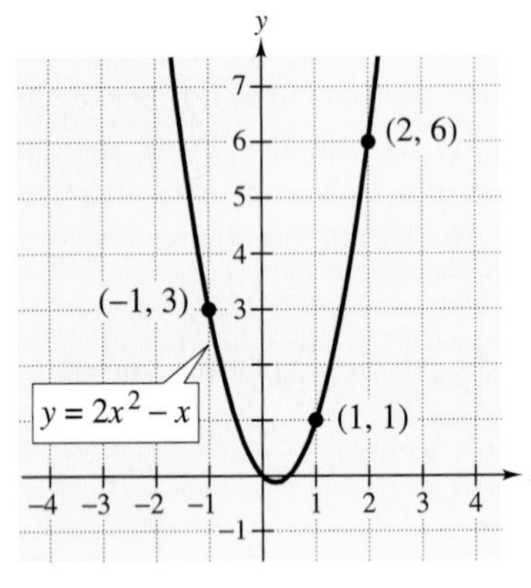

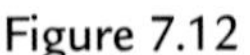
Figure 7.12

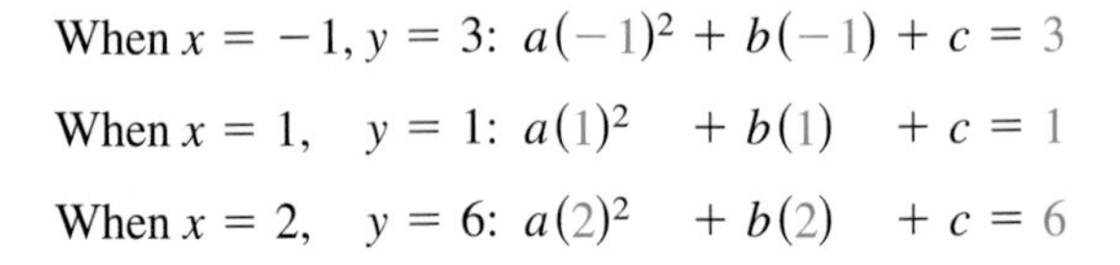

When $x = -1, y = 3$: $a(-1)^2 + b(-1) + c = 3$

When $x = 1, \quad y = 1$: $a(1)^2 \quad + b(1) \quad + c = 1$

When $x = 2, \quad y = 6$: $a(2)^2 \quad + b(2) \quad + c = 6$

This produces the following system of linear equations.

$$a - b + c = 3 \qquad \text{Equation 1}$$

$$a + b + c = 1 \qquad \text{Equation 2}$$

$$4a + 2b + c = 6 \qquad \text{Equation 3}$$

The solution of this system is $a = 2$, $b = -1$, and $c = 0$. So, the equation of the parabola is

$$y = 2x^2 - x$$

as shown in Figure 7.12.

Discussing the Concept — Fitting a Quadratic Model

The data in the table represent the United States government's annual net receipts y (in billions of dollars) from individual income taxes for the year x from 1994 through 1996, where $x = 4$ corresponds to 1994. (Source: U.S. Office of Management and Budget)

x	4	5	6
y	543	590	656

Use a system of three linear equations to find a quadratic model that fits the data. According to your model, what were the annual net receipts from individual income taxes in 1997? The actual annual net receipts for 1997 were \$737 billion. How does the value obtained from your quadratic model compare? Suppose you had been involved in planning the 1997 federal budget and had used this model to estimate how much federal income could be expected from 1997 individual income taxes. When you review the actual 1997 tax receipts and see that the model wasn't completely accurate, how do you evaluate the model's prediction performance? Are you satisfied with it? Why or why not?

7.3 Exercises

Integrated Review *Concepts, Skills, and Problem Solving*

Keep mathematically in shape by doing these exercises *before* the problems of this section.

Properties and Definitions

1. A linear equation of the form $2x + 8 = 7$ has how many solutions?

2. What is the usual first step in solving an equation such as

$$\frac{t}{6} + \frac{5}{8} = \frac{7}{4}?$$

Simplifying Expressions

In Exercises 3–6, simplify the expression. (Assume all variables are positive.)

3. $4x^2(x^3)^2$

4. $(2x^2y)^3(xy^3)^4$

5. $\dfrac{8x^{-4}}{2x^7}$

6. $\left(\dfrac{t^4}{3}\right)^{-1}$

Solving Equations

7. $|2x - 4| = 6$

8. $\frac{1}{4}(5 - 2x) = 9x - 7x$

Models and Graphs

9. The speed of a ship is 15 knots. Write the distance d the ship travels as a function of time t. Graph the model.

10. The length of each edge of a cube is s inches. Write the volume V of the cube as a function of s.

11. Express the area A of a circle as a function of its circumference C.

Developing Skills

In Exercises 1 and 2, determine whether each ordered triple is a solution of the system of linear equations.

1.
$$\begin{aligned} x + 3y + 2z &= 1 \\ 5x - y + 3z &= 16 \\ -3x + 7y + z &= -14 \end{aligned}$$

(a) $(0, 3, -2)$ (b) $(12, 5, -13)$

(c) $(1, -2, 3)$ (d) $(-2, 5, -3)$

2.
$$\begin{aligned} 3x - y + 4z &= -10 \\ -x + y + 2z &= 6 \\ 2x - y + z &= -8 \end{aligned}$$

(a) $(-2, 4, 0)$ (b) $(0, -3, 10)$

(c) $(1, -1, 5)$ (d) $(7, 19, -3)$

In Exercises 3–6, use back-substitution to solve the system of linear equations. See Example 1.

3.
$$\begin{aligned} x - 2y + 4z &= 4 \\ 3y - z &= 2 \\ z &= -5 \end{aligned}$$

4.
$$\begin{aligned} 5x + 4y - z &= 0 \\ 10y - 3z &= 11 \\ z &= 3 \end{aligned}$$

5.
$$\begin{aligned} x - 2y + 4z &= 4 \\ y &= 3 \\ y + z &= 2 \end{aligned}$$

6.
$$\begin{aligned} x &= 10 \\ 3x + 2y &= 2 \\ x + y + 2z &= 0 \end{aligned}$$

In Exercises 7 and 8, determine whether the two systems of linear equations are equivalent. Give reasons for your answer.

7.
$$\begin{aligned} x + 3y - z &= 6 \\ 2x - y + 2z &= 1 \\ 3x + 2y - z &= 2 \end{aligned} \qquad \begin{aligned} x + 3y - z &= 6 \\ -7y + 4z &= 1 \\ -7y - 4z &= -16 \end{aligned}$$

8.
$$\begin{aligned} x - 2y + 3z &= 9 \\ -x + 3y &= -4 \\ 2x - 5y + 5z &= 17 \end{aligned} \qquad \begin{aligned} x - 2y + 3z &= 9 \\ y + 3z &= 5 \\ -y - z &= -1 \end{aligned}$$

In Exercises 9–12, perform the row operation and write the equivalent system of linear equations. See Example 2.

9. Add Equation 1 to Equation 2.

$$\begin{aligned} x - 2y &= 8 && \text{Equation 1} \\ -x + 3y &= 6 && \text{Equation 2} \end{aligned}$$

What did this operation accomplish?

10. Add -2 times Equation 1 to Equation 2.

$$\begin{aligned} 2x + 3y &= 7 && \text{Equation 1} \\ 4x - 2y &= -2 && \text{Equation 2} \end{aligned}$$

What did this operation accomplish?

11. Add Equation 1 to Equation 2.

$$\begin{aligned} x - 2y + 3z &= 5 && \text{Equation 1} \\ -x + y + 5z &= 4 && \text{Equation 2} \\ 2x - 3z &= 0 && \text{Equation 3} \end{aligned}$$

What did this operation accomplish?

12. Add -2 times Equation 1 to Equation 3.

$$\begin{aligned} x - 2y + 3z &= 5 && \text{Equation 1} \\ -x + y + 5z &= 4 && \text{Equation 2} \\ 2x - 3z &= 0 && \text{Equation 3} \end{aligned}$$

What did this operation accomplish?

In Exercises 13–40, solve the system of linear equations. See Examples 3–6.

13. $$\begin{aligned} x + z &= 4 \\ y &= 2 \\ 4x + z &= 7 \end{aligned}$$

14. $$\begin{aligned} x &= 3 \\ -x + 3y &= 3 \\ y + 2z &= 4 \end{aligned}$$

15. $$\begin{aligned} x + y + z &= 6 \\ 2x - y + z &= 3 \\ 3x - z &= 0 \end{aligned}$$

16. $$\begin{aligned} x + y + z &= 2 \\ -x + 3y + 2z &= 8 \\ 4x + y &= 4 \end{aligned}$$

17. $$\begin{aligned} x + y + z &= -3 \\ 4x + y - 3z &= 11 \\ 2x - 3y + 2z &= 9 \end{aligned}$$

18. $$\begin{aligned} x - y + 2z &= -4 \\ 3x + y - 4z &= -6 \\ 2x + 3y - 4z &= 4 \end{aligned}$$

19. $$\begin{aligned} x + 2y + 6z &= 5 \\ -x + y - 2z &= 3 \\ x - 4y - 2z &= 1 \end{aligned}$$

20. $$\begin{aligned} x + 6y + 2z &= 9 \\ 3x - 2y + 3z &= -1 \\ 5x - 5y + 2z &= 7 \end{aligned}$$

21. $$\begin{aligned} 2x + 2z &= 2 \\ 5x + 3y &= 4 \\ 3y - 4z &= 4 \end{aligned}$$

22. $$\begin{aligned} 6y + 4z &= -12 \\ 3x + 3y &= 9 \\ 2x - 3z &= 10 \end{aligned}$$

23. $$\begin{aligned} x + y + 8z &= 3 \\ 2x + y + 11z &= 4 \\ x + 3z &= 0 \end{aligned}$$

24. $$\begin{aligned} 2x - 4y + z &= 0 \\ 3x + 2z &= -1 \\ -6x + 3y + 2z &= -10 \end{aligned}$$

25. $$\begin{aligned} 2x + y + 3z &= 1 \\ 2x + 6y + 8z &= 3 \\ 6x + 8y + 18z &= 5 \end{aligned}$$

26. $$\begin{aligned} 3x - y - 2z &= 5 \\ 2x + y + 3z &= 6 \\ 6x - y - 4z &= 9 \end{aligned}$$

27. $$\begin{aligned} y + z &= 5 \\ 2x + 4z &= 4 \\ 2x - 3y &= -14 \end{aligned}$$

28. $$\begin{aligned} 5x + 2y &= -8 \\ z &= 5 \\ 3x - y + z &= 9 \end{aligned}$$

29. $$\begin{aligned} 2x + 6y - 4z &= 8 \\ 3x + 10y - 7z &= 12 \\ -2x - 6y + 5z &= -3 \end{aligned}$$

30. $$\begin{aligned} x + 2y - 2z &= 4 \\ 2x + 5y - 7z &= 5 \\ 3x + 7y - 9z &= 10 \end{aligned}$$

31. $$\begin{aligned} 2x + z &= 1 \\ 5y - 3z &= 2 \\ 6x + 20y - 9z &= 11 \end{aligned}$$

32. $$\begin{aligned} 2x + y - z &= 4 \\ y + 3z &= 2 \\ 3x + 2y &= 4 \end{aligned}$$

33. $$\begin{aligned} 3x + y + z &= 2 \\ 4x + 2z &= 1 \\ 5x - y + 3z &= 0 \end{aligned}$$

34. $$\begin{aligned} 2x + 3z &= 4 \\ 5x + y + z &= 2 \\ 11x + 3y - 3z &= 0 \end{aligned}$$

35. $$\begin{aligned} 0.2x + 1.3y + 0.6z &= 0.1 \\ 0.1x + 0.3z &= 0.7 \\ 2x + 10y + 8z &= 8 \end{aligned}$$

36. $$\begin{aligned} 0.3x - 0.1y + 0.2z &= 0.35 \\ 2x + y - 2z &= -1 \\ 2x + 4y + 3z &= 10.5 \end{aligned}$$

37. $$\begin{aligned} x + 4y - 2z &= 2 \\ -3x + y + z &= -2 \\ 5x + 7y - 5z &= 6 \end{aligned}$$

38. $$\begin{aligned} x - 2y - z &= 3 \\ 2x + y - 3z &= 1 \\ x + 8y - 3z &= -7 \end{aligned}$$

39. $$\begin{aligned} -4x + y + 0.2z &= 6 \\ 6x - 3y + 0.5z &= -4 \\ -8x + 2y + 0.6z &= 14 \end{aligned}$$

40. $$\begin{aligned} x + 6y + 2z &= 9 \\ 3x - 2y + 3z &= -1 \\ 5x - 5y + 2z &= 7 \end{aligned}$$

In Exercises 41 and 42, find a system of linear equations in three variables that has the given point as a solution. (*Note:* There are many correct answers.)

41. $(4, -3, 2)$

42. $(5, 7, -10)$

Solving Problems

Vertical Motion In Exercises 43–46, find the position equation $s = \frac{1}{2}at^2 + v_0t + s_0$ for an object that has the indicated heights at the specified times.

43. $s = 128$ feet at $t = 1$ second
$s = 80$ feet at $t = 2$ seconds
$s = 0$ feet at $t = 3$ seconds

44. $s = 48$ feet at $t = 1$ second
$s = 64$ feet at $t = 2$ seconds
$s = 48$ feet at $t = 3$ seconds

45. $s = 32$ feet at $t = 1$ second
$s = 32$ feet at $t = 2$ seconds
$s = 0$ feet at $t = 3$ seconds

46. $s = 10$ feet at $t = 0$ second
$s = 54$ feet at $t = 1$ second
$s = 46$ feet at $t = 3$ seconds

Curve-Fitting In Exercises 47–52, find a quadratic equation $y = ax^2 + bx + c$ whose graph passes through the points.

47. $(0, -4), (1, 1), (2, 10)$ **48.** $(0, 5), (1, 6), (2, 5)$

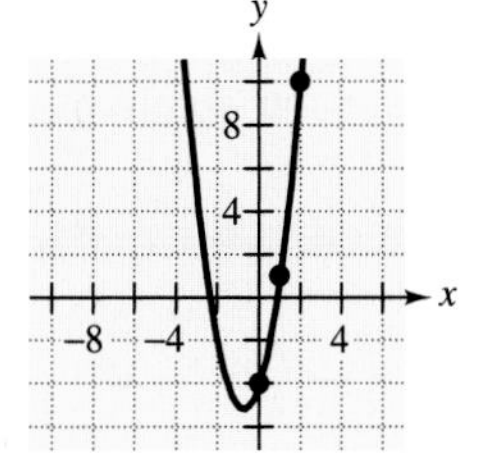

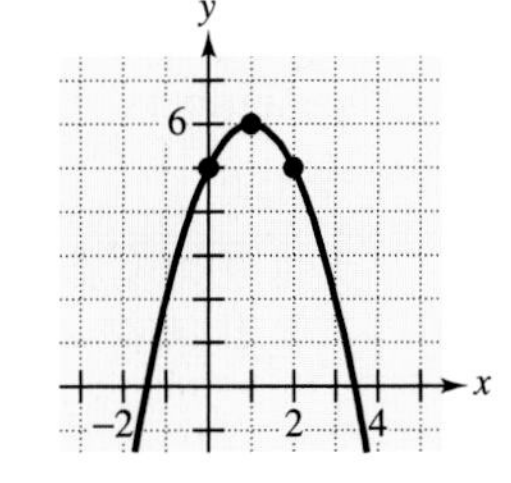

49. $(1, 0), (2, -1), (3, 0)$ **50.** $(1, 2), (2, 1), (3, -4)$

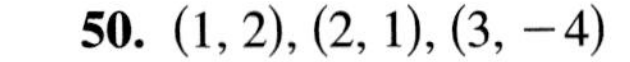

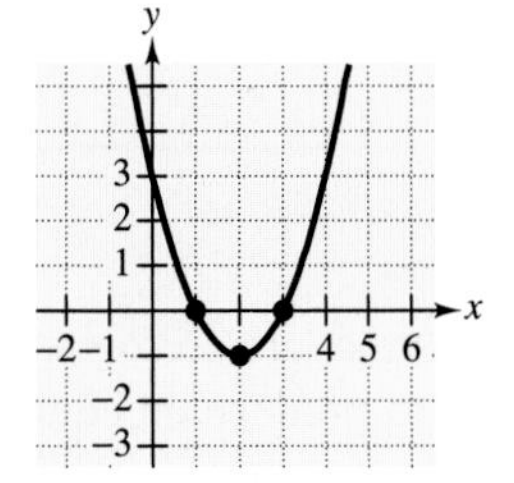

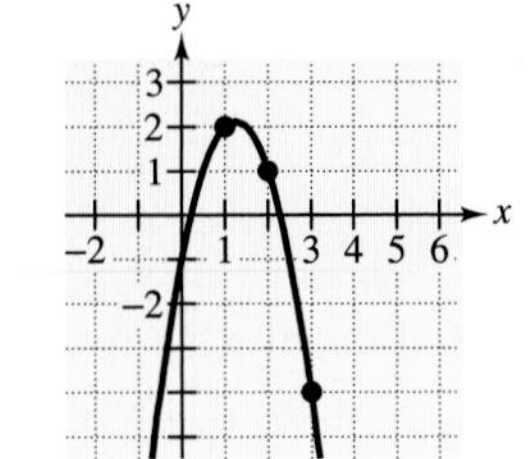

51. $(-1, -3), (1, 1), (2, 0)$

52. $(-1, -1), (1, 1), (2, -4)$

53. *Diagonals of a Polygon* The total numbers of sides and diagonals of regular polygons with three, four, and five sides are 3, 6, and 10, as shown in the figure. Find a quadratic function $y = ax^2 + bx + c$, where x is the number of sides in the polygon, that fits these data. Does it give the correct answer for a polygon with six sides?

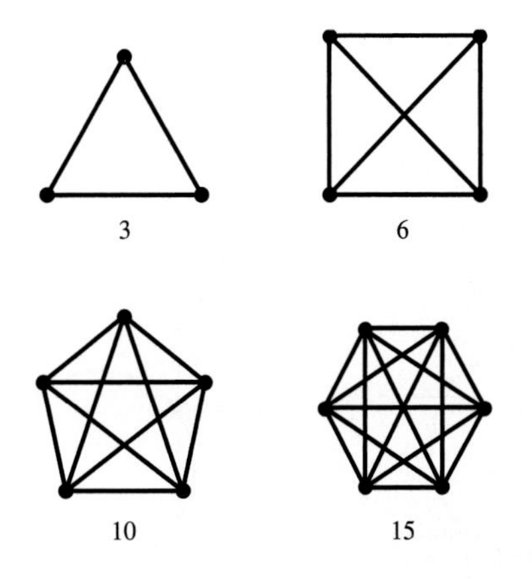

54. *Graphical Estimation* The table gives the numbers y of metric tons of newsprint (in thousands) produced in the years 1993 through 1995 in the United States. (Source: American Forest and Paper Association)

Year	1993	1994	1995
y	6412	6336	6352

(a) Find a quadratic equation $y = at^2 + bt + c$ whose graph passes through the three points, letting $t = 0$ correspond to 1990.

(b) Use a graphing utility to graph the model found in part (a).

(c) Use the model in part (a) to predict newsprint production in the year 2000 if the trend continues.

Curve-Fitting In Exercises 55–60, find the equation of the circle $x^2 + y^2 + Dx + Ey + F = 0$ that passes through the points.

55. $(0, 0), (2, -2), (4, 0)$

56. $(0, 0), (0, 6), (-3, 3)$

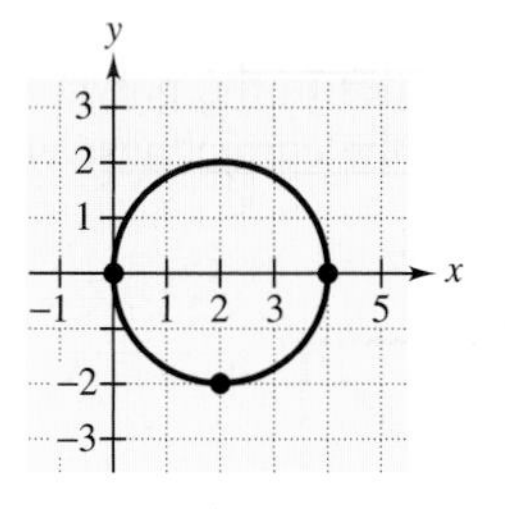

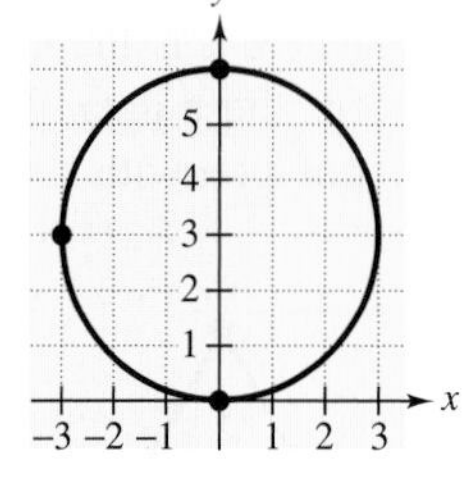

57. $(3, -1), (-2, 4), (6, 8)$

58. $(0, 0), (0, 2), (3, 0)$

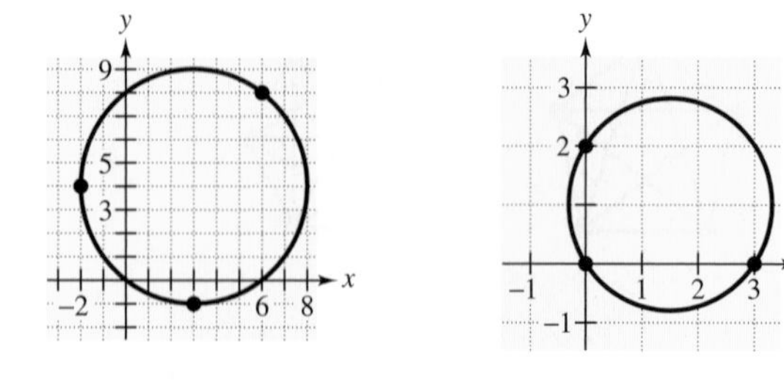

59. $(-3, 5), (4, 6), (5, 5)$

60. $(5, 13), (17, 5), (10, 12)$

61. *Crop Spraying* A mixture of 12 gallons of chemical A, 16 gallons of chemical B, and 26 gallons of chemical C is required to kill a certain destructive crop insect. Commercial spray X contains 1, 2, and 2 parts, respectively, of these chemicals. Spray Y contains only chemical C. Spray Z contains only chemicals A and B in equal amounts. How much of each type of commercial spray is needed to get the desired mixture?

62. *Chemistry* A chemist needs 10 liters of a 25% acid solution. It is mixed from three solutions whose concentrations are 10%, 20%, and 50%. How many liters of each solution will satisfy the following?

(a) Use 2 liters of the 50% solution.

(b) Use as little as possible of the 50% solution.

(c) Use as much as possible of the 50% solution.

63. *School Orchestra* The table shows the percents of each section of the North High School orchestra that were chosen to participate in the city orchestra, the county orchestra, and the state orchestra. Thirty members of the city orchestra, 17 members of the county orchestra, and 10 members of the state orchestra are from North High. How many members are in each section of North High's orchestra?

Orchestra	String	Wind	Percussion
City orchestra	40%	30%	50%
County orchestra	20%	25%	25%
State orchestra	10%	15%	25%

64. *Rewriting a Fraction* The fraction $1/(x^3 - x)$ can be written as a sum of three fractions as follows.

$$\frac{1}{x^3 - x} = \frac{A}{x} + \frac{B}{x + 1} + \frac{C}{x - 1}.$$

The numbers A, B, and C are the solutions of the system

$$\begin{aligned} A + B + C &= 0 \\ -B + C &= 0 \\ -A \qquad\quad &= 1. \end{aligned}$$

Solve the system and write the expression as the sum of three fractions.

Explaining Concepts

65. Answer parts (d)–(f) of Motivating the Chapter on page 373.

66. Give an example of a system of linear equations that is in row-echelon form.

67. Show how to use back-substitution to solve the system you found in Exercise 66.

68. Describe the row operations that are performed on a system of linear equations to produce an equivalent system of equations.

69. Write a system of four linear equations in four unknowns, and solve it by elimination.

Mid-Chapter Quiz

Take this quiz as you would take a quiz in class. After you are done, check your work against the answers given in the back of the book.

1. Which is the solution of the system $5x - 12y = 2$ and $2x + 1.5y = 26$: $(1, -2)$ or $(10, 4)$? Explain your reasoning.

In Exercises 2–4, graph the equations in the system. Use the graphs to determine the number of solutions of the system.

2. $\begin{aligned} -6x + 9y &= 9 \\ 2x - 3y &= 6 \end{aligned}$

3. $\begin{aligned} x - 2y &= -4 \\ 3x - 2y &= 4 \end{aligned}$

4. $\begin{aligned} y &= x - 1 \\ y &= 1 + 2x - x^2 \end{aligned}$

In Exercises 5–8, solve the system of equations graphically.

5. $\begin{aligned} x \quad &= 4 \\ 2x - y &= 6 \end{aligned}$

6. $\begin{aligned} y &= \tfrac{1}{3}(1 - 2x) \\ y &= \tfrac{1}{3}(5x - 13) \end{aligned}$

7. $\begin{aligned} 2x + 7y &= 16 \\ 3x + 2y &= 24 \end{aligned}$

8. $\begin{aligned} 7x - 17y &= -169 \\ x^2 + y^2 &= 169 \end{aligned}$

In Exercises 9–12, use substitution to solve the system.

9. $\begin{aligned} 2x - 3y &= 4 \\ y &= 2 \end{aligned}$

10. $\begin{aligned} y &= 5 - x^2 \\ y &= 2(x + 1) \end{aligned}$

11. $\begin{aligned} 5x - y &= 32 \\ 6x - 9y &= 18 \end{aligned}$

12. $\begin{aligned} 0.2x + 0.7y &= 8 \\ -x + 2y &= 15 \end{aligned}$

In Exercises 13–16, use Gaussian elimination to solve the linear system.

13. $\begin{aligned} x + 10y &= 18 \\ 5x + 2y &= 42 \end{aligned}$

14. $\begin{aligned} 3x + 11y &= 38 \\ 7x - 5y &= -34 \end{aligned}$

15. $\begin{aligned} a + b + c &= 1 \\ 4a + 2b + c &= 2 \\ 9a + 3b + c &= 4 \end{aligned}$

16. $\begin{aligned} x \qquad + 4z &= 17 \\ -3x + 2y - z &= -20 \\ x - 5y + 3z &= 19 \end{aligned}$

In Exercises 17 and 18, find a system of linear equations that has the unique solution. (There are many correct answers.)

17. $(10, -12)$

18. $(1, 3, -7)$

19. Twenty gallons of a 30% brine solution is obtained by mixing a 20% solution with a 50% solution. Let x represent the number of gallons of the 20% solution and let y represent the number of gallons of the 50% solution. Write a system of equations that models this problem and solve the system.

20. Find the equation of the parabola $y = ax^2 + bx + c$ that passes through the points $(1, 2)$, $(-1, -4)$, and $(2, 8)$.

7.4 Matrices and Linear Systems

Objectives

1. Form a coefficient and an augmented matrix and form a linear system from the augmented matrix.
2. Perform elementary row operations to solve a system of linear equations.
3. Use matrices and Gaussian elimination with back-substitution to solve a system of linear equations.

1 Form a coefficient and an augmented matrix and form a linear system from the augmented matrix.

Matrices

In this section, you will study a streamlined technique for solving systems of linear equations. This technique involves the use of a rectangular array of real numbers called a **matrix.** (The plural of matrix is *matrices.*) Here is an example of a matrix.

$$\begin{array}{c c} & \begin{array}{cccc} \textit{Column 1} & \textit{Column 2} & \textit{Column 3} & \textit{Column 4} \end{array} \\ \begin{array}{c} \textit{Row 1} \\ \textit{Row 2} \\ \textit{Row 3} \end{array} & \begin{bmatrix} 3 & -2 & 4 & 1 \\ 0 & 1 & -1 & 2 \\ 2 & 0 & -3 & 0 \end{bmatrix} \end{array}$$

This matrix has three rows and four columns, which means that its **order** is 3×4, which is read as "3 by 4." Each number in the matrix is an **entry** of the matrix.

Study Tip

The order of a matrix is always given as *row by column.*

Example 1 Examples of Matrices

The following matrices have the indicated orders.

a. Order: 2×3

$$\begin{bmatrix} 1 & -2 & 4 \\ 0 & 1 & -2 \end{bmatrix}$$

b. Order: 2×2

$$\begin{bmatrix} 0 & 0 \\ 0 & 0 \end{bmatrix}$$

c. Order: 3×2

$$\begin{bmatrix} 1 & -3 \\ -2 & 0 \\ 4 & -2 \end{bmatrix}$$

A matrix with the same number of rows as columns is called a **square matrix.** For instance, the 2×2 matrix in part (b) is square.

Study Tip

Note the use of 0 for the missing y-variable in the third equation, and also note the fourth column of constant terms in the augmented matrix.

A matrix derived from a system of linear equations (each written in standard form) is the **augmented matrix** of the system. Moreover, the matrix derived from the coefficients of the system (but that does not include the constant terms) is the **coefficient matrix** of the system. Here is an example.

System

$$\begin{cases} x - 4y + 3z = 5 \\ -x + 3y - z = -3 \\ 2x - 4z = 6 \end{cases}$$

Coefficient Matrix

$$\begin{bmatrix} 1 & -4 & 3 \\ -1 & 3 & -1 \\ 2 & 0 & -4 \end{bmatrix}$$

Augmented Matrix

$$\left[\begin{array}{ccc|c} 1 & -4 & 3 & 5 \\ -1 & 3 & -1 & -3 \\ 2 & 0 & -4 & 6 \end{array}\right]$$

When forming either the coefficient matrix or the augmented matrix of a system, you should begin by vertically aligning the variables in the equations.

Given System	*Align Variables*	*Form Augmented Matrix*
$\begin{aligned} x + 3y &= 9 \\ -y + 4z &= -2 \\ x - 5z &= 0 \end{aligned}$	$\begin{array}{rrrrl} x & + 3y & & = & 9 \\ & -y & + 4z & = & -2 \\ x & & - 5z & = & 0 \end{array}$	$\begin{bmatrix} 1 & 3 & 0 & \vdots & 9 \\ 0 & -1 & 4 & \vdots & -2 \\ 1 & 0 & -5 & \vdots & 0 \end{bmatrix}$

Example 2 Forming Coefficient and Augmented Matrices

Form the coefficient matrix and the augmented matrix for each system of linear equations.

a. $\begin{aligned} -x + 5y &= 2 \\ 7x - 2y &= -6 \end{aligned}$ **b.** $\begin{aligned} 3x + 2y - z &= 1 \\ x + 2z &= -3 \\ -2x - y &= 4 \end{aligned}$ **c.** $\begin{aligned} x &= 3y - 1 \\ 2y - 5 &= 9x \end{aligned}$

Solution

	System	*Coefficient Matrix*	*Augmented Matrix*
a.	$\begin{aligned} -x + 5y &= 2 \\ 7x - 2y &= -6 \end{aligned}$	$\begin{bmatrix} -1 & 5 \\ 7 & -2 \end{bmatrix}$	$\begin{bmatrix} -1 & 5 & \vdots & 2 \\ 7 & -2 & \vdots & -6 \end{bmatrix}$
b.	$\begin{array}{rrrrl} 3x & + 2y & - z & = & 1 \\ x & & + 2z & = & -3 \\ -2x & - y & & = & 4 \end{array}$	$\begin{bmatrix} 3 & 2 & -1 \\ 1 & 0 & 2 \\ -2 & -1 & 0 \end{bmatrix}$	$\begin{bmatrix} 3 & 2 & -1 & \vdots & 1 \\ 1 & 0 & 2 & \vdots & -3 \\ -2 & -1 & 0 & \vdots & 4 \end{bmatrix}$
c.	$\begin{aligned} x - 3y &= -1 \\ -9x + 2y &= 5 \end{aligned}$	$\begin{bmatrix} 1 & -3 \\ -9 & 2 \end{bmatrix}$	$\begin{bmatrix} 1 & -3 & \vdots & -1 \\ -9 & 2 & \vdots & 5 \end{bmatrix}$

Example 3 Forming Linear Systems from Their Matrices

Write systems of linear equations that are represented by the following matrices.

a. $\begin{bmatrix} 3 & -5 & \vdots & 4 \\ -1 & 2 & \vdots & 0 \end{bmatrix}$ **b.** $\begin{bmatrix} 1 & 3 & \vdots & 2 \\ 0 & 1 & \vdots & -3 \end{bmatrix}$ **c.** $\begin{bmatrix} 2 & 0 & -8 & \vdots & 1 \\ -1 & 1 & 1 & \vdots & 2 \\ 5 & -1 & 7 & \vdots & 3 \end{bmatrix}$

Solution

a. $\begin{aligned} 3x - 5y &= 4 \\ -x + 2y &= 0 \end{aligned}$

b. $\begin{aligned} x + 3y &= 2 \\ y &= -3 \end{aligned}$

c. $\begin{array}{rrrrl} 2x & & - 8z & = & 1 \\ -x & + y & + z & = & 2 \\ 5x & - y & + 7z & = & 3 \end{array}$

2 Perform elementary row operations to solve a system of linear equations.

Elementary Row Operations

In Section 7.3, you studied three operations that can be used on a system of linear equations to produce an equivalent system: (1) interchange two rows, (2) multiply a row by a nonzero constant, and (3) add a multiple of a row to another row. In matrix terminology, these three operations correspond to **elementary row operations.**

Study Tip

Although elementary row operations are simple to perform, they involve a lot of arithmetic. Because it is easy to make a mistake, we suggest that you get in the habit of noting the elementary row operations performed in each step so that you can go back and check your work. People use different schemes to denote which elementary row operations have been performed. The scheme we use is to write an abbreviated version of the row operation to the left of the row that has been changed.

▶ Elementary Row Operations

Any of the following **elementary row operations** performed on an augmented matrix will produce a matrix that is row-equivalent to the original matrix. Two matrices are **row-equivalent** if one can be obtained from the other by a sequence of elementary row operations.

1. Interchange two rows.
2. Multiply a row by a nonzero constant.
3. Add a multiple of a row to another row.

Example 4 Elementary Row Operations

a. Interchange the first and second rows.

Original Matrix

$$\begin{bmatrix} 0 & 1 & 3 & 4 \\ -1 & 2 & 0 & 3 \\ 2 & -3 & 4 & 1 \end{bmatrix}$$

New Row-Equivalent Matrix

$$\begin{matrix} R_2 \\ R_1 \\ \\ \end{matrix} \begin{bmatrix} -1 & 2 & 0 & 3 \\ 0 & 1 & 3 & 4 \\ 2 & -3 & 4 & 1 \end{bmatrix}$$

b. Multiply the first row by $\frac{1}{2}$.

Original Matrix

$$\begin{bmatrix} 2 & -4 & 6 & -2 \\ 1 & 3 & -3 & 0 \\ 5 & -2 & 1 & 2 \end{bmatrix}$$

New Row-Equivalent Matrix

$$\begin{matrix} \frac{1}{2}R_1 \rightarrow \\ \\ \\ \end{matrix} \begin{bmatrix} 1 & -2 & 3 & -1 \\ 1 & 3 & -3 & 0 \\ 5 & -2 & 1 & 2 \end{bmatrix}$$

c. Add -2 times the first row to the third row.

Original Matrix

$$\begin{bmatrix} 1 & 2 & -4 & 3 \\ 0 & 3 & -2 & -1 \\ 2 & 1 & 5 & -2 \end{bmatrix}$$

New Row-Equivalent Matrix

$$\begin{matrix} \\ \\ -2R_1 + R_3 \rightarrow \end{matrix} \begin{bmatrix} 1 & 2 & -4 & 3 \\ 0 & 3 & -2 & -1 \\ 0 & -3 & 13 & -8 \end{bmatrix}$$

d. Add 6 times the first row to the second row.

Original Matrix

$$\begin{bmatrix} 1 & 2 & 2 & -4 \\ -6 & -11 & 3 & 18 \\ 0 & 0 & 4 & 7 \end{bmatrix}$$

New Row-Equivalent Matrix

$$\begin{matrix} \\ 6R_1 + R_2 \rightarrow \\ \\ \end{matrix} \begin{bmatrix} 1 & 2 & 2 & -4 \\ 0 & 1 & 15 & -6 \\ 0 & 0 & 4 & 7 \end{bmatrix}$$

In Section 7.3, Gaussian elimination was used with back-substitution to solve a system of linear equations. Example 5 demonstrates the matrix version of Gaussian elimination. The two methods are essentially the same. The basic difference is that with matrices you do not need to keep writing the variables.

Technology: Tip

Most graphing utilities are capable of performing row operations on matrices. Some graphing utilities have a function that will return the reduced row-echelon form of a matrix. Consult the user's guide of your graphing utility to learn how to perform elementary row operations. Most graphing utilities store the resulting matrix of each step into an answer variable. We suggest that you store the results of each operation into a matrix variable.

Enter the matrix from Example 5 into your graphing utility and perform the indicated row operations.

Example 5 Solving a System of Linear Equations

Linear System

$$\begin{aligned} x - 2y + 2z &= 9 \\ -x + 3y \quad\quad &= -4 \\ 2x - 5y + z &= 10 \end{aligned}$$

Associated Augmented Matrix

$$\begin{bmatrix} 1 & -2 & 2 & \vdots & 9 \\ -1 & 3 & 0 & \vdots & -4 \\ 2 & -5 & 1 & \vdots & 10 \end{bmatrix}$$

Add the first equation to the second equation.

$$\begin{aligned} x - 2y + 2z &= 9 \\ y + 2z &= 5 \\ 2x - 5y + z &= 10 \end{aligned}$$

Add the first row to the second row ($R_1 + R_2$).

$$R_1 + R_2 \rightarrow \begin{bmatrix} 1 & -2 & 2 & \vdots & 9 \\ 0 & 1 & 2 & \vdots & 5 \\ 2 & -5 & 1 & \vdots & 10 \end{bmatrix}$$

Add -2 times the first equation to the third equation.

$$\begin{aligned} x - 2y + 2z &= 9 \\ y + 2z &= 5 \\ -y - 3z &= -8 \end{aligned}$$

Add -2 times the first row to the third row ($-2R_1 + R_3$).

$$-2R_1 + R_3 \rightarrow \begin{bmatrix} 1 & -2 & 2 & \vdots & 9 \\ 0 & 1 & 2 & \vdots & 5 \\ 0 & -1 & -3 & \vdots & -8 \end{bmatrix}$$

Add the second equation to the third equation.

$$\begin{aligned} x - 2y + 2z &= 9 \\ y + 2z &= 5 \\ -z &= -3 \end{aligned}$$

Add the second row to the third row ($R_2 + R_3$).

$$R_2 + R_3 \rightarrow \begin{bmatrix} 1 & -2 & 2 & \vdots & 9 \\ 0 & 1 & 2 & \vdots & 5 \\ 0 & 0 & -1 & \vdots & -3 \end{bmatrix}$$

Multiply the third equation by -1.

$$\begin{aligned} x - 2y + 2z &= 9 \\ y + 2z &= 5 \\ z &= 3 \end{aligned}$$

Multiply the third row by -1.

$$-R_3 \rightarrow \begin{bmatrix} 1 & -2 & 2 & \vdots & 9 \\ 0 & 1 & 2 & \vdots & 5 \\ 0 & 0 & 1 & \vdots & 3 \end{bmatrix}$$

At this point, you can use back-substitution to find that the solution is

$$x = 1, \quad y = -1, \quad \text{and} \quad z = 3.$$

Check this in the original system as follows.

$(1) - 2(-1) + 2(3) = 9$ Substitute in Equation 1. ✓

$-(1) + 3(-1) = -4$ Substitute in Equation 2. ✓

$2(1) - 5(-1) + (3) = 10$ Substitute in Equation 3. ✓

The last matrix in Example 5 is in **row-echelon form.** The term *echelon* refers to the stair-step pattern formed by the nonzero elements of the matrix.

3 Use matrices and Gaussian elimination with back-substitution to solve a system of linear equations.

Solving a System of Linear Equations

▶ Gaussian Elimination with Back-Substitution

To use matrices and Gaussian elimination to solve a system of linear equations, use the following steps.

1. Write the augmented matrix of the system of linear equations.
2. Use elementary row operations to rewrite the augmented matrix in row-echelon form.
3. Write the system of linear equations corresponding to the matrix in row-echelon form, and use back-substitution to find the solution.

When you perform Gaussian elimination with back-substitution, we suggest that you operate from *left to right by columns,* using elementary row operations to obtain zeros in all entries directly below the leading 1s.

Example 6 Gaussian Elimination with Back-Substitution

Solve the system of linear equations.

$$\begin{aligned} 2x - 3y &= -2 \\ x + 2y &= 13 \end{aligned}$$

Solution

$$\begin{bmatrix} 2 & -3 & \vdots & -2 \\ 1 & 2 & \vdots & 13 \end{bmatrix}$$ Augmented matrix for system of linear equations

$$\begin{matrix} R_2 \\ R_1 \end{matrix} \begin{bmatrix} 1 & 2 & \vdots & 13 \\ 2 & -3 & \vdots & -2 \end{bmatrix}$$ First column has leading 1 in upper left corner.

$$-2R_1 + R_2 \rightarrow \begin{bmatrix} 1 & 2 & \vdots & 13 \\ 0 & -7 & \vdots & -28 \end{bmatrix}$$ First column has a zero under its leading 1.

$$-\tfrac{1}{7}R_2 \rightarrow \begin{bmatrix} 1 & 2 & \vdots & 13 \\ 0 & 1 & \vdots & 4 \end{bmatrix}$$ Second column has leading 1 in second row.

The system of linear equations that corresponds to the (row-echelon) matrix is

$$\begin{aligned} x + 2y &= 13 \\ y &= 4. \end{aligned}$$

Using back-substitution, you can find that the solution of the system is

$x = 5$ and $y = 4$.

Check this solution in the original system, as follows.

$2(5) - 3(4) = -2$ Substitute in Equation 1. ✓

$5 - 2(4) = 13$ Substitute in Equation 2. ✓

Example 7 Gaussian Elimination with Back-Substitution

Solve the system of linear equations.

$$\begin{aligned} 3x + 3y \qquad &= 9 \\ 2x \qquad - 3z &= 10 \\ 6y + 4z &= -12 \end{aligned}$$

Solution

$$\left[\begin{array}{ccc:c} 3 & 3 & 0 & 9 \\ 2 & 0 & -3 & 10 \\ 0 & 6 & 4 & -12 \end{array}\right]$$ Augmented matrix for system of linear equations

$$\tfrac{1}{3}R_1 \rightarrow \left[\begin{array}{ccc:c} 1 & 1 & 0 & 3 \\ 2 & 0 & -3 & 10 \\ 0 & 6 & 4 & -12 \end{array}\right]$$ First column has leading 1 in upper left corner.

$$-2R_1 + R_2 \rightarrow \left[\begin{array}{ccc:c} 1 & 1 & 0 & 3 \\ 0 & -2 & -3 & 4 \\ 0 & 6 & 4 & -12 \end{array}\right]$$ First column has zeros under its leading 1.

$$-\tfrac{1}{2}R_2 \rightarrow \left[\begin{array}{ccc:c} 1 & 1 & 0 & 3 \\ 0 & 1 & \frac{3}{2} & -2 \\ 0 & 6 & 4 & -12 \end{array}\right]$$ Second column has leading 1 in second row.

$$-6R_2 + R_3 \rightarrow \left[\begin{array}{ccc:c} 1 & 1 & 0 & 3 \\ 0 & 1 & \frac{3}{2} & -2 \\ 0 & 0 & -5 & 0 \end{array}\right]$$ Second column has zero under its leading 1.

$$-\tfrac{1}{5}R_3 \rightarrow \left[\begin{array}{ccc:c} 1 & 1 & 0 & 3 \\ 0 & 1 & \frac{3}{2} & -2 \\ 0 & 0 & 1 & 0 \end{array}\right]$$ Third column has leading 1 in third row.

The system of linear equations that corresponds to this (row-echelon) matrix is

$$\begin{aligned} x + y \qquad &= 3 \\ y + \frac{3}{2}z &= -2 \\ z &= 0. \end{aligned}$$

Using back-substitution, you can find that the solution is

$x = 5, \quad y = -2, \quad$ and $\quad z = 0.$

Check this in the original system, as follows.

$3(5) + 3(-2) = 9$ Substitute in Equation 1. ✓

$2(5) - 3(0) = 10$ Substitute in Equation 2. ✓

$6(-2) + 4(0) = -12$ Substitute in Equation 3. ✓

Example 8 A System with No Solution

Solve the system of linear equations.

$$6x - 10y = -4$$
$$9x - 15y = 5$$

Solution

$$\begin{bmatrix} 6 & -10 & \vdots & -4 \\ 9 & -15 & \vdots & 5 \end{bmatrix}$$ Augmented matrix for system of linear equations

$$\tfrac{1}{6}R_1 \rightarrow \begin{bmatrix} 1 & -\frac{5}{3} & \vdots & -\frac{2}{3} \\ 9 & -15 & \vdots & 5 \end{bmatrix}$$ First column has leading 1 in upper left corner.

$$-9R_1 + R_2 \rightarrow \begin{bmatrix} 1 & -\frac{5}{3} & \vdots & -\frac{2}{3} \\ 0 & 0 & \vdots & 11 \end{bmatrix}$$ First column has a zero under its leading 1.

The "equation" that corresponds to the second row of this matrix is $0 = 11$. Because this is a false statement, the system of equations has no solution.

Example 9 A System with Infinitely Many Solutions

Solve the system of linear equations.

$$12x - 6y = -3$$
$$-8x + 4y = 2$$

Solution

$$\begin{bmatrix} 12 & -6 & \vdots & -3 \\ -8 & 4 & \vdots & 2 \end{bmatrix}$$ Augmented matrix for system of linear equations

$$\tfrac{1}{12}R_1 \rightarrow \begin{bmatrix} 1 & -\frac{1}{2} & \vdots & -\frac{1}{4} \\ -8 & 4 & \vdots & 2 \end{bmatrix}$$ First column has leading 1 in upper left corner.

$$8R_1 + R_2 \rightarrow \begin{bmatrix} 1 & -\frac{1}{2} & \vdots & -\frac{1}{4} \\ 0 & 0 & \vdots & 0 \end{bmatrix}$$ First column has a zero under its leading 1.

Because the second row of the matrix is all zeros, you can conclude that the system of equations has an infinite number of solutions, represented by all points (x, y) on the line

$$x - \frac{1}{2}y = -\frac{1}{4}.$$

Because this line can be written as

$$x = -\frac{1}{4} + \frac{1}{2}y$$

you can write the solution set as

$$\left(-\frac{1}{4} + \frac{1}{2}a, a\right), \quad \text{where } a \text{ is any real number.}$$

Example 10 An Investment Portfolio

You have a portfolio totaling \$219,000 and want to invest in municipal bonds, blue-chip stocks, and growth or speculative stocks. The municipal bonds pay 6% annually. Over a 5-year period, you expect blue-chip stocks to return 10% annually and growth stocks to return 15% annually. You want a combined annual return of 8%, and you also want to have only one-fourth of the portfolio invested in stocks. How much should be allocated to each type of investment?

Solution

To solve this problem, let M represent municipal bonds, B represent blue-chip stocks, and G represent growth stocks. These three equations make up the following system.

$$M + B + G = 219{,}000$$ Equation 1: total investment is \$219,000.

$$0.06M + 0.10B + 0.15G = 17{,}520$$ Equation 2: combined annual return is 8%.

$$B + G = 54{,}750$$ Equation 3: $\frac{1}{4}$ of investment is allocated to stocks.

The augmented matrix for this system is

$$\begin{bmatrix} 1 & 1 & 1 & \vdots & 219{,}000 \\ 0.06 & 0.10 & 0.15 & \vdots & 17{,}520 \\ 0 & 1 & 1 & \vdots & 54{,}750 \end{bmatrix}.$$

Using elementary row operations, the reduced row-echelon form of this matrix is

$$\begin{bmatrix} 1 & 1 & 1 & \vdots & 219{,}000 \\ 0 & 1 & 2.25 & \vdots & 109{,}500 \\ 0 & 0 & 1 & \vdots & 43{,}800 \end{bmatrix}.$$

From the row-echelon form, you can see that $G = 43{,}800$. By back-substituting G into the revised second equation, you can determine that $B = 10{,}950$. By back-substituting B and G into the first equation, you can determine that $M = 164{,}250$. So, you should invest \$164,250 in municipal bonds, \$10,950 in blue-chip stocks, and \$43,800 in growth or speculative stocks. Check your solution by substituting these values into the original equations of the system.

Discussing the Concept — Analyzing Solutions to Systems of Equations

Use a graphing utility to graph each system of equations given in Example 6, Example 8, and Example 9. Verify the solution given in each example and explain how you can reach the same conclusion by using the graph. Summarize how you can conclude that a system has a unique solution, no solution, or infinitely many solutions when you use Gaussian elimination.

7.4 Exercises

Integrated Review *Concepts, Skills, and Problem Solving*

Keep mathematically in shape by doing these exercises *before* the problems of this section.

Properties and Definitions

In Exercises 1–4, name the property illustrated.

1. $2ab - 2ab = 0$ **2.** $8t \cdot 1 = 8t$

3. $b + 3a = 3a + b$ **4.** $3(2x) = (3 \cdot 2)x$

Algebraic Operations

In Exercises 5–10, plot the points on the rectangular coordinate system. Find the slope of the line passing through the points. If not possible, state why.

5. $(-3, 2), \left(-\frac{3}{2}, -2\right)$ **6.** $(0, -6), (8, 0)$

7. $\left(\frac{5}{2}, \frac{7}{2}\right), \left(\frac{5}{2}, 4\right)$ **8.** $\left(-\frac{5}{8}, -\frac{3}{4}\right), \left(1, -\frac{9}{2}\right)$

9. $(3, 1.2), (-3, 2.1)$ **10.** $(12, 8), (6, 8)$

Problem Solving

11. Through a membership drive, the membership for a public television station was increased by 10%. The current number of members is 8415. How many members did the station have before the membership drive?

12. A sales representative indicates that if a customer waits another month for a new car that currently costs $23,500, the price will increase by 4%. The customer has a certificate of deposit that comes due in 1 month and will pay a penalty for early withdrawal if the money is withdrawn before the due date. Determine the maximum penalty for early withdrawal that would equal the cost increase of waiting to buy the car.

Developing Skills

In Exercises 1–6, determine the order of the matrix. See Example 1.

1. $\begin{bmatrix} 3 & -2 \\ -4 & 0 \\ 2 & -7 \\ -1 & -3 \end{bmatrix}$

2. $\begin{bmatrix} 4 & 0 & -5 \\ -1 & 8 & 9 \\ 0 & -3 & 4 \end{bmatrix}$

3. $\begin{bmatrix} 5 & -8 & 32 \\ 7 & 15 & 28 \end{bmatrix}$

4. $\begin{bmatrix} -2 & 5 \\ 0 & -1 \end{bmatrix}$

5. $\begin{bmatrix} 4 \\ -2 \\ 0 \\ 1 \end{bmatrix}$

6. $\begin{bmatrix} 1 & -1 & 2 & 3 \end{bmatrix}$

In Exercises 7–12, form the augmented matrix for the system of linear equations. See Example 2.

7. $\begin{aligned} 4x - 5y &= -2 \\ -x + 8y &= 10 \end{aligned}$

8. $\begin{aligned} 8x + 3y &= 25 \\ 3x - 9y &= 12 \end{aligned}$

9. $\begin{aligned} x + 10y - 3z &= 2 \\ 5x - 3y + 4z &= 0 \\ 2x + 4y &= 6 \end{aligned}$

10. $\begin{aligned} 9x - 3y + z &= 13 \\ 12x - 8z &= 5 \\ 3x + 4y - z &= 6 \end{aligned}$

11. $\begin{aligned} 5x + y - 3z &= 7 \\ 2y + 4z &= 12 \end{aligned}$

12. $\begin{aligned} 10x + 6y - 8z &= -4 \\ -4x - 7y &= 9 \end{aligned}$

In Exercises 13–18, write the system of linear equations represented by the augmented matrix. (Use variables *x*, *y*, *z*, and *w*.) See Example 3.

13. $\left[\begin{array}{cc:c} 4 & 3 & 8 \\ 1 & -2 & 3 \end{array}\right]$

14. $\left[\begin{array}{cc:c} 9 & -4 & 0 \\ 6 & 1 & -4 \end{array}\right]$

15. $\left[\begin{array}{ccc:c} 1 & 0 & 2 & -10 \\ 0 & 3 & -1 & 5 \\ 4 & 2 & 0 & 3 \end{array}\right]$

16. $\left[\begin{array}{ccc:c} 4 & -1 & 3 & 5 \\ 2 & 0 & -2 & -1 \\ -1 & 6 & 0 & 3 \end{array}\right]$

17. $\left[\begin{array}{cccc:c} 5 & 8 & 2 & 0 & -1 \\ -2 & 15 & 5 & 1 & 9 \\ 1 & 6 & -7 & 0 & -3 \end{array}\right]$

18. $\begin{bmatrix} 7 & 3 & -2 & 4 & \vdots & 2 \\ -1 & 0 & 4 & -1 & \vdots & 6 \\ 8 & 3 & 0 & 0 & \vdots & -4 \\ 0 & 2 & -4 & 3 & \vdots & 12 \end{bmatrix}$

In Exercises 19–24, fill in the blank(s) by using elementary row operations to form a row-equivalent matrix. See Examples 4 and 5.

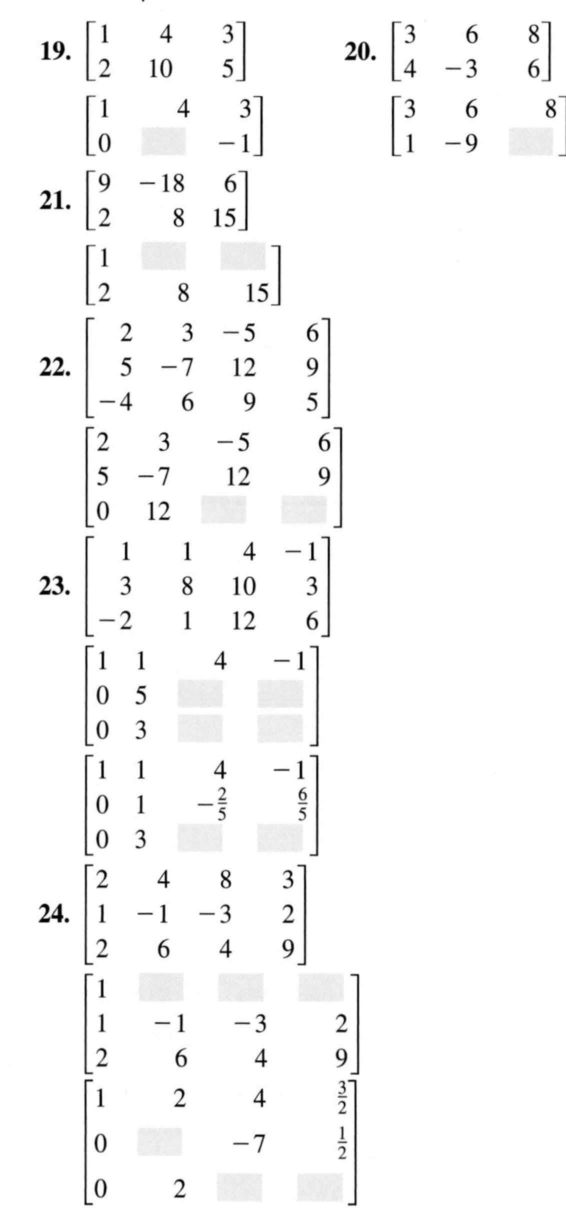

In Exercises 25–30, convert the matrix to row-echelon form. (*Note:* There is more than one correct answer.)

25. $\begin{bmatrix} 1 & 2 & 3 \\ 2 & -1 & -4 \end{bmatrix}$ **26.** $\begin{bmatrix} 1 & 3 & 6 \\ -4 & -9 & 3 \end{bmatrix}$

27. $\begin{bmatrix} 4 & 6 & 1 \\ -2 & 2 & 5 \end{bmatrix}$ **28.** $\begin{bmatrix} 3 & 2 & 6 \\ 2 & 3 & -3 \end{bmatrix}$

29. $\begin{bmatrix} 1 & 1 & 0 & 5 \\ -2 & -1 & 2 & -10 \\ 3 & 6 & 7 & 14 \end{bmatrix}$ **30.** $\begin{bmatrix} 1 & 2 & -1 & 3 \\ 3 & 7 & -5 & 14 \\ -2 & -1 & -3 & 8 \end{bmatrix}$

In Exercises 31–34, use the matrix capabilities of a graphing utility to write the matrix in row-echelon form. (*Note:* There is more than one correct answer.)

31. $\begin{bmatrix} 1 & -1 & -1 & 1 \\ 4 & -4 & 1 & 8 \\ -6 & 8 & 18 & 0 \end{bmatrix}$

32. $\begin{bmatrix} 1 & -3 & 0 & -7 \\ -3 & 10 & 1 & 23 \\ 4 & -10 & 2 & -24 \end{bmatrix}$

33. $\begin{bmatrix} 1 & 1 & -1 & 3 \\ 2 & 1 & 2 & 5 \\ 3 & 2 & 1 & 8 \end{bmatrix}$

34. $\begin{bmatrix} 1 & -3 & -2 & -8 \\ 1 & 3 & -2 & 17 \\ 1 & 2 & -2 & -5 \end{bmatrix}$

In Exercises 35–40, write the system of linear equations represented by the augmented matrix. Then use back-substitution to find the solution. (Use variables x, y, and z.)

35. $\begin{bmatrix} 1 & -2 & \vdots & 4 \\ 0 & 1 & \vdots & -3 \end{bmatrix}$ **36.** $\begin{bmatrix} 1 & 5 & \vdots & 0 \\ 0 & 1 & \vdots & -1 \end{bmatrix}$

37. $\begin{bmatrix} 1 & 5 & \vdots & 3 \\ 0 & 1 & \vdots & -2 \end{bmatrix}$ **38.** $\begin{bmatrix} 1 & 5 & -3 & \vdots & 0 \\ 0 & 1 & 0 & \vdots & 6 \\ 0 & 0 & 1 & \vdots & -5 \end{bmatrix}$

39. $\begin{bmatrix} 1 & -1 & 2 & \vdots & 4 \\ 0 & 1 & -1 & \vdots & 2 \\ 0 & 0 & 1 & \vdots & -2 \end{bmatrix}$

40. $\begin{bmatrix} 1 & 2 & -2 & \vdots & -1 \\ 0 & 1 & 1 & \vdots & 9 \\ 0 & 0 & 1 & \vdots & -3 \end{bmatrix}$

In Exercises 41–62, use matrices to solve the system of linear equations. See Examples 5–9.

41. $\begin{aligned} x + 2y &= 7 \\ 3x + y &= 8 \end{aligned}$

42. $\begin{aligned} 2x + 6y &= 16 \\ 2x + 3y &= 7 \end{aligned}$

43. $\begin{aligned} 6x - 4y &= 2 \\ 5x + 2y &= 7 \end{aligned}$

44. $\begin{aligned} x - 3y &= 5 \\ -2x + 6y &= -10 \end{aligned}$

45. $\begin{aligned} -x + 2y &= 1.5 \\ 2x - 4y &= 3 \end{aligned}$

46. $\begin{aligned} 2x - y &= -0.1 \\ 3x + 2y &= 1.6 \end{aligned}$

47. $\begin{aligned} x - 2y - z &= 6 \\ y + 4z &= 5 \\ 4x + 2y + 3z &= 8 \end{aligned}$

48. $\begin{aligned} x - 3z &= -2 \\ 3x + y - 2z &= 5 \\ 2x + 2y + z &= 4 \end{aligned}$

49. $\begin{aligned} x + y - 5z &= 3 \\ x - 2z &= 1 \\ 2x - y - z &= 0 \end{aligned}$

50. $\begin{aligned} 2y + z &= 3 \\ -4y - 2z &= 0 \\ x + y + z &= 2 \end{aligned}$

51. $\begin{aligned} 2x + 4y &= 10 \\ 2x + 2y + 3z &= 3 \\ -3x + y + 2z &= -3 \end{aligned}$

52. $\begin{aligned} 2x - y + 3z &= 24 \\ 2y - z &= 14 \\ 7x - 5y &= 6 \end{aligned}$

53. $\begin{aligned} x - 3y + 2z &= 8 \\ 2y - z &= -4 \\ x + z &= 3 \end{aligned}$

54. $\begin{aligned} 2x + 3z &= 3 \\ 4x - 3y + 7z &= 5 \\ 8x - 9y + 15z &= 9 \end{aligned}$

55. $\begin{aligned} -2x - 2y - 15z &= 0 \\ x + 2y + 2z &= 18 \\ 3x + 3y + 22z &= 2 \end{aligned}$

56. $\begin{aligned} 2x + 4y + 5z &= 5 \\ x + 3y + 3z &= 2 \\ 2x + 4y + 4z &= 2 \end{aligned}$

57. $\begin{aligned} 2x + 4z &= 1 \\ x + y + 3z &= 0 \\ x + 3y + 5z &= 0 \end{aligned}$

58. $\begin{aligned} 3x + y - 2z &= 2 \\ 6x + 2y - 4z &= 1 \\ -3x - y + 2z &= 1 \end{aligned}$

59. $\begin{aligned} x + 3y &= 2 \\ 2x + 6y &= 4 \\ 2x + 5y + 4z &= 3 \end{aligned}$

60. $\begin{aligned} 2x + 2y + z &= 8 \\ 2x + 3y + z &= 7 \\ 6x + 8y + 3z &= 22 \end{aligned}$

61. $\begin{aligned} 2x + y - 2z &= 4 \\ 3x - 2y + 4z &= 6 \\ -4x + y + 6z &= 12 \end{aligned}$

62. $\begin{aligned} 3x + 3y + z &= 4 \\ 2x + 6y + z &= 5 \\ -x - 3y + 2z &= -5 \end{aligned}$

Solving Problems

63. *Simple Interest* A corporation borrowed \$1,500,000 to expand its product line. Some of the money was borrowed at 8%, some at 9%, and the remainder at 12%. The annual interest payment to the lenders was \$133,000. If the amount borrowed at 8% was 4 times the amount borrowed at 12%, how much was borrowed at each rate?

64. *Investments* An inheritance of \$16,000 was divided among three investments yielding a total of \$990 in simple interest per year. The interest rates for the three investments were 5%, 6%, and 7%. Find the amount placed in each investment if the 5% and 6% investments were \$3000 and \$2000 less than the 7% investment, respectively.

Investment Portfolio In Exercises 65 and 66, consider an investor with a portfolio totaling $500,000 that is to be allocated among the following types of investments: certificates of deposit, municipal bonds, blue-chip stocks, and growth or speculative stocks. How much should be allocated to each type of investment?

65. The certificates of deposit pay 10% annually, and the municipal bonds pay 8% annually. Over a 5-year period, the investor expects the blue-chip stocks to return 12% annually and the growth stocks to return 13% annually. The investor wants a combined annual return of 10% and also wants to have only one-fourth of the portfolio invested in stocks.

66. The certificates of deposit pay 9% annually, and the municipal bonds pay 5% annually. Over a 5-year period, the investor expects the blue-chip stocks to return 12% annually and the growth stocks to return 14% annually. The investor wants a combined annual return of 10% and also wants to have only one-fourth of the portfolio invested in stocks.

67. ***Nut Mixture*** A grocer wishes to mix three kinds of nuts costing $3.50, $4.50, and $6.00 per pound to obtain 50 pounds of a mixture priced at $4.95 per pound. How many pounds of each variety should the grocer use if half the mixture is composed of the two cheapest varieties?

68. ***Nut Mixture*** A grocer wishes to mix three kinds of nuts costing $3.00, $4.00, and $6.00 per pound to obtain 50 pounds of a mixture priced at $4.10 per pound. How many pounds of each variety should the grocer use if three-quarters of the mixture is composed of the two cheapest varieties?

69. ***Number Problem*** The sum of three positive numbers is 33. The second number is 3 greater than the first, and the third is 4 times the first. Find the three numbers.

70. ***Number Problem*** The sum of three positive numbers is 24. The second number is 4 greater than the first, and the third is 3 times the first. Find the three numbers.

Curve-Fitting In Exercises 71–74, find a quadratic equation $y = ax^2 + bx + c$ whose graph passes through the given points.

71. **72.**

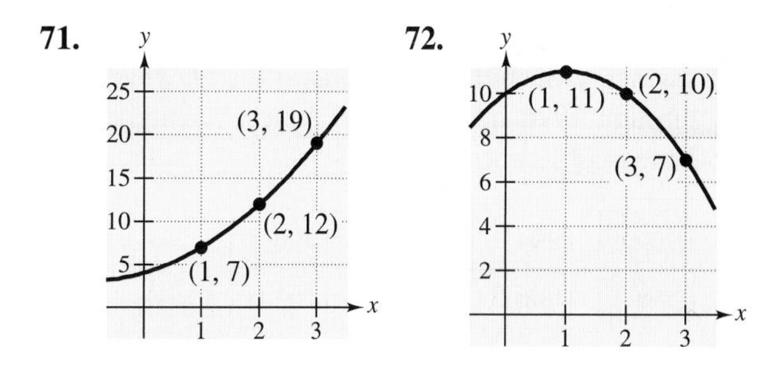

73. $(1, 8), (2, 2), (3, -25)$

74. $(1, 1), (-3, 17), \left(2, -\frac{1}{2}\right)$

Curve-Fitting In Exercises 75 and 76, find the equation of the circle $x^2 + y^2 + Dx + Ey + F = 0$ that passes through the points.

75. $(1, 1), (3, 3), (4, 2)$

76. $(-1, 2), (2, 3), (3, 2)$

77. ***Mathematical Modeling*** A videotape of the path of a ball thrown by a baseball player was analyzed on a television set with a grid covering the screen. The tape was paused three times and the coordinates of the ball were measured each time. The coordinates were approximately $(0, 6)$, $(25, 18.5)$, and $(50, 26)$ (see figure). The x-coordinate was the horizontal distance in feet from the player and the y-coordinate was the height in feet of the ball above the ground.

(a) Find the equation $y = ax^2 + bx + c$ of the graph that passes through the three points.

(b) Use a graphing utility to graph the model in part (a). Use the graph to approximate the maximum height of the ball and the point at which the ball struck the ground.

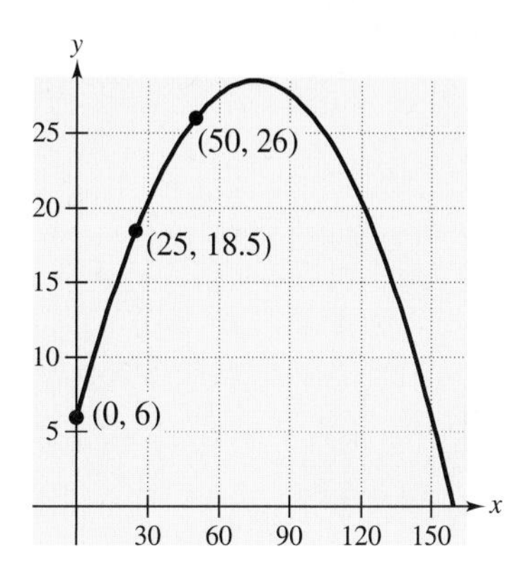

78. *Data Analysis* The table gives the gross private savings y (in billions of dollars) for the years 1994 through 1996 in the United States. (Source: U.S. Bureau of Economic Analysis)

Year	1994	1995	1996
y	1006.3	1072.3	1161.0

(a) Create a bar graph of the data.

(b) Find a quadratic equation $y = at^2 + bt + c$ whose graph passes through the three points, with $t = 0$ corresponding to 1990.

(c) Use a graphing utility to graph the model in part (b).

(d) Use the model in part (b) to predict gross private savings in the year 2000 if the trend continues.

79. *Rewriting a Fraction* The fraction

$$\frac{2x^2 - 9x}{(x - 2)^3}$$

can be written as a sum of three fractions, as follows.

$$\frac{2x^2 - 9x}{(x - 2)^3} = \frac{A}{x - 2} + \frac{B}{(x - 2)^2} + \frac{C}{(x - 2)^3}.$$

The numbers A, B, and C are the solutions of the system

$$\begin{aligned} 4A - 2B + C &= 0 \\ -4A + B \qquad &= -9 \\ A \qquad\qquad &= 2. \end{aligned}$$

Write the expression as the sum of three fractions.

80. *Rewriting a Fraction* The fraction

$$\frac{x + 1}{x(x^2 + 1)}$$

can be written as a sum of two fractions, as follows.

$$\frac{x + 1}{x(x^2 + 1)} = \frac{A}{x} + \frac{Bx + C}{x^2 + 1}$$

The numbers A, B, and C are the solutions of the system

$$\begin{aligned} 2A + B + C &= 2 \\ 2A + B - C &= 0 \\ 5A + 4B + 2C &= 3. \end{aligned}$$

Solve the system and verify that the sum of the two resulting fractions is the original fraction.

Explaining Concepts

81. Describe the three elementary row operations that can be performed on an augmented matrix.

82. What is the relationship between the three elementary row operations on an augmented matrix and the row operations on a system of linear equations?

83. What is meant by saying that two augmented matrices are *row-equivalent*?

84. Give an example of a matrix in *row-echelon form.*

85. Describe the row-echelon form of an augmented matrix that corresponds to a system of linear equations that is inconsistent.

86. Describe the row-echelon form of an augmented matrix that corresponds to a system of linear equations that has an infinite number of solutions.

7.5 Determinants and Linear Systems

Objectives

1. Find the determinants of a 2×2 matrix and a 3×3 matrix.
2. Use determinants and Cramer's Rule to solve a system of linear equations.
3. Use the determinant to find the area of a triangle, to test for collinear points, and to find the equation of a line.

1 Find the determinants of a 2×2 matrix and a 3×3 matrix.

The Determinant of a Matrix

Associated with each square matrix is a real number called its **determinant.** The use of determinants arose from special number patterns that occur during the solution of systems of linear equations. For instance, the system

$$a_1x + b_1y = c_1$$
$$a_2x + b_2y = c_2$$

has a solution given by

$$x = \frac{c_1b_2 - c_2b_1}{a_1b_2 - a_2b_1} \quad \text{and} \quad y = \frac{a_1c_2 - a_2c_1}{a_1b_2 - a_2b_1}$$

provided that $a_1b_2 - a_2b_1 \neq 0$. Note that the denominator of each fraction is the same. This denominator is called the **determinant** of the coefficient matrix of the system.

Coefficient Matrix	*Determinant*
$A = \begin{bmatrix} a_1 & b_1 \\ a_2 & b_2 \end{bmatrix}$	$\det(A) = a_1b_2 - a_2b_1$

The determinant of the matrix A can also be denoted by vertical bars on both sides of the matrix, as indicated in the following definition.

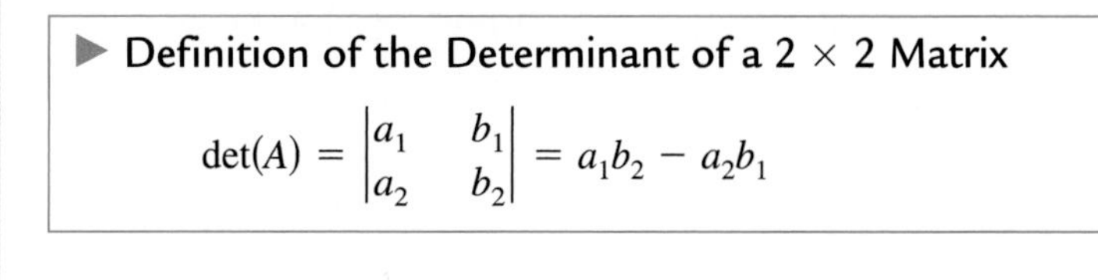

Definition of the Determinant of a 2×2 Matrix

$$\det(A) = \begin{vmatrix} a_1 & b_1 \\ a_2 & b_2 \end{vmatrix} = a_1b_2 - a_2b_1$$

The Granger Collection

Arthur Cayley
(1821–1895)

Cayley is credited with creating the theory of matrices. Determinants had been studied as rectangular arrays of numbers since the middle of the 18th century. So, the use and basic properties of matrices were well established when Cayley first published articles introducing matrices as distinct entities.

A convenient method for remembering the formula for the determinant of a 2×2 matrix is shown in the following diagram.

$$\det(A) = \begin{vmatrix} a_1 & b_1 \\ a_2 & b_2 \end{vmatrix} = a_1b_2 - a_2b_1$$

Note that the determinant is given by the difference of the products of the two diagonals of the matrix.

Example 1 The Determinant of a 2 × 2 Matrix

Find the determinant of each matrix.

a. $A = \begin{bmatrix} 2 & -3 \\ 1 & 4 \end{bmatrix}$ **b.** $B = \begin{bmatrix} -1 & 2 \\ 2 & -4 \end{bmatrix}$ **c.** $C = \begin{bmatrix} 1 & 3 \\ 2 & 5 \end{bmatrix}$

Solution

a. $\det(A) = \begin{vmatrix} 2 & -3 \\ 1 & 4 \end{vmatrix} = 2(4) - 1(-3) = 8 + 3 = 11$

b. $\det(B) = \begin{vmatrix} -1 & 2 \\ 2 & -4 \end{vmatrix} = (-1)(-4) - 2(2) = 4 - 4 = 0$

c. $\det(C) = \begin{vmatrix} 1 & 3 \\ 2 & 5 \end{vmatrix} = 1(5) - 2(3) = 5 - 6 = -1$

A graphing utility with matrix capabilities can be used to evaluate the determinant of a square matrix. Consult the user's guide for your graphing utility to learn how to evaluate a determinant. Use the graphing utility to check the result in Example 1(a). Then try to evaluate the determinant of the 3 × 3 matrix at the right using a graphing utility. Finish the evaluation of the determinant by expanding by minors to check the result.

Notice in Example 1 that the determinant of a matrix can be positive, zero, or negative.

One way to evaluate the determinant of a 3 × 3 matrix, called **expanding by minors,** allows you to write the determinant of a 3 × 3 matrix in terms of three 2 × 2 determinants. The **minor** of an entry in a 3 × 3 matrix is the determinant of the 2 × 2 matrix that remains after deletion of the row and column in which the entry occurs. Here are two examples.

Given Determinant	*Entry*	*Minor of Entry*	*Value of Minor*
$\begin{vmatrix} 1 & -1 & 3 \\ 0 & 2 & 5 \\ -2 & 4 & -7 \end{vmatrix}$	1	$\begin{vmatrix} 2 & 5 \\ 4 & -7 \end{vmatrix}$	$2(-7) - 4(5) = -34$
$\begin{vmatrix} 1 & -1 & 3 \\ 0 & 2 & 5 \\ -2 & 4 & -7 \end{vmatrix}$	-1	$\begin{vmatrix} 0 & 5 \\ -2 & -7 \end{vmatrix}$	$0(-7) - (-2)(5) = 10$

▶ Expanding by Minors

$$\det(A) = \begin{vmatrix} a_1 & b_1 & c_1 \\ a_2 & b_2 & c_2 \\ a_3 & b_3 & c_3 \end{vmatrix}$$

$$= a_1(\text{minor of } a_1) - b_1(\text{minor of } b_1) + c_1(\text{minor of } c_1)$$

$$= a_1\begin{vmatrix} b_2 & c_2 \\ b_3 & c_3 \end{vmatrix} - b_1\begin{vmatrix} a_2 & c_2 \\ a_3 & c_3 \end{vmatrix} + c_1\begin{vmatrix} a_2 & b_2 \\ a_3 & b_3 \end{vmatrix}$$

This pattern is called **expanding by minors** along the first row. A similar pattern can be used to expand by minors along any row or column.

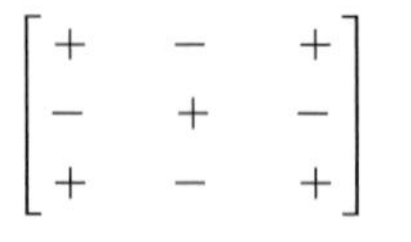

Figure 7.13 *Sign Pattern for a 3 × 3 Matrix*

The *signs* of the terms used in expanding by minors follow the alternating pattern shown in Figure 7.13. For instance, the signs used to expand by minors along the second row are −, +, −, as shown at the top of page 427.

$$\det(A) = \begin{vmatrix} a_1 & b_1 & c_1 \\ a_2 & b_2 & c_2 \\ a_3 & b_3 & c_3 \end{vmatrix}$$
$$= -a_2(\text{minor of } a_2) + b_2(\text{minor of } b_2) - c_2(\text{minor of } c_2)$$

Example 2 Finding the Determinant of a 3 × 3 Matrix

Find the determinant of $A = \begin{bmatrix} -1 & 1 & 2 \\ 0 & 2 & 3 \\ 3 & 4 & 2 \end{bmatrix}$.

Solution

By expanding by minors along the *first column*, you obtain the following.

$$\det(A) = \begin{vmatrix} -1 & 1 & 2 \\ 0 & 2 & 3 \\ 3 & 4 & 2 \end{vmatrix}$$
$$= (-1)\begin{vmatrix} 2 & 3 \\ 4 & 2 \end{vmatrix} - (0)\begin{vmatrix} 1 & 2 \\ 4 & 2 \end{vmatrix} + (3)\begin{vmatrix} 1 & 2 \\ 2 & 3 \end{vmatrix}$$
$$= (-1)(4 - 12) - (0)(2 - 8) + (3)(3 - 4)$$
$$= 8 - 0 - 3$$
$$= 5$$

Example 3 Finding the Determinant of a 3 × 3 Matrix

Find the determinant of $A = \begin{bmatrix} 1 & 2 & 1 \\ 3 & 0 & 2 \\ 4 & 0 & -1 \end{bmatrix}$.

Solution

By expanding by minors along the *second column*, you obtain the following.

$$\det(A) = \begin{vmatrix} 1 & 2 & 1 \\ 3 & 0 & 2 \\ 4 & 0 & -1 \end{vmatrix}$$
$$= -(2)\begin{vmatrix} 3 & 2 \\ 4 & -1 \end{vmatrix} + (0)\begin{vmatrix} 1 & 1 \\ 4 & -1 \end{vmatrix} - (0)\begin{vmatrix} 1 & 1 \\ 3 & 2 \end{vmatrix}$$
$$= -(2)(-3 - 8) + 0 - 0$$
$$= 22$$

Note in the expansion in Examples 2 and 3 that a zero entry will always yield a zero term when expanding by minors. Thus, when you are evaluating the determinant of a matrix, you should choose to expand along the row or column that has the most zero entries.

2 Use determinants and Cramer's Rule to solve a system of linear equations.

Cramer's Rule

So far in this chapter, you have studied three methods for solving a system of linear equations: substitution, elimination (with equations), and elimination (with matrices). We now look at one more method, called **Cramer's Rule,** which is named after Gabriel Cramer (1704–1752). This rule uses determinants to write the solution of a system of linear equations.

▶ Cramer's Rule

1. For the system of linear equations

$$a_1x + b_1y = c_1$$
$$a_2x + b_2y = c_2$$

the solution is given by

$$x = \frac{D_x}{D} = \frac{\begin{vmatrix} c_1 & b_1 \\ c_2 & b_2 \end{vmatrix}}{\begin{vmatrix} a_1 & b_1 \\ a_2 & b_2 \end{vmatrix}}, \qquad y = \frac{D_y}{D} = \frac{\begin{vmatrix} a_1 & c_1 \\ a_2 & c_2 \end{vmatrix}}{\begin{vmatrix} a_1 & b_1 \\ a_2 & b_2 \end{vmatrix}}$$

provided that $D \neq 0$.

2. For the system of linear equations

$$a_1x + b_1y + c_1z = d_1$$
$$a_2x + b_2y + c_2z = d_2$$
$$a_3x + b_3y + c_3z = d_3$$

the solution is given by

$$x = \frac{D_x}{D} = \frac{\begin{vmatrix} d_1 & b_1 & c_1 \\ d_2 & b_2 & c_2 \\ d_3 & b_3 & c_3 \end{vmatrix}}{\begin{vmatrix} a_1 & b_1 & c_1 \\ a_2 & b_2 & c_2 \\ a_3 & b_3 & c_3 \end{vmatrix}},$$

$$y = \frac{D_y}{D} = \frac{\begin{vmatrix} a_1 & d_1 & c_1 \\ a_2 & d_2 & c_2 \\ a_3 & d_3 & c_3 \end{vmatrix}}{\begin{vmatrix} a_1 & b_1 & c_1 \\ a_2 & b_2 & c_2 \\ a_3 & b_3 & c_3 \end{vmatrix}},$$

$$z = \frac{D_z}{D} = \frac{\begin{vmatrix} a_1 & b_1 & d_1 \\ a_2 & b_2 & d_2 \\ a_3 & b_3 & d_3 \end{vmatrix}}{\begin{vmatrix} a_1 & b_1 & c_1 \\ a_2 & b_2 & c_2 \\ a_3 & b_3 & c_3 \end{vmatrix}}, \quad D \neq 0$$

Example 4 Using Cramer's Rule for a 2 × 2 System

Use Cramer's Rule to solve the system of linear equations.

$$4x - 2y = 10$$
$$3x - 5y = 11$$

Solution

Begin by finding the determinant of the coefficient matrix, $D = -14$.

$$x = \frac{D_x}{D} = \frac{\begin{vmatrix} 10 & -2 \\ 11 & -5 \end{vmatrix}}{-14} = \frac{(-50) - (-22)}{-14} = \frac{-28}{-14} = 2$$

$$y = \frac{D_y}{D} = \frac{\begin{vmatrix} 4 & 10 \\ 3 & 11 \end{vmatrix}}{-14} = \frac{44 - 30}{-14} = \frac{14}{-14} = -1$$

The solution is $(2, -1)$. Check this in the original system of equations.

Example 5 Using Cramer's Rule for a 3 × 3 System

Use Cramer's Rule to solve the system of linear equations.

$$-x + 2y - 3z = 1$$
$$2x + \qquad z = 0$$
$$3x - 4y + 4z = 2$$

Solution

The determinant of the coefficient matrix is $D = 10$.

$$x = \frac{D_x}{D} = \frac{\begin{vmatrix} 1 & 2 & -3 \\ 0 & 0 & 1 \\ 2 & -4 & 4 \end{vmatrix}}{10} = \frac{8}{10} = \frac{4}{5}$$

$$y = \frac{D_y}{D} = \frac{\begin{vmatrix} -1 & 1 & -3 \\ 2 & 0 & 1 \\ 3 & 2 & 4 \end{vmatrix}}{10} = \frac{-15}{10} = -\frac{3}{2}$$

$$z = \frac{D_z}{D} = \frac{\begin{vmatrix} -1 & 2 & 1 \\ 2 & 0 & 0 \\ 3 & -4 & 2 \end{vmatrix}}{10} = \frac{-16}{10} = -\frac{8}{5}$$

The solution is $\left(\frac{4}{5}, -\frac{3}{2}, -\frac{8}{5}\right)$. Check this in the original system of equations.

When using Cramer's Rule, remember that the method *does not* apply if the determinant of the coefficient matrix is zero.

3 Use the determinant to find the area of a triangle, to test for collinear points, and to find the equation of a line.

Applications of Determinants

In addition to Cramer's Rule, determinants have many other practical applications. For instance, you can use a determinant to find the area of a triangle whose vertices are given by three points on a rectangular coordinate system.

▶ Area of a Triangle

The area of a triangle with vertices (x_1, y_1), (x_2, y_2), and (x_3, y_3) is

$$\text{Area} = \pm\frac{1}{2}\begin{vmatrix} x_1 & y_1 & 1 \\ x_2 & y_2 & 1 \\ x_3 & y_3 & 1 \end{vmatrix}$$

where the symbol $(\pm)$ indicates that the appropriate sign should be chosen to yield a positive area.

Example 6 Finding the Area of a Triangle

Find the area of the triangle whose vertices are $(2, 0)$, $(1, 3)$, and $(3, 2)$, as shown in Figure 7.14.

Solution

Choose $(x_1, y_1) = (2, 0)$, $(x_2, y_2) = (1, 3)$, and $(x_3, y_3) = (3, 2)$. To find the area of the triangle, evaluate the determinant

$$\begin{vmatrix} x_1 & y_1 & 1 \\ x_2 & y_2 & 1 \\ x_3 & y_3 & 1 \end{vmatrix} = \begin{vmatrix} 2 & 0 & 1 \\ 1 & 3 & 1 \\ 3 & 2 & 1 \end{vmatrix}$$

$$= 2\begin{vmatrix} 3 & 1 \\ 2 & 1 \end{vmatrix} - 0\begin{vmatrix} 1 & 1 \\ 3 & 1 \end{vmatrix} + 1\begin{vmatrix} 1 & 3 \\ 3 & 2 \end{vmatrix}$$

$$= 2(1) - 0 + 1(-7)$$

$$= -5.$$

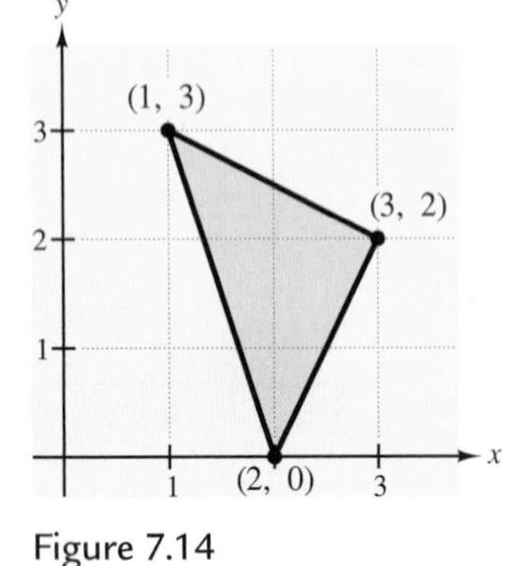

Figure 7.14

Using this value, you can conclude that the area of the triangle is

$$\text{Area} = -\frac{1}{2}\begin{vmatrix} 2 & 0 & 1 \\ 1 & 3 & 1 \\ 3 & 2 & 1 \end{vmatrix}$$

$$= -\frac{1}{2}(-5)$$

$$= \frac{5}{2}.$$

To see the benefit of the "determinant formula," try finding the area of the triangle in Example 6 using the standard formula:

$$\text{Area} = \tfrac{1}{2}(\text{base})(\text{height}).$$

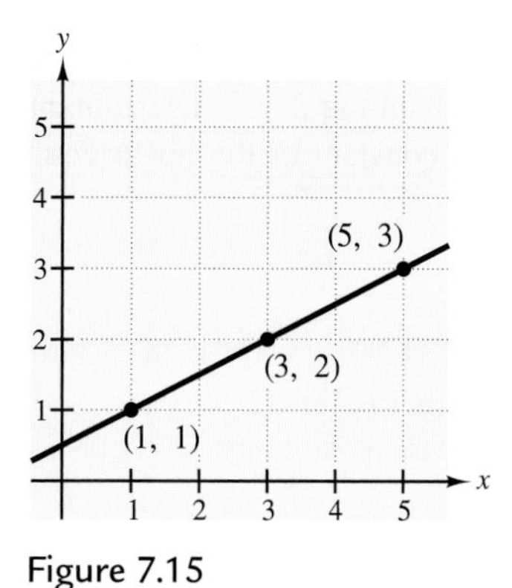

Figure 7.15

Suppose the three points in Example 6 had been on the same line. What would have happened had we applied the area formula to three such points? The answer is that the determinant would have been zero. Consider, for instance, the three collinear points (1, 1), (3, 2), and (5, 3), as shown in Figure 7.15. The area of the "triangle" that has these three points as vertices is

$$\frac{1}{2}\begin{vmatrix} 1 & 1 & 1 \\ 3 & 2 & 1 \\ 5 & 3 & 1 \end{vmatrix} = \frac{1}{2}\left(1\begin{vmatrix} 2 & 1 \\ 3 & 1 \end{vmatrix} - 1\begin{vmatrix} 3 & 1 \\ 5 & 1 \end{vmatrix} + 1\begin{vmatrix} 3 & 2 \\ 5 & 3 \end{vmatrix}\right)$$

$$= \frac{1}{2}[-1 - (-2) + (-1)]$$

$$= 0.$$

This result is generalized as follows.

> **▶ Test for Collinear Points**
>
> Three points (x_1, y_1), (x_2, y_2), and (x_3, y_3) are collinear (lie on the same line) if and only if
>
> $$\begin{vmatrix} x_1 & y_1 & 1 \\ x_2 & y_2 & 1 \\ x_3 & y_3 & 1 \end{vmatrix} = 0.$$

Example 7 Testing for Collinear Points

Determine whether the points $(-2, -2)$, $(1, 1)$, and $(7, 5)$ lie on the same line. (See Figure 7.16.)

Solution

Letting $(x_1, y_1) = (-2, -2)$, $(x_2, y_2) = (1, 1)$, and $(x_3, y_3) = (7, 5)$, you have

$$\begin{vmatrix} x_1 & y_1 & 1 \\ x_2 & y_2 & 1 \\ x_3 & y_3 & 1 \end{vmatrix} = \begin{vmatrix} -2 & -2 & 1 \\ 1 & 1 & 1 \\ 7 & 5 & 1 \end{vmatrix}$$

$$= -2\begin{vmatrix} 1 & 1 \\ 5 & 1 \end{vmatrix} - (-2)\begin{vmatrix} 1 & 1 \\ 7 & 1 \end{vmatrix} + 1\begin{vmatrix} 1 & 1 \\ 7 & 5 \end{vmatrix}$$

$$= -2(-4) - (-2)(-6) + 1(-2)$$

$$= -6.$$

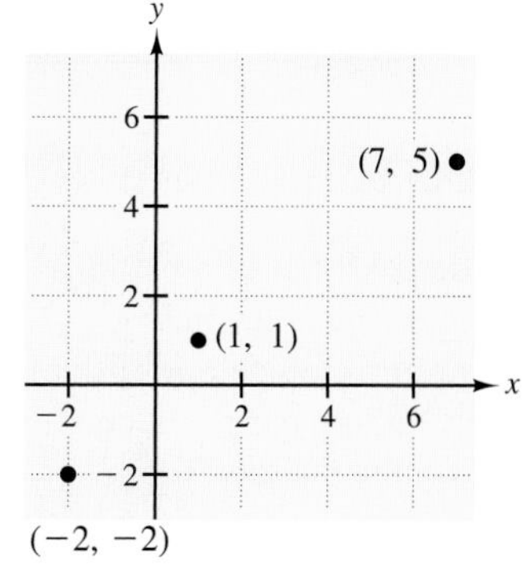

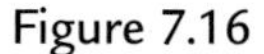

Figure 7.16

Because the value of this determinant *is not* zero, you can conclude that the three points *do not* lie on the same line.

As a good review, look at how the slope can be used to verify the result in Example 7. Label the points $A(-2, -2)$, $B(1, 1)$, and $C(7, 5)$. Because the slopes from A to B and from A to C are different, the points are not collinear.

You can also use determinants to find the equation of a line through two points. In this case the first row consists of the variables x and y and the number 1. By expanding by minors along the first row, the resulting 2×2 determinants are the coefficients of the variables x and y and the constant of the linear equation, as shown in Example 8.

Two-Point Form of the Equation of a Line

An equation of the line passing through the distinct points (x_1, y_1) and (x_2, y_2) is given by

$$\begin{vmatrix} x & y & 1 \\ x_1 & y_1 & 1 \\ x_2 & y_2 & 1 \end{vmatrix} = 0.$$

Example 8 Finding an Equation of a Line

Find an equation of the line passing through $(-2, 1)$ and $(3, -2)$.

Solution

$$\begin{vmatrix} x & y & 1 \\ -2 & 1 & 1 \\ 3 & -2 & 1 \end{vmatrix} = 0$$

$$x\begin{vmatrix} 1 & 1 \\ -2 & 1 \end{vmatrix} - y\begin{vmatrix} -2 & 1 \\ 3 & 1 \end{vmatrix} + 1\begin{vmatrix} -2 & 1 \\ 3 & -2 \end{vmatrix} = 0$$

$$3x + 5y + 1 = 0$$

So, an equation of the line is $3x + 5y + 1 = 0$.

Discussing the Concept — Determinant of a 3 × 3 Matrix

There is an alternative method for evaluating the determinant of a 3×3 matrix A. (This method works *only* for 3×3 matrices.) To apply this method, copy the first and second columns of A to form fourth and fifth columns. The determinant of A is then obtained by adding the products of three diagonals and subtracting the products of three diagonals.

$$\begin{array}{ccccc} & & -4 & 0 & 6 \\ 0 & 2 & 1 & 0 & 2 \\ 3 & -1 & 2 & 3 & -1 \\ 4 & -4 & 1 & 4 & -4 \\ & & 0 & 16 & -12 \end{array}$$

$$|A| = [0 + 16 - 12] - [(-4) + 0 + 6] = 2$$

Try using this technique to find the determinants of the matrices in Examples 2 and 3. Do you think this method is easier than expanding by minors?

7.5 Exercises

Integrated Review *Concepts, Skills, and Problem Solving*

Keep mathematically in shape by doing these exercises *before* the problems of this section.

Properties and Definitions

In Exercises 1–4, use $(px + m)(qx + n) = ax^2 + bx + c$.

1. $a =$

2. $b =$

3. $c =$

4. If $a = 1$, must $p = 1$ and $q = 1$? Explain.

Solving Equations

In Exercises 5–10, solve the equation.

5. $3x^2 + 9x - 12 = 0$

6. $x^2 - x - 6 = 0$

7. $4x^2 - 20x + 25 = 0$

8. $x^2 - 16 = 0$

9. $x^3 + 64 = 0$

10. $3x^3 - 6x^2 + 4x - 8 = 0$

Models

In Exercises 11 and 12, translate the phrase into an algebraic expression.

11. The time to travel 320 miles if the average speed is r miles per hour

12. The perimeter of a triangle if the sides are $x + 1, \frac{1}{2}x + 5$, and $3x + 1$

Developing Skills

In Exercises 1–12, find the determinant of the matrix. See Example 1.

1. $\begin{bmatrix} 2 & 1 \\ 3 & 4 \end{bmatrix}$

2. $\begin{bmatrix} -3 & 1 \\ 5 & 2 \end{bmatrix}$

3. $\begin{bmatrix} 5 & 2 \\ -6 & 3 \end{bmatrix}$

4. $\begin{bmatrix} 2 & -2 \\ 4 & 3 \end{bmatrix}$

5. $\begin{bmatrix} 5 & -4 \\ -10 & 8 \end{bmatrix}$

6. $\begin{bmatrix} 4 & -3 \\ 0 & 0 \end{bmatrix}$

7. $\begin{bmatrix} 2 & 6 \\ 0 & 3 \end{bmatrix}$

8. $\begin{bmatrix} -2 & 3 \\ 6 & -9 \end{bmatrix}$

9. $\begin{bmatrix} -7 & 6 \\ \frac{1}{2} & 3 \end{bmatrix}$

10. $\begin{bmatrix} \frac{2}{3} & \frac{5}{6} \\ 14 & -2 \end{bmatrix}$

11. $\begin{bmatrix} 0.3 & 0.5 \\ 0.5 & 0.3 \end{bmatrix}$

12. $\begin{bmatrix} -1.2 & 4.5 \\ 0.4 & -0.9 \end{bmatrix}$

In Exercises 13–32, evaluate the determinant of the matrix. Expand by minors along the row or column that appears to make the computation easiest. See Examples 2 and 3.

13. $\begin{bmatrix} 2 & 3 & -1 \\ 6 & 0 & 0 \\ 4 & 1 & 1 \end{bmatrix}$

14. $\begin{bmatrix} 10 & 2 & -4 \\ 8 & 0 & -2 \\ 4 & 0 & 2 \end{bmatrix}$

15. $\begin{bmatrix} 1 & 1 & 2 \\ 3 & 1 & 0 \\ -2 & 0 & 3 \end{bmatrix}$

16. $\begin{bmatrix} 2 & 1 & 3 \\ 1 & 4 & 4 \\ 1 & 0 & 2 \end{bmatrix}$

17. $\begin{bmatrix} 2 & 4 & 6 \\ 0 & 3 & 1 \\ 0 & 0 & -5 \end{bmatrix}$

18. $\begin{bmatrix} 2 & 3 & 1 \\ 0 & 5 & -2 \\ 0 & 0 & -2 \end{bmatrix}$

19. $\begin{bmatrix} -2 & 2 & 3 \\ 1 & -1 & 0 \\ 0 & 1 & 4 \end{bmatrix}$

20. $\begin{bmatrix} 3 & 2 & 2 \\ 2 & 2 & 2 \\ -4 & 4 & 3 \end{bmatrix}$

21. $\begin{bmatrix} 1 & 4 & -2 \\ 3 & 6 & -6 \\ -2 & 1 & 4 \end{bmatrix}$

22. $\begin{bmatrix} 2 & -1 & 0 \\ 4 & 2 & 1 \\ 4 & 2 & 1 \end{bmatrix}$

23. $\begin{bmatrix} -3 & 2 & 1 \\ 4 & 5 & 6 \\ 2 & -3 & 1 \end{bmatrix}$

24. $\begin{bmatrix} -3 & 4 & 2 \\ 6 & 3 & 1 \\ 4 & -7 & -8 \end{bmatrix}$

25. $\begin{bmatrix} 1 & 4 & -2 \\ 3 & 2 & 0 \\ -1 & 4 & 3 \end{bmatrix}$

26. $\begin{bmatrix} 6 & 8 & -7 \\ 0 & 0 & 0 \\ 4 & -6 & 22 \end{bmatrix}$

27. $\begin{bmatrix} 2 & -5 & 0 \\ 4 & 7 & 0 \\ -7 & 25 & 3 \end{bmatrix}$

28. $\begin{bmatrix} 8 & 7 & 6 \\ -4 & 0 & 0 \\ 5 & 1 & 4 \end{bmatrix}$

29. $\begin{bmatrix} 0.1 & 0.2 & 0.3 \\ -0.3 & 0.2 & 0.2 \\ 5 & 4 & 4 \end{bmatrix}$

30. $\begin{bmatrix} -0.4 & 0.4 & 0.3 \\ 0.2 & 0.2 & 0.2 \\ 0.3 & 0.2 & 0.2 \end{bmatrix}$

31. $\begin{bmatrix} x & y & 1 \\ 3 & 1 & 1 \\ -2 & 0 & 1 \end{bmatrix}$

32. $\begin{bmatrix} x & y & 1 \\ -2 & -2 & 1 \\ 1 & 5 & 1 \end{bmatrix}$

In Exercises 33–38, use a graphing utility to evaluate the determinant of the matrix.

33. $\begin{bmatrix} 5 & -3 & 2 \\ 7 & 5 & -7 \\ 0 & 6 & -1 \end{bmatrix}$

34. $\begin{bmatrix} -\frac{1}{2} & -1 & 6 \\ 8 & -\frac{1}{4} & -4 \\ 1 & 2 & 1 \end{bmatrix}$

35. $\begin{bmatrix} 3 & -1 & 2 \\ 1 & -1 & 2 \\ -2 & 3 & 10 \end{bmatrix}$

36. $\begin{bmatrix} \frac{1}{2} & \frac{3}{2} & \frac{1}{2} \\ 4 & 8 & 10 \\ -2 & -6 & 12 \end{bmatrix}$

37. $\begin{bmatrix} 0.2 & 0.8 & -0.3 \\ 0.1 & 0.8 & 0.6 \\ -10 & -5 & 1 \end{bmatrix}$

38. $\begin{bmatrix} 0.4 & 0.3 & 0.3 \\ -0.2 & 0.6 & 0.6 \\ 3 & 1 & 1 \end{bmatrix}$

In Exercises 39–56, use Cramer's Rule to solve the system of linear equations. (If not possible, state the reason.) See Examples 4 and 5.

39. $\begin{aligned} x + 2y &= 5 \\ -x + y &= 1 \end{aligned}$

40. $\begin{aligned} 2x - y &= -10 \\ 3x + 2y &= -1 \end{aligned}$

41. $\begin{aligned} 3x + 4y &= -2 \\ 5x + 3y &= 4 \end{aligned}$

42. $\begin{aligned} 18x + 12y &= 13 \\ 30x + 24y &= 23 \end{aligned}$

43. $\begin{aligned} 20x + 8y &= 11 \\ 12x - 24y &= 21 \end{aligned}$

44. $\begin{aligned} 13x - 6y &= 17 \\ 26x - 12y &= 8 \end{aligned}$

45. $\begin{aligned} -0.4x + 0.8y &= 1.6 \\ 2x - 4y &= 5 \end{aligned}$

46. $\begin{aligned} -0.4x + 0.8y &= 1.6 \\ 0.2x + 0.3y &= 2.2 \end{aligned}$

47. $\begin{aligned} 3u + 6v &= 5 \\ 6u + 14v &= 11 \end{aligned}$

48. $\begin{aligned} 3x_1 + 2x_2 &= 1 \\ 2x_1 + 10x_2 &= 6 \end{aligned}$

49. $\begin{aligned} 4x - y + z &= -5 \\ 2x + 2y + 3z &= 10 \\ 5x - 2y + 6z &= 1 \end{aligned}$

50. $\begin{aligned} 4x - 2y + 3z &= -2 \\ 2x + 2y + 5z &= 16 \\ 8x - 5y - 2z &= 4 \end{aligned}$

51. $\begin{aligned} 3x + 4y + 4z &= 11 \\ 4x - 4y + 6z &= 11 \\ 6x - 6y &= 3 \end{aligned}$

52. $\begin{aligned} 14x_1 - 21x_2 - 7x_3 &= 10 \\ -4x_1 + 2x_2 - 2x_3 &= 4 \\ 56x_1 - 21x_2 + 7x_3 &= 5 \end{aligned}$

53. $\begin{aligned} 3a + 3b + 4c &= 1 \\ 3a + 5b + 9c &= 2 \\ 5a + 9b + 17c &= 4 \end{aligned}$

54. $\begin{aligned} 2x + 3y + 5z &= 4 \\ 3x + 5y + 9z &= 7 \\ 5x + 9y + 17z &= 13 \end{aligned}$

55. $\begin{aligned} 5x - 3y + 2z &= 2 \\ 2x + 2y - 3z &= 3 \\ x - 7y + 8z &= -4 \end{aligned}$

56. $\begin{aligned} 3x + 2y + 5z &= 4 \\ 4x - 3y - 4z &= 1 \\ -8x + 2y + 3z &= 0 \end{aligned}$

In Exercises 57–60, solve the system of linear equations using a graphing utility and Cramer's Rule.

57. $\begin{aligned} -3x + 10y &= 22 \\ 9x - 3y &= 0 \end{aligned}$

58. $\begin{aligned} 3x + 7y &= 3 \\ 7x + 25y &= 11 \end{aligned}$

59. $\begin{aligned} 3x - 2y + 3z &= 8 \\ x + 3y + 6z &= -3 \\ x + 2y + 9z &= -5 \end{aligned}$

60. $\begin{aligned} 6x + 4y - 8z &= -22 \\ -2x + 2y + 3z &= 13 \\ -2x + 2y - z &= 5 \end{aligned}$

In Exercises 61 and 62, solve the equation.

61. $\begin{vmatrix} 5 - x & 4 \\ 1 & 2 - x \end{vmatrix} = 0$

62. $\begin{vmatrix} 4 - x & -2 \\ 1 & 1 - x \end{vmatrix} = 0$

Solving Problems

Area of a Triangle In Exercises 63–70, use a determinant to find the area of the triangle with the given vertices.

63. $(0, 3), (4, 0), (8, 5)$

64. $(2, 0), (0, 5), (6, 3)$

65. $(0, 0), (3, 1), (1, 5)$

66. $(-2, -3), (2, -3), (0, 4)$

67. $(-2, 1), (3, -1), (1, 6)$

68. $(-4, 2), (1, 5), (4, -4)$

69. $\left(0, \frac{1}{2}\right), \left(\frac{5}{2}, 0\right), (4, 3)$

70. $\left(\frac{1}{4}, 0\right), \left(0, \frac{3}{4}\right), (8, -2)$

Area of a Region In Exercises 71–74, find the area of the shaded region of the figure.

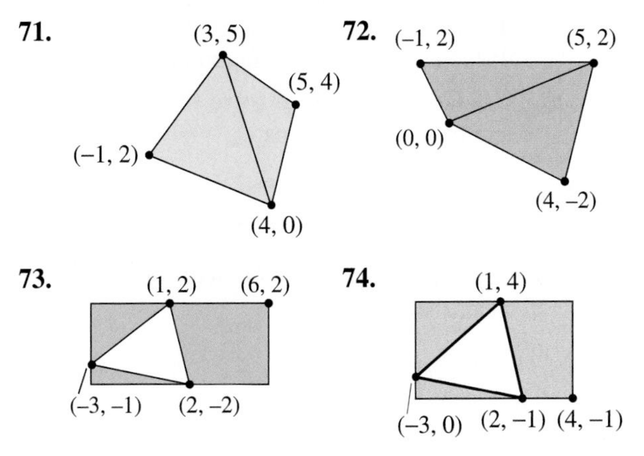

75. *Area of a Region* A large region of forest has been infested with gypsy moths. The region is roughly triangular, as shown in the figure. Approximate the number of square miles in this region.

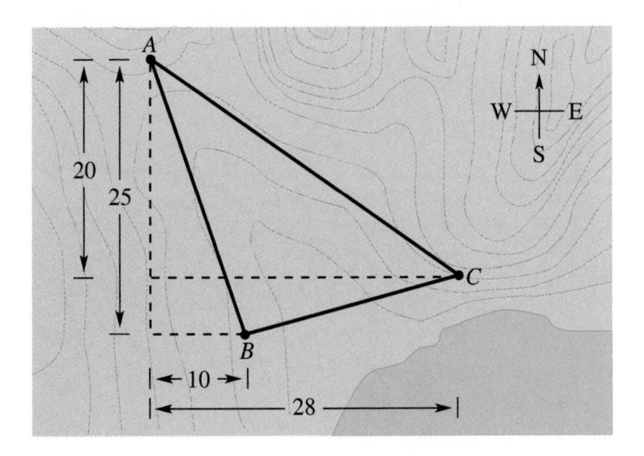

76. *Area of a Region* You have purchased a triangular tract of land, as shown in the figure. How many square feet are there in the tract of land?

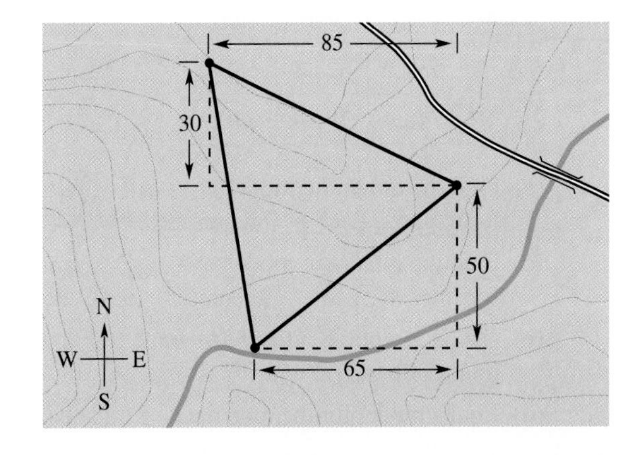

Collinear Points In Exercises 77–82, determine whether the points are collinear.

77. $(-1, 11), (0, 8), (2, 2)$

78. $(-1, -1), (1, 9), (2, 13)$

79. $(-1, -5), (1, -1), (4, 5)$

80. $(-1, 8), (1, 2), (2, 0)$

81. $\left(-2, \frac{1}{3}\right), (2, 1), \left(3, \frac{1}{5}\right)$

82. $\left(0, \frac{1}{2}\right), \left(1, \frac{7}{6}\right), \left(9, \frac{13}{2}\right)$

Equation of a Line In Exercises 83–90, use a determinant to find the equation of the line through the points.

83. $(0, 0), (5, 3)$

84. $(-4, 3), (2, 1)$

85. $(10, 7), (-2, -7)$

86. $(-8, 3), (4, 6)$

87. $\left(-2, \frac{3}{2}\right), (3, -3)$

88. $\left(-\frac{1}{2}, 3\right), \left(\frac{5}{2}, 1\right)$

89. $(2, 3.6), (8, 10)$

90. $(3, 1.6), (5, -2.2)$

Curve-Fitting In Exercises 91–96, use Cramer's Rule to find a quadratic equation $y = ax^2 + bx + c$ whose graph passes through the points. Use a graphing utility to plot the points and graph the model.

91. $(0, 1), (1, -3), (-2, 21)$

92. $(-1, 0), (1, 4), (4, -5)$

93. $(-2, 6), (2, -2), (4, 0)$

94. $(-2, 6), (1, 9), (3, 1)$

95. $(1, -1), (-1, -5), \left(\frac{1}{2}, \frac{1}{4}\right)$

96. $(2, 3), \left(-1, \frac{9}{2}\right), (-2, 9)$

97. *Mathematical Modeling* The table gives the merchandise exports y_1 and the merchandise imports y_2 (in billions of dollars) for the years 1995 through 1997 in the United States. (Source: U.S. International Trade Administration)

Year	1995	1996	1997
y_1	584.7	624.8	689.2
y_2	743.4	791.4	870.7

(a) Find the quadratic model $y_1 = a_1t^2 + b_1t + c_1$ for exports. Let $t = 0$ represent 1990.

(b) Find the quadratic model $y_2 = a_2t^2 + b_2t + c_2$ for imports. Let $t = 0$ represent 1990.

(c) Use a graphing utility to graph the models found in parts (a) and (b).

(d) Find a model for the merchandise trade balance $y_1 - y_2$.

(e) Use a graphing utility to graph the model for the merchandise trade balance. What does the graph show concerning this balance?

98. *Mathematical Modeling* The table gives the agricultural products exports y_1 and the agricultural products imports y_2 (in billions of dollars) for the years 1995 through 1997 in the United States. (Source: U.S. International Trade Administration)

Year	1995	1996	1997
y_1	56.0	60.6	57.1
y_2	29.3	32.6	35.2

Table for 98

(a) Find the quadratic model $y_1 = a_1t^2 + b_1t + c_1$ for the exports. Let $t = 0$ represent 1990.

(b) Find the quadratic model $y_2 = a_2t^2 + b_2t + c_2$ for the imports. Let $t = 0$ represent 1990.

(c) Use a graphing utility to graph the models found in parts (a) and (b).

(d) Find a model for the agricultural products trade balance $y_1 - y_2$.

(e) Use a graphing utility to graph the model for the agricultural products trade balance. What does the graph show concerning this balance?

99. (a) Use Cramer's Rule to solve the following system of linear equations.

$$\begin{aligned} kx + (1 - k)y &= 1 \\ (1 - k)x + \qquad ky &= 3 \end{aligned}$$

(b) For what value(s) of k will the system be inconsistent?

Explaining Concepts

100. Answer parts (g) and (h) of Motivating the Chapter on page 373.

101. Explain the difference between a square matrix and its determinant.

102. Is it possible to find the determinant of a 2×3 matrix? Explain.

103. What is meant by the *minor* of an entry of a square matrix?

104. What conditions must be met in order to use Cramer's Rule to solve a system of linear equations?

CHAPTER SUMMARY

Key Terms

system of equations, *p. 374*
solution of a system of equations, *p. 374*
points of intersection, *p. 375*
consistent system, *p. 376*
dependent system, *p. 376*
inconsistent system, *p. 376*
row-echelon form, *p. 399*
equivalent systems, *p. 400*
Gaussian elimination, *p. 400*
row operations, *p. 400*
matrix, *p. 412*
matrix order, *p. 412*
square matrix, *p. 412*
augmented matrix, *p. 412*
coefficient matrix, *p. 412*
row-equivalent matrices, *p. 414*
minor (of an entry), *p. 426*

Key Concepts

7.1 The method of substitution

1. Solve one of the equations for one variable in terms of the other.
2. Substitute the expression found in Step 1 into the other equation to obtain an equation in one variable.
3. Solve the equation obtained in Step 2.
4. Back-substitute the solution from Step 3 into the expression obtained in Step 1 to find the value of the other variable.
5. Check the solution in the original system.

7.2 The method of elimination

1. Obtain coefficients for x (or y) that differ only in sign by multiplying all of the terms of one or both equations by suitably chosen constants.
2. Add the equations to eliminate one variable and solve the resulting equation.
3. Back-substitute the value obtained in Step 2 into either of the original equations and solve for the other variable.
4. Check your solution in both of the original equations.

7.3 Operations that produce equivalent systems

Each of the following row operations on a system of linear equations produces an equivalent system of linear equations.

1. Interchange two equations.
2. Multiply one of the equations by a nonzero constant.
3. Add a multiple of one of the equations to another equation to replace the latter equation.

7.4 Elementary row operations for matrices

Two matrices are row-equivalent if one can be obtained from the other by a sequence of elementary row operations.

1. Interchange two rows.
2. Multiply a row by a nonzero constant.
3. Add a multiple of a row to another row.

7.4 Gaussian elimination with back-substitution

To use matrices and Gaussian elimination to solve a system of linear equations, use the following steps.

1. Write the augmented matrix of the system of linear equations.
2. Use elementary row operations to rewrite the augmented matrix in row-echelon form.
3. Write the system of linear equations corresponding to the matrix in row-echelon form, and use back-substitution to find the solution.

7.5 Determinant of a matrix

$$\det(A) = \begin{vmatrix} a_1 & b_1 \\ a_2 & b_2 \end{vmatrix} = a_1b_2 - a_2b_1$$

7.5 Expanding by minors

The minor of an entry in a 3×3 matrix is the determinant of the 2×2 matrix that remains after the deletion of the row and column in which the entry occurs.

$$\det(A) = \begin{vmatrix} a_1 & b_1 & c_1 \\ a_2 & b_2 & c_2 \\ a_3 & b_3 & c_3 \end{vmatrix}$$

$$= a_1(\text{minor of } a_1) - b_1(\text{minor of } b_1) + c_1(\text{minor of } c_1)$$

$$= a_1\begin{vmatrix} b_2 & c_2 \\ b_3 & c_3 \end{vmatrix} - b_1\begin{vmatrix} a_2 & c_2 \\ a_3 & c_3 \end{vmatrix} + c_1\begin{vmatrix} a_2 & b_2 \\ a_3 & b_3 \end{vmatrix}$$

REVIEW EXERCISES

Reviewing Skills

7.1 In Exercises 1–4, determine whether each ordered pair is a solution of the system of equations.

1. $3x + 7y = 2$
$5x + 6y = 9$
(a) $(3, 4)$
(b) $(3, -1)$

2. $-2x + 5y = 21$
$9x - y = 13$
(a) $(2, 5)$
(b) $(-2, 4)$

3. $x^2 + y^2 = 41$
$20x + 10y = 30$
(a) $(4, -5)$
(b) $(7, 12)$

4. $x^2 + 2y = 1$
$2x - 5y = 26$
(a) $(3, -4)$
(b) $(2, 8)$

In Exercises 5–14, solve the system graphically.

5. $x + y = 2$
$x - y = 0$

6. $2x = 3(y - 1)$
$y = x$

7. $x - y = 3$
$-x + y = 1$

8. $x + y = -1$
$3x + 2y = 0$

9. $2x - y = 0$
$-x + y = 4$

10. $x = y + 3$
$x = y + 1$

11. $2x + y = 4$
$-4x - 2y = -8$

12. $3x - 2y = 6$
$-6x + 4y = 12$

13. $3x - 2y = -2$
$-5x + 2y = 2$

14. $2x - y = 4$
$-3x + 4y = -11$

In Exercises 15–18, use a graphing utility to solve the system.

15. $5x - 3y = 3$
$2x + 2y = 14$

16. $8x + 5y = 1$
$3x - 4y = 18$

17. $y = x^2 - 4$
$2x - 3y = 11$

18. $y = 9 - x^2$
$2x + y = 6$

In Exercises 19–28, use substitution to solve the system.

19. $2x + 3y = 1$
$x + 4y = -2$

20. $3x - 7y = 10$
$-2x + y = -14$

21. $-5x + 2y = 4$
$10x - 4y = 7$

22. $5x + 2y = 3$
$2x + 3y = 10$

23. $3x - 7y = 5$
$5x - 9y = -5$

24. $24x - 4y = 20$
$6x - y = 5$

25. $y = 5x^2$
$y = -15x - 10$

26. $y^2 = 16x$
$x - y = -4$

27. $x^2 + y^2 = 1$
$x + y = -1$

28. $x^2 + y^2 = 32$
$x + y = 0$

7.2 In Exercises 29–34, use elimination to solve the system in two variables.

29. $x + y = 0$
$2x + y = 0$

30. $4x + y = 1$
$x - y = 4$

31. $2x - y = 2$
$6x + 8y = 39$

32. $3x + 2y = 11$
$x - 3y = -11$

33. $0.2x + 0.3y = 0.14$
$0.4x + 0.5y = 0.20$

34. $0.1x + 0.5y = -0.17$
$-0.3x - 0.2y = -0.01$

7.3 In Exercises 35–40, use elimination to solve the system in three variables.

35. $-x + y + 2z = 1$
$2x + 3y + z = -2$
$5x + 4y + 2z = 4$

36. $2x + 3y + z = 10$
$2x - 3y - 3z = 22$
$4x - 2y + 3z = -2$

37. $x - y - z = 1$
$-2x + y + 3z = -5$
$3x + 4y - z = 6$

38. $-3x + y + 2z = -13$
$-x - y + z = 0$
$2x + 2y - 3z = -1$

39. $\begin{aligned} x \quad\quad - 4z &= 17 \\ -2x + 4y + 3z &= -14 \\ 5x - y + 2z &= -3 \end{aligned}$

40. $\begin{aligned} 2x + 3y - 5z &= 3 \\ -x + 2y \quad\quad &= 3 \\ 3x + 5y + 2z &= 15 \end{aligned}$

7.4 In Exercises 41–48, use matrices and elementary row operations to solve the system.

41. $\begin{aligned} 5x + 4y &= 2 \\ -x + y &= -22 \end{aligned}$

42. $\begin{aligned} 2x - 5y &= 2 \\ 3x - 7y &= 1 \end{aligned}$

43. $\begin{aligned} 0.2x - 0.1y &= 0.07 \\ 0.4x - 0.5y &= -0.01 \end{aligned}$

44. $\begin{aligned} 2x + y &= 0.3 \\ 3x - y &= -1.3 \end{aligned}$

45. $\begin{aligned} x + 2y + 6z &= 4 \\ -3x + 2y - z &= -4 \\ 4x + \quad\quad 2z &= 16 \end{aligned}$

46. $\begin{aligned} -x + 3y - z &= -4 \\ 2x \quad\quad + 6z &= 14 \\ -3x - y + z &= 10 \end{aligned}$

47. $\begin{aligned} 2x_1 + 3x_2 + 3x_3 &= 3 \\ 6x_1 + 6x_2 + 12x_3 &= 13 \\ 12x_1 + 9x_2 - x_3 &= 2 \end{aligned}$

48. $\begin{aligned} -x_1 + 2x_2 + 3x_3 &= 4 \\ 2x_1 - 4x_2 - x_3 &= -13 \\ 3x_1 + 2x_2 - 4x_3 &= -1 \end{aligned}$

7.5 In Exercises 49–54, find the determinant of the matrix.

49. $\begin{bmatrix} 7 & 10 \\ 10 & 15 \end{bmatrix}$

50. $\begin{bmatrix} -3.4 & 1.2 \\ -5 & 2.5 \end{bmatrix}$

51. $\begin{bmatrix} 8 & 6 & 3 \\ 6 & 3 & 0 \\ 3 & 0 & 2 \end{bmatrix}$

52. $\begin{bmatrix} 7 & -1 & 10 \\ -3 & 0 & -2 \\ 12 & 1 & 1 \end{bmatrix}$

53. $\begin{bmatrix} 8 & 3 & 2 \\ 1 & -2 & 4 \\ 6 & 0 & 5 \end{bmatrix}$

54. $\begin{bmatrix} 4 & 0 & 10 \\ 0 & 10 & 0 \\ 10 & 0 & 34 \end{bmatrix}$

In Exercises 55–60, solve the system of linear equations by using Cramer's Rule. (If not possible, state the reason.)

55. $\begin{aligned} 7x + 12y &= 63 \\ 2x + 3y &= 15 \end{aligned}$

56. $\begin{aligned} 12x + 42y &= -17 \\ 30x - 18y &= 19 \end{aligned}$

57. $\begin{aligned} 3x - 2y &= 16 \\ 12x - 8y &= -5 \end{aligned}$

58. $\begin{aligned} 4x + 24y &= 20 \\ -3x + 12y &= -5 \end{aligned}$

59. $\begin{aligned} -x + y + 2z &= 1 \\ 2x + 3y + z &= -2 \\ 5x + 4y + 2z &= 4 \end{aligned}$

60. $\begin{aligned} 2x_1 + x_2 + 2x_3 &= 4 \\ 2x_1 + 2x_2 \quad\quad &= 5 \\ 2x_1 - x_2 + 6x_3 &= 2 \end{aligned}$

In Exercises 61 and 62, create a system of equations having the given solution. (Each problem has many correct answers.)

61. $\left(\frac{2}{3}, -4\right)$

62. $(-10, 12)$

Solving Problems

63. ***Break-Even Analysis*** A small business invests \$25,000 in equipment to produce a product. Each unit of the product costs \$3.75 to produce and is sold for \$5.25. How many units must be sold before the business breaks even?

64. ***Break-Even Analysis*** A small business invests \$33,000 in equipment to produce a product. Each unit of the product costs \$1.70 to produce and is sold for \$5.00. How many units must be sold before the business breaks even?

65. ***Acid Mixture*** One hundred gallons of a 60% acid solution is obtained by mixing a 75% solution with a 50% solution. How many gallons of each solution must be used to obtain the desired mixture?

66. ***Alcohol Mixture*** Fifty gallons of a 90% alcohol solution is obtained by mixing a 100% solution with a 75% solution. How many gallons of each solution must be used to obtain the desired mixture?

67. *Geometry* The perimeter of a rectangle is 480 meters and its length is 150% of its width. Find the dimensions of the rectangle.

68. *Rope Length* Suppose that you must cut a rope that is 128 inches long into two pieces such that one piece is three times as long as the other. Find the length of each piece.

69. *Cassette Tape Sales* You are the manager of a music store and are going over receipts for the previous week's sales. Six hundred and fifty cassette tapes of two different types were sold. One type of cassette sold for \$9.95 and the other sold for \$14.95. The total cassette receipts were \$7717.50. The cash register that was supposed to record the number of each type of cassette sold malfunctioned. Can you recover the information? If so, how many of each type of cassette were sold?

70. *Flying Speeds* Two planes leave Pittsburgh and Philadelphia at the same time, each going to the other city. Because of the wind, one plane flies 25 miles per hour faster than the other. Find the ground speed of each plane if the cities are 275 miles apart and the planes pass one another (at different altitudes) after 40 minutes of flying time.

71. *Flying Speeds* One plane flies 450 miles from City A to City B while a second plane, leaving at the same time, flies from City B to City A. Because of the wind, one plane travels 40 miles per hour faster than the other. Find the ground speed of each plane, if the planes pass one another (at different altitudes) after 50 minutes of flying time.

72. *Investments* An inheritance of \$20,000 is divided among three investments yielding \$1780 in interest per year. The interest rates for the three investments are 7%, 9%, and 11%. Find the amount placed in each investment if the second and third are \$3000 and \$1000 less than the first, respectively.

73. *Number Problem* The sum of three positive numbers is 68. The second number is four greater than the first, and the third is twice the first. Find the three numbers.

Curve-Fitting In Exercises 74 and 75, find a quadratic equation $y = ax^2 + bx + c$ whose graph passes through the given points.

74. $(0, -6), (1, -3), (2, 4)$

75. $(-5, 0), (1, -6), (2, 14)$

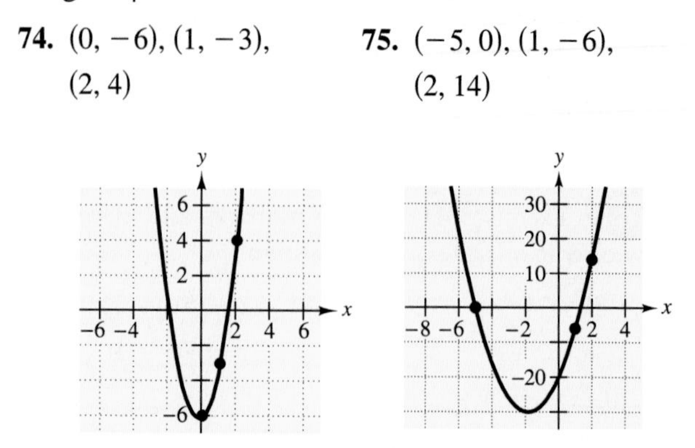

76. *Mathematical Modeling* A child throws a softball over a garage. The location of the eaves and the peak of the roof are given by $(0, 10)$, $(15, 15)$, and $(30, 10)$.

(a) Find the equation $y = ax^2 + bx + c$ for the path of the ball if the ball follows a path 1 foot over the eaves and the peak of the roof.

(b) Use a graphing utility to graph the path of the ball in part (a).

(c) From the graph, estimate how far from the edge of the garage the child was standing if the ball was at a height of 5 feet when it left his hand.

In Exercises 77–80, use a determinant to find the area of the triangle with the given vertices.

77. $(1, 0), (5, 0), (5, 8)$

78. $(-4, 0), (4, 0), (0, 6)$

79. $(1, 2), (4, -5), (3, 2)$

80. $\left(\frac{3}{2}, 1\right), \left(4, -\frac{1}{2}\right), (4, 2)$

In Exercises 81–84, use a determinant to find the equation of the line through the points.

81. $(-4, 0), (4, 4)$

82. $(2, 5), (6, -1)$

83. $\left(-\frac{5}{2}, 3\right), \left(\frac{7}{2}, 1\right)$

84. $(-0.8, 0.2), (0.7, 3.2)$

Chapter Test

Take this test as you would take a test in class. After you are done, check your work against the answers given in the back of the book.

$$2x - 2y = 1$$
$$-x + 2y = 0$$

System for 1

1. Which ordered pair is the solution of the system at the left: $(3, -4)$ or $\left(1, \frac{1}{2}\right)$?

In Exercises 2–13, use the indicated method to solve the system.

2. *Substitution:* $\begin{aligned} 5x - y &= 6 \\ 4x - 3y &= -4 \end{aligned}$

3. *Substitution:* $\begin{aligned} x + y &= 8 \\ x^2 + y &= 10 \end{aligned}$

4. *Graphical:* $\begin{aligned} x - 2y &= -1 \\ 2x + 3y &= 12 \end{aligned}$

5. *Elimination:* $\begin{aligned} 3x - 4y &= -14 \\ -3x + y &= 8 \end{aligned}$

6. *Elimination:* $\begin{aligned} 8x + 3y &= 3 \\ 4x - 6y &= -1 \end{aligned}$

7. *Elimination:* $\begin{aligned} x + 2y - 4z &= 0 \\ 3x + y - 2z &= 5 \\ 3x - y + 2z &= 7 \end{aligned}$

8. *Matrices:* $\begin{aligned} x \qquad\quad - 3z &= -10 \\ -2y + 2z &= 0 \\ x - 2y \qquad &= -7 \end{aligned}$

9. *Matrices:* $\begin{aligned} x - 3y + z &= -3 \\ 3x + 2y - 5z &= 18 \\ y + z &= -1 \end{aligned}$

10. *Cramer's Rule:* $\begin{aligned} 2x - 7y &= 7 \\ 3x + 7y &= 13 \end{aligned}$

11. *Graphical:* $\begin{aligned} x - 2y &= -3 \\ 2x + 3y &= 22 \end{aligned}$

12. *Any Method:* $\begin{aligned} 3x - 2y + z &= 12 \\ x - 3y \qquad &= 2 \\ -3x \qquad - 9z &= -6 \end{aligned}$

13. *Any Method:* $\begin{aligned} 4x + y + 2z &= -4 \\ 3y + z &= 8 \\ -3x + y - 3z &= 5 \end{aligned}$

$$A = \begin{bmatrix} 3 & -2 & 0 \\ -1 & 5 & 3 \\ 2 & 7 & 1 \end{bmatrix}$$

Matrix for 15

$$5x - 8y = 3$$
$$3x + ay = 0$$

System for 16

14. Describe the types of possible solutions of a system of linear equations.

15. Evaluate the determinant of A, as shown at the left.

16. Find the value of a such that the system at the left is inconsistent.

17. Find a system of linear equations with integer coefficients that has the solution $(5, -3)$. (The problem has many correct answers.)

18. Two people share the driving on a 200-mile trip. One person drives four times as far as the other. Write a system of linear equations that models the problem. Find the distance each person drives.

19. Find a quadratic equation $y = ax^2 + bx + c$ whose graph passes through the points $(0, 4)$, $(1, 3)$, and $(2, 6)$.

20. An inheritance of \$25,000 is divided among three investments yielding \$1275 in interest per year. The interest rates for the three investments are 4.5%, 5%, and 8%. Find the amount placed in each investment if the second and third investments are \$4000 and \$10,000 less than the first, respectively.

21. Find the area of the triangle with vertices $(0, 0)$, $(5, 4)$, and $(6, 0)$.

8 Rational Expressions, Equations, and Functions

Daemmrich/The Image Works

In 1997, approximately 17% (103,600) of all boats sold in the United States were canoes. (Source: National Marine Manufacturers Association)

Motivating the Chapter

A Canoe Trip

You and a friend are planning a canoe trip on a river. You want to travel 10 miles upstream and 10 miles back downstream during daylight hours. You know that in still water you are able to paddle the canoe at an average speed of 5 miles per hour. While traveling upstream your average speed will be 5 miles per hour minus the speed of the current, and while traveling downstream your average speed will be 5 miles per hour plus the speed of the current.

See Section 8.3, Exercise 109

a. Write an expression that represents the time it will take to travel upstream in terms of the speed, x (in miles per hour), of the current. Write an expression that represents the time it will take to travel downstream in terms of the speed of the current.

b. Write a function f for the entire time (in hours) of the trip in terms of x.

c. Write the rational function f as a single fraction.

See Section 8.4, Exercise 95

d. What is the speed of the current if the time of the trip is $6\frac{1}{4}$ hours? Explain.

e. If the speed of the current is 4 miles per hour, can you and your friend make the trip during 12 hours of daylight? Explain.

See Section 8.5, Exercise 81

f. Sketch a graph of f. In the context of the problem, what is the domain of f?

g. Does the graph of f have an x-intercept? What would an x-intercept mean in the context of this problem?

h. Find any vertical asymptotes of the graph of f. Interpret the vertical asymptotes in the context of this problem.

8.1 Rational Expressions and Functions

Objectives

1. Determine the domain of a rational function.
2. Simplify a rational expression using the Cancellation Rule.

1 Determine the domain of a rational function.

The Domain of a Rational Function

Algebra may seem simpler when you realize that it consists primarily of a basic set of operations, including addition, subtraction, multiplication, division, factoring, and simplifying, that are applied to different types of algebraic expressions. For instance, in Chapters 5 and 6 you applied these operations to *polynomials*. In this chapter, you will apply these operations to *rational expressions*. Like polynomials, rational expressions can be used to describe functions. Such functions are called **rational functions.**

Study Tip

Every polynomial is also a rational expression because you can consider the denominator to be 1. The domain of every polynomial is the set of all real numbers. That is,

$$\text{Domain} = (-\infty, \infty).$$

▶ Definition of a Rational Function

Let $u(x)$ and $v(x)$ be polynomials. The function

$$f(x) = \frac{u(x)}{v(x)}$$

is a **rational function.** The **domain** of f is the set of all real numbers for which $v(x) \neq 0$.

Technology: Discovery

Use a graphing utility to graph the equation

$$y = \frac{4}{x - 2}.$$

Then use the trace or table feature of the utility to determine the behavior of the graph near $x = 2$. Graph the equations that correspond to parts (b) and (c) of Example 1. How does each of these graphs differ from the graph of $y = 4/(x - 2)$?

Example 1 Finding the Domain of a Rational Function

Find the domains of the following rational functions.

a. $f(x) = \dfrac{4}{x - 2}$ **b.** $g(x) = \dfrac{2x + 5}{8}$ **c.** $h(x) = \dfrac{3x - 1}{x^2 - 2x - 3}$

Solution

a. The denominator is zero when $x - 2 = 0$ or $x = 2$. So, the domain is all real values of x such that $x \neq 2$. In interval notation, you can write the domain as

$$\text{Domain} = (-\infty, 2) \cup (2, \infty).$$

b. The denominator, 8, is never zero, so the domain is the set of *all* real numbers. In interval notation, you can write the domain as

$$\text{Domain} = (-\infty, \infty).$$

c. The denominator is zero when $x^2 - 2x - 3 = 0$. Solving this equation by factoring, you find that the denominator is zero when $x = 3$ or when $x = -1$. So, the domain is all real values of x such that $x \neq 3$ and $x \neq -1$. In interval notation, you can write the domain as

$$\text{Domain} = (-\infty, -1) \cup (-1, 3) \cup (3, \infty).$$

In applications involving rational functions, it is often necessary to place restrictions on the domain besides those that make the denominator zero. To indicate such a restriction, write the domain to the right of the fraction. For instance, the domain of the rational function

$$f(x) = \frac{x^2 + 20}{x + 4}, \qquad x > 0$$

is the set of positive real numbers, as indicated by the inequality $x > 0$. Note that the normal domain of this function would be all real values of x such that $x \neq -4$. However, because "$x > 0$" is listed to the right of the function, the domain is restricted by this inequality.

Study Tip

When a rational function is written, the domain is usually not listed with the function. It is *implied* that the real numbers that make the denominator zero are excluded from the function. For instance, you know to exclude $x = 2$ and $x = -2$ from the function

$$f(x) = \frac{3x + 2}{x^2 - 4}$$

without having to list this information with the function.

Example 2 An Application Involving a Restricted Domain

You have started a small manufacturing business. The initial investment for the business is \$120,000. The cost of each unit that you manufacture is \$15. So, your total cost of producing x units is

$$C = 15x + 120{,}000. \qquad \text{Cost function}$$

Your average cost per unit depends on the number of units produced. For instance, the average cost per unit $\overline{C}$ for producing 100 units is

$$\overline{C} = \frac{15(100) + 120{,}000}{100} \qquad \text{Substitute 100 for } x.$$

$$= \$1215. \qquad \text{Average cost per unit for 100 units}$$

The average cost per unit decreases as the number of units increases. For instance, the average cost per unit $\overline{C}$ for producing 1000 units is

$$\overline{C} = \frac{15(1000) + 120{,}000}{1000} \qquad \text{Substitute 1000 for } x.$$

$$= \$135. \qquad \text{Average cost per unit for 1000 units}$$

In general, the average cost of producing x units is

$$\overline{C} = \frac{15x + 120{,}000}{x}. \qquad \text{Average cost per unit for } x \text{ units}$$

What is the domain of this rational function?

Solution

If you were considering this function from only a mathematical point of view, you would say that the domain is all real values of x such that $x \neq 0$. However, because this function is a mathematical model representing a real-life situation, you must consider which values of x make sense in real life. For this model, the variable x represents the number of units that you produce. Assuming that you cannot produce a fractional number of units, you conclude that the domain is the set of positive integers. That is,

$$\text{Domain} = \{1, 2, 3, 4, . . .\}.$$

2 Simplify a rational expression using the Cancellation Rule.

Simplifying Rational Expressions

As with numerical fractions, a rational expression is said to be **simplified** or **in reduced form** if its numerator and denominator have no factors in common (other than ± 1). To simplify fractions, you can apply the following rule.

The Granger Collection

Karl Weierstrass
(1815–1897)

In the 19th century, mathematicians were expanding their knowledge of calculus and laying the foundation for complex numbers. At that time, the properties of real numbers had not been finalized. Teaching at the University of Berlin, Weierstrass recognized the need for a logical foundation for the real number system. His work contributed much to the formal real number system that forms the foundation of modern algebra.

> ### ▶ Cancellation Rule for Fractions
>
> Let u, v, and w represent numbers, variables, or algebraic expressions such that $v \neq 0$ and $w \neq 0$. Then the following Cancellation Rule is valid.
>
> $$\frac{u\cancel{w}}{v\cancel{w}} = \frac{u}{v}$$

Be sure you see that this Cancellation Rule allows you to cancel only factors, not terms. For instance, consider the following.

$$\frac{\cancel{2} \cdot 2}{\cancel{2}(x + 5)}$$ You can cancel common factor 2.

$$\frac{3 + x}{3 + 2x}$$ You cannot cancel common term 3.

Using the Cancellation Rule to simplify a rational expression requires two steps: (1) completely factor the numerator and denominator and (2) apply the Cancellation Rule to cancel any *factors* that are common to both the numerator and denominator. So, your success in simplifying rational expressions actually lies in your ability to *completely factor* the polynomials in both the numerator and denominator.

Example 3 Simplifying a Rational Expression

Simplify $\dfrac{2x^3 - 6x}{6x^2}$.

Solution

First note that the domain of the rational expression is all real values of x such that $x \neq 0$. Then, completely factor both the numerator and denominator.

$$\frac{2x^3 - 6x}{6x^2} = \frac{2x(x^2 - 3)}{2x(3x)}$$ Factor numerator and denominator.

$$= \frac{\cancel{2x}(x^2 - 3)}{\cancel{2x}(3x)}$$ Cancel common factor $2x$.

$$= \frac{x^2 - 3}{3x}$$ Simplified form

In simplified form, the domain of the rational expression is the same as that of the original expression, all real values of x such that $x \neq 0$.

Example 4 Simplifying a Rational Expression

Simplify $\frac{x^2 + 2x - 15}{3x - 9}$.

Solution

The domain of the rational expression is all real values of x such that $x \neq 3$.

$$\frac{x^2 + 2x - 15}{3x - 9} = \frac{(x + 5)(x - 3)}{3(x - 3)} \qquad \text{Factor numerator and denominator.}$$

$$= \frac{(x + 5)\cancel{(x - 3)}}{3\cancel{(x - 3)}} \qquad \text{Cancel common factor } (x - 3).$$

$$= \frac{x + 5}{3}, \quad x \neq 3 \qquad \text{Simplified form}$$

Technology: Tip

Use the table feature of your graphing utility to compare the two functions in Example 4.

$$y_1 = \frac{x^2 + 2x - 15}{3x - 9}$$

$$y_2 = \frac{x + 5}{3}$$

Set the increment value of the table to 1 and compare the values at $x = 0, 1, 2, 3, 4$, and 5. Next set the increment value to 0.1 and compare the values at $x = 2.8, 2.9, 3.0, 3.1$, and 3.2. From the table you can see that the functions differ only at $x = 3$. This shows why $x \neq 3$ must be written as part of the reduced form of the original expression.

Canceling common factors from the numerator and denominator of a rational expression can change its domain. For instance, in Example 4 the domain of the original expression is all real values of x such that $x \neq 3$. So, the original expression is equal to the simplified expression for all real numbers *except* 3.

Example 5 Simplifying a Rational Expression

Simplify $\frac{x^3 - 16x}{x^2 - 2x - 8}$.

Solution

The domain of the rational expression is all real values of x such that $x \neq -2$ and $x \neq 4$.

$$\frac{x^3 - 16x}{x^2 - 2x - 8} = \frac{x(x^2 - 16)}{(x + 2)(x - 4)} \qquad \text{Partially factor.}$$

$$= \frac{x(x + 4)(x - 4)}{(x + 2)(x - 4)} \qquad \text{Factor completely.}$$

$$= \frac{x(x + 4)\cancel{(x - 4)}}{(x + 2)\cancel{(x - 4)}} \qquad \text{Cancel common factor } (x - 4).$$

$$= \frac{x(x + 4)}{x + 2}, \quad x \neq 4 \qquad \text{Simplified form}$$

In this text, when simplifying a rational expression, we follow the convention of listing *by the simplified expression* all values of x that must be specifically excluded from the domain in order to make the domains of the simplified and original expressions agree. For instance, in Example 5 the restriction $x \neq 4$ must be listed with the simplified expression in order to make the two domains agree. Note that the value of -2 is excluded from both domains, so it is not necessary to list this value.

Example 6 Simplification Involving a Change of Sign

Simplify $\frac{2x^2 - 9x + 4}{12 + x - x^2}$.

Study Tip

Be sure to *completely* factor the numerator and denominator of a rational expression before concluding that there is no common factor. This may involve a change in sign to see if further reduction is possible. Note that the Distributive Property allows you to write $(b - a)$ as $-(a - b)$. Watch for this in Example 6.

Solution

The domain of the rational expression is all real values of x such that $x \neq -3$ and $x \neq 4$.

$$\frac{2x^2 - 9x + 4}{12 + x - x^2} = \frac{(2x - 1)(x - 4)}{(4 - x)(3 + x)}$$ Factor numerator and denominator.

$$= \frac{(2x - 1)(x - 4)}{-(x - 4)(3 + x)}$$ $(4 - x) = -(x - 4)$

$$= \frac{(2x - 1)\cancel{(x - 4)}}{-\cancel{(x - 4)}(3 + x)}$$ Cancel common factor $(x - 4)$.

$$= -\frac{2x - 1}{3 + x}, \quad x \neq 4$$ Simplified form

The simplified form is equivalent to the original expression for all values of x except 4. Note that -3 is excluded from the domains of both the original and simplified expressions.

In Example 6, be sure you see that when dividing the numerator and denominator by the common factor of $(x - 4)$, you keep the negative sign. In the simplified form of the fraction, we usually like to move the negative sign out in front of the fraction. However, this is a personal preference. All of the following forms are legitimate.

$$-\frac{2x - 1}{3 + x} = \frac{-(2x - 1)}{3 + x} = \frac{2x - 1}{-3 - x} = \frac{2x - 1}{-(3 + x)}$$

In the next three examples, the Cancellation Rule is used to simplify rational expressions that involve more than one variable.

Example 7 A Rational Expression Involving Two Variables

Simplify $\frac{3xy + y^2}{2y}$.

Solution

The domain of the rational expression is all real values of y such that $y \neq 0$.

$$\frac{3xy + y^2}{2y} = \frac{y(3x + y)}{2y}$$ Factor numerator and denominator.

$$= \frac{\cancel{y}(3x + y)}{2\cancel{y}}$$ Cancel common factor y.

$$= \frac{3x + y}{2}, \quad y \neq 0$$ Simplified form

Example 8 A Rational Expression Involving Two Variables

Simplify $\dfrac{2x^2 + 2xy - 4y^2}{5x^3 - 5xy^2}$.

Solution

The domain of the rational expression is all real numbers such that $x \neq 0$ and $x \neq \pm y$.

$$\frac{2x^2 + 2xy - 4y^2}{5x^3 - 5xy^2} = \frac{2(x - y)(x + 2y)}{5x(x - y)(x + y)} \qquad \text{Factor numerator and denominator.}$$

$$= \frac{2\cancel{(x - y)}(x + 2y)}{5x\cancel{(x - y)}(x + y)} \qquad \text{Cancel common factor } (x - y).$$

$$= \frac{2(x + 2y)}{5x(x + y)}, \quad x \neq y \qquad \text{Simplified form}$$

Example 9 A Rational Expression Involving Two Variables

Simplify $\dfrac{4x^2y - y^3}{2x^2y - xy^2}$.

Solution

The domain of the rational expression is all real numbers such that $x \neq 0$, $y \neq 0$, and $y \neq 2x$.

$$\frac{4x^2y - y^3}{2x^2y - xy^2} = \frac{(2x - y)(2x + y)y}{(2x - y)xy} \qquad \text{Factor numerator and denominator.}$$

$$= \frac{\cancel{(2x - y)}(2x + y)\cancel{y}}{\cancel{(2x - y)}x\cancel{y}} \qquad \text{Cancel common factors } (2x - y) \text{ and } y.$$

$$= \frac{2x + y}{x}, \quad y \neq 0,\ y \neq 2x \qquad \text{Simplified form}$$

As you study the examples and work the exercises in this chapter, keep in mind that you are *rewriting expressions in simpler forms.* You are not solving equations. Equal signs are used in the steps of the simplification process only to indicate that the new form of the expression is *equivalent* to the previous one.

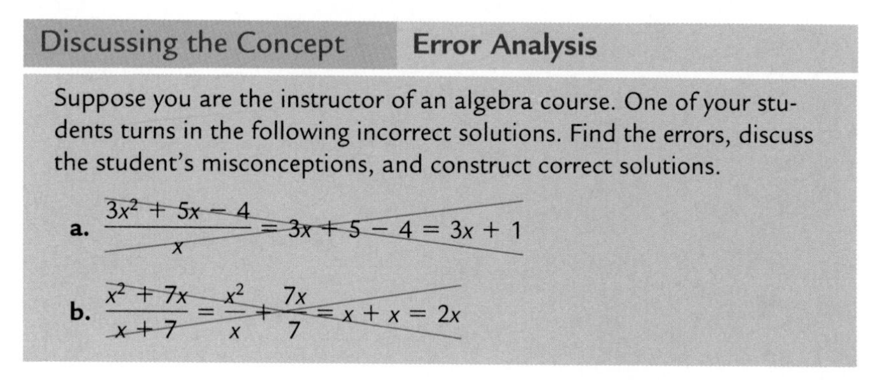

Discussing the Concept **Error Analysis**

Suppose you are the instructor of an algebra course. One of your students turns in the following incorrect solutions. Find the errors, discuss the student's misconceptions, and construct correct solutions.

a. $\dfrac{3x^2 + 5x - 4}{x} = 3x + 5 - 4 = 3x + 1$

b. $\dfrac{x^2 + 7x}{x + 7} = \dfrac{x^2}{x} + \dfrac{7x}{7} = x + x = 2x$

8.1 Exercises

Integrated Review *Concepts, Skills, and Problem Solving*

Keep mathematically in shape by doing these exercises *before* the problems of this section.

Properties and Definitions

1. Define the slope of the line through the points (x_1, y_1) and (x_2, y_2).
2. Make a statement about the slope m of the line for each of the following.
   (a) The line rises from left to right.
   (b) The line falls from left to right.
   (c) The line is horizontal.
   (d) The line is vertical.

Simplifying Expressions

In Exercises 3–8, simplify the expression.

3. $2(x + 5) - 3 - (2x - 3)$
4. $3(y + 4) + 5 - (3y + 5)$
5. $4 - 2[3 + 4(x + 1)]$
6. $5x + x[3 - 2(x - 3)]$
7. $\left(\frac{5}{x^2}\right)^2$
8. $-\frac{(2u^2v)^2}{-3uv^2}$

Problem Solving

9. Determine the number of gallons of a 30% solution that must be mixed with a 60% solution to obtain 20 gallons of a 40% solution.
10. A suit sells for \$375 during a 25% storewide clearance sale. What was the original price of the suit?

Developing Skills

In Exercises 1–20, find the domain of the expression. See Example 1.

1. $\frac{5}{x - 8}$
2. $\frac{9}{x - 13}$
3. $\frac{7x}{x + 4}$
4. $\frac{2y}{6 - y}$
5. $\frac{x^2 + 9}{4}$
6. $\frac{y^2 - 3}{7}$
7. $x^4 - 2x^2 - 5$
8. $t^3 - 4t^2 + 1$
9. $\frac{x}{x^2 + 4}$
10. $\frac{4x}{x^2 + 16}$
11. $\frac{y - 4}{y(y + 3)}$
12. $\frac{z + 2}{z(z - 4)}$
13. $\frac{5t}{t^2 - 16}$
14. $\frac{x}{x^2 - 4}$
15. $\frac{y + 5}{y^2 - 3y}$
16. $\frac{t - 6}{t^2 + 5t}$
17. $\frac{8x}{x^2 - 5x + 6}$
18. $\frac{3t}{t^2 - 2t - 3}$
19. $\frac{u^2}{3u^2 - 2u - 5}$
20. $\frac{y + 5}{4y^2 - 5y - 6}$

In Exercises 21–26, evaluate the function as indicated. If not possible, state the reason.

21. $f(x) = \frac{4x}{x + 3}$
    (a) $f(1)$ (b) $f(-2)$
    (c) $f(-3)$ (d) $f(0)$
22. $f(x) = \frac{x - 10}{4x}$
    (a) $f(10)$ (b) $f(0)$
    (c) $f(-2)$ (d) $f(12)$
23. $g(x) = \frac{x^2 - 4x}{x^2 - 9}$
    (a) $g(0)$ (b) $g(4)$
    (c) $g(3)$ (d) $g(-3)$
24. $g(t) = \frac{t - 2}{2t - 5}$
    (a) $g(2)$ (b) $g\left(\frac{5}{2}\right)$
    (c) $g(-2)$ (d) $g(0)$
25. $h(s) = \frac{s^2}{s^2 - s - 2}$
    (a) $h(10)$ (b) $h(0)$
    (c) $h(-1)$ (d) $h(2)$

26. $f(x) = \dfrac{x^3 + 1}{x^2 - 6x + 9}$

(a) $f(-1)$ (b) $f(3)$
(c) $f(-2)$ (d) $f(2)$

In Exercises 27–32, describe the domain. See Example 2.

27. *Geometry* A rectangle of length x inches has an area of 500 square inches. The perimeter P of the rectangle is given by

$$P = 2\left(x + \frac{500}{x}\right).$$

28. *Cost* The cost C in millions of dollars for the government to seize $p\%$ of a certain illegal drug as it enters the country is given by

$$C = \frac{528p}{100 - p}.$$

29. *Inventory Cost* The inventory cost I when x units of a product are ordered from a supplier is given by

$$I = \frac{0.25x + 2000}{x}.$$

30. *Average Cost* The average cost $\overline{C}$ for a manufacturer to produce x units of a product is given by

$$\overline{C} = \frac{1.35x + 4570}{x}.$$

31. *Pollution Removal* The cost C in dollars of removing $p\%$ of the air pollutants in the stack emission of a utility company is given by the rational function

$$C = \frac{80{,}000p}{100 - p}.$$

32. *Video Rental* The average cost of a movie video rental $\overline{M}$ when you consider the cost of purchasing a video cassette recorder and renting x movie videos at \$2.49 per movie is

$$\overline{M} = \frac{150 + 2.49x}{x}.$$

In Exercises 33–40, complete the statement.

33. $\dfrac{5(\quad)}{6(x + 3)} = \dfrac{5}{6}, \quad x \neq -3$

34. $\dfrac{7(\quad)}{15(x - 10)} = \dfrac{7}{15}, \quad x \neq 10$

35. $\dfrac{3x(x + 16)^2}{2(\quad)} = \dfrac{x}{2}, \quad x \neq -16$

36. $\dfrac{25x^2(x - 10)}{12(\quad)} = \dfrac{5x}{12}, \quad x \neq 10, \quad x \neq 0$

37. $\dfrac{(x + 5)(\quad)}{3x^2(x - 2)} = \dfrac{x + 5}{3x}, \quad x \neq 2$

38. $\dfrac{(3y - 7)(\quad)}{y^2 - 4} = \dfrac{3y - 7}{y + 2}, \quad y \neq 2$

39. $\dfrac{8x(\quad)}{x^2 - 3x - 10} = \dfrac{8x}{x - 5}, \quad x \neq -2$

40. $\dfrac{(3 - z)(\quad)}{z^3 + 2z^2} = \dfrac{3 - z}{z^2}, \quad z \neq -2$

In Exercises 41–78, simplify the expression. See Examples 3–9.

41. $\dfrac{5x}{25}$

42. $\dfrac{32y}{24}$

43. $\dfrac{12y^2}{2y}$

44. $\dfrac{15z^3}{15z^3}$

45. $\dfrac{18x^2y}{15xy^4}$

46. $\dfrac{16y^2z^2}{60y^5z}$

47. $\dfrac{3x^2 - 9x}{12x^2}$

48. $\dfrac{8x^3 + 4x^2}{20x}$

49. $\dfrac{x^2(x - 8)}{x(x - 8)}$

50. $\dfrac{a^2b(b - 3)}{b^3(b - 3)^2}$

51. $\dfrac{2x - 3}{4x - 6}$

52. $\dfrac{y^2 - 81}{2y - 18}$

53. $\dfrac{5 - x}{3x - 15}$

54. $\dfrac{x^2 - 36}{6 - x}$

55. $\dfrac{a + 3}{a^2 + 6a + 9}$

56. $\dfrac{u^2 - 12u + 36}{u - 6}$

57. $\dfrac{x^2 - 7x}{x^2 - 14x + 49}$

58. $\dfrac{z^2 + 22z + 121}{3z + 33}$

59. $\dfrac{y^3 - 4y}{y^2 + 4y - 12}$

60. $\dfrac{x^2 - 7x}{x^2 - 4x - 21}$

61. $\dfrac{x^3 - 4x}{x^2 - 5x + 6}$

62. $\dfrac{x^4 - 25x^2}{x^2 + 2x - 15}$

63. $\dfrac{3x^2 - 7x - 20}{12 + x - x^2}$

64. $\dfrac{2x^2 + 3x - 5}{7 - 6x - x^2}$

65. $\dfrac{2x^2 + 19x + 24}{2x^2 - 3x - 9}$

66. $\dfrac{2y^2 + 13y + 20}{2y^2 + 17y + 30}$

67. $\dfrac{15x^2 + 7x - 4}{25x^2 - 16}$

68. $\dfrac{56z^2 - 3z - 20}{49z^2 - 16}$

69. $\dfrac{3xy^2}{xy^2 + x}$

70. $\dfrac{x + 3x^2y}{3xy + 1}$

71. $\dfrac{y^2 - 64x^2}{5(3y + 24x)}$

72. $\dfrac{x^2 - 25z^2}{x + 5z}$

73. $\dfrac{5xy + 3x^2y^2}{xy^3}$

74. $\dfrac{4u^2v - 12uv^2}{18uv}$

75. $\dfrac{u^2 - 4v^2}{u^2 + uv - 2v^2}$

76. $\dfrac{x^2 + 4xy}{x^2 - 16y^2}$

77. $\dfrac{3m^2 - 12n^2}{m^2 + 4mn + 4n^2}$

78. $\dfrac{x^2 + xy - 2y^2}{x^2 + 3xy + 2y^2}$

In Exercises 79–82, explain how you can show that the two expressions are not equivalent.

79. $\dfrac{x - 4}{4} \neq x - 1$

80. $\dfrac{x - 4}{x} \neq -4$

81. $\dfrac{3x + 2}{4x + 2} \neq \dfrac{3}{4}$

82. $\dfrac{1 - x}{2 - x} \neq \dfrac{1}{2}$

In Exercises 83 and 84, complete the table. What can you conclude?

83.

x	-2	-1	0	1	2	3	4
$\dfrac{x^2 - x - 2}{x - 2}$							
$x + 1$							

84.

x	-2	-1	0	1	2	3	4
$\dfrac{x^2 + 5x}{x}$							
$x + 5$							

Solving Problems

Geometry In Exercises 85 and 86, find the ratio of the area of the shaded portion to the total area of the figure.

85.

86.

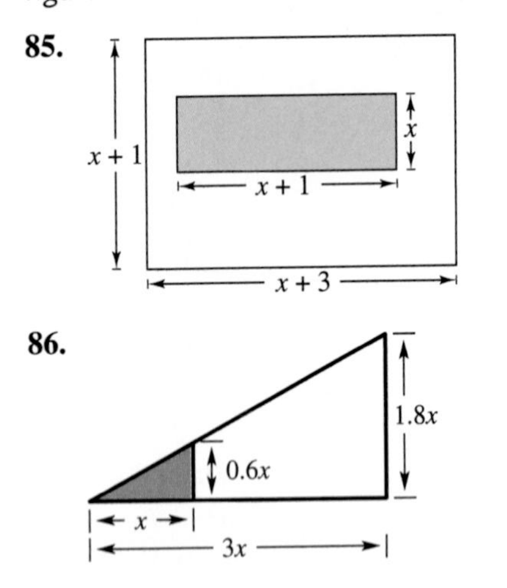

87. *Average Cost* A machine shop has a setup cost of \$2500 for the production of a new product. The cost for labor and material in producing each unit is \$9.25.

(a) Write the total cost C as a function of x, the number of units produced.

(b) Write the average cost per unit $\overline{C} = C/x$ as a function of x, the number of units produced.

(c) Determine the domain of the function in part (b).

(d) Find the value of $\overline{C}(100)$.

88. *Average Cost* A greeting card company has an initial investment of \$60,000. The cost of producing one dozen cards is \$6.50.

(a) Write the total cost C as a function of x, the number of cards in dozens produced.

(b) Write the average cost per dozen $\overline{C} = C/x$ as a function of x, the number of cards in dozens produced.

(c) Determine the domain of the function in part (b).

(d) Find the value of $\overline{C}(11{,}000)$.

89. *Distance Traveled* A van starts on a trip and travels at an average speed of 45 miles per hour. Three hours later, a car starts on the same trip and travels at an average speed of 60 miles per hour.

(a) Find the distance each vehicle has traveled when the car has been on the road for t hours.

(b) Use the result of part (a) to write the distance between the van and the car as a function of t.

(c) Write the ratio of the distance the car has traveled to the distance the van has traveled as a function of t.

90. *Distance Traveled* A car starts on a trip and travels at an average speed of 55 miles per hour. Two hours later, a second car starts on the same trip and travels at an average speed of 65 miles per hour.

(a) Find the distance each vehicle has traveled when the second car has been on the road for t hours.

(b) Use the result of part (a) to write the distance between the first car and the second car as a function of t.

(c) Write the ratio of the distance the second car has traveled to the distance the first car has traveled as a function of t.

91. *Geometry* One swimming pool is circular and another is rectangular. The rectangular pool's width is three times its depth. Its length is 6 feet more than its width. The circular pool has a diameter that is twice the width of the rectangular pool, and it is 2 feet deeper. Find the ratio of the circular pool's volume to the rectangular pool's volume.

92. *Geometry* A circular pool has a radius five times its depth. A rectangular pool has the same depth as the circular pool. Its width is 4 feet more than three times its depth and its length is 2 feet less than six times its depth. Find the ratio of the rectangular pool's volume to the circular pool's volume.

Cost of Medicare In Exercises 93 and 94, use the following polynomial models, which give the total annual cost of Medicare C (in billions of dollars) and the U.S. population enrolled in Medicare P (in millions) from 1990 through 1996 (see figures).

$C = 107.30 + 15.09t$

$P = 34.26 + 0.65t$

In these models, t represents the year, with $t = 0$ corresponding to 1990. (Source: U.S. Health Care Financing Administration)

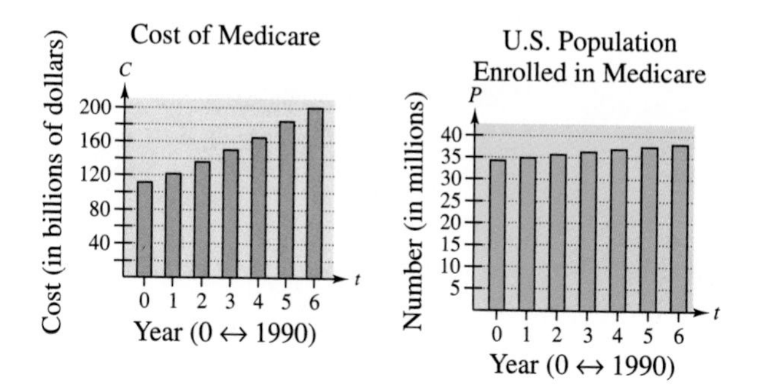

93. Find a rational model that represents the average cost of Medicare per person enrolled during the years 1990 to 1996.

94. Use the model found in Exercise 93 to complete the table showing the average cost of Medicare per person enrolled.

Year	1990	1991	1992	1993
Average cost				

Year	1994	1995	1996
Average cost			

Explaining Concepts

95. Define the term *rational expression.*

96. Give an example of a rational function whose domain is the set of all real numbers.

97. How do you determine whether a rational expression is in simplified form?

98. Can you cancel common terms from the numerator and denominator of a rational expression? Explain.

99. Explain the error in the following.

$$\frac{2x^2}{x^2+4} = \frac{2x^2}{x^2+4} = \frac{2}{1+4} = \frac{2}{5}$$

100. Is the following statement true? Explain.

$$\frac{6x-5}{5-6x} = -1$$

8.2 Multiplying and Dividing Rational Expressions

Objectives

1 Multiply rational expressions and simplify.

2 Divide rational expressions and simplify.

3 Simplify a complex fraction.

1 Multiply rational expressions and simplify.

Multiplying Rational Expressions

The rule for multiplying rational expressions is the same as the rule for multiplying numerical fractions.

$$\frac{3}{4} \cdot \frac{7}{6} = \frac{21}{24} = \frac{\cancel{3} \cdot 7}{\cancel{3} \cdot 8} = \frac{7}{8}$$

That is, you *multiply numerators, multiply denominators, and write the new fraction in simplified form.*

▶ Multiplying Rational Expressions

Let u, v, w, and z be real numbers, variables, or algebraic expressions such that $v \neq 0$ and $z \neq 0$. Then the product of u/v and w/z is given by

$$\frac{u}{v} \cdot \frac{w}{z} = \frac{uw}{vz}.$$

In order to recognize common factors, write the numerators and denominators in factored form, as demonstrated in Example 1.

Example 1 Multiplying Rational Expressions

Multiply the rational expressions.

$$\frac{4x^3y}{3xy^4} \cdot \frac{-6x^2y^2}{10x^4}$$

Solution

$$\frac{4x^3y}{3xy^4} \cdot \frac{-6x^2y^2}{10x^4} = \frac{(4x^3y) \cdot (-6x^2y^2)}{(3xy^4) \cdot (10x^4)}$$ Multiply numerators and denominators.

$$= \frac{-24x^5y^3}{30x^5y^4}$$ Simplify.

$$= \frac{-4(\cancel{6})(\cancel{x^5})(\cancel{y^3})}{5(\cancel{6})(\cancel{x^5})(\cancel{y^3})(y)}$$ Factor and cancel.

$$= -\frac{4}{5y}, \quad x \neq 0$$ Simplified form

Example 2 Multiplying Rational Expressions

Multiply the rational expressions.

$$\frac{x}{5x^2 - 20x} \cdot \frac{x - 4}{2x^2 + x - 3}$$

Solution

$$\frac{x}{5x^2 - 20x} \cdot \frac{x - 4}{2x^2 + x - 3}$$

$$= \frac{x \cdot (x - 4)}{(5x^2 - 20x) \cdot (2x^2 + x - 3)}$$ Multiply numerators and denominators.

$$= \frac{x(x - 4)}{5x(x - 4)(x - 1)(2x + 3)}$$ Factor.

$$= \frac{\cancel{x}\cancel{(x - 4)}}{5\cancel{x}\cancel{(x - 4)}(x - 1)(2x + 3)}$$ Cancel common factors.

$$= \frac{1}{5(x - 1)(2x + 3)}, \; x \neq 0, \; x \neq 4$$ Simplified form

Technology: Tip

You can use a graphing utility to check your results when multiplying rational expressions. For instance, in Example 3, try graphing the equations

$$y_1 = \frac{4x^2 - 4x}{x^2 + 2x - 3} \cdot \frac{x^2 + x - 6}{4x}$$

and

$$y_2 = x - 2$$

on the same screen. If the two graphs coincide, as shown below, you can conclude that the two functions are equivalent.

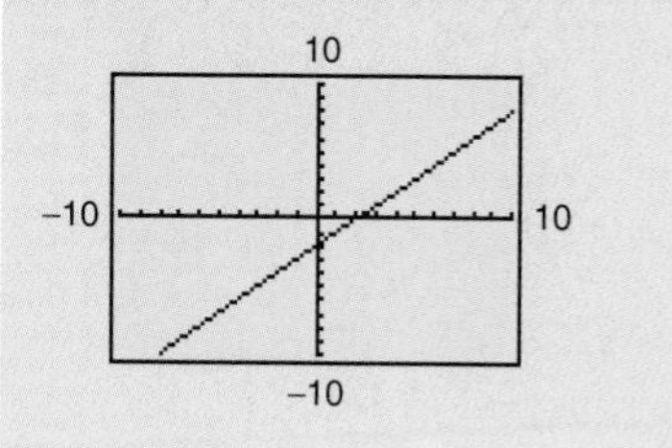

Example 3 Multiplying Rational Expressions

Multiply the rational expressions.

$$\frac{4x^2 - 4x}{x^2 + 2x - 3} \cdot \frac{x^2 + x - 6}{4x}$$

Solution

$$\frac{4x^2 - 4x}{x^2 + 2x - 3} \cdot \frac{x^2 + x - 6}{4x}$$

$$= \frac{4x(x - 1)(x + 3)(x - 2)}{(x - 1)(x + 3)(4x)}$$ Factor and multiply.

$$= \frac{\cancel{4x}\cancel{(x - 1)}\cancel{(x + 3)}(x - 2)}{\cancel{(x - 1)}\cancel{(x + 3)}\cancel{(4x)}}$$ Cancel common factors.

$$= x - 2, \; x \neq 0, \; x \neq 1, \; x \neq -3$$ Simplified form

The rule for multiplying fractions can be extended to cover products involving expressions that are not in fractional form. To do this, rewrite the nonfractional expression as a fraction whose denominator is 1. Here is a simple example.

$$\frac{x + 3}{x - 2} \cdot (5x) = \frac{x + 3}{x - 2} \cdot \frac{5x}{1}$$

$$= \frac{(x + 3)(5x)}{x - 2}$$

$$= \frac{5x(x + 3)}{x - 2}$$

In the next example, note how to divide out a factor that differs only in sign. The Distributive Property is used in the step in which $(y - x)$ is rewritten as $(-1)(x - y)$.

Example 4 Multiplying Rational Expressions

Multiply the rational expressions.

$$\frac{x - y}{y^2 - x^2} \cdot \frac{x^2 - xy - 2y^2}{3x - 6y}$$

Solution

$$\frac{x - y}{y^2 - x^2} \cdot \frac{x^2 - xy - 2y^2}{3x - 6y}$$

$$= \frac{x - y}{(y + x)(y - x)} \cdot \frac{(x - 2y)(x + y)}{3(x - 2y)} \qquad \text{Factor.}$$

$$= \frac{x - y}{(y + x)(-1)(x - y)} \cdot \frac{(x - 2y)(x + y)}{3(x - 2y)} \qquad \text{Factor: } (y - x) = -1(x - y).$$

$$= \frac{(x - y)(x - 2y)(x + y)}{(y + x)(-1)(x - y)(3)(x - 2y)} \qquad \text{Multiply.}$$

$$= \frac{\cancel{(x - y)}\cancel{(x - 2y)}\cancel{(x + y)}}{\cancel{(x + y)}(-1)\cancel{(x - y)}(3)\cancel{(x - 2y)}} \qquad \text{Cancel common factors.}$$

$$= -\frac{1}{3}, \; x \neq y, \; x \neq -y, \; x \neq 2y \qquad \text{Simplified form}$$

The rule for multiplying rational expressions can be extended to cover products of three or more fractions, as shown in Example 5.

Example 5 Multiplying Three Rational Expressions

Multiply the rational expressions.

$$\frac{x^2 - 3x + 2}{x + 2} \cdot \frac{3x}{x - 2} \cdot \frac{2x + 4}{x^2 - 5x}$$

Solution

$$\frac{x^2 - 3x + 2}{x + 2} \cdot \frac{3x}{x - 2} \cdot \frac{2x + 4}{x^2 - 5x}$$

$$= \frac{(x - 1)(x - 2)(3)(x)(2)(x + 2)}{(x + 2)(x - 2)(x)(x - 5)} \qquad \text{Factor and multiply.}$$

$$= \frac{(x - 1)\cancel{(x - 2)}(3)\cancel{(x)}(2)\cancel{(x + 2)}}{\cancel{(x + 2)}\cancel{(x - 2)}\cancel{(x)}(x - 5)} \qquad \text{Cancel common factors.}$$

$$= \frac{6(x - 1)}{x - 5}, \; x \neq 0, \; x \neq 2, \; x \neq -2 \qquad \text{Simplified form}$$

---

2 Divide rational expressions and simplify.

Dividing Rational Expressions

To divide two rational expressions, multiply the first fraction by the *reciprocal* of the second. That is, simply *invert the divisor and multiply.* For instance, to perform the following division

$$\frac{x}{x+3} \div \frac{4}{x-1}$$

invert the fraction $4/(x-1)$ and multiply, as follows.

$$\frac{x}{x+3} \div \frac{4}{x-1} = \frac{x}{x+3} \cdot \frac{x-1}{4}$$ Invert divisor and multiply.

$$= \frac{x(x-1)}{(x+3)(4)}$$ Multiply numerators and denominators.

$$= \frac{x(x-1)}{4(x+3)}$$ Simplify.

> ### ▶ Dividing Rational Expressions
>
> Let u, v, w, and z be real numbers, variables, or algebraic expressions such that $v \neq 0$, $w \neq 0$, and $z \neq 0$. The quotient of u/v and w/z is
>
> $$\frac{u}{v} \div \frac{w}{z} = \frac{u}{v} \cdot \frac{z}{w} = \frac{uz}{vw}.$$

Example 6 Dividing Rational Expressions

Perform the division.

$$\frac{2x}{3x-12} \div \frac{x^2-2x}{x^2-6x+8}$$

Solution

$$\frac{2x}{3x-12} \div \frac{x^2-2x}{x^2-6x+8}$$

$$= \frac{2x}{3x-12} \cdot \frac{x^2-6x+8}{x^2-2x}$$ Invert divisor and multiply.

$$= \frac{(2)(x)(x-2)(x-4)}{(3)(x-4)(x)(x-2)}$$ Factor and multiply.

$$= \frac{(2)(\cancel{x})(\cancel{x-2})(\cancel{x-4})}{(3)(\cancel{x-4})(\cancel{x})(\cancel{x-2})}$$ Cancel common factors.

$$= \frac{2}{3}, \quad x \neq 0, \; x \neq 2, \; x \neq 4$$ Simplified form

Remember the original expression is equivalent to $2/3$ except for $x = 0$, $x = 2$, and $x = 4$.

3 Simplify a complex fraction.

Complex Fractions

Problems involving division of two rational expressions are sometimes written as **complex fractions.** A complex fraction is one that has a fraction in its numerator or denominator, or both. The rules for dividing fractions still apply in such cases.

Example 7 Simplifying a Complex Fraction

$$\frac{\left(\frac{x^2+2x-3}{x-3}\right)}{4x+12} = \frac{\left(\frac{x^2+2x-3}{x-3}\right)}{\left(\frac{4x+12}{1}\right)}$$ Rewrite denominator.

$$= \frac{x^2+2x-3}{x-3} \cdot \frac{1}{4x+12}$$ Invert divisor and multiply.

$$= \frac{(x-1)(x+3)}{(x-3)(4)(x+3)}$$ Factor.

$$= \frac{(x-1)\cancel{(x+3)}}{(x-3)(4)\cancel{(x+3)}}$$ Cancel common factor.

$$= \frac{x-1}{4(x-3)},\quad x \neq -3$$ Simplified form

Note that in Example 7 the domain of the complex fraction is restricted by the two denominators in the expression, $x - 3$ and $4x + 12$. So, the domain of the original expression is all real values of x such that $x \neq 3$ and $x \neq -3$.

Discussing the Concept Using a Table

Complete the following table for the given values of x.

x	60	100	1000	10,000	100,000	1,000,000
$\frac{x-10}{x+10}$						
$\frac{x+50}{x-50}$						
$\frac{x-10}{x+10} \cdot \frac{x+50}{x-50}$						

What kind of pattern do you see? Try to explain what is going on. Can you see why?

8.2 Exercises

Integrated Review — *Concepts, Skills, and Problem Solving*

Keep mathematically in shape by doing these exercises *before* the problems of this section.

Properties and Definitions

1. Explain how to factor the difference of two squares $9t^2 - 4$.
2. Explain how to factor the perfect square trinomial $4x^2 - 12x + 9$.
3. Explain how to factor the sum of two cubes $8x^3 + 64$.
4. Factor $3x^2 + 13x - 10$, and explain how you can prove that your answer is correct.

Algebraic Operations

In Exercises 5–10, factor the expression completely.

5. $5x - 20x^2$ **6.** $64 - (x - 6)^2$

7. $15x^2 - 16x - 15$ **8.** $16t^2 + 8t + 1$

9. $y^3 - 64$ **10.** $8x^3 + 1$

Graphs

In Exercises 11 and 12, sketch the graphs of the lines through the given point with the indicated slopes. Make the sketches on the same set of coordinate axes.

	Point	*Slopes*	
11.	$(2, -3)$	(a) 0	(b) undefined
		(c) 2	(d) $-\frac{1}{3}$
12.	$(-1, 4)$	(a) 2	(b) -1
		(c) $\frac{1}{2}$	(d) undefined

Developing Skills

In Exercises 1 and 2, evaluate the function as indicated. If not possible, state the reason.

	Expression	*Values*	
1.	$f(x) = \dfrac{x - 10}{4x}$	(a) $f(10)$	(b) $f(0)$
		(c) $f(-2)$	(d) $f(12)$
2.	$g(x) = \dfrac{x^2 - 4x}{x^2 - 9}$	(a) $g(0)$	(b) $g(4)$
		(c) $g(3)$	(d) $g(-3)$

In Exercises 3–10, complete the statement.

3. $\dfrac{7x^2}{3y(\quad)} = \dfrac{7}{3y}, \quad x \neq 0$

4. $\dfrac{14x(x - 3)^2}{(x - 3)(\quad)} = \dfrac{2x}{x - 3}, \quad x \neq 3$

5. $\dfrac{3x(x + 2)^2}{(x - 4)(\quad)} = \dfrac{3x}{x - 4}, \quad x \neq -2$

6. $\dfrac{(x + 1)^3}{x(\quad)} = \dfrac{x + 1}{x}, \quad x \neq -1$

7. $\dfrac{3u(\quad)}{7v(u + 1)} = \dfrac{3u}{7v}, \quad u \neq -1$

8. $\dfrac{(3t + 5)(\quad)}{5t^2(3t - 5)} = \dfrac{3t + 5}{t}, \quad t \neq \dfrac{5}{3}$

9. $\dfrac{13x(\quad)}{4 - x^2} = \dfrac{13x}{x - 2}, \quad x \neq -2$

10. $\dfrac{x^2(\quad)}{x^2 - 10x} = \dfrac{x^2}{10 - x}, \quad x \neq 0$

In Exercises 11–40, multiply and simplify. See Examples 1–5.

11. $\frac{45}{28} \cdot \frac{77}{60}$ **12.** $24\left(-\frac{7}{18}\right)$

13. $7x \cdot \dfrac{9}{14x}$ **14.** $\dfrac{6}{5a} \cdot (25a)$

15. $\dfrac{8s^3}{9s} \cdot \dfrac{6s^2}{32s}$ **16.** $\dfrac{3x^4}{7x} \cdot \dfrac{8x^2}{9}$

17. $16u^4 \cdot \dfrac{12}{8u^2}$ **18.** $25x^3 \cdot \dfrac{8}{35x}$

19. $\dfrac{8}{3 + 4x} \cdot (9 + 12x)$ **20.** $(6 - 4x) \cdot \dfrac{10}{3 - 2x}$

21. $\dfrac{8u^2v}{3u + v} \cdot \dfrac{u + v}{12u}$ **22.** $\dfrac{1 - 3xy}{4x^2y} \cdot \dfrac{46x^4y^2}{15 - 45xy}$

23. $\dfrac{12 - r}{3} \cdot \dfrac{3}{r - 12}$

24. $\dfrac{8 - z}{8 + z} \cdot \dfrac{z + 8}{z - 8}$

25. $\dfrac{(2x - 3)(x + 8)}{x^3} \cdot \dfrac{x}{3 - 2x}$

26. $\dfrac{x + 14}{x^3(10 - x)} \cdot \dfrac{x(x - 10)}{5}$

27. $\dfrac{4r - 12}{r - 2} \cdot \dfrac{r^2 - 4}{r - 3}$

28. $\dfrac{5y - 20}{5y + 15} \cdot \dfrac{2y + 6}{y - 4}$

29. $\dfrac{2t^2 - t - 15}{t + 2} \cdot \dfrac{t^2 - t - 6}{t^2 - 6t + 9}$

30. $\dfrac{y^2 - 16}{y^2 + 8y + 16} \cdot \dfrac{3y^2 - 5y - 2}{y^2 - 6y + 8}$

31. $(x^2 - 4y^2) \cdot \dfrac{xy}{(x - 2y)^2}$

32. $(u - 2v)^2 \cdot \dfrac{u + 2v}{u - 2v}$

33. $\dfrac{x^2 + 2xy - 3y^2}{(x + y)^2} \cdot \dfrac{x^2 - y^2}{x + 3y}$

34. $\dfrac{(x - 2y)^2}{x + 2y} \cdot \dfrac{x^2 + 7xy + 10y^2}{x^2 - 4y^2}$

35. $\dfrac{x + 5}{x - 5} \cdot \dfrac{2x^2 - 9x - 5}{3x^2 + x - 2} \cdot \dfrac{x^2 - 1}{x^2 + 7x + 10}$

36. $\dfrac{t^2 + 4t + 3}{2t^2 - t - 10} \cdot \dfrac{t}{t^2 + 3t + 2} \cdot \dfrac{2t^2 + 4t^3}{t^2 + 3t}$

37. $\dfrac{9 - x^2}{2x + 3} \cdot \dfrac{4x^2 + 8x - 5}{4x^2 - 8x + 3} \cdot \dfrac{6x^4 - 2x^3}{8x^2 + 4x}$

38. $\dfrac{16x^2 - 1}{4x^2 + 9x + 5} \cdot \dfrac{5x^2 - 9x - 18}{x^2 - 12x + 36} \cdot \dfrac{12 + 4x - x^2}{4x^2 - 13x + 3}$

39. $\dfrac{x^3 + 3x^2 - 4x - 12}{x^3 - 3x^2 - 4x + 12} \cdot \dfrac{x^2 - 9}{x}$

40. $\dfrac{xu - yu + xv - yv}{xu + yu - xv - yv} \cdot \dfrac{xu + yu + xv + yv}{xu - yu - xv + yv}$

In Exercises 41–60, divide and simplify. See Examples 6 and 7.

41. $-\dfrac{5}{12} \div \dfrac{45}{32}$

42. $-\dfrac{7}{15} \div \left(-\dfrac{14}{25}\right)$

43. $x^2 \div \dfrac{3x}{4}$

44. $\dfrac{u}{10} \div u^2$

45. $\dfrac{7xy^2}{10u^2v} \div \dfrac{21x^3}{45uv}$

46. $\dfrac{25x^2y}{60x^3y^2} \div \dfrac{5x^4y^3}{16x^2y}$

47. $\dfrac{3(a + b)}{4} \div \dfrac{(a + b)^2}{2}$

48. $\dfrac{x^2 + 9}{5(x + 2)} \div \dfrac{x + 3}{5(x^2 - 4)}$

49. $\dfrac{(x^3y)^2}{(x + 2y)^2} \div \dfrac{x^2y}{(x + 2y)^3}$

50. $\dfrac{x^2 - y^2}{2x^2 - 8x} \div \dfrac{(x - y)^2}{2xy}$

51. $\dfrac{\left(\dfrac{x^2}{12}\right)}{\left(\dfrac{5x}{18}\right)}$

52. $\dfrac{\left(\dfrac{3u^2}{6v^3}\right)}{\left(\dfrac{u}{3v}\right)}$

53. $\dfrac{\left(\dfrac{25x^2}{x - 5}\right)}{\left(\dfrac{10x}{5 + 4x - x^2}\right)}$

54. $\dfrac{\left(\dfrac{5x}{x + 7}\right)}{\left(\dfrac{10}{x^2 + 8x + 7}\right)}$

55. $\dfrac{16x^2 + 8x + 1}{3x^2 + 8x - 3} \div \dfrac{4x^2 - 3x - 1}{x^2 + 6x + 9}$

56. $\dfrac{9x^2 - 24x + 16}{x^2 + 10x + 25} \div \dfrac{6x^2 - 5x - 4}{2x^2 + 3x - 35}$

57. $\dfrac{x^2 + 3x - 2x - 6}{x^2 - 4} \div \dfrac{x + 3}{x^2 + 4x + 4}$

58. $\dfrac{t^3 + t^2 - 9t - 9}{t^2 - 5t + 6} \div \dfrac{t^2 + 6t + 9}{t - 2}$

59. $\dfrac{\left(\dfrac{x^2 - 3x - 10}{x^2 - 4x + 4}\right)}{\left(\dfrac{21 + 4x - x^2}{x^2 - 5x - 14}\right)}$

60. $\dfrac{\left(\dfrac{x^2 + 5x + 6}{4x^2 - 20x + 25}\right)}{\left(\dfrac{x^2 - 5x - 24}{4x^2 - 25}\right)}$

In Exercises 61–68, perform the operations and simplify. (In Exercises 67 and 68, n is a positive integer.)

61. $\left[\dfrac{x^2}{9} \cdot \dfrac{3(x + 4)}{x^2 + 2x}\right] \div \dfrac{x}{x + 2}$

62. $\left(\dfrac{x^2 + 6x + 9}{x^2} \cdot \dfrac{2x + 1}{x^2 - 9}\right) \div \dfrac{4x^2 + 4x + 1}{x^2 - 3x}$

63. $\left[\dfrac{xy + y}{4x} \div (3x + 3)\right] \div \dfrac{y}{3x}$

64. $\dfrac{3u^2 - u - 4}{u^2} \div \dfrac{3u^2 + 12u + 4}{u^4 - 3u^3}$

65. $\dfrac{2x^2 + 5x - 25}{3x^2 + 5x + 2} \cdot \dfrac{3x^2 + 2x}{x + 5} \div \left(\dfrac{x}{x + 1}\right)^2$

66. $\dfrac{t^2 - 100}{4t^2} \cdot \dfrac{t^3 - 5t^2 - 50t}{t^4 + 10t^3} \div \dfrac{(t - 10)^2}{5t}$

67. $x^3 \cdot \dfrac{x^{2n} - 9}{x^{2n} + 4x^n + 3} \div \dfrac{x^{2n} - 2x^n - 3}{x}$

68. $\dfrac{x^{n+1} - 8x}{x^{2n} + 2x^n + 1} \cdot \dfrac{x^{2n} - 4x^n - 5}{x} \div x^n$

In Exercises 69–72, use a graphing utility to graph the two equations on the same screen. Use the graphs to verify that the expressions are equivalent. Verify the results algebraically.

69. $y_1 = \dfrac{3x + 2}{x} \cdot \dfrac{x^2}{9x^2 - 4}$

$y_2 = \dfrac{x}{3x - 2}, \quad x \neq 0, \quad x \neq -\dfrac{2}{3}$

70. $y_1 = \dfrac{x^2 - 10x + 25}{x^2 - 25} \cdot \dfrac{x + 5}{2}$

$y_2 = \dfrac{x - 5}{2}, \quad x \neq \pm 5$

71. $y_1 = \dfrac{3x + 15}{x^4} \div \dfrac{x + 5}{x^2}$

$y_2 = \dfrac{3}{x^2}, \quad x \neq -5$

72. $y_1 = (x^2 + 6x + 9) \cdot \dfrac{3}{2x(x + 3)}$

$y_2 = \dfrac{3(x + 3)}{2x}, \quad x \neq -3$

Solving Problems

Geometry In Exercises 73 and 74, write an expression for the area of the shaded region. Then simplify.

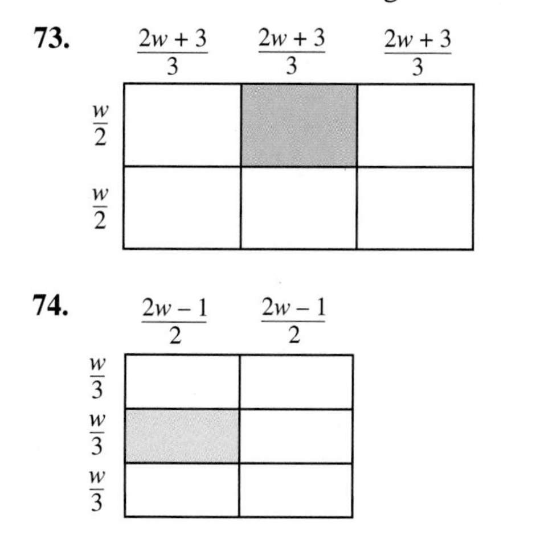

Probability In Exercises 75–78, consider an experiment in which a marble is tossed into a rectangular box with dimensions x centimeters by $2x + 1$ centimeters. The probability that the marble will come to rest in the unshaded portion of the box is equal to the ratio of the unshaded area to the total area of the figure. Find the probability in simplified form.

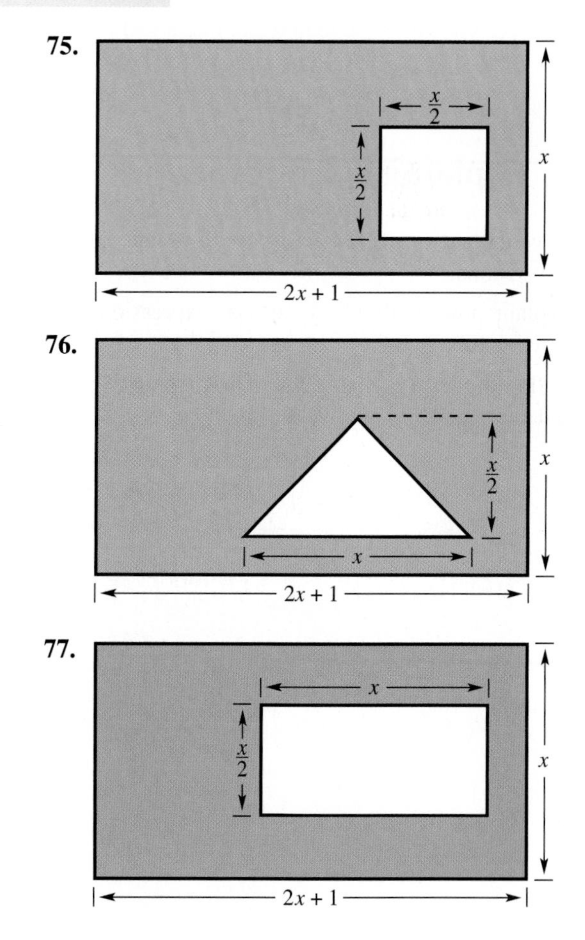

78.

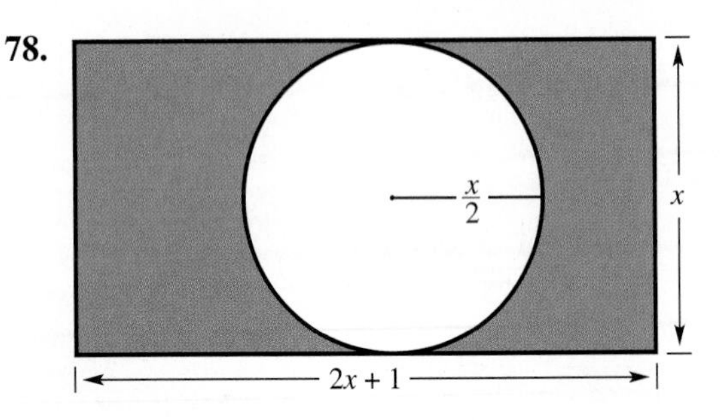

79. *Photocopy Rate* A photocopier produces copies at a rate of 20 pages per minute.

(a) Determine the time required to copy 1 page.

(b) Determine the time required to copy x pages.

(c) Determine the time required to copy 35 pages.

80. *Pumping Rate* The rate for a pump is 15 gallons per minute.

(a) Determine the time required to pump 1 gallon.

(b) Determine the time required to pump x gallons.

(c) Determine the time required to pump 130 gallons.

81. *Analyzing Data* The number N (in thousands) of subscribers to a cellular telephone service and the annual revenue R (in millions of dollars) generated by subscribers in the United States for the period 1990 through 1996 can be modeled by

$N = 6357 + 1070t^2$ and $R = 6115.2 + 590.7t^2$

where t is time in years, with $t = 0$ representing 1990. (Source: Cellular Telecommunications Industry Association)

(a) Use a graphing utility to graph the two models.

(b) Find a model for the average monthly bill per subscriber. (*Note:* Modify the revenue function from years to months.)

(c) Use the model in part (b) to complete the table.

Year, t	0	2	4	6
Monthly bill				

(d) The number of subscribers and the revenue were increasing over the last few years, and yet the average monthly bill was decreasing. Explain how this is possible.

Explaining Concepts

82. In your own words, explain how to divide rational expressions.

83. Explain how to divide a rational expression by a polynomial.

84. Define the term *complex fraction*. Give an example and show how to simplify the fraction.

85. *Error Analysis* Describe the error.

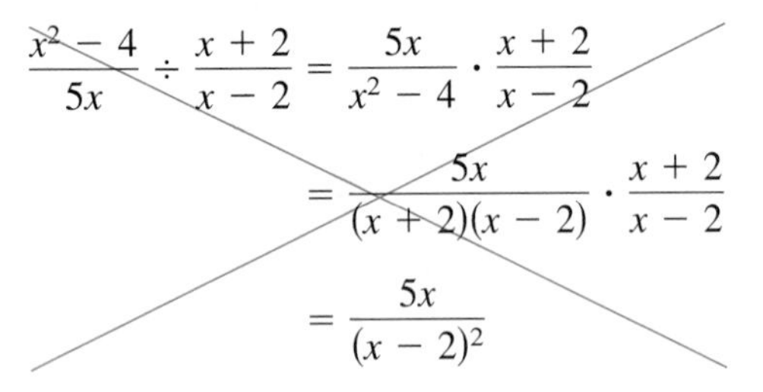

8.3 Adding and Subtracting Rational Expressions

Objectives

1. Add or subtract rational expressions with like denominators and simplify.
2. Add or subtract rational expressions with unlike denominators and simplify.
3. Simplify a complex fraction.

1 Add or subtract rational expressions with like denominators and simplify.

Adding or Subtracting with Like Denominators

As with numerical fractions, the procedure used to add or subtract two rational expressions depends on whether the expressions have *like* or *unlike* denominators. To add or subtract two rational expressions with *like* denominators, simply combine their numerators and place the result over the common denominator.

▶ Adding or Subtracting with Like Denominators

If u, v, and w are real numbers, variables, or variable expressions, and $w \neq 0$, the following rules are valid.

1. $\dfrac{u}{w} + \dfrac{v}{w} = \dfrac{u+v}{w}$ Add fractions with like denominators.

2. $\dfrac{u}{w} - \dfrac{v}{w} = \dfrac{u-v}{w}$ Subtract fractions with like denominators.

Example 1 Adding and Subtracting with Like Denominators

a. $\dfrac{x}{4} + \dfrac{5-x}{4} = \dfrac{x + (5-x)}{4} = \dfrac{5}{4}$

b. $\dfrac{7}{2x-3} - \dfrac{3x}{2x-3} = \dfrac{7-3x}{2x-3}$

Example 2 Subtracting Rational Expressions and Simplifying

$$\frac{x}{x^2-2x-3} - \frac{3}{x^2-2x-3} = \frac{x-3}{x^2-2x-3}$$ Subtract.

$$= \frac{x-3}{(x-3)(x+1)}$$ Factor.

$$= \frac{\cancel{(x-3)}(1)}{\cancel{(x-3)}(x+1)}$$ Cancel common factor.

$$= \frac{1}{x+1}, \quad x \neq 3$$ Simplified form

Study Tip

After adding or subtracting two (or more) rational expressions, check the resulting fraction to see if it can be simplified, as illustrated in Example 2.

The rules for adding and subtracting rational expressions with like denominators can be extended to cover sums and differences involving three or more rational expressions, as illustrated in Example 3.

Example 3 Combining Three Rational Expressions

$$\frac{x^2 - 26}{x - 5} - \frac{2x + 4}{x - 5} + \frac{10 + x}{x - 5}$$

$$= \frac{(x^2 - 26) - (2x + 4) + (10 + x)}{x - 5}$$ Write numerator over common denominator.

$$= \frac{x^2 - 26 - 2x - 4 + 10 + x}{x - 5}$$ Distributive Property

$$= \frac{x^2 - x - 20}{x - 5}$$ Simplify.

$$= \frac{\cancel{(x - 5)}(x + 4)}{\cancel{x - 5}}$$ Factor and cancel common factor.

$$= x + 4, \quad x \neq 5$$ Simplified form

2 Add or subtract rational expressions with unlike denominators and simplify.

Adding or Subtracting with Unlike Denominators

To add or subtract rational expressions with *unlike* denominators, you must first rewrite each expression using the **least common multiple (LCM)** of the denominators of the individual expressions. The least common multiple of two (or more) polynomials is the simplest polynomial that is a multiple of each of the original polynomials. This means that the LCM must contain all the *different* factors in the polynomials and each of these factors must be repeated the maximum number of times it occurs in any one of the polynomials.

Example 4 Finding Least Common Multiples

a. The least common multiple of

$$6x = 2 \cdot 3 \cdot x, \quad 2x^2 = 2 \cdot x \cdot x, \quad \text{and} \quad 9x^3 = 3 \cdot 3 \cdot x \cdot x \cdot x$$

is $2 \cdot 3 \cdot 3 \cdot x \cdot x \cdot x = 18x^3$.

b. The least common multiple of

$$x^2 - x = x(x - 1) \quad \text{and} \quad 2x - 2 = 2(x - 1)$$

is $2x(x - 1)$.

c. The least common multiple of

$$3x^2 + 6x = 3x(x + 2) \quad \text{and} \quad x^2 + 4x + 4 = (x + 2)^2$$

is $3x(x + 2)^2$.

To add or subtract rational expressions with *unlike* denominators, you must first rewrite the rational expressions so that they have *like* denominators. The like denominator that you use is the least common multiple of the original denominators and is called the **least common denominator (LCD)** of the original rational expressions. Once the rational expressions have been written with like denominators, you can simply add or subtract these rational expressions using the rules given at the beginning of this section.

Technology: Tip

You can use a graphing utility to check your results when adding or subtracting rational expressions. For instance, in Example 5, try graphing the equations

$$y_1 = \frac{7}{6x} + \frac{5}{8x}$$

and

$$y_2 = \frac{43}{24x}$$

on the same screen. If the two graphs coincide, as shown below, you can conclude that the two functions are equivalent.

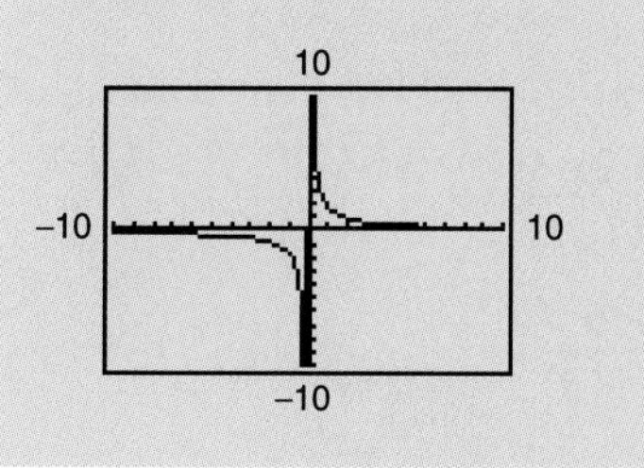

Example 5 Adding with Unlike Denominators

Add the rational expressions: $\frac{7}{6x} + \frac{5}{8x}$.

Solution

The least common denominator of $6x$ and $8x$ is $24x$, so the first step is to rewrite each fraction with this denominator.

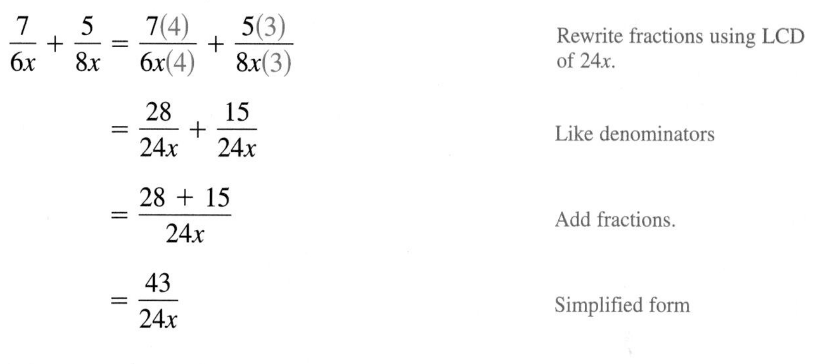

$$\frac{7}{6x} + \frac{5}{8x} = \frac{7(4)}{6x(4)} + \frac{5(3)}{8x(3)}$$ Rewrite fractions using LCD of $24x$.

$$= \frac{28}{24x} + \frac{15}{24x}$$ Like denominators

$$= \frac{28 + 15}{24x}$$ Add fractions.

$$= \frac{43}{24x}$$ Simplified form

Example 6 Subtracting with Unlike Denominators

Subtract the rational expressions: $\frac{3}{x - 3} - \frac{5}{x + 2}$.

Solution

The least common denominator is $(x - 3)(x + 2)$.

$$\frac{3}{x - 3} - \frac{5}{x + 2}$$

$$= \frac{3(x + 2)}{(x - 3)(x + 2)} - \frac{5(x - 3)}{(x - 3)(x + 2)}$$ Rewrite fractions using LCD of $(x - 3)(x + 2)$.

$$= \frac{3x + 6}{(x - 3)(x + 2)} - \frac{5x - 15}{(x - 3)(x + 2)}$$ Distributive Property

$$= \frac{(3x + 6) - (5x - 15)}{(x - 3)(x + 2)}$$ Subtract fractions.

$$= \frac{3x + 6 - 5x + 15}{(x - 3)(x + 2)}$$ Distributive Property

$$= \frac{-2x + 21}{(x - 3)(x + 2)}$$ Simplified form

Study Tip

In Example 7, the factors in the denominator are $x^2 - 4 = (x+2)(x-2)$ and $2 - x$. Because

$$(2 - x) = (-1)(x - 2),$$

the original addition problem can be written as a subtraction problem.

Example 7 Adding with Unlike Denominators

$$\frac{6x}{x^2 - 4} + \frac{3}{(2 - x)}$$

$$= \frac{6x}{(x+2)(x-2)} + \frac{3}{(-1)(x-2)} \quad \text{Factor.}$$

$$= \frac{6x}{(x+2)(x-2)} - \frac{3(x+2)}{(x+2)(x-2)} \quad \text{Rewrite fractions using LCD of } (x+2)(x-2).$$

$$= \frac{6x}{(x+2)(x-2)} - \frac{3x+6}{(x+2)(x-2)} \quad \text{Distributive Property}$$

$$= \frac{6x - (3x+6)}{(x+2)(x-2)} \quad \text{Subtract.}$$

$$= \frac{6x - 3x - 6}{(x+2)(x-2)} \quad \text{Distributive Property}$$

$$= \frac{3x - 6}{(x+2)(x-2)} \quad \text{Simplify.}$$

$$= \frac{3\cancel{(x-2)}}{(x+2)\cancel{(x-2)}} \quad \text{Factor and cancel common factor.}$$

$$= \frac{3}{x+2}, \quad x \neq 2 \quad \text{Simplified form}$$

Example 8 Combining Three Rational Expressions

$$\frac{2x - 5}{6x + 9} - \frac{4}{2x^2 + 3x} + \frac{1}{x}$$

$$= \frac{(2x-5)(x)}{3(2x+3)(x)} - \frac{(4)(3)}{x(2x+3)(3)} + \frac{3(2x+3)}{(x)(3)(2x+3)} \quad \text{Rewrite fractions using LCD of } 3x(2x+3).$$

$$= \frac{2x^2 - 5x}{3x(2x+3)} - \frac{12}{3x(2x+3)} + \frac{6x+9}{3x(2x+3)} \quad \text{Distributive Property}$$

$$= \frac{2x^2 - 5x - 12 + 6x + 9}{3x(2x+3)} \quad \text{Combine numerators.}$$

$$= \frac{2x^2 + x - 3}{3x(2x+3)} \quad \text{Simplify.}$$

$$= \frac{(x-1)(2x+3)}{3x(2x+3)} \quad \text{Factor.}$$

$$= \frac{(x-1)\cancel{(2x+3)}}{3x\cancel{(2x+3)}} \quad \text{Cancel common factor.}$$

$$= \frac{x-1}{3x}, \quad x \neq -\frac{3}{2} \quad \text{Simplified form}$$

3 Simplify a complex fraction.

Complex Fractions

Complex fractions can have numerators or denominators that are the sums or differences of fractions. To simplify a complex fraction, first combine its numerator and its denominator into single fractions. Then divide by inverting the divisor and multiplying.

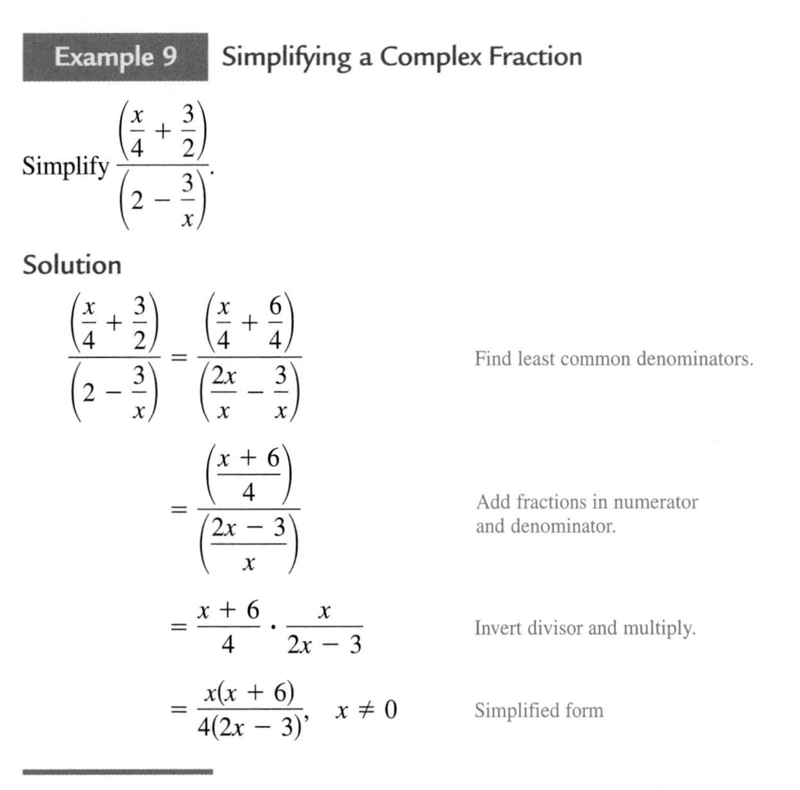

Example 9 Simplifying a Complex Fraction

Simplify $\dfrac{\left(\dfrac{x}{4}+\dfrac{3}{2}\right)}{\left(2-\dfrac{3}{x}\right)}$.

Solution

$$\frac{\left(\frac{x}{4}+\frac{3}{2}\right)}{\left(2-\frac{3}{x}\right)}=\frac{\left(\frac{x}{4}+\frac{6}{4}\right)}{\left(\frac{2x}{x}-\frac{3}{x}\right)}$$ Find least common denominators.

$$=\frac{\left(\frac{x+6}{4}\right)}{\left(\frac{2x-3}{x}\right)}$$ Add fractions in numerator and denominator.

$$=\frac{x+6}{4}\cdot\frac{x}{2x-3}$$ Invert divisor and multiply.

$$=\frac{x(x+6)}{4(2x-3)},\quad x\neq 0$$ Simplified form

Another way to simplify the complex fraction given in Example 9 is to multiply the numerator and denominator by the least common denominator of *every* fraction in the numerator and denominator. For this fraction, notice what happens when we multiply the numerator and denominator by $4x$.

$$\frac{\left(\frac{x}{4}+\frac{3}{2}\right)}{\left(2-\frac{3}{x}\right)}=\frac{\left(\frac{x}{4}+\frac{3}{2}\right)}{\left(2-\frac{3}{x}\right)}\cdot\frac{4x}{4x}$$ Multiply numerator and denominator by LCD of $4x$.

$$=\frac{\frac{x}{4}(4x)+\frac{3}{2}(4x)}{2(4x)-\frac{3}{x}(4x)}$$ Distributive Property

$$=\frac{x^2+6x}{8x-12}$$ Simplify.

$$=\frac{x(x+6)}{4(2x-3)},\quad x\neq 0$$ Simplified form

Example 10 Simplifying a Complex Fraction

Simplify $\dfrac{\left(\dfrac{2}{x+2}\right)}{\left(\dfrac{1}{x+2}+\dfrac{2}{x}\right)}$.

Solution

The least common denominator of the fractions is

$x(x + 2)$.

Multiplying each fraction by this LCD yields

$$\frac{\left(\dfrac{2}{x+2}\right)}{\left(\dfrac{1}{x+2}+\dfrac{2}{x}\right)} = \frac{\left(\dfrac{2}{x+2}\right)(x)(x+2)}{\dfrac{1}{x+2}(x)(x+2)+\dfrac{2}{x}(x)(x+2)}$$

Multiply numerator and denominator by LCD of $x(x + 2)$.

$$= \frac{2x}{x+2(x+2)}$$

Simplify.

$$= \frac{2x}{3x+4}, \quad x \neq -2, x \neq 0.$$

Simplified form

The factors of the LCD cannot be equal to zero in order for the expression to be defined. So, for the reduced form of the expression you must specify that $x \neq -2$ and $x \neq 0$.

Discussing the Concept **Comparing Two Methods**

Evaluate each of the following expressions at the given value of the variable in two different ways: (1) combine and simplify the rational expressions first and then evaluate the simplified expression at the given variable value, and (2) substitute the given value of the variable first and then simplify the resulting expression. Do you get the same result with each method? Discuss which method you prefer and why. List any advantages and/or disadvantages of each method.

a. $\dfrac{1}{m-4} - \dfrac{1}{m+4} + \dfrac{3m}{m^2-16}, \quad m = 2$

b. $\dfrac{x-2}{x^2-9} + \dfrac{3x+2}{x^2-5x+6}, \quad x = 4$

c. $\dfrac{3y^2+16y-8}{y^2+2y-8} - \dfrac{y-1}{y-2} + \dfrac{y}{y+4}, \quad y = 3$

8.3 Exercises

Integrated Review *Concepts, Skills, and Problem Solving*

Keep mathematically in shape by doing these exercises *before* the problems of this section.

Properties and Definitions

1. Write the equation $5y - 3x - 4 = 0$ in the following forms.
   (a) Slope-intercept form
   (b) Point-slope form (many correct answers)
2. Explain how you can visually determine the sign of the slope of a line by observing its graph.

Simplifying Expressions

In Exercises 3–10, perform the multiplication and simplify.

3. $-6x(10 - 7x)$
4. $(2 - y)(3 + 2y)$
5. $(11 - x)(11 + x)$
6. $(4 - 5z)(4 + 5z)$
7. $(x + 1)^2$
8. $t(t^2 + 1) - t(t^2 - 1)$
9. $(x - 2)(x^2 + 2x + 4)$
10. $t(t - 4)(2t + 3)$

Creating Expressions

In Exercises 11 and 12, find expressions for the perimeter and area of the region. Simplify the expressions.

11. 12.

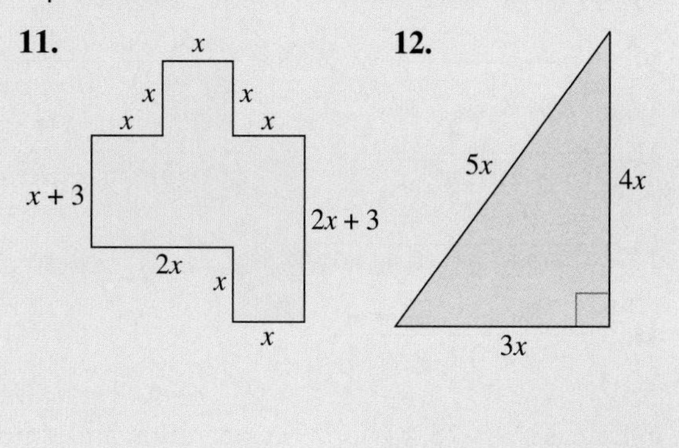

Developing Skills

In Exercises 1–18, combine and simplify. See Examples 1–3.

1. $\frac{5}{8} + \frac{7}{8}$
2. $\frac{7}{12} - \frac{5}{12}$
3. $\frac{5x}{8} - \frac{7x}{8}$
4. $\frac{7y}{12} + \frac{9y}{12}$
5. $\frac{2}{3a} - \frac{11}{3a}$
6. $\frac{6x}{13} - \frac{7x}{13}$
7. $\frac{x}{9} - \frac{x + 2}{9}$
8. $\frac{4 - y}{4} + \frac{3y}{4}$
9. $\frac{z^2}{3} + \frac{z^2 - 2}{3}$
10. $\frac{10x^2 + 1}{3} - \frac{10x^2}{3}$
11. $\frac{2x + 5}{3} + \frac{1 - x}{3}$
12. $\frac{16 + z}{5z} - \frac{11 - z}{5z}$
13. $\frac{3y}{3} - \frac{3y - 3}{3} - \frac{7}{3}$
14. $\frac{-16u}{9} - \frac{27 - 16u}{9} + \frac{2}{9}$
15. $\frac{3y - 22}{y - 6} - \frac{2y - 16}{y - 6}$
16. $\frac{5x - 1}{x + 4} + \frac{5 - 4x}{x + 4}$
17. $\frac{2x - 1}{x(x - 3)} + \frac{1 - x}{x(x - 3)}$
18. $\frac{7s - 5}{2s + 5} + \frac{3(s + 10)}{2s + 5}$

In Exercises 19–30, find the least common multiple of the expressions. See Example 4.

19. $5x^2, 20x^3$
20. $14t^2, 42t^5$
21. $9y^3, 12y$
22. $44m^2, 10m$
23. $15x^2, 3(x + 5)$
24. $6x^2, 15x(x - 1)$
25. $63z^2(z + 1), 14(z + 1)^4$
26. $18y^3, 27y(y - 3)^2$
27. $8t(t + 2), 14(t^2 - 4)$
28. $2y^2 + y - 1, 4y^2 - 2y$
29. $6(x^2 - 4), 2x(x + 2)$
30. $t^3 + 3t^2 + 9t, 2t^2(t^2 - 9)$

In Exercises 31–36, find the missing algebraic expression that makes the two fractions equivalent.

31. $\frac{7x^2}{4a(\quad)} = \frac{7}{4a}, \quad x \neq 0$
32. $\frac{3y(x - 3)^2}{(x - 3)(\quad)} = \frac{21y}{x - 3}$
33. $\frac{5r(\quad)}{3v(u + 1)} = \frac{5r}{3v}, \quad u \neq -1$

34. $\dfrac{(3t+5)(\quad)}{10t^2(3t-5)} = \dfrac{3t+5}{2t},\quad t \neq \dfrac{5}{3}$

35. $\dfrac{7y(\quad)}{4-x^2} = \dfrac{7y}{x-2},\quad x \neq -2$

36. $\dfrac{4x^2(\quad)}{x^2-10x} = \dfrac{4x^2}{10-x},\quad x \neq 0$

In Exercises 37–44, find the least common denominator of the two fractions and rewrite each fraction using the least common denominator.

37. $\dfrac{n+8}{3n-12}, \dfrac{10}{6n^2}$

38. $\dfrac{8s}{(s+2)^2}, \dfrac{3}{s^3+s^2-2s}$

39. $\dfrac{2}{x^2(x-3)}, \dfrac{5}{x(x+3)}$

40. $\dfrac{5t}{2t(t-3)^2}, \dfrac{4}{t(t-3)}$

41. $\dfrac{v}{2v^2+2v}, \dfrac{4}{3v^2}$

42. $\dfrac{4x}{(x+5)^2}, \dfrac{x-2}{x^2-25}$

43. $\dfrac{x-8}{x^2-25}, \dfrac{9x}{x^2-10x+25}$

44. $\dfrac{3y}{y^2-y-12}, \dfrac{y-4}{y^2+3y}$

In Exercises 45–78, perform the operation and simplify. See Examples 5–8.

45. $\dfrac{5}{4x} - \dfrac{3}{5}$

46. $\dfrac{10}{b} + \dfrac{1}{10b}$

47. $\dfrac{7}{a} + \dfrac{14}{a^2}$

48. $\dfrac{1}{6u^2} - \dfrac{2}{9u}$

49. $\dfrac{20}{x-4} + \dfrac{20}{4-x}$

50. $\dfrac{15}{2-t} - \dfrac{7}{t-2}$

51. $\dfrac{3x}{x-8} - \dfrac{6}{8-x}$

52. $\dfrac{1}{y-6} + \dfrac{y}{6-y}$

53. $25 + \dfrac{10}{x+4}$

54. $\dfrac{100}{x-10} - 8$

55. $\dfrac{3x}{3x-2} + \dfrac{2}{2-3x}$

56. $\dfrac{y}{5y-3} - \dfrac{3}{3-5y}$

57. $-\dfrac{1}{6x} + \dfrac{1}{6(x-3)}$

58. $\dfrac{3}{t(t+1)} + \dfrac{4}{t}$

59. $\dfrac{x}{x+3} - \dfrac{5}{x-2}$

60. $\dfrac{1}{x+4} - \dfrac{1}{x+2}$

61. $\dfrac{3}{x+1} - \dfrac{2}{x}$

62. $\dfrac{5}{x-4} - \dfrac{3}{x}$

63. $\dfrac{3}{x-5} + \dfrac{2}{x+5}$

64. $\dfrac{7}{2x-3} + \dfrac{3}{2x+3}$

65. $\dfrac{4}{x^2} - \dfrac{4}{x^2+1}$

66. $\dfrac{2}{y^2+2} + \dfrac{1}{2y^2}$

67. $\dfrac{x}{x^2-9} + \dfrac{3}{x^2-5x+6}$

68. $\dfrac{x}{x^2-x-30} - \dfrac{1}{x+5}$

69. $\dfrac{4}{x-4} + \dfrac{16}{(x-4)^2}$

70. $\dfrac{3}{x-2} - \dfrac{1}{(x-2)^2}$

71. $\dfrac{y}{x^2+xy} - \dfrac{x}{xy+y^2}$

72. $\dfrac{5}{x+y} + \dfrac{5}{x^2-y^2}$

73. $\dfrac{4}{x} - \dfrac{2}{x^2} + \dfrac{4}{x+3}$

74. $\dfrac{5}{2} - \dfrac{1}{2x} - \dfrac{3}{x+1}$

75. $\dfrac{3u}{u^2-2uv+v^2} + \dfrac{2}{u-v} - \dfrac{u}{u-v}$

76. $\dfrac{1}{x-y} - \dfrac{3}{x+y} + \dfrac{3x-y}{x^2-y^2}$

77. $\dfrac{x+2}{x-1} - \dfrac{2}{x+6} - \dfrac{14}{x^2+5x-6}$

78. $\dfrac{x}{x^2+15x+50} + \dfrac{7}{x+10} - \dfrac{x-1}{x+5}$

In Exercises 79 and 80, use a graphing utility to graph the two equations on the same screen. Use the graphs to verify that the expressions are equivalent. Verify the results algebraically.

79. $y_1 = \dfrac{2}{x} + \dfrac{4}{x-2},\quad y_2 = \dfrac{6x-4}{x(x-2)}$

80. $y_1 = 3 - \dfrac{1}{x-1},\quad y_2 = \dfrac{3x-4}{x-1}$

In Exercises 81–96, simplify the complex fraction. See Examples 9 and 10.

81. $\dfrac{\left(\dfrac{1}{2}\right)}{\left(3+\dfrac{1}{x}\right)}$

82. $\dfrac{\left(\dfrac{2}{3}\right)}{\left(4-\dfrac{1}{x}\right)}$

83. $\dfrac{\left(\dfrac{4}{x}+3\right)}{\left(\dfrac{4}{x}-3\right)}$

84. $\dfrac{\left(\dfrac{1}{t}-1\right)}{\left(\dfrac{1}{t}+1\right)}$

85. $\dfrac{\left(16x-\dfrac{1}{x}\right)}{\left(\dfrac{1}{x}-4\right)}$

86. $\dfrac{\left(\dfrac{36}{y}-y\right)}{6+y}$

87. $\dfrac{\left(3+\dfrac{9}{x-3}\right)}{\left(4+\dfrac{12}{x-3}\right)}$

88. $\dfrac{\left(x+\dfrac{2}{x-3}\right)}{\left(x+\dfrac{6}{x-3}\right)}$

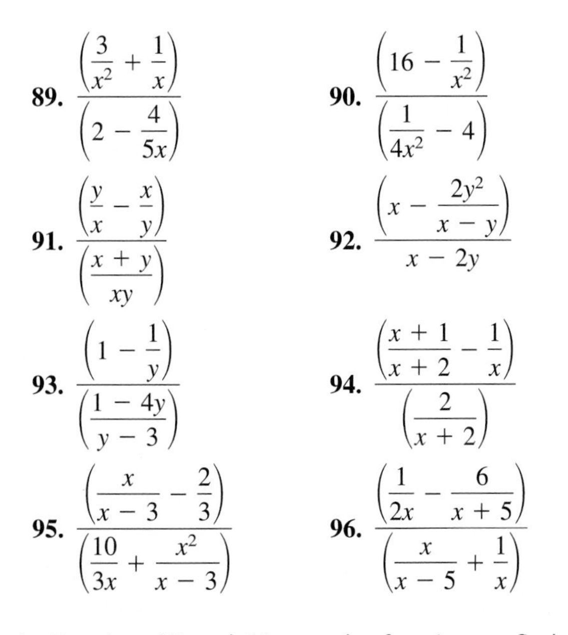

89. $\dfrac{\left(\dfrac{3}{x^2} + \dfrac{1}{x}\right)}{\left(2 - \dfrac{4}{5x}\right)}$

90. $\dfrac{\left(16 - \dfrac{1}{x^2}\right)}{\left(\dfrac{1}{4x^2} - 4\right)}$

91. $\dfrac{\left(\dfrac{y}{x} - \dfrac{x}{y}\right)}{\left(\dfrac{x + y}{xy}\right)}$

92. $\dfrac{\left(x - \dfrac{2y^2}{x - y}\right)}{x - 2y}$

93. $\dfrac{\left(1 - \dfrac{1}{y}\right)}{\left(\dfrac{1 - 4y}{y - 3}\right)}$

94. $\dfrac{\left(\dfrac{x + 1}{x + 2} - \dfrac{1}{x}\right)}{\left(\dfrac{2}{x + 2}\right)}$

95. $\dfrac{\left(\dfrac{x}{x - 3} - \dfrac{2}{3}\right)}{\left(\dfrac{10}{3x} + \dfrac{x^2}{x - 3}\right)}$

96. $\dfrac{\left(\dfrac{1}{2x} - \dfrac{6}{x + 5}\right)}{\left(\dfrac{x}{x - 5} + \dfrac{1}{x}\right)}$

In Exercises 97 and 98, use the function to find and simplify the expression for

$$\frac{f(2 + h) - f(2)}{h}.$$

97. $f(x) = \dfrac{1}{x}$

98. $f(x) = \dfrac{x}{x - 1}$

In Exercises 99 and 100, use a graphing utility to complete the table. Comment on the domains and equivalence of the expressions.

99.

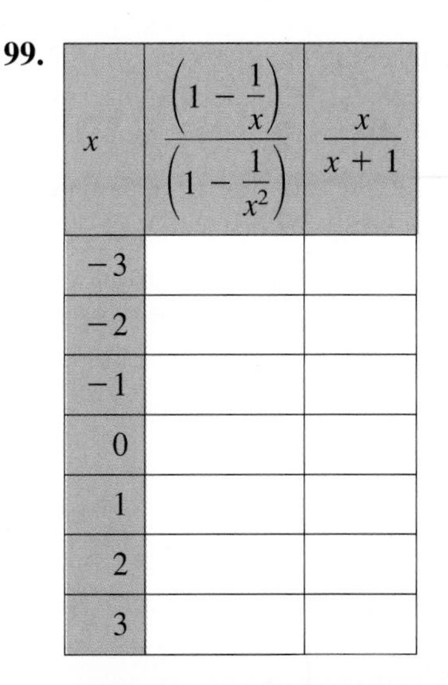

x	$\dfrac{\left(1 - \dfrac{1}{x}\right)}{\left(1 - \dfrac{1}{x^2}\right)}$	$\dfrac{x}{x + 1}$
-3		
-2		
-1		
0		
1		
2		
3		

100.

x	$\dfrac{\left(1 + \dfrac{4}{x} + \dfrac{4}{x^2}\right)}{\left(1 - \dfrac{4}{x^2}\right)}$	$\dfrac{x + 2}{x - 2}$
-3		
-2		
-1		
0		
1		
2		
3		

Solving Problems

101. *Work Rate* After working together for t hours on a common task, two workers have completed fractional parts of the job equal to $t/4$ and $t/6$. What fractional part of the task has been completed?

102. *Work Rate* After working together for t hours on a common task, two workers have completed fractional parts of the job equal to $t/3$ and $t/5$. What fractional part of the task has been completed?

103. *Average of Two Numbers* Determine the average of the two real numbers $x/4$ and $x/6$.

104. *Average of Three Numbers* Determine the average of the three real numbers x, $x/2$, and $x/3$.

105. *Equal Parts* Find two real numbers that divide the real number line between $x/5$ and $x/3$ into three equal parts (see figure).

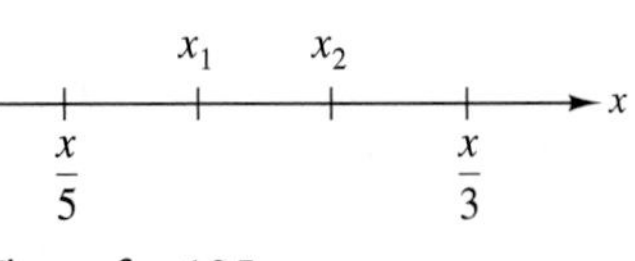

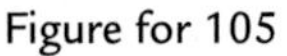

Figure for 105

106. *Monthly Payment* The approximate annual percentage rate r of a monthly installment loan is

$$r = \frac{\left[\dfrac{24(NM - P)}{N}\right]}{\left(P + \dfrac{MN}{12}\right)}$$

where N is the total number of payments, M is the monthly payment, and P is the amount financed.

(a) Approximate the annual percentage rate for a 4-year car loan of \$10,000 that has monthly payments of \$300.

(b) Simplify the expression for the annual percentage rate r, and then rework part (a).

107. *Parallel Resistance* When two resistors are connected in parallel (see figure), the total resistance is

$$\frac{1}{\left(\frac{1}{R_1} + \frac{1}{R_2}\right)}.$$

Simplify this complex fraction.

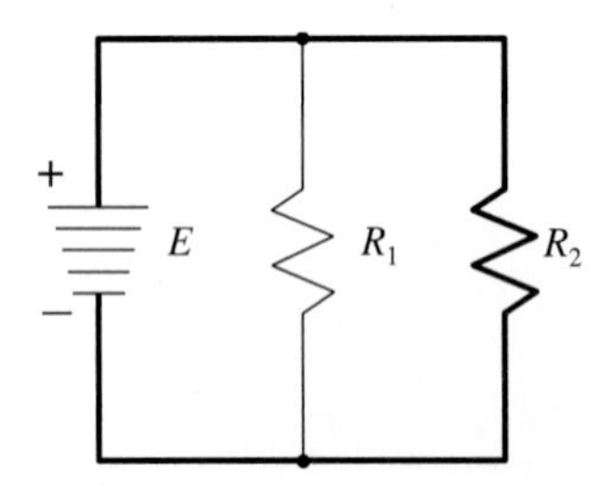

108. *Using Two Models* From 1990 through 1996, the circulations of morning and evening daily newspapers M and E (in millions) can be approximated by

$$M = 41.1 + 0.61t \quad \text{and} \quad E = \frac{20.675 - 1.675t}{1 - 0.023t}$$

where $t = 0$ represents 1990. (Source: Editor & Publisher Co.)

(a) Write an expression for the total daily circulation T. Simplify the result.

(b) Use a graphing utility to graph the functions M, E, and T on the same screen.

(c) Morning circulation is increasing over the time period, whereas evening circulation is decreasing. Use the graphs in part (b) to discuss the change in total circulation.

(d) Use the result of part (a) to approximate the total circulation in the United States in 1991.

Explaining Concepts

109. Answer parts (a)–(c) of Motivating the Chapter on page 443.

110. In your own words, describe how to add or subtract rational expressions with like denominators.

111. In your own words, describe how to add or subtract rational expressions with unlike denominators.

112. Is it possible for the least common denominator of two fractions to be the same as one of the fraction's denominators? If so, give an example.

113. *Error Analysis* Describe the error.

$$\frac{x-1}{x+4} - \frac{4x-11}{x+4} = \frac{x-1-4x-11}{x+4}$$

$$= \frac{-3x-12}{x+4} = \frac{-3(x+4)}{x+4}$$

$$= -3$$

114. *Error Analysis* Describe the error.

$$\frac{2}{x} - \frac{3}{x+1} + \frac{x+1}{x^2}$$

$$= \frac{2x(x+1) - 3x^2 + (x+1)^2}{x^2(x+1)}$$

$$= \frac{2x^2 + x - 3x^2 + x^2 + 1}{x^2(x+1)}$$

$$= \frac{x+1}{x^2(x+1)} = \frac{1}{x^2}$$

Mid-Chapter Quiz

Take this quiz as you would take a quiz in class. After you are done, check your work against the answers given in the back of the book.

1. Determine the domain of $\dfrac{y+2}{y(y-4)}$.

2. Evaluate $h(x) = (x^2 - 9)/(x^2 - x - 2)$ for the indicated values of x. If it is not possible, state the reason.

(a) $h(-3)$ (b) $h(0)$ (c) $h(-1)$ (d) $h(5)$

In Exercises 3–8, write the expression in reduced form.

3. $\dfrac{9y^2}{6y}$ **4.** $\dfrac{8u^3v^2}{36uv^3}$ **5.** $\dfrac{4x^2 - 1}{x - 2x^2}$

6. $\dfrac{(z+3)^2}{2z^2 + 5z - 3}$ **7.** $\dfrac{7ab + 3a^2b^2}{a^2b}$ **8.** $\dfrac{2mn^2 - n^3}{2m^2 + mn - n^2}$

In Exercises 9–18, perform the operations and simplify your answer.

9. $\dfrac{11t^2}{6} \cdot \dfrac{9}{33t}$ **10.** $(x^2 + 2x) \cdot \dfrac{5}{x^2 - 4}$

11. $\dfrac{4}{3(x-1)} \cdot \dfrac{12x}{6(x^2 + 2x - 3)}$ **12.** $\dfrac{5u}{3(u+v)} \cdot \dfrac{2(u^2 - v^2)}{3v} \div \dfrac{25u^2}{18(u-v)}$

13. $\dfrac{\left(\dfrac{9t^2}{3-t}\right)}{\left(\dfrac{6t}{t-3}\right)}$ **14.** $\dfrac{\left(\dfrac{10}{x^2 + 2x}\right)}{\left(\dfrac{15}{x^2 + 3x + 2}\right)}$ **15.** $\dfrac{4x}{x+5} - \dfrac{3x}{4}$

16. $4 + \dfrac{x}{x^2 - 4} - \dfrac{2}{x^2}$ **17.** $\dfrac{\left(1 - \dfrac{2}{x}\right)}{\left(\dfrac{3}{x} - \dfrac{4}{5}\right)}$ **18.** $\dfrac{\left(\dfrac{3}{x} + \dfrac{x}{3}\right)}{\left(\dfrac{x+3}{6x}\right)}$

19. You start a business with a setup cost of \$6000. The cost of material for producing each unit of your product is \$10.50.

(a) Write an algebraic fraction that gives the average cost per unit when x units are produced. Explain your reasoning.

(b) Find the average cost per unit when $x = 500$ units are produced.

20. Find the ratio of the shaded portion of the figure to the total area of the figure.

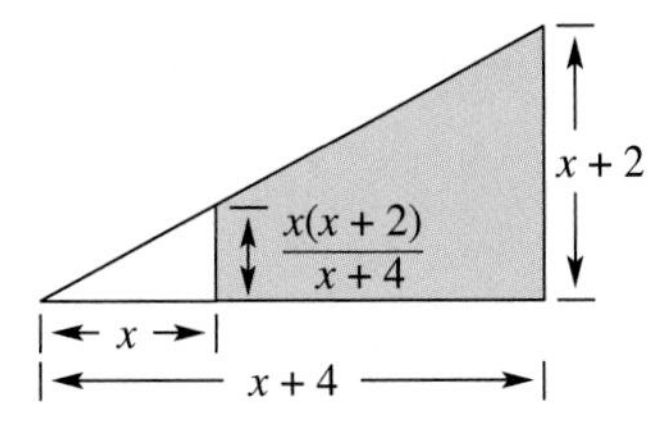

8.4 Solving Rational Equations

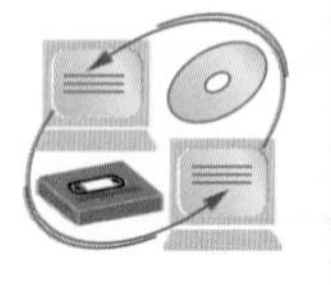

Objectives

1. Solve a rational equation containing constant denominators.
2. Solve a rational equation containing variable denominators.
3. Solve an application problem involving a rational equation.

1 Solve a rational equation containing constant denominators.

Equations Containing Constant Denominators

In Section 3.2, you studied a strategy for solving equations that contain fractions with *constant* denominators. We review that procedure here because it is the basis for solving more general equations involving fractions. Recall from Section 3.2 that you can "clear an equation of fractions" by multiplying both sides of the equation by the least common denominator (LCD) of the fractions in the equation. Note how this is done in the next three examples.

Example 1 An Equation Containing Constant Denominators

Solve $\frac{3}{5} = \frac{x}{2} + 1$.

Solution

The least common denominator of the two fractions is 10, so begin by multiplying both sides of the equation by 10.

$\frac{3}{5} = \frac{x}{2} + 1$	Original equation
$10\left(\frac{3}{5}\right) = 10\left(\frac{x}{2} + 1\right)$	Multiply both sides by LCD of 10.
$6 = 5x + 10$	Simplify.
$-4 = 5x$	Subtract 10 from both sides.
$-\frac{4}{5} = x$	Divide both sides by 5.

The solution is $-\frac{4}{5}$. You can check this as follows.

Check

$\frac{3}{5} \stackrel{?}{=} \frac{-4/5}{2} + 1$	Substitute $-\frac{4}{5}$ for x in the original equation.
$\frac{3}{5} \stackrel{?}{=} -\frac{4}{5} \cdot \frac{1}{2} + 1$	Invert and multiply.
$\frac{3}{5} = -\frac{2}{5} + 1$	Solution checks. ✓

Example 2 An Equation Containing Constant Denominators

Solve $\frac{x-3}{6} = 7 - \frac{x}{12}$.

Solution

$$\frac{x-3}{6} = 7 - \frac{x}{12}$$ Original equation

$$12\left(\frac{x-3}{6}\right) = 12\left(7 - \frac{x}{12}\right)$$ Multiply both sides by LCD of 12.

$$2x - 6 = 84 - x$$ Distribute and simplify.

$$3x - 6 = 84$$ Add x to both sides.

$$3x = 90$$ Add 6 to both sides.

$$x = 30$$ Divide both sides by 3.

Check

$$\frac{30-3}{6} \stackrel{?}{=} 7 - \frac{30}{12}$$ Substitute 30 for x in the original equation.

$$\frac{27}{6} = \frac{42}{6} - \frac{15}{6}$$ Solution checks. ✓

Example 3 An Equation Containing Constant Denominators

Solve $\frac{x+2}{6} - \frac{x-4}{8} = \frac{2}{3}$.

Solution

$$\frac{x+2}{6} - \frac{x-4}{8} = \frac{2}{3}$$ Original equation

$$24\left(\frac{x+2}{6} - \frac{x-4}{8}\right) = 24\left(\frac{2}{3}\right)$$ Multiply both sides by LCD of 24.

$$4(x+2) - 3(x-4) = 8(2)$$ Distribute and simplify.

$$4x + 8 - 3x + 12 = 16$$ Distributive Property

$$x + 20 = 16$$ Combine like terms.

$$x = -4$$ Subtract 20 from both sides.

Check

$$\frac{-4+2}{6} - \frac{-4-4}{8} \stackrel{?}{=} \frac{2}{3}$$ Substitute -4 for x in the original equation.

$$-\frac{1}{3} + 1 = \frac{2}{3}$$ Solution checks. ✓

2 Solve a rational equation containing variable denominators.

Equations Containing Variable Denominators

Remember that you always *exclude* those values of a variable that make the denominator of a rational expression zero. This is especially critical for solving equations that contain variable denominators. You will see why in the examples that follow.

Technology: Tip

You can use a graphing utility to estimate the solution of the equation in Example 4. To do this, graph the left side of the equation and the right side of the equation on the same screen.

$$y_1 = \frac{7}{x} - \frac{1}{3x} \text{ and } y_2 = \frac{8}{3}$$

The solution of the equation is the x-coordinate of the point at which the two graphs intersect, as shown below.

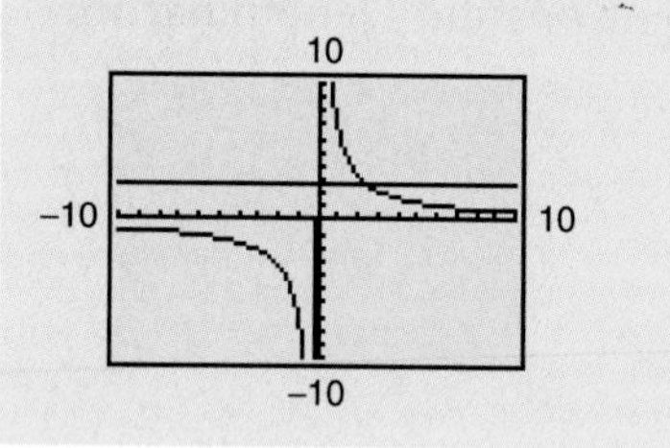

Example 4 An Equation Containing Variable Denominators

Solve the equation.

$$\frac{7}{x} - \frac{1}{3x} = \frac{8}{3}$$

Solution

For this equation, the least common denominator is $3x$. So, begin by multiplying both sides of the equation by $3x$.

$$\frac{7}{x} - \frac{1}{3x} = \frac{8}{3}$$ Original equation

$$3x\left(\frac{7}{x} - \frac{1}{3x}\right) = 3x\left(\frac{8}{3}\right)$$ Multiply both sides by LCD of $3x$.

$$\frac{21x}{x} - \frac{3x}{3x} = \frac{24x}{3}$$ Distributive Property

$$21 - 1 = 8x$$ Simplify.

$$\frac{20}{8} = x$$ Combine like terms and divide both sides by 8.

$$x = \frac{5}{2}$$ Simplify.

The solution is $\frac{5}{2}$. You can check this as follows.

Check

$$\frac{7}{x} - \frac{1}{3x} = \frac{8}{3}$$ Original equation

$$\frac{7}{5/2} - \frac{1}{3(5/2)} \stackrel{?}{=} \frac{8}{3}$$ Substitute $\frac{5}{2}$ for x.

$$7\left(\frac{2}{5}\right) - \frac{2}{15} \stackrel{?}{=} \frac{8}{3}$$ Invert and multiply.

$$\frac{14}{5} - \frac{2}{15} \stackrel{?}{=} \frac{8}{3}$$ Simplify.

$$\frac{40}{15} \stackrel{?}{=} \frac{8}{3}$$ Combine like terms.

$$\frac{8}{3} = \frac{8}{3}$$ Solution checks. ✓

Throughout the text, we have emphasized the importance of checking solutions. Up to this point, the main reason for checking has been to make sure that you did not make errors in the solution process. In the next example you will see that there is another reason for checking solutions in the *original* equation. That is, even with no mistakes in the solution process, it can happen that a "trial solution" does not satisfy the original equation. This type of "solution" is called **extraneous.** An extraneous solution of an equation does not, by definition, satisfy the original equation, and therefore *must not* be listed as an actual solution.

Example 5 An Equation with No Solution

Solve the equation.

$$\frac{5x}{x-2} = 7 + \frac{10}{x-2}$$

Solution

The least common denominator for this equation is $x - 2$. So, begin by multiplying both sides of the equation by $x - 2$.

$$\frac{5x}{x-2} = 7 + \frac{10}{x-2}$$ Original equation

$$(x-2)\left(\frac{5x}{x-2}\right) = (x-2)\left(7 + \frac{10}{x-2}\right)$$ Multiply both sides by $x - 2$.

$$5x = 7(x-2) + 10, \quad x \neq 2$$ Distribute and simplify.

$$5x = 7x - 14 + 10$$ Distributive Property

$$5x = 7x - 4$$ Combine like terms.

$$-2x = -4$$ Subtract $7x$ from both sides.

$$x = 2$$ Divide both sides by -2.

At this point, the solution appears to be 2. However, by performing the following check, you will see that this "trial solution" is extraneous.

Check

$$\frac{5x}{x-2} = 7 + \frac{10}{x-2}$$ Original equation

$$\frac{5(2)}{2-2} \stackrel{?}{=} 7 + \frac{10}{2-2}$$ Substitute 2 for x.

$$\frac{10}{0} \stackrel{?}{=} 7 + \frac{10}{0}$$ Solution does not check. ✗

Because the check results in *division by zero*, 2 is extraneous. Therefore, the original equation has no solution.

Notice that 2 is excluded from the domains of the two fractions in the original equation in Example 5. You may find it helpful when solving these types of equations to list the domain restrictions before beginning the solution process.

Example 6 An Equation Containing Variable Denominators

Solve $\frac{4}{x-2} + \frac{3x}{x+1} = 3$.

Solution

The domain is all real values of x such that $x \neq 2$ and $x \neq -1$. The least common denominator is $(x-2)(x+1)$.

$$\frac{4}{x-2} + \frac{3x}{x+1} = 3$$

$$(x-2)(x+1)\left(\frac{4}{x-2} + \frac{3x}{x+1}\right) = 3(x-2)(x+1)$$

$$4(x+1) + 3x(x-2) = 3(x^2 - x - 2), \quad x \neq 2, x \neq -1$$

$$4x + 4 + 3x^2 - 6x = 3x^2 - 3x - 6$$

$$3x^2 - 2x + 4 = 3x^2 - 3x - 6$$

$$x = -10$$

The solution is -10. Check this in the original equation.

Technology: Discovery

Use a graphing utility to graph the equation

$$y = \frac{3x}{x+1} - \frac{12}{x^2-1} - 2.$$

Then use the zoom and trace features of the utility to determine the x-intercepts. How do the x-intercepts compare with the solutions to Example 7? What can you conclude?

So far in this section, each of the equations has had one solution or no solution. The equation in the next example has two solutions.

Example 7 An Equation That Has Two Solutions

Solve $\frac{3x}{x+1} = \frac{12}{x^2-1} + 2$.

Solution

The domain is all real values of x such that $x \neq 1$ and $x \neq -1$. The least common denominator is $(x+1)(x-1) = x^2 - 1$.

$$\frac{3x}{x+1} = \frac{12}{x^2-1} + 2$$ Original equation

$$(x^2-1)\left(\frac{3x}{x+1}\right) = (x^2-1)\left(\frac{12}{x^2-1} + 2\right)$$ Multiply both sides by LCD of $x^2 - 1$.

$$(x-1)(3x) = 12 + 2(x^2-1), \quad x \neq \pm 1$$ Simplify.

$$3x^2 - 3x = 12 + 2x^2 - 2$$ Distributive Property

$$x^2 - 3x - 10 = 0$$ Subtract $2x^2$ and 10 from both sides.

$$(x+2)(x-5) = 0$$ Factor.

$$x + 2 = 0 \quad \Rightarrow \quad x = -2$$ Set $x + 2$ equal to 0.

$$x - 5 = 0 \quad \Rightarrow \quad x = 5$$ Set $x - 5$ equal to 0.

The solutions are -2 and 5. Check these in the original equation.

3 Solve an application problem involving a rational equation.

Applications

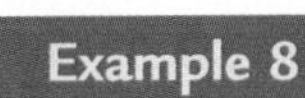

Example 8 Average Cost

A manufacturing plant can produce x units of a certain item for \$26 per unit *plus* an initial investment of \$80,000. How many units must be produced to have an average cost of \$30 per unit?

Solution

Verbal Model: [Average cost per unit] = [Total cost] ÷ [Number of units]

Labels:
Number of units $= x$ (units)
Average cost per unit $= 30$ (dollars per unit)
Total cost $= 26x + 80{,}000$ (dollars)

Equation:

$$30 = \frac{26x + 80{,}000}{x}$$

$$30x = 26x + 80{,}000, \quad x \neq 0$$

$$4x = 80{,}000$$

$$x = 20{,}000$$

The plant should produce 20,000 units.

Example 9 A Work-Rate Problem

With only the cold water valve open, it takes 8 minutes to fill the tub of a washer. With both the hot and cold water valves open, it takes only 5 minutes. How long will it take the tub to fill with only the hot water valve open?

Solution

Verbal Model: [Rate for cold water] + [Rate for hot water] = [Rate for warm water]

Labels:
Rate for cold water $= \frac{1}{8}$ (tub per minute)
Rate for hot water $= \frac{1}{t}$ (tub per minute)
Rate for warm water $= \frac{1}{5}$ (tub per minute)

Equation:

$$\frac{1}{8} + \frac{1}{t} = \frac{1}{5}$$

$$5t + 40 = 8t$$

$$40 = 3t$$

$$\frac{40}{3} = t$$

So, it takes $13\frac{1}{3}$ minutes to fill the tub with hot water.

Example 10 Batting Average

In this year's playing season, a baseball player has been at bat 140 times and has hit the ball safely 35 times. So, the "batting average" for the player is $35/140 = .250$. How many consecutive times must the player hit safely to obtain a batting average of .300?

Solution

Verbal Model: Batting average = Total hits ÷ Total times at bat

Labels: Current times at bat $= 140$
Current hits $= 35$
Additional consecutive hits $= x$

Equation:

$$.300 = \frac{x + 35}{x + 140}$$

$$.300(x + 140) = x + 35$$

$$.3x + 42 = x + 35$$

$$7 = 0.7x$$

$$10 = x$$

The player must hit safely the next 10 times at bat. After that, the batting average will be $45/150 = .300$.

Examples 8, 9, and 10 are types of application problems that you have seen earlier in the text. The difference now is that the variable appears in the denominator of a rational expression. When determining the domain of a real-life problem you must also consider the context of the problem. For instance, in Example 10 the additional times at bat could not be a negative number. The problem implies that the domain be all real numbers greater than or equal to 0.

Discussing the Concept — Interpreting Average Cost

You buy sand in bulk for a construction project. You find that the total cost of your order depends on the weight of the order. The total cost C in dollars is given by $C = 100 + 50x - 0.2x^2$, $1 \le x \le 50$, where x is the weight in thousands of pounds. Construct a rational function representing the average cost per thousand pounds. Because of cost constraints, you can proceed with the project only if the average cost of the sand is less than \$50 per thousand pounds. What is the smallest order you can place and still proceed with the project?

8.4 Exercises

Integrated Review *Concepts, Skills, and Problem Solving*

Keep mathematically in shape by doing these exercises *before* the problems of this section.

Properties and Definitions

In Exercises 1 and 2, determine the quadrants in which the point must be located.

1. $(-2, y)$, y is a real number.

2. $(x, 3)$, x is a real number.

3. Give the positions of points whose y-coordinates are 0.

4. Find the coordinates of the point 9 units to the right of the y-axis and 6 units below the x-axis.

Solving Inequalities

In Exercises 5–10, solve the inequality.

5. $7 - 3x > 4 - x$ **6.** $2(x + 6) - 20 < 2$

7. $|x - 3| < 2$ **8.** $|x - 5| > 3$

9. $\left|\frac{1}{4}x - 1\right| \geq 3$ **10.** $\left|2 - \frac{1}{3}x\right| \leq 10$

Problem Solving

11. A jogger leaves a given point on a fitness trail running at a rate of 6 miles per hour. Five minutes later a second jogger leaves from the same location running at 8 miles per hour. How long will it take the second runner to overtake the first, and how far will each have run at that point?

12. An inheritance of \$24,000 is invested in two bonds that pay 7.5% and 9% simple interest. The annual interest is \$1935. How much is invested in each bond?

Developing Skills

In Exercises 1–4, determine whether the values of x are solutions to the equation.

Equation	Values	
1. $\frac{x}{3} - \frac{x}{5} = \frac{4}{3}$	(a) $x = 0$	(b) $x = -1$
	(c) $x = \frac{1}{8}$	(d) $x = 10$
2. $x = 4 + \frac{21}{x}$	(a) $x = 0$	(b) $x = -3$
	(c) $x = 7$	(d) $x = -1$
3. $\frac{x}{4} + \frac{3}{4x} = 1$	(a) $x = -1$	(b) $x = 1$
	(c) $x = 3$	(d) $x = -3$
4. $5 - \frac{1}{x - 3} = 2$	(a) $x = \frac{10}{3}$	(b) $x = -\frac{1}{3}$
	(c) $x = 0$	(d) $x = 1$

In Exercises 5–18, solve the equation. See Examples 1–3.

5. $\frac{x}{6} - 1 = \frac{2}{3}$ **6.** $\frac{y}{8} + 7 = -\frac{1}{2}$

7. $\frac{z + 2}{3} = 4 - \frac{z}{12}$ **8.** $\frac{x - 5}{5} + 3 = -\frac{x}{4}$

9. $\frac{2y - 9}{6} = 3y - \frac{3}{4}$ **10.** $\frac{4x - 2}{7} - \frac{5}{14} = 2x$

11. $\frac{4t}{3} = 15 - \frac{t}{6}$ **12.** $\frac{x}{3} + \frac{x}{6} = 10$

13. $\frac{5y - 1}{12} + \frac{y}{3} = -\frac{1}{4}$ **14.** $\frac{z - 4}{9} - \frac{3z + 1}{18} = \frac{3}{2}$

15. $\frac{h + 2}{5} - \frac{h - 1}{9} = \frac{2}{3}$ **16.** $\frac{u - 2}{6} + \frac{2u + 5}{15} = 3$

17. $\frac{x + 5}{4} - \frac{3x - 8}{3} = \frac{4 - x}{12}$

18. $\frac{2x - 7}{10} - \frac{3x + 1}{5} = \frac{6 - x}{5}$

In Exercises 19–58, solve the equation. (Check for extraneous solutions.) See Examples 4–7.

19. $\frac{9}{25 - y} = -\frac{1}{4}$ **20.** $\frac{2}{u + 4} = \frac{5}{8}$

21. $5 - \frac{12}{a} = \frac{5}{3}$ **22.** $\frac{6}{b} + 22 = 24$

23. $\dfrac{4}{x} - \dfrac{7}{5x} = -\dfrac{1}{2}$

24. $\dfrac{5}{3} = \dfrac{6}{7x} + \dfrac{2}{x}$

25. $\dfrac{12}{y+5} + \dfrac{1}{2} = 2$

26. $\dfrac{7}{8} - \dfrac{16}{t-2} = \dfrac{3}{4}$

27. $\dfrac{5}{x} = \dfrac{25}{3(x+2)}$

28. $\dfrac{10}{x+4} = \dfrac{15}{4(x+1)}$

29. $\dfrac{8}{3x+5} = \dfrac{1}{x+2}$

30. $\dfrac{500}{3x+5} = \dfrac{50}{x-3}$

31. $\dfrac{3}{x+2} - \dfrac{1}{x} = \dfrac{1}{5x}$

32. $\dfrac{12}{x+5} + \dfrac{5}{x} = \dfrac{20}{x}$

33. $\dfrac{1}{2} = \dfrac{18}{x^2}$

34. $\dfrac{1}{4} = \dfrac{16}{z^2}$

35. $\dfrac{32}{t} = 2t$

36. $\dfrac{20}{u} = \dfrac{u}{5}$

37. $x + 1 = \dfrac{72}{x}$

38. $\dfrac{48}{x} = x - 2$

39. $1 = \dfrac{16}{y} - \dfrac{39}{y^2}$

40. $x - \dfrac{24}{x} = 5$

41. $\dfrac{2x}{3x-10} - \dfrac{5}{x} = 0$

42. $\dfrac{x+42}{x} = x$

43. $\dfrac{2x}{5} = \dfrac{x^2-5x}{5x}$

44. $\dfrac{3x}{4} = \dfrac{x^2+3x}{8x}$

45. $\dfrac{2}{6q+5} - \dfrac{3}{4(6q+5)} = \dfrac{1}{28}$

46. $\dfrac{10}{x(x-2)} + \dfrac{4}{x} = \dfrac{5}{x-2}$

47. $\dfrac{4}{2x+3} + \dfrac{17}{5x-3} = 3$

48. $\dfrac{1}{x-1} + \dfrac{3}{x+1} = 2$

49. $\dfrac{2}{x-10} - \dfrac{3}{x-2} = \dfrac{6}{x^2-12x+20}$

50. $\dfrac{5}{x+2} + \dfrac{2}{x^2-6x-16} = \dfrac{-4}{x-8}$

51. $\dfrac{x+3}{x^2-9} + \dfrac{4}{3-x} - 2 = 0$

52. $1 - \dfrac{6}{4-x} = \dfrac{x+2}{x^2-16}$

53. $\dfrac{x}{x-2} + \dfrac{3x}{x-4} = \dfrac{-2(x-6)}{x^2-6x+8}$

54. $\dfrac{2(x+1)}{x^2-4x+3} + \dfrac{6x}{x-3} = \dfrac{3x}{x-1}$

55. $\dfrac{2(x+7)}{x+4} - 2 = \dfrac{2x+20}{2x+8}$

56. $\dfrac{2x^2-5}{x^2-4} + \dfrac{6}{x+2} = \dfrac{4x-7}{x-2}$

57. $\dfrac{x}{2} = \dfrac{2 - \frac{3}{x}}{1 - \frac{1}{x}}$

58. $\dfrac{2x}{3} = \dfrac{1 + \frac{2}{x}}{1 + \frac{1}{x}}$

In Exercises 59–62, (a) use the graph to determine any x-intercepts of the graph, and (b) set $y = 0$ and solve the resulting rational equation to confirm the result of part (a).

59. $y = \dfrac{x+2}{x-2}$

60. $y = \dfrac{2x}{x+4}$

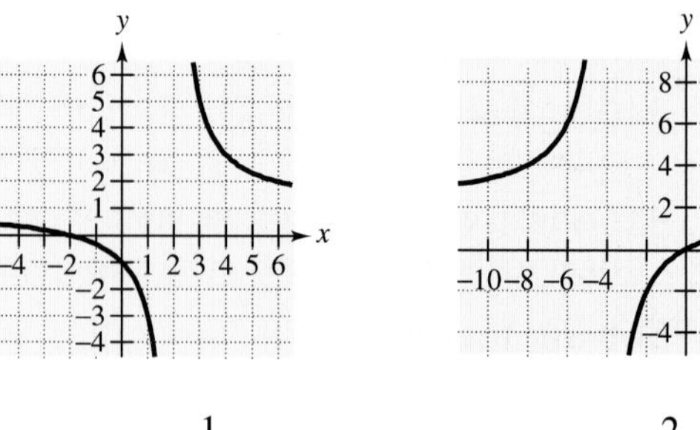

61. $y = x - \dfrac{1}{x}$

62. $y = x - \dfrac{2}{x} - 1$

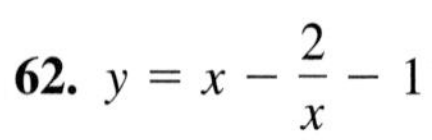

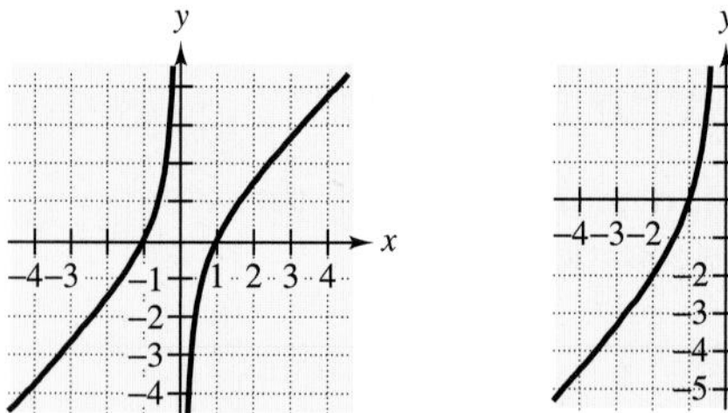

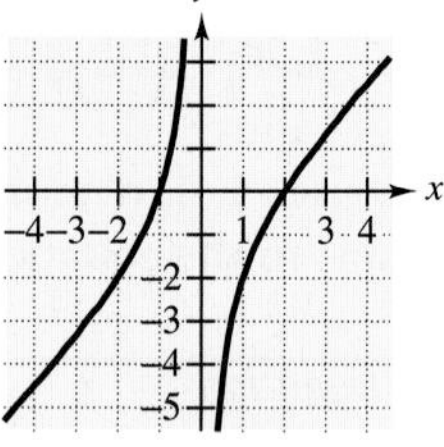

In Exercises 63–70, (a) use a graphing utility to graph the equation and determine any x-intercepts of the graph, and (b) set $y = 0$ and solve the resulting rational equation to confirm the result of part (a).

63. $y = \dfrac{x-4}{x+5}$

64. $y = \dfrac{1}{x} - \dfrac{3}{x+4}$

65. $y = \dfrac{1}{x} + \dfrac{4}{x-5}$

66. $y = 20\left(\dfrac{2}{x} - \dfrac{3}{x-1}\right)$

67. $y = (x+1) - \dfrac{6}{x}$

68. $y = \dfrac{x^2-4}{x}$

69. $y = (x-1) - \dfrac{12}{x}$

70. $y = \dfrac{x}{2} - \dfrac{4}{x} - 1$

Solving Problems

71. *Number Problem* Find a number such that the sum of the number and its reciprocal is $\frac{65}{8}$.

72. *Number Problem* Find a number such that the sum of two times the number and three times its reciprocal is $\frac{97}{4}$.

73. *Wind Speed* A plane has a speed of 300 miles per hour in still air. Find the speed of the wind if the plane travels a distance of 680 miles with a tail wind in the same time it takes to travel 520 miles into a head wind.

74. *Average Speed* During the first part of a 6-hour trip, you travel 240 miles at an average speed of r miles per hour. For the next 72 miles of the trip, you increase your speed by 10 miles per hour. What are your two average speeds?

75. *Speed* One person runs 2 miles per hour faster than a second person. The first person runs 5 miles in the same time the second person runs 4 miles. Find the speed of each person.

76. *Speed* The speed of a commuter plane is 150 miles per hour slower than that of a passenger jet. The commuter plane travels 450 miles in the same time the jet travels 1150 miles. Find the speed of each plane.

77. *Speed* A boat travels at a speed of 20 miles per hour in still water. It travels 48 miles upstream and then returns to the starting point in a total of 5 hours. Find the speed of the current.

78. *Speed* You traveled 72 miles in a certain time period. If you had traveled 6 miles per hour faster, the trip would have taken 10 minutes less time. What was your speed?

79. *Partnership Costs* A group plans to start a new business that will require $240,000 for start-up capital. The individuals in the group will share the cost equally. If two additional people joined the group, the cost per person would decrease by $4000. How many people are presently in the group?

80. *Partnership Costs* A group of people agree to share equally in the cost of a $150,000 endowment to a college. If they could find four more people to join the group, each person's share of the cost would decrease by $6250. How many people are presently in the group?

81. *Partnership Costs* Some partners buy a piece of property for $78,000 by sharing the cost equally. To ease the financial burden, they look for three additional partners to reduce the cost per person by $1300. How many partners are presently in the group?

82. *Population Growth* A biologist introduces 100 insects into a culture. The population P of the culture is approximated by the model

$$P = \frac{500(1 + 3t)}{5 + t}$$

where t is the time in hours. Find the time required for the population to increase to 1000 insects.

83. *Pollution Removal* The cost C in dollars of removing $p\%$ of the air pollutants in the stack emission of a utility company is modeled by

$$C = \frac{120{,}000p}{100 - p}.$$

(a) Use a graphing utility to graph the model. Use the result to estimate graphically the percent of stack emission that can be removed for $680,000.

(b) Use the model to determine algebraically the percent of stack emission that can be removed for $680,000.

84. *Average Cost* The average cost for producing x units of a product is given by

$$\text{Average cost} = 1.50 + \frac{4200}{x}.$$

Determine the number of units that must be produced to have an average cost of $2.90.

Work-Rate Problem In Exercises 85 and 86, complete the table by finding the time required for two individuals to complete a task. The first two columns in the table give the times required for the two individuals to complete the task working alone. (Assume that when they work together their individual rates do not change.)

85.

Person #1	Person #2	Together
6 hours	6 hours	
3 minutes	5 minutes	
5 hours	$2\frac{1}{2}$ hours	

86.

Person #1	Person #2	Together
4 days	4 days	
$5\frac{1}{2}$ hours	3 hours	
a days	b days	

87. *Work-Rate Problem* One landscaper works $1\frac{1}{2}$ times as fast as another landscaper. Find their individual times if it takes them 9 hours working together to complete a certain job.

88. *Work-Rate Problem* Assume that the slower of the two landscapers in Exercise 87 is given another job after 4 hours. The faster of the two must work an additional 10 hours to complete the job. Find the individual times.

89. *Swimming Pool* The flow rate for one pipe is $1\frac{1}{4}$ times that of another pipe. A swimming pool can be filled in 5 hours using both pipes. Find the time required to fill the pool using only the pipe with the slower flow rate.

90. *Swimming Pool* Assume the pipe with the faster flow rate in Exercise 89 is shut off after 1 hour and it takes an additional 10 hours to fill the pool. Find the filling time for each pipe.

Computer and Data Processing Services In Exercises 91 and 92, use the following model, which approximates the total revenue y (in billions of dollars) from computer and data services in the United States for the years 1990 through 1995.

$$y = \frac{87{,}709 - 1236t}{1000 - 93t}, \quad 0 \le t \le 5$$

In this model, $t = 0$ represents 1990. (Source: Current Business Reports)

91. Evaluate the model for each year and compare the results with the actual data shown in the table.

Year	1990	1991	1992	1993	1994	1995
Revenue	88.3	94.4	104.7	116.8	133.1	152.2

92. (a) Use a graphing utility to graph the model. Use the graph to forecast when revenue will exceed $250 billion.

(b) Explain why the model fails after the year 2000.

93. *Construction* The unit for determining the size of a nail is the *penny.* For example, 8d represents an 8-penny nail. The number N of finishing nails per pound can be modeled by

$$N = 43.4 + \frac{9353}{x^2}$$

where x is the size of the nail. (Source: Standard Handbook for Mechanical Engineers)

(a) What is the domain of the function?

(b) Use a graphing utility to graph the function.

(c) Use the graph to determine the size of the finishing nail if there are 135 nails per pound.

94. *Think About It* It is important to distinguish between equations and expressions. In parts (a) through (d), if the exercise is an equation, solve it; if it is an expression, simplify it.

(a) $\dfrac{16}{x^2 - 16} + \dfrac{x}{2x - 8} = \dfrac{1}{2}$

(b) $\dfrac{16}{x^2 - 16} + \dfrac{x}{2x - 8} + \dfrac{1}{2}$

(c) $\dfrac{5}{x + 3} + \dfrac{5}{3} + 3$

(d) $\dfrac{5}{x + 3} + \dfrac{5}{3} = 3$

Explaining Concepts

95. Answer parts (d) and (e) of Motivating the Chapter on page 443.

96. (a) Explain the difference between an equation and an expression.

(b) Compare the use of a common denominator in solving rational equations with its use in adding or subtracting rational expressions.

97. Describe how to solve a rational equation.

98. Define the term *extraneous solution.* How do you identify an extraneous solution?

99. Describe the steps that can be used to transform an equation into an equivalent equation.

100. Explain how you can use a graphing utility to estimate the solution of a rational equation.

101. When can you use cross-multiplication to solve a rational equation? Explain.

8.5 Graphs of Rational Functions

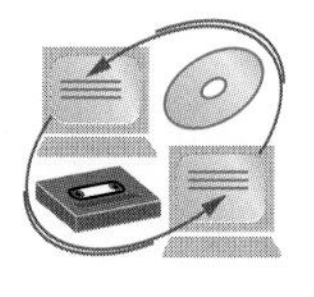

Objectives

1. Use a table of values to sketch the graph of a rational function.
2. Determine horizontal and vertical asymptotes of a rational function.
3. Use asymptotes and intercepts to sketch the graph of a rational function.
4. Use the graph of a rational function to solve an application problem.

1 Use a table of values to sketch the graph of a rational function.

Introduction

Recall that the domain of a rational function consists of all values of x for which the denominator is not zero. For instance, the domain of

$$f(x) = \frac{x+2}{x-1}$$

is all real numbers except $x = 1$. When graphing a rational function, pay special attention to the shape of the graph near x-values that are not in the domain.

Example 1 Sketching the Graph of a Rational Function

Sketch the graph of $f(x) = \dfrac{x+2}{x-1}$.

Solution

Begin by noticing that the domain is all real numbers except $x = 1$. Next, construct a table of values, including x-values that are close to 1 on the left *and* the right.

x-Values to the Left of 1

x	-3	-2	-1	0	0.5	0.9
$f(x)$	0.25	0	-0.5	-2	-5	-29

x-Values to the Right of 1

x	1.1	1.5	2	3	4	5
$f(x)$	31	7	4	2.5	2	1.75

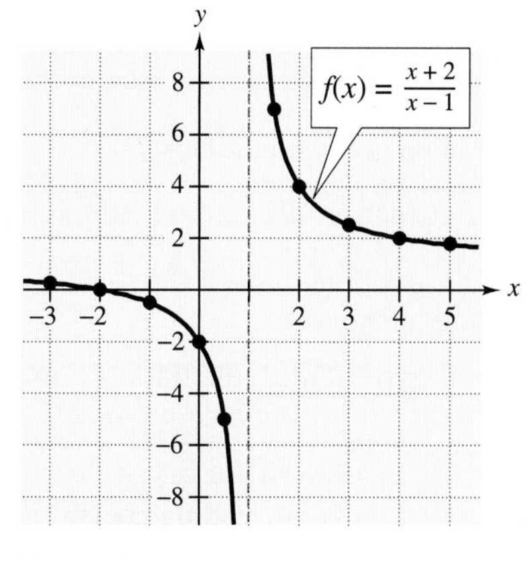

Figure 8.1

Plot the points to the left of 1 and connect them with a smooth curve, as shown in Figure 8.1. Do the same for the points to the right of 1. *Do not* connect the two portions of the graph, which are called its **branches.**

In Figure 8.1, as x approaches 1 from the left, the values of $f(x)$ approach negative infinity, and as x approaches 1 from the right, the values of $f(x)$ approach positive infinity.

2 Determine horizontal and vertical asymptotes of a rational function.

Horizontal and Vertical Asymptotes

An **asymptote** of a graph is a line to which the graph becomes arbitrarily close as $|x|$ or $|y|$ increases without bound. In other words, if a graph has an asymptote, it is possible to move far enough out on the graph so that there is almost no difference between the graph and the asymptote.

The graph in Figure 8.2(a) has two asymptotes: the line $x = -2$ is a **vertical asymptote,** and the line $y = 1$ is a **horizontal asymptote.** The graph in Figure 8.2(b) has three asymptotes: the lines $x = -2$ and $x = 2$ are vertical asymptotes, and the line $y = 2$ is a horizontal asymptote.

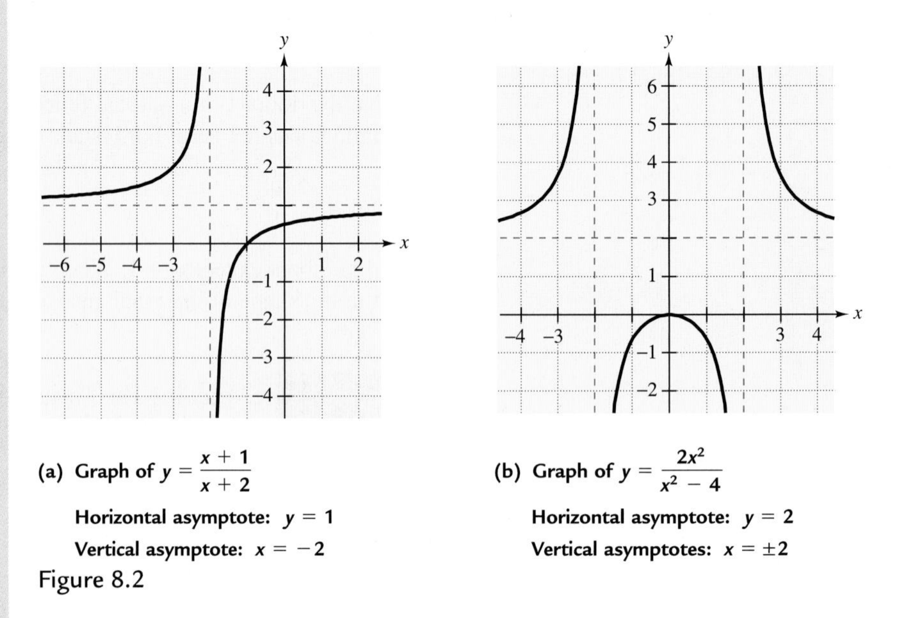

(a) Graph of $y = \frac{x+1}{x+2}$

Horizontal asymptote: $y = 1$

Vertical asymptote: $x = -2$

(b) Graph of $y = \frac{2x^2}{x^2 - 4}$

Horizontal asymptote: $y = 2$

Vertical asymptotes: $x = \pm 2$

Figure 8.2

The graph of a rational function may have no horizontal or vertical asymptotes, or it may have several.

▶ Guidelines for Finding Asymptotes

Let $f(x) = p(x)/q(x)$, where $p(x)$ and $q(x)$ have no common factors.

1. The graph of f has a vertical asymptote at each x-value for which the denominator is zero.
2. The graph of f has at most one horizontal asymptote.
   (a) If the degree of $p(x)$ is less than the degree of $q(x)$, the line $y = 0$ is a horizontal asymptote.
   (b) If the degree of $p(x)$ is equal to the degree of $q(x)$, the line $y = a/b$ is a horizontal asymptote, where a is the leading coefficient of $p(x)$ and b is the leading coefficient of $q(x)$.
   (c) If the degree of $p(x)$ is greater than the degree of $q(x)$, the graph has no horizontal asymptote.

Technology: Tip

A graphing utility can help you sketch the graph of a rational function. With most graphing utilities, however, there are problems with graphs of rational functions. If you use *connected mode*, the graphing utility will try to connect any branches of the graph. If you use *dot mode*, the graphing utility will draw a dotted (rather than a solid) graph. Both of these options are shown below for the graph of $y = (x - 1)/(x - 3)$.

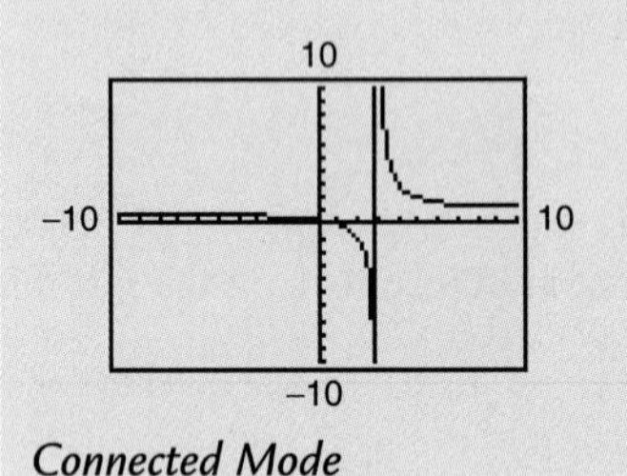

Connected Mode

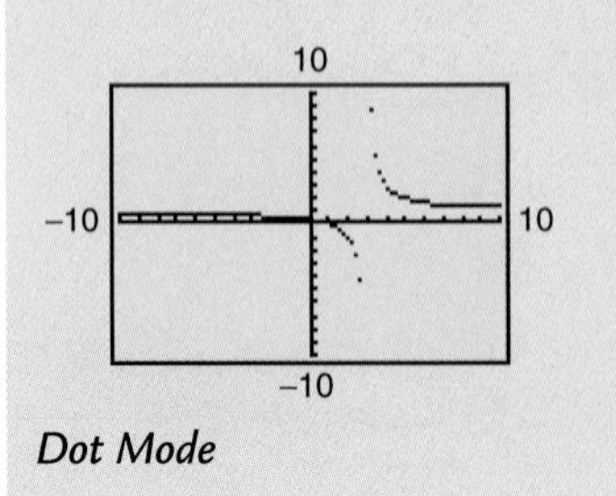

Dot Mode

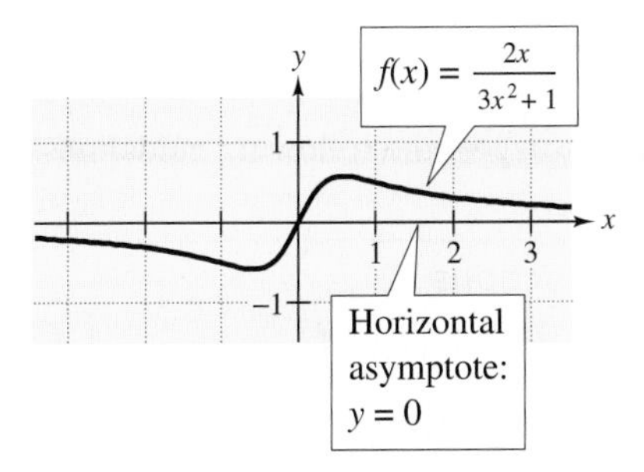

Figure 8.3

Example 2 Finding Horizontal and Vertical Asymptotes

Find all horizontal and vertical asymptotes of the graph of

$$f(x) = \frac{2x}{3x^2 + 1}.$$

Solution

For this rational function, the degree of the numerator is less than the degree of the denominator. This implies that the graph has the line

$$y = 0 \qquad \text{Horizontal asymptote}$$

as a horizontal asymptote, as shown in Figure 8.3. To find any vertical asymptotes, set the denominator equal to zero and solve the resulting equation for x.

$$3x^2 + 1 = 0 \qquad \text{Set denominator equal to zero.}$$

Because this equation has no real solution, you can conclude that the graph has no vertical asymptote.

Remember that the graph of a rational function can have at most one horizontal asymptote, but it can have several vertical asymptotes. For instance, the graph in Example 3 has two vertical asymptotes.

Example 3 Finding Horizontal and Vertical Asymptotes

Find all horizontal and vertical asymptotes of the graph of

$$f(x) = \frac{2x^2}{x^2 - 1}.$$

Solution

For this rational function, the degree of the numerator is equal to the degree of the denominator. The leading coefficient of the numerator is 2, and the leading coefficient of the denominator is 1. So, the graph has the line

$$y = \frac{2}{1} = 2 \qquad \text{Horizontal asymptote}$$

as a horizontal asymptote, as shown in Figure 8.4. To find any vertical asymptotes, set the denominator equal to zero and solve the resulting equation for x.

$$x^2 - 1 = 0 \qquad \text{Set denominator equal to zero.}$$

$$(x + 1)(x - 1) = 0 \qquad \text{Factor.}$$

$$x + 1 = 0 \implies x = -1 \qquad \text{Set 1st factor equal to 0.}$$

$$x - 1 = 0 \implies x = 1 \qquad \text{Set 2nd factor equal to 0.}$$

This equation has two real solutions: -1 and 1. So, the graph has two vertical asymptotes: the lines $x = -1$ and $x = 1$.

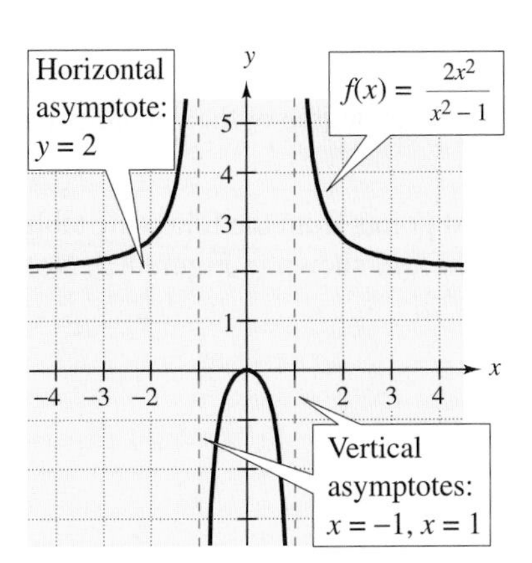

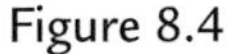
Figure 8.4

3 Use asymptotes and intercepts to sketch the graph of a rational function.

Graphing Rational Functions

To sketch the graph of a rational function, we suggest the following guidelines.

> **Guidelines for Graphing Rational Functions**
>
> Let $f(x) = p(x)/q(x)$, where $p(x)$ and $q(x)$ have no common factors.
>
> 1. Find and plot the y-intercept (if any) by evaluating $f(0)$.
> 2. Set the numerator equal to zero and solve the equation for x. The real solutions represent the x-intercepts of the graph. Plot these intercepts.
> 3. Find and sketch the horizontal and vertical asymptotes of the graph.
> 4. Plot at least one point both between and beyond each x-intercept and vertical asymptote.
> 5. Use smooth curves to complete the graph between and beyond the vertical asymptotes.
> 6. If $p(x)$ and $q(x)$ have a common factor $x - a$, then the graph of $p(x)/q(x)$ has a hole at $x = a$.

Technology: Discovery

The rational function

$$f(x) = \frac{x+2}{x^2-4}$$

has a common factor $x + 2$ in its numerator and denominator. Use the table feature of your graphing utility to examine the values of the function *near* and *at* $x = -2$. What happens at $x = -2$? Is $x = -2$ a vertical asymptote? Explain. Graph this function and use the trace feature to verify your explanation.

Example 4 Sketching the Graph of a Rational Function

Sketch the graph of $f(x) = \dfrac{2}{x-3}$.

Solution

Begin by noting that the numerator and denominator have no common factors. Following the guidelines above produces the following.

- Because $f(0) = \dfrac{2}{0-3} = -\dfrac{2}{3}$, the y-intercept is $\left(0, -\dfrac{2}{3}\right)$.
- Because the numerator is never zero, there are no x-intercepts.
- Because the denominator is zero when $x - 3 = 0$ or $x = 3$, the line $x = 3$ is a vertical asymptote.
- Because the degree of the numerator is less than the degree of the denominator, the line $y = 0$ is a horizontal asymptote.

Plot the intercepts, asymptotes, and the additional points from the following table. Then complete the graph by drawing two branches, as shown in Figure 8.5. Note that the two branches are not connected.

x	1	2	4	5	6
$f(x)$	-1	-2	2	1	$\frac{2}{3}$

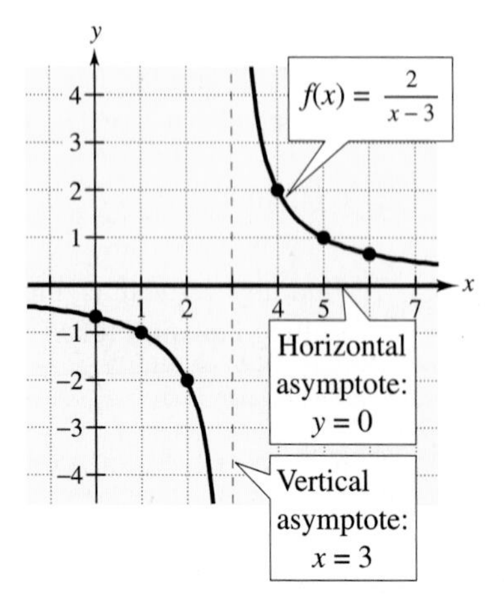

Figure 8.5

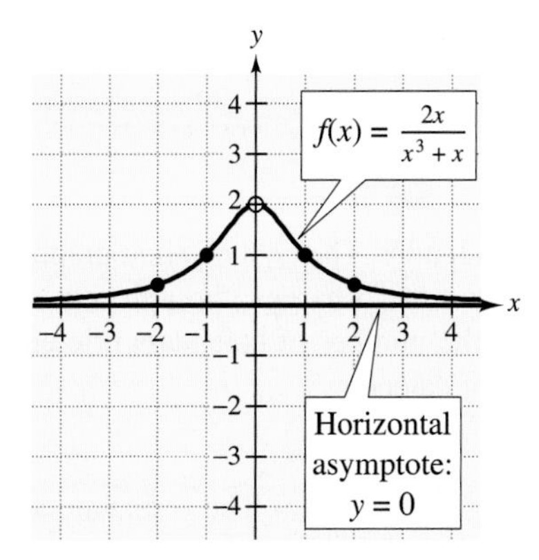

Figure 8.6

Example 5 A Rational Function with a Hole in Its Graph

Sketch the graph of $f(x) = \dfrac{2x}{x^3 + x}$.

Solution

Begin by noting that the numerator and denominator have a common factor x, and so the graph has a hole at $x = 0$.

- The simplified form of the equation is $f(x) = \dfrac{2}{x^2 + 1}$.
- Because the numerator of the simplified form is never zero, there are no x-intercepts.
- Because the denominator is never zero, there are no vertical asymptotes.
- Because the degree of the numerator is less than the degree of the denominator, the line $y = 0$ is a horizontal asymptote.

By identifying the hole, sketching the asymptotes, and plotting the additional points from the following table, you can obtain the graph shown in Figure 8.6.

x	-2	-1	0	1	2
$f(x)$	$\frac{2}{5}$	1	Undefined	1	$\frac{2}{5}$

Example 6 Sketching the Graph of a Rational Function

Sketch the graph of $f(x) = \dfrac{4x^2}{x^2 - 4}$.

Solution

Begin by noting that the numerator and denominator have no common factors.

- Because $f(0) = \dfrac{4(0)^2}{(0)^2 - 4} = 0$, the y-intercept is $(0, 0)$.
- Because the numerator is zero when $4x^2 = 0$ or $x = 0$, the x-intercept is $(0, 0)$.
- Because the denominator is zero when $x^2 - 4 = 0$ or $(x - 2)(x + 2) = 0$, the lines $x = 2$ and $x = -2$ are vertical asymptotes.
- Because the degree of the numerator equals the degree of the denominator, the line $y = a/b = 4/1 = 4$ is a horizontal asymptote.

By plotting the intercept, sketching the asymptotes, and plotting the additional points from the following table, you can obtain the graph shown in Figure 8.7.

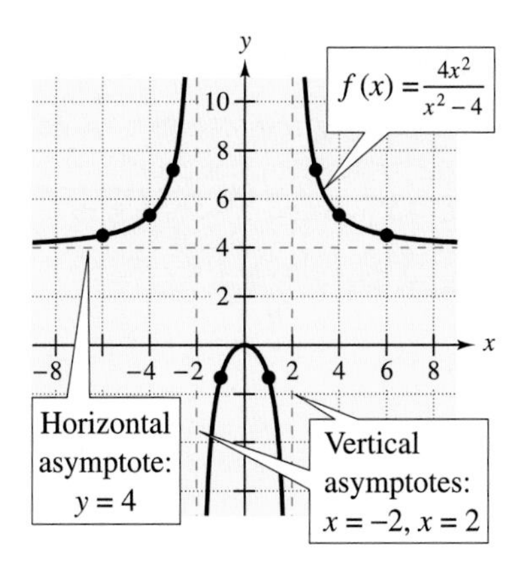

Figure 8.7

x	-6	-4	-3	-1	1	3	4	6
$f(x)$	$\frac{9}{2}$	$\frac{16}{3}$	$\frac{36}{5}$	$-\frac{4}{3}$	$-\frac{4}{3}$	$\frac{36}{5}$	$\frac{16}{3}$	$\frac{9}{2}$

4 Use the graph of a rational function to solve an application problem.

Application

Example 7 Finding the Average Cost 

As a fundraising project, a club is publishing a calendar. The cost of photography and typesetting is \$850. In addition to these "one-time" charges, the unit cost of printing each calendar is \$3.25. Let x represent the number of calendars printed. Write a model that represents the average cost per calendar.

Solution

To begin, you need to find the total cost of printing the calendars. The verbal model for the total cost is

$$\boxed{\text{Total cost}} = \boxed{\text{Unit cost}} \times \boxed{\text{Number of calendars}} + \boxed{\text{Cost of photography and typesetting}}.$$

The total cost C of printing x calendars is

$$C = 3.25x + 850. \qquad \text{Total cost function}$$

The verbal model for the average cost per calendar is

$$\boxed{\text{Average cost}} = \frac{\boxed{\text{Total cost}}}{\boxed{\text{Number of calendars}}}.$$

The average cost per calendar $\overline{C}$ for printing x calendars is

$$\overline{C} = \frac{3.25x + 850}{x}. \qquad \text{Average cost function}$$

From the graph shown in Figure 8.8, notice that the average cost decreases as the number of calendars increases.

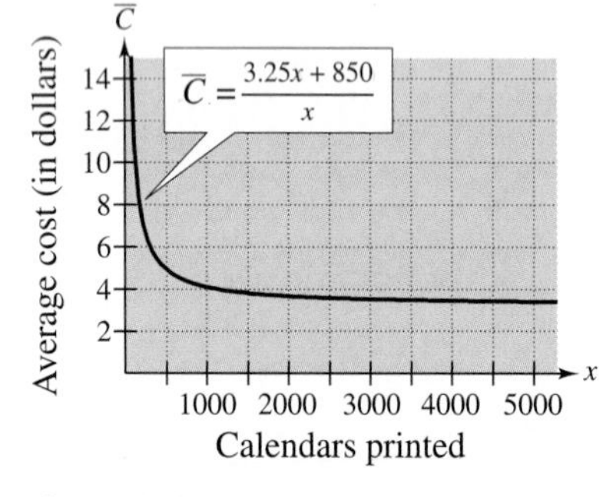

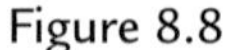
Figure 8.8

Discussing the Concept — More About the Average Cost

In Example 7, what is the horizontal asymptote of the graph of the average cost function? What is the significance of this asymptote in the problem? Is it possible to sell enough calendars to obtain an average cost of \$3.00 per calendar? Explain your reasoning.

8.5 Exercises

Integrated Review *Concepts, Skills, and Problem Solving*

Keep mathematically in shape by doing these exercises *before* the problems of this section.

Properties and Definitions

1. Identify the leading coefficient in $7x^2 + 3x - 4$. Explain.

2. State the degree of the product $(x^4 + 3)(x - 4)$. Explain.

3. Sketch a graph for which y is not a function of x. Explain.

4. Sketch a graph for which y is a function of x. Explain.

Multiplying Expressions

In Exercises 5–10, find the product.

5. $-2x^5(5x^3)$

6. $3x(5 - 2x)$

7. $(2x - 15)^2$

8. $(3x + 2)(7x - 10)$

9. $[(x + 1) - y][(x + 1) + y]$

10. $(x + 3)(x^2 - 3x + 9)$

Problem Solving

11. The height of a triangle is 12 meters less than its base. Find the base and height of the triangle if its area is 80 square meters.

12. An open box with a square base is to be constructed from 825 square inches of material. What should be the dimensions of the base if the height of the box is to be 10 inches?

Developing Skills

Numerical Analysis In Exercises 1 and 2, (a) complete each table, (b) use the tables to sketch the graph, and (c) determine the domain of the function. See Example 1.

x	0	0.5	0.9	0.99	0.999
y					

x	2	1.5	1.1	1.01	1.001
y					

x	2	5	10	100	1000
y					

1. $f(x) = \dfrac{4}{x - 1}$

2. $f(x) = \dfrac{2x}{x - 1}$

Numerical Analysis In Exercises 3–6, (a) complete each table, (b) use the tables to sketch the graph, and (c) determine the domain of the function.

x	2	2.5	2.9	2.99	2.999
y					

x	4	3.5	3.1	3.01	3.001
y					

x	4	5	10	100	1000
y					

3. $f(x) = 2 + \dfrac{1}{x - 3}$

4. $f(x) = \dfrac{2}{x - 3}$

5. $f(x) = \dfrac{3x}{x^2 - 9}$

6. $f(x) = \dfrac{5x^2}{x^2 - 9}$

In Exercises 7–24, find the domain of the function and identify any horizontal and vertical asymptotes. See Examples 2 and 3.

7. $f(x) = \dfrac{5}{x^2}$

8. $g(x) = \dfrac{3}{x}$

9. $f(x) = \dfrac{x}{x + 8}$

10. $f(u) = \dfrac{u^2}{u - 10}$

11. $g(t) = \dfrac{2t - 5}{3t - 9}$

12. $h(x) = \dfrac{4x - 3}{2x + 5}$

13. $y = \dfrac{3 - 5x}{1 - 3x}$

14. $y = \dfrac{3x + 2}{2x - 1}$

15. $g(t) = \dfrac{3}{t(t - 1)}$

16. $h(s) = \dfrac{2s + 1}{s(s + 3)}$

17. $y = \dfrac{2x^2}{x^2 + 1}$

18. $g(t) = \dfrac{3t^3}{t^2 + 1}$

19. $y = \dfrac{x^2 - 4}{x^2 - 1}$

20. $y = \dfrac{x^2 - 9}{x^2 - 2x - 8}$

21. $g(z) = 1 - \dfrac{2}{z}$

22. $h(v) = \dfrac{3}{v} - 2$

23. $g(x) = 2x + \dfrac{4}{x}$

24. $f(t) = \dfrac{5}{t} - 4t$

In Exercises 25–28, identify the horizontal and vertical asymptotes of the function. Use the asymptotes to match the rational function with its graph. [The graphs are labeled (a), (b), (c), and (d).]

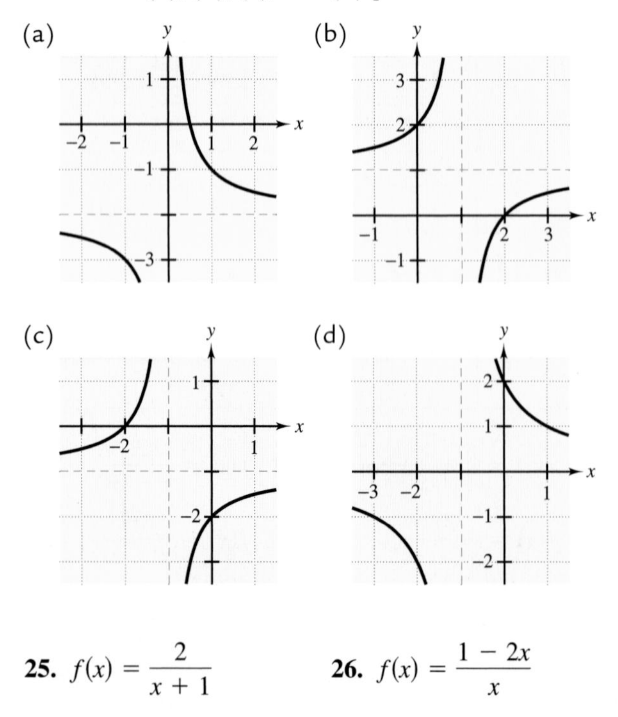

25. $f(x) = \dfrac{2}{x + 1}$

26. $f(x) = \dfrac{1 - 2x}{x}$

27. $f(x) = \dfrac{x - 2}{x - 1}$

28. $f(x) = -\dfrac{x + 2}{x + 1}$

In Exercises 29–32, match the rational function with its graph. [The graphs are labeled (a), (b), (c), and (d).]

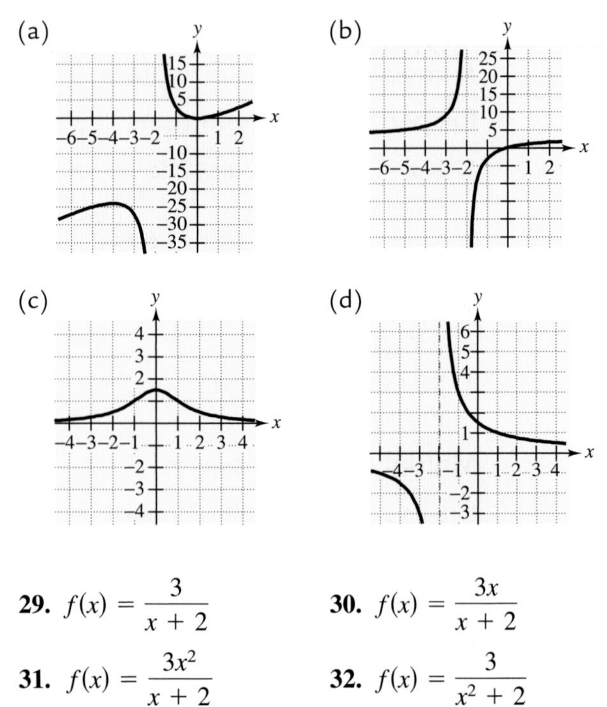

29. $f(x) = \dfrac{3}{x + 2}$

30. $f(x) = \dfrac{3x}{x + 2}$

31. $f(x) = \dfrac{3x^2}{x + 2}$

32. $f(x) = \dfrac{3}{x^2 + 2}$

In Exercises 33–58, sketch the graph of the function. As sketching aids, check for intercepts, vertical asymptotes, and horizontal asymptotes. See Examples 4–6.

33. $g(x) = \dfrac{5}{x}$

34. $f(x) = \dfrac{5}{x^2}$

35. $g(x) = \dfrac{5}{x - 4}$

36. $f(x) = \dfrac{5}{(x - 4)^2}$

37. $f(x) = \dfrac{1}{x - 2}$

38. $f(x) = \dfrac{3}{x + 1}$

39. $g(x) = \dfrac{1}{2 - x}$

40. $g(x) = \dfrac{-3}{x + 1}$

41. $y = \dfrac{3x}{x^2 + 4x}$

42. $y = \dfrac{2x}{x^2 + 4x}$

43. $h(u) = \dfrac{3u^2}{u^2 - 3u}$

44. $g(v) = \dfrac{2v^2}{v^2 + v}$

45. $y = \dfrac{2x + 4}{x}$

46. $y = \dfrac{x - 2}{x}$

47. $y = \dfrac{2x^2}{x^2 + 1}$

48. $y = \dfrac{10}{x^2 + 2}$

49. $y = \dfrac{4}{x^2 + 1}$

50. $y = \dfrac{4x^2}{x^2 + 1}$

51. $g(t) = 3 - \dfrac{2}{t}$

52. $f(x) = \dfrac{4}{x} + 2$

53. $y = -\dfrac{x}{x^2 - 4}$

54. $y = \dfrac{4x + 6}{x^2 - 9}$

55. $f(x) = \dfrac{3x^2}{x^2 - x - 2}$

56. $g(x) = \dfrac{2x^2}{x^2 + 2x - 3}$

57. $f(x) = \dfrac{x^2 - 4}{x^2 - 3x - 10}$

58. $g(t) = \dfrac{t^2 - 9}{t^2 + 6t + 9}$

In Exercises 59–68, use a graphing utility to graph the function. Give the domain of the function and identify any horizontal or vertical asymptotes.

59. $f(x) = \dfrac{3}{x + 2}$

60. $f(x) = \dfrac{3x}{x + 2}$

61. $h(x) = \dfrac{x - 3}{x - 1}$

62. $h(x) = \dfrac{x^2}{x - 2}$

63. $f(t) = \dfrac{6}{t^2 + 1}$

64. $g(t) = 2 + \dfrac{3}{t + 1}$

65. $y = \dfrac{2(x^2 + 1)}{x^2}$

66. $y = \dfrac{2(x^2 - 1)}{x^2}$

67. $y = \dfrac{3}{x} + \dfrac{1}{x - 2}$

68. $y = \dfrac{x}{2} - \dfrac{2}{x}$

Think About It In Exercises 69 and 70, use a graphing utility to graph the function. Explain why there is no vertical asymptote when a superficial examination of the function seems to indicate that there should be one.

69. $g(x) = \dfrac{4 - 2x}{x - 2}$

70. $h(x) = \dfrac{x^2 - 9}{x + 3}$

Solving Problems

71. *Average Cost* The cost of producing x units of a product is $C = 2500 + 0.50x$, $x > 0$.

(a) Write the average cost $\overline{C}$ as a function of x.

(b) Find the average cost of producing $x = 1000$ and $x = 10{,}000$ units.

(c) Use a graphing utility to graph the average cost function. Determine the horizontal asymptote of the graph. Interpret the result.

72. *Average Cost* The cost of producing x units of a product is $C = 30{,}000 + 1.25x$, $x > 0$.

(a) Write the average cost $\overline{C}$ as a function of x.

(b) Find the average cost of producing $x = 10{,}000$ and $x = 100{,}000$ units.

(c) Use a graphing utility to graph the average cost function. Determine the horizontal asymptote of the graph. Interpret the result.

73. *Medicine* The concentration of a certain chemical in the bloodstream t hours after injection into the muscle tissue is given by

$$C = \frac{2t}{4t^2 + 25}, \quad t \geq 0.$$

(a) Determine the horizontal asymptote of the function and interpret its meaning in the context of the problem.

(b) Use a graphing utility to graph the function. Approximate the time when the concentration is the greatest.

74. *Concentration of a Mixture* A 25-liter container contains 5 liters of a 25% brine solution. You add x liters of a 75% brine solution to the container. The concentration C of the resulting mixture is

$$C = \frac{3x + 5}{4(x + 5)}.$$

(a) Determine the domain of the rational function based on the physical constraints of the problem.

(b) Use a graphing utility to graph the function. As the container is filled, what does the concentration of the brine appear to approach?

75. *Geometry* A rectangular region of length x and width y has an area of 400 square meters.

(a) Sketch a figure that gives a visual representation of the problem.

(b) Verify that the perimeter P is given by

$$P = 2\left(x + \frac{400}{x}\right).$$

(c) Determine the domain of the function based on the physical constraints of the problem.

(d) Use a graphing utility to graph the function. Approximate the dimensions of the rectangle that has a minimum perimeter.

76. *Sales* The cumulative number N (in thousands) of units of a product sold over a period of t years is modeled by

$$N = \frac{150t(1 + 4t)}{1 + 0.15t^2}, \quad t \geq 0.$$

(a) Estimate N when $t = 1$, $t = 2$, and $t = 4$.

(b) Use a graphing utility to graph the function. Determine the horizontal asymptote.

(c) Explain the meaning of the horizontal asymptote in the context of the problem.

Think About It In Exercises 77–80, write a rational function satisfying each of the criteria. Use a graphing utility to verify the results.

77. Vertical asymptote: $x = 3$
Horizontal asymptote: $y = 2$
Zero of the function: $x = -1$

78. Vertical asymptote: $x = -2$
Horizontal asymptote: $y = 0$
Zero of the function: $x = 3$

79. Vertical asymptotes: $x = 4$ and $x = -2$
Horizontal asymptote: $y = 0$
Zero of the function: $x = 6$

80. Vertical asymptotes: $x = 1$ and $x = -1$
Horizontal asymptote: $y = 1$
Zero of the function: $x = 0$

Explaining Concepts

81. Answer parts (f)–(h) of Motivating the Chapter on page 443.

82. In your own words, describe how to determine the domain of a rational function. Give an example of a rational function whose domain is all real numbers except 2.

83. In your own words, describe what is meant by an *asymptote* of a graph.

84. *True or False?* If the graph of a rational function f has a vertical asymptote at $x = 3$, it is possible to sketch the graph without lifting your pencil from the paper. Explain.

85. Does every rational function have a vertical asymptote? Explain.

8.6 Variation

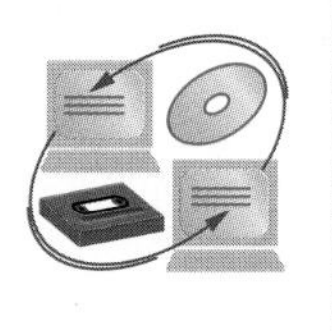

Objectives

1. Solve an application problem involving direct variation.
2. Solve an application problem involving inverse variation.
3. Solve an application problem involving joint variation.

1 Solve an application problem involving direct variation.

Direct Variation

In the mathematical model for **direct variation,** y is a *linear* function of x. Specifically, $y = kx$.

Study Tip

To use this mathematical model in applications involving direct variation, you are usually given specific values of x and y, which then enable you to find the value of the constant k.

> **Direct Variation**
>
> The following statements are equivalent. The number k is the **constant of proportionality.**
>
> 1. y **varies directly** as x.
> 2. y is **directly proportional** to x.
> 3. $y = kx$ for some constant k.

Example 1 Direct Variation 

Assume that the total revenue R (in dollars) obtained from selling x units of a product is directly proportional to the number of units sold. When 10,000 units are sold, the total revenue is \$142,500.

a. Find a model that relates the total revenue R to the number of units sold x.

b. Find the total revenue obtained from selling 12,000 units.

Solution

a. Because the total revenue is directly proportional to the number of units sold, the linear model is $R = kx$. To find the value of the constant k, substitute 142,500 for R and 10,000 for x

$$142{,}500 = k(10{,}000) \quad \text{Substitute for } R \text{ and } x.$$

which implies that $k = 142{,}500/10{,}000 = 14.25$. So, the equation relating the total revenue to the total number of units sold is

$$R = 14.25x. \quad \text{Direct variation model}$$

The graph of this equation is shown in Figure 8.9.

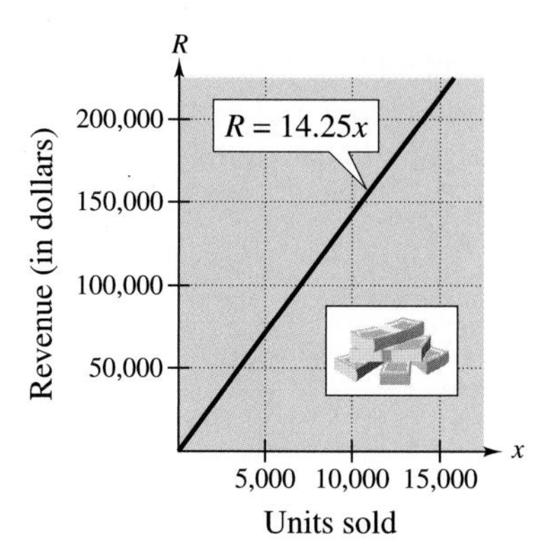

Figure 8.9

b. When $x = 12{,}000$, the total revenue is

$$R = 14.25(12{,}000) = \$171{,}000.$$

Example 2 Direct Variation

Hooke's Law for springs states that the distance a spring is stretched (or compressed) is proportional to the force on the spring. A force of 20 pounds stretches a particular spring 5 inches.

a. Find a mathematical model that relates the distance the spring is stretched to the force applied to the spring.

b. How far will a force of 30 pounds stretch the spring?

Solution

a. For this problem, let d represent the distance (in inches) that the spring is stretched and let F represent the force (in pounds) that is applied to the spring. Because the distance d is proportional to the force F, the model is

$$d = kF.$$

To find the value of the constant k, use the fact that $d = 5$ when $F = 20$. Substituting these values into the given model produces

$$5 = k(20) \qquad \text{Substitute 5 for } d \text{ and 20 for } F.$$

$$\frac{5}{20} = k \qquad \text{Divide both sides by 20.}$$

$$\frac{1}{4} = k. \qquad \text{Simplify.}$$

So, the equation relating distance and force is

$$d = \frac{1}{4}F. \qquad \text{Direct variation model}$$

b. When $F = 30$, the distance is

$$d = \frac{1}{4}(30) = 7.5 \text{ inches.} \qquad \text{See Figure 8.10.}$$

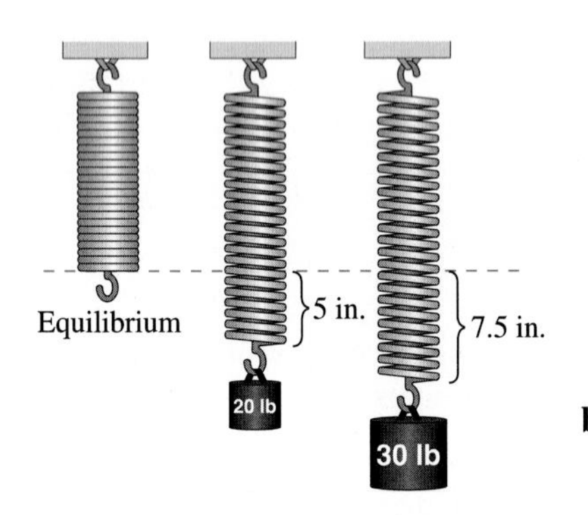

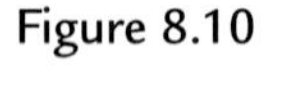
Figure 8.10

In Example 2, you can get a clearer understanding of Hooke's Law by using the model $d = \frac{1}{4}F$ to create a table or a graph (see Figure 8.11). From the table or from the graph, you can see what it means for the distance to be "proportional to the force."

Force, F	10 lb	20 lb	30 lb	40 lb	50 lb	60 lb
Distance, d	2.5 in.	5.0 in.	7.5 in.	10.0 in.	12.5 in.	15.0 in.

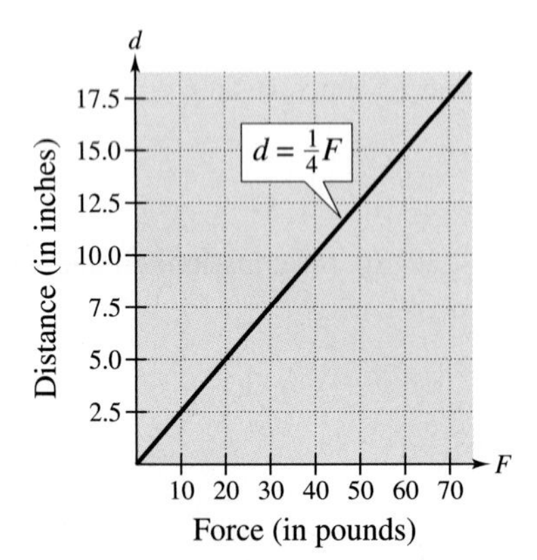

Figure 8.11

In Examples 1 and 2, the direct variations are such that an *increase* in one variable corresponds to an *increase* in the other variable. There are, however, other applications of direct variation in which an increase in one variable corresponds to a *decrease* in the other variable. For instance, in the model $y = -2x$, an increase in x will yield a decrease in y.

A second type of direct variation relates one variable to a *power* of another.

▶ Direct Variation as *n*th Power

The following statements are equivalent.

1. y **varies directly as the *n*th power** of x.
2. y is **directly proportional to the *n*th power** of x.
3. $y = kx^n$ for some constant k.

Example 3 Direction Variation as a Power

The distance a ball rolls down an inclined plane is directly proportional to the square of the time it rolls. Assume that, during the first second, a ball rolls down a particular plane a distance of 6 feet.

a. Find a mathematical model that relates the distance traveled to the time.

b. How far will the ball roll during the first 2 seconds?

Solution

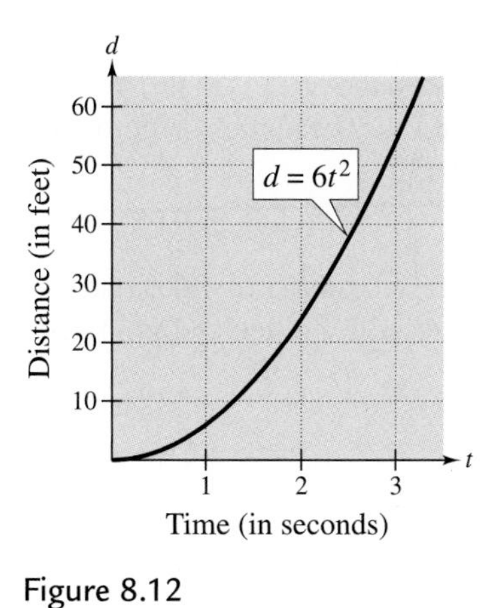

Figure 8.12

a. Letting d be the distance (in feet) that the ball rolls and letting t be the time (in seconds), you obtain the model

$$d = kt^2.$$

Because $d = 6$ when $t = 1$, you obtain

$$d = kt^2 \qquad \text{Original equation}$$

$$6 = k(1)^2 \qquad \text{Substitute 6 for } d \text{ and 1 for } t.$$

$$6 = k. \qquad \text{Simplify.}$$

So, the equation relating distance to time is

$$d = 6t^2. \qquad \text{Direct variation as 2nd power model}$$

The graph of this equation is shown in Figure 8.12.

b. When $t = 2$, the distance traveled is

$$d = 6(2)^2 = 6(4) = 24 \text{ feet.} \qquad \text{See Figure 8.13.}$$

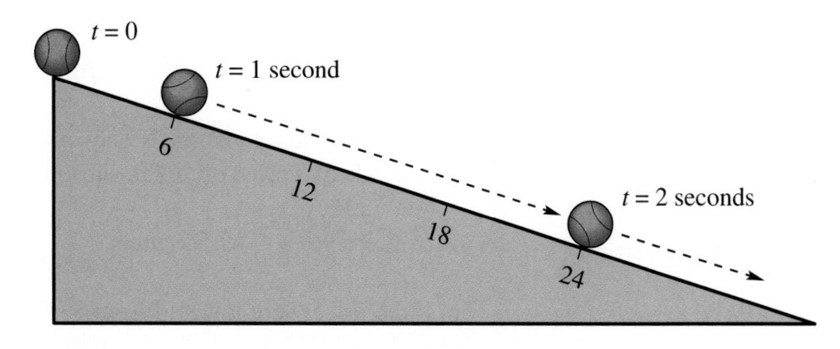

Figure 8.13

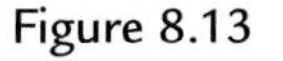

2 Solve an application problem involving inverse variation.

Inverse Variation

Another type of variation is called **inverse variation.** For this type of variation, we say that one of the variables is inversely proportional to the other variable.

> **Inverse Variation**
>
> **1.** The following three statements are equivalent.
>
> **a.** y **varies inversely** as x.
>
> **b.** y is **inversely proportional** to x.
>
> **c.** $y = \frac{k}{x}$ for some constant k.
>
> **2.** If $y = \frac{k}{x^n}$, then y is inversely proportional to the nth power of x.

Example 4 Inverse Variation 

The marketing department of a large company has found that the demand for one of its products varies inversely as the price of the product. (When the price is low, more people are willing to buy the product than when the price is high.) When the price of the product is \$7.50, the monthly demand is 50,000 units. Approximate the monthly demand if the price is reduced to \$6.00.

Solution

Let x represent the number of units that are sold each month (the demand), and let p represent the price per unit (in dollars). Because the demand is inversely proportional to the price, the model is

$$x = \frac{k}{p}.$$

By substituting $x = 50{,}000$ when $p = 7.50$, you obtain

$$50{,}000 = \frac{k}{7.50} \qquad \text{Substitute 50,000 for } x \text{ and 7.50 for } p.$$

$$375{,}000 = k. \qquad \text{Multiply both sides by 7.50.}$$

So, the model is

$$x = \frac{375{,}000}{p}. \qquad \text{Inverse variation model}$$

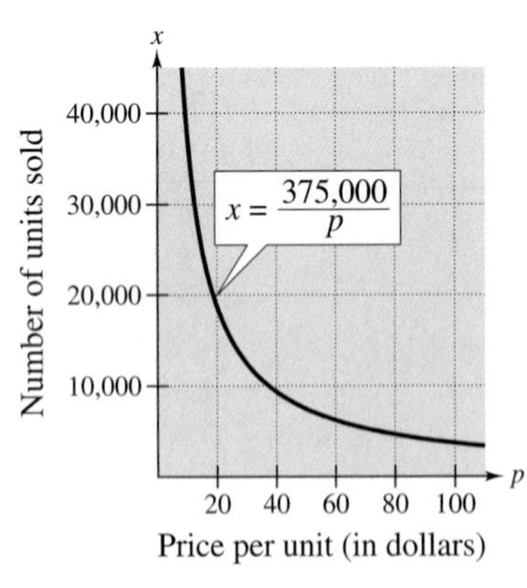

Figure 8.14

The graph of this equation is shown in Figure 8.14. To find the demand that corresponds to a price of \$6.00, substitute $p = 6$ into the equation and obtain

$$x = \frac{375{,}000}{6} = 62{,}500 \text{ units.}$$

So, if the price were lowered from \$7.50 per unit to \$6.00 per unit, the monthly demand could be expected to increase from 50,000 units to 62,500 units.

Some applications of variation involve problems with *both* direct and inverse variation in the same model.

Example 5 Direct and Inverse Variation 

A company determines that the demand for one of its products is directly proportional to the amount spent on advertising and inversely proportional to the price of the product. When \$40,000 is spent on advertising and the price per unit is \$20, the monthly demand is 10,000 units.

a. If the amount of advertising were increased to \$50,000, how much could the price be increased to maintain a monthly demand of 10,000 units?

b. If you were in charge of the advertising department, would you recommend this increased expense in advertising?

Solution

a. Let x represent the number of units that are sold each month (the demand), let a represent the amount spent on advertising (in dollars), and let p represent the price per unit (in dollars). Because the demand is directly proportional to the advertising and inversely proportional to the price, the model is

$$x = \frac{ka}{p}.$$

By substituting $x = 10{,}000$ when $a = 40{,}000$ and $p = 20$, you obtain

$$10{,}000 = \frac{k(40{,}000)}{20}$$ Substitute 10,000 for x, 40,000 for a, and 20 for p.

$$200{,}000 = 40{,}000k$$ Multiply both sides by 20.

$$5 = k.$$ Divide both sides by 40,000.

So, the model is

$$x = \frac{5a}{p}.$$ Direct and inverse variation model

To find the price that corresponds to a demand of 10,000 and an advertising expense of \$50,000, substitute $x = 10{,}000$ and $a = 50{,}000$ into the model and solve for p.

$$10{,}000 = \frac{5(50{,}000)}{p} \quad \Longrightarrow \quad p = \frac{5(50{,}000)}{10{,}000} = \$25$$

So, the price increase would be $\$25 - \$20 = \$5$.

b. The total revenue from selling 10,000 units at \$20 each is \$200,000, and the revenue from selling 10,000 units at \$25 each is \$250,000. So, increasing the advertising expense from \$40,000 to \$50,000 would increase the revenue by \$50,000. This implies that you should recommend the increased expense in advertising.

Amount of Advertising	*Price*	*Revenue*
\$40,000	\$20.00	$10{,}000 \times 20 = \$200{,}000$
\$50,000	\$25.00	$10{,}000 \times 25 = \$250{,}000$

3 Solve an application problem involving joint variation.

Joint Variation

A third type of variation is called **joint variation.** For this type of variation, we say that one variable varies directly with the product of two variables.

> **Joint Variation**
>
> 1. The following three statements are equivalent.
>   a. z **varies jointly** as x and y.
>   b. z is **jointly proportional** to x and y.
>   c. $z = kxy$ for some constant k.
> 2. If $z = kx^n y^m$, then z is jointly proportional to the nth power of x and the mth power of y.

Example 6 Joint Variation

The *simple interest* for a certain savings account is jointly proportional to the time and the principal. After one quarter (3 months), the interest for a principal of \$6000 is \$120. How much interest would a principal of \$7500 earn in 5 months?

Solution

To begin, let I represent the interest earned (in dollars), let P represent the principal (in dollars), and let t represent the time (in years). Because the interest is jointly proportional to the time and the principal, the model is

$$I = ktP.$$

Because $I = 120$ when $P = 6000$ and $t = \frac{1}{4}$, it follows that

$$k = 120/\left(6000 \cdot \tfrac{1}{4}\right) = 0.08.$$

So, the model that relates interest to time and principal is

$$I = 0.08tP. \qquad \text{Joint variation model}$$

To find the interest earned on a principal of \$7500 over a 5-month period of time, substitute $P = 7500$ and $t = \frac{5}{12}$ into the model and obtain an interest of

$$I = 0.08\left(\frac{5}{12}\right)(7500) = \$250.$$

Discussing the Concept Creating Variation Models

For each type of variation, create a problem for which $k = 24$. Sketch a graph of each model and discuss how the graphs are the same and how they differ.

Direct variation: $y = kx$ *Inverse variation:* $y = \frac{k}{x}$ *Joint variation:* $z = kxy$

8.6 Exercises

Integrated Review *Concepts, Skills, and Problem Solving*

Keep mathematically in shape by doing these exercises *before* the problems of this section.

Properties and Definitions

1. Sketch a curve on the rectangular coordinate system such that y *is not* a function of x. Explain.
2. Sketch a curve on the rectangular coordinate system such that y *is* a function of x. Explain.
3. Determine the domain of $f(x) = x^2 - 4x + 9$.
4. Determine the domain of $h(x) = \dfrac{x - 1}{x^2(x^2 + 1)}$.

Functions

In Exercises 5–8, consider the function

$$f(x) = 2x^3 - 3x^2 - 18x + 27$$
$$= (2x - 3)(x + 3)(x - 3).$$

5. Use a graphing utility to graph both expressions for the function. Are the graphs the same?
6. Verify the factorization by multiplying the polynomials in the factored form of f.
7. Verify the factorization by performing the long division

$$\frac{2x^3 - 3x^2 - 18x + 27}{2x - 3}$$

and then factoring the quotient.

8. Verify the factorization by performing the long division

$$\frac{2x^3 - 3x^2 - 18x + 27}{x^2 - 9}.$$

In Exercises 9 and 10, use the function to find and simplify the expression for

$$\frac{f(2 + h) - f(2)}{h}.$$

9. $f(x) = x^2 - 3$
10. $f(x) = \dfrac{3}{x + 5}$

Modeling

11. The inventor of a new game believes that the variable cost for producing the game is $5.75 per unit and the fixed costs are $12,000. If x is the number of games produced, express the total cost C as a function of x.
12. The length of a rectangle is one and one-half times its width. Express the perimeter P of the rectangle as a function of the rectangle's width w.

Developing Skills

In Exercises 1–14, write a model for the statement.

1. I varies directly as V.
2. C varies directly as r.
3. V is directly proportional to t.
4. s varies directly as the cube of t.
5. u is directly proportional to the square of v.
6. V varies directly as the cube root of x.
7. p varies inversely as d.
8. S is inversely proportional to the square of v.
9. P is inversely proportional to the square root of $1 + r$.
10. A varies inversely as the fourth power of t.
11. A varies jointly as l and w.
12. V varies jointly as h and the square of r.
13. *Boyle's Law* If the temperature of a gas is not allowed to change, its absolute pressure P is inversely proportional to its volume V.

14. *Newton's Law of Universal Gravitation* The gravitational attraction F between two particles of masses m_1 and m_2 is directly proportional to the product of the masses and inversely proportional to the square of the distance r between the particles.

In Exercises 15–22, write a verbal sentence using variation terminology to describe the formula.

15. *Area of a Triangle:* $A = \frac{1}{2}bh$

16. *Area of a Circle:* $A = \pi r^2$

17. *Area of a Rectangle:* $A = lw$

18. *Surface Area of a Sphere:* $A = 4\pi r^2$

19. *Volume of a Right Circular Cylinder:* $V = \pi r^2 h$

20. *Volume of a Sphere:* $V = \frac{4}{3}\pi r^3$

21. *Average Speed:* $r = \dfrac{d}{t}$

22. *Height of a Cylinder:* $h = \dfrac{V}{\pi r^2}$

In Exercises 23–36, find the constant of proportionality and write an equation that relates the variables.

23. s varies directly as t, and $s = 20$ when $t = 4$.

24. h is directly proportional to r, and $h = 28$ when $r = 12$.

25. F is directly proportional to the square of x, and $F = 500$ when $x = 40$.

26. v varies directly as the square root of s, and $v = 24$ when $s = 16$.

27. H is directly proportional to u, and $H = 100$ when $u = 40$.

28. M varies directly as the cube of n, and $M = 0.012$ when $n = 0.2$.

29. n varies inversely as m, and $n = 32$ when $m = 1.5$.

30. q is inversely proportional to p, and $q = \frac{3}{2}$ when $p = 50$.

31. g varies inversely as the square root of z, and $g = \frac{4}{5}$ when $z = 25$.

32. u varies inversely as the square of v, and $u = 40$ when $v = \frac{1}{2}$.

33. F varies jointly as x and y, and $F = 500$ when $x = 15$ and $y = 8$.

34. V varies jointly as h and the square of b, and $V = 288$ when $h = 6$ and $b = 12$.

35. d varies directly as the square of x and inversely with r, and $d = 3000$ when $x = 10$ and $r = 4$.

36. z is directly proportional to x and inversely proportional to the square root of y, and $z = 720$ when $x = 48$ and $y = 81$.

Solving Problems

37. *Revenue* The total revenue R is directly proportional to the number of units sold x. When 500 units are sold, the revenue is \$3875.

(a) Find the revenue when 635 units are sold.

(b) Interpret the constant of proportionality.

38. *Revenue* The total revenue R is directly proportional to the number of units sold x. When 25 units are sold, the revenue is \$300.

(a) Find the revenue when 42 units are sold.

(b) Interpret the constant of proportionality.

39. *Hooke's Law* A force of 50 pounds stretches a spring 5 inches (see figure).

(a) How far will a force of 20 pounds stretch the spring?

(b) What force is required to stretch the spring 1.5 inches?

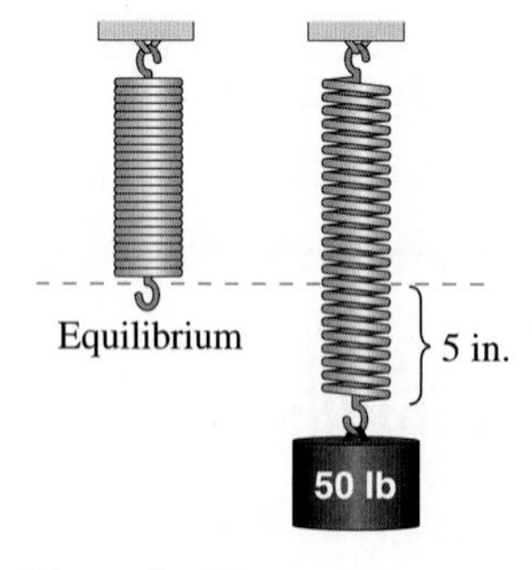

Figure for 39

40. *Hooke's Law* A force of 50 pounds stretches a spring 3 inches.

(a) How far will a force of 20 pounds stretch the spring?

(b) What force is required to stretch the spring 1.5 inches?

41. *Hooke's Law* A baby weighing $10\frac{1}{2}$ pounds compresses the spring of a baby scale 7 millimeters (see figure). Determine the weight of a baby that will compress the spring 12 millimeters.

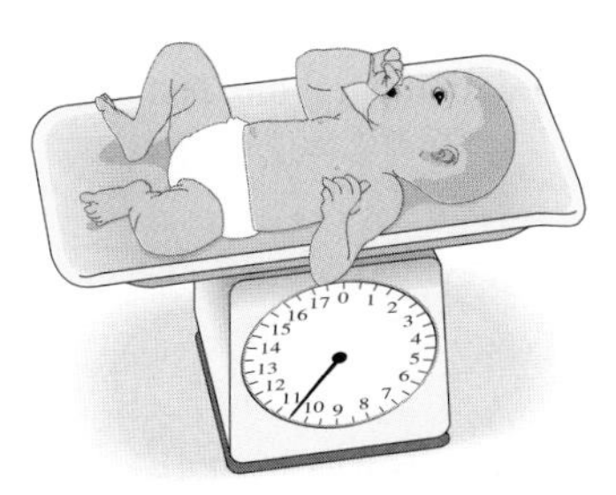

42. *Hooke's Law* A force of 50 pounds stretches the spring of a scale 1.5 inches.

(a) Write the force F as a function of the distance x the spring is stretched.

(b) Graph the function in part (a) where $0 \leq x \leq 5$. Identify the graph.

43. *Free-Falling Object* The velocity v of a free-falling object is proportional to the time that the object has fallen. The constant of proportionality is the acceleration due to gravity. Find the acceleration due to gravity if the velocity of a falling object is 96 feet per second after the object has fallen for 3 seconds.

44. *Free-Falling Object* Neglecting air resistance, the distance d that an object falls varies directly as the square of the time t it has fallen. If an object falls 64 feet in 2 seconds, determine the distance it will fall in 6 seconds.

45. *Stopping Distance* The stopping distance d of an automobile is directly proportional to the square of its speed s. On a certain road surface, a car requires 75 feet to stop when its speed is 30 miles per hour. Estimate the stopping distance if the brakes are applied when the car is traveling at 50 miles per hour under similar road conditions.

46. *Velocity of a Stream* The diameter d of a particle that can be moved by a stream is directly proportional to the square of the velocity v of the stream. A stream with a velocity of $\frac{1}{4}$ mile per hour can move coarse sand particles of about 0.02 inch diameter. What must the velocity be to carry particles with a diameter of 0.12 inch?

47. *Frictional Force* The frictional force F between the tires and the road that is required to keep a car on a curved section of a highway is directly proportional to the square of the speed s of the car. If the speed of the car is doubled, the force will change by what factor?

48. *Power Generation* The power P generated by a wind turbine varies directly as the cube of the wind speed w. The turbine generates 750 watts of power in a 25-mile-per-hour wind. Find the power it generates in a 40-mile-per-hour wind.

49. *Best Buy* The prices of 9-inch, 12-inch, and 15-inch diameter pizzas at a certain pizza shop are \$6.78, \$9.78, and \$12.18, respectively. One would expect that the price of a certain size pizza would be directly proportional to its surface area. Is that the case for this pizza shop? If not, which size pizza is the best buy?

50. *Travel Time* The travel time between two cities is inversely proportional to the average speed. If a train travels between two cities in 3 hours at an average speed of 65 miles per hour, how long would it take at an average speed of 80 miles per hour? What does the constant of proportionality measure in this problem?

51. *Demand Function* A company has found that the daily demand x for its product is inversely proportional to the price p. When the price is \$5, the demand is 800 units. Approximate the demand if the price is increased to \$6.

52. *Predator-Prey* The number N of prey t months after a natural predator is introduced into a test area is inversely proportional to $t + 1$. If $N = 500$ when $t = 0$, find N when $t = 4$.

53. *Weight of an Astronaut* A person's weight on the moon varies directly with his or her weight on earth. Neil Armstrong, the first man on the moon, weighed 360 pounds on earth, including his heavy equipment. On the moon he weighed only 60 pounds with the equipment. If the first woman in space, Valentina V. Tereshkova, had landed on the moon and weighed 54 pounds with equipment, how much would she have weighed on earth with her equipment?

54. *Weight of an Astronaut* The gravitational force F with which an object is attracted to the earth is inversely proportional to the square of its distance r from the center of the earth. If an astronaut weighs 190 pounds on the surface of the earth ($r \approx 4000$ miles), what will the astronaut weigh 1000 miles above the earth's surface?

55. *Amount of Illumination* The illumination I from a light source varies inversely as the square of the distance d from the light source. If you raise a study lamp from 18 inches to 36 inches above your desk (see figure), the illumination will change by what factor?

56. *Snowshoes* When a person walks, the pressure P on each sole varies inversely with the area A of the sole. Denise is trudging through deep snow, wearing boots that have a sole area of 29 square inches each. The sole pressure is 4 pounds per square inch. If Denise were wearing snowshoes, each with an area 11 times that of her boot soles, what would be the pressure on each snowshoe? The constant of variation in this problem is Denise's weight. How much does she weigh?

57. *Oil Spill* The graph shows the percent p of oil that remained in Chedabucto Bay, Nova Scotia, after an oil spill. The cleaning of the spill was left primarily to natural actions such as wave motion, evaporation, photochemical decomposition, and bacterial decomposition. After about a year, the percent that remained varied inversely as time. Find a model that relates p and t, where t is the number of years since the spill. Then use it to find the percent of oil that remained $6\frac{1}{2}$ years after the spill, and compare the result with the graph.

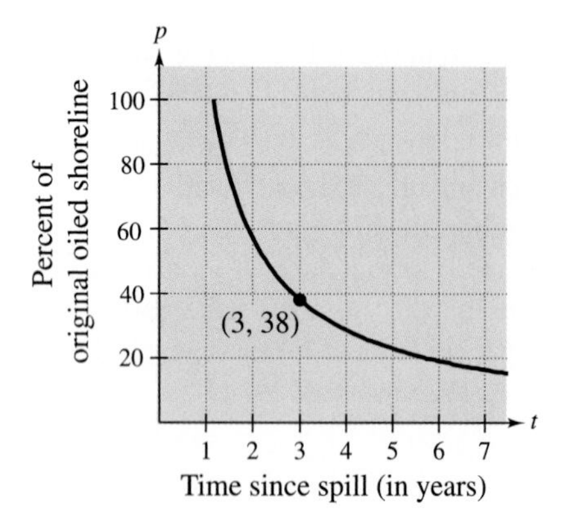

58. *Ocean Temperatures* The graph shows the temperature of the water in the north central Pacific Ocean. At depths greater than 900 meters, the water temperature varies inversely with the water depth. Find a model that relates the temperature T to the depth d. Then use it to find the water temperature at a depth of 4385 meters, and compare the result with the graph.

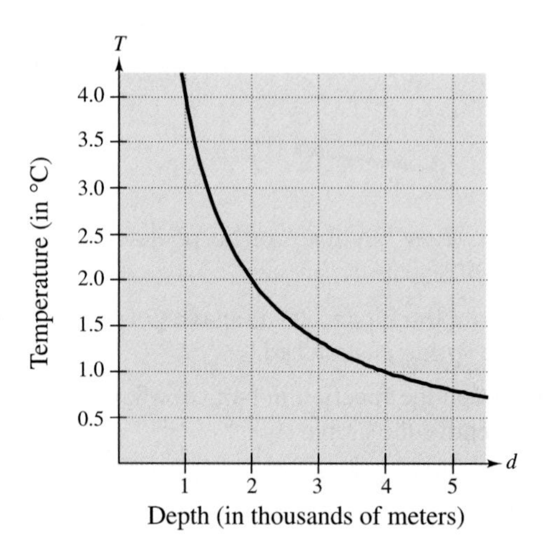

59. *Simple Interest* Simple interest varies jointly as the product of the interest rate and the time. An investment at 9% for 3 years earns \$202.50. How much will the investment earn in 4 years? What does the constant of proportionality measure in this problem?

60. *Engineering* The load P that can be safely supported by a horizontal beam varies jointly as the product of the width W of the beam and the square of the depth D and inversely as the length L.

(a) Write a model for the statement.

(b) How does P change when the width and length of the beam are both doubled?

(c) How does P change when the width and depth of the beam are doubled?

(d) How does P change when all three of the dimensions are doubled?

(e) How does P change when the depth of the beam is cut in half?

(f) A beam with width 3 inches, depth 8 inches, and length 10 feet can safely support 2000 pounds. Determine the safe load of a beam made from the same material if its depth is increased to 10 inches.

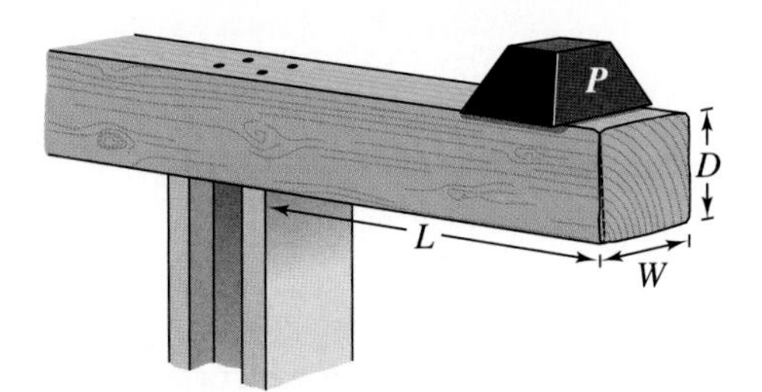

Figure for 60

In Exercises 61–64, complete the table and plot the resulting points.

x	2	4	6	8	10
$y = kx^2$					

61. $k = 1$

62. $k = 2$

63. $k = \frac{1}{2}$

64. $k = \frac{1}{4}$

In Exercises 65–68, complete the table and plot the resulting points.

x	2	4	6	8	10
$y = \frac{k}{x^2}$					

65. $k = 2$

66. $k = 5$

67. $k = 10$

68. $k = 20$

In Exercises 69–72, determine whether the variation model is of the form $y = kx$ or $y = k/x$, and find k.

69.

x	10	20	30	40	50
y	$\frac{2}{5}$	$\frac{1}{5}$	$\frac{2}{15}$	$\frac{1}{10}$	$\frac{2}{25}$

70.

x	10	20	30	40	50
y	2	4	6	8	10

71.

x	10	20	30	40	50
y	-3	-6	-9	-12	-15

72.

x	10	20	30	40	50
y	60	30	20	15	12

Explaining Concepts

73. Suppose the constant of proportionality is positive and y varies directly as x. If one of the variables increases, how will the other change? Explain.

74. Suppose the constant of proportionality is positive and y varies inversely as x. If one of the variables increases, how will the other change? Explain.

75. If y varies directly as the square of x and x is doubled, how will y change? Use the properties of exponents to explain your answer.

76. If y varies inversely as the square of x and x is doubled, how will y change? Use the properties of exponents to explain your answer.

CHAPTER SUMMARY

Key Terms

rational function, *p. 444*
domain of a rational function, *p. 444*
simplified form, *p. 446*
complex fraction, *p. 458*
least common multiple, *p. 464*
least common denominator, *p. 465*
extraneous solution, *p. 477*
asymptote, *p. 486*
constant of proportionality, *p. 495*
direct variation, *p. 495*
directly proportional, *p. 495*
varies directly as the nth power, *p. 497*
directly proportional to the nth power, *p. 497*
inverse variation, *p. 498*
inversely proportional, *p. 498*
joint variation, *p. 500*
jointly proportional, *p. 500*

Key Concepts

8.2 Multiplying rational expressions

To multiply two rational expressions, multiply numerators, multiply denominators, and write the new fraction in simplified form.

8.2 Dividing rational expressions

To divide two rational expressions, multiply the first fraction by the reciprocal of the second fraction, and write the new fraction in simplified form.

8.3 Adding or subtracting with like denominators

To add or subtract two rational expressions with like denominators, simply combine their numerators and place the result over the common denominator.

8.3 Adding or subtracting with unlike denominators

Rewrite the rational expressions with like denominators by finding the least common denominator. Then add or subtract as with like denominators.

8.4 Solving rational equations

1. To remove a fraction in a rational equation, multiply both sides by the least common denominator of all fractions in the equation.
2. Exclude those values of a variable that make the denominator of a rational expression zero.
3. Check your solutions to determine if any are extraneous solutions.

8.5 Guidelines for graphing rational functions

Let $f(x) = p(x)/q(x)$, where $p(x)$ and $q(x)$ have no common factors.

1. Find and plot the y-intercept.
2. Find and plot the x-intercept(s).
3. Find and sketch the horizontal and vertical asymptotes of the graph.
4. Plot at least one point between and beyond each x-intercept and vertical asymptote.
5. Use smooth curves to complete the graph between and beyond the vertical asymptotes.
6. If $p(x)$ and $q(x)$ have a common factor $x - a$, then the graph of $p(x)/q(x)$ has a hole at $x = a$.

8.6 Direct variation

The following statements are equivalent. The number k is the constant of proportionality.

1. y varies directly as x.
2. y is directly proportional to x.
3. $y = kx$ for some constant k.

8.6 Direct variation as nth power

The following statements are equivalent.

1. y varies directly as the nth power of x.
2. y is directly proportional to the nth power of x.
3. $y = kx^n$ for some constant k.

8.6 Inverse variation

1. The following three statements are equivalent.
   a. y varies inversely as x.
   b. y is inversely proportional to x.
   c. $y = k/x$ for some constant k.
2. If $y = k/x^n$, then y is inversely proportional to the nth power of x.

8.6 Joint variation

1. The following three statements are equivalent.
   a. z varies jointly as x and y.
   b. z is jointly proportional to x and y.
   c. $z = kxy$ for some constant k.
2. If $z = kx^ny^m$, then z is jointly proportional to the nth power of x and the mth power of y.

REVIEW EXERCISES

Reviewing Skills

8.1 In Exercises 1–4, determine the domain of the rational function.

1. $f(y) = \dfrac{3y}{y - 8}$

2. $g(t) = \dfrac{t + 4}{t + 12}$

3. $g(u) = \dfrac{u}{u^2 - 7u + 6}$

4. $f(x) = \dfrac{x - 12}{x(x^2 - 16)}$

In Exercises 5–12, simplify the rational expression using the Cancellation Rule.

5. $\dfrac{6x^4y^2}{15xy^2}$

6. $\dfrac{2(y^3z)^2}{28(yz^2)^2}$

7. $\dfrac{5b - 15}{30b - 120}$

8. $\dfrac{4a}{10a^2 + 26a}$

9. $\dfrac{9x - 9y}{y - x}$

10. $\dfrac{x + 3}{x^2 - x - 12}$

11. $\dfrac{x^2 - 5x}{2x^2 - 50}$

12. $\dfrac{x^2 + 3x + 9}{x^3 - 27}$

8.2 In Exercises 13–28, multiply or divide the rational expressions.

13. $3x(x^2y)^2$

14. $2b(-3b)^3$

15. $\dfrac{24x^4}{15x}$

16. $\dfrac{8u^2v}{6v}$

17. $\dfrac{7}{8} \cdot \dfrac{2x}{y} \cdot \dfrac{y^2}{14x^2}$

18. $\dfrac{15(x^2y)^3}{3y^3} \cdot \dfrac{12y}{x}$

19. $\dfrac{60z}{z + 6} \cdot \dfrac{z^2 - 36}{5}$

20. $\dfrac{1}{6}(x^2 - 16) \cdot \dfrac{3}{x^2 - 8x + 16}$

21. $\dfrac{u}{u - 3} \cdot \dfrac{3u - u^2}{4u^2}$

22. $x^2 \cdot \dfrac{x + 1}{x^2 - x} \cdot \dfrac{(5x - 5)^2}{x^2 + 6x + 5}$

23. $\dfrac{\left(\dfrac{6}{x}\right)}{\left(\dfrac{2}{x^3}\right)}$

24. $\dfrac{0}{\left(\dfrac{5x^2}{2y}\right)}$

25. $25y^2 \div \dfrac{xy}{5}$

26. $\dfrac{6}{z^2} \div 4z^2$

27. $\dfrac{x^2 - 7x}{x + 1} \div \dfrac{x^2 - 14x + 49}{x^2 - 1}$

28. $\left(\dfrac{6x}{y^2}\right)^2 \div \left(\dfrac{3x}{y}\right)^3$

In Exercises 29 and 30, simplify the complex fraction.

29. $\dfrac{\left(\dfrac{6x^2}{x^2 + 2x - 35}\right)}{\left(\dfrac{x^3}{x^2 - 25}\right)}$

30. $\dfrac{\left[\dfrac{24 - 18x}{(2 - x)^2}\right]}{\left(\dfrac{60 - 45x}{x^2 - 4x - 4}\right)}$

8.3 In Exercises 31–42, add or subtract the rational expressions and simplify.

31. $\frac{4}{9} - \frac{11}{9}$

32. $-\frac{3}{8} + \frac{7}{6} - \frac{1}{12}$

33. $\frac{15}{16} - \frac{5}{24} - 1$

34. $\dfrac{2(3y + 4)}{2y + 1} + \dfrac{3 - y}{2y + 1}$

35. $\dfrac{1}{x + 5} + \dfrac{3}{x - 12}$

36. $\dfrac{2}{x - 10} + \dfrac{3}{4 - x}$

37. $5x + \dfrac{2}{x - 3} - \dfrac{3}{x + 2}$

38. $4 - \dfrac{4x}{x + 6} + \dfrac{7}{x - 5}$

39. $\dfrac{6}{x} - \dfrac{6x - 1}{x^2 + 4}$

40. $\dfrac{5}{x + 2} + \dfrac{25 - x}{x^2 - 3x - 10}$

41. $\dfrac{5}{x + 3} - \dfrac{4x}{(x + 3)^2} - \dfrac{1}{x - 3}$

42. $\dfrac{8}{y} - \dfrac{3}{y + 5} + \dfrac{4}{y - 2}$

In Exercises 43–46, simplify the complex fraction.

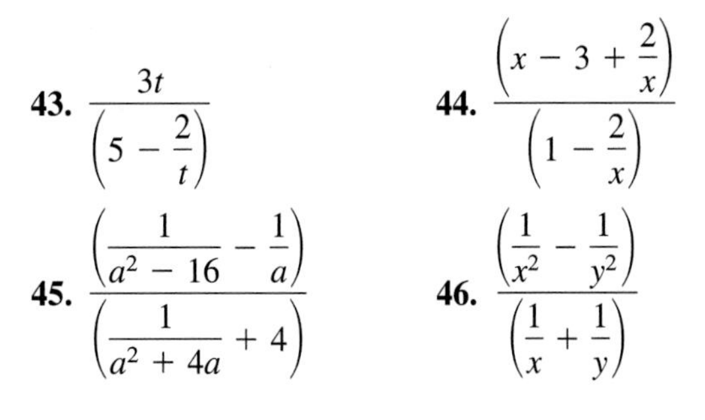

43. $\dfrac{3t}{\left(5 - \dfrac{2}{t}\right)}$

44. $\dfrac{\left(x - 3 + \dfrac{2}{x}\right)}{\left(1 - \dfrac{2}{x}\right)}$

45. $\dfrac{\left(\dfrac{1}{a^2 - 16} - \dfrac{1}{a}\right)}{\left(\dfrac{1}{a^2 + 4a} + 4\right)}$

46. $\dfrac{\left(\dfrac{1}{x^2} - \dfrac{1}{y^2}\right)}{\left(\dfrac{1}{x} + \dfrac{1}{y}\right)}$

In Exercises 47–50, use a graphing utility to graph the two functions on the same screen. Use the graphs to verify that the expressions are equivalent. Verify the results algebraically.

47. $y_1 = \dfrac{x^2 + 6x + 9}{x^2} \cdot \dfrac{x^2 - 3x}{x + 3}$

$y_2 = \dfrac{x^2 - 9}{x}, \quad x \neq -3$

48. $y_1 = \dfrac{1}{x} - \dfrac{3}{x + 3}, \quad y_2 = \dfrac{3 - 2x}{x(x + 3)}$

49. $y_1 = \dfrac{\left(\dfrac{1}{x} - \dfrac{1}{2}\right)}{2x}, \quad y_2 = \dfrac{2 - x}{4x^2}$

50. $y_1 = \dfrac{x^3 - 2x^2 - 7}{x - 2}, \quad y_2 = x^2 - \dfrac{7}{x - 2}$

8.4 In Exercises 51–66, solve the equation.

51. $\dfrac{3x}{8} = -15 + \dfrac{x}{4}$

52. $\dfrac{t + 1}{6} = \dfrac{1}{2} - 2t$

53. $8 - \dfrac{12}{t} = \dfrac{1}{3}$

54. $5 + \dfrac{2}{x} = \dfrac{1}{4}$

55. $\dfrac{2}{y} - \dfrac{1}{3y} = \dfrac{1}{3}$

56. $\dfrac{7}{4x} - \dfrac{6}{8x} = 1$

57. $r = 2 + \dfrac{24}{r}$

58. $\dfrac{2}{x} - \dfrac{x}{6} = \dfrac{2}{3}$

59. $8\left(\dfrac{6}{x} - \dfrac{1}{x + 5}\right) = 15$

60. $\dfrac{3}{y + 1} - \dfrac{8}{y} = 1$

61. $\dfrac{4x}{x - 5} + \dfrac{2}{x} = -\dfrac{4}{x - 5}$

62. $\dfrac{2x}{x + 3} - \dfrac{3}{x} = 0$

63. $\dfrac{12}{x^2 + x - 12} - \dfrac{1}{x - 3} = -1$

64. $\dfrac{3}{x - 1} + \dfrac{6}{x^2 - 3x + 2} = 2$

65. $\dfrac{5}{x^2 - 4} - \dfrac{6}{x - 2} = -5$

66. $\dfrac{3}{x^2 - 9} + \dfrac{4}{x + 3} = 1$

In Exercises 67 and 68, (a) use a graphing utility to determine any x-intercepts of the graph of the equation, and (b) set $y = 0$ and solve the resulting equation to confirm the result algebraically.

67. $y = \dfrac{1}{x} - \dfrac{1}{2x + 3}$

68. $y = \dfrac{x}{4} - \dfrac{2}{x} - \dfrac{1}{2}$

8.5 In Exercises 69–72, match the function with its graph. [The graphs are labeled (a), (b), (c), and (d).]

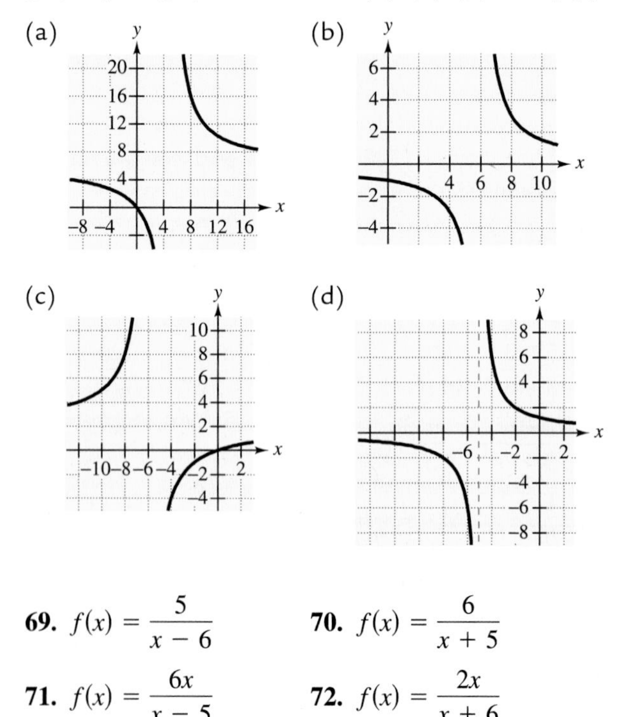

69. $f(x) = \dfrac{5}{x - 6}$

70. $f(x) = \dfrac{6}{x + 5}$

71. $f(x) = \dfrac{6x}{x - 5}$

72. $f(x) = \dfrac{2x}{x + 6}$

In Exercises 73–88, use asymptotes and intercepts to graph the rational function.

73. $f(x) = -\dfrac{5}{x^2}$

74. $f(x) = \dfrac{4}{x}$

75. $P(x) = \dfrac{3x + 6}{x - 2}$

76. $s(x) = \dfrac{2x - 6}{x + 4}$

77. $g(x) = \dfrac{2 + x}{1 - x}$

78. $h(x) = \dfrac{x - 3}{x - 2}$

79. $f(x) = \dfrac{x}{x^2 + 1}$

80. $f(x) = \dfrac{2x}{x^2 + 4}$

81. $h(x) = \dfrac{4}{(x - 1)^2}$

82. $g(x) = \dfrac{-2}{(x + 3)^2}$

83. $y = \dfrac{x}{x^2 - 1}$

84. $y = \dfrac{2x}{x^2 - 4}$

85. $y = \dfrac{2x^2}{x^2 - 4x}$

86. $y = \dfrac{2x}{x^2 + 3x}$

87. $y = \dfrac{x - 4}{x^2 - 3x - 4}$

88. $y = \dfrac{2x + 1}{2x^2 - 5x - 3}$

Think About It In Exercises 89 and 90, write a rational function satisfying each of the criteria. Use a graphing utility to verify the results.

89. Vertical asymptote: $x = 4$
Horizontal asymptote: $y = 3$
Zero of the function: $x = 0$

90. Vertical asymptote: $x = -3$
Horizontal asymptote: $y = 0$
Zero of the function: $x = 2$

8.6 In Exercises 91–94, find the constant of proportionality and write an equation that relates the variables.

91. y varies directly as the cube root of x, and $y = 12$ when $x = 8$.

92. r varies inversely as s, and $r = 45$ when $s = \frac{3}{5}$.

93. T varies jointly as r and the square of s, and $T = 5000$ when $r = 0.09$ and $s = 1000$.

94. D is directly proportional to the cube of x and inversely proportional to y, and $D = 810$ when $x = 3$ and $y = 25$.

Solving Problems

95. *Geometry* A rectangle with a width of w inches has an area of 36 square inches. The perimeter of the rectangle is given by

$$P = 2\left(w + \frac{36}{w}\right).$$

Describe the domain of the function.

96. *Average Cost* The average cost $\overline{C}$ for a manufacturer to produce x units of a product is given by

$$\overline{C} = \frac{15{,}000 + 0.75x}{x}.$$

Describe the domain of the function.

97. *Average Speed* You drive 56 miles on a service call for your company. On the return trip, which takes 10 minutes less than the original trip, your average speed is 8 miles per hour faster. What is your average speed on the return trip?

98. *Average Speed* You drive 220 miles to see a friend. On the return trip, which takes 20 minutes less than the original trip, your average speed is 5 miles per hour faster. What is your average speed on the return trip?

99. *Batting Average* In this year's playing season, a baseball player has been at bat 150 times and has hit the ball safely 45 times. Thus, the batting average for the player is $45/150 = 0.300$. How many consecutive times must the player hit safely to obtain a batting average of 0.400?

100. *Batting Average* In this year's playing season, a softball player has been at bat 75 times and has hit the ball safely 23 times. So, the batting average for this player is $23/75 \approx 0.307$. How many consecutive times must the player hit safely to obtain a batting average of 0.350?

101. *Forming a Partnership* A group of people agree to share equally in the cost of a \$60,000 piece of machinery. If they could find two more people to join the group, each person's share of the cost would decrease by \$5000. How many people are presently in the group?

102. *Forming a Partnership* An individual is planning to start a small business that will require $28,000 before any income can be generated. Because it is difficult to borrow for new ventures, the individual wants a group of friends to divide the cost equally for a future share of the profit. The person has found some investors, but three more are needed so that the price per person will be $1200 less. How many investors are currently in the group?

103. *Work-Rate Problem* Suppose that in 12 minutes your supervisor can complete a task that you need 15 minutes to complete. Determine the time required to complete the task if you work together.

104. *Work-Rate Problem* Suppose that in 21 minutes your supervisor can complete a task that you need 24 minutes to complete. Determine the time required to complete the task if you work together.

105. *Population of Fish* The Parks and Wildlife Commission introduces 80,000 fish into a large lake. The population (in thousands) is

$$N = \frac{20(4 + 3t)}{1 + 0.05t}, \quad t > 0$$

where t is time in years.

(a) Find the population when t is 5, 10, and 25 years.

(b) In how many years will the population reach 752,000?

106. *Hooke's Law* A force of 100 pounds stretches a spring 4 inches. Find the force required to stretch the spring 6 inches.

107. *Stopping Distance* The stopping distance d of an automobile is directly proportional to the square of its speed s. How will the stopping distance be changed by doubling the speed of the car?

108. *Demand Function* A company has found that the daily demand x for its product varies inversely as the square root of the price p. When the price is $25 the demand is approximately 1000 units. Approximate the demand if the price is increased to $28.

109. *Weight of an Astronaut* The gravitational force F with which an object is attracted to the earth is inversely proportional to the square of its distance r from the center of the earth. If an astronaut weighs 200 pounds on the surface of the earth ($r \approx 4000$ miles), what will the astronaut weigh 500 miles above the earth's surface?

Chapter Test

Take this test as you would take a test in class. After you are done, check your work against the answers given in the back of the book.

1. Find the domain of $\dfrac{3y}{y^2 - 25}$.

2. Find the least common denominator of $\dfrac{3}{x^2}, \dfrac{x}{x-3}$, and $\dfrac{2x}{x^3(x+3)}$.

3. Simplify the rational expression. (a) $\dfrac{2-x}{3x-6}$ (b) $\dfrac{2a^2 - 5a - 12}{5a - 20}$

In Exercises 4–12, perform the operation and simplify.

4. $\dfrac{4z^3}{5} \cdot \dfrac{25}{12z^2}$

5. $\dfrac{y^2 + 8y + 16}{2(y-2)} \cdot \dfrac{8y - 16}{(y+4)^3}$

6. $(4x^2 - 9) \cdot \dfrac{2x + 3}{2x^2 - x - 3}$

7. $\dfrac{(2xy^2)^3}{15} \div \dfrac{12x^3}{21}$

8. $\dfrac{\left(\dfrac{3x}{x+2}\right)}{\left(\dfrac{12}{x^3 + 2x^2}\right)}$

9. $\dfrac{\left(9x - \dfrac{1}{x}\right)}{\left(\dfrac{1}{x} - 3\right)}$

10. $2x + \dfrac{1 - 4x^2}{x + 1}$

11. $\dfrac{5x}{x+2} - \dfrac{2}{x^2 - x - 6}$

12. $\dfrac{3}{x} - \dfrac{5}{x^2} + \dfrac{2x}{x^2 + 2x + 1}$

13. Graph each function.

(a) $f(x) = \dfrac{3}{x-3}$ (b) $g(x) = \dfrac{3x}{x-3}$

In Exercises 14 and 15, solve the equation.

14. $\dfrac{3}{h+2} = \dfrac{1}{8}$

15. $\dfrac{2}{x+5} - \dfrac{3}{x+3} = \dfrac{1}{x}$

16. Graph the rational function $f(x) = \dfrac{x-4}{x^2 - x - 12}$.

17. One painter works $1\frac{1}{2}$ times as fast as another. Find their individual times for painting a room if it takes them 4 hours working together.

18. Write a mathematical model for the statement, "S varies directly as the square of x and inversely as y."

19. Find the constant of proportionality if v varies directly as the square root of u and $v = \frac{3}{2}$ when $u = 36$.

20. If the temperature of a gas is not allowed to change, its absolute pressure P is inversely proportional to its volume V, according to Boyle's Law. A large balloon is filled with 180 cubic meters of helium at atmospheric pressure (1 atm) at sea level. What is the volume of the helium if the balloon rises to an altitude at which the atmospheric pressure is 0.75 atm? (Assume that the temperature does not change.)

9 Radicals and Complex Numbers

William Taufic/The Stock Market

Greenhouses represented 57% (approximately 466 million square feet) of the total covered growing area used in floriculture crop production in 1997. (Source: U.S. Department of Agriculture)

Motivating the Chapter

Building a Greenhouse

You are building a greenhouse in the form of a half cylinder. The volume of the greenhouse is to be approximately 35,350 cubic feet.

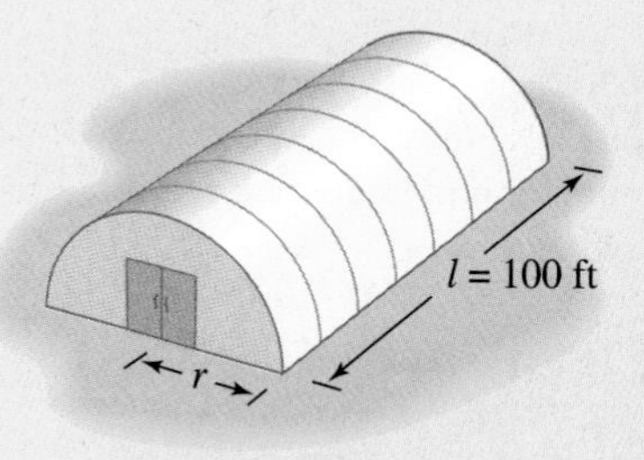

See Section 9.1, Exercise 151

a. The formula for the radius (in feet) of a half cylinder is

$$r = \sqrt{\frac{2V}{\pi l}}$$

where V is the volume (in cubic feet) and l is the length (in feet). Find the radius of the greenhouse and round your result to the nearest whole number. Use this value of r in parts (b)–(d).

b. Beams to hold the sprinkler system are to be placed across the building. The formula for the height h at which the beams are to be placed is

$$h = \sqrt{r^2 - \left(\frac{a}{2}\right)^2}$$

where a is the length of the beam. Rewrite h as a function of a.

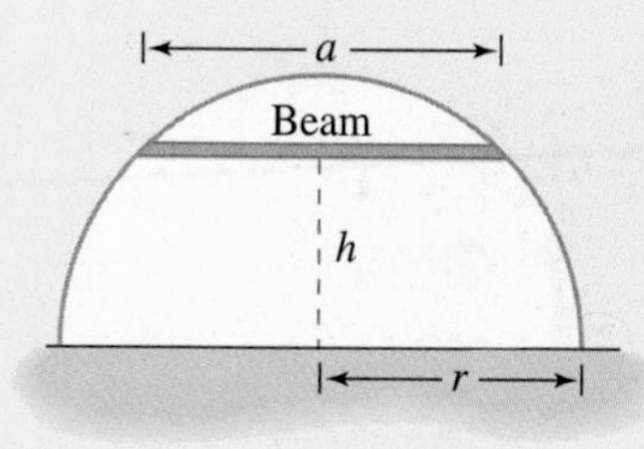

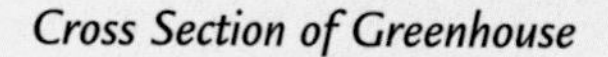
Cross Section of Greenhouse

c. Suppose the length of each beam is $a = 25$ feet. Find the height h at which the beams should be placed.

d. The equation from part (b) can be rewritten as

$$a = 2\sqrt{r^2 - h^2}.$$

Suppose that the height is $h = 8$ feet. What is the length a of each beam?

See Section 9.4, Exercise 103

e. The cost of building the greenhouse is estimated to be \$25,000. The money to pay for the greenhouse was invested in an interest-bearing account 10 years ago at an annual percentage rate of 7%. The amount of money earned can be found using the formula

$$r = \left(\frac{A}{P}\right)^{1/n} - 1,$$

where r is the annual percentage rate (in decimal form), A is the amount in the account after 10 years, P is the initial deposit, and n is the number of years. What initial deposit P would have generated enough money to cover the building cost of \$25,000?

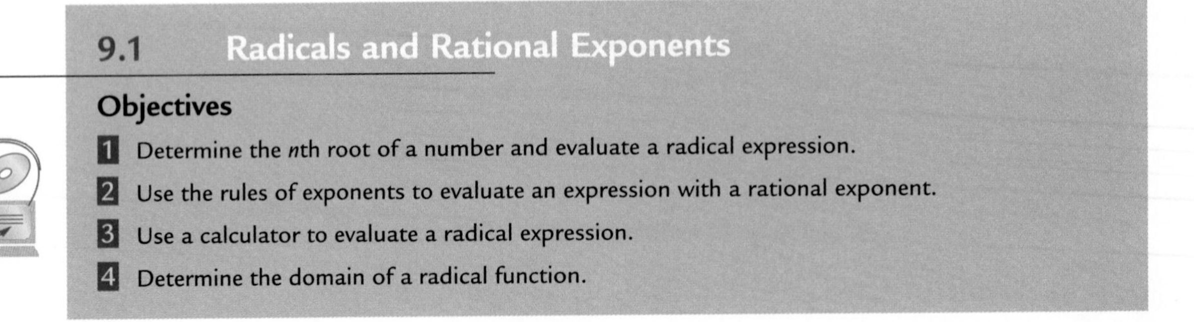

9.1 Radicals and Rational Exponents

Objectives

1. Determine the *n*th root of a number and evaluate a radical expression.
2. Use the rules of exponents to evaluate an expression with a rational exponent.
3. Use a calculator to evaluate a radical expression.
4. Determine the domain of a radical function.

1 Determine the *n*th root of a number and evaluate a radical expression.

Roots and Radicals

The **square root** of a number is defined as one of its two equal factors. For example, 5 is a square root of 25 because 5 is one of the two equal factors of 25. In a similar way, a **cube root** of a number is one of its three equal factors.

Number	*Equal Factors*	*Root*	*Type*
$9 = 3^2$	$3 \cdot 3$	3	Square root
$25 = (-5)^2$	$(-5)(-5)$	-5	Square root
$-27 = (-3)^3$	$(-3)(-3)(-3)$	-3	Cube root
$64 = (4)^3$	$4 \cdot 4 \cdot 4$	4	Cube root
$16 = 2^4$	$2 \cdot 2 \cdot 2 \cdot 2$	2	Fourth root

▶ Definition of *n*th Root of a Number

Let a and b be real numbers and let n be an integer such that $n \geq 2$. If

$$a = b^n$$

then b is an ***n*th root of *a*.** If $n = 2$, the root is a **square root,** and if $n = 3$, the root is a **cube root.**

Some numbers have more than one nth root. For example, both 5 and -5 are square roots of 25 because $25 = 5^2$ and $25 = (-5)^2$. To avoid ambiguity about which root of a number you are talking about, the **principal *n*th root** of a number is defined in terms of a radical symbol $\sqrt[n]{\ }$.

Study Tip

"Having the same sign as a" means that the principal nth root of a is positive if a is positive and negative if a is negative. For example, $\sqrt{4} = 2$ and $\sqrt[3]{-8} = -2$.

▶ Principal *n*th Root of a Number

Let a be a real number that has at least one (real number) nth root. The **principal *n*th root of *a*** is the nth root that has the same sign as a, and it is denoted by the **radical symbol**

$$\sqrt[n]{a}. \qquad \text{Principal } n\text{th root}$$

The positive integer n is the **index** of the radical, and the number a is the **radicand.** If $n = 2$, omit the index and write $\sqrt{a}$ rather than $\sqrt[2]{a}$.

Example 1 Finding Roots of a Number

Find the roots.

a. $\sqrt{36}$ **b.** $-\sqrt{36}$ **c.** $\sqrt{-4}$ **d.** $\sqrt[3]{8}$ **e.** $\sqrt[3]{-8}$

Solution

a. $\sqrt{36} = 6$ because $6 \cdot 6 = 6^2 = 36$.

b. $-\sqrt{36} = -6$ because $6 \cdot 6 = 6^2 = 36$.

c. $\sqrt{-4}$ is not real because there is no real number that when multiplied by itself yields -4.

d. $\sqrt[3]{8} = 2$ because $2 \cdot 2 \cdot 2 = 2^3 = 8$.

e. $\sqrt[3]{-8} = -2$ because $(-2)(-2)(-2) = (-2)^3 = -8$.

Study Tip

To remember the properties of nth roots in which a is negative, consider the following visual scheme.

$\sqrt[\text{odd}]{\text{negative number}}$ is negative.

$\sqrt[\text{even}]{\text{negative number}}$ is *not* a real number.

▶ Properties of nth Roots

Property	*Example*
1. If a is a positive real number and n is *even*, then a has exactly two (real) nth roots, which are denoted by $\sqrt[n]{a}$ and $-\sqrt[n]{a}$.	The two real square roots of 81 are $\sqrt{81} = 9$ and $-\sqrt{81} = -9$.
2. If a is any real number and n is *odd*, then a has only one (real) nth root, which is denoted by $\sqrt[n]{a}$.	$\sqrt[3]{27} = 3$ $\sqrt[3]{-64} = -4$
3. If a is a negative real number and n is *even*, then a has no (real) nth root.	$\sqrt{-64}$ is not a real number.

Integers such as 1, 4, 9, 16, 49, and 81 are called **perfect squares** because they have integer square roots. Similarly, integers such as 1, 8, 27, 64, and 125 are called **perfect cubes** because they have integer cube roots.

Example 2 Classifying Perfect Squares and Perfect Cubes

State whether the number is a perfect square, a perfect cube, both, or neither.

a. 81 **b.** -125 **c.** 64 **d.** 32

Solution

a. 81 is a perfect square because $9^2 = 81$. It is not a perfect cube.

b. -125 is a perfect cube because $(-5)^3 = -125$. It is not a perfect square.

c. 64 is a perfect square because $8^2 = 64$, and it is also a perfect cube because $4^3 = 64$.

d. 32 is not a perfect square or a perfect cube. (It is a perfect 5th power because $2^5 = 32$.)

Raising a number to the nth power and taking the principal nth root of a number can be thought of as *inverse* operations. Here are some examples.

$$(\sqrt{4})^2 = (2)^2 = 4 \quad \text{and} \quad \sqrt{2^2} = \sqrt{4} = 2$$

$$(\sqrt[3]{27})^3 = (3)^3 = 27 \quad \text{and} \quad \sqrt[3]{3^3} = \sqrt[3]{27} = 3$$

$$(\sqrt[4]{16})^4 = (2)^4 = 16 \quad \text{and} \quad \sqrt[4]{2^4} = \sqrt[4]{16} = 2$$

$$(\sqrt[5]{-243})^5 = (-3)^5 = -243 \quad \text{and} \quad \sqrt[5]{(-3)^5} = \sqrt[5]{-243} = -3$$

▶ Inverse Properties of nth Powers and nth Roots

Let a be a real number, and let n be an integer such that $n \geq 2$.

Property	*Example*
1. If a has a principal nth root, then $(\sqrt[n]{a})^n = a$.	$(\sqrt{5})^2 = 5$
2. If n is *odd*, then $\sqrt[n]{a^n} = a$.	$\sqrt[3]{5^3} = 5$
If n is *even*, then $\sqrt[n]{a^n} = \lvert a \rvert$.	$\sqrt{(-5)^2} = \lvert -5 \rvert = 5$

Example 3 Evaluating nth Roots of nth Powers

Evaluate each radical expression.

a. $\sqrt[3]{4^3}$ **b.** $\sqrt[3]{(-2)^3}$ **c.** $(\sqrt{7})^2$

d. $\sqrt{(-3)^2}$ **e.** $\sqrt{-3^2}$

Solution

a. Because the index of the radical is odd, you can write

$$\sqrt[3]{4^3} = 4.$$

b. Because the index of the radical is odd, you can write

$$\sqrt[3]{(-2)^3} = -2.$$

c. Using the inverse property of powers and roots, you can write

$$(\sqrt{7})^2 = 7.$$

d. Because the index of the radical is even, you must include absolute value signs, and write

$$\sqrt{(-3)^2} = \lvert -3 \rvert = 3.$$

e. Because $\sqrt{-3^2} = \sqrt{-9}$ is an even root of a negative number, its value is not a real number.

2 Use the rules of exponents to evaluate an expression with a rational exponent.

Rational Exponents

> **▶ Definition of Rational Exponents**
>
> Let a be a real number, and let n be an integer such that $n \geq 2$. If the principal nth root of a exists, we define $a^{1/n}$ to be
>
> $$a^{1/n} = \sqrt[n]{a}.$$
>
> If m is a positive integer that has no common factor with n, then
>
> $$a^{m/n} = (a^{1/n})^m = \left(\sqrt[n]{a}\right)^m \quad \text{and} \quad a^{m/n} = (a^m)^{1/n} = \sqrt[n]{a^m}.$$

Study Tip

The numerator of a rational exponent denotes the *power* to which the base is raised, and the denominator denotes the *root* to be taken.

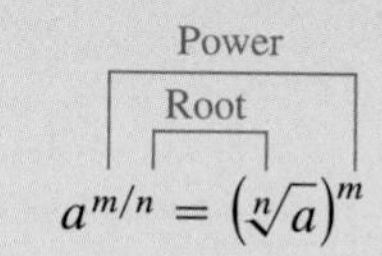

$$a^{m/n} = \left(\sqrt[n]{a}\right)^m$$

It does not matter in which order the two operations are performed, provided the nth root exists. Here is an example.

$$8^{2/3} = \left(\sqrt[3]{8}\right)^2 = 2^2 = 4 \qquad \text{Cube root, then second power}$$

$$8^{2/3} = \sqrt[3]{8^2} = \sqrt[3]{64} = 4 \qquad \text{Second power, then cube root}$$

The rules of exponents that we listed in Section 5.3 also apply to rational exponents (provided the roots indicated by the denominators exist). We relist those rules here, with different examples.

> **▶ Summary of Rules of Exponents**
>
> Let r and s be rational numbers, and let a and b be real numbers, variables, or algebraic expressions.

Product and Quotient Rules	*Example*
1. $a^r \cdot a^s = a^{r+s}$	$4^{1/2}(4^{1/3}) = 4^{5/6}$
2. $\dfrac{a^r}{a^s} = a^{r-s}, \quad a \neq 0$	$\dfrac{x^2}{x^{1/2}} = x^{2-(1/2)} = x^{3/2}$
Power Rules	
3. $(ab)^r = a^r \cdot b^r$	$(2x)^{1/2} = 2^{1/2}(x^{1/2})$
4. $\left(\dfrac{a}{b}\right)^r = \dfrac{a^r}{b^r}, \quad b \neq 0$	$\left(\dfrac{x}{3}\right)^{2/3} = \dfrac{x^{2/3}}{3^{2/3}}$
5. $(a^r)^s = a^{rs}$	$(x^3)^{1/2} = x^{3/2}$
Zero and Negative Exponent Rules	
6. $a^{-r} = \dfrac{1}{a^r}, \quad a \neq 0$	$4^{-3/2} = \dfrac{1}{4^{3/2}} = \dfrac{1}{(2)^3} = \dfrac{1}{8}$
7. $a^0 = 1$	$(3x)^0 = 1$
8. $\left(\dfrac{a}{b}\right)^{-r} = \left(\dfrac{b}{a}\right)^r, \quad a \neq 0,\ b \neq 0$	$\left(\dfrac{x}{4}\right)^{-1/2} = \left(\dfrac{4}{x}\right)^{1/2} = \dfrac{2}{x^{1/2}}$

Technology: Discovery

Use a calculator to evaluate the expressions below.

$$\frac{3.4^{4.6}}{3.4^{3.1}} \quad \text{and} \quad 3.4^{1.5}$$

How are these two expressions related? Use your calculator to verify some of the other rules of exponents.

Example 4 Evaluating Expressions with Rational Exponents

Use rules of exponents to rewrite each expression in simpler form.

a. $8^{4/3}$ **b.** $(4^2)^{3/2}$ **c.** $25^{-3/2}$

d. $\left(\frac{64}{125}\right)^{2/3}$ **e.** $-9^{1/2}$ **f.** $(-9)^{1/2}$

Solution

a. $8^{4/3} = (8^{1/3})^4 = \left(\sqrt[3]{8}\right)^4 = 2^4 = 16$ Root is 3. Power is 4.

b. $(4^2)^{3/2} = 4^{2\cdot(3/2)} = 4^{6/2} = 4^3 = 64$ Root is 2. Power is 3.

c. $25^{-3/2} = \dfrac{1}{25^{3/2}} = \dfrac{1}{\left(\sqrt{25}\right)^3} = \dfrac{1}{5^3} = \dfrac{1}{125}$ Root is 2. Power is 3.

d. $\left(\dfrac{64}{125}\right)^{2/3} = \dfrac{64^{2/3}}{125^{2/3}} = \dfrac{\left(\sqrt[3]{64}\right)^2}{\left(\sqrt[3]{125}\right)^2} = \dfrac{4^2}{5^2} = \dfrac{16}{25}$ Root is 3. Power is 2.

e. $-9^{1/2} = -\sqrt{9} = -(3) = -3$ Root is 2. Power is 1.

f. $(-9)^{1/2} = \sqrt{-9}$ is not a real number. Root is 2. Power is 1.

In parts (e) and (f) of Example 4, be sure that you see the distinction between the expressions $-9^{1/2}$ and $(-9)^{1/2}$.

Example 5 Using Rules of Exponents

Rewrite each expression using rational exponents.

a. $x\sqrt[4]{x^3}$ **b.** $\dfrac{\sqrt[3]{x^2}}{\sqrt{x^3}}$ **c.** $\sqrt[3]{x^2y}$

Solution

a. $x\sqrt[4]{x^3} = x(x^{3/4}) = x^{1+(3/4)} = x^{7/4}$

b. $\dfrac{\sqrt[3]{x^2}}{\sqrt{x^3}} = \dfrac{x^{2/3}}{x^{3/2}} = x^{(2/3)-(3/2)} = x^{-5/6} = \dfrac{1}{x^{5/6}}$

c. $\sqrt[3]{x^2y} = (x^2y)^{1/3} = (x^2)^{1/3}y^{1/3} = x^{2/3}y^{1/3}$

Example 6 Using Rules of Exponents

Use rules of exponents to rewrite each expression in simpler form.

a. $\sqrt{\sqrt[3]{x}}$ **b.** $\dfrac{(2x-1)^{4/3}}{\sqrt[3]{2x-1}}$

Solution

a. $\sqrt{\sqrt[3]{x}} = \sqrt{x^{1/3}} = (x^{1/3})^{1/2} = x^{(1/3)(1/2)} = x^{1/6}$

b. $\dfrac{(2x-1)^{4/3}}{\sqrt[3]{2x-1}} = \dfrac{(2x-1)^{4/3}}{(2x-1)^{1/3}} = (2x-1)^{(4/3)-(1/3)} = (2x-1)^{3/3} = 2x-1$

3 Use a calculator to evaluate a radical expression.

Radicals and Calculators

There are two methods of evaluating radicals on most calculators. For square roots, you can use the *square root key* [√]. For other roots, you should first convert the radical to exponential form and then use the *exponential key* [y^x] or [^].

Technology: Tip

Some graphing utilities have cube root functions $\sqrt[3]{}$ and xth root functions $\sqrt[x]{}$ that can be used to evaluate roots other than square roots. Consult the user's guide of your graphing utility for keystrokes.

Example 7 Evaluating Roots with a Calculator

Evaluate the following. Round the result to three decimal places.

a. $\sqrt{5}$ **b.** $\sqrt[5]{25}$ **c.** $\sqrt[3]{-4}$ **d.** $(8)^{1/2}$ **e.** $(1.4)^{-2/5}$

Solution

a. 5 [√] Scientific

[√] 5 [ENTER] Graphing

The display is 2.236068. Rounded to three decimal places, $\sqrt{5} \approx 2.236$.

b. First rewrite the expression as $\sqrt[5]{25} = 25^{1/5}$. Then use one of the following keystroke sequences.

25 [y^x] [(] 1 [÷] 5 [)] [=] Scientific

25 [^] [(] 1 [÷] 5 [)] [ENTER] Graphing

The display is 1.9036539. Rounded to three decimal places, $\sqrt[5]{25} \approx 1.904$.

c. If your calculator does not have a cube root key, use the fact that

$$\sqrt[3]{-4} = \sqrt[3]{(-1)(4)} = \sqrt[3]{-1}\sqrt[3]{4} = -\sqrt[3]{4}$$

and attach the negative sign of the radicand as the last keystroke.

4 [y^x] [(] 1 [÷] 3 [)] [=] [+/−] Scientific

[$\sqrt[3]{}$] [(−)] 4 [ENTER] Graphing

The display is −1.5874011. Rounded to three decimal places, $\sqrt[3]{-4} \approx -1.587$.

d. 8 [y^x] [(] 1 [÷] 2 [)] [=] Scientific

8 [^] [(] 1 [÷] 2 [)] [ENTER] Graphing

The display is 2.8284271. Rounded to three decimal places, $(8)^{1/2} \approx 2.828$.

e. 1.4 [y^x] [(] 2 [÷] 5 [+/−] [)] [=] Scientific

1.4 [^] [(] [(−)] 2 [÷] 5 [)] [ENTER] Graphing

The display is 0.8740752. Rounded to three decimal places, $(1.4)^{-2/5} \approx 0.874$.

Some calculators have a cube root key or submenu command. If your calculator does, try using it to evaluate the expression in Example 7(c).

4 Determine the domain of a radical function.

Radical Functions

The **domain** of the radical function $f(x) = \sqrt[n]{x}$ is the set of all real numbers such that x has a principal nth root.

Technology: Discovery

Consider the function $f(x) = x^{2/3}$.

a. What is the domain of the function?

b. Use your graphing utility to graph the following, in order.

$y_1 = x^{(2 \div 3)}$

$y_2 = (x^2)^{1/3}$ Power, then root

$y_3 = (x^{1/3})^2$ Root, then power

c. Are the graphs all the same? Are their domains all the same?

d. On your graphing utility, which of the forms properly represent the function $f(x) = x^{m/n}$?

$y_1 = x^{(m \div n)}$

$y_2 = (x^m)^{1/n}$

$y_3 = (x^{1/n})^m$

e. Explain how the domains of $f(x) = x^{2/3}$ and $g(x) = x^{-2/3}$ differ.

▶ Domain of a Radical Function

Let n be an integer that is greater than or equal to 2.

1. If n is odd, the domain of $f(x) = \sqrt[n]{x}$ is the set of all real numbers.
2. If n is even, the domain of $f(x) = \sqrt[n]{x}$ is the set of all nonnegative real numbers.

Example 8 Finding the Domain of a Radical Function

Describe the domain of each function.

a. $f(x) = \sqrt{x}$

b. $f(x) = \sqrt[3]{x}$

c. $f(x) = \sqrt{x^2}$

d. $f(x) = \sqrt{x^3}$

Solution

a. The domain of $f(x) = \sqrt{x}$ is the set of all nonnegative real numbers. For instance, 2 is in the domain, but -2 is not because $\sqrt{-2}$ is not a real number.

b. The domain of $f(x) = \sqrt[3]{x}$ is the set of all real numbers because for any real number x, the expression $\sqrt[3]{x}$ is a real number.

c. The domain of $f(x) = \sqrt{x^2}$ is the set of all real numbers because for any real number x, the expression x^2 is a nonnegative real number.

d. The domain of $f(x) = \sqrt{x^3}$ is the set of all nonnegative real numbers. For instance, 1 is in the domain, but -1 is not because $\sqrt{(-1)^3} = \sqrt{-1}$ is not a real number.

Discussing the Concept — Describing Domains and Ranges

Discuss the domain and range of each of the following functions. Use a graphing utility to verify your conclusions.

a. $y = x^{3/2}$ **b.** $y = x^2$

c. $y = x^{1/3}$ **d.** $y = (\sqrt{x})^2$

e. $y = x^{-4/5}$ **f.** $y = \sqrt[3]{x^2}$

9.1 Exercises

Integrated Review *Concepts, Skills, and Problem Solving*

Keep mathematically in shape by doing these exercises *before* the problems of this section.

Properties and Definitions

In Exercises 1–4, complete the property of exponents.

1. $a^m \cdot a^n =$ **2.** $(ab)^m =$

3. $(a^m)^n =$ **4.** $\dfrac{a^m}{a^n} =$, if $m > n$

Solving Equations

In Exercises 5–10, solve for *y*.

5. $3x + y = 4$ **6.** $2x + 3y = 2$

7. $x^2 + 3y = 4$ **8.** $x^2 + y - 4 = 0$

9. $2\sqrt{x} - 3y = 15$ **10.** $6|x| - 5y + 10 = 0$

Problem Solving

11. You can mow a lawn in 4 hours and your friend can mow it in 6 hours. Working together, how long will it take to mow the lawn?

12. A truck driver traveled at an average speed of 54 miles per hour on a 90-mile trip to pick up a load of freight. On the return trip, the average speed was 42 miles per hour. Find the average speed for the round trip.

Developing Skills

In Exercises 1–8, find the root if it exists. See Example 1.

1. $\sqrt{64}$ **2.** $-\sqrt{100}$

3. $-\sqrt{49}$ **4.** $\sqrt{-25}$

5. $\sqrt[3]{-8}$ **6.** $\sqrt[3]{-64}$

7. $\sqrt{-1}$ **8.** $-\sqrt[3]{1}$

In Exercises 9–14, complete the statement. See Example 2.

9. Because $7^2 = 49$, ______ is a square root of 49.

10. Because $24.5^2 = 600.25$, ______ is a square root of 600.25.

11. Because $4.2^3 = 74.088$, ______ is a cube root of 74.088.

12. Because $6^4 = 1296$, ______ is a fourth root of 1296.

13. Because $45^2 = 2025$, 45 is called the ______ of 2025.

14. Because $12^3 = 1728$, 12 is called the ______ of 1728.

In Exercises 15–44, evaluate each radical expression without using a calculator. If not possible, state the reason. See Example 3.

15. $\sqrt{8^2}$ **16.** $-\sqrt{10^2}$

17. $\sqrt{(-10)^2}$ **18.** $\sqrt{(-12)^2}$

19. $\sqrt{-9^2}$ **20.** $\sqrt{-12^2}$

21. $-\sqrt{\left(\frac{2}{3}\right)^2}$ **22.** $\sqrt{\left(\frac{3}{4}\right)^2}$

23. $\sqrt{-\left(\frac{3}{10}\right)^2}$ **24.** $\sqrt{\left(-\frac{3}{5}\right)^2}$

25. $\left(\sqrt{5}\right)^2$ **26.** $-\left(\sqrt{10}\right)^2$

27. $-\left(\sqrt{23}\right)^2$ **28.** $\left(-\sqrt{18}\right)^2$

29. $\sqrt[3]{(5)^3}$ **30.** $\sqrt[3]{(-2)^3}$

31. $\sqrt[3]{10^3}$ **32.** $\sqrt[3]{4^3}$

33. $-\sqrt[3]{(-6)^3}$ **34.** $-\sqrt[3]{9^3}$

35. $\sqrt[3]{\left(-\frac{1}{4}\right)^3}$ **36.** $-\sqrt[3]{\left(\frac{1}{5}\right)^3}$

37. $\left(\sqrt[3]{11}\right)^3$ **38.** $\left(\sqrt[3]{-6}\right)^3$

39. $\left(-\sqrt[3]{24}\right)^3$ **40.** $\left(\sqrt[3]{21}\right)^3$

41. $\sqrt[4]{3^4}$ **42.** $\sqrt[5]{(-2)^5}$

43. $-\sqrt[4]{-5^4}$ **44.** $-\sqrt[4]{2^4}$

In Exercises 45–48, determine whether the square root is a rational or irrational number.

45. $\sqrt{6}$ **46.** $\sqrt{\frac{9}{16}}$

47. $\sqrt{900}$ **48.** $\sqrt{72}$

In Exercises 49–54, fill in the missing description.

	Radical Form	*Rational Exponent Form*
49.	$\sqrt{16} = 4$	
50.	$\sqrt[4]{81} = 3$	
51.	$\sqrt[3]{27^2} = 9$	
52.		$125^{1/3} = 5$
53.		$256^{3/4} = 64$
54.		$27^{2/3} = 9$

In Exercises 55–74, evaluate without using a calculator. See Example 4.

55. $25^{1/2}$
56. $49^{1/2}$
57. $-36^{1/2}$
58. $-121^{1/2}$
59. $-(16)^{3/4}$
60. $-(125)^{2/3}$
61. $32^{-2/5}$
62. $81^{-3/4}$
63. $(-27)^{-2/3}$
64. $(-243)^{-3/5}$
65. $\left(\frac{8}{27}\right)^{2/3}$
66. $\left(\frac{256}{625}\right)^{1/4}$
67. $\left(\frac{121}{9}\right)^{-1/2}$
68. $\left(\frac{27}{1000}\right)^{-4/3}$
69. $(3^3)^{2/3}$
70. $(8^2)^{3/2}$
71. $-(4^4)^{3/4}$
72. $(-2^3)^{5/3}$
73. $\left(\dfrac{1}{5^3}\right)^{-2/3}$
74. $\left(\dfrac{4}{6^2}\right)^{-3/2}$

In Exercises 75–94, rewrite the expression using rational exponents. See Example 5.

75. $\sqrt{t}$
76. $\sqrt[3]{x}$
77. $x\sqrt[4]{x^3}$
78. $t\sqrt[5]{t^2}$
79. $u^2\sqrt[3]{u}$
80. $y\sqrt[4]{y^2}$
81. $s^4\sqrt{s^5}$
82. $n^3\sqrt[4]{n^6}$
83. $\dfrac{\sqrt{x}}{\sqrt{x^3}}$
84. $\dfrac{\sqrt[3]{x^2}}{\sqrt[3]{x^4}}$
85. $\dfrac{\sqrt[4]{t}}{\sqrt{t^5}}$
86. $\dfrac{\sqrt[3]{x^4}}{\sqrt{x^3}}$
87. $\sqrt[3]{x^2} \cdot \sqrt[3]{x^7}$
88. $\sqrt[5]{z^3} \cdot \sqrt[5]{z^2}$
89. $\sqrt[4]{y^3} \cdot \sqrt[3]{y}$
90. $\sqrt[6]{x^5} \cdot \sqrt[3]{x^4}$
91. $\sqrt[4]{x^3y}$
92. $\sqrt[3]{u^4v^2}$
93. $z^2\sqrt{y^5z^4}$
94. $x^2\sqrt[3]{xy^4}$

In Exercises 95–116, simplify the expression. See Example 6.

95. $3^{1/4} \cdot 3^{3/4}$
96. $2^{2/5} \cdot 2^{3/5}$
97. $(2^{1/2})^{2/3}$
98. $(4^{1/3})^{9/4}$
99. $\dfrac{2^{1/5}}{2^{6/5}}$
100. $\dfrac{5^{-3/4}}{5}$
101. $(c^{3/2})^{1/3}$
102. $(k^{-1/3})^{3/2}$
103. $\dfrac{18y^{4/3}z^{-1/3}}{24y^{-2/3}z}$
104. $\dfrac{a^{3/4} \cdot a^{1/2}}{a^{5/2}}$
105. $(3x^{-1/3}y^{3/4})^2$
106. $(-2u^{3/5}v^{-1/5})^3$
107. $\left(\dfrac{x^{1/4}}{x^{1/6}}\right)^3$
108. $\left(\dfrac{3m^{1/6}n^{1/3}}{4n^{-2/3}}\right)^2$
109. $\sqrt{\sqrt[4]{y}}$
110. $\sqrt[3]{\sqrt{2x}}$
111. $\sqrt[4]{\sqrt{x^3}}$
112. $\sqrt[5]{\sqrt[3]{y^4}}$
113. $\dfrac{(x+y)^{3/4}}{\sqrt[4]{x+y}}$
114. $\dfrac{(a-b)^{1/3}}{\sqrt[3]{a-b}}$
115. $\dfrac{(3u-2v)^{2/3}}{\sqrt{(3u-2v)^3}}$
116. $\dfrac{\sqrt[4]{2x+y}}{(2x+y)^{3/2}}$

In Exercises 117–130, use a calculator to approximate the quantity accurate to four decimal places. If not possible, state the reason. See Example 7.

117. $\sqrt{73}$
118. $\sqrt{-532}$
119. $315^{2/5}$
120. $962^{2/3}$
121. $1698^{-3/4}$
122. $382.5^{-3/2}$
123. $\sqrt[4]{342}$
124. $\sqrt[3]{159}$
125. $\sqrt[3]{545^2}$
126. $\sqrt[5]{-35^3}$
127. $\dfrac{8-\sqrt{35}}{2}$
128. $\dfrac{-5+\sqrt{3215}}{10}$
129. $\dfrac{3+\sqrt{17}}{9}$
130. $\dfrac{7-\sqrt{241}}{12}$

In Exercises 131–136, describe the domain of the function. See Example 8.

131. $f(x) = 3\sqrt{x}$
132. $h(x) = \sqrt[4]{x}$
133. $g(x) = \dfrac{2}{\sqrt[4]{x}}$
134. $g(x) = \dfrac{10}{\sqrt[3]{x}}$
135. $f(x) = \sqrt{-x}$
136. $f(x) = \sqrt[3]{x^4}$

In Exercises 137–140, describe the domain of the function algebraically. Use a graphing utility to graph the function. Did the graphing utility omit part of the domain? If so, complete the graph by hand.

137. $y = \dfrac{5}{\sqrt[4]{x^3}}$
138. $y = 4\sqrt[3]{x}$
139. $g(x) = 2x^{3/5}$
140. $h(x) = 5x^{2/3}$

In Exercises 141–144, perform the multiplication. Use a graphing utility to confirm your result.

141. $x^{1/2}(2x - 3)$

142. $x^{4/3}(3x^2 - 4x + 5)$

143. $y^{-1/3}(y^{1/3} + 5y^{4/3})$

144. $(x^{1/2} - 3)(x^{1/2} + 3)$

Solving Problems

Mathematical Modeling In Exercises 145 and 146, use the formula for the *declining balances method*

$$r = 1 - \left(\frac{S}{C}\right)^{1/n}$$

to find the depreciation rate r. In the formula, n is the useful life of the item (in years), S is the salvage value (in dollars), and C is the original cost (in dollars).

145. A \$75,000 truck depreciates over an 8-year period, as shown in the graph. Find r.

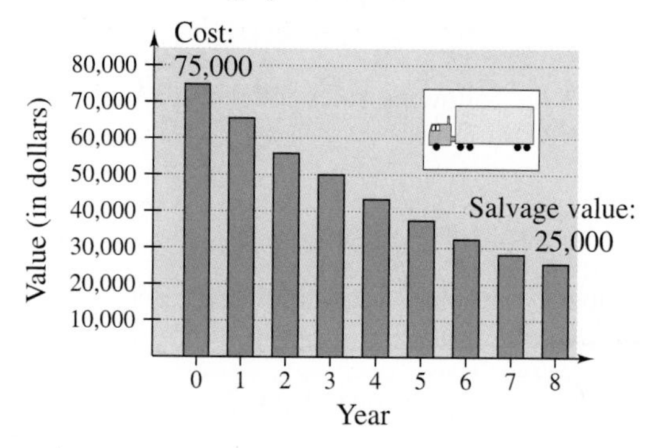

146. A \$125,000 printing press depreciates over a 10-year period, as shown in the graph. Find r.

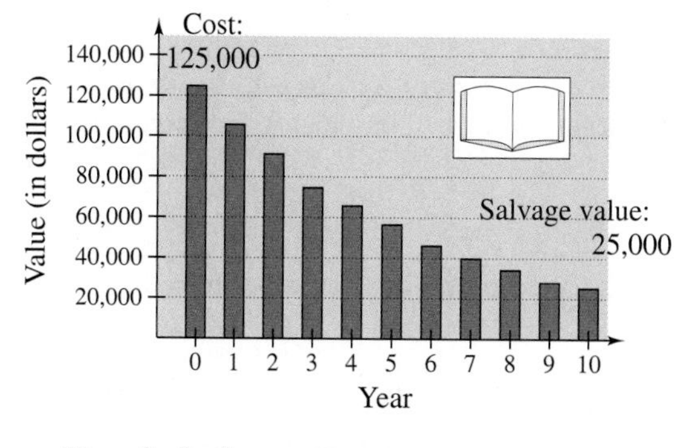

147. *Geometry* Find the dimensions of a piece of carpet for a classroom with 529 square feet of floor space, assuming the floor is square.

148. *Geometry* Find the dimensions of a square mirror with an area of 1024 square inches.

149. *Geometry* The length of a diagonal of a rectangular solid of length l, width w, and height h is

$\sqrt{l^2 + w^2 + h^2}$.

Approximate to two decimal places the length of the diagonal of the solid shown in the figure.

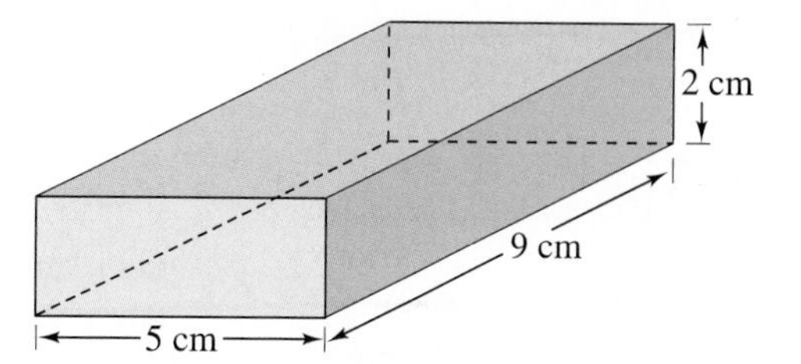

150. *Velocity of a Stream* A stream of water moving at a rate of v feet per second can carry particles of size $0.03\sqrt{v}$ inches.

(a) Find the particle size that can be carried by a stream flowing at the rate of $\frac{3}{4}$ foot per second. Approximate to three decimal places.

(b) Find the particle size that can be carried by a stream flowing at the rate of $\frac{3}{16}$ foot per second. Approximate to three decimal places.

Explaining Concepts

151. Answer parts (a)–(d) of Motivating the Chapter on page 513.

152. In your own words, define the nth root of a number.

153. Define the *radicand* and the *index* of a radical.

154. If n is even, what must be true about the radicand for the nth root to be a real number? Explain.

155. Is it true that $\sqrt{2} = 1.414$? Explain.

156. Given a real number x, state the conditions on n for each of the following.

(a) $\sqrt[n]{x^n} = x$ (b) $\sqrt[n]{x^n} = |x|$

157. *Investigation* Find all possible "last digits" of perfect squares. (For instance, the last digit of 81 is 1 and the last digit of 64 is 4.) Is it possible that 4,322,788,986 is a perfect square?

9.2 Simplifying Radical Expressions

Objectives

1. Use the Multiplication and Division Properties of Radicals to simplify a radical expression.
2. Use rationalization techniques to simplify a radical expression.
3. Use the Distributive Property to add and subtract like radicals.
4. Use the Pythagorean Theorem in an application problem.

1 Use the Multiplication and Division Properties of Radicals to simplify a radical expression.

Simplifying Radicals

In this section, you will study ways to simplify and combine radicals. For instance, the expression $\sqrt{12}$ can be simplified as

$$\sqrt{12} = \sqrt{4 \cdot 3} = \sqrt{4}\sqrt{3} = 2\sqrt{3}.$$

This rewritten form is based on the following rules for multiplying and dividing radicals.

Study Tip

The Multiplication and Division Properties of Radicals can be shown to be true by converting the radicals to exponential form and using the rules of exponents on page 517.

Using Property 3

$$\sqrt[n]{uv} = (uv)^{1/n}$$
$$= u^{1/n}v^{1/n}$$
$$= \sqrt[n]{u}\sqrt[n]{v}$$

Using Property 4

$$\sqrt[n]{\frac{u}{v}} = \left(\frac{u}{v}\right)^{1/n}$$
$$= \frac{u^{1/n}}{v^{1/n}} = \frac{\sqrt[n]{u}}{\sqrt[n]{v}}$$

▶ Multiplying and Dividing Radicals

Let u and v be real numbers, variables, or algebraic expressions. If the nth roots of u and v are real, the following properties are true.

1. $\sqrt[n]{uv} = \sqrt[n]{u}\sqrt[n]{v}$ — Multiplication Property of Radicals
2. $\sqrt[n]{\dfrac{u}{v}} = \dfrac{\sqrt[n]{u}}{\sqrt[n]{v}}, \quad v \neq 0$ — Division Property of Radicals

You can use the Multiplication Property of Radicals to *simplify* square root expressions by finding the largest perfect square factor and removing it from the radical, as follows.

$$\sqrt{48} = \sqrt{16 \cdot 3} = \sqrt{16}\sqrt{3} = 4\sqrt{3}$$

This simplification process is called **removing perfect square factors from the radical.**

Example 1 Removing Constant Factors from Radicals

Simplify each radical by removing as many factors as possible.

a. $\sqrt{75}$ **b.** $\sqrt{72}$ **c.** $\sqrt{162}$

Solution

a. $\sqrt{75} = \sqrt{25 \cdot 3} = \sqrt{25}\sqrt{3} = 5\sqrt{3}$ — 25 is a perfect square factor of 75.

b. $\sqrt{72} = \sqrt{36 \cdot 2} = \sqrt{36}\sqrt{2} = 6\sqrt{2}$ — 36 is a perfect square factor of 72.

c. $\sqrt{162} = \sqrt{81 \cdot 2} = \sqrt{81}\sqrt{2} = 9\sqrt{2}$ — 81 is a perfect square factor of 162.

When removing *variable* factors from a square root radical, remember that it is not valid to write $\sqrt{x^2} = x$ *unless* you happen to know that x is nonnegative. Without knowing anything about x, the only way you can simplify $\sqrt{x^2}$ is to include absolute value signs when you remove x from the radical.

$\sqrt{x^2} = |x|$ Restricted by absolute value signs

When simplifying the expression $\sqrt{x^3}$, it is not necessary to include absolute value signs because the domain of this expression does not include negative numbers.

$\sqrt{x^3} = \sqrt{x^2(x)} = x\sqrt{x}$ Restricted by domain of radical

Example 2 Removing Variable Factors from Radicals

Simplify the radical expression.

a. $\sqrt{25x^2}$ **b.** $\sqrt{12x^3}$, $x \ge 0$ **c.** $\sqrt{144x^4}$ **d.** $\sqrt{72x^3y^2}$

Solution

a. $\sqrt{25x^2} = \sqrt{5^2x^2} = \sqrt{5^2}\sqrt{x^2} = 5|x|$ $\sqrt{x^2} = |x|$

b. $\sqrt{12x^3} = \sqrt{2^2x^2(3x)} = 2x\sqrt{3x}$ $\sqrt{2^2}\sqrt{x^2} = 2x,\ x \ge 0$

c. $\sqrt{144x^4} = \sqrt{12^2(x^2)^2} = 12x^2$ $\sqrt{12^2}\sqrt{(x^2)^2} = 12|x^2| = 12x^2$

d. $\sqrt{72x^3y^2} = \sqrt{6^2x^2y^2(2x)}$ $\sqrt{6^2}\sqrt{x^2}\sqrt{y^2} = 6|x||y|$

$= 6|x||y|\sqrt{2x}$

In the same way that perfect squares can be removed from square root radicals, perfect nth powers can be removed from nth root radicals.

Study Tip

To find the perfect nth root factor of 486 in Example 3(c), you can write the prime factorization of the number.

$486 = 2 \cdot 3 \cdot 3 \cdot 3 \cdot 3 \cdot 3$

$= 2 \cdot 3^5$

From its prime factorization you can see that 3^5 is a fifth root factor of 486.

$\sqrt[5]{486} = \sqrt[5]{2 \cdot 3^5}$

$= \sqrt[5]{3^5}\sqrt[5]{2}$

$= 3\sqrt[5]{2}$

Example 3 Removing Factors from Radicals

Simplify the radical expressions.

a. $\sqrt[3]{40}$ **b.** $\sqrt[4]{x^5}$, $x \ge 0$ **c.** $\sqrt[5]{486x^7}$ **d.** $\sqrt[3]{128x^3y^5}$

Solution

a. $\sqrt[3]{40} = \sqrt[3]{8(5)}$ $\sqrt[3]{2^3} = 2$

$= \sqrt[3]{2^3(5)}$

$= 2\sqrt[3]{5}$

b. $\sqrt[4]{x^5} = \sqrt[4]{x^4(x)}$ $\sqrt[4]{x^4} = x,\ x \ge 0$

$= x\sqrt[4]{x}$

c. $\sqrt[5]{486x^7} = \sqrt[5]{243x^5(2x^2)}$

$= \sqrt[5]{3^5x^5(2x^2)}$

$= 3x\sqrt[5]{2x^2}$ $\sqrt[5]{3^5}\sqrt[5]{x^5} = 3x$

d. $\sqrt[3]{128x^3y^5} = \sqrt[3]{64x^3y^3(2y^2)}$

$= \sqrt[3]{4^3x^3y^3(2y^2)}$

$= 4xy\sqrt[3]{2y^2}$ $\sqrt[3]{4^3}\sqrt[3]{x^3}\sqrt[3]{y^3} = 4xy$

2 Use rationalization techniques to simplify a radical expression.

Rationalization Techniques

Removing factors from radicals is only one of three techniques used to simplify radicals. The three techniques are summarized as follows.

> **Simplifying Radical Expressions**
>
> A radical expression is said to be in simplest form if all three of the following are true.
>
> **1.** All possible nth powered factors have been removed from each radical.
>
> **2.** No radical contains a fraction.
>
> **3.** No denominator of a fraction contains a radical.

To meet the last two conditions, you can use a technique called **rationalizing the denominator.** This involves multiplying both the numerator and denominator by a *rationalizing factor* that creates a perfect nth power in the denominator.

Example 4 Rationalizing the Denominator

Simplify the expression.

a. $\sqrt{\dfrac{3}{5}}$ **b.** $\dfrac{4}{\sqrt[3]{9}}$ **c.** $\dfrac{8}{3\sqrt{18}}$

Study Tip

When rationalizing a denominator, remember that for square roots you want a perfect square in the denominator, for cube roots you want a perfect cube, and so on.

Solution

a. $\sqrt{\dfrac{3}{5}} = \dfrac{\sqrt{3}}{\sqrt{5}} = \dfrac{\sqrt{3}}{\sqrt{5}} \cdot \dfrac{\sqrt{5}}{\sqrt{5}} = \dfrac{\sqrt{15}}{\sqrt{25}} = \dfrac{\sqrt{15}}{5}$ Multiply by $\sqrt{5}/\sqrt{5}$ to create a perfect square in the denominator.

b. $\dfrac{4}{\sqrt[3]{9}} = \dfrac{4}{\sqrt[3]{9}} \cdot \dfrac{\sqrt[3]{3}}{\sqrt[3]{3}} = \dfrac{4\sqrt[3]{3}}{\sqrt[3]{27}} = \dfrac{4\sqrt[3]{3}}{3}$ Multiply by $\sqrt[3]{3}/\sqrt[3]{3}$ to create a perfect cube in the denominator.

c. $\dfrac{8}{3\sqrt{18}} = \dfrac{8}{3\sqrt{18}} \cdot \dfrac{\sqrt{2}}{\sqrt{2}} = \dfrac{8\sqrt{2}}{3\sqrt{36}} = \dfrac{8\sqrt{2}}{3(6)} = \dfrac{4\sqrt{2}}{9}$

Example 5 Rationalizing the Denominator

Simplify the expression.

a. $\sqrt{\dfrac{8x}{12y^5}}$ **b.** $\sqrt[3]{\dfrac{54x^6y^3}{5z^2}}$

Solution

a. $\sqrt{\dfrac{8x}{12y^5}} = \sqrt{\dfrac{(4)(2)x}{(4)(3)y^5}} = \sqrt{\dfrac{2x}{3y^5}} = \dfrac{\sqrt{2x}}{\sqrt{3y^5}} \cdot \dfrac{\sqrt{3y}}{\sqrt{3y}} = \dfrac{\sqrt{6xy}}{\sqrt{9y^6}} = \dfrac{\sqrt{6xy}}{3y^3}$

b. $\sqrt[3]{\dfrac{54x^6y^3}{5z^2}} = \dfrac{\sqrt[3]{(3^3)(2)(x^6)(y^3)}}{\sqrt[3]{5z^2}} \cdot \dfrac{\sqrt[3]{25z}}{\sqrt[3]{25z}} = \dfrac{3x^2y\sqrt[3]{50z}}{\sqrt[3]{5^3z^3}} = \dfrac{3x^2y\sqrt[3]{50z}}{5z}$

3 Use the Distributive Property to add and subtract like radicals.

Adding and Subtracting Radicals

Two or more radical expressions are *alike* if they have the same radicand and the same index. For instance, $\sqrt{2}$ and $3\sqrt{2}$ are alike, but $\sqrt{3}$ and $\sqrt[3]{3}$ are not alike. Two radical expressions that are alike can be added or subtracted by adding or subtracting their coefficients. *Before* concluding that two radicals cannot be combined, you should first rewrite them in simplest form. This is illustrated in Example 6(c).

Example 6 Combining Radicals

a. $\sqrt{7} + 5\sqrt{7} - 2\sqrt{7} = (1 + 5 - 2)\sqrt{7}$ Distributive Property

$= 4\sqrt{7}$ Simplify.

b. $6\sqrt{x} - \sqrt[3]{4} - 5\sqrt{x} + 2\sqrt[3]{4} = 6\sqrt{x} - 5\sqrt{x} - \sqrt[3]{4} + 2\sqrt[3]{4}$ Group like terms.

$= (6 - 5)\sqrt{x} + (-1 + 2)\sqrt[3]{4}$ Distributive Property

$= \sqrt{x} + \sqrt[3]{4}$ Simplify.

c. $3\sqrt[3]{x} + \sqrt[3]{8x} = 3\sqrt[3]{x} + 2\sqrt[3]{x}$ $\sqrt[3]{8} = \sqrt[3]{2^3} = 2$

$= (3 + 2)\sqrt[3]{x}$ Distributive Property

$= 5\sqrt[3]{x}$ Simplify.

Example 7 Simplifying Before Combining Radicals

a. $\sqrt{45x} + 3\sqrt{20x} = 3\sqrt{5x} + 6\sqrt{5x} = 9\sqrt{5x}$

b. $5\sqrt{x^3} - x\sqrt{4x} = 5x\sqrt{x} - 2x\sqrt{x} = 3x\sqrt{x}$

c. $\sqrt[3]{54y^5} + 4\sqrt[3]{2y^2} = 3y\sqrt[3]{2y^2} + 4\sqrt[3]{2y^2} = (3y + 4)\sqrt[3]{2y^2}$

In some instances, it may be necessary to rationalize denominators before combining radicals.

Example 8 Rationalizing Denominators Before Simplifying

$\sqrt{7} - \frac{5}{\sqrt{7}} = \sqrt{7} - \left(\frac{5}{\sqrt{7}} \cdot \frac{\sqrt{7}}{\sqrt{7}}\right)$ Multiply by $\sqrt{7}/\sqrt{7}$ to create a perfect square in the denominator.

$= \sqrt{7} - \frac{5\sqrt{7}}{7}$ Simplify.

$= \left(1 - \frac{5}{7}\right)\sqrt{7}$ Distributive Property

$= \frac{2}{7}\sqrt{7}$ Simplify.

4 Use the Pythagorean Theorem in an application problem.

Application of Radicals

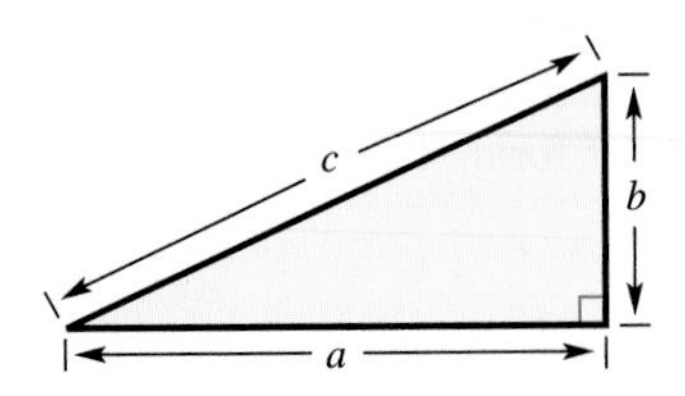

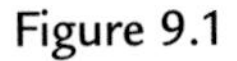

Figure 9.1

A common use of radicals occurs in applications involving right triangles. Recall that a right triangle is one that contains a right (or 90°) angle, as shown in Figure 9.1. The relationship among the three sides of a right triangle is described by the **Pythagorean Theorem,** which says that if a and b are the lengths of the legs and c is the length of the hypotenuse, then

$$c = \sqrt{a^2 + b^2}. \qquad \text{Pythagorean Theorem}$$

For instance, if $a = 6$ and $b = 9$, then

$$c = \sqrt{6^2 + 9^2} = \sqrt{117} = \sqrt{9}\sqrt{13} = 3\sqrt{13}.$$

Example 9 An Application of the Pythagorean Theorem

A softball diamond has the shape of a square with 60-foot sides (see Figure 9.2). The catcher is 5 feet behind home plate. How far does the catcher have to throw to reach second base?

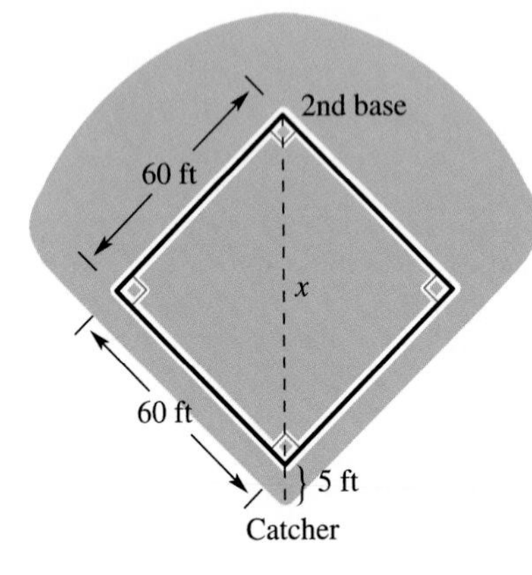

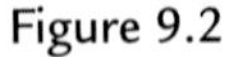

Figure 9.2

Solution

In Figure 9.2, let x be the hypotenuse of a right triangle with 60-foot legs. So, by the Pythagorean Theorem, you have the following.

$x = \sqrt{60^2 + 60^2}$ — Pythagorean Theorem

$x = \sqrt{7200}$ — Simplify.

$x = \sqrt{3600 \cdot 2}$ — 3600 is a perfect square factor of 7200.

$x = 60\sqrt{2}$ — Simplify.

$x \approx 84.9$ feet

So, the distance from home plate to second base is approximately 84.9 feet. Because the catcher is 5 feet behind home plate, the catcher must make a throw of

$$x + 5 \approx 84.9 + 5 = 89.9 \text{ feet.}$$

Discussing the Concept **Error Analysis**

Suppose you are an algebra instructor and one of your students hands in the following work. Find and correct the errors, and discuss how you can help your student avoid such errors in the future.

a. $7\sqrt{3} + 4\sqrt{2} = 11\sqrt{5}$

b. $3\sqrt[3]{k} - 6\sqrt{k} = -3\sqrt{k}$

9.2 Exercises

Integrated Review *Concepts, Skills, and Problem Solving*

Keep mathematically in shape by doing these exercises *before* the problems of this section.

Properties and Definitions

1. Explain how to determine the half-plane satisfying $x - y > -3$.
2. Describe the difference between the graphs of $3x + 4y \leq 4$ and $3x + 4y < 4$.

In Exercises 3–8, factor the expression completely.

3. $-x^3 + 3x^2 - x + 3$
4. $4t^2 - 169$
5. $x^2 - 3x + 2$
6. $2x^2 + 5x - 7$
7. $11x^2 + 6x - 5$
8. $4x^2 - 28x + 49$

Problem Solving

9. Twelve hundred tickets were sold for a theater production and the receipts for the performance were $21,120. The tickets for adults and students sold for $20 and $12.50, respectively. How many of each kind of ticket were sold?
10. A quality control engineer for a certain buyer found two defective units in a sample of 75. At that rate, what is the expected number of defective units in a shipment of 10,000 units?

Developing Skills

In Exercises 1–20, simplify the radical. See Example 1.

1. $\sqrt{20}$
2. $\sqrt{27}$
3. $\sqrt{50}$
4. $\sqrt{125}$
5. $\sqrt{96}$
6. $\sqrt{84}$
7. $\sqrt{216}$
8. $\sqrt{147}$
9. $\sqrt{1183}$
10. $\sqrt{1176}$
11. $\sqrt{0.04}$
12. $\sqrt{0.25}$
13. $\sqrt{0.0072}$
14. $\sqrt{0.0027}$
15. $\sqrt{2.42}$
16. $\sqrt{9.8}$
17. $\sqrt{\frac{15}{4}}$
18. $\sqrt{\frac{5}{36}}$
19. $\sqrt{\frac{13}{25}}$
20. $\sqrt{\frac{15}{36}}$

In Exercises 21–56, simplify the expression. See Examples 2 and 3.

21. $\sqrt{9x^5}$
22. $\sqrt{64x^3}$
23. $\sqrt{48y^4}$
24. $\sqrt{32x}$
25. $\sqrt{117y^6}$
26. $\sqrt{160x^8}$
27. $\sqrt{120x^2y^3}$
28. $\sqrt{125u^4v^6}$
29. $\sqrt{192a^5b^7}$
30. $\sqrt{363x^{10}y^9}$
31. $\sqrt[3]{48}$
32. $\sqrt[3]{81}$
33. $\sqrt[3]{112}$
34. $\sqrt[4]{112}$
35. $\sqrt[3]{40x^5}$
36. $\sqrt[3]{54z^7}$
37. $\sqrt[4]{324y^6}$
38. $\sqrt[5]{160x^8}$
39. $\sqrt[3]{x^4y^3}$
40. $\sqrt[3]{a^5b^6}$
41. $\sqrt[4]{3x^4y^2}$
42. $\sqrt[4]{128u^4v^7}$
43. $\sqrt[5]{32x^5y^6}$
44. $\sqrt[3]{16x^4y^5}$
45. $\sqrt[3]{\frac{35}{64}}$
46. $\sqrt[4]{\frac{5}{16}}$
47. $\sqrt[5]{\frac{15}{243}}$
48. $\sqrt[3]{\frac{1}{1000}}$
49. $\sqrt[5]{\frac{32x^2}{y^5}}$
50. $\sqrt[3]{\frac{16z^3}{y^6}}$
51. $\sqrt[3]{\frac{54a^4}{b^9}}$
52. $\sqrt[4]{\frac{3u^2}{16v^8}}$
53. $\sqrt{\frac{32a^4}{b^2}}$
54. $\sqrt{\frac{18x^2}{z^6}}$
55. $\sqrt[4]{(3x^2)^4}$
56. $\sqrt[5]{96x^5}$

In Exercises 57–78, rationalize the denominator and simplify further, if possible. See Examples 4 and 5.

57. $\sqrt{\frac{1}{3}}$
58. $\sqrt{\frac{1}{5}}$
59. $\frac{1}{\sqrt{7}}$
60. $\frac{1}{\sqrt{15}}$
61. $\frac{12}{\sqrt{3}}$
62. $\frac{5}{\sqrt{10}}$

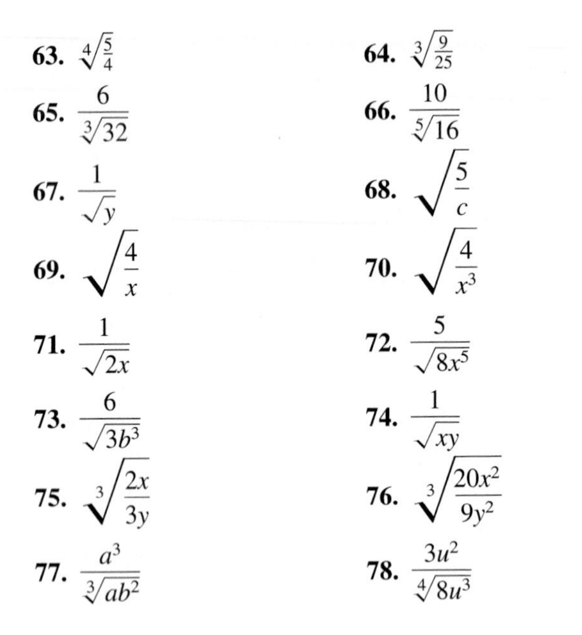

63. $\sqrt[4]{\frac{5}{4}}$ **64.** $\sqrt[3]{\frac{9}{25}}$

65. $\dfrac{6}{\sqrt[3]{32}}$ **66.** $\dfrac{10}{\sqrt[5]{16}}$

67. $\dfrac{1}{\sqrt{y}}$ **68.** $\sqrt{\dfrac{5}{c}}$

69. $\sqrt{\dfrac{4}{x}}$ **70.** $\sqrt{\dfrac{4}{x^3}}$

71. $\dfrac{1}{\sqrt{2x}}$ **72.** $\dfrac{5}{\sqrt{8x^5}}$

73. $\dfrac{6}{\sqrt{3b^3}}$ **74.** $\dfrac{1}{\sqrt{xy}}$

75. $\sqrt[3]{\dfrac{2x}{3y}}$ **76.** $\sqrt[3]{\dfrac{20x^2}{9y^2}}$

77. $\dfrac{a^3}{\sqrt[3]{ab^2}}$ **78.** $\dfrac{3u^2}{\sqrt[4]{8u^3}}$

In Exercises 79–92, combine the radical expressions, if possible. See Examples 6 and 7.

79. $3\sqrt{2} - \sqrt{2}$ **80.** $6\sqrt{5} - 2\sqrt{5}$

81. $12\sqrt{8} - 3\sqrt[3]{8}$ **82.** $4\sqrt{32} + 7\sqrt{32}$

83. $\sqrt[4]{3} - 5\sqrt[4]{7} - 12\sqrt[4]{3}$

84. $9\sqrt[3]{17} + 7\sqrt[3]{2} - 4\sqrt[3]{17} + \sqrt[3]{2}$

85. $2\sqrt[3]{54} + 12\sqrt[3]{16}$ **86.** $4\sqrt[4]{48} - \sqrt[4]{243}$

87. $5\sqrt{9x} - 3\sqrt{x}$

88. $3\sqrt{x+1} + 10\sqrt{x+1}$

89. $\sqrt{25y} + \sqrt{64y}$ **90.** $\sqrt[3]{16t^4} - \sqrt[3]{54t^4}$

91. $10\sqrt[3]{z} - \sqrt[3]{z^4}$ **92.** $5\sqrt[3]{24u^2} + 2\sqrt[3]{81u^5}$

In Exercises 93–98, perform the addition or subtraction and simplify your answer. See Example 8.

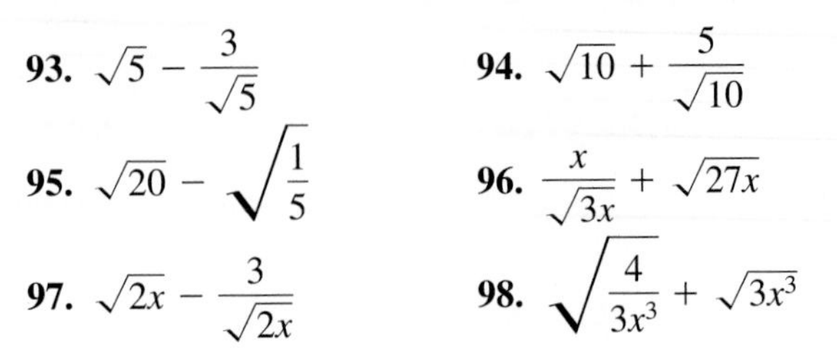

93. $\sqrt{5} - \dfrac{3}{\sqrt{5}}$ **94.** $\sqrt{10} + \dfrac{5}{\sqrt{10}}$

95. $\sqrt{20} - \sqrt{\dfrac{1}{5}}$ **96.** $\dfrac{x}{\sqrt{3x}} + \sqrt{27x}$

97. $\sqrt{2x} - \dfrac{3}{\sqrt{2x}}$ **98.** $\sqrt{\dfrac{4}{3x^3}} + \sqrt{3x^3}$

In Exercises 99–102, place the correct inequality symbol (<, >, or =) between the numbers.

99. $\sqrt{7} + \sqrt{18}$ ▢ $\sqrt{7 + 18}$

100. $\sqrt{10} - \sqrt{6}$ ▢ $\sqrt{10 - 6}$

101. 5 ▢ $\sqrt{3^2 + 2^2}$

102. 5 ▢ $\sqrt{3^2 + 4^2}$

In Exercises 103–106, find the length of the hypotenuse of the right triangle. See Example 9.

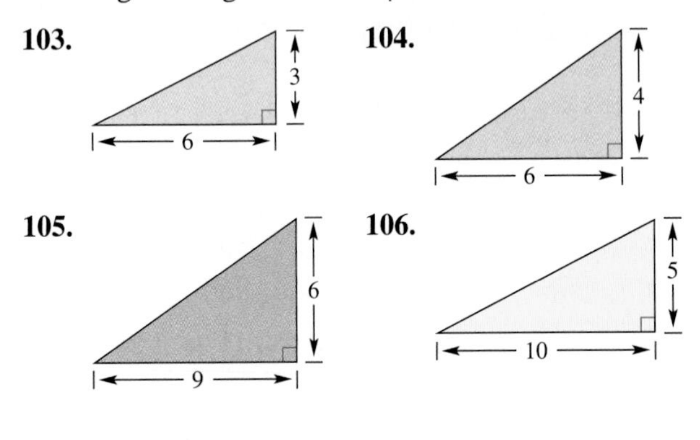

Solving Problems

107. *Geometry* The foundation of a house is 40 feet long and 30 feet wide. The height of the attic is 5 feet (see figure).

(a) Use the Pythagorean Theorem to find the length of the hypotenuse of the right triangle formed by the roof line.

(b) Use the result of part (a) to determine the total area of the roof.

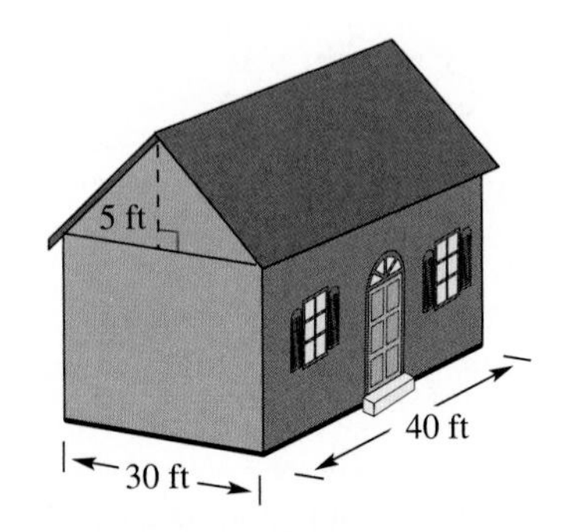

Figure for 107

108. *Geometry* The four corners are cut from a 4-foot-by-8-foot sheet of plywood, as shown in the figure. Find the perimeter of the remaining piece of plywood.

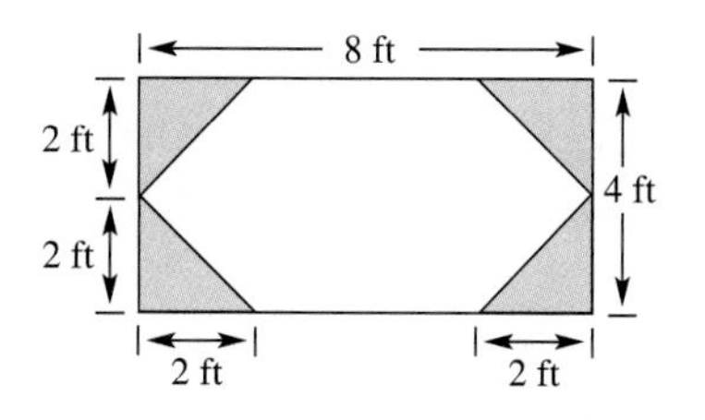

109. *Vibrating String* The frequency f in cycles per second of a vibrating string is given by

$$f = \frac{1}{100}\sqrt{\frac{400 \times 10^6}{5}}.$$

Use a calculator to approximate this number. (Round the result to two decimal places.)

110. *Period of a Pendulum* The period T in seconds of a pendulum (see figure) is given by

$$T = 2\pi\sqrt{\frac{L}{32}}$$

where L is the length of the pendulum in feet. Find the period of a pendulum whose length is 4 feet. (Round the result to two decimal places.)

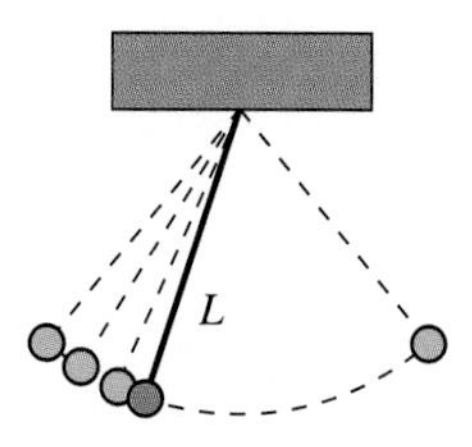

111. *Mathematical Modeling* The average salary S (in thousands of dollars) of public school teachers for the years 1992 through 1997 in the United States is modeled by

$$S = 34.7 + 1.6t - 2.4\sqrt{t}, \quad 2 \le t \le 7$$

where t is years, with $t = 0$ corresponding to 1990. (Source: Educational Research Service)

(a) Use a graphing utility to graph the model over the specified domain.

(b) Use a graphing utility to predict the year when the average salary will reach $48,000.

112. *Calculator Experiment* Enter any positive real number into your calculator and find its square root. Then repeatedly take the square root of the result.

$$\sqrt{x}, \quad \sqrt{\sqrt{x}}, \quad \sqrt{\sqrt{\sqrt{x}}}, \ . \ . \ .$$

What real number does the display appear to be approaching?

113. Square the real number $5/\sqrt{3}$ and note that the radical is eliminated from the denominator. Is this equivalent to rationalizing the denominator? Why or why not?

114. *The Square Root Spiral* The square root spiral (see figure) is formed by a sequence of right triangles, each with a side whose length is 1. Let r_n be the length of the hypotenuse of the nth triangle.

(a) Each leg of the first triangle has a length of 1 unit. Use the Pythagorean Theorem to show that $r_1 = \sqrt{2}$.

(b) Find r_2, r_3, r_4, r_5, and r_6.

(c) What can you conclude about r_n?

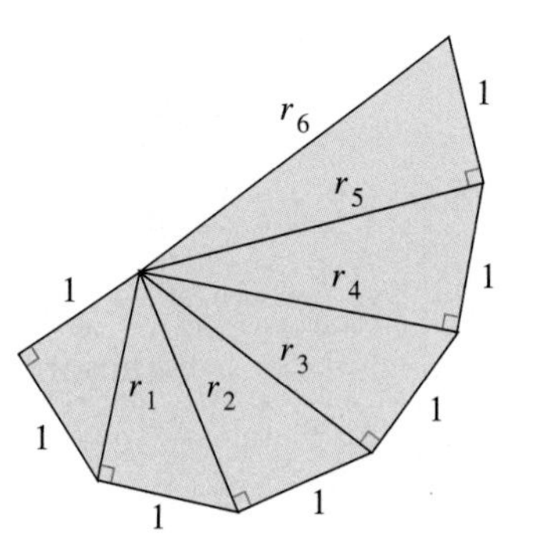

Explaining Concepts

115. Give an example of multiplying two radicals.

116. Describe the three conditions that characterize a simplified radical expression.

117. Is $\sqrt{2} + \sqrt{18}$ in simplest form? Explain.

118. Describe the steps you would use to simplify $1/\sqrt{3}$.

119. For what values of x is $\sqrt{x^2} \neq x$? Explain.

120. Explain what it means for two radical expressions to be alike.

9.3 Multiplying and Dividing Radical Expressions

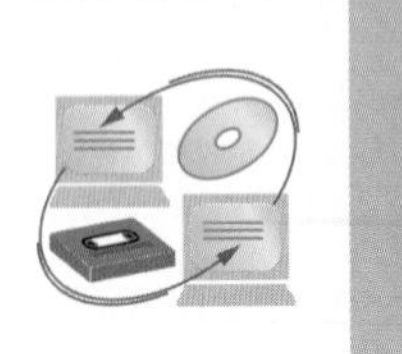

Objectives

1 Use the Distributive Property or the FOIL Method to multiply radical expressions.

2 Determine the product of conjugates.

3 Simplify a quotient involving radicals by rationalizing the denominator.

1 Use the Distributive Property or the FOIL Method to multiply radical expressions.

Multiplying Radical Expressions

You can multiply radical expressions by using the Distributive Property or the FOIL Method. In both procedures, you also make use of the Multiplication Property of Radicals from Section 9.2,

$$\sqrt[n]{a}\sqrt[n]{b} = \sqrt[n]{ab},$$

where a and b are real numbers whose nth roots are also real numbers.

Example 1 Multiplying Radical Expressions

Find the products and simplify.

a. $\sqrt{6} \cdot \sqrt{3}$ **b.** $\sqrt[3]{5} \cdot \sqrt[3]{16}$

Solution

a. $\sqrt{6} \cdot \sqrt{3} = \sqrt{6 \cdot 3} = \sqrt{18} = \sqrt{9 \cdot 2} = 3\sqrt{2}$

b. $\sqrt[3]{5} \cdot \sqrt[3]{16} = \sqrt[3]{5 \cdot 16} = \sqrt[3]{80} = \sqrt[3]{8 \cdot 10} = 2\sqrt[3]{10}$

Example 2 Multiplying Radical Expressions

Find the products and simplify.

a. $\sqrt{3}\left(2 + \sqrt{5}\right)$ **b.** $\sqrt{2}\left(4 - \sqrt{8}\right)$ **c.** $\sqrt{6}\left(\sqrt{12} - \sqrt{3}\right)$

Solution

a. $\sqrt{3}\left(2 + \sqrt{5}\right) = 2\sqrt{3} + \sqrt{3}\sqrt{5}$ Distributive Property

$= 2\sqrt{3} + \sqrt{15}$ Multiplication Property of Radicals

b. $\sqrt{2}\left(4 - \sqrt{8}\right) = 4\sqrt{2} - \sqrt{2}\sqrt{8}$ Distributive Property

$= 4\sqrt{2} - \sqrt{16}$ Multiplication Property of Radicals

$= 4\sqrt{2} - 4$ Simplify.

c. $\sqrt{6}\left(\sqrt{12} - \sqrt{3}\right) = \sqrt{6}\sqrt{12} - \sqrt{6}\sqrt{3}$ Distributive Property

$= \sqrt{72} - \sqrt{18}$ Multiplication Property of Radicals

$= 6\sqrt{2} - 3\sqrt{2}$ Find perfect square factors.

$= 3\sqrt{2}$ Simplify.

In Example 2, the Distributive Property was used to multiply radical expressions. In Example 3, note how the FOIL Method can be used to multiply binomial radical expressions.

Example 3 Using the FOIL Method

a. $(2\sqrt{7} - 4)(\sqrt{7} + 1) = \overbrace{2(\sqrt{7})^2}^{F} + \overbrace{2\sqrt{7}}^{O} - \overbrace{4\sqrt{7}}^{I} - \overbrace{4}^{L}$ FOIL Method

$= 2(7) + (2 - 4)\sqrt{7} - 4$ Combine like radicals.

$= 10 - 2\sqrt{7}$ Combine like terms.

b. $(3 - \sqrt{x})(1 + \sqrt{x}) = 3 + 3\sqrt{x} - \sqrt{x} - (\sqrt{x})^2$

$= 3 + 2\sqrt{x} - x, \quad x \geq 0$ Combine like radicals.

2 Determine the product of conjugates.

Conjugates

The expressions $3 + \sqrt{6}$ and $3 - \sqrt{6}$ are called **conjugates** of each other. Notice that they differ only by the sign between the terms. The product of two conjugates is the difference of two squares, which is given by the special product formula $(a + b)(a - b) = a^2 - b^2$. Now, in addition to the FOIL Method, you can also use special product formulas to multiply certain binomial radical expressions.

Example 4 Using a Special Product Formula

Find the conjugate of the expression and multiply the number by its conjugate.

a. $(2 - \sqrt{5})$ **b.** $(\sqrt{x} - 2)$ **c.** $(\sqrt{3} + \sqrt{x})$

Solution

a. The conjugate of $(2 - \sqrt{5})$ is $(2 + \sqrt{5})$.

$(2 - \sqrt{5})(2 + \sqrt{5}) = 2^2 - (\sqrt{5})^2$ Special product formula

$= 4 - 5$

$= -1$

b. The conjugate of $(\sqrt{x} - 2)$ is $(\sqrt{x} + 2)$.

$(\sqrt{x} - 2)(\sqrt{x} + 2) = (\sqrt{x})^2 - (2)^2$ Special product formula

$= x - 4, \quad x \geq 0$

c. The conjugate of $(\sqrt{3} + \sqrt{x})$ is $(\sqrt{3} - \sqrt{x})$.

$(\sqrt{3} + \sqrt{x})(\sqrt{3} - \sqrt{x}) = (\sqrt{3})^2 - (\sqrt{x})^2$ Special product formula

$= 3 - x, \quad x \geq 0$

3 Simplify a quotient involving radicals by rationalizing the denominator.

Dividing Radical Expressions

To simplify a *quotient* involving radicals, we rationalize the denominator. For single-term denominators, you can use the rationalizing process described in Section 9.2. To rationalize a denominator involving two terms, multiply both the numerator and denominator by the *conjugate of the denominator.*

Example 5 Simplifying Quotients Involving Radicals

a. $\dfrac{\sqrt{3}}{1-\sqrt{5}} = \dfrac{\sqrt{3}}{1-\sqrt{5}} \cdot \dfrac{1+\sqrt{5}}{1+\sqrt{5}}$ Multiply numerator and denominator by conjugate of denominator.

$= \dfrac{\sqrt{3}(1+\sqrt{5})}{1^2-(\sqrt{5})^2}$ Special product formula

$= \dfrac{\sqrt{3}+\sqrt{15}}{1-5}$ Simplify.

$= \dfrac{\sqrt{3}+\sqrt{15}}{-4}$ Simplify.

b. $\dfrac{4}{2-\sqrt{3}} = \dfrac{4}{2-\sqrt{3}} \cdot \dfrac{2+\sqrt{3}}{2+\sqrt{3}}$ Multiply numerator and denominator by conjugate of denominator.

$= \dfrac{4(2+\sqrt{3})}{2^2-(\sqrt{3})^2}$ Special product formula

$= \dfrac{8+4\sqrt{3}}{4-3}$ Simplify.

$= 8+4\sqrt{3}$ Simplify.

Example 6 Simplifying Quotients Involving Radicals

$\dfrac{5\sqrt{2}}{\sqrt{7}+\sqrt{2}} = \dfrac{5\sqrt{2}}{\sqrt{7}+\sqrt{2}} \cdot \dfrac{\sqrt{7}-\sqrt{2}}{\sqrt{7}-\sqrt{2}}$ Multiply numerator and denominator by conjugate of denominator.

$= \dfrac{5(\sqrt{14}-\sqrt{4})}{(\sqrt{7})^2-(\sqrt{2})^2}$ Special product formula

$= \dfrac{5(\sqrt{14}-2)}{7-2}$ Simplify.

$= \dfrac{\cancel{5}(\sqrt{14}-2)}{\cancel{5}}$ Cancel common factor.

$= \sqrt{14}-2$ Simplest form

Example 7 Dividing Radical Expressions

Perform the division and simplify.

a. $6 \div \left(\sqrt{x} - 2\right)$ **b.** $\left(2 - \sqrt{3}\right) \div \left(\sqrt{6} + \sqrt{2}\right)$

Solution

a. $\dfrac{6}{\sqrt{x} - 2} = \dfrac{6}{\sqrt{x} - 2} \cdot \dfrac{\sqrt{x} + 2}{\sqrt{x} + 2}$ Multiply numerator and denominator by conjugate of denominator.

$= \dfrac{6\left(\sqrt{x} + 2\right)}{\left(\sqrt{x}\right)^2 - 2^2}$ Special product formula

$= \dfrac{6\sqrt{x} + 12}{x - 4}, \quad x \geq 0$ Simplify.

b. $\dfrac{2 - \sqrt{3}}{\sqrt{6} + \sqrt{2}} = \dfrac{2 - \sqrt{3}}{\sqrt{6} + \sqrt{2}} \cdot \dfrac{\sqrt{6} - \sqrt{2}}{\sqrt{6} - \sqrt{2}}$ Multiply numerator and denominator by conjugate of denominator.

$= \dfrac{2\sqrt{6} - 2\sqrt{2} - \sqrt{18} + \sqrt{6}}{\left(\sqrt{6}\right)^2 - \left(\sqrt{2}\right)^2}$ Special product formula

$= \dfrac{3\sqrt{6} - 2\sqrt{2} - 3\sqrt{2}}{6 - 2}$ Simplify.

$= \dfrac{3\sqrt{6} - 5\sqrt{2}}{4}$ Simplify.

Discussing the Concept The Golden Section

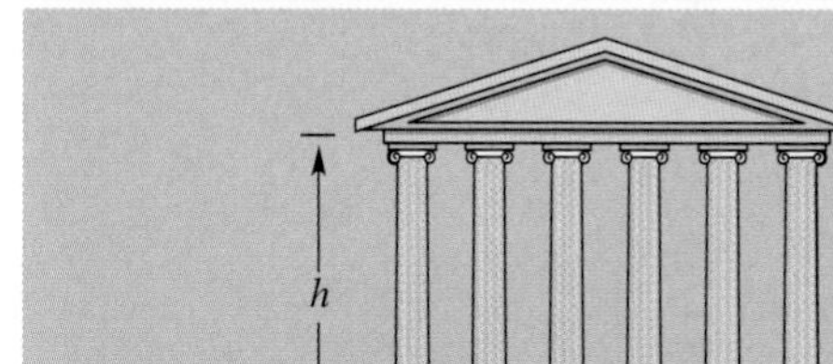

The ratio of the width of the Temple of Hephaestus to its height is approximately

$$\frac{w}{h} \approx \frac{2}{\sqrt{5} - 1}.$$

This number is called the **golden section.** Early Greeks believed that the most aesthetically pleasing rectangles were those whose sides had this ratio.

a. Rationalize the denominator for this number. Approximate your answer, rounded to two decimal places.

b. Use the Pythagorean Theorem, a straightedge, and a compass to construct a rectangle whose sides have the golden section as their ratio.

9.3 Exercises

Integrated Review *Concepts, Skills, and Problem Solving*

Keep mathematically in shape by doing these exercises *before* the problems of this section.

Properties and Definitions

In Exercises 1–4, use $x^2 + bx + c = (x + m)(x + n)$.

1. $mn =$ ▭

2. If $c > 0$, then what must be true about the signs of m and n?

3. If $c < 0$, then what must be true about the signs of m and n?

4. If m and n have like signs, then $m + n =$ ▭.

Equations of Lines

In Exercises 5–10, find an equation of the line through the two points.

5. $(-1, -2), (3, 6)$

6. $(1, 5), (6, 0)$

7. $(6, 3), (10, 3)$

8. $(4, -2), (4, 5)$

9. $\left(\frac{4}{3}, 8\right), (5, 6)$

10. $(7, 4), (10, 1)$

Models

In Exercises 11 and 12, translate the phrase into an algebraic expression.

11. The time to travel 360 miles if the average speed is r miles per hour

12. The perimeter of a rectangle of length L and width $L/3$

Developing Skills

In Exercises 1–42, multiply and simplify. See Examples 1–3.

1. $\sqrt{2} \cdot \sqrt{8}$

2. $\sqrt{6} \cdot \sqrt{18}$

3. $\sqrt{3} \cdot \sqrt{6}$

4. $\sqrt{5} \cdot \sqrt{10}$

5. $\sqrt[3]{12} \cdot \sqrt[3]{6}$

6. $\sqrt[3]{9} \cdot \sqrt[3]{9}$

7. $\sqrt[4]{8} \cdot \sqrt[4]{6}$

8. $\sqrt[4]{54} \cdot \sqrt[4]{3}$

9. $\sqrt{5}(2 - \sqrt{3})$

10. $\sqrt{11}(\sqrt{5} - 3)$

11. $\sqrt{2}(\sqrt{20} + 8)$

12. $\sqrt{7}(\sqrt{14} + 3)$

13. $\sqrt{6}(\sqrt{12} - \sqrt{3})$

14. $\sqrt{10}(\sqrt{5} + \sqrt{6})$

15. $\sqrt{2}(\sqrt{18} - \sqrt{10})$

16. $\sqrt{5}(\sqrt{15} + \sqrt{5})$

17. $\sqrt{y}(\sqrt{y} + 4)$

18. $\sqrt{x}(5 - \sqrt{x})$

19. $\sqrt{a}(4 - \sqrt{a})$

20. $\sqrt{z}(\sqrt{z} + 5)$

21. $\sqrt[3]{4}(\sqrt[3]{2} - 7)$

22. $\sqrt[3]{9}(\sqrt[3]{3} + 2)$

23. $(\sqrt{3} + 2)(\sqrt{3} - 2)$

24. $(3 - \sqrt{5})(3 + \sqrt{5})$

25. $(\sqrt{5} + 3)(\sqrt{3} - 5)$

26. $(\sqrt{7} + 6)(\sqrt{2} + 6)$

27. $(\sqrt{20} + 2)^2$

28. $(4 - \sqrt{20})^2$

29. $(\sqrt[3]{6} - 3)(\sqrt[3]{4} + 3)$

30. $(\sqrt[3]{9} + 5)(\sqrt[3]{5} - 5)$

31. $(10 + \sqrt{2x})^2$

32. $(5 - \sqrt{3v})^2$

33. $(9\sqrt{x} + 2)(5\sqrt{x} - 3)$

34. $(16\sqrt{u} - 3)(\sqrt{u} - 1)$

35. $(3\sqrt{x} - 5)(3\sqrt{x} + 5)$

36. $(7 - 3\sqrt{3t})(7 + 3\sqrt{3t})$

37. $(\sqrt[3]{2x} + 5)^2$

38. $(\sqrt[3]{3x} - 4)^2$

39. $(\sqrt[3]{y} + 2)(\sqrt[3]{y^2} - 5)$

40. $(\sqrt[3]{2y} + 10)(\sqrt[3]{4y^2} - 10)$

41. $(\sqrt[3]{t} + 1)(\sqrt[3]{t^2} + 4\sqrt[3]{t} - 3)$

42. $(\sqrt{x} - 2)(\sqrt{x^3} - 2\sqrt{x^2} + 1)$

In Exercises 43–48, complete the statement.

43. $5x\sqrt{3} + 15\sqrt{3} = 5\sqrt{3}(\quad)$

44. $x\sqrt{7} - x^2\sqrt{7} = x\sqrt{7}(\quad)$

45. $4\sqrt{12} - 2x\sqrt{27} = 2\sqrt{3}(\quad)$

46. $5\sqrt{50} + 10y\sqrt{8} = 5\sqrt{2}(\quad)$

47. $6u^2 + \sqrt{18u^3} = 3u(\quad)$

48. $12s^3 - \sqrt{32s^4} = 4s^2(\quad)$

In Exercises 49–62, find the conjugate of the expression. Then multiply the expression by its conjugate. See Example 4.

49. $2 + \sqrt{5}$

50. $\sqrt{2} - 9$

51. $\sqrt{11} - \sqrt{3}$

52. $\sqrt{10} + \sqrt{7}$

53. $\sqrt{15} + 3$

54. $\sqrt{11} + 3$

55. $\sqrt{x} - 3$

56. $\sqrt{t} + 7$

57. $\sqrt{2u} - \sqrt{3}$

58. $\sqrt{5a} + \sqrt{2}$

59. $2\sqrt{2} + \sqrt{4}$

60. $4\sqrt{3} + \sqrt{2}$

61. $\sqrt{x} + \sqrt{y}$

62. $3\sqrt{u} + \sqrt{3v}$

In Exercises 63–66, simplify the expression.

63. $\dfrac{4 - 8\sqrt{x}}{12}$

64. $\dfrac{-3 + 27\sqrt{2y}}{18}$

65. $\dfrac{-2y + \sqrt{12y^3}}{8y}$

66. $\dfrac{-t^2 - \sqrt{2t^3}}{3t}$

In Exercises 67–70, evaluate the function.

67. $f(x) = x^2 - 6x + 1$

(a) $f(2 - \sqrt{3})$ (b) $f(3 - 2\sqrt{2})$

68. $g(x) = x^2 + 8x + 11$

(a) $g(-4 + \sqrt{5})$ (b) $g(-4\sqrt{2})$

69. $f(x) = x^2 - 2x - 1$

(a) $f(1 + \sqrt{2})$ (b) $f(\sqrt{4})$

70. $g(x) = x^2 - 4x + 1$

(a) $g(1 + \sqrt{5})$ (b) $g(2 - \sqrt{3})$

In Exercises 71–94, rationalize the denominator of the expression and simplify. See Examples 5–7.

71. $\dfrac{6}{\sqrt{2} - 2}$

72. $\dfrac{8}{\sqrt{7} + 3}$

73. $\dfrac{7}{\sqrt{3} + 5}$

74. $\dfrac{5}{9 - \sqrt{6}}$

75. $\dfrac{3}{2\sqrt{10} - 5}$

76. $\dfrac{4}{3\sqrt{5} - 1}$

77. $\dfrac{2}{\sqrt{6} + \sqrt{2}}$

78. $\dfrac{10}{\sqrt{9} + \sqrt{5}}$

79. $\dfrac{9}{\sqrt{3} - \sqrt{7}}$

80. $\dfrac{12}{\sqrt{5} + \sqrt{8}}$

81. $(\sqrt{7} + 2) \div (\sqrt{7} - 2)$

82. $(5 - \sqrt{3}) \div (3 + \sqrt{3})$

83. $(\sqrt{x} - 5) \div (2\sqrt{x} - 1)$

84. $(2\sqrt{t} + 1) \div (2\sqrt{t} - 1)$

85. $\dfrac{3x}{\sqrt{15} - \sqrt{3}}$

86. $\dfrac{5y}{\sqrt{12} + \sqrt{10}}$

87. $\dfrac{2t^2}{\sqrt{5} - \sqrt{t}}$

88. $\dfrac{5x}{\sqrt{x} - \sqrt{2}}$

89. $\dfrac{8a}{\sqrt{3a} + \sqrt{a}}$

90. $\dfrac{7z}{\sqrt{5z} - \sqrt{z}}$

91. $\dfrac{3(x - 4)}{x^2 - \sqrt{x}}$

92. $\dfrac{6(y + 1)}{y^2 + \sqrt{y}}$

93. $\dfrac{\sqrt{u + v}}{\sqrt{u - v} - \sqrt{u}}$

94. $\dfrac{z}{\sqrt{u + z} - \sqrt{u}}$

In Exercises 95–98, use a graphing utility to graph the functions on the same screen. Use the graphs to verify that the functions are equivalent. Verify your results algebraically.

95. $y_1 = \dfrac{10}{\sqrt{x} + 1}$

$y_2 = \dfrac{10(\sqrt{x} - 1)}{x - 1}$

96. $y_1 = \dfrac{4x}{\sqrt{x} + 4}$

$y_2 = \dfrac{4x(\sqrt{x} - 4)}{x - 16}$

97. $y_1 = \dfrac{2\sqrt{x}}{2 - \sqrt{x}}$

$y_2 = \dfrac{2(2\sqrt{x} + x)}{4 - x}$

98. $y_1 = \dfrac{\sqrt{2x} + 6}{\sqrt{2x} - 2}$

$y_2 = \dfrac{x + 6 + 4\sqrt{2x}}{x - 2}$

Rationalizing Numerators In the study of calculus, students sometimes rewrite an expression by rationalizing the numerator. (*Note:* The results will not be in simplest radical form.) In Exercises 99–102, rationalize the numerator.

99. $\dfrac{\sqrt{2}}{7}$

100. $\dfrac{\sqrt{10}}{5}$

101. $\dfrac{\sqrt{7} + \sqrt{3}}{5}$

102. $\dfrac{\sqrt{x} + 6}{\sqrt{2}}$

Solving Problems

103. *Strength of a Wooden Beam* The rectangular cross section of a wooden beam cut from a log of diameter 24 inches (see figure) will have maximum strength if its width w and height h are given by

$$w = 8\sqrt{3} \quad \text{and} \quad h = \sqrt{24^2 - \left(8\sqrt{3}\right)^2}.$$

Find the area of the rectangular cross section and express the area in simplest form.

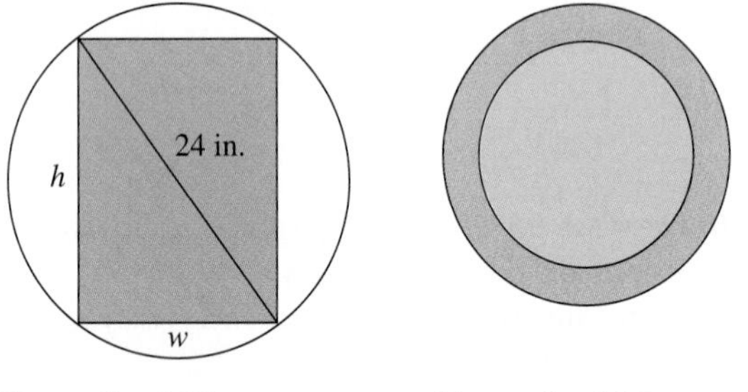

Figure for 103 Figure for 104

104. *Geometry* The areas of the circles in the figure are 15 square centimeters and 20 square centimeters. Find the ratio of the radius of the small circle to the radius of the large circle.

105. *Force to Move a Block* The force required to slide a steel block weighing 500 pounds across a milling machine is

$$\frac{500k}{\dfrac{1}{\sqrt{k^2+1}} + \dfrac{k^2}{\sqrt{k^2+1}}}$$

where k is the friction constant (see figure). Simplify this expression.

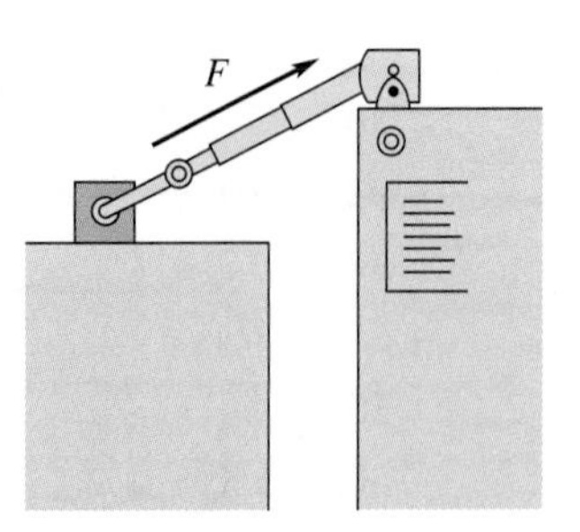

Figure for 105

106. *Wind Chill* The term "wind chill" goes back to the Antarctic explorer Paul A. Siple, who coined it in a 1939 dissertation, *Adaptation of the Explorer to the Climate of Antarctica.* During the 1940s, Siple and Charles F. Passel conducted experiments on the time needed to freeze water in a plastic cylinder that was exposed to the elements. They found that the time depends on the initial temperature of the water, the outside temperature, and the wind speed. The National Weather Service uses the following formula for determining "wind chill."

$$T_{wc} = 0.0817\left(3.71\sqrt{v} + 5.81 - 0.25v\right)(T - 91.4) + 91.4$$

where T is the air temperature in degrees Fahrenheit, T_{wc} is the wind chill, and v is the wind speed in miles per hour. Use the formula to determine the wind chill for each combination of air temperature and wind speed given in the table.

v \ T	0°	5°	10°	15°	20°	25°
10 mi/hr						
20 mi/hr						
30 mi/hr						
40 mi/hr						

Explaining Concepts

107. Multiply $\sqrt{3}\left(1 - \sqrt{6}\right)$. State an algebraic property to justify each step.

108. Describe the differences and similarities of using the FOIL Method with polynomial expressions and with radical expressions.

109. Multiply $3 - \sqrt{2}$ by its conjugate. Explain why the result has no radicals.

110. Is the number $3/\left(1 + \sqrt{5}\right)$ in simplest form? If not, explain the steps for writing it in simplest form.

Mid-Chapter Quiz

Take this quiz as you would take a quiz in class. After you are done, check your work against the answers given in the back of the book.

In Exercises 1–4, evaluate the expression.

1. $\sqrt{225}$ **2.** $\sqrt[4]{\frac{81}{16}}$ **3.** $64^{1/2}$ **4.** $(-27)^{2/3}$

In Exercises 5–8, simplify the expression.

5. $\sqrt{27x^2}$ **6.** $\sqrt[4]{81x^6}$ **7.** $\sqrt{\dfrac{4u^3}{9}}$ **8.** $\sqrt[3]{\dfrac{16}{u^6}}$

In Exercises 9 and 10, combine the radical expressions, if possible.

9. $\sqrt{200y} - 3\sqrt{8y}$ **10.** $6x\sqrt[3]{5x^2} + 2\sqrt[3]{40x^4}$

In Exercises 11–18, simplify the radical expression.

11. $\sqrt{8}(3 + \sqrt{32})$

12. $(\sqrt{50} - 4)\sqrt{2}$

13. $(\sqrt{6} + 3)(4\sqrt{6} - 7)$

14. $(9 + 2\sqrt{3})(2 + 7\sqrt{3})$

15. $\dfrac{\sqrt{7}}{1 + \sqrt{3}}$

16. $\dfrac{6\sqrt{2}}{2\sqrt{2} - 4}$

17. $4 \div (\sqrt{6} + 3)$

18. $(4\sqrt{2} - 2\sqrt{3}) \div (\sqrt{2} + \sqrt{6})$

In Exercises 19 and 20, write the conjugate of the number. Find the product of the number and its conjugate.

19. $1 + \sqrt{4}$ **20.** $\sqrt{10} - 5$

21. Explain why $\sqrt{5^2 + 12^2} \neq 17$. Determine the correct value of the radical.

22. The four corners are cut from an $8\frac{1}{2}$-inch-by-11-inch sheet of paper, as shown in the figure at the left. Find the perimeter of the remaining piece of paper.

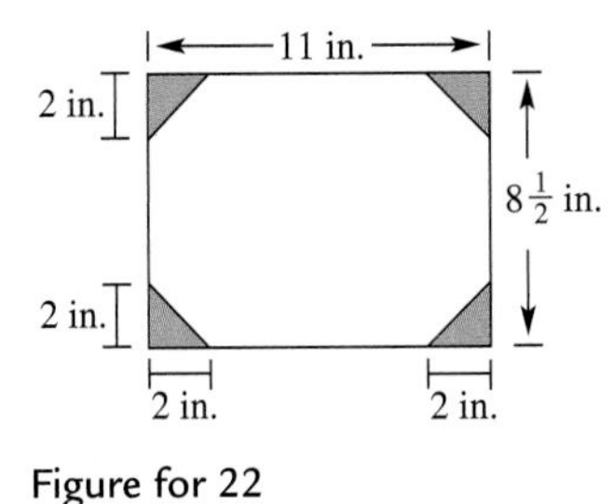

Figure for 22

9.4 Solving Radical Equations

Objectives

1 Solve a radical equation by raising both sides to the nth power.

2 Solve an application problem involving a radical equation.

1 Solve a radical equation by raising both sides to the nth power.

Solving Radical Equations

Solving equations involving radicals is somewhat like solving equations that contain fractions—you try to get rid of the radicals and obtain a polynomial equation. Then you solve the polynomial equation using the standard procedures. The following property plays a key role.

Raising Both Sides of an Equation to the nth Power

Let u and v be real numbers, variables, or algebraic expressions, and let n be a positive integer. If $u = v$, then it follows that

$$u^n = v^n.$$

This is called **raising both sides of an equation to the nth power.**

To use this property to solve an equation, first try to isolate one of the radicals on one side of the equation.

Technology: Tip

To use a graphing utility to check the solution in Example 1, sketch the graph of

$$y = \sqrt{x} - 8$$

as shown below. Notice that the graph crosses the x-axis when $x = 64$, which confirms the solution that was obtained algebraically.

1

60 70

−1

Example 1 Solving an Equation Having One Radical

Solve $\sqrt{x} - 8 = 0$.

Solution

$\sqrt{x} - 8 = 0$ Original equation

$\sqrt{x} = 8$ Isolate radical.

$(\sqrt{x})^2 = 8^2$ Square both sides.

$x = 64$ Simplify.

Check

$\sqrt{x} - 8 = 0$ Original equation

$\sqrt{64} - 8 \stackrel{?}{=} 0$ Substitute 64 for x.

$8 - 8 = 0$ Solution checks. ✓

So, the equation has one solution: $x = 64$.

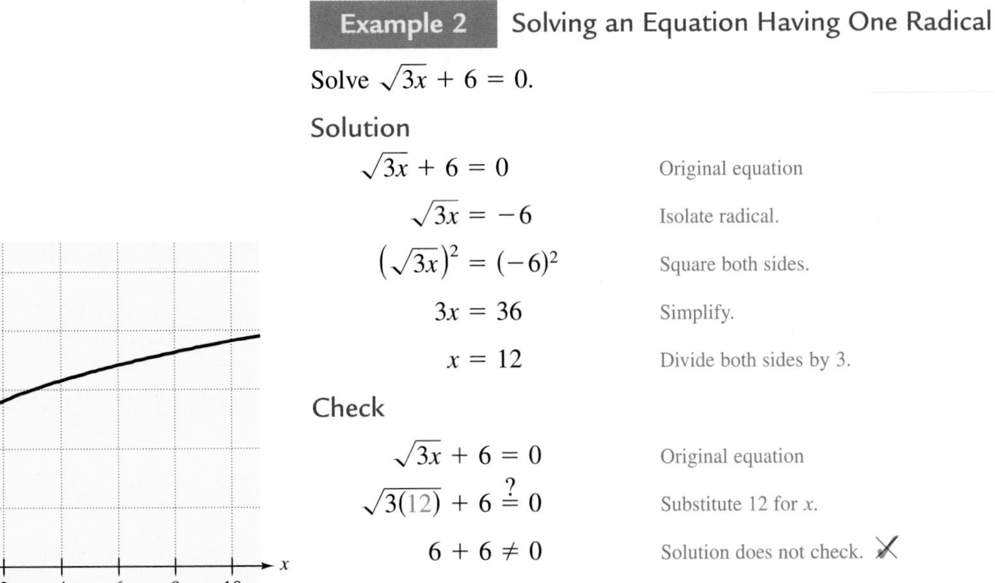

Example 2 Solving an Equation Having One Radical

Solve $\sqrt{3x} + 6 = 0$.

Solution

$$\sqrt{3x} + 6 = 0 \qquad \text{Original equation}$$

$$\sqrt{3x} = -6 \qquad \text{Isolate radical.}$$

$$\left(\sqrt{3x}\right)^2 = (-6)^2 \qquad \text{Square both sides.}$$

$$3x = 36 \qquad \text{Simplify.}$$

$$x = 12 \qquad \text{Divide both sides by 3.}$$

Check

$$\sqrt{3x} + 6 = 0 \qquad \text{Original equation}$$

$$\sqrt{3(12)} + 6 \stackrel{?}{=} 0 \qquad \text{Substitute 12 for } x.$$

$$6 + 6 \neq 0 \qquad \text{Solution does not check. ✗}$$

The solution $x = 12$ is extraneous. So, the equation has no solution. You can also check this graphically, as shown in Figure 9.3.

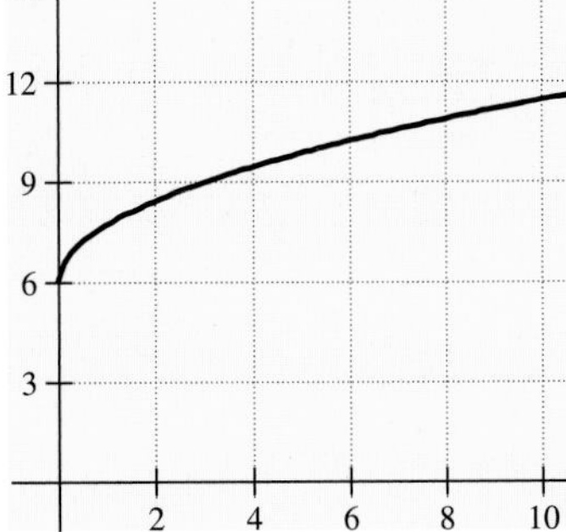

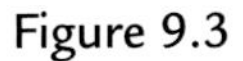
Figure 9.3

As you can see from Example 2, checking solutions of a radical equation is especially important because raising both sides of an equation to the nth power to remove the radical often introduces *extraneous* solutions.

Example 3 Solving an Equation Having One Radical

Solve $\sqrt[3]{2x + 1} - 2 = 3$.

Solution

$$\sqrt[3]{2x + 1} - 2 = 3 \qquad \text{Original equation}$$

$$\sqrt[3]{2x + 1} = 5 \qquad \text{Isolate radical.}$$

$$\left(\sqrt[3]{2x + 1}\right)^3 = 5^3 \qquad \text{Cube both sides.}$$

$$2x + 1 = 125 \qquad \text{Simplify.}$$

$$2x = 124 \qquad \text{Subtract 1 from both sides.}$$

$$x = 62 \qquad \text{Divide both sides by 2.}$$

Check

$$\sqrt[3]{2x + 1} - 2 = 3 \qquad \text{Original equation}$$

$$\sqrt[3]{2(62) + 1} - 2 \stackrel{?}{=} 3 \qquad \text{Substitute 62 for } x.$$

$$\sqrt[3]{125} - 2 \stackrel{?}{=} 3 \qquad \text{Simplify.}$$

$$5 - 2 = 3 \qquad \text{Solution checks. ✓}$$

So, the equation has one solution: $x = 62$. You can also check the solution graphically, as shown in Figure 9.4.

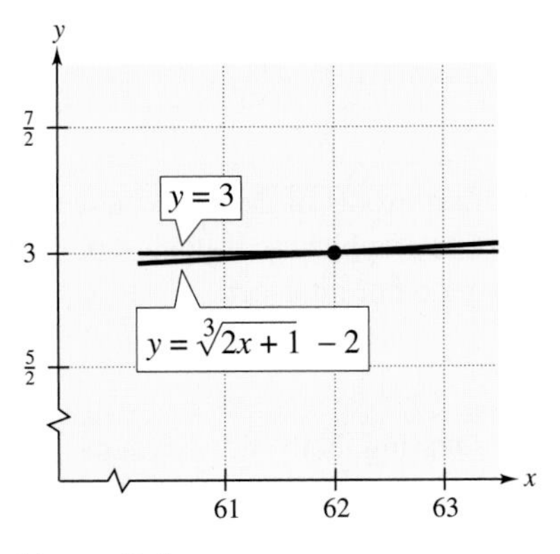

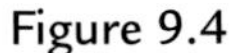
Figure 9.4

Technology: Tip

In Example 4, you can graphically check the solution of the equation by "graphing the left side and right side on the same screen." That is, by graphing the equations

$$y = \sqrt{5x + 3}$$

and

$$y = \sqrt{x + 11}$$

on the same screen, as shown below, you can see that the two graphs intersect when $x = 2$.

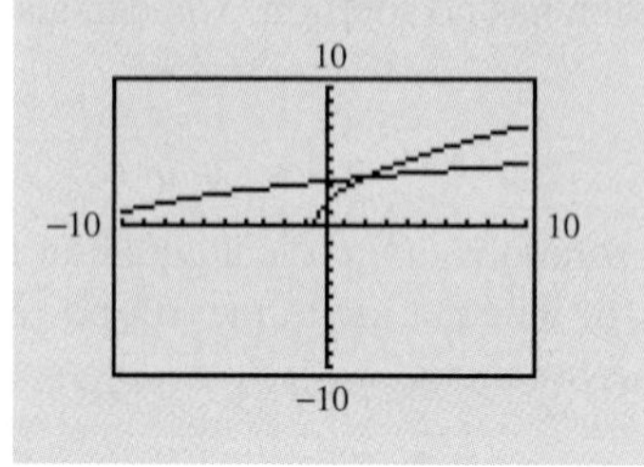

Example 4 Solving an Equation Having Two Radicals

Solve $\sqrt{5x + 3} = \sqrt{x + 11}$.

Solution

$\sqrt{5x + 3} = \sqrt{x + 11}$	Original equation
$(\sqrt{5x + 3})^2 = (\sqrt{x + 11})^2$	Square both sides.
$5x + 3 = x + 11$	Simplify.
$5x = x + 8$	Subtract 3 from both sides.
$4x = 8$	Subtract x from both sides.
$x = 2$	Divide both sides by 4.

Check

$\sqrt{5x + 3} = \sqrt{x + 11}$	Original equation
$\sqrt{5(2) + 3} \stackrel{?}{=} \sqrt{2 + 11}$	Substitute 2 for x.
$\sqrt{13} = \sqrt{13}$	Solution checks. ✓

So, the equation has one solution: $x = 2$.

Example 5 Solving an Equation Having Two Radicals

Solve $\sqrt[4]{3x} + \sqrt[4]{2x - 5} = 0$.

Solution

$\sqrt[4]{3x} + \sqrt[4]{2x - 5} = 0$	Original equation
$\sqrt[4]{3x} = -\sqrt[4]{2x - 5}$	Isolate radicals.
$(\sqrt[4]{3x})^4 = (-\sqrt[4]{2x - 5})^4$	Raise both sides to 4th power.
$3x = 2x - 5$	Simplify.
$x = -5$	Subtract $2x$ from both sides.

Check

$\sqrt[4]{3x} + \sqrt[4]{2x - 5} = 0$	Original equation
$\sqrt[4]{3(-5)} + \sqrt[4]{2(-5) - 5} \stackrel{?}{=} 0$	Substitute -5 for x.
$\sqrt[4]{-15} + \sqrt[4]{-15} \neq 0$	Solution does not check. ✗

The solution does not check because it yields fourth roots of negative radicands. So, this equation has no solution. Try checking this graphically. If you graph both sides of the equation, you will discover that they do not intersect.

In the next example you will see that squaring both sides results in a quadratic equation. Remember that you must check the solutions in the *original* radical equation.

Example 6 An Equation That Converts to a Quadratic Equation

Solve $\sqrt{x} + 2 = x$.

Solution

$\sqrt{x} + 2 = x$	Original equation
$\sqrt{x} = x - 2$	Isolate radical.
$(\sqrt{x})^2 = (x - 2)^2$	Square both sides.
$x = x^2 - 4x + 4$	Simplify.
$-x^2 + 5x - 4 = 0$	General form
$(-1)(x - 4)(x - 1) = 0$	Factor.
$x - 4 = 0 \quad \Rightarrow \quad x = 4$	Set 1st factor equal to 0.
$x - 1 = 0 \quad \Rightarrow \quad x = 1$	Set 2nd factor equal to 0.

Check

$\sqrt{x} + 2 = x$	Original equation	$\sqrt{x} + 2 = x$	Original equation
$\sqrt{4} + 2 \stackrel{?}{=} 4$	Substitute 4 for x.	$\sqrt{1} + 2 \stackrel{?}{=} 1$	Substitute 1 for x.
$2 + 2 \stackrel{?}{=} 4$	Simplify.	$1 + 2 \stackrel{?}{=} 1$	Simplify.
$4 = 4$	Solution checks. ✓	$3 = 1$	Solution does not check. ✗

From the check you can see that $x = 1$ is an extraneous solution. So, the only solution to the equation is $x = 4$.

When an equation contains two radicals, it may not be possible to isolate both. In such cases, you may have to raise both sides of the equation to a power at *two* different stages in the solution.

Example 7 Repeatedly Squaring Both Sides of an Equation

Solve $\sqrt{3t + 1} = 2 - \sqrt{3t}$.

Solution

$\sqrt{3t + 1} = 2 - \sqrt{3t}$	Original equation
$(\sqrt{3t + 1})^2 = (2 - \sqrt{3t})^2$	Square both sides (1st time).
$3t + 1 = 4 - 4\sqrt{3t} + 3t$	Simplify.
$-3 = -4\sqrt{3t}$	Isolate radical.
$(-3)^2 = (-4\sqrt{3t})^2$	Square both sides (2nd time).
$9 = 16(3t)$	Simplify.
$\frac{3}{16} = t$	Divide both sides by 48.

The solution is $t = \frac{3}{16}$. Check this in the original equation.

2 Solve an application problem involving a radical equation.

Applications

Example 8 An Application Involving Electricity

The amount of power consumed by an electrical appliance is given by

$$I = \sqrt{\frac{P}{R}}$$

where I is the current measured in amps, R is the resistance measured in ohms, and P is the power measured in watts. Find the power used by an electric heater for which $I = 10$ amps and $R = 16$ ohms.

Solution

$I = \sqrt{\frac{P}{R}}$	Original equation
$10 = \sqrt{\frac{P}{16}}$	Substitute for I and R.
$10^2 = \left(\sqrt{\frac{P}{16}}\right)^2$	Square both sides.
$100 = \frac{P}{16}$	Simplify.
$1600 = P$	Multiply both sides by 16.

Check

$10 \stackrel{?}{=} \sqrt{\frac{1600}{16}}$	Substitute 10 for I, 16 for R, and 1600 for P in the original equation.
$10 \stackrel{?}{=} \sqrt{100}$	Simplify.
$10 = 10$	Solution checks. ✓

So, the solution is $P = 1600$ watts.

An alternative way to solve the problem in Example 8 would be first to solve the equation for P.

$I = \sqrt{\frac{P}{R}}$	Original equation
$I^2 = \left(\sqrt{\frac{P}{R}}\right)^2$	Square both sides.
$I^2 = \frac{P}{R}$	Simplify.
$I^2R = P$	Multiply both sides by R.

At this stage, you can substitute the known values of I and R to obtain

$$P = (10)^2 16 = 1600.$$

Example 9 The Velocity of a Falling Object

The velocity of a free-falling object can be determined from the equation

$$v = \sqrt{2gh}$$

where v is the velocity measured in feet per second, $g = 32$ feet per second per second, and h is the distance (in feet) the object has fallen. Find the height from which a rock has been dropped if it strikes the ground with a velocity of 50 feet per second.

Solution

$$v = \sqrt{2gh} \qquad \text{Original equation}$$

$$50 = \sqrt{2(32)h} \qquad \text{Substitute for } v \text{ and } g.$$

$$50^2 = \left(\sqrt{64h}\right)^2 \qquad \text{Square both sides.}$$

$$2500 = 64h \qquad \text{Simplify.}$$

$$39 \approx h \qquad \text{Divide both sides by 64.}$$

Check

Because the value of h was rounded in the solution, the check will not result in an equality. The expressions on each side of the equal sign will be approximately equal to each other.

$$v = \sqrt{2gh} \qquad \text{Original equation}$$

$$50 \stackrel{?}{\approx} \sqrt{2(32)(39)} \qquad \text{Substitute 50 for } v\text{, 32 for } g\text{, and 39 for } h.$$

$$50 \stackrel{?}{\approx} \sqrt{2496} \qquad \text{Simplify.}$$

$$50 \approx 49.96 \qquad \text{Solution checks. ✓}$$

So, the height from which the rock has been dropped is approximately 39 feet.

Discussing the Concept — An Experiment

Without using a stopwatch, you can find the length of time an object has been falling by using the following equation from physics

$$t = \sqrt{\frac{h}{384}}$$

where t is the length of time in seconds and h is the distance in inches the object has fallen. How far does an object fall in 0.25 second? in 0.10 second?

Use this equation to test how long it takes members of your group to catch a falling ruler. Hold the ruler vertically while another student holds his or her hands near the lower end of the ruler ready to catch it. Before releasing the ruler, record the mark on the ruler closest to the top of the catcher's hands. Release the ruler. After it has been caught, again note the mark closest to the top of the catcher's hands. (The difference between these two measurements is h.) Which member of your class reacts most quickly?

9.4 Exercises

Integrated Review *Concepts, Skills, and Problem Solving*

Keep mathematically in shape by doing these exercises *before* the problems of this section.

Properties and Definitions

1. Explain how you determine the domain of the function

$$f(x) = \frac{4}{(x+2)(x-3)}.$$

2. Explain the excluded value $(x \neq -3)$ in the following equation.

$$\frac{2x^2+5x-3}{x^2-9} = \frac{2x-1}{x-3}, \quad x \neq 3, \quad x \neq -3$$

Simplifying Expressions

In Exercises 3–6, simplify the expression. (Assume that no variable is equal to 0.)

3. $(-3x^2y^3)^2 \cdot (4xy^2)$

4. $(x^2 - 3xy)^0$

5. $\dfrac{64r^2s^4}{16rs^2}$

6. $\left(\dfrac{3x}{4y^3}\right)^2$

In Exercises 7–10, perform the operation and simplify.

7. $\dfrac{x+13}{x^3(3-x)} \cdot \dfrac{x(x-3)}{5}$

8. $\dfrac{x+2}{5x+15} \cdot \dfrac{x-2}{5(x-3)}$

9. $\dfrac{2x}{x-5} - \dfrac{5}{5-x}$

10. $\dfrac{3}{x-1} - 5$

Graphs

In Exercises 11 and 12, graph the function and identify any intercepts.

11. $y = 2x - 3$

12. $y = -\frac{3}{4}x + 2$

Developing Skills

In Exercises 1–4, determine whether the values of x are solutions of the radical equation.

Equation	Values of x	
1. $\sqrt{x} - 10 = 0$	(a) $x = -4$	(b) $x = -100$
	(c) $x = \sqrt{10}$	(d) $x = 100$
2. $\sqrt{3x} - 6 = 0$	(a) $x = \frac{2}{3}$	(b) $x = 2$
	(c) $x = 12$	(d) $x = -\frac{1}{3}\sqrt{6}$
3. $\sqrt[3]{x-4} = 4$	(a) $x = -60$	(b) $x = 68$
	(c) $x = 20$	(d) $x = 0$
4. $\sqrt[4]{2x} + 2 = 6$	(a) $x = 128$	(b) $x = 2$
	(c) $x = -2$	(d) $x = 0$

In Exercises 5–52, solve the equation and check your solution(s) in the original equation. (Some of the equations have no solution.) See Examples 1–7.

5. $\sqrt{x} = 20$

6. $\sqrt{x} = 5$

7. $\sqrt{x} = 3$

8. $\sqrt{t} = 4$

9. $\sqrt[3]{z} = 3$

10. $\sqrt[4]{x} = 2$

11. $\sqrt{y} - 7 = 0$

12. $\sqrt{t} - 13 = 0$

13. $\sqrt{u} + 13 = 0$

14. $\sqrt{y} + 15 = 0$

15. $\sqrt{x} - 8 = 0$

16. $\sqrt{x} - 10 = 0$

17. $\sqrt{10x} = 30$

18. $\sqrt{8x} = 6$

19. $\sqrt{-3x} = 9$

20. $\sqrt{-4y} = 4$

21. $\sqrt{5t} - 2 = 0$

22. $6 - \sqrt{8x} = 0$

23. $\sqrt{3y+1} = 4$

24. $\sqrt{3-2x} = 2$

25. $\sqrt{4-5x} = -3$

26. $\sqrt{2t-7} = -5$

27. $\sqrt{3y+5} - 3 = 4$

28. $\sqrt{5z-2} + 7 = 10$

29. $5\sqrt{x+2} = 8$

30. $2\sqrt{x+4} = 7$

31. $\sqrt{3x+2} + 5 = 0$

32. $\sqrt{1-x} + 10 = 4$

33. $\sqrt{x+3} = \sqrt{2x-1}$

34. $\sqrt{3t+1} = \sqrt{t+15}$

35. $\sqrt{3y-5} - 3\sqrt{y} = 0$

36. $\sqrt{2u+10} - 2\sqrt{u} = 0$

37. $\sqrt[3]{3x-4} = \sqrt[3]{x+10}$

38. $2\sqrt[3]{10 - 3x} = \sqrt[3]{2 - x}$

39. $\sqrt[3]{2x + 15} - \sqrt[3]{x} = 0$

40. $\sqrt[4]{2x} + \sqrt[4]{x + 3} = 0$

41. $\sqrt{x^2 + 5} = x + 3$ **42.** $\sqrt{x^2 - 4} = x - 2$

43. $\sqrt{2x} = x - 4$ **44.** $\sqrt{x} = x - 6$

45. $\sqrt{8x + 1} = x + 2$ **46.** $\sqrt{3x + 7} = x + 3$

47. $\sqrt{z + 2} = 1 + \sqrt{z}$

48. $\sqrt{2x + 5} = 7 - \sqrt{2x}$

49. $\sqrt{2t + 3} = 3 - \sqrt{2t}$

50. $\sqrt{x} + \sqrt{x + 2} = 2$

51. $\sqrt{x + 5} - \sqrt{x} = 1$

52. $\sqrt{x + 3} - \sqrt{x - 1} = 1$

In Exercises 53–60, solve the equation and check your solution(s) in the original equation.

53. $t^{3/2} = 8$ **54.** $v^{2/3} = 25$

55. $3y^{1/3} = 18$ **56.** $2x^{3/4} = 54$

57. $(x + 4)^{2/3} = 4$ **58.** $(u - 2)^{4/3} = 81$

59. $(2x + 5)^{1/3} + 3 = 0$ **60.** $(x - 6)^{3/2} - 27 = 0$

In Exercises 61–70, use a graphing utility to graph each side of the equation on the same screen. Use the graphs to approximate the solution(s).

61. $\sqrt{x} = 2(2 - x)$ **62.** $\sqrt{2x + 3} = 4x - 3$

63. $\sqrt{x^2 + 1} = 5 - 2x$ **64.** $\sqrt{8 - 3x} = x$

65. $\sqrt{x + 3} = 5 - \sqrt{x}$ **66.** $\sqrt[3]{5x - 8} = 4 - \sqrt[3]{x}$

67. $4\sqrt[3]{x} = 7 - x$ **68.** $\sqrt[3]{x + 4} = \sqrt{6 - x}$

69. $\sqrt{15 - 4x} = 2x$ **70.** $\dfrac{4}{\sqrt{x}} = 3\sqrt{x} - 4$

In Exercises 71–76, match the function with its graph. [The graphs are labeled (a), (b), (c), (d), (e), and (f).]

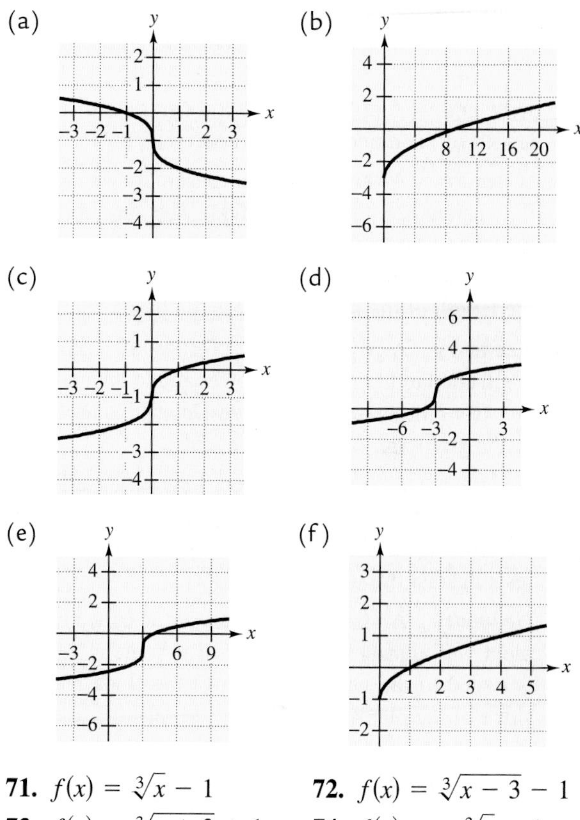

71. $f(x) = \sqrt[3]{x} - 1$ **72.** $f(x) = \sqrt[3]{x - 3} - 1$

73. $f(x) = \sqrt[3]{x + 3} + 1$ **74.** $f(x) = -\sqrt[3]{x} - 1$

75. $f(x) = \sqrt{x} - 1$ **76.** $f(x) = \sqrt{x} - 3$

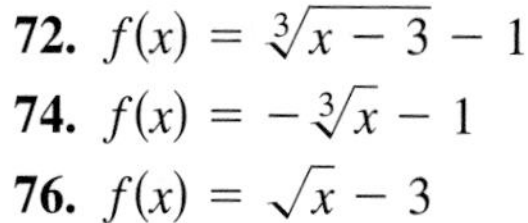

Solving Problems

Geometry In Exercises 77–80, find the length x of the unknown side of the right triangle. (Round your answer to two decimal places.)

77. **78.**

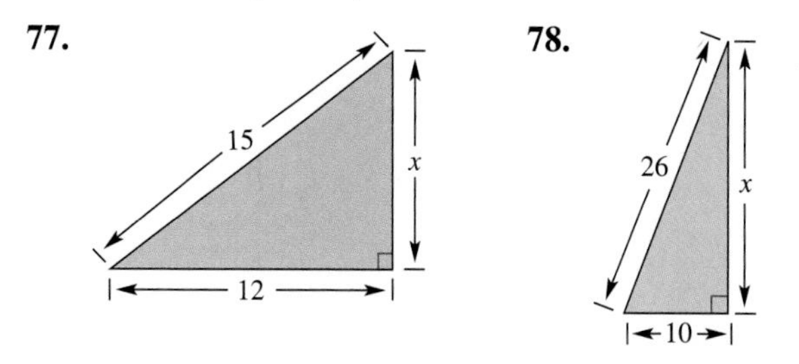

79.

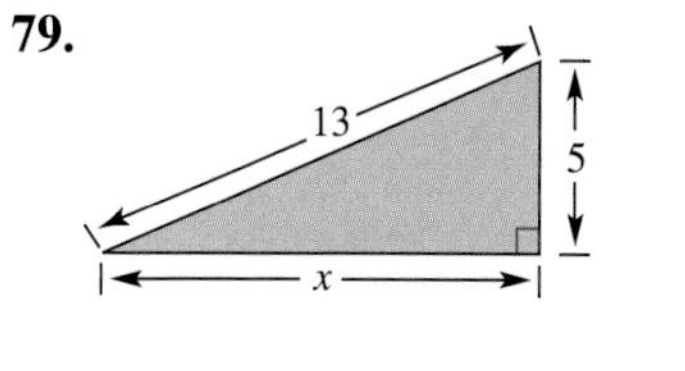

80.

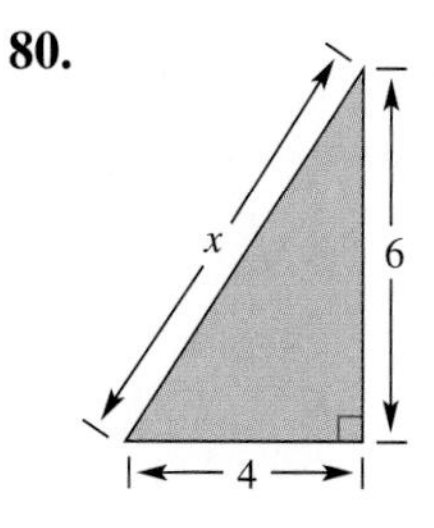

81. *Drawing a Diagram* The screen of a computer monitor has a diagonal of 13.75 inches and a width of 8.25 inches. Draw a diagram of the computer monitor and find the length of the screen.

82. *Drawing a Diagram* A basketball court is 50 feet wide and 94 feet long. Draw a diagram and find the length of the diagonal of the court.

83. *Geometry* A house has a basement floor with dimensions 26 feet by 32 feet. The gas hot water heater and furnace are diagonally across the basement from where the natural gas line enters the house. Find the length of the gas line across the basement.

84. *Geometry* A guy wire on a 100-foot-tall radio tower is attached to the top of the tower and to an anchor 50 feet from the base of the tower. Determine the length of the guy wire.

85. *Geometry* A ladder is 17 feet long and the bottom of the ladder is 8 feet from the wall of a house. Determine the height at which the top of the ladder rests against the wall.

86. *Geometry* A 10-foot plank is used to brace a basement wall during construction of a home. The plank is nailed to the wall 6 feet above the floor. Find the slope of the plank.

87. *Geometry* Determine the length and width of a rectangle with a perimeter of 92 inches and a diagonal of 34 inches.

88. *Geometry* Determine the length and width of a rectangle with a perimeter of 68 inches and a diagonal of 26 inches.

89. *Geometry* The surface area of a cone is given by

$$S = \pi r\sqrt{r^2 + h^2}$$

as shown in the figure. Solve the equation for h.

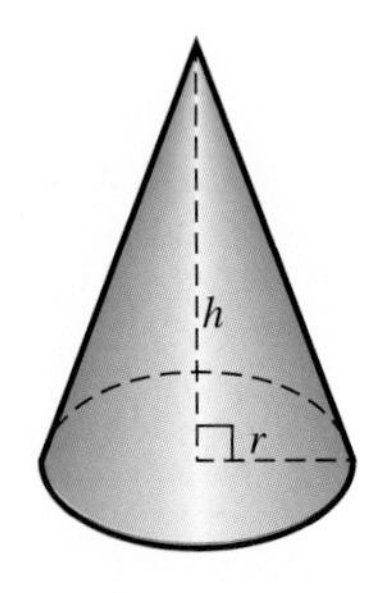

90. *Geometry* Write a function that gives the radius of a circle in terms of the circle's area A. Use a graphing utility to graph this function.

Height of an Object In Exercises 91 and 92, use the formula $t = \sqrt{d/16}$, which gives the time t in seconds for a free-falling object to fall d feet.

91. A construction worker drops a nail and observes it strike a water puddle after approximately 2 seconds. Estimate the height from which the nail was dropped.

92. A construction worker drops a nail and observes it strike a water puddle after approximately 3 seconds. Estimate the height from which the nail was dropped.

Free-Falling Object In Exercises 93–96, use the equation for the velocity of a free-falling object ($v = \sqrt{2gh}$), as described in Example 9.

93. An object is dropped from a height of 50 feet. Find the velocity of the object when it strikes the ground.

94. An object is dropped from a height of 200 feet. Find the velocity of the object when it strikes the ground.

95. An object that was dropped strikes the ground with a velocity of 60 feet per second. Find the height from which the object was dropped.

96. An object that was dropped strikes the ground with a velocity of 120 feet per second. Find the height from which the object was dropped.

Length of a Pendulum In Exercises 97 and 98, use the equation of the time t in seconds for a pendulum of length L feet to go through one complete cycle (its period). The equation is $t = 2\pi\sqrt{L/32}$.

97. How long is the pendulum of a grandfather clock with a period of 1.5 seconds (see figure)?

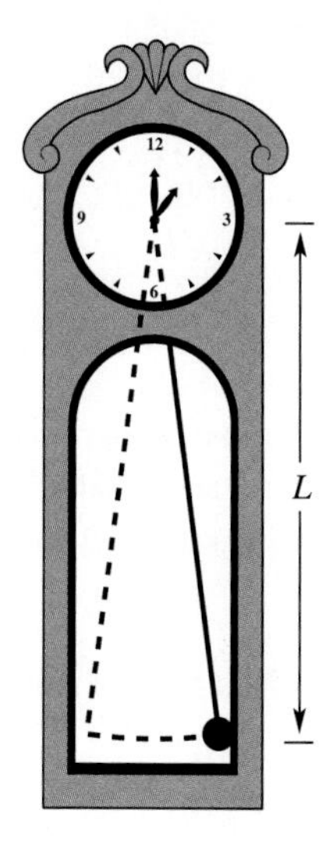

98. How long is the pendulum of a mantle clock with a period of 0.75 second?

99. *Demand for a Product* The demand equation for a certain product is

$$p = 50 - \sqrt{0.8(x - 1)}$$

where x is the number of units demanded per day and p is the price per unit. Find the demand if the price is \$30.02.

100. *Airline Passengers* An airline offers daily flights between Chicago and Denver. The total monthly cost of the flights is

$$C = \sqrt{0.2x + 1}, \quad 0 \le x$$

where C is measured in millions of dollars and x is measured in thousands of passengers (see figure). The total cost of the flights for a certain month is 2.5 million dollars. Approximately how many passengers flew that month?

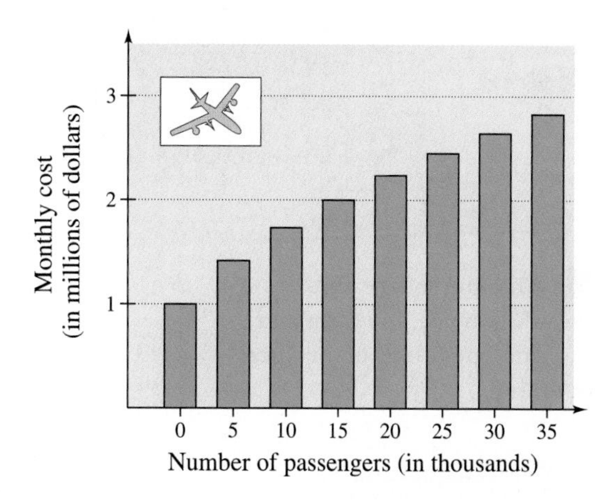

101. *Federal Grants-in-Aid* The total amount spent on federal grants-in-aid F (in billions of dollars) in the United States for the years 1990 through 1997 is modeled by

$$F = 133.5 + 9.3t + 18.0\sqrt{t}, \quad 0 \le t \le 7$$

where t is time in years, with $t = 0$ corresponding to 1990. (Source: U.S. Office of Management and Budget)

(a) Use a graphing utility to graph the function.

(b) In what year did the grants total approximately 225 billion dollars?

102. *Exploration* The solution of the equation $x + \sqrt{x - a} = b$ is $x = 20$. Find a and b. (There are many correct answers.)

Explaining Concepts

103. Answer part (e) of Motivating the Chapter on page 513.

104. In your own words, describe the steps that can be used to solve a radical equation.

105. Does raising both sides of an equation to the nth power always yield an equivalent equation? Explain.

106. One reason for checking a solution in the original equation is to discover errors that were made when solving the equation. Describe another reason.

107. *Error Analysis* Describe the error.

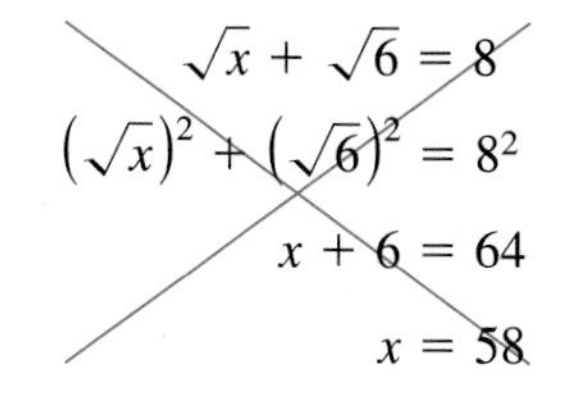

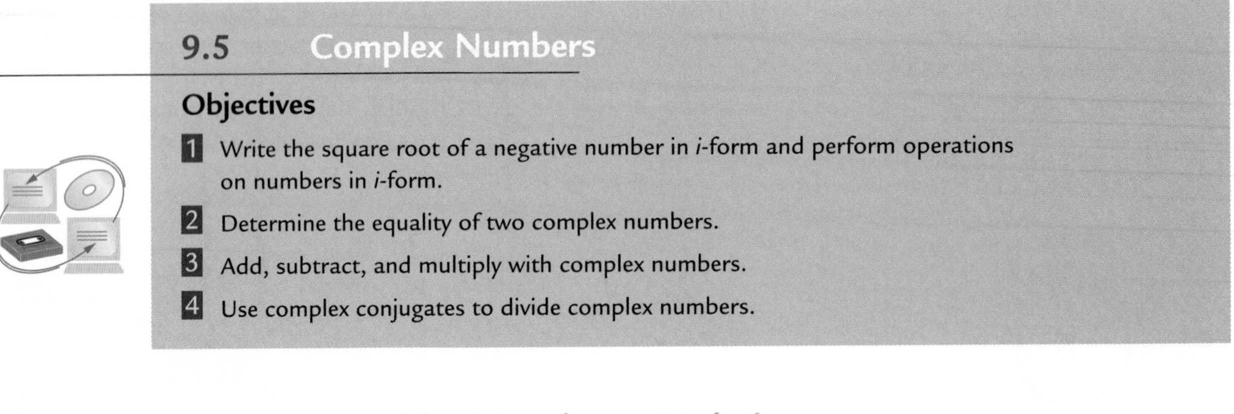

9.5 Complex Numbers

Objectives

1. Write the square root of a negative number in *i*-form and perform operations on numbers in *i*-form.
2. Determine the equality of two complex numbers.
3. Add, subtract, and multiply with complex numbers.
4. Use complex conjugates to divide complex numbers.

1 Write the square root of a negative number in *i*-form and perform operations on numbers in *i*-form.

The Imaginary Unit *i*

In Section 9.1, you learned that a negative number has no *real* square root. For instance, $\sqrt{-1}$ is not real because there is no real number x such that $x^2 = -1$. Thus, as long as you are dealing only with real numbers, the equation

$$x^2 = -1$$

has no solution. To overcome this deficiency, mathematicians have expanded the set of numbers, using the **imaginary unit *i*,** defined as

$$i = \sqrt{-1}. \qquad \text{Imaginary unit}$$

This number has the property that $i^2 = -1$. So, the imaginary unit i is a solution of the equation $x^2 = -1$.

The Square Root of a Negative Number

Let c be a positive real number. Then the square root of $-c$ is given by

$$\sqrt{-c} = \sqrt{c(-1)} = \sqrt{c}\sqrt{-1} = \sqrt{c}\,i.$$

When writing $\sqrt{-c}$ in the ***i*-form,** $\sqrt{c}\,i$, note that i is outside the radical.

Technology: Discovery

Use a calculator to evaluate the following radicals. Does one result in an error message? Explain why.

a. $\sqrt{121}$

b. $\sqrt{-121}$

c. $-\sqrt{121}$

Example 1 Writing Numbers in *i*-Form

a. $\sqrt{-36} = \sqrt{36(-1)} = \sqrt{36}\sqrt{-1} = 6i$

b. $\sqrt{-\dfrac{16}{25}} = \sqrt{\dfrac{16}{25}(-1)} = \sqrt{\dfrac{16}{25}}\sqrt{-1} = \dfrac{4}{5}i$

c. $\sqrt{-5} = \sqrt{5(-1)} = \sqrt{5}\sqrt{-1} = \sqrt{5}\,i$

d. $\sqrt{-54} = \sqrt{54(-1)} = \sqrt{54}\sqrt{-1} = 3\sqrt{6}\,i$

e. $\dfrac{\sqrt{-48}}{\sqrt{-3}} = \dfrac{\sqrt{48}\sqrt{-1}}{\sqrt{3}\sqrt{-1}} = \dfrac{\sqrt{48}\,i}{\sqrt{3}\,i} = \sqrt{\dfrac{48}{3}} = \sqrt{16} = 4$

f. $\dfrac{\sqrt{-18}}{\sqrt{2}} = \dfrac{\sqrt{18}\sqrt{-1}}{\sqrt{2}} = \dfrac{\sqrt{18}\,i}{\sqrt{2}} = \sqrt{\dfrac{18}{2}}\,i = \sqrt{9}\,i = 3i$

To perform operations with square roots of negative numbers, you must *first* write the numbers in i-form. Once the numbers are written in i-form, you add, subtract, and multiply as follows.

$ai + bi = (a + b)i$ — Addition

$ai - bi = (a - b)i$ — Subtraction

$(ai)(bi) = ab(i^2) = ab(-1) = -ab$ — Multiplication

Example 2 Operations with Square Roots of Negative Numbers

Perform the operation.

a. $\sqrt{-9} + \sqrt{-49}$ **b.** $\sqrt{-32} - 2\sqrt{-2}$

Solution

a. $\sqrt{-9} + \sqrt{-49} = \sqrt{9}\sqrt{-1} + \sqrt{49}\sqrt{-1}$ — Property of radicals

$= 3i + 7i$ — Write in i-form.

$= 10i$ — Simplify.

b. $\sqrt{-32} - 2\sqrt{-2} = \sqrt{32}\sqrt{-1} - 2\sqrt{2}\sqrt{-1}$ — Property of radicals

$= 4\sqrt{2}i - 2\sqrt{2}i$ — Write in i-form.

$= 2\sqrt{2}i$ — Simplify.

Study Tip

When performing operations with numbers in i-form, you sometimes need to be able to evaluate powers of the imaginary unit i. The first several powers of i are as follows.

$i^1 = i$

$i^2 = -1$

$i^3 = i(i^2) = i(-1) = -i$

$i^4 = (i^2)(i^2) = (-1)(-1) = 1$

$i^5 = i(i^4) = i(1) = i$

$i^6 = (i^2)(i^4) = (-1)(1) = -1$

$i^7 = (i^3)(i^4) = (-i)(1) = -i$

$i^8 = (i^4)(i^4) = (1)(1) = 1$

Note how the pattern of values $i, -1, -i$, and 1 repeats itself for powers greater than 4.

Example 3 Multiplying Square Roots of Negative Numbers

Find each product.

a. $\sqrt{-15}\sqrt{-15}$ **b.** $\sqrt{-5}(\sqrt{-45} - \sqrt{-4})$

Solution

a. $\sqrt{-15}\sqrt{-15} = (\sqrt{15}i)(\sqrt{15}i)$ — Write in i-form.

$= (\sqrt{15})^2 i^2$ — Multiply.

$= 15(-1)$ — Definition of i

$= -15$ — Simplify.

b. $\sqrt{-5}(\sqrt{-45} - \sqrt{-4}) = \sqrt{5}i(3\sqrt{5}i - 2i)$ — Write in i-form.

$= (\sqrt{5}i)(3\sqrt{5}i) - (\sqrt{5}i)(2i)$ — Distributive Property

$= 3(5)(-1) - 2\sqrt{5}(-1)$ — Multiply.

$= -15 + 2\sqrt{5}$ — Simplify.

When multiplying square roots of negative numbers, be sure to write them in i-form *before multiplying.* If you do not, you can obtain incorrect answers. For instance, in Example 3(a) be sure you see that

$$\sqrt{-15}\sqrt{-15} \neq \sqrt{(-15)(-15)} = \sqrt{225} = 15.$$

2 Determine the equality of two complex numbers.

Complex Numbers

A number of the form $a + bi$, where a and b are real numbers, is called a **complex number.**

> ▶ **Definition of a Complex Number**
>
> If a and b are real numbers, the number $a + bi$ is a **complex number,** and it is said to be written in **standard form.** If $b = 0$, the number $a + bi = a$ is a real number. If $b \neq 0$, the number $a + bi$ is called an **imaginary number.** A number of the form bi, where $b \neq 0$, is called a **pure imaginary number.**

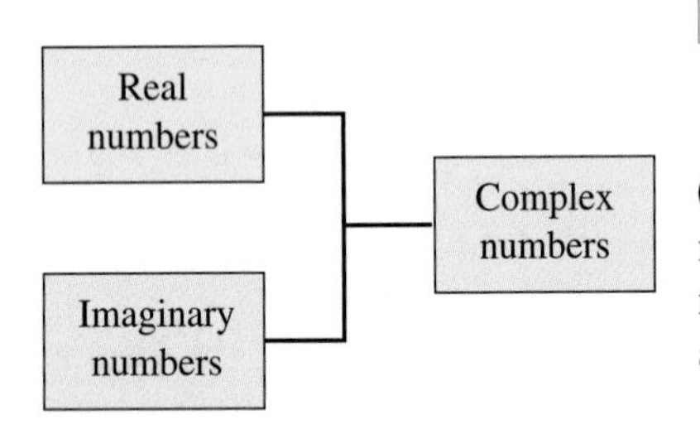

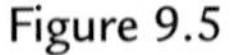

Figure 9.5

A number cannot be both real and imaginary. For instance, the numbers -2, 0, 1, $\frac{1}{2}$, and $\sqrt{2}$ are real numbers (but they are *not* imaginary numbers), and the numbers $-3i$, $2 + 4i$, and $-1 + i$ are imaginary numbers (but they are *not* real numbers). The diagram shown in Figure 9.5 further illustrates the relationship among real, complex, and imaginary numbers.

Two complex numbers $a + bi$ and $c + di$, in standard form, are equal if and only if $a = c$ and $b = d$.

Example 4 Equality of Two Complex Numbers

a. Are the complex numbers $\sqrt{9} + \sqrt{-48}$ and $3 - 4\sqrt{3}i$ equal?

b. Find values of x and y such that the equation is valid.

$$3x - \sqrt{-25} = -6 + 3yi$$

Solution

a. Begin by writing the first number in standard form.

$$\sqrt{9} + \sqrt{-48} = \sqrt{3^2} + \sqrt{4^2(3)(-1)} \qquad \text{Factor.}$$

$$= 3 + 4\sqrt{3}i \qquad \text{Write in } i\text{-form.}$$

From this form, you can see that the two numbers are not equal because they have imaginary parts that differ in sign.

b. Begin by writing the left side of the equation in standard form.

$$3x - \sqrt{-25} = -6 + 3yi \qquad \text{Original equation}$$

$$3x - 5i = -6 + 3yi \qquad \text{Both sides in standard form}$$

For these two numbers to be equal, their real parts must be equal to each other and their imaginary parts must be equal to each other.

Real Parts	*Imaginary Parts*
$3x = -6$	$3y = -5$
$x = -2$	$y = -\frac{5}{3}$

So, $x = -2$ and $y = -\frac{5}{3}$.

3 Add, subtract, and multiply with complex numbers.

Operations with Complex Numbers

The real number a is called the **real part** of the complex number $a + bi$, and the number bi is called the **imaginary part** of the complex number. To add or subtract two complex numbers, we add (or subtract) the real and imaginary parts separately. This is similar to combining like terms of a polynomial.

$$(a + bi) + (c + di) = (a + c) + (b + d)i \quad \text{Addition of complex numbers}$$

$$(a + bi) - (c + di) = (a - c) + (b - d)i \quad \text{Subtraction of complex numbers}$$

Benoit Mandelbrot

(1924–)

Until very recently it was thought that shapes in nature, such as clouds, coastlines, and mountain ranges, could not be described in mathematical terms. In the 1970s, Mandelbrot discovered that many of these shapes do have patterns in their irregularity—they are made up of smaller parts that are scaled-down versions of the shapes themselves. Computers using mathematical terms with complex numbers are able to generate the larger images. Mandelbrot coined the term *fractals* for these shapes and for the geometry used to describe them.

Example 5 Adding and Subtracting Complex Numbers

a. $(3 - i) + (-2 + 4i) = (3 - 2) + (-1 + 4)i = 1 + 3i$

b. $3i + (5 - 3i) = 5 + (3 - 3)i = 5$

c. $4 - (-1 + 5i) + (7 + 2i) = [4 - (-1) + 7] + (-5 + 2)i = 12 - 3i$

d. $(6 + 3i) + \left(2 - \sqrt{-8}\right) - \sqrt{-4} = (6 + 3i) + \left(2 - 2\sqrt{2}i\right) - 2i$

$$= (6 + 2) + \left(3 - 2\sqrt{2} - 2\right)i$$

$$= 8 + \left(1 - 2\sqrt{2}\right)i$$

Note in part (b) that the sum of two complex numbers can be a real number.

The Commutative, Associative, and Distributive Properties of real numbers are also valid for complex numbers.

Example 6 Multiplying Complex Numbers

a. $(7i)(-3i) = -21i^2$ — Multiply.

$= -21(-1) = 21$ — $i^2 = -1$

b. $(1 - i)\left(\sqrt{-9}\right) = (1 - i)(3i)$ — Write in i-form.

$= (1)(3i) - (i)(3i)$ — Distributive Property

$= 3i - 3(i^2)$ — Simplify.

$= 3i - 3(-1) = 3 + 3i$ — $i^2 = -1$

c. $(2 - i)(4 + 3i) = 8 + 6i - 4i - 3i^2$ — FOIL Method

$= 8 + 6i - 4i - 3(-1)$ — $i^2 = -1$

$= 11 + 2i$ — Combine like terms.

d. $(3 + 2i)(3 - 2i) = 9 - 6i + 6i - 4i^2$ — FOIL Method

$= 9 - 6i + 6i - 4(-1)$ — $i^2 = -1$

$= 9 + 4 = 13$ — Combine like terms.

4 Use complex conjugates to divide complex numbers.

Complex Conjugates

In Example 6(d), note that the product of two complex numbers can be a real number. This occurs with pairs of complex numbers of the form $a + bi$ and $a - bi$, called **complex conjugates.** In general, the product of complex conjugates has the following form.

$$\begin{aligned}(a + bi)(a - bi) &= a^2 - (bi)^2\\ &= a^2 - b^2i^2\\ &= a^2 - b^2(-1)\\ &= a^2 + b^2\end{aligned}$$

Here are some examples.

Complex Number	*Complex Conjugate*	*Product*
$4 - 5i$	$4 + 5i$	$4^2 + 5^2 = 41$
$3 + 2i$	$3 - 2i$	$3^2 + 2^2 = 13$
$-2 = -2 + 0i$	$-2 = -2 - 0i$	$(-2)^2 + 0^2 = 4$
$i = 0 + i$	$-i = 0 - i$	$0^2 + 1^2 = 1$

Complex conjugates are used to divide one complex number by another. To do this, multiply the numerator and denominator by the *complex conjugate of the denominator,* as shown in Example 7.

Example 7 Division of Complex Numbers

a. $\dfrac{2 - i}{4i} = \dfrac{2 - i}{4i} \cdot \dfrac{(-i)}{(-i)}$ Multiply numerator and denominator by complex conjugate of denominator.

$= \dfrac{-2i + i^2}{-4i^2}$ Multiply fractions.

$= \dfrac{-2i + (-1)}{-4(-1)}$ $i^2 = -1$

$= -\dfrac{2i + 1}{4}$ Simplify.

b. $\dfrac{5}{3 - 2i} = \dfrac{5}{3 - 2i} \cdot \dfrac{3 + 2i}{3 + 2i}$ Multiply numerator and denominator by complex conjugate of denominator.

$= \dfrac{5(3 + 2i)}{(3 - 2i)(3 + 2i)}$ Multiply fractions.

$= \dfrac{5(3 + 2i)}{3^2 + 2^2}$ Product of complex conjugates

$= \dfrac{15 + 10i}{13}$ Simplify.

$= \dfrac{15}{13} + \dfrac{10}{13}i$ Standard form

Example 8 Division of Complex Numbers

Divide $2 + 3i$ by $4 - 2i$.

Solution

$$\frac{2+3i}{4-2i} = \frac{2+3i}{4-2i} \cdot \frac{4+2i}{4+2i}$$ Multiply numerator and denominator by complex conjugate of denominator.

$$= \frac{8+16i+6i^2}{4^2+2^2}$$ Multiply fractions.

$$= \frac{8+16i+6(-1)}{20}$$ $i^2 = -1$

$$= \frac{2+16i}{20}$$ Combine like terms.

$$= \frac{2}{20} + \frac{16i}{20}$$ Standard form

$$= \frac{1}{10} + \frac{4}{5}i$$ Simplify.

Example 9 Verifying a Complex Solution of an Equation

Show that $x = 2 + i$ is a solution of the equation $x^2 - 4x + 5 = 0$.

Solution

$$x^2 - 4x + 5 = 0$$ Original equation

$$(2+i)^2 - 4(2+i) + 5 \stackrel{?}{=} 0$$ Substitute $2 + i$ for x.

$$4 + 4i + i^2 - 8 - 4i + 5 \stackrel{?}{=} 0$$ Expand.

$$i^2 + 1 \stackrel{?}{=} 0$$ Combine like terms.

$$(-1) + 1 \stackrel{?}{=} 0$$ $i^2 = -1$

$$0 = 0$$ Solution checks. ✓

So, $2 + i$ is a solution of the original equation.

Discussing the Concept **Prime Polynomials**

The polynomial $x^2 + 1$ is prime *with respect to the integers.* It is not, however, prime *with respect to the complex numbers.* Show how $x^2 + 1$ can be factored using complex numbers.

9.5 Exercises

Integrated Review *Concepts, Skills, and Problem Solving*

Keep mathematically in shape by doing these exercises *before* the problems of this section.

Properties and Definitions

1. In your own words, describe how to multiply

$$\frac{3t}{5} \cdot \frac{8t^2}{15}.$$

2. In your own words, describe how to divide

$$\frac{3t}{5} \div \frac{8t^2}{15}.$$

3. In your own words, describe how to add

$$\frac{3t}{5} + \frac{8t^2}{15}.$$

4. What is the value of $\frac{t-5}{5-t}$? Explain.

Simplifying Expressions

In Exercises 5–10, simplify the expression.

5. $\frac{x^2}{2x+3} \div \frac{5x}{2x+3}$ **6.** $\frac{x-y}{5x} \div \frac{x^2-y^2}{x^2}$

7. $\frac{\frac{9}{x}}{\left(\frac{6}{x}+2\right)}$ **8.** $\frac{\left(1+\frac{2}{x}\right)}{\left(x-\frac{4}{x}\right)}$

9. $\frac{\left(\frac{4}{x^2-9}+\frac{2}{x-2}\right)}{\left(\frac{1}{x+3}+\frac{1}{x-3}\right)}$ **10.** $\frac{\left(\frac{1}{x+1}+\frac{1}{2}\right)}{\left(\frac{3}{2x^2+4x+2}\right)}$

Problem Solving

11. Find two real numbers that divide the real number line between $x/2$ and $4x/3$ into three equal parts.

12. When two capacitors with capacitances C_1 and C_2, respectively, are connected in series, the equivalent capacitance is given by

$$\frac{1}{\left(\frac{1}{C_1}+\frac{1}{C_2}\right)}.$$

Simplify this complex fraction.

Developing Skills

In Exercises 1–20, write the number in *i*-form. See Example 1.

1. $\sqrt{-4}$ **2.** $\sqrt{-9}$

3. $-\sqrt{-144}$ **4.** $\sqrt{-49}$

5. $\sqrt{-\frac{4}{25}}$ **6.** $-\sqrt{-\frac{36}{121}}$

7. $\sqrt{-0.09}$ **8.** $\sqrt{-0.0004}$

9. $\sqrt{-8}$ **10.** $\sqrt{-75}$

11. $\sqrt{-27}$ **12.** $\sqrt{-80}$

13. $\sqrt{-7}$ **14.** $\sqrt{-15}$

15. $\frac{\sqrt{-12}}{\sqrt{-3}}$ **16.** $\frac{\sqrt{-45}}{\sqrt{-5}}$

17. $\frac{\sqrt{-20}}{\sqrt{4}}$ **18.** $\frac{\sqrt{72}}{\sqrt{-2}}$

19. $\sqrt{-\frac{18}{64}}$ **20.** $\sqrt{-\frac{8}{25}}$

In Exercises 21–42, perform the operation(s) and write the result in standard form. See Examples 2 and 3.

21. $\sqrt{-16}+\sqrt{-36}$ **22.** $\sqrt{-25}-\sqrt{-9}$

23. $\sqrt{-50}-\sqrt{-8}$ **24.** $\sqrt{-500}+\sqrt{-45}$

25. $\sqrt{-48}+\sqrt{-12}-\sqrt{-27}$

26. $\sqrt{-32}-\sqrt{-18}+\sqrt{-50}$

27. $\sqrt{-8}\sqrt{-2}$ **28.** $\sqrt{-25}\sqrt{-6}$

29. $\sqrt{-18}\sqrt{-3}$ **30.** $\sqrt{-7}\sqrt{-7}$

31. $\sqrt{-0.16}\sqrt{-1.21}$ **32.** $\sqrt{-0.49}\sqrt{-1.44}$

33. $\sqrt{-3}\left(\sqrt{-3}+\sqrt{-4}\right)$

34. $\sqrt{-12}\left(\sqrt{-3}-\sqrt{-12}\right)$

35. $\sqrt{-5}\left(\sqrt{-16} - \sqrt{-10}\right)$

36. $\sqrt{-24}\left(\sqrt{-9} + \sqrt{-4}\right)$

37. $\sqrt{-2}\left(3 - \sqrt{-8}\right)$

38. $\sqrt{-9}\left(1 + \sqrt{-16}\right)$

39. $\left(\sqrt{-16}\right)^2$

40. $\left(\sqrt{-2}\right)^2$

41. $\left(\sqrt{-4}\right)^3$

42. $\left(\sqrt{-5}\right)^3$

In Exercises 43–50, determine a and b. See Example 4.

43. $3 - 4i = a + bi$

44. $-8 + 6i = a + bi$

45. $5 - 4i = (a + 3) + (b - 1)i$

46. $-10 + 12i = 2a + (5b - 3)i$

47. $-4 - \sqrt{-8} = a + bi$

48. $\sqrt{-36} - 3 = a + bi$

49. $(a + 5) + (b - 1)i = 7 - 3i$

50. $(2a + 1) + (2b + 3)i = 5 + 12i$

In Exercises 51–66, perform the operation(s) and write the result in standard form. See Example 5.

51. $(4 - 3i) + (6 + 7i)$

52. $(-10 + 2i) + (4 - 7i)$

53. $(-4 - 7i) + (-10 - 33i)$

54. $(15 + 10i) - (2 + 10i)$

55. $13i - (14 - 7i)$

56. $(-21 - 50i) + (21 - 20i)$

57. $(30 - i) - (18 + 6i) + 3i^2$

58. $(4 + 6i) + (15 + 24i) - (1 - i)$

59. $6 - (3 - 4i) + 2i$

60. $22 + (-5 + 8i) + 10i$

61. $\left(\frac{4}{3} + \frac{1}{3}i\right) + \left(\frac{5}{6} + \frac{7}{6}i\right)$

62. $(0.05 + 2.50i) - (6.2 + 11.8i)$

63. $15i - (3 - 25i) + \sqrt{-81}$

64. $(-1 + i) - \sqrt{2} - \sqrt{-2}$

65. $8 - \left(5 - \sqrt{-63}\right) + (4 - 5i)$

66. $\left(7 - \sqrt{-96}\right) - (-8 + 10i) - 3i$

In Exercises 67–96, perform the operation and write the result in standard form. See Example 6.

67. $(3i)(12i)$

68. $(-5i)(4i)$

69. $(3i)(-8i)$

70. $(-2i)(-10i)$

71. $(-6i)(-i)(6i)$

72. $(10i)(12i)(-3i)$

73. $(-3i)^3$

74. $(8i)^2$

75. $(-3i)^2$

76. $(2i)^4$

77. $-5(13 + 2i)$

78. $10(8 - 6i)$

79. $4i(-3 - 5i)$

80. $-3i(10 - 15i)$

81. $(9 - 2i)\left(\sqrt{-4}\right)$

82. $(11 + 3i)\left(\sqrt{-25}\right)$

83. $\sqrt{-20}\left(6 + 2\sqrt{5}i\right)$

84. $\sqrt{-24}\left(-3\sqrt{6} - 4i\right)$

85. $(4 + 3i)(-7 + 4i)$

86. $(3 + 5i)(2 + 15i)$

87. $(-7 + 7i)(4 - 2i)$

88. $(3 + 5i)(2 - 15i)$

89. $\left(-2 + \sqrt{-5}\right)\left(-2 - \sqrt{-5}\right)$

90. $\left(-3 - \sqrt{-12}\right)\left(4 - \sqrt{-12}\right)$

91. $(3 - 4i)^2$

92. $(7 + i)^2$

93. $(2 + 5i)^2$

94. $(8 - 3i)^2$

95. $(2 + i)^3$

96. $(3 - 2i)^3$

In Exercises 97–108, multiply the number by its conjugate.

97. $2 + i$

98. $3 + 2i$

99. $-2 - 8i$

100. $10 - 3i$

101. $5 - \sqrt{6}i$

102. $-4 + \sqrt{2}i$

103. $10i$

104. 20

105. $1 + \sqrt{-3}$

106. $-3 - \sqrt{-5}$

107. $1.5 + \sqrt{-0.25}$

108. $3.2 - \sqrt{-0.04}$

In Exercises 109–118, perform the operation and write the result in standard form. See Examples 7 and 8.

109. $\dfrac{20}{2i}$

110. $\dfrac{1 + i}{3i}$

111. $\dfrac{4}{1 - i}$

112. $\dfrac{20}{3 + i}$

113. $\dfrac{-12}{2 + 7i}$

114. $\dfrac{15}{2(1 - i)}$

115. $\dfrac{4i}{1 - 3i}$

116. $\dfrac{17i}{5 + 3i}$

117. $\dfrac{2 + 3i}{1 + 2i}$

118. $\dfrac{4 - 5i}{4 + 5i}$

In Exercises 119–122, perform the operation and write the result in standard form.

119. $\dfrac{1}{1 - 2i} + \dfrac{4}{1 + 2i}$

120. $\dfrac{3i}{1 + i} + \dfrac{2}{2 + 3i}$

121. $\dfrac{i}{4 - 3i} - \dfrac{5}{2 + i}$

122. $\dfrac{1 + i}{i} - \dfrac{3}{5 - 2i}$

In Exercises 123–126, determine whether each number is a solution of the equation. See Example 9.

123. $x^2 + 2x + 5 = 0$

(a) $x = -1 + 2i$ (b) $x = -1 - 2i$

124. $x^2 - 4x + 13 = 0$

(a) $x = 2 - 3i$ (b) $x = 2 + 3i$

125. $x^3 + 4x^2 + 9x + 36 = 0$

(a) $x = -4$ (b) $x = -3i$

126. $x^3 - 8x^2 + 25x - 26 = 0$

(a) $x = 2$ (b) $x = 3 - 2i$

127. *Cube Roots* The principal cube root of 125, $\sqrt[3]{125}$, is 5. Evaluate the expression x^3 for each of the following values of x.

(a) $\dfrac{-5 + 5\sqrt{3}i}{2}$ (b) $\dfrac{-5 - 5\sqrt{3}i}{2}$

128. *Cube Roots* The principal cube root of 27, $\sqrt[3]{27}$, is 3. Evaluate the expression x^3 for each of the following values of x.

(a) $\dfrac{-3 + 3\sqrt{3}i}{2}$ (b) $\dfrac{-3 - 3\sqrt{3}i}{2}$

129. *Pattern Recognition* Use the results of Exercises 127 and 128 to list possible cube roots of (a) 1, (b) 8, and (c) 64. Verify your results algebraically.

130. *Algebraic Properties* Consider the complex number $1 + 5i$.

(a) Find the additive inverse of the number.

(b) Find the multiplicative inverse of the number.

In Exercises 131–134, perform the operations.

131. $(a + bi) + (a - bi)$ **132.** $(a + bi)(a - bi)$

133. $(a + bi) - (a - bi)$

134. $(a + bi)^2 + (a - bi)^2$

Explaining Concepts

135. Define the imaginary unit i.

136. Explain why the equation $x^2 = -1$ does not have real number solutions.

137. *Error Analysis* Describe the error.

$$\sqrt{-3}\sqrt{-3} = \sqrt{(-3)(-3)} = \sqrt{9} = 3$$

138. *True or False?* Some numbers are both real and imaginary. Explain.

139. Find the product of $3 - 2i$ and its complex conjugate.

140. Describe the methods for adding, subtracting, multiplying, and dividing complex numbers.

CHAPTER SUMMARY

Key Terms

square root, *p. 514*
cube root, *p. 514*
*n*th root of *a*, *p. 514*
principal *n*th root of *a*, *p. 514*
radical symbol, *p. 514*
index, *p. 514*
radicand, *p. 514*
rationalizing the denominator, *p. 526*
Pythagorean Theorem, *p. 528*
conjugates, *p. 533*
imaginary unit *i*, *p. 550*
i-form, *p. 550*
complex number, *p. 552*
imaginary number, *p. 552*
real part, *p. 553*
imaginary part, *p. 553*
complex conjugates, *p. 554*

Key Concepts

9.1 Properties of *n*th roots

1. If a is a positive real number and n is even, then a has exactly two (real) nth roots, which are denoted by $\sqrt[n]{a}$ and $-\sqrt[n]{a}$.
2. If a is any real number and n is odd, then a has only one (real) nth root, which is denoted by $\sqrt[n]{a}$.
3. If a is a negative real number and n is even, then a has no (real) nth root.

9.1 Inverse properties of *n*th powers and *n*th roots

Let a be a real number, and let n be an integer such that $n \geq 2$.

1. If a has a principal nth root, then $\left(\sqrt[n]{a}\right)^n = a$.
2. If n is odd, then $\sqrt[n]{a^n} = a$.
   If n is even, then $\sqrt[n]{a^n} = |a|$.

9.1 Domain of a radical function

Let n be an integer that is greater than or equal to 2.

1. If n is odd, the domain of $f(x) = \sqrt[n]{x}$ is the set of all real numbers.
2. If n is even, the domain of $f(x) = \sqrt[n]{x}$ is the set of all nonnegative real numbers.

9.1 Rules of exponents

Let r and s be rational numbers, and let a and b be real numbers, variables, or algebraic expressions.

1. $a^r \cdot a^s = a^{r+s}$
2. $\dfrac{a^r}{a^s} = a^{r-s}, \quad a \neq 0$
3. $(ab)^r = a^r \cdot b^r$
4. $\left(\dfrac{a}{b}\right)^r = \dfrac{a^r}{b^r}, \quad b \neq 0$
5. $(a^r)^s = a^{rs}$
6. $a^{-r} = \dfrac{1}{a^r}, \quad a \neq 0$
7. $a^0 = 1$
8. $\left(\dfrac{a}{b}\right)^{-r} = \left(\dfrac{b}{a}\right)^r, \quad a \neq 0, b \neq 0$

9.2 Multiplying and dividing radicals

Let u and v be real numbers, variables, or algebraic expressions. If the nth roots of u and v are real, the following properties are true.

1. $\sqrt[n]{uv} = \sqrt[n]{u}\,\sqrt[n]{v}$
2. $\sqrt[n]{\dfrac{u}{v}} = \dfrac{\sqrt[n]{u}}{\sqrt[n]{v}}, \quad v \neq 0$

9.2 Simplifying radical expressions

A radical expression is said to be in simplest form if all three of the following are true.

1. All possible nth powered factors have been removed from each radical.
2. No radical contains a fraction.
3. No denominator of a fraction contains a radical.

9.4 Raising both sides of an equation to the *n*th power

Let u and v be real numbers, variables, or algebraic expressions, and let n be a positive integer. If $u = v$, then it follows that $u^n = v^n$. This is called raising both sides of an equation to the nth power.

9.5 Square root of a negative number

Let c be a positive real number. Then the square root of $-c$ is given by $\sqrt{-c} = \sqrt{c(-1)} = \sqrt{c}\sqrt{-1} = \sqrt{c}\,i$. When writing $\sqrt{-c}$ in the i-form, $\sqrt{c}\,i$, note that the i is outside the radical.

REVIEW EXERCISES

Reviewing Skills

9.1 In Exercises 1–16, evaluate the radical expression.

1. $\sqrt{49}$
2. $\sqrt{64}$
3. $-\sqrt{81}$
4. $\sqrt{-16}$
5. $\sqrt[3]{-8}$
6. $\sqrt[3]{-27}$
7. $-\sqrt[3]{64}$
8. $-\sqrt[3]{125}$
9. $\sqrt{(1.2)^2}$
10. $\sqrt{(0.4)^2}$
11. $\sqrt{\left(\frac{5}{6}\right)^2}$
12. $\sqrt{\left(\frac{8}{15}\right)^2}$
13. $\sqrt[3]{-\left(\frac{1}{5}\right)^3}$
14. $-\sqrt[3]{\left(-\frac{27}{64}\right)^3}$
15. $\sqrt{-2^2}$
16. $\sqrt{-4^2}$

In Exercises 17–20, fill in the missing description.

	Radical Form	*Rational Exponent Form*
17.	$\sqrt{49} = 7$	
18.	$\sqrt[3]{0.125} = 0.5$	
19.		$216^{1/3} = 6$
20.		$16^{1/4} = 2$

In Exercises 21–28, use the rules of exponents to evaluate the expression.

21. $27^{4/3}$
22. $16^{3/4}$
23. $-(5^2)^{3/2}$
24. $(-9)^{5/2}$
25. $8^{-4/3}$
26. $243^{-2/5}$
27. $-\left(\frac{27}{64}\right)^{2/3}$
28. $\left(-\frac{8}{125}\right)^{1/3}$

In Exercises 29–40, use the rules of exponents to simplify the expression.

29. $x^{3/4} \cdot x^{-1/6}$
30. $a^{2/3} \cdot a^{3/5}$
31. $z\sqrt[3]{z^2}$
32. $x^2\sqrt[4]{x^3}$
33. $\dfrac{\sqrt[4]{x^3}}{\sqrt{x^4}}$
34. $\dfrac{\sqrt{x^3}}{\sqrt[3]{x^2}}$
35. $\sqrt[3]{a^3b^2}$
36. $\sqrt[5]{x^6y^2}$
37. $\sqrt[4]{\sqrt{x}}$
38. $\sqrt{\sqrt[3]{x^4}}$
39. $\dfrac{(3x+2)^{2/3}}{\sqrt[3]{3x+2}}$
40. $\dfrac{\sqrt[5]{3x+6}}{(3x+6)^{4/5}}$

In Exercises 41–44, evaluate the expression. Round the result to two decimal places.

41. $75^{-3/4}$
42. $510^{5/3}$
43. $\sqrt{13^2 - 4(2)(7)}$
44. $\dfrac{-3.7 + \sqrt{15.8}}{2(2.3)}$

In Exercises 45–48, use a graphing utility to graph the function. Give the domain of the function.

45. $y = 3\sqrt[3]{2x}$
46. $y = \dfrac{10}{\sqrt[4]{x^2+1}}$
47. $g(x) = 4x^{3/4}$
48. $h(x) = \frac{1}{2}x^{4/3}$

9.2 In Exercises 49–58, simplify the radical expression.

49. $\sqrt{360}$
50. $\sqrt{\frac{50}{9}}$
51. $\sqrt{75u^5v^4}$
52. $\sqrt{24x^3y^4}$
53. $\sqrt{0.25x^4y}$
54. $\sqrt{0.16s^6t^3}$
55. $\sqrt[4]{64a^2b^5}$
56. $\sqrt{36x^3y^2}$
57. $\sqrt[3]{48a^3b^4}$
58. $\sqrt[4]{32u^4v^5}$

In Exercises 59–64, rationalize the denominator and simplify further, if possible.

59. $\sqrt{\frac{5}{6}}$
60. $\sqrt{\frac{3}{20}}$
61. $\dfrac{3}{\sqrt{12x}}$
62. $\dfrac{4y}{\sqrt{10z}}$
63. $\dfrac{2}{\sqrt[3]{2x}}$
64. $\sqrt[3]{\dfrac{16t}{s^2}}$

In Exercises 65–76, perform the operations and simplify.

65. $2\sqrt{7} - 5\sqrt{7} + 4\sqrt{7}$
66. $3\sqrt{5} - 7\sqrt{5} + 2\sqrt{5}$
67. $3\sqrt{40} - 10\sqrt{90}$
68. $9\sqrt{50} - 5\sqrt{8} + \sqrt{48}$
69. $5\sqrt{x} - \sqrt[3]{x} + 9\sqrt{x} - 8\sqrt[3]{x}$
70. $\sqrt{3x} - \sqrt[4]{6x^2} + 2\sqrt[4]{6x^2} - 4\sqrt{3x}$
71. $10\sqrt[4]{y+3} - 3\sqrt[4]{y+3}$
72. $5\sqrt[3]{x-3} + 4\sqrt[3]{x-3}$

73. $\sqrt{25x} + \sqrt{49x} - \sqrt[3]{8x}$

74. $\sqrt[3]{81x^4} + \sqrt[3]{24x^4} - \sqrt{3x}$

75. $\sqrt{5} - \dfrac{3}{\sqrt{5}}$

76. $\dfrac{4}{\sqrt{2}} + 3\sqrt{2}$

9.3 In Exercises 77–88, multiply the radical expressions and simplify.

77. $\sqrt{15} \cdot \sqrt{20}$

78. $\sqrt{42} \cdot \sqrt{21}$

79. $\sqrt{5}(\sqrt{10} + 3)$

80. $\sqrt{6}(\sqrt{24} - 8)$

81. $\sqrt{10}(\sqrt{2} + \sqrt{5})$

82. $\sqrt{12}(\sqrt{6} - \sqrt{8})$

83. $(2\sqrt{3} + 7)(\sqrt{6} - 2)$

84. $(2 - 4\sqrt{3})(7 + \sqrt{3})$

85. $(\sqrt{5} + 6)^2$

86. $(4 - 3\sqrt{2})^2$

87. $(\sqrt{3} - \sqrt{x})(\sqrt{3} + \sqrt{x})$

88. $(2 + 3\sqrt{5})(2 - 3\sqrt{5})$

In Exercises 89–96, simplify the quotient.

89. $\dfrac{3}{1 - \sqrt{2}}$

90. $\dfrac{\sqrt{5}}{\sqrt{10} + 3}$

91. $\dfrac{3\sqrt{8}}{2\sqrt{2} + \sqrt{3}}$

92. $\dfrac{7\sqrt{6}}{\sqrt{3} - 4\sqrt{2}}$

93. $\dfrac{\sqrt{2} - 1}{\sqrt{3} - 4}$

94. $\dfrac{3 + \sqrt{3}}{5 - \sqrt{3}}$

95. $(\sqrt{x} + 10) \div (\sqrt{x} - 10)$

96. $(3\sqrt{s} + 4) \div (\sqrt{s} + 2)$

In Exercises 97–100, use a graphing utility to graph the functions on the same screen. Use the graphs to verify that the expressions are equivalent. Verify the results algebraically.

97. $y_1 = \sqrt{\dfrac{5}{2x}}$, $y_2 = \dfrac{\sqrt{10x}}{2x}$

98. $y_1 = \dfrac{x}{1 + \sqrt{x}}$, $y_2 = \dfrac{x(1 - \sqrt{x})}{1 - x}$

99. $y_1 = 5\sqrt{x} - 2\sqrt{x}$, $y_2 = 3\sqrt{x}$

100. $y_1 = -2\sqrt{9x} + 10\sqrt{x}$, $y_2 = 4\sqrt{x}$

9.4 In Exercises 101–116, solve the given equation.

101. $\sqrt{y} = 15$

102. $\sqrt{x} - 3 = 0$

103. $\sqrt{3x} + 9 = 0$

104. $\sqrt{4x} + 6 = 9$

105. $\sqrt{2(a - 7)} = 14$

106. $\sqrt{5(4 - 3x)} = 10$

107. $\sqrt[3]{5x - 7} - 3 = -1$

108. $\sqrt[4]{2x + 3} + 4 = 5$

109. $\sqrt[3]{5x + 2} - \sqrt[3]{7x - 8} = 0$

110. $\sqrt[4]{9x - 2} - \sqrt[4]{8x} = 0$

111. $\sqrt{2(x + 5)} = x + 5$

112. $y - 2 = \sqrt{y + 4}$

113. $\sqrt{v - 6} = 6 - v$

114. $\sqrt{5t} = 1 + \sqrt{5(t - 1)}$

115. $\sqrt{1 + 6x} = 2 - \sqrt{6x}$

116. $\sqrt{2 + 9b} + 1 = 3\sqrt{b}$

9.5 In Exercises 117–122, write the complex number in *i*-form.

117. $\sqrt{-48}$

118. $\sqrt{-0.16}$

119. $10 - 3\sqrt{-27}$

120. $3 + 2\sqrt{-500}$

121. $\frac{3}{4} - 5\sqrt{-\frac{3}{25}}$

122. $-0.5 + 3\sqrt{-1.21}$

In Exercises 123–130, perform the operation and write the result in standard form.

123. $\sqrt{-81} + \sqrt{-36}$

124. $\sqrt{-49} + \sqrt{-1}$

125. $\sqrt{-121} - \sqrt{-84}$

126. $\sqrt{-169} - \sqrt{-4}$

127. $\sqrt{-5}\sqrt{-5}$

128. $\sqrt{-24}\sqrt{-6}$

129. $\sqrt{-10}(\sqrt{-4} - \sqrt{-7})$

130. $\sqrt{-5}(\sqrt{-10} + \sqrt{-15})$

In Exercises 131–134, find *x* and *y* such that the two complex numbers are equal.

131. $4x - \sqrt{-36} = 8 - 2yi$

132. $5x + \sqrt{-81} = 25 + 3yi$

133. $24 + \sqrt{-5y} = 6x + 25i$

134. $10 - \sqrt{-4y} = 2x - 16i$

In Exercises 135–142, add, subtract, or multiply the complex numbers.

135. $(-4 + 5i) - (-12 + 8i)$

136. $(-8 + 3i) - (6 + 7i)$

137. $(3 - 8i) + (5 + 12i)$

138. $(-6 + 3i) + (-1 + i)$

139. $(4 - 3i)(4 + 3i)$ **140.** $(12 - 5i)(2 + 7i)$

141. $(6 - 5i)^2$ **142.** $(2 - 9i)^2$

In Exercises 143–148, use complex conjugates to perform the division.

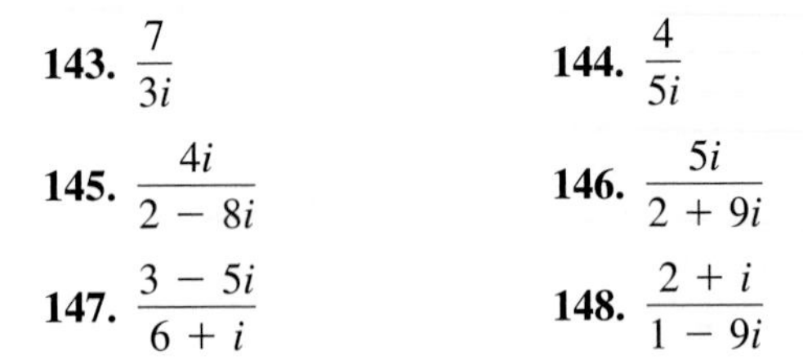

143. $\dfrac{7}{3i}$ **144.** $\dfrac{4}{5i}$

145. $\dfrac{4i}{2 - 8i}$ **146.** $\dfrac{5i}{2 + 9i}$

147. $\dfrac{3 - 5i}{6 + i}$ **148.** $\dfrac{2 + i}{1 - 9i}$

Solving Problems

149. *Perimeter* The four corners are cut from an $8\frac{1}{2}$-inch-by-14-inch sheet of paper (see figure). Find the perimeter of the remaining piece of paper.

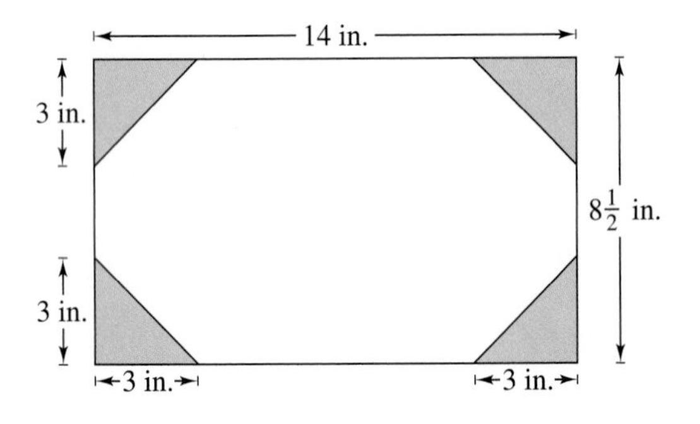

150. *Geometry* Determine the length and width of a rectangle with a perimeter of 84 inches and a diagonal of 30 inches.

151. *Length of a Pendulum* The time t in seconds for a pendulum of length L in feet to go through one complete cycle (its period) is

$$t = 2\pi\sqrt{\frac{L}{32}}.$$

How long is the pendulum of a grandfather clock with a period of 1.3 seconds?

152. *Height of a Bridge* The time t in seconds for a free-falling object to fall d feet is given by

$$t = \sqrt{\frac{d}{16}}.$$

A child drops a pebble from a bridge and observes it strike the water after approximately 4 seconds. Estimate the height of the bridge.

Power In Exercises 153–156, use the equation

$$I = \sqrt{\frac{P}{R}}$$

to find the amount of power P (in watts) consumed by an electrical device that operates with a current I (in amps) and a resistance R (in ohms).

153. $I = 5$ amps, $R = 20$ ohms

154. $I = 10$ amps, $R = 20$ ohms

155. $I = 15$ amps, $R = 40$ ohms

156. $I = 15$ amps, $R = 20$ ohms

157. *Velocity of a Falling Object* The velocity of a free-falling object can be determined from the equation

$$v = \sqrt{2gh}$$

where v is the velocity (in feet per second), $g = 32$ feet per second per second, and h is the distance (in feet) the object has fallen. Find the height from which a rock has been dropped if it strikes the ground with a velocity of 25 feet per second.

Chapter Test

Take this test as you would take a test in class. After you are done, check your work against the answers given in the back of the book.

In Exercises 1 and 2, evaluate the expressions without using a calculator.

1. (a) $16^{3/2}$
(b) $\sqrt{5}\sqrt{20}$

2. (a) $27^{-2/3}$
(b) $\sqrt{2}\sqrt{18}$

In Exercises 3–5, simplify the expressions.

3. (a) $\left(\dfrac{x^{1/2}}{x^{1/3}}\right)^2$
(b) $5^{1/4} \cdot 5^{7/4}$

4. (a) $\sqrt{\dfrac{32}{9}}$
(b) $\sqrt[3]{24}$

5. (a) $\sqrt{24x^3}$
(b) $\sqrt[4]{16x^5y^8}$

6. In your own words, explain the meaning of "rationalize" and demonstrate by rationalizing the denominator in the expression $\dfrac{3}{\sqrt{6}}$.

7. Combine: $5\sqrt{3x} - 3\sqrt{75x}$

8. Multiply and simplify: $\sqrt{5}(\sqrt{15x} + 3)$

9. Expand: $(4 - \sqrt{2x})^2$

10. Factor: $7\sqrt{27} + 14y\sqrt{12} = 7\sqrt{3}(\quad)$

In Exercises 11–13, solve the equation.

11. $\sqrt{3y} - 6 = 3$

12. $\sqrt{x^2 - 1} = x - 2$

13. $\sqrt{x} - x + 6 = 0$

In Exercises 14 and 15, find x and y such that the two complex numbers are equal.

14. $3x + \sqrt{-4y} = 12 + 40i$

15. $27 - \sqrt{-16y} = 9x - 4i$

In Exercises 16–19, perform the operation and simplify.

16. $(2 + 3i) - \sqrt{-25}$

17. $(2 - 3i)^2$

18. $\sqrt{-16}(1 + \sqrt{-4})$

19. $(3 - 2i)(1 + 5i)$

20. Divide $5 - 2i$ by i. Write the result in standard form.

21. The velocity v (in feet per second) of an object is given by $v = \sqrt{2gh}$, where $g = 32$ feet per second per second and h is the distance (in feet) the object has fallen. Find the height from which a rock has been dropped if it strikes the ground with a velocity of 80 feet per second.

Cumulative Test: Chapters 7–9

Take this test as you would take a test in class. After you are done, check your work against the answers given in the back of the book.

In Exercises 1–6, match the system of equations with its graph. [The graphs are labeled (a), (b), (c), (d), (e), and (f).]

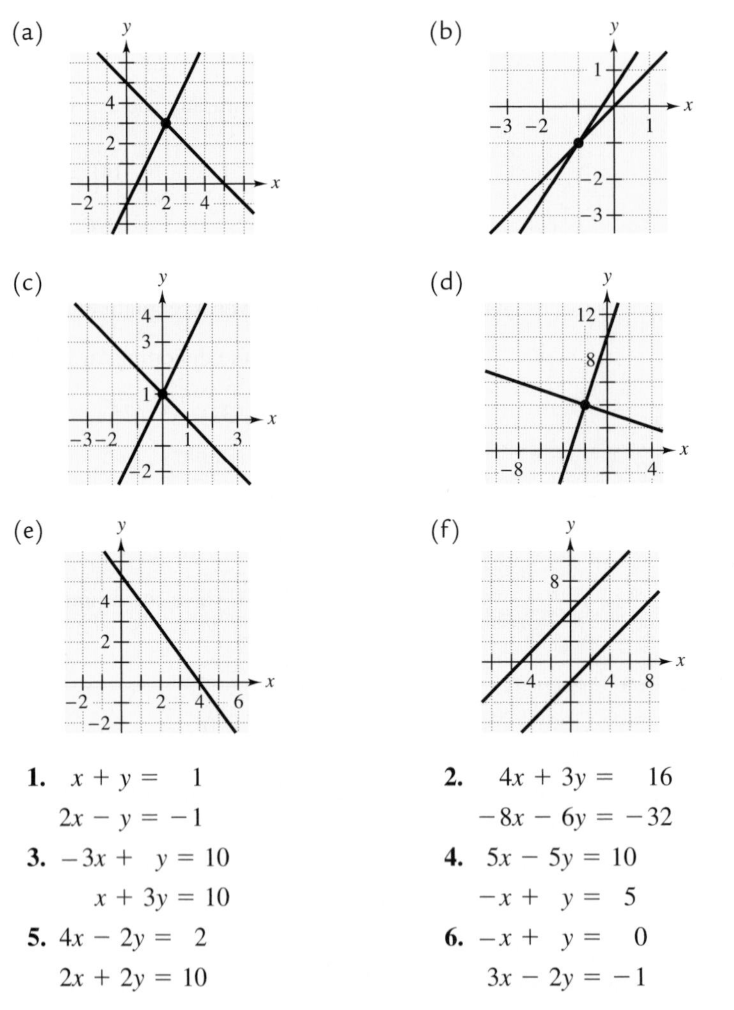

1. $x + y = 1$
$2x - y = -1$

2. $4x + 3y = 16$
$-8x - 6y = -32$

3. $-3x + y = 10$
$x + 3y = 10$

4. $5x - 5y = 10$
$-x + y = 5$

5. $4x - 2y = 2$
$2x + 2y = 10$

6. $-x + y = 0$
$3x - 2y = -1$

In Exercises 7–10, solve the system of equations by the specified method.

7. Graphical: $x - y = 1$
$2x + y = 5$

8. Substitution: $4x + 2y = 8$
$x - 5y = 13$

9. Elimination: $4x - 3y = 8$
$-2x + y = -6$

10. Cramer's Rule: $2x - y = 4$
$3x + y = -5$

In Exercises 11–20, perform the operation(s) and/or simplify.

11. $\dfrac{x^2 + 8x + 16}{18x^2} \cdot \dfrac{2x^4 + 4x^3}{x^2 - 16}$

12. $\dfrac{2}{x} - \dfrac{x}{x^3 + 3x^2} + \dfrac{1}{x + 3}$

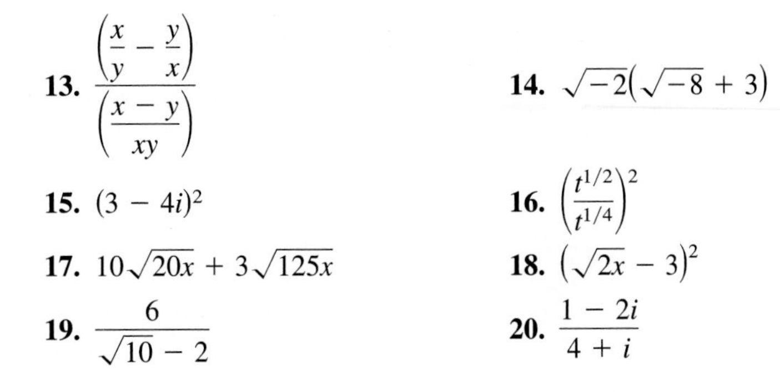

13. $\dfrac{\left(\dfrac{x}{y} - \dfrac{y}{x}\right)}{\left(\dfrac{x - y}{xy}\right)}$

14. $\sqrt{-2}\left(\sqrt{-8} + 3\right)$

15. $(3 - 4i)^2$

16. $\left(\dfrac{t^{1/2}}{t^{1/4}}\right)^2$

17. $10\sqrt{20x} + 3\sqrt{125x}$

18. $\left(\sqrt{2x} - 3\right)^2$

19. $\dfrac{6}{\sqrt{10} - 2}$

20. $\dfrac{1 - 2i}{4 + i}$

In Exercises 21–24, solve the equation.

21. $\dfrac{1}{x} + \dfrac{4}{10 - x} = 1$

22. $\dfrac{x - 3}{x} + 1 = \dfrac{x - 4}{x - 6}$

23. $\sqrt{x} - x + 12 = 0$

24. $\sqrt{5 - x} + 10 = 11$

In Exercises 25–28, match the function with its graph. [The graphs are labeled (a), (b), (c), and (d).]

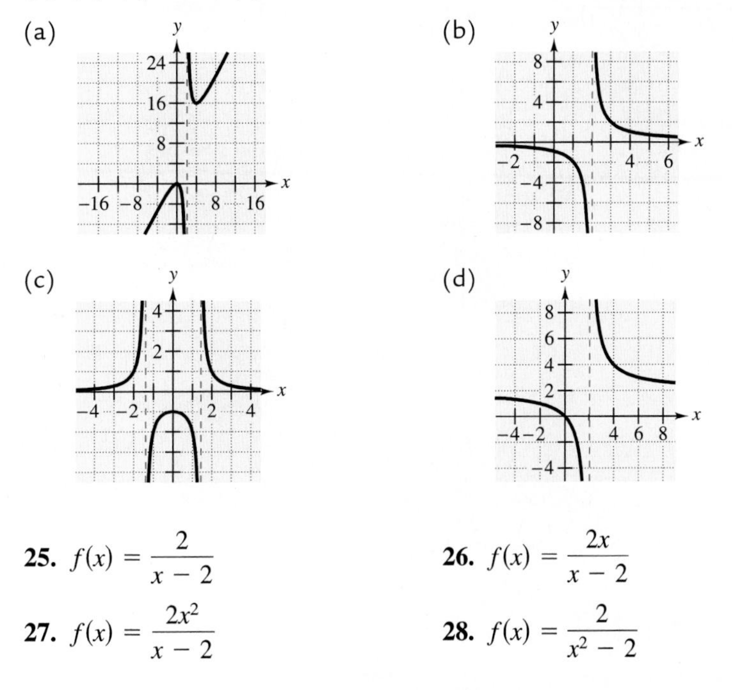

25. $f(x) = \dfrac{2}{x - 2}$

26. $f(x) = \dfrac{2x}{x - 2}$

27. $f(x) = \dfrac{2x^2}{x - 2}$

28. $f(x) = \dfrac{2}{x^2 - 2}$

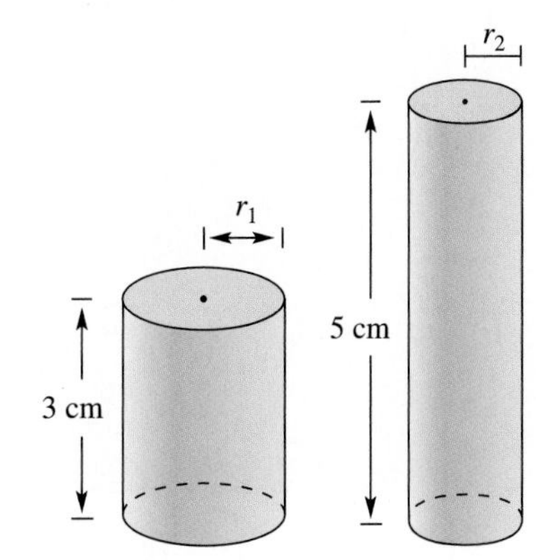

Figure for 31

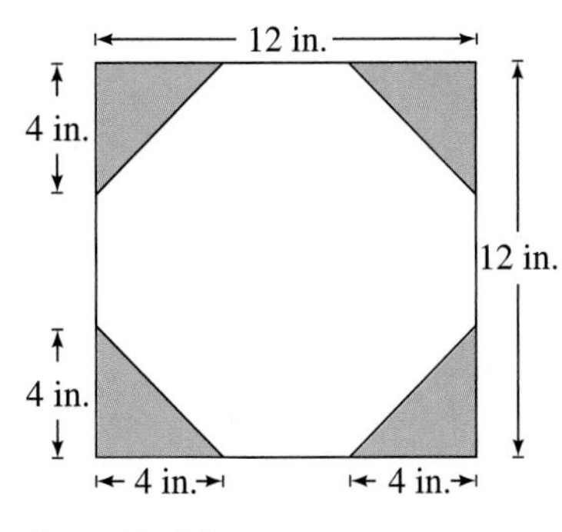

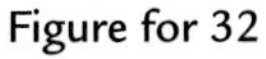

Figure for 32

29. The stopping distance d of a car is directly proportional to the square of its speed s. On a certain type of pavement, a car requires 50 feet to stop when its speed is 25 miles per hour. Estimate the stopping distance when the speed of the car is 40 miles per hour. Explain your reasoning.

30. The number N of prey t months after a predator is introduced into an area is inversely proportional to $t + 1$. If $N = 300$ when $t = 0$, find N when $t = 5$.

31. The volume V of a right circular cylinder is $V = \pi r^2 h$. The two cylinders in the figure have equal volumes. Write r_2 as a function of r_1.

32. The four corners are cut from a 12-inch-by-12-inch piece of glass, as shown in the figure. Find the perimeter of the remaining piece of glass.

10 Quadratic Equations and Inequalities

Tom Bean/The Stock Market

The footbridge over the dells of the Eau Claire River is part of the Ice Age Trail, a 1000-mile state scenic trail in Marathon County, Wisconsin.

Motivating the Chapter

Height of a Falling Object

You drop a rock from a bridge 100 feet above a river. The height h (in feet) of the rock at any time t (in seconds) is

$$h = -16t^2 + v_0t + h_0$$

where v_0 is the initial velocity (in feet per second) of the rock and h_0 is the initial height.

See Section 10.1, Exercise 137

a. Suppose you drop the rock ($v_0 = 0$ ft/sec). How long will it take to hit the water? What method did you use to solve the quadratic equation? Explain why you used that method.

b. Suppose you throw the rock straight up with an initial velocity of 32 feet per second. Find the time(s) when h is 100 feet. What method did you use to solve this quadratic equation? Explain why you used this method.

See Section 10.3, Exercise 103

c. Suppose you throw the rock straight up with an initial velocity of 32 feet per second. Find the time when h is 50 feet. What method did you use to solve this quadratic equation? Explain why you used this method.

d. You move to a lookout point that is 84 feet above the river. If you throw the rock straight up at the same rate as when you were 100 feet above the river, would you expect it to reach the water in less time? Verify your conclusion algebraically.

See Section 10.4, Exercise 106

e. Find the maximum height reached by the rock when it is thrown straight up with an initial velocity of 32 feet per second from a height of 100 feet.

f. Sketch a graph of the function in part (e). Describe the relationship between the vertex of the graph and the maximum height reached by the rock.

See Section 10.6, Exercise 116

g. You throw a rock straight up with an initial velocity of 32 feet per second from a height of 100 feet. During what interval of time is the height greater than 52 feet?

10.1 Factoring and Extracting Square Roots

Objectives

1. Solve a quadratic equation by factoring.
2. Solve a quadratic equation by extracting square roots.
3. Solve a quadratic equation with complex solutions by extracting complex square roots.
4. Use substitution to solve an equation of quadratic form.

1 Solve a quadratic equation by factoring.

Corbis-Bettmann

Pierre de Fermat
(1601–1665)

Fermat's Last Theorem states that the equation $x^n + y^n = z^n$ has no solution when x, y, and z are nonzero integers and $n > 2$. In 1637, Fermat wrote in the margin of a book that he had discovered a proof of this theorem; however, his proof has never been found. On June 23, 1993, 356 years later, a 200-page proof was presented at a gathering of mathematicians at Cambridge University in England by an American mathematician, Andrew Wiles.

Solving Quadratic Equations by Factoring

In this chapter, you will study methods for solving quadratic equations and equations of quadratic form. To begin, let's review the method of factoring that you studied in Section 6.5.

Remember that the first step in solving a quadratic equation by factoring is to write the equation in general form. Next, factor the left side. Finally, set each factor equal to zero and solve for x. It is important to check each solution in the original equation.

Example 1 Solving Quadratic Equations by Factoring

a.

$x^2 + 5x = 24$		Original equation
$x^2 + 5x - 24 = 0$		General form
$(x + 8)(x - 3) = 0$		Factor.
$x + 8 = 0$	$\Rightarrow$ $x = -8$	Set 1st factor equal to 0.
$x - 3 = 0$	$\Rightarrow$ $x = 3$	Set 2nd factor equal to 0.

b.

$3x^2 = 4 - 11x$		Original equation
$3x^2 + 11x - 4 = 0$		General form
$(3x - 1)(x + 4) = 0$		Factor.
$3x - 1 = 0$	$\Rightarrow$ $x = \frac{1}{3}$	Set 1st factor equal to 0.
$x + 4 = 0$	$\Rightarrow$ $x = -4$	Set 2nd factor equal to 0.

c.

$9x^2 + 12 = 3 + 12x + 5x^2$		Original equation
$4x^2 - 12x + 9 = 0$		General form
$(2x - 3)(2x - 3) = 0$		Repeated factor
$2x - 3 = 0$	$\Rightarrow$ $x = \dfrac{3}{2}$	Set factor equal to 0.

When the two solutions of a quadratic equation are identical, they are called a **double** or **repeated solution.** This occurred in Example 1(c).

2 Solve a quadratic equation by extracting square roots.

Extracting Square Roots

Consider the following equation, where $d > 0$ and u is an algebraic expression.

$u^2 = d$		Original equation
$u^2 - d = 0$		General form
$(u + \sqrt{d})(u - \sqrt{d}) = 0$		Factor.
$u + \sqrt{d} = 0$	⇒ $u = -\sqrt{d}$	Set 1st factor equal to 0.
$u - \sqrt{d} = 0$	⇒ $u = \sqrt{d}$	Set 2nd factor equal to 0.

Because the solutions differ only in sign, they can be written together using a "plus or minus sign"

$$u = \pm\sqrt{d}.$$

This form of the solution is read as "u is equal to plus or minus the square root of d." Solving an equation of the form $u^2 = d$ *without* going through the steps of factoring is called **extracting square roots.**

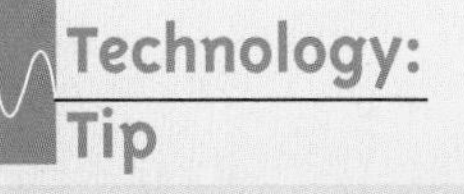

Technology: Tip

To graphically check solutions of an equation written in general form, graph the left side of the equation and locate its x-intercepts. For instance, in Example 2(b), write the equation as

$$(x - 2)^2 - 10 = 0$$

and then sketch the graph of

$$y = (x - 2)^2 - 10$$

as shown below. From the graph, you can determine that the x-intercepts are approximately 5.16 and -1.16.

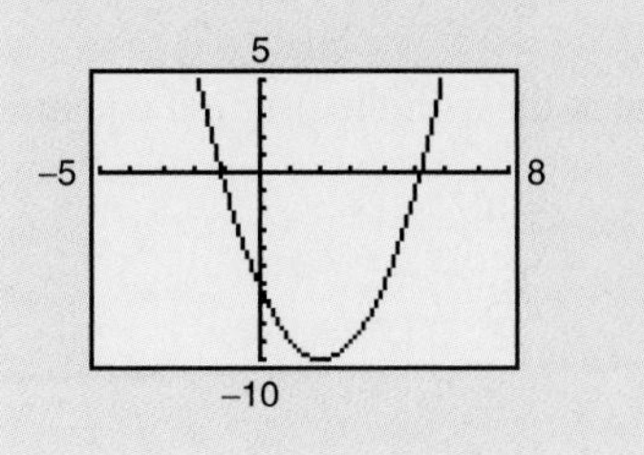

▶ Extracting Square Roots

The equation $u^2 = d$, where $d > 0$, has exactly two solutions:

$$u = \sqrt{d} \quad \text{and} \quad u = -\sqrt{d}.$$

These solutions can also be written as $u = \pm\sqrt{d}$.

Example 2 Extracting Square Roots

a. $3x^2 = 15$ — Original equation

$x^2 = 5$ — Divide both sides by 3.

$x = \pm\sqrt{5}$ — Extract square roots.

The solutions are $\sqrt{5}$ and $-\sqrt{5}$. Check these in the original equation.

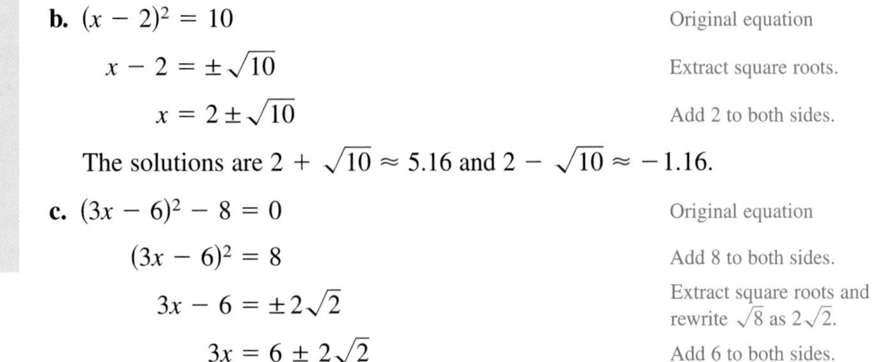

b. $(x - 2)^2 = 10$ — Original equation

$x - 2 = \pm\sqrt{10}$ — Extract square roots.

$x = 2 \pm \sqrt{10}$ — Add 2 to both sides.

The solutions are $2 + \sqrt{10} \approx 5.16$ and $2 - \sqrt{10} \approx -1.16$.

c. $(3x - 6)^2 - 8 = 0$ — Original equation

$(3x - 6)^2 = 8$ — Add 8 to both sides.

$3x - 6 = \pm 2\sqrt{2}$ — Extract square roots and rewrite $\sqrt{8}$ as $2\sqrt{2}$.

$3x = 6 \pm 2\sqrt{2}$ — Add 6 to both sides.

$x = \dfrac{6 \pm 2\sqrt{2}}{3}$ — Divide both sides by 3.

The solutions are $(6 + 2\sqrt{2})/3 \approx 2.94$ and $(6 - 2\sqrt{2})/3 \approx 1.06$.

3 Solve a quadratic equation with complex solutions by extracting complex square roots.

Quadratic Equations with Complex Solutions

Prior to Section 9.5, the only solutions to find were real numbers. But now that you have studied complex numbers, it makes sense to look for other types of solutions. For instance, although the quadratic equation $x^2 + 1 = 0$ has no solutions that are real numbers, it does have two solutions that are complex numbers: i and $-i$. To check this, substitute i and $-i$ for x.

$$(i)^2 + 1 = -1 + 1 = 0 \qquad \text{Solution checks. } \checkmark$$

$$(-i)^2 + 1 = -1 + 1 = 0 \qquad \text{Solution checks. } \checkmark$$

One way to find complex solutions of a quadratic equation is to extend the *extracting square roots* technique to cover the case where d is a negative number.

> ▶ **Extracting Complex Square Roots**
>
> The equation $u^2 = d$, where $d < 0$, has exactly two solutions:
>
> $$u = \sqrt{|d|}i \quad \text{and} \quad u = -\sqrt{|d|}i.$$
>
> These solutions can also be written as $u = \pm\sqrt{|d|}i$.

Example 3 Extracting Complex Square Roots

a. $x^2 + 8 = 0$ — Original equation

$x^2 = -8$ — Subtract 8 from both sides.

$x = \pm\sqrt{8}i$ — Extract complex square roots.

$x = \pm 2\sqrt{2}i$ — Simplify.

The solutions are $2\sqrt{2}i$ and $-2\sqrt{2}i$. Check these in the original equation.

b. $(x - 4)^2 = -3$ — Original equation

$x - 4 = \pm\sqrt{-3}$ — Extract complex square roots.

$x - 4 = \pm\sqrt{3}i$ — Write in i-form.

$x = 4 \pm \sqrt{3}i$ — Add 4 to both sides.

The solutions are $4 + \sqrt{3}i$ and $4 - \sqrt{3}i$. Check these in the original equation.

c. $2(3x - 5)^2 + 32 = 0$ — Original equation

$2(3x - 5)^2 = -32$ — Subtract 32 from both sides.

$(3x - 5)^2 = -16$ — Divide both sides by 2.

$3x - 5 = \pm 4i$ — Extract complex square roots.

$3x = 5 \pm 4i$ — Add 5 to both sides.

$x = \dfrac{5 \pm 4i}{3}$ — Divide both sides by 3.

The solutions are $(5 + 4i)/3$ and $(5 - 4i)/3$. Check these in the original equation.

Technology: Tip

When graphically checking solutions, it is important to realize that only the real solutions appear as x-intercepts—the complex solutions cannot be estimated from the graph. For instance, in Example 3(a), the graph of

$$y = x^2 + 8$$

has no x-intercepts. This agrees with the fact that both of its solutions are complex numbers.

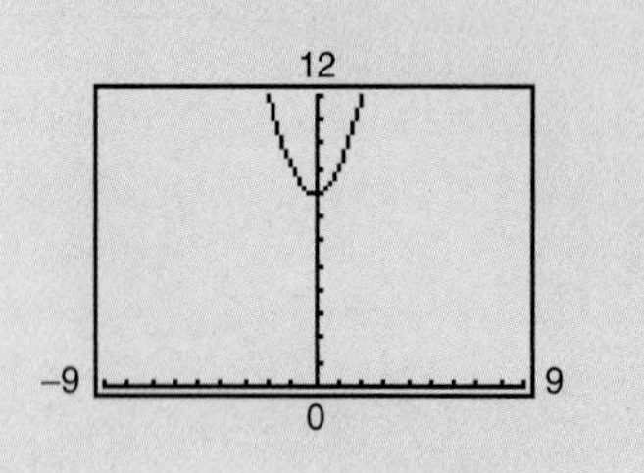

4 Use substitution to solve an equation of quadratic form.

Equations of Quadratic Form

Both the factoring and extraction of square roots methods can be applied to nonquadratic equations that are of **quadratic form.** An equation is said to be of quadratic form if it has the form

$$au^2 + bu + c = 0$$

where u is an algebraic expression. Here are some examples.

Equation	*Written in Quadratic Form*
$x^4 + 5x^2 + 4 = 0$	$(x^2)^2 + 5(x^2) + 4 = 0$
$x - 5\sqrt{x} + 6 = 0$	$(\sqrt{x})^2 - 5(\sqrt{x}) + 6 = 0$
$2x^{2/3} + 5x^{1/3} - 3 = 0$	$2(\sqrt[3]{x})^2 + 5(\sqrt[3]{x}) - 3 = 0$
$18 + 2x^2 + (x^2 + 9)^2 = 8$	$(x^2 + 9)^2 + 2(x^2 + 9) - 8 = 0$

To solve an equation of quadratic form, it helps to make a substitution and rewrite the equation in terms of u, as demonstrated in Examples 4 and 5.

Example 4 Solving an Equation of Quadratic Form

Solve $x^4 - 13x^2 + 36 = 0$.

Solution

Begin by writing the original equation in quadratic form, as follows.

$$x^4 - 13x^2 + 36 = 0 \qquad \text{Original equation}$$

$$(x^2)^2 - 13(x^2) + 36 = 0 \qquad \text{Write in quadratic form.}$$

Next, let $u = x^2$ and substitute u into the equation written in quadratic form. Then, factor and solve the equation.

$$u^2 - 13u + 36 = 0 \qquad \text{Substitute } u \text{ for } x^2.$$

$$(u - 4)(u - 9) = 0 \qquad \text{Factor.}$$

$$u - 4 = 0 \quad \Rightarrow \quad u = 4 \qquad \text{Set 1st factor equal to 0.}$$

$$u - 9 = 0 \quad \Rightarrow \quad u = 9 \qquad \text{Set 2nd factor equal to 0.}$$

At this point you have found the "u-solutions." To find the "x-solutions," replace u by x^2 and solve for x.

$$u = 4 \quad \Rightarrow \quad x^2 = 4 \quad \Rightarrow \quad x = \pm 2$$

$$u = 9 \quad \Rightarrow \quad x^2 = 9 \quad \Rightarrow \quad x = \pm 3$$

The solutions are 2, -2, 3, and -3. Check these in the original equation.

Technology: Tip

You may find it helpful to graph the equation with a graphing utility before you begin. The graph will indicate the number of real solutions an equation has. For instance, the graph shown below is from the equation in Example 4. You can see from the graph that there are four real solutions.

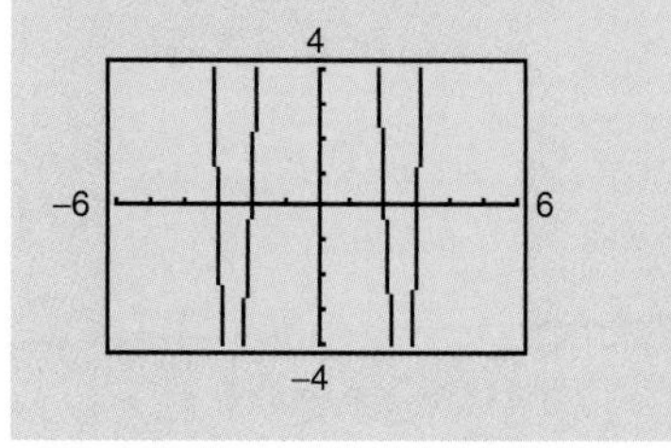

Be sure you see in Example 4 that the u-solutions of 4 and 9 represent only a temporary step. They are not solutions of the original equation and cannot be substituted into the original equation.

Example 5 Solving an Equation of Quadratic Form

Solve $x - 5\sqrt{x} + 6 = 0$.

Solution

This equation is of quadratic form with $u = \sqrt{x}$.

$x - 5\sqrt{x} + 6 = 0$ Original equation

$(\sqrt{x})^2 - 5(\sqrt{x}) + 6 = 0$ Write in quadratic form.

$u^2 - 5u + 6 = 0$ Substitute u for $\sqrt{x}$.

$(u - 2)(u - 3) = 0$ Factor.

$u - 2 = 0 \implies u = 2$ Set 1st factor equal to 0.

$u - 3 = 0 \implies u = 3$ Set 2nd factor equal to 0.

Now, using the u-solutions of 2 and 3, you obtain the following x-solutions.

$u = 2 \implies \sqrt{x} = 2 \implies x = 4$

$u = 3 \implies \sqrt{x} = 3 \implies x = 9$

Example 6 Surface Area of a Sphere

The surface area of a sphere of radius r is given by $S = 4\pi r^2$. If the surface area of a softball is $144/\pi$ square inches, find the diameter d of the softball.

Solution

$\frac{144}{\pi} = 4\pi r^2$ Substitute $144/\pi$ for S.

$\frac{36}{\pi^2} = r^2 \implies \pm\sqrt{\frac{36}{\pi^2}} = r$ Divide both sides by 4π and extract square roots.

Choosing the positive root, you get $r = 6/\pi$, and so the diameter of the softball is

$$d = 2r = 2\left(\frac{6}{\pi}\right) = \frac{12}{\pi} \approx 3.82 \text{ inches.}$$

Discussing the Concept — Analyzing Solutions of Quadratic Equations

Use a graphing utility to graph each of the following equations. How many times does the graph of each equation cross the x-axis?

a. $y = 2x^2 + x - 15$ **b.** $y = (3x - 1)^2 + 3$

Now set each equation equal to zero and solve the resulting equations. How many of each type of solution (real or complex) does each equation have? Summarize the relationship between the number of x-intercepts in the graph of a quadratic equation and the number and type of roots found algebraically.

10.1 Exercises

Integrated Review *Concepts, Skills, and Problem Solving*

Keep mathematically in shape by doing these exercises *before* the problems of this section.

Properties and Definitions

1. Identify the leading coefficient in $5t - 3t^3 + 7t^2$. Explain.

2. State the degree of the product $(y^2 - 2)(y^3 + 7)$. Explain.

3. Sketch a graph for which y is not a function of x. Explain why it is not a function.

4. Sketch a graph for which y is a function of x. Explain why it is a function.

Simplifying Expressions

In Exercises 5–10, simplify the expression.

5. $(x^3 \cdot x^{-2})^{-3}$

6. $(5x^{-4}y^5)(-3x^2y^{-1})$

7. $\left(\dfrac{2x}{3y}\right)^{-2}$

8. $\left(\dfrac{7u^{-4}}{3v^{-2}}\right)\left(\dfrac{14u}{6v^2}\right)^{-1}$

9. $\dfrac{6u^2v^{-3}}{27uv^3}$

10. $\dfrac{-14r^4s^2}{-98rs^2}$

Problem Solving

11. The number N of prey t months after a natural predator is introduced into the test area is inversely proportional to the square root of $t + 1$. If $N = 300$ when $t = 0$, find N when $t = 8$.

12. The travel time between two cities is inversely proportional to the average speed. If a train travels between two cities in 2 hours at an average speed of 58 miles per hour, how long would it take at an average speed of 72 miles per hour? What does the constant of proportionality measure in this problem?

Developing Skills

In Exercises 1–20, solve the equation by factoring. See Example 1.

1. $x^2 - 12x + 35 = 0$

2. $x^2 + 15x + 44 = 0$

3. $x^2 + x - 72 = 0$

4. $x^2 - 2x - 48 = 0$

5. $x^2 + 4x = 45$

6. $x^2 - 7x = 18$

7. $x^2 - 12x + 36 = 0$

8. $x^2 + 60x + 900 = 0$

9. $9x^2 + 24x + 16 = 0$

10. $8x^2 - 10x + 3 = 0$

11. $4x^2 - 12x = 0$

12. $25y^2 - 75y = 0$

13. $u(u - 9) - 12(u - 9) = 0$

14. $16x(x - 8) - 12(x - 8) = 0$

15. $3x(x - 6) - 5(x - 6) = 0$

16. $3(4 - x) - 2x(4 - x) = 0$

17. $(y - 4)(y - 3) = 6$

18. $(6 + u)(1 - u) = 10$

19. $2x(3x + 2) = 5 - 6x^2$

20. $(2z + 1)(2z - 1) = -4z^2 - 5z + 2$

In Exercises 21–42, solve the quadratic equation by extracting square roots. See Example 2.

21. $x^2 = 64$

22. $z^2 = 169$

23. $6x^2 = 54$

24. $5t^2 = 125$

25. $25x^2 = 16$

26. $9z^2 = 121$

27. $\dfrac{y^2}{2} = 32$

28. $\dfrac{x^2}{6} = 24$

29. $4x^2 - 25 = 0$

30. $16y^2 - 121 = 0$

31. $4u^2 - 225 = 0$

32. $16x^2 - 1 = 0$

33. $(x + 4)^2 = 169$

34. $(y - 20)^2 = 625$

35. $(x - 3)^2 = 0.25$

36. $(x + 2)^2 = 0.81$

37. $(x - 2)^2 = 7$

38. $(x + 8)^2 = 28$

39. $(2x + 1)^2 = 50$

40. $(3x - 5)^2 = 48$

41. $(4x - 3)^2 - 98 = 0$

42. $(5x + 11)^2 - 300 = 0$

In Exercises 43-64, solve the equation by extracting complex square roots. See Example 3.

43. $z^2 = -36$

44. $x^2 = -9$

45. $x^2 + 4 = 0$
46. $y^2 + 16 = 0$
47. $9u^2 + 17 = 0$
48. $4v^2 + 9 = 0$
49. $(t - 3)^2 = -25$
50. $(x + 5)^2 = -81$
51. $(3z + 4)^2 + 144 = 0$
52. $(2y - 3)^2 + 25 = 0$
53. $(2x + 3)^2 = -54$
54. $(6y - 5)^2 = -8$
55. $9(x + 6)^2 = -121$
56. $4(x - 4)^2 = -169$
57. $(x - 1)^2 = -27$
58. $(2x + 3)^2 = -54$
59. $(x + 1)^2 + 0.04 = 0$
60. $(x - 3)^2 + 2.25 = 0$
61. $\left(c - \frac{2}{3}\right)^2 + \frac{1}{9} = 0$
62. $\left(u + \frac{5}{8}\right)^2 + \frac{49}{16} = 0$
63. $\left(x + \frac{7}{3}\right)^2 = -\frac{38}{9}$
64. $\left(y - \frac{5}{6}\right)^2 = -\frac{4}{5}$

In Exercises 65–80, find all real and complex solutions of the equation.

65. $2x^2 - 5x = 0$
66. $3t^2 + 6t = 0$
67. $2x^2 + 5x - 12 = 0$
68. $3x^2 + 8x - 16 = 0$
69. $x^2 - 900 = 0$
70. $y^2 - 225 = 0$
71. $x^2 + 900 = 0$
72. $y^2 + 225 = 0$
73. $\frac{2}{3}x^2 = 6$
74. $\frac{1}{3}x^2 = 4$
75. $(x - 5)^2 - 100 = 0$
76. $(y + 12)^2 - 400 = 0$
77. $(x - 5)^2 + 100 = 0$
78. $(y + 12)^2 + 400 = 0$
79. $(x + 2)^2 + 18 = 0$
80. $(x + 2)^2 - 18 = 0$

In Exercises 81–90, use a graphing utility to graph the function. Use the graph to approximate any x-intercepts. Set $y = 0$ and solve the resulting equation. Compare the result with the x-intercepts of the graph.

81. $y = x^2 - 9$
82. $y = 5x - x^2$
83. $y = x^2 - 2x - 15$
84. $y = 9 - 4(x - 3)^2$
85. $y = 4 - (x - 3)^2$
86. $y = 4(x + 1)^2 - 9$
87. $y = 2x^2 - x - 6$
88. $y = 4x^2 - x - 14$
89. $y = 3x^2 - 8x - 16$
90. $y = 5x^2 + 9x - 18$

In Exercises 91–96, use a graphing utility to graph the function and observe that the graph has no x-intercepts. Set $y = 0$ and solve the resulting equation. Identify the type of roots of the equation.

91. $y = x^2 + 7$
92. $y = x^2 + 5$
93. $y = (x - 1)^2 + 1$
94. $y = (x + 2)^2 + 3$
95. $y = (x + 3)^2 + 5$
96. $y = (x - 2)^2 + 3$

In Exercises 97–100, solve for y in terms of x. Let f and g be those functions where f represents the positive square root and g the negative square root. Use a graphing utility to sketch the graphs of f and g in the same viewing rectangle.

97. $x^2 + y^2 = 4$
98. $x^2 - y^2 = 4$
99. $x^2 + 4y^2 = 4$
100. $x - y^2 = 0$

In Exercises 101–120, solve the equation of quadratic form. (Find all real *and* complex solutions.) See Examples 4 and 5.

101. $x^4 - 5x^2 + 4 = 0$
102. $x^4 - 10x^2 + 25 = 0$
103. $x^4 - 5x^2 + 6 = 0$
104. $x^4 - 11x^2 + 30 = 0$
105. $x^4 - 3x^2 - 4 = 0$
106. $x^4 - x^2 - 6 = 0$
107. $(x^2 - 4)^2 + 2(x^2 - 4) - 3 = 0$
108. $(x^2 - 1)^2 + (x^2 - 1) - 6 = 0$
109. $x - 7\sqrt{x} + 10 = 0$
110. $x - 11\sqrt{x} + 24 = 0$
111. $x^{2/3} - x^{1/3} - 6 = 0$
112. $x^{2/3} + 3x^{1/3} - 10 = 0$
113. $2x^{2/3} - 7x^{1/3} + 5 = 0$
114. $3x^{2/3} + 8x^{1/3} + 5 = 0$
115. $x^{2/5} - 3x^{1/5} + 2 = 0$
116. $x^{2/5} + 5x^{1/5} + 6 = 0$
117. $2x^{2/5} - 7x^{1/5} + 3 = 0$
118. $2x^{2/5} + 3x^{1/5} + 1 = 0$
119. $\frac{1}{x^2} - \frac{3}{x} + 2 = 0$
120. $3\left(\frac{x}{x + 1}\right)^2 + 7\left(\frac{x}{x + 1}\right) - 6 = 0$

Think About It In Exercises 121–126, find a quadratic equation having the given solutions.

121. $5, -2$
122. $-2, 3$
123. $1 + \sqrt{2}, 1 - \sqrt{2}$
124. $-3 + \sqrt{5}, -3 - \sqrt{5}$
125. $5i, -5i$
126. $2i, -2i$

Solving Problems

Free-Falling Object In Exercises 127–130, find the time required for an object to reach the ground when it is dropped from a height of s_0 feet. The height h (in feet) is given by

$$h = -16t^2 + s_0$$

where t measures time in seconds from the time when the object is released.

127. $s_0 = 256$

128. $s_0 = 48$

129. $s_0 = 128$

130. $s_0 = 500$

131. *Free-Falling Object* The height h (in feet) of an object thrown vertically upward from a tower 144 feet tall is given by

$$h = 144 + 128t - 16t^2$$

where t measures the time in seconds from the time when the object is released. How long does it take for the object to reach the ground?

132. *Revenue* The revenue R (in dollars) when x units of a product are sold is given by

$$R = x\left(120 - \frac{1}{2}x\right).$$

Determine the number of units that must be sold to produce a revenue of \$7000.

Compound Interest The amount A after 2 years when a principal of P dollars is invested at percentage rate r compounded annually is given by

$$A = P(1 + r)^2.$$

In Exercises 133 and 134, find r.

133. $P = \$1500,\ A = \$1685.40$

134. $P = \$5000,\ A = \$5724.50$

National Health Expenditures In Exercises 135 and 136, use the following model, which gives the national expenditures for health care in the United States from 1990 through 1996.

$$y = (26.6 + t)^2, \quad 0 \le t \le 6$$

In this model, y represents the expenditures (in billions of dollars) and t represents the year, with $t = 0$ corresponding to 1990 (see figure). (Source: U.S. Health Care Financing Administration)

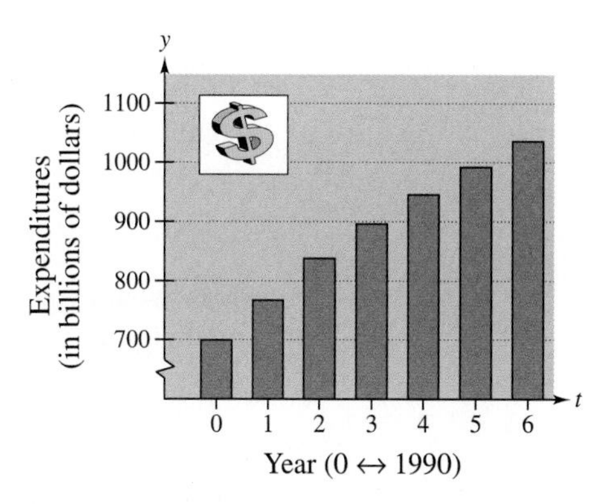

135. Analytically determine the year when expenditures were approximately \$892 billion. Graphically confirm the result.

136. Analytically determine the year when expenditures were approximately \$1000 billion. Graphically confirm the result.

Explaining Concepts

137. Answer parts (a) and (b) of Motivating the Chapter on page 567.

138. For a quadratic equation $ax^2 + bx + c = 0$, where a, b, and c are real numbers with $a \neq 0$, explain why b and c can equal 0, but a cannot.

139. Explain the Zero-Factor Property and how it can be used to solve a quadratic equation.

140. Is it possible for a quadratic equation to have only one solution? If so, give an example.

141. *True or False?* The only solution of the equation $x^2 = 25$ is $x = 5$. Explain.

142. Describe the steps in solving a quadratic equation by extracting square roots.

143. Describe the procedure for solving an equation in quadratic form. Give an example.

10.2 Completing the Square

Objectives

1 Rewrite a quadratic equation in completed square form.

2 Solve a quadratic equation by completing the square.

1 Rewrite a quadratic equation in completed square form.

Constructing Perfect Square Trinomials

Consider the quadratic equation

$(x - 2)^2 = 10.$ Completed square form

You know from Example 2(b) in the preceding section that this equation has two solutions: $2 + \sqrt{10}$ and $2 - \sqrt{10}$. Suppose you had been given the equation in its general form

$x^2 - 4x - 6 = 0.$ General form

How would you solve this equation if you were given only the general form? You could try factoring, but after attempting to do so you would find that the left side of the equation is not factorable (using integer coefficients).

In this section, you will study a technique for rewriting an equation in a completed square form. This technique is called **completing the square.** Note that prior to completing the square, the coefficient of the second-degree term must be 1.

Completing the Square

To **complete the square** for the expression $x^2 + bx$, add $(b/2)^2$, which is the square of half the coefficient of x. Consequently,

$$x^2 + bx + \left(\frac{b}{2}\right)^2 = \left(x + \frac{b}{2}\right)^2.$$

Example 1 Creating a Perfect Square Trinomial

What term should be added to $x^2 - 8x$ so that it becomes a perfect square trinomial?

Solution

For this expression, the coefficient of the x-term is -8. Divide this term by 2, and square the result to obtain $(-4)^2 = 16$. This is the term that should be added to the expression to make it a perfect square trinomial.

$x^2 - 8x + 16 = x^2 - 8x + (-4)^2$ Add 16 to the expression.

$= (x - 4)^2$ Completed square form

2 Solve a quadratic equation by completing the square.

Solving Equations by Completing the Square

When completing the square to solve an equation, remember that it is essential to *preserve the equality*. Thus, when you add a constant term to one side of the equation, you must be sure to add the same constant to the other side of the equation.

Study Tip

Completing the square can be used to solve *any* quadratic equation. However, sometimes it is easier to factor an equation than to complete the square. For instance, the equation in Example 2 could easily be factored as $x(x + 12) = 0$. But remember, not all equations are factorable. Don't spend a lot of time trying to factor when you know that completing the square will work.

Example 2 Completing the Square: Leading Coefficient Is 1

Solve $x^2 + 12x = 0$.

Solution

$$x^2 + 12x = 0 \qquad \text{Original equation}$$

$$x^2 + 12x + (6)^2 = 36 \qquad \text{Add } \left(\tfrac{12}{2}\right)^2 = 36 \text{ to both sides.}$$

(half)^2

$$(x + 6)^2 = 36 \qquad \text{Completed square form}$$

$$x + 6 = \pm\sqrt{36} \qquad \text{Extract square roots.}$$

$$x = -6 \pm 6 \qquad \text{Subtract 6 from both sides.}$$

$$x = -6 + 6 \text{ or } x = -6 - 6 \qquad \text{Separate solutions.}$$

$$x = 0 \qquad x = -12 \qquad \text{Solutions}$$

The solutions are 0 and -12. Check these in the original equation.

Example 3 Completing the Square: Leading Coefficient Is 1

Solve $x^2 - 6x + 7 = 0$.

Solution

$$x^2 - 6x + 7 = 0 \qquad \text{Original equation}$$

$$x^2 - 6x = -7 \qquad \text{Subtract 7 from both sides.}$$

$$x^2 - 6x + (-3)^2 = -7 + 9 \qquad \text{Add } \left(-\frac{6}{2}\right)^2 = 9 \text{ to both sides.}$$

(half)^2

$$(x - 3)^2 = 2 \qquad \text{Completed square form}$$

$$x - 3 = \pm\sqrt{2} \qquad \text{Extract square roots.}$$

$$x = 3 \pm \sqrt{2} \qquad \text{Add 3 to both sides.}$$

$$x = 3 + \sqrt{2} \text{ or } x = 3 - \sqrt{2} \qquad \text{Separate solutions.}$$

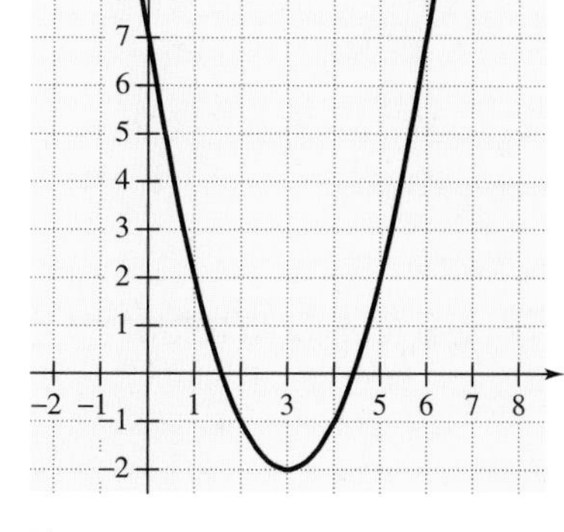

Figure 10.1

The solutions are $3 + \sqrt{2}$ and $3 - \sqrt{2}$. Check these in the original equation. Also try checking the solutions graphically, as shown in Figure 10.1.

If the leading coefficient of a quadratic expression is not 1, you must divide both sides of the equation by this coefficient *before* completing the square. This process is demonstrated in Example 4.

Example 4 A Leading Coefficient That Is Not 1

Solve $2x^2 - x - 2 = 0$.

Solution

$$2x^2 - x - 2 = 0 \qquad \text{Original equation}$$

$$2x^2 - x = 2 \qquad \text{Add 2 to both sides.}$$

$$x^2 - \frac{1}{2}x = 1 \qquad \text{Divide both sides by 2.}$$

$$x^2 - \frac{1}{2}x + \left(-\frac{1}{4}\right)^2 = 1 + \frac{1}{16} \qquad \text{Add } \left(-\tfrac{1}{4}\right)^2 = \tfrac{1}{16} \text{ to both sides.}$$

$$\left(x - \frac{1}{4}\right)^2 = \frac{17}{16} \qquad \text{Completed square form}$$

$$x - \frac{1}{4} = \pm\frac{\sqrt{17}}{4} \qquad \text{Extract square roots.}$$

$$x = \frac{1}{4} \pm \frac{\sqrt{17}}{4} \qquad \text{Add } \tfrac{1}{4} \text{ to both sides.}$$

The solutions are $\frac{1}{4}(1 + \sqrt{17})$ and $\frac{1}{4}(1 - \sqrt{17})$.

Example 5 A Quadratic Equation with Complex Solutions

Solve $x^2 - 4x + 8 = 0$.

Solution

$$x^2 - 4x + 8 = 0 \qquad \text{Original equation}$$

$$x^2 - 4x = -8 \qquad \text{Subtract 8 from both sides.}$$

$$x^2 - 4x + (-2)^2 = -8 + 4 \qquad \text{Add } (-2)^2 = 4 \text{ to both sides.}$$

$$(x - 2)^2 = -4 \qquad \text{Completed square form}$$

$$x - 2 = \pm 2i \qquad \text{Extract complex square roots.}$$

$$x = 2 \pm 2i \qquad \text{Add 2 to both sides.}$$

The solutions are $2 + 2i$ and $2 - 2i$. The first of these is checked below.

Check

$$x^2 - 4x + 8 = 0 \qquad \text{Original equation}$$

$$(2 + 2i)^2 - 4(2 + 2i) + 8 \stackrel{?}{=} 0 \qquad \text{Substitute } 2 + 2i \text{ for } x.$$

$$4 + 8i - 4 - 8 - 8i + 8 \stackrel{?}{=} 0 \qquad \text{Simplify.}$$

$$0 = 0 \qquad \text{Solution checks. ✓}$$

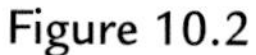
Figure 10.2

Example 6 Dimensions of a Cereal Box 

A cereal box has a volume of 441 cubic inches. Its height is 12 inches and its base has the dimensions x by $x + 7$. (See Figure 10.2.) Find the dimensions of the base in inches.

Solution

$$lwh = V$$ Formula for volume of a rectangular box

$$(x + 7)(x)(12) = 441$$ Substitute 441 for V, $x + 7$ for length, x for width, and 12 for height.

$$12x^2 + 84x = 441$$ Multiply factors.

$$x^2 + 7x = \frac{441}{12}$$ Divide both sides by 12.

$$x^2 + 7x + \left(\frac{7}{2}\right)^2 = \frac{147}{4} + \frac{49}{4}$$ Add $\left(\frac{7}{2}\right)^2 = \frac{49}{4}$ to both sides.

$$\left(x + \frac{7}{2}\right)^2 = \frac{196}{4}$$ Completed square form

$$x + \frac{7}{2} = \pm\sqrt{49}$$ Extract square roots.

$$x = -\frac{7}{2} \pm 7$$ Subtract $\frac{7}{2}$ from both sides.

Choosing the positive root, you get

$$x = -\frac{7}{2} + 7 = \frac{7}{2} = 3.5 \text{ inches}$$ Width of base

and

$$x + 7 = \frac{7}{2} + 7 = \frac{21}{2} = 10.5 \text{ inches.}$$ Length of base

Discussing the Concept **Error Analysis**

Suppose you teach an algebra class and one of your students hands in the following solution. Find and correct the error(s). Discuss how to explain the error(s) to your student.

1. Solve $x^2 + 6x - 13 = 0$ by completing the square.

$$x^2 + 6x = 13$$

$$x^2 + 6x + \left(\frac{6}{2}\right)^2 = 13$$

$$(x + 3)^2 = 13$$

$$x + 3 = \pm\sqrt{13}$$

$$x = -3 \pm \sqrt{13}$$

10.2 Exercises

Integrated Review *Concepts, Skills, and Problem Solving*

Keep mathematically in shape by doing these exercises *before* the problems of this section.

Properties and Definitions

In Exercises 1–4, complete the property of exponents and/or simplify.

1. $(ab)^4 =$

2. $(a^r)^s =$

3. $\left(\dfrac{a}{b}\right)^{-r} =$ $, a \neq 0, b \neq 0$

4. $a^{-r} =$ $, a \neq 0$

Solving Equations

In Exercises 5–8, solve the equation.

5. $\dfrac{4}{x} - \dfrac{2}{3} = 0$

6. $2x - 3[1 + (4 - x)] = 0$

7. $3x^2 - 13x - 10 = 0$

8. $x(x - 3) = 40$

Graphing

In Exercises 9–12, graph the function.

9. $g(x) = \frac{2}{3}x - 5$

10. $h(x) = 5 - \sqrt{x}$

11. $f(x) = \dfrac{4}{x + 2}$

12. $f(x) = 2x + |x - 1|$

Developing Skills

In Exercises 1–16, add a term to the expression so that it becomes a perfect square trinomial. See Example 1.

1. $x^2 + 8x +$

2. $x^2 + 12x +$

3. $y^2 - 20y +$

4. $y^2 - 2y +$

5. $x^2 - 16x +$

6. $x^2 + 18x +$

7. $t^2 + 5t +$

8. $u^2 + 7u +$

9. $x^2 - 9x +$

10. $y^2 - 11y +$

11. $a^2 - \frac{1}{3}a +$

12. $y^2 + \frac{4}{3}y +$

13. $y^2 - \frac{3}{5}y +$

14. $x^2 - \frac{6}{5}x +$

15. $r^2 - 0.4r +$

16. $s^2 + 4.6s +$

In Exercises 17–34, solve the quadratic equation (a) by completing the square and (b) by factoring. See Examples 2–4.

17. $x^2 - 20x = 0$

18. $x^2 + 32x = 0$

19. $x^2 + 6x = 0$

20. $t^2 - 10t = 0$

21. $y^2 - 5y = 0$

22. $t^2 - 9t = 0$

23. $t^2 - 8t + 7 = 0$

24. $y^2 - 8y + 12 = 0$

25. $x^2 + 2x - 24 = 0$

26. $x^2 + 12x + 27 = 0$

27. $x^2 + 7x + 12 = 0$

28. $z^2 + 3z - 10 = 0$

29. $x^2 - 3x - 18 = 0$

30. $t^2 - 5t - 36 = 0$

31. $2x^2 - 14x + 12 = 0$

32. $3x^2 - 3x - 6 = 0$

33. $4x^2 + 4x - 15 = 0$

34. $3x^2 - 13x + 12 = 0$

In Exercises 35–72, solve the quadratic equation by completing the square. Give the solutions in exact form and in decimal form rounded to two decimal places. (The solutions may be complex numbers.) See Examples 2–5.

35. $x^2 - 4x - 3 = 0$

36. $x^2 - 6x + 7 = 0$

37. $x^2 + 4x - 3 = 0$

38. $x^2 + 6x + 7 = 0$

39. $u^2 - 4u + 1 = 0$

40. $a^2 - 10a - 15 = 0$

41. $x^2 + 2x + 3 = 0$

42. $x^2 - 6x + 12 = 0$

43. $x^2 - 10x - 2 = 0$

44. $x^2 + 8x - 4 = 0$

45. $y^2 + 20y + 10 = 0$

46. $y^2 + 6y - 24 = 0$

47. $t^2 + 5t + 3 = 0$

48. $u^2 - 9u - 1 = 0$

49. $v^2 + 3v - 2 = 0$

50. $z^2 - 7z + 9 = 0$

51. $-x^2 + x - 1 = 0$

52. $1 - x - x^2 = 0$

53. $x^2 - 7x + 12 = 0$

54. $y^2 + 5y + 9 = 0$

55. $x^2 - \frac{2}{3}x - 3 = 0$

56. $x^2 + \frac{4}{5}x - 1 = 0$

57. $v^2 + \frac{3}{4}v - 2 = 0$

58. $u^2 - \frac{2}{3}u + 5 = 0$

59. $2x^2 + 8x + 3 = 0$

60. $3x^2 - 24x - 5 = 0$

61. $3x^2 + 9x + 5 = 0$

62. $5x^2 - 15x + 7 = 0$

63. $4y^2 + 4y - 9 = 0$

64. $4z^2 - 3z + 2 = 0$

65. $5x^2 - 3x + 10 = 0$

66. $7x^2 + 4x + 3 = 0$

67. $x(x - 7) = 2$

68. $2x\left(x + \frac{4}{3}\right) = 5$

69. $0.5t^2 + t + 2 = 0$

70. $0.1x^2 + 0.5x = -0.2$

71. $0.1x^2 + 0.2x + 0.5 = 0$

72. $0.02x^2 + 0.10x - 0.05 = 0$

In Exercises 73–78, find the real solutions.

73. $\frac{x}{2} - \frac{1}{x} = 1$

74. $\frac{x}{2} + \frac{5}{x} = 4$

75. $\frac{x^2}{4} = \frac{x + 1}{2}$

76. $\frac{x^2 + 2}{24} = \frac{x - 1}{3}$

77. $\sqrt{2x + 1} = x - 3$

78. $\sqrt{3x - 2} = x - 2$

In Exercises 79–86, use a graphing utility to graph the function. Use the graph to approximate any x-intercepts of the graph. Set $y = 0$ and solve the resulting equation. Compare the result with the x-intercepts of the graph.

79. $y = x^2 + 4x - 1$

80. $y = x^2 + 6x - 4$

81. $y = x^2 - 2x - 5$

82. $y = 2x^2 - 6x - 5$

83. $y = \frac{1}{3}x^2 + 2x - 6$

84. $y = \frac{1}{2}x^2 - 3x + 1$

85. $y = -x^2 - x + 3$

86. $y = \sqrt{x} - x + 2$

Solving Problems

87. *Geometric Modeling*

(a) Find the area of the two adjoining rectangles and large square in the figure.

(b) Find the area of the small square region in the lower right-hand corner of the figure and add it to the area found in part (a).

(c) Find the dimensions and the area of the entire figure after adjoining the small square in the lower right-hand corner of the figure. Note that you have shown geometrically the technique of completing the square.

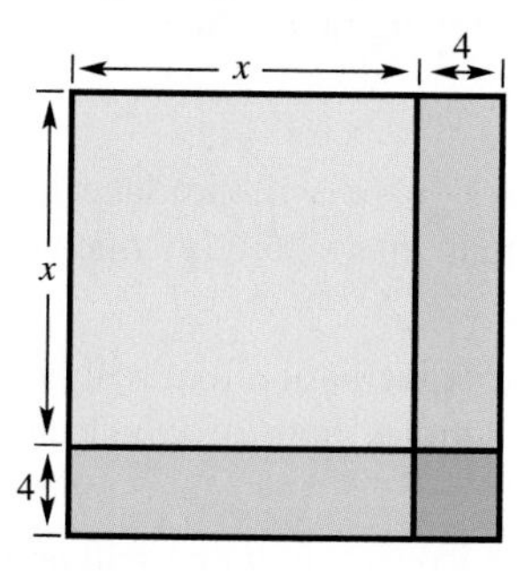

88. *Completing the Square* Repeat Exercise 87 for the model shown below.

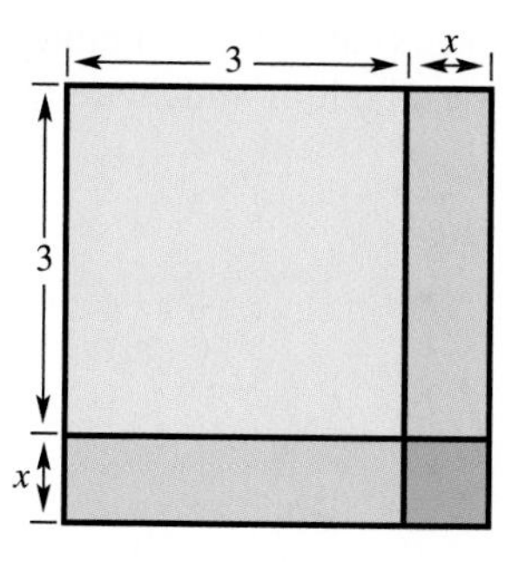

89. *Geometry* Find the dimensions of the triangle in the figure if its area is 12 square centimeters.

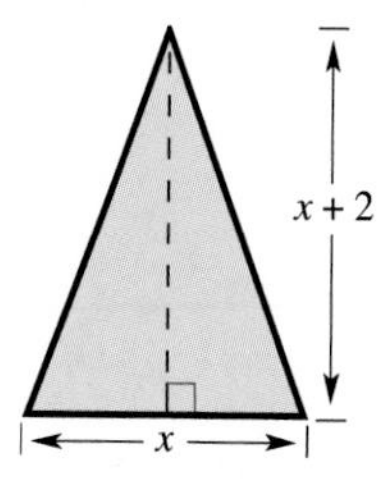

90. *Geometry* The area of the rectangle in the figure is 160 square feet. Find the rectangle's dimensions.

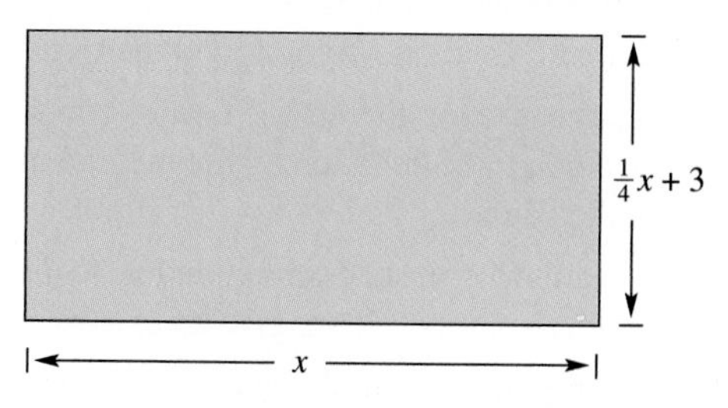

91. *Fencing In a Corral* You have 200 meters of fencing to enclose two adjacent rectangular corrals (see figure). The total area of the enclosed region is 1400 square meters. What are the dimensions of each corral? (The corrals are the same size.)

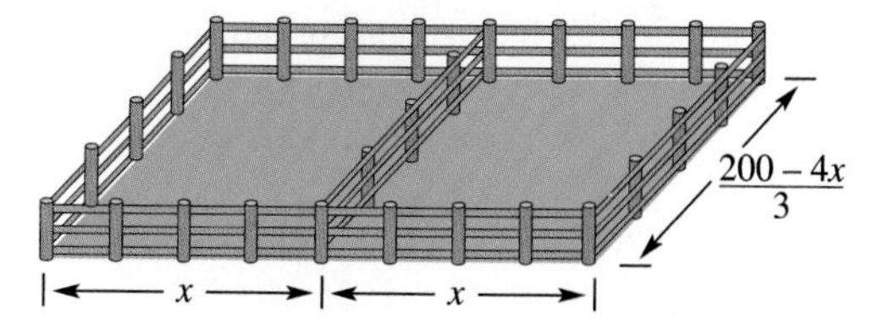

92. *Geometry* An open box with a rectangular base of x inches by $x + 4$ inches has a height of 6 inches (see figure). Find the dimensions of the box if its volume is 840 cubic inches.

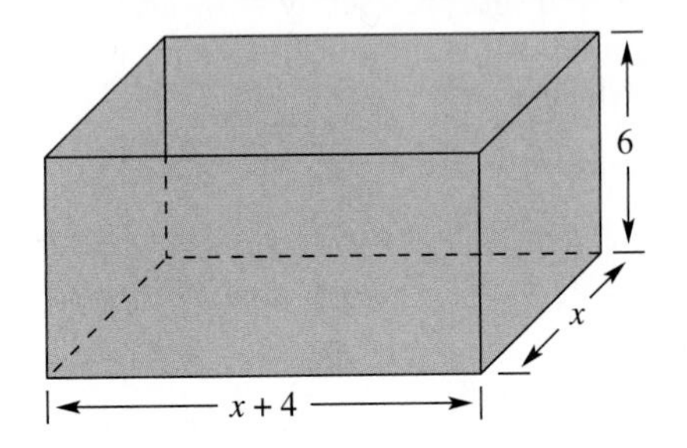

93. *Cutting Across the Lawn* On the sidewalk, the distance from the dormitory to the cafeteria is 400 meters (see figure). By cutting across the lawn, the walking distance is shortened to 300 meters. How long is each part of the L-shaped sidewalk?

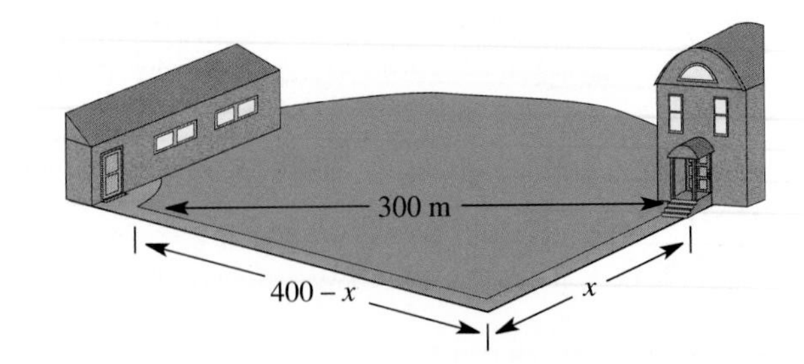

Figure for 93

94. *Revenue* The revenue R from selling x units of a certain product is

$$R = x\left(50 - \frac{1}{2}x\right).$$

Find the number of units that must be sold to produce a revenue of $1218.

95. *Revenue* The revenue R from selling x units of a certain product is

$$R = x\left(100 - \frac{1}{10}x\right).$$

Find the number of units that must be sold to produce a revenue of $12,000.

Explaining Concepts

96. What is a perfect square trinomial?

97. What term must be added to $x^2 + 5x$ to complete the square? Explain how you found the term.

98. Explain the use of extracting square roots when solving a quadratic equation by the method of completing the square.

99. Is it possible for a quadratic equation to have no real number solution? If so, give an example.

100. When using the method of completing the square to solve a quadratic equation, what is the first step if the leading coefficient is not 1? Is the resulting equation equivalent to the given equation? Explain.

101. *True or False?* If you solve a quadratic equation by completing the square and obtain solutions that are rational numbers, then you could have solved the equation by factoring. Explain.

102. Consider the following quadratic equation.

$$(x - 1)^2 = d$$

(a) What value(s) of d will produce a quadratic equation that has exactly one (repeated) solution?

(b) Describe the value(s) of d that will produce two different solutions, both of which are rational numbers.

(c) Describe the value(s) of d that will produce two different solutions, both of which are irrational numbers.

(d) Describe the value(s) of d that will produce two different solutions, both of which are complex numbers.

10.3 The Quadratic Formula

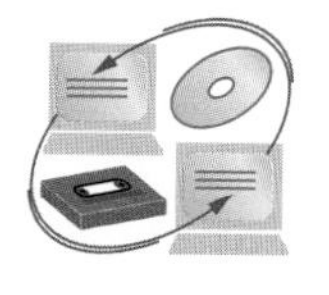

Objectives

1. Derive the Quadratic Formula by completing the square for a general quadratic equation.
2. Use the Quadratic Formula to solve a quadratic equation.
3. Determine the type of solution to a quadratic equation using the discriminant.

1 Derive the Quadratic Formula by completing the square for a general quadratic equation.

The Quadratic Formula

A fourth technique for solving a quadratic equation involves the **Quadratic Formula.** This formula is derived by completing the square for a general quadratic equation.

$ax^2 + bx + c = 0$	General form, $a \neq 0$
$ax^2 + bx = -c$	Subtract c from both sides.
$x^2 + \frac{b}{a}x = -\frac{c}{a}$	Divide both sides by a.
$x^2 + \frac{b}{a}x + \left(\frac{b}{2a}\right)^2 = -\frac{c}{a} + \left(\frac{b}{2a}\right)^2$	Add $\left(\frac{b}{2a}\right)^2$ to both sides.
$\left(x + \frac{b}{2a}\right)^2 = \frac{b^2 - 4ac}{4a^2}$	Simplify.
$x + \frac{b}{2a} = \pm\sqrt{\frac{b^2 - 4ac}{4a^2}}$	Extract square roots.
$x = -\frac{b}{2a} \pm \frac{\sqrt{b^2 - 4ac}}{2\|a\|}$	Subtract $\frac{b}{2a}$ from both sides.
$x = \frac{-b \pm \sqrt{b^2 - 4ac}}{2a}$	Simplify.

Study Tip

The Quadratic Formula is one of the most important formulas in algebra, and you should memorize it. We have found that it helps to try to memorize a verbal statement of the rule. For instance, you might try to remember the following verbal statement of the Quadratic Formula: "Minus b, plus or minus the square root of b squared minus $4ac$, all divided by $2a$."

▶ The Quadratic Formula

The solutions of $ax^2 + bx + c = 0$, $a \neq 0$, are given by the **Quadratic Formula**

$$x = \frac{-b \pm \sqrt{b^2 - 4ac}}{2a}.$$

The expression inside the radical, $b^2 - 4ac$, is called the **discriminant.**

1. If $b^2 - 4ac > 0$, the equation has two real solutions.
2. If $b^2 - 4ac = 0$, the equation has one (repeated) real solution.
3. If $b^2 - 4ac < 0$, the equation has no real solutions.

2 Use the Quadratic Formula to solve a quadratic equation.

Solving Equations by the Quadratic Formula

When using the Quadratic Formula, remember that *before* the formula can be applied, you must first write the quadratic equation in general form in order to determine the values of a, b, and c.

Example 1 The Quadratic Formula: Two Distinct Solutions

$$x^2 + 6x = 16 \qquad \text{Original equation}$$

$$x^2 + 6x - 16 = 0 \qquad \text{Write in general form.}$$

$$x = \frac{-b \pm \sqrt{b^2 - 4ac}}{2a} \qquad \text{Quadratic Formula}$$

$$x = \frac{-6 \pm \sqrt{6^2 - 4(1)(-16)}}{2(1)} \qquad \text{Substitute: } a = 1, b = 6, c = -16.$$

$$x = \frac{-6 \pm \sqrt{100}}{2} \qquad \text{Simplify.}$$

$$x = \frac{-6 \pm 10}{2} \qquad \text{Simplify.}$$

$$x = 2 \quad \text{or} \quad x = -8 \qquad \text{Solutions}$$

The solutions are 2 and -8. Check these in the original equation.

Study Tip

In Example 1, the solutions are rational numbers, which means that the equation could have been solved by factoring. Try solving the equation by factoring.

Example 2 The Quadratic Formula: Two Distinct Solutions

$$-x^2 - 4x + 8 = 0 \qquad \text{Leading coefficient is negative.}$$

$$x^2 + 4x - 8 = 0 \qquad \text{Multiply both sides by } -1.$$

$$x = \frac{-b \pm \sqrt{b^2 - 4ac}}{2a} \qquad \text{Quadratic Formula}$$

$$x = \frac{-4 \pm \sqrt{4^2 - 4(1)(-8)}}{2(1)} \qquad \text{Substitute: } a = 1, b = 4, c = -8.$$

$$x = \frac{-4 \pm \sqrt{48}}{2} \qquad \text{Simplify.}$$

$$x = \frac{-4 \pm 4\sqrt{3}}{2} \qquad \text{Simplify.}$$

$$x = \frac{2(-2 \pm 2\sqrt{3})}{2} \qquad \text{Factor numerator.}$$

$$x = \frac{\cancel{2}(-2 \pm 2\sqrt{3})}{\cancel{2}} \qquad \text{Cancel common factor.}$$

$$x = -2 \pm 2\sqrt{3} \qquad \text{Solutions}$$

The solutions are $-2 + 2\sqrt{3}$ and $-2 - 2\sqrt{3}$. Check these in the original equation.

Study Tip

If the leading coefficient of a quadratic equation is negative, we suggest that you begin by multiplying both sides of the equation by -1, as shown in Example 2. This will produce a positive leading coefficient, which is less cumbersome to work with.

Study Tip

Example 3 could have been solved as follows, without dividing both sides by 2 in the first step.

$$x = \frac{-(-24) \pm \sqrt{(-24)^2 - 4(18)(8)}}{2(18)}$$

$$x = \frac{24 \pm \sqrt{576 - 576}}{36}$$

$$x = \frac{24 \pm 0}{36}$$

$$x = \frac{2}{3}$$

While the result is the same, dividing both sides by 2 simplifies the equation before the Quadratic Formula is applied and so allows you to work with smaller numbers.

Example 3 The Quadratic Formula: One Repeated Solution

$18x^2 - 24x + 8 = 0$ Original equation

$9x^2 - 12x + 4 = 0$ Divide both sides by 2.

$$x = \frac{-b \pm \sqrt{b^2 - 4ac}}{2a}$$ Quadratic Formula

$$x = \frac{-(-12) \pm \sqrt{(-12)^2 - 4(9)(4)}}{2(9)}$$ Substitute 9 for a, -12 for b, and 4 for c.

$$x = \frac{12 \pm \sqrt{144 - 144}}{18}$$ Simplify.

$$x = \frac{12 \pm \sqrt{0}}{18}$$ Simplify.

$$x = \frac{2}{3}$$ Solution

The only solution is $\frac{2}{3}$. Check this in the original equation.

Note in the next example how the Quadratic Formula can be used to solve a quadratic equation that has complex solutions.

Example 4 The Quadratic Formula: Complex Solutions

$2x^2 - 4x + 5 = 0$ Original equation

$$x = \frac{-b \pm \sqrt{b^2 - 4ac}}{2a}$$ Quadratic Formula

$$x = \frac{-(-4) \pm \sqrt{(-4)^2 - 4(2)(5)}}{2(2)}$$ Substitute 2 for a, -4 for b, and 5 for c.

$$x = \frac{4 \pm \sqrt{-24}}{4}$$ Simplify.

$$x = \frac{4 \pm 2\sqrt{6}i}{4}$$ Write in i-form.

$$x = \frac{2(2 \pm \sqrt{6}i)}{2 \cdot 2}$$ Factor numerator and denominator.

$$x = \frac{\cancel{2}(2 \pm \sqrt{6}i)}{\cancel{2} \cdot 2}$$ Cancel common factor.

$$x = \frac{2 \pm \sqrt{6}i}{2}$$ Solutions

The solutions are $\frac{1}{2}(2 + \sqrt{6}i)$ and $\frac{1}{2}(2 - \sqrt{6}i)$. Check these in the original equation.

3 Determine the type of solution to a quadratic equation using the discriminant.

The Discriminant

The radicand in the Quadratic Formula, $b^2 - 4ac$, is called the discriminant because it allows you to "discriminate" among different types of solutions.

Study Tip

From Examples 1–4, you can see that equations with rational or repeated solutions could have been solved by factoring. A quick calculation of the discriminant will help you decide which solution method to use to solve a quadratic equation.

1. Use factoring if

$$b^2 - 4ac \text{ is } \begin{cases} \text{zero} \\ \text{or a} \\ \text{perfect square} \end{cases}.$$

2. Use completing the square or the Quadratic Formula if

$$b^2 - 4ac \text{ is } \begin{cases} \text{negative} \\ \text{or not} \\ \text{a perfect square} \end{cases}.$$

▶ Using the Discriminant

Let a, b, and c be rational numbers such that $a \neq 0$. The discriminant of the quadratic equation $ax^2 + bx + c = 0$ is given by $b^2 - 4ac$, and can be used to classify the solutions of the equation as follows.

Discriminant: $b^2 - 4ac$	*Solution Types*
1. Perfect square	Two distinct rational solutions (Example 1)
2. Positive nonperfect square	Two distinct irrational solutions (Example 2)
3. Zero	One repeated rational solution (Example 3)
4. Negative number	Two distinct imaginary solutions (Example 4)

Example 5 Using the Discriminant

Determine the type of solution for each quadratic equation.

a. $x^2 - x + 2 = 0$ **b.** $2x^2 - 3x - 2 = 0$

c. $x^2 - 2x + 1 = 0$ **d.** $x^2 - 2x - 1 = 9$

Solution

Equation	*Discriminant*	*Solution Types*
a. $x^2 - x + 2 = 0$	$b^2 - 4ac = (-1)^2 - 4(1)(2)$ $= 1 - 8$ $= -7$	Two distinct imaginary solutions
b. $2x^2 - 3x - 2 = 0$	$b^2 - 4ac = (-3)^2 - 4(2)(-2)$ $= 9 + 16$ $= 25$	Two distinct rational solutions
c. $x^2 - 2x + 1 = 0$	$b^2 - 4ac = (-2)^2 - 4(1)(1)$ $= 4 - 4$ $= 0$	One repeated rational solution
d. $x^2 - 2x - 1 = 9$	$b^2 - 4ac = (-2)^2 - 4(1)(-10)$ $= 4 + 40$ $= 44$	Two distinct irrational solutions

Technology: Discovery

Use a graphing utility to graph the equations below.

a. $y = x^2 - x + 2$

b. $y = 2x^2 - 3x - 2$

c. $y = x^2 - 2x + 1$

d. $y = x^2 - 2x - 10$

Describe the solution type of each equation and check your results with those shown in Example 5. Why do you think the discriminant is used to determine solution types?

▶ Summary of Methods for Solving Quadratic Equations

Method	*Example*
1. Factoring	$3x^2 + x = 0$
	$x(3x + 1) = 0 \implies x = 0 \text{ and } 3x + 1 = 0$
2. Extracting square roots	$(x + 2)^2 = 9$
	$x + 2 = \pm 3 \implies x = -2 + 3 = 1 \text{ and } x = -2 - 3 = -5$
3. Completing the square	$x^2 + 6x = 3$
	$x^2 + 6x + \left(\frac{1}{2} \cdot 6\right)^2 = 3 + \left(\frac{1}{2} \cdot 6\right)^2$
	$(x + 3)^2 = 12 \implies x = -3 \pm \sqrt{12}$
4. Using the Quadratic Formula	$3x^2 - 2x + 2 = 0 \implies x = \dfrac{-(-2) \pm \sqrt{(-2)^2 - 4(3)(2)}}{2(3)} = \dfrac{1 \pm \sqrt{5}i}{3}$

Technology: Tip

A graphing utility program that uses the Quadratic Formula to solve quadratic equations can be found at our website, *www.hmco.com*. Programs are available for several current models of graphing utilities.

Example 6 Using a Calculator with the Quadratic Formula

Solve $1.2x^2 - 17.8x + 8.05 = 0$.

Solution

Using the Quadratic Formula, you can write

$$x = \frac{-(-17.8) \pm \sqrt{(-17.8)^2 - 4(1.2)(8.05)}}{2(1.2)}.$$

To evaluate these solutions, begin by calculating the square root.

17.8 [+/−] [x^2] [−] 4 [×] 1.2 [×] 8.05 [=] [√] Scientific

[√] [(] [(] [(−)] 17.8 [)] [x^2] [−] 4 [×] 1.2 Graphing

[×] 8.05 [)] [ENTER]

The display for either of these keystroke sequences should be 16.67932852. Storing this result and using the recall key, we find the following two solutions.

$$x \approx \frac{17.8 + 16.67932852}{2.4} \approx 14.366 \qquad \text{Add stored value.}$$

$$x \approx \frac{17.8 - 16.67932852}{2.4} \approx 0.467 \qquad \text{Subtract stored value.}$$

Discussing the Concept **Problem Posing**

Suppose you are writing a quiz that covers quadratic equations. Write four quadratic equations, including one with solutions $x = \frac{5}{3}$ and $x = -2$ and one with solutions $x = 4 \pm \sqrt{3}$, and instruct students to use any of the four solution methods: factoring, extracting square roots, completing the square, and using the Quadratic Formula. Trade quizzes with a class member and check each other's work.

10.3 Exercises

Integrated Review *Concepts, Skills, and Problem Solving*

Keep mathematically in shape by doing these exercises *before* the problems of this section.

Properties and Definitions

In Exercises 1 and 2, rewrite the expression using the specified property, where a and b are nonnegative real numbers.

1. Multiplication Property: $\sqrt{ab} =$ ▭

2. Division Property: $\sqrt{\dfrac{a}{b}} =$ ▭

3. Is $\sqrt{72}$ in simplest form? Explain.

4. Is $10/\sqrt{5}$ in simplest form? Explain.

Simplifying Expressions

In Exercises 5–10, perform the operation and simplify the expression.

5. $\sqrt{128} + 3\sqrt{50}$ **6.** $3\sqrt{5}\sqrt{500}$

7. $(3 + \sqrt{2})(3 - \sqrt{2})$ **8.** $(3 + \sqrt{2})^2$

9. $\dfrac{8}{\sqrt{10}}$ **10.** $\dfrac{5}{\sqrt{12} - 2}$

Problem Solving

11. Determine the length and width of a rectangle with a perimeter of 50 inches and a diagonal of $5\sqrt{13}$ inches.

12. The demand equation for a certain product is given by

$$p = 75 - \sqrt{1.2(x - 10)}$$

where x is the number of units demanded per day and p is the price per unit. Find the demand if the price is \$59.90.

Developing Skills

In Exercises 1–4, write in general form.

1. $2x^2 = 7 - 2x$ **2.** $7x^2 + 15x = 5$

3. $x(10 - x) = 5$ **4.** $x(3x + 8) = 15$

In Exercises 5–16, solve the quadratic equation (a) by using the Quadratic Formula and (b) by factoring. See Examples 1–4.

5. $x^2 - 11x + 28 = 0$ **6.** $x^2 - 12x + 27 = 0$

7. $x^2 + 6x + 8 = 0$ **8.** $x^2 + 9x + 14 = 0$

9. $4x^2 + 4x + 1 = 0$ **10.** $9x^2 + 12x + 4 = 0$

11. $4x^2 + 12x + 9 = 0$ **12.** $9x^2 - 30x + 25 = 0$

13. $6x^2 - x - 2 = 0$ **14.** $10x^2 - 11x + 3 = 0$

15. $x^2 - 5x - 300 = 0$ **16.** $x^2 + 20x - 300 = 0$

In Exercises 17–46, solve the quadratic equation by using the Quadratic Formula. (Find all real *and* complex solutions.) See Examples 1–4.

17. $x^2 - 2x - 4 = 0$ **18.** $x^2 - 2x - 6 = 0$

19. $t^2 + 4t + 1 = 0$ **20.** $y^2 + 6y + 4 = 0$

21. $x^2 + 6x - 3 = 0$ **22.** $x^2 + 8x - 4 = 0$

23. $x^2 - 10x + 23 = 0$ **24.** $u^2 - 12u + 29 = 0$

25. $2x^2 + 3x + 3 = 0$ **26.** $2x^2 - x + 1 = 0$

27. $3v^2 - 2v - 1 = 0$ **28.** $4x^2 + 6x + 1 = 0$

29. $2x^2 + 4x - 3 = 0$ **30.** $2x^2 + 3x + 3 = 0$

31. $9z^2 + 6z - 4 = 0$ **32.** $8y^2 - 8y - 1 = 0$

33. $-4x^2 - 6x + 3 = 0$

34. $-5x^2 - 15x + 10 = 0$

35. $8x^2 - 6x + 2 = 0$ **36.** $6x^2 + 3x - 9 = 0$

37. $-4x^2 + 10x + 12 = 0$

38. $-15x^2 - 10x + 25 = 0$

39. $9x^2 = 1 + 9x$ **40.** $7x^2 = 3 - 5x$

41. $3x - 2x^2 = 4 - 5x^2$

42. $x - x^2 = 1 - 6x^2$ **43.** $x^2 - 0.4x - 0.16 = 0$

44. $x^2 + 0.6x - 0.41 = 0$

45. $2.5x^2 + x - 0.9 = 0$

46. $0.09x^2 - 0.12x - 0.26 = 0$

In Exercises 47–56, use the discriminant to determine the type of solutions of the quadratic equation. See Example 5.

47. $x^2 + x + 1 = 0$

48. $x^2 + x - 1 = 0$

49. $2x^2 - 5x - 4 = 0$

50. $10x^2 + 5x + 1 = 0$

51. $5x^2 + 7x + 3 = 0$

52. $3x^2 - 2x - 5 = 0$

53. $4x^2 - 12x + 9 = 0$

54. $2x^2 + 10x + 6 = 0$

55. $3x^2 - x + 2 = 0$

56. $9x^2 - 24x + 16 = 0$

In Exercises 57–74, solve the quadratic equation by the most convenient method. (Find all real *and* complex solutions.)

57. $z^2 - 169 = 0$

58. $t^2 = 144$

59. $5y^2 + 15y = 0$

60. $7u^2 + 49u = 0$

61. $25(x - 3)^2 - 36 = 0$

62. $9(x + 4)^2 + 16 = 0$

63. $2y(y - 18) + 3(y - 18) = 0$

64. $4y(y + 7) - 5(y + 7) = 0$

65. $x^2 + 8x + 25 = 0$

66. $x^2 - 3x - 4 = 0$

67. $x^2 - 24x + 128 = 0$

68. $y^2 + 21y + 108 = 0$

69. $3x^2 - 13x + 169 = 0$

70. $2x^2 - 15x + 225 = 0$

71. $18x^2 + 15x - 50 = 0$

72. $14x^2 + 11x - 40 = 0$

73. $1.2x^2 - 0.8x - 5.5 = 0$

74. $2x^2 + 8x + 4.5 = 0$

In Exercises 75–82, use a graphing utility to graph the function. Use the graph to approximate any x-intercepts of the graph. Set $y = 0$ and solve the resulting equation. Compare the result with the x-intercepts of the graph.

75. $y = 3x^2 - 6x + 1$

76. $y = x^2 + x + 1$

77. $y = -(4x^2 - 20x + 25)$

78. $y = x^2 - 4x + 3$

79. $y = 5x^2 - 18x + 6$

80. $y = 15x^2 + 3x - 105$

81. $y = -0.04x^2 + 4x - 0.8$

82. $y = 3.7x^2 - 10.2x + 3.2$

In Exercises 83–86, use a graphing utility to determine the number of real solutions of the quadratic equation. Verify your answer using the discriminant.

83. $2x^2 - 5x + 5 = 0$

84. $2x^2 - x - 1 = 0$

85. $\frac{1}{5}x^2 + \frac{6}{5}x - 8 = 0$

86. $\frac{1}{3}x^2 - 5x + 25 = 0$

In Exercises 87–90, solve the equation.

87. $\dfrac{2x^2}{5} - \dfrac{x}{2} = 1$

88. $\dfrac{x^2 - 9x}{6} = \dfrac{x - 1}{2}$

89. $\sqrt{x + 3} = x - 1$

90. $\sqrt{2x - 3} = x - 2$

Think About It In Exercises 91–94, describe the values of c such that the equation has (a) two real number solutions, (b) one real number solution, and (c) two complex number solutions.

91. $x^2 - 6x + c = 0$

92. $x^2 - 12x + c = 0$

93. $x^2 + 8x + c = 0$

94. $x^2 + 2x + c = 0$

Solving Problems

95. *Geometry* A rectangle has a width of x inches, a length of $x + 6.3$ inches, and an area of 58.14 square inches. Find its dimensions.

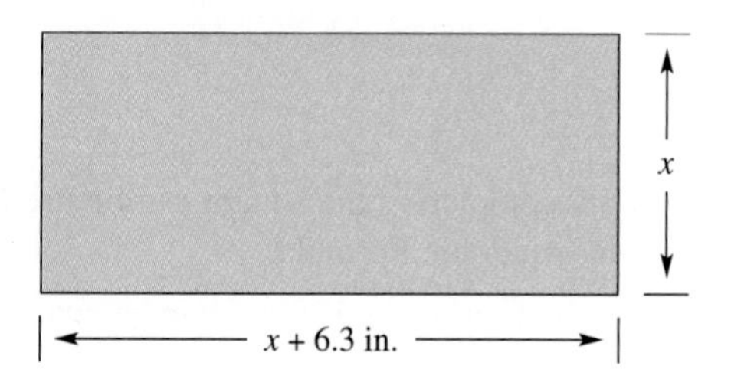

96. *Geometry* A rectangle has a length of $x + 1.5$ inches, a width of x inches, and an area of 18.36 square inches. Find its dimensions.

97. *Free-Falling Object* A ball is thrown vertically upward at a velocity of 40 feet per second from a bridge that is 50 feet above the level of the water (see figure). The height h (in feet) of the ball at time t (in seconds) after it is thrown is

$$h = -16t^2 + 40t + 50.$$

(a) Find the time when the ball is again 50 feet above the water.

(b) Find the time when the ball strikes the water.

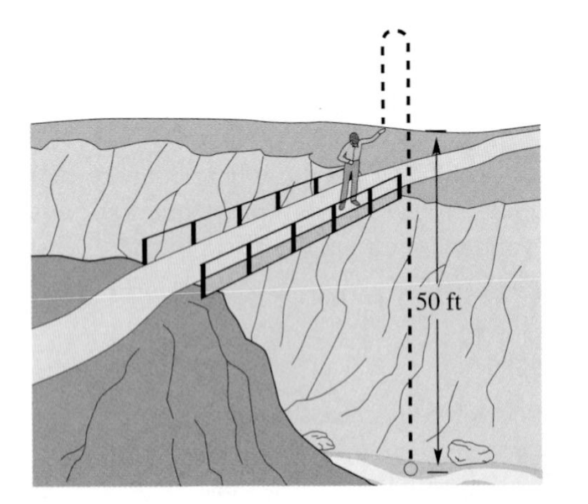

98. *Free-Falling Object* A ball is thrown vertically upward at a velocity of 20 feet per second from a bridge that is 40 feet above the level of the water. The height h (in feet) of the ball at time t (in seconds) after it is thrown is

$$h = -16t^2 + 20t + 40.$$

(a) Find the time when the ball is again 40 feet above the water.

(b) Find the time when the ball strikes the water.

99. *Aerospace Employment* The following model approximates the number of people employed in the aerospace industry in the United States from 1990 through 1996.

$$y = 831.3 - 85.71t + 3.452t^2, \quad 0 \le t \le 6$$

In this model, y represents the number employed in the aerospace industry (in thousands) and t represents the year, with $t = 0$ corresponding to 1990. (Source: U.S. Department of Commerce)

(a) Use a graphing utility to graph the model.

(b) Use the graph in part (a) to find the year in which there were approximately 750,000 employed in the aerospace industry in the United States. Verify your answer algebraically.

(c) Use the model to estimate the number employed in the aerospace industry in 1997.

100. *Cellular Phone Subscribers* The numbers of cellular phone subscribers (in millions) in the United States for the years 1989 through 1996 can be modeled by

$$s = 0.84t^2 + 1.51t + 4.70, \quad -1 \le t \le 6$$

where $t = 0$ corresponds to 1990. (Source: Cellular Telecommunications Industry Association)

(a) Use a graphing utility to graph the model.

(b) Use the model to determine the year in which the cellular phone companies had 10 million subscribers. Verify your answer algebraically.

101. *Exploration* Determine the solutions x_1 and x_2 of each quadratic equation. Use the values of x_1 and x_2 to fill in the boxes.

Equation	x_1, x_2	$x_1 + x_2$	x_1x_2
(a) $x^2 - x - 6 = 0$			
(b) $2x^2 + 5x - 3 = 0$			
(c) $4x^2 - 9 = 0$			
(d) $x^2 - 10x + 34 = 0$			

102. *Think About It* Consider a general quadratic equation $ax^2 + bx + c = 0$ whose solutions are x_1 and x_2. Use the results of Exercise 101 to determine a relationship among the coefficients a, b, and c, and the sum $(x_1 + x_2)$ and product (x_1x_2) of the solutions.

Explaining Concepts

103. Answer parts (c) and (d) of Motivating the Chapter on page 567.

104. State the Quadratic Formula *in words.*

105. What is the discriminant of $ax^2 + bx + c = 0$? How is the discriminant related to the number and type of solutions of the equation?

106. Explain how completing the square can be used to develop the Quadratic Formula.

107. Summarize the four methods for solving a quadratic equation.

Mid-Chapter Quiz

Take this quiz as you would take a quiz in class. After you are done, check your work against the answers given in the back of the book.

In Exercises 1–8, solve the quadratic equation by the specified method.

1. Factoring:

$2x^2 - 72 = 0$

2. Factoring:

$2x^2 + 3x - 20 = 0$

3. Extracting square roots:

$t^2 = 12$

4. Extracting square roots:

$(u - 3)^2 - 16 = 0$

5. Completing the square:

$s^2 + 10s + 1 = 0$

6. Completing the square:

$2y^2 + 6y - 5 = 0$

7. Quadratic Formula:

$x^2 + 4x - 6 = 0$

8. Quadratic Formula:

$6v^2 - 3v - 4 = 0$

In Exercises 9–16, solve the equation by the most convenient method. (Find all the real *and* complex solutions.)

9. $x^2 + 5x + 7 = 0$

10. $36 - (t - 4)^2 = 0$

11. $x(x - 10) + 3(x - 10) = 0$

12. $x(x - 3) = 10$

13. $4b^2 - 12b + 9 = 0$

14. $3m^2 + 10m + 5 = 0$

15. $x - 2\sqrt{x} - 24 = 0$

16. $x^4 + 7x^2 + 12 = 0$

In Exercises 17 and 18, use a graphing utility to graph the function. Use the graph to approximate any *x*-intercepts of the graph. Set $y = 0$ and solve the resulting equation. Write a paragraph comparing the results of your algebraic and graphical solutions.

17. $y = \frac{1}{2}x^2 - 3x - 1$

18. $y = x^2 + 0.45x - 4$

19. The revenue R from selling x units of a certain product is given by

$R = x(20 - 0.2x)$.

Find the number of units that must be sold to produce a revenue of \$500.

20. The perimeter of a rectangle with sides x and $100 - x$ is 200 meters. Its area A is given by $A = x(100 - x)$. Determine the dimensions of the rectangle if its area is 2275 square meters.

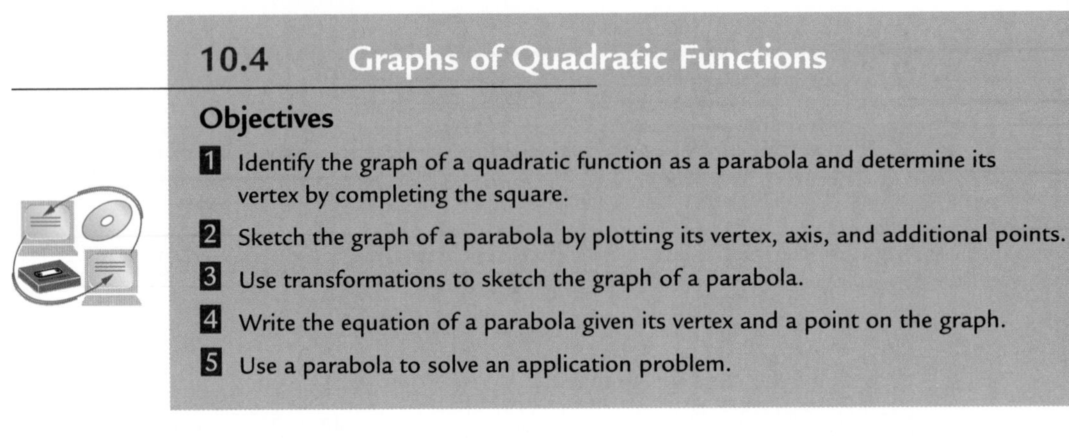

10.4 Graphs of Quadratic Functions

Objectives

1. Identify the graph of a quadratic function as a parabola and determine its vertex by completing the square.
2. Sketch the graph of a parabola by plotting its vertex, axis, and additional points.
3. Use transformations to sketch the graph of a parabola.
4. Write the equation of a parabola given its vertex and a point on the graph.
5. Use a parabola to solve an application problem.

1 Identify the graph of a quadratic function as a parabola and determine its vertex by completing the square.

Graphs of Quadratic Functions

In this section, you will study graphs of quadratic functions.

$$f(x) = ax^2 + bx + c \qquad \text{Quadratic function}$$

Figure 10.3 shows the graph of a simple quadratic function, $f(x) = x^2$.

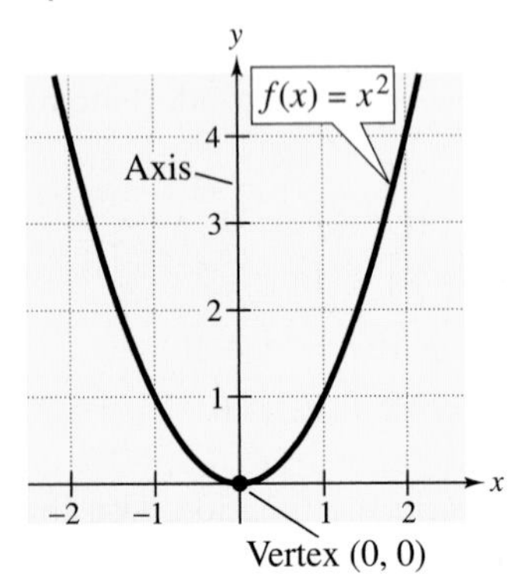

Figure 10.3

> **Graphs of Quadratic Functions**
>
> The graph of $f(x) = ax^2 + bx + c$, $a \neq 0$, is a **parabola.** The completed-square form
>
> $$f(x) = a(x - h)^2 + k \qquad \text{Standard form}$$
>
> is the **standard form** of the function. The **vertex** of the parabola occurs at the point (h, k), and the vertical line passing through the vertex is the **axis** of the parabola.

Every parabola is *symmetric* about its axis, which means that if it were folded along its axis, the two parts would match.

If a is positive, the graph of $f(x) = ax^2 + bx + c$ opens up, and if a is negative, the graph opens down, as shown in Figure 10.4. Observe in Figure 10.4 that the y-coordinate of the vertex identifies the minimum function value if $a > 0$ and the maximum function value if $a < 0$.

Technology: Discovery

You can use a graphing utility to discover a rule for determining the appearance of a parabola. Graph the equations below.

$$y_1 = x^2 - 3x - 5$$
$$y_2 = 7 - 3x^2$$
$$y_3 = -4 + 6x^2$$
$$y_4 = -x^2 - 6x$$

In your own words, write a rule for determining whether the graph of a parabola opens up or down by just looking at the equation. Does $y = 8 - 2x - 2x^2$ open up or down?

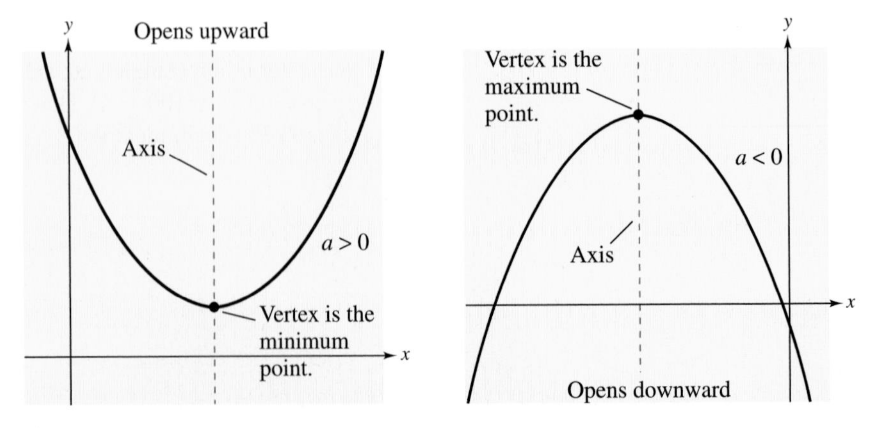

Figure 10.4

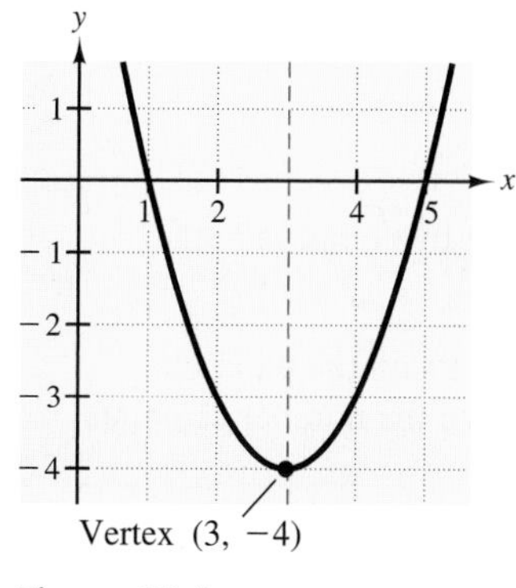

Figure 10.5

Study Tip

When a number is added to a function and then that same number is subtracted from the function, the value of the function remains unchanged. Notice in Example 1 that $(-3)^2$ is added to the function to complete the square and then $(-3)^2$ is subtracted from the function so that the value of the function remains the same.

Example 1 Finding the Vertex by Completing the Square

Find the vertex of the graph of $f(x) = x^2 - 6x + 5$.

Solution

Begin by writing the function in standard form.

$f(x) = x^2 - 6x + 5$	Original function
$f(x) = x^2 - 6x + (-3)^2 - (-3)^2 + 5$	Add and subtract $(-3)^2$.
$f(x) = (x^2 - 6x + 9) - 9 + 5$	Regroup terms.
$f(x) = (x - 3)^2 - 4$	Standard form

From the standard form, you can see that the vertex of the parabola occurs at the point $(3, -4)$, as shown in Figure 10.5. The minimum value of the function is $f(3) = -4$.

In Example 1, the vertex of the graph was found by *completing the square*. Another approach to finding the vertex is to complete the square once for a general function and then use the resulting formula for the vertex.

$f(x) = ax^2 + bx + c$	Quadratic function
$= ax^2 + bx + \frac{b^2}{4a} - \frac{b^2}{4a} + c$	Add and subtract $\frac{b^2}{4a}$.
$= a\left[x^2 + \frac{b}{a}x + \left(\frac{b}{2a}\right)^2\right] + c - \frac{b^2}{4a}$	Group terms.
$= a\left(x + \frac{b}{2a}\right)^2 + c - \frac{b^2}{4a}$	Standard form

From this form you can see that the vertex occurs when $x = -b/2a$.

Example 2 Finding the Vertex with a Formula

Find the vertex of the graph of $f(x) = x^2 + x$.

Solution

From the given function, it follows that $a = 1$ and $b = 1$. So, the x-coordinate of the vertex is

$$x = \frac{-b}{2a} = \frac{-1}{2(1)} = -\frac{1}{2}$$

and the y-coordinate is

$$f\left(-\frac{b}{2a}\right) = f\left(-\frac{1}{2}\right) = \left(-\frac{1}{2}\right)^2 + \left(-\frac{1}{2}\right) = \frac{1}{4} - \frac{1}{2} = -\frac{1}{4}.$$

So, the vertex of the parabola is $\left(-\frac{1}{2}, -\frac{1}{4}\right)$, the minimum value of the function is $f\left(-\frac{1}{2}\right) = -\frac{1}{4}$, and the parabola opens upward, as shown in Figure 10.6.

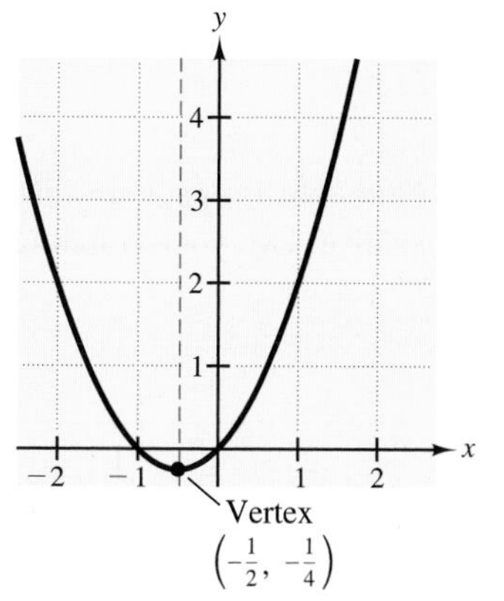

Figure 10.6

2 Sketch the graph of a parabola by plotting its vertex, axis, and additional points.

Sketching a Parabola by Point-Plotting

To obtain an accurate sketch of a parabola, the following guidelines are useful. Remember that the intercepts are convenient points to plot as well.

▶ Sketching a Parabola

1. Determine the vertex and axis of the parabola by completing the square or by formula.
2. Plot the vertex, axis, and a few additional points on the parabola. (Using the symmetry about the axis can reduce the number of points you need to plot.)
3. Use the fact that the parabola opens upward if $a > 0$ and opens downward if $a < 0$ to complete the sketch.

Study Tip

The x- and y-intercepts are useful points to plot. Another convenient fact is that the x-coordinate of the vertex lies halfway between the x-intercepts. Keep this in mind as you study the examples and do the exercises in this section.

Example 3 Sketching a Parabola

Sketch the graph of $x^2 - y + 6x + 8 = 0$.

Solution

Begin by writing the equation in standard form.

$$x^2 - y + 6x + 8 = 0 \qquad \text{Original equation}$$

$$-y = -x^2 - 6x - 8 \qquad \text{Subtract } x^2 + 6x + 8 \text{ from both sides.}$$

$$y = x^2 + 6x + 8 \qquad \text{Multiply both sides by } -1.$$

$$y = (x^2 + 6x + 3^2 - 3^2) + 8 \qquad \text{Add and subtract } 3^2.$$

$$y = (x^2 + 6x + 9) - 9 + 8 \qquad \text{Regroup terms.}$$

$$y = (x + 3)^2 - 1 \qquad \text{Standard form}$$

The vertex occurs at the point $(-3, -1)$ and the axis is given by the line $x = -3$. After plotting this information, calculate a few additional points on the parabola, as shown in the table. Note that the y-intercept is $(0, 8)$ and the x-intercepts are solutions to the equation

$$x^2 + 6x + 8 = (x + 4)(x + 2) = 0.$$

The graph of the parabola is shown in Figure 10.7. Note that it opens upward because the leading coefficient (in standard form) is positive. Use your graphing utility to verify the graph shown in Figure 10.7.

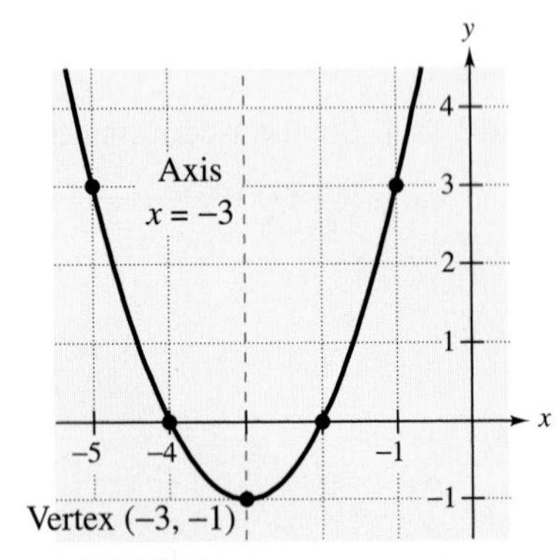

Figure 10.7

x-Value	-5	-4	-3	-2	-1
y-Value	3	0	-1	0	3
Solution point	$(-5, 3)$	$(-4, 0)$	$(-3, -1)$	$(-2, 0)$	$(-1, 3)$

3 Use transformations to sketch the graph of a parabola.

Transformations of Graphs of Functions

Many functions have graphs that are simple **transformations** of a basic graph. The following list summarizes the various types of **horizontal** and **vertical shifts** of the graphs of functions.

Use a graphing utility to display the graphs of $y = x^2 + c$, where c is equal to -2, 0, 2, and 4. What conclusions can you make?

Vertical and Horizontal Shifts

Let c be a positive real number. **Vertical** and **horizontal shifts** of the graph of the function $y = f(x)$ are represented as follows.

1. Vertical shift c units **upward:** $h(x) = f(x) + c$
2. Vertical shift c units **downward:** $h(x) = f(x) - c$
3. Horizontal shift c units to the **right:** $h(x) = f(x - c)$
4. Horizontal shift c units to the **left:** $h(x) = f(x + c)$

The graph of a quadratic function can be viewed as a transformation of the graph $f(x) = x^2$. The transformation can be read easily from the completed-square form of the quadratic function

$$g(x) = (x - h)^2 + k,$$

where the horizontal shift is h units and the vertical shift is k units.

Example 4 Shifts of the Graphs of Functions

Use the graph of $f(x) = x^2$ to sketch the graph of each function.

a. $g(x) = x^2 - 2$ **b.** $h(x) = (x + 3)^2$

Solution

a. Relative to the graph of $f(x) = x^2$, the graph of $g(x) = x^2 - 2$ represents a *downward shift* of two units, as shown in Figure 10.8.

b. Relative to the graph of $f(x) = x^2$, the graph of $h(x) = (x + 3)^2$ represents a *left shift* of three units, as shown in Figure 10.9.

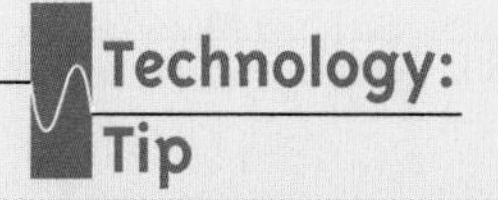

A program called *parabola* can be found at our website *www.hmco.com*. This program will give you practice in working with reflections, horizontal shifts, and vertical shifts for a variety of graphing calculator models. The program will sketch a graph of the function

$$y = R(x + H)^2 + V$$

where R is ± 1, H is an integer between -6 and 6, and V is an integer between -3 and 3. After you determine the values for R, H, and V, the program will list the answers.

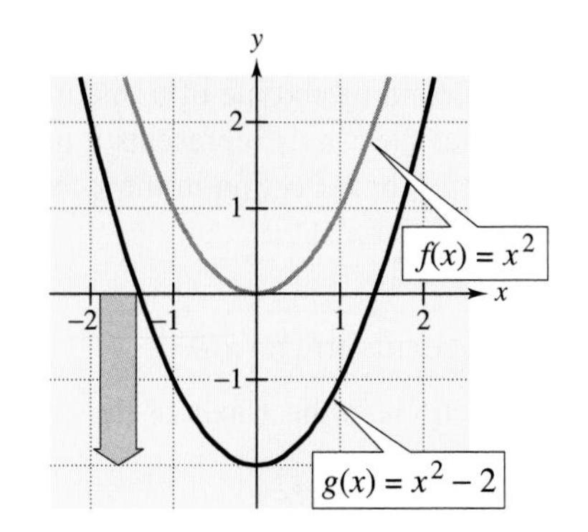

Vertical Shift: Two Units Down
Figure 10.8

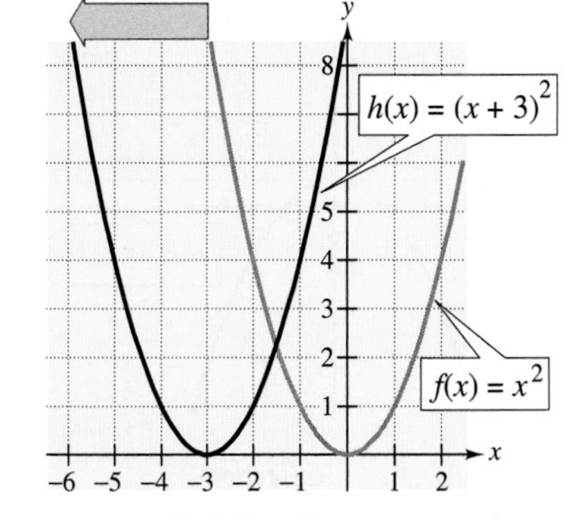

Horizontal Shift: Three Units Left
Figure 10.9

Some graphs can be obtained from combinations of vertical and horizontal shifts, as shown in the next example. The order in which the shifts are carried out does not matter. Try to visualize this in the graphs shown in the next example.

Technology: Discovery

Use a graphing utility to display the graph of $y = (x - h)^2 + 2$, where $h = -2, 0, 2$, and 4. Explain the transformations.

Use a graphing utility to display the graph of $y = ax^2$, where $a = -1$ and 1. Explain the transformation.

Example 5 Shifts of the Graphs of Functions

Use the graph of $f(x) = x^2$ to sketch the graph of each function.

a. $g(x) = (x + 3)^2 - 1$

b. $h(x) = (x - 1)^2 + 2$

Solution

a. Relative to the graph of $f(x) = x^2$, the graph of $g(x) = (x + 3)^2 - 1$ represents a *left shift* of three units, followed by a *downward shift* of one unit, as shown in Figure 10.10.

b. Relative to the graph of $f(x) = x^2$, the graph of $h(x) = (x - 1)^2 + 2$ represents a *right shift* of one unit, followed by an *upward shift* of two units, as shown in Figure 10.11.

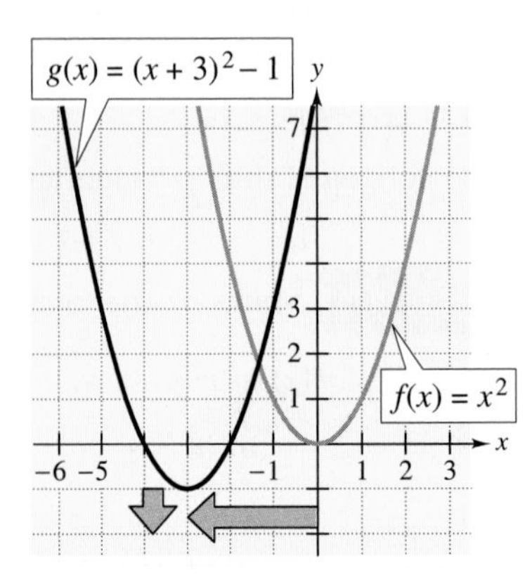

Horizontal Shift: Three Units Left
Vertical Shift: One Unit Down
Figure 10.10

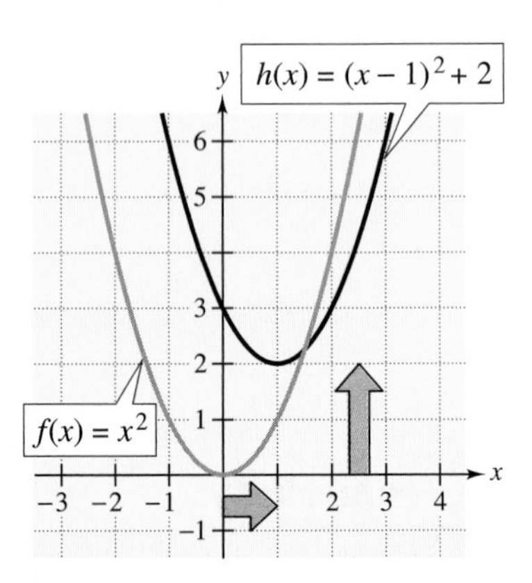

Horizontal Shift: One Unit Right
Vertical Shift: Two Units Up
Figure 10.11

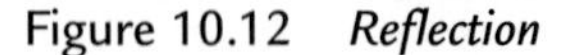

The second basic type of transformation is a **reflection.** For instance, if you imagine that the x-axis represents a mirror, then the graph of $h(x) = -x^2$ is the mirror image (or reflection in the x-axis) of the graph of $f(x) = x^2$, as shown in Figure 10.12.

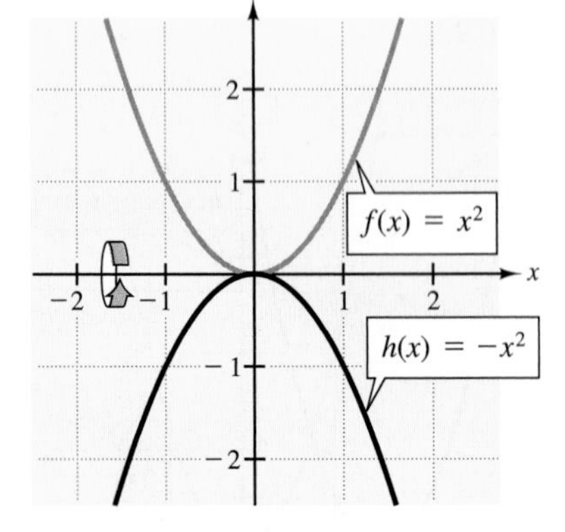

Figure 10.12 *Reflection*

▶ Reflections in the x-Axis

Reflections in the x-axis of the graph of $y = f(x)$ are represented by

$h(x) = -f(x)$. Reflection in the x-axis

4 Write the equation of a parabola given its vertex and a point on the graph.

Writing the Equation of a Parabola

To write the equation of a parabola with a vertical axis, use the fact that its standard equation has the form $y = a(x - h)^2 + k$, where (h, k) is the vertex.

Example 6 Writing the Equation of a Parabola

Write the equation of the parabola whose vertex is $(-2, 1)$ and whose y-intercept is $(0, -3)$, as shown in Figure 10.13.

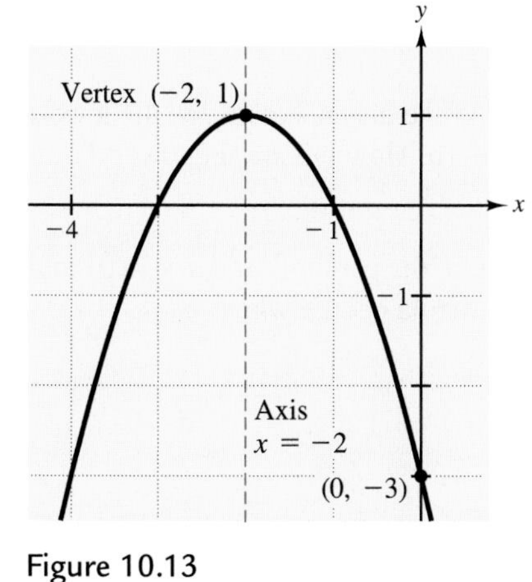

Figure 10.13

Solution

Because the vertex occurs at $(h, k) = (-2, 1)$, you can write the following.

$y = a(x - h)^2 + k$ — Standard form

$y = a[x - (-2)]^2 + 1$ — Substitute -2 for h and 1 for k.

$y = a(x + 2)^2 + 1$ — Simplify.

To find the value of a, use the fact that the y-intercept is $(0, -3)$.

$y = a(x + 2)^2 + 1$ — Standard form

$-3 = a(0 + 2)^2 + 1$ — Substitute 0 for x and -3 for y.

$-1 = a$ — Simplify.

This implies that the standard form of the equation of the parabola is

$$y = -(x + 2)^2 + 1.$$

Example 7 Writing the Equation of a Parabola

Write the equation of the parabola that has a vertex of $(3, -4)$ and contains the point $(5, -2)$, as shown in Figure 10.14.

Solution

Because the vertex occurs at $(h, k) = (3, -4)$, you can write the following.

$y = a(x - h)^2 + k$ — Standard form

$y = a(x - 3)^2 + (-4)$ — Substitute 3 for h and -4 for k.

$y = a(x - 3)^2 - 4$ — Simplify.

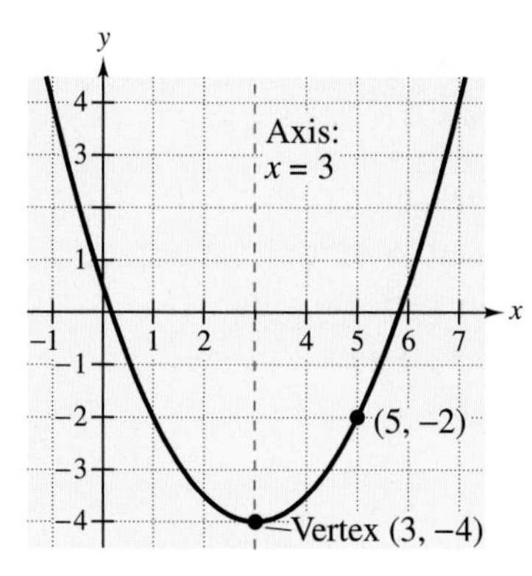

Figure 10.14

To find the value of a, use the fact that $(5, -2)$ is a point on the parabola.

$y = a(x - 3)^2 - 4$ — Standard form

$-2 = a(5 - 3)^2 - 4$ — Substitute 5 for x and -2 for y.

$\frac{1}{2} = a$ — Simplify.

This implies that the standard form of the equation is

$$y = \frac{1}{2}(x - 3)^2 - 4.$$

5 Use a parabola to solve an application problem.

Application

Example 8 An Application Involving a Minimum Point

A suspension bridge is 100 feet long, as shown in Figure 10.15(a). The bridge is supported by cables attached at the tops of the towers at each end of the bridge. Each cable hangs in the shape of a parabola (see Figure 10.15(b)) given by

$$y = 0.01x^2 - x + 35$$

where x and y are both measured in feet. (a) Find the distance between the lowest point of the cable and the roadbed of the bridge. (b) How tall are the towers?

100 ft

(a)

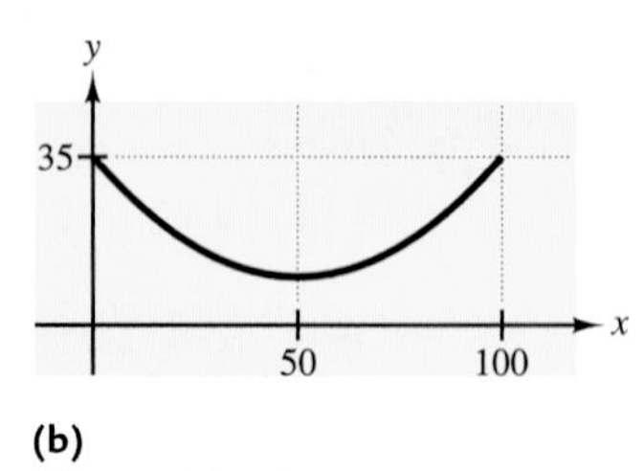

(b)

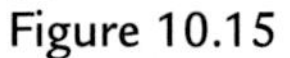

Figure 10.15

Solution

a. Because the lowest point occurs at the vertex of the parabola and $a = 0.01$ and $b = -1$, it follows that the vertex of the parabola occurs when

$$x = \frac{-b}{2a} = \frac{1}{0.02} = 50.$$

At this x-value, the value of y is

$$y = 0.01(50)^2 - 50 + 35 = 10.$$

Thus, the minimum distance between the cable and the roadbed is 10 feet.

b. Because the vertex of the parabola occurs at the midpoint of the bridge, the two towers are located at the points where $x = 0$ and $x = 100$. Substituting an x-value of 0, you can find that the corresponding y-value is

$$y = 0.01(0)^2 - 0 + 35 = 35.$$

So, the towers are each 35 feet high. (Try substituting $x = 100$ in the equation to see that you obtain the same y-value.)

Discussing the Concept Quadratic Modeling

The data in the table represent the average monthly temperature y in degrees Fahrenheit in Savannah, Georgia for the month x, with $x = 1$ corresponding to November. (Source: National Climate Data Center) Plot the data. Find a quadratic model for the data and use it to find the average temperatures for December and February. The actual average temperature for both December and February is 52°F. How well do you think the model fits the data? Use the model to predict the average temperature for June. How useful do you think the model would be for the whole year?

x	1	3	5
y	59	49	59

10.4 Exercises

Integrated Review — *Concepts, Skills, and Problem Solving*

Keep mathematically in shape by doing these exercises *before* the problems of this section.

Properties and Definitions

1. Fill in the blanks: $(x + b)^2 = x^2 + \square x + \square$.

2. Fill in the blank so that the expression is a perfect square trinomial. Explain how the constant is determined.

$x^2 + 5x + \square$

Simplifying Expressions

In Exercises 3–10, simplify the expression.

3. $(4x + 3y) - 3(5x + y)$

4. $(-15u + 4v) + 5(3u - 9v)$

5. $2x^2 + (2x - 3)^2 + 12x$

6. $y^2 - (y + 2)^2 + 4y$

7. $\sqrt{24x^2y^3}$

8. $\sqrt[3]{9} \cdot \sqrt[3]{15}$

9. $(12a^{-4}b^6)^{1/2}$

10. $(16^{1/3})^{3/4}$

Problem Solving

In Exercises 11 and 12, find the time required for an object to reach the ground when it is dropped from a height of s_0 feet. The height h (in feet) is given by $h = -16t^2 + s_0$, where t measures time in seconds from when the object is released.

11. $s_0 = 80$

12. $s_0 = 150$

Developing Skills

In Exercises 1–6, match the equation with its graph. [The graphs are labeled (a), (b), (c), (d), (e), and (f).]

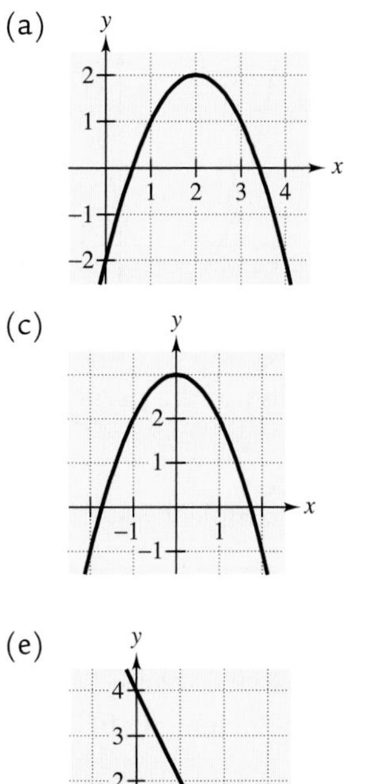

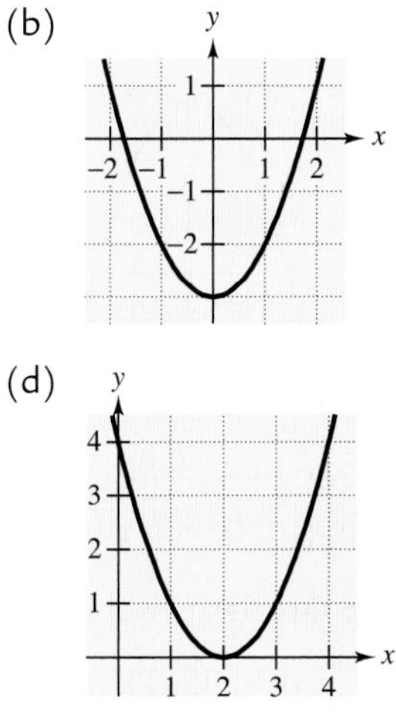

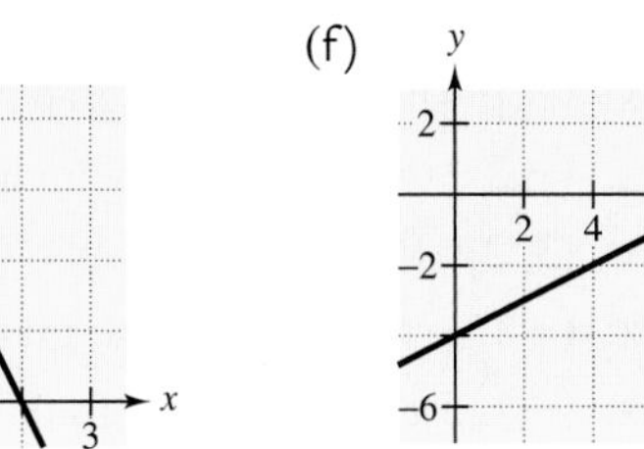

1. $y = 4 - 2x$

2. $y = \frac{1}{2}x - 4$

3. $y = x^2 - 3$

4. $y = -x^2 + 3$

5. $y = (x - 2)^2$

6. $y = 2 - (x - 2)^2$

In Exercises 7–18, write the equation in standard form and find the vertex of its graph. See Example 1.

7. $y = x^2 + 2$

8. $y = x^2 + 2x$

9. $y = x^2 - 4x + 7$

10. $y = x^2 + 6x - 5$

11. $y = x^2 + 6x + 5$

12. $y = x^2 - 4x + 5$

13. $y = -x^2 + 6x - 10$

14. $y = 4 - 8x - x^2$

15. $y = -x^2 + 2x - 7$

16. $y = -x^2 - 10x + 10$

17. $y = 2x^2 + 6x + 2$

18. $y = 3x^2 - 3x - 9$

In Exercises 19–24, find the vertex of the graph of the function by formula. See Example 2.

19. $f(x) = x^2 - 8x + 15$

20. $f(x) = x^2 + 4x + 1$

21. $g(x) = -x^2 - 2x + 1$

22. $h(x) = -x^2 + 14x - 14$

23. $y = 4x^2 + 4x + 4$

24. $y = 9x^2 - 12x$

In Exercises 25–32, state whether the graph opens upward or downward and find the vertex.

25. $y = 2(x - 0)^2 + 2$
26. $y = -3(x + 5)^2 - 3$
27. $y = 4 - (x - 10)^2$
28. $y = 2(x - 12)^2 + 3$
29. $y = x^2 - 6$
30. $y = -(x + 1)^2$
31. $y = -(x - 3)^2$
32. $y = x^2 - 6x$

In Exercises 33–40, find the x- and y-intercepts of the graph.

33. $y = 25 - x^2$
34. $y = x^2 - 49$
35. $y = x^2 - 9x$
36. $y = x^2 + 4x$
37. $y = 4x^2 - 12x + 9$
38. $y = 10 - x - 2x^2$
39. $y = x^2 - 3x + 3$
40. $y = x^2 - 3x - 10$

In Exercises 41–64, sketch the graph of the function. Identify the vertex and any x-intercepts. Use a graphing utility to verify your results. See Example 3.

41. $g(x) = x^2 - 4$
42. $h(x) = x^2 - 9$
43. $f(x) = -x^2 + 4$
44. $f(x) = -x^2 + 9$
45. $f(x) = x^2 - 3x$
46. $g(x) = x^2 - 4x$
47. $y = -x^2 + 3x$
48. $y = -x^2 + 4x$
49. $y = (x - 4)^2$
50. $y = -(x + 4)^2$
51. $y = x^2 - 8x + 15$
52. $y = x^2 + 4x + 2$
53. $y = -(x^2 + 6x + 5)$
54. $y = -x^2 + 2x + 8$
55. $q(x) = -x^2 + 6x - 7$
56. $g(x) = x^2 + 4x + 7$
57. $y = 2(x^2 + 6x + 8)$
58. $y = 3x^2 - 6x + 4$
59. $y = \frac{1}{2}(x^2 - 2x - 3)$
60. $y = -\frac{1}{2}(x^2 - 6x + 7)$
61. $y = \frac{1}{5}(3x^2 - 24x + 38)$
62. $y = \frac{1}{5}(2x^2 - 4x + 7)$
63. $f(x) = 5 - \frac{1}{3}x^2$
64. $f(x) = \frac{1}{3}x^2 - 2$

In Exercises 65–72, identify the transformation of the graph of $f(x) = x^2$ and sketch a graph of h. See Examples 4 and 5.

65. $h(x) = x^2 + 2$
66. $h(x) = x^2 - 4$
67. $h(x) = (x + 2)^2$
68. $h(x) = (x - 4)^2$
69. $h(x) = (x - 1)^2 + 3$
70. $h(x) = (x + 2)^2 - 1$
71. $h(x) = (x + 3)^2 + 1$
72. $h(x) = (x - 3)^2 - 2$

In Exercises 73–76, use a graphing utility to approximate the vertex of the graph. Check the result algebraically.

73. $y = \frac{1}{6}(2x^2 - 8x + 11)$
74. $y = -\frac{1}{4}(4x^2 - 20x + 13)$
75. $y = -0.7x^2 - 2.7x + 2.3$
76. $y = 0.75x^2 - 7.50x + 23.00$

In Exercises 77–82, write an equation of the parabola. See Example 6.

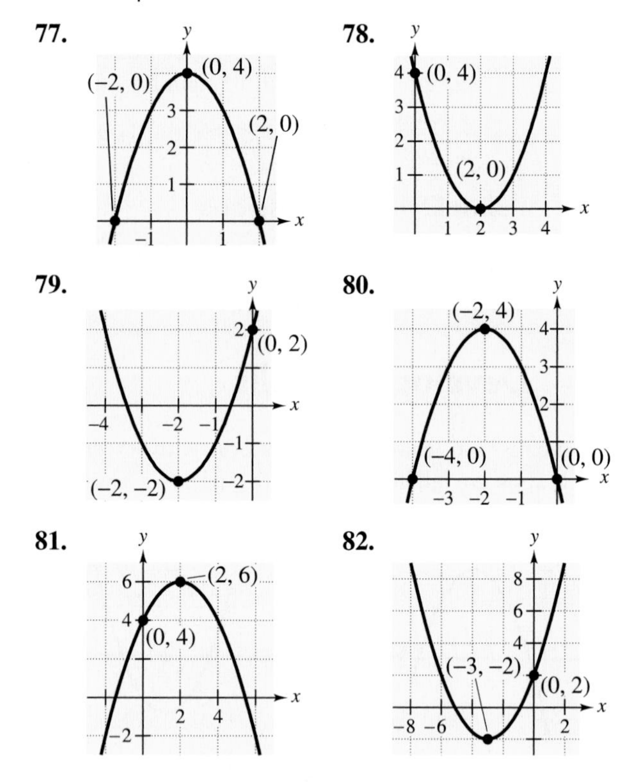

In Exercises 83–90, write an equation of the parabola

$y = a(x - h)^2 + k$

that satisfies the conditions. See Example 7.

83. Vertex: $(2, 1)$; $a = 1$
84. Vertex: $(-3, -3)$; $a = 1$
85. Vertex: $(2, -4)$; Point on the graph: $(0, 0)$
86. Vertex: $(-2, -4)$; Point on the graph: $(0, 0)$
87. Vertex: $(3, 2)$; Point on the graph: $(1, 4)$
88. Vertex: $(-1, -1)$; Point on the graph: $(0, 4)$
89. Vertex: $(-1, 5)$; Point on the graph: $(0, 1)$
90. Vertex: $(5, 2)$; Point on the graph: $(10, 3)$

In Exercises 91–94, identify the transformation of $y = x^2$ that will produce the given graph.

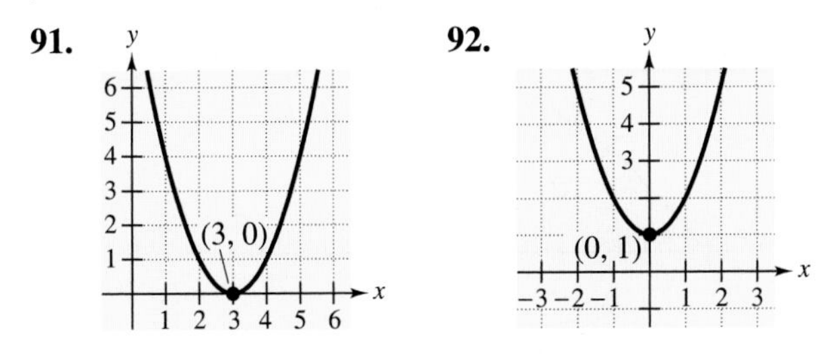

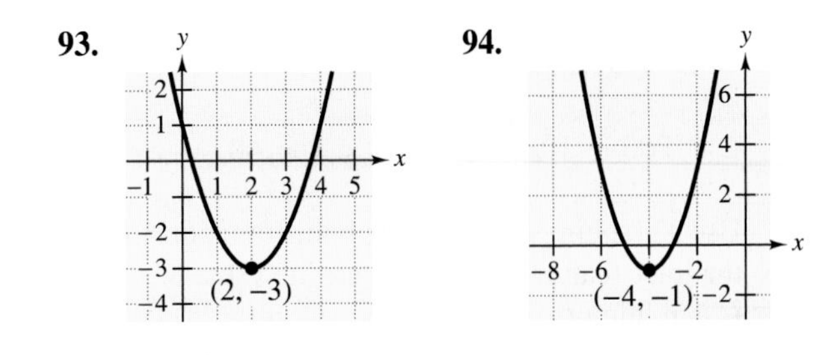

Solving Problems

95. *Path of a Ball* The height y (in feet) of a ball thrown by a child is given by

$$y = -\frac{1}{12}x^2 + 2x + 4$$

where x is the horizontal distance (in feet) from where the ball was thrown.

(a) How high was the ball when it left the child's hand?

(b) How high was the ball when it reached its maximum height?

(c) How far from the child did the ball strike the ground?

96. *Path of a Ball* Repeat Exercise 95 if the path of the ball is modeled by $y = -\frac{1}{16}x^2 + 2x + 5$.

97. *Maximum Height of a Diver* The path of a diver is given by $y = -\frac{4}{9}x^2 + \frac{24}{9}x + 10$ where y is the height in feet and x is the horizontal distance from the end of the diving board in feet. What is the maximum height of the diver?

98. *Maximum Height of a Diver* Repeat Exercise 97 if the path of the diver is modeled by

$$y = -\frac{4}{3}x^2 + \frac{10}{3}x + 10.$$

99. *Graphical Estimation* The number N (in thousands) of personnel in the Marine Corps reserves in the United States for the years 1990 through 1996 is approximated by the model

$$N = 83.64 + 14.89t - 2.04t^2, \quad 0 \le t \le 6.$$

In this model, t is time in years, with $t = 0$ corresponding to 1990. (Source: U.S. Department of Defense)

(a) Use a graphing utility to graph the function.

(b) Determine the year when the number in the Marine Corps reserves was greatest. Approximate the number that year.

100. *Graphical Estimation* The profit (in thousands of dollars) for a company is given by

$$P = 230 + 20s - \frac{1}{2}s^2$$

where s is the amount (in hundreds of dollars) spent on advertising. Use a graphing utility to graph the profit function and approximate the amount of advertising that yields a maximum profit. Verify the maximum profit algebraically.

101. *Graphical Interpretation* A company manufactures radios that cost the company \$60 each. For buyers who purchase 100 or fewer radios, the purchase price is \$90 per radio. To encourage large orders, the company will reduce the price *per radio* for orders over 100, as follows. If 101 radios are purchased, the price is \$89.85 per unit. If 102 radios are purchased, the price is \$89.70 per unit. If $(100 + x)$ radios are purchased, the price per unit is

$$p = 90 - x(0.15)$$

where x is the amount over 100 in the order.

(a) Show algebraically that the profit P for x orders over 100 is

$$P = (100 + x)[90 - x(0.15)] - (100 + x)60$$
$$= 3000 + 15x - \frac{3}{20}x^2.$$

(b) Find the vertex of the profit curve and determine the order size for maximum profit.

(c) Would you recommend this pricing scheme? Explain.

102. *Graphical Estimation* The cost of producing x units of a product is given by

$$C = 800 - 10x + \frac{1}{4}x^2, \quad 0 < x < 40.$$

Use a graphing utility to graph this function and use the trace feature to approximate the value of x when C is minimum.

103. *Geometry* The area of a rectangle is given by the function

$$A = \frac{2}{\pi}(100x - x^2), \quad 0 < x < 100$$

where x is the length of the base of the rectangle in feet. Use a graphing utility to graph the function and use the trace feature to approximate the value of x when A is maximum.

104. *Bridge Design* A bridge is to be constructed over a gorge with the main supporting arch being a parabola (see figure). The equation of the parabola is

$$y = 4\left(100 - \frac{x^2}{2500}\right)$$

where x and y are measured in feet.

(a) Find the length of the road across the gorge.

(b) Find the height of the parabolic arch at the center of the span.

(c) Find the lengths of the vertical girders at intervals of 100 feet from the center of the bridge.

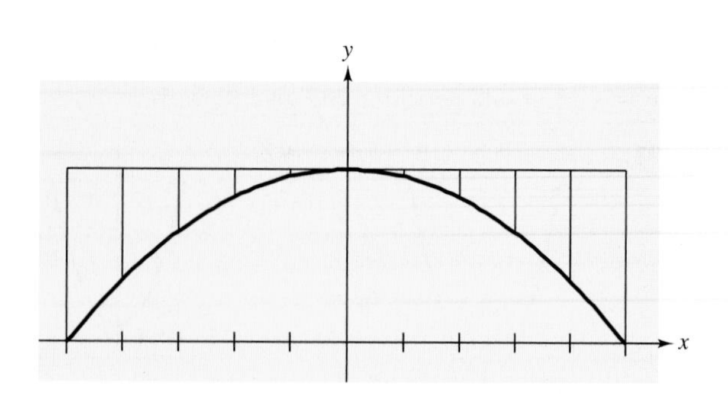

Figure for 104

105. *Highway Design* A highway department engineer must design a parabolic arc to create a turn in a freeway around a city. The vertex of the parabola is placed at the origin, and the parabola must connect with roads represented by the equations

$$y = -0.4x - 100, \quad x < -500$$

and

$$y = 0.4x - 100, \quad x > 500$$

(see figure). Find an equation for the parabolic arc.

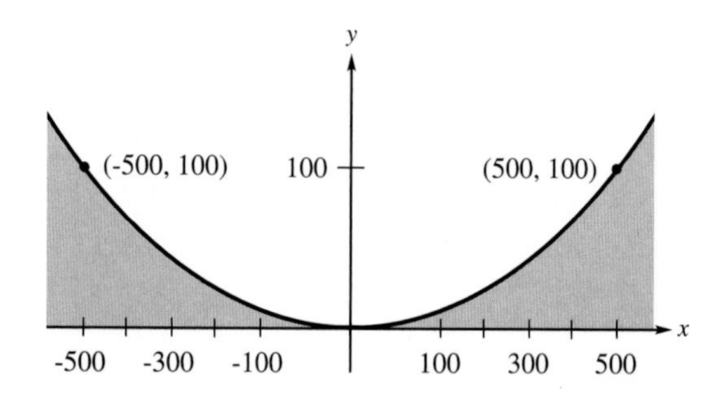

Explaining Concepts

106. Answer parts (e) and (f) of Motivating the Chapter on page 567.

107. In your own words, describe the graph of the quadratic function $f(x) = ax^2 + bx + c$.

108. Explain how to find the vertex of the graph of a quadratic function.

109. Explain how to find any x- or y-intercepts of the graph of a quadratic function.

110. Explain how to determine whether the graph of a quadratic function opens up or down.

111. How is the discriminant related to the graph of a quadratic function?

112. Is it possible for the graph of a quadratic function to have two y-intercepts? Explain.

113. Explain how to determine the maximum (or minimum) value of a quadratic function.

10.5 Applications of Quadratic Equations

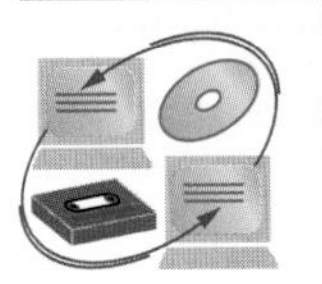

Objectives

1 Use a quadratic equation to solve an application problem.

1 Use a quadratic equation to solve an application problem.

Applications of Quadratic Equations

Example 1 An Investment Problem

A car dealer bought a fleet of cars from a car rental agency for a total of \$120,000. By the time the dealer had sold all but four of the cars, at an average profit of \$2500 each, the original investment of \$120,000 had been regained. How many cars did the dealer sell, and what was the average price per car?

Solution

Although this problem is stated in terms of average price and average profit per car, we can use a model that assumes that each car sold for the same price.

Verbal Model: Selling price per car = Cost per car + Profit per car

Labels:
Number of cars sold $= x$ (cars)
Number of cars bought $= x + 4$ (cars)
Selling price per car $= \dfrac{120{,}000}{x}$ (dollars per car)
Cost per car $= \dfrac{120{,}000}{x + 4}$ (dollars per car)
Profit per car $= 2500$ (dollars per car)

Equation:

$$\frac{120{,}000}{x} = \frac{120{,}000}{x + 4} + 2500$$

$$120{,}000(x + 4) = 120{,}000x + 2500x(x + 4), \quad x \neq 0, x \neq -4$$

$$120{,}000x + 480{,}000 = 120{,}000x + 2500x^2 + 10{,}000x$$

$$0 = 2500x^2 + 10{,}000x - 480{,}000$$

$$0 = x^2 + 4x - 192$$

$$0 = (x - 12)(x + 16)$$

$$x - 12 = 0 \Rightarrow x = 12$$

$$x + 16 = 0 \Rightarrow x = -16$$

Choosing the positive value, it follows that the dealer sold 12 cars at an average price of $\frac{1}{12}(120{,}000) = \$10{,}000$ per car. Check this result in the original statement of the problem.

Example 2 Geometry

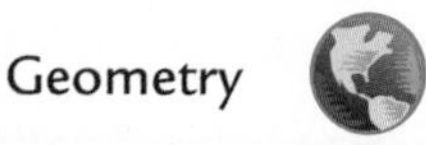

A picture is 6 inches taller than it is wide and has an area of 216 square inches. What are the dimensions of the picture?

Solution

Begin by drawing a diagram, as shown in Figure 10.16.

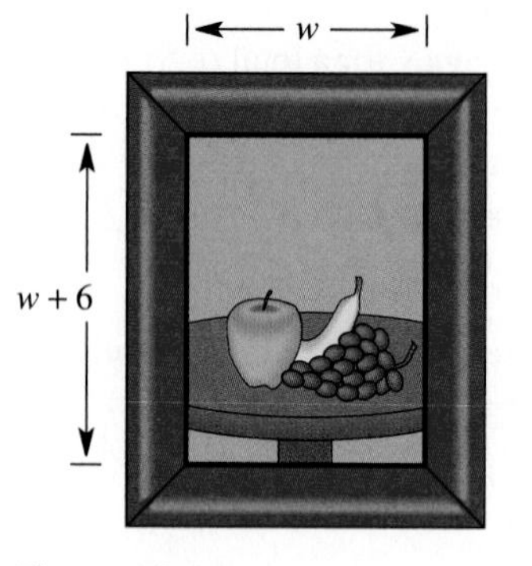

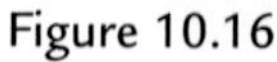
Figure 10.16

Verbal Model: Area of picture = Width · Height

Labels: Picture width $= w$ (inches)
Picture height $= w + 6$ (inches)
Area $= 216$ (square inches)

Equation:

$$216 = w(w + 6)$$
$$0 = w^2 + 6w - 216$$
$$0 = (w + 18)(w - 12)$$
$$w + 18 = 0 \Rightarrow w = -18$$
$$w - 12 = 0 \Rightarrow w = 12$$

Of the two possible solutions, choose the positive value of w and conclude that the picture is $w = 12$ inches wide and $w + 6$ or 18 inches tall. Check these dimensions in the original statement of the problem.

Example 3 An Interest Problem

The formula

$$A = P(1 + r)^2$$

represents the amount of money A in an account in which P dollars is deposited for 2 years at an annual interest rate of r (in decimal form). Find the interest rate if a deposit of \$6000 increases to \$6933.75 over a 2-year period.

Solution

$A = P(1 + r)^2$	Given formula
$6933.75 = 6000(1 + r)^2$	Substitute for A and P.
$1.155625 = (1 + r)^2$	Divide both sides by 6000.
$\pm 1.075 = 1 + r$	Extract square roots.
$0.075 = r$	Choose positive solution.

The annual interest rate is $r = 0.075 = 7.5\%$.

Check

$A = P(1 + r)^2$	Given formula
$6933.75 \stackrel{?}{=} 6000(1 + 0.075)^2$	Substitute 6933.75 for A, 6000 for P, and 0.075 for r.
$6933.75 \stackrel{?}{=} 6000(1.155625)$	Simplify.
$6933.75 = 6933.75$	Solution checks. ✓

Example 4 Reduced Rates

A ski club chartered a bus for a ski trip at a cost of \$720. In an attempt to lower the bus fare per skier, the club invited nonmembers to go along. When four nonmembers joined the trip, the fare per skier decreased by \$6.00. How many club members are going on the trip?

Solution

Verbal Model: [Cost per skier] · [Number of skiers] = [\$720]

Labels:
Number of ski club members $= x$ (people)
Number of skiers $= x + 4$ (people)
Original cost per skier $= \dfrac{720}{x}$ (dollars)
New cost per skier $= \dfrac{720}{x} - 6.00$ (dollars)

Equation:

$$\left(\frac{720}{x} - 6.00\right)(x + 4) = 720$$

$$\left(\frac{720 - 6x}{x}\right)(x + 4) = 720$$

$$(720 - 6x)(x + 4) = 720x, \quad x \neq 0$$

$$720x - 6x^2 - 24x + 2880 = 720x$$

$$-6x^2 - 24x + 2880 = 0$$

$$x^2 + 4x - 480 = 0$$

$$(x + 24)(x - 20) = 0$$

$$x + 24 = 0 \Rightarrow x = -24$$

$$x - 20 = 0 \Rightarrow x = 20$$

Choosing the positive value of x implies that there are 20 ski club members. Check this solution in the original equation, as follows.

Check

Number of Skiers	*Cost per Skier*
20	$\dfrac{720}{20} = \$36.00$
24	$\dfrac{720}{24} = \$30.00$

From these two calculations, you can see that the difference in cost per skier is $\$36.00 - \$30.00 = \$6.00$.

Example 5 An Application Involving the Pythagorean Theorem

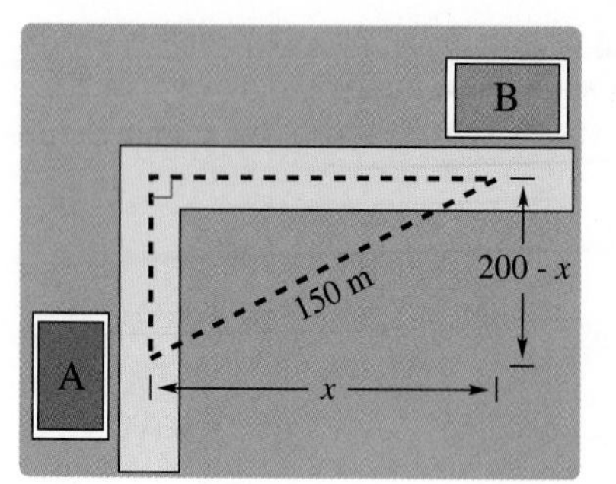

Figure 10.17

An L-shaped sidewalk from building A to building B on a college campus is 200 meters long, as shown in Figure 10.17. By cutting diagonally across the grass, students shorten the walking distance to 150 meters. What are the lengths of the two legs of the sidewalk?

Solution

Verbal Model: $a^2 + b^2 = c^2$ Pythagorean Theorem

Labels: Length of one leg $= x$ (meters)
Length of other leg $= 200 - x$ (meters)
Length of diagonal $= 150$ (meters)

Equation:

$$x^2 + (200 - x)^2 = (150)^2$$
$$2x^2 - 400x + 40{,}000 = 22{,}500$$
$$2x^2 - 400x + 17{,}500 = 0$$
$$x^2 - 200x + 8750 = 0$$

By the Quadratic Formula, you can find the solutions as follows.

$$x = \frac{200 \pm \sqrt{(-200)^2 - 4(1)(8750)}}{2(1)}$$
$$= \frac{200 \pm \sqrt{5000}}{2}$$
$$= \frac{200 \pm 50\sqrt{2}}{2}$$
$$= \frac{2(100 \pm 25\sqrt{2})}{2}$$
$$= 100 \pm 25\sqrt{2}$$

Both solutions are positive, and it does not matter which one you choose. If you let

$$x = 100 + 25\sqrt{2} \approx 135.4 \text{ meters,}$$

the length of the other leg is

$$200 - x \approx 200 - 135.4$$
$$\approx 64.6 \text{ meters.}$$

In Example 5, notice that you obtain the same dimensions if you choose the other value of x. That is, if the length of one leg is

$$x = 100 - 25\sqrt{2} \approx 64.6 \text{ meters,}$$

the length of the other leg is

$$200 - x \approx 200 - 64.6$$
$$\approx 135.4 \text{ meters.}$$

Example 6 Work-Rate Problem

An office contains two copy machines. Machine B is known to take 12 minutes longer than Machine A to copy the company's monthly report. Using both machines together, it takes 8 minutes to reproduce the report. How long would it take each machine alone to reproduce the report?

Solution

Verbal Model: [Work done by machine A] + [Work done by machine B] = [1 complete job]

[Rate for A] · [Time for both] + [Rate for B] · [Time for both] = [1]

Labels: Time for machine A $= t$ (minutes)

Rate for machine A $= \frac{1}{t}$ (job per minute)

Time for machine B $= t + 12$ (minutes)

Rate for machine B $= \frac{1}{t + 12}$ (job per minute)

Time for both machines $= 8$ (minutes)

Rate for both machines $= \frac{1}{8}$ (job per minute)

Equation:

$$\frac{1}{t}(8) + \frac{1}{t + 12}(8) = 1$$

$$8\left(\frac{1}{t} + \frac{1}{t + 12}\right) = 1$$

$$8\left[\frac{t + 12 + t}{t(t + 12)}\right] = 1$$

$$8t(t + 12)\left[\frac{2t + 12}{t(t + 12)}\right] = t(t + 12)$$

$$8(2t + 12) = t^2 + 12t$$

$$16t + 96 = t^2 + 12t$$

$$0 = t^2 - 4t - 96$$

$$0 = (t - 12)(t + 8)$$

$$t - 12 = 0 \quad \Rightarrow \quad t = 12$$

$$t + 8 = 0 \quad \Rightarrow \quad t = -8$$

Choose the positive value for t and find that

Time for machine A $= t = 12$ minutes

Time for machine B $= t + 12 = 24$ minutes.

Check these solutions in the original equation.

Example 7 The Height of a Model Rocket

A model rocket is projected straight upward from ground level according to the height equation $h = -16t^2 + 192t$, $t \geq 0$, where h is the height in feet and t is the time in seconds. (a) After how many seconds will the height be 432 feet? (b) When will the rocket hit the ground?

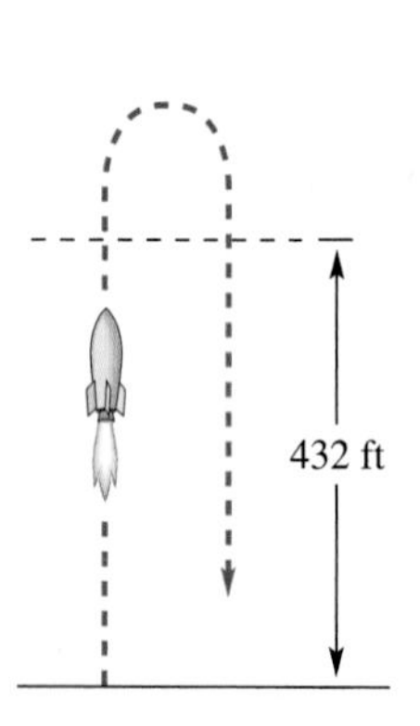

Figure 10.18

Solution

a.

$$h = -16t^2 + 192t \quad \text{Original equation}$$

$$432 = -16t^2 + 192t \quad \text{Substitute 432 for } h.$$

$$16t^2 - 192t + 432 = 0 \quad \text{Standard form}$$

$$t^2 - 12t + 27 = 0 \quad \text{Divide both sides by 16.}$$

$$(t - 3)(t - 9) = 0 \quad \text{Factor.}$$

$$t - 3 = 0 \Rightarrow t = 3 \quad \text{Set 1st factor equal to 0.}$$

$$t - 9 = 0 \Rightarrow t = 9 \quad \text{Set 2nd factor equal to 0.}$$

The rocket attains a height of 432 feet at two different times—once (going up) after 3 seconds, and again (coming down) after 9 seconds. (See Figure 10.18.)

b. To find the time it takes for the rocket to hit the ground, let the height be 0.

$$0 = -16t^2 + 192t$$

$$0 = t^2 - 12t$$

$$0 = t(t - 12)$$

$$t = 0 \quad \text{or} \quad t = 12$$

The rocket will hit the ground after 12 seconds. (Note that the time of $t = 0$ seconds corresponds to the time of lift-off.)

Discussing the Concept Analyzing Quadratic Functions

Use a graphing utility to graph $y_1 = 3x^2 + 2x - 1$ and $y_2 = -x^2 + 5x + 4$. For each function, use the zoom and trace features to find either the maximum or minimum function value. Discuss other methods that you could use to find these values.

10.5 Exercises

Integrated Review *Concepts, Skills, and Problem Solving*

Keep mathematically in shape by doing these exercises *before* the problems of this section.

Properties and Definitions

1. Define the slope of the line through the points (x_1, y_1) and (x_2, y_2).

2. Give the following forms of an equation of a line.

(a) Slope-intercept form

(b) Point-slope form

(c) General form

(d) Horizontal line

Equations of Lines

In Exercises 3–10, find the general form of the equation of the line through the two points.

3. $(0, 0), (4, -2)$ **4.** $(0, 0), (100, 75)$

5. $(-1, -2), (3, 6)$ **6.** $(1, 5), (6, 0)$

7. $\left(\frac{3}{2}, 8\right), \left(\frac{11}{2}, \frac{5}{2}\right)$ **8.** $(0, 2), (7.3, 15.4)$

9. $(0, 8), (5, 8)$ **10.** $(-3, 2), (-3, 5)$

Problem Solving

11. A group of people agree to share equally in the cost of a $250,000 endowment to a college. If they could find two more people to join the group, each person's share of the cost would decrease by $6250. How many people are presently in the group?

12. A boat travels at a speed of 18 miles per hour in still water. It travels 35 miles upstream and then returns to the starting point in a total of 4 hours. Find the speed of the current.

Solving Problems

1. *Selling Price* A store owner bought a case of eggs for $21.60. By the time all but 6 dozen of the eggs had been sold at a profit of $0.30 per dozen, the original investment of $21.60 had been regained. How many dozen eggs did the owner sell, and what was the selling price per dozen?

2. *Selling Price* A manager of a computer store bought several computers of the same model for $27,000. When all but three of the computers had been sold at a profit of $750 per computer, the original investment of $27,000 had been regained. How many computers were sold, and what was the selling price of each?

3. *Selling Price* A storeowner bought a case of video games for $480. By the time he had sold all but eight of them at a profit of $10 each, the original investment of $480 had been regained. How many video games were sold, and what was the selling price of each game?

4. *Selling Price* A math club bought a case of sweatshirts for $850 to sell as a fundraiser. By the time all but 16 sweatshirts had been sold at a profit of $8 per sweatshirt, the original investment of $850 had been regained. How many sweatshirts were sold, and what was the selling price of each sweatshirt?

Dimensions of a Rectangle In Exercises 5–14, complete the table of widths, lengths, perimeters, and areas of rectangles.

	Width	Length	Perimeter	Area
5.	$0.75l$	l	42 in.	
6.	w	$1.5w$	40 m	
7.	w	$2.5w$		250 ft^2
8.	w	$1.5w$		216 cm^2
9.	$\frac{1}{3}l$	l		192 in.^{2}
10.	$\frac{3}{4}l$	l		2700 in.^{2}
11.	w	$w + 3$	54 km	
12.	$l - 6$	l	108 ft	
13.	$l - 20$	l		12,000 m^2
14.	w	$w + 5$		500 ft^2

15. *Geometry* A picture frame is 4 inches taller than it is wide and has an area of 192 square inches. What are the dimensions of the picture frame?

16. *Geometry* The top of a coffee table is 3 feet longer than it is wide and has an area of 10 square feet. What are the dimensions of the top of the coffee table?

17. *Geometry* The height of a triangle is 8 inches less than its base. The area of the triangle is 192 square inches. Find the dimensions of the triangle.

18. *Geometry* The height of a triangle is 25 inches greater than its base. The area of the triangle is 625 square inches. Find the dimensions of the triangle.

19. *Lumber Storage Area* A retail lumberyard plans to store lumber in a rectangular region adjoining the sales office (see figure). The region will be fenced on three sides and the fourth side will be bounded by the wall of the office building. Find the dimensions of the region if 350 feet of fencing is available and the area of the region is 12,500 square feet.

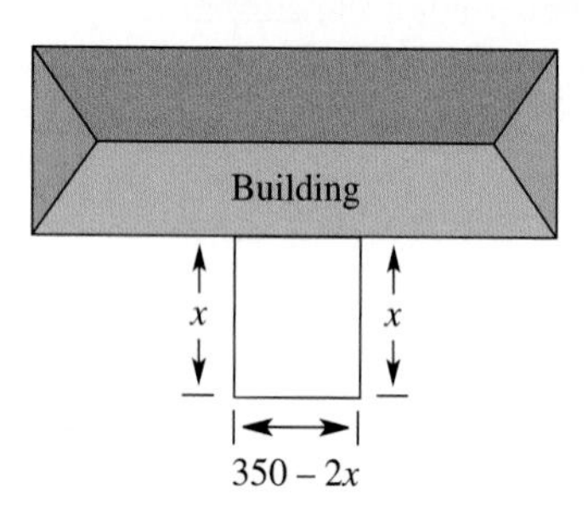

20. *Fencing the Yard* You have 100 feet of fencing. Do you have enough to enclose a rectangular region whose area is 630 square feet? Is there enough to enclose a circular area of 630 square feet? Explain.

21. *Fencing the Yard* A family built a fence around three sides of their property (see figure). In total, they used 550 feet of fencing. By their calculations, the lot is 1 acre (43,560 square feet). Is this correct? Explain your answer.

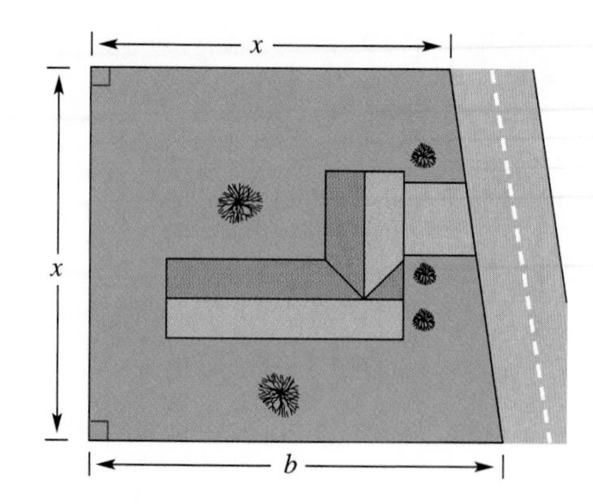

Figure for 21

22. *Geometry* Your home is on a square lot. To add more space to your yard, you purchase an additional 20 feet along the side of the property (see figure). The area of the lot is now 25,500 square feet. What are the dimensions of the new lot?

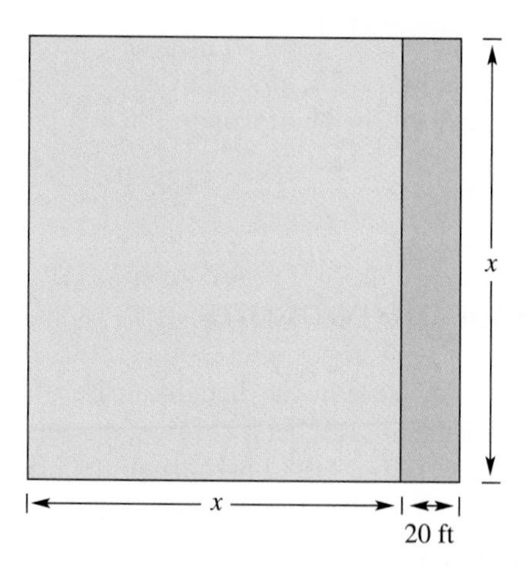

23. *Open Conduit* An open-topped rectangular conduit for carrying water in a manufacturing process is made by folding up the edges of a sheet of aluminum 48 inches wide (see figure). A cross section of the conduit must have an area of 288 square inches. Find the width and height of the conduit.

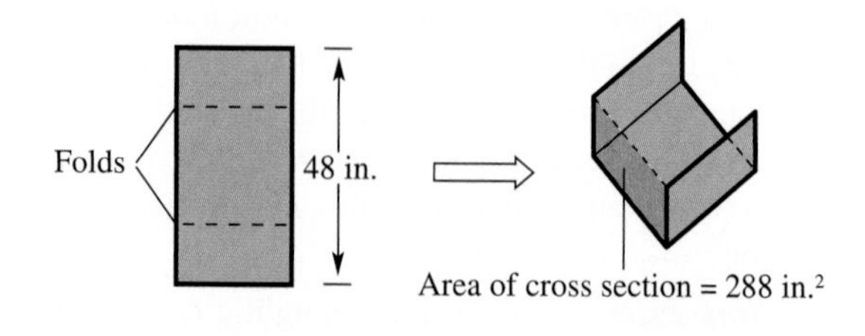

Compound Interest In Exercises 24–29, find the interest rate r. Use the formula $A = P(1 + r)^2$, where A is the amount after 2 years in an account earning r percent (in decimal form) compounded annually, and P is the original investment. See Example 3.

24. $P = \$10{,}000$ $A = \$11{,}990.25$

25. $P = \$3000$ $A = \$3499.20$

26. $P = \$500$ $A = \$572.45$

27. $P = \$250$ $A = \$280.90$

28. $P = \$6500$ $A = \$7370.46$

29. $P = \$8000$ $A = \$8420.20$

30. *Reduced Ticket Price* A service organization paid \$210 for a block of tickets to a ball game. The block contained three more tickets than the organization needed for its members. By inviting three more people to attend (and share in the cost), the organization lowered the price per ticket by \$3.50. How many people are going to the game?

31. *Reduced Ticket Price* A service organization buys a block of tickets for a ball game for \$240. After eight more people decide to go to the game, the price per ticket is decreased by \$1. How many people are going to the game?

32. *Reduced Fare* A science club charters a bus to attend a science fair at a cost of \$480. In an attempt to lower the bus fare per person, the club invites nonmembers to go along. When two nonmembers join the trip, the fare per person is decreased by \$1. How many people are going on the excursion?

33. *Venture Capital* Eighty thousand dollars is needed to begin a small business. The cost will be divided equally among the investors. Some have made a commitment to invest. If three more investors are found, the amount required from each would decrease by \$6000. How many have made a commitment to invest in the business?

34. *Dimensions of a Rectangle* The perimeter of a rectangle is 102 inches and the length of the diagonal is 39 inches. Find the dimensions of the rectangle.

35. *Delivery Route* You are asked to deliver pizza to offices B and C in your city (see figure), and you are required to keep a log of all the mileages between stops. You forget to look at the odometer at stop B, but after getting to stop C you record the total distance traveled from the pizza shop as 18 miles. The return distance from C to A is 16 miles. If the route approximates a right triangle, estimate the distance from A to B.

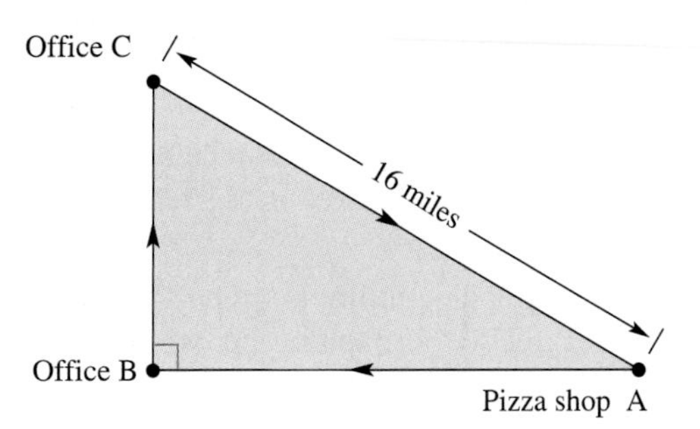

Figure for 35

36. *Shortcut* An L-shaped sidewalk from building A to building B on a high school campus is 100 yards long, as shown in the figure. By cutting diagonally across the grass, students shorten the walking distance to 80 yards. What are the lengths of the two legs of the sidewalk?

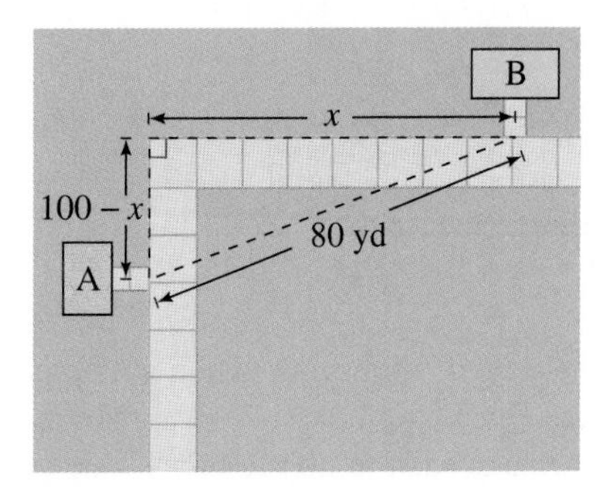

37. *Dimensions of a Rectangle* An adjustable rectangular form has minimum dimensions of 3 meters by 4 meters. The length and width can be expanded by equal amounts x (see figure).

(a) Write the length d of the diagonal as a function of x. Use a graphing utility to graph the function. Use the graph to approximate the value of x when $d = 10$ meters.

(b) Find x algebraically when $d = 10$.

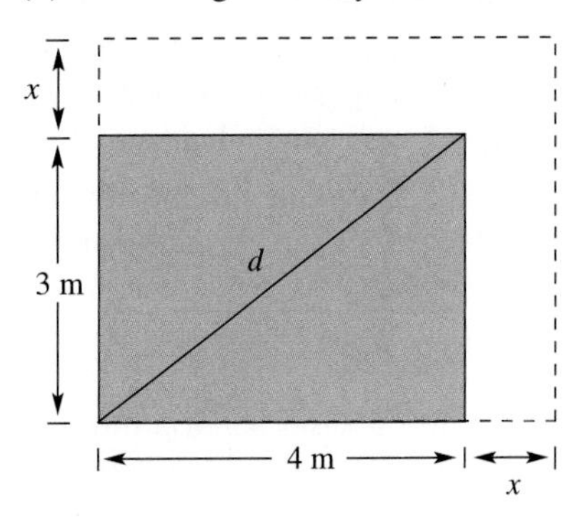

38. *Solving Graphically and Numerically* A meteorologist is positioned 100 feet from the point where a weather balloon is launched (see figure). The instrument package lifted vertically by the balloon transmits data to the meteorologist.

(a) Write the distance d between the balloon and the meteorologist as a function of the height h of the balloon.

(b) Use a graphing utility to graph the function in part (a). Use the graph to approximate the value of h when $d = 200$ feet.

(c) Complete the following table.

h	0	100	200	300
d				

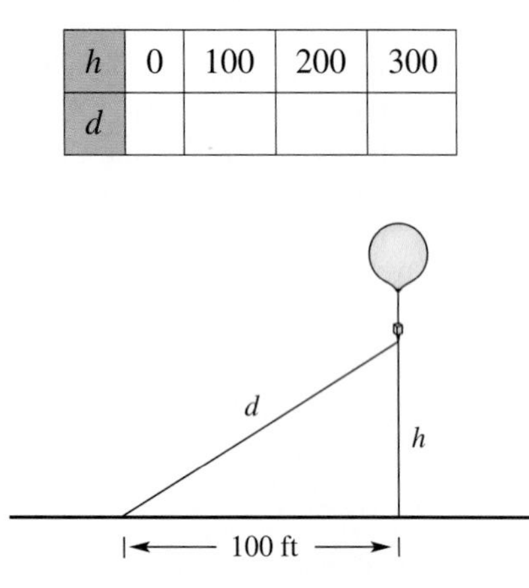

39. *Work-Rate Problem* Working together, two people can complete a task in 5 hours. Working alone, how long would it take each to do the task if one person took 2 hours longer than the other?

40. *Work-Rate Problem* An office contains two printers. Machine B is known to take 3 minutes longer than Machine A to produce the company's monthly financial report. Using both machines together, it takes 6 minutes to produce the report. How long would it take each machine to produce the report?

41. *Work-Rate Problem* A builder works with two plumbing companies. Company A is known to take 3 days longer than Company B to do the plumbing in a particular style of house. Using both companies, it takes 4 days. How long would it take to do the plumbing using each company individually?

42. *Work-Rate Problem* Working together, two people can complete a task in 6 hours. Working alone, one person takes 2 hours longer than the other. How long would it take each to do the task alone?

Free-Falling Object In Exercises 43–46, find the time necessary for an object to fall to ground level from an initial height of h_0 feet if its height h at any time t (in seconds) is given by

$$h = h_0 - 16t^2.$$

43. $h_0 = 144$

44. $h_0 = 625$

45. $h_0 = 1454$ (height of the Sears Tower)

46. $h_0 = 984$ (height of the Eiffel Tower)

47. *Height of a Baseball* The height h in feet of a baseball hit 3 feet above the ground is given by

$$h = 3 + 75t - 16t^2$$

where t is time in seconds. Find the time when the ball hits the ground in the outfield.

48. *Hitting Baseballs* You are hitting baseballs. When tossing the ball into the air, your hand is 5 feet above the ground (see figure). You hit the ball when it falls back to a height of 4 feet. If you toss the ball with an initial velocity of 25 feet per second, the height h of the ball t seconds after leaving your hand is given by

$$h = 5 + 25t - 16t^2.$$

How much time will pass before you hit the ball?

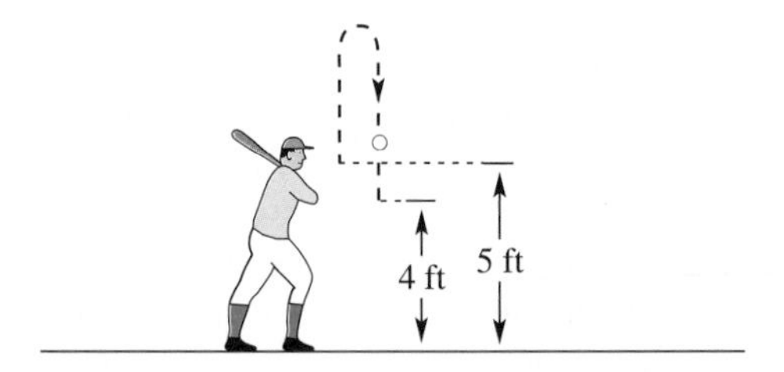

49. *Height of a Model Rocket* A model rocket is projected straight upward from ground level according to the height equation $h = -16t^2 + 160t$, where h is the height of the rocket in feet and t is the time in seconds.

(a) After how many seconds will the height be 336 feet?

(b) When will the rocket hit the ground?

50. *Height of a Tennis Ball* A tennis ball is tossed vertically upward from a height of 5 feet according to the height equation $h = -16t^2 + 21t + 5$, where h is the height of the tennis ball in feet and t is the time in seconds.

(a) After how many seconds will the height be 11 feet?

(b) When will the tennis ball hit the ground?

Number Problems In Exercises 51–56, find two positive integers that satisfy the given requirement.

51. The product of two consecutive integers is 240.

52. The product of two consecutive integers is 1122.

53. The product of two consecutive even integers is 224.

54. The product of two consecutive even integers is 528.

55. The product of two consecutive odd integers is 483.

56. The product of two consecutive odd integers is 255.

57. *Air Speed* An airline runs a commuter flight between two cities that are 720 miles apart. If the average speed of the planes could be increased by 40 miles per hour, the travel time would be decreased by 12 minutes. What air speed is required to obtain this decrease in travel time?

58. *Average Speed* A truck traveled the first 100 miles of a trip at one speed and the last 135 miles at an average speed of 5 miles per hour less. If the entire trip took 5 hours, what was the average speed for the first part of the trip?

59. *Speed* A small business uses a minivan to make deliveries. The cost per hour for fuel for the van is $C = v^2/600$, where v is the speed in miles per hour. The driver is paid \$5 per hour. Find the speed if the cost for wages and fuel for a 110-mile trip is \$20.39.

60. *Distance* Find any points on the line $y = 14$ that are 13 units from the point (1, 2).

61. *Geometry* The area of an ellipse is given by $A = \pi ab$ (see figure). For a certain ellipse, it is required that $a + b = 20$.

(a) Show that $A = \pi a(20 - a)$.

(b) Complete the table.

a	4	7	10	13	16
A					

(c) Find two values of a such that $A = 300$.

(d) Use a graphing utility to graph the area function.

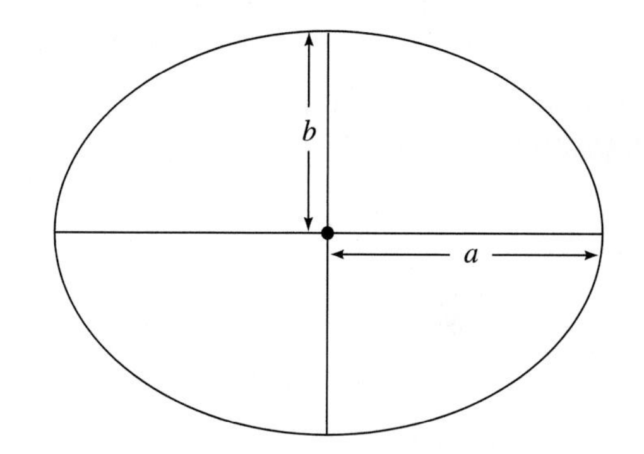

Figure for 61

62. *Data Analysis* For the years 1990 through 1996, the sales s (in millions of dollars) of snowmobiles in the United States can be approximated by the model

$$s = 12.88t^2 + 43.86t + 300.83, \; 0 \le t \le 6$$

where t is time in years, with $t = 0$ corresponding to 1990. (Source: National Sporting Goods Association)

(a) Use a graphing utility to graph the model over the specified domain.

(b) In which year were sales approximately \$400 million?

Explaining Concepts

63. In your own words, describe guidelines for solving word problems.

64. Describe the strategies that can be used to solve a quadratic equation.

65. *Unit Analysis* Describe the units of the product.

$$\frac{9 \text{ dollars}}{\text{hour}} \cdot (20 \text{ hours})$$

66. *Unit Analysis* Describe the units of the product.

$$\frac{20 \text{ feet}}{\text{minute}} \cdot \frac{1 \text{ minute}}{60 \text{ seconds}} \cdot (45 \text{ seconds})$$

67. Give an example of a quadratic equation that has only one repeated solution.

68. Give an example of a quadratic equation that has two imaginary solutions.

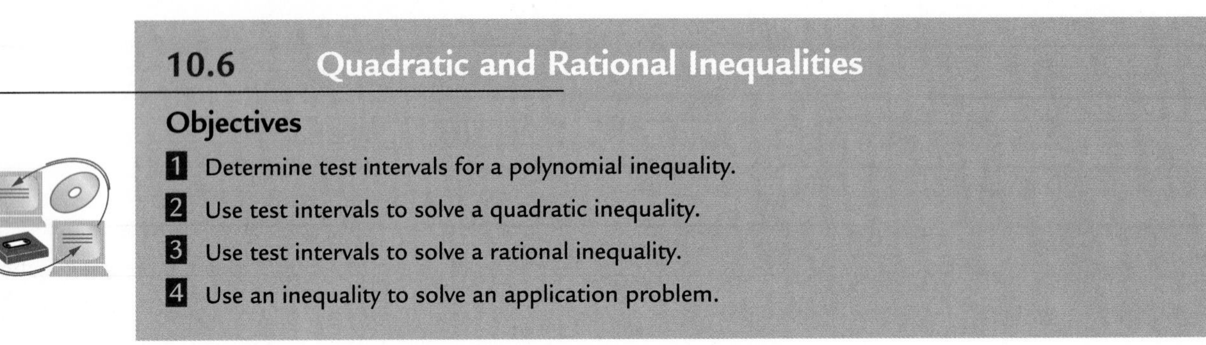

10.6 Quadratic and Rational Inequalities

Objectives

1 Determine test intervals for a polynomial inequality.

2 Use test intervals to solve a quadratic inequality.

3 Use test intervals to solve a rational inequality.

4 Use an inequality to solve an application problem.

1 Determine test intervals for a polynomial inequality.

Finding Test Intervals

When working with polynomial inequalities, it is important to realize that the value of a polynomial can change sign only at its **zeros.** That is, a polynomial can change signs only at the x-values that make the value of the polynomial zero. For instance, the first-degree polynomial $x + 2$ has a zero at -2, and it changes sign at that zero. You can picture this result on the real number line, as shown in Figure 10.19.

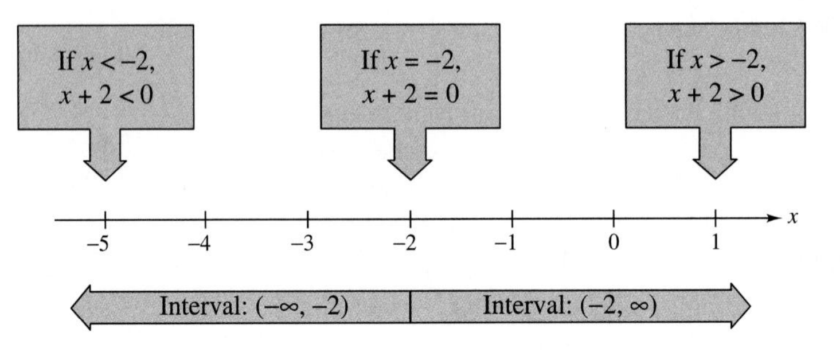

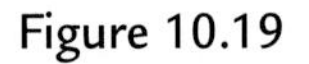
Figure 10.19

Note in Figure 10.19 that the zero of the polynomial partitions the real number line into two **test intervals.** The value of the polynomial is negative for every x-value in the first test interval $(-\infty, -2)$, and it is positive for every x-value in the second test interval $(-2, \infty)$. You can use the same basic approach to determine the test intervals for any polynomial.

▶ Finding Test Intervals for a Polynomial

1. Find all real zeros of the polynomial, and arrange the zeros in increasing order. The zeros of a polynomial are called its **critical numbers.**

2. Use the critical numbers of the polynomial to determine its test intervals.

3. Choose a representative x-value in each test interval and evaluate the polynomial at that value. If the value of the polynomial is negative, the polynomial will have negative values for *every* x-value in the interval. If the value of the polynomial is positive, the polynomial will have positive values for *every* x-value in the interval.

2 Use test intervals to solve a quadratic inequality.

Quadratic Inequalities

The concepts of critical numbers and test intervals can be used to solve nonlinear inequalities, as demonstrated in Examples 1, 2, and 4.

Example 1 Solving a Quadratic Inequality

Solve the inequality $x^2 - 5x < 0$.

Solution

First find the *critical numbers* for $x^2 - 5x < 0$ by finding the solutions of the equation $x^2 - 5x = 0$.

$$x^2 - 5x = 0 \qquad \text{Corresponding equation}$$

$$x(x - 5) = 0 \qquad \text{Factor.}$$

$$x = 0,\ x = 5 \qquad \text{Critical numbers}$$

This implies that the test intervals are

$$(-\infty, 0), \quad (0, 5), \quad \text{and} \quad (5, \infty).$$

To test an interval, choose a convenient number in the interval and compute the sign of $x^2 - 5x$.

Interval	*x-Value*	*Polynomial Value*	*Conclusion*
$(-\infty, 0)$	$x = -1$	$(-1)^2 - 5(-1) = 6$	Positive
$(0, 5)$	$x = 1$	$(1)^2 - 5(1) = -4$	Negative
$(5, \infty)$	$x = 6$	$(6)^2 - 5(6) = 6$	Positive

From this you can conclude that the value of the polynomial is positive for all x-values in $(-\infty, 0)$ and $(5, \infty)$, and negative for all x-values in $(0, 5)$. This implies that the solution of the inequality $x^2 - 5x < 0$ is the interval $(0, 5)$, as shown in Figure 10.20.

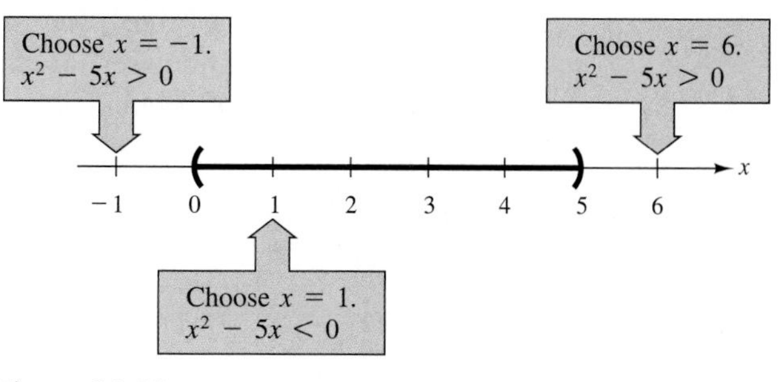

Figure 10.20

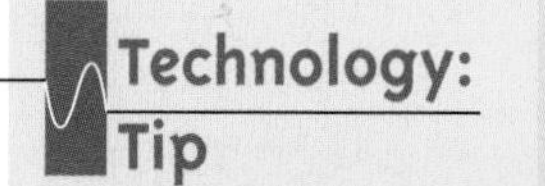

Most graphing utilities can sketch the graph of the solution set of a quadratic inequality. Consult the user's guide for your graphing utility. Notice that the solution set for the quadratic inequality

$$x^2 - 5x < 0$$

shown below appears as a horizontal line above the x-axis.

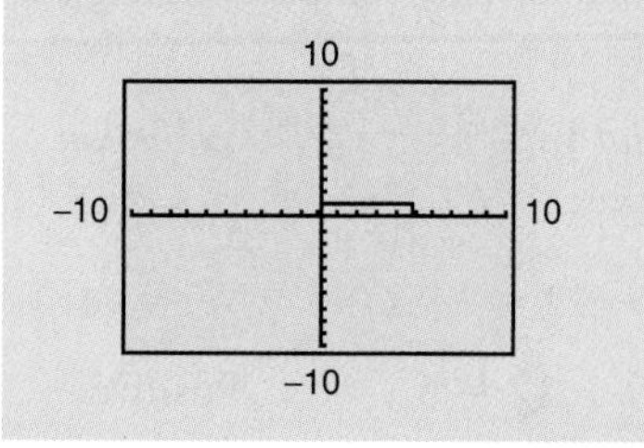

In Example 1, note that you would have used the same basic procedure if the inequality symbol had been $\leq$, $>$, or $\geq$. For instance, from Figure 10.20, you can see that the solution set of the inequality $x^2 - 5x \geq 0$ consists of the union of the half-open intervals $(-\infty, 0]$ and $[5, \infty)$, which is written as $(-\infty, 0] \cup [5, \infty)$.

Just as in solving quadratic *equations*, the first step in solving a quadratic *inequality* is to write the inequality in **general form,** with the polynomial on the left and zero on the right. Factorization of the polynomial then shows the critical numbers, as demonstrated in Example 2.

Study Tip

In Examples 1 and 2, the critical numbers are found by factoring. With quadratic polynomials that do not factor, you can use the Quadratic Formula to find the critical numbers. For instance, to solve the inequality

$$x^2 - 2x - 1 \le 0$$

you can use the Quadratic Formula to determine that the critical numbers are

$$1 - \sqrt{2} \approx -0.414$$

and

$$1 + \sqrt{2} \approx 2.414.$$

Example 2 Solving a Quadratic Inequality

Solve the inequality $2x^2 + 5x \ge 12$.

Solution

Begin by writing the inequality in general form.

$$2x^2 + 5x \ge 12 \qquad \text{Original inequality}$$

$$2x^2 + 5x - 12 \ge 0 \qquad \text{Write in general form.}$$

Next, find the critical numbers for $2x^2 + 5x - 12 \ge 0$ by finding the solutions to the equation $2x^2 + 5x - 12 = 0$.

$$2x^2 + 5x - 12 = 0 \qquad \text{Original equation}$$

$$(x + 4)(2x - 3) = 0 \qquad \text{Factor.}$$

$$x = -4, x = \frac{3}{2} \qquad \text{Critical numbers}$$

This implies that the test intervals are $(-\infty, -4)$, $\left(-4, \frac{3}{2}\right)$, and $\left(\frac{3}{2}, \infty\right)$. To test an interval, choose a convenient number in the interval and compute the sign of $2x^2 + 5x - 12$.

Interval	*x-Value*	*Polynomial Value*	*Conclusion*
$(-\infty, -4)$	$x = -5$	$2(-5)^2 + 5(-5) - 12 = 13$	Positive
$\left(-4, \frac{3}{2}\right)$	$x = 0$	$2(0)^2 + 5(0) - 12 = -12$	Negative
$\left(\frac{3}{2}, \infty\right)$	$x = 2$	$2(2)^2 + 5(2) - 12 = 6$	Positive

From this you can conclude that the value of the polynomial is positive for all x-values in $(-\infty, -4)$ and $\left(\frac{3}{2}, \infty\right)$, and negative for all x-values in $\left(-4, \frac{3}{2}\right)$. This implies that the solution set of the inequality $2x^2 + 5x - 12 \ge 0$ is $(-\infty, -4] \cup \left[\frac{3}{2}, \infty\right)$, as shown in Figure 10.21.

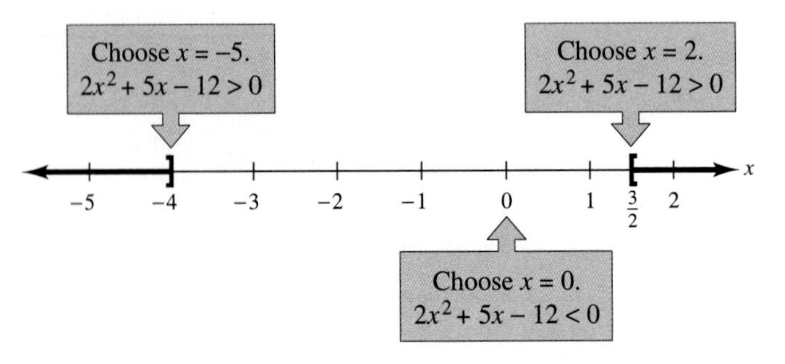

Figure 10.21

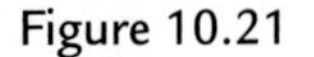

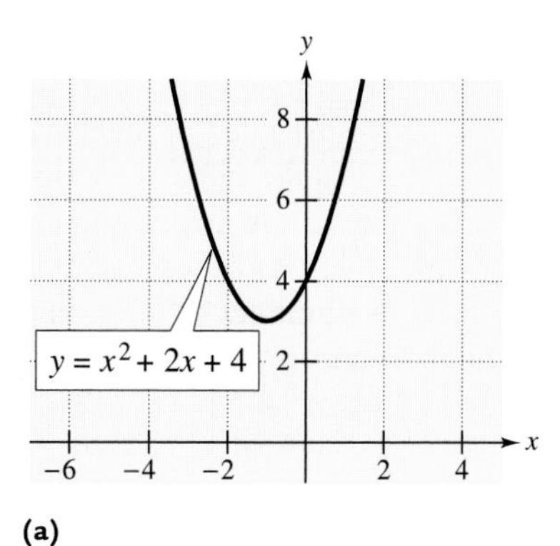

(a)

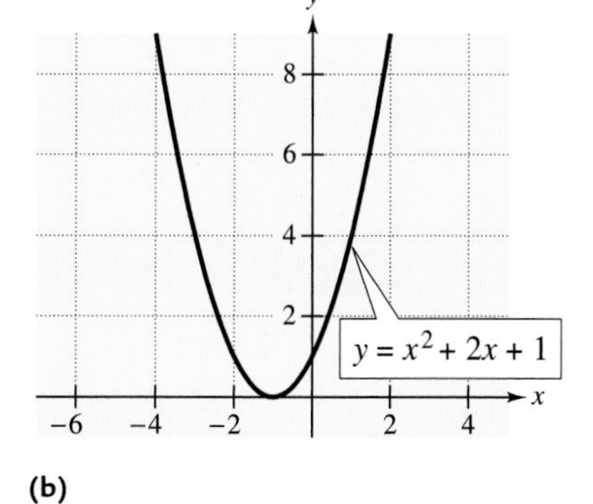

(b)

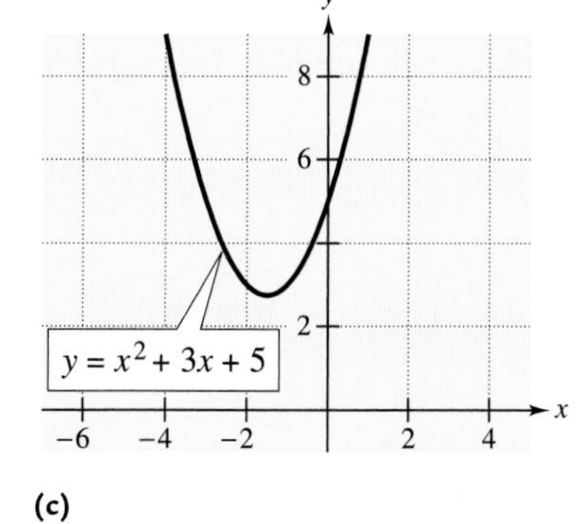

(c)

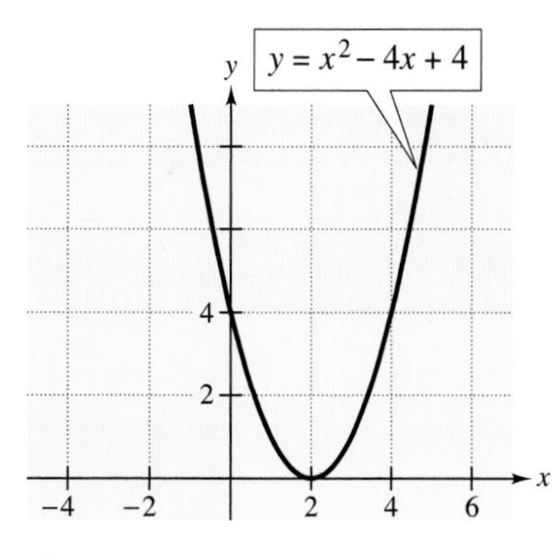

(d)
Figure 10.22

The solutions of the quadratic inequalities in Examples 1 and 2 consist, respectively, of a single interval and the union of two intervals. When solving the exercises for this section, you should be on the watch for some unusual solution sets, as illustrated in Example 3.

Example 3 Unusual Solution Sets

Solve each inequality to verify that the given solution set is correct.

a. The solution set of the quadratic inequality

$$x^2 + 2x + 4 > 0$$

consists of the entire set of real numbers, $(-\infty, \infty)$. This is true because the value of the quadratic $x^2 + 2x + 4$ is positive for every real value of x, as shown in Figure 10.22(a).

b. The solution set of the quadratic inequality

$$x^2 + 2x + 1 \le 0$$

consists of the single number $\{-1\}$. This is true because $x^2 + 2x + 1 = (x + 1)^2$ has just one critical number, $x = -1$, and it is the only value that satisfies the inequality. [See Figure 10.22(b).]

c. The solution set of the quadratic inequality

$$x^2 + 3x + 5 < 0$$

is empty. This is true because the value of the quadratic $x^2 + 3x + 5$ is not less than zero for any value of x, as shown in Figure 10.22(c).

d. The solution set of the quadratic inequality

$$x^2 - 4x + 4 > 0$$

consists of all real numbers *except* the number 2. In interval notation, this solution set can be written as $(-\infty, 2) \cup (2, \infty)$. [See Figure 10.22(d).]

Remember that checking the solution set of an inequality is not as straightforward as checking the solutions of an equation, because inequalities tend to have infinitely many solutions. Even so, we suggest that you check several x-values in your solution set to confirm that they satisfy the inequality. Also try checking x-values that are not in the solution set to verify that they do not satisfy the inequality.

For instance, the solution of $x^2 - 5x < 0$ is $(0, 5)$. Try checking some numbers in this interval to verify that they satisfy the inequality. Then check some numbers outside the interval to verify that they do not satisfy the inequality.

3 Use test intervals to solve a rational inequality.

Rational Inequalities

The concepts of critical numbers and test intervals can be extended to inequalities involving rational expressions. To do this, use the fact that the value of a rational expression can change sign only at its *zeros* (the x-values for which its numerator is zero) and its *undefined values* (the x-values for which its denominator is zero). These two types of numbers make up the **critical numbers** of a rational inequality. For instance, the critical numbers of the inequality

$$\frac{x-2}{(x-1)(x+3)} < 0$$

are $x = 2$ (the numerator is zero), and $x = 1$ and $x = -3$ (the denominator is zero). From these three critical numbers you can see that the inequality has *four* test intervals.

$$(-\infty, -3), \quad (-3, 1), \quad (1, 2), \quad \text{and} \quad (2, \infty)$$

Study Tip

When solving a rational inequality, you should begin by writing the inequality in general form, with the rational expression (as a single fraction) on the left and zero on the right. For instance, the first step in solving

$$\frac{2x}{x+3} < 4$$

is to write it as

$$\frac{2x}{x+3} < 4$$

$$\frac{2x}{x+3} - 4 < 0$$

$$\frac{2x - 4(x+3)}{x+3} < 0$$

$$\frac{-2x - 12}{x+3} < 0.$$

Try solving this inequality. You should find that the solution set is $(-\infty, -6) \cup (-3, \infty)$.

Example 4 Solving a Rational Inequality

Solve the inequality $\dfrac{x}{x-2} > 0$.

Solution

The numerator is zero when $x = 0$ and the denominator is zero when $x = 2$. So, the two critical numbers are 0 and 2, which implies that the test intervals are

$$(-\infty, 0), \quad (0, 2), \quad \text{and} \quad (2, \infty).$$

To test an interval, choose a convenient number and compute the sign of $x/(x-2)$.

Interval	*x-Value*	*Rational Expression Value*	*Conclusion*
$(-\infty, 0)$	$x = -1$	$(-1)/(-1-2) = \frac{1}{3}$	Positive
$(0, 2)$	$x = 1$	$(1)/(1-2) = -1$	Negative
$(2, \infty)$	$x = 3$	$(3)/(3-2) = 3$	Positive

From this you can conclude that the value of the rational expression is positive for all x-values in $(-\infty, 0)$ and $(2, \infty)$, and negative for all x-values in $(0, 2)$. This implies that the solution set of the inequality $x/(x-2) > 0$ is $(-\infty, 0) \cup (2, \infty)$, as shown in Figure 10.23.

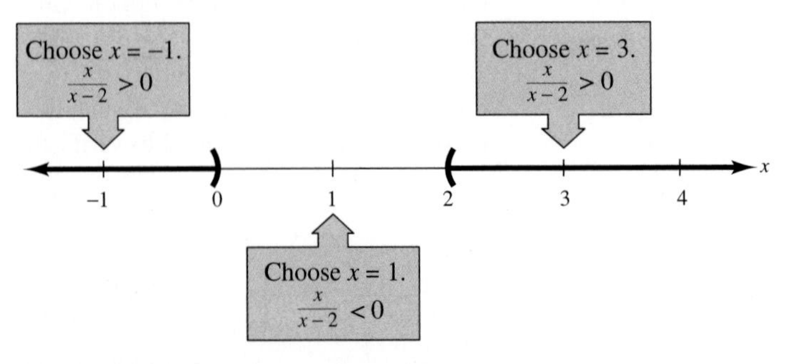

Figure 10.23

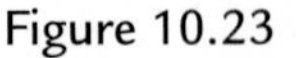

4 Use an inequality to solve an application problem.

Application

Example 5 The Height of a Projectile 

A projectile is fired straight up from ground level with an initial velocity of 256 feet per second, as shown in Figure 10.24, so that its height at any time t is given by

$$h = -16t^2 + 256t$$

where the height h is measured in feet and the time t is measured in seconds. During what interval of time will the height of the projectile exceed 960 feet?

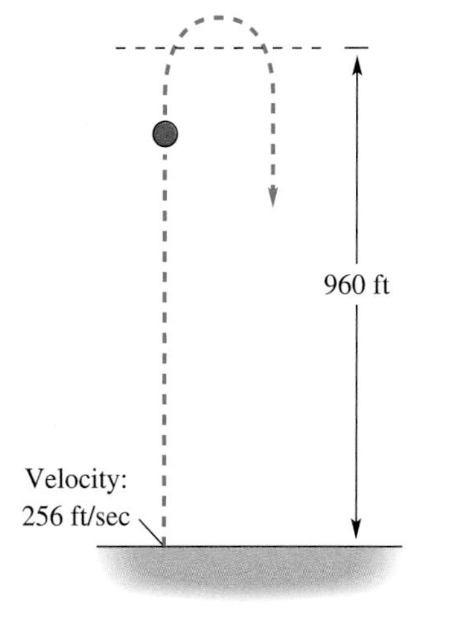

Figure 10.24

Solution

To solve this problem, find the values of t for which h is greater than 960.

$-16t^2 + 256t > 960$ — Original inequality

$-16t^2 + 256t - 960 > 0$ — Subtract 960 from both sides.

$t^2 - 16t + 60 < 0$ — Divide both sides by -16 and reverse the inequality.

$(t - 6)(t - 10) < 0$ — Factor.

So, the critical numbers are $t = 6$ and $t = 10$. A test of the intervals $(-\infty, 6)$, $(6, 10)$, and $(10, \infty)$ shows that the solution interval is $(6, 10)$. So, the height of the object will exceed 960 feet for values of t that are greater than 6 seconds and less than 10 seconds.

Discussing the Concept — Graphing an Inequality

You can use a graph on a rectangular coordinate system as an alternative method for solving an inequality. For instance, to solve the inequality in Example 1

$$x^2 - 5x < 0$$

you can sketch the graph of $y = x^2 - 5x$. Using a graphing utility, you can obtain the graph shown below. From the graph, you can see that the only part of the curve that lies below the x-axis is the portion for which $0 < x < 5$. So, the solution of $x^2 - 5x < 0$ is $0 < x < 5$. Try using this graphing approach to solve the following inequalities.

a. $x^2 + 4x > 0$ **b.** $x^2 - 16 \le 0$

c. $x^2 - 3x - 18 \le 0$ **d.** $3x^2 + 5x + 4 \ge 0$

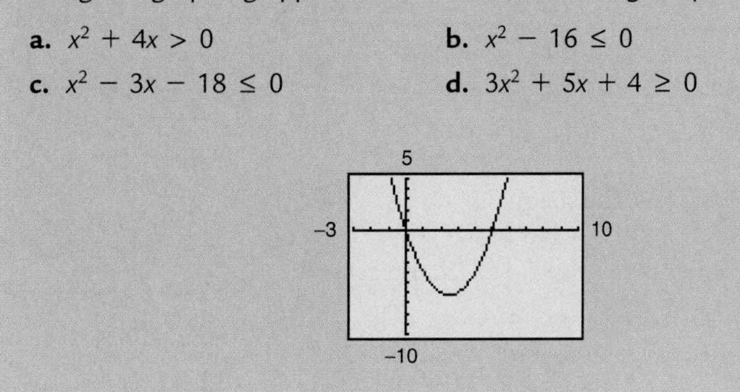

10.6 Exercises

Integrated Review *Concepts, Skills, and Problem Solving*

Keep mathematically in shape by doing these exercises *before* the problems of this section.

Properties and Definitions

1. Is 36.82×10^8 written in scientific notation? Explain.

2. The numbers $n_1 \times 10^2$ and $n_2 \times 10^4$ are written in scientific notation. The product of these two numbers must lie in what interval? Explain.

Simplifying Expressions

In Exercises 3–8, factor the expression.

3. $6u^2v - 192v^2$

4. $5x^{2/3} - 10x^{1/3}$

5. $x(x - 10) - 4(x - 10)$

6. $x^3 + 3x^2 - 4x - 12$

7. $16x^2 - 121$

8. $4x^3 - 12x^2 + 16x$

Mathematical Modeling

In Exercises 9–12, find a mathematical model for the area of the figure.

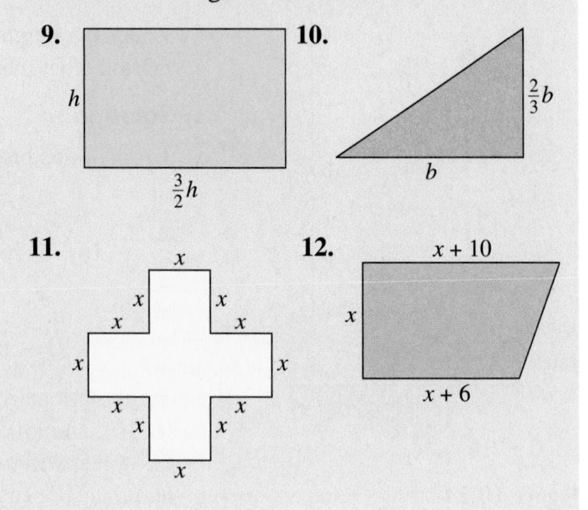

Developing Skills

In Exercises 1–10, find the critical numbers.

1. $x(2x - 5)$

2. $5x(x - 3)$

3. $4x^2 - 81$

4. $9y^2 - 16$

5. $x(x + 3) - 5(x + 3)$

6. $y(y - 4) - 3(y - 4)$

7. $x^2 - 4x + 3$

8. $3x^2 - 2x - 8$

9. $4x^2 - 20x + 25$

10. $4x^2 - 4x - 3$

In Exercises 11–20, determine the intervals for which the polynomial is entirely negative and entirely positive.

11. $x - 4$

12. $3 - x$

13. $3 - \frac{1}{2}x$

14. $\frac{2}{3}x - 8$

15. $2x(x - 4)$

16. $7x(3 - x)$

17. $4 - x^2$

18. $x^2 - 9$

19. $x^2 - 4x - 5$

20. $2x^2 - 4x - 3$

In Exercises 21–60, solve the inequality and sketch the graph of the solution on the real number line. (Some of the inequalities have no solution.) See Examples 1–3.

21. $2x + 6 \ge 0$

22. $5x - 20 < 0$

23. $-\frac{3}{4}x + 6 < 0$

24. $3x - 2 \ge 0$

25. $3x(x - 2) < 0$

26. $2x(x - 6) > 0$

27. $3x(2 - x) \ge 0$

28. $2x(6 - x) > 0$

29. $x^2 > 4$

30. $z^2 \le 9$

31. $x^2 + 3x \le 10$

32. $t^2 - 4t > 12$

33. $u^2 + 2u - 2 > 1$

34. $t^2 - 15t + 50 < 0$

35. $x^2 + 4x + 5 < 0$

36. $x^2 + 6x + 10 > 0$

37. $x^2 + 2x + 1 \ge 0$

38. $y^2 - 5y + 6 > 0$

39. $x^2 - 4x + 2 > 0$

40. $-x^2 + 8x - 11 \le 0$

41. $x^2 - 6x + 9 \ge 0$

42. $x^2 + 8x + 16 < 0$

43. $u^2 - 10u + 25 < 0$

44. $y^2 + 16y + 64 \le 0$

45. $3x^2 + 2x - 8 \le 0$

46. $2t^2 - 3t - 20 \ge 0$

47. $-6u^2 + 19u - 10 > 0$

48. $4x^2 - 4x - 63 < 0$

49. $-2u^2 + 7u + 4 < 0$

50. $-3x^2 - 4x + 4 \le 0$

51. $4x^2 + 28x + 49 \le 0$

52. $9x^2 - 24x + 16 \ge 0$

53. $(x - 5)^2 < 0$

54. $(y + 3)^2 \ge 0$

55. $6 - (x - 5)^2 < 0$

56. $(y + 3)^2 - 6 \ge 0$

57. $16 \le (u + 5)^2$

58. $25 \ge (x - 3)^2$

59. $x(x - 2)(x + 2) > 0$

60. $x^2(x - 2) \le 0$

In Exercises 61–68, use a graphing utility to graph the function and solve the inequality.

61. $y = x^2 - 6x, \quad y < 0$

62. $y = 2x^2 + 5x, \quad y > 0$

63. $y = 0.5x^2 + 1.25x - 3, \quad y > 0$

64. $y = \frac{1}{3}x^2 - 3x, \quad y < 0$

65. $y = x^2 + 4x + 4, \quad y \geq 9$

66. $y = x^2 - 6x + 9, \quad y < 16$

67. $y = 9 - 0.2(x - 2)^2, \quad y < 4$

68. $y = 8x - x^2, \quad y > 12$

In Exercises 69–72, determine the critical numbers of the rational expression and locate them on the real number line.

69. $\frac{5}{x - 3}$

70. $\frac{-6}{x + 2}$

71. $\frac{2x}{x + 5}$

72. $\frac{x - 2}{x - 10}$

In Exercises 73–94, solve the rational inequality. As part of your solution, include a graph that shows the test intervals. See Example 4.

73. $\frac{5}{x - 3} > 0$

74. $\frac{3}{4 - x} > 0$

75. $\frac{-5}{x - 3} > 0$

76. $\frac{-3}{4 - x} > 0$

77. $\frac{x}{x - 3} < 0$

78. $\frac{x}{2 - x} < 0$

79. $\frac{x + 3}{x - 4} \leq 0$

80. $\frac{z - 1}{z + 3} < 0$

81. $\frac{y - 4}{y + 6} < 0$

82. $\frac{u + 3}{u + 7} \leq 0$

83. $\frac{y - 3}{2y - 11} \geq 0$

84. $\frac{x + 5}{3x + 2} \geq 0$

85. $\frac{x + 2}{4x + 6} \leq 0$

86. $\frac{u - 6}{3u - 5} \leq 0$

87. $\frac{3(u - 3)}{u + 1} < 0$

88. $\frac{2(4 - t)}{4 + t} > 0$

89. $\frac{6}{x - 4} > 2$

90. $\frac{1}{x + 2} > -3$

91. $\frac{4x}{x + 2} < -1$

92. $\frac{6x}{x - 4} < 5$

93. $\frac{x - 1}{x - 3} \leq 2$

94. $\frac{x + 4}{x - 5} \geq 10$

In Exercises 95–102, use a graphing utility to solve the rational inequality.

95. $\frac{1}{x} - x > 0$

96. $\frac{1}{x} - 4 < 0$

97. $\frac{x + 6}{x + 1} - 2 < 0$

98. $\frac{x + 12}{x + 2} - 3 \geq 0$

99. $\frac{6x - 3}{x + 5} < 2$

100. $\frac{3x - 4}{x - 4} < -5$

101. $x + \frac{1}{x} > 3$

102. $4 - \frac{1}{x^2} > 1$

Graphical Analysis In Exercises 103–106, use a graphing utility to graph the function. Use the graph to approximate the values of x that satisfy the specified inequalities.

	Equation	*Inequalities*	
103.	$y = \frac{3x}{x - 2}$	(a) $y \leq 0$	(b) $y \geq 6$
104.	$y = \frac{2(x - 2)}{x + 1}$	(a) $y \leq 0$	(b) $y \geq 8$
105.	$y = \frac{2x^2}{x^2 + 4}$	(a) $y \geq 1$	(b) $y \leq 2$
106.	$y = \frac{5x}{x^2 + 4}$	(a) $y \geq 1$	(b) $y \geq 0$

Solving Problems

107. *Height of a Projectile* A projectile is fired vertically upward from ground level with an initial velocity of 128 feet per second, so that its height at any time t is given by $h = -16t^2 + 128t$ where the height h is measured in feet and the time t is measured in seconds. During what interval of time will the height of the projectile exceed 240 feet?

108. *Height of a Projectile* A projectile is fired vertically upward from ground level with an initial velocity of 88 feet per second, so that its height at any time t is given by $h = -16t^2 + 88t$ where the height h is measured in feet and the time t is measured in seconds. During what interval of time will the height of the projectile exceed 50 feet?

109. *Annual Interest Rate* You are investing \$1000 in a certificate of deposit for 2 years and you want the interest for that time period to exceed \$150. The interest is compounded annually. What interest rate should you have? [*Hint:* Solve the inequality $1000(1 + r)^2 > 1150$.]

110. *Annual Interest Rate* You are investing \$500 in a certificate of deposit for 2 years and you want the interest for that time to exceed \$50. The interest is compounded annually. What interest rate should you have? [*Hint:* Solve the inequality $500(1 + r)^2 > 550$.]

111. *Company Profits* The revenue and cost equations for a product are given by

$$R = x(50 - 0.0002x)$$

$$C = 12x + 150{,}000$$

where R and C are measured in dollars and x represents the number of units sold (see figure). How many units must be sold to obtain a profit of at least \$1,650,000?

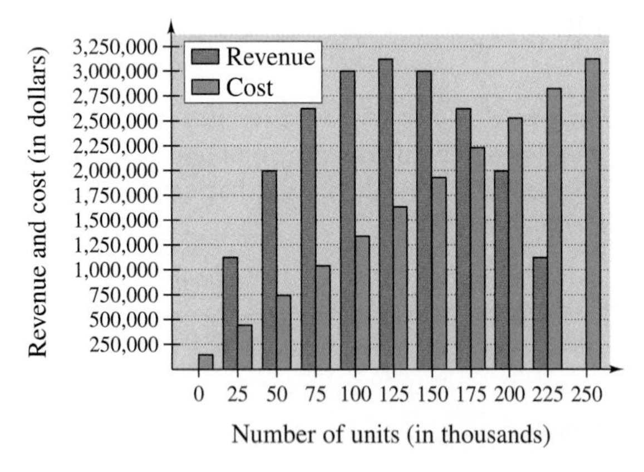

112. *Geometry* A rectangular playing field with a perimeter of 100 meters is to have an area of at least 500 square meters. Within what bounds must the length of the field lie?

113. *Geometry* You have 64 feet of fencing to enclose a rectangular region. Determine the interval for the length such that the area will exceed 240 square feet.

114. *Average Cost* The cost of producing x units of a product is $C = 3000 + 0.75x,\ x > 0$.

(a) Write the average cost $\overline{C} = C/x$ as a function of x.

(b) Use a graphing utility to graph the average cost function in part (a). Determine the horizontal asymptote of the graph.

(c) How many units must be produced if the average cost per unit is to be less than \$2?

115. *Data Analysis* The temperature T (in degrees Fahrenheit) of a metal in a laboratory experiment was recorded every 2 minutes for a period of 16 minutes. The table gives the experimental data, where t is the time in minutes.

t	0	2	4	6	8
T	250	290	338	410	498

t	10	12	14	16
T	560	530	370	160

A model for these data is

$$T = \frac{244.20 - 13.23t}{1 - 0.13t + 0.005t^2}.$$

(a) Use a graphing utility to plot the data and graph the model.

(b) Use the graph to approximate the times when the temperature was at least 400°F.

Explaining Concepts

116. Answer part (g) of Motivating the Chapter on page 567.

117. Explain the change in an inequality when both sides are multiplied by a negative real number.

118. Give a verbal description of the intervals $(-\infty, 5] \cup (10, \infty)$.

119. Define the term *critical number* and explain its use in solving quadratic inequalities.

120. In your own words, describe the procedure for solving quadratic inequalities.

121. Give an example of a quadratic inequality that has no real solution.

122. Explain the distinction between the critical numbers of a quadratic inequality and those of a rational inequality.

CHAPTER SUMMARY

Key Terms

double or repeated solution, *p. 568*
quadratic form, *p. 571*
discriminant, *p. 583*
parabola, *p. 592*
vertex, *p. 592*
axis, *p. 592*
transformations, *p. 595*
reflection, *p. 596*
test intervals, *p. 614*
critical numbers, *p. 614*

Key Concepts

10.1 Extracting square roots

The equation $u^2 = d$, where $d > 0$, has exactly two solutions: $u = \sqrt{d}$ and $u = -\sqrt{d}$.

10.1 Extracting complex square roots

The equation $u^2 = d$, where $d < 0$, has exactly two solutions: $u = \sqrt{|d|}\,i$ and $u = -\sqrt{|d|}\,i$.

10.2 Completing the square

To complete the square for the expression $x^2 + bx$, add $(b/2)^2$, which is the square of half the coefficient of x. Consequently, $x^2 + bx + (b/2)^2 = (x + b/2)^2$.

10.3 The Quadratic Formula

The solutions of $ax^2 + bx + c = 0$, $a \neq 0$, are given by the Quadratic Formula $x = \dfrac{-b \pm \sqrt{b^2 - 4ac}}{2a}$.

The expression inside the radical, $b^2 - 4ac$, is called the discriminant.

1. If $b^2 - 4ac > 0$, the equation has two real solutions.
2. If $b^2 - 4ac = 0$, the equation has one (repeated) real solution.
3. If $b^2 - 4ac < 0$, the equation has no real solutions.

10.3 Using the discriminant

The discriminant of the quadratic equation $ax^2 + bx + c = 0$, $a \neq 0$, can be used to classify the solutions of the equation as follows.

Discriminant	*Solution Types*
1. Perfect square	Two distinct rational solutions
2. Positive nonperfect square	Two distinct irrational solutions
3. Zero	One repeated rational solution
4. Negative number	Two distinct imaginary solutions

10.4 Graphs of quadratic functions

The graph of $f(x) = ax^2 + bx + c$, $a \neq 0$, is a parabola.

The completed-square form $f(x) = a(x - h)^2 + k$ is the standard form of the function. The vertex of the parabola occurs at the point (h, k).

10.4 Sketching a parabola

1. Determine the vertex and axis of the parabola by completing the square or by formula.
2. Plot the vertex, axis, and a few additional points on the parabola.
3. Use the fact that the parabola opens upward if $a > 0$ and opens downward if $a < 0$ to complete the sketch.

10.4 Vertical and horizontal shifts

Let c be a positive real number. Shifts of the graph of the function $y = f(x)$ are represented as follows.

1. Vertical shift c units upward: $h(x) = f(x) + c$
2. Vertical shift c units downward: $h(x) = f(x) - c$
3. Horizontal shift c units to the right: $h(x) = f(x - c)$
4. Horizontal shift c units to the left: $h(x) = f(x + c)$

10.6 Finding test intervals for inequalities

1. For a polynomial function, find all the real zeros. For a rational function, find all the real zeros and those x-values for which the function is undefined.
2. Arrange the numbers found in part 1 in increasing order. These numbers are called critical numbers.
3. Use the critical numbers to determine the test intervals.
4. Choose a representative x-value in each test interval and evaluate the function at that value. If the value of the function is negative, the function will have negative values for every x-value in the interval. If the value of the function is positive, the function will have positive values for every x-value in the interval.

REVIEW EXERCISES

Reviewing Skills

10.1 In Exercises 1–10, solve the quadratic equation by factoring.

1. $x^2 + 12x = 0$
2. $u^2 - 18u = 0$
3. $4y^2 - 1 = 0$
4. $2z^2 - 72 = 0$
5. $4y^2 + 20y + 25 = 0$
6. $x^2 + \frac{8}{3}x + \frac{16}{9} = 0$
7. $2x^2 - 2x - 180 = 0$
8. $15x^2 - 30x - 45 = 0$
9. $6x^2 - 12x = 4x^2 - 3x + 18$
10. $10x - 8 = 3x^2 - 9x + 12$

In Exercises 11–22, solve the quadratic equation by extracting square roots. Find all real and complex solutions.

11. $4x^2 = 10{,}000$
12. $2x^2 = 98$
13. $y^2 - 12 = 0$
14. $y^2 - 8 = 0$
15. $(x - 16)^2 = 400$
16. $(x + 3)^2 = 900$
17. $z^2 = -121$
18. $u^2 = -36$
19. $y^2 + 50 = 0$
20. $x^2 + 48 = 0$
21. $(y + 4)^2 + 18 = 0$
22. $(x - 2)^2 + 24 = 0$

In Exercises 23–30, solve the equation of quadratic form.

23. $x^4 - 4x^2 - 5 = 0$
24. $x^4 - 10x^2 + 9 = 0$
25. $x - 4\sqrt{x} + 3 = 0$
26. $x + 2\sqrt{x} - 3 = 0$
27. $(x^2 - 2x)^2 - 4(x^2 - 2x) - 5 = 0$
28. $(\sqrt{x} - 2)^2 + 2(\sqrt{x} - 2) - 3 = 0$
29. $x^{2/3} + 3x^{1/3} - 28 = 0$
30. $x^{2/5} + 4x^{1/5} + 3 = 0$

10.2 In Exercises 31–38, solve the equation by completing the square. (Find all real and complex solutions.)

31. $x^2 - 6x - 3 = 0$
32. $x^2 + 12x + 6 = 0$
33. $x^2 - 3x + 3 = 0$
34. $u^2 - 5u + 6 = 0$
35. $y^2 - \frac{2}{3}y + 2 = 0$
36. $t^2 + \frac{1}{2}t - 1 = 0$
37. $2y^2 + 10y + 3 = 0$
38. $3x^2 - 2x + 2 = 0$

10.3 In Exercises 39–46, use the Quadratic Formula to solve the equation. Find all real and complex solutions.

39. $y^2 + y - 30 = 0$
40. $x^2 - x - 72 = 0$
41. $2y^2 + y - 21 = 0$
42. $2x^2 - 3x - 20 = 0$
43. $5x^2 - 16x + 2 = 0$
44. $3x^2 + 12x + 4 = 0$
45. $0.3t^2 - 2t + 5 = 0$
46. $-u^2 + 2.5u + 3 = 0$

In Exercises 47–54, determine the type of solution of the quadratic equation using the discriminant.

47. $x^2 + 4x + 4 = 0$
48. $y^2 - 26y + 169 = 0$
49. $s^2 - s - 20 = 0$
50. $r^2 - 5r - 45 = 0$
51. $3t^2 + 17t + 10 = 0$
52. $7x^2 + 3x - 18 = 0$
53. $v^2 - 6v + 21 = 0$
54. $9y^2 + 1 = 0$

10.4 In Exercises 55–58, determine the vertex of the graph of the quadratic function.

55. $f(x) = x^2 - 8x + 3$
56. $g(x) = x^2 + 12x - 9$
57. $h(u) = 2u^2 - u + 3$
58. $f(t) = 3t^2 + 2t - 6$

In Exercises 59–62, sketch the graph of the function. Identify the vertex and any x-intercepts. Use a graphing utility to verify your results.

59. $y = x^2 + 8x$
60. $y = -x^2 + 3x$
61. $y = x^2 - 6x + 5$
62. $y = x^2 + 3x - 10$

In Exercises 63–66, identify the transformation of the graph of $f(x) = x^2$ and sketch a graph of h.

63. $h(x) = x^2 + 3$
64. $h(x) = x^2 - 1$
65. $h(x) = (x + 2)^2 - 3$
66. $h(x) = (x - 2)^2 + 4$

In Exercises 67–72, write the standard form of an equation of the parabola that satisfies the conditions.

67. Vertex: $(3, 5)$; $a = -2$
68. Vertex: $(-2, 3)$; $a = 3$
69. Vertex: $(2, -5)$; y-intercept: $(0, 3)$
70. Vertex: $(-4, 0)$; y-intercept: $(0, -6)$
71. Vertex: $(5, 0)$; passes through the point $(1, 1)$
72. Vertex: $(-2, 5)$; passes through the point $(0, 1)$

10.6 In Exercises 73–82, solve the inequality and graph its solution on the real number line. (Some of the inequalities have no solution.)

73. $5x(7 - x) > 0$

74. $-2x(x - 10) \leq 0$

75. $16 - (x - 2)^2 \leq 0$

76. $(x - 5)^2 - 36 > 0$

77. $2x^2 + 3x - 20 < 0$

78. $3x^2 - 2x - 8 > 0$

79. $\dfrac{x + 3}{2x - 7} \geq 0$

80. $\dfrac{2x - 9}{x - 1} \leq 0$

81. $\dfrac{2x - 2}{x + 6} + 2 < 0$

82. $\dfrac{3x + 1}{x - 2} > 4$

Solving Problems

83. *Selling Price* A car dealer bought a fleet of cars from a car rental agency for a total of $80,000. By the time the dealer had sold all but four of the cars, at an average profit of $1000 each, the original investment of $80,000 had been regained. How many cars did the dealer sell, and what was the average price per car?

84. *Selling Price* A manager of a computer store bought several computers of the same model for $27,000. When all but five of the computers had been sold at a profit of $900 per computer, the original investment of $27,000 had been regained. How many computers were sold, and what was the selling price of each computer?

85. *Geometry* The length of a rectangle is 12 inches greater than its width. The area of the rectangle is 108 square inches. Find the dimensions of the rectangle.

86. *Geometry* Find the dimensions of a triangle if its height is 4 centimeters less than its base and its area is 480 square centimeters.

87. *Compound Interest* You want to invest $20,000 for 2 years at an annual interest rate of r (in decimal form). Interest on the account is compounded annually. Find the interest rate if a deposit of $20,000 increases to $21,424.50 over a 2-year period.

88. *Compound Interest* You want to invest $35,000 for 2 years at an annual interest rate of r (in decimal form). Interest on the account is compounded annually. Find the interest rate if a deposit of $35,000 increases to $38,955.88 over a 2-year period.

89. *Reduced Fare* A college wind ensemble charters a bus at a cost of $360 to attend a concert. In an attempt to lower the bus fare per person, the ensemble invites nonmembers to go along. When eight nonmembers join the trip, the fare is decreased by $1.50 per person. How many people are going on the excursion?

90. *Think About It* When six nonmembers go along on the excursion described in Exercise 89, the fare is decreased by $16. Describe how it is possible to have fewer nonmembers and a greater decrease in the fare.

91. *Decreased Price* A Little League baseball team obtains a block of tickets for a ball game for $96. After three more people decide to go to the game, the price per ticket is decreased by $1.60. How many people are going to the game?

92. *Shortcut* A corner lot has an L-shaped sidewalk along its sides. The total length of the sidewalk is 51 feet. By cutting diagonally across the lot, the walking distance is shortened to 39 feet. What are the lengths of the two legs of the sidewalk?

93. *Shortcut* Two buildings are connected by an L-shaped protected walkway. The distance between buildings via the walkway is 140 feet. By cutting diagonally across the grass, the walking distance is shortened to 100 feet. What are the lengths of the two legs of the walkway?

94. *Work-Rate Problem* Working together, two people can complete a task in 6 hours. Working alone, how long would it take each to do the task if one person takes 3 hours longer than the other?

95. *Work-Rate Problem* Working together, two people can complete a task in 10 hours. Working alone, how long would it take each to do the task if one person takes 2 hours longer than the other?

96. *Vertical Motion* The height h in feet of an object above the ground is

$$h = 200 - 16t^2, \quad t \geq 0$$

where t is time in seconds.

(a) After how many seconds will the height be 164 feet?

(b) Find the time when the object strikes the ground.

97. *Vertical Motion* The height h in feet of an object above the ground is

$$h = -16t^2 + 64t + 192, \quad t > 0$$

where t is time in seconds.

(a) After how many seconds will the height be 256 feet?

(b) Find the time when the object strikes the ground.

98. *Path of a Projectile* The path y of a projectile is

$$y = -\frac{1}{16}x^2 + 5x$$

where y is the height (in feet) and x is the horizontal distance (in feet) from where the projectile is launched.

(a) Use a graphing utility to graph the path of the projectile.

(b) How high is the projectile when it is at its maximum height?

(c) How far from the launch point does the projectile strike the ground?

99. *Graphical Estimation* The height y (in feet) of a ball thrown by a child is given by

$$y = -\frac{1}{10}x^2 + 3x + 6$$

where x is the horizontal distance (in feet) from where the ball is thrown.

(a) Use a graphing utility to graph the path of the ball.

(b) How high is the ball when it leaves the child's hand?

(c) What is the maximum height of the ball?

(d) How far from the child does the ball hit the ground?

100. *Graphical Estimation* The enrollment E (in millions) in public schools grades 1 through 8 in the United States for the years 1970 through 1995 is approximated by the model

$$E = 32.931 - 0.816t + 0.032t^2$$

where t represents the calendar year, with $t = 0$ corresponding to 1970. (Source: U.S. National Center for Education Statistics)

(a) Use a graphing utility to graph the enrollment function.

(b) Use the graph in part (a) to approximate the year when enrollment was minimum.

(c) Use the model to predict the enrollment in the year 2000.

101. *Graphical Estimation* The number N of orders for U.S. civil jet transport aircraft for the years 1990 through 1996 is approximated by the model

$$N = 653.50 - 379.89t + 62.75t^2$$

where t represents the calendar year, with $t = 0$ corresponding to 1990. (Source: Aerospace Industries Association of America)

(a) Use a graphing utility to graph the function N.

(b) Use the graph in part (a) to approximate the year when the number of orders was minimum.

(c) Use the model to predict the number of orders in the year 2000.

102. *Average Cost* The cost of producing x units of a product is $C = 100{,}000 + 0.9x$, and therefore the average cost per unit is $\overline{C} = C/x$. Find the number of units that must be produced if $\overline{C} < 2$.

103. *Average Cost* The cost of producing x units of a product is $C = 50{,}000 + 1.2x$, and so the average cost per unit is $\overline{C} = C/x$. Find the number of units that must be produced if $\overline{C} < 5$.

104. *Annual Interest Rate* You are investing \$3000 in a certificate of deposit for 2 years and you want the interest for that time to exceed \$370. The interest is compounded annually. What interest rate should you have? [*Hint:* Solve the inequality $3000(1 + r)^2 > 3370$.]

105. *Height of a Projectile* A projectile is fired straight up from ground level with an initial velocity of 312 feet per second. Its height at any time t is given by $h = -16t^2 + 312t$, where the height h is measured in feet and the time t is measured in seconds. During what interval of time will the height of the projectile exceed 1200 feet?

Chapter Test

Take this test as you would take a test in class. After you are done, check your work against the answers given in the back of the book.

In Exercises 1–6, solve the equation by the specified method.

1. Factoring:
$x(x + 5) - 10(x + 5) = 0$

2. Factoring:
$8x^2 - 21x - 9 = 0$

3. Extracting square roots:
$(x - 2)^2 = 0.09$

4. Extracting square roots:
$(x + 3)^2 + 81 = 0$

5. Completing the square:
$2x^2 - 6x + 3 = 0$

6. Quadratic Formula:
$2y(y - 2) = 7$

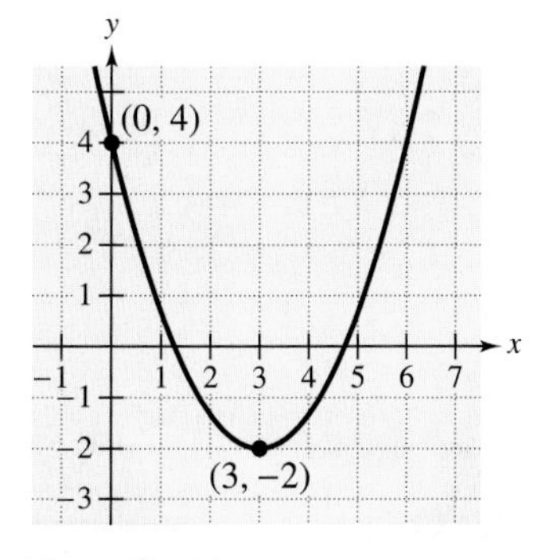

Figure for 12

In Exercises 7 and 8, solve the equation of quadratic form.

7. $x - 5\sqrt{x} + 4 = 0$

8. $x^4 + 6x^2 - 16 = 0$

9. Find the discriminant and explain how it can be used to determine the type of solutions of the quadratic equation $5x^2 - 12x + 10 = 0$.

10. Find a quadratic equation having solutions -4 and 5.

11. Sketch the graph of $f(x) = -2(x - 2)^2 + 8$. Label its vertex and intercepts.

12. Write the standard form of an equation for the parabola shown at the left.

13. Write the standard form of an equation of the parabola that has a vertex of $(3, -2)$ and that contains the point $(-1, 6)$.

In Exercises 14–17, solve the inequality and sketch its solution.

14. $16 \leq (x - 2)^2$

15. $2x(x - 3) < 0$

16. $\dfrac{3u + 2}{u - 3} \leq 2$

17. $\dfrac{3}{x - 2} > 4$

18. The width of a rectangle is 8 feet less than its length. The area of the rectangle is 240 square feet. Find the dimensions of the rectangle.

19. An English club chartered a bus for a trip to a Shakespearean festival. The cost of the bus was \$1250. To lower the per person cost of the bus, nonmembers were invited. When 10 nonmembers joined the trip, the fare per person decreased by \$6.25. How many club members were going on the trip?

20. An object is dropped from a height of 75 feet. Its height h (in feet) at any time t is given by $h = -16t^2 + 75$, where the time t is measured in seconds. Find the time at which the object has fallen to a height of 35 feet.

21. The revenue R for a chartered bus trip is given by $R = -\frac{1}{20}(n^2 - 240n)$, where n is the number of passengers and $80 \leq n \leq 160$. How many passengers will produce a maximum revenue? Explain your reasoning.

22. A projectile is fired straight up from ground level with an initial velocity of 288 feet per second. Its height at any time t is given by $h = -16t^2 + 288t$, where the height h is measured in feet and the time t is measured in seconds. During what time interval will the height of the projectile exceed 1040 feet?

11 Exponential and Logarithmic Functions

Bruce Ayers/Tony Stone Images

In 1997, the total financial assets held by households in the United States was $27 trillion. Assets included deposits in savings and money market accounts, government securities, corporate equities, pension funds, and mutual fund shares. (Source: Board of Governors of the Federal Reserve System)

Motivating the Chapter

Choosing the Best Investment

You receive an inheritance of $5000 and want to invest it.

See Section 11.1, Exercise 104

a. Complete the table by finding the amount A of the $5000 investment after 3 years with an annual interest rate of $r = 6\%$. Which form of compounding gives you the greatest balance?

Compounding	Amount A
Annually	
Quarterly	
Monthly	
Daily	
Hourly	
Continuously	

b. You are considering two different investment options. The first investment option has an interest rate of 7% compounded continuously. The second investment option has an interest rate of 8% compounded quarterly. Which investment would you choose? Explain.

See Section 11.5, Exercise 139

c. What annual percentage rate is needed to obtain a balance of $6200 in 3 years if the interest is compounded monthly?

d. If $r = 6\%$ and the interest is compounded continuously, how long will it take for your inheritance to grow to $7500?

e. What is the *effective yield* on your investment if the interest rate is 8% compounded quarterly?

f. With an interest rate of 6%, compounded continuously, how long will it take your inheritance to double? How long will it take your inheritance to quadruple (reach four times the original amount)?

11.1 Exponential Functions

Objectives

1. Evaluate exponential functions.
2. Use the point-plotting method or a graphing utility to graph an exponential function.
3. Evaluate the natural base e and graph the natural exponential function.
4. Use an exponential function to solve an application problem.

1 Evaluate exponential functions.

Exponential Functions

In this section, you will study a new type of function called an **exponential function.** Whereas polynomial and rational functions have terms with variable bases and constant exponents, exponential functions have terms with *constant bases* and *variable exponents.* Here are some examples.

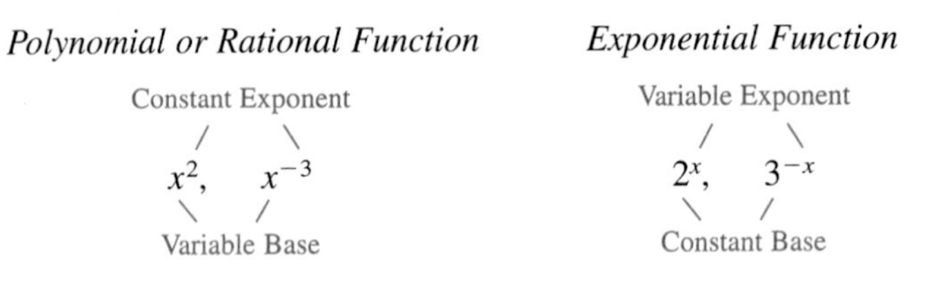

▶ **Definition of Exponential Function**

The **exponential function f with base a** is denoted by

$$f(x) = a^x$$

where $a > 0$, $a \neq 1$, and x is any real number.

The base $a = 1$ is excluded from exponential functions because $f(x) = 1^x = 1$ is a constant function, *not* an exponential function.

In Chapter 9, you learned to evaluate a^x for integer and rational values of x. For example, you know that

$$a^3 = a \cdot a \cdot a, \quad a^{-4} = \frac{1}{a^4}, \quad \text{and} \quad a^{5/3} = \left(\sqrt[3]{a}\right)^5.$$

However, to evaluate a^x for any real number x, you need to interpret forms with *irrational* exponents, such as $a^{\sqrt{2}}$ or a^{π}. For the purpose of this text, it is sufficient to think of a number such as

$$a^{\sqrt{2}}$$

where $\sqrt{2} \approx 1.414214$, as the number that has the successively closer approximations

$$a^{1.4}, a^{1.41}, a^{1.414}, a^{1.4142}, a^{1.41421}, a^{1.414214}, \ldots .$$

The rules of exponents that were discussed in Section 9.1 can be extended to cover exponential functions, as described on page 631.

▶ Rules of Exponential Functions

1. $a^x \cdot a^y = a^{x+y}$ — Product Rule
2. $\dfrac{a^x}{a^y} = a^{x-y}$ — Quotient Rule
3. $(a^x)^y = a^{xy}$ — Power Rule
4. $a^{-x} = \dfrac{1}{a^x} = \left(\dfrac{1}{a}\right)^x$ — Negative Exponent Rule

To evaluate exponential functions with a calculator, you can use the exponential key [y^x] (where y is the base and x is the exponent) or [^]. For example, to evaluate $3^{-1.3}$, you can use the following keystrokes.

Keystrokes	*Display*	
3 [y^x] 1.3 [+/−] [=]	0.239741	Scientific
3 [^] [(] [(−)] 1.3 [)] [ENTER]	0.239741	Graphing

Example 1 Evaluating Exponential Functions

Evaluate each function at the indicated values of x. Use a calculator only if it is necessary or more efficient.

Function	*Values*
a. $f(x) = 2^x$	$x = 3, x = -4, x = \pi$
b. $g(x) = 12^x$	$x = 3, x = -0.1, x = \frac{5}{7}$
c. $j(x) = 200(1.04)^{2x}$	$x = 1, x = -2, x = \sqrt{2}$

Solution

Evaluation	*Comment*
a. $f(3) = 2^3 = 8$	Calculator is not necessary.
$f(-4) = 2^{-4} = \dfrac{1}{2^4} = \dfrac{1}{16}$	Calculator is not necessary.
$f(\pi) = 2^{\pi} \approx 8.825$	Calculator is necessary.
b. $g(3) = 12^3 = 1728$	Calculator is more efficient.
$g(-0.1) = 12^{-0.1} \approx 0.7800$	Calculator is necessary.
$g\left(\dfrac{5}{7}\right) = 12^{5/7} \approx 5.900$	Calculator is necessary.
c. $j(1) = 200(1.04)^{2(1)} = 216.32$	Calculator is more efficient.
$j(-2) = 200(1.04)^{2(-2)} \approx 170.961$	Calculator is more efficient.
$j(\sqrt{2}) = 200(1.04)^{2\sqrt{2}} \approx 223.464$	Calculator is necessary.

2 Use the point-plotting method or a graphing utility to graph an exponential function.

Graphs of Exponential Functions

The basic nature of the graph of an exponential function can be determined by the point-plotting method or by using a graphing utility.

Example 2 The Graphs of Exponential Functions

In the same coordinate plane, sketch the graphs of the following functions. Determine the domains and ranges.

a. $f(x) = 2^x$ **b.** $g(x) = 4^x$

Solution

The table lists some values of each function, and Figure 11.1 shows their graphs. From the graphs, you can see that the domain of each function is the set of all real numbers and that the range of each function is the set of all positive real numbers.

x	-2	-1	0	1	2	3
2^x	$\frac{1}{4}$	$\frac{1}{2}$	1	2	4	8
4^x	$\frac{1}{16}$	$\frac{1}{4}$	1	4	16	64

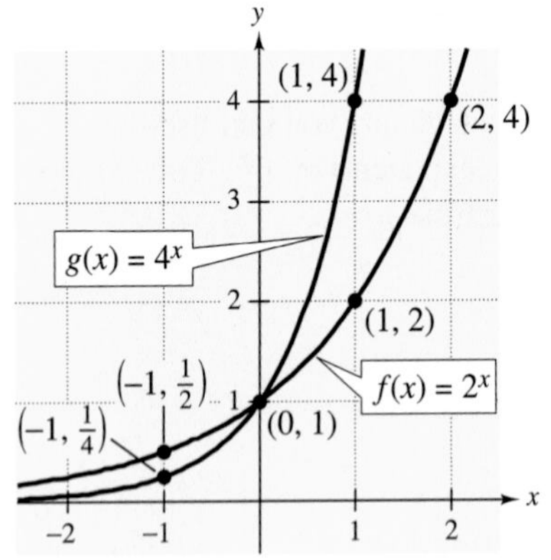

Figure 11.1

Note in the next example that a graph of the form $f(x) = a^x$ (as shown in Example 2) is a reflection in the y-axis of a graph of the form $g(x) = a^{-x}$.

Example 3 The Graphs of Exponential Functions

In the same coordinate plane, sketch the graph of each function.

a. $f(x) = 2^{-x}$ **b.** $g(x) = 4^{-x}$

Solution

The table lists some values of each function, and Figure 11.2 shows their graphs.

x	-3	-2	-1	0	1	2
2^{-x}	8	4	2	1	$\frac{1}{2}$	$\frac{1}{4}$
4^{-x}	64	16	4	1	$\frac{1}{4}$	$\frac{1}{16}$

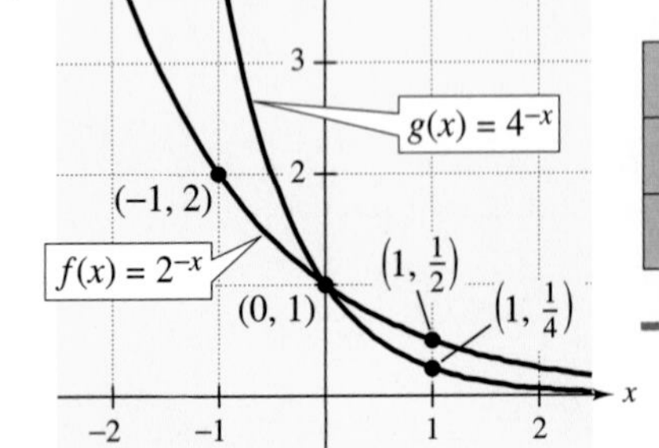

Figure 11.2

Examples 2 and 3 suggest that for $a > 1$, the graph of $y = a^x$ increases and the graph of $y = a^{-x}$ decreases. Remember that $a^{-x} = (1/a)^x$. So, increasing the power on a fraction $(1/a) < 1$ yields smaller and smaller values. The graphs shown in Figure 11.3 are typical of the graphs of exponential functions. Note that each has a y-intercept at $(0, 1)$ and a horizontal asymptote of $y = 0$.

Graph of $y = a^x$

- Domain: $(-\infty, \infty)$
- Range: $(0, \infty)$
- Intercept: $(0, 1)$
- Increasing

Graph of $y = a^{-x} = \left(\frac{1}{a}\right)^x$

- Domain: $(-\infty, \infty)$
- Range: $(0, \infty)$
- Intercept: $(0, 1)$
- Decreasing

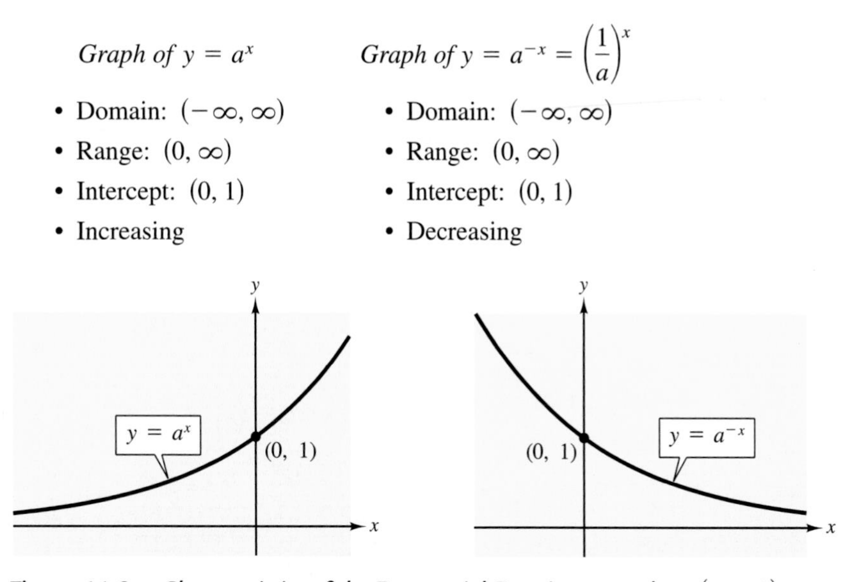

Figure 11.3 *Characteristics of the Exponential Functions* a^x *and* a^{-x} $(a > 1)$

In the next two examples, notice how the graph of $y = a^x$ can be used to sketch the graphs of functions of the form $f(x) = b \pm a^{x+c}$. Also note that the transformation in Example 4(a) keeps the x-axis as a horizontal asymptote, but the transformation in Example 4(b) yields a new horizontal asymptote of $y = -2$. Also, be sure to note how the y-intercept is affected by each transformation.

Example 4 Sketching Graphs of Exponential Functions

Each of the following graphs is a transformation of the graph of $f(x) = 3^x$, as shown in Figure 11.4.

a. Because $g(x) = 3^{x+1} = f(x + 1)$, the graph of g can be obtained by shifting the graph of f 1 unit to the left.

b. Because $h(x) = 3^x - 2 = f(x) - 2$, the graph of h can be obtained by shifting the graph of f down 2 units.

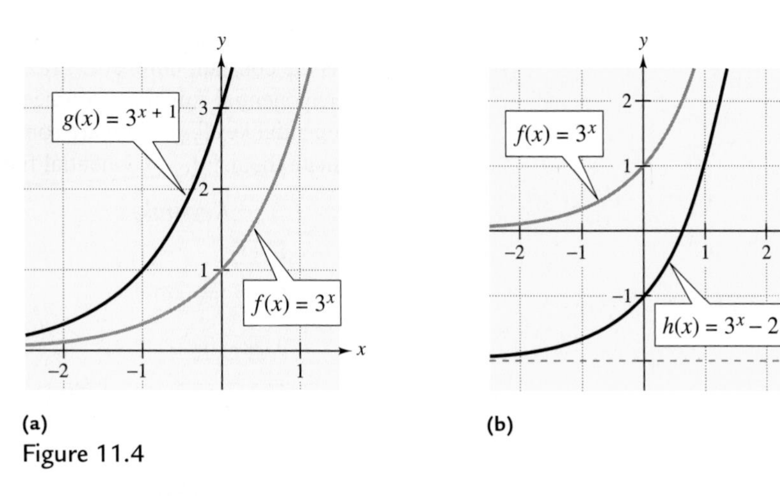

(a) (b)

Figure 11.4

Example 5 Reflections of Exponential Functions

Each of the following graphs is a reflection of the graph of $f(x) = 3^x$, as shown in Figure 11.5.

a. Because $k(x) = -3^x = -f(x)$, the graph of k can be obtained by reflecting the graph of f in the x-axis.

b. Because $j(x) = 3^{-x} = f(-x)$, the graph of j can be obtained by reflecting the graph of f in the y-axis.

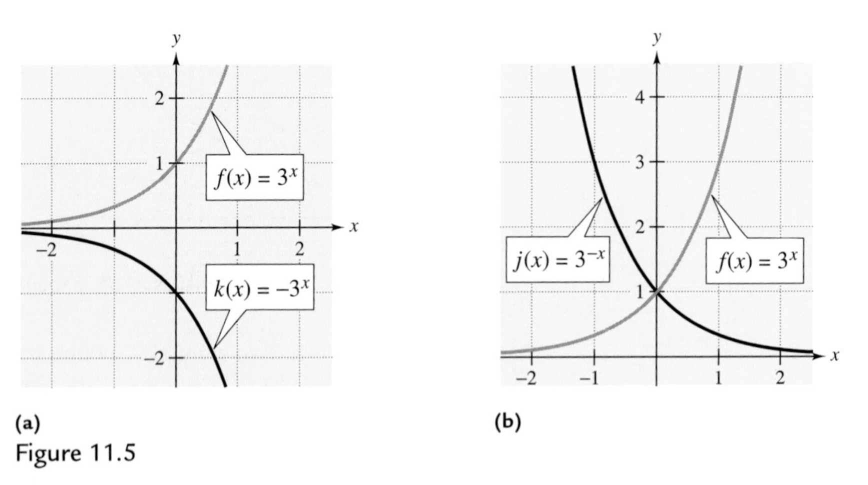

(a) (b)
Figure 11.5

The Natural Exponential Function

3 Evaluate the natural base e and graph the natural exponential function.

So far, we have used integers or rational numbers as bases of exponential functions. In many applications of exponential functions, the convenient choice for a base is the irrational number, denoted by the letter "e."

$e \approx 2.71828\ .\ .\ .$ Natural base

This number is called the **natural base.** The function

$f(x) = e^x$ Natural exponential function

is called the **natural exponential function.** Be sure you understand that for this function, e is the constant number 2.71828 . . . , and x is a variable. To evaluate the natural exponential function, you need a calculator, preferably one having a natural exponential key [e^x]. Here are some examples of how to use such a calculator to evaluate the natural exponential function.

Value	*Keystrokes*	*Display*	
e^2	2 [e^x]	7.3890561	Scientific
e^2	[e^x] 2 [ENTER]	7.3890561	Graphing
e^{-3}	3 [+/−] [e^x]	0.049787	Scientific
e^{-3}	[e^x] [(] [(−)] 3 [)] [ENTER]	0.049787	Graphing
$e^{0.32}$	.32 [e^x]	1.3771278	Scientific
$e^{0.32}$	[e^x] .32 [ENTER]	1.3771278	Graphing

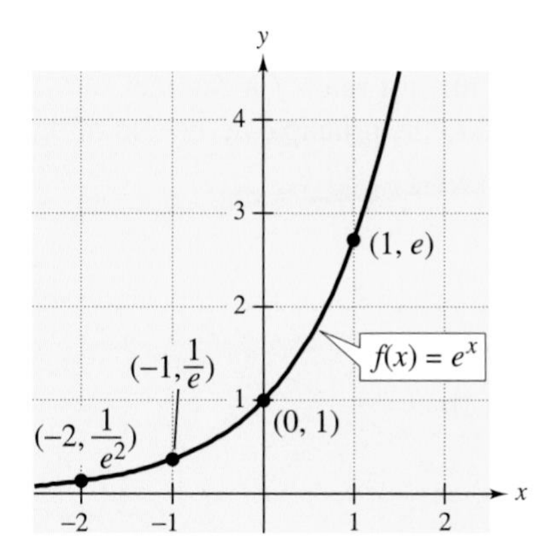

Figure 11.6

Some calculators do not have a key labeled e^x. If your calculator does not have this key, but does have a key labeled $\ln x$, you will have to use the two-keystroke sequence INV $\ln x$ in place of e^x.

After evaluating the natural exponential function at several values, as shown in the table, you can sketch its graph, as shown in Figure 11.6.

x	-1.5	-1.0	-0.5	0.0	0.5	1.0	1.5
$f(x) = e^x$	0.223	0.368	0.607	1.000	1.649	2.718	4.482

From the graph, notice the following properties of the natural exponential function.

- The domain is the set of all real numbers.
- The range is the set of positive real numbers.
- The y-intercept is $(0, 1)$.

4 Use an exponential function to solve an application problem.

Applications

A common scientific application of exponential functions is that of **radioactive decay.**

Example 6 Radioactive Decay

Let y represent the mass of a particular radioactive element whose half-life is 25 years. The initial mass is 10 grams. After t years, the mass (in grams) is given by

$$y = 10\left(\frac{1}{2}\right)^{t/25}, \quad t \geq 0.$$

How much of the initial mass remains after 120 years?

Solution

When $t = 120$, the mass is given by

$$y = 10\left(\frac{1}{2}\right)^{120/25} \qquad \text{Substitute 120 for } t.$$

$$= 10\left(\frac{1}{2}\right)^{4.8} \qquad \text{Simplify.}$$

$$\approx 0.359 \text{ gram.} \qquad \text{Use a calculator.}$$

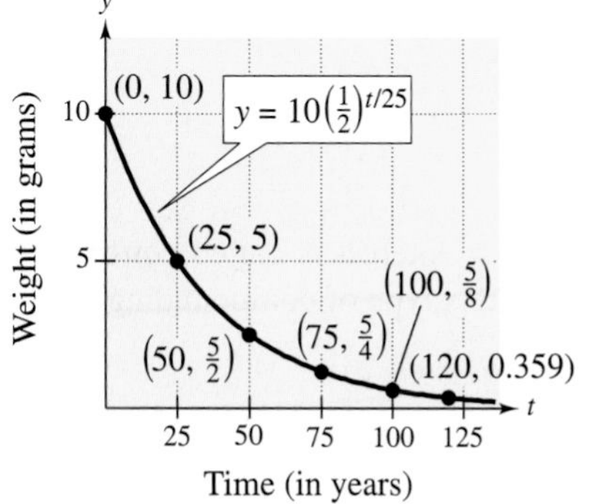

Figure 11.7

So, after 120 years, the mass has decayed from an initial amount of 10 grams to only 0.359 gram. Note in Figure 11.7 that the graph of the function shows the 25-year half-life. That is, after 25 years the mass is 5 grams (half of the original), after another 25 years the mass is 2.5 grams, and so on.

One of the most familiar uses of exponential functions involves **compound interest.** A principal P is invested at an annual interest rate r (in decimal form), compounded once a year. If the interest is added to the principal at the end of the year, the balance is

$$A = P + Pr = P(1 + r).$$

This pattern of multiplying the previous principal by $(1 + r)$ is then repeated each successive year, as shown below.

Time in Years	*Balance at Given Time*
0	$A = P$
1	$A = P(1 + r)$
2	$A = P(1 + r)(1 + r) = P(1 + r)^2$
3	$A = P(1 + r)^2(1 + r) = P(1 + r)^3$
$\vdots$	$\vdots$
t	$A = P(1 + r)^t$

To account for more frequent compounding of interest (such as quarterly or monthly compounding), let n be the number of compoundings per year and let t be the number of years. Then the rate per compounding is r/n and the account balance after t years is

$$A = P\left(1 + \frac{r}{n}\right)^{nt}.$$

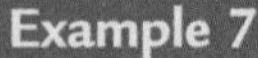

Example 7 Finding the Balance for Compound Interest

A sum of \$10,000 is invested at an annual interest rate of 7.5%, compounded monthly. Find the balance in the account after 10 years.

Solution

Using the formula for compound interest, with $P = 10{,}000$, $r = 0.075$, $n = 12$ (for monthly compounding), and $t = 10$, you obtain the following balance.

$$A = 10{,}000\left(1 + \frac{0.075}{12}\right)^{12(10)} \approx \$21{,}120.65$$

A second method that banks use to compute interest is called **continuous compounding.** The formula for the balance for this type of compounding is

$$A = Pe^{rt}.$$

The formulas for both types of compounding are summarized on the next page.

Formulas for Compound Interest

After t years, the balance A in an account with principal P and annual interest rate r (in decimal form) is given by the following formulas.

1. For n compoundings per year: $A = P\left(1 + \frac{r}{n}\right)^{nt}$
2. For continuous compounding: $A = Pe^{rt}$

Example 8 Comparing Three Types of Compounding

A total of \$15,000 is invested at an annual interest rate of 8%. Find the balance after 6 years if the interest is compounded

a. quarterly, **b.** monthly, and **c.** continuously.

Technology: Discovery

Use a graphing utility to evaluate

$$A = 15{,}000\left(1 + \frac{0.08}{n}\right)^{n(6)}$$

for $n = 1000$, 10,000, and 100,000. Compare these values with those found in parts (a) and (b) of Example 8.

As n gets larger and larger, do you think that the value of A will ever exceed the value found in Example 8(c)? Explain.

Solution

a. Letting $P = 15{,}000$, $r = 0.08$, $n = 4$, and $t = 6$, the balance after 6 years at quarterly compounding is

$$A = 15{,}000\left(1 + \frac{0.08}{4}\right)^{4(6)}$$
$$= \$24{,}126.56.$$

b. Letting $P = 15{,}000$, $r = 0.08$, $n = 12$, and $t = 6$, the balance after 6 years at monthly compounding is

$$A = 15{,}000\left(1 + \frac{0.08}{12}\right)^{12(6)}$$
$$= \$24{,}202.53.$$

c. Letting $P = 15{,}000$, $r = 0.08$, and $t = 6$, the balance after 6 years at continuous compounding is

$$A = 15{,}000e^{0.08(6)}$$
$$= \$24{,}241.12.$$

Note that the balance is greater with continuous compounding than with quarterly and monthly compounding.

Example 8 illustrates the following general rule. For a given principal, interest rate, and time, the more often the interest is compounded per year, the greater the balance will be. Moreover, the balance obtained by continuous compounding is larger than the balance obtained by compounding n times per year.

Discussing the Concept — Finding a Pattern

Use a graphing utility to investigate the function $f(x) = k^x$ for $0 < k < 1$, $k = 1$, and $k > 1$. Discuss the effect that k has on the shape of the graph.

11.1 Exercises

Integrated Review *Concepts, Skills, and Problem Solving*

Keep mathematically in shape by doing these exercises *before* the problems of this section.

Properties and Definitions

1. Explain how to determine the half-plane satisfying $x + y < 5$.

2. Describe the difference between the graphs of $3x - 5y \le 15$ and $3x - 5y < 15$.

Graphing Inequalities

In Exercises 3–10, graph the inequality.

3. $y > x - 2$

4. $y \le 5 - \frac{3}{2}x$

5. $y < \frac{2}{3}x - 1$

6. $x > 6 - y$

7. $y \le -2$

8. $x > 7$

9. $2x + 3y \ge 6$

10. $5x - 2y < 5$

Problem Solving

11. Working together, two people can complete a task in 10 hours. Working alone, one person takes 3 hours longer than the other. How long would it take each to do the task alone?

12. A family is setting up the boundaries for a backyard volleyball court. The court is to be 60 feet long and 30 feet wide. To be assured that the court is rectangular, someone suggests that they measure the diagonals of the court. What should be the length of each diagonal?

Developing Skills

In Exercises 1–8, simplify the expression.

1. $2^x \cdot 2^{x-1}$

2. $10e^{2x} \cdot e^{-x}$

3. $\dfrac{e^{x+2}}{e^x}$

4. $\dfrac{3^{2x+3}}{3^{x+1}}$

5. $(2e^x)^3$

6. $-4e^{-2x}$

7. $\sqrt[3]{-8e^{3x}}$

8. $\sqrt{4e^{6x}}$

In Exercises 9–16, approximate the expression to three decimal places.

9. $4^{\sqrt{3}}$

10. $6^{-\pi}$

11. $e^{1/3}$

12. $e^{-1/3}$

13. $4(3e^4)^{1/2}$

14. $(9e^2)^{3/2}$

15. $\dfrac{4e^3}{12e^2}$

16. $\dfrac{6e^5}{10e^7}$

In Exercises 17–30, evaluate the function as indicated. Use a calculator only if it is more efficient or necessary. (Round to three decimal places.) See Example 1.

17. $f(x) = 3^x$
(a) $x = -2$
(b) $x = 0$
(c) $x = 1$

18. $F(x) = 3^{-x}$
(a) $x = -2$
(b) $x = 0$
(c) $x = 1$

19. $g(x) = 1.07^x$
(a) $x = -1$
(b) $x = 3$
(c) $x = \sqrt{5}$

20. $G(x) = 2.04^{-x}$
(a) $x = -1$
(b) $x = 1$
(c) $x = \sqrt{3}$

21. $f(t) = 500\left(\frac{1}{2}\right)^t$
(a) $t = 0$
(b) $t = 1$
(c) $t = \pi$

22. $g(s) = 1200\left(\frac{2}{3}\right)^s$
(a) $s = 0$
(b) $s = 2$
(c) $s = \sqrt{2}$

23. $f(x) = 1000(1.05)^{2x}$
(a) $x = 0$
(b) $x = 5$
(c) $x = 10$

24. $g(t) = 10{,}000(1.03)^{4t}$
(a) $t = 1$
(b) $t = 3$
(c) $t = 5.5$

25. $h(x) = \dfrac{5000}{(1.06)^{8x}}$
(a) $x = 5$
(b) $x = 10$
(c) $x = 20$

26. $P(t) = \dfrac{10{,}000}{(1.01)^{12t}}$
(a) $t = 2$
(b) $t = 10$
(c) $t = 20$

27. $g(x) = 10e^{-0.5x}$
(a) $x = -4$
(b) $x = 4$
(c) $x = 8$

28. $A(t) = 200e^{0.1t}$
(a) $t = 10$
(b) $t = 20$
(c) $t = 40$

29. $g(x) = \dfrac{1000}{2 + e^{-0.12x}}$
(a) $x = 0$
(b) $x = 10$
(c) $x = 50$

30. $f(z) = \dfrac{100}{1 + e^{-0.05z}}$
(a) $z = 0$
(b) $z = 10$
(c) $z = 20$

In Exercises 31–50, graph the function. See Examples 2 and 3.

31. $f(x) = 3^x$
32. $f(x) = 3^{-x} = \left(\frac{1}{3}\right)^x$
33. $h(x) = \frac{1}{2}(3^x)$
34. $h(x) = \frac{1}{2}(3^{-x})$
35. $g(x) = 3^x - 2$
36. $g(x) = 3^x + 1$
37. $f(x) = 4^{x-5}$
38. $f(x) = 4^{x+1}$
39. $g(x) = 4^x - 5$
40. $g(x) = 4^x + 1$
41. $f(t) = 2^{-t^2}$
42. $f(t) = 2^{t^2}$
43. $f(x) = -2^{0.5x}$
44. $h(t) = -2^{-0.5t}$
45. $h(x) = 2^{0.5x}$
46. $g(x) = 2^{-0.5x}$
47. $f(x) = -\left(\frac{1}{3}\right)^x$
48. $f(x) = \left(\frac{3}{4}\right)^x + 1$
49. $g(t) = 200\left(\frac{1}{2}\right)^t$
50. $h(y) = 27\left(\frac{2}{3}\right)^y$

In Exercises 51–58, match the function with its graph. [The graphs are labeled (a), (b), (c), (d), (e), (f), (g), and (h).]

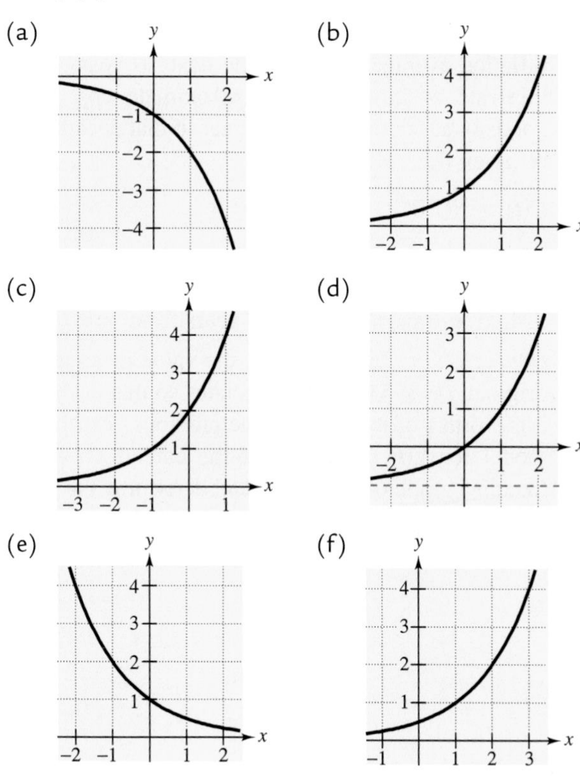
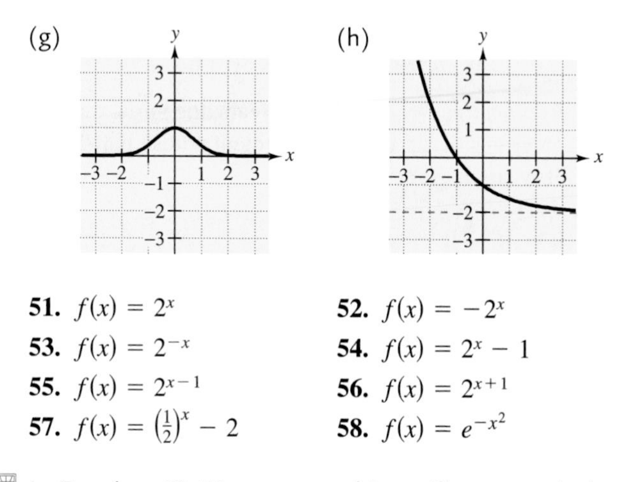

51. $f(x) = 2^x$
52. $f(x) = -2^x$
53. $f(x) = 2^{-x}$
54. $f(x) = 2^x - 1$
55. $f(x) = 2^{x-1}$
56. $f(x) = 2^{x+1}$
57. $f(x) = \left(\frac{1}{2}\right)^x - 2$
58. $f(x) = e^{-x^2}$

In Exercises 59–70, use a graphing utility to graph the function.

59. $y = 5^{x/3}$
60. $y = 5^{-x/3}$
61. $y = 5^{(x-2)/3}$
62. $y = 5^{-x/3} + 2$
63. $y = 500(1.06)^t$
64. $y = 100(1.06)^{-t}$
65. $y = 3e^{0.2x}$
66. $y = 50e^{-0.05x}$
67. $P(t) = 100e^{-0.1t}$
68. $A(t) = 1000e^{0.08t}$
69. $y = 6e^{-x^2/3}$
70. $g(x) = 7e^{(x+1)/2}$

In Exercises 71–76, identify the transformation of the graph of $f(x) = 4^x$ and sketch a graph of h. See Examples 4 and 5.

71. $h(x) = 4^x - 1$
72. $h(x) = 4^x + 2$
73. $h(x) = 4^{x+2}$
74. $h(x) = 4^{x-4}$
75. $h(x) = -4^x$
76. $h(x) = -4^x + 2$

77. ***Think About It*** What type of function does each equation represent?
(a) $f(x) = 2x$
(b) $f(x) = \sqrt{2x}$
(c) $f(x) = 2^x$
(d) $f(x) = 2x^2$

78. ***Identifying Graphs*** Identify the graphs of

$y_1 = e^{0.2x}$, $y_2 = e^{0.5x}$, and $y_3 = e^x$

in the figure. Describe the effect on the graph of $y = e^{kx}$ when $k > 0$ is changed.

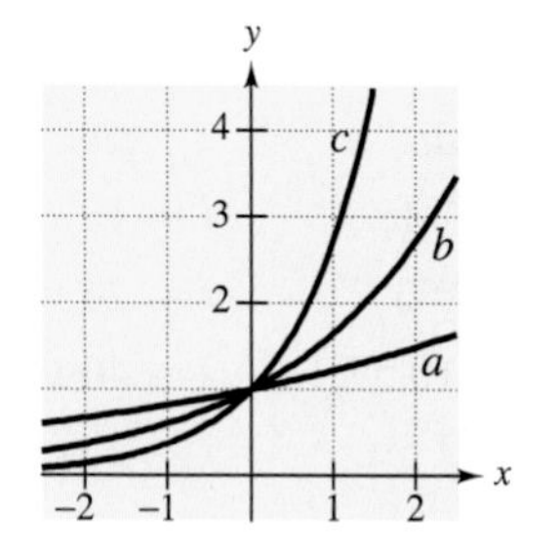

Solving Problems

79. *Radioactive Decay* After t years, the initial mass of 16 grams of a radioactive element whose half-life is 30 years is given by

$$y = 16\left(\frac{1}{2}\right)^{t/30}, \quad t \geq 0.$$

How much of the initial mass remains after 80 years?

80. *Radioactive Decay* After t years, the initial mass of 23 grams of a radioactive element whose half-life is 45 years is given by

$$y = 23\left(\frac{1}{2}\right)^{t/45}, \quad t \geq 0.$$

How much of the initial mass remains after 150 years?

In Exercises 81–86, complete the table to determine the balance A for P dollars invested at rate r for t years, compounded n times per year.

n	1	4	12	365	Continuous compounding
A					

	Principal	*Rate*	*Time*
81.	$P = \$100$	$r = 8\%$	$t = 20$ years
82.	$P = \$400$	$r = 8\%$	$t = 50$ years
83.	$P = \$2000$	$r = 9\%$	$t = 10$ years
84.	$P = \$1500$	$r = 7\%$	$t = 2$ years
85.	$P = \$5000$	$r = 10\%$	$t = 40$ years
86.	$P = \$10{,}000$	$r = 9.5\%$	$t = 30$ years

In Exercises 87–90, complete the table to determine the principal P that will yield a balance of A dollars when invested at rate r for t years, compounded n times per year.

n	1	4	12	365	Continuous compounding
P					

	Balance	*Rate*	*Time*
87.	$A = \$5000$	$r = 7\%$	$t = 10$ years
88.	$A = \$100{,}000$	$r = 9\%$	$t = 20$ years
89.	$A = \$1{,}000{,}000$	$r = 10.5\%$	$t = 40$ years
90.	$A = \$2500$	$r = 7.5\%$	$t = 2$ years

91. *Price and Demand* The daily demand x and the price p for a certain product are related by

$$p = 25 - 0.4e^{0.02x}.$$

Find the prices for demands of (a) $x = 100$ units and (b) $x = 125$ units.

92. *Population Growth* The population of the United States (in recent years) can be approximated by the exponential function

$$P(t) = 205.7(1.0098)^t$$

where P is the population (in millions) and t is the time in years, with $t = 0$ corresponding to 1970. Use the model to estimate the population in the years (a) 2000 and (b) 2010.

93. *Property Value* Suppose that the value of a piece of property doubles every 15 years. If you buy the property for $64,000, its value t years after the date of purchase should be

$$V(t) = 64{,}000(2)^{t/15}.$$

Use the model to approximate the value of the property (a) 5 years and (b) 20 years after it is purchased.

94. *Inflation Rate* Suppose that the annual rate of inflation averages 5% over the next 10 years. With this rate of inflation, the approximate cost C of goods or services during any year in that decade will be given by

$$C(t) = P(1.05)^t, \quad 0 \leq t \leq 10$$

where t is time in years and P is the present cost. If the price of an oil change for your car is presently $24.95, estimate the price 10 years from now.

95. *Depreciation* After t years, the value of a car that originally cost $16,000 depreciates so that each year it is worth $\frac{3}{4}$ of its value for the previous year. Find a model for $V(t)$, the value of the car after t years. Sketch a graph of the model and determine the value of the car 2 years after it was purchased.

96. *Depreciation* Suppose straight-line depreciation is used to determine the value of the car in Exercise 95. Assume that the car depreciates $3000 per year.

(a) Write a linear equation for $V(t)$, the value of the car for year t.

(b) Sketch the graph of the model in part (a) on the same coordinate axes used for the graph in Exercise 95.

(c) If you were selling the car after owning it for 2 years, which depreciation model would you prefer?

(d) If you were selling the car after 4 years, which model would be to your advantage?

97. *Graphical Interpretation* An investment of $500 in two different accounts with respective interest rates of 6% and 8% is compounded continuously.

(a) Write an exponential function to represent the balance after t years for each account.

(b) Use a graphing utility to graph each of the models in the same viewing rectangle.

(c) Use a graphing utility to graph the function $A_2 - A_1$ on the same screen used for the graphs in part (b).

(d) Use the graphs to discuss the rates of increase of the balances in the two accounts.

98. *Savings Plan* Suppose you decide to start saving pennies according to the following pattern. You save 1 penny the first day, 2 pennies the second day, 4 the third day, 8 the fourth day, and so on. Each day you save twice the number of pennies as the previous day. Write an exponential function that models this problem. How many pennies do you save on the thirtieth day? (In the next chapter you will learn how to find the total number saved.)

99. *Parachute Drop* A parachutist jumps from a plane and opens the parachute at a height of 2000 feet (see figure). The distance between the parachutist and the ground is

$$h = 1950 + 50e^{-1.6t} - 20t$$

where h is the distance in feet and t is the time in seconds. (The time $t = 0$ corresponds to the time when the parachute is opened.)

(a) Use a graphing utility to graph the function.

(b) Find the distance between the parachutist and the ground when $t = 0, 25, 50$, and 75.

(c) Approximate the time when the parachutist reaches the ground.

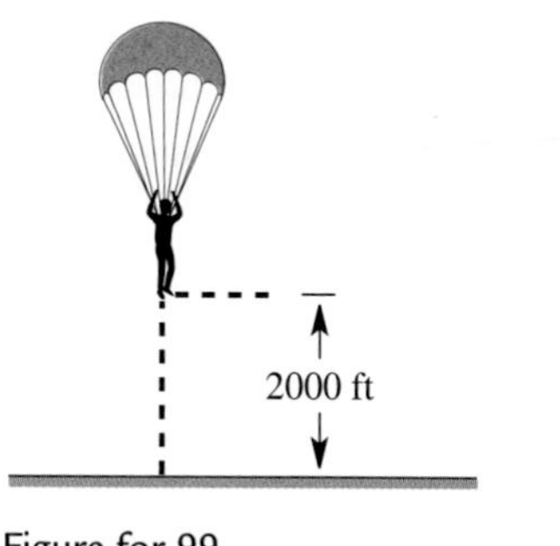

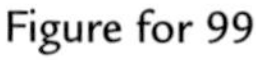
Figure for 99

100. *Parachute Drop* A parachutist jumps from a plane and opens the parachute at a height of 3000 feet. The distance between the parachutist and the ground is

$$h = 2940 + 60e^{-1.7t} - 22t$$

where h is the distance in feet and t is the time in seconds. (The time $t = 0$ corresponds to the time when the parachute is opened.)

(a) Use a graphing utility to graph the function.

(b) Find the distance between the parachutist and the ground when $t = 0, 50$, and 100.

(c) Approximate the time when the parachutist reaches the ground.

101. *Data Analysis* A meteorologist measures the atmospheric pressure P (in kilograms per square meter) at altitude h (in kilometers). The data are shown in the table.

h	0	5	10	15	20
P	10,332	5583	2376	1240	517

(a) Use a graphing utility to plot the data points.

(b) A model for the data is given by

$$P = 10{,}958e^{-0.15h}.$$

Use a graphing utility to graph the model in the same viewing rectangle as in part (a). How well does the model fit the data?

(c) Use a graphing utility to create a table comparing the model with the data points.

(d) Estimate the atmospheric pressure at an altitude of 8 kilometers.

(e) Use the graph to estimate the altitude at which the atmospheric pressure is 2000 kilograms per square meter.

102. *Data Analysis* For the years 1991 through 1996, the median price of a one-family home sold in the United States is given in the table. (Source: National Association of Realtors)

Year	1991	1992	1993
Price	\$100,300	\$103,700	\$106,800

Year	1994	1995	1996
Price	\$109,900	\$113,100	\$118,200

A model for these data is given by

$y = 97{,}107e^{0.0317t}$

where t is time in years, with $t = 0$ representing 1990.

(a) Use the model to complete the table and compare the results with the actual data.

Year	1991	1992	1993	1994	1995	1996
Price						

(b) Use a graphing utility to graph the model.

(c) If the model were used to predict home prices in the years ahead, would the predictions be increasing at a faster rate or a slower rate with increasing t?

(d) Compare the model with the continuous compound interest model. What does the coefficient of t represent?

103. *Calculator Experiment*

(a) Use a calculator to complete the table.

x	1	10	100	1000	10,000
$\left(1 + \frac{1}{x}\right)^x$					

(b) Use the table to sketch the graph of the function

$$f(x) = \left(1 + \frac{1}{x}\right)^x.$$

Does this graph appear to be approaching a horizontal asymptote?

(c) From parts (a) and (b), what conclusions can you make about the value of

$$\left(1 + \frac{1}{x}\right)^x$$

as x gets larger and larger?

Explaining Concepts

104. Answer parts (a) and (b) of Motivating the Chapter on page 629.

105. Describe the differences between exponential functions and polynomial or rational functions.

106. Explain why 1^x is not an exponential function.

107. Compare the graphs of $f(x) = 3^x$ and $g(x) = \left(\frac{1}{3}\right)^x$.

108. Describe some applications of the exponential functions $f(x) = a^x$ and $g(x) = a^{-x}$.

109. *True or False?* $e = \dfrac{271{,}801}{99{,}990}$. Explain.

110. Without using a calculator, how do you know that $2^{\sqrt{2}}$ is greater than 2, but less than 4?

11.2 Inverse Functions

Objectives

1. Form the composition of two functions and find the domain of the composite function.
2. Use the Horizontal Line Test to determine if a function has an inverse.
3. Find the inverse of a function algebraically.
4. Graphically show that two functions are inverses of each other.

1 Form the composition of two functions and find the domain of the composite function.

Composition of Functions

Two functions can be combined to form another function called the **composition** of the two functions. For instance, if $f(x) = 2x^2$ and $g(x) = x - 1$, the composition of f with g is denoted by $f \circ g$ and is evaluated as

$$f(g(x)) = f(x - 1) = 2(x - 1)^2.$$

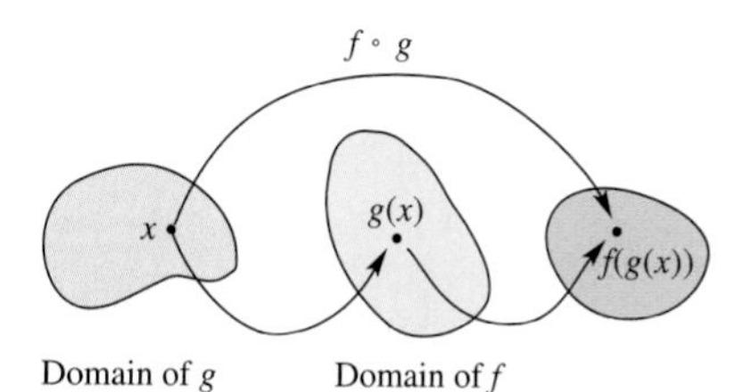

Figure 11.8

Definition of Composition of Two Functions

The **composition** of the functions f and g is given by

$$(f \circ g)(x) = f(g(x)).$$

The domain of the **composite function** $(f \circ g)$ is the set of all x in the domain of g such that $g(x)$ is in the domain of f. (See Figure 11.8.)

Study Tip

A composite function can be viewed as a *function within a function,* where the composition

$$(f \circ g)(x) = f(g(x))$$

has f as the "outer" function and g as the "inner" function. This is reversed in the composition

$$(g \circ f)(x) = g(f(x)).$$

Example 1 Forming the Composition of Two Functions

Find the composition of f with g. Evaluate the composite function when $x = 1$ and when $x = -3$.

$$f(x) = 2x + 4 \quad \text{and} \quad g(x) = 3x - 1$$

Solution

The composition of f with g is given by

$$\begin{aligned}
(f \circ g)(x) &= f(g(x)) && \text{Composite form} \\
&= f(3x - 1) && g(x) = 3x - 1 \text{ is the inner function.} \\
&= 2(3x - 1) + 4 && \text{Input } 3x - 1 \text{ into the outer function } f. \\
&= 6x - 2 + 4 && \text{Simplify.} \\
&= 6x + 2. && \text{Simplify.}
\end{aligned}$$

When $x = 1$, the value of this function is

$$(f \circ g)(1) = 6(1) + 2 = 8.$$

When $x = -3$, the value of this function is

$$(f \circ g)(-3) = 6(-3) + 2 = -16.$$

The composition of f with g is generally *not* the same as the composition of g with f. This is illustrated in Example 2.

Example 2 Comparing the Compositions of Functions

Given $f(x) = 2x - 3$ and $g(x) = x^2 + 1$, find each of the following.

a. $(f \circ g)(x)$ **b.** $(g \circ f)(x)$

Solution

a. The composition of f with g is as follows.

$$
\begin{aligned}
(f \circ g)(x) &= f(g(x)) && \text{Composite form} \\
&= f(x^2 + 1) && g(x) = x^2 + 1 \text{ is the inner function.} \\
&= 2(x^2 + 1) - 3 && \text{Input } x^2 + 1 \text{ into the outer function } f. \\
&= 2x^2 + 2 - 3 && \text{Distributive Property} \\
&= 2x^2 - 1 && \text{Simplify.}
\end{aligned}
$$

b. The composition of g with f is as follows.

$$
\begin{aligned}
(g \circ f)(x) &= g(f(x)) && \text{Composite form} \\
&= g(2x - 3) && f(x) = 2x - 3 \text{ is the inner function.} \\
&= (2x - 3)^2 + 1 && \text{Input } 2x - 3 \text{ into the outer function } g. \\
&= 4x^2 - 12x + 9 + 1 && \text{Expand } (2x - 3)^2. \\
&= 4x^2 - 12x + 10 && \text{Simplify.}
\end{aligned}
$$

Note that $(f \circ g)(x) \neq (g \circ f)(x)$.

Study Tip

To determine the domain of a composite function, first write the composite function in simplest form. Then use the fact that its domain *either is equal to or is a restriction of the domain of the "inner" function.* Check to see if the domain of $g \circ f$ in Example 3 is the same as or is a restriction of the domain of f.

Example 3 Finding the Domain of a Composite Function

Find the domain of the composition of $(f \circ g)(x)$ when $f(x) = x^2$ and $g(x) = \sqrt{x}$.

Solution

The composition of f with g is given by

$$
\begin{aligned}
(f \circ g)(x) &= f(g(x)) && \text{Composite form} \\
&= f(\sqrt{x}) && g(x) = \sqrt{x} \text{ is the inner function.} \\
&= (\sqrt{x})^2 && \text{Input } \sqrt{x} \text{ into the outer function } f. \\
&= x, \quad x \geq 0. && \text{Domain of } f \circ g \text{ is all } x \geq 0.
\end{aligned}
$$

The domain of the inner function $g(x) = \sqrt{x}$ is the interval $[0, \infty)$. The simplified form of $f \circ g$ has no restriction on this set of numbers. So, the restriction $x \geq 0$ must be added to the composition of this function. The domain of $f \circ g$ is $[0, \infty)$.

2 Use the Horizontal Line Test to determine if a function has an inverse.

Inverse and One-to-One Functions

In Section 4.3, you learned that a function can be represented by a set of ordered pairs. For instance, the function $f(x) = x + 2$ from the set $A = \{1, 2, 3, 4\}$ to the set $B = \{3, 4, 5, 6\}$ can be written as follows.

$$f(x) = x + 2: \quad \{(1, 3), (2, 4), (3, 5), (4, 6)\}$$

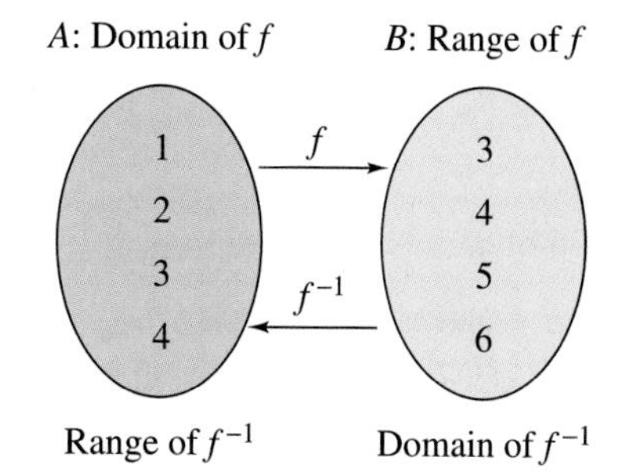

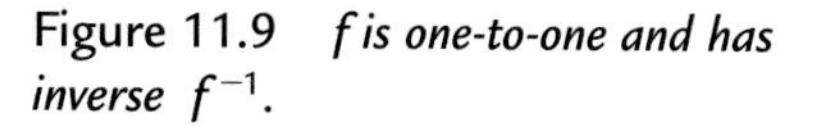

Figure 11.9 *f is one-to-one and has inverse* f^{-1}.

By interchanging the first and second coordinates of each of these ordered pairs, you can form another function that is called the **inverse function** of f. The inverse function is denoted by f^{-1}. It is a function from the set B to the set A, and can be written as follows.

$$f^{-1}(x) = x - 2: \quad \{(3, 1), (4, 2), (5, 3), (6, 4)\}$$

Interchanging the ordered pairs for a function f will only produce another function when f is one-to-one. A function f is **one-to-one** if each value of the dependent variable corresponds to exactly one value of the independent variable. Figure 11.9 shows that the domain of f is the range of f^{-1} and the range of f is the domain of f^{-1}.

> **Horizontal Line Test for Inverse Functions**
>
> A function f has an inverse function if and only if f is one-to-one. Graphically, a function f has an inverse function if and only if no *horizontal* line intersects the graph of f at more than one point.

Example 4 Applying the Horizontal Line Test

a. The function $f(x) = x^3 - 1$ *has* an inverse because no horizontal line intersects its graph at more than one point, as shown in Figure 11.10.

b. The function $f(x) = x^2 - 1$ *does not have* an inverse because it is possible to find a horizontal line that intersects the graph of f at more than one point, as shown in Figure 11.11.

c. The function f: $\{(1, 3), (2, 0), (3, -1), (4, 3)\}$ *does not have* an inverse function because the horizontal line $y = 3$ intersects two points of the graph of f, as shown in Figure 11.12.

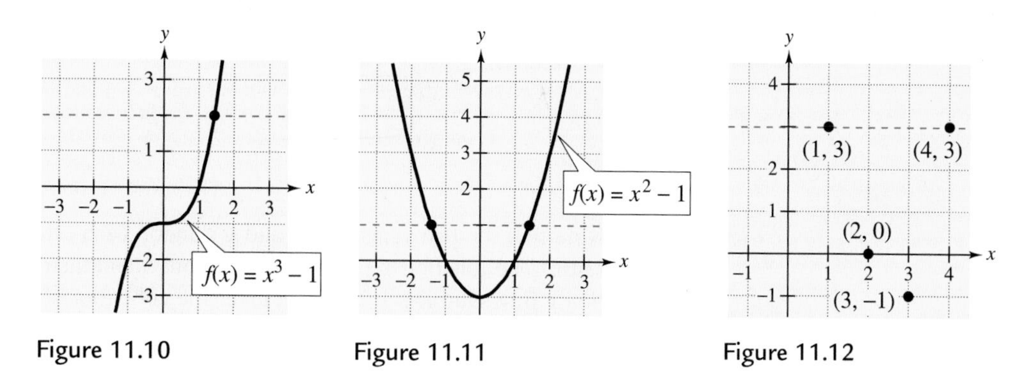

Figure 11.10

Figure 11.11

Figure 11.12

Don't be confused by the use of -1 to denote the inverse function f^{-1}. Whenever we write f^{-1}, we will *always* be referring to the inverse of the function f and *not* the reciprocal of $f(x)$.

The formal definition of the inverse of a function is given as follows.

▶ Definition of the Inverse of a Function

Let f and g be two functions such that

$$f(g(x)) = x \quad \text{for every } x \text{ in the domain of } g$$

and

$$g(f(x)) = x \quad \text{for every } x \text{ in the domain of } f.$$

The function g is the **inverse** of the function f, and is denoted by f^{-1} (read "f-inverse"). So, $f(f^{-1}(x)) = x$ and $f^{-1}(f(x)) = x$. The domain of f is equal to the range of f^{-1}, and vice versa.

If the function g is the inverse of the function f, it must also be true that the function f is the inverse of the function g. For this reason, you can refer to the functions f and g as being *inverses of each other.*

Example 5 Verifying Inverse Functions

Show that each function is an inverse of the other.

$$f(x) = x^3 + 1 \quad \text{and} \quad g(x) = \sqrt[3]{x - 1}$$

Solution

Begin by noting that the domain and range of both functions are the entire set of real numbers. To show that f and g are inverses of each other, you need to show that $f(g(x)) = x$ and $g(f(x)) = x$, as follows.

$f(g(x)) = f\left(\sqrt[3]{x-1}\right)$ — $g(x) = \sqrt[3]{x-1}$ is the inner function.

$= \left(\sqrt[3]{x-1}\right)^3 + 1$ — Input $\sqrt[3]{x-1}$ into the outer function f.

$= (x - 1) + 1$ — Simplify.

$= x$ — Simplify.

$g(f(x)) = g(x^3 + 1)$ — $f(x) = x^3 + 1$ is the inner function.

$= \sqrt[3]{(x^3 + 1) - 1}$ — Input $x^3 + 1$ into the outer function g.

$= \sqrt[3]{x^3}$ — Simplify.

$= x$ — Simplify.

Note that the two functions f and g "undo" each other in the following verbal sense. The function f first cubes the input x and then adds 1, whereas the function g first subtracts 1, and then takes the cube root of the result.

3 Find the inverse of a function algebraically.

Finding Inverse Functions

You can find the inverse of a simple function by inspection. For instance, the inverse of $f(x) = 10x$ is $f^{-1}(x) = x/10$. For more complicated functions, however, it is best to use the following steps for finding the inverse of a function. The key step in these guidelines is switching the roles of x and y. This step corresponds to the fact that inverse functions have ordered pairs with the coordinates reversed.

Study Tip

You can graph a function and use the Horizontal Line Test to see if the function is one-to-one before trying to find its inverse.

▶ Finding the Inverse of a Function

1. In the equation for $f(x)$, replace $f(x)$ by y.
2. Interchange the roles of x and y, and solve for y.
3. If the new equation does not represent y as a function of x, the function f does not have an inverse function.
4. If the new equation represents y as a function of x, replace y by $f^{-1}(x)$.

Technology: Discovery

Use a graphing utility to graph $f(x) = x^3 + 1$, $f^{-1}(x) = \sqrt[3]{x - 1}$, and $y = x$ in the same viewing window.

a. Relative to the line $y = x$, how do the graphs of f and f^{-1} compare?

b. For the graph of f, complete the table.

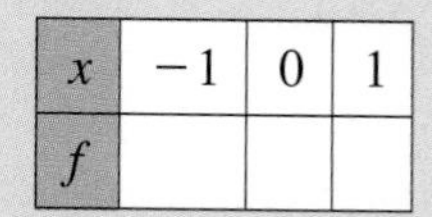

x	-1	0	1
f			

For the graph of f^{-1}, complete the table.

x	0	1	2
f^{-1}			

What can you conclude about the coordinates of the points on the graph of f compared with those on the graph of f^{-1}?

Example 6 Finding the Inverse of a Function

Find the inverse, if it exists.

a. $f(x) = 2x + 3$ **b.** $f(x) = x^2$ **c.** $f(x) = x^3 + 3$

Solution

a.

$f(x) = 2x + 3$	Original function
$y = 2x + 3$	Replace $f(x)$ by y.
$x = 2y + 3$	Interchange x and y.
$y = \dfrac{x - 3}{2}$	Solve for y.
$f^{-1}(x) = \dfrac{x - 3}{2}$	Replace y by $f^{-1}(x)$.

b.

$f(x) = x^2$	Original function
$y = x^2$	Replace $f(x)$ by y.
$x = y^2$	Interchange x and y.
$\pm\sqrt{x} = y$	Solve for y.

Because y is not a function of x, the original function f has no inverse.

c.

$f(x) = x^3 + 3$	Original function
$y = x^3 + 3$	Replace $f(x)$ by y.
$x = y^3 + 3$	Interchange x and y.
$\sqrt[3]{x - 3} = y$	Solve for y.
$f^{-1}(x) = \sqrt[3]{x - 3}$	Replace y by $f^{-1}(x)$.

4 Graphically show that two functions are inverses of each other.

Graphs of Inverse Functions

The graphs of a function f and its inverse f^{-1} are related to each other in the following way. If the point (a, b) lies on the graph of f, the point (b, a) must lie on the graph of f^{-1}, and vice versa. This means that the graph of f^{-1} is a reflection of the graph of f in the line $y = x$, as shown in Figure 11.13.

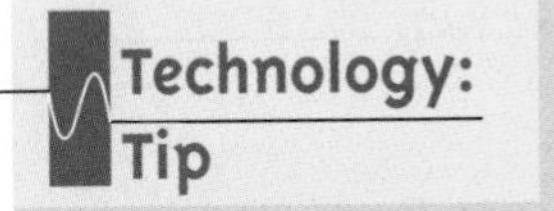

A graphing utility program that graphs the function f and its reflection in the line $y = x$ can be found at our website *www.hmco.com*. Programs are available for several current models of graphing utilities.

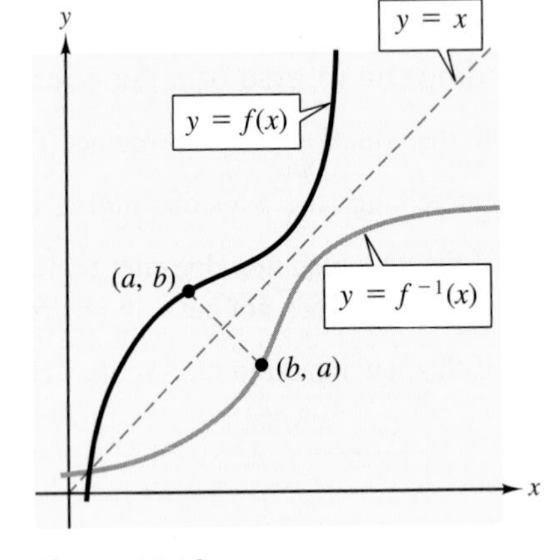

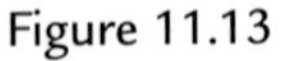
Figure 11.13

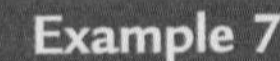

Example 7 The Graphs of f and f^{-1}

Sketch the graphs of the inverse functions $f(x) = 2x - 3$ and $f^{-1}(x) = \frac{1}{2}(x + 3)$ on the same rectangular coordinate system and show that the graphs are reflections of each other in the line $y = x$.

Solution

The graphs of f and f^{-1} are shown in Figure 11.14. Visually, it appears that the graphs are reflections of each other. You can further verify this by testing a few points on each graph. Note in the following list that if the point (a, b) is on the graph of f, the point (b, a) is on the graph of f^{-1}.

$f(x) = 2x - 3$	$f^{-1}(x) = \frac{1}{2}(x + 3)$
$(-1, -5)$	$(-5, -1)$
$(0, -3)$	$(-3, 0)$
$(1, -1)$	$(-1, 1)$
$(2, 1)$	$(1, 2)$
$(3, 3)$	$(3, 3)$

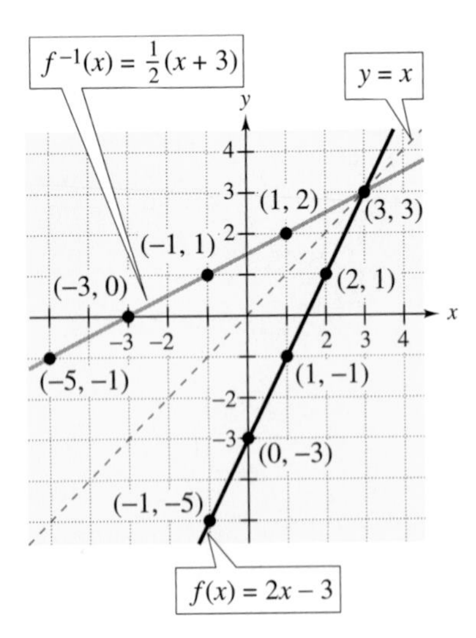

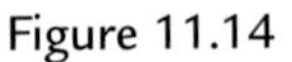
Figure 11.14

You can sketch the graph of the inverse of a function without knowing the equation of the inverse function. Simply find the coordinates of points that lie on the original function. By interchanging the x- and y-coordinates you have points that lie on the graph of the inverse function. Plot these points and sketch the graph of the inverse.

In Example 6(b) we said that the function $f(x) = x^2$ has no inverse. What we mean is that *assuming the domain of f is the entire real line*, the function $f(x) = x^2$ has no inverse. If, however, we restrict the domain of f to the nonnegative real numbers, f does have an inverse.

Example 8 The Graphs of f and f^{-1}

Graphically show that each function is an inverse of the other.

$$f(x) = x^2, \quad x \geq 0, \quad \text{and} \quad f^{-1}(x) = \sqrt{x}$$

Solution

The graphs of f and f^{-1} are shown in Figure 11.15. Visually, it appears that the graphs are reflections of each other in the line $y = x$. You can further verify this by testing a few points on each graph. Note in the following list that if the point (a, b) is on the graph of f, the point (b, a) is on the graph of f^{-1}.

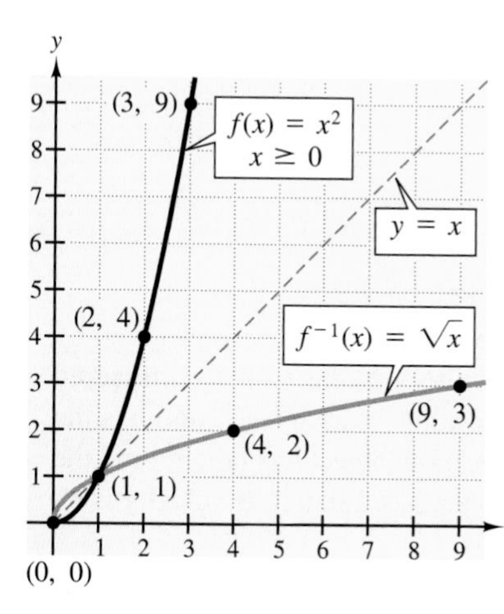

Figure 11.15

$f(x) = x^2, \quad x \geq 0$	$f^{-1}(x) = \sqrt{x}$
(0, 0)	(0, 0)
(1, 1)	(1, 1)
(2, 4)	(4, 2)
(3, 9)	(9, 3)

Discussing the Concept — Error Analysis

Suppose you are an algebra instructor and one of your students hands in the following solutions. Find and correct the errors and discuss how you can help your student avoid such errors in the future.

a. If $f(x) = 2x - 1$ and $g(x) = x^3 + 1$, find $(f \circ g)(2)$.

$$\begin{aligned}(f \circ g)(2) &= (2 \cdot 2 - 1)(2^3 + 1)\\ &= (4 - 1)(8 + 1)\\ &= (3)(9)\\ &= 27\end{aligned}$$

b. If $f(x) = 3x^2 + x$ and $g(x) = x - 2$, find $(f \circ g)(1)$.

$$\begin{aligned}(f \circ g)(1) &= f(1) - 2\\ &= [3(1)^2 + 1] - 2\\ &= (3 + 1) - 2\\ &= 2\end{aligned}$$

11.2 Exercises

Integrated Review — *Concepts, Skills, and Problem Solving*

Keep mathematically in shape by doing these exercises *before* the problems of this section.

Properties and Definitions

1. Decide whether $x - y^2 = 0$ represents y as a function of x. Explain.

2. Decide whether $|x| - 2y = 4$ represents y as a function of x. Explain.

3. Explain why the domains of f and g are not the same.

$$f(x) = \sqrt{4 - x^2} \qquad g(x) = \frac{6}{\sqrt{4 - x^2}}$$

4. Determine the range of $h(x) = 8 - \sqrt{x}$ over the domain $\{0, 4, 9, 16\}$.

Simplifying Expressions

In Exercises 5–10, perform the operations and simplify.

5. $-(5x^2 - 1) + (3x^2 - 5)$

6. $(-2x)(-5x)(3x + 4)$

7. $(u - 4v)(u + 4v)$

8. $(3a - 2b)^2$

9. $(t - 2)^3$

10. $\dfrac{6x^3 - 3x^2}{12x}$

Problem Solving

11. The velocity of a free-falling body is given by $v = \sqrt{2gh}$, where v is the velocity measured in feet per second, $g = 32$ ft/sec^2, and h is the distance (in feet) the object has fallen. Find the distance an object has fallen if its velocity is 80 feet per second.

12. The cost for a long-distance telephone call is \$0.95 for the first minute and \$0.35 for each additional minute. The total cost of a call is \$5.15. Find the length of the call.

Developing Skills

In Exercises 1–10, find the compositions. See Examples 1 and 2.

1. $f(x) = x - 3, \quad g(x) = 2x - 4$
(a) $(f \circ g)(x)$ (b) $(g \circ f)(x)$
(c) $(f \circ g)(4)$ (d) $(g \circ f)(7)$

2. $f(x) = x + 1, \quad g(x) = 5 - 3x$
(a) $(f \circ g)(x)$ (b) $(g \circ f)(x)$
(c) $(f \circ g)(3)$ (d) $(g \circ f)(3)$

3. $f(x) = x + 5, \quad g(x) = 2x^2 - 6$
(a) $(f \circ g)(x)$ (b) $(g \circ f)(x)$
(c) $(f \circ g)(2)$ (d) $(g \circ f)(-3)$

4. $f(x) = x^2 - 3x, \quad g(x) = 3 - 2x$
(a) $(f \circ g)(x)$ (b) $(g \circ f)(x)$
(c) $(f \circ g)(-1)$ (d) $(g \circ f)(3)$

5. $f(x) = |x - 3|, \quad g(x) = 3x$
(a) $(f \circ g)(x)$ (b) $(g \circ f)(x)$
(c) $(f \circ g)(1)$ (d) $(g \circ f)(2)$

6. $f(x) = |x|, \quad g(x) = 2x + 5$
(a) $(f \circ g)(x)$ (b) $(g \circ f)(x)$
(c) $(f \circ g)(-2)$ (d) $(g \circ f)(-4)$

7. $f(x) = \sqrt{x - 4}, \quad g(x) = x + 5$
(a) $(f \circ g)(x)$ (b) $(g \circ f)(x)$
(c) $(f \circ g)(3)$ (d) $(g \circ f)(8)$

8. $f(x) = \sqrt{x + 6}, \quad g(x) = 2x - 3$
(a) $(f \circ g)(x)$ (b) $(g \circ f)(x)$
(c) $(f \circ g)(3)$ (d) $(g \circ f)(-2)$

9. $f(x) = \frac{1}{x - 3}$, $g(x) = \frac{2}{x^2}$

(a) $(f \circ g)(x)$ (b) $(g \circ f)(x)$

(c) $(f \circ g)(-1)$ (d) $(g \circ f)(2)$

10. $f(x) = \frac{4}{x^2 - 4}$, $g(x) = \frac{1}{x}$

(a) $(f \circ g)(x)$ (b) $(g \circ f)(x)$

(c) $(f \circ g)(-2)$ (d) $(g \circ f)(1)$

In Exercises 11–14, use the functions f and g to find the indicated values.

$f = \{(-2, 3), (-1, 1), (0, 0), (1, -1), (2, -3)\}$,

$g = \{(-3, 1), (-1, -2), (0, 2), (2, 2), (3, 1)\}$

11. (a) $f(1)$ (b) $g(-1)$ (c) $(g \circ f)(1)$

12. (a) $g(0)$ (b) $f(2)$ (c) $(f \circ g)(0)$

13. (a) $(f \circ g)(-3)$ (b) $(g \circ f)(-2)$

14. (a) $(f \circ g)(2)$ (b) $(g \circ f)(2)$

In Exercises 15–18, use the functions f and g to find the indicated values.

$f = \{(0, 1), (1, 2), (2, 5), (3, 10), (4, 17)\}$,

$g = \{(5, 4), (10, 1), (2, 3), (17, 0), (1, 2)\}$

15. (a) $f(3)$ (b) $g(10)$ (c) $(g \circ f)(3)$

16. (a) $g(2)$ (b) $f(0)$ (c) $(f \circ g)(10)$

17. (a) $(g \circ f)(4)$ (b) $(f \circ g)(2)$

18. (a) $(f \circ g)(1)$ (b) $(g \circ f)(0)$

In Exercises 19–26, find the domain of the composition (a) $f \circ g$ and (b) $g \circ f$. See Example 3.

19. $f(x) = x + 1$
$g(x) = 2x - 5$

20. $f(x) = 2 - 3x$
$g(x) = 5x + 3$

21. $f(x) = \sqrt{x}$
$g(x) = x - 2$

22. $f(x) = \sqrt{x - 5}$
$g(x) = x + 3$

23. $f(x) = x^2 - 1$
$g(x) = \sqrt{x + 3}$

24. $f(x) = \sqrt{2x - 1}$
$g(x) = x^2 + 1$

25. $f(x) = \frac{x}{x + 5}$
$g(x) = \sqrt{x - 1}$

26. $f(x) = \frac{x}{x - 4}$
$g(x) = \sqrt{x}$

In Exercises 27–32, use the Horizontal Line Test to determine whether the function has an inverse. See Example 4.

27. $f(x) = x^2 - 2$

28. $f(x) = \frac{1}{5}x$

29. $f(x) = x^2, \quad x \geq 0$

30. $f(x) = \sqrt{-x}$

31. $g(x) = \sqrt{25 - x^2}$

32. $g(x) = |x - 4|$

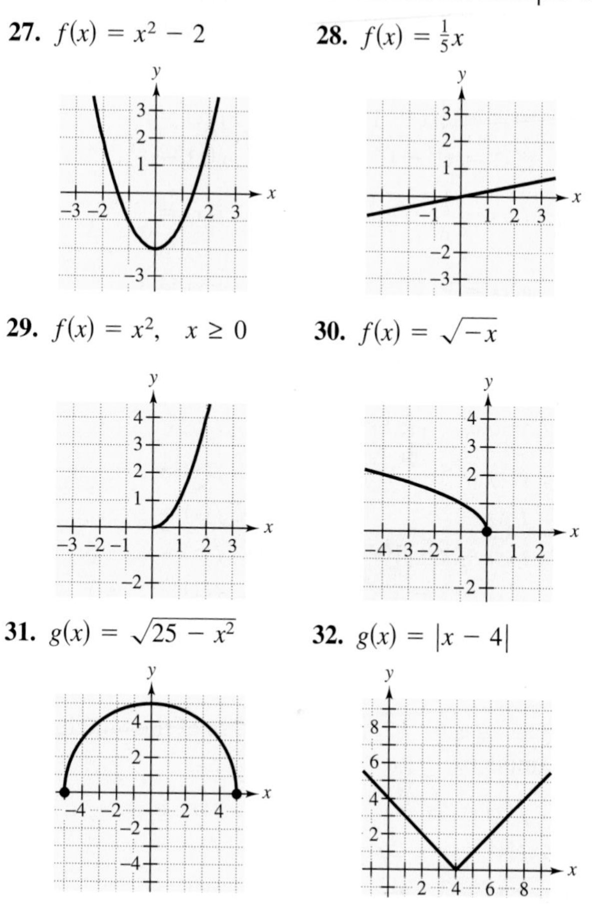

In Exercises 33–42, use a graphing utility to graph the function and determine whether the function is one-to-one.

33. $f(x) = x^3 - 1$

34. $f(x) = (2 - x)^3$

35. $f(t) = \sqrt[3]{5 - t}$

36. $h(t) = 4 - \sqrt[3]{t}$

37. $g(x) = x^4 - 6$

38. $f(x) = (x + 2)^5$

39. $h(t) = \frac{5}{t}$

40. $g(t) = \frac{5}{t^2}$

41. $f(s) = \frac{4}{s^2 + 1}$

42. $f(x) = \frac{1}{x^2 - 2}$

In Exercises 43–54, verify algebraically that the functions f and g are inverses of each other. See Example 5.

43. $f(x) = 10x$
$g(x) = \frac{1}{10}x$

44. $f(x) = \frac{2}{3}x$
$g(x) = \frac{3}{2}x$

45. $f(x) = x + 15$
$g(x) = x - 15$

46. $f(x) = 3 - x$
$g(x) = 3 - x$

47. $f(x) = 1 - 2x$
$g(x) = \frac{1}{2}(1 - x)$

48. $f(x) = 2x - 1$
$g(x) = \frac{1}{2}(x + 1)$

49. $f(x) = 2 - 3x$
$g(x) = \frac{1}{3}(2 - x)$

50. $f(x) = -\frac{1}{4}x + 3$
$g(x) = -4(x - 3)$

51. $f(x) = \sqrt[3]{x + 1}$
$g(x) = x^3 - 1$

52. $f(x) = x^7$
$g(x) = \sqrt[7]{x}$

53. $f(x) = \dfrac{1}{x}$
$g(x) = \dfrac{1}{x}$

54. $f(x) = \dfrac{1}{x - 3}$
$g(x) = 3 + \dfrac{1}{x}$

In Exercises 55–66, find the inverse of the function f. Verify that $f(f^{-1}(x))$ and $f^{-1}(f(x))$ are equal to the identity function. See Example 6.

55. $f(x) = 5x$
56. $f(x) = 2x$
57. $f(x) = \frac{1}{2}x$
58. $f(x) = \frac{1}{3}x$
59. $f(x) = x + 10$
60. $f(x) = x - 5$
61. $f(x) = 3 - x$
62. $f(x) = 8 - x$
63. $f(x) = x^7$
64. $f(x) = x^5$
65. $f(x) = \sqrt[3]{x}$
66. $f(x) = x^{1/5}$

In Exercises 67–82, find the inverse of the function. See Example 6.

67. $f(x) = 8x$
68. $f(x) = \frac{1}{10}x$
69. $g(x) = x + 25$
70. $f(x) = 7 - x$
71. $g(x) = 3 - 4x$
72. $g(t) = 6t + 1$
73. $g(t) = \frac{1}{4}t + 2$
74. $h(s) = 5 - \frac{3}{2}s$
75. $h(x) = \sqrt{x}$
76. $h(x) = \sqrt{x + 5}$
77. $f(t) = t^3 - 1$
78. $h(t) = t^5 + 8$
79. $g(s) = \dfrac{5}{s + 4}$
80. $f(s) = \dfrac{2}{3 - s}$
81. $f(x) = \sqrt{x + 3}, \quad x \ge -3$
82. $f(x) = \sqrt{x^2 - 4}, \quad x \ge 2$

In Exercises 83–88, sketch the graphs of f and f^{-1} on the same rectangular coordinate system. Show that the graphs are reflections of each other in the line $y = x$. See Example 7.

83. $f(x) = x + 4,\ f^{-1}(x) = x - 4$
84. $f(x) = x - 7,\ f^{-1}(x) = x + 7$
85. $f(x) = 3x - 1,\ f^{-1}(x) = \frac{1}{3}(x + 1)$
86. $f(x) = 5 - 4x,\ f^{-1}(x) = -\frac{1}{4}(x - 5)$
87. $f(x) = x^2 - 1,\ x \ge 0,$
$f^{-1}(x) = \sqrt{x + 1}$
88. $f(x) = (x + 2)^2,\ x \ge -2,$
$f^{-1}(x) = \sqrt{x} - 2$

In Exercises 89–92, match the graph with the graph of its inverse. [The graphs of the inverse functions are labeled (a), (b), (c), and (d).]

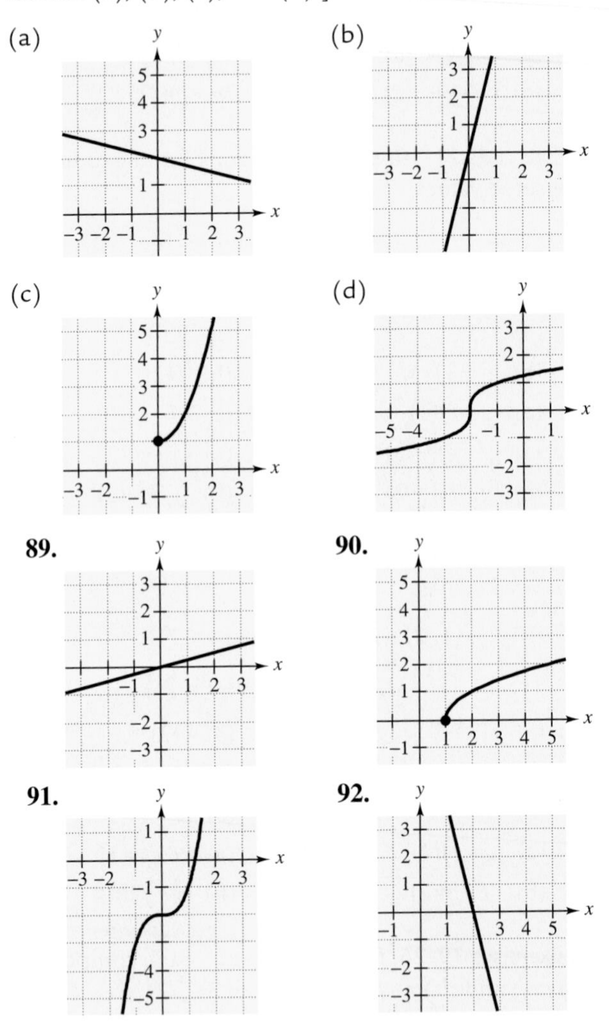

In Exercises 93–100, use a graphing utility to verify that the functions are inverses of each other.

93. $f(x) = \frac{1}{3}x$
$g(x) = 3x$

94. $f(x) = \frac{1}{5}x - 1$
$g(x) = 5x + 5$

95. $f(x) = \sqrt{x + 1}$
$g(x) = x^2 - 1,\ x \ge 0$

96. $f(x) = \sqrt{4 - x}$
$g(x) = 4 - x^2,\ x \ge 0$

97. $f(x) = \frac{1}{8}x^3$
$g(x) = 2\sqrt[3]{x}$

98. $f(x) = \sqrt[3]{x + 2}$
$g(x) = x^3 - 2$

99. $f(x) = 3x + 4$
$g(x) = \frac{1}{3}(x - 4)$

100. $f(x) = |x - 2|,\ x \ge 2$
$g(x) = x + 2,\ x \ge 0$

In Exercises 101–104, delete part of the graph of the function so that the remaining part is one-to-one. Find the inverse of the remaining part and give the domain of the inverse. (*Note:* There is more than one correct answer.) See Example 8.

101. $f(x) = (x - 2)^2$

102. $f(x) = 9 - x^2$

103. $f(x) = |x| + 1$

104. $f(x) = |x - 2|$

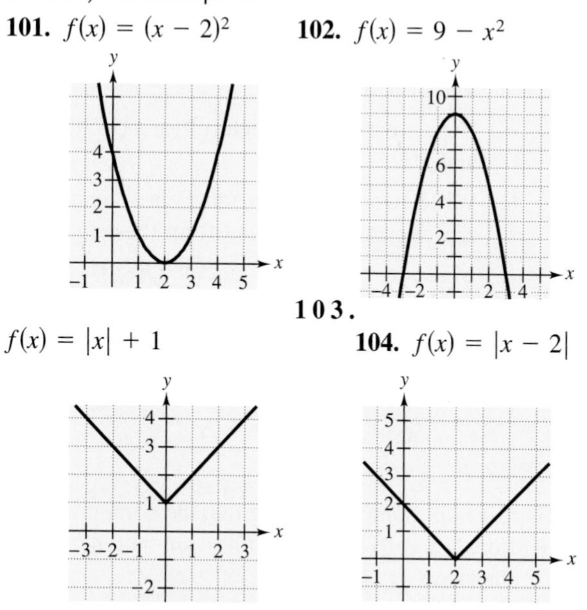

In Exercises 105–108, use the graph of f to sketch the graph of f^{-1}.

105. **106.** **107.** **108.**

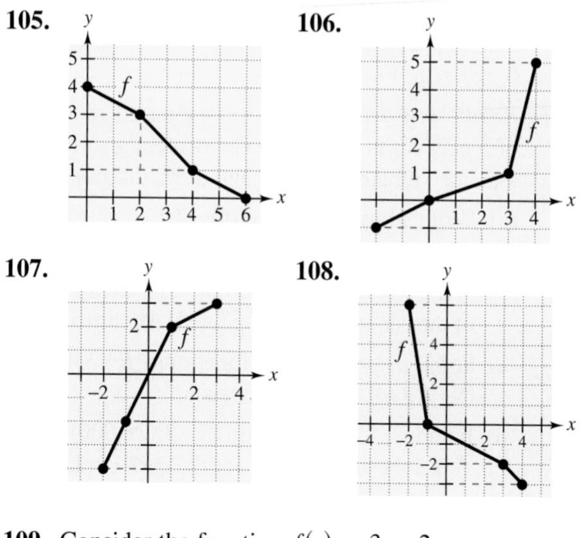

109. Consider the function $f(x) = 3 - 2x$.

(a) Find $f^{-1}(x)$.

(b) Find $(f^{-1})^{-1}(x)$.

Solving Problems

110. *Hourly Wage* Your wage is \$9.00 per hour plus \$0.65 for each unit produced per hour. Thus, your hourly wage y in terms of the number of units produced is $y = 9 + 0.65x$.

(a) Determine the inverse of the function.

(b) What does each variable represent in the inverse function?

(c) Determine the number of units produced when your hourly wage averages \$14.20.

111. *Cost* You need 100 pounds of two commodities that cost \$0.50 and \$0.75 per pound.

(a) Verify that your total cost is $y = 0.50x + 0.75(100 - x)$, where x is the number of pounds of the less expensive commodity.

(b) Find the inverse of the function. What does each variable represent in the inverse function?

(c) Use the context of the problem to determine the domain of the inverse function.

(d) Determine the number of pounds of the less expensive commodity purchased if the total cost is \$60.

112. *Geometry* You are standing on a bridge over a calm pond and drop a pebble, causing ripples of concentric circles in the water (see figure). The radius (in feet) of the outer ripple is given by $r(t) = 0.6t$, where t is time in seconds after the pebble hits the water. The area of the circle is given by the function $A(r) = \pi r^2$. Find an equation for the composition $A(r(t))$. What are the input and output of this composite function?

113. *Sales Bonus* You are a sales representative for a clothing manufacturer. You are paid an annual salary plus a bonus of 2% of your sales over \$200,000. Consider the two functions $f(x) = x - 200{,}000$ and $g(x) = 0.02x$. If x is greater than \$200,000, which of the following represents your bonus? Explain.

(a) $f(g(x))$

(b) $g(f(x))$

114. *Daily Production Cost* The daily cost of producing x units in a manufacturing process is $C(x) = 8.5x + 300$. The number of units produced in t hours during a day is given by $x(t) = 12t$, $0 \le t \le 8$. Find, simplify, and interpret $(C \circ x)(t)$.

115. *Rebate and Discount* The suggested retail price of a new car is p dollars. The dealership advertised a factory rebate of \$2000 and a 5% discount.

(a) Write a function R in terms of p, giving the cost of the car after receiving the factory rebate.

(b) Write a function S in terms of p, giving the cost of the car after receiving the dealership discount.

(c) Form the composite functions $(R \circ S)(p)$ and $(S \circ R)(p)$ and interpret each.

(d) Find $(R \circ S)(26{,}000)$ and $(S \circ R)(26{,}000)$. Which yields the smaller cost for the car? Explain.

116. *Exploration and Conjecture* Consider the functions $f(x) = 4x$ and $g(x) = x + 6$.

(a) Find $(f \circ g)(x)$.

(b) Find $(f \circ g)^{-1}(x)$.

(c) Find $f^{-1}(x)$ and $g^{-1}(x)$.

(d) Find $(g^{-1} \circ f^{-1})(x)$ and compare the result with that of part (b).

(e) Repeat parts (a) through (d) for $f(x) = x^3 + 1$ and $g(x) = 2x$.

(f) Write two one-to-one functions f and g, and repeat parts (a) through (d) for these functions.

(g) Make a conjecture about $(f \circ g)^{-1}(x)$ and $(g^{-1} \circ f^{-1})(x)$.

Explaining Concepts

True or False? In Exercises 117–120, decide whether the statement is true or false. If true, explain your reasoning. If false, give an example.

117. If the inverse of f exists, the y-intercept of f is an x-intercept of f^{-1}. Explain.

118. There exists no function f such that $f = f^{-1}$.

119. If the inverse of f exists, the domains of f and f^{-1} are the same.

120. If the inverse of f exists and its graph passes through the point $(2, 2)$, the graph of f^{-1} also passes through the point $(2, 2)$.

121. Give an example showing that the composite functions $(f \circ g)(x)$ and $(g \circ f)(x)$ are not necessarily the same.

122. Describe how to find the inverse of a function given by a set of ordered pairs. Give an example.

123. Describe how to find the inverse of a function given by an equation in x and y. Give an example.

124. Give an example of a function that does not have an inverse.

125. Explain the Horizontal Line Test. What is the relationship between this test and a function being one-to-one?

126. Describe the relationship between the graph of a function and its inverse.

11.3 Logarithmic Functions

Objectives

1. Evaluate a logarithmic function.
2. Sketch the graph of a logarithmic function using its inverse exponential function.
3. Use the natural base e to define the natural logarithmic function.
4. Use the change-of-base formula to evaluate logarithms.

1 Evaluate a logarithmic function.

Logarithmic Functions

In Section 11.2, you were introduced to the concept of the inverse of a function. Moreover, you saw that if a function has the property that no horizontal line intersects the graph of the function more than once, the function must have an inverse. By looking back at the graphs of the exponential functions introduced in Section 11.1, you will see that every function of the form

$f(x) = a^x$ Exponential functions have inverses.

passes the Horizontal Line Test, and so must have an inverse. To describe the inverse of $f(x) = a^x$, we follow the steps used in Section 11.2.

$y = a^x$ Replace $f(x)$ by y.

$x = a^y$ Interchange x and y.

At this point we have no way to solve for y. A verbal description of y in the equation $x = a^y$ is "y equals the exponent needed on base a to get x." This inverse of $f(x) = a^x$ is denoted by

$$f^{-1}(x) = \log_a x.$$

Definition of Logarithmic Function

Let a and x be positive real numbers such that $a \neq 1$. The **logarithm of *x* with base *a*** is denoted by $\log_a x$ and is defined as follows.

$$y = \log_a x \quad \text{if and only if} \quad a^y = x$$

The function $f(x) = \log_a x$ is the **logarithmic function with base *a*.**

From the definition it is clear that

Logarithmic Equation		*Exponential Equation*
$y = \log_a x$	is equivalent to	$a^y = x$.

So, to find the value of $\log_a x$, *think*

"$\log_a x =$ the exponent needed on base a to get x."

For instance,

$y = \log_2 8$ Think: "The exponent needed on 2 to get 8."

$y = 3.$

That is,

$3 = \log_2 8.$ This is equivalent to $2^3 = 8$.

By now it should be clear that *a logarithm is an exponent.*

John Napier
(1550–1617)

John Napier, a Scottish mathematician, developed logarithms as a way to simplify some of the tedious calculations of his day. Beginning in 1594, Napier worked about 20 years on logarithms, but he was only partially successful in his quest. Nonetheless, the development of logarithms was a step forward and received immediate recognition.

Example 1 Evaluating Logarithms

Evaluate each logarithm.

a. $\log_2 16$ **b.** $\log_3 9$ **c.** $\log_4 2$

Solution

In each case you should answer the question, "To what power must the base be raised to obtain the given number?"

a. The power to which 2 must be raised to obtain 16 is 4. That is,

$$2^4 = 16 \quad \Rightarrow \quad \log_2 16 = 4.$$

b. The power to which 3 must be raised to obtain 9 is 2. That is,

$$3^2 = 9 \quad \Rightarrow \quad \log_3 9 = 2.$$

c. The power to which 4 must be raised to obtain 2 is $\frac{1}{2}$. That is,

$$4^{1/2} = 2 \quad \Rightarrow \quad \log_4 2 = \frac{1}{2}.$$

Example 2 Evaluating Logarithms

Evaluate each logarithm.

a. $\log_5 1$ **b.** $\log_{10} \frac{1}{10}$ **c.** $\log_3(-1)$ **d.** $\log_4 0$

Solution

a. The power to which 5 must be raised to obtain 1 is 0. That is,

$$5^0 = 1 \quad \Rightarrow \quad \log_5 1 = 0.$$

b. The power to which 10 must be raised to obtain $\frac{1}{10}$ is -1. That is,

$$10^{-1} = \frac{1}{10} \quad \Rightarrow \quad \log_{10} \frac{1}{10} = -1.$$

c. There is no power to which 3 can be raised to obtain -1. The reason for this is that for any value of x, 3^x is a positive number. So, $\log_3(-1)$ is undefined.

d. There is no power to which 4 can be raised to obtain 0. So, $\log_4 0$ is undefined.

Study Tip

Study the results in Example 2 carefully. Each of the logarithms illustrates an important special property of logarithms that you should know.

The following properties of logarithms follow directly from the definition of the logarithmic function with base a.

▶ Properties of Logarithms

Let a and x be positive real numbers such that $a \neq 1$. Then the following properties are true.

1. $\log_a 1 = 0$ because $a^0 = 1$.

2. $\log_a a = 1$ because $a^1 = a$.

3. $\log_a a^x = x$ because $a^x = a^x$.

The logarithmic function with base 10 is called the **common logarithmic function.** On most calculators, this function can be evaluated with the common logarithmic key [LOG], as illustrated in the next example.

Example 3 Evaluating Common Logarithms

Evaluate each logarithm. Use a calculator only if necessary.

a. $\log_{10} 100$

b. $\log_{10} 0.01$

c. $\log_{10} 5$

Solution

a. Because $10^2 = 100$, it follows that

$$\log_{10} 100 = 2.$$

b. Because $10^{-2} = \frac{1}{100} = 0.01$, it follows that

$$\log_{10} 0.01 = -2.$$

c. There is no simple power to which 10 can be raised to obtain 5, so you should use a calculator to evaluate $\log_{10} 5$.

Keystrokes	*Display*	
5 [LOG]	0.69897	Scientific
[LOG] 5 [ENTER]	0.69897	Graphing

So, rounded to three decimal places, $\log_{10} 5 \approx 0.699$.

Be sure you see that the value of a logarithm can be zero or negative, as in Example 3(b), *but* you cannot take the logarithm of zero or a negative number. This means that the logarithms $\log_{10}(-10)$ and $\log_5 0$ are not valid.

2 Sketch the graph of a logarithmic function using its inverse exponential function.

Graphs of Logarithmic Functions

To sketch the graph of $y = \log_a x$, we can use the fact that the graphs of inverse functions are reflections of each other in the line $y = x$.

Example 4 Graphs of Exponential and Logarithmic Functions

On the same rectangular coordinate system, sketch the graphs of the following.

a. $f(x) = 2^x$ **b.** $g(x) = \log_2 x$

Solution

a. Begin by making a table of values for $f(x) = 2^x$.

x	-2	-1	0	1	2	3
$f(x) = 2^x$	$\frac{1}{4}$	$\frac{1}{2}$	1	2	4	8

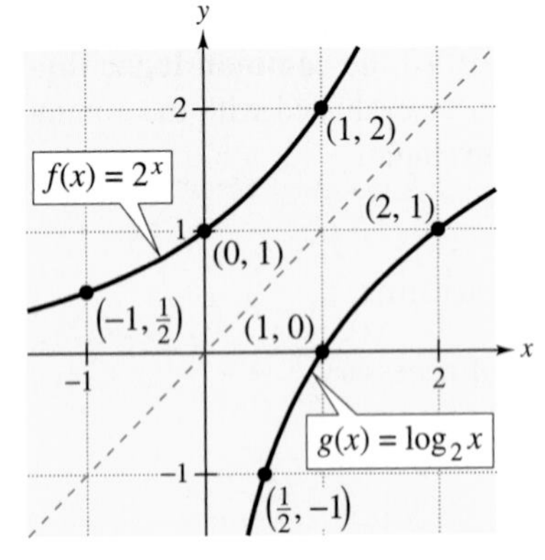

Figure 11.16 *Inverse Functions*

By plotting these points and connecting them with a smooth curve, you obtain the graph shown in Figure 11.16.

b. Because $g(x) = \log_2 x$ is the inverse function of $f(x) = 2^x$, the graph of g is obtained by reflecting the graph of f in the line $y = x$, as shown in Figure 11.16.

Study Tip

In Example 4, the inverse nature of logarithmic functions is used to sketch the graph of $g(x) = \log_2 x$. You could also use a standard point-plotting approach or a graphing utility.

Notice from the graph of $g(x) = \log_2 x$, shown in Figure 11.16, that the domain of the function is the set of positive numbers and the range is the set of all real numbers. The basic characteristics of the graph of a logarithmic function are summarized in Figure 11.17. In this figure, note that the graph has one x-intercept at $(1, 0)$. Also note that the y-axis is a vertical asymptote of the graph.

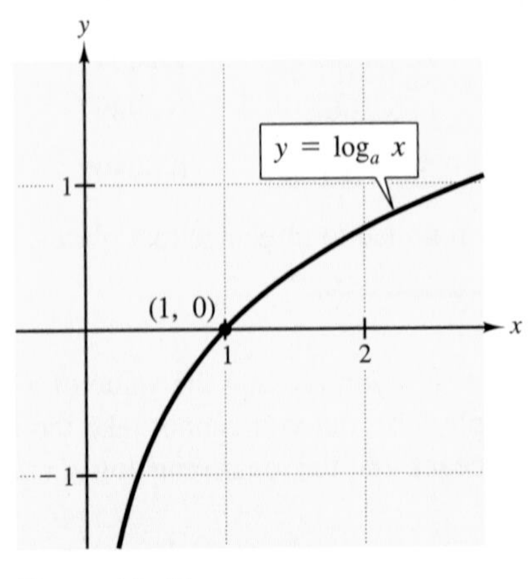

Figure 11.17

Graph of $y = \log_a x$, $a > 1$

- Domain: $(0, \infty)$
- Range: $(-\infty, \infty)$
- Intercept: $(1, 0)$
- Vertical asymptote: $x = 0$
- Increasing

In the following example, the graph of $\log_a x$ is used to sketch the graphs of functions of the form $y = b \pm \log_a(x + c)$. Notice how each transformation affects the vertical asymptote.

Example 5 Sketching the Graphs of Logarithmic Functions

The graph of each of the following functions is similar to the graph of $f(x) = \log_{10} x$, as shown in Figure 11.18. From the graph you can determine the domain of the function.

a. Because $g(x) = \log_{10}(x - 1) = f(x - 1)$, the graph of g can be obtained by shifting the graph of f 1 unit to the right. The vertical asymptote of the graph of g is $x = 1$. The domain of g is $x > 1$.

b. Because $h(x) = 2 + \log_{10} x = 2 + f(x)$, the graph of h can be obtained by shifting the graph of f 2 units up. The vertical asymptote of the graph of h is $x = 0$. The domain of h is $x > 0$.

c. Because $k(x) = -\log_{10} x = -f(x)$, the graph of k can be obtained by reflecting the graph of f in the x-axis. The vertical asymptote of the graph of k is $x = 0$. The domain of k is $x > 0$.

d. Because $j(x) = \log_{10}(-x) = f(-x)$, the graph of j can be obtained by reflecting the graph of f in the y-axis. The vertical asymptote of the graph of j is $x = 0$. The domain of j is $x < 0$.

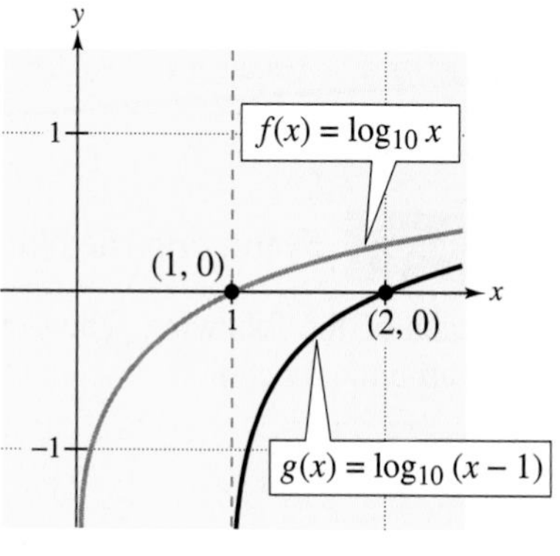

(a)

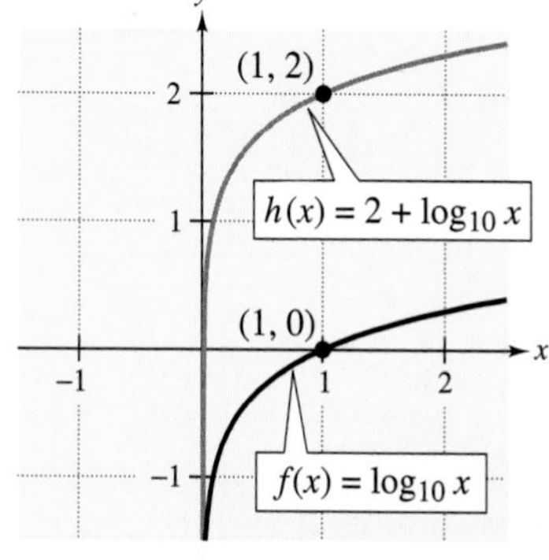

(b)

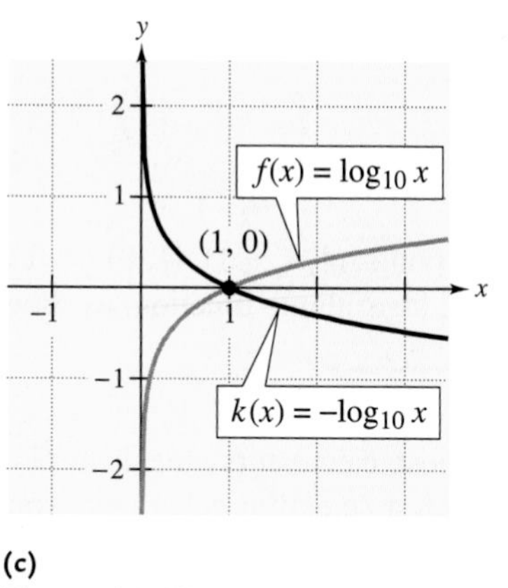

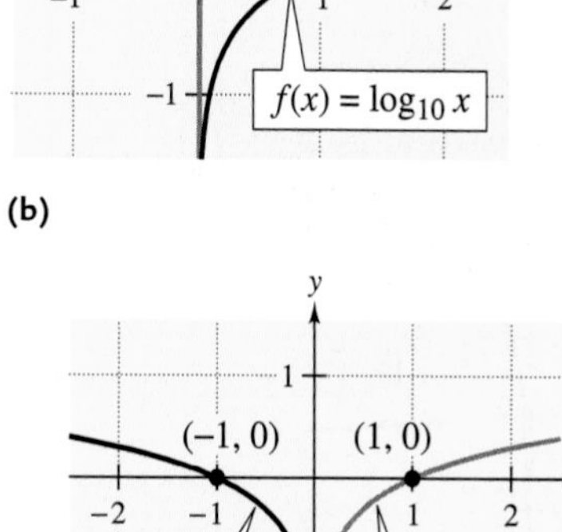

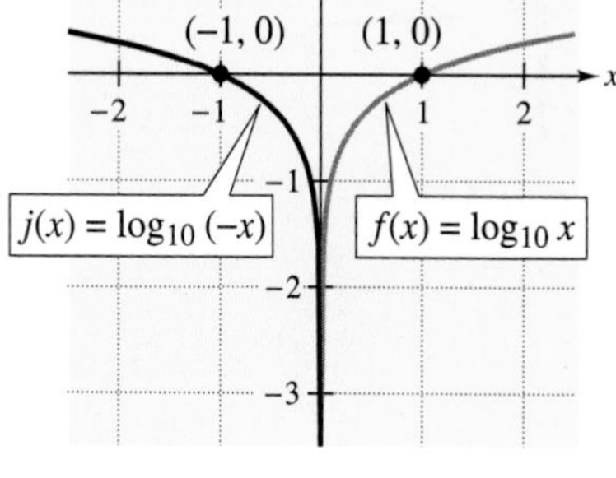

(c)

(d)

Figure 11.18

3 Use the natural base e to define the natural logarithmic function.

The Natural Logarithmic Function

As with exponential functions, the most widely used base for logarithmic functions is the number e. The logarithmic function with base e is the **natural logarithmic function** and is denoted by the special symbol $\ln x$, which is read as "el en of x."

> **The Natural Logarithmic Function**
>
> The function defined by
>
> $$f(x) = \log_e x = \ln x$$
>
> where $x > 0$, is called the **natural logarithmic function.**

The three properties of logarithms listed earlier in this section are also valid for natural logarithms.

> **Properties of Natural Logarithms**
>
> Let x be a positive real number. Then the following properties are true.
>
> **1.** $\ln 1 = 0$ because $e^0 = 1$.
>
> **2.** $\ln e = 1$ because $e^1 = e$.
>
> **3.** $\ln e^x = x$ because $e^x = e^x$.

Example 6 Evaluating the Natural Logarithmic Function

Evaluate each of the following. Then incorporate the results into a graph of the natural logarithmic function.

a. $\ln e^2$ **b.** $\ln \dfrac{1}{e}$

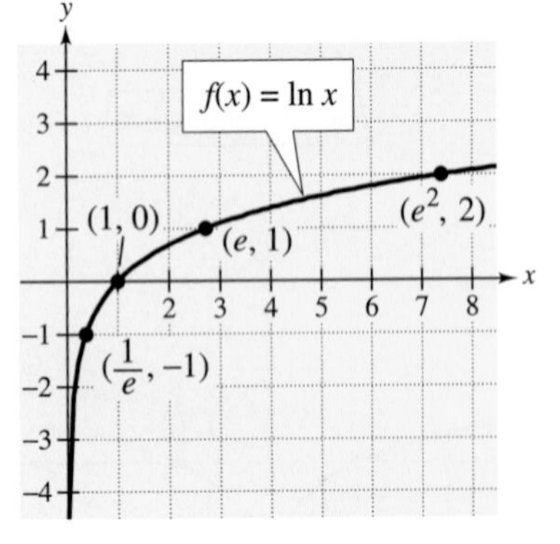

Figure 11.19

Solution

Using the property that $\ln e^x = x$, you obtain the following.

a. $\ln e^2 = 2$

b. $\ln \dfrac{1}{e} = \ln e^{-1} = -1$

Using the points $(1/e, -1)$, $(1, 0)$, $(e, 1)$, and $(e^2, 2)$, you can sketch the graph of the natural logarithmic function, as shown in Figure 11.19.

On most calculators, the natural logarithm key is denoted by LN. For instance, on a scientific calculator, you can evaluated ln 2 as 2 LN and on a graphing calculator, you can evaluate it as LN 2 ENTER. In either case, you should obtain a display of 0.6931472.

4 Use the change-of-base formula to evaluate logarithms.

Change of Base

Although 10 and e are the most frequently used bases, you occasionally need to evaluate logarithms with other bases. In such cases the following **change-of-base formula** is useful.

Technology: Tip

You can use a graphing utility to graph logarithmic functions that do not have a base of 10 by using the change-of-base formula. Use the change-of-base formula to rewrite $g(x) = \log_2 x$ in Example 4 on page 658. Use the trace feature to estimate $g(x) = \log_2 x$ when $x = 3$. Verify your estimate arithmetically using a calculator.

▶ Change-of-Base Formula

Let a, b, and x be positive real numbers such that $a \neq 1$ and $b \neq 1$. Then $\log_a x$ is given as follows.

$$\log_a x = \frac{\log_b x}{\log_b a} \qquad \text{or} \qquad \log_a x = \frac{\ln x}{\ln a}$$

The usefulness of this change-of-base formula is that you can use a calculator that has only the common logarithm key [LOG] and the natural logarithm key [LN] to evaluate logarithms to any base.

Example 7 Changing the Base to Evaluate Logarithms

a. Use *common* logarithms to evaluate $\log_3 5$.

b. Use *natural* logarithms to evaluate $\log_6 2$.

Solution

Using the change-of-base formula, you can convert to common and natural logarithms by writing

$$\log_3 5 = \frac{\log_{10} 5}{\log_{10} 3} \qquad \text{and} \qquad \log_6 2 = \frac{\ln 2}{\ln 6}.$$

Now, use the following keystrokes.

a.

Keystrokes	*Display*	
5 [LOG] [÷] 3 [LOG] [=]	1.4649735	Scientific
[LOG] 5 [÷] [LOG] 3 [ENTER]	1.4649735	Graphing

So, $\log_3 5 \approx 1.465$.

b.

Keystrokes	*Display*	
2 [LN] [÷] 6 [LN] [=]	0.3868528	Scientific
[LN] 2 [÷] [LN] 6 [ENTER]	0.3868528	Graphing

So, $\log_6 2 \approx 0.387$.

In Example 7(a), $\log_3 5$ could have been evaluated using natural logarithms in the change-of-base formula.

$$\log_3 5 = \frac{\ln 5}{\ln 3} \approx 1.465$$

Notice that you get the same answer whether you use natural logarithms or common logarithms in the change-of-base formula.

At this point, you have been introduced to all the basic types of functions that are covered in this course: polynomial functions, radical functions, rational functions, exponential functions, and logarithmic functions. The only other common types of functions are *trigonometric functions*, which you will study if you go on to take a course in trigonometry or precalculus.

Discussing the Concept **Comparing Models**

Suppose you work for a research and development firm that deals with a wide variety of disciplines. Your supervisor has asked you to give a presentation to your department on four basic kinds of mathematical models. Identify each of the models shown below. Develop a presentation describing the types of data sets that each model would best represent. Include distinctions in domain, range, intercepts, and a discussion of the types of applications to which each model is suited.

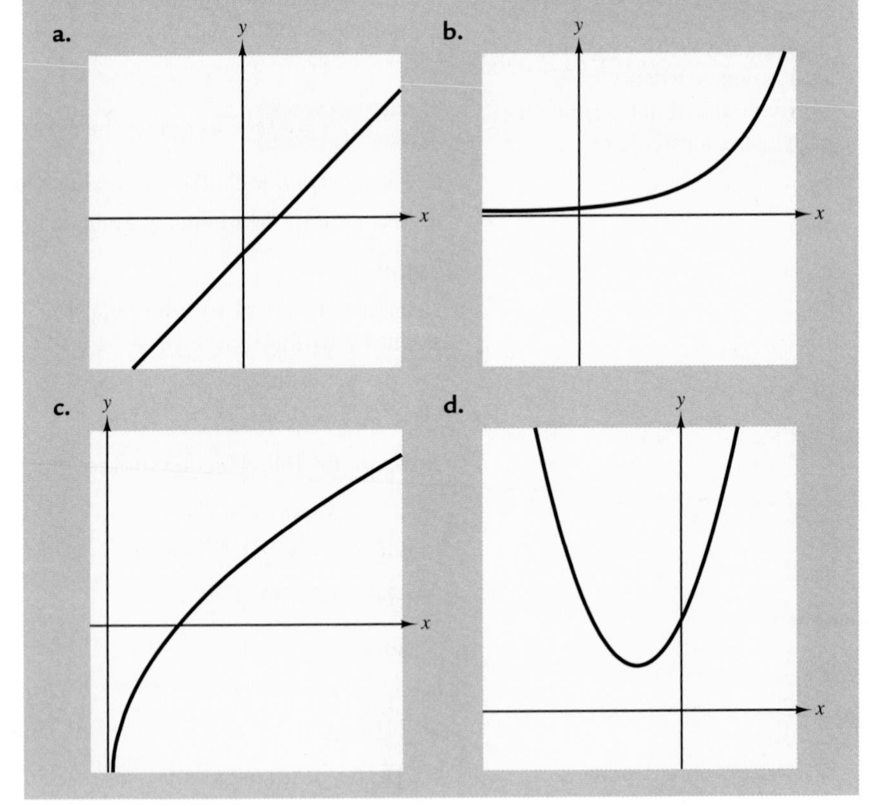

11.3 Exercises

Integrated Review *Concepts, Skills, and Problem Solving*

Keep mathematically in shape by doing these exercises *before* the problems of this section.

Properties and Definitions

In Exercises 1-4, identify the transformation of $f(x) = x^2$ needed to sketch the graph of the function.

1. $g(x) = (x - 4)^2$ **2.** $h(x) = -x^2$

3. $j(x) = x^2 + 1$ **4.** $k(x) = (x + 3)^2 - 5$

Factoring

In Exercises 5-8, factor the expression completely.

5. $2x^3 - 6x$ **6.** $16 - (y + 2)^2$

7. $t^2 + 10t + 25$ **8.** $5 - u + 5u^2 - u^3$

Graphing

In Exercises 9-12, graph the equation.

9. $y = 3 - \frac{1}{2}x$ **10.** $3x - 4y = 6$

11. $y = x^2 - 6x + 5$ **12.** $y = -(x - 2)^2 + 1$

Developing Skills

In Exercises 1-12, write the logarithmic equation in exponential form.

1. $\log_5 25 = 2$ **2.** $\log_6 36 = 2$

3. $\log_4 \frac{1}{16} = -2$ **4.** $\log_8 \frac{1}{8} = -1$

5. $\log_3 \frac{1}{243} = -5$ **6.** $\log_{10} 10{,}000 = 4$

7. $\log_{36} 6 = \frac{1}{2}$ **8.** $\log_{32} 4 = \frac{2}{5}$

9. $\log_8 4 = \frac{2}{3}$ **10.** $\log_{16} 8 = \frac{3}{4}$

11. $\log_2 2.462 \approx 1.3$ **12.** $\log_3 1.179 \approx 0.15$

In Exercises 13-24, write the exponential equation in logarithmic form.

13. $7^2 = 49$ **14.** $6^4 = 1296$

15. $3^{-2} = \frac{1}{9}$ **16.** $5^{-4} = \frac{1}{625}$

17. $8^{2/3} = 4$ **18.** $81^{3/4} = 27$

19. $25^{-1/2} = \frac{1}{5}$ **20.** $6^{-3} = \frac{1}{216}$

21. $4^0 = 1$ **22.** $6^1 = 6$

23. $5^{1.4} \approx 9.518$ **24.** $10^{0.12} \approx 1.318$

In Exercises 25-48, evaluate the logarithm without using a calculator. (If not possible, state the reason.) See Examples 1 and 2.

25. $\log_2 8$ **26.** $\log_3 27$

27. $\log_{10} 10$ **28.** $\log_8 8$

29. $\log_{10} 1000$ **30.** $\log_{10} 0.00001$

31. $\log_2 \frac{1}{4}$ **32.** $\log_3 \frac{1}{9}$

33. $\log_4 \frac{1}{64}$ **34.** $\log_5 \frac{1}{125}$

35. $\log_{10} \frac{1}{10{,}000}$ **36.** $\log_{10} \frac{1}{100}$

37. $\log_2(-3)$ **38.** $\log_4(-4)$

39. $\log_4 1$ **40.** $\log_3 1$

41. $\log_5(-6)$ **42.** $\log_2 0$

43. $\log_9 3$ **44.** $\log_{25} 125$

45. $\log_{16} 8$ **46.** $\log_{144} 12$

47. $\log_7 7^4$ **48.** $\log_5 5^3$

In Exercises 49-54, use a calculator to evaluate the common logarithm. Round to four decimal places. See Example 3.

49. $\log_{10} 31$ **50.** $\log_{10} 5310$

51. $\log_{10} 0.85$ **52.** $\log_{10} 0.345$

53. $\log_{10}(\sqrt{2} + 4)$ **54.** $\log_{10} \frac{\sqrt{3}}{2}$

In Exercises 55–58, sketch the graphs of f and g on the same set of coordinate axes. What can you conclude about the relationship between f and g? See Example 4.

55. $f(x) = \log_3 x$
$g(x) = 3^x$

56. $f(x) = \log_4 x$
$g(x) = 4^x$

57. $f(x) = \log_6 x$
$g(x) = 6^x$

58. $f(x) = \log_{1/2} x$
$g(x) = \left(\frac{1}{2}\right)^x$

In Exercises 59–62, state the relationship between the functions f and g.

59. $f(x) = 7^x$
$g(x) = \log_7 x$

60. $f(x) = 5^x$
$g(x) = \log_5 x$

61. $f(x) = e^x$
$g(x) = \ln x$

62. $f(x) = 10^x$
$g(x) = \log_{10} x$

In Exercises 63–68, identify the transformation of the graph of $f(x) = \log_2 x$ and sketch the graph of h. See Example 5.

63. $h(x) = 3 + \log_2 x$ **64.** $h(x) = -4 + \log_2 x$

65. $h(x) = \log_2(x - 2)$ **66.** $h(x) = \log_2(x + 4)$

67. $h(x) = \log_2(-x)$ **68.** $h(x) = -\log_2(x)$

In Exercises 69–74, match the function with its graph. [The graphs are labeled (a), (b), (c), (d), (e), and (f).]

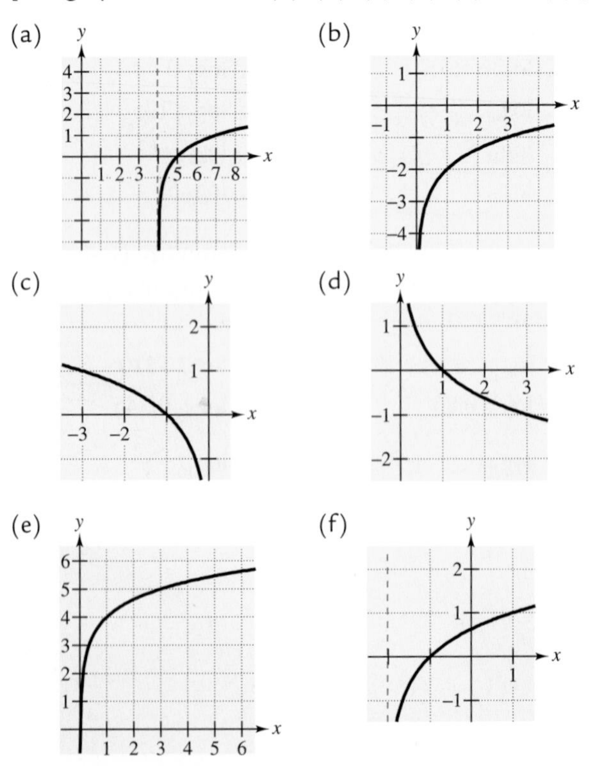

69. $f(x) = 4 + \log_3 x$ **70.** $f(x) = -2 + \log_3 x$

71. $f(x) = -\log_3 x$ **72.** $f(x) = \log_3(-x)$

73. $f(x) = \log_3(x - 4)$ **74.** $f(x) = \log_3(x + 2)$

In Exercises 75–84, sketch the graph of the function.

75. $f(x) = \log_5 x$ **76.** $g(x) = \log_8 x$

77. $g(t) = -\log_2 t$ **78.** $h(s) = -2\log_3 s$

79. $f(x) = 3 + \log_2 x$ **80.** $f(x) = -2 + \log_3 x$

81. $g(x) = \log_2(x - 3)$ **82.** $h(x) = \log_3(x + 1)$

83. $f(x) = \log_{10}(10x)$ **84.** $g(x) = \log_4(4x)$

In Exercises 85–90, find the domain and vertical asymptote of the logarithmic function. Sketch its graph.

85. $f(x) = \log_4 x$ **86.** $g(x) = \log_6 x$

87. $h(x) = \log_4(x - 3)$ **88.** $f(x) = -\log_6(x + 2)$

89. $y = -\log_3 x + 2$ **90.** $y = \log_5(x - 1) + 4$

In Exercises 91–96, use a graphing utility to graph the function. Determine the domain and identify any vertical asymptotes.

91. $y = 5\log_{10} x$ **92.** $y = 5\log_{10}(x - 3)$

93. $y = -3 + 5\log_{10} x$ **94.** $y = 5\log_{10}(3x)$

95. $y = \log_{10}\left(\frac{x}{5}\right)$ **96.** $y = \log_{10}(-x)$

In Exercises 97–102, use a calculator to evaluate the natural logarithm. Round to four decimal places. See Example 6.

97. $\ln 25$ **98.** $\ln 6.57$

99. $\ln 0.75$ **100.** $\ln(\sqrt{3} + 1)$

101. $\ln\left(\frac{1 + \sqrt{5}}{3}\right)$ **102.** $\ln\left(1 + \frac{0.10}{12}\right)$

In Exercises 103–108, match the function with its graph. [The graphs are labeled (a), (b), (c), (d), (e), and (f).]

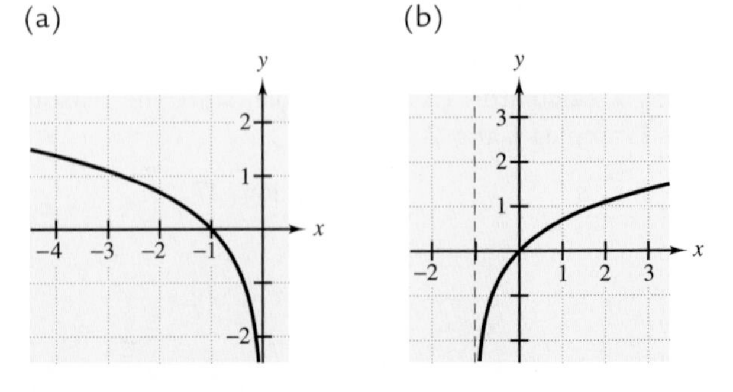

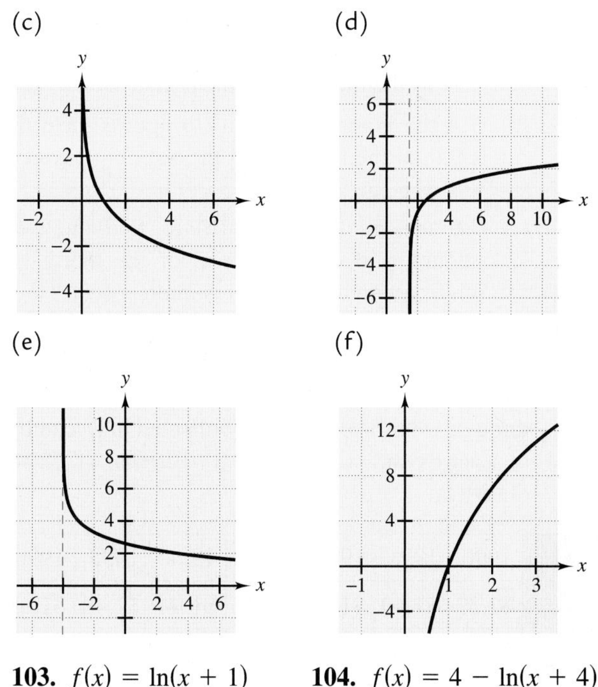

103. $f(x) = \ln(x + 1)$ **104.** $f(x) = 4 - \ln(x + 4)$

105. $f(x) = \ln\left(x - \frac{3}{2}\right)$ **106.** $f(x) = -\frac{3}{2}\ln x$

107. $f(x) = 10\ln x$ **108.** $f(x) = \ln(-x)$

In Exercises 109–116, sketch the graph of the function.

109. $f(x) = -\ln x$ **110.** $f(x) = -2\ln x$

111. $f(x) = 3\ln x$ **112.** $h(t) = 4\ln t$

113. $f(x) = 1 + \ln x$ **114.** $h(x) = 2 + \ln x$

115. $g(t) = 2\ln(t - 4)$ **116.** $g(x) = -3\ln(x + 3)$

In Exercises 117–120, use a graphing utility to graph the function. Determine the domain and identify any vertical asymptotes.

117. $g(x) = \ln(x + 6)$ **118.** $h(x) = -\ln(x - 2)$

119. $f(t) = 3 + 2\ln t$ **120.** $g(t) = \ln(3 - t)$

In Exercises 121–134, use a calculator to evaluate the logarithm by means of the change-of-base formula. Use (a) the common logarithm key and (b) the natural logarithm key. See Example 7.

121. $\log_8 132$ **122.** $\log_5 510$

123. $\log_3 7$ **124.** $\log_7 4$

125. $\log_2 0.72$ **126.** $\log_{12} 0.6$

127. $\log_{15} 1250$ **128.** $\log_{20} 125$

129. $\log_{1/2} 4$ **130.** $\log_{1/3} 18$

131. $\log_4 \sqrt{42}$ **132.** $\log_3 \sqrt{26}$

133. $\log_2(1 + e)$ **134.** $\log_4(2 + e^3)$

Solving Problems

135. *American Elk* The antler spread a (in inches) and shoulder height h (in inches) of an adult male American elk are related by the model

$$h = 116\log_{10}(a + 40) - 176.$$

Approximate the shoulder height of a male American elk with an antler spread of 55 inches.

136. *Intensity of Sound* The relationship between the number of decibels B and the intensity of a sound I in watts per centimeter squared is given by

$$B = 10\log_{10}\left(\frac{I}{10^{-16}}\right).$$

Determine the number of decibels of a sound with an intensity of 10^{-4} watts per centimeter squared.

137. *Creating a Table* The time t in years for an investment to double in value when compounded continuously at the rate r is given by

$$t = \frac{\ln 2}{r}.$$

Complete the following table, which shows the "doubling times" for several annual percentage rates.

r	0.07	0.08	0.09	0.10	0.11	0.12
t						

138. *Tornadoes* Most tornadoes last less than 1 hour and travel less than 20 miles. The speed of the wind S (in miles per hour) near the center of the tornado is related to the distance the tornado travels d (in miles) by the model

$$S = 93\log_{10} d + 65.$$

On March 18, 1925, a large tornado struck portions of Missouri, Illinois, and Indiana, covering a distance of 220 miles. Approximate the speed of the wind near the center of this tornado.

139. *Tractrix* A person walking along a dock (the y-axis) drags a boat by a 10-foot rope (see figure). The boat travels along a path known as a *tractrix*. The equation of the path is

$$y = 10 \ln\left(\frac{10 + \sqrt{100 - x^2}}{x}\right) - \sqrt{100 - x^2}.$$

(a) Use a graphing utility to graph the function. What is the domain of the function?

(b) Identify any asymptotes of the graph.

(c) Determine the position of the person when the x-coordinate of the position of the boat is $x = 2$.

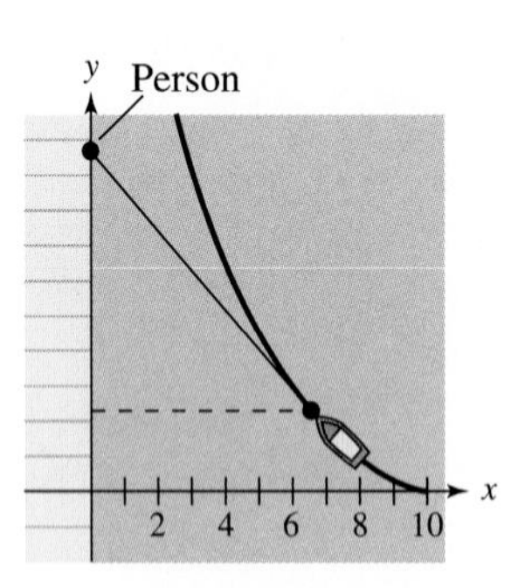

140. *Home Mortgage* The model

$$t = 10.042 \ln\left(\frac{x}{x - 1250}\right), \quad 1250 < x$$

approximates the length t (in years) of a home mortgage of \$150,000 at 10% in terms of the monthly payment x.

(a) Use a graphing utility to graph the model. Describe the change in the length of the mortgage as the monthly payment increases.

(b) Use the graph in part (a) to approximate the length of the mortgage if the monthly payment is \$1316.35.

(c) Use the result of part (b) to find the total amount paid over the term of the mortgage. What amount of the total is interest costs?

141. *Data Analysis* The table gives the prices x of a half-gallon of milk and the prices y of a half-gallon of ice cream in the United States for the years 1990 through 1996. The data were collected in December of each year. (Source: U.S. Bureau of Labor Statistics)

Year	1990	1991	1992	1993	1994	1995	1996
x	\$1.39	\$1.40	\$1.39	\$1.43	\$1.44	\$1.48	\$1.65
y	\$2.54	\$2.63	\$2.49	\$2.59	\$2.62	\$2.68	\$2.94

A model for the data is

$$y = 435.33 - 527.72x + 396.68 \ln x + 88.05x^2.$$

(a) Use a graphing utility to plot the data and graph the model on the same screen.

(b) Use the model to estimate the price of a half-gallon of ice cream if the price of a half-gallon of milk was \$1.59.

Explaining Concepts

142. Write "logarithm of x with base 5" symbolically.

143. Explain the relationship between the functions $f(x) = 2^x$ and $g(x) = \log_2 x$.

144. Explain why $\log_a a = 1$.

145. Explain why $\log_a a^x = x$.

146. What are common logarithms and natural logarithms?

147. Describe how to use a calculator to find the logarithm of a number if the base is not 10 or e.

Think About It In Exercises 148–153, answer the question for the function $f(x) = \log_{10} x$. (Do not use a calculator.)

148. What is the domain of f?

149. Find the inverse function of f.

150. Describe the values of $f(x)$ for $1000 \le x \le 10{,}000$.

151. Describe the values of x, given that $f(x)$ is negative.

152. By what amount will x increase, given that $f(x)$ is increased by 1 unit?

153. Find the ratio of a to b, given that $f(a) = 3 + f(b)$.

Mid-Chapter Quiz

Take this quiz as you would take a quiz in class. After you are done, check your work against the answers given in the back of the book.

1. Given $f(x) = \left(\frac{4}{3}\right)^x$, find (a) $f(2)$, (b) $f(0)$, (c) $f(-1)$, and (d) $f(1.5)$.

2. Find the domain and range of $g(x) = 2^{-0.5x}$.

In Exercises 3–6, sketch the graph of the function.

3. $y = \frac{1}{2}(4^x)$

4. $y = 5(2^{-x})$

5. $f(t) = 12e^{-0.4t}$

6. $g(x) = 100(1.08)^x$

7. You deposit \$750 at $7\frac{1}{2}\%$ interest, compounded n times per year or continuously. Complete the table, which shows the balance A after 20 years for several types of compounding.

n	1	4	12	365	Continuous compounding
A					

8. A gallon of milk costs \$2.23 now. If the price increases by 4% each year, what will the price be after 5 years?

9. Given $f(x) = 2x - 3$ and $g(x) = x^3$, find the indicated composition.

(a) $(f \circ g)(x)$ (b) $(g \circ f)(x)$ (c) $(f \circ g)(-2)$ (d) $(g \circ f)(4)$

10. Verify algebraically and graphically that $f(x) = 3 - 5x$ and $g(x) = \frac{1}{5}(3 - x)$ are inverses of each other.

In Exercises 11 and 12, find the inverse of the function.

11. $h(x) = 10x + 3$

12. $g(t) = \frac{1}{2}t^3 + 2$

13. Write the logarithmic equation $\log_4\left(\frac{1}{16}\right) = -2$ in exponential form.

14. Write the exponential equation $3^4 = 81$ in logarithmic form.

15. Evaluate $\log_5 125$ without the aid of a calculator.

16. Write a paragraph comparing the graphs of $f(x) = \log_5 x$ and $g(x) = 5^x$.

In Exercises 17 and 18, use a graphing utility to sketch the graph of the function.

17. $f(t) = \frac{1}{2} \ln t$

18. $h(x) = 3 - \ln x$

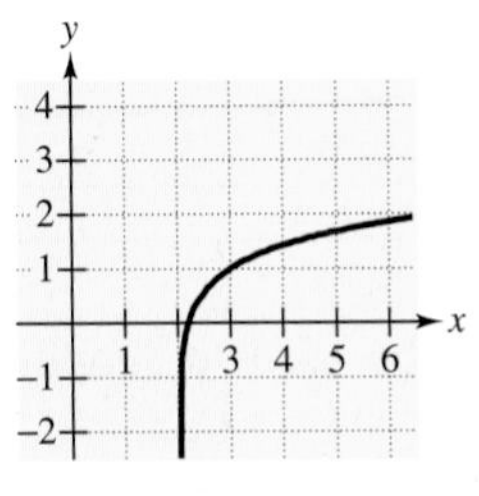

Figure for 19

19. Use the graph of f at the left to determine h and k if $f(x) = \log_5(x - h) + k$.

20. Use a calculator and the change-of-base formula to evaluate $\log_6 450$.

11.4 Properties of Logarithms

Objectives

1. Use the properties of logarithms to evaluate a logarithm.
2. Use the properties of logarithms to rewrite a logarithmic expression.
3. Use the properties of logarithms to solve an application problem.

1 Use the properties of logarithms to evaluate a logarithm.

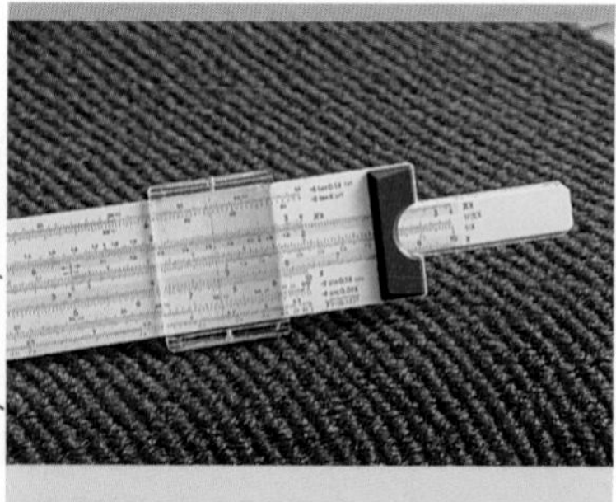

Tony Freeman/PhotoEdit

Slide Rule

Before electronic hand-held calculators became available in the 1970's, mathematicians, engineers, and scientists relied on a tool called the slide rule. Created by Edmund Gunter, (1581–1626) the slide rule uses logarithms to quickly multiply and divide numbers.

Properties of Logarithms

You know from the preceding section that the logarithmic function with base a is the *inverse* of the exponential function with base a. So, it makes sense that each property of exponents should have a corresponding property of logarithms. For instance, the exponential property

$a^0 = 1$ Exponential property

has the corresponding logarithmic property

$\log_a 1 = 0.$ Corresponding logarithmic property

In this section you will study the logarithmic properties that correspond to the following three exponential properties:

	Base a	*Natural Base*	
1.	$a^n a^m = a^{n+m}$	$e^n e^m = e^{n+m}$	Product Rule
2.	$\dfrac{a^n}{a^m} = a^{n-m}$	$\dfrac{e^n}{e^m} = e^{n-m}$	Quotient Rule
3.	$(a^n)^m = a^{nm}$	$(e^n)^m = e^{nm}$	Power Rule

▶ Properties of Logarithms

Let a be a positive real number such that $a \neq 1$, and let n be a real number. If u and v are real numbers, variables, or algebraic expressions such that $u > 0$ and $v > 0$, the following properties are true.

	Logarithm with Base a	*Natural Logarithm*	
1.	$\log_a(uv) = \log_a u + \log_a v$	$\ln(uv) = \ln u + \ln v$	Product Rule
2.	$\log_a \dfrac{u}{v} = \log_a u - \log_a v$	$\ln \dfrac{u}{v} = \ln u - \ln v$	Quotient Rule
3.	$\log_a u^n = n \log_a u$	$\ln u^n = n \ln u$	Power Rule

There is no general property of logarithms that can be used to simplify $\log_a(u + v)$. Specifically,

$\log_a(u + v)$ *does not equal* $\log_a u + \log_a v$.

Example 1 Using Properties of Logarithms

Use the fact that $\ln 2 \approx 0.693$, $\ln 3 \approx 1.099$, and $\ln 5 \approx 1.609$ to approximate each of the following.

a. $\ln \frac{2}{3}$ **b.** $\ln 10$ **c.** $\ln 30$

Solution

a. $\ln \frac{2}{3} = \ln 2 - \ln 3$ Quotient Rule

$\approx 0.693 - 1.099$ Substitute for ln 2 and ln 3.

$= -0.406$ Simplify.

b. $\ln 10 = \ln(2 \cdot 5)$ Factor.

$= \ln 2 + \ln 5$ Product Rule

$\approx 0.693 + 1.609$ Substitute for ln 2 and ln 5.

$= 2.302$ Simplify.

c. $\ln 30 = \ln(2 \cdot 3 \cdot 5)$ Factor.

$= \ln 2 + \ln 3 + \ln 5$ Product Rule

$\approx 0.693 + 1.099 + 1.609$ Substitute for ln 2, ln 3, and ln 5.

$= 3.401$ Simplify.

When using the properties of logarithms, it helps to state the properties *verbally*. For instance, the verbal form of the Product Property $\ln(uv) = \ln u + \ln v$ is: *The log of a product is the sum of the logs of the factors.* Similarly, the verbal form of the Quotient Property $\ln\left(\frac{u}{v}\right) = \ln u - \ln v$ is: *The log of a quotient is the difference of the logs of the numerator and denominator.*

Study Tip

Remember that you can verify results such as those given in Example 2 with a calculator.

Example 2 Using Properties of Logarithms

Use the properties of logarithms to verify that $-\ln 2 = \ln \frac{1}{2}$.

Solution

Using the Power Rule, you can write the following.

$-\ln 2 = (-1) \ln 2$ Rewrite coefficient as -1.

$= \ln 2^{-1}$ Power Rule

$= \ln \frac{1}{2}$ Rewrite 2^{-1} as $\frac{1}{2}$.

2 Use the properties of logarithms to rewrite a logarithmic expression.

Rewriting Logarithmic Expressions

In Examples 1 and 2, the properties of logarithms were used to rewrite logarithmic expressions involving the log of a *constant.* A more common use of the properties is to rewrite the log of a *variable expression.*

Example 3 Rewriting the Logarithm of a Product

Use the properties of logarithms to rewrite $\log_{10} 7x^3$.

Solution

$$\log_{10} 7x^3 = \log_{10} 7 + \log_{10} x^3 \qquad \text{Product Rule}$$

$$= \log_{10} 7 + 3 \log_{10} x \qquad \text{Power Rule}$$

When you rewrite a logarithmic expression as in Example 3, you are **expanding** the expression. The reverse procedure is demonstrated in Example 4, and is called **condensing** a logarithmic expression.

Example 4 Condensing a Logarithmic Expression

Use the properties of logarithms to condense each expression.

a. $\ln x - \ln 3$

b. $2 \log_3 x + \log_3 5$

Solution

a. Using the Quotient Rule, you can write

$$\ln x - \ln 3 = \ln \frac{x}{3}. \qquad \text{Quotient Rule}$$

b. $2 \log_3 x + \log_3 5 = \log_3 x^2 + \log_3 5$ Power Rule

$$= \log_3 5x^2 \qquad \text{Product Rule}$$

Confirm these results by graphing both sides of the equation on a graphing utility.

Example 5 Expanding a Logarithmic Expression

Expand the logarithmic expression.

$$\log_2 3xy^2, \quad x > 0, \ y > 0$$

Solution

$$\log_2 3xy^2 = \log_2 3 + \log_2 x + \log_2 y^2 \qquad \text{Product Rule}$$

$$= \log_2 3 + \log_2 x + 2 \log_2 y \qquad \text{Power Rule}$$

Sometimes expanding or condensing logarithmic expressions involves several steps. In the next example, be sure that you can justify each step in the solution. Also, notice how different the expanded expression is from the original.

Example 6 Expanding a Logarithmic Expression

Expand the logarithmic expression.

$\ln\sqrt{x^2-1}, \quad x > 1$

Solution

$$\ln\sqrt{x^2-1} = \ln(x^2-1)^{1/2} \qquad \text{Rewrite using fractional exponent.}$$

$$= \tfrac{1}{2}\ln(x^2-1) \qquad \text{Power Rule}$$

$$= \tfrac{1}{2}\ln[(x-1)(x+1)] \qquad \text{Factor.}$$

$$= \tfrac{1}{2}[\ln(x-1)+\ln(x+1)] \qquad \text{Product Rule}$$

$$= \tfrac{1}{2}\ln(x-1)+\tfrac{1}{2}\ln(x+1) \qquad \text{Distributive Property}$$

Example 7 Condensing a Logarithmic Expression

Use the properties of logarithms to condense the expression.

a. $\ln 2 - 2\ln x$ **b.** $3(\ln 4 + \ln x)$

Solution

a. $$\ln 2 - 2\ln x = \ln 2 - \ln x^2, \quad x > 0 \qquad \text{Power Rule}$$

$$= \ln\frac{2}{x^2}, \quad x > 0 \qquad \text{Quotient Rule}$$

b. $$3(\ln 4 + \ln x) = 3(\ln 4x) \qquad \text{Product Rule}$$

$$= \ln(4x)^3 \qquad \text{Power Rule}$$

$$= \ln 64x^3, \quad x \ge 0 \qquad \text{Simplify.}$$

When you expand or condense a logarithmic expression, it is possible to change the domain of the expression. For instance, the domain of the function

$$f(x) = 2\ln x \qquad \text{Domain is the set of positive real numbers.}$$

is the set of positive real numbers, whereas the domain of

$$g(x) = \ln x^2 \qquad \text{Domain is the set of nonzero real numbers.}$$

is the set of nonzero real numbers. So, when you expand or condense a logarithmic expression, you should check to see whether the rewriting has changed the domain of the expression. In such cases, you should restrict the domain appropriately. For instance, you can write

$$f(x) = 2\ln x = \ln x^2, \quad x > 0.$$

3 Use the properties of logarithms to solve an application problem.

Application

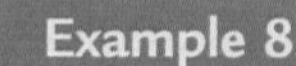

Example 8 Human Memory Model

Students participating in a psychological experiment attended several lectures on a subject. Every month for a year after that, the students were tested to see how much of the material they remembered. The average scores for the group are given by the **human memory model**

$$f(t) = 80 - \ln(t + 1)^9, \quad 0 \le t \le 12$$

where t is the time in months. Find the average scores for the group after 2 months and 8 months.

Solution

To make the calculations easier, rewrite the model as

$$f(t) = 80 - 9\ln(t + 1), \quad 0 \le t \le 12.$$

After 2 months, the average score will be

$$f(2) = 80 - 9\ln 3 \approx 70.1 \qquad \text{Average score after 2 months}$$

and after 8 months, the average score will be

$$f(8) = 80 - 9\ln 9 \approx 60.2. \qquad \text{Average score after 8 months}$$

The graph of the function is shown in Figure 11.20.

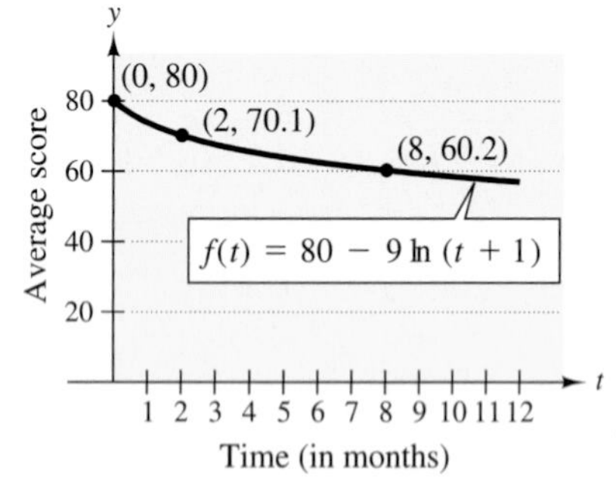

Figure 11.20 *Human Memory Model*

Discussing the Concept Mathematical Modeling

The data in the table represent the annual number y of new AIDS cases in males reported in the United States for the year x from 1993 through 1997, with $x = 3$ corresponding to 1993. (Source: U.S. Centers for Disease Control and Prevention)

x	3	4	5	6	7
y	85,781	63,357	57,512	53,009	45,738

a. Plot the data. Would a linear model fit the points well?

b. Add a third row to the table giving the values of $\ln y$.

c. Plot the coordinate pairs $(x, \ln y)$. Would a linear model fit these points well? If so, draw the best-fitting line and find its equation.

d. Describe the shapes of the two scatter plots. Using your knowledge of logarithms, explain why the second scatter plot is so different from the first.

11.4 Exercises

Integrated Review *Concepts, Skills, and Problem Solving*

Keep mathematically in shape by doing these exercises *before* the problems of this section.

Properties and Definitions

In Exercises 1 and 2, use the property of radicals to fill in the blank.

1. Multiplication Property: $\sqrt[n]{u}\,\sqrt[n]{v} = $ ▭

2. Division Property: $\dfrac{\sqrt[n]{u}}{\sqrt[n]{v}} = $ ▭

3. Explain why the radicals $\sqrt{2x}$ and $\sqrt[3]{2x}$ cannot be added.

4. Is $1/\sqrt{2x}$ in simplest form? Explain.

Simplifying Expressions

In Exercises 5–10, perform the operations and simplify. (Assume all variables are positive.)

5. $25\sqrt{3x} - 3\sqrt{12x}$

6. $(\sqrt{x} + 3)(\sqrt{x} - 3)$

7. $\sqrt{u}(\sqrt{20} - \sqrt{5})$

8. $(2\sqrt{t} + 3)^2$

9. $\dfrac{50x}{\sqrt{2}}$

10. $\dfrac{12}{\sqrt{t+2} + \sqrt{t}}$

Problem Solving

11. The demand equation for a product is given by $p = 30 - \sqrt{0.5(x-1)}$, where x is the number of units demanded per day and p is the price per unit. Find the demand if the price is \$26.76.

12. The sale price of a computer is \$1955. The discount is 15% of the list price. Find the list price.

Developing Skills

In Exercises 1–24, use properties of logarithms to evaluate the expression without using a calculator. (If not possible, state the reason.)

1. $\log_5 5^2$

2. $\log_3 9$

3. $\log_2\left(\frac{1}{8}\right)^3$

4. $\log_8\left(\frac{1}{64}\right)^5$

5. $\log_6 \sqrt{6}$

6. $\ln \sqrt[3]{e}$

7. $\ln 8^0$

8. $\log_4 4^2$

9. $\ln e^4$

10. $\ln e^{-4}$

11. $\log_4 8 + \log_4 2$

12. $\log_6 2 + \log_6 3$

13. $\log_8 4 + \log_8 16$

14. $\log_{10} 5 + \log_{10} 20$

15. $\log_4 8 - \log_4 2$

16. $\log_5 50 - \log_5 2$

17. $\log_6 72 - \log_6 2$

18. $\log_3 324 - \log_3 4$

19. $\log_2 5 - \log_2 40$

20. $\log_3\left(\frac{2}{3}\right) + \log_3\left(\frac{1}{2}\right)$

21. $\ln e^8 + \ln e^4$

22. $\ln e^5 - \ln e^2$

23. $\ln \dfrac{e^3}{e^2}$

24. $\ln(e^2 \cdot e^4)$

In Exercises 25–36, use $\log_4 2 = 0.5000$, $\log_4 3 \approx 0.7925$, and the properties of logarithms to approximate the value of the logarithm. Do not use a calculator. See Example 1.

25. $\log_4 4$

26. $\log_4 8$

27. $\log_4 6$

28. $\log_4 24$

29. $\log_4 \frac{3}{2}$

30. $\log_4 \frac{9}{2}$

31. $\log_4 \sqrt{2}$

32. $\log_4 \sqrt[3]{9}$

33. $\log_4(3 \cdot 2^4)$

34. $\log_4 \sqrt{3 \cdot 2^5}$

35. $\log_4 3^0$

36. $\log_4 4^3$

In Exercises 37–42, use $\log_{10} 3 \approx 0.477$, $\log_{10} 12 \approx 1.079$, and the properties of logarithms to approximate the value of the logarithm. Use a calculator to verify your result.

37. $\log_{10} 9$

38. $\log_{10} \frac{1}{4}$

39. $\log_{10} 36$

40. $\log_{10} 144$

41. $\log_{10} \sqrt{36}$

42. $\log_{10} 5^0$

In Exercises 43–76, use the properties of logarithms to expand the given expression. See Examples 3, 5, and 6.

43. $\log_3 11x$ **44.** $\log_2 3x$

45. $\log_7 x^2$ **46.** $\log_3 x^3$

47. $\log_5 x^{-2}$ **48.** $\log_2 s^{-4}$

49. $\log_4 \sqrt{3x}$ **50.** $\log_3 \sqrt[3]{5y}$

51. $\ln 3y$ **52.** $\ln 5x$

53. $\log_2 \dfrac{z}{17}$ **54.** $\log_{10} \dfrac{7}{y}$

55. $\ln \dfrac{5}{x-2}$ **56.** $\log_4 \dfrac{1}{\sqrt{t}}$

57. $\ln x^2(y-2)$ **58.** $\ln y(y-1)^2$

59. $\log_4[x^6(x-7)^2]$ **60.** $\log_8[(x-y)^4z^6]$

61. $\log_3 \sqrt[3]{x+1}$ **62.** $\log_5 \sqrt{xy}$

63. $\ln \sqrt{x(x+2)}$ **64.** $\ln \sqrt[3]{x(x+5)}$

65. $\ln\left(\dfrac{x+1}{x-1}\right)^2$ **66.** $\log_2\left(\dfrac{x^2}{x-3}\right)^3$

67. $\ln \sqrt[3]{\dfrac{x^2}{x+1}}$ **68.** $\ln \sqrt{\dfrac{3x}{x-5}}$

69. $\ln \dfrac{a^3(b-4)}{c^2}$ **70.** $\log_3 \dfrac{x^2y}{z^7}$

71. $\ln \dfrac{x\sqrt[3]{y}}{(wz)^4}$ **72.** $\log_4 \dfrac{\sqrt[3]{a+1}}{(ab)^4}$

73. $\log_6\left[a\sqrt{b}(c-d)^3\right]$ **74.** $\ln[(xy)^2(x+3)^4]$

75. $\ln\left[(x+y)\dfrac{\sqrt[5]{w+2}}{3t}\right]$ **76.** $\ln\left[(u-v)\dfrac{\sqrt[3]{u-4}}{3v}\right]$

In Exercises 77–108, use the properties of logarithms to condense the expression. See Examples 4 and 7.

77. $\log_{12} x - \log_{12} 3$ **78.** $\log_6 12 - \log_6 y$

79. $\log_2 3 + \log_2 x$ **80.** $\log_5 2x + \log_5 3y$

81. $\log_{10} 4 - \log_{10} x$ **82.** $\ln 10x - \ln z$

83. $4 \ln b$ **84.** $10 \log_4 z$

85. $-2 \log_5 2x$ **86.** $-5 \ln(x+3)$

87. $\frac{1}{3} \ln(2x+1)$ **88.** $-\frac{1}{2} \log_3 5y$

89. $\log_3 2 + \frac{1}{2} \log_3 y$ **90.** $\ln 6 - 3 \ln z$

91. $2 \ln x + 3 \ln y - \ln z$ **92.** $4 \ln 3 - 2 \ln x - \ln y$

93. $5 \ln 2 - \ln x + 3 \ln y$

94. $4 \ln 2 + 2 \ln x - \frac{1}{2} \ln y$

95. $4(\ln x + \ln y)$ **96.** $\frac{1}{2}(\ln 8 + \ln 2x)$

97. $2[\ln x - \ln(x+1)]$ **98.** $5\left[\ln x - \frac{1}{2} \ln(x+4)\right]$

99. $\log_4(x+8) - 3 \log_4 x$

100. $5 \log_3 x + \log_3(x-6)$

101. $\frac{1}{2} \log_5(x+2) - \log_5(x-3)$

102. $\frac{1}{4} \log_6(x+1) - 5 \log_6(x-4)$

103. $5 \log_6(c+d) - \frac{1}{2} \log_6(m-n)$

104. $2 \log_5(x+y) + 3 \log_5 w$

105. $\frac{1}{5}(3 \log_2 x - 4 \log_2 y)$

106. $\frac{1}{3}[\ln(x-6) - 4 \ln y - 2 \ln z]$

107. $\frac{1}{5} \log_6(x-3) - 2 \log_6 x - 3 \log_6(x+1)$

108. $3\left[\frac{1}{2} \log_9(a+6) - 2 \log_9(a-1)\right]$

In Exercises 109–114, simplify the expression.

109. $\ln 3e^2$ **110.** $\log_3(3^2 \cdot 4)$

111. $\log_5 \sqrt{50}$ **112.** $\log_2 \sqrt{22}$

113. $\log_4 \dfrac{4}{x^2}$ **114.** $\ln \dfrac{6}{e^5}$

In Exercises 115–118, use a graphing utility to graph the two equations on the same screen. Use the graphs to verify that the expressions are equivalent. Assume $x > 0$.

115. $y_1 = \ln\left(\dfrac{10}{x^2+1}\right)^2$

$y_2 = 2[\ln 10 - \ln(x^2+1)]$

116. $y_1 = \ln \sqrt{x(x+1)}$

$y_2 = \frac{1}{2}[\ln x + \ln(x+1)]$

117. $y_1 = \ln[x^2(x+2)]$

$y_2 = 2 \ln x + \ln(x+2)$

118. $y_1 = \ln\left(\dfrac{\sqrt{x}}{x-3}\right)$

$y_2 = \frac{1}{2} \ln x - \ln(x-3)$

119. *Think About It* Explain how you can show that

$$\frac{\ln x}{\ln y} \neq \ln \frac{x}{y}.$$

120. *Think About It* Without a calculator, approximate the natural logarithms of as many integers as possible between 1 and 20 using $\ln 2 \approx 0.6931$, $\ln 3 \approx 1.0986$, $\ln 5 \approx 1.6094$, and $\ln 7 \approx 1.9459$. Explain the method you used. Then verify your results with a calculator and explain any differences in the results.

Solving Problems

121. *Intensity of Sound* The relationship between the number of decibels B and the intensity of a sound I in watts per centimeter squared is given by

$$B = 10 \log_{10}\left(\frac{I}{10^{-16}}\right).$$

Use properties of logarithms to write the formula in simpler form, and determine the number of decibels of a sound with an intensity of 10^{-10} watts per centimeter squared.

122. *Human Memory Model* Students participating in a psychological experiment attended several lectures on a subject. Every month for a year after that, the students were tested to see how much of the material they remembered. The average scores for the group are given by the human memory model

$$f(t) = 80 - \log_{10}(t + 1)^{12}, \quad 0 \le t \le 12$$

where t is the time in months.

(a) Find the average scores for the group after 2 months and 8 months.

(b) Use a graphing utility to graph the function.

Biology In Exercises 123 and 124, use the following information. The energy E (in kilocalories per gram molecule) required to transport a substance from the outside to the inside of a living cell is given by

$$E = 1.4(\log_{10} C_2 - \log_{10} C_1)$$

where C_1 and C_2 are the concentrations of the substance outside and inside the cell, respectively.

123. Condense the equation.

124. The concentration of a particular substance inside a cell is twice the concentration outside the cell. How much energy is required to transport the substance from outside to inside the cell?

Explaining Concepts

True or False? In Exercises 125–132, use properties of logarithms to determine whether the equation is true or false. If false, state why or give an example to show that it is false.

125. $\ln e^{2-x} = 2 - x$

126. $\log_2 8x = 3 + \log_2 x$

127. $\log_8 4 + \log_8 16 = 2$

128. $\log_3(u + v) = \log_3 u + \log_3 v$

129. $\log_3(u + v) = \log_3 u \cdot \log_3 v$

130. $\dfrac{\log_6 10}{\log_6 3} = \log_6 10 - \log_6 3$

131. If $f(x) = \log_a x$, then $f(ax) = 1 + f(x)$.

132. If $f(x) = \log_a x$, then $f(a^n) = n$.

True or False? In Exercises 133–138, determine whether the statement is true or false given that $f(x) = \ln x$. If false, state why or give an example to show that the statement is false.

133. $f(0) = 0$

134. $f(2x) = \ln 2 + \ln x$

135. $f(x - 3) = \ln x - \ln 3, \quad x > 3$

136. $\sqrt{f(x)} = \frac{1}{2} \ln x$

137. If $f(u) = 2f(v)$, then $v = u^2$.

138. If $f(x) > 0$, then $x > 1$.

11.5 Solving Exponential and Logarithmic Equations

Objectives

1. Solve an exponential or a logarithmic equation in which both sides have the same base.
2. Use inverse properties to solve an exponential equation.
3. Use inverse properties to solve a logarithmic equation.
4. Use an exponential or a logarithmic equation to solve an application problem.

1 Solve an exponential or a logarithmic equation in which both sides have the same base.

Exponential and Logarithmic Equations

In this section, you will study procedures for *solving equations* that involve exponential or logarithmic expressions. As a simple example, consider the exponential equation $2^x = 16$. By rewriting this equation in the form $2^x = 2^4$, you can see that the solution is $x = 4$. To solve this equation, you can use one of the following properties, which result from the fact that exponential and logarithmic functions are one-to-one.

▶ Properties of Exponential and Logarithmic Equations

Let a be a positive real number such that $a \neq 1$, and let x and y be real numbers. Then the following properties are true.

1. $a^x = a^y$ if and only if $x = y$.

2. $\log_a x = \log_a y$ if and only if $x = y$ $(x > 0, y > 0)$.

Example 1 Solving Exponential and Logarithmic Equations

Solve each equation.

a. $4^{x+2} = 64$ **b.** $\ln(2x - 3) = \ln 11$

Solution

a.

$4^{x+2} = 64$	Original equation
$4^{x+2} = 4^3$	Rewrite with like bases.
$x + 2 = 3$	Property of exponential equations
$x = 1$	Subtract 2 from both sides.

The solution is 1. Check this in the original equation.

b.

$\ln(2x - 3) = \ln 11$	Original equation
$2x - 3 = 11$	Property of logarithmic equations
$2x = 14$	Add 3 to both sides.
$x = 7$	Divide both sides by 2.

The solution is 7. Check this in the original equation.

2 Use inverse properties to solve an exponential equation.

Solving Exponential Equations

In Example 1(a), you were able to solve the given equation because both sides of the equation could be written in exponential form (with the same base). However, if only one side of the equation can be written in exponential form or if both sides cannot be written with the same base, it is more difficult to solve the equation. For example, how would you solve the following equation?

$$2^x = 7$$

To solve this equation, you must find the power to which 2 can be raised to obtain 7. To do this, rewrite the exponential form into logarithmic form by taking the logarithm of both sides and use one of the following inverse properties of exponents and logarithms.

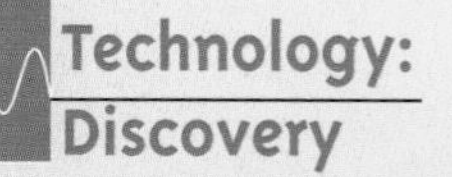

Use a graphing utility to graph both sides of each equation. What does this tell you about the inverse properties of exponents and logarithms?

1. (a) $\log_{10}(10^x) = x$
   (b) $10^{(\log_{10} x)} = x$
2. (a) $\ln(e^x) = x$
   (b) $e^{(\ln x)} = x$

▶ Inverse Properties of Exponents and Logarithms

	Base a	*Natural Base e*
1.	$\log_a(a^x) = x$	$\ln(e^x) = x$
2.	$a^{(\log_a x)} = x$	$e^{(\ln x)} = x$

Example 2 Solving an Exponential Equation

Solve the exponential equation.

a. $2^x = 7$ **b.** $4^{x-3} = 9$ **c.** $2e^x = 10$

Solution

a. To isolate the x, take the $\log_2$ of both sides of the equation, as follows.

$2^x = 7$ Original equation

$\log_2 2^x = \log_2 7$ Take the logarithm of both sides.

$x = \log_2 7$ Inverse property

The solution is $x = \log_2 7 \approx 2.807$. Check this in the original equation.

b. To isolate the x, take the $\log_4$ of both sides of the equation, as follows.

$4^{x-3} = 9$ Original equation

$\log_4 4^{x-3} = \log_4 9$ Take the logarithm of both sides.

$x - 3 = \log_4 9$ Inverse property

$x = \log_4 9 + 3$ Add 3 to both sides.

The solution is $x = \log_4 9 + 3 \approx 4.585$. Check this in the original equation.

c. $2e^x = 10$ Original equation

$e^x = 5$ Divide both sides by 2.

$\ln e^x = \ln 5$ Take the logarithm of both sides.

$x = \ln 5$ Inverse property

The solution is $x = \ln 5 \approx 1.609$. Check this in the original equation.

Study Tip

Remember that to evaluate a logarithm like $\log_2 7$ you need to use the change of base formula.

$$\log_2 7 = \frac{\ln 7}{\ln 2} \approx 2.807$$

Similarly,

$$\log_4 9 + 3 = \frac{\ln 9}{\ln 4} + 3 \approx$$

$$1.585 + 3 \approx 4.585$$

Graphical Check of Solutions
Remember that you can use a graphing utility to solve equations graphically or check solutions that are obtained algebraically. For instance, to check the solutions in Examples 2(a) and 2(c), graph both sides of the equations, as shown below.

Graph $y = 2^x$ and $y = 7$. Then approximate the intersection of the two graphs to be $x \approx 2.807$.

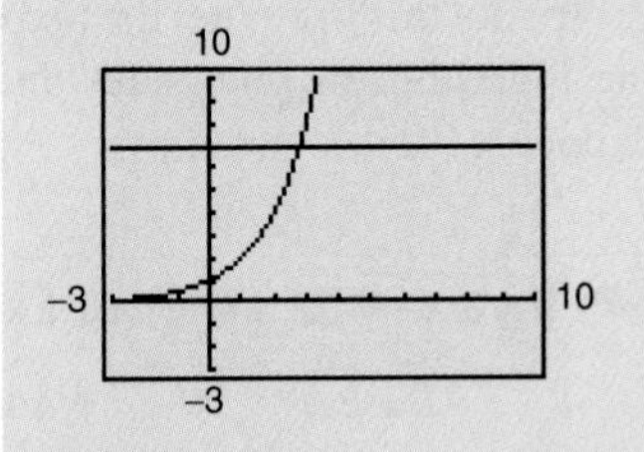

Graph $y = 2e^x$ and $y = 10$. Then approximate the intersection of the two graphs to be $x \approx 1.609$.

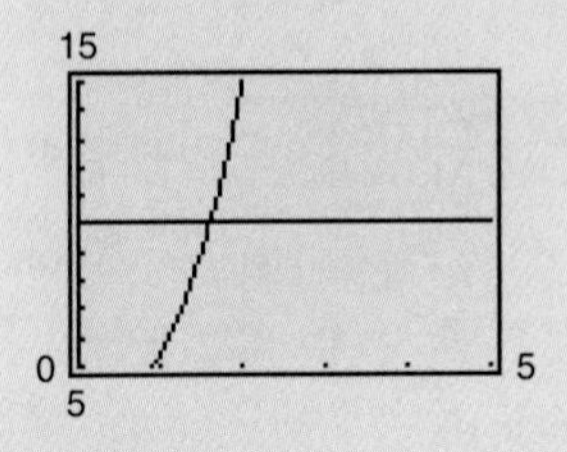

Example 3 Solving an Exponential Equation

Solve $5 + e^{x+1} = 20$.

Solution

$5 + e^{x+1} = 20$	Original equation
$e^{x+1} = 15$	Subtract 5 from both sides.
$\ln e^{x+1} = \ln 15$	Take the logarithm of both sides.
$x + 1 = \ln 15$	Inverse property
$x = -1 + \ln 15$	Subtract 1 from both sides.

The solution is $x = -1 + \ln 15 \approx 1.708$.

Check

$5 + e^{x+1} = 20$	Original equation
$5 + e^{1.708+1} \stackrel{?}{=} 20$	Substitute 1.708 for x.
$5 + e^{2.708} \stackrel{?}{=} 20$	Simplify.
$5 + 14.999 \approx 20$	Solution checks. ✓

3 Use inverse properties to solve a logarithmic equation.

Solving Logarithmic Equations

You know how to solve an exponential equation by *taking the logarithms of both sides.* To solve a logarithmic equation, you need to **exponentiate** both sides. For instance, to solve a logarithmic equation such as

$$\ln x = 2$$

you can exponentiate both sides of the equation as follows.

$\ln x = 2$ Original equation

$e^{\ln x} = e^2$ Exponentiate both sides.

$x = e^2$ Inverse property

This procedure is demonstrated in the next three examples. We suggest the following guidelines for solving exponential and logarithmic equations.

▶ Solving Exponential and Logarithmic Equations

1. To solve an exponential equation, first isolate the exponential expression, then **take the logarithm of both sides of the equation** and solve for the variable.
2. To solve a logarithmic equation, first isolate the logarithmic expression, then **exponentiate both sides of the equation** and solve for the variable.

Example 4 Solving a Logarithmic Equation

Solve $2 \log_4 x = 5$.

Solution

$2 \log_4 x = 5$ Original equation

$\log_4 x = \dfrac{5}{2}$ Divide both sides by 2.

$4^{\log_4 x} = 4^{5/2}$ Exponentiate both sides.

$x = 4^{5/2}$ Inverse property

$x = 32$ Simplify.

The solution is $x = 32$. Check this in the original equation, as follows.

Check

$2 \log_4 x = 5$ Original equation

$2 \log_4(32) \stackrel{?}{=} 5$ Substitute 32 for x.

$2(2.5) \stackrel{?}{=} 5$ Use a calculator.

$5 = 5$ Solution checks. ✓

Example 5 Solving a Logarithmic Equation

Solve $3 \log_{10} x = 6$.

Solution

$3 \log_{10} x = 6$ Original equation

$\log_{10} x = 2$ Divide both sides by 3.

$10^{\log_{10} x} = 10^2$ Exponentiate both sides.

$x = 100$ Inverse property

The solution is $x = 100$. Check this in the original equation.

Example 6 Solving a Logarithmic Equation

Solve $20 \ln 0.2x = 30$.

Solution

$20 \ln 0.2x = 30$ Original equation

$\ln 0.2x = 1.5$ Divide both sides by 20.

$e^{\ln 0.2x} = e^{1.5}$ Exponentiate both sides.

$0.2x = e^{1.5}$ Inverse property

$x = 5e^{1.5}$ Divide both sides by 0.2.

The solution is $x = 5e^{1.5} \approx 22.408$. Check this in the original equation.

The next example uses logarithmic properties as part of the solution.

Example 7 Solving a Logarithmic Equation

Solve $\log_3 2x - \log_3(x - 3) = 1$.

Solution

$\log_3 2x - \log_3(x - 3) = 1$ Original equation

$\log_3 \frac{2x}{x - 3} = 1$ Condense the left side.

$3^{\log[2x/(x-3)]} = 3^1$ Exponentiate both sides.

$\frac{2x}{x - 3} = 3$ Inverse property

$2x = 3x - 9$ Multiply both sides by $x - 3$.

$-x = -9$ Subtract $3x$ from both sides.

$x = 9$ Divide both sides by -1.

The solution is $x = 9$. Check this in the original equation.

4 Use an exponential or a logarithmic equation to solve an application problem.

Application

Example 8 Compound Interest

A deposit of \$5000 is placed in a savings account for 2 years. The interest on the account is compounded continuously. At the end of 2 years, the balance in the account is \$5867.55. What is the annual interest rate for this account?

Solution

Using the formula for continuously compounded interest, $A = Pe^{rt}$, you have the following solution.

Formula: $A = Pe^{rt}$

Labels:
Principal $= P = 5000$ (dollars)
Amount $= A = 5867.55$ (dollars)
Time $= t = 2$ (years)
Annual interest rate $= r$ (percent in decimal form)

Equation: $5867.55 = 5000e^{2r}$

$\frac{5867.55}{5000} = e^{2r}$ Divide both sides by 5000.

$1.1735 \approx e^{2r}$ Simplify.

$\ln 1.1735 \approx \ln(e^{2r})$ Logarithmic form

$0.16 \approx 2r$ Inverse property

$0.08 \approx r$ Divide both sides by 2.

The rate is 8%. Check this in the original statement of the problem, as follows.

Check

$A = Pe^{rt}$ Original equation

$A = 5000e^{(0.08)(2)}$ Substitute 5000 for P, 0.08 for r, and 2 for t.

$A = \$5867.55$ Solution checks. ✓

Discussing the Concept **Solving Equations**

Solve each equation.

a. $x^2 - 5x - 14 = 0$

b. $e^{2x} - 5e^x - 14 = 0$

c. $(\ln x)^2 - 5 \ln x - 14 = 0$

Explain your strategy. What are the similarities among the three equations? One of the equations has only one solution. Explain why.

11.5 Exercises

Integrated Review *Concepts, Skills, and Problem Solving*

Keep mathematically in shape by doing these exercises *before* the problems of this section.

Properties and Definitions

1. Is it possible for the system

$$7x - 2y = 8$$
$$x + y = 4$$

to have exactly two solutions? Explain.

2. Explain why the following system has no solution.

$$8x - 4y = 5$$
$$-2x + y = 1$$

Solving Equations

In Exercises 3–8, solve the equation.

3. $\frac{2}{3}x + \frac{2}{3} = 4x - 6$

4. $x^2 - 10x + 17 = 0$

5. $\frac{5}{2x} - \frac{4}{x} = 3$

6. $\frac{1}{x} + \frac{2}{x-5} = 0$

7. $|x - 4| = 3$

8. $\sqrt{x + 2} = 7$

Models and Graphing

9. A train is traveling at 73 miles per hour. Write the distance d the train travels as a function of the time t. Graph the function.

10. The diameter of a right circular cylinder is 10 centimeters. Write the volume V of the cylinder as a function of its height h if the formula for its volume is $V = \pi r^2 h$. Graph the model.

11. The height of a right circular cylinder is 10 centimeters. Write the volume V of the cylinder as a function of its radius r if the formula for its volume is $V = \pi r^2 h$. Graph the model.

12. A force of 100 pounds stretches a spring 4 inches. Write the force F as a function of the distance x that the spring is stretched. Graph the model.

Developing Skills

In Exercises 1–6, determine whether the x-values are solutions of the equation.

1. $3^{2x-5} = 27$
(a) $x = 1$
(b) $x = 4$

2. $4^{x+3} = 16$
(a) $x = -1$
(b) $x = 0$

3. $e^{x+5} = 45$
(a) $x = -5 + \ln 45$
(b) $x = -5 + e^{45}$

4. $2^{3x-1} = 324$
(a) $x \approx 3.1133$
(b) $x \approx 2.4327$

5. $\log_9(6x) = \frac{3}{2}$
(a) $x = 27$
(b) $x = \frac{9}{2}$

6. $\ln(x + 3) = 2.5$
(a) $x = -3 + e^{2.5}$
(b) $x \approx 9.1825$

In Exercises 7–34, solve the equation. (Do not use a calculator.) See Example 1.

7. $2^x = 2^5$

8. $5^x = 5^3$

9. $3^{x+4} = 3^{12}$

10. $10^{1-x} = 10^4$

11. $3^{x-1} = 3^7$

12. $4^{x+4} = 4^3$

13. $4^{3x} = 16$

14. $3^{2x} = 81$

15. $6^{2x-1} = 216$

16. $5^{3-2x} = 625$

17. $5^x = \frac{1}{125}$

18. $3^x = \frac{1}{243}$

19. $2^{x+2} = \frac{1}{16}$

20. $3^{2-x} = 9$

21. $4^{x+3} = 32^x$

22. $9^{x-2} = 243^{x+1}$

23. $\ln 5x = \ln 22$

24. $\ln 3x = \ln 24$

25. $\log_6 3x = \log_6 18$

26. $\log_5 2x = \log_5 36$

27. $\ln(2x - 3) = \ln 15$

28. $\ln(2x - 3) = \ln 17$

29. $\log_2(x + 3) = \log_2 7$

30. $\log_4(x - 4) = \log_4 12$

31. $\log_5(2x - 3) = \log_5(4x - 5)$

32. $\log_3(4 - 3x) = \log_3(2x + 9)$

33. $\log_3(2 - x) = 2$

34. $\log_2(3x - 1) = 5$

In Exercises 35–38, simplify the expression.

35. $\ln e^{2x-1}$

36. $\log_3 3^{x^2}$

37. $10^{\log_{10} 2x}$

38. $e^{\ln(x+1)}$

In Exercises 39–82, solve the exponential equation. Round the result to two decimal places. See Examples 2 and 3.

39. $2^x = 45$ **40.** $5^x = 21$

41. $3^x = 3.6$ **42.** $2^x = 1.5$

43. $10^{2y} = 52$ **44.** $8^{4x} = 20$

45. $7^{3y} = 126$ **46.** $5^{5y} = 305$

47. $3^{x+4} = 6$ **48.** $5^{3-x} = 15$

49. $10^{x+6} = 250$ **50.** $12^{x-1} = 324$

51. $3e^x = 42$ **52.** $6e^{-x} = 3$

53. $\frac{1}{4}e^x = 5$ **54.** $\frac{2}{3}e^x = 1$

55. $\frac{1}{2}e^{3x} = 20$ **56.** $4e^{-3x} = 6$

57. $250(1.04)^x = 1000$ **58.** $32(1.5)^x = 640$

59. $300e^{x/2} = 9000$ **60.** $6000e^{-2t} = 1200$

61. $1000^{0.12x} = 25{,}000$ **62.** $10{,}000e^{-0.1t} = 4000$

63. $\frac{1}{5}(4^{x+2}) = 300$ **64.** $3(2^{t+4}) = 350$

65. $6 + 2^{x-1} = 1$ **66.** $5^{x+6} - 4 = 12$

67. $7 + e^{2-x} = 28$ **68.** $9 + e^{5-x} = 32$

69. $8 - 12e^{-x} = 7$ **70.** $4 - 2e^x = -23$

71. $4 + e^{2x} = 10$ **72.** $10 + e^{4x} = 18$

73. $32 + e^{7x} = 46$ **74.** $50 - e^{x/2} = 35$

75. $23 - 5e^{x+1} = 3$ **76.** $2e^x + 5 = 115$

77. $4(1 + e^{x/3}) = 84$ **78.** $50(3 - e^{2x}) = 125$

79. $\dfrac{8000}{(1.03)^t} = 6000$ **80.** $\dfrac{5000}{(1.05)^x} = 250$

81. $\dfrac{300}{2 - e^{-0.15t}} = 200$ **82.** $\dfrac{500}{1 + e^{-0.1t}} = 400$

In Exercises 83–118, solve the logarithmic equation. Round the result to two decimal places. See Examples 4–7.

83. $\log_{10} x = 3$ **84.** $\log_{10} x = -2$

85. $\log_2 x = 4.5$ **86.** $\log_4 x = 2.1$

87. $4 \log_3 x = 28$ **88.** $6 \log_2 x = 18$

89. $16 \ln x = 30$ **90.** $12 \ln x = 20$

91. $\log_{10} 4x = 2$ **92.** $\log_3 6x = 4$

93. $\ln 2x = 3$ **94.** $\ln(0.5t) = \frac{1}{4}$

95. $\ln x^2 = 6$ **96.** $\ln \sqrt{x} = 6.5$

97. $2 \log_4(x + 5) = 3$ **98.** $5 \log_{10}(x + 2) = 15$

99. $2 \log_8(x + 3) = 3$ **100.** $\frac{2}{3} \ln(x + 1) = -1$

101. $1 - 2 \ln x = -4$ **102.** $5 - 4 \log_2 x = 2$

103. $-1 + 3 \log_{10} \dfrac{x}{2} = 8$ **104.** $-5 + 2 \ln 3x = 5$

105. $\log_4 x + \log_4 5 = 2$ **106.** $\log_5 x - \log_5 4 = 2$

107. $\log_6(x + 8) + \log_6 3 = 2$

108. $\log_7(x - 1) - \log_7 4 = 1$

109. $\log_5(x + 3) - \log_5 x = 1$

110. $\log_3(x - 2) + \log_3 5 = 3$

111. $\log_{10} x + \log_{10}(x - 3) = 1$

112. $\log_{10} x + \log_{10}(x + 1) = 0$

113. $\log_2(x - 1) + \log_2(x + 3) = 3$

114. $\log_6(x - 5) + \log_6 x = 2$

115. $\log_4 3x + \log_4(x - 2) = \frac{1}{2}$

116. $\log_{10}(25x) - \log_{10}(x - 1) = 2$

117. $\log_2 x + \log_2(x + 2) - \log_2 3 = 4$

118. $\log_3 2x + \log_3(x - 1) - \log_3 4 = 1$

In Exercises 119–122, use a graphing utility to approximate the x-intercept of the graph.

119. $y = 10^{x/2} - 5$ **120.** $y = 2e^x - 21$

121. $y = 6 \ln(0.4x) - 13$

122. $y = 5 \log_{10}(x + 1) - 3$

In Exercises 123–126, use a graphing utility to approximate the point of intersection of the graphs.

123. $y_1 = 2$, $y_2 = e^x$ **124.** $y_1 = 2$, $y_2 = \ln x$

125. $y_1 = 3$, $y_2 = 2 \ln(x + 3)$ **126.** $y_1 = 200$, $y_2 = 1000e^{-x/2}$

Solving Problems

127. *Compound Interest* A deposit of \$10,000 is placed in a savings account for 2 years. The interest for the account is compounded continuously. At the end of 2 years, the balance in the account is \$11,972.17. What is the annual interest rate for this account?

128. *Compound Interest* A deposit of \$2500 is placed in a savings account for 2 years. The interest for the account is compounded continuously. At the end of 2 years, the balance in the account is \$2847.07. What is the annual interest rate for this account?

129. *Doubling Time* Solve the exponential equation

$$5000 = 2500e^{0.09t}$$

for t to determine the number of years for an investment of \$2500 to double in value when compounded continuously at the rate of 9%.

130. *Doubling Rate* Solve the exponential equation

$$10{,}000 = 5000e^{10r}$$

for r to determine the interest rate required for an investment of \$5000 to double in value when compounded continuously for 10 years.

131. *Intensity of Sound* The relationship between the number of decibels B and the intensity of a sound I in watts per centimeter squared is given by

$$B = 10 \log_{10}\left(\frac{I}{10^{-16}}\right).$$

Determine the intensity of a sound I if it registers 75 decibels on a decibel meter.

132. *Intensity of Sound* The relationship between the number of decibels B and the intensity of a sound I in watts per centimeter squared is given by

$$B = 10 \log_{10}\left(\frac{I}{10^{-16}}\right).$$

Determine the intensity of a sound I if it registers 90 decibels on a decibel meter.

133. *Muon Decay* A muon is an elementary particle that is similar to an electron, but much heavier. Muons are unstable—they quickly decay to form electrons and other particles. In an experiment conducted in 1943, the number of muon decays m (of an original 5000 muons) was related to the time T (in microseconds) by the model

$$T = 15.7 - 2.48 \ln m.$$

How many decays were recorded when $T = 2.5$?

134. *Friction* In order to restrain an untrained horse, a person partially wraps the rope around a cylindrical post in a corral (see figure). If the horse is pulling on the rope with a force of 200 pounds, the force F in pounds required by the person is

$$F = 200e^{-0.5\pi\theta/180}$$

where *theta* is the angle of wrap in degrees. Find the smallest value of *theta* if F cannot exceed 80 pounds.

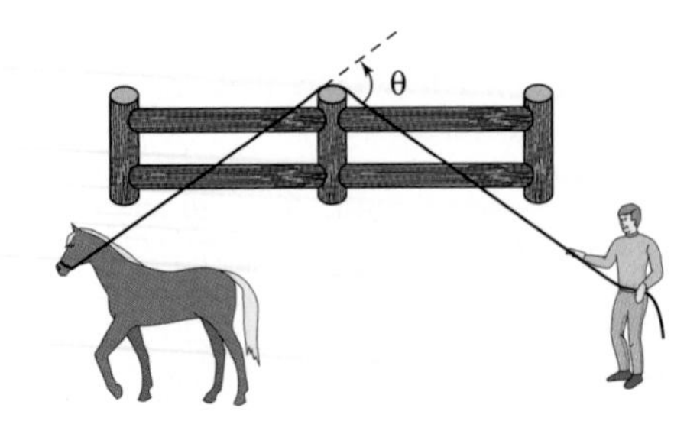

Figure for 134

135. *Human Memory Model* The average score A for a group of students who took a test t months after the completion of a course is given by the memory model

$$A = 80 - \log_{10}(t + 1)^{12}.$$

How long after completing the course will the average score fall to $A = 72$?

(a) Answer the question algebraically by letting $A = 72$ and solving the resulting equation.

(b) Answer the question graphically by using a graphing utility to graph the equations $y_1 = 80 - \log_{10}(t + 1)^{12}$ and $y_2 = 72$, and finding their point(s) of intersection.

(c) Which strategy works better for this problem? Explain.

136. *Military Personnel* The number N (in thousands) of United States military personnel on active duty in foreign countries for the years 1990 through 1996 is modeled by the equation

$$N = 273.1 + 355.8e^{-t}, \quad 0 \le t \le 6$$

where t is time in years, with $t = 0$ corresponding to 1990. (Source: U.S. Department of Defense)

(a) Use a graphing utility to graph the equation over the specified domain.

(b) Use the graph in part (a) to estimate the value of t when $N = 500$.

137. *Making Ice Cubes* You place a tray of 60°F water into a freezer that is set at 0°F. The water cools according to Newton's Law of Cooling

$$kt = \ln \frac{T - S}{T_0 - S}$$

where T is the temperature of the water (in °F), t is the number of hours the tray is in the freezer, S is the temperature of the surrounding air, and T_0 is the original temperature of the water.

(a) If the water freezes in 4 hours, what is the constant k? (*Hint:* Water freezes at 32°F.)

(b) Suppose you lower the temperature in the freezer to -10°F. At this temperature, how long will it take for the ice cubes to form?

(c) Suppose the initial temperature of the water is 50°F. If the freezer temperature is 0°F, how long will it take for the ice cubes to form?

138. *Oceanography* Oceanographers use the density d (in grams per cubic centimeter) of seawater to obtain information about the circulation of water masses and the rates at which waters of different densities mix. For water with a salinity of 30%, the water temperature T (in °C) is related to the density by

$$T = 7.9 \ln(1.0245 - d) + 61.84.$$

Find the densities of the subantarctic water and the antarctic bottom water shown in the figure.

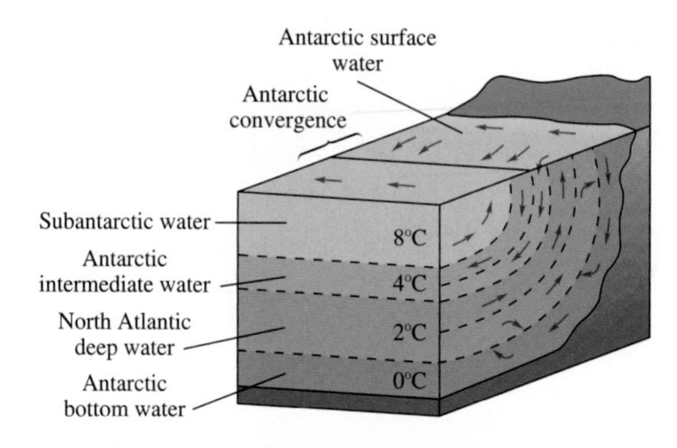

Figure for 138 *This cross section shows complex currents at various depths in the South Atlantic Ocean off Antarctica.*

Explaining Concepts

139. Answer parts (c)–(f) of Motivating the Chapter on page 629.

140. State the three basic properties of logarithms.

141. Which equation requires logarithms for its solution?

$2^{x-1} = 32$ or $2^{x-1} = 30$?

142. Explain how to solve $10^{2x-1} = 5316$.

143. In your own words, state the guidelines for solving exponential and logarithmic equations.

11.6 Applications

Objectives

1. Use an exponential equation to solve a compound interest problem.
2. Use an exponential equation to solve a growth or decay problem.
3. Use a logarithmic equation to solve an intensity problem.

1 Use an exponential equation to solve a compound interest problem.

Compound Interest

In Section 11.1, you were introduced to the following two exponential formulas for compound interest. In these formulas, A is the balance, P is the principal, r is the annual interest rate (in decimal form), and t is the time in years.

n Compoundings per Year	*Continuous Compounding*
$A = P\left(1 + \dfrac{r}{n}\right)^{nt}$	$A = Pe^{rt}$

Example 1 Finding the Annual Interest Rate 

An investment of \$50,000 is made in an account that compounds interest quarterly. After 4 years, the balance in the account is \$71,381.07. What is the annual interest rate for this account?

Solution

Formula: $A = P\left(1 + \dfrac{r}{n}\right)^{nt}$

Labels:
Principal $= P = 50{,}000$ (dollars)
Amount $= A = 71{,}381.07$ (dollars)
Time $= t = 4$ (years)
Number of compoundings per year $= n = 4$
Annual interest rate $= r$ (percent in decimal form)

Equation: $71{,}381.07 = 50{,}000\left(1 + \dfrac{r}{4}\right)^{(4)(4)}$

$\dfrac{71{,}381.07}{50{,}000} = \left(1 + \dfrac{r}{4}\right)^{16}$ Divide both sides by 50,000.

$1.42762 \approx \left(1 + \dfrac{r}{4}\right)^{16}$ Simplify.

$(1.42762)^{1/16} \approx 1 + \dfrac{r}{4}$ Raise both sides to $\frac{1}{16}$ power.

$1.0225 \approx 1 + \dfrac{r}{4}$ Simplify.

$0.09 \approx r$ Subtract 1 and then multiply both sides by 4.

The annual interest rate is approximately 9%. Check this in the original problem.

Study Tip

Solving an exponential equation often requires "getting rid of" the exponent on the variable expression. This can be accomplished by raising both sides of the equation to the reciprocal power. For instance, in Example 1 the variable expression had power 16, so both sides were raised to the reciprocal power $\frac{1}{16}$.

Example 2 Doubling Time for Continuous Compounding

An investment is made in a trust fund at an annual interest rate of 8.75%, compounded continuously. How long will it take for the investment to double?

Study Tip

In "doubling time" problems you do not need to know the value of the principal P to find the doubling time. As shown in Example 2, the factor P divides out of the equation and so does not affect the doubling time.

Solution

$A = Pe^{rt}$ Formula for continuous compounding

$2P = Pe^{0.0875t}$ Substitute known values.

$2 = e^{0.0875t}$ Divide both sides by P.

$\ln 2 = 0.0875t$ Inverse property

$\dfrac{\ln 2}{0.0875} = t$ Divide both sides by 0.0875.

$7.92 \approx t$

It will take approximately 7.92 years for the investment to double.

Check

$A = Pe^{rt}$ Formula for continuous compounding

$2P \stackrel{?}{=} Pe^{0.0875(7.92)}$ Substitute $2P$ for A, 0.0875 for r, and 7.92 for t.

$2P \stackrel{?}{=} Pe^{0.693}$ Simplify.

$2P \approx 1.9997P$ Solution checks. ✓

Example 3 Finding the Type of Compounding

You deposit \$1000 in an account. At the end of 1 year your balance is \$1077.63. If the bank tells you that the annual interest rate for the account is 7.5%, how was the interest compounded?

Solution

If the interest had been compounded continuously at 7.5%, the balance would have been

$$A = 1000e^{(0.075)(1)} = \$1077.88.$$

Because the actual balance is slightly less than this, you should use the formula for interest that is compounded n times per year.

$$A = 1000\left(1 + \frac{0.075}{n}\right)^n = 1077.63$$

At this point, it is not clear what you should do to solve the equation for n. However, by completing a table, you can see that $n = 12$. So, the interest was compounded monthly.

n	1	4	12	365
$\left(1 + \dfrac{0.075}{n}\right)^n$	1.075	1.07714	1.07763	1.07788

In Example 3, notice that an investment of \$1000 compounded monthly produced a balance of \$1077.63 at the end of 1 year. Because \$77.63 of this amount is interest, the **effective yield** for the investment is

$$\text{Effective yield} = \frac{\text{year's interest}}{\text{amount invested}} = \frac{77.63}{1000} = 0.07763 = 7.763\%.$$

In other words, the effective yield for an investment collecting compound interest is the *simple interest rate* that would yield the same balance at the end of 1 year.

Example 4 Finding the Effective Yield 

An investment is made in an account that pays 6.75% interest, compounded continuously. What is the effective yield for this investment?

Solution

Notice that you do not have to know the principal or the time that the money will be left in the account. Instead, you can choose an arbitrary principal, such as \$1000. Then, because effective yield is based on the balance at the end of 1 year, you can use the following formula.

$$\begin{aligned} A &= Pe^{rt} \\ &= 1000e^{0.0675(1)} \\ &= 1069.83 \end{aligned}$$

Now, because the account would earn \$69.83 in interest after 1 year for a principal of \$1000, you can conclude that the effective yield is

$$\begin{aligned} \text{Effective yield} &= \frac{69.83}{1000} \\ &= 0.06983 \\ &= 6.983\%. \end{aligned}$$

2 Use an exponential equation to solve a growth or decay problem.

Growth and Decay

The balance in an account earning *continuously* compounded interest is one example of a quantity that increases over time according to the **exponential growth model** $y = Ce^{kt}$.

▶ Exponential Growth and Decay

The mathematical model for exponential growth or decay is given by

$$y = Ce^{kt}.$$

For this model, t is the time, C is the original amount of the quantity, and y is the amount after time t. The number k is a constant that is determined by the rate of growth. If $k > 0$, the model represents **exponential growth,** and if $k < 0$, it represents **exponential decay.**

One common application of exponential growth is in modeling the growth of a population. Example 5 illustrates the use of the growth model

$$y = Ce^{kt}, \quad k > 0.$$

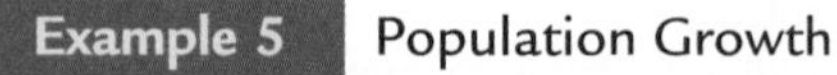

Example 5 Population Growth

A country's population was 2 million in 1990 and 3 million in 2000. What would you predict the population of the country to be in 2010?

Solution

If you assumed a *linear growth model*, you would simply predict the population in the year 2010 to be 4 million. If, however, you assumed an *exponential growth model*, the model would have the form

$$y = Ce^{kt}.$$

In this model, let $t = 0$ represent the year 1990. The given information about the population can be described by the following table.

t (years)	0	10	20
Ce^{kt} (million)	$Ce^{k(0)} = 2$	$Ce^{k(10)} = 3$	$Ce^{k(20)} = ?$

To find the population when $t = 20$, you must first find the values of C and k. From the table, you can use the fact that $Ce^{k(0)} = Ce^0 = 2$ to conclude that $C = 2$. Then, using this value of C, you can solve for k as follows.

$$Ce^{k(10)} = 3 \qquad \text{From table}$$

$$2e^{10k} = 3 \qquad \text{Substitute value of } C.$$

$$e^{10k} = \frac{3}{2} \qquad \text{Divide both sides by 2.}$$

$$10k = \ln\frac{3}{2} \qquad \text{Inverse property}$$

$$k = \frac{1}{10}\ln\frac{3}{2} \qquad \text{Divide both sides by 10.}$$

$$k \approx 0.0405 \qquad \text{Simplify.}$$

Finally, you can use this value of k to conclude that the population in the year 2010 is given by

$$2e^{0.0405(20)} \approx 2(2.25) = 4.5 \text{ million.}$$

Figure 11.21 graphically compares the exponential growth model with a linear growth model.

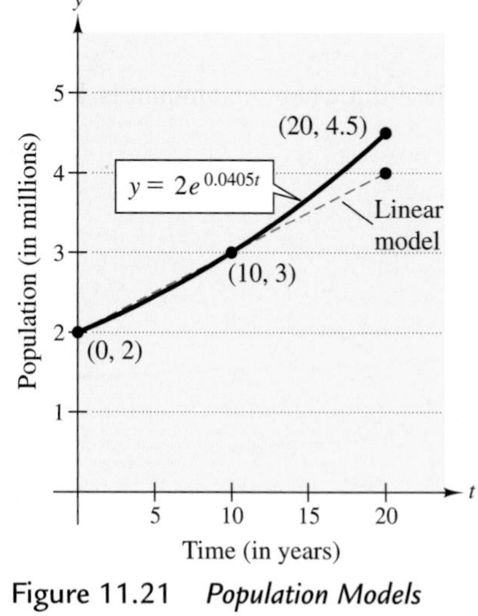

Figure 11.21 *Population Models*

Example 6 Radioactive Decay 

Radioactive iodine is a by-product of some types of nuclear reactors. Its **half-life** is 60 days. That is, after 60 days, a given amount of radioactive iodine will have decayed to half the original amount. Suppose a nuclear accident occurs and releases 20 grams of radioactive iodine. How long will it take for the radioactive iodine to decay to a level of 1 gram?

Solution

To solve this problem, use the model for exponential decay.

$$y = Ce^{kt}$$

Next, use the information given in the problem to set up the following table.

t (days)	0	60	?
Ce^{kt} (grams)	$Ce^{k(0)} = 20$	$Ce^{k(60)} = 10$	$Ce^{k(t)} = 1$

Because $Ce^{k(0)} = Ce^0 = 20$, you can conclude that $C = 20$. Then, using this value of C, you can solve for k, as follows.

$$Ce^{k(60)} = 10 \qquad \text{From table}$$

$$20e^{60k} = 10 \qquad \text{Substitute value of } C.$$

$$e^{60k} = \frac{1}{2} \qquad \text{Divide both sides by 20.}$$

$$60k = \ln\frac{1}{2} \qquad \text{Inverse property}$$

$$k = \frac{1}{60}\ln\frac{1}{2} \qquad \text{Divide both sides by 60.}$$

$$\approx -0.01155 \qquad \text{Simplify.}$$

Finally, you can use this value of k to find the time when the amount is 1 gram, as follows.

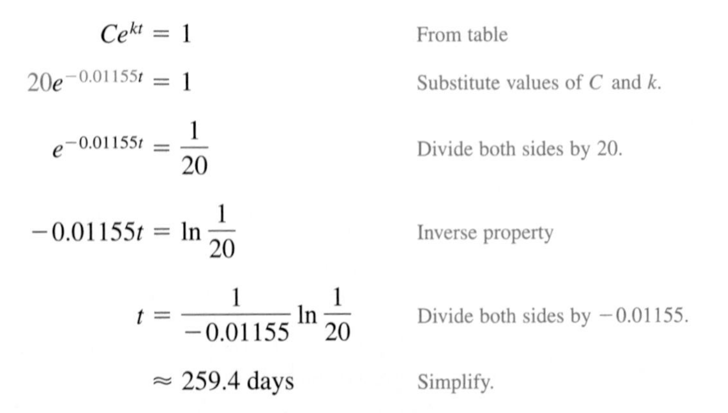

$$Ce^{kt} = 1 \qquad \text{From table}$$

$$20e^{-0.01155t} = 1 \qquad \text{Substitute values of } C \text{ and } k.$$

$$e^{-0.01155t} = \frac{1}{20} \qquad \text{Divide both sides by 20.}$$

$$-0.01155t = \ln\frac{1}{20} \qquad \text{Inverse property}$$

$$t = \frac{1}{-0.01155}\ln\frac{1}{20} \qquad \text{Divide both sides by } -0.01155.$$

$$\approx 259.4 \text{ days} \qquad \text{Simplify.}$$

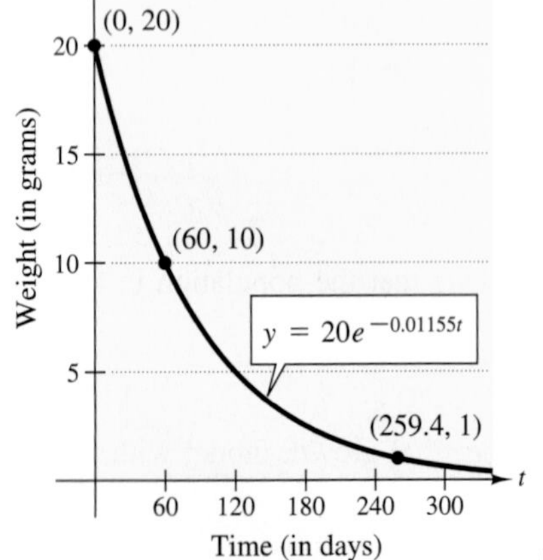

Figure 11.22 *Radioactive Decay*

So, 20 grams of radioactive iodine will have decayed to 1 gram after about 259.4 days. This solution is shown graphically in Figure 11.22.

3 Use a logarithmic equation to solve an intensity problem.

Intensity Models

On the **Richter scale,** the magnitude R of an earthquake can be measured by the **intensity model**

$$R = \log_{10} I$$

where I is the intensity of the shock wave.

Peter Menzel/Stock Boston

Earthquakes take place along faults in the earth's crust. The 1989 earthquake in California took place along the San Andreas Fault.

Example 7 Earthquake Intensity

In 1906, San Francisco experienced an earthquake that measured 8.6 on the Richter scale. In 1989, another earthquake, which measured 7.7 on the Richter scale, struck the same area. Compare the intensities of these two earthquakes.

Solution

The intensity of the 1906 earthquake is given as follows.

$$8.6 = \log_{10} I \qquad \text{Given}$$

$$10^{8.6} = I \qquad \text{Inverse property}$$

The intensity of the 1989 earthquake can be found in a similar way.

$$7.7 = \log_{10} I \qquad \text{Given}$$

$$10^{7.7} = I \qquad \text{Inverse property}$$

The ratio of these two intensities is

$$\frac{I \text{ for } 1906}{I \text{ for } 1989} = \frac{10^{8.6}}{10^{7.7}}$$

$$= 10^{8.6-7.7}$$

$$= 10^{0.9}$$

$$\approx 7.94.$$

Thus, the 1906 earthquake had an intensity that was about eight times greater than the 1989 earthquake.

Discussing the Concept **Problem Posing**

Write a problem that could be answered by investigating the exponential growth model

$$y = 10e^{0.08t}$$

or the exponential decay model

$$y = 5e^{-0.25t}.$$

Exchange your problem for that of another class member, and solve one another's problems.

11.6 Exercises

Integrated Review *Concepts, Skills, and Problem Solving*

Keep mathematically in shape by doing these exercises *before* the problems of this section.

Properties and Definitions

In Exercises 1–4, identify the type of variation given in the model.

1. $y = kx^2$

2. $y = \dfrac{k}{x}$

3. $z = kxy$

4. $z = \dfrac{kx}{y}$

Simplifying Expressions

In Exercises 5–10, solve the system of equations.

5. $\begin{aligned} x - y &= 0 \\ x + 2y &= 9 \end{aligned}$

6. $\begin{aligned} 2x + 5y &= 15 \\ 3x + 6y &= 20 \end{aligned}$

7. $\begin{aligned} y &= x^2 \\ -3x + 2y &= 2 \end{aligned}$

8. $\begin{aligned} x - y^3 &= 0 \\ x - 2y^2 &= 0 \end{aligned}$

9. $\begin{aligned} x - y \phantom{{}- 2z} &= -1 \\ x + 2y - 2z &= 3 \\ 3x - y + 2z &= 3 \end{aligned}$

10. $\begin{aligned} 2x + y - 2z &= 1 \\ x \phantom{{}+ y} - z &= 1 \\ 3x + 3y + z &= 12 \end{aligned}$

Graphs

In Exercises 11 and 12, use the function $y = -x^2 + 4x$.

11. (a) Does the graph open up or down? Explain.

(b) Find the x-intercepts algebraically.

(c) Find the coordinates of the vertex of the parabola.

12. Use a graphing utility to graph the function and geometrically verify the results of Exercise 11.

Solving Problems

Annual Interest Rate In Exercises 1–8, find the annual interest rate. See Example 1.

	Principal	*Balance*	*Time*	*Compounding*
1.	\$500	\$1004.83	10 years	Monthly
2.	\$3000	\$21,628.70	20 years	Quarterly
3.	\$1000	\$36,581.00	40 years	Daily
4.	\$200	\$314.85	5 years	Yearly
5.	\$750	\$8267.38	30 years	Continuously
6.	\$2000	\$4234.00	10 years	Continuously
7.	\$5000	\$22,405.68	25 years	Daily
8.	\$10,000	\$110,202.78	30 years	Daily

Doubling Time In Exercises 9–16, find the time for an investment to double. Use a graphing utility to check the result graphically. See Example 2.

	Principal	*Rate*	*Compounding*
9.	\$6000	8%	Quarterly
10.	\$500	$5\frac{1}{4}$%	Monthly
11.	\$2000	10.5%	Daily
12.	\$10,000	9.5%	Yearly
13.	\$1500	7.5%	Continuously
14.	\$100	6%	Continuously
15.	\$300	5%	Yearly
16.	\$12,000	4%	Continuously

Type of Compounding In Exercises 17–20, determine the type of compounding. Solve the problem by trying the common types of compounding. See Example 3.

	Principal	*Balance*	*Time*	*Rate*
17.	\$750	\$1587.75	10 years	7.5%
18.	\$10,000	\$73,890.56	20 years	10%
19.	\$100	\$141.48	5 years	7%
20.	\$4000	\$4788.76	2 years	9%

Effective Yield In Exercises 21–28, find the effective yield. See Example 4.

	Rate	*Compounding*
21.	8%	Continuously
22.	9.5%	Daily
23.	7%	Monthly
24.	8%	Yearly
25.	6%	Quarterly
26.	9%	Quarterly
27.	8%	Monthly
28.	$5\frac{1}{4}\%$	Daily

29. ***Think About It*** Is it necessary to know the principal P to find the doubling time in Exercises 9–16? Explain.

30. ***Effective Yield***

(a) Is it necessary to know the principal P to find the effective yield in Exercises 21–28? Explain.

(b) When the interest is compounded more frequently, what inference can you make about the difference between the effective yield and the stated annual percentage rate?

Principal In Exercises 31–38, find the principal that must be deposited in an account to obtain the given balance.

	Balance	*Rate*	*Time*	*Compounding*
31.	\$10,000	9%	20 years	Continuously
32.	\$5000	8%	5 years	Continuously
33.	\$750	6%	3 years	Daily
34.	\$3000	7%	10 years	Monthly
35.	\$25,000	7%	30 years	Monthly
36.	\$8000	6%	2 years	Monthly
37.	\$1000	5%	1 year	Daily
38.	\$100,000	9%	40 years	Daily

Balance After Monthly Deposits In Exercises 39–42, you make monthly deposits of P dollars into a savings account at an annual interest rate r, compounded continuously. Find the balance A after t years given that

$$A = \frac{P(e^{rt} - 1)}{e^{r/12} - 1}.$$

	Principal	*Rate*	*Time*
39.	$P = 30$	$r = 8\%$	$t = 10$ years
40.	$P = 100$	$r = 9\%$	$t = 30$ years
41.	$P = 50$	$r = 10\%$	$t = 40$ years
42.	$P = 20$	$r = 7\%$	$t = 20$ years

Balance After Monthly Deposits In Exercises 43 and 44, you make monthly deposits of \$30 into a savings account at an annual interest rate of 8%, compounded continuously (see figure).

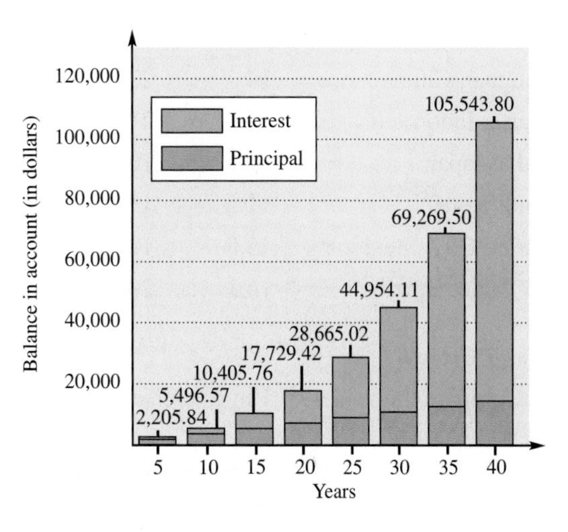

43. Find the total amount that has been deposited into the account in 20 years and the total interest earned.

44. Find the total amount that has been deposited into the account in 40 years and the total interest earned.

Exponential Growth and Decay In Exercises 45–48, find the constant k such that the graph of $y = Ce^{kt}$ passes through the given points.

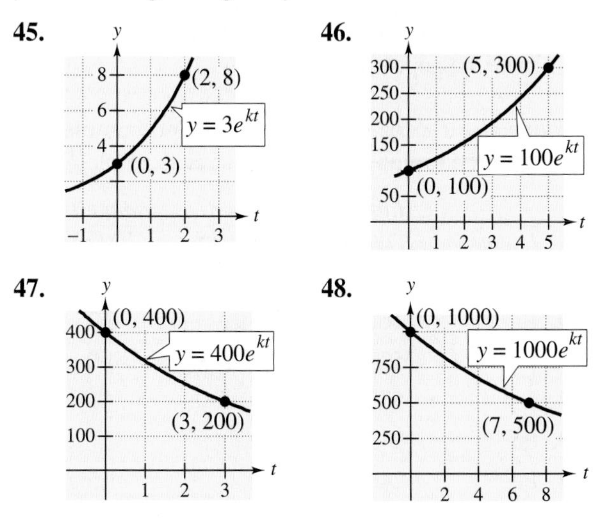

Population of a Region In Exercises 49–56, the population (in millions) of an urban region for 1994 and the predicted population (in millions) for the year 2015 are given. Find the constants C and k to obtain the exponential growth model $y = Ce^{kt}$ for the population. (Let $t = 0$ correspond to the year 1994.) Use the model to predict the population of the region in the year 2020. See Example 5. (Source: United Nations)

Region	*1994*	*2015*
49. Los Angeles	12.2	14.3
50. New York	16.3	17.6
51. Shanghai, China	14.7	23.4
52. Jakarta, Indonesia	11.0	21.2
53. Osaka, Japan	10.5	10.6
54. Seoul, Korea	11.5	13.1
55. Mexico City, Mexico	15.5	18.8
56. Sao Paulo, Brazil	16.1	20.8

57. *Rate of Growth*

(a) Compare the values of k in Exercises 51 and 53. Which is larger? Explain.

(b) What variable in the continuous compound interest formula is equivalent to k in the model for population growth? Use your answer to give an interpretation of k.

58. *World Population* The figure in the next column shows the population P (in billions) of the world as projected by the Population Reference Bureau. The bureau's projection can be modeled by

$$P = \frac{11.14}{1 + 1.101e^{-0.051t}}$$

where $t = 0$ represents 1990. Use the model to estimate the population in 2020.

Radioactive Decay In Exercises 59–64, complete the table for the radioactive isotopes. See Example 6.

Isotope	*Half-Life (Years)*	*Initial Quantity*	*Amount After 1000 Years*
59. Ra^{226}	1620	6 g	
60. Ra^{226}	1620		0.25 g
61. C^{14}	5730		4.0 g
62. C^{14}	5730	10 g	
63. Pu^{230}	24,360	4.2 g	
64. Pu^{230}	24,360		1.5 g

65. *Radioactive Decay* Radioactive radium (Ra^{226}) has a half-life of 1620 years. If you start with 5 grams of the isotope, how much remains after 1000 years?

66. *Radioactive Decay* The isotope Pu^{230} has a half-life of 24,360 years. If you start with 10 grams of the isotope, how much remains after 10,000 years?

67. *Radioactive Decay* Carbon 14 (C^{14}) has a half-life of 5730 years. If you start with 5 grams of this isotope, how much remains after 1000 years?

68. *Carbon 14 Dating* C^{14} dating assumes that the carbon dioxide on earth today has the same radioactive content as it did centuries ago. If this is true, the amount of C^{14} absorbed by a tree that grew several centuries ago should be the same as the amount of C^{14} absorbed by a tree growing today. A piece of ancient charcoal contains only 15% as much of the radioactive carbon as a piece of modern charcoal. How long ago did the tree burn to make the ancient charcoal if the half-life of C^{14} is 5730 years? (Round your answer to the nearest 100 years.)

69. *Depreciation* A car that cost \$22,000 new has a depreciated value of \$16,500 after 1 year. Find the value of the car when it is 3 years old by using the exponential model $y = Ce^{kt}$.

70. *Depreciation* After x years, the value y of a truck that cost \$32,000 new is given by

$y = 32{,}000(0.8)^x$.

(a) Use a graphing utility to graph the model.

(b) Graphically approximate the value of the truck after 1 year.

(c) Graphically approximate the time when the truck's value will be \$16,000.

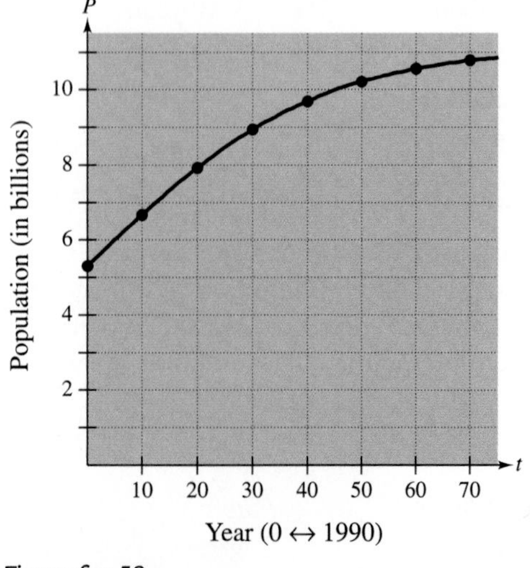

Figure for 58

Earthquake Intensity In Exercises 71–74, compare the intensities of the two earthquakes. See Example 7.

	Location	*Date*	*Magnitude*
71.	Alaska	3/27/64	8.4
	San Fernando Valley	2/9/71	6.6
72.	Long Beach, California	3/10/33	6.2
	Morocco	2/29/60	5.8
73.	Mexico City, Mexico	9/19/85	8.1
	Nepal	8/20/88	6.5
74.	Chile	8/16/06	8.6
	Armenia, USSR	12/7/88	6.8

Acidity Model In Exercises 75–78, use the acidity model

$$\text{pH} = -\log_{10}[\text{H}^+]$$

where acidity (pH) is a measure of the hydrogen ion concentration $[\text{H}^+]$ (measured in moles of hydrogen per liter) of a solution.

75. Find the pH of a solution that has a hydrogen ion concentration of 9.2×10^{-8}.

76. Compute the hydrogen ion concentration if the pH of a solution is 4.7.

77. A certain fruit has a pH of 2.5 and an antacid tablet has a pH of 9.5. The hydrogen ion concentration of the fruit is how many times the concentration of the tablet?

78. If the pH of a solution is decreased by 1 unit, the hydrogen ion concentration is increased by what factor?

79. *Population Growth* The population p of a certain species t years after it is introduced into a new habitat is given by

$$p(t) = \frac{5000}{1 + 4e^{-t/6}}.$$

(a) Use a graphing utility to graph the population function.

(b) Determine the population size that was introduced into the habitat.

(c) Determine the population size after 9 years.

(d) After how many years will the population be 2000?

80. *Sales Growth* Annual sales y of a product x years after it is introduced are approximated by

$$y = \frac{2000}{1 + 4e^{-x/2}}.$$

(a) Use a graphing utility to graph the model.

(b) Use the graph in part (a) to approximate annual sales when $x = 4$.

(c) Use the graph in part (a) to approximate the time when annual sales are $y = 1100$ units.

(d) Use the graph in part (a) to estimate the maximum level that annual sales will approach.

81. *Advertising Effect* The sales S (in thousands of units) of a product after spending x hundred dollars in advertising are given by

$$S = 10(1 - e^{kx}).$$

(a) Find S as a function of x if 2500 units are sold when \$500 is spent on advertising.

(b) How many units will be sold if advertising expenditures are raised to \$700?

Explaining Concepts

82. If the equation $y = Ce^{kt}$ models exponential growth, what must be true about k?

83. If the equation $y = Ce^{kt}$ models exponential decay, what must be true about k?

84. The formulas for periodic and continuous compounding have the four variables $A, P, r,$ and t in common. Explain what each variable measures.

85. What is meant by the effective yield of an investment? Explain how it is computed.

86. In your own words, explain what is meant by the half-life of a radioactive isotope.

87. If the reading on the Richter scale is increased by 1, the intensity of the earthquake is increased by what factor?

CHAPTER SUMMARY

Key Terms

exponential function, p. 630
natural base e, p. 634
natural exponential function, p. 634
composition, p. 643
inverse function, p. 645
one-to-one, p. 645
logarithmic function, p. 655
common logarithmic function, p. 657
natural logarithmic function, p. 660
exponentiate, p. 679
exponential growth, p. 688
exponential decay, p. 688

Key Concepts

11.1 Rules of exponential functions

1. $a^x \cdot a^y = a^{x+y}$
2. $\dfrac{a^x}{a^y} = a^{x-y}$
3. $(a^x)^y = a^{xy}$
4. $a^{-x} = \dfrac{1}{a^x} = \left(\dfrac{1}{a}\right)^x$

11.2 Composition of two functions

The composition of two functions f and g is given by $(f \circ g)(x) = f(g(x))$. The domain of the composite function $(f \circ g)$ is the set of all x in the domain of g such that $g(x)$ is in the domain of f.

11.2 Horizontal Line Test for inverse functions

A function f has an inverse function if and only if f is one-to-one. Graphically, a function f has an inverse if and only if no horizontal line intersects the graph of f at more than one point.

11.2 Finding the inverse of a function

1. In the equation for $f(x)$, replace $f(x)$ by y.
2. Interchange the roles of x and y, and solve for y.
3. If the new equation does not represent y as a function of x, the function f does not have an inverse function.
4. If the new equation represents y as a function of x, replace y by $f^{-1}(x)$.

11.3 Properties of logarithms and natural logarithms

Let a and x be positive real numbers such that $a \neq 1$. Then the following properties are true.

1. $\log_a 1 = 0$ because $a^0 = 1$.
   $\ln 1 = 0$ because $e^0 = 1$.
2. $\log_a a = 1$ because $a^1 = a$.
   $\ln e = 1$ because $e^1 = e$.
3. $\log_a a^x = x$ because $a^x = a^x$.
   $\ln e^x = x$ because $e^x = e^x$.

11.3 Change-of-base formula

Let a, b, and x be positive real numbers such that $a \neq 1$ and $b \neq 1$. Then $\log_a x = \dfrac{\log_b x}{\log_b a}$ or $\log_a x = \dfrac{\ln x}{\ln a}$.

11.4 Properties of logarithms

Let a be a positive real number such that $a \neq 1$, and let n be a real number. If u and v are real numbers, variables, or algebraic expressions such that $u > 0$ and $v > 0$, the following properties are true.

	Logarithm with base a	*Natural logarithm*
1.	$\log_a(uv) = \log_a u + \log_a v$	$\ln(uv) = \ln u + \ln v$
2.	$\log_a \dfrac{u}{v} = \log_a u - \log_a v$	$\ln \dfrac{u}{v} = \ln u - \ln v$
3.	$\log_a u^n = n \log_a u$	$\ln u^n = n \ln u$

11.5 Properties of exponential and logarithmic equations

Let a be a positive real number such that $a \neq 1$, and let x and y be real numbers. Then the following properties are true.

1. $a^x = a^y$ if and only if $x = y$.
2. $\log_a x = \log_a y$ if and only if $x = y$ $(x > 0, y > 0)$.

11.5 Inverse properties of exponents and logarithms

	Base a	*Natural base e*
1.	$\log_a(a^x) = x$	$\ln(e^x) = x$
2.	$a^{(\log_a x)} = x$	$e^{(\ln x)} = x$

11.5 Solving exponential and logarithmic equations

1. To solve an exponential equation, first isolate the exponential expression, then take the logarithm of both sides of the equation and solve for the variable.
2. To solve a logarithmic equation, first isolate the logarithmic expression, then exponentiate both sides of the equation and solve for the variable.

REVIEW EXERCISES

Reviewing Skills

11.1 In Exercises 1–4, evaluate the exponential function as indicated.

1. $f(x) = 2^x$
 (a) $x = -3$ (b) $x = 1$ (c) $x = 2$

2. $g(x) = 2^{-x}$
 (a) $x = -2$ (b) $x = 0$ (c) $x = 2$

3. $g(t) = e^{-t/3}$
 (a) $t = -3$ (b) $t = \pi$ (c) $t = 6$

4. $h(s) = 1 - e^{0.2s}$
 (a) $s = 0$ (b) $s = 2$ (c) $s = \sqrt{10}$

In Exercises 5–8, match the function with the sketch of its graph. [The graphs are labeled (a), (b), (c), and (d).]

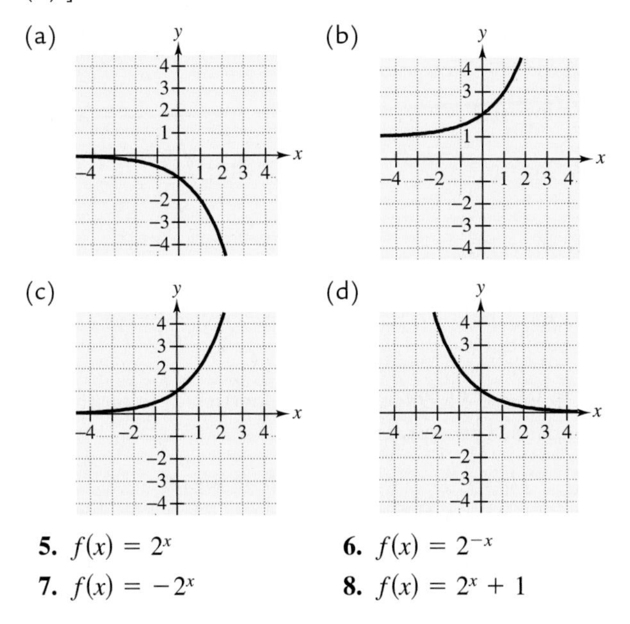

5. $f(x) = 2^x$ **6.** $f(x) = 2^{-x}$

7. $f(x) = -2^x$ **8.** $f(x) = 2^x + 1$

In Exercises 9–18, use the point-plotting method to sketch the graph of the exponential function.

9. $f(x) = 3^x$ **10.** $f(x) = 3^{-x}$

11. $f(x) = 3^x - 1$ **12.** $f(x) = 3^x + 2$

13. $f(x) = 3^{(x+1)}$ **14.** $f(x) = 3^{(x-1)}$

15. $y = 3^{x/2}$ **16.** $f(x) = 3^{-x/2}$

17. $y = 3^{x/2} - 2$ **18.** $f(x) = 3^{x/2} + 3$

In Exercises 19–22, use a graphing utility to graph the exponential function.

19. $y = 5e^{-x/4}$

20. $y = 6 - e^{x/2}$

21. $f(x) = e^{x+2}$

22. $h(t) = \dfrac{8}{1 + e^{-t/5}}$

11.2 In Exercises 23–26, form $f \circ g$ and $g \circ f$ and evaluate the composite functions.

23. $f(x) = x + 2,\ g(x) = x^2$
 (a) $(f \circ g)(2)$ (b) $(g \circ f)(-1)$

24. $f(x) = \sqrt[3]{x},\ g(x) = x + 2$
 (a) $(f \circ g)(6)$ (b) $(g \circ f)(64)$

25. $f(x) = \sqrt{x + 1},\ g(x) = x^2 - 1$
 (a) $(f \circ g)(5)$ (b) $(g \circ f)(-1)$

26. $f(x) = \dfrac{1}{x - 5},\ g(x) = \dfrac{5x + 1}{x}$
 (a) $(f \circ g)(1)$ (b) $(g \circ f)\left(\dfrac{1}{5}\right)$

In Exercises 27 and 28, form the compositions (a) $f \circ g$ and (b) $g \circ f$, and find the domains of the composites.

27. $f(x) = \sqrt{x - 4},\ g(x) = 2x$

28. $f(x) = \dfrac{2}{x - 4},\ g(x) = x^2$

In Exercises 29–32, use the Horizontal Line Test to determine whether the function has an inverse.

29. $f(x) = x^2 - 25$ **30.** $f(x) = \frac{1}{4}x^3$

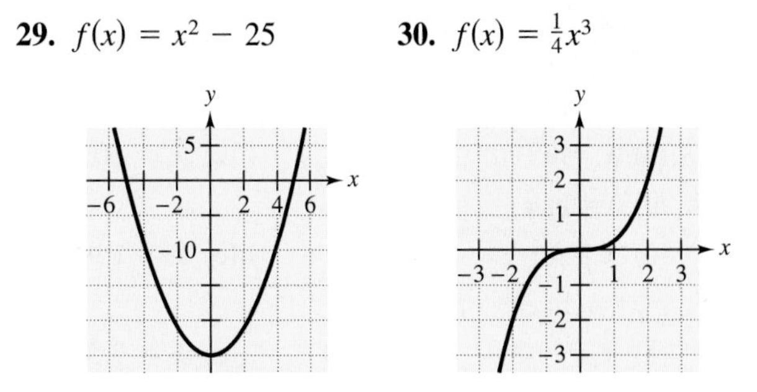

31. $h(x) = 4\sqrt[3]{x}$

32. $g(x) = \sqrt{9 - x^2}$

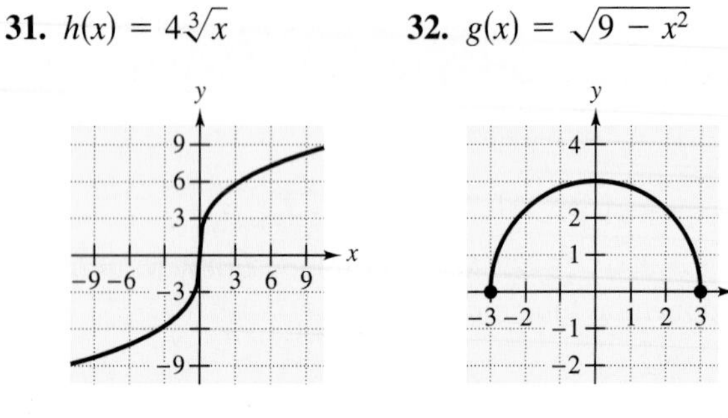

In Exercises 33–38, find the inverse of the function algebraically. (If not possible, state the reason.)

33. $f(x) = 3x + 4$

34. $f(x) = 2x - 3$

35. $h(x) = \sqrt{x}$

36. $g(x) = x^2 + 2,\ x \ge 0$

37. $f(t) = t^3 + 4$

38. $h(t) = \sqrt[3]{t - 1}$

11.3 In Exercises 39 and 40, write the exponential equation in logarithmic form.

39. $4^3 = 64$

40. $25^{3/2} = 125$

In Exercises 41 and 42, write the logarithmic equation in exponential form.

41. $\ln e = 1$

42. $\log_3 \frac{1}{9} = -2$

In Exercises 43–50, evaluate the logarithm.

43. $\log_{10} 1000$

44. $\log_9 3$

45. $\log_3 \frac{1}{9}$

46. $\log_4 \frac{1}{16}$

47. $\ln e^7$

48. $\log_a \dfrac{1}{a}$

49. $\ln 1$

50. $\ln e^{-3}$

In Exercises 51–56, evaluate the logarithmic function as indicated.

51. $f(x) = \log_3 x$

(a) $x = 1$ (b) $x = 27$ (c) $x = 0.5$

52. $g(x) = \log_{10} x$

(a) $x = 0.01$ (b) $x = 0.1$ (c) $x = 30$

53. $f(x) = \ln x$

(a) $x = e$ (b) $x = \frac{1}{3}$ (c) $x = 10$

54. $h(x) = \ln x$

(a) $x = e^2$ (b) $x = \frac{5}{4}$ (c) $x = 1200$

55. $g(x) = \ln e^{3x}$

(a) $x = -2$ (b) $x = 0$ (c) $x = 7.5$

56. $f(x) = \log_2 \sqrt{x}$

(a) $x = 4$ (b) $x = 64$ (c) $x = 5.2$

In Exercises 57–66, use the point-plotting method to graph the logarithmic function.

57. $f(x) = \log_3 x$

58. $f(x) = -\log_3 x$

59. $f(x) = -2 + \log_3 x$

60. $f(x) = 2 + \log_3 x$

61. $y = \log_2(x - 4)$

62. $y = \log_4(x + 1)$

63. $y = \ln(x - 3)$

64. $y = -\ln(x + 2)$

65. $y = 5 - \ln x$

66. $y = 3 + \ln x$

In Exercises 67–70, use the change-of-base formula to evaluate the logarithm. Round the result to three decimal places.

67. $\log_4 9$

68. $\log_{1/2} 5$

69. $\log_{12} 200$

70. $\log_3 0.28$

11.4 In Exercises 71–76, approximate the logarithm given that $\log_5 2 \approx 0.43068$ and $\log_5 3 \approx 0.68261$.

71. $\log_5 18$

72. $\log_5 \sqrt{6}$

73. $\log_5 \frac{1}{2}$

74. $\log_5 \frac{2}{3}$

75. $\log_5(12)^{2/3}$

76. $\log_5(5^2 \cdot 6)$

In Exercises 77–84, use the properties of logarithms to expand the expression.

77. $\log_4 6x^4$

78. $\log_{10} 2x^{-3}$

79. $\log_5 \sqrt{x + 2}$

80. $\ln \sqrt[3]{\dfrac{x}{5}}$

81. $\ln \dfrac{x + 2}{x - 2}$

82. $\ln x(x - 3)^2$

83. $\ln\left[\sqrt{2x}(x + 3)^5\right]$

84. $\log_3 \dfrac{a^2\sqrt{b}}{cd^5}$

In Exercises 85–94, use the properties of logarithms to condense the expression.

85. $-\frac{2}{3} \ln 3y$

86. $5 \log_2 y$

87. $\log_8 16x + \log_8 2x^2$

88. $\log_4 6x - \log_4 10$

89. $-2(\ln 2x - \ln 3)$

90. $4(1 + \ln x + \ln x)$

91. $4[\log_2 k - \log_2(k - t)]$

92. $\frac{1}{3}(\log_8 a + 2 \log_8 b)$

93. $3 \ln x + 4 \ln y + \ln z$

94. $\ln(x + 4) - 3 \ln x - \ln y$

True or False? In Exercises 95–100, use the properties of logarithms to determine whether the equation is true or false. If false, state why or give an example to show that it is false.

95. $\log_2 4x = 2 \log_2 x$

96. $\dfrac{\ln 5x}{\ln 10x} = \ln \dfrac{1}{2}$

97. $\log_{10} 10^{2x} = 2x$

98. $e^{\ln t} = t$

99. $\log_4 \dfrac{16}{x} = 2 - \log_4 x$

100. $6 \ln x + 6 \ln y = \ln(xy)^6$

11.5 In Exercises 101–110, solve the equation.

101. $2^x = 64$

102. $5^x = 25$

103. $4^{x-3} = \frac{1}{16}$

104. $3^{x-2} = 81$

105. $\log_3 x = 5$

106. $\log_4 x = 3$

107. $\log_2 2x = \log_2 100$

108. $\ln(x + 4) = \ln 7$

109. $\log_3(2x + 1) = 2$

110. $\log_5(x - 10) = 2$

In Exercises 111–124, solve the equation.

111. $3^x = 500$

112. $8^x = 1000$

113. $\ln x = 7.25$

114. $\ln x = -0.5$

115. $2e^{0.5x} = 45$

116. $100e^{-0.6x} = 20$

117. $12(1 - 4^x) = 18$

118. $25(1 - e^t) = 12$

119. $\log_{10} 2x = 1.5$

120. $\log_2 2x = -0.65$

121. $\frac{1}{3} \log_2 x + 5 = 7$

122. $4 \log_5(x + 1) = 4.8$

123. $\log_2 x + \log_2 3 = 3$

124. $2 \log_4 x - \log_4(x - 1) = 1$

11.6 *Annual Interest Rate* In Exercises 125–130, find the annual interest rate.

	Principal	*Balance*	*Time*	*Compounding*
125.	\$250	\$410.90	10 years	Quarterly
126.	\$1000	\$1348.85	5 years	Monthly
127.	\$5000	\$15,399.30	15 years	Daily
128.	\$10,000	\$35,236.45	20 years	Yearly
129.	\$1500	\$24,666.97	40 years	Continuously
130.	\$7500	\$15,877.50	15 years	Continuously

Effective Yield In Exercises 131–136, find the effective yield.

	Rate	*Compounding*
131.	5.5%	Daily
132.	6%	Monthly
133.	7.5%	Quarterly
134.	8%	Yearly
135.	7.5%	Continuously
136.	4%	Continuously

Radioactive Decay In Exercises 137–142, complete the table for the radioactive isotopes.

	Isotope	*Half-Life (Years)*	*Initial Quantity*	*Amount After 1000 Years*
137.	Ra^{226}	1620	3.5 g	
138.	Ra^{226}	1620		0.5 g
139.	C^{14}	5730		2.6 g
140.	C^{14}	5730	10 g	
141.	Pu^{230}	24,360	5 g	
142.	Pu^{230}	24,360		2.5 g

Solving Problems

143. *Inflation Rate* If the annual rate of inflation averages 5% over the next 10 years, the approximate cost C of goods or services during any year in that decade will be given by

$$C(t) = P(1.05)^t, \quad 0 \le t \le 10$$

where t is time in years and P is the present cost. If the price of an oil change for your car is presently \$24.95, when will it cost \$30.00?

144. *Doubling Time* Find the time for an investment of \$1000 to double in value when invested at 8% compounded monthly.

145. *Doubling Time* Find the time for an investment of \$750 to double in value when invested at 5.5% compounded continuously.

146. *Product Demand* The daily demand x and price p for a product are related by

$$p = 25 - 0.4e^{0.02x}.$$

Approximate the demand if the price is \$16.97.

147. *Sound Intensity* The relationship between the number of decibels B and the intensity of a sound I in watts per centimeter squared is given by

$$B = 10 \log_{10}\left(\frac{I}{10^{-16}}\right).$$

Determine the intensity of a sound in watts per centimeter squared if the decibel level is 125.

148. *Sound Intensity* The relationship between the number of decibels B and the intensity of a sound I in watts per centimeter squared is given by

$$B = 10 \log_{10}\left(\frac{I}{10^{-16}}\right).$$

Determine the intensity of a sound in watts per centimeter squared if the decibel level is 150.

149. *Population Limit* The population p of a certain species t years after it is introduced into a new habitat is given by

$$p(t) = \frac{600}{1 + 2e^{-0.2t}}.$$

Use a graphing utility to graph the function. Use the graph to determine the limiting size of the population in this habitat.

150. *Deer Herd* The state Parks and Wildlife Department releases 100 deer into a wilderness area. The population P of the herd can be modeled by

$$P = \frac{500}{1 + 4e^{-0.36t}}$$

where t is measured in years.

(a) Find the population after 5 years.

(b) After how many years will the population be 250?

151. *Ventilation* The table gives the required ventilation rate V (in cubic feet per minute per person) for a room in a public building with an air space of x cubic feet per person.

x	100	200	300	400
V	25	17	12	9

A model for the data is $V = 78.56 - 11.6314 \ln x$.

(a) Use a graphing utility to plot the data and graph the model.

(b) Use the model to determine V if $x = 250$.

152. *Comparing Models* The figure gives the assets (in billions of dollars) in client accounts for the years 1993 through 1997 for Merrill Lynch. A list of models ($t = 3$ represents 1993) for the data is also given. For each of the models, (a) use a graphing utility to obtain its graph, and (b) use the graphs of part (a) to determine which model "best fits" the data. (Source: Merrill Lynch 1997 Annual Report)

Linear: $A = 156.5t - 8.3$

Quadratic: $A = 50.6t^2 - 349.9t + 1156.5$

Exponential: $A = 282.4e^{0.193t}$

Logarithmic: $A = 1133.3 + 620.8t - 2210.9 \ln t$

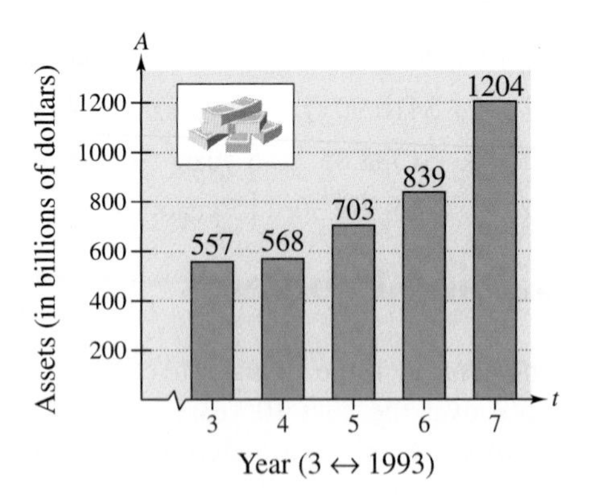

Chapter Test

Take this test as you would take a test in class. After you are done, check your work against the answers given in the back of the book.

1. Evaluate $f(t) = 54\left(\frac{2}{3}\right)^t$ when $t = -1, 0, \frac{1}{2}$, and 2.

2. Sketch a graph of the function $f(x) = 2^{x/3}$.

3. Form the composition of (a) $f \circ g$ and (b) $g \circ f$, and find the domains of the composites.

$f(x) = 3x - 4, \qquad g(x) = x^2 + 1$

4. Find the inverse of the function $f(x) = 5x + 6$.

5. Verify algebraically that the two functions are inverses of each other.

$f(x) = -\frac{1}{2}x + 3, \qquad g(x) = -2x + 6$

6. Describe the relationship between the graphs of $f(x) = \log_5 x$ and $g(x) = 5^x$.

7. Use the properties of logarithms to expand $\log_4\left(5x^2/\sqrt{y}\right)$.

8. Use the properties of logarithms to condense $\ln x - 4 \ln y$.

9. Simplify $\log_5 5^3 \cdot 6$.

In Exercises 10–17, solve the equation.

10. $\log_4 x = 3$

11. $10^{3y} = 832$

12. $400e^{0.08t} = 1200$

13. $3 \ln(2x - 3) = 10$

14. $8(2 - 3^x) = -56$

15. $\log_2 x + \log_2 4 = 5$

16. $\ln x - \ln 2 = 4$

17. $30(e^x + 9) = 300$

18. Determine the balance after 20 years if \$2000 is invested at 7% compounded (a) quarterly and (b) continuously.

19. Determine the principal that will yield \$100,000 when invested at 9% compounded quarterly for 25 years.

20. A principal of \$500 yields a balance of \$1006.88 in 10 years when the interest is compounded continuously. What is the annual interest rate?

21. A car that cost \$18,000 new has a depreciated value of \$14,000 after 1 year. Find the value of the car when it is 3 years old by using the exponential model $y = Ce^{kt}$.

In Exercises 22–24, the population of a certain species t years after it is introduced into a new habitat is given by

$$p(t) = \frac{2400}{1 + 3e^{-t/4}}.$$

22. Determine the population size that was introduced into the habitat.

23. Determine the population after 4 years.

24. After how many years will the population be 1200?

12 Sequences, Series, and Probability

Lawrence Migdale/Stock Boston

Interest in genealogy, which is the study of family history and ancestry, is growing. Today there are many resources and organizations available to assist both the amateur and professional genealogist.

Motivating the Chapter

Ancestors and Descendants

See Section 12.3, Exercise 127

a. Your ancestors consist of your two parents (first generation), your four grandparents (second generation), your eight great-grandparents (third generation), and so on. Write a geometric sequence that will describe the number of ancestors for each generation.

b. If your ancestry could be traced back 66 generations (approximately 2000 years), how many different ancestors would you have?

c. A common ancestor is one to whom you are related in more than one way. (See figure.) From the results of part (b), do you think that you have had no common ancestors in the past 2000 years?

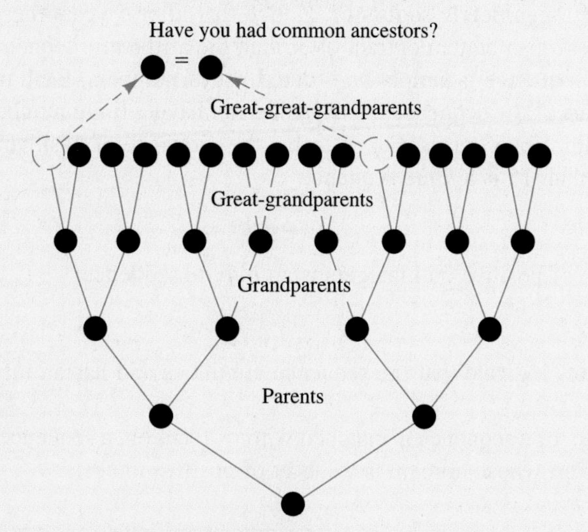

See Section 12.6, Exercise 55

d. One set of your great-grandparents had eight children (of which one is your grandparent). Each of those children had five children (of which one is your parent). And each of those children had three children. How many direct descendants do your great-grandparents have?

e. One hundred fifty people were able to attend your family reunion, to which everyone was asked to bring an exchange gift. All the names were put into a bowl to be drawn randomly to determine the order for receiving a gift. Will the drawing be done *with* or *without* replacement? Explain.

f. What is the probability that your name will be chosen first to receive a gift? After 45 names are drawn, your name is still in the bowl. What is the probability that your name will be chosen next?

12.1 Sequences and Series

Objectives

1. Write the terms of a sequence given its nth term.
2. Write the terms of a sequence involving factorials.
3. Find the apparent nth term of a sequence.
4. Sum the terms of a sequence to obtain a series.

1 Write the terms of a sequence given its nth term.

Sequences

Suppose you were given the following choice of contract offers for the next 5 years of employment.

Contract A $20,000 the first year and a $2200 raise each year

Contract B $20,000 the first year and a 10% raise each year

Which contract offers the greater salary over the 5-year period? The salaries for each contract are shown at the left. The salaries for contract A represent the first five terms of an **arithmetic sequence,** and the salaries for contract B represent the first five terms of a **geometric sequence.** Notice that after 5 years the geometric sequence represents a better contract offer than the arithmetic sequence.

Year	*Contract A*	*Contract B*
1	$20,000	$20,000
2	$22,200	$22,000
3	$24,400	$24,200
4	$26,600	$26,620
5	$28,800	$29,282
Total	$122,000	$122,102

A mathematical **sequence** is simply an ordered list of numbers. Each number in the list is a **term** of the sequence. A sequence can have a finite number of terms or an infinite number of terms. For instance, the sequence of positive odd integers that are less than 15 is a *finite* sequence

$$1, 3, 5, 7, 9, 11, 13$$ Finite sequence

whereas the sequence of positive odd integers is an *infinite* sequence.

$$1, 3, 5, 7, 9, 11, 13, . . .$$ Infinite sequence

Note that the three dots indicate that the sequence continues and has an infinite number of terms.

Because each term of a sequence is matched with its location, a sequence can be defined as a **function** whose domain is a subset of positive integers.

> **Sequences**
>
> An **infinite sequence** $a_1, a_2, a_3, . . . , a_n, . . .$ is a function whose domain is the set of positive integers.
>
> A **finite sequence** $a_1, a_2, a_3, . . . , a_n$ is a function whose domain is the finite set $\{1, 2, 3, . . ., n\}$.

In some cases it is convenient to begin subscripting a sequence with 0 instead of 1. Then the domain of the infinite sequence is the set of nonnegative integers and the domain of the finite sequence is the set $\{0, 1, 2, . . . , n\}$.

$a_{(\ )} = 2(\ ) + 1$

$a_{(1)} = 2(1) + 1 = 3$

$a_{(2)} = 2(2) + 1 = 5$

$\vdots$

$a_{(51)} = 2(51) + 1 = 103$

The subscripts of a sequence are used in place of function notation. For instance, if parentheses replaced the n in $a_n = 2n + 1$, the notation would be similar to function notation, as shown at the left.

Most graphing utilities have a "sequence graphing mode" that allows you to plot the terms of a sequence as points on a rectangular coordinate system. For instance, the graph of the first six terms of the sequence given by

$$a_n = n^2 - 1$$

is shown below.

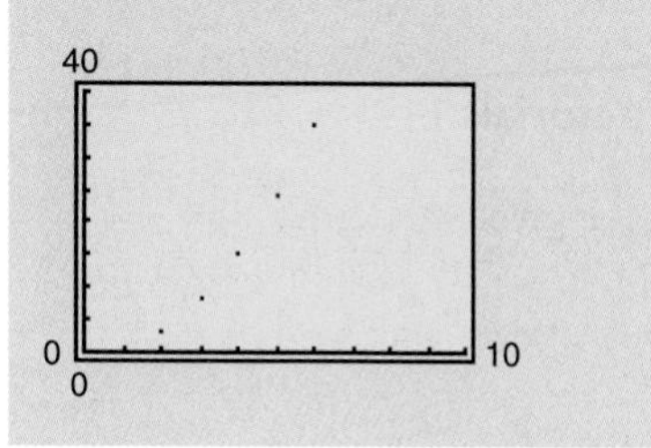

Example 1 Finding the Terms of a Sequence

Write the first six terms of the sequence whose nth term is

$$a_n = n^2 - 1. \qquad \text{Begin sequence with } n = 1.$$

Solution

$a_1 = (1)^2 - 1 = 0 \qquad a_2 = (2)^2 - 1 = 3 \qquad a_3 = (3)^2 - 1 = 8$

$a_4 = (4)^2 - 1 = 15 \qquad a_5 = (5)^2 - 1 = 24 \qquad a_6 = (6)^2 - 1 = 35$

To represent the entire sequence, you can write the following.

$$0, 3, 8, 15, 24, 35, \ldots, n^2 - 1, \ldots$$

Example 2 Finding the Terms of a Sequence

Write the first six terms of the sequence whose nth term is

$$a_n = 3(2^n). \qquad \text{Begin sequence with } n = 0.$$

Solution

$a_0 = 3(2^0) = 3 \cdot 1 = 3 \qquad a_1 = 3(2^1) = 3 \cdot 2 = 6$

$a_2 = 3(2^2) = 3 \cdot 4 = 12 \qquad a_3 = 3(2^3) = 3 \cdot 8 = 24$

$a_4 = 3(2^4) = 3 \cdot 16 = 48 \qquad a_5 = 3(2^5) = 3 \cdot 32 = 96$

The entire sequence can be written as follows.

$$3, 6, 12, 24, 48, 96, \ldots, 3(2^n), \ldots$$

Example 3 A Sequence Whose Terms Alternate in Sign

Write the first six terms of the sequence whose nth term is

$$a_n = \frac{(-1)^n}{2n - 1}. \qquad \text{Begin sequence with } n = 1.$$

Solution

$a_1 = \frac{(-1)^1}{2(1) - 1} = -\frac{1}{1} \qquad a_2 = \frac{(-1)^2}{2(2) - 1} = \frac{1}{3} \qquad a_3 = \frac{(-1)^3}{2(3) - 1} = -\frac{1}{5}$

$a_4 = \frac{(-1)^4}{2(4) - 1} = \frac{1}{7} \qquad a_5 = \frac{(-1)^5}{2(5) - 1} = -\frac{1}{9} \qquad a_6 = \frac{(-1)^6}{2(6) - 1} = \frac{1}{11}$

The entire sequence can be written as follows.

$$-1, \frac{1}{3}, -\frac{1}{5}, \frac{1}{7}, -\frac{1}{9}, \frac{1}{11}, \ldots, \frac{(-1)^n}{2n - 1}, \ldots$$

2 Write the terms of a sequence involving factorials.

Factorial Notation

Some very important sequences in mathematics involve terms that are defined with special types of products called **factorials.**

> **Definition of Factorial**
>
> If n is a positive integer, ***n* factorial** is defined as
>
> $$n! = 1 \cdot 2 \cdot 3 \cdot 4 \cdot \cdot \cdot \cdot \cdot (n-1) \cdot n.$$
>
> As a special case, zero factorial is defined as $0! = 1$.

The first several factorial values are as follows.

$0! = 1$ $\qquad$ $1! = 1$

$2! = 1 \cdot 2 = 2$ $\qquad$ $3! = 1 \cdot 2 \cdot 3 = 6$

$4! = 1 \cdot 2 \cdot 3 \cdot 4 = 24$ $\qquad$ $5! = 1 \cdot 2 \cdot 3 \cdot 4 \cdot 5 = 120$

Many calculators have a factorial key, denoted by [n!]. If your calculator has such a key, try using it to evaluate $n!$ for several values of n. You will see that the value of n does not have to be very large before the value of $n!$ is huge. For instance,

$$10! = 3{,}628{,}800.$$

Example 4 A Sequence Involving Factorials

Write the first six terms of the sequence with the given nth term.

a. $a_n = \dfrac{1}{n!}$ $\qquad$ **b.** $a_n = \dfrac{2^n}{n!}$ $\qquad$ Begin both sequences with $n = 0$.

Solution

a. $a_0 = \dfrac{1}{0!} = \dfrac{1}{1} = 1$ $\qquad$ $a_1 = \dfrac{1}{1!} = \dfrac{1}{1} = 1$

$a_2 = \dfrac{1}{2!} = \dfrac{1}{2}$ $\qquad$ $a_3 = \dfrac{1}{3!} = \dfrac{1}{1 \cdot 2 \cdot 3} = \dfrac{1}{6}$

$a_4 = \dfrac{1}{4!} = \dfrac{1}{1 \cdot 2 \cdot 3 \cdot 4} = \dfrac{1}{24}$ $\qquad$ $a_5 = \dfrac{1}{5!} = \dfrac{1}{1 \cdot 2 \cdot 3 \cdot 4 \cdot 5} = \dfrac{1}{120}$

b. $a_0 = \dfrac{2^0}{0!} = \dfrac{1}{1} = 1$ $\qquad$ $a_1 = \dfrac{2^1}{1!} = \dfrac{2}{1} = 2$

$a_2 = \dfrac{2^2}{2!} = \dfrac{4}{2} = 2$ $\qquad$ $a_3 = \dfrac{2^3}{3!} = \dfrac{8}{1 \cdot 2 \cdot 3} = \dfrac{8}{6} = \dfrac{4}{3}$

$a_4 = \dfrac{2^4}{4!} = \dfrac{2 \cdot 2 \cdot 2 \cdot 2}{1 \cdot 2 \cdot 3 \cdot 4} = \dfrac{2}{3}$ $\qquad$ $a_5 = \dfrac{2^5}{5!} = \dfrac{2 \cdot 2 \cdot 2 \cdot 2 \cdot 2}{1 \cdot 2 \cdot 3 \cdot 4 \cdot 5} = \dfrac{4}{15}$

3 Find the apparent nth term of a sequence.

Finding the nth Term of a Sequence

Sometimes you will have the first several terms of a sequence and need to find a formula (the nth term) that will generate those terms. *Pattern recognition* is crucial in finding a form for the nth term.

Study Tip

Simply listing the first few terms is not sufficient to define a unique sequence—the nth term *must be given.* Consider the sequence

$$\frac{1}{2}, \frac{1}{4}, \frac{1}{8}, \frac{1}{15}, \ldots$$

The first three terms are identical to the first three terms of the sequence in Example 5(a). However, the nth term of this sequence is defined as

$$a_n = \frac{6}{(n+1)(n^2 - n + 6)}.$$

Example 5 Finding the nth Term of a Sequence

Write an expression for the nth term of each sequence.

a. $\frac{1}{2}, \frac{1}{4}, \frac{1}{8}, \frac{1}{16}, \frac{1}{32}, \ldots$ **b.** $1, -4, 9, -16, 25, \ldots$

Solution

a.

n:	1	2	3	4	5	. . .	n
Terms:	$\frac{1}{2}$	$\frac{1}{4}$	$\frac{1}{8}$	$\frac{1}{16}$	$\frac{1}{32}$	. . .	a_n

Pattern: The numerator is 1 and each denominator is an increasing power of 2.

$$a_n = \frac{1}{2^n}$$

b.

n:	1	2	3	4	5	. . .	n
Terms:	1	-4	9	-16	25	. . .	a_n

Pattern: The terms have alternating signs with those in the even positions being negative. Each term is the square of n.

$$a_n = (-1)^{n+1}n^2$$

4 Sum the terms of a sequence to obtain a series.

Series

In the opening illustration of this section, the terms of the finite sequence were *added.* If you add all the terms of an infinite sequence, you obtain a **series.**

Definition of a Series

For an infinite sequence

$$a_1, a_2, a_3, \ldots, a_n, \ldots$$

1. the sum of all the terms

$$a_1 + a_2 + a_3 + \cdots + a_n + \cdots$$

is called an **infinite series,** or simply a **series,** and

2. the sum of the first n terms

$$S_n = a_1 + a_2 + a_3 + \cdots + a_n$$

is called a **partial sum.**

Example 6 Finding Partial Sums

Find the indicated partial sums for each sequence.

a. Find S_1, S_2, and S_5 for $a_n = 3n - 1$.

b. Find S_2, S_3, and S_4 for $a_n = \dfrac{(-1)^n}{n + 1}$.

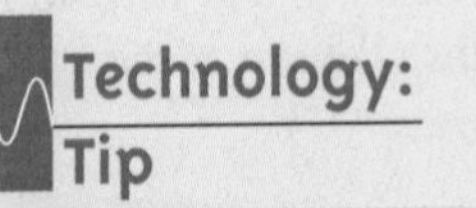

Most graphing utilities have a built-in program that will calculate the partial sum of a sequence. Consult the user's guide for your graphing utility.

Solution

a. The first five terms of the sequence $a_n = 3n - 1$ are

$$a_1 = 2,\ a_2 = 5,\ a_3 = 8,\ a_4 = 11, \text{ and } a_5 = 14.$$

So, the partial sums are

$$S_1 = 2,\ S_2 = 2 + 5 = 7, \text{ and } S_5 = 2 + 5 + 8 + 11 + 14 = 40.$$

b. The first four terms of the sequence $a_n = \dfrac{(-1)^n}{n + 1}$ are

$$a_1 = -\frac{1}{2},\ a_2 = \frac{1}{3},\ a_3 = -\frac{1}{4}, \text{ and } a_4 = \frac{1}{5}.$$

So, the partial sums are

$$S_2 = -\frac{1}{2} + \frac{1}{3} = -\frac{1}{6},\ S_3 = -\frac{1}{2} + \frac{1}{3} - \frac{1}{4} = -\frac{5}{12}, \text{ and}$$

$$S_4 = -\frac{1}{2} + \frac{1}{3} - \frac{1}{4} + \frac{1}{5} = -\frac{13}{60}.$$

A convenient shorthand notation for denoting a partial sum is called **sigma notation.** This name comes from the use of the uppercase Greek letter sigma, written as Σ.

Study Tip

If a_i is a constant in the partial sum

$$\sum_{i=m}^{n} a_i$$

then the partial sum is the product $(n - m + 1)(a_i)$.

For instance,

$$\sum_{i=1}^{4} 5 = 5 + 5 + 5 + 5$$
$$= 4(5)$$
$$= 20.$$

▶ Definition of Sigma Notation

The sum of the first n terms of the sequence whose nth term is a_n is

$$\sum_{i=1}^{n} a_i = a_1 + a_2 + a_3 + a_4 + \cdot \cdot \cdot + a_n$$

where i is the **index of summation,** n is the **upper limit of summation,** and 1 is the **lower limit of summation.**

Example 7 Sigma Notation for Sums

$$\sum_{i=1}^{6} 2i = 2(1) + 2(2) + 2(3) + 2(4) + 2(5) + 2(6)$$
$$= 2 + 4 + 6 + 8 + 10 + 12$$
$$= 42$$

Study Tip

In Example 7, the index of summation is i and the summation begins with $i = 1$. Any letter can be used as the index of summation, and the summation can begin with any integer. For instance, in Example 8, the index of summation is k and the summation begins with $k = 0$.

Example 8 Sigma Notation for Sums

Find the sum $\displaystyle\sum_{k=0}^{8} \frac{1}{k!}$.

Solution

$$\sum_{k=0}^{8} \frac{1}{k!} = \frac{1}{0!} + \frac{1}{1!} + \frac{1}{2!} + \frac{1}{3!} + \frac{1}{4!} + \frac{1}{5!} + \frac{1}{6!} + \frac{1}{7!} + \frac{1}{8!}$$

$$= 1 + 1 + \frac{1}{2} + \frac{1}{6} + \frac{1}{24} + \frac{1}{120} + \frac{1}{720} + \frac{1}{5040} + \frac{1}{40{,}320}$$

$$\approx 2.71828$$

Note that this sum is approximately $e = 2.71828. . . .$

Example 9 Writing a Sum in Sigma Notation

Write the following sums in sigma notation.

a. $\frac{2}{2} + \frac{2}{3} + \frac{2}{4} + \frac{2}{5} + \frac{2}{6}$ **b.** $1 - \frac{1}{3} + \frac{1}{9} - \frac{1}{27} + \frac{1}{81}$

Solution

a. The pattern has numerators of 2 and denominators that range over the integers from 2 to 6. So, one possible sigma notation is

$$\sum_{i=1}^{5} \frac{2}{i+1} = \frac{2}{2} + \frac{2}{3} + \frac{2}{4} + \frac{2}{5} + \frac{2}{6}.$$

b. The numerators alternate in sign and the denominators are integer powers of 3, starting with 3^0 and ending with 3^4. So, one possible sigma notation is

$$\sum_{i=0}^{4} \frac{(-1)^i}{3^i} = \frac{1}{3^0} + \frac{-1}{3^1} + \frac{1}{3^2} + \frac{-1}{3^3} + \frac{1}{3^4}.$$

Discussing the Concept — Finding a Pattern

You learned in this section that a sequence is an ordered list of numbers. Study the following sequence and see if you can guess what its next term should be.

Z, O, T, T, F, F, S, S, E, N, T, E, T, . . .

(*Hint:* You might try to figure out what numbers the letters represent.) Construct another sequence with letters. Can the other members of your class guess the next term?

12.1 Exercises

Integrated Review *Concepts, Skills, and Problem Solving*

Keep mathematically in shape by doing these exercises *before* the problems of this section.

Properties and Definitions

1. Demonstrate the Multiplicative Property of Equality for the equation $-7x = 35$.

2. Demonstrate the Additive Property of Equality for the equation $7x + 63 = 35$.

3. How do you determine whether $t = -3$ is a solution of the equation $t^2 + 4t + 3 = 0$?

4. What is the usual first step in solving an equation such as

$$\frac{3}{x} - \frac{1}{x+1} = 10?$$

Simplifying Expressions

In Exercises 5–10, simplify the expression.

5. $(x + 10)^{-2}$

6. $\dfrac{18(x-3)^5}{(x-3)^2}$

7. $(a^2)^{-4}$

8. $(8x^3)^{1/3}$

9. $\sqrt{128x^3}$

10. $\dfrac{5}{\sqrt{x}-2}$

Graphs and Models

In Exercises 11 and 12, (a) write a function for the area of the region, (b) use a graphing utility to graph the function, and (c) approximate the value of x if the area of the region is 200 square units.

11. **12.**

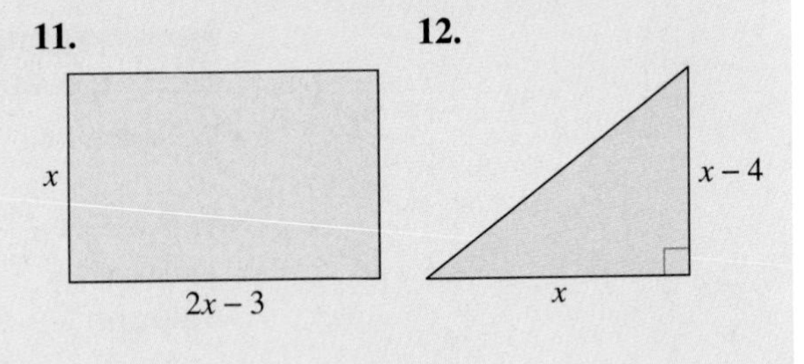

Developing Skills

In Exercises 1–22, write the first five terms of the sequence. (Begin with $n = 1$.) See Examples 1–4.

1. $a_n = 2n$

2. $a_n = 3n$

3. $a_n = (-1)^n 2n$

4. $a_n = (-1)^{n+1} 3n$

5. $a_n = \left(\frac{1}{2}\right)^n$

6. $a_n = \left(\frac{1}{3}\right)^n$

7. $a_n = \left(-\frac{1}{2}\right)^{n+1}$

8. $a_n = \left(\frac{2}{3}\right)^{n-1}$

9. $a_n = (-0.2)^{n-1}$

10. $a_n = \left(-\frac{2}{3}\right)^{n-1}$

11. $a_n = \dfrac{1}{n+1}$

12. $a_n = \dfrac{3}{2n+1}$

13. $a_n = \dfrac{2n}{3n+2}$

14. $a_n = \dfrac{5n}{4n+3}$

15. $a_n = \dfrac{(-1)^n}{n^2}$

16. $a_n = \dfrac{1}{\sqrt{n}}$

17. $a_n = 5 - \dfrac{1}{2^n}$

18. $a_n = 7 + \dfrac{1}{3^n}$

19. $a_n = \dfrac{(n+1)!}{n!}$

20. $a_n = \dfrac{n!}{(n-1)!}$

21. $a_n = \dfrac{2 + (-2)^n}{n!}$

22. $a_n = \dfrac{1 + (-1)^n}{n^2}$

In Exercises 23–26, find the indicated term of the sequence.

23. $a_n = (-1)^n(5n - 3)$
$a_{15} = \square$

24. $a_n = (-1)^{n-1}(2n + 4)$
$a_{14} = \square$

25. $a_n = \dfrac{n^2 - 2}{(n-1)!}$
$a_8 = \square$

26. $a_n = \dfrac{n^2}{n!}$
$a_{12} = \square$

In Exercises 27–38, simplify the expression.

27. $\dfrac{5!}{4!}$

28. $\dfrac{18!}{17!}$

29. $\dfrac{10!}{12!}$

30. $\dfrac{5!}{8!}$

31. $\dfrac{25!}{20!5!}$

32. $\dfrac{20!}{15! \cdot 5!}$

33. $\dfrac{n!}{(n+1)!}$

34. $\dfrac{(n+2)!}{n!}$

35. $\dfrac{(2n+1)!}{(n-1)!}$

36. $\dfrac{(3n)!}{(3n+2)!}$

37. $\dfrac{(2n)!}{(2n-1)!}$

38. $\dfrac{(2n+2)!}{(2n)!}$

In Exercises 39–42, match the sequence with the graph of its first 10 terms. [The graphs are labeled (a), (b), (c), and (d).]

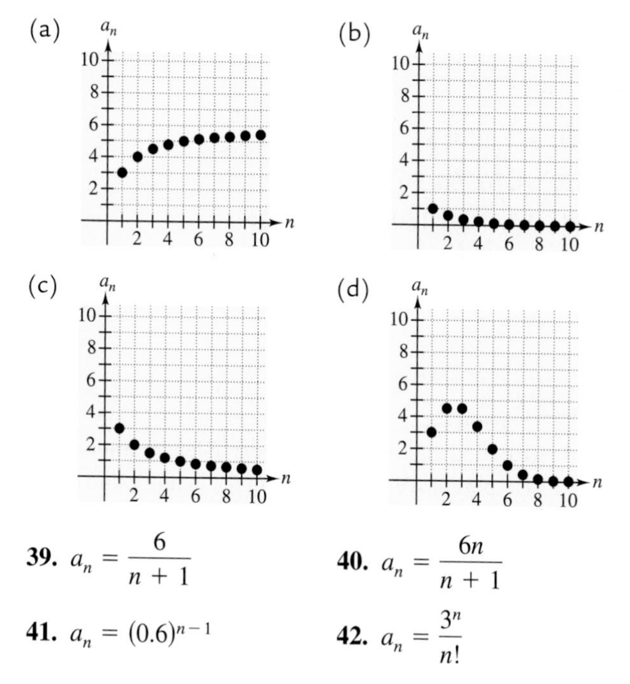

39. $a_n = \dfrac{6}{n+1}$

40. $a_n = \dfrac{6n}{n+1}$

41. $a_n = (0.6)^{n-1}$

42. $a_n = \dfrac{3^n}{n!}$

In Exercises 43–48, use a graphing utility to graph the first 10 terms of the sequence.

43. $a_n = (-0.8)^{n-1}$

44. $a_n = \dfrac{2n^2}{n^2+1}$

45. $a_n = \dfrac{1}{2}n$

46. $a_n = \dfrac{n+2}{n}$

47. $a_n = 3 - \dfrac{4}{n}$

48. $a_n = 10\left(\dfrac{3}{4}\right)^{n-1}$

In Exercises 49–66, write an expression for the nth term of the sequence. (Assume that n begins with 1.) See Example 5.

49. 3, 6, 9, 12, 15, . . .

50. 5, 10, 15, 20, 25, . . .

51. 1, 4, 7, 10, 13, . . .

52. 3, 7, 11, 15, 19, . . .

53. 0, 3, 8, 15, 24, . . .

54. 1, 8, 27, 64, 125, . . .

55. 2, −4, 6, −8, 10, . . .

56. 1, −1, 1, −1, 1, . . .

57. $\frac{2}{3}, \frac{3}{4}, \frac{4}{5}, \frac{5}{6}, \frac{6}{7}, \ldots$

58. $\frac{2}{1}, \frac{3}{3}, \frac{4}{5}, \frac{5}{7}, \frac{6}{9}, \ldots$

59. $\frac{1}{2}, \frac{-1}{4}, \frac{1}{8}, \frac{-1}{16}, \ldots$

60. $1, \frac{1}{4}, \frac{1}{9}, \frac{1}{16}, \frac{1}{25}, \ldots$

61. $1, \frac{1}{2}, \frac{1}{4}, \frac{1}{8}, \ldots$

62. $\frac{1}{3}, \frac{2}{9}, \frac{4}{27}, \frac{8}{81}, \ldots$

63. $1 + \frac{1}{1}, 1 + \frac{1}{2}, 1 + \frac{1}{3}, 1 + \frac{1}{4}, 1 + \frac{1}{5}, \ldots$

64. $1 + \frac{1}{2}, 1 + \frac{3}{4}, 1 + \frac{7}{8}, 1 + \frac{15}{16}, 1 + \frac{31}{32}, \ldots$

65. $1, \frac{1}{2}, \frac{1}{6}, \frac{1}{24}, \frac{1}{120}, \ldots$

66. $1, 2, \dfrac{2^2}{2}, \dfrac{2^3}{6}, \dfrac{2^4}{24}, \dfrac{2^5}{120}, \ldots$

In Exercises 67–82, find the partial sum. See Examples 6–8.

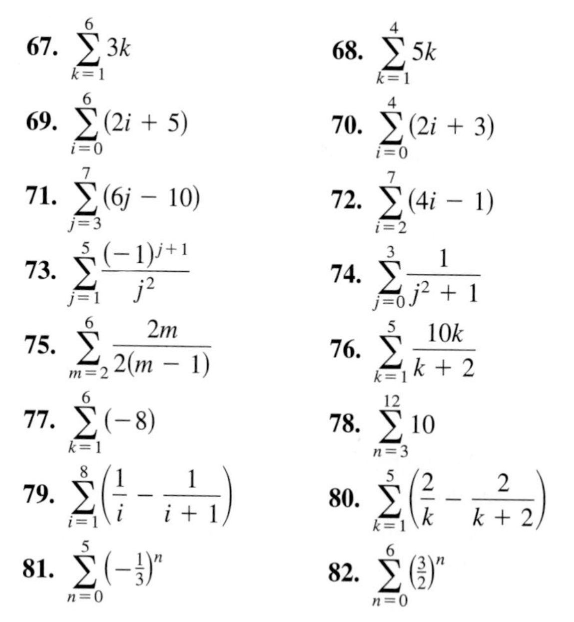

67. $\sum_{k=1}^{6} 3k$

68. $\sum_{k=1}^{4} 5k$

69. $\sum_{i=0}^{6} (2i+5)$

70. $\sum_{i=0}^{4} (2i+3)$

71. $\sum_{j=3}^{7} (6j-10)$

72. $\sum_{i=2}^{7} (4i-1)$

73. $\sum_{j=1}^{5} \dfrac{(-1)^{j+1}}{j^2}$

74. $\sum_{j=0}^{3} \dfrac{1}{j^2+1}$

75. $\sum_{m=2}^{6} \dfrac{2m}{2(m-1)}$

76. $\sum_{k=1}^{5} \dfrac{10k}{k+2}$

77. $\sum_{k=1}^{6} (-8)$

78. $\sum_{n=3}^{12} 10$

79. $\sum_{i=1}^{8} \left(\dfrac{1}{i} - \dfrac{1}{i+1}\right)$

80. $\sum_{k=1}^{5} \left(\dfrac{2}{k} - \dfrac{2}{k+2}\right)$

81. $\sum_{n=0}^{5} \left(-\frac{1}{3}\right)^n$

82. $\sum_{n=0}^{6} \left(\frac{3}{2}\right)^n$

In Exercises 83–90, use a graphing utility to find the sum.

83. $\sum_{n=1}^{6} 3n^2$

84. $\sum_{n=0}^{5} 2n^2$

85. $\sum_{j=2}^{6} (j! - j)$

86. $\sum_{i=0}^{4} (i! + 4)$

87. $\sum_{j=0}^{4} \dfrac{6}{j!}$

88. $\sum_{k=1}^{6} \left(\dfrac{1}{2k} - \dfrac{1}{2k-1}\right)$

89. $\sum_{k=1}^{6} \ln k$

90. $\sum_{k=2}^{4} \dfrac{k}{\ln k}$

In Exercises 91-108, write the sum using sigma notation. (Begin with $k = 0$ or $k = 1$.) See Example 9.

91. $1 + 2 + 3 + 4 + 5$

92. $8 + 9 + 10 + 12 + 13 + 14$

93. $2 + 4 + 6 + 8 + 10$

94. $24 + 30 + 36 + 42$

95. $\frac{1}{2(1)} + \frac{1}{2(2)} + \frac{1}{2(3)} + \frac{1}{2(4)} + \cdots + \frac{1}{2(10)}$

96. $\frac{3}{1+1} + \frac{3}{1+2} + \frac{3}{1+3} + \cdots + \frac{3}{1+50}$

97. $\frac{1}{1^2} + \frac{1}{2^2} + \frac{1}{3^2} + \frac{1}{4^2} + \cdots + \frac{1}{20^2}$

98. $\frac{1}{2^0} + \frac{1}{2^1} + \frac{1}{2^2} + \frac{1}{2^3} + \cdots + \frac{1}{2^{12}}$

99. $\frac{1}{3^0} - \frac{1}{3^1} + \frac{1}{3^2} - \frac{1}{3^3} + \cdots - \frac{1}{3^9}$

100. $\left(-\frac{2}{3}\right)^0 + \left(-\frac{2}{3}\right)^1 + \left(-\frac{2}{3}\right)^2 + \cdots + \left(-\frac{2}{3}\right)^{20}$

101. $\frac{4}{1+3} + \frac{4}{2+3} + \frac{4}{3+3} + \cdots + \frac{4}{20+3}$

102. $\frac{1}{2^3} - \frac{1}{4^3} + \frac{1}{6^3} - \frac{1}{8^3} + \cdots + \frac{1}{14^3}$

103. $\frac{1}{2} + \frac{2}{3} + \frac{3}{4} + \frac{4}{5} + \frac{5}{6} + \cdots + \frac{11}{12}$

104. $\frac{2}{4} + \frac{4}{7} + \frac{6}{10} + \frac{8}{13} + \frac{10}{16} + \cdots + \frac{20}{31}$

105. $\frac{2}{4} + \frac{4}{5} + \frac{6}{6} + \frac{8}{7} + \cdots + \frac{40}{23}$

106. $\left(2 + \frac{1}{1}\right) + \left(2 + \frac{1}{2}\right) + \left(2 + \frac{1}{3}\right) + \cdots + \left(2 + \frac{1}{25}\right)$

107. $1 + 1 + 2 + 6 + 24 + 120 + 720$

108. $1 + 1 + \frac{1}{2} + \frac{1}{6} + \frac{1}{24} + \frac{1}{120} + \frac{1}{720}$

Arithmetic Mean In Exercises 109–112, find the arithmetic mean of the set. The *arithmetic mean* of a set of n measurements $x_1, x_2, x_3, \ldots, x_n$ is

$$\bar{x} = \frac{1}{n}\sum_{i=1}^{n} x_i$$

109. 3, 7, 2, 1, 5

110. 84, 69, 66, 96

111. 0.5, 0.8, 1.1, 0.8, 0.7, 0.7, 1.0

112. -1.0, 4.2, 5.4, -3.2, 3.6

Solving Problems

113. *Compound Interest* A deposit of \$500 is made in an account that earns 7% interest compounded yearly. The balance in the account after N years is given by

$$A_N = 500(1 + 0.07)^N, \quad N = 1, 2, 3, \ldots.$$

(a) Compute the first eight terms of the sequence.

(b) Find the balance in this account after 40 years by computing A_{40}.

(c) Use a graphing utility to graph the first 40 terms of the sequence.

(d) The terms are increasing. Is the rate of growth of the terms increasing? Explain.

114. *Depreciation* At the end of each year, the value of a car with an initial cost of \$26,000 is three-fourths what it was at the beginning of the year. Thus, after n years, its value is given by

$$a_n = 26{,}000\left(\frac{3}{4}\right)^n, \quad n = 1, 2, 3, \ldots.$$

(a) Find the value of the car 3 years after it was purchased by computing a_3.

(b) Find the value of the car 6 years after it was purchased by computing a_6. Is this value half of what it was after 3 years? Explain.

115. *Soccer Ball* The number of degrees a_n in each angle of a regular n-sided polygon is

$$a_n = \frac{180(n-2)}{n}, \quad n \geq 3.$$

The surface of a soccer ball is made of regular hexagons and pentagons. If a soccer ball is taken apart and flattened, as shown in the figure, the sides don't meet each other. Use the terms a_5 and a_6 to explain why there are gaps between adjacent hexagons.

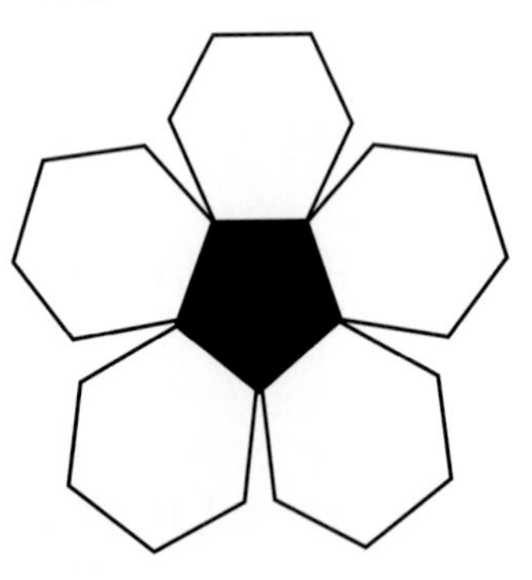

116. *Stars* The number of degrees d_n in each tip of each n-pointed star in the figure is given by

$$d_n = \frac{180(n - 4)}{n}, \quad n \geq 5.$$

Write the first six terms of this sequence.

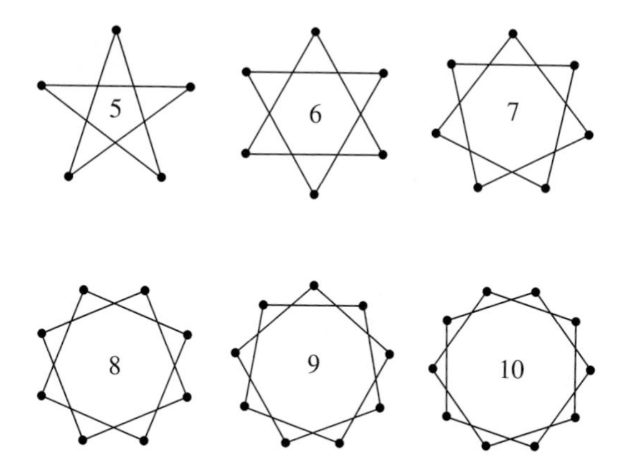

117. *Stars* The stars in Exercise 116 were formed by placing n equally spaced points on a circle and connecting each point with the second point from it on the circle. The stars in the figure for this exercise were formed in a similar way except that each point was connected with the third point from it. For these stars, the number of degrees in a tip is given by

$$d_n = \frac{180(n - 6)}{n}, \quad n \geq 7.$$

Write the first five terms of this sequence.

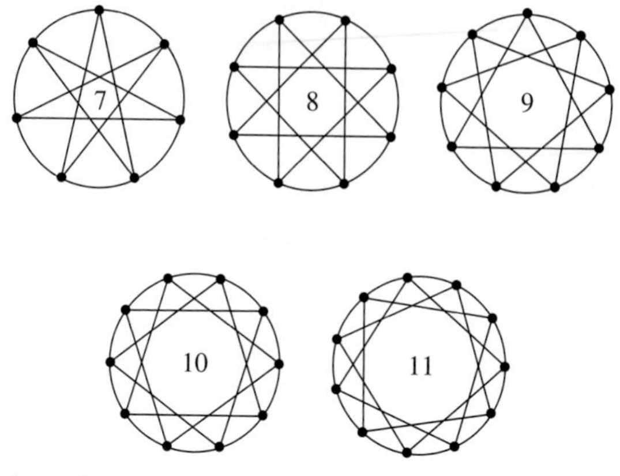

Figure for 117

118. *Outpatient Surgery* The number a_n (in thousands) of outpatient surgeries performed in hospitals in the United States for the years 1990 through 1995 is modeled by

$$a_n = 11{,}791 + 436n, \quad n = 1, 2, \ldots, 5$$

where n is the year, with $n = 0$ corresponding to 1990. Find the terms of this finite sequence and use a graphing utility to construct a bar graph that represents the sequence. (Source: American Hospital Association)

Explaining Concepts

119. Give an example of an infinite sequence.

120. State the definition of n factorial.

121. The nth term of a sequence is $a_n = (-1)^n n$. Which terms of the sequence are negative?

In Exercises 122–124, decide whether the statement is true. Explain your reasoning.

122. $\sum_{i=1}^{4} (i^2 + 2i) = \sum_{i=1}^{4} i^2 + \sum_{i=1}^{4} 2i$

123. $\sum_{k=1}^{4} 3k = 3\sum_{k=1}^{4} k$

124. $\sum_{j=1}^{4} 2^j = \sum_{j=3}^{6} 2^{j-2}$

12.2 Arithmetic Sequences

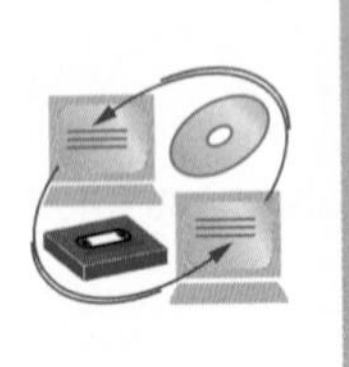

Objectives

1. Find the common difference and the nth term of an arithmetic sequence.
2. Find the nth partial sum of an arithmetic sequence.
3. Use an arithmetic sequence to solve an application problem.

1 Find the common difference and the nth term of an arithmetic sequence.

Arithmetic Sequences

A sequence whose consecutive terms have a common difference is called an **arithmetic sequence.**

> **Definition of an Arithmetic Sequence**
>
> A sequence is called **arithmetic** if the differences between consecutive terms are the same. So, the sequence
>
> $$a_1, a_2, a_3, a_4, \ldots, a_n, \ldots$$
>
> is arithmetic if there is a number d such that
>
> $$a_2 - a_1 = d, \quad a_3 - a_2 = d, \quad a_4 - a_3 = d$$
>
> and so on. The number d is the **common difference** of the sequence.

Example 1 Examples of Arithmetic Sequences

a. The sequence whose nth term is $3n + 2$ is arithmetic. For this sequence, the common difference between consecutive terms is 3.

$$\underbrace{5, 8}_{8-5=3}, 11, 14, \ldots, 3n + 2, \ldots$$

b. The sequence whose nth term is $7 - 5n$ is arithmetic. For this sequence, the common difference between consecutive terms is -5.

$$\underbrace{2, -3}_{-3-2=-5}, -8, -13, \ldots, 7 - 5n, \ldots$$

c. The sequence whose nth term is $\frac{1}{4}(n + 3)$ is arithmetic. For this sequence, the common difference between consecutive terms is $\frac{1}{4}$.

$$\underbrace{1, \frac{5}{4}}_{\frac{5}{4}-1=\frac{1}{4}}, \frac{3}{2}, \frac{7}{4}, \ldots, \frac{n+3}{4}, \ldots$$

Study Tip

The nth term of an arithmetic sequence can be derived from the following pattern.

$a_1 = a_1$ 1st term

$a_2 = a_1 + d$ 2nd term

$a_3 = a_1 + 2d$ 3rd term

$a_4 = a_1 + 3d$ 4th term

$a_5 = a_1 + 4d$ 5th term

1 less

⋮ ⋮

$a_n = a_1 + (n - 1)d$ nth term

1 less

▶ The nth Term of an Arithmetic Sequence

The nth term of an arithmetic sequence has the form

$$a_n = a_1 + (n - 1)d$$

where d is the common difference between the terms of the sequence, and a_1 is the first term.

Example 2 Finding the nth Term of an Arithmetic Sequence

Find a formula for the nth term of the arithmetic sequence whose common difference is 2 and whose first term is 5.

Solution

You know that the formula for the nth term is of the form $a_n = a_1 + (n - 1)d$. Moreover, because the common difference is $d = 2$, and the first term is $a_1 = 5$, the formula must have the form

$$a_n = 5 + 2(n - 1).$$

So, the formula for the nth term is

$$a_n = 2n + 3.$$

The sequence therefore has the following form.

$$5, 7, 9, 11, 13, \ldots, 2n + 3, \ldots$$

If you know the nth term and the common difference of an arithmetic sequence, you can find the $(n + 1)$th term by using the following **recursion formula.**

$$a_{n+1} = a_n + d$$

Example 3 Using a Recursion Formula

The 12th term of an arithmetic sequence is 52 and the common difference is 3.

a. What is the 13th term of the sequence? **b.** What is the first term?

Solution

a. Using the recursion formula $a_{13} = a_{12} + d$, you know that $a_{12} = 52$ and $d = 3$. So, the 13th term of the sequence is

$$a_{13} = 52 + 3 = 55.$$

b. Using $n = 12$, $d = 3$, and $a_{12} = 52$ in the formula $a_n = a_1 + (n - 1)d$ yields

$$52 = a_1 + (12 - 1)(3)$$

$$19 = a_1.$$

2 Find the nth partial sum of an arithmetic sequence.

The Partial Sum of an Arithmetic Sequence

The sum of the first n terms of an arithmetic sequence is called the ***n*th partial sum** of the sequence. For instance, the 5th partial sum of the arithmetic sequence whose nth term is $3n + 4$ is

$$\sum_{i=1}^{5} (3i + 4) = 7 + 10 + 13 + 16 + 19 = 65.$$

To find a formula for the nth partial sum S_n of an arithmetic sequence, write out S_n forwards and backwards and then add the two forms, as follows.

$$S_n = a_1 + (a_1 + d) + (a_1 + 2d) + \cdots + [a_1 + (n - 1)d] \qquad \text{Forwards}$$

$$S_n = a_n + (a_n - d) + (a_n - 2d) + \cdots + [a_n - (n - 1)d] \qquad \text{Backwards}$$

$$2S_n = (a_1 + a_n) + (a_1 + a_n) + (a_1 + a_n) + \cdots + [a_1 + a_n] \qquad \text{Sum of two equations}$$

$$= n(a_1 + a_n) \qquad n \text{ groups of } (a_1 + a_n)$$

Dividing both sides by 2 yields the following formula.

Study Tip

You can use the formula for the nth partial sum of an arithmetic sequence to find the sum of consecutive numbers. For instance, the sum of the integers from 1 to 100 is

$$\sum_{i=1}^{100} i = 100\left(\frac{1 + 100}{2}\right)$$

$$= 50(101)$$

$$= 5050.$$

▶ The nth Partial Sum of an Arithmetic Sequence

The nth partial sum of the arithmetic sequence whose nth term is a_n is

$$\sum_{i=1}^{n} a_i = a_1 + a_2 + a_3 + a_4 + \cdots + a_n$$

$$= n\left(\frac{a_1 + a_n}{2}\right).$$

In other words, to find the sum of the first n terms of an arithmetic sequence, find the average of the first and nth terms, and multiply by n.

Example 4 Finding the nth Partial Sum

Find the sum of the first 20 terms of the arithmetic sequence whose nth term is $4n + 1$.

Solution

The first term of this sequence is $a_1 = 4(1) + 1 = 5$ and the 20th term is $a_{20} = 4(20) + 1 = 81$. So, the sum of the first 20 terms is given by

$$\sum_{i=1}^{n} a_i = n\left(\frac{a_1 + a_n}{2}\right) \qquad n\text{th partial sum formula}$$

$$\sum_{i=1}^{20} (4i + 1) = 20\left(\frac{a_1 + a_{20}}{2}\right) \qquad \text{Substitute 20 for } n.$$

$$= 10(5 + 81) \qquad \text{Substitute 5 for } a_1 \text{ and 81 for } a_{20}.$$

$$= 10(86) \qquad \text{Simplify.}$$

$$= 860. \qquad n\text{th partial sum}$$

3 Use an arithmetic sequence to solve an application problem.

Application

Example 5 Total Sales

Your business sells \$100,000 worth of products during its first year. You have a goal of increasing annual sales by \$25,000 each year for 9 years. If you meet this goal, how much will you sell during your first 10 years of business?

Solution

The annual sales during the first 10 years form the following arithmetic sequence.

\$100,000, \$125,000, \$150,000, \$175,000, \$200,000,
\$225,000, \$250,000, \$275,000, \$300,000, \$325,000

Using the formula for the nth partial sum of an arithmetic sequence, you find the total sales during the first 10 years as follows.

$$\text{Total sales} = n\left(\frac{a_1 + a_n}{2}\right) \qquad n\text{th partial sum formula}$$

$$= 10\left(\frac{100{,}000 + 325{,}000}{2}\right) \qquad \text{Substitute for } n, a_1, \text{ and } a_n.$$

$$= 10(212{,}500) \qquad \text{Simplify.}$$

$$= \$2{,}125{,}000 \qquad \text{Simplify.}$$

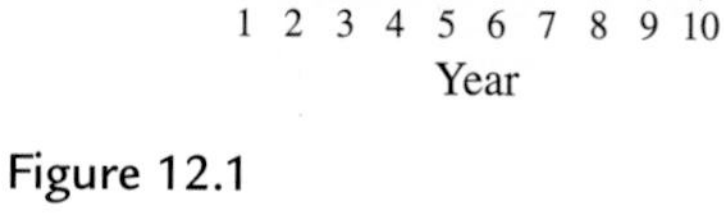

Figure 12.1

From the bar graph shown in Figure 12.1, notice that the annual sales for this company follows a *linear growth* pattern. In other words, saying that a quantity increases arithmetically is the same as saying that it increases linearly.

Discussing the Concept **Using Arithmetic Sequences**

A magic square is a square table of positive integers in which each row, column, and diagonal adds up to the same number. One example is shown below. In addition, the values in the middle row, in the middle column, and along both diagonals form arithmetic sequences. See if you can complete the following magic squares.

6	1	8
7	5	3
2	9	4

a.

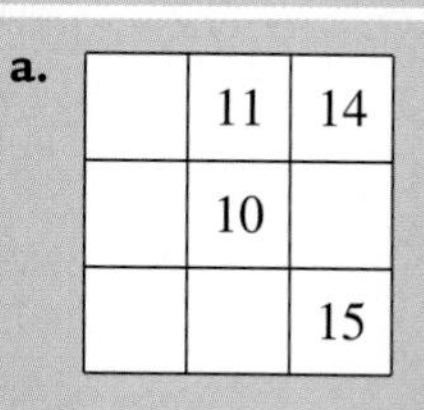

	11	14
	10	
		15

b.

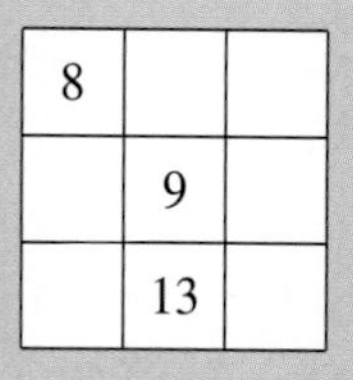

8		
		9
	13	

c.

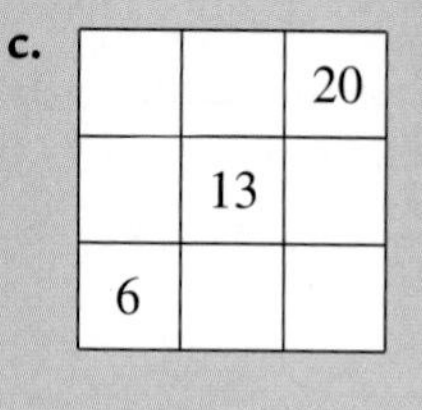

		20
	13	
6		

12.2 Exercises

Integrated Review *Concepts, Skills, and Problem Solving*

Keep mathematically in shape by doing these exercises *before* the problems of this section.

Properties and Definitions

1. In your own words, state the definition of an algebraic expression.
2. State the definition of a term of an algebraic expression.
3. Write a trinomial of degree 3.
4. Write a monomial of degree 4.

Domain

In Exercises 5–10, find the domain of the function.

5. $f(x) = x^3 - 2x$
6. $g(x) = \sqrt[3]{x}$
7. $h(x) = \sqrt{16 - x^2}$
8. $A(x) = \dfrac{3}{36 - x^2}$
9. $g(t) = \ln(t - 2)$
10. $f(s) = 630e^{-0.2s}$

Problem Solving

11. Determine the balance when \$10,000 is invested at $7\frac{1}{2}\%$ compounded daily for 15 years.
12. Determine the amount after 5 years if \$4000 is invested in an account earning 6% compounded monthly.

Developing Skills

In Exercises 1–10, find the common difference of the arithmetic sequence. See Example 1.

1. $2, 5, 8, 11, \ldots$
2. $-8, 0, 8, 16, \ldots$
3. $100, 94, 88, 82, \ldots$
4. $3200, 2800, 2400, 2000, \ldots$
5. $10, -2, -14, -26, -38, \ldots$
6. $4, \frac{9}{2}, 5, \frac{11}{2}, 6, \ldots$
7. $1, \frac{5}{3}, \frac{7}{3}, 3, \ldots$
8. $\frac{1}{2}, \frac{5}{4}, 2, \frac{11}{4}, \ldots$
9. $\frac{7}{2}, \frac{9}{4}, 1, -\frac{1}{4}, -\frac{3}{2}, \ldots$
10. $\frac{5}{2}, \frac{11}{6}, \frac{7}{6}, \frac{1}{2}, -\frac{1}{6}, \ldots$

In Exercises 11–26, determine whether the sequence is arithmetic. If so, find the common difference.

11. $2, 4, 6, 8, \ldots$
12. $1, 2, 4, 8, 16, \ldots$
13. $10, 8, 6, 4, 2, \ldots$
14. $2, 6, 10, 14, \ldots$
15. $32, 16, 0, -16, \ldots$
16. $32, 16, 8, 4, \ldots$
17. $3.2, 4, 4.8, 5.6, \ldots$
18. $8, 4, 2, 1, 0.5, 0.25, \ldots$
19. $2, \frac{7}{2}, 5, \frac{13}{2}, \ldots$
20. $3, \frac{5}{2}, 2, \frac{3}{2}, 1, \ldots$
21. $\frac{1}{3}, \frac{2}{3}, \frac{4}{3}, \frac{8}{3}, \frac{16}{3}, \ldots$
22. $\frac{9}{4}, 2, \frac{7}{4}, \frac{3}{2}, \frac{5}{4}, \ldots$
23. $1, \sqrt{2}, \sqrt{3}, 2, \sqrt{5}, \ldots$
24. $1, 4, 9, 16, 25, \ldots$
25. $\ln 4, \ln 8, \ln 12, \ln 16, \ldots$
26. $e, e^2, e^3, e^4, \ldots$

In Exercises 27-36, write the first five terms of the arithmetic sequence. (Begin with $n = 1$.)

27. $a_n = 3n + 4$
28. $a_n = 5n - 4$
29. $a_n = -2n + 8$
30. $a_n = -10n + 100$
31. $a_n = \frac{5}{2}n - 1$
32. $a_n = \frac{2}{3}n + 2$
33. $a_n = \frac{3}{5}n + 1$
34. $a_n = \frac{3}{4}n - 2$
35. $a_n = -\frac{1}{4}(n - 1) + 4$
36. $a_n = 4(n + 2) + 24$

In Exercises 37–54, find a formula for the *n*th term of the arithmetic sequence. See Example 2.

37. $a_1 = 3, \quad d = \frac{1}{2}$

38. $a_1 = -1, \quad d = 1.2$

39. $a_1 = 1000, \quad d = -25$

40. $a_1 = 64, \quad d = -8$

41. $a_3 = 20, \quad d = -4$

42. $a_1 = 12, \quad d = -3$

43. $a_1 = 3, \quad d = \frac{3}{2}$

44. $a_6 = 5, \quad d = \frac{3}{2}$

45. $a_1 = 5, \quad a_5 = 15$

46. $a_2 = 93, \quad a_6 = 65$

47. $a_3 = 16, \quad a_4 = 20$

48. $a_5 = 30, \quad a_4 = 25$

49. $a_1 = 50, \quad a_3 = 30$

50. $a_{10} = 32, \quad a_{12} = 48$

51. $a_2 = 10, \quad a_6 = 8$

52. $a_7 = 8, \quad a_{13} = 6$

53. $a_1 = 0.35, \quad a_2 = 0.30$

54. $a_1 = 0.08, \quad a_2 = 0.082$

In Exercises 55–62, write the first five terms of the arithmetic sequence defined recursively. See Example 3.

55. $a_1 = 25$
$a_{k+1} = a_k + 3$

56. $a_1 = 12$
$a_{k+1} = a_k - 6$

57. $a_1 = 9$
$a_{k+1} = a_k - 3$

58. $a_1 = 8$
$a_{k+1} = a_k + 7$

59. $a_1 = -10$
$a_{k+1} = a_k + 6$

60. $a_1 = -20$
$a_{k+1} = a_k - 4$

61. $a_1 = 100$
$a_{k+1} = a_k - 20$

62. $a_1 = 4.2$
$a_{k+1} = a_k + 0.4$

In Exercises 63–72, find the *n*th partial sum. See Example 4.

63. $\sum_{k=1}^{20} k$

64. $\sum_{k=1}^{30} 4k$

65. $\sum_{k=1}^{50} (k + 3)$

66. $\sum_{n=1}^{30} (n + 2)$

67. $\sum_{k=1}^{10} (5k - 2)$

68. $\sum_{k=1}^{100} (4k - 1)$

69. $\sum_{n=1}^{500} \frac{n}{2}$

70. $\sum_{n=1}^{600} \frac{2n}{3}$

71. $\sum_{n=1}^{30} \left(\frac{1}{3}n - 4\right)$

72. $\sum_{n=1}^{75} (0.3n + 5)$

In Exercises 73–84, find the *n*th partial sum of the arithmetic sequence.

73. $5, 12, 19, 26, 33, \ldots, \quad n = 12$

74. $2, 12, 22, 32, 42, \ldots, \quad n = 20$

75. $2, 8, 14, 20, \ldots, \quad n = 25$

76. $500, 480, 460, 440, \ldots, \quad n = 20$

77. $200, 175, 150, 125, 100, \ldots, \quad n = 8$

78. $800, 785, 770, 755, 740, \ldots, \quad n = 25$

79. $-50, -38, -26, -14, -2, \ldots, \quad n = 50$

80. $-16, -8, 0, 8, 16, \ldots, \quad n = 30$

81. $1, 4.5, 8, 11.5, 15, \ldots, \quad n = 12$

82. $2.2, 2.8, 3.4, 4.0, 4.6, \ldots, \quad n = 12$

83. $a_1 = 0.5, \ a_4 = 1.7, \ldots, \quad n = 10$

84. $a_1 = 15, \ a_{100} = 307, \ldots, \quad n = 100$

In Exercises 85–90, match the arithmetic sequence with its graph. [The graphs are labeled (a), (b), (c), (d), (e), and (f).]

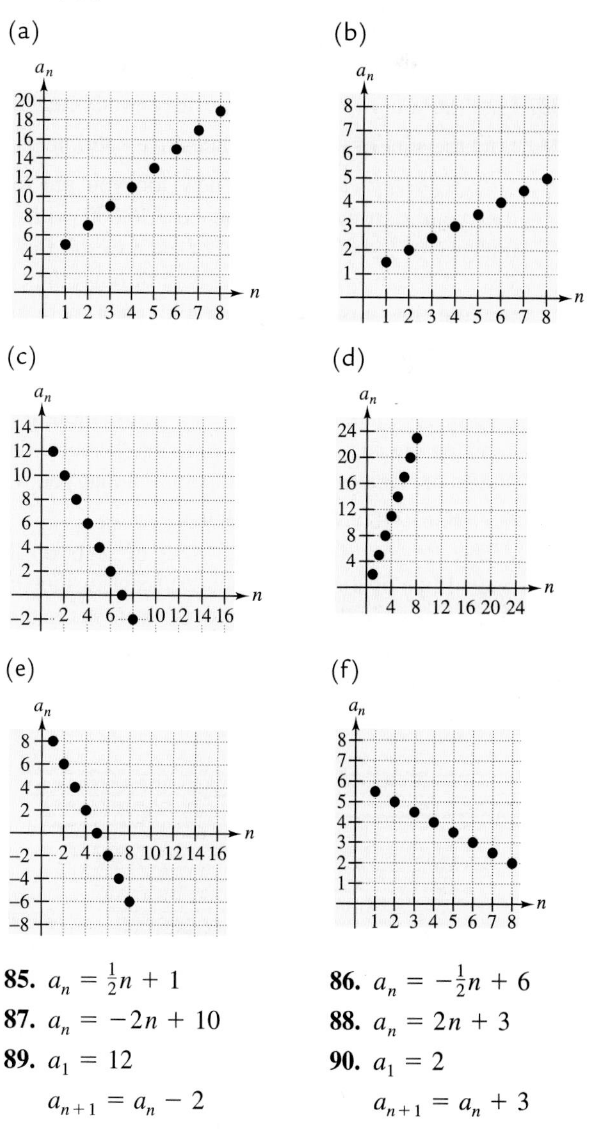

85. $a_n = \frac{1}{2}n + 1$

86. $a_n = -\frac{1}{2}n + 6$

87. $a_n = -2n + 10$

88. $a_n = 2n + 3$

89. $a_1 = 12$
$a_{n+1} = a_n - 2$

90. $a_1 = 2$
$a_{n+1} = a_n + 3$

In Exercises 91–96, use a graphing utility to graph the first 10 terms of the sequence.

91. $a_n = -2n + 21$

92. $a_n = -25n + 500$

93. $a_n = \frac{3}{5}n + \frac{3}{2}$

94. $a_n = \frac{3}{2}n + 1$

95. $a_n = 2.5n - 8$

96. $a_n = 6.2n + 3$

In Exercises 97–102, use a graphing utility to find the sum.

97. $\sum_{j=1}^{25} (750 - 30j)$

98. $\sum_{n=1}^{40} (1000 - 25n)$

99. $\sum_{i=1}^{60} \left(300 - \frac{8}{3}i\right)$

100. $\sum_{n=1}^{20} \left(500 - \frac{1}{10}n\right)$

101. $\sum_{n=1}^{50} (2.15n + 5.4)$

102. $\sum_{n=1}^{60} (200 - 3.4n)$

Solving Problems

103. Find the sum of the first 75 positive integers.

104. Find the sum of the integers from 35 to 100.

105. Find the sum of the first 50 positive even integers.

106. Find the sum of the first 100 positive odd integers.

107. *Salary Increases* In your new job you are told that your starting salary will be $36,000 with an increase of $2000 at the end of each of the first 5 years. How much will you be paid through the end of your first six years of employment with the company?

108. *Would You Accept This Job?* Suppose that you receive 25 cents on the first day of the month, 50 cents on the second day, 75 cents on the third day, and so on. Determine the total amount that you will receive during a 30-day month.

109. *Ticket Prices* There are 20 rows of seats on the main floor of a concert hall: 20 seats in the first row, 21 seats in the second row, 22 seats in the third row, and so on (see figure). How much should you charge per ticket in order to obtain $15,000 for the sale of all the seats on the main floor?

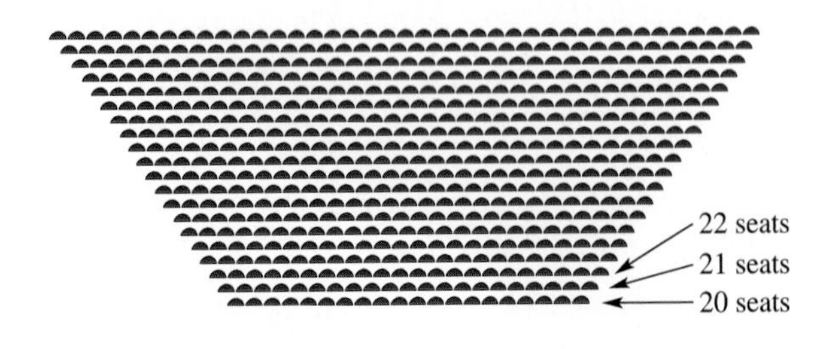

110. *Pile of Logs* Logs are stacked in a pile as shown in the figure. The top row has 15 logs and the bottom row has 21 logs. How many logs are in the pile?

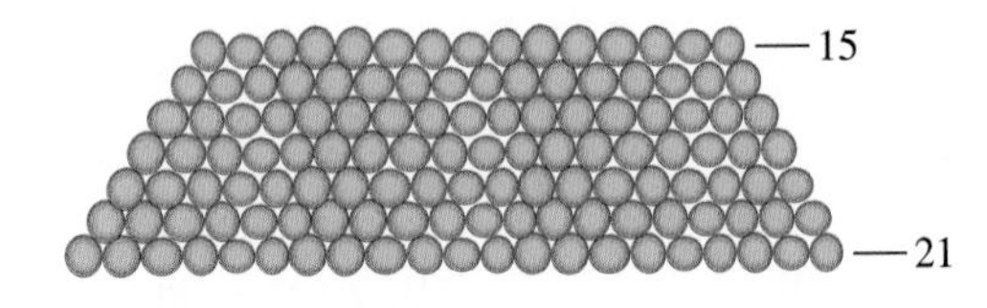

111. *Baling Hay* In the first two trips baling hay around a large field (see figure), a farmer obtains 93 bales and 89 bales, respectively. The farmer estimates that the same pattern will continue. Estimate the total number of bales made if there are another six trips around the field.

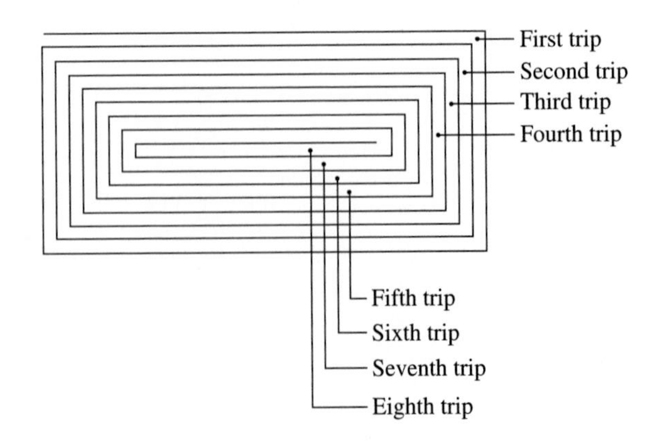

112. *Baling Hay* In the first two trips baling hay around a field (see figure), a farmer obtains 64 bales and 60 bales, respectively. The farmer estimates that the same pattern will continue. Estimate the total number of bales made if there are another four trips around the field.

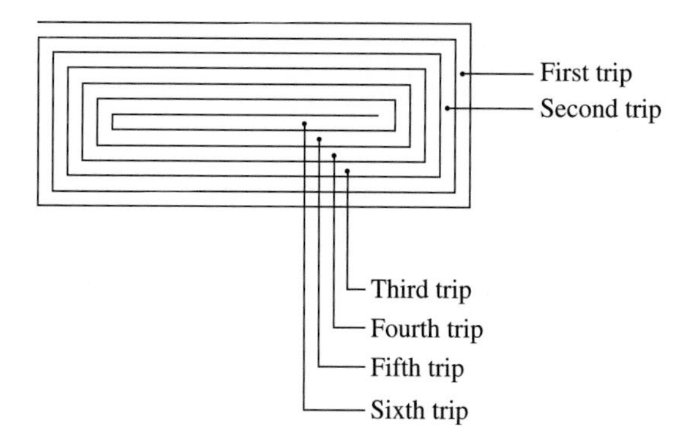

113. *Clock Chimes* A clock chimes once at 1:00, twice at 2:00, three times at 3:00, and so on. The clock also chimes once at 15-minute intervals that are not on the hour. How many times does the clock chime in a 12-hour period?

114. *Clock Chimes* A clock chimes once at 1:00, twice at 2:00, three times at 3:00, and so on. The clock also chimes once on the half-hour. How many times does the clock chime in a 12-hour period?

115. *Free-Falling Object* A free-falling object will fall 16 feet during the first second, 48 more feet during the second, 80 more feet during the third, and so on. What is the total distance the object will fall in 8 seconds if this pattern continues?

116. *Free-Falling Object* A free-falling object will fall 4.9 meters during the first second, 14.7 more meters during the second, 24.5 more meters during the third, and so on. What is the total distance the object will fall in 5 seconds if this pattern continues?

117. *Pattern Recognition*

(a) Compute the sums of positive odd integers.

$1 + 3 = \square$

$1 + 3 + 5 = \square$

$1 + 3 + 5 + 7 = \square$

$1 + 3 + 5 + 7 + 9 = \square$

$1 + 3 + 5 + 7 + 9 + 11 = \square$

(b) Use the sums in part (a) to make a conjecture about the sums of positive odd integers. Check your conjecture for the sum

$1 + 3 + 5 + 7 + 9 + 11 + 13 = \square.$

(c) Verify your conjecture in part (b) analytically.

Explaining Concepts

118. In your own words, explain what makes a sequence arithmetic.

119. The second and third terms of an arithmetic sequence are 12 and 15, respectively. What is the first term?

120. Explain how the first two terms of an arithmetic sequence can be used to find the nth term.

121. Explain what is meant by a recursion formula.

122. Explain what is meant by the nth partial sum of a sequence.

123. Explain how to find the sum of the integers from 100 to 200.

124. Each term of an arithmetic sequence is multiplied by a constant C. Is the resulting sequence arithmetic? If so, how does the common difference compare with the common difference of the original sequence?

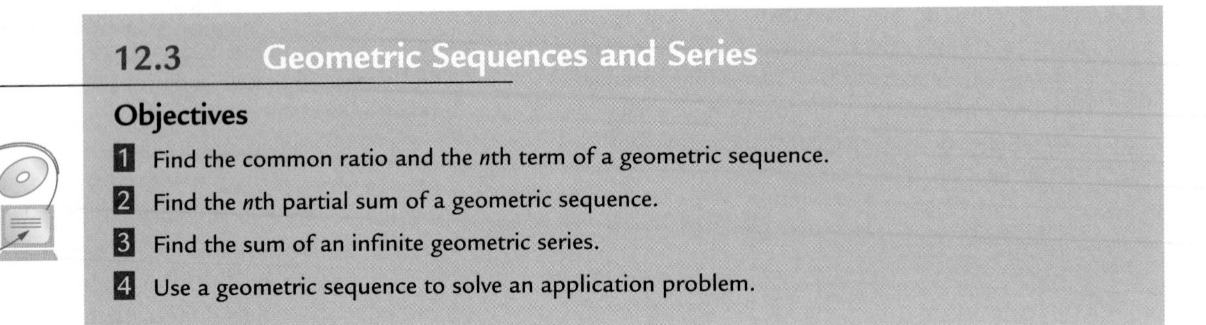

12.3 Geometric Sequences and Series

Objectives

1. Find the common ratio and the nth term of a geometric sequence.
2. Find the nth partial sum of a geometric sequence.
3. Find the sum of an infinite geometric series.
4. Use a geometric sequence to solve an application problem.

1 Find the common ratio and the nth term of a geometric sequence.

Geometric Sequences

In Section 12.2, you studied sequences whose consecutive terms have a common *difference*. In this section, you will study sequences whose consecutive terms have a common *ratio*.

Definition of a Geometric Sequence

A sequence is called **geometric** if the ratios of consecutive terms are the same. So, the sequence $a_1, a_2, a_3, a_4, \ldots, a_n, \ldots$ is geometric if there is a number r, $r \neq 0$, such that

$$\frac{a_2}{a_1} = r, \quad \frac{a_3}{a_2} = r, \quad \frac{a_4}{a_3} = r$$

and so on. The number r is the **common ratio** of the sequence.

Example 1 Examples of Geometric Sequences

a. The sequence whose nth term is 2^n is geometric. For this sequence, the common ratio between consecutive terms is 2.

$$\underbrace{2, 4}_{\frac{4}{2} = 2}, 8, 16, \ldots, 2^n, \ldots$$

b. The sequence whose nth term is $4(3^n)$ is geometric. For this sequence, the common ratio between consecutive terms is 3.

$$\underbrace{12, 36}_{\frac{36}{12} = 3}, 108, 324, \ldots, 4(3^n), \ldots$$

c. The sequence whose nth term is $\left(-\frac{1}{3}\right)^n$ is geometric. For this sequence, the common ratio between consecutive terms is $-\frac{1}{3}$.

$$\underbrace{-\frac{1}{3}, \frac{1}{9}}_{\frac{1/9}{-1/3} = -\frac{1}{3}}, -\frac{1}{27}, \frac{1}{81}, \ldots, \left(-\frac{1}{3}\right)^n, \ldots$$

Study Tip

If you know the nth term of a geometric sequence, the $(n + 1)$th term can be found by multiplying by r. That is,

$$a_{n+1} = ra_n.$$

▶ The nth Term of a Geometric Sequence

The nth term of a geometric sequence has the form

$$a_n = a_1 r^{n-1}$$

where r is the common ratio of consecutive terms of the sequence. So, every geometric sequence can be written in the following form.

$$a_1, a_1 r, a_1 r^2, a_1 r^3, a_1 r^4, \ldots, a_1 r^{n-1}, \ldots$$

Example 2 Finding the nth Term of a Geometric Sequence

Find a formula for the nth term of the geometric sequence whose common ratio is 3 and whose first term is 1. What is the eighth term of this sequence?

Solution

The formula for the nth term is of the form $a_1 r^{n-1}$. Moreover, because the common ratio is $r = 3$, and the first term is $a_1 = 1$, the formula must have the form

$$a_n = a_1 r^{n-1} \quad \text{Formula for geometric sequence}$$

$$= (1)(3)^{n-1} \quad \text{Substitute 1 for } a_1 \text{ and 3 for } r.$$

$$= 3^{n-1}. \quad \text{Simplify.}$$

The sequence therefore has the following form.

$$1, 3, 9, 27, 81, \ldots, 3^{n-1}, \ldots$$

The eighth term of the sequence is $a_8 = 3^{8-1} = 3^7 = 2187$.

Example 3 Finding the nth Term of a Geometric Sequence

Find a formula for the nth term of the geometric sequence whose first two terms are 4 and 2.

Solution

Because $a_1 = 4$ and $a_2 = 2$, the common ratio is

$$\frac{a_2}{a_1} = \frac{2}{4} = \frac{1}{2} \quad \text{So, } r = \tfrac{1}{2}.$$

the formula for the nth term must be

$$a_n = a_1 r^{n-1} \quad \text{Formula for geometric sequence}$$

$$= 4\left(\frac{1}{2}\right)^{n-1}. \quad \text{Substitute 4 for } a_1 \text{ and } \tfrac{1}{2} \text{ for } r.$$

The sequence therefore has the following form.

$$4, 2, 1, \frac{1}{2}, \frac{1}{4}, \ldots, 4\left(\frac{1}{2}\right)^{n-1}, \ldots$$

2 Find the nth partial sum of a geometric sequence.

The Partial Sum of a Geometric Sequence

> **The nth Partial Sum of a Geometric Sequence**
>
> The nth partial sum of the geometric sequence whose nth term is $a_n = a_1 r^{n-1}$ is given by
>
> $$\sum_{i=1}^{n} a_1 r^{i-1} = a_1 + a_1 r + a_1 r^2 + a_1 r^3 + \cdot \cdot \cdot + a_1 r^{n-1} = a_1\left(\frac{r^n - 1}{r - 1}\right).$$

Joseph Fourier

(1768–1830)

Some of the early work in representing functions by series was done by the French mathematician Joseph Fourier. Fourier's work is important in the history of calculus, partly because it forced 18th-century mathematicians to question the then-prevailing narrow concept of a function. Both Cauchy and Dirichlet were motivated by Fourier's work in series, and in 1837 Dirichlet published the general definition of a function that is used today.

Example 4 Finding the nth Partial Sum

Find the sum $1 + 2 + 4 + 8 + 16 + 32 + 64 + 128$.

Solution

This is a geometric sequence whose common ratio is $r = 2$. Because the first term of the sequence is $a_1 = 1$, it follows that the sum is

$$\sum_{i=1}^{8} 2^{i-1} = (1)\left(\frac{2^8 - 1}{2 - 1}\right) = \frac{256 - 1}{2 - 1} = 255.$$ Substitute 1 for a_1 and 2 for r.

Example 5 Finding the nth Partial Sum

Find the sum of the first five terms of the geometric sequence whose nth term is $a_n = \left(\frac{2}{3}\right)^n$.

Solution

$$\sum_{i=1}^{5} \left(\frac{2}{3}\right)^i = \frac{2}{3}\left[\frac{(2/3)^5 - 1}{(2/3) - 1}\right]$$ Substitute $\frac{2}{3}$ for a_1 and $\frac{2}{3}$ for r.

$$= \frac{2}{3}\left[\frac{(32/243) - 1}{-1/3}\right]$$ Simplify.

$$= \frac{2}{3}\left(-\frac{211}{243}\right)(-3)$$ Simplify.

$$= \frac{422}{243}$$ Simplify.

$$\approx 1.737$$

3 Find the sum of an infinite geometric series.

Geometric Series

Suppose that in Example 5, you were to find the sum of all the terms of the *infinite* geometric sequence

$$\frac{2}{3}, \frac{4}{9}, \frac{8}{27}, \frac{16}{81}, \ldots, \left(\frac{2}{3}\right)^n, \ldots.$$

A summation of all the terms of an infinite geometric sequence is called an **infinite geometric series,** or simply a **geometric series.**

Evaluate $\left(\frac{1}{2}\right)^n$ for $n = 1, 10, 100,$ and 1000. What happens to the value of $\left(\frac{1}{2}\right)^n$ as n increases? Make a conjecture about the value of $\left(\frac{1}{2}\right)^n$ as n approaches infinity.

In your mind, would this sum be infinitely large or would it be a finite number? Consider the formula for the nth partial sum of a geometric sequence.

$$S_n = a_1\left(\frac{r^n - 1}{r - 1}\right) = a_1\left(\frac{1 - r^n}{1 - r}\right)$$

Suppose that $|r| < 1$ and you let n become larger and larger. It follows that r^n gets closer and closer to 0, so that the term r^n drops out of the formula above. You then get the sum

$$S = a_1\left(\frac{1}{1 - r}\right) = \frac{a_1}{1 - r}.$$

Notice that this sum is not dependent on the nth term of the sequence. In the case of Example 5, $r = \left(\frac{2}{3}\right) < 1$, and so the sum of the infinite geometric sequence is

$$S = \sum_{i=1}^{\infty}\left(\frac{2}{3}\right)^i = \frac{a_1}{1 - r} = \frac{2/3}{1 - (2/3)} = \frac{2/3}{1/3} = 2.$$

Sum of an Infinite Geometric Series

If $a_1, a_1r, a_1r^2, \ldots, a_1r^n, \ldots$ is an infinite geometric sequence, then for $|r| < 1$, the sum of the terms is

$$S = \sum_{i=0}^{\infty} a_1r^i = \frac{a_1}{1 - r}.$$

Example 6 Evaluating a Geometric Series

Find the value of each sum.

a. $\displaystyle\sum_{i=1}^{\infty} 5\left(\frac{3}{4}\right)^{i-1}$ **b.** $\displaystyle\sum_{n=0}^{\infty} 4\left(\frac{3}{10}\right)^n$ **c.** $\displaystyle\sum_{i=0}^{\infty}\left(-\frac{3}{5}\right)^i$

Solution

a. The series is geometric with $a_1 = 5\left(\frac{3}{4}\right)^{1-1} = 5$ and $r = \frac{3}{4}$. So,

$$\sum_{i=1}^{\infty} 5\left(\frac{3}{4}\right)^{i-1} = \frac{5}{1 - (3/4)}$$

$$= \frac{5}{1/4} = 20.$$

b. The series is geometric with $a_1 = 4\left(\frac{3}{10}\right)^0 = 4$ and $r = \frac{3}{10}$. So,

$$\sum_{n=0}^{\infty} 4\left(\frac{3}{10}\right)^n = \frac{4}{1 - (3/10)}$$

$$= \frac{4}{7/10} = \frac{40}{7}.$$

c. The series is geometric with $a_1 = \left(-\frac{3}{5}\right)^0 = 1$ and $r = \frac{3}{5}$. So,

$$\sum_{i=0}^{\infty}\left(-\frac{3}{5}\right)^i = \frac{1}{1 - (-3/5)} = \frac{1}{1 + (3/5)} = \frac{5}{8}$$

4 Use a geometric sequence to solve an application problem.

Applications

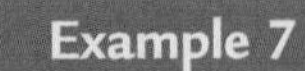
Example 7 A Lifetime Salary 

You have accepted a job that pays a salary of $28,000 the first year. During the next 39 years, suppose you receive a 6% raise each year. What will your total salary be over the 40-year period?

Solution

Using a geometric sequence, your salary during the first year will be

$$a_1 = 28{,}000.$$

Then, with a 6% raise, your salary during the next 2 years will be as follows.

$$\begin{aligned} a_2 &= 28{,}000 + 28{,}000(0.06) \\ &= 28{,}000(1.06)^1 \\ a_3 &= 28{,}000(1.06) + 28{,}000(1.06)(0.06) \\ &= 28{,}000(1.06)^2 \end{aligned}$$

From this pattern, you can see that the common ratio of the geometric sequence is $r = 1.06$. Using the formula for the nth partial sum of a geometric sequence, you will find that the total salary over the 40-year period is given by

$$\begin{aligned} \text{Total salary} &= \sum_{i=1}^{n} a_1 r^{i-1} \\ &= a_1\left(\frac{r_n - 1}{r - 1}\right) \\ &= 28{,}000\left[\frac{(1.06)^{40} - 1}{1.06 - 1}\right] \\ &= 28{,}000\left[\frac{(1.06)^{40} - 1}{0.06}\right] \\ &\approx \$4{,}333{,}335. \end{aligned}$$

The bar graph in Figure 12.2 illustrates your salary during the 40-year period.

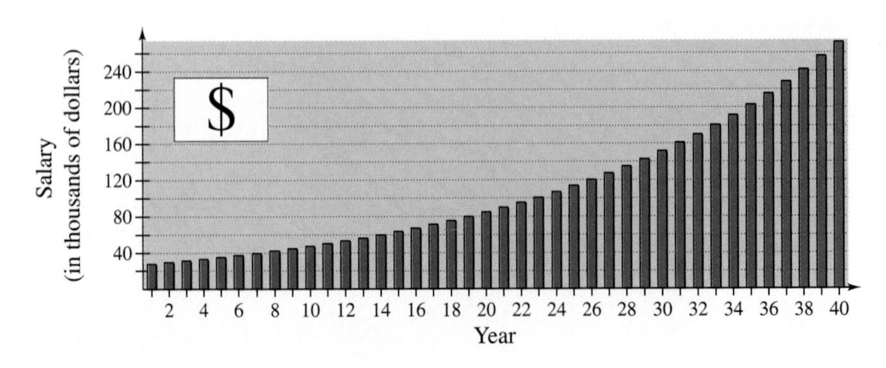

Figure 12.2

Example 8 Increasing Annuity

You deposit \$100 in an account each month for 2 years. The account pays an annual interest rate of 9%, compounded monthly. What is your balance at the end of 2 years? (This type of savings plan is called an **increasing annuity.**)

Solution

The first deposit would earn interest for the full 24 months, the second deposit would earn interest for 23 months, the third deposit would earn interest for 22 months, and so on. Using the formula for compound interest,

$$a_n = P\left(1 + \frac{r}{12}\right)^n \qquad n \text{ is months}$$

$$= 100\left(1 + \frac{0.09}{12}\right)^n$$

$$= 100(1 + 0.0075)^n$$

you can see that the total of the 24 deposits would be

$$\text{Total} = a_1 + a_2 + \cdot \cdot \cdot + a_{24}$$

$$= 100(1.0075)^1 + 100(1.0075)^2 + \cdot \cdot \cdot + 100(1.0075)^{24}$$

$$= 100(1.0075)\left(\frac{1.0075^{24} - 1}{1.0075 - 1}\right) \qquad a_1\left(\frac{r^n - 1}{r - 1}\right)$$

$$= \$2638.49.$$

Discussing the Concept **Annual Revenue**

The two bar graphs below show the annual revenues for two companies. One company's revenue grew at an arithmetic rate, whereas the other grew at a geometric rate. Which company had the greatest revenue during the 10-year period? Which company would you rather own? Explain.

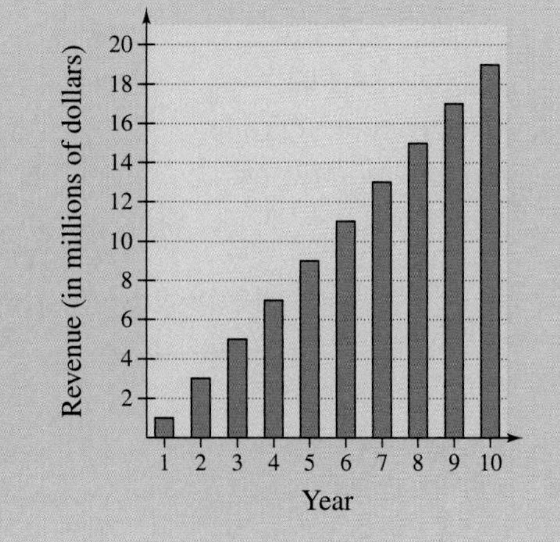

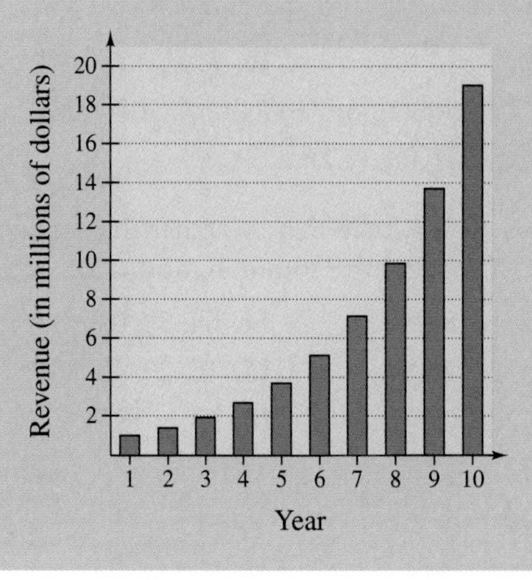

12.3 Exercises

Integrated Review

Concepts, Skills, and Problem Solving

Keep mathematically in shape by doing these exercises *before* the problems of this section.

Properties and Definitions

1. Relative to the x- and y-axes, explain the meaning of each coordinate of the point $(-6, 4)$.

2. A point lies 5 units from the x-axis and 10 units from the y-axis. Give the ordered pair for such a point in each quadrant.

3. In your own words, define the graph of the function $y = f(x)$.

4. Describe the procedure for finding the x- and y-intercepts of the graph of $f(x) = 2\sqrt{x + 4}$.

Solving Inequalities

In Exercises 5–10, solve the inequality.

5. $3x - 5 > 0$

6. $\frac{3}{2}y + 11 < 20$

7. $100 < 2x + 30 < 150$

8. $-5 < -\frac{x}{6} < 2$

9. $2x^2 - 7x + 5 > 0$

10. $2x - \frac{5}{x} > 3$

Problem Solving

11. A television set is advertised as having a 19-inch screen. Determine the dimensions of the square screen if its diagonal is 19 inches.

12. A construction worker is building the forms for the rectangular foundation of a home that is 25 feet wide and 40 feet long. To make sure that the corners are square the worker measures the diagonal of the foundation. What should that measurement be?

Developing Skills

In Exercises 1–12, find the common ratio of the geometric sequence. See Example 1.

1. $2, 6, 18, 54, \ldots$

2. $5, -10, 20, -40, \ldots$

3. $1, -3, 9, -27, \ldots$

4. $54, 18, 6, 2, \ldots$

5. $12, -6, 3, -\frac{3}{2}, \ldots$

6. $9, 6, 4, \frac{8}{3}, \ldots$

7. $1, -\frac{3}{2}, \frac{9}{4}, -\frac{27}{8}, \ldots$

8. $5, -\frac{5}{2}, \frac{5}{4}, -\frac{5}{8}, \ldots$

9. $1, \pi, \pi^2, \pi^3, \ldots$

10. $e, e^2, e^3, e^4, \ldots$

11. $500(1.06), 500(1.06)^2, 500(1.06)^3, 500(1.06)^4, \ldots$

12. $1.1, (1.1)^2, (1.1)^3, (1.1)^4, \ldots$

In Exercises 13–24, determine whether the sequence is geometric. If so, find the common ratio.

13. $64, 32, 16, 8, \ldots$

14. $64, 32, 0, -32, \ldots$

15. $10, 15, 20, 25, \ldots$

16. $10, 20, 40, 80, \ldots$

17. $5, 10, 20, 40, \ldots$

18. $54, -18, 6, -2, \ldots$

19. $1, 8, 27, 64, 125, \ldots$

20. $12, 7, 2, -3, -8, \ldots$

21. $1, -\frac{2}{3}, \frac{4}{9}, -\frac{8}{27}, \ldots$

22. $\frac{1}{3}, -\frac{2}{3}, \frac{4}{3}, -\frac{8}{3}, \ldots$

23. $10(1 + 0.02), 10(1 + 0.02)^2, 10(1 + 0.02)^3, \ldots$

24. $1, 0.2, 0.04, 0.008, \ldots$

In Exercises 25–38, write the first five terms of the geometric sequence.

25. $a_1 = 4, \quad r = 2$

26. $a_1 = 3, \quad r = 4$

27. $a_1 = 6, \quad r = \frac{1}{3}$

28. $a_1 = 4, \quad r = \frac{1}{2}$

29. $a_1 = 1, \quad r = -\frac{1}{2}$

30. $a_1 = 32, \quad r = -\frac{3}{4}$

31. $a_1 = 4, \quad r = -\frac{1}{2}$

32. $a_1 = 4, \quad r = \frac{3}{2}$

33. $a_1 = 1000, \quad r = 1.01$

34. $a_1 = 200, \quad r = 1.07$

35. $a_1 = 4000, \quad r = 1/1.01$

36. $a_1 = 1000, \quad r = 1/1.05$

37. $a_1 = 10, \quad r = \frac{3}{5}$

38. $a_1 = 36, \quad r = \frac{2}{3}$

In Exercises 39–52, find the specified nth term of the geometric sequence.

39. $a_1 = 6, \quad r = \frac{1}{2}, \quad a_{10} = \square$

40. $a_1 = 8, \quad r = \frac{3}{4}, \quad a_8 = \square$

41. $a_1 = 3, \quad r = \sqrt{2}, \quad a_{10} = \square$

42. $a_1 = 5, \quad r = \sqrt{3}, \quad a_9 = \square$

43. $a_1 = 200, \quad r = 1.2, \quad a_{12} = \square$

44. $a_1 = 500, \quad r = 1.06, \quad a_{40} = \square$

45. $a_1 = 120, \quad r = -\frac{1}{3}, \quad a_{10} = $ ▭

46. $a_1 = 240, \quad r = -\frac{1}{4}, \quad a_{13} = $ ▭

47. $a_1 = 4, \quad a_2 = 3, \quad a_5 = $ ▭

48. $a_1 = 1, \quad a_2 = 9, \quad a_7 = $ ▭

49. $a_1 = 1, \quad a_3 = \frac{9}{4}, \quad a_6 = $ ▭

50. $a_3 = 6, \quad a_5 = \frac{8}{3}, \quad a_6 = $ ▭

51. $a_2 = 12, \quad a_3 = 16, \quad a_4 = $ ▭

52. $a_4 = 100, \quad a_5 = -25, \quad a_7 = $ ▭

In Exercises 53–66, find the formula for the *n*th term of the geometric sequence. (Begin with $n = 1$.) See Examples 2 and 3.

53. $a_1 = 2, \quad r = 3$

54. $a_1 = 5, \quad r = 4$

55. $a_1 = 1, \quad r = 2$

56. $a_1 = 25, \quad r = 4$

57. $a_1 = 1, \quad r = -\frac{1}{5}$

58. $a_1 = 12, \quad r = -\frac{4}{3}$

59. $a_1 = 4, \quad r = -\frac{1}{2}$

60. $a_1 = 9, \quad r = \frac{2}{3}$

61. $a_1 = 8, \quad a_2 = 2$

62. $a_1 = 18, \quad a_2 = 8$

63. $a_1 = 14, \quad a_2 = \frac{21}{2}$

64. $a_1 = 36, \quad a_2 = \frac{27}{2}$

65. $4, -6, 9, -\frac{27}{2}, \ldots$

66. $1, \frac{3}{2}, \frac{9}{4}, \frac{27}{8}, \ldots$

In Exercises 67–70, match the geometric sequence with its graph. [The graphs are labeled (a), (b), (c), and (d).]

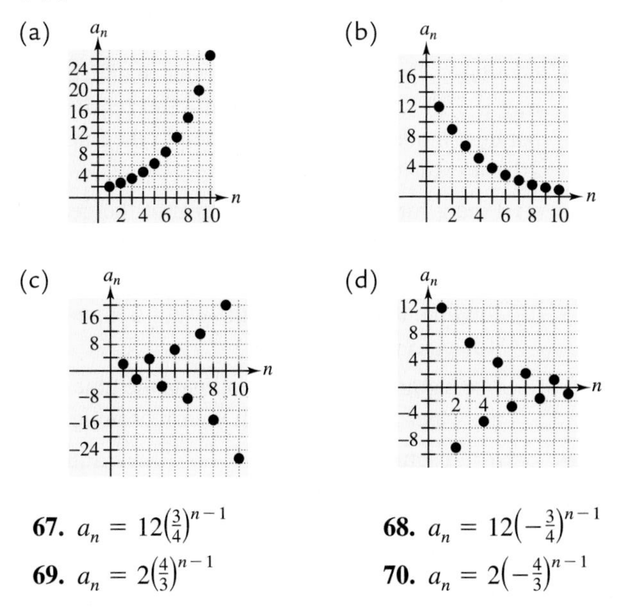

67. $a_n = 12\left(\frac{3}{4}\right)^{n-1}$

68. $a_n = 12\left(-\frac{3}{4}\right)^{n-1}$

69. $a_n = 2\left(\frac{4}{3}\right)^{n-1}$

70. $a_n = 2\left(-\frac{4}{3}\right)^{n-1}$

In Exercises 71–80, find the *n*th partial sum. See Examples 4 and 5.

71. $\sum_{i=1}^{10} 2^{i-1}$

72. $\sum_{i=1}^{6} 3^{i-1}$

73. $\sum_{i=1}^{12} 3\left(\frac{3}{2}\right)^{i-1}$

74. $\sum_{i=1}^{20} 12\left(\frac{2}{3}\right)^{i-1}$

75. $\sum_{i=1}^{15} 3\left(-\frac{1}{3}\right)^{i-1}$

76. $\sum_{i=1}^{8} 8\left(-\frac{1}{4}\right)^{i-1}$

77. $\sum_{i=1}^{12} 4(-2)^{i-1}$

78. $\sum_{i=1}^{20} 16\left(\frac{1}{2}\right)^{i-1}$

79. $\sum_{i=1}^{8} 6(0.1)^{i-1}$

80. $\sum_{i=1}^{24} 1000(1.06)^{i-1}$

In Exercises 81–92, find the *n*th partial sum of the geometric sequence.

81. $1, -3, 9, -27, 81, \ldots, n = 10$

82. $3, -6, 12, -24, 48, \ldots, n = 12$

83. $8, 4, 2, 1, \frac{1}{2}, \ldots, n = 15$

84. $9, 6, 4, \frac{8}{3}, \frac{16}{9}, \ldots, n = 10$

85. $4, 12, 36, 108, \ldots, n = 8$

86. $\frac{1}{36}, -\frac{1}{12}, \frac{1}{4}, -\frac{3}{4}, \ldots, n = 20$

87. $60, -15, \frac{15}{4}, -\frac{15}{16}, \ldots, n = 12$

88. $40, -10, \frac{5}{2}, -\frac{5}{8}, \frac{5}{32}, \ldots, n = 10$

89. $30, 30(1.06), 30(1.06)^2, 30(1.06)^3, \ldots, n = 20$

90. $100, 100(1.08), 100(1.08)^2, 100(1.08)^3, \ldots, n = 40$

91. $500, 500(1.04), 500(1.04)^2, 500(1.04)^3, \ldots, n = 18$

92. $1, \sqrt{2}, 2, 2\sqrt{2}, 4, \ldots, n = 12$

In Exercises 93–100, find the sum. See Example 6.

93. $\sum_{n=0}^{\infty} \left(\frac{1}{2}\right)^n$

94. $\sum_{n=0}^{\infty} 2\left(\frac{2}{3}\right)^n$

95. $\sum_{n=0}^{\infty} \left(-\frac{1}{2}\right)^n$

96. $\sum_{n=0}^{\infty} \left(\frac{1}{10}\right)^n$

97. $\sum_{n=0}^{\infty} 2\left(-\frac{2}{3}\right)^n$

98. $\sum_{n=0}^{\infty} 4\left(\frac{1}{4}\right)^n$

99. $8 + 6 + \frac{9}{2} + \frac{27}{8} + \cdots$

100. $3 - 1 + \frac{1}{3} - \frac{1}{9} + \cdots$

In Exercises 101–104, use a graphing utility to graph the first 10 terms of the sequence.

101. $a_n = 20(-0.6)^{n-1}$

102. $a_n = 4(1.4)^{n-1}$

103. $a_n = 15(0.6)^{n-1}$

104. $a_n = 8(-0.6)^{n-1}$

Solving Problems

105. *Depreciation* A company buys a machine for \$250,000. During the next 5 years, the machine depreciates at the rate of 25% per year. (That is, at the end of each year, the depreciated value is 75% of what it was at the beginning of the year.)

(a) Find a formula for the nth term of the geometric sequence that gives the value of the machine n full years after it was purchased.

(b) Find the depreciated value of the machine at the end of 5 full years.

(c) During which year did the machine depreciate the most?

106. *Population Increase* A city of 500,000 people is growing at the rate of 1% per year. (That is, at the end of each year, the population is 1.01 times the population at the beginning of the year.)

(a) Find a formula for the nth term of the geometric sequence that gives the population n years from now.

(b) Estimate the population 20 years from now.

107. *Salary Increases* You accept a job that pays a salary of \$30,000 the first year. During the next 39 years, you receive a 5% raise each year. What would your total salary be over the 40-year period?

108. *Salary Increases* You accept a job that pays a salary of \$30,000 the first year. During the next 39 years, you receive a 5.5% raise each year.

(a) What would your total salary be over the 40-year period?

(b) How much more income did the extra 0.5% provide than the result in Exercise 107?

Increasing Annuity In Exercises 109–114, find the balance A in an increasing annuity in which a principal of P dollars is invested each month for t years, compounded monthly at rate r.

109.	$P = \$100$	$t = 10$ years	$r = 9\%$
110.	$P = \$50$	$t = 5$ years	$r = 7\%$
111.	$P = \$30$	$t = 40$ years	$r = 8\%$
112.	$P = \$200$	$t = 30$ years	$r = 10\%$
113.	$P = \$75$	$t = 30$ years	$r = 6\%$
114.	$P = \$100$	$t = 25$ years	$r = 8\%$

115. *Would You Accept This Job?* You start work at a company that pays \$0.01 for the first day, \$0.02 for the second day, \$0.04 for the third day, and so on. If the daily wage keeps doubling, what would your total income be for working (a) 29 days and (b) 30 days?

116. *Would You Accept This Job?* You start work at a company that pays \$0.01 for the first day, \$0.03 for the second day, \$0.09 for the third day, and so on. If the daily wage keeps tripling, what would your total income be for working (a) 25 days and (b) 26 days?

117. *Power Supply* The electrical power for an implanted medical device decreases by 0.1% each day.

(a) Find a formula for the nth term of the geometric sequence that gives the percent of the initial power n days after the device is implanted.

(b) What percent of the initial power is still available 1 year after the device is implanted?

(c) The power supply needs to be changed when half the power is depleted. Use a graphing utility to graph the first 750 terms of the sequence and estimate when the power source should be changed.

118. *Cooling* The temperature of water in an ice cube tray is 70°F when it is placed in a freezer. Its temperature n hours after being placed in the freezer is 20% less than 1 hour earlier.

(a) Find a formula for the nth term of the geometric sequence that gives the temperature of the water n hours after being placed in the freezer.

(b) Find the temperature of the water 6 hours after it is placed in the freezer.

(c) Use a graphing utility to estimate the time when the water freezes. Explain your reasoning.

119. *Area* A square has 12-inch sides. A new square is formed by connecting the midpoints of the sides of the square. Then two of the triangles are shaded (see figure). This process is repeated five more times. What is the total area of the shaded region?

120. *Area* A square has 12-inch sides. The square is divided into nine smaller squares and the center square is shaded (see figure). Each of the eight unshaded squares is then divided into nine smaller squares and each center square is shaded. This process is repeated four more times. What is the total area of the shaded region?

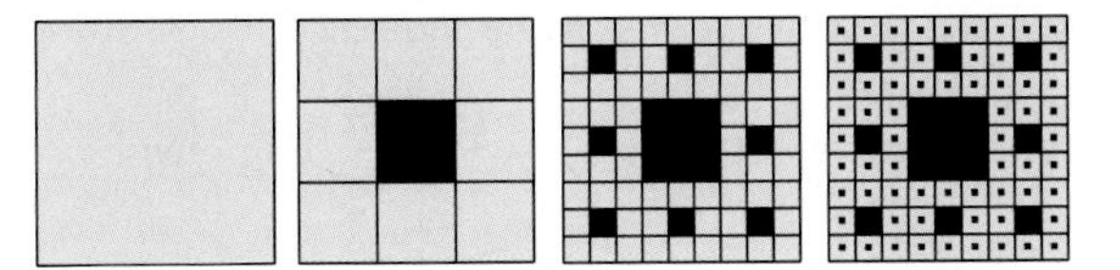

121. *Bungee Jumping* A bungee jumper jumps from a bridge and stretches a cord 100 feet. Successive bounces stretch the cord 75% of each previous length (see figure). Find the total distance traveled by the bungee jumper during 10 bounces.

$100 + 2(100)(0.75) + \cdot \cdot \cdot + 2(100)(0.75)^{10}$

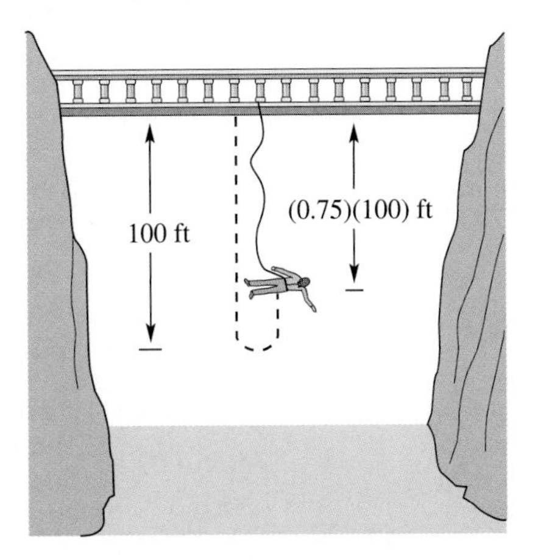

122. *Distance* A ball is dropped from a height of 16 feet. Each time it drops h feet, it rebounds $0.81h$ feet.

(a) Find the total distance traveled by the ball.

(b) The ball takes the following time for each fall.

$$s_1 = -16t^2 + 16, \qquad s_1 = 0 \text{ if } t = 1$$
$$s_2 = -16t^2 + 16(0.81), \qquad s_2 = 0 \text{ if } t = 0.9$$
$$s_3 = -16t^2 + 16(0.81)^2, \qquad s_3 = 0 \text{ if } t = (0.9)^2$$
$$s_4 = -16t^2 + 16(0.81)^3, \qquad s_4 = 0 \text{ if } t = (0.9)^3$$
$$\vdots \qquad\qquad\qquad \vdots$$
$$s_n = -16t^2 + 16(0.81)^{n-1}, s_n = 0 \text{ if } t = (0.9)^{n-1}$$

Beginning with s_2, the ball takes the same amount of time to bounce up as it does to fall, and thus the total time elapsed before it comes to rest is

$$t = 1 + 2\sum_{n=1}^{\infty} (0.9)^n.$$

Find this total.

Explaining Concepts

123. Answer parts (a)–(c) of Motivating the Chapter on page 703.

124. In your own words, explain what makes a sequence geometric.

125. What is the general formula for the nth term of a geometric sequence?

126. The second and third terms of a geometric sequence are 6 and 3, respectively. What is the first term?

127. Give an example of a geometric sequence whose terms alternate in sign.

128. Explain why the terms of a geometric sequence decrease when $a_1 > 0$ and $0 < r < 1$.

129. In your own words, describe an increasing annuity.

130. Explain what is meant by the nth partial sum of a sequence.

Mid-Chapter Quiz

Take this quiz as you would take a quiz in class. After you are done, check your work against the answers given in the back of the book.

In Exercises 1 and 2, write the first five terms of the sequence.

1. $a_n = 32\left(\frac{1}{4}\right)^{n-1}$ (Begin with $n = 1$.) **2.** $a_n = \frac{(-3)^n n}{n+4}$ (Begin with $n = 1$.)

In Exercises 3–6, find the sum.

3. $\sum_{k=1}^{4} 10k$ **4.** $\sum_{i=1}^{10} 4$ **5.** $\sum_{j=1}^{5} \frac{60}{j+1}$ **6.** $\sum_{n=1}^{8} 8\left(-\frac{1}{2}\right)$

In Exercises 7 and 8, write the sum using sigma notation.

7. $\frac{2}{3(1)} + \frac{2}{3(2)} + \frac{2}{3(3)} + \cdots + \frac{2}{3(20)}$ **8.** $\frac{1}{1^3} - \frac{1}{2^3} + \frac{1}{3^3} - \cdots + \frac{1}{25^3}$

In Exercises 9 and 10, find the common difference of the arithmetic sequence.

9. $1, \frac{3}{2}, 2, \frac{5}{2}, 3, \ldots$ **10.** $100, 94, 88, 82, 76, \ldots$

In Exercises 11 and 12, find the common ratio of the geometric sequence.

11. $2, 6, 18, 54, 162, \ldots$ **12.** $2, 1, \frac{1}{2}, \frac{1}{4}, \frac{1}{8}, \ldots$

In Exercises 13 and 14, find a formula for a_n.

13. *Arithmetic*, $a_1 = 20,\ a_4 = 11$ **14.** *Geometric*, $a_1 = 32,\ r = -\frac{1}{4}$

In Exercises 15–20, find the sum.

15. $\sum_{n=1}^{50} (3n + 5)$ **16.** $\sum_{n=1}^{300} \frac{n}{5}$ **17.** $\sum_{i=1}^{8} 9\left(\frac{2}{3}\right)^{i-1}$

18. $\sum_{j=1}^{20} 500(1.06)^{j-1}$ **19.** $\sum_{i=0}^{\infty} 3\left(\frac{2}{3}\right)^{i}$ **20.** $\sum_{i=0}^{\infty} \frac{4}{5}\left(\frac{1}{4}\right)^{i}$

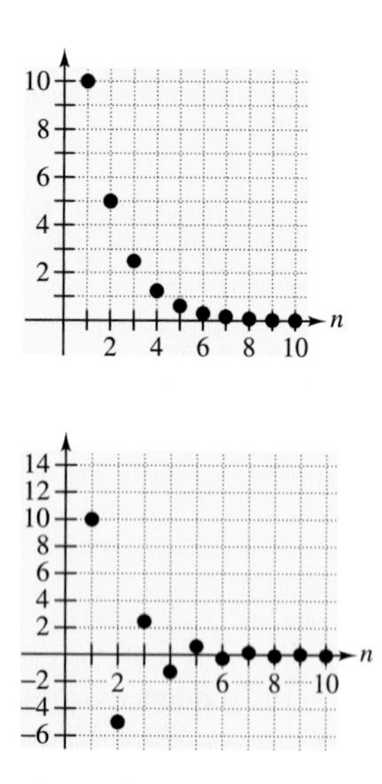

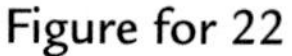

Figure for 22

21. Find the 12th term of $625, -250, 100, -40, \ldots$.

22. Match $a_n = 10\left(\frac{1}{2}\right)^{n-1}$ and $b_n = 10\left(-\frac{1}{2}\right)^{n-1}$ with the graphs at the left.

23. The temperature of a coolant decreases by 25.75°F the first hour. For each subsequent hour, the temperature decreases by 2.25°F less than it decreased the previous hour. How much does the temperature decrease during the 10th hour?

24. The sequence given by $a_n = 2^{n-1}$ is geometric. Describe the sequence given by $b_n = \ln a_n$.

12.4 The Binomial Theorem

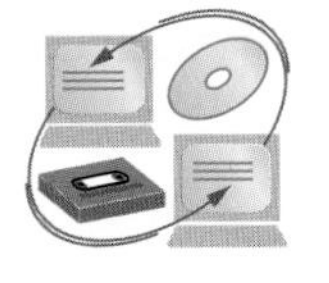

Objectives

1. Determine the coefficients of a binomial raised to a power.
2. Arrange the binomial coefficients in a triangular pattern known as Pascal's Triangle.
3. Expand a binomial raised to a power using Pascal's Triangle and the Binomial Theorem.

1 Determine the coefficients of a binomial raised to a power.

Binomial Coefficients

Recall that a **binomial** is a polynomial that has two terms. In this section, you will study a formula that provides a quick method of raising a binomial to a power. To begin, let's look at the expansion of $(x + y)^n$ for several values of n.

$$(x + y)^0 = 1$$

$$(x + y)^1 = x + y$$

$$(x + y)^2 = x^2 + 2xy + y^2$$

$$(x + y)^3 = x^3 + 3x^2y + 3xy^2 + y^3$$

$$(x + y)^4 = x^4 + 4x^3y + 6x^2y^2 + 4xy^3 + y^4$$

$$(x + y)^5 = x^5 + 5x^4y + 10x^3y^2 + 10x^2y^3 + 5xy^4 + y^5$$

There are several observations you can make about these expansions.

1. In each expansion, there are $n + 1$ terms.
2. In each expansion, x and y have symmetrical roles. The powers of x decrease by 1 in successive terms, whereas the powers of y increase by 1.
3. The sum of the powers of each term is n. For instance, in the expansion of $(x + y)^5$, the sum of the powers of each term is 5.

$$(x + y)^5 = x^5 + \overbrace{5x^4y^1}^{4+1=5} + \overbrace{10x^3y^2}^{3+2=5} + 10x^2y^3 + 5xy^4 + y^5$$

4. The coefficients increase and then decrease in a symmetric pattern.

The coefficients of a binomial expansion are called **binomial coefficients.** To find them, you can use the following theorem.

> **The Binomial Theorem**
>
> In the expansion of $(x + y)^n$
>
> $$(x + y)^n = x^n + nx^{n-1}y + \cdot \cdot \cdot + {}_nC_rx^{n-r}y^r + \cdot \cdot \cdot + nxy^{n-1} + y^n$$
>
> the coefficient of $x^{n-r}y^r$ is given by
>
> $${}_nC_r = \frac{n!}{(n-r)!r!}.$$

Example 1 Finding Binomial Coefficients

Find each binomial coefficient.

a. ${}_8C_2$ **b.** ${}_{10}C_3$ **c.** ${}_7C_0$ **d.** ${}_8C_8$ **e.** ${}_9C_6$

Solution

a. $${}_8C_2 = \frac{8!}{6! \cdot 2!} = \frac{(8 \cdot 7) \cdot \cancel{6!}}{\cancel{6!} \cdot 2!} = \frac{8 \cdot 7}{2 \cdot 1} = 28$$

b. $${}_{10}C_3 = \frac{10!}{7! \cdot 3!} = \frac{(10 \cdot 9 \cdot 8) \cdot \cancel{7!}}{\cancel{7!} \cdot 3!} = \frac{10 \cdot 9 \cdot 8}{3 \cdot 2 \cdot 1} = 120$$

c. $${}_7C_0 = \frac{7!}{7! \cdot 0!} = 1$$

d. $${}_8C_8 = \frac{8!}{0! \cdot 8!} = 1$$

e. $${}_9C_6 = \frac{9!}{3! \cdot 6!} = \frac{(9 \cdot 8 \cdot 7) \cdot \cancel{6!}}{3! \cdot \cancel{6!}} = \frac{9 \cdot 8 \cdot 7}{3 \cdot 2 \cdot 1} = 84$$

Technology: Tip

As you will learn later in this text, the formula for the binomial coefficient is the same as the formula for combinations in the study of probability. Most graphing utilities have a function called *combination* that can be used to evaluate a binomial coefficient. Consult the user's guide for your graphing utility.

When $r \neq 0$ and $r \neq n$, as in parts (a) and (b) above, there is a simple pattern for evaluating binomial coefficients.

$${}_8C_2 = \frac{\overbrace{8 \cdot 7}^{\text{2 factors}}}{\underbrace{2 \cdot 1}_{\text{2 factorial}}} \quad \text{and} \quad {}_{10}C_3 = \frac{\overbrace{10 \cdot 9 \cdot 8}^{\text{3 factors}}}{\underbrace{3 \cdot 2 \cdot 1}_{\text{3 factorial}}}$$

Example 2 Finding Binomial Coefficients

Find each binomial coefficient.

a. ${}_7C_3$ **b.** ${}_7C_4$ **c.** ${}_{12}C_1$ **d.** ${}_{12}C_{11}$

Solution

a. $${}_7C_3 = \frac{7 \cdot 6 \cdot 5}{3 \cdot 2 \cdot 1} = 35$$

b. $${}_7C_4 = \frac{7 \cdot 6 \cdot 5 \cdot 4}{4 \cdot 3 \cdot 2 \cdot 1} = 35$$

c. $${}_{12}C_1 = \frac{12!}{11!1!} = \frac{(12) \cdot \cancel{11!}}{\cancel{11!} \cdot 1!} = \frac{12}{1} = 12$$

d. $${}_{12}C_{11} = \frac{12!}{1! \cdot 11!} = \frac{(12) \cdot \cancel{11!}}{1! \cdot \cancel{11!}} = \frac{12}{1} = 12$$

In Example 2, it is not a coincidence that the answers to parts (a) and (b) are the same, and that those in parts (c) and (d) are the same. In general, it is true that

$${}_nC_r = {}_nC_{n-r}.$$

This shows the symmetric property of binomial coefficients.

2 Arrange the binomial coefficients in a triangular pattern known as Pascal's Triangle.

Pascal's Triangle

There is a convenient way to remember a pattern for binomial coefficients. By arranging the coefficients in a triangular pattern, you obtain the following array, which is called **Pascal's Triangle.** This triangle is named after the famous French mathematician Blaise Pascal (1623–1662).

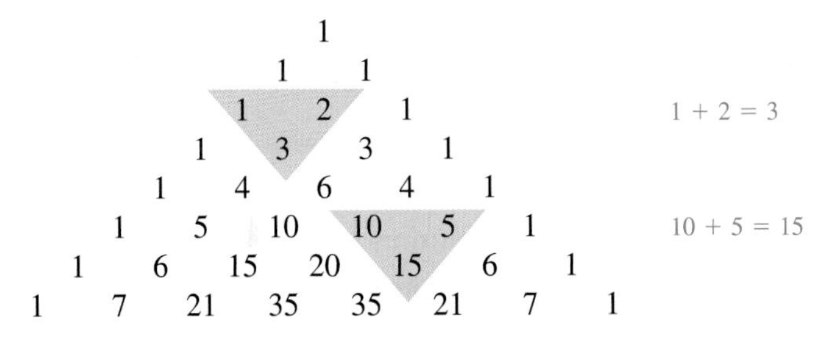

Study Tip

The top row in Pascal's Triangle is called the *zero row* because it corresponds to the binomial expansion

$$(x + y)^0 = 1.$$

Similarly, the next row is called the *first row* because it corresponds to the binomial expansion

$$(x + y)^1 = 1(x) + 1(y).$$

In general, the *nth row* in Pascal's Triangle gives the coefficients of $(x + y)^n$.

The first and last numbers in each row of Pascal's Triangle are 1. As shown above, every other number in each row is formed by adding the two numbers immediately above the number. Pascal noticed that numbers in this triangle are precisely the same numbers that are the coefficients of binomial expansions.

$$(x + y)^0 = 1 \qquad \text{0th row}$$
$$(x + y)^1 = 1x + 1y \qquad \text{1st row}$$
$$(x + y)^2 = 1x^2 + 2xy + 1y^2 \qquad \text{2nd row}$$
$$(x + y)^3 = 1x^3 + 3x^2y + 3xy^2 + 1y^3 \qquad \text{3rd row}$$
$$(x + y)^4 = 1x^4 + 4x^3y + 6x^2y^2 + 4xy^3 + 1y^4 \qquad \vdots$$
$$(x + y)^5 = 1x^5 + 5x^4y + 10x^3y^2 + 10x^2y^3 + 5xy^4 + 1y^5$$
$$(x + y)^6 = 1x^6 + 6x^5y + 15x^4y^2 + 20x^3y^3 + 15x^2y^4 + 6xy^5 + 1y^6$$
$$(x + y)^7 = 1x^7 + 7x^6y + 21x^5y^2 + 35x^4y^3 + 35x^3y^4 + 21x^2y^5 + 7xy^6 + 1y^7$$

You can use the seventh row of Pascal's Triangle to find the eighth row.

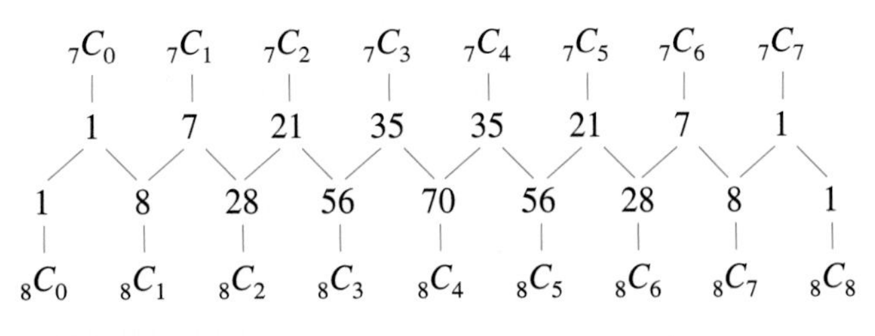

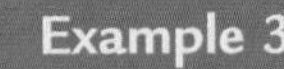

Example 3 Using Pascal's Triangle

To evaluate $_5C_2$, use the fifth row of Pascal's Triangle

1	5	10	10	5	1
$_5C_0$	$_5C_1$	$_5C_2$	$_5C_3$	$_5C_4$	$_5C_5$

to obtain $_5C_2 = 10$.

3 Expand a binomial raised to a power using Pascal's Triangle and the Binomial Theorem.

Binomial Expansions

As mentioned at the beginning of this section, when you write out the coefficients for a binomial that is raised to a power, you are **expanding a binomial.** The formulas for binomial coefficients give you an easy way to expand binomials, as demonstrated in the next three examples.

Example 4 Expanding a Binomial

Write the expansion for the expression.

$$(x + 1)^5$$

Solution

The binomial coefficients from the fifth row of Pascal's Triangle are

1, 5, 10, 10, 5, 1.

So, the expansion is as follows.

$$\begin{aligned}(x + 1)^5 &= (1)x^5 + (5)x^4(1) + (10)x^3(1^2) + (10)x^2(1^3) + (5)x(1^4) + (1)(1^5)\\ &= x^5 + 5x^4 + 10x^3 + 10x^2 + 5x + 1\end{aligned}$$

To expand binomials representing *differences*, rather than sums, you alternate signs. Here are two examples.

$$(x - 1)^3 = x^3 - 3x^2 + 3x - 1$$

$$(x - 1)^4 = x^4 - 4x^3 + 6x^2 - 4x + 1$$

Example 5 Expanding a Binomial

Write the expansion for the expression.

a. $(x - 3)^4$ **b.** $(2x - 1)^3$

Solution

a. The binomial coefficients from the fourth row of Pascal's Triangle are

1, 4, 6, 4, 1.

So, the expansion is as follows.

$$\begin{aligned}(x - 3)^4 &= (1)x^4 - (4)x^3(3) + (6)x^2(3^2) - (4)x(3^3) + (1)(3^4)\\ &= x^4 - 12x^3 + 54x^2 - 108x + 81\end{aligned}$$

b. The binomial coefficients from the third row of Pascal's Triangle are

1, 3, 3, 1.

So, the expansion is as follows.

$$\begin{aligned}(2x - 1)^3 &= (1)(2x)^3 - (3)(2x)^2(1) + (3)(2x)(1^2) - (1)(1^3)\\ &= 8x^3 - 12x^2 + 6x - 1\end{aligned}$$

Example 6 Expanding a Binomial

Write the expansion for $(x - 2y)^4$.

Solution

Use the fourth row of Pascal's Triangle, as follows.

$$\begin{aligned}(x - 2y)^4 &= (1)x^4 - (4)x^3(2y) + (6)x^2(2y)^2 - (4)x(2y)^3 + (1)(2y)^4 \\ &= x^4 - 8x^3y + 24x^2y^2 - 32xy^3 + 16y^4\end{aligned}$$

Example 7 Finding a Term in the Binomial Expansion

a. Find the sixth term of $(a + 2b)^8$. **b.** Find the 12th term of $(2a - b)^{15}$.

Solution

a. From the Binomial Theorem, you can see that the $(r + 1)$th term is ${}_nC_r x^{n-r}y^r$. So in this case, $6 = r + 1$ means that $r = 5$. Because $n = 8$, $x = a$, and $y = 2b$, the sixth term in the binomial expansion is

$$\begin{aligned}{}_8C_5 a^{8-5}(2b)^5 &= 56 \cdot a^3 \cdot (2b)^5 \\ &= 56(2^5)a^3b^5 \\ &= 1792a^3b^5.\end{aligned}$$

b. From the Binomial Theorem, you can see that the $(r + 1)$th term is ${}_nC_r x^{n-r}y^r$. So in this case, $12 = r + 1$ means that $r = 11$. Because $n = 15$, $x = 2a$, and $y = -b$, the 12th term in the binomial expansion is

$$\begin{aligned}{}_{15}C_{11}(2a)^{15-11}(-b)^{11} &= 1365 \cdot (2a)^4 \cdot (-b)^{11} \\ &= 1365(2^4)(-1)^{11}a^4b^{11} \\ &= -21{,}840a^4b^{11}.\end{aligned}$$

Discussing the Concept **Finding a Pattern**

By adding the terms in each of the rows of Pascal's Triangle, you obtain the following.

Row 0: $1 = 1$

Row 1: $1 + 1 = 2$

Row 2: $1 + 2 + 1 = 4$

Row 3: $1 + 3 + 3 + 1 = 8$

Row 4: $1 + 4 + 6 + 4 + 1 = 16$

Find a pattern for this sequence. Then use the pattern to find the sum of the terms in the 10th row of Pascal's Triangle. Finally, check your answer by actually adding the terms of the 10th row.

12.4 Exercises

Integrated Review *Concepts, Skills, and Problem Solving*

Keep mathematically in shape by doing these exercises *before* the problems of this section.

Properties and Definitions

1. Is it possible to find the determinant of the following matrix? Explain.

$$\begin{bmatrix} 3 & 2 & 6 \\ 1 & -4 & 7 \end{bmatrix}$$

2. State the three elementary row operations that can be used to transform a matrix into a second row-equivalent matrix.

3. Is the following matrix in row-echelon form? Explain.

$$\begin{bmatrix} 1 & 2 & 6 \\ 0 & 1 & 7 \end{bmatrix}$$

Determinants

In Exercises 4–7, evaluate the determinant of the matrix.

4. $\begin{bmatrix} 10 & 25 \\ 6 & -5 \end{bmatrix}$

5. $\begin{bmatrix} 3 & 7 \\ -2 & 6 \end{bmatrix}$

6. $\begin{bmatrix} 3 & -2 & 1 \\ 0 & 5 & 3 \\ 6 & 1 & 1 \end{bmatrix}$

7. $\begin{bmatrix} 4 & 3 & 5 \\ 3 & 2 & -2 \\ 5 & -2 & 0 \end{bmatrix}$

Problem Solving

8. Use a determinant to find the area of the triangle with vertices $(-5, 8)$, $(10, 0)$, and $(3, -4)$.

9. The path of a ball passes through the points $(0, 2)$, $(10, 8)$, and $(20, 0)$. Use Cramer's Rule to find the equation of the parabola that models the path of the ball.

10. Use determinants to find the equation of the line through $(2, -1)$ and $(4, 7)$.

Developing Skills

In Exercises 1–12, evaluate the binomial coefficient ${}_nC_r$. See Examples 1 and 2.

1. ${}_6C_4$ **2.** ${}_7C_3$

3. ${}_{10}C_5$ **4.** ${}_{12}C_9$

5. ${}_{20}C_{20}$ **6.** ${}_{15}C_0$

7. ${}_{18}C_{18}$ **8.** ${}_{200}C_1$

9. ${}_{50}C_{48}$ **10.** ${}_{75}C_1$

11. ${}_{25}C_4$ **12.** ${}_{18}C_5$

In Exercises 13–22, use a graphing utility to evaluate ${}_nC_r$.

13. ${}_{30}C_6$ **14.** ${}_{25}C_{10}$

15. ${}_{12}C_7$ **16.** ${}_{40}C_5$

17. ${}_{52}C_5$ **18.** ${}_{100}C_6$

19. ${}_{200}C_{195}$ **20.** ${}_{500}C_4$

21. ${}_{25}C_{12}$ **22.** ${}_{1000}C_2$

In Exercises 23–28, use Pascal's Triangle to evaluate ${}_nC_r$. See Example 3.

23. ${}_6C_2$ **24.** ${}_9C_3$

25. ${}_7C_3$ **26.** ${}_9C_5$

27. ${}_8C_4$ **28.** ${}_{10}C_6$

In Exercises 29–38, use Pascal's Triangle to expand the expression. See Examples 4–6.

29. $(a + 2)^3$ **30.** $(x + 3)^5$

31. $(x + y)^8$ **32.** $(r - s)^7$

33. $(2x - 1)^5$ **34.** $(4 - 3y)^3$

35. $(2y + z)^6$ **36.** $(2t - s)^5$

37. $(x^2 + 2)^4$ **38.** $(3 - y^4)^5$

In Exercises 39–50, use the Binomial Theorem to expand the expression.

39. $(x + 3)^6$ **40.** $(x - 5)^4$

41. $(x - 4)^6$ **42.** $(x - 8)^4$

43. $(x + y)^4$ **44.** $(u + v)^6$

45. $(u - 2v)^3$ **46.** $(2x + y)^5$

47. $(3a + 2b)^4$ **48.** $(4u - 3v)^3$

49. $(2x^2 - y)^5$ **50.** $(x - 4y^3)^4$

In Exercises 51–60, find the coefficient of the given term of the expression.

	Expression	*Term*
51.	$(x + 1)^{10}$	x^7
52.	$(x + 3)^{12}$	x^9
53.	$(x - y)^{15}$	x^4y^{11}
54.	$(x - 3y)^{14}$	x^3y^{11}
55.	$(2x + y)^{12}$	x^3y^9
56.	$(x + y)^{10}$	x^7y^3
57.	$(x^2 - 3)^4$	x^4
58.	$(3 - y^3)^5$	y^9
59.	$(\sqrt{x} + 1)^8$	x^2
60.	$\left(\frac{1}{u} + 2\right)^6$	$\frac{1}{u^4}$

In Exercises 61–64, use the Binomial Theorem to approximate the quantity accurate to three decimal places. For example,

$(1.02)^{10} \approx 1 + 10(0.02) + 45(0.02)^2.$

61. $(1.02)^8$ **62.** $(2.005)^{10}$

63. $(2.99)^{12}$ **64.** $(1.98)^9$

Solving Problems

Probability In Exercises 65–68, use the Binomial Theorem to expand the expression. In the study of probability, it is sometimes necessary to use the expansion $(p + q)^n$, where $p + q = 1$.

65. $\left(\frac{1}{2} + \frac{1}{2}\right)^5$ **66.** $\left(\frac{2}{3} + \frac{1}{3}\right)^4$

67. $\left(\frac{1}{4} + \frac{3}{4}\right)^4$ **68.** $\left(\frac{2}{5} + \frac{3}{5}\right)^3$

69. *Patterns in Pascal's Triangle* Describe the pattern.

1
1 1
1 2 1
1 3 3 1
1 4 6 4 1
1 5 10 10 5 1
1 6 15 20 15 6 1

70. *Patterns in Pascal's Triangle* Use each encircled group of numbers to form a 2×2 matrix. Find the determinant of each matrix. Describe the pattern.

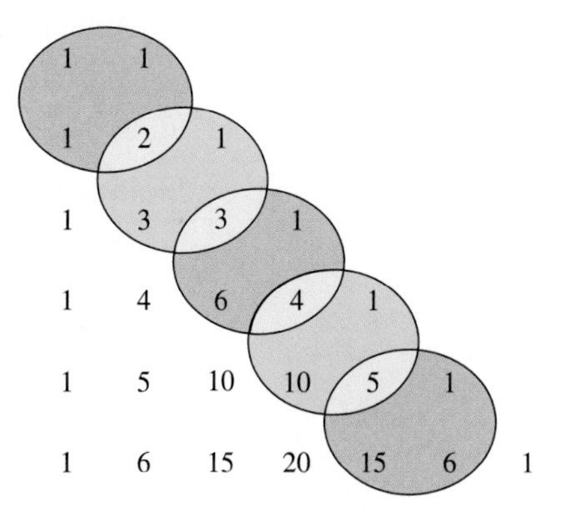

Explaining Concepts

71. How many terms are in the expansion of $(x + y)^n$?

72. Describe the pattern for the exponents with base x in the expansion of $(x + y)^n$.

73. How do the expansions of $(x + y)^n$ and $(x - y)^n$ differ?

74. Which of the following is equal to ${}_{11}C_5$? Explain.

(a) $\dfrac{11 \cdot 10 \cdot 9 \cdot 8 \cdot 7}{5 \cdot 4 \cdot 3 \cdot 2 \cdot 1}$ (b) $\dfrac{11 \cdot 10 \cdot 9 \cdot 8 \cdot 7}{6 \cdot 5 \cdot 4 \cdot 3 \cdot 2 \cdot 1}$

75. What is the relationship between ${}_nC_r$ and ${}_nC_{n-r}$? Explain.

76. In your own words, explain how to form the rows in Pascal's Triangle.

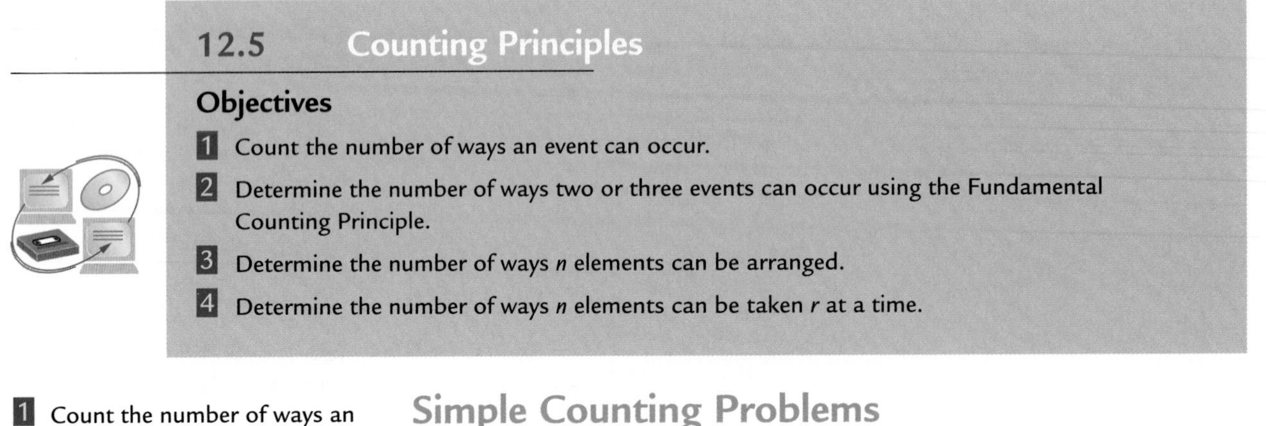

12.5 Counting Principles

Objectives

1. Count the number of ways an event can occur.
2. Determine the number of ways two or three events can occur using the Fundamental Counting Principle.
3. Determine the number of ways n elements can be arranged.
4. Determine the number of ways n elements can be taken r at a time.

1 Count the number of ways an event can occur.

Simple Counting Problems

The last two sections of this chapter contain a brief introduction to some of the basic counting principles and their application to probability. In the next section, you will see that much of probability has to do with counting the number of ways an event can occur. Examples 1, 2, and 3 describe some simple cases.

Example 1 A Random Number Generator

A random number generator (on a computer) selects an integer from 1 to 30. Find the number of ways each event can occur.

a. An even integer is selected.

b. A number that is less than 12 is selected.

c. A prime number is selected.

d. A perfect square is selected.

Solution

a. Because half of the numbers from 1 to 30 are even, this event can occur in 15 different ways.

b. The positive integers that are less than 12 are as follows.

$$\{1, 2, 3, 4, 5, 6, 7, 8, 9, 10, 11\}$$

Because this set has 11 members, you can conclude that there are 11 different ways this event can happen.

c. The prime numbers between 1 and 30 are as follows.

$$\{2, 3, 5, 7, 11, 13, 17, 19, 23, 29\}$$

Because this set has 10 members, you can conclude that there are 10 different ways this event can happen.

d. The perfect square numbers between 1 and 30 are as follows.

$$\{1, 4, 9, 16, 25\}$$

Because this set has five members, you can conclude that there are five different ways this event can happen.

Example 2 Selecting Pairs of Numbers at Random

Eight pieces of paper are numbered from 1 to 8 and placed in a box. One piece of paper is drawn from the box, its number is written down, and the piece of paper is replaced in the box. Then, a second piece of paper is drawn from the box, and its number is written down. Finally, the two numbers are added together. How many different ways can a total of 12 be obtained?

Solution

To solve this problem, count the different ways that a total of 12 can be obtained using two numbers between 1 and 8.

First number + Second number = 12

After considering the various possibilities, you can see that this equation can be satisfied in the following five ways.

First Number:	4	5	6	7	8
Second Number:	8	7	6	5	4

So, a total of 12 can be obtained in *five* different ways.

Solving counting problems can be tricky. Often, seemingly minor changes in the statement of a problem can affect the answer. For instance, compare the counting problem in the next example with that given in Example 2.

Example 3 Selecting Pairs of Numbers at Random

Eight pieces of paper are numbered from 1 to 8 and placed in a box. Two pieces of paper are drawn from the box, and the numbers on them are written down and totaled. How many different ways can a total of 12 be obtained?

Solution

To solve this problem, count the different ways that a total of 12 can be obtained *using two different numbers* between 1 and 8.

First number + Second number = 12

After considering the various possibilities, you can see that this equation can be satisfied in the following four ways.

First Number:	4	5	7	8
Second Number:	8	7	5	4

So, a total of 12 can be obtained in *four* different ways.

Examples 2 and 3 differ in that the random selection in Example 2 occurs *with replacement*, whereas the random selection in Example 3 occurs *without replacement*, which eliminates the possibility of choosing two 6s. When doing such exercises, be sure to note if selection is *with* or *without* replacement.

2 Determine the number of ways two or three events can occur using the Fundamental Counting Principle.

Counting Principles

The first three examples in this section are considered simple counting problems in which you can *list* each possible way that an event can occur. When it is possible, this is always the best way to solve a counting problem. However, some events can occur in so many different ways that it is not feasible to write out the entire list. In such cases, you must rely on formulas and counting principles. The most important of these is called the **Fundamental Counting Principle.**

Fundamental Counting Principle

Let E_1 and E_2 be two events. The first event E_1 can occur in m_1 different ways. After E_1 has occurred, E_2 can occur in m_2 different ways. The number of ways that the two events can occur is

$$m_1 \cdot m_2.$$

The Fundamental Counting Principle can be extended to three or more events. For instance, the number of ways that three events E_1, E_2, and E_3 can occur is

$$m_1 \cdot m_2 \cdot m_3.$$

Example 4 Applying the Fundamental Counting Principle

How many "two-letter words" can be made from the English alphabet?

Solution

The English alphabet contains 26 letters. So, the number of possible "two-letter words" is $26 \cdot 26 = 676$.

Example 5 Applying the Fundamental Counting Principle

Telephone numbers in the United States have 10 digits. The first three are the *area code* and the next seven are the *local telephone number*. How many different telephone numbers are possible within each area code? (A telephone number cannot have 0 or 1 as its first or second digit.)

Peter Fownes/Photo Network

In 1996, there were 178 million active telephone numbers in use in the United States.

Solution

There are only eight choices for the first and second digits because neither can be 0 or 1. For each of the other digits, there are 10 choices.

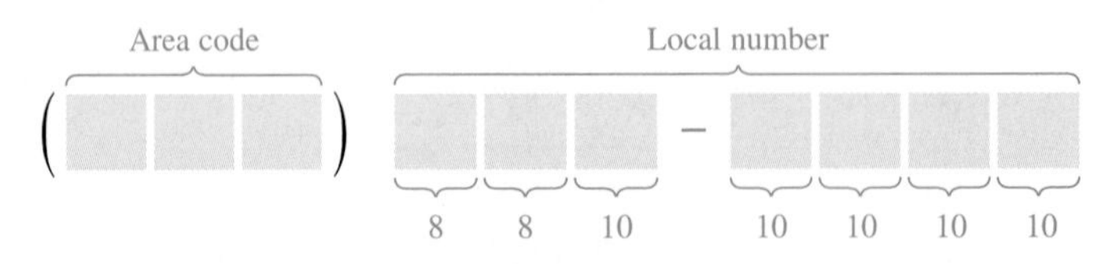

So, by the Fundamental Counting Principle, the number of local telephone numbers that are possible within each area code is

$$8 \cdot 8 \cdot 10 \cdot 10 \cdot 10 \cdot 10 \cdot 10 = 6{,}400{,}000.$$

3 Determine the number of ways n elements can be arranged.

Permutations

One important application of the Fundamental Counting Principle is in determining the number of ways that n elements can be arranged (in order). An ordering of n elements is called a **permutation** of the elements.

> **Definition of Permutation**
>
> A **permutation** of n different elements is an ordering of the elements such that one element is first, one is second, one is third, and so on.

Example 6 Listing Permutations

How many permutations are possible for the letters A, B, and C?

Solution

The possible permutations of the letters A, B, and C are as follows.

A, B, C B, A, C C, A, B

A, C, B B, C, A C, B, A

So, six permutations are possible.

Example 7 Finding the Number of Permutations of n Elements

How many permutations are possible for the letters A, B, C, D, E, and F?

Solution

1st position: Any of the *six* letters.
2nd position: Any of the remaining *five* letters.
3rd position: Any of the remaining *four* letters.
4th position: Any of the remaining *three* letters.
5th position: Any of the remaining *two* letters.
6th position: The *one* remaining letter.

So, the number of choices for the six positions are as follows.

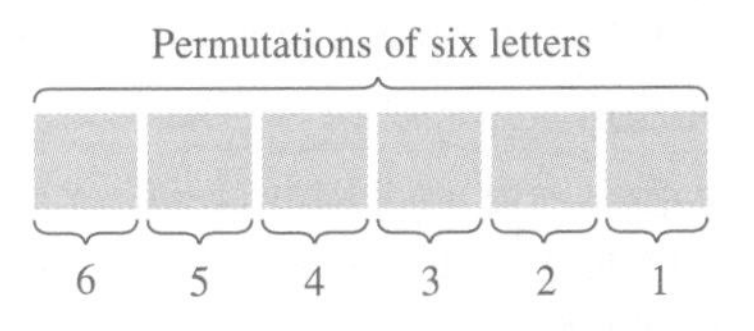

By the Fundamental Counting Principle, the total number of permutations of the six letters is

$$6 \cdot 5 \cdot 4 \cdot 3 \cdot 2 \cdot 1 = 6!$$
$$= 720.$$

The result obtained in Example 7 is generalized below.

> **Number of Permutations of n Elements**
>
> The number of permutations of n elements is given by
>
> $$n \cdot (n - 1) \cdot \cdot \cdot \cdot \cdot 4 \cdot 3 \cdot 2 \cdot 1 = n!.$$
>
> So, there are $n!$ different ways that n elements can be ordered.

Example 8 Finding the Number of Permutations

How many ways can you form a four-digit number using each of the digits 1, 3, 5, and 7 exactly once?

Solution

One way to solve this problem is simply to list the number of ways.

1357, 1375, 1537, 1573, 1735, 1753
3157, 3175, 3517, 3571, 3715, 3751
5137, 5173, 5317, 5371, 5713, 5731
7135, 7153, 7315, 7351, 7513, 7531

Another way to solve the problem is to use the formula for the number of permutations of four elements. By that formula, there are $4! = 24$ permutations.

Example 9 Finding the Number of Permutations

How many ways can you form a six-digit number using each of the digits 1, 2, 3, 4, 5, and 6 exactly once?

Solution

By the formula for the number of permutations of six elements, there are $6! = 720$ permutations.

Example 10 Finding the Number of Permutations

You are a supervisor for 11 different employees. One of your responsibilities is to perform an annual evaluation for each employee, and then rank the 11 different performances. How many different rankings are possible?

Solution

Because there are 11 different employees, you have 11 choices for first ranking. After choosing the first ranking, you can choose any of the remaining 10 for second ranking, and so on.

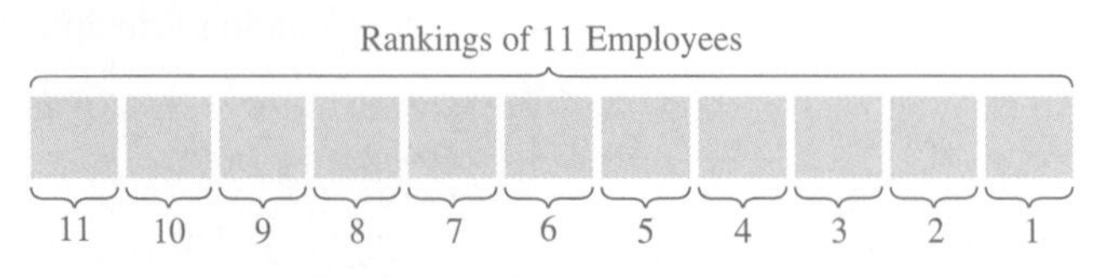

So, the number of different rankings is $11! = 39{,}916{,}800$.

4 Determine the number of ways n elements can be taken r at a time.

Combinations

When counting the number of possible permutations of a set of elements, order is important. The final topic in this section is a method of selecting subsets of a larger set in which order is *not important*. Such subsets are called **combinations of n elements taken r at a time.** For instance, the combinations

$$\{A, B, C\} \quad \text{and} \quad \{B, A, C\}$$

are equivalent because both sets contain the same three elements, and the order in which the elements are listed is *not important*. So, you would count only one of the two sets. A common example of how a combination occurs is a card game in which the player is free to reorder the cards after they have been dealt.

Do you remember this riddle? How many different ways can two of the three (wolf, goat, and cabbage) be left on the shore?

> *A farmer wants to take a wolf, a goat, and cabbage across a river. Unattended, the wolf will eat the goat, and the goat will eat the cabbage. The farmer can only take one of the three on each trip across the river. How can the farmer take all three safely across the river?*

There are three possible combinations that can be left on the shore.

{wolf, goat}, {wolf, cabbage}, {goat, cabbage}

Of the three combinations, the farmer can only leave the wolf alone with the cabbage when he crosses the river.

Example 11 Combination of n Elements Taken r at a Time

In how many different ways can three letters be chosen from the letters A, B, C, D, and E? (The order of the three letters is not important.)

Solution

The following subsets represent the different combinations of three letters that can be chosen from five letters.

{A, B, C} {A, B, D} {A, B, E} {A, C, D} {A, C, E}
{A, D, E} {B, C, D} {B, C, E} {B, D, E} {C, D, E}

From this list, you can conclude that there are 10 different ways that three letters can be chosen from five letters. Because order is not important, the set {B, C, A} is not chosen. It is represented by the set {A, B, C}.

The formula for the number of combinations of n elements taken r at a time is as follows.

Number of Combinations of n Elements Taken r at a Time

The number of combinations of n elements taken r at a time is

$${}_nC_r = \frac{n!}{(n-r)!r!}.$$

Study Tip

When solving problems involving counting principles, you need to be able to distinguish among the various counting principles in order to determine which is necessary to solve the problem correctly. To do this, ask yourself the following questions.

1. Is the order of the elements important? *Permutation*
2. Are the chosen elements a subset of a larger set in which order is not important? *Combination*
3. Does the problem involve two or more separate events? *Fundamental Counting Principle*

Note that the formula for ${}_nC_r$ is the same one given for binomial coefficients. To see how this formula is used, consider the counting problem given in Example 11. In that problem, you need to find the number of combinations of five elements taken three at a time. Thus, $n = 5$, $r = 3$, and the number of combinations is

$$
\begin{aligned}
{}_5C_3 &= \frac{5!}{2!3!} \\
&= \frac{5 \cdot 4 \cdot 3}{3 \cdot 2 \cdot 1} \\
&= 10
\end{aligned}
$$

which is the same as the answer obtained in Example 11.

Example 12 Combinations of n Elements Taken r at a Time

A standard poker hand consists of five cards dealt from a deck of 52. How many different poker hands are possible? (After the cards are dealt, the player may reorder them, and therefore order is not important.)

Solution

Use the formula for the number of combinations of 52 elements taken five at a time, as follows.

$$
\begin{aligned}
{}_{52}C_5 &= \frac{52!}{47!5!} \\
&= \frac{52 \cdot 51 \cdot 50 \cdot 49 \cdot 48}{5 \cdot 4 \cdot 3 \cdot 2 \cdot 1} \\
&= 2{,}598{,}960
\end{aligned}
$$

So, there are almost 2.6 million different hands.

Discussing the Concept Applying Counting Methods

The Boston Market restaurant chain offers individual rotisserie chicken meals with two side items. Customers can choose either dark or white meat and can select side items from a list of 15. How many different meals are available if two different side items are to be ordered? Of the 15 side items, nine are hot and six are cold. How many different meals are available if a customer wishes to order one hot and one cold side item? (Source: Boston Market, Inc.)

12.5 Exercises

Integrated Review *Concepts, Skills, and Problem Solving*

Keep mathematically in shape by doing these exercises *before* the problems of this section.

Properties and Definitions

1. Which of the following functions are exponential? Explain.

$f(x) = 5x^2 \quad g(x) = 2(5^x)$

2. Explain why $e^{2-x^2} = e^2 \cdot e^{-x^2}$.

Logarithms and Exponents

In Exercises 3–6, rewrite the equation in exponential form.

3. $\log_4 64 = 3$

4. $\log_3 \frac{1}{81} = -4$

5. $\ln 1 = 0$

6. $\ln 5 \approx 1.6094 \ldots$

In Exercises 7–10, solve the equation. (Round the result to two decimal places.)

7. $3^x = 50$

8. $e^{x/2} = 8$

9. $\log_2(x - 5) = 6$

10. $\ln(x + 3) = 10$

Problem Solving

11. After t years, the value of a car that cost \$22,000 is given by

$V(t) = 22{,}000(0.8)^t.$

Sketch a graph of the function and determine when the value of the car is \$15,000.

12. Carbon 14 has a half-life of 5730 years. If you start with 10 grams of this isotope, how much remains after 3000 years?

Solving Problems

Random Selection In Exercises 1–6, find the number of ways the specified event can occur when one or more marbles are selected from a bowl containing 10 marbles numbered 0 through 9. See Examples 1–3.

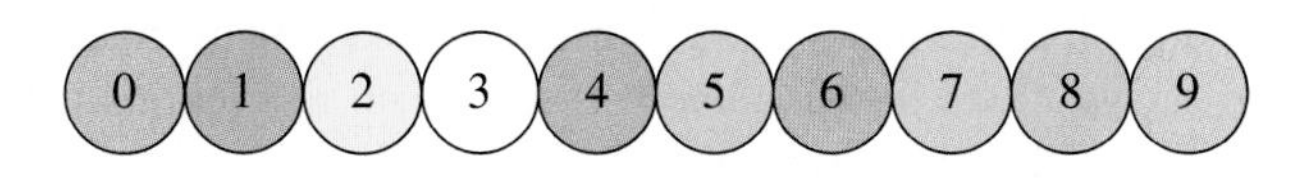

1. One marble is drawn and its number is even.

2. One marble is drawn and its number is prime.

3. Two marbles are drawn one after the other. The first is replaced before the second is drawn. The sum of the numbers is 10.

4. Two marbles are drawn one after the other. The first is replaced before the second is drawn. The sum of the numbers is 7.

5. Two marbles are drawn without replacement. The sum of the numbers is 10.

6. Two marbles are drawn without replacement. The sum of the numbers is 7.

Random Selection In Exercises 7–16, find the number of ways the specified event can occur when one or more marbles are selected from a bowl containing 20 marbles numbered 1 through 20.

7. One marble is drawn and its number is odd.

8. One marble is drawn and its number is even.

9. One marble is drawn and its number is prime.

10. One marble is drawn and its number is greater than 12.

11. One marble is drawn and its number is divisible by 3.

12. One marble is drawn and its number is divisible by 6.

13. Two marbles are drawn one after the other. The first is replaced before the second is drawn. The sum of the numbers is 8.

14. Two marbles are drawn one after the other. The first is replaced before the second is drawn. The sum of the numbers is 15.

15. Two marbles are drawn without replacement. The sum of the numbers is 8.

16. Three marbles are drawn one after the other. Each marble is replaced before the next is drawn. The sum of the numbers is 15.

17. *Staffing Choices* A small grocery store needs to open another checkout line. Three people who can run the cash register are available and two people are available to bag groceries. In how many different ways can the additional checkout line be staffed?

18. *Computer System* You are in the process of purchasing a new computer system. You must choose one of three monitors, one of two computers, and one of two keyboards. How many different configurations of the system are available to you?

19. *Identification Numbers* In a statistical study, each participant was given an identification label consisting of a letter of the alphabet followed by a single digit (0 is a digit). How many distinct identification labels can be made in this way?

20. *Identification Numbers* How many identification labels (see Exercise 19) can be made by one letter of the alphabet followed by a two-digit number?

21. *License Plates* How many distinct automobile license plates can be formed by using a four-digit number followed by two letters?

22. *License Plates* How many distinct license plates can be formed by using three letters followed by a three-digit number?

23. *Three-Digit Numbers* How many three-digit numbers can be formed in each of the following situations?

(a) The hundreds digit cannot be 0.

(b) No repetition of digits is allowed.

(c) The number cannot be greater than 400.

24. *Toboggan Ride* Five people line up on a toboggan at the top of a hill. In how many ways can they be seated if only two of the five are willing to sit in the front seat?

25. *Task Assignment* Four people are assigned to four different tasks. In how many ways can the assignments be made if one of the four is not qualified for the first task?

26. *Taking a Trip* Five people are taking a long trip in a car. Two sit in the front seat and three in the back seat. Three of the people agree to share the driving. In how many different arrangements can the five people sit?

27. *Aircraft Boarding* Eight people are boarding an aircraft. Three have tickets for first class and board before those in economy class. In how many different ways can the eight people board the aircraft?

28. *Permutations* List all the permutations of the letters X, Y, and Z.

29. *Permutations* List all the permutations of the letters A, B, C, and D.

30. *Permutations* List all the permutations of two letters selected from the letters A, B, and C.

31. *Permutations* List all the permutations of two letters selected from the letters A, B, C, and D.

32. *Seating Arrangement* In how many ways can five children be seated in a single row of five chairs?

33. *Seating Arrangement* In how many ways can six people be seated in a six-passenger car?

34. *Posing for a Photograph* In how many ways can four children line up in one row to have their picture taken?

35. *Combination Lock* A combination lock will open when the right choice of three numbers (from 1 to 40, inclusive) is selected. How many different lock combinations are possible?

36. *Access Code* An access code will unlock a door when the right choice of five numbers (from 0 to 9, inclusive) is selected. How many different access codes are possible?

37. *Work Assignments* Eight workers are assigned to eight different tasks. In how many ways can this be done assuming there are no restrictions in making the assignments?

38. *Work Assignments* Out of eight workers, five are selected and assigned to different tasks. In how many ways can this be done if there are no restrictions in making the assignments?

39. *Choosing Offers* From a pool of 10 candidates, the offices of president, vice-president, secretary, and treasurer will be filled. In how many ways can the offices be filled, if each of the 10 candidates can hold any office?

40. *Time Management Study* There are eight steps in accomplishing a certain task and these steps can be performed in any order. Management wants to test each possible order to determine which is the least time-consuming.

(a) How many different orders will have to be tested?

(b) How many different orders will have to be tested if one step in accomplishing the task must be done first? (The other seven steps can be performed in any order.)

41. *Number of Subsets* List all the subsets with two elements that can be formed from the set of letters

$\{A, B, C, D, E, F\}.$

42. *Number of Subsets* List all the subsets with three elements that can be formed from the set of letters

$\{A, B, C, D, E, F\}.$

43. *Committee Selection* Three students are selected from a class of twenty to form a fundraising committee. In how many ways can the committee be formed?

44. *Committee Selection* In how many ways can a committee of five be formed from a group of 30 people?

45. *Menu Selection* A group of four people go out to dinner at a restaurant. There are nine entrees on the menu and the four people decide that no two will order the same thing. In how many ways can the four people order from the nine entrees?

46. *Menu Selection* A group of six people go out to dinner at a restaurant. There are 12 entrees on the menu and the six people decide that no two will order the same thing. In how many ways can the six people order from the 12 entrees?

47. *Test Questions* A student may answer any nine questions from the 12 questions on an exam. Determine the number of ways the student can select the nine questions.

48. *Test Questions* A student may answer any three questions from the 10 questions on an exam. Determine the number of ways the student can select the three questions.

49. *Basketball Lineup* A high school basketball team has 15 players. Use a graphing utility to determine the number of ways the coach can choose 5 players in the starting lineup. (Assume each player can play each position.)

50. *Softball League* Six churches form a softball league. If each team must play every other team twice during the season, what is the total number of league games played?

51. *Job Applicants* An employer interviews six people for four openings in the company. Four of the six people are women. If all six are qualified, in how many ways could the employer fill the four positions if

(a) the selection is random?

(b) exactly two women are selected?

52. *Defective Units* A shipment of 10 microwave ovens contains two defective units. In how many ways can a vending company purchase three of these units and receive

(a) all good units? (b) two good units?
(c) one good unit?

53. *Group Selection* Four people are to be selected from four couples. In how many ways can this be done if

(a) there are no restrictions?

(b) one person from each couple must be selected?

54. *Geometry* Eight points are located in the coordinate plane such that no three lie on the same line. How many different triangles can be formed having their vertices as three of the eight points?

55. *Geometry* Three points that are not on the same line determine three lines. How many lines are determined by seven points, no three of which are on a line?

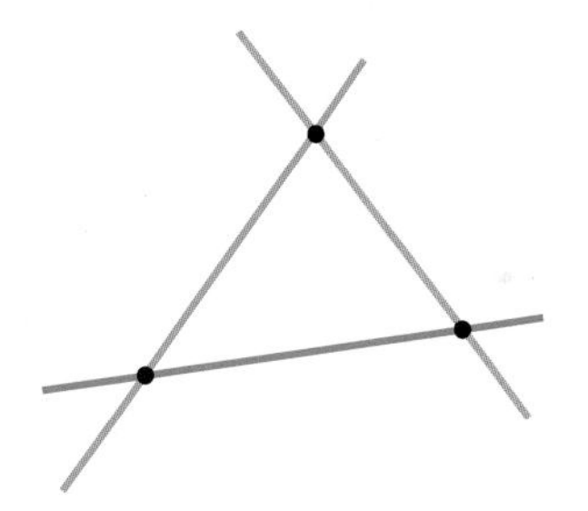

Geometry In Exercises 56–59, find the number of diagonals of the polygon. (A line segment connecting any two nonadjacent vertices of a polygon is called a *diagonal* of the polygon.)

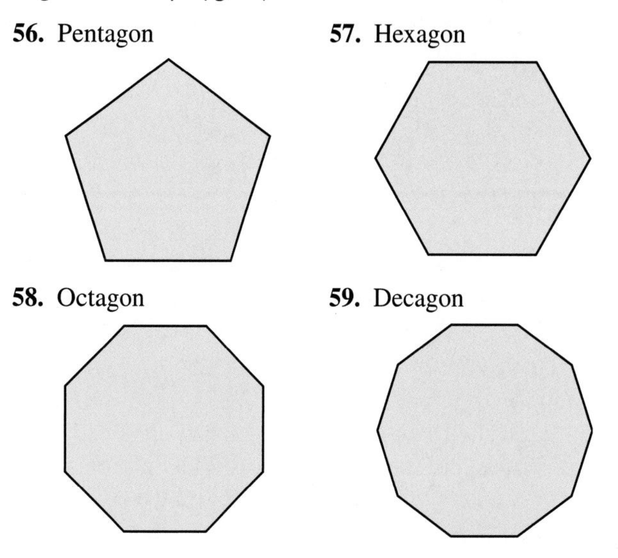

60. *Relationships* As the size of a group increases, the number of relationships increases dramatically (see figure). Determine the number of two-person relationships in a group that has the following numbers.

(a) 3 (b) 4 (c) 6 (d) 8 (e) 10 (f) 12

Explaining Concepts

61. State the Fundamental Counting Principle.

62. When you use the Fundamental Counting Principle, what are you counting?

63. Give examples of a permutation and a combination.

64. Without calculating the numbers, determine which is greater: the combination of 10 elements taken six at a time or the permutation of 10 elements taken six at a time. Explain.

12.6 Probability

Objectives

1. Determine the probability that an event will occur.
2. Use counting principles to determine the probability that an event will occur.

1 Determine the probability that an event will occur.

The Probability of an Event

The **probability of an event** is a number from 0 to 1 that indicates the likelihood that the event will occur. An event that is certain to occur has a probability of 1. An event that cannot occur has a probability of 0. An event that is equally likely to occur or not occur has a probability of $\frac{1}{2}$ or 0.5.

Probability of 0: Event cannot occur. — 0

Probability of 0.5: Event is equally likely to occur or not occur. — 0.5

Probability of 1: Event must occur. — 1

▶ The Probability of an Event

Consider a **sample space** S that is composed of a finite number of outcomes, each of which is equally likely to occur. A subset E of the sample space is an **event.** The probability that an outcome E will occur is

$$P = \frac{\text{number of outcomes in event}}{\text{number of outcomes in sample space}}.$$

Example 1 Finding the Probability of an Event 

a. You select a card from a standard deck of 52 cards. What is the probability that the card is an ace?

$$P = \frac{\text{number of aces in the deck}}{\text{number of cards in the deck}} = \frac{4}{52} = \frac{1}{13}$$

b. You are dialing a friend's phone number but cannot remember the last digit. If you choose a digit at random, the probability that it is correct is

$$P = \frac{\text{number of correct digits}}{\text{number of possible digits}} = \frac{1}{10}.$$

c. On a multiple-choice test, you know that the answer to a question is not (a) or (d), but you are not sure about (b), (c), and (e). If you guess, the probability that you are wrong is

$$P = \frac{\text{number of wrong answers}}{\text{number of possible answers}} = \frac{2}{3}.$$

Example 2 Conducting a Poll

The Centers for Disease Control took a survey of 11,631 high school students. The students were asked whether they considered themselves to be a good weight, underweight, or overweight. The results are shown in Figure 12.3.

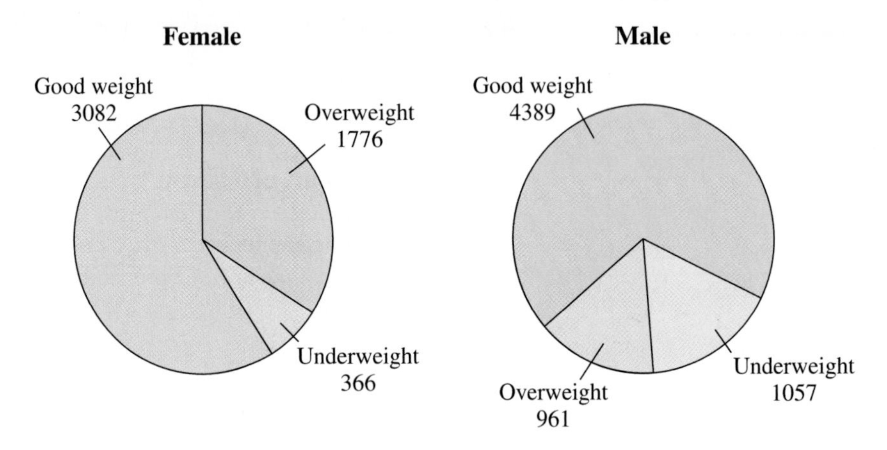

Figure 12.3

a. If you choose a female at random from those surveyed, the probability that she said she was underweight is

$$P = \frac{\text{number of females who answered "underweight"}}{\text{number of females in survey}}$$

$$= \frac{366}{3082 + 366 + 1776}$$

$$= \frac{366}{5224}$$

$$\approx 0.07.$$

b. If you choose a person who answered "underweight" from those surveyed, the probability that the person is female is

$$P = \frac{\text{number of females who answered "underweight"}}{\text{number in survey who answered "underweight"}}$$

$$= \frac{366}{366 + 1057}$$

$$= \frac{366}{1423}$$

$$\approx 0.26.$$

Polls such as the one described in Example 2 are often used to make inferences about a population that is larger than the sample. For instance, from Example 2, you might infer that 7% of *all* high school girls consider themselves to be underweight. When you make such an inference, it is important that those surveyed are representative of the entire population.

Example 3 Using Area to Find Probability 

You have just stepped into the tub to take a shower when one of your contact lenses falls out. (You have not yet turned on the water.) Assuming the lens is equally likely to land anywhere on the bottom of the tub, what is the probability that it lands in the drain? Use the dimensions in Figure 12.4 to answer the question.

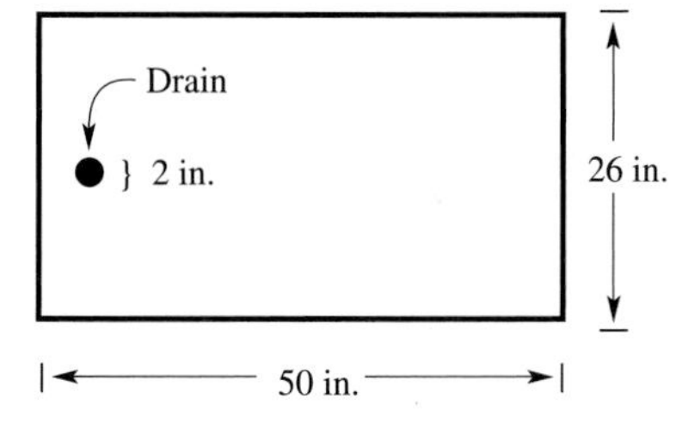

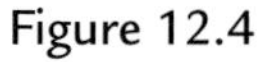
Figure 12.4

Solution

Because the area of the tub bottom is $(26)(50) = 1300$ square inches and the area of the drain is

$$\pi(1^2) = \pi \qquad \text{Area of drain}$$

square inches, the probability that the lens lands in the drain is about

$$P = \frac{\pi}{1300} \approx 0.0024.$$

Example 4 The Probability of Inheriting Certain Genes 

Common parakeets have genes that can produce any one of four feather colors:

Green: BBCC, BBCc, BbCC, BbCc
Blue: BBcc, Bbcc
Yellow: bbCC, bbCc
White: bbcc

	BC	Bc	bC	bc
BC	BBCC	BBCc	BbCC	BbCc
Bc	BBCc	BBcc	BbCc	Bbcc
bC	BbCC	BbCc	bbCC	bbCc
bc	BbCc	Bbcc	bbCc	bbcc

Figure 12.5

Use the *Punnett square* in Figure 12.5 to find the probability that an offspring to two green parents (both with BbCc feather genes) will be yellow. Note that each parent passes along a B or b gene and a C or c gene.

Solution

The probability that an offspring will be yellow is

$$P = \frac{\text{number of yellow possibilities}}{\text{number of possibilities}}$$

$$= \frac{3}{16}.$$

2 Use counting principles to determine the probability that an event will occur.

Using Counting Methods to Find Probabilities

In the following examples, you will see how the basic counting principles you learned in Section 12.5 are used to determine the probability that an event will occur.

Standard 52-Card Deck

A♠	A♥	A♦	A♣
K♠	K♥	K♦	K♣
Q♠	Q♥	Q♦	Q♣
J♠	J♥	J♦	J♣
10♠	10♥	10♦	10♣
9♠	9♥	9♦	9♣
8♠	8♥	8♦	8♣
7♠	7♥	7♦	7♣
6♠	6♥	6♦	6♣
5♠	5♥	5♦	5♣
4♠	4♥	4♦	4♣
3♠	3♥	3♦	3♣
2♠	2♥	2♦	2♣

Figure 12.6

Example 5 The Probability of a Royal Flush

Five cards are dealt at random from a standard deck of 52 playing cards (see Figure 12.6). What is the probability that the cards are 10-J-Q-K-A of the same suit?

Solution

On page 746, you saw that the number of possible five-card hands from a deck of 52 cards is ${}_{52}C_5 = 2{,}598{,}960$. Because only four of these five-card hands are 10-J-Q-K-A of the same suit, the probability that the hand contains these cards is

$$P = \frac{4}{2{,}598{,}960} = \frac{1}{649{,}740}.$$

Example 6 Conducting a Survey

A survey was conducted of 500 adults who had worn Halloween costumes. Each person was asked how he or she acquired a Halloween costume: created it, rented it, bought it, or borrowed it. The results are shown in Figure 12.7. What is the probability that the first four people who were polled all created their costumes?

Solution

To answer this question, you need to use the formula for the number of combinations *twice*. First, find the number of ways to choose four people from 360 who created their own costumes.

$${}_{360}C_4 = \frac{360 \cdot 359 \cdot 358 \cdot 357}{4 \cdot 3 \cdot 2 \cdot 1} = 688{,}235{,}310$$

Next, find the number of ways to choose four people from the 500 who were surveyed.

$${}_{500}C_4 = \frac{500 \cdot 499 \cdot 498 \cdot 497}{4 \cdot 3 \cdot 2 \cdot 1} = 2{,}573{,}031{,}125$$

The probability that all of the first four people surveyed created their own costumes is the ratio of these two numbers.

$$P = \frac{\text{number of ways to choose 4 from 360}}{\text{number of ways to choose 4 from 500}}$$

$$= \frac{688{,}235{,}310}{2{,}573{,}031{,}125}$$

$$\approx 0.267$$

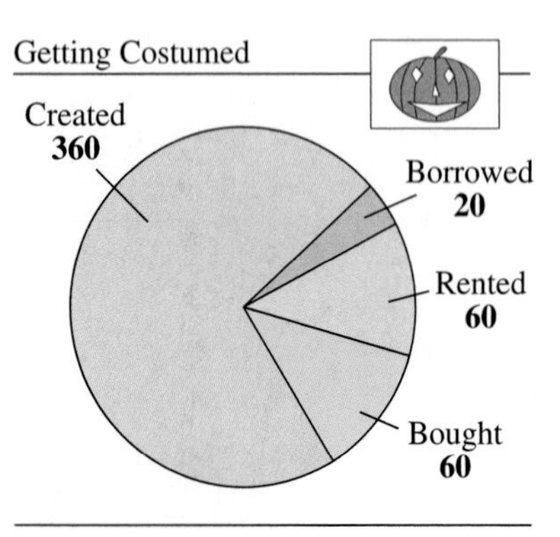

Figure 12.7

Example 7 Forming a Committee

To obtain input from 200 company employees, the management of a company selected a committee of five. Of the 200 employees, 56 were from minority groups. None of the 56, however, was selected to be on the committee. Does this indicate that the management's selection was biased?

Solution

Part of the solution is similar to that of Example 6. If the five committee members were selected at random, the probability that all five would be nonminority is

$$\begin{aligned}
P &= \frac{\text{number of ways to choose 5 from 144 nonminority employees}}{\text{number of ways to choose 5 from 200 employees}} \\
&= \frac{{}_{144}C_5}{{}_{200}C_5} \\
&= \frac{481{,}008{,}528}{2{,}535{,}650{,}040} \\
&\approx 0.19.
\end{aligned}$$

So, if the committee were chosen at random (that is, without bias), the likelihood that it would have no minority members is about 0.19. Although this does not *prove* that there was bias, it does suggest it.

Discussing the Concept — Probability of Guessing Correctly

You are taking a chemistry test and are asked to arrange the first 10 elements in the order in which they appear on the periodic table of elements. Suppose that you have no idea of the correct order and simply guess. Does the following computation represent the probability that you guess correctly? Explain.

Solution

You have 10 choices for the first element, nine choices for the second, eight choices for the third, and so on. The number of different orders is $10! = 3{,}628{,}800$, which means that your probability of guessing correctly is

$$P = \frac{1}{3{,}628{,}800}.$$

12.6 Exercises

Integrated Review *Concepts, Skills, and Problem Solving*

Keep mathematically in shape by doing these exercises *before* the problems of this section.

Properties and Definitions

In Exercises 1–6, complete the property of logarithms.

1. $\log_a 1 =$ 　　**2.** $\log_a a =$

3. $\log_a a^x =$ 　　**4.** $\log_a(uv) =$

5. $\log_a \dfrac{u}{v} =$ 　　**6.** $\log_a u^n =$

Rewriting Expressions

In Exercises 7–10, use the properties of logarithms to expand the expression as a sum, difference, or multiple of logarithms.

7. $\log_2(x^2y)$ 　　**8.** $\log_2\sqrt{x^2+1}$

9. $\ln\dfrac{7}{x-3}$ 　　**10.** $\ln\left(\dfrac{u+2}{u-2}\right)^2$

Graphs and Models

11. Annual sales y of a product x years after it is introduced are approximated by

$$y = \frac{10{,}000}{1 + 4e^{-x/3}}.$$

(a) Use a graphing utility to graph the equation.

(b) Use the graph in part (a) to approximate the time when annual sales are $y = 5000$ units.

(c) Use the graph in part (a) to estimate the maximum level of annual sales.

12. An investment is made in an account that pays 5.5% interest, compounded continuously. What is the effective yield for this investment?

Solving Problems

Sample Space In Exercises 1–4, determine the number of outcomes in the sample space for the experiment.

1. One letter from the alphabet is chosen.

2. A six-sided die is tossed twice and the sum is recorded.

3. Two county supervisors are selected from five supervisors, A, B, C, D, and E, to study a recycling plan.

4. A salesperson makes a presentation about a product in three homes per day. In each home there may be a sale (denote by Y) or there may be no sale (denote by N).

Sample Space In Exercises 5–8, list the outcomes in the sample space for the experiment.

5. A taste tester must taste and rank three brands of yogurt, A, B, and C, according to preference.

6. A coin and a die are tossed.

7. A basketball tournament between two teams consists of three games. For each game, your team may win (denote by W) or lose (denote by L).

8. Two students are randomly selected from four students, A, B, C, and D.

In Exercises 9 and 10, you are given the probability that an event *will* occur. Find the probability that the event *will not* occur.

9. $p = 0.35$ 　　**10.** $p = 0.8$

In Exercises 11 and 12, you are given the probability that an event *will not* occur. Find the probability that the event *will* occur.

11. $p = 0.82$ 　　**12.** $p = 0.13$

Coin Tossing In Exercises 13–16, a coin is tossed three times. Find the probability of the specified event. Use the sample space

$S = \{HHH, HHT, HTH, HTT, THH, THT, TTH, TTT\}$.

13. The event of getting exactly two heads

14. The event of getting a tail on the second toss

15. The event of getting at least one head

16. The event of getting no more than two heads

Playing Cards In Exercises 17–20, a card is drawn from a standard deck of 52 playing cards. Find the probability of drawing the indicated event.

17. A red card

18. A queen

19. A face card

20. A black face card

Tossing a Die In Exercises 21–24, a six-sided die is tossed. Find the probability of the specified event.

21. The number is a 5.

22. The number is a 7.

23. The number is no more than 5.

24. The number is at least 1.

United States Blood Types In Exercise 25 and 26, use the circle graph, which shows the percent of people in the United States in 1996 with each blood type. See Example 2. (Source: America's Blood Centers)

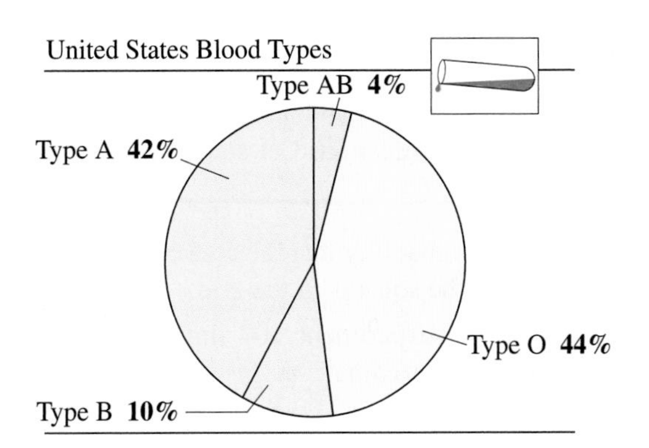

25. A person is selected at random from the United States population. What is the probability that the person *does not* have blood type B?

26. What is the probability that a person selected at random from the United States population *does* have blood type B? How is this probability related to the probability found in Exercise 25?

Charitable Giving In Exercises 27–30, use the circle graph, which shows the percent of households in the United States by dollar amount given for the year 1995. Answer the question for a household selected at random from the population. (Source: U.S. Bureau of the Census)

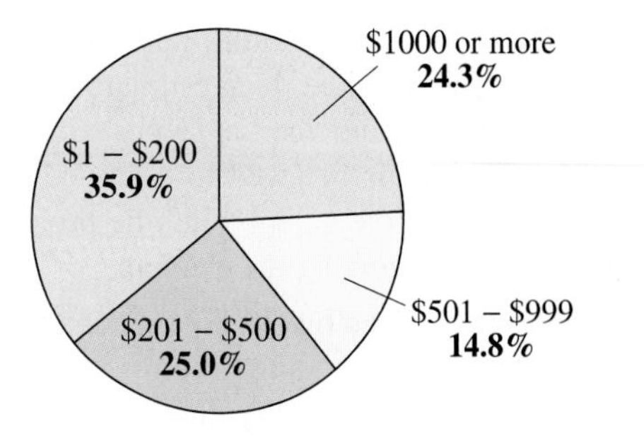

Figure for 27–30

27. What is the probability that the household gave at least $1000?

28. What is the probability that the household gave no more than $200?

29. What is the probability that the household gave no more than $500?

30. What is the probability that the household gave more than $500?

31. *Multiple-Choice Test* A student takes a multiple-choice test in which there are five choices for each question. Find the probability that the first question is answered correctly given the following conditions.

(a) The student has no idea of the answer and guesses at random.

(b) The student can eliminate two of the choices and guesses from the remaining choices.

(c) The student knows the answer.

32. *Multiple-Choice Test* A student takes a multiple-choice test in which there are four choices for each question. Find the probability that the first question is answered correctly given the following conditions.

(a) The student has no idea of the answer and guesses at random.

(b) The student can eliminate two of the choices and guesses from the remaining choices.

(c) The student knows the answer.

33. *Class Election* Three people are running for class president. It is estimated that the probability that Candidate A will win is 0.5 and the probability that Candidate B will win is 0.3.

(a) What is the probability that *either* Candidate A *or* Candidate B will win?

(b) What is the probability that the third candidate will win?

34. *Winning an Election* Jones, Smith, and Thomas are candidates for public office. It is estimated that Jones and Smith have about the same probability of winning, and Thomas is believed to be twice as likely to win as either of the others. Find the probability of each candidate winning the election.

35. *Continuing Education* In a high school graduating class of 325 students, 255 are going to continue their education. What is the probability that a student selected at random from the class will not be furthering his or her education?

36. *Study Questions* An instructor gives the class a list of four study questions for the next exam. Two of the four study questions will be on the exam. Find the probability that a student who only knows the material relating to three of the four questions will be able to correctly answer both questions selected for the exam.

Geometry In Exercises 37–40, the specified probability is the ratio of two areas. See Example 3.

37. *Meteorites* The largest meteorite in the United States was found in the Willamette Valley of Oregon in 1902. Earth contains 57,510,000 square miles of land and 139,440,000 square miles of water. What is the probability that a meteorite that hits the earth will fall onto land? What is the probability that a meteorite that hits the earth will fall into water?

38. *Game* A child uses a spring-loaded device to shoot a marble into the square box shown in the figure. The base of the square is horizontal and the marble has an equal likelihood of coming to rest at any point on the base. In parts (a)–(d), find the probability that the marble comes to rest in the specified region.

(a) The red center (b) The blue ring

(c) The purple border (d) Not in the yellow ring

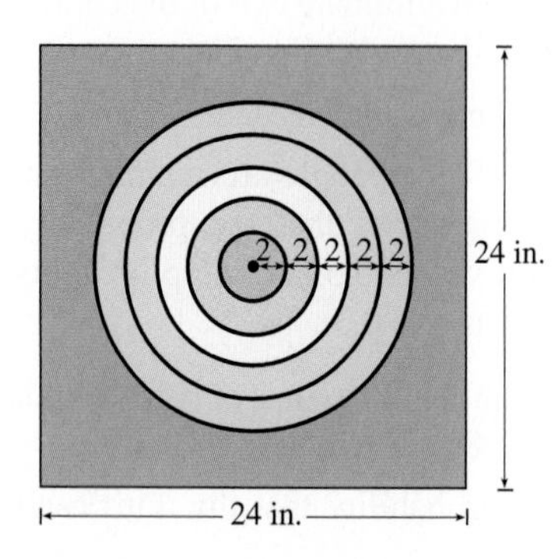

39. *Meeting Time* You and a friend agree to meet at a favorite restaurant between 5:00 and 6:00 P.M. The one who arrives first will wait 15 minutes for the other, after which the first person will leave (see figure). What is the probability that the two of you actually meet, assuming that your arrival times are random within the hour?

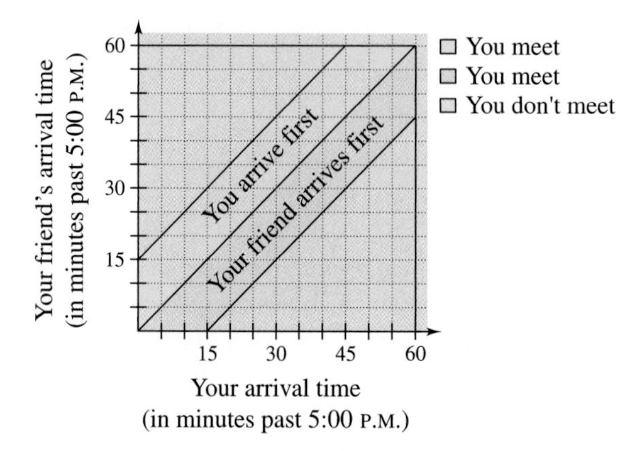

40. *Estimating* π A coin of diameter d is dropped onto a paper that contains a grid of squares d units on a side (see figure).

(a) Find the probability that the coin covers a vertex of one of the squares in the grid.

(b) Repeat the experiment 100 times and use the results to approximate π.

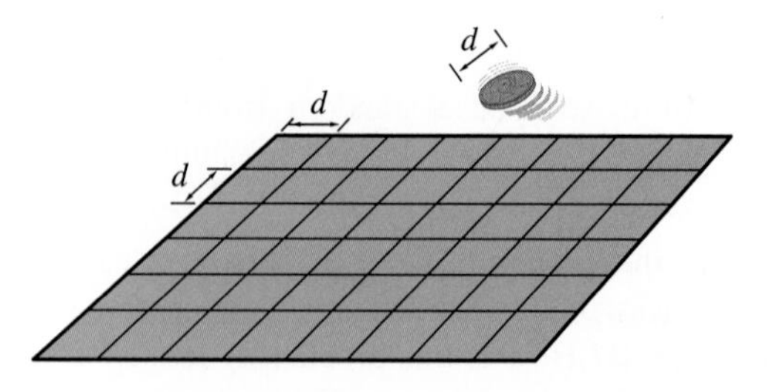

In Exercises 41 and 42, complete the Punnett square and answer the questions. See Example 4.

41. *Girl or Boy?* The genes that determine the sex of humans are denoted by XX (female) and XY (male). (See figure.)

(a) What is the probability that a newborn will be a girl? a boy?

(b) Explain why it is equally likely that a newborn baby will be a boy or a girl.

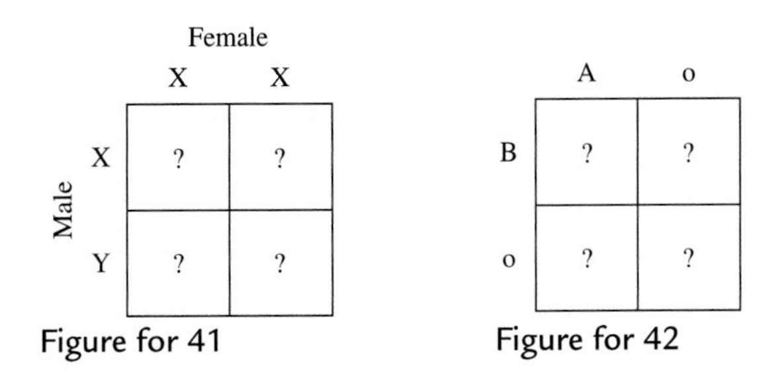

Figure for 41

Figure for 42

42. *Blood Types* There are four basic human blood types: A (AA or Ao), B (BB or Bo), AB (AB), and O (oo) (see figure).

(a) What is the blood type of each parent?

(b) What is the probability that their offspring will have blood type A? B? AB? O?

In Exercises 43–54, some of the sample spaces are large. Therefore, you should use the counting principles discussed in Section 12.5. See Examples 5 and 6.

43. *Game Show* On a game show, you are given five digits to arrange in the proper order for the price of a car. If you arrange them correctly, you win the car. Find the probability of winning if you know the correct position of only one digit and must guess the positions of the other digits.

44. *Game Show* On a game show you are given four digits to arrange in the proper order for the price of a grandfather clock. What is the probability of winning given the following conditions?

(a) You guess the position of each digit.

(b) You know the first digit, but must guess the remaining three.

45. *Lottery* You buy a lottery ticket inscribed with a five-digit number. On the designated day, five digits (from 0 to 9 inclusive) are randomly selected. What is the probability that you have a winning ticket?

46. *Shelving Books* A parent instructs a young child to place a five-volume set of books on a bookshelf. Find the probability that the books are in the correct order if the child places them at random.

47. *Preparing for a Test* An instructor gives her class a list of 10 study problems, from which she will select eight to be answered on an exam. If you know how to solve eight of the problems, what is the probability that you will be able to correctly answer all eight questions on the exam?

48. *Committee Selection* A committee of three students is to be selected from a group of three girls and five boys. Find the probability that the committee is composed entirely of girls.

49. *Defective Units* A shipment of 10 food processors to a certain store contains two defective units. If you purchase two of these food processors as birthday gifts for friends, determine the probability that you get both defective units.

50. *Defective Units* A shipment of 12 compact disc players contains two defective units. A husband and wife buy three of these compact disc players to give to their children as Christmas gifts.

(a) What is the probability that none of the three units is defective?

(b) What is the probability that at least one of the units is defective?

51. *Book Selection* Four books are selected at random from a shelf containing six novels and four autobiographies. Find the probability that the four autobiographies are selected.

52. *Card Selection* Five cards are selected from a standard deck of 52 cards. Find the probability that four aces are selected.

53. *Card Selection* Five cards are selected from a standard deck of 52 cards. Find the probability that all hearts are selected.

54. *Card Selection* Five cards are selected from a standard deck of 52 cards. Find the probability that two aces and three queens are selected.

Explaining Concepts

55. Answer parts (d)–(f) of Motivating the Chapter on page 703.

56. The probability of an event must be a real number in what interval? Is the interval open or closed?

57. The probability of an event is $\frac{3}{4}$. What is the probability that the event *does not* occur? Explain.

58. What is the sum of the probabilities of all the occurrences of outcomes in a sample space? Explain.

59. The weather forecast indicates that the probability of rain is 40%. Explain what this means.

CHAPTER SUMMARY

Key Terms

arithmetic sequence, *pp. 704, 714*
geometric sequence, *pp. 704, 722*
sequence, *p. 704*
term of a sequence, *p. 704*
infinite sequence, *p. 704*
finite sequence, *p. 704*
factorials, *p. 706*
series, *p. 707*
infinite series, *p. 707*
partial sum, *p. 707*
sigma notation, *p. 708*
index of summation, *p. 708*
upper limit of summation, *p. 708*
lower limit of summation, *p. 708*
common difference, *p. 714*
recursion formula, *p. 715*
*n*th partial sum, *p. 716*
common ratio, *p. 722*
infinite geometric series, *p. 724*
binomial coefficients, *p. 733*
Pascal's Triangle, *p. 735*
expanding a binomial, *p. 736*
permutation, *p. 743*
combination, *p. 745*
probability of an event, *p. 751*
sample space, *p. 751*
event, *p. 751*

Key Concepts

12.2 The *n*th term of an arithmetic sequence

The nth term of an arithmetic sequence has the form $a_n = a_1 + (n-1)d$, where d is the common difference of the sequence, and a_1 is the first term.

12.2 The *n*th partial sum of an arithmetic sequence

The nth partial sum of the arithmetic sequence whose nth term is a_n is

$$\sum_{i=1}^{n} a_i = a_1 + a_2 + a_3 + a_4 + \cdots + a_n = n\left(\frac{a_1 + a_n}{2}\right).$$

12.3 The *n*th term of a geometric sequence

The nth term of a geometric sequence has the form $a_n = a_1 r^{n-1}$, where r is the common ratio of consecutive terms of the sequence. So, every geometric sequence can be written in the following form.

$$a_1, a_1r, a_1r^2, a_1r^3, a_1r^4, \ldots, a_1r^{n-1}, \ldots$$

12.3 The *n*th partial sum of a geometric sequence

The nth partial sum of the geometric sequence whose nth term is $a_n = a_1r^{n-1}$ is given by

$$\sum_{i=1}^{n} a_1r^{i-1} = a_1 + a_1r + a_1r^2 + a_1r^3 + \cdots + a_1r^{n-1}$$
$$= a_1\left(\frac{r^n - 1}{r - 1}\right).$$

12.3 Sum of an infinite geometric series

If $a_1, a_1r, a_1r^2, \ldots, a_1r^n, \ldots$ is an infinite geometric sequence, then for $|r| < 1$, the sum of the terms is

$$\sum_{i=1}^{\infty} a_1r^i = \frac{a_1}{1 - r}.$$

12.4 The Binomial Theorem

In the expansion of $(x + y)^n$

$$(x + y)^n = x^n + nx^{n-1}y + \cdots + {}_nC_r x^{n-r}y^r + \cdots + nxy^{n-1} + y^n$$

the coefficient of $x^{n-r}y^r$ is given by

$${}_nC_r = \frac{n!}{(n-r)!r!}.$$

12.5 Fundamental Counting Principle

Let E_1 and E_2 be two events. The first event E_1 can occur in m_1 different ways. After E_1 has occurred, E_2 can occur in m_2 different ways. The number of ways that the two events can occur is $m_1 \cdot m_2$.

12.5 Number of permutations of *n* elements

The number of permutations of n elements is given by $n \cdot (n-1) \cdot \cdots \cdot 4 \cdot 3 \cdot 2 \cdot 1 = n!$.

12.5 Number of combinations of *n* elements taken *r* at a time

The number of combinations of n elements taken r at a time is ${}_nC_r = \dfrac{n!}{(n-r)!r!}$.

12.6 The probability of an event

The probability that an outcome E will occur is

$$P = \frac{\text{number of outcomes in event}}{\text{number of outcomes in sample space}}.$$

REVIEW EXERCISES

Reviewing Skills

12.1 In Exercises 1–4, write the first five terms of the sequence. (Begin with $n = 1$.)

1. $a_n = 3n + 5$

2. $a_n = \frac{1}{2}n - 4$

3. $a_n = \dfrac{1}{2^n} + \dfrac{1}{2}$

4. $a_n = (n + 1)!$

In Exercises 5–8, find the nth term of the sequence.

5. $1, 3, 5, 7, 9, \ldots$

6. $3, -6, 9, -12, 15, \ldots$

7. $\frac{1}{4}, \frac{2}{9}, \frac{3}{16}, \frac{4}{25}, \frac{5}{36}, \ldots$

8. $\frac{0}{2}, \frac{1}{3}, \frac{2}{4}, \frac{3}{5}, \frac{4}{6}, \ldots$

In Exercises 9–14, match the sequence with its graph. [The graphs are labeled (a), (b), (c), (d), (e), and (f).]

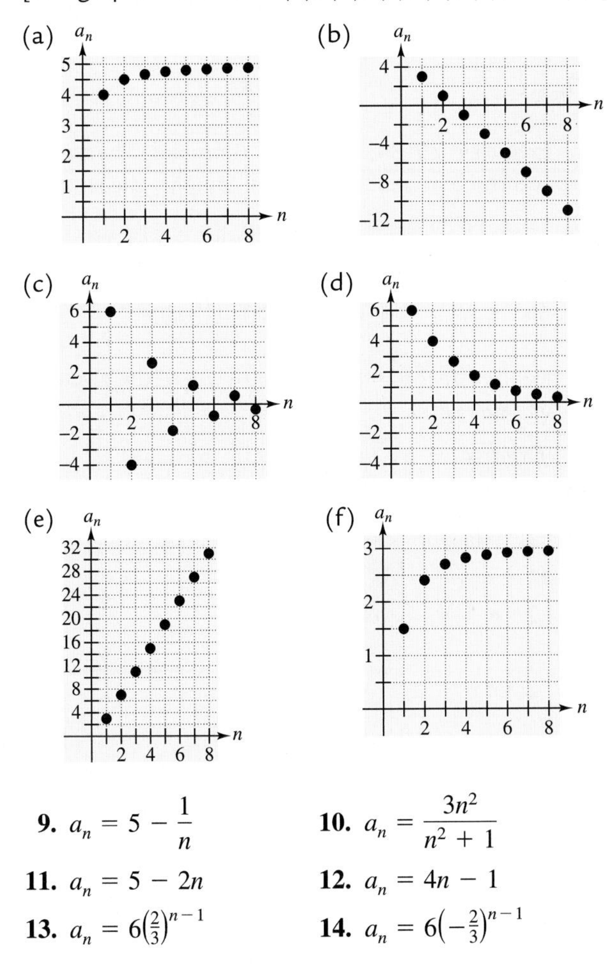

9. $a_n = 5 - \dfrac{1}{n}$

10. $a_n = \dfrac{3n^2}{n^2 + 1}$

11. $a_n = 5 - 2n$

12. $a_n = 4n - 1$

13. $a_n = 6\left(\frac{2}{3}\right)^{n-1}$

14. $a_n = 6\left(-\frac{2}{3}\right)^{n-1}$

In Exercises 15–18, evaluate the sum.

15. $\displaystyle\sum_{k=1}^{4} 7$

16. $\displaystyle\sum_{k=1}^{4} \frac{(-1)^k}{k}$

17. $\displaystyle\sum_{n=1}^{4} \left(\frac{1}{n} - \frac{1}{n+1}\right)$

18. $\displaystyle\sum_{n=1}^{4} \left(\frac{1}{n} - \frac{1}{n+2}\right)$

In Exercises 19–22, use sigma notation to write the sum.

19. $[5(1) - 3] + [5(2) - 3] + [5(3) - 3] + [5(4) - 3]$

20. $[9 - 10(1)] + [9 - 10(2)] + [9 - 10(3)] + [9 - 10(4)]$

21. $\dfrac{1}{3(1)} + \dfrac{1}{3(2)} + \dfrac{1}{3(3)} + \dfrac{1}{3(4)} + \dfrac{1}{3(5)} + \dfrac{1}{3(6)}$

22. $\left(-\frac{1}{3}\right)^0 + \left(-\frac{1}{3}\right)^1 + \left(-\frac{1}{3}\right)^2 + \left(-\frac{1}{3}\right)^3 + \left(-\frac{1}{3}\right)^4$

12.2 In Exercises 23 and 24, find the common difference of the arithmetic sequence.

23. $30, 27.5, 25, 22.5, 20, \ldots$

24. $9, 12, 15, 18, 21, \ldots$

In Exercises 25–32, write the first five terms of the arithmetic sequence. (Begin with $n = 1$.)

25. $a_n = 132 - 5n$

26. $a_n = 2n + 3$

27. $a_n = \frac{3}{4}n + \frac{1}{2}$

28. $a_n = -\frac{3}{5}n + 1$

29. $a_1 = 5$, $a_{k+1} = a_k + 3$

30. $a_1 = 12$, $a_{k+1} = a_k + 1.5$

31. $a_1 = 80$, $a_{k+1} = a_k - \frac{5}{2}$

32. $a_1 = 25$, $a_{k+1} = a_k - 6$

In Exercises 33–36, find a formula for the nth term of the arithmetic sequence.

33. $a_1 = 10, \quad d = 4$

34. $a_1 = 32, \quad d = -2$

35. $a_1 = 1000, \quad a_2 = 950$

36. $a_1 = 12, \quad a_2 = 20$

In Exercises 37–40, find the nth partial sum of the arithmetic sequence.

37. $\sum_{k=1}^{12} (7k - 5)$

38. $\sum_{k=1}^{10} (100 - 10k)$

39. $\sum_{j=1}^{100} \frac{j}{4}$

40. $\sum_{j=1}^{50} \frac{3j}{2}$

In Exercises 41 and 42, use a graphing utility to evaluate the sum.

41. $\sum_{i=1}^{60} (1.25i + 4)$

42. $\sum_{i=1}^{100} (5000 - 3.5i)$

12.3 In Exercises 43 and 44, find the common ratio of the geometric sequence.

43. $8, 12, 18, 27, \frac{81}{2}, \ldots$

44. $27, -18, 12, -8, \frac{16}{3}, \ldots$

In Exercises 45–50, write the first five terms of the geometric sequence.

45. $a_1 = 10, \quad r = 3$

46. $a_1 = 2, \quad r = -5$

47. $a_1 = 100, \quad r = -\frac{1}{2}$

48. $a_1 = 12, \quad r = \frac{1}{6}$

49. $a_1 = 3$
$a_{k+1} = 2a_k$

50. $a_1 = 36$
$a_{k+1} = \frac{1}{2}a_k$

In Exercises 51–56, find a formula for the nth term of the geometric sequence.

51. $a_1 = 1, \quad r = -\frac{2}{3}$

52. $a_1 = 100, \quad r = 1.07$

53. $a_1 = 24, \quad a_2 = 48$

54. $a_1 = 16, \quad a_2 = -4$

55. $a_1 = 12, \quad a_4 = -\frac{3}{2}$

56. $a_2 = 1, \quad a_3 = \frac{1}{3}$

In Exercises 57–64, find the nth partial sum of the geometric sequence.

57. $\sum_{n=1}^{12} 2^n$

58. $\sum_{n=1}^{12} (-2)^n$

59. $\sum_{k=1}^{8} 5\left(-\frac{3}{4}\right)^k$

60. $\sum_{k=1}^{10} 4\left(\frac{3}{2}\right)^k$

61. $\sum_{i=1}^{8} (1.25)^{i-1}$

62. $\sum_{i=1}^{8} (-1.25)^{i-1}$

63. $\sum_{n=1}^{120} 500(1.01)^n$

64. $\sum_{n=1}^{40} 1000(1.1)^n$

In Exercises 65–68, find the sum of the infinite geometric series.

65. $\sum_{i=1}^{\infty} \left(\frac{7}{8}\right)^{i-1}$

66. $\sum_{i=1}^{\infty} \left(\frac{1}{3}\right)^{i-1}$

67. $\sum_{k=1}^{\infty} 4\left(\frac{2}{3}\right)^{k-1}$

68. $\sum_{k=1}^{\infty} 1.3\left(\frac{1}{10}\right)^{k-1}$

In Exercises 69 and 70, use a graphing utility to evaluate the sum.

69. $\sum_{k=1}^{50} 50(1.2)^{k-1}$

70. $\sum_{j=1}^{60} 25(0.9)^{j-1}$

12.4 In Exercises 71–74, find the binomial coefficient.

71. ${}_8C_3$

72. ${}_{12}C_2$

73. ${}_{12}C_0$

74. ${}_{100}C_1$

In Exercises 75–78, use a graphing utility to evaluate ${}_nC_r$.

75. ${}_{40}C_4$

76. ${}_{15}C_9$

77. ${}_{25}C_6$

78. ${}_{32}C_2$

In Exercises 79–84, use the Binomial Theorem to expand the binomial expression. Simplify the result.

79. $(x + 1)^{10}$

80. $(y - 2)^6$

81. $(3x - 2y)^4$

82. $(2u + 5v)^4$

83. $(u^2 + v^3)^9$

84. $(x^4 - y^5)^8$

In Exercises 85–88, find the coefficient of the given term of the expression.

	Expression	*Term*
85.	$(x - 3)^{10}$	x^5
86.	$(x + 4)^9$	x^6
87.	$(x + 2y)^7$	x^4y^3
88.	$(2x - 3y)^5$	x^2y^3

Solving Problems

89. Find the sum of the first 50 positive integers that are multiples of 4.

90. Find the sum of the integers from 225 to 300.

91. *Auditorium Seating* Each row in a small auditorium has three more seats than the preceding row. Find the seating capacity of the auditorium if the front row seats 22 people and there are 12 rows of seats.

92. *Depreciation* A company pays $120,000 for a machine. During the next 5 years, the machine depreciates at the rate of 30% per year. (That is, at the end of each year, the depreciated value is 70% of what it was at the beginning of the year.)

(a) Find a formula for the nth term of the geometric sequence that gives the value of the machine n full years after it was purchased.

(b) Find the depreciated value of the machine at the end of 5 full years.

93. *Population Increase* A city of 85,000 people is growing at the rate of 1.2% per year. (That is, at the end of each year, the population is 1.012 times what it was at the beginning of the year.)

(a) Find a formula for the nth term of the geometric sequence that gives the population n years from now.

(b) Estimate the population 50 years from now.

94. *Salary Increase* You accept a job that pays a salary of $32,000 the first year. During the next 39 years, you receive a 5.5% raise each year. What would your total salary be over the 40-year period?

95. *Morse Code* In Morse code, all characters are transmitted using a sequence of *dots* and *dashes*. How many different characters can be formed by using a sequence of three dots and dashes? (These can be repeated. For example, dash-dot-dot represents the letter *d*.)

96. *Random Selection* Ten marbles numbered 0 through 9 are placed in a bag. List the ways in which two marbles having a sum of 8 can be drawn without replacement.

97. *Committee Selection* Determine the number of ways a committee of five people can be formed from a group of 15 people.

98. *Program Listing* There are seven participants in a piano recital. In how many orders can their names be listed in the program?

99. *Rolling a Die* Find the probability of obtaining a number greater than 4 when a six-sided die is rolled.

100. *Coin Tossing* Find the probability of obtaining at least one head when a coin is tossed four times.

101. *Book Selection* A child who does not know how to read carries a four-volume set of books to a bookshelf. Find the probability that the child will put the books on the shelf in the correct order.

102. *Rolling a Die* Are the chances of rolling a 3 with one six-sided die the same as the chances of rolling a total of 6 with two six-sided dice? If not, which has the greater probability of occurring?

103. *Hospital Inspection* As part of a monthly inspection at a hospital, the inspection team randomly selects reports from eight of the 84 nurses who are on duty. What is the probability that none of the reports selected will be from the ten most experienced nurses?

104. *Target Shooting* An archer shoots an arrow at the target shown in the figure. Suppose that the arrow is equally likely to hit any point on the target. What is the probability that the arrow hits the bull's-eye? What is the probability that the arrow hits the blue ring?

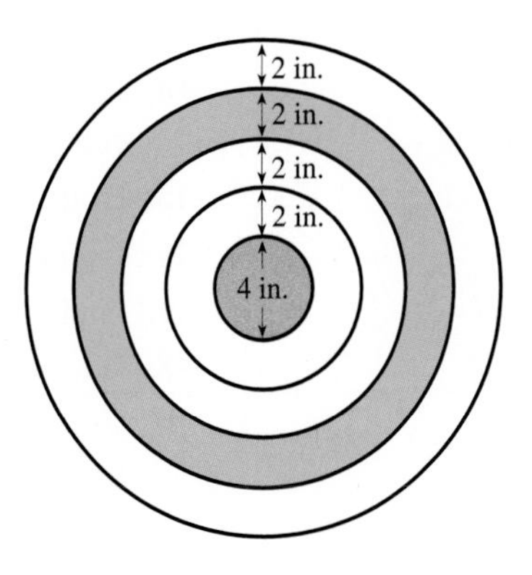

Chapter Test

Take this test as you would take a test in class. After you are done, check your work against the answers given in the back of the book.

1. Write the first five terms of the sequence $a_n = \left(-\frac{2}{3}\right)^{n-1}$. (Begin with $n = 1$.)
2. Evaluate: $\sum_{j=0}^{4} (3j + 1)$
3. Evaluate: $\sum_{n=1}^{5} (3 - 4n)$
4. Use sigma notation to write $\frac{2}{3(1) + 1} + \frac{2}{3(2) + 1} + \cdots + \frac{2}{3(12) + 1}$.
5. Write the first five terms of the arithmetic sequence whose first term is $a_1 = 12$ and whose common difference is $d = 4$.
6. Find a formula for the nth term of the arithmetic sequence whose first term is $a_1 = 5000$ and whose common difference is $d = -100$.
7. Find the sum of the first 50 positive integers that are multiples of 3.
8. Find the common ratio of the geometric sequence: $2, -3, \frac{9}{2}, -\frac{27}{4}, \ldots$.
9. Find a formula for the nth term of the geometric sequence whose first term is $a_1 = 4$ and whose common ratio is $r = \frac{1}{2}$.
10. Evaluate: $\sum_{n=1}^{8} 2(2^n)$
11. Evaluate: $\sum_{n=1}^{10} 3\left(\frac{1}{2}\right)^n$
12. Evaluate: $\sum_{i=1}^{\infty} \left(\frac{1}{2}\right)^i$
13. Evaluate: $\sum_{i=1}^{\infty} 4\left(\frac{2}{3}\right)^{i-1}$
14. Fifty dollars is deposited each month in an increasing annuity that pays 8%, compounded monthly. What is the balance after 25 years?
15. Evaluate: ${}_{20}C_3$
16. Explain how to use Pascal's Triangle to expand $(x - 2)^5$.
17. Find the coefficient of the term x^3y^5 in the expansion of $(x + y)^8$.
18. How many license plates can consist of one letter followed by three digits?
19. Four students are randomly selected from a class of 25 to answer questions from a reading assignment. In how many ways can the four be selected?
20. The weather report indicates that the probability of snow tomorrow is 0.75. What is the probability that it will not snow?
21. A card is drawn from a standard deck of playing cards. Find the probability that it is a red face card.
22. Suppose two spark plugs require replacement in a four-cylinder engine. If the mechanic randomly removes two plugs, find the probability that they are the two defective plugs.

Cumulative Test: Chapters 10–12

Take this test as you would take a test in class. After you are done, check your work against the answers given in the back of the book.

In Exercises 1–4, solve the equation.

1. $(x - 5)^2 + 50 = 0$

2. $3x^2 + 6x + 2 = 0$

3. $x + \dfrac{4}{x} = 4$

4. $\sqrt{x + 10} = x - 2$

5. Find a quadratic function for the parabola that has vertex $(2, 3)$ and passes through the point $(1, 1)$.

In Exercises 6 and 7, solve the inequality.

6. $3x(2x - 7) \geq 0$

7. $\dfrac{1}{x + 1} + \dfrac{2x}{x - 1} < 1$

In Exercises 8 and 9, graph the function.

8. $y = 2^{-x/2}$

9. $y = 4e^{-x^2/4}$

In Exercises 10 and 11, solve for x.

10. $2^{x-3} = 32$

11. $8^{3x-1} = 64^{x+2}$

In Exercises 12 and 13, evaluate the logarithm.

12. $\ln 1$

13. $\log_3 \frac{1}{81}$

14. Describe the relationship between the graphs of $f(x) = e^x$ and $g(x) = \ln x$.

15. Use the properties of logarithms to condense $3(\log_2 x + \log_2 y) - \log_2 z$.

In Exercises 16–19, solve the equation.

16. $4 \log_2 x = 10$

17. $3(1 + e^{2x}) = 20$

18. $e^{3x+1} - 4 = 7$

19. $2 \log_3(x - 1) = \log_3(2x + 6)$

In Exercises 20 and 21, find the most apparent nth term of the sequence.

20. Arithmetic: $8, 6, 4, 2, 0, \ldots$

21. Geometric: $-1, \frac{2}{3}, -\frac{4}{9}, \frac{8}{27}, -\frac{16}{81}, \ldots$

22. Find the sum of the first 20 terms of the arithmetic sequence $8, 12, 16, 20, \ldots$.

23. Find the sum: $\displaystyle\sum_{i=0}^{\infty} 3\left(\tfrac{1}{2}\right)^i$

24. Use the Binomial Theorem to expand and simplify $(z - 3)^4$.

25. After t years, the value of a car that cost \$25,000 new is given by $V(t) = 25{,}000(0.8)^t$. Sketch a graph of the function and determine when the value of the car is \$15,000.

26. Three thousand dollars is deposited in an interest-paying account that compounds quarterly. What is the interest rate if the balance in the account is \$3750 after 3 years?

27. The salary for the first year of a job is \$32,000. During the next 9 years there is a 5% raise each year. Determine the total compensation over the 10-year period.

28. A personnel manager has 10 applicants to fill three different positions. In how many ways can this be done, assuming that all the applicants are qualified for any of the three positions?

29. On a game show, the digits 3, 4, and 5 must be arranged in the proper order to form the price of an appliance. If they are arranged correctly, the contestant wins the appliance. What is the probability of winning if the contestant knows that the price is at least \$400?

Appendix A
Conic Sections

Introduction to Conic Sections ■ Circles ■ Ellipses ■ Hyperbolas ■ Application

Introduction to Conic Sections

In Section 10.4, you learned that the graph of a second-degree equation of the form

$$y = ax^2 + bx + c$$

is a parabola. A parabola is one of four types of **conics** or **conic sections.** The other three types are circles, ellipses, and hyperbolas. All four types have equations that are of second degree. As indicated in Figure A.1, the name "conic" relates to the fact that each of these figures can be obtained by intersecting a plane with a double-napped cone.

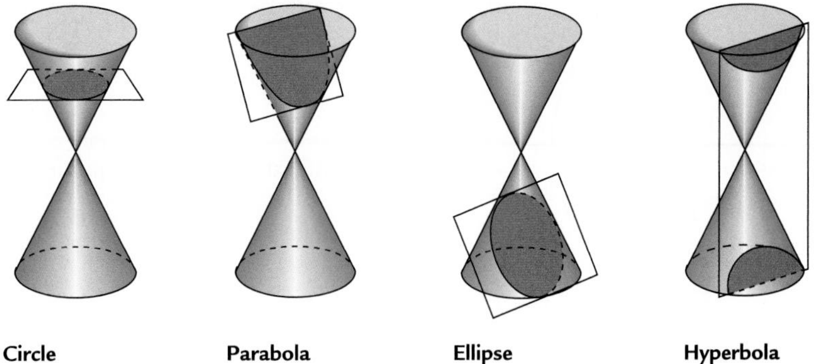

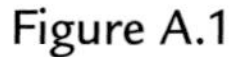
Figure A.1

Conic sections were discovered during the classical Greek period, which lasted from 600 to 300 B.C. By the beginning of the Alexandrian period, enough was known of conics for Apollonius (262–190 B.C.) to produce an eight-volume work on the subject.

This early Greek study was largely concerned with the geometrical properties of conics. It was not until the early seventeenth century that the broad applicability of conics became apparent. For instance, reflective surfaces in satellite dishes, flashlights, and telescopes often are of parabolic shape. The orbits of planets are elliptical, and the orbits of comets are usually elliptical or hyperbolic. Ellipses and parabolas are also used in building archways and bridges.

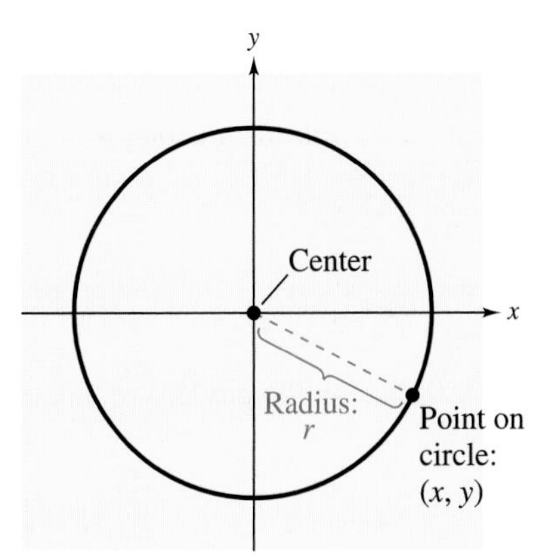

Figure A.2

Circles

A **circle** in the rectangular coordinate plane consists of all points (x, y) that are a given positive distance r from a fixed point, called the **center** of the circle. The positive distance r is the **radius** of the circle. If the center of the circle is the origin, as shown in Figure A.2, the relationship between the coordinates of any point (x, y) on the circle and the radius r is given by

$$\text{Radius} = r = \sqrt{(x-0)^2 + (y-0)^2} \qquad \text{Distance Formula, center at } (0, 0)$$
$$= \sqrt{x^2 + y^2}.$$

If the center of the circle is translated to the point (h, k), the relationship between the coordinates of any point (x, y) on the circle and the radius r is given by

$$r = \sqrt{(x-h)^2 + (y-k)^2}. \qquad \text{Distance Formula, center at } (h, k)$$

By squaring both sides of this equation, you obtain the **standard form of the equation of a circle.**

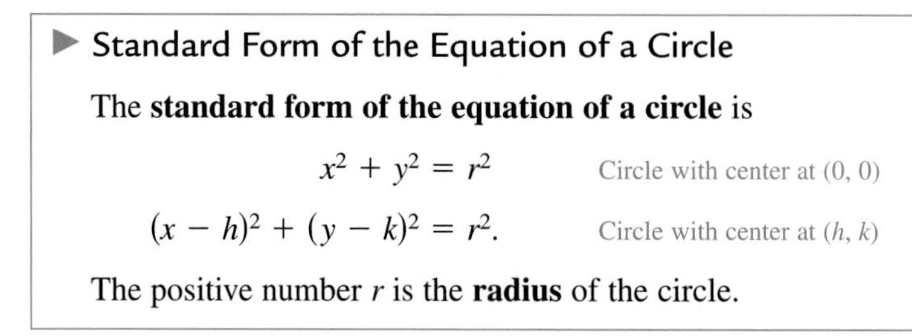

▶ Standard Form of the Equation of a Circle

The **standard form of the equation of a circle** is

$$x^2 + y^2 = r^2 \qquad \text{Circle with center at } (0, 0)$$

$$(x-h)^2 + (y-k)^2 = r^2. \qquad \text{Circle with center at } (h, k)$$

The positive number r is the **radius** of the circle.

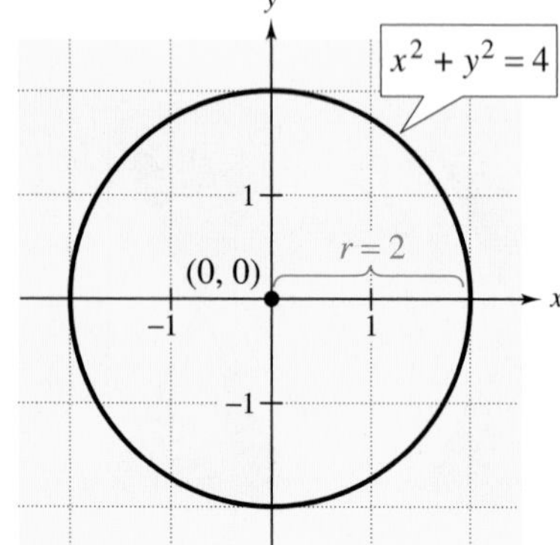

Figure A.3

Example 1 Finding an Equation of a Circle

Find an equation of the circle whose center is at (0, 0) and whose radius is 2.

Solution

Use the standard form of the equation of a circle with center at the origin.

$$x^2 + y^2 = r^2 \qquad \text{Standard form with center at } (0, 0)$$

$$x^2 + y^2 = 2^2 \qquad \text{Substitute 2 for } r.$$

$$x^2 + y^2 = 4 \qquad \text{Equation of circle}$$

The circle given by this equation is shown in Figure A.3.

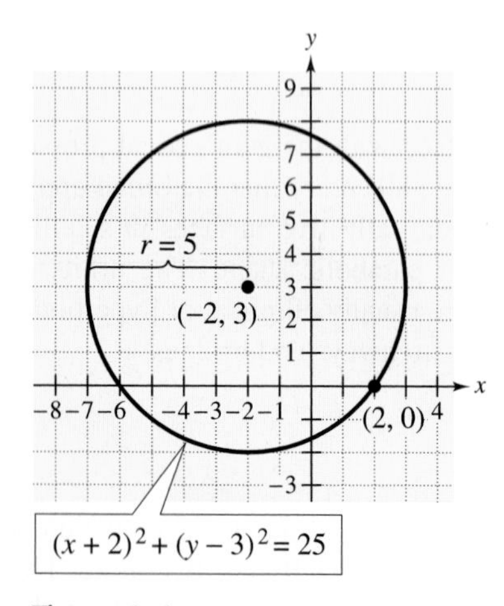

Figure A.4

Example 2 Finding an Equation of a Circle

Find an equation of the circle whose center is at $(-2, 3)$ and that contains the point $(2, 0)$.

Solution

Begin by using the standard form of the equation of a circle with center at (h, k).

$$(x-h)^2 + (y-k)^2 = r^2 \qquad \text{Standard form with center at } (h, k)$$

$$[x - (-2)]^2 + (y-3)^2 = r^2 \qquad \text{Substitute } -2 \text{ for } h \text{ and 3 for } k.$$

$$(x+2)^2 + (y-3)^2 = r^2 \qquad \text{Simplify.}$$

To find r^2, use the information that the point $(2, 0)$ lies on the circle.

$$(2+2)^2 + (0-3)^2 = r^2 \qquad \text{Substitute 2 for } x \text{ and 0 for } y.$$

$$(4)^2 + (-3)^2 = r^2 \qquad \text{Simplify.}$$

$$16 + 9 = r^2 \qquad \text{Simplify.}$$

$$25 = r^2 \qquad \text{Simplify.}$$

So, an equation of this circle is $(x + 2)^2 + (y - 3)^2 = 25$. The graph of the circle is shown in Figure A.4 on the previous page.

To sketch the circle for a given equation, write the equation in standard form. From the standard form, you can identify the center and radius.

Example 3 Finding the Center and Radius of a Circle

Identify the center and radius of the circle given by the equation, and sketch the circle.

$$x^2 + y^2 + 2x - 6y + 1 = 0$$

Solution

$$x^2 + y^2 + 2x - 6y + 1 = 0 \qquad \text{Original equation}$$

$$(x^2 + 2x) + (y^2 - 6y) = -1 \qquad \text{Group terms.}$$

$$(x^2 + 2x + 1) + (y^2 - 6y + 9) = -1 + 1 + 9 \qquad \text{Complete each square.}$$

$$(x + 1)^2 + (y - 3)^2 = 9 \qquad \text{Standard form}$$

From this standard form, you can see that the graph of the equation is a circle whose center is at $(-1, 3)$ and whose radius is 3, as shown in Figure A.5.

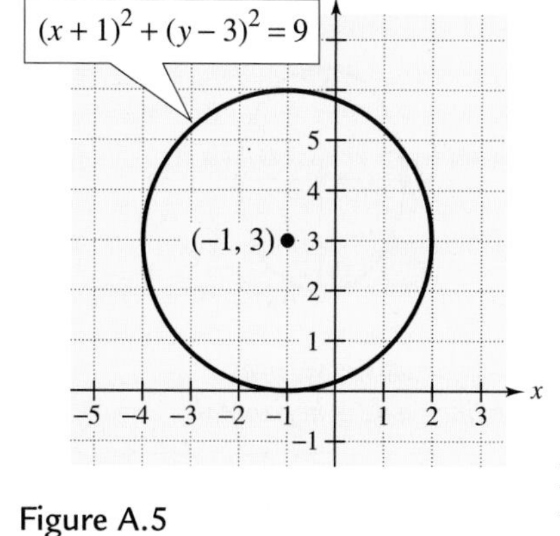

Figure A.5

Ellipses

An **ellipse** is the set of all points (x, y) such that the sum of the distances between (x, y) and two distinct fixed points is a constant. As shown in Figure A.6(a), each of the two fixed points is a **focus** of the ellipse. (The plural of focus is *foci.*) In this text, we restrict the study of ellipses to those whose centers are at the origin.

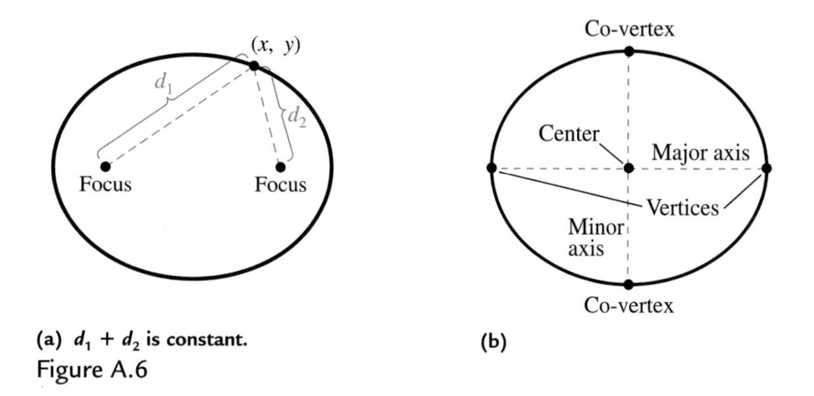

(a) $d_1 + d_2$ is constant. **(b)**

Figure A.6

The line through the foci intersects the ellipse at the **vertices,** as shown in Figure A.6(b). The line segment joining the vertices is the **major axis,** and its midpoint is the **center** of the ellipse. The line segment perpendicular to the major axis at the center is the **minor axis** of the ellipse, and the points at which the minor axis intersects the ellipse are **co-vertices.**

Study Tip

If the equation of an ellipse is of the form $\frac{x^2}{a^2} + \frac{y^2}{b^2} = 1$, its major axis is horizontal. Because a is greater than b and its square is the denominator of the x^2-term, you can conclude that the major axis lies along the x-axis. Similarly, if the equation of an ellipse is of the form $\frac{x^2}{b^2} + \frac{y^2}{a^2} = 1$, its major axis is vertical.

▶ Standard Form of the Equation of an Ellipse

The **standard form of the equation of an ellipse** with center at the origin and major and minor axes of lengths $2a$ and $2b$, respectively, is

$$\frac{x^2}{a^2} + \frac{y^2}{b^2} = 1 \quad \text{or} \quad \frac{x^2}{b^2} + \frac{y^2}{a^2} = 1, \qquad 0 < b < a.$$

The vertices lie on the major axis, a units from the center, and the co-vertices lie on the minor axis, b units from the center.

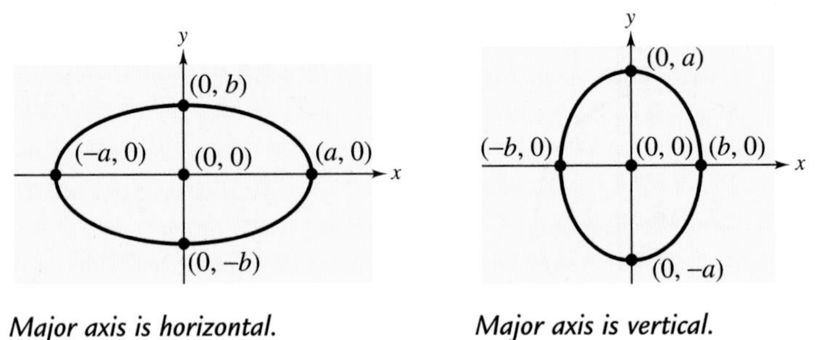

Major axis is horizontal.
Minor axis is vertical.

Major axis is vertical.
Minor axis is horizontal.

Example 4 Finding an Equation of an Ellipse

Find an equation of the ellipse whose vertices are $(-3, 0)$ and $(3, 0)$ and whose co-vertices are $(0, -2)$ and $(0, 2)$.

Solution

Begin by plotting the vertices and co-vertices, as shown in Figure A.7. The center of the ellipse is (0, 0), because it is the point that lies halfway between the vertices (and halfway between the co-vertices). Thus, the equation of the ellipse has the form

$$\frac{x^2}{a^2} + \frac{y^2}{b^2} = 1.$$

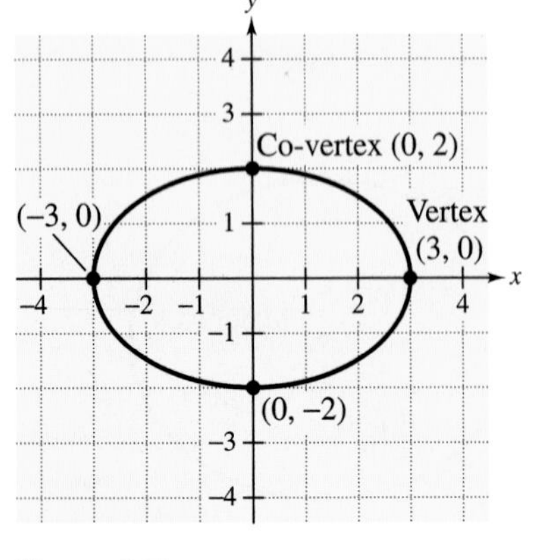

Figure A.7

For this ellipse, the major axis is horizontal. Thus, a is the distance between the center and either vertex, which implies that $a = 3$. Similarly, b is the distance between the center and either co-vertex, which implies that $b = 2$. Thus, the standard equation of the ellipse is

$$\frac{x^2}{3^2} + \frac{y^2}{2^2} = 1.$$

As an aid in sketching an ellipse, first write its equation in standard form.

Example 5 Sketching an Ellipse

Sketch the ellipse $4x^2 + y^2 = 36$, and identify the vertices and co-vertices.

Solution

$$4x^2 + y^2 = 36 \qquad \text{Original equation}$$

$$\frac{4x^2}{36} + \frac{y^2}{36} = \frac{36}{36} \qquad \text{Divide both sides by 36.}$$

$$\frac{x^2}{9} + \frac{y^2}{36} = 1 \qquad \text{Simplify.}$$

$$\frac{x^2}{3^2} + \frac{y^2}{6^2} = 1 \qquad \text{Standard form}$$

Because the denominator of the y^2-term is larger than the denominator of the x^2-term, the major axis is vertical. Because $a = 6$, the vertices are $(0, -6)$ and $(0, 6)$. Finally, because $b = 3$, the co-vertices are $(-3, 0)$ and $(3, 0)$. (See Figure A.8.)

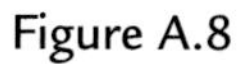
Figure A.8

Hyperbolas

A **hyperbola** on the rectangular coordinate system consists of all points (x, y) such that the *difference* of the distances between (x, y) and two fixed points is a constant, as shown in Figure A.9. The two fixed points are called the **foci** of the hyperbola. As with ellipses, we will consider only equations of hyperbolas whose foci lie on the x-axis or on the y-axis (whose centers are at the origin). The line on which the foci lie is called the **transverse axis** of the hyperbola.

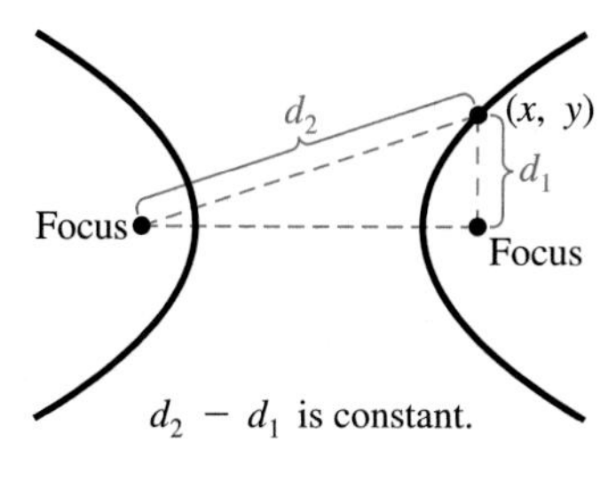

Figure A.9

Standard Form of the Equation of a Hyperbola

The **standard form of the equation of a hyperbola** whose center is at the origin is given by

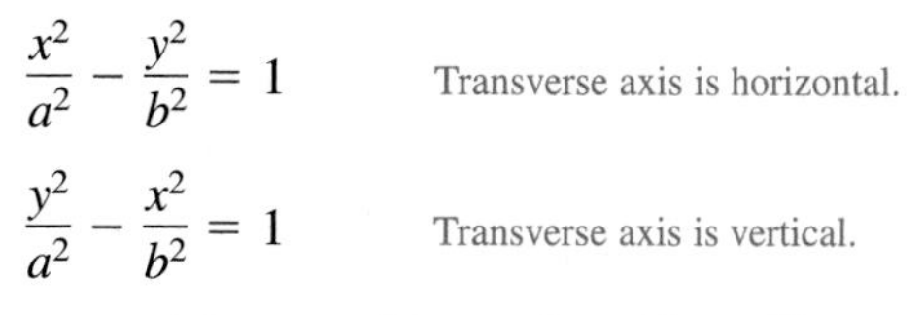

$$\frac{x^2}{a^2} - \frac{y^2}{b^2} = 1 \qquad \text{Transverse axis is horizontal.}$$

$$\frac{y^2}{a^2} - \frac{x^2}{b^2} = 1 \qquad \text{Transverse axis is vertical.}$$

where a and b are positive real numbers. The **vertices** of the hyperbola lie on the transverse axis, a units from the center.

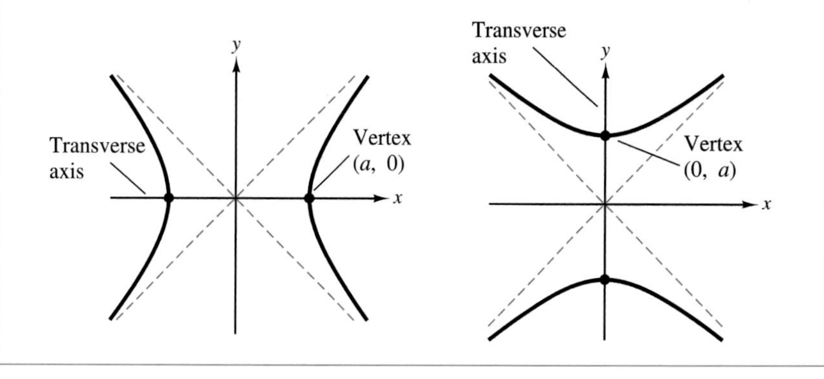

A hyperbola has two disconnected parts, each of which is a **branch** of the hyperbola. The two branches approach a pair of intersecting straight lines called **asymptotes** of the hyperbola. The two asymptotes intersect at the center of the hyperbola.

To sketch a hyperbola, form a **central rectangle** whose center is the origin and whose width and height are $2a$ and $2b$. Note in Figure A.10 that the asymptotes pass through the corners of the central rectangle and the vertices of the hyperbola lie at the centers of opposite sides of the central rectangle.

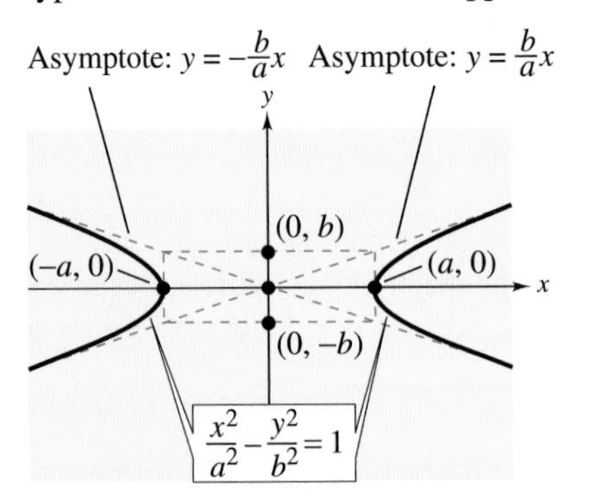

Transverse axis is horizontal.

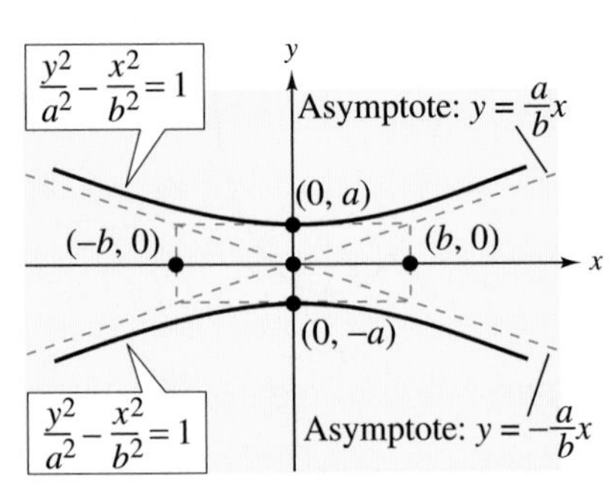

Transverse axis is vertical.

Figure A.10

Example 6 Sketching a Hyperbola

Sketch the hyperbola given by $\frac{x^2}{36} - \frac{y^2}{16} = 1.$

Solution

From the standard form of the equation $\frac{x^2}{6^2} - \frac{y^2}{4^2} = 1$ you can see that the center of the hyperbola is the origin and the transverse axis is horizontal. Therefore, the vertices lie 6 units to the left and right of the center at the points $(-6, 0)$ and $(6, 0)$. Because $a = 6$ and $b = 4$, you can sketch the hyperbola by first drawing a central rectangle whose width is $2a = 12$ and whose height is $2b = 8$, as shown in Figure A.11(a). Next, draw the asymptotes of the hyperbola through the corners of the central rectangle and plot the vertices. Finally, draw the hyperbola, as shown in Figure A.11(b).

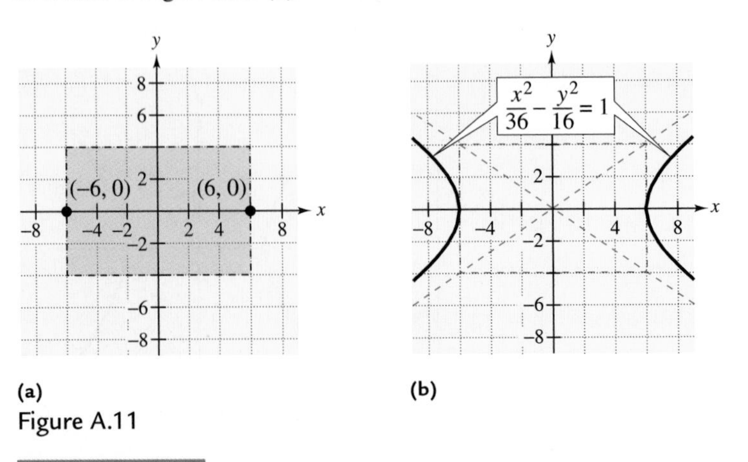

(a) (b)

Figure A.11

Finding an equation of a hyperbola is a little more difficult than finding equations of the other three types of conics. However, if you know the vertices and the asymptotes, you can find the values of a and b, which enable you to write the equation. Notice in Example 7, that the key to this procedure is knowing that the central rectangle has a width of $2b$ and a height of $2a$.

Example 7 Finding the Equation of a Hyperbola

Find an equation of the hyperbola with a vertical transverse axis whose vertices are $(0, 3)$ and $(0, -3)$ and whose asymptotes are given by $y = \frac{3}{5}x$ and $y = -\frac{3}{5}x$.

Solution

To begin, sketch the lines that represent the asymptotes, as shown in Figure A.12(a). Note that these two lines intersect at the origin, which implies that the center of the hyperbola is $(0, 0)$. Next, plot the two vertices at the points $(0, 3)$ and $(0, -3)$. Because you know where the vertices are located, you can sketch the central rectangle of the hyperbola, as shown in Figure A.12(a). Note that the corners of the central rectangle occur at the points

$$(-5, 3), \quad (5, 3), \quad (-5, -3), \quad \text{and} \quad (5, -3).$$

Because the width of the central rectangle is $2b = 10$, it follows that $b = 5$. Similarly, because the height of the central rectangle is $2a = 6$, it follows that $a = 3$. Now that you know the values of a and b, you can conclude that the standard form of the equation of the hyperbola is

$$\frac{y^2}{3^2} - \frac{x^2}{5^2} = 1.$$

The graph is shown in Figure A.12(b).

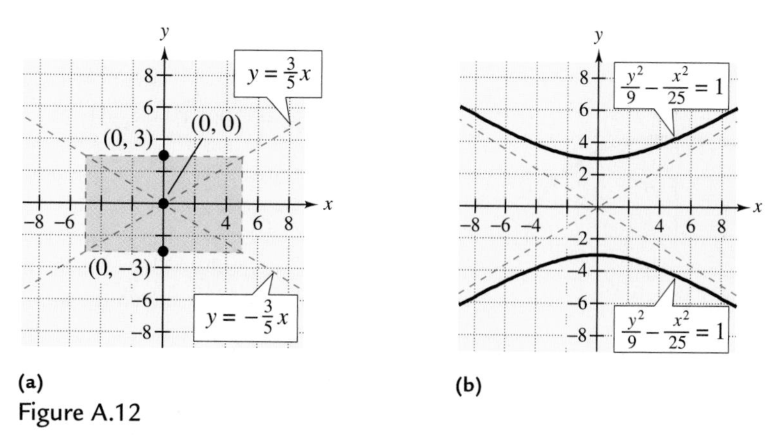

(a) **(b)**

Figure A.12

Application

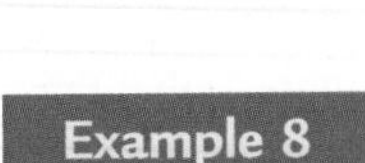

Example 8 An Application Involving an Ellipse

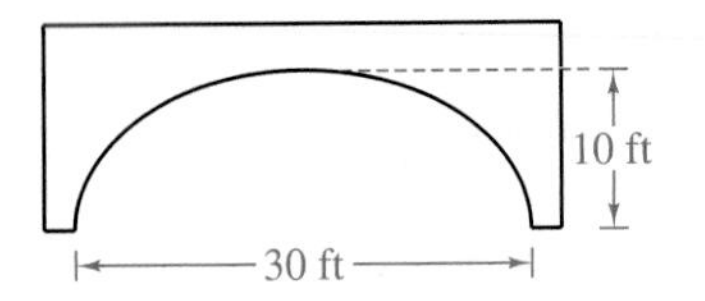

Figure A.13

You are responsible for designing a semielliptical archway, as shown in Figure A.13. The height of the archway is 10 feet and its width is 30 feet. Find an equation of the ellipse and use the equation to sketch an accurate diagram of the archway.

Solution

To make the equation simple, place the origin at the center of the ellipse. This means that the standard form of the equation is

$$\frac{x^2}{a^2} + \frac{y^2}{b^2} = 1.$$

Because the major axis is horizontal, it follows that $a = 15$ and $b = 10$, which implies that the equation is

$$\frac{x^2}{15^2} + \frac{y^2}{10^2} = 1.$$

In order to make an accurate sketch of the semiellipse, it is helpful to solve this equation for y as follows.

$$\frac{x^2}{15^2} + \frac{y^2}{10^2} = 1$$

$$\frac{x^2}{225} + \frac{y^2}{100} = 1$$

$$\frac{y^2}{100} = 1 - \frac{x^2}{225}$$

$$y^2 = 100\left(1 - \frac{x^2}{225}\right)$$

$$y = 10\sqrt{1 - \frac{x^2}{225}}$$

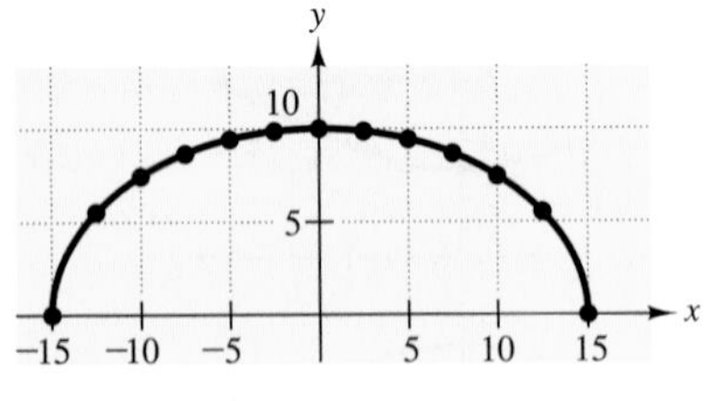

Figure A.14

Next, calculate several y-values for the archway, as shown in the table. Then use the values in the table to sketch the archway, as shown in Figure A.14.

x-Value	± 15	± 12.5	± 10	± 7.5	± 5	± 2.5	0
y-Value	0	5.53	7.45	8.66	9.43	9.86	10

Appendix A Exercises

In Exercises 1–8, find an equation of the circle that satisfies the given criteria.

1. Center: $(0, 0)$; radius: 5

2. Center: $(0, 0)$; radius: $\frac{5}{2}$

3. Center: $(-4, 1)$; radius: 3

4. Center: $(5, 8)$; radius: $\sqrt{10}$

5. Center: $(0, 0)$; passes through the point $(5, 2)$

6. Center: $(0, 0)$; passes through the point $(-1, -4)$

7. Center: $(7, -5)$; passes through the point $(1, 3)$

8. Center: $(-4, -1)$; passes through the point $(-3, 0)$

In Exercises 9–16, identify the center and radius of the circle and sketch its graph.

9. $x^2 + y^2 = 16$ **10.** $x^2 + y^2 = 25$

11. $(x - 2)^2 + (y - 3)^2 = 4$

12. $(x + 4)^2 + (y - 3)^2 = 25$

13. $25x^2 + 25y^2 - 144 = 0$

14. $\dfrac{x^2}{4} + \dfrac{y^2}{4} - 1 = 0$

15. $x^2 + y^2 - 4x - 2y + 1 = 0$

16. $x^2 + y^2 + 6x - 4y - 3 = 0$

In Exercises 17–20, write the standard form of the equation of the ellipse, centered at the origin.

	Vertices	*Co-vertices*
17.	$(-4, 0), (4, 0)$	$(0, -3), (0, 3)$
18.	$(-6, 0), (6, 0)$	$(0, -1), (0, 1)$
19.	$(0, -2), (0, 2)$	$(-5, 0), (5, 0)$
20.	$(0, -5), (0, 5)$	$(-2, 0), (2, 0)$

21. Write the standard form of the equation of the ellipse, centered at the origin, with a horizontal major axis of length 20 and a vertical minor axis of length 12.

22. Write the standard form of the equation of the ellipse, centered at the origin, with a horizontal minor axis of length 30 and a vertical major axis of length 50.

In Exercises 23–28, sketch the ellipse. Identify its vertices and co-vertices.

23. $\dfrac{x^2}{16} + \dfrac{y^2}{4} = 1$ **24.** $\dfrac{x^2}{9} + \dfrac{y^2}{25} = 1$

25. $\dfrac{x^2}{\frac{25}{9}} + \dfrac{y^2}{\frac{16}{9}} = 1$ **26.** $\dfrac{x^2}{1} + \dfrac{y^2}{\frac{1}{4}} = 1$

27. $4x^2 + y^2 - 4 = 0$ **28.** $4x^2 + 9y^2 - 36 = 0$

In Exercises 29–36, sketch the hyperbola. Identify its vertices and asymptotes.

29. $\dfrac{x^2}{9} - \dfrac{y^2}{25} = 1$ **30.** $\dfrac{x^2}{4} - \dfrac{y^2}{9} = 1$

31. $\dfrac{y^2}{9} - \dfrac{x^2}{25} = 1$ **32.** $\dfrac{y^2}{4} - \dfrac{x^2}{9} = 1$

33. $y^2 - x^2 = 9$ **34.** $y^2 - x^2 = 1$

35. $4y^2 - x^2 + 16 = 0$ **36.** $4y^2 - 9x^2 - 36 = 0$

In Exercises 37–40, find an equation of the hyperbola centered at the origin.

	Vertices	*Asymptotes*	
37.	$(-4, 0), (4, 0)$	$y = 2x$	$y = -2x$
38.	$(-2, 0), (2, 0)$	$y = \frac{1}{3}x$	$y = -\frac{1}{3}x$
39.	$(-9, 0), (9, 0)$	$y = \frac{3}{2}x$	$y = -\frac{3}{2}x$
40.	$(0, -5), (0, 5)$	$y = x$	$y = -x$

In Exercises 41–46, match the equation with its graph. [The graphs are labeled (a), (b), (c), (d), (e), and (f).]

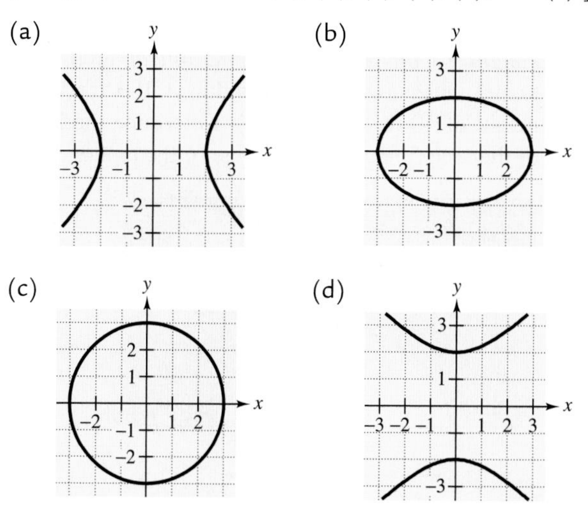

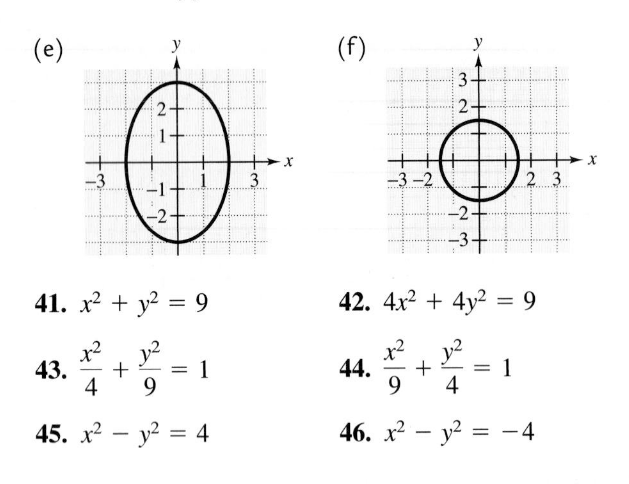

41. $x^2 + y^2 = 9$

42. $4x^2 + 4y^2 = 9$

43. $\dfrac{x^2}{4} + \dfrac{y^2}{9} = 1$

44. $\dfrac{x^2}{9} + \dfrac{y^2}{4} = 1$

45. $x^2 - y^2 = 4$

46. $x^2 - y^2 = -4$

47. *Satellite Orbit* Find an equation of the circular orbit of a satellite 500 miles above the surface of the earth. Place the origin of the rectangular coordinate system at the center of the earth and assume the radius of the earth to be 4000 miles.

48. *Architecture* The top portion of a stained-glass window is in the form of a pointed Gothic arch (see figure). Each side of the arch is an arc of a circle that has a radius of 12 feet and a center at the base of the opposite arch. Find an equation of one of the circles and use it to determine the height of the point of the arch above the horizontal base of the window.

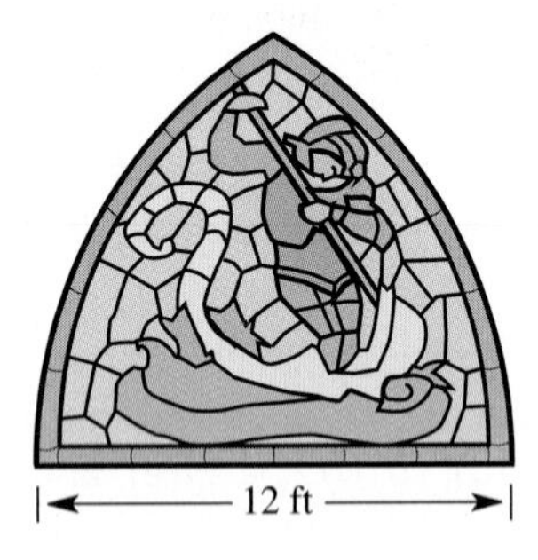

49. *Height of an Arch* A *semielliptical* arch for a tunnel under a river has a width of 100 feet and a height of 40 feet (see figure). Determine the height of the arch 5 feet from the edge of the tunnel.

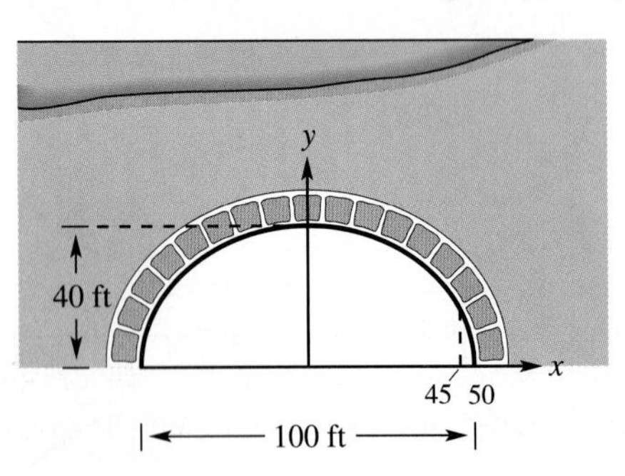

50. *Height of an Arch* A *semicircular* arch for a tunnel under a river has a diameter of 100 feet (see figure). Determine the height of the arch 5 feet from the edge of the tunnel.

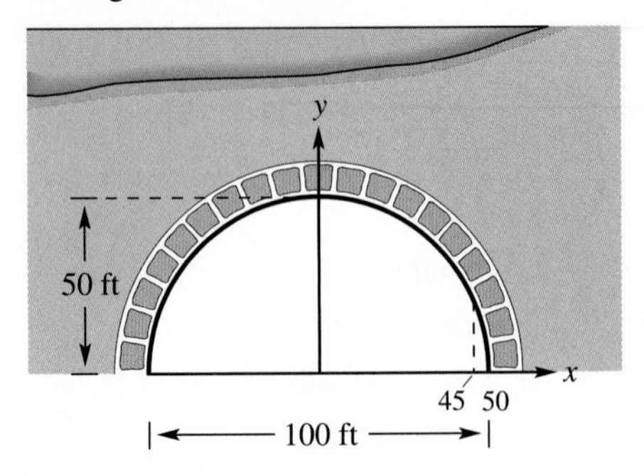

51. *Geometry* A rectangle centered at the origin with sides parallel to the coordinate axes is placed in a circle of radius 25 inches centered at the origin (see figure). The length of the rectangle is $2x$ inches.

(a) Show that the width and area of the rectangle are given by $2\sqrt{625 - x^2}$ and $4x\sqrt{625 - x^2}$, respectively.

(b) Use a graphing utility to graph the area function. Approximate the value of x for which the area is maximum.

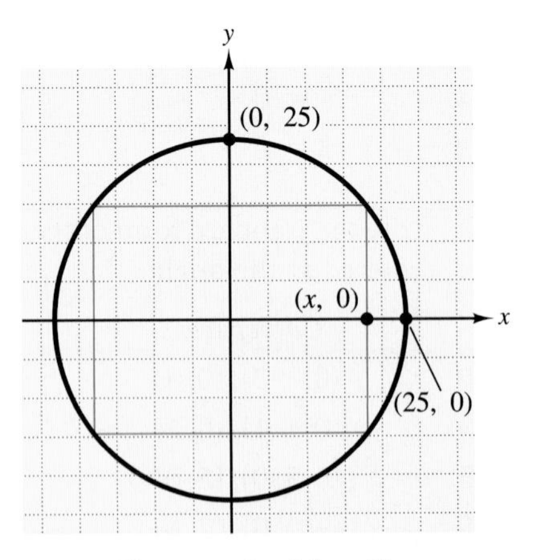

52. *Area* The area A of the ellipse

$$\frac{x^2}{a^2} + \frac{y^2}{b^2} = 1$$

is given by $A = \pi ab$. Find an equation of the ellipse with area 301.59 square units and $a + b = 20$.

53. *Hyperbolic Mirror* A hyperbolic mirror (used in some telescopes) has the property that a light ray directed at the first focus will be reflected to the second focus. The foci of a hyperbolic mirror are $(\pm 12, 0)$. Find the vertex of the mirror if its mount has coordinates (12, 12).

Appendix B
Introduction to Graphing Utilities

Introduction ■ Using a Graphing Utility ■ Using Special Features of a Graphing Utility

Introduction

In Section 4.2 you studied the point-plotting method for sketching the graph of an equation. One of the disadvantages of the point-plotting method is that to get a good idea about the shape of a graph you need to plot *many* points. By plotting only a few points, you can badly misrepresent the graph.

For instance, consider the equation $y = x^3$. To graph this equation, suppose you calculated only the following three points.

x	-1	0	1
$y = x^3$	-1	0	1

By plotting these three points, as shown in Figure B.1(a), you might assume that the graph of the equation is a straight line. This, however, is not correct. By plotting several more points, as shown in Figure B.1(b), you can see that the actual graph is not straight at all.

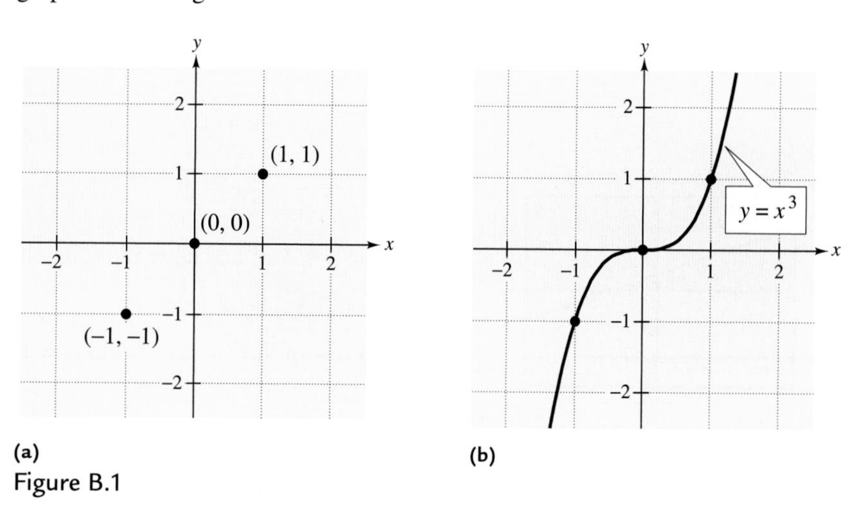

(a) (b)
Figure B.1

Thus, the point-plotting method leaves you with a dilemma. On the one hand, the method can be very inaccurate if only a few points are plotted. But, on the other hand, it is very time-consuming to plot a dozen (or more) points. Technology can help you solve this dilemma. Plotting several points (or even hundreds of points) on a rectangular coordinate system is something that a computer or graphing calculator can do easily.

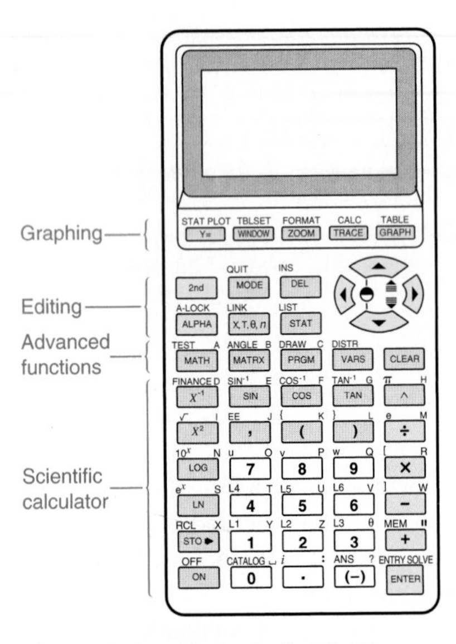

Figure B.2 *Keypad of a TI-83 Graphics Calculator*

Using a Graphing Utility

There are many different graphing utilities: some are graphing packages for computers and some are hand-held graphing calculators. In this section we describe the steps used to graph an equation with a *TI-83* graphing utility. (See Figure B.2.) We will often give keystroke sequences for illustration; however, these may not agree precisely with the steps required by *your* calculator.*

▶ Graphing an Equation with a *TI-83* Graphing Calculator

Before performing the following steps, set your calculator so that all of the standard defaults are active. For instance, all of the options at the left of the [MODE] screen should be highlighted.

1. Set the viewing window for the graph. (See Example 3.) To set the standard viewing window, press [ZOOM] 6.
2. Rewrite the equation so that y is isolated on the left side of the equation.
3. Press the [Y=] key. Then enter the right side of the equation on the first line of the display. (The first line is labeled $Y_1 =$.)
4. Press the [GRAPH] key.

Example 1 Graphing a Linear Equation

Sketch the graph of $2y + x = 4$.

Solution

To begin, solve the given equation for y in terms of x.

$$2y + x = 4 \qquad \text{Original equation}$$

$$2y = -x + 4 \qquad \text{Subtract } x \text{ from both sides.}$$

$$y = -\frac{1}{2}x + 2 \qquad \text{Divide both sides by 2.}$$

Press the [Y=] key, and enter the following keystrokes.

[(−)] [X,T,θ,n] [÷] 2 [+] 2

The top row of the display should now be as follows.

Y_1 = -X/2 + 2

Press the [GRAPH] key, and the screen should look like that shown in Figure B.3.

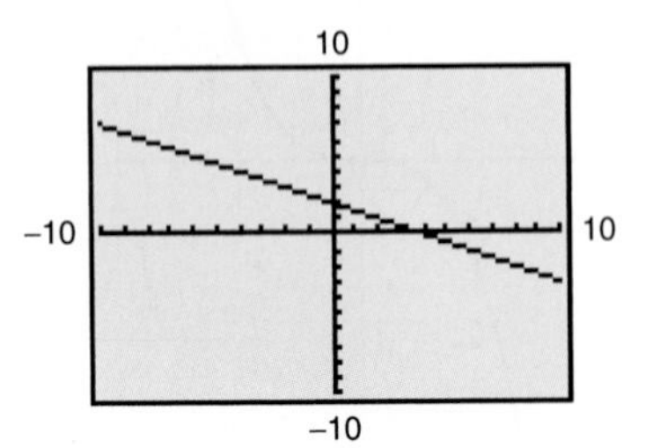

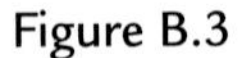
Figure B.3

*The graphing calculator keystrokes given in this section correspond to the *TI-83* graphing utility by Texas Instruments. For other graphing utilities, the keystrokes may differ. Consult your user's guide.

In Figure B.3, notice that the calculator screen does not label the tick marks on the x-axis or the y-axis. To see what the tick marks represent, you can press [WINDOW]. If you set your calculator to the standard graphing defaults before working Example 1, the screen should show the following values.

Xmin = -10	The minimum x-value is -10.
Xmax = 10	The maximum x-value is 10.
Xscl = 1	The x-scale is 1 unit per tick mark.
Ymin = -10	The minimum y-value is -10.
Ymax = 10	The maximum y-value is 10.
Yscl = 1	The y-scale is 1 unit per tick mark.
Xres = 1	Sets the pixel resolution.

These settings are summarized visually in Figure B.4.

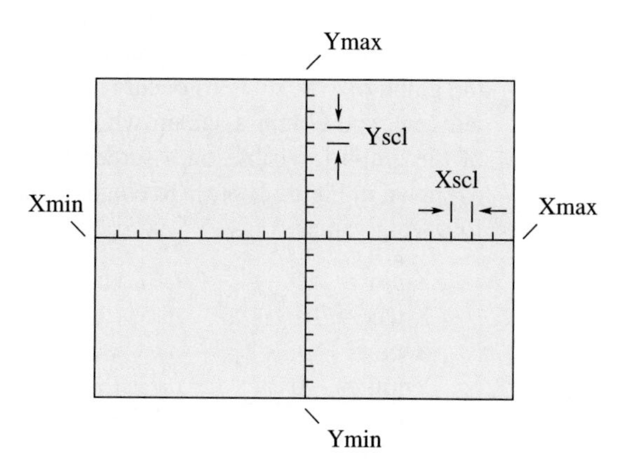

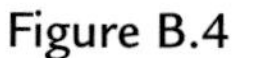
Figure B.4

Example 2 Graphing an Equation Involving Absolute Value

Sketch the graph of $y = |x - 3|$.

Solution

This equation is already written so that y is isolated on the left side of the equation. Press the [Y=] key, and enter the following keystrokes.

[ABS] [(] [X,T,θ,n] [−] 3 [)]

The top row of the display should now be as follows.

$Y_1 = \text{abs}(X - 3)$

Press the [GRAPH] key, and the screen should look like that shown in Figure B.5.

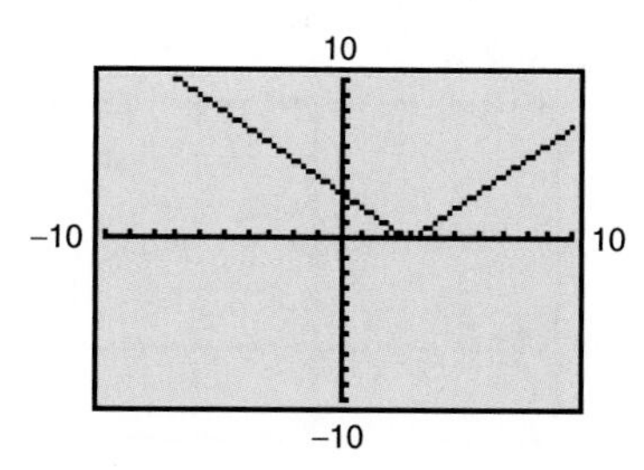

Figure B.5

Using the Special Features of a Graphing Utility

To use your graphing utility to its best advantage, you must learn to set the viewing window, as illustrated in the next example.

Example 3 Setting the Viewing Window

Sketch the graph of $y = x^2 + 12$.

Solution

Press [Y=] and enter $x^2 + 12$ on the first line.

[X,T,θ,n] [x^2] [+] 12

Press the [GRAPH] key. If your calculator is set to the standard viewing window, nothing will appear on the screen. The reason for this is that the lowest point on the graph of $y = x^2 + 12$ occurs at the point (0, 12). Using the standard viewing window, you obtain a screen whose largest y-value is 10. In other words, none of the graph is visible on a screen whose y-values vary between -10 and 10, as shown in Figure B.6(a). To change these settings, press [WINDOW] and enter the following values.

Xmin = -10	The minimum x-value is -10.
Xmax = 10	The maximum x-value is 10.
Xscl = 1	The x-scale is 1 unit per tick mark.
Ymin = -10	The minimum y-value is -10.
Ymax = 30	The maximum y-value is 30.
Yscl = 5	The y-scale is 5 units per tick mark.
Xres = 1	Sets the pixel resolution.

Press [GRAPH] and you will obtain the graph shown in Figure B.6(b). On this graph, note that each tick mark on the y-axis represents 5 units because you changed the y-scale to 5. Also note that the highest point on the y-axis is now 30 because you changed the maximum value of y to 30.

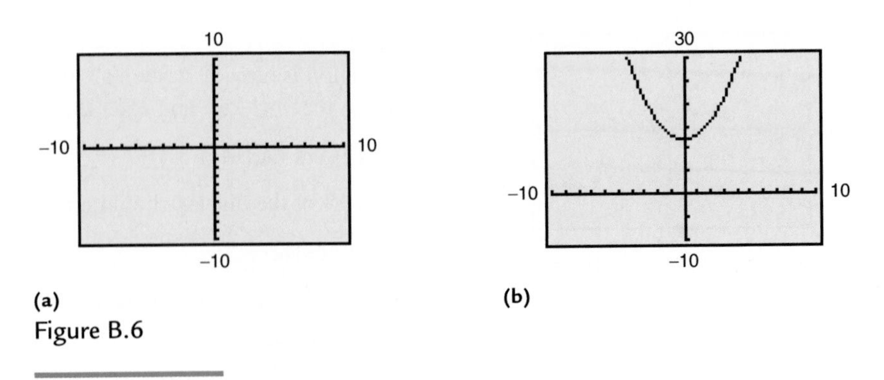

(a) (b)

Figure B.6

If you changed the y-maximum and y-scale on your utility as indicated in Example 3, you should return to the standard settings before working Example 4. To do this, press [ZOOM] 6.

Example 4 Using a Square Setting

Sketch the graph of $y = x$. The graph of this equation is a straight line that makes a 45° angle with the x-axis and with the y-axis. From the graph on your utility, does the angle appear to be 45°?

Solution

Press [Y=] and enter x on the first line.

$$Y_1 = X$$

Press the [GRAPH] key and you will obtain the graph shown in Figure B.7(a). Notice that the angle the line makes with the x-axis doesn't appear to be 45°. The reason for this is that the screen is wider than it is tall. This makes the tick marks on the x-axis farther apart than the tick marks on the y-axis. To obtain the same distance between tick marks on both axes, you can change the graphing settings from "standard" to "square." To do this, press the following keys.

[ZOOM] 5 Square setting

The screen should look like that shown in Figure B.7(b). Note in this figure that the square setting has changed the viewing window so that the x-values vary between -15 and 15.

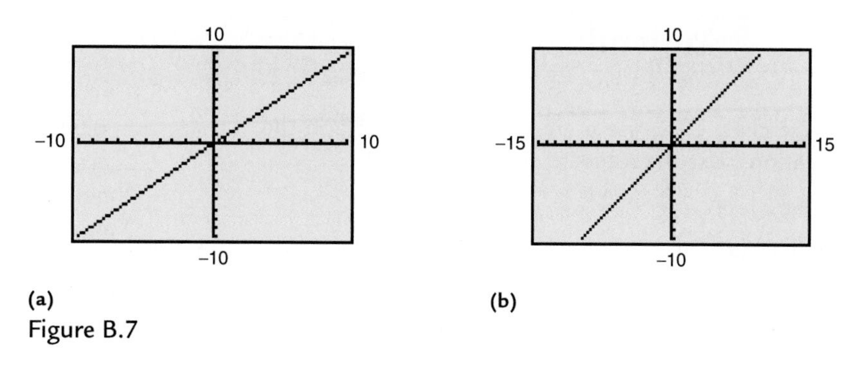

(a) (b)
Figure B.7

There are many possible square settings on a graphing utility. To create a square setting, you need the following ratio to be $\frac{2}{3}$.

$$\frac{\text{Ymax} - \text{Ymin}}{\text{Xmax} - \text{Xmin}}$$

For instance, the setting in Example 4 is square because (Ymax $-$ Ymin) $= 20$ and (Xmax $-$ Xmin) $= 30$.

Example 5 Sketching More than One Graph on the Same Screen

Sketch the graphs of the following equations on the same screen.

$$y = -x + 4, \quad y = -x, \quad \text{and} \quad y = -x - 4$$

Solution

To begin, press [Y=] and enter all three equations on the first three lines. The display should now be as follows.

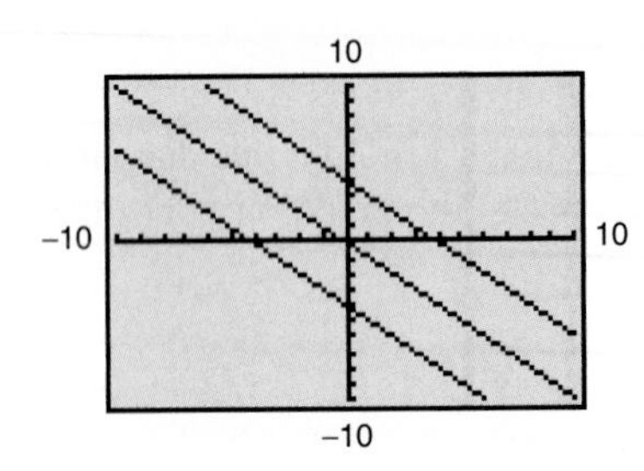

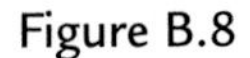
Figure B.8

$Y_1 = -X + 4$ (−) X,T,θ,n + 4

$Y_2 = -X$ (−) X,T,θ,n

$Y_3 = -X - 4$ (−) X,T,θ,n − 4

Press the GRAPH key and you will obtain the graph shown in Figure B.8. Note that the graph of each equation is a straight line, and that the lines are parallel to each other.

Another special feature of a graphing utility is the trace feature. This feature is used to find solution points of an equation. For example, you can approximate the x- and y-intercepts of $y = 3x + 6$ by first graphing the equation, then pressing the TRACE key, and finally pressing the ◄ ► keys. To get a better approximation of a solution point, you can use the following keystrokes repeatedly.

ZOOM 2 ENTER

Check to see that you get an x-intercept of $(-2, 0)$ and a y-intercept of $(0, 6)$. Use the trace feature to find the x- and y-intercepts of $y = \frac{1}{2}x - 4$.

Appendix B Exercises

In Exercises 1–12, use a graphing utility to graph the equation. (Use the standard setting.)

1. $y = -3x$ **2.** $y = x - 4$

3. $y = \frac{3}{4}x - 6$ **4.** $y = -3x + 2$

5. $y = \frac{1}{2}x^2$ **6.** $y = -\frac{2}{3}x^2$

7. $y = x^2 - 4x + 2$ **8.** $y = -0.5x^2 - 2x + 2$

9. $y = |x - 3|$ **10.** $y = |x + 4|$

11. $y = |x^2 - 4|$ **12.** $y = |x - 2| - 5$

In Exercises 13–16, use a graphing utility to graph the equation using the given window settings.

13. $y = 27x + 100$

Xmin = 0
Xmax = 5
Xscl = .5
Ymin = 75
Ymax = 250
Yscl = 25
Xres = 1

14. $y = 50{,}000 - 6000x$

Xmin = 0
Xmax = 7
Xscl = .5
Ymin = 0
Ymax = 50000
Yscl = 5000
Xres = 1

15. $y = 0.001x^2 + 0.5x$

Xmin = -500
Xmax = 200
Xscl = 50
Ymin = -100
Ymax = 100
Yscl = 20
Xres = 1

16. $y = 100 - 0.5|x|$

Xmin = -300
Xmax = 300
Xscl = 60
Ymin = -100
Ymax = 100
Yscl = 20
Xres = 1

In Exercises 17–20, find a viewing window that shows the important characteristics of the graph.

17. $y = 15 + |x - 12|$ **18.** $y = 15 + (x - 12)^2$

19. $y = -15 + |x + 12|$ **20.** $y = -15 + (x + 12)^2$

In Exercises 21–24, graph both equations on the same screen. Are the graphs identical? If so, what rule of algebra is being illustrated?

21. $y_1 = 2x + (x + 1)$
$y_2 = (2x + x) + 1$

22. $y_1 = \frac{1}{2}(3 - 2x)$
$y_2 = \frac{3}{2} - x$

23. $y_1 = 2\left(\frac{1}{2}\right)$
$y_2 = 1$

24. $y_1 = x(0.5x)$
$y_2 = (0.5x)x$

In Exercises 25–32, use the trace feature of a graphing utility to approximate the x- and y-intercepts of the graph.

25. $y = 9 - x^2$

26. $y = 3x^2 - 2x - 5$

27. $y = 6 - |x + 2|$

28. $y = |x - 2|^2 - 3$

29. $y = 2x - 5$

30. $y = 4 - |x|$

31. $y = x^2 + 1.5x - 1$

32. $y = x^3 - 4x$

Geometry In Exercises 33–36, graph the equations on the same display. Using a "square setting," determine the geometrical shape bounded by the graphs.

33. $y = -4, \quad y = -|x|$

34. $y = |x|, \quad y = 5$

35. $y = |x| - 8, \quad y = -|x| + 8$

36. $y = -\frac{1}{2}x + 7, \quad y = \frac{8}{3}(x + 5), \quad y = \frac{2}{7}(3x - 4)$

Modeling Data In Exercises 37 and 38, use the following models, which give the number of pieces of first-class mail and the number of periodicals handled by the U.S. Postal Service.

First Class

$y = 0.07x^2 + 1.06x + 88.97, \quad 0 \le x \le 7$

Periodicals

$y = 0.02x^2 - 0.23x + 10.70, \quad 0 \le x \le 7$

In these models, y is the number of pieces handled (in billions) and x is the year, with $x = 0$ corresponding to 1990. (Source: U.S. Postal Service)

37. Use the following setting to graph both models on the same display of a graphing utility.

Xmin = 0
Xmax = 7
Xscl = 1
Ymin = -5
Ymax = 115
Yscl = 10
Xres = 1

38. (a) Were the numbers of pieces of first-class mail and periodicals increasing or decreasing over time?

(b) Is the distance between the graphs increasing or decreasing over time? What does this mean to the U.S. Postal Service?

Appendix C
Further Concepts in Geometry

C.1 Exploring Congruence and Similarity

Identifying Congruent Figures ■ Identifying Similar Figures ■
Reading and Using Definitions ■ Congruent Triangles ■ Classifying Triangles

Identifying Congruent Figures

Two figures are *congruent* if they have the same shape and the same size. Each of the triangles in Figure C.1 is congruent to each of the other triangles. The triangles in Figure C.2 are not congruent to each other.

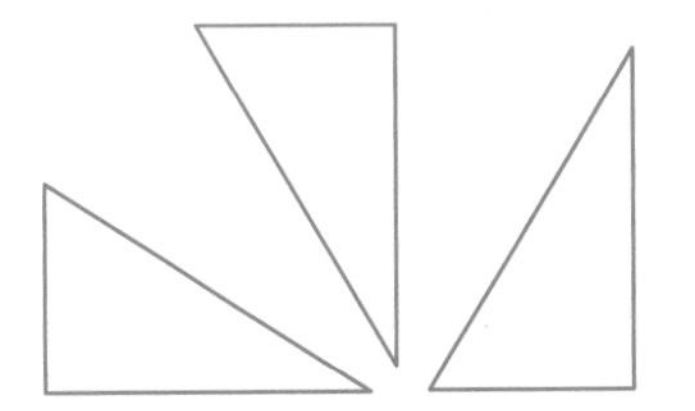

Congruent
Figure C.1

Not Congruent
Figure C.2

Notice that two figures can be congruent without having the same orientation. If two figures are congruent, then either one can be moved (and turned or flipped if necessary) so that it coincides with the other figure.

Example 1 Dividing Regions into Congruent Parts

Divide the region into two congruent parts.

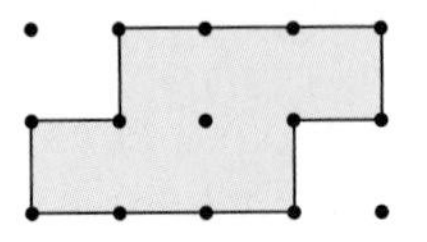

Solution

There are many solutions to this problem. Some of the solutions are shown in Figure C.3. Can you think of others?

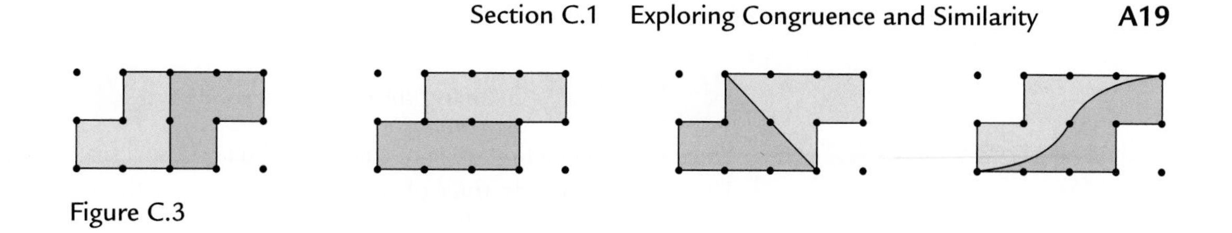

Figure C.3

Identifying Similar Figures

Two figures are *similar* if they have the same shape. (They may or may not have the same size.) Each of the quadrilaterals in Figure C.4 is similar to the others. The quadrilaterals in Figure C.5 are not similar to each other.

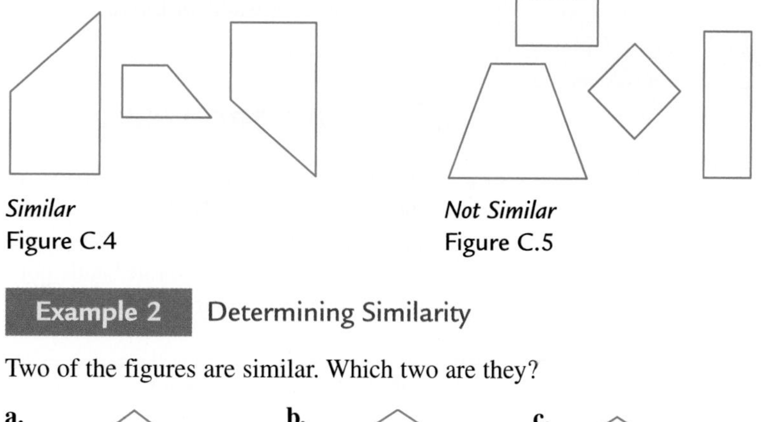

Similar
Figure C.4

Not Similar
Figure C.5

Example 2 Determining Similarity

Two of the figures are similar. Which two are they?

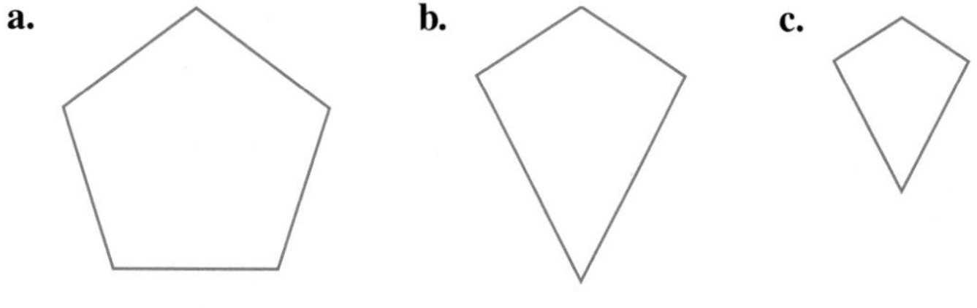

Solution

The first figure has five sides and the other two figures have four sides. Because similar figures must have the same shape, the first figure is not similar to either of the others. Because you are told that two figures are similar, it follows that the second and third figures are similar.

Example 3 Determining Similarity

You wrote an essay on Euclid, the Greek mathematician who is famous for writing a geometry book titled *Elements of Geometry*. You are making a copy of the essay using a photocopier that is set at 75% reduction. Is each image on the copied pages similar to its original?

Solution

Every image on a copied page *is* similar to its original. The copied pages are smaller, but that doesn't matter because similar figures do not have to be the same size.

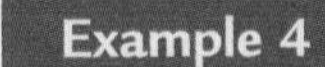

Example 4 Drawing an Object to Scale

You are drawing a floor plan of a building. You choose a scale of $\frac{1}{8}$ inch to 1 foot. That is, $\frac{1}{8}$ inch of the floor plan represents 1 foot of the actual building. What dimensions should you draw for a room that is 12 feet wide and 18 feet long?

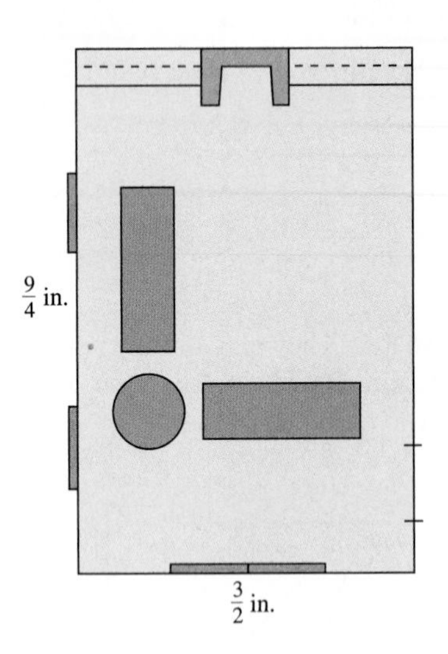

Figure C.6

Solution

Because each foot is represented as $\frac{1}{8}$ inch, the width of the room should be

$$12\left(\tfrac{1}{8}\right) = \tfrac{12}{8} = \tfrac{3}{2} = 1\tfrac{1}{2}$$

and the length of the room should be

$$18\left(\tfrac{1}{8}\right) = \tfrac{18}{8} = \tfrac{9}{4} = 2\tfrac{1}{4}.$$

The scale dimensions of the room are $1\frac{1}{2}$ inches by $2\frac{1}{4}$ inches. See Figure C.6.

Reading and Using Definitions

A definition uses *known* words to describe a *new* word. If no words were known, then no new words could be defined. Hence, some words such as **point, line,** and **plane** must be commonly understood without being defined. Some statements such as "a point lies on a line" and "point C lies between points A and B" are also not defined. See Figure C.7.

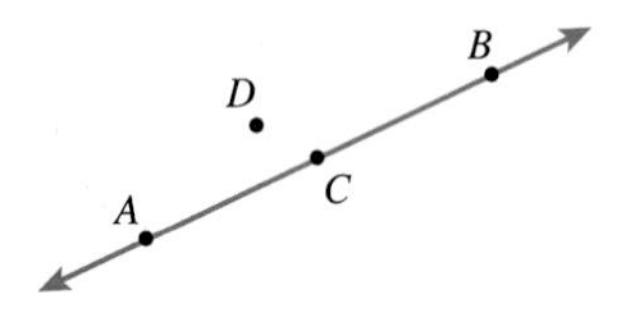

C is between A and B. D is not between A and B.
Figure C.7

▶ Segments and Rays

Consider the line $\overleftrightarrow{AB}$ that contains the points A and B. (In geometry, the word *line* means a *straight line*.)

The **line segment** (or simply **segment**) $\overline{AB}$ consists of the *endpoints* A and B and all points on the line $\overleftrightarrow{AB}$ that lie between A and B.

The **ray** $\overrightarrow{AB}$ consists of the *initial point* A and all points on the line $\overleftrightarrow{AB}$ that lie on the same side of A as B lies. If C is between A and B, then $\overrightarrow{CA}$ and $\overrightarrow{CB}$ are **opposite** rays.

Points, segments, or rays that lie on the same line are **collinear.**

Lines are drawn with two arrowheads, line segments are drawn with no arrowhead, and rays are drawn with a single arrowhead. See Figure C.8.

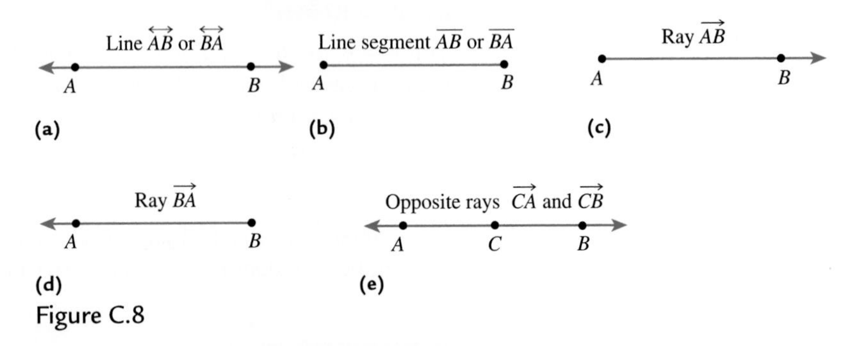

Figure C.8

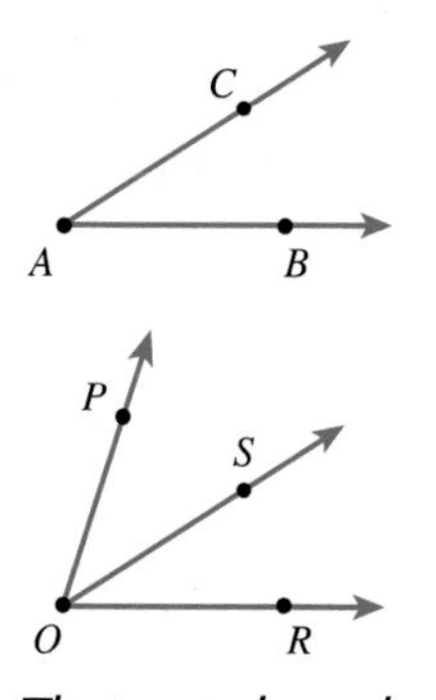

The top angle can be denoted by $\angle A$ or by $\angle BAC$. In the lower figure, the angle $\angle ROS$ should not be denoted by $\angle O$ because the figure contains three angles whose vertex is O.
Figure C.9

It follows that $\overline{AB}$ and $\overline{BA}$ denote the same segment, but $\overrightarrow{AB}$ and $\overrightarrow{BA}$ do not denote the same ray. No length is given to lines or rays because each is infinitely long. The *length* of the line segment $\overline{AB}$ is denoted by AB.

▶ Angles

An **angle** consists of two different rays that have the same initial point. The rays are the *sides* of the angle. The angle that consists of the rays $\overrightarrow{AB}$ and $\overrightarrow{AC}$ is denoted by $\angle BAC$, $\angle CAB$, or $\angle A$. The point A is the **vertex** of the angle. See Figure C.9.

The measure of $\angle A$ is denoted by $m\angle A$. Angles are classified as **acute, right, obtuse,** and **straight.**

Acute	$0° < m\angle A < 90°$
Right	$m\angle A = 90°$
Obtuse	$90° < m\angle A < 180°$
Straight	$m\angle A = 180°$

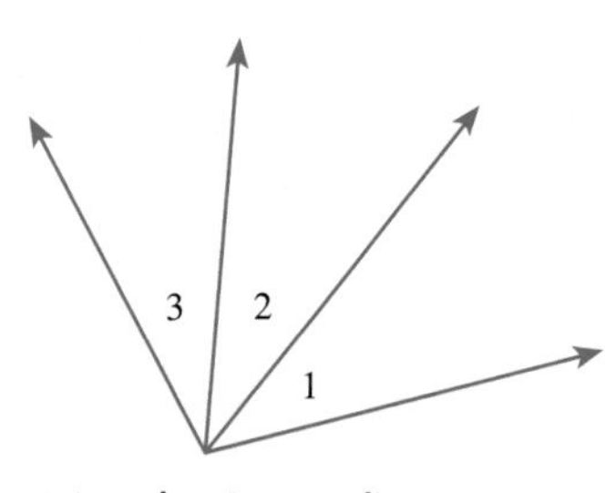

$\angle 1$ and $\angle 2$ are adjacent.
$\angle 1$ and $\angle 3$ are not adjacent.
Figure C.10

In geometry, *unless specifically stated otherwise*, angles are assumed to have a measure that is greater than 0° and less than or equal to 180°.

Every nonstraight angle has an **interior** and an **exterior.** A point D is in the interior of $\angle A$ if it is between points that lie on each side of the angle. Two angles (such as $\angle ROS$ and $\angle SOP$ shown in Figure C.9) are **adjacent** if they share a common vertex and side, but have no common interior points. In Figure C.10, $\angle 1$ and $\angle 3$ share a vertex, but not a common side, so $\angle 1$ and $\angle 3$ are not adjacent.

▶ Segment and Angle Congruence

Two segments are **congruent,** $\overline{AB} \cong \overline{CD}$, if they have the same length. Two angles are **congruent,** $\angle P \cong \angle Q$, if they have the same measure.

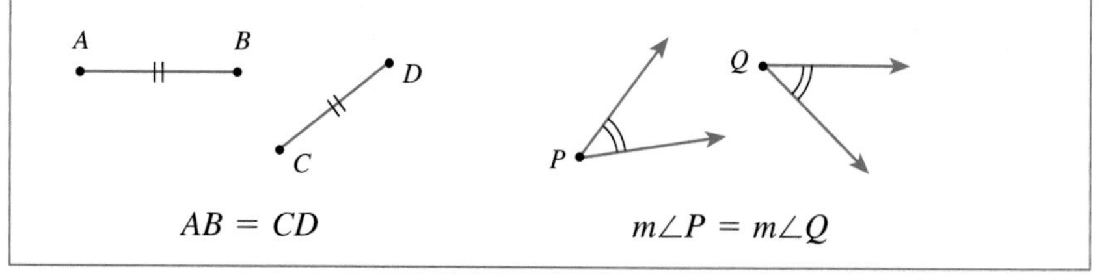

Definitions can always be interpreted "forward" and "backward." For instance, the definition of congruent segments means (1) if two segments have the same measure, then they are congruent, and (2) if two segments are congruent, then they have the same measure. You learned that two figures are congruent if they have the same shape and size.

Congruent Triangles

If $\triangle ABC$ is **congruent** to $\triangle PQR$, then there is a correspondence between their angles and sides such that corresponding angles are congruent and corresponding sides are congruent. The notation $\triangle ABC \cong \triangle PQR$ indicates the congruence *and* the correspondence, as shown in Figure C.11. If two triangles are congruent, then you know that they share many properties.

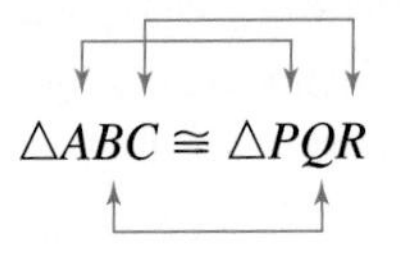

Corresponding angles are	*Corresponding sides are*
$\angle A \cong \angle P$	$\overline{AB} \cong \overline{PQ}$
$\angle B \cong \angle Q$	$\overline{BC} \cong \overline{QR}$
$\angle C \cong \angle R$	$\overline{CA} \cong \overline{RP}$

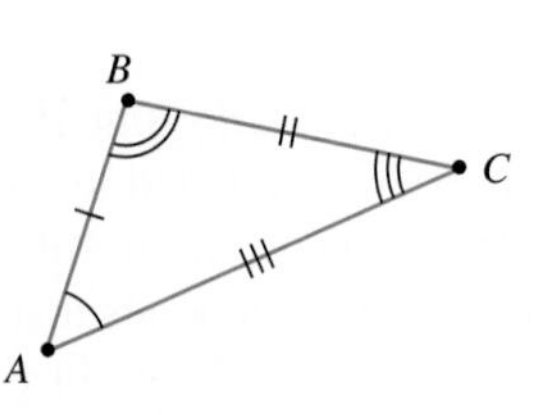

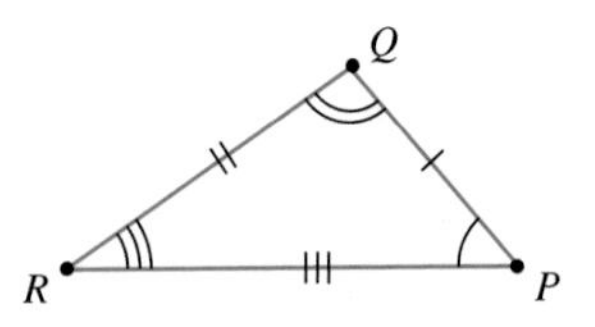

Figure C.11

Naming Congruent Parts

You and a friend have identical drafting triangles, as shown in Figure C.12. Name all congruent parts.

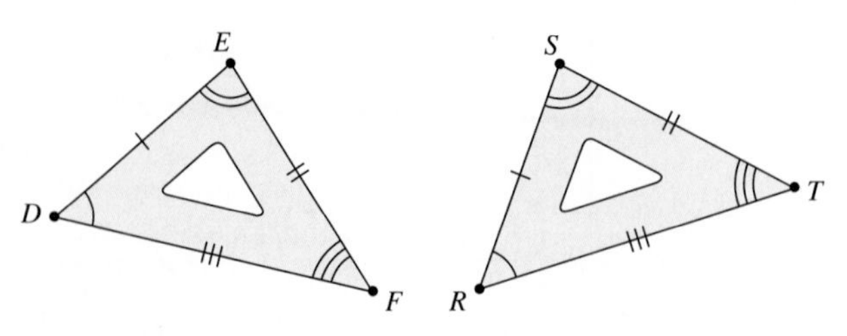

Figure C.12

Solution

Given that $\triangle DEF \cong \triangle RST$, the congruent angles and sides are as follows.

Angles: $\angle D \cong \angle R$, $\angle E \cong \angle S$, $\angle F \cong \angle T$

Sides: $\overline{DE} \cong \overline{RS}$, $\overline{EF} \cong \overline{ST}$, $\overline{FD} \cong \overline{TR}$

Classifying Triangles

A triangle can be classified by relationships among its sides or among its angles, as shown in the following definitions.

▶ Classification by Sides

1. An **equilateral triangle** has three congruent sides.
2. An **isosceles triangle** has at least two congruent sides.
3. A **scalene triangle** has no sides congruent.

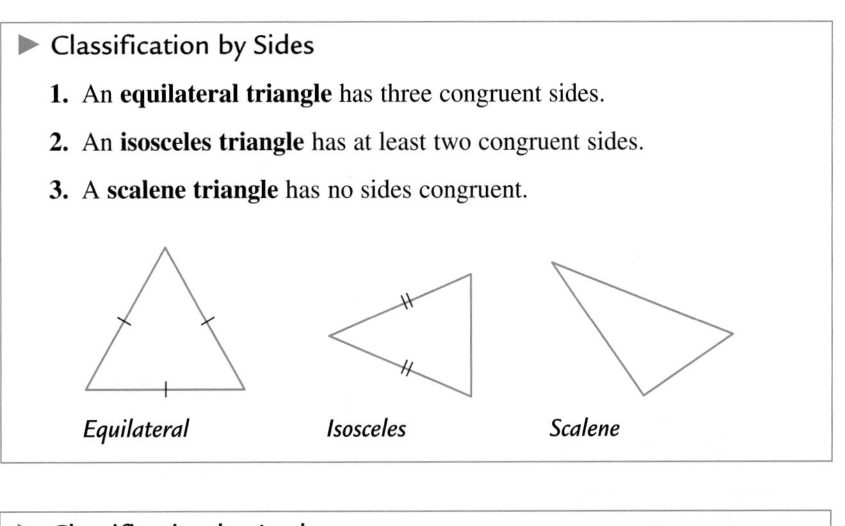

▶ Classification by Angles

1. An **acute triangle** has three acute angles. If these angles are all congruent, then the triangle is also **equiangular.**
2. A **right triangle** has exactly one right angle.
3. An **obtuse triangle** has exactly one obtuse angle.

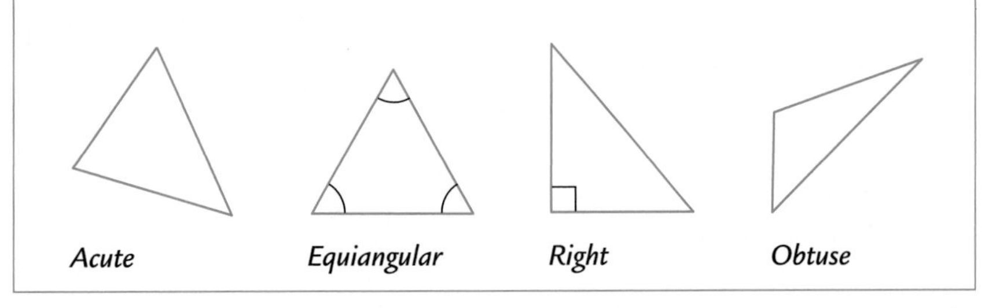

In $\triangle ABC$, each of the points A, B, and C is a **vertex** of the triangle. (The plural of vertex is *vertices*.) The side $\overline{BC}$ is the side *opposite* $\angle A$. Two sides that share a common vertex are *adjacent sides* (see Figure C.13).

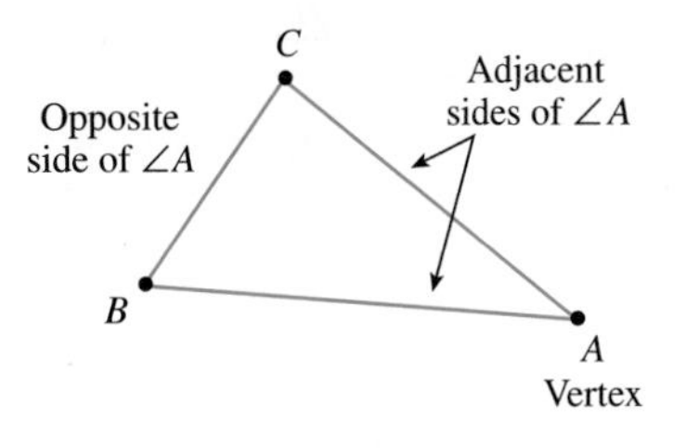

Figure C.13

The sides of right triangles and isosceles triangles are given special names. In a right triangle, the sides adjacent to the right angle are the **legs** of the triangle. The side opposite the right angle is the **hypotenuse** of the triangle (see Figure C.14).

Figure C.14

Figure C.15

An isosceles triangle can have three congruent sides. If it has only two, then the two congruent sides are the **legs** of the triangle. The third side is the **base** of the triangle (see Figure C.15).

C.1 Exercises

In Exercises 1–3, copy the region on a piece of dot paper. Then divide the region into two congruent parts. How many different ways can you do this?

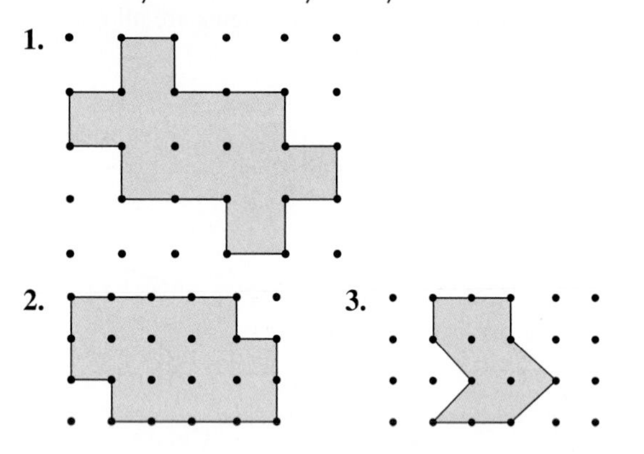

4. Two of the figures are congruent. Which are they?

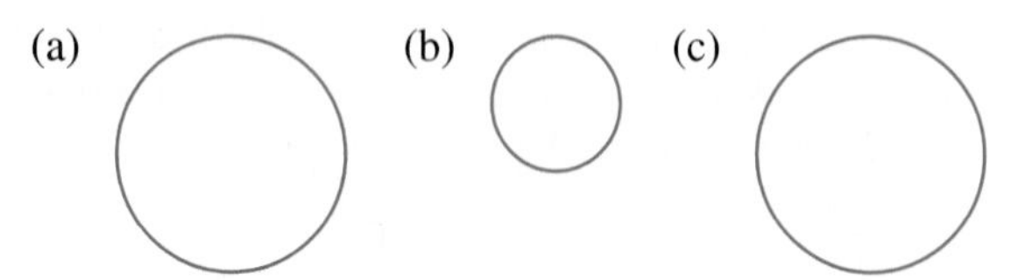

5. Two of the figures are similar. Which are they?

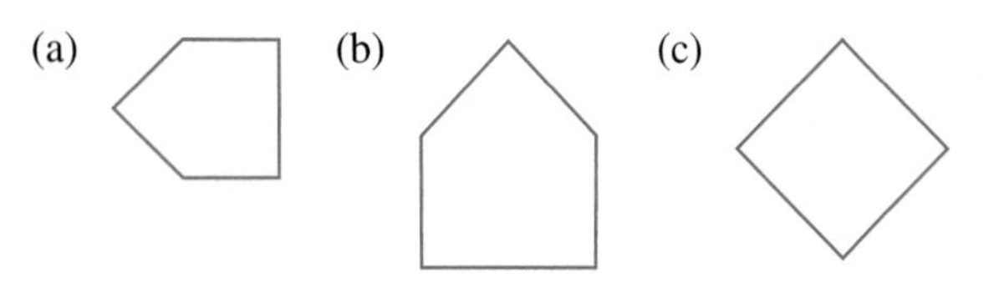

In Exercises 6 and 7, copy the region on a piece of paper. Then divide the region into four congruent parts.

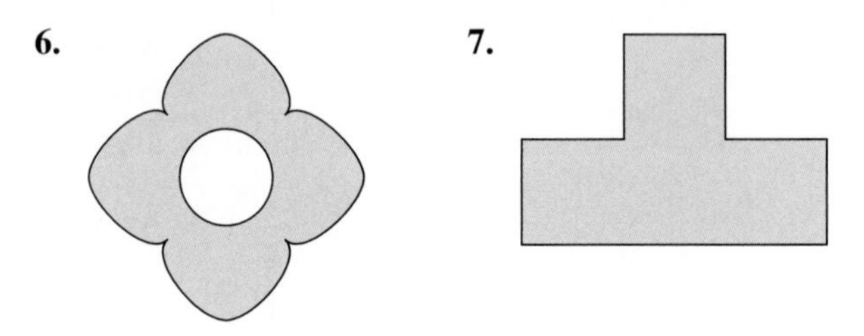

In Exercises 8–11, use the triangular grid below. In the grid, each small triangle has sides of 1 unit.

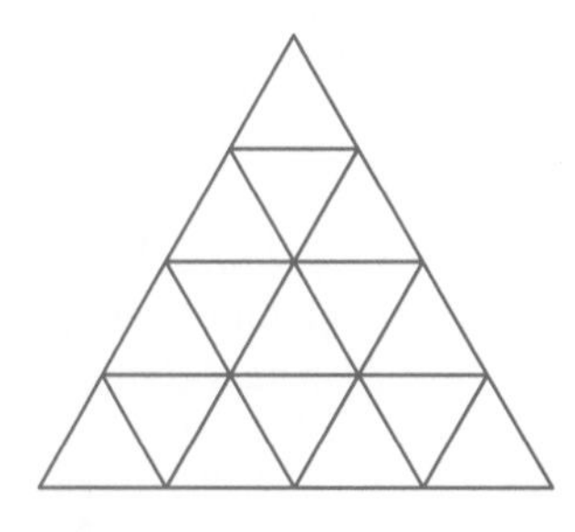

8. How many congruent triangles with 1-unit sides are in the grid?

9. How many congruent triangles with 2-unit sides are in the grid?

10. How many congruent triangles with 3-unit sides are in the grid?

11. Does the grid contain triangles that are not similar to each other?

12. *True or False?* If two figures are congruent, then they are similar.

13. *True or False?* If two figures are similar, then they are congruent.

14. *True or False?* A triangle can be similar to a square.

15. *True or False?* Any two squares are similar.

In Exercises 16–19, match the description with its correct notation.

(a) $\overline{PQ}$ (b) PQ (c) $\overleftrightarrow{PQ}$ (d) $\overrightarrow{PQ}$

16. The line through P and Q

17. The ray from P through Q

18. The segment between P and Q

19. The length of the segment between P and Q

20. The point R is between points S and T. Which of the following are true?

(a) R, S, and T are collinear.
(b) $\overrightarrow{SR}$ is the same as $\overrightarrow{ST}$.
(c) $\overline{ST}$ is the same as $\overline{TS}$.
(d) $\overrightarrow{ST}$ is the same as $\overrightarrow{TS}$.

In Exercises 21–23, use the figure below.

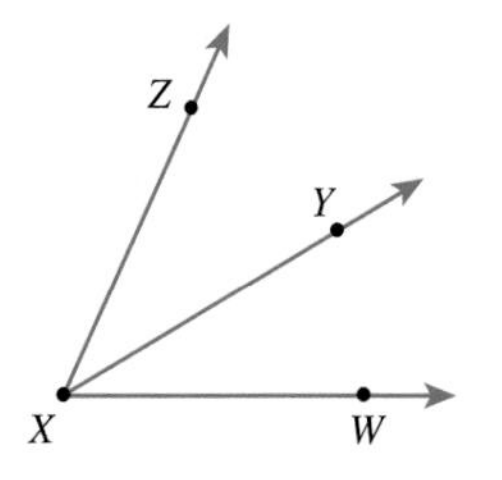

21. The figure shows three angles whose vertex is X. Write two names for each angle. Which two angles are adjacent?

22. Is Y in the interior or exterior of $\angle WXZ$?

23. Which is the best estimate for $m\angle WXY$?

(a) 15° (b) 30° (c) 45°

In Exercises 24–29, match the triangle with its name.

(a) Equilateral (b) Scalene (c) Obtuse
(d) Equiangular (e) Isosceles (f) Right

24. Side lengths: 2 cm, 3 cm, 4 cm

25. Angle measures: 60°, 60°, 60°

26. Side lengths: 3 cm, 2 cm, 3 cm

27. Angle measures: 30°, 60°, 90°

28. Side lengths: 4 cm, 4 cm, 4 cm

29. Angle measures: 20°, 145°, 15°

In Exercises 30–32, use the figure, in which $\triangle LMP \cong \triangle ONQ$.

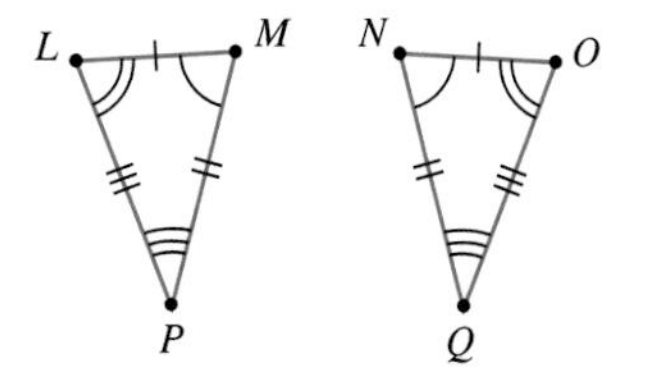

30. Name three pairs of congruent angles.

31. Name three pairs of congruent sides.

32. If $\triangle LMP$ is isosceles, explain why $\triangle ONQ$ must be isosceles.

In Exercises 33–36, use the definition of congruence to complete the statement.

33. If $\triangle ABC \cong \triangle TUV$, then $m\angle C =$ ▢.

34. If $\triangle PQR \cong \triangle XYZ$, then $\angle P \cong$ ▢.

35. If $\triangle LMN \cong \triangle TUV$, then $\overline{LN} \cong$ ▢.

36. If $\triangle DEF \cong \triangle NOP$, then $DE =$ ▢.

37. Copy and complete the table. Write *Yes* if it is possible to sketch a triangle with both characteristics. Write *No* if it is not possible. Illustrate your results with sketches. (The first is done for you.)

	Scalene	Isosceles	Equilateral
Acute	Yes	?	?
Obtuse	?	?	?
Right	?	?	?

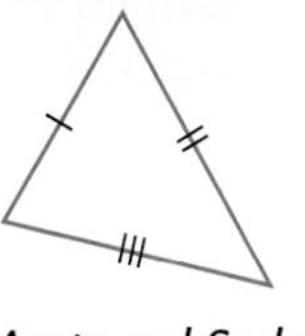

Acute and Scalene

In Exercises 38–41, $\triangle ABC$ is isosceles with $\overline{AC} \cong \overline{BC}$. Solve for x. Then decide whether the triangle is equilateral. (The figures are not necessarily drawn to scale.)

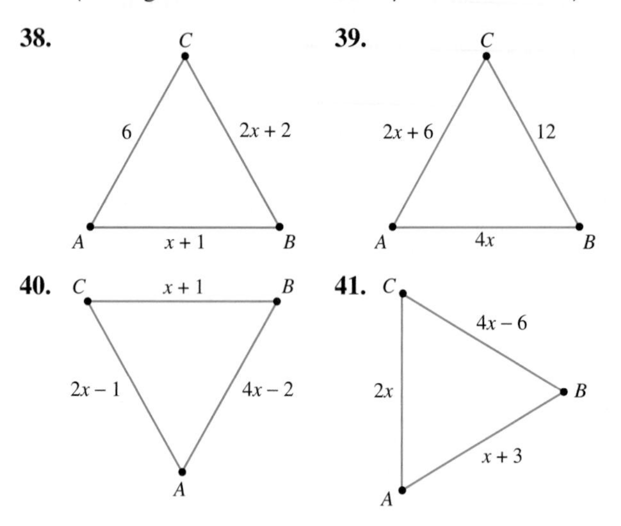

42. *Landscape Design* You are designing a patio. Your plans use a scale of $\frac{1}{8}$ inch to 1 foot. The patio is 24 feet by 36 feet. What are its dimensions on the plans?

43. *Architecture* The Pentagon, near Washington D.C., covers a region that is about 1200 feet by 1200 feet. About how large would a $\frac{1}{8}$-inch to 1-foot scale drawing of the Pentagon be? Would such a scale be reasonable?

Coordinate Geometry In Exercises 44 and 45, use the following figure.

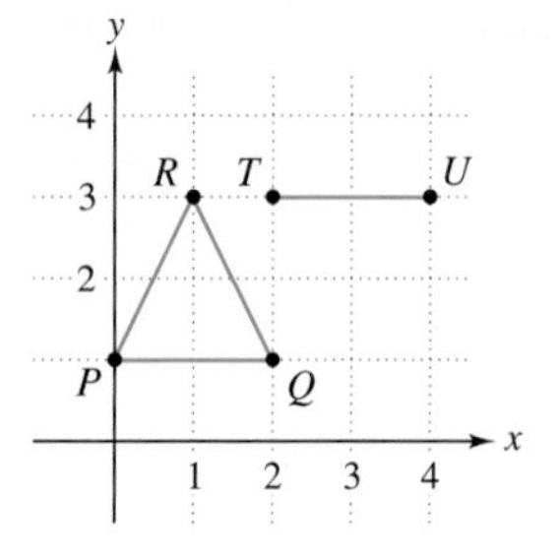

44. Find a location of S such that $\triangle PQR \cong \triangle PQS$.

45. Find two locations of V such that $\triangle PQR \cong \triangle TUV$.

46. *Logical Reasoning* Arrange 16 toothpicks as shown below. What is the least number of toothpicks you must remove to create four congruent triangles? (Each toothpick must be the side of at least one triangle.) Sketch your result.

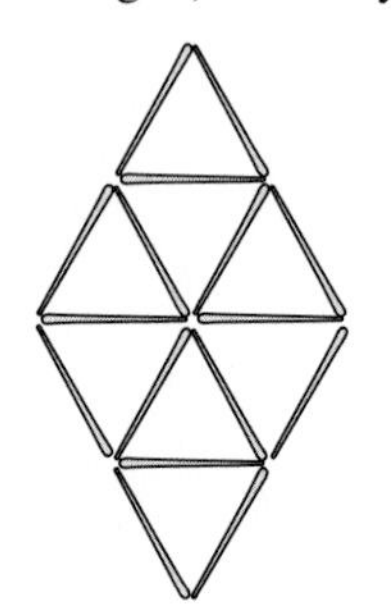

47. *Logical Reasoning* Show how you could arrange six toothpicks to form four congruent triangles. Each triangle has one toothpick for each side, and you cannot bend, break, or overlap the toothpicks. (*Hint:* The figure can be three-dimensional.)

C.2 Angles

Identifying Special Pairs of Angles ■ Angles Formed by a Transversal ■ Angles of a Triangle

Identifying Special Pairs of Angles

You have been introduced to several definitions concerning angles. For instance, you know that two angles are *adjacent* if they share a common vertex and side but have no common interior points. Here are some other definitions for pairs of angles. See Figure C.16.

2
1

∠1 and ∠2 are complementary angles.
Figure C.16

▶ **Definitions for Pairs of Angles**

Two angles are **vertical angles** if their sides form two pairs of opposite rays.

Two adjacent angles are a **linear pair** if their noncommon sides are opposite rays.

Two angles are **complementary** if the sum of their measures is 90°. Each angle is the *complement* of the other.

Two angles are **supplementary** if the sum of their measures is 180°. Each angle is the *supplement* of the other.

Example 1 Identifying Special Pairs of Angles

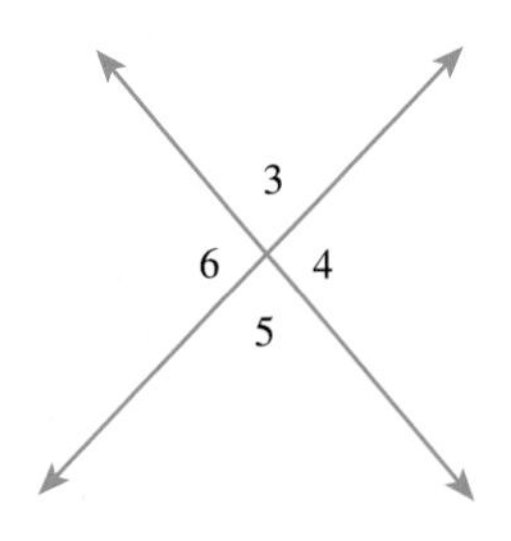

Figure C.17

Use the terms defined above to describe relationships between the labeled angles in Figure C.17.

Solution

a. $\angle 3$ and $\angle 5$ are vertical angles. So are $\angle 4$ and $\angle 6$.

b. There are four sets of linear pairs:

$\angle 3$ and $\angle 4$, $\angle 4$ and $\angle 5$, $\angle 5$ and $\angle 6$, and $\angle 3$ and $\angle 6$.

The angles in each of these pairs are also supplementary angles.

In Example 1(b), note that the linear pairs are also supplementary. This result is stated in the following postulate.

▶ **Linear Pair Postulate**

If two angles form a linear pair, then they are supplementary—i.e., the sum of their measures is 180°.

The relationship between vertical angles is stated in the following theorem.

▶ **Vertical Angles Theorem**

If two angles are vertical angles, then they are congruent.

Angles Formed by a Transversal

A **transversal** is a line that intersects two or more coplanar lines at different points. The angles that are formed when the transversal intersects the lines have the following names.

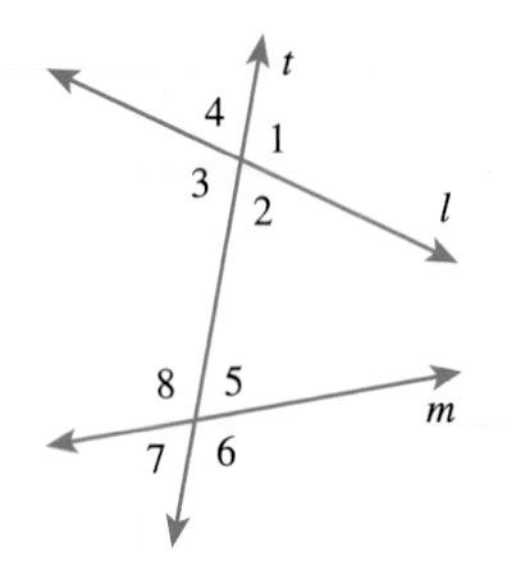

Figure C.18

> **▶ Angles Formed by a Transversal**
>
> In Figure C.18, the transversal t intersects the lines l and m.
>
> Two angles are **corresponding angles** if they occupy corresponding positions, such as $\angle 1$ and $\angle 5$.
>
> Two angles are **alternate interior angles** if they lie between l and m on opposite sides of t, such as $\angle 2$ and $\angle 8$.
>
> Two angles are **alternate exterior angles** if they lie outside l and m on opposite sides of t, such as $\angle 1$ and $\angle 7$.
>
> Two angles are **consecutive interior angles** if they lie between l and m on the same side of t, such as $\angle 2$ and $\angle 5$.

Example 2 Naming Pairs of Angles

In Figure C.19, how is $\angle 9$ related to the other angles?

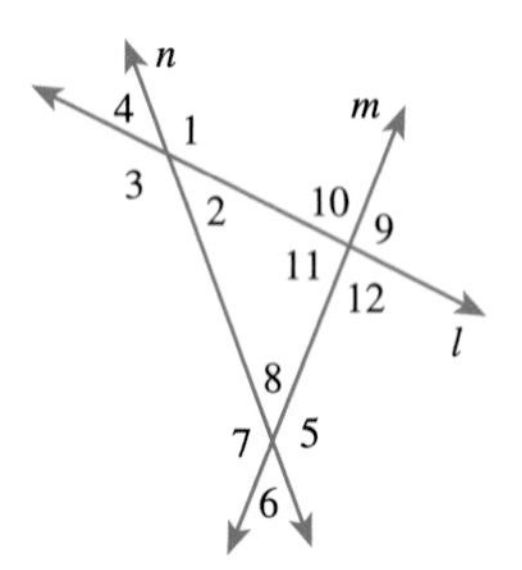

Figure C.19

Solution

You can consider that $\angle 9$ is formed by the transversal l as it intersects m and n, or you can consider $\angle 9$ to be formed by the transversal m as it intersects l and n. Considering one or the other of these, you have the following.

a. $\angle 9$ and $\angle 10$ are a linear pair. So are $\angle 9$ and $\angle 12$.

b. $\angle 9$ and $\angle 11$ are vertical angles.

c. $\angle 9$ and $\angle 7$ are alternate exterior angles. So are $\angle 9$ and $\angle 3$.

d. $\angle 9$ and $\angle 5$ are corresponding angles. So are $\angle 9$ and $\angle 1$.

To help build understanding regarding angles formed by a transversal, consider relationships between two lines. **Parallel lines** are coplanar lines that do not intersect. (Recall from Section 4.4 that two nonvertical lines are parallel if and only if they have the same slope.) **Intersecting lines** are coplanar and have exactly one point in common. If intersecting lines meet at right angles, they are perpendicular; otherwise, they are **oblique.** See Figure C.20.

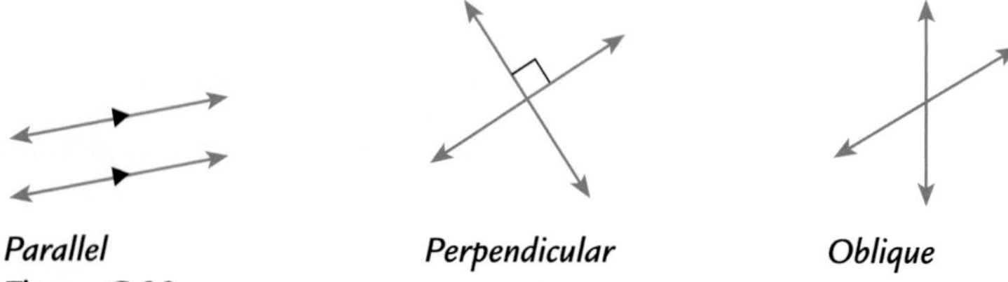

Figure C.20

Many of the angles formed by a transversal that intersects *parallel* lines are congruent. The following postulate and theorems list useful results.

> **▶ Corresponding Angles Postulate**
>
> If two parallel lines are cut by a transversal, then the pairs of corresponding angles are congruent.

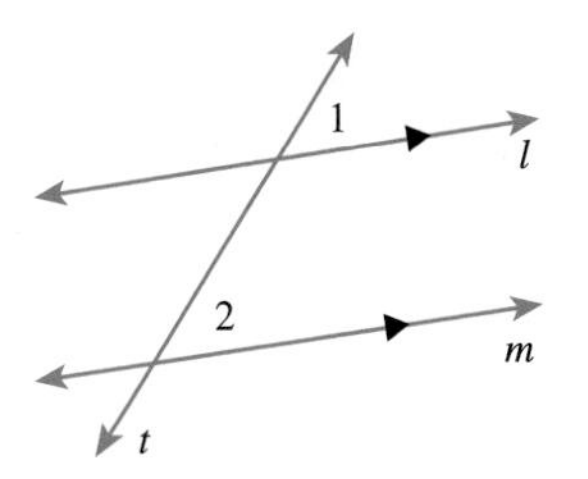

$l \parallel m$, $\angle 1 \cong \angle 2$
Figure C.21

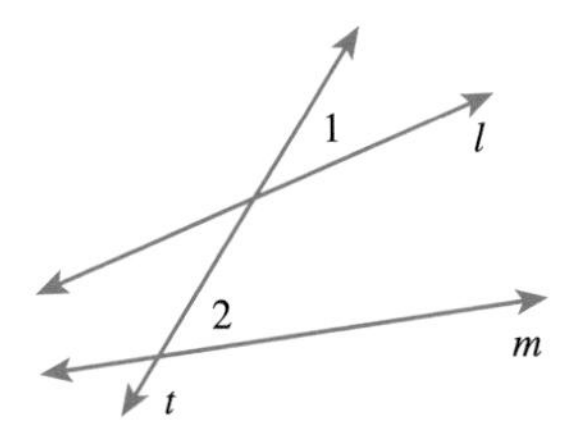

$l \nparallel m$, $\angle 1 \ncong \angle 2$
Figure C.22

Note that the hypothesis of this postulate states that the lines must be parallel, as shown in Figure C.21. If the lines are not parallel, then the corresponding angles are not congruent, as shown in Figure C.22.

> **Angle Theorems**
>
> **Alternate Interior Angles Theorem** If two parallel lines are cut by a transversal, then the pairs of alternate interior angles are congruent.
>
> **Consecutive Interior Angles Theorem** If two parallel lines are cut by a transversal, then the pairs of consecutive interior angles are supplementary.
>
> **Alternate Exterior Angles Theorem** If two parallel lines are cut by a transversal, then the pairs of alternate exterior angles are congruent.
>
> **Perpendicular Transversal Theorem** If a transversal is perpendicular to one of two parallel lines, then it is perpendicular to the second.

Example 3 Using Properties of Parallel Lines

In Figure C.23, lines r and s are parallel lines cut by a transversal, l. Find the measure of each labeled angle.

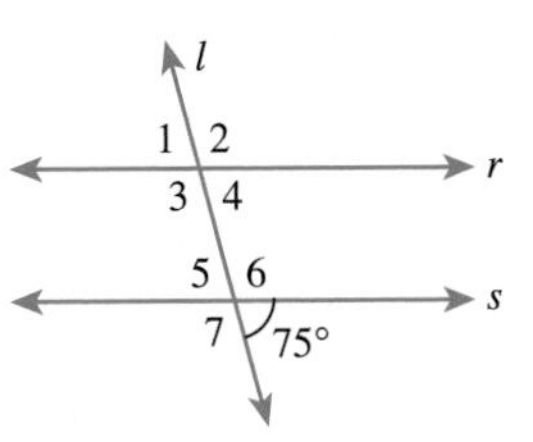

Figure C.23

Solution

$\angle 1$ and the given angle are alternate exterior angles and are congruent. So $m\angle 1 = 75°$. $\angle 5$ and the given angle are vertical angles. Because vertical angles are congruent, they have the same measure. So, $m\angle 5 = 75°$. Similarly, $\angle 1 \cong \angle 4$ and $m\angle 1 = m\angle 4 = 75°$. There are several sets of linear pairs, including

$\angle 1$ and $\angle 2$; $\angle 3$ and $\angle 4$; $\angle 5$ and $\angle 6$; $\angle 5$ and $\angle 7$.

The angles in each of these pairs are also supplementary angles; the sum of the measures of each pair of angles is 180°. Because one angle of each pair measures 75°, the supplements each measure 105°. So $\angle 2$, $\angle 3$, $\angle 6$, and $\angle 7$ each measure 105°.

Angles of a Triangle

The word "triangle" means "three angles." When the sides of a triangle are extended, however, other angles are formed. The original three angles of the triangle are the **interior angles.** The angles that are adjacent to the interior angles are the **exterior angles** of the triangle. Each vertex has a pair of exterior angles, as shown in Figure C.24 on the following page.

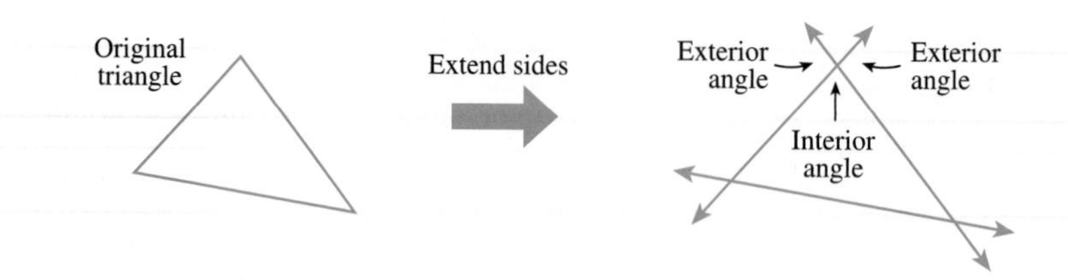

Figure C.24

You could cut a triangle out of a piece of paper. Tear off the three angles and place them adjacent to each other, as shown in Figure C.25. What do you observe? (You could perform this investigation by measuring with a protractor or using a computer drawing program.) You should arrive at the conclusion given in the following theorem.

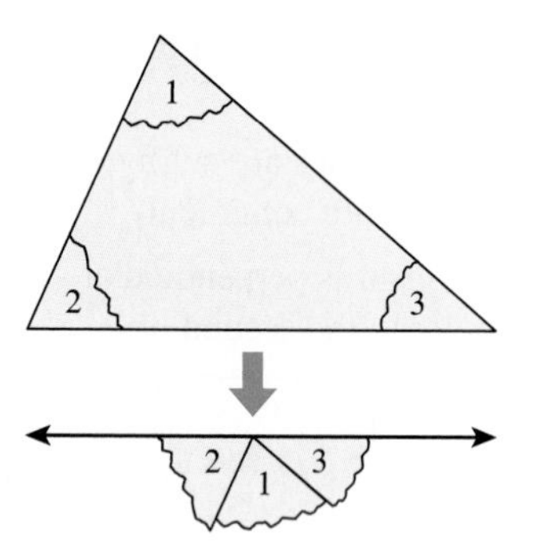

Figure C.25

> **Triangle Sum Theorem**
>
> The sum of the measures of the interior angles of a triangle is 180°.

Example 4 Using the Triangle Sum Theorem

In the triangle in Figure C.26, find $m\angle 1$, $m\angle 2$, and $m\angle 3$.

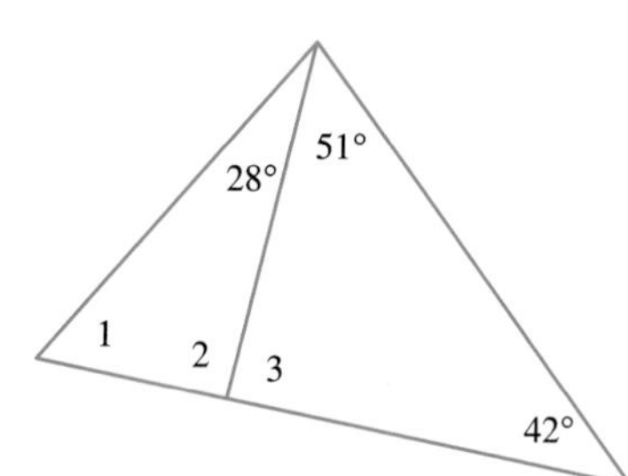

Figure C.26

Solution

To find the measure of $\angle 3$, use the Triangle Sum Theorem, as follows.

$$m\angle 3 = 180° - (51° + 42°) = 87°$$

Knowing the measure of $\angle 3$, you can use the Linear Pair Postulate to write $m\angle 2 = 180° - 87° = 93°$. Using the Triangle Sum Theorem, you have

$$m\angle 1 = 180° - (28° + 93°) = 59°.$$

The next theorem is one that you might have anticipated from the investigation earlier. As shown in Figure C.27, if you had torn only two of the angles from the paper triangle, you could put them together to cover exactly one of the exterior angles.

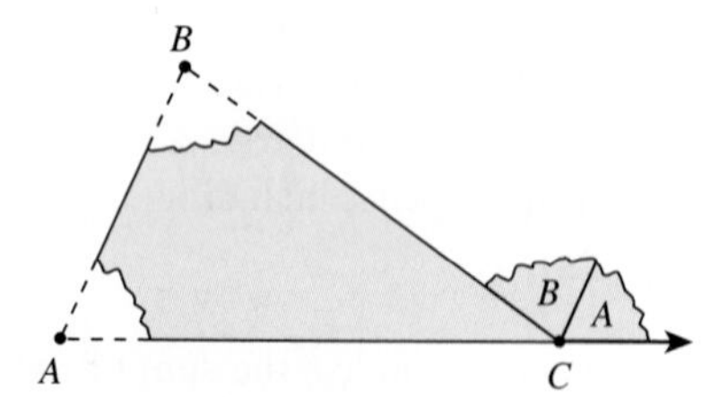

Figure C.27

> **Exterior Angle Theorem**
>
> The measure of an exterior angle of a triangle is equal to the sum of the measures of the two remote (nonadjacent) interior angles. (See Figure C.28.)

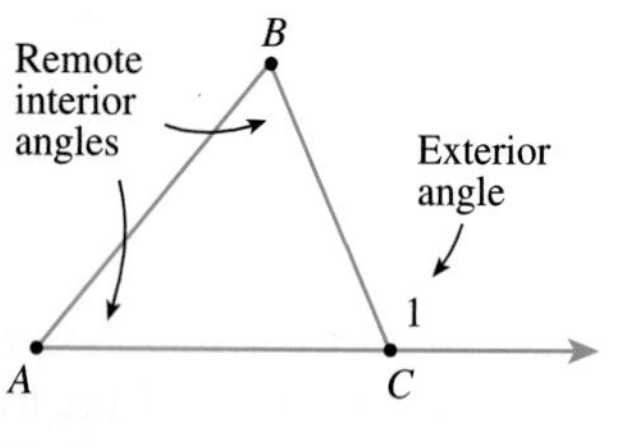

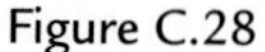
Figure C.28

C.2 Exercises

In Exercises 1–6, sketch a pair of angles that fits the description. Label the angles as $\angle 1$ and $\angle 2$.

1. A linear pair of angles
2. Supplementary angles for which $\angle 1$ is acute
3. Acute vertical angles
4. Adjacent congruent complementary angles
5. Obtuse vertical angles
6. Adjacent congruent supplementary angles

In Exercises 7–12, use the figure to determine relationships among the given angles.

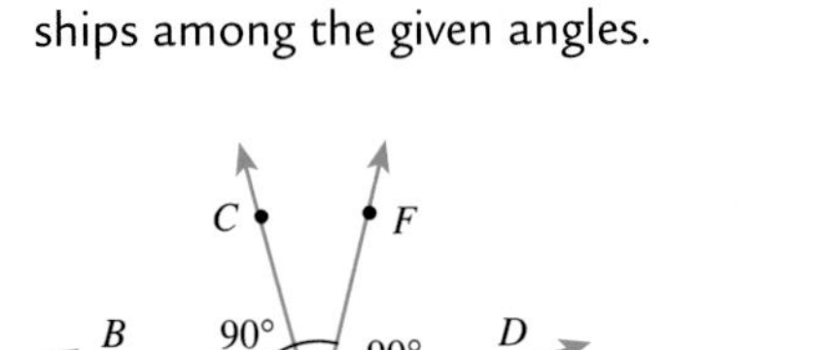

7. $\angle AOC$ and $\angle COD$
8. $\angle AOB$ and $\angle BOC$
9. $\angle BOC$ and $\angle COE$
10. $\angle AOB$ and $\angle EOD$
11. $\angle BOC$ and $\angle COF$
12. $\angle AOB$ and $\angle AOE$

In Exercises 13–18, use the following information to decide whether the statement is true or false. (*Hint*: Make a sketch.)

Vertical angles: $\angle 1$ and $\angle 2$;
Linear pairs: $\angle 1$ and $\angle 3$, $\angle 1$ and $\angle 4$.

13. If $m\angle 3 = 30°$, then $m\angle 4 = 150°$.
14. If $m\angle 1 = 150°$, then $m\angle 4 = 30°$.
15. $\angle 2$ and $\angle 3$ are congruent.
16. $m\angle 3 + m\angle 1 = m\angle 4 + m\angle 2$
17. $\angle 3 \cong \angle 4$
18. $m\angle 3 = 180° - m\angle 2$

In Exercises 19–24, find the value of x.

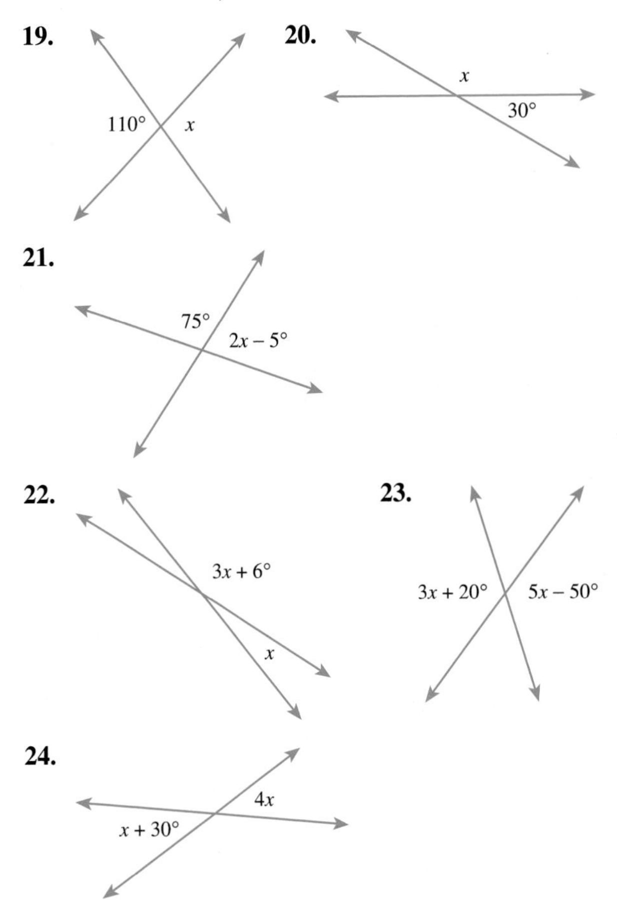

25. In the figure, P, S, and T are collinear. If $m\angle P = 40°$ and $m\angle QST = 110°$, what is $m\angle Q$? (*Hint:* $m\angle P + m\angle Q + m\angle PSQ = 180°$)

(a) 40° (b) 55° (c) 70° (d) 110° (e) 140°

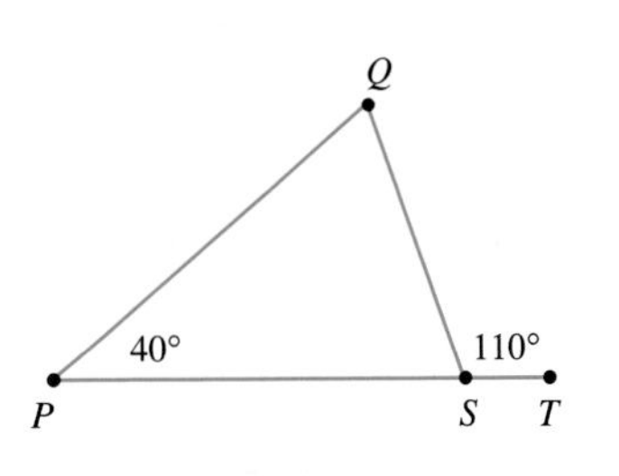

In Exercises 26–29, use the figure below.

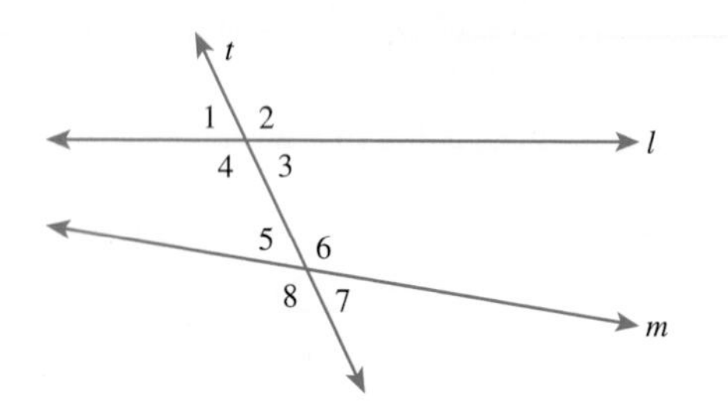

26. Name two corresponding angles.

27. Name two alternate interior angles.

28. Name two alternate exterior angles.

29. Name two consecutive interior angles.

In Exercises 30–33, $l_1 \parallel l_2$. Find the measures of $\angle 1$ and $\angle 2$. Explain your reasoning.

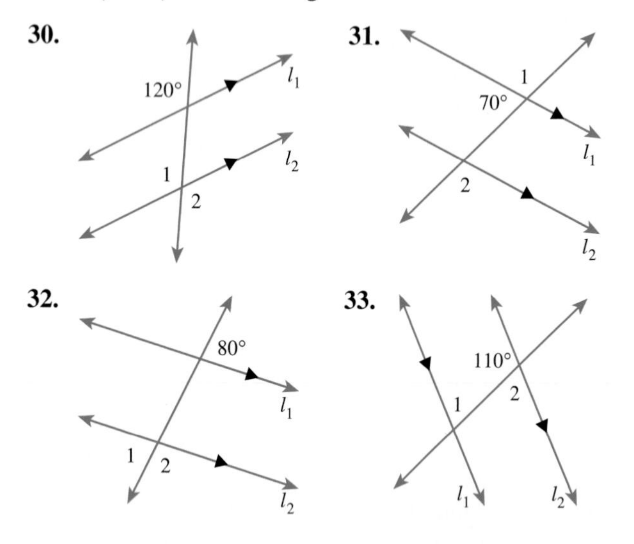

In Exercises 34–36, $m \parallel n$ and $k \parallel l$. Determine the values of a and b.

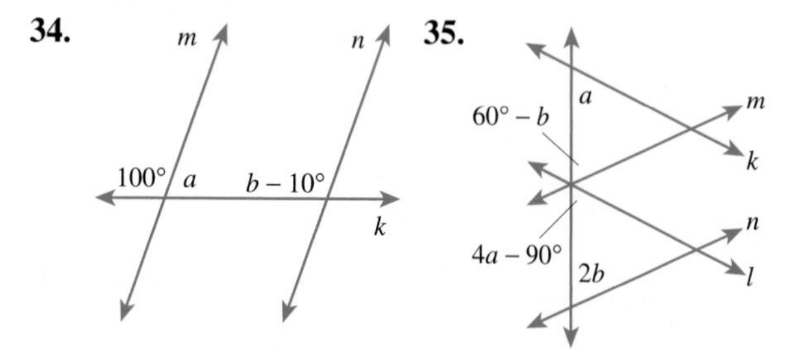

36.

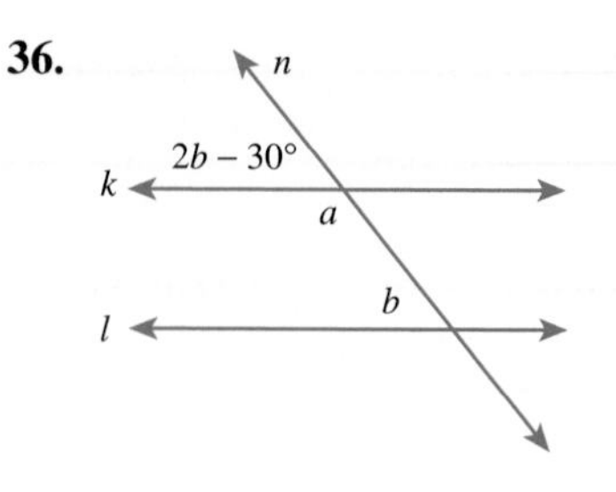

In Exercises 37–39, use the following figure.

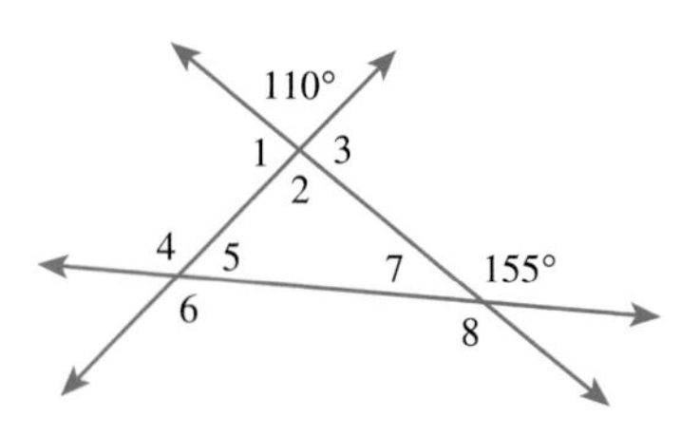

37. Name the interior angles of the triangle.

38. Name the exterior angles of the triangle.

39. Two angle measures are given in the figure. Find the measure of the eight labeled angles.

In Exercises 40–43, use the following figure, in which $\triangle ABC \cong \triangle DEF$.

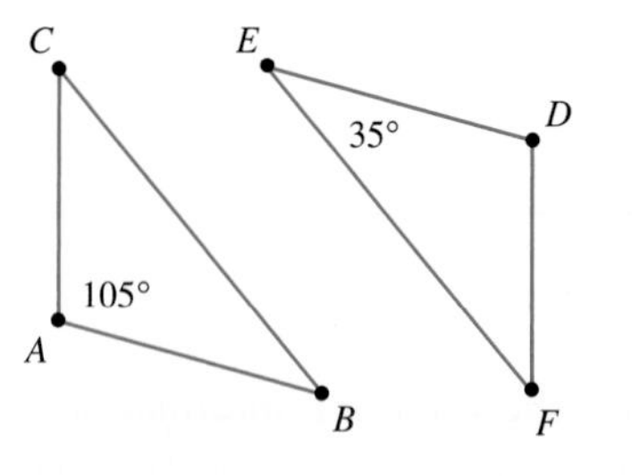

40. What is the measure of $\angle D$?

41. What is the measure of $\angle B$?

42. What is the measure of $\angle C$?

43. What is the measure of $\angle F$?

44. *True or False?* A right triangle can have an obtuse angle.

45. *True or False?* A triangle that has two 60° angles must be equiangular.

46. *True or False?* If a right triangle has two congruent angles, then it must have two 45° angles.

In Exercises 47 and 48, find the measure of each labeled angle.

47.

48.

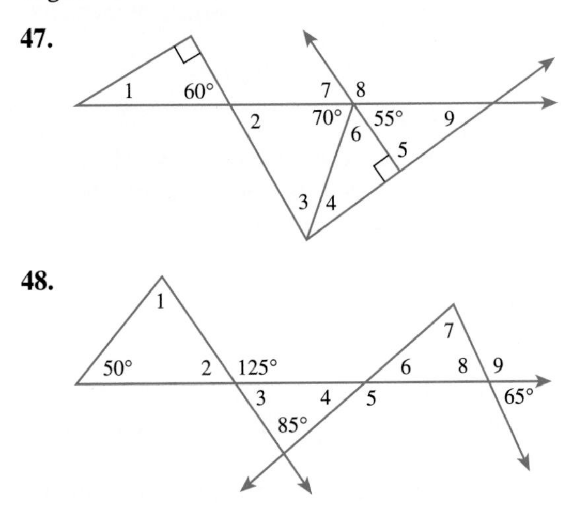

In Exercises 49 and 50, draw and label a right triangle, $\triangle ABC$, for which the right angle is $\angle C$. What is $m\angle B$?

49. $m\angle A = 13°$ **50.** $m\angle A = 47°$

In Exercises 51 and 52, draw two noncongruent, isosceles triangles that have an exterior angle with the given measure.

51. 130° **52.** 145°

In Exercises 53–56, find the measures of the interior angles.

53.

54.

55.

56.

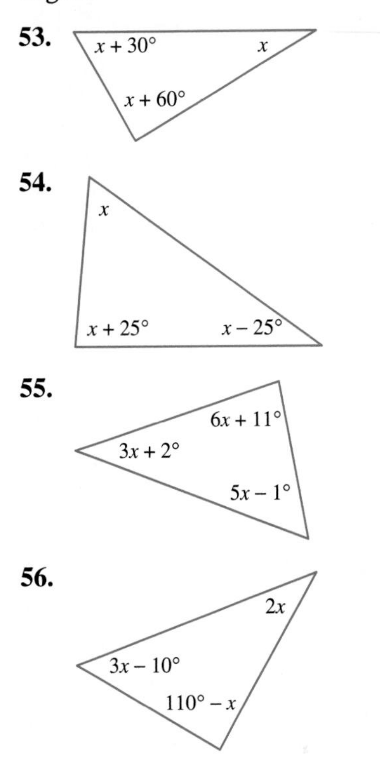

Appendix D
Further Concepts in Statistics

Stem-and-Leaf Plots ■ Histograms and Frequency Distributions ■ Line Graphs ■ Choosing an Appropriate Graph ■ Scatter Plots ■ Fitting a Line to Data ■ Measures of Central Tendency

Stem-and-Leaf Plots

Statistics is the branch of mathematics that studies techniques for collecting, organizing, and interpreting data. In this section, you will study several ways to organize and interpret data.

One type of plot that can be used to organize sets of numbers by hand is a **stem-and-leaf plot.** A set of test scores and the corresponding stem-and-leaf plot are shown below.

Test Scores

93, 70, 76, 58, 86, 93, 82, 78, 83, 86,
64, 78, 76, 66, 83, 83, 96, 74, 69, 76,
64, 74, 79, 76, 88, 76, 81, 82, 74, 70

Stems	*Leaves*
5	8
6	4 4 6 9
7	0 0 4 4 4 6 6 6 6 6 8 8 9
8	1 2 2 3 3 3 6 6 8
9	3 3 6

Note that the *leaves* represent the units digits of the numbers and the *stems* represent the tens digits. Stem-and-leaf plots can also be used to compare two sets of data, as shown in the following example.

Example 1 Comparing Two Sets of Data

Use a stem-and-leaf plot to compare the test scores given above with the following test scores. Which set of test scores is better?

90, 81, 70, 62, 64, 73, 81, 92, 73, 81, 92, 93, 83, 75, 76,
83, 94, 96, 86, 77, 77, 86, 96, 86, 77, 86, 87, 87, 79, 88

Solution

Begin by ordering the second set of scores.

62, 64, 70, 73, 73, 75, 76, 77, 77, 77, 79, 81, 81, 81, 83,
83, 86, 86, 86, 86, 87, 87, 88, 90, 92, 92, 93, 94, 96, 96

Now that the data have been ordered, you can construct a *double* stem-and-leaf plot by letting the leaves to the right of the stems represent the units digits for the first group of test scores and letting the leaves to the left of the stems represent the units digits for the second group of test scores.

Leaves (2nd Group)	*Stems*	*Leaves (1st Group)*
	5	8
4 2	6	4 4 6 9
9 7 7 7 6 5 3 3 0	7	0 0 4 4 4 6 6 6 6 6 8 8 9
8 7 7 6 6 6 6 3 3 1 1 1	8	1 2 2 3 3 3 6 6 8
6 6 4 3 2 2 0	9	3 3 6

By comparing the two sets of leaves, you can see that the second group of test scores is better than the first group.

Example 2 Using a Stem-and-Leaf Plot

The table below shows the percent of the population of each state and the District of Columbia that was at least 65 years old in 1997. Use a stem-and-leaf plot to organize the data. (Source: U.S. Bureau of the Census)

AK	5.3	AL	13.0	AR	14.3	AZ	13.2	CA	11.1
CO	10.1	CT	14.4	DC	13.9	DE	12.9	FL	18.5
GA	9.9	HI	13.2	IA	15.0	ID	11.3	IL	12.5
IN	12.5	KS	13.5	KY	12.5	LA	11.4	MA	14.1
MD	11.5	ME	13.9	MI	12.4	MN	12.3	MO	13.7
MS	12.2	MT	13.2	NC	12.5	ND	14.4	NE	13.7
NH	12.1	NJ	13.7	NM	11.2	NV	11.5	NY	13.4
OH	13.4	OK	13.4	OR	13.3	PA	15.8	RI	15.8
SC	12.1	SD	14.3	TN	12.5	TX	10.1	UT	8.7
VA	11.2	VT	12.3	WA	11.5	WI	13.2	WV	15.1
WY	11.3								

Solution

Begin by ordering the numbers, as shown below.

5.3, 8.7, 9.9, 10.1, 10.1, 11.1, 11.2, 11.2, 11.3, 11.3, 11.4, 11.5, 11.5, 11.5, 12.1, 12.1, 12.2, 12.3, 12.3, 12.4, 12.5, 12.5, 12.5, 12.5, 12.5, 12.9, 13.0, 13.2, 13.2, 13.2, 13.2, 13.3, 13.4, 13.4, 13.4, 13.5, 13.7, 13.7, 13.7, 13.9, 13.9, 14.1, 14.3, 14.3, 14.4, 14.4, 15.0, 15.1, 15.8, 15.8, 18.5

Next construct the stem-and-leaf plot using the leaves to represent the digits to the right of the decimal points.

Stems	*Leaves*	
5.	3	Alaska has the lowest percent.
6.		
7.		
8.	7	
9.	9	
10.	1 1	
11.	1 2 2 3 3 4 5 5 5	
12.	1 1 2 3 3 4 5 5 5 5 5 9	
13.	0 2 2 2 2 3 4 4 4 5 7 7 7 9 9	
14.	1 3 3 4 4	
15.	0 1 8 8	
16.		
17.		
18.	5	Florida has the highest percent.

Histograms and Frequency Distributions

With data such as those given in Example 2, it is useful to group the numbers into intervals and plot the frequency of the data in each interval. For instance, the **frequency distribution** and **histogram** shown in Figure D.1 represent the data given in Example 2.

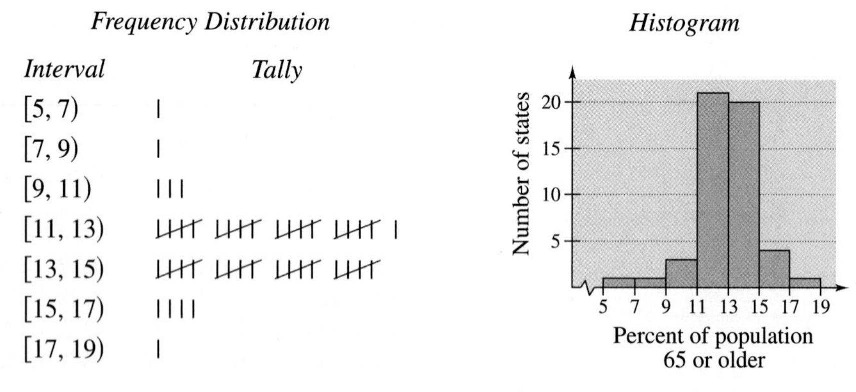

Frequency Distribution

Interval	*Tally*
[5, 7)	|
[7, 9)	|
[9, 11)	|||
[11, 13)	卌 卌 卌 卌 |
[13, 15)	卌 卌 卌 卌
[15, 17)	||||
[17, 19)	|

Figure D.1

A histogram has a portion of a real number line as its horizontal axis. A **bar graph** is similar to a histogram, except that the rectangles (bars) can be either horizontal or vertical and the labels of the bars are not necessarily numbers.

Another difference between a bar graph and a histogram is that the bars in a bar graph are usually separated by spaces, whereas the bars in a histogram are not separated by spaces.

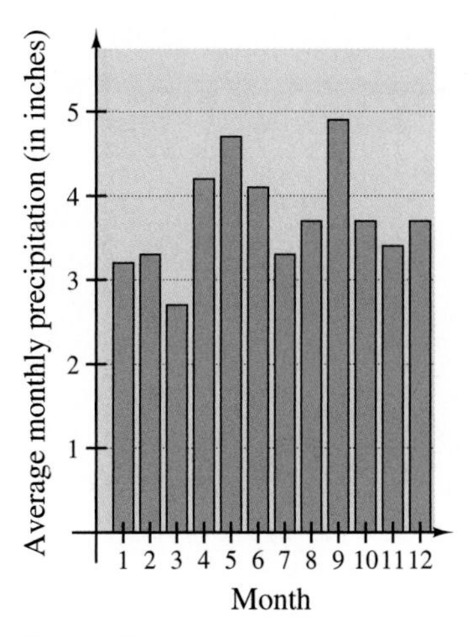

Figure D.2

Example 3 Constructing a Bar Graph

The data below show the average monthly precipitation (in inches) in Houston, Texas. Construct a bar graph for these data. What can you conclude? (Source: PC USA)

January	3.2	February	3.3	March	2.7
April	4.2	May	4.7	June	4.1
July	3.3	August	3.7	September	4.9
October	3.7	November	3.4	December	3.7

Solution

To create a bar graph, begin by drawing a vertical axis to represent the precipitation and a horizontal axis to represent the months. The bar graph is shown in Figure D.2. From the graph, you can see that Houston receives a fairly consistent amount of rain throughout the year—the driest month tends to be March and the wettest month tends to be September.

Line Graphs

A **line graph** is similar to a standard coordinate graph. Line graphs are usually used to show trends over periods of time.

Example 4 Constructing a Line Graph

The following data show the number of immigrants (in thousands) to the United States for the years 1970 through 1996. Construct a line graph of the data. What can you conclude? (Source: U.S. Immigration and Naturalization Service)

Year	*Number*	*Year*	*Number*	*Year*	*Number*
1970	373	1971	370	1972	385
1973	400	1974	395	1975	386
1976	399	1977	462	1978	601
1979	460	1980	531	1981	597
1982	594	1983	560	1984	544
1985	570	1986	602	1987	602
1988	643	1989	1091	1990	1536
1991	1827	1992	974	1993	904
1994	804	1995	720	1996	916

Solution

Begin by drawing a vertical axis to represent the number of immigrants in thousands. Then label the horizontal axis with years and plot the points shown in the table. Finally, connect the points with line segments, as shown on the next page in Figure D.3. From the line graph, you can see that the number of immigrants steadily increased until 1989, when there was a sharp increase followed by a sudden decrease in 1992.

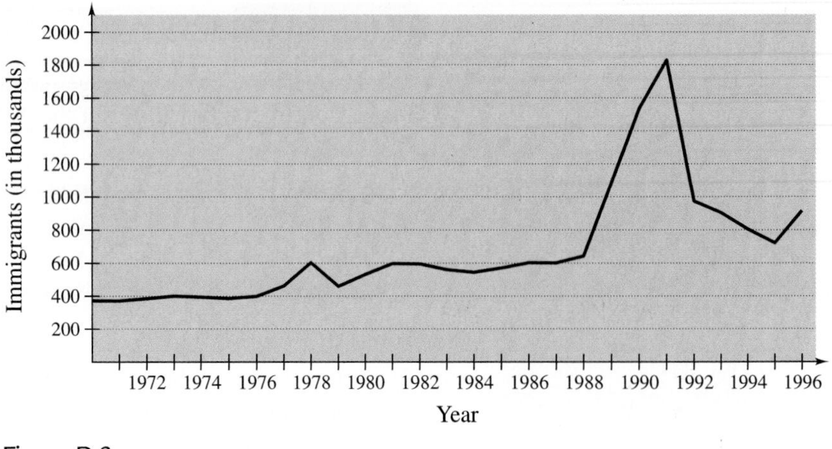

Figure D.3

Choosing an Appropriate Graph

Line graphs and bar graph are commonly used for displaying data. When you are using a graph to organize and present data, you must first decide which type of graph to use.

Example 5 Organizing Data with a Graph

Listed below are the daily average numbers of miles walked by people while working at their jobs. Organize the data graphically. (Source: American Podiatry Association)

Occupation	*Miles Walked per Day*
Mail Carrier	4.4
Medical Doctor	3.5
Nurse	3.9
Police Officer	6.8
Television Reporter	4.2

Solution

You can use a bar graph because the data fall into distinct categories, and it would be useful to compare totals. The bar graph shown in Figure D.4 is horizontal. This makes it easier to label each bar. Also notice that the occupations are listed in order of the number of miles walked.

Study Tip

Here are some guidelines to use when you must decide which type of graph to use.

1. Use a bar graph when the data fall into distinct categories and you want to compare totals.
2. Use a line graph when you want to show the relationship between consecutive amounts or data over time.

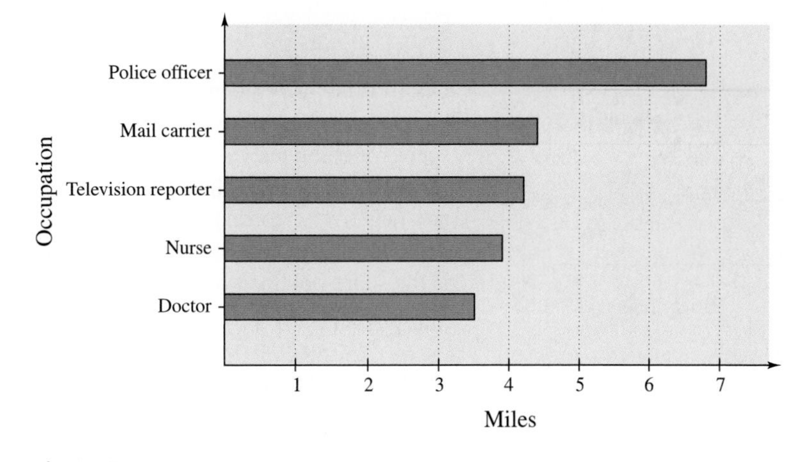

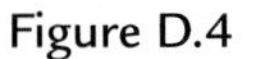
Figure D.4

Scatter Plots

Many real-life situations involve finding relationships between two variables, such as the year and the number of people in the labor force. In a typical situation, data are collected and written as a set of ordered pairs. The graph of such a set is called a **scatter plot.**

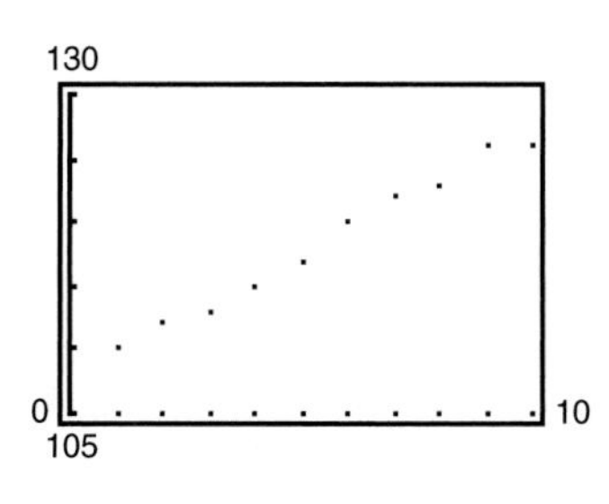

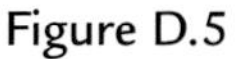
Figure D.5

From the scatter plot in Figure D.5 that relates the year t with the number of people in the labor force P, it appears that the points describe a relationship that is nearly linear. (The relationship is not *exactly* linear because the labor force did not increase by precisely the same amount each year.) A mathematical equation that approximates the relationship between t and P is called a *mathematical model.* When developing a mathematical model, you strive for two (often conflicting) goals—accuracy and simplicity.

Consider a collection of ordered pairs of the form (x, y). If y tends to increase as x increases, the collection is said to have a **positive correlation.** If y tends to decrease as x increases, the collection is said to have a **negative correlation.** Figure D.6 shows three examples: one with a positive correlation, one with a negative correlation, and one with no (discernible) correlation.

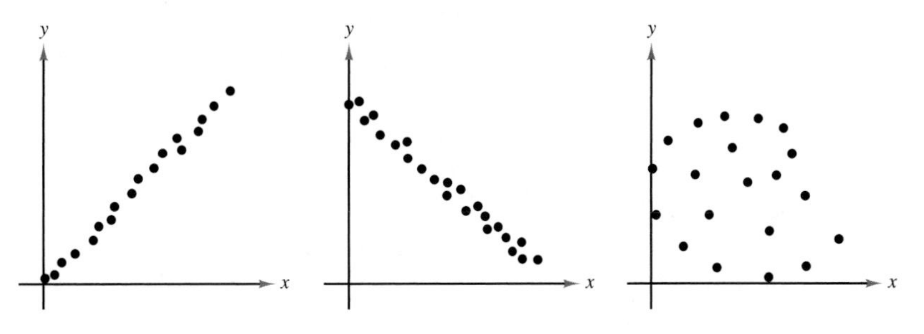

Figure D.6

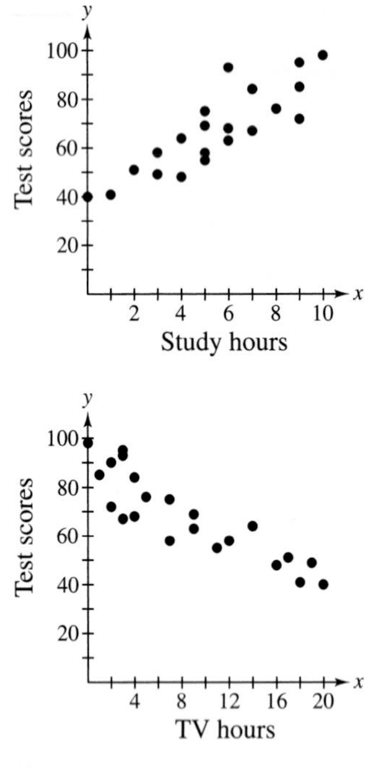

Figure D.7

Example 6 Interpreting Scatter Plots

On a Friday, 22 students in a class were asked to keep track of the numbers of hours they spent studying for a test on Monday and the numbers of hours they spent watching television. The numbers are shown below. Construct a scatter plot for each set of data. Then determine whether the points are positively correlated, are negatively correlated, or have no discernible correlation. What can you conclude? (The first coordinate is the number of hours and the second coordinate is the score obtained on Monday's test.)

Study Hours: (0, 40), (1, 41), (2, 51), (3, 58), (3, 49), (4, 48), (4, 64), (5, 55), (5, 69), (5, 58), (5, 75), (6, 68), (6, 63), (6, 93), (7, 84), (7, 67), (8, 90), (8, 76), (9, 95), (9, 72), (9, 85), (10, 98)

TV Hours: (0, 98), (1, 85), (2, 72), (2, 90), (3, 67), (3, 93), (3, 95), (4, 68), (4, 84), (5, 76), (7, 75), (7, 58), (9, 63), (9, 69), (11, 55), (12, 58), (14, 64), (16, 48), (17, 51), (18, 41), (19, 49), (20, 40)

Solution

Scatter plots for the two sets of data are shown in Figure D.7. The scatter plot relating study hours and test scores has a positive correlation. This means that the more a student studied, the higher his or her score tended to be. The scatter plot relating television hours and test scores has a negative correlation. This means that the more time a student spent watching television, the lower his or her score tended to be.

Fitting a Line to Data

Finding a linear model that represents the relationship described by a scatter plot is called **fitting a line to data.** You can do this graphically by simply sketching the line that appears to fit the points, finding two points on the line, and then finding the equation of the line that passes through the two points.

Example 7 Fitting a Line to Data

Find a linear model that relates the year and the number of people P (in millions) who were part of the United States labor force from 1987 through 1997. In the table below, t represents the year, with $t = 0$ corresponding to 1987. (Source: U.S. Bureau of Labor Statistics)

t	0	1	2	3	4	5	6	7	8	9	10
P	120	122	124	126	126	128	129	131	132	134	136

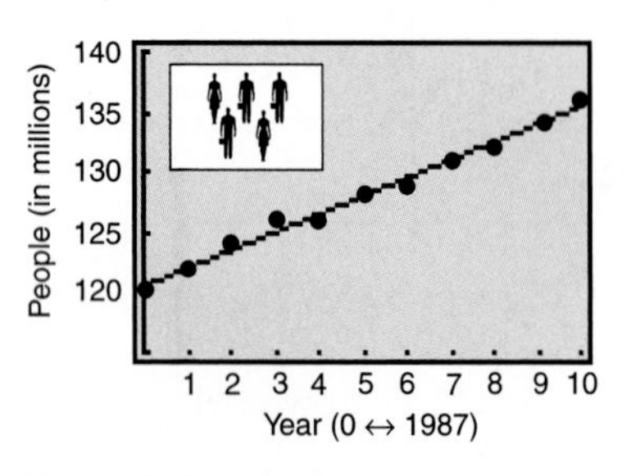

Figure D.8

Solution

After plotting the data from the table, draw the line that you think best represents the data, as shown in Figure D.8. Two points that lie on this line are (0, 120) and (9, 134). Using the point-slope form, you can find the equation of the line to be $P = 14t/9 + 120$.

Once you have found a model, you can measure how well the model fits the data by comparing the actual values with the values given by the model, as shown in the following table for the data and model in Example 7.

t	0	1	2	3	4	5	6	7	8	9	10
Actual P	120	122	124	126	126	128	129	131	132	134	136
Model P	120	121.6	123.1	124.7	126.2	127.8	129.3	130.9	132.4	134	135.6

The sum of the squares of the differences between the actual values and the model's values is the **sum of the squared differences.** The model that has the least sum is called the **least squares regression line** for the data. For the model in Example 7, the sum of the squared differences is 3.16. The least squares regression line for the data is

$$P = 1.5t + 120.5. \qquad \text{Best-fitting linear model}$$

The sum of the squared differences is 2.5.

Many graphing utilities have "built-in" least squares regression programs. If your graphing utility has such a program, enter the data in the table and use it to find the least squares regression line.

Measures of Central Tendency

In many real-life situations, it is helpful to describe data by a single number that is most representative of the entire collection of numbers. Such a number is called a **measure of central tendency.** The most commonly used measures are as follows.

1. The **mean,** or **average,** of n numbers is the sum of the numbers divided by n.
2. The **median** of n numbers is the middle number when the numbers are written in order. If n is even, the median is the average of the two middle numbers.
3. The **mode** of n numbers is the number that occurs most frequently. If two numbers tie for most frequent occurrence, the collection has two modes and is called **bimodal.**

Example 8 Comparing Measures of Central Tendency

You are interviewing for a job. The interviewer tells you that the average income of the company's 25 employees is \$60,849. The actual annual incomes of the 25 employees are shown below. What are the mean, median, and mode of the incomes? Was the person telling you the truth?

\$17,305,	\$478,320,	\$45,678,	\$18,980,	\$17,408,
\$25,676,	\$28,906,	\$12,500,	\$24,540,	\$33,450,
\$12,500,	\$33,855,	\$37,450,	\$20,432,	\$28,956,
\$34,983,	\$36,540,	\$250,921,	\$36,853,	\$16,430,
\$32,654,	\$98,213,	\$48,980,	\$94,024,	\$35,671

Solution

The mean of the incomes is

$$\text{Mean} = \frac{17{,}305 + 478{,}320 + 45{,}678 + 18{,}980 + \cdot \cdot \cdot + 35{,}671}{25}$$

$$= \frac{1{,}521{,}225}{25} = \$60{,}849.$$

To find the median, order the incomes as follows.

$12,500,	$12,500,	$16,430,	$17,305,	$17,408,
$18,980,	$20,432,	$24,540,	$25,676,	$28,906,
$28,956,	$32,654,	$33,450,	$33,855,	$34,983,
$35,671,	$36,540,	$36,853,	$37,450,	$45,678,
$48,980,	$94,024,	$98,213,	$250,921,	$478,320

From this list, you can see that the median (the middle number) is $33,450. From the same list, you can see that $12,500 is the only income that occurs more than once. Thus, the mode is $12,500. Technically, the person was telling the truth because the average is (generally) defined to be the mean. However, of the three measures of central tendency

Mean: $60,849 *Median:* $33,450 *Mode:* $12,500

it seems clear that the median is most representative. The mean is inflated by the two highest salaries.

Which of the three measures of central tendency is the most representative? The answer is that it depends on the distribution of the data *and* the way in which you plan to use the data.

For instance, in Example 8, the mean salary of $60,849 does not seem very representative to a potential employee. To a city income tax collector who wants to estimate 1% of the total income of the 25 employees, however, the mean is precisely the right measure.

Example 9 Choosing a Measure of Central Tendency

Which measure of central tendency is the most representative of the data given in each of the following frequency distributions?

a.

Number	*Tally*
1	7
2	20
3	15
4	11
5	8
6	3
7	2
8	0
9	15

b.

Number	*Tally*
1	9
2	8
3	7
4	6
5	5
6	6
7	7
8	8
9	9

c.

Number	*Tally*
1	6
2	1
3	2
4	3
5	5
6	5
7	4
8	3
9	0

Solution

a. For these data, the mean is 4.23, the median is 3, and the mode is 2. Of these, the mode is probably the most representative.

b. For these data, the mean and median are each 5 and the modes are 1 and 9 (the distribution is bimodal). Of these, the mean or median is the most representative.

c. For these data, the mean is 4.59, the median is 5, and the mode is 1. Of these, the mean or median is the most representative.

Appendix D Exercises

1. Construct a stem-and-leaf plot for the following exam scores for a class of 30 students. The scores are for a 100-point exam.

77, 100, 77, 70, 83, 89, 87, 85, 81, 84, 81, 78, 89, 78, 88, 85, 90, 92, 75, 81, 85, 100, 98, 81, 78, 75, 85, 89, 82, 75

2. *Insurance Coverage* The following table shows the total number of persons (in thousands) without health insurance coverage in the 50 states and the District of Columbia in 1996. Use a stem-and-leaf plot to organize the data. (Source: U.S. Bureau of the Census)

AK	89	AL	550	AR	566	AZ	1159	CA	6514
CO	644	CT	368	DC	80	DE	98	FL	2722
GA	1319	HI	101	IA	335	ID	196	IL	1337
IN	600	KS	292	KY	601	LA	890	MA	766
MD	581	ME	146	MI	857	MN	480	MO	700
MS	518	MT	124	NC	1160	ND	62	NE	190
NH	109	NJ	1317	NM	412	NV	255	NY	3132
OH	1292	OK	570	OR	496	PA	1133	RI	93
SC	634	SD	67	TN	841	TX	4680	UT	240
VA	811	VT	65	WA	761	WI	438	WV	261
WY	66								

In Exercises 3 and 4, use the following set of data, which lists students' scores on a 100-point exam.

93, 84, 100, 92, 66, 89, 78, 52, 71, 85, 83, 95, 98, 99, 93, 81, 80, 79, 67, 59, 90, 55, 77, 62, 90, 78, 66, 63, 93, 87, 74, 96, 72, 100, 70, 73

3. Use a stem-and-leaf plot to organize the data.

4. Draw a histogram to represent the data.

5. Complete the following frequency distribution table and draw a histogram to represent the data.

44, 33, 17, 23, 16, 18, 44, 47, 18, 20, 25, 27, 18, 29, 29, 28, 27, 18, 36, 22, 32, 38, 33, 41, 49, 48, 45, 38, 49, 15

Interval	*Tally*
[15, 22)	
[22, 29)	
[29, 36)	
[36, 43)	
[43, 50)	

6. *Snowfall* The data below show the seasonal snowfall (in inches) in Lincoln, Nebraska for the years 1968 through 1997 (the amounts are listed in order by year). How would you organize these data? Explain your reasoning. (Source: University of Nebraska-Lincoln)

39.8, 26.2, 49.0, 21.6, 29.2, 33.6, 42.1, 21.1, 21.8, 31.0, 34.4, 23.3, 13.0, 32.3, 38.0, 47.5, 21.5, 18.9, 15.7, 13.0, 19.1, 18.7, 25.8, 23.8, 32.1, 21.3, 21.8, 30.7, 29.0, 44.6

7. *Travel to the United States* The data below give the places of origin and the numbers of travelers (in millions) to the United States in 1995. Construct a bar graph for these data. (Source: U.S. Department of Commerce)

Canada	14.7	Mexico	8.0
Europe	8.8	Far East	6.6
Other	5.2		

8. *Fruit Crops* The data below show the cash receipts (in millions of dollars) from fruit crops of farmers in 1996. Construct a bar graph for these data. (Source: U.S. Department of Agriculture)

Apples	1846	Peaches	380
Cherries	264	Pears	292
Grapes	2334	Plums and Prunes	295
Lemons	228	Strawberries	770
Oranges	1798		

Handling Garbage In Exercises 9–14, use the line graph given below. (Source: Franklin Associates)

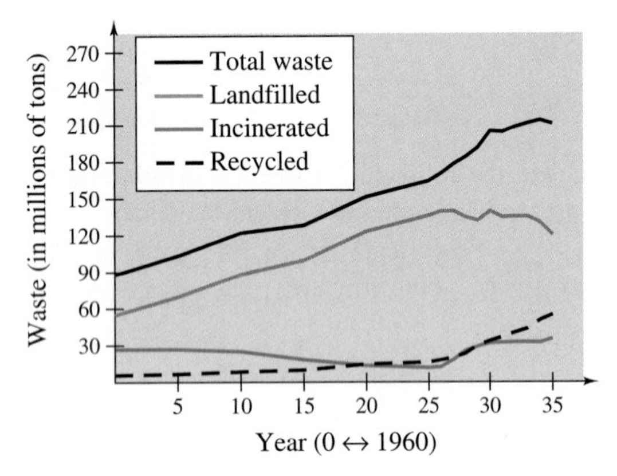

9. Estimate the total waste in 1985 and 1995.

10. Estimate the amount of incinerated garbage in 1990.

11. Which quantities increased every year?

12. During which time period did the amount of incinerated garbage decrease?

13. What is the relationship among the four quantities in the line graph?

14. Why do you think landfill garbage is decreasing?

15. *College Attendance* The following table shows the enrollment in a liberal arts college. Construct a line graph for the data.

Year	1993	1994	1995	1996
Enrollment	1675	1704	1710	1768

Year	1997	1998	1999	2000
Enrollment	1833	1918	1967	1972

16. *Oil Imports* The table shows the crude oil imports into the United States in millions of barrels for the years 1988 through 1997. Construct a line graph for the data and state what information it reveals. (Source: U.S. Energy Information Administration)

Year	1988	1989	1990	1991	1992
Oil imports	1864	2133	2151	2110	2220

Year	1993	1994	1995	1996	1997
Oil imports	2477	2578	2643	2748	2918

Table for 16

17. *Stock Market* The list below shows stock prices for selected companies in April of 1999. Draw a graph that best represents the data. Explain why you chose that type of graph. (Source: Value Line)

Company	*Stock Price*
Sears, Roebuck	$46
Wal-Mart Stores	$98
JC Penney	$40
K Mart Corp.	$16
The Gap, Inc.	$68

18. *Net Profit* The table shows the net profits (in millions of dollars) of Callaway Golf Co. for the years 1992 through 1997. Draw a graph that best represents the net profit and explain why you chose that type of graph. (Source: Value Line)

Year	1992	1993	1994	1995	1996	1997
Net profit	19.3	41.2	78.0	97.7	122.3	139.9

19. *Camcorders* The factory sales (in millions of dollars) of camcorders for the years 1990 through 1996 are given in the table. Organize the data graphically. Explain your reasoning. (Source: Electronic Industries Association)

Year	1990	1991	1992	1993	1994	1995	1996
Number	2260	2013	1841	1958	1985	2135	2084

20. *Owning Cats* The average numbers (out of 100) of cat owners who state various reasons for owning a cat are listed below. Organize the data graphically. Explain your reasoning. (Source: Gallup Poll)

Reason for Owning a Cat	*Number*
Have a pet to play with	93
Companionship	84
Help children learn responsibility	78
Have a pet to communicate with	62
Security	51

Interpreting a Scatter Plot In Exercises 21–24, use the scatter plot shown. The scatter plot compares the number of hits x made by 30 softball players during the first half of the season with the number of runs batted in y.

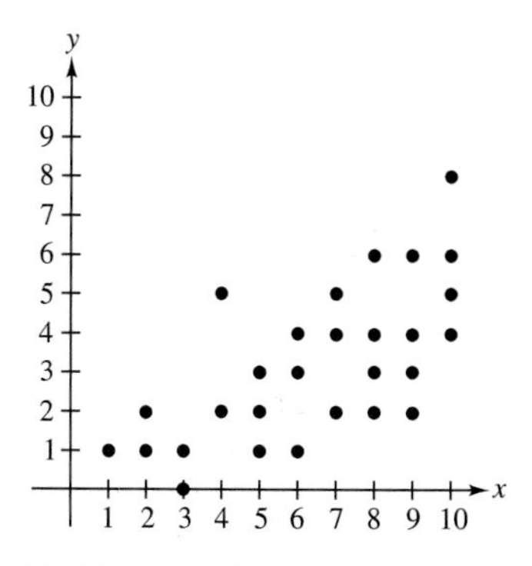

21. Do x and y have a positive correlation, a negative correlation, or no correlation?

22. Why does the scatter plot show only 28 points?

23. From the scatter plot, does it appear that players with more hits tend to have more runs batted in?

24. Can a player have more runs batted in than hits? Explain.

In Exercises 25–28, decide whether a scatter plot relating the two quantities would tend to have a positive, a negative, or no correlation. Explain.

25. The age and value of a car

26. A student's study time and test scores

27. The height and age of a pine tree

28. A student's height and test scores

In Exercises 29–32, use the data in the table, which shows the relationship between the altitude A (in thousands of feet) and the air pressure P (in pounds per square foot).

A	0	5	10	15	20	25
P	14.7	12.3	10.2	8.4	6.8	5.4

A	30	35	40	45	50
P	4.5	3.5	2.8	2.1	1.8

29. Sketch a scatter plot of the data.

30. How are A and P related?

31. Estimate the air pressure at 42,500 feet.

32. Estimate the altitude at which the air pressure is 5.0 pounds per square foot.

Crop Yield In Exercises 33–36, use the data in the table, where x is the number of units of fertilizer applied to sample plots and y is the yield (in bushels) of a crop.

x	0	1	2	3	4	5	6	7	8
y	58	60	59	61	63	66	65	67	70

33. Sketch a scatter plot of the data.

34. Determine whether the points are positively correlated, are negatively correlated, or have no discernible correlation.

35. Sketch a linear model that you think best represents the data. Find an equation of the line you sketched. Use the line to predict the yield if 10 units of fertilizer are used.

36. Can the model found in Exercise 35 be used to predict yields for arbitrarily large values of x? Explain.

Speed of Sound In Exercises 37–40, use the data in the table, where h is altitude in thousands of feet and v is the speed of sound in feet per second.

h	0	5	10	15	20	25	30	35
v	1116	1097	1077	1057	1036	1015	995	973

37. Sketch a scatter plot of the data.

38. Determine whether the points are positively correlated, are negatively correlated, or have no discernible correlation.

39. Sketch a linear model that you think best represents the data. Find an equation of the line you sketched. Use the line to predict the speed of sound at an altitude of 27,000 feet.

40. The speed of sound at an altitude of 70,000 feet is approximately 971 feet per second. What does this suggest about the validity of using the model in Exercise 39 to extrapolate beyond the data given in the table?

In Exercises 41–44, use a graphing utility to find the least squares regression line for the data. Sketch a scatter plot and the regression line.

41. (0, 23), (1, 20), (2, 19), (3, 17), (4, 15), (5, 11), (6, 10)

42. (4, 52.8), (5, 54.7), (6, 55.7), (7, 57.8), (8, 60.2), (9, 63.1), (10, 66.5)

43. (−10, 5.1), (−5, 9.8), (0, 17.5), (2, 25.4), (4, 32.8), (6, 38.7), (8, 44.2), (10, 50.5)

44. $(-10, 213.5)$, $(-5, 174.9)$, $(0, 141.7)$, $(5, 119.7)$, $(8, 102.4)$, $(10, 87.6)$

45. *School Enrollment* The table gives the preprimary school enrollments y (in millions) for the years 1990 through 1995, where $t = 0$ corresponds to 1990. (Source: U.S. Bureau of the Census)

t	0	1	2	3	4	5
y	11.21	11.37	11.54	11.95	12.33	12.52

(a) Use a graphing utility to find the least squares regression line. Use the equation to estimate enrollment in 1996.

(b) Make a scatter plot of the data and sketch the graph of the regression line.

(c) Use a graphing utility to determine the correlation coefficient.

46. *Advertising* The management of a department store ran an experiment to determine if a relationship existed between sales S (in thousands of dollars) and the amount spent on advertising x (in thousands of dollars). The following data were collected.

x	1	2	3	4	5	6	7	8
S	405	423	455	466	492	510	525	559

(a) Use a graphing utility to find the least squares regression line. Use the equation to estimate sales if $4500 is spent on advertising.

(b) Make a scatter plot of the data and sketch the graph of the regression line.

(c) Use a graphing utility to determine the correlation coefficient.

In Exercises 47–52, find the mean, median, and mode of the set of measurements.

47. 5, 12, 7, 14, 8, 9, 7

48. 30, 37, 32, 39, 33, 34, 32

49. 5, 12, 7, 24, 8, 9, 7

50. 20, 37, 32, 39, 33, 34, 32

51. *Electric Bills* A person had the following monthly bills for electricity. What are the mean and median of the collection of bills?

Jan.	$67.92	Feb.	$59.84	Mar.	$52.00
Apr.	$52.50	May	$57.99	June	$65.35
July	$81.76	Aug.	$74.98	Sept.	$87.82
Oct.	$83.18	Nov.	$65.35	Dec.	$57.00

52. *Car Rental* A car rental company kept the following record of the numbers of miles driven by a car that was rented. What are the mean, median, and mode of these data?

Monday	410	Tuesday	260
Wednesday	320	Thursday	320
Friday	460	Saturday	150

53. *Six-Child Families* A study was done on families having six children. The table gives the number of families in the study with the indicated number of girls. Determine the mean, median, and mode of this set of data.

Number of girls	0	1	2	3	4	5	6
Frequency	1	24	45	54	50	19	7

54. *Baseball* A baseball fan examined the records of a favorite baseball player's performance during his last 50 games. The number of games in which the player had 0, 1, 2, 3, and 4 hits are recorded in the table.

Number of hits	0	1	2	3	4
Frequency	14	26	7	2	1

(a) Determine the average number of hits per game.

(b) Determine the player's batting average if he had 200 at bats during the 50-game series.

55. *Think About It* Construct a collection of numbers that has the following properties. If this is not possible, explain why it is not.

Mean $= 6$, median $= 4$, mode $= 4$

56. *Think About It* Construct a collection of numbers that has the following properties. If this is not possible, explain why it is not.

Mean $= 6$, median $= 6$, mode $= 4$

57. *Test Scores* A professor records the following scores for a 100-point exam.

99, 64, 80, 77, 59, 72, 87, 79, 92, 88,
90, 42, 20, 89, 42, 100, 98, 84, 78, 91

Which measure of central tendency best describes these test scores?

58. *Shoe Sales* A salesperson sold eight pairs of a certain style of men's shoes. The sizes of the eight pairs were as follows: $10\frac{1}{2}$, 8, 12, $10\frac{1}{2}$, 10, $9\frac{1}{2}$, 11, and $10\frac{1}{2}$. Which measure (or measures) of central tendency best describes the typical shoe size for these data?

Appendix E
Introduction to Logic

E.1 Statements and Truth Tables

Statements ■ Truth Tables

Statements

In everyday speech and in mathematics we make inferences that adhere to common **laws of logic.** These laws (or methods of reasoning) allow us to build an algebra of statements by using logical operations to form compound statements from simpler ones. One of the primary goals of logic is to determine the truth value (true or false) of a compound statement knowing the truth values of its simpler component statements. For instance, the compound statement "The temperature is below freezing and it is snowing" is true only if both component statements are true.

> **Definition of a Statement**
>
> 1. A **statement** (or proposition) is a sentence to which only one truth value (either true or false) can be meaningfully assigned.
> 2. An **open statement** is a sentence that contains one or more variables and becomes a statement when each variable is replaced by a specific item from a designated set.

Example 1 Statements, Nonstatements, and Open Statements

Statement	*Truth Value*
A square is a rectangle.	T
-3 is less than -5.	F
Nonstatement	*Truth Value*
Do your homework.	No truth value can be meaningfully assigned.
Did you call the police?	No truth value can be meaningfully assigned.
Open Statement	*Truth Value*
x is an irrational number.	We need a value of x.
She is a computer science major.	We need a specific person.

Symbolically, statements are represented by lowercase letters p, q, r, and so on. Statements can be changed or combined to form **compound statements** by means of the three logical operations **and, or,** and **not,** which are represented by $\wedge$ (and), $\vee$ (or), and ~ (not). In logic, the word *or* is used in the *inclusive* sense (meaning "and/or" in everyday language). That is, the statement "p or q" is true if p is true, q is true, or both p and q are true. The following list summarizes the terms and symbols used with these three operations of logic.

▶ Operations of Logic

Operation	*Verbal Statement*	*Symbolic Form*	*Name of Operation*
~	not p	$\sim p$	**Negation**
$\wedge$	p and q	$p \wedge q$	**Conjunction**
$\vee$	p or q	$p \vee q$	**Disjunction**

Compound statements can be formed using more than one logical operation, as demonstrated in Example 2.

Example 2 Forming Negations and Compound Statements

The statements p and q are as follows.

p: The temperature is below freezing.
q: It is snowing.

Write the verbal form for each of the following.

a. $p \wedge q$ **b.** $\sim p$ **c.** $\sim(p \vee q)$ **d.** $\sim p \wedge \sim q$

Solution

a. The temperature is below freezing and it is snowing.

b. The temperature is not below freezing.

c. It is not true that the temperature is below freezing or it is snowing.

d. The temperature is not below freezing and it is not snowing.

Example 3 Forming Compound Statements

The statements p and q are as follows.

p: The temperature is below freezing.
q: It is snowing.

a. Write the symbolic form for: *The temperature is not below freezing or it is not snowing.*

b. Write the symbolic form for: *It is not true that the temperature is below freezing and it is snowing.*

Solution

a. The symbolic form is: $\sim p \vee \sim q$ **b.** The symbolic form is: $\sim(p \wedge q)$

Truth Tables

To determine the truth value of a compound statement, it is helpful to construct charts called **truth tables.** The following tables represent the three basic logical operations.

Negation

p	q	$\sim p$	$\sim q$
T	T	F	F
T	F	F	T
F	T	T	F
F	F	T	T

Conjunction

p	q	$p \wedge q$
T	T	T
T	F	F
F	T	F
F	F	F

Disjunction

p	q	$p \vee q$
T	T	T
T	F	T
F	T	T
F	F	F

For the sake of uniformity, all truth tables with two component statements will have **T** and **F** values for p and q assigned in the order shown in the first columns of these three tables. Truth tables for several operations can be combined into one chart by using the same two first columns. For each operation, a new column is added. Such an arrangement is especially useful with compound statements that involve more than one logical operation and for showing that two statements are logically equivalent.

▶ Logical Equivalence

Two compound statements are **logically equivalent** if they have identical truth tables. Symbolically, the equivalence of the statements p and q is denoted by writing $p \equiv q$.

Example 4 Logical Equivalence

Use a truth table to show the logical equivalence of the statements $\sim p \wedge \sim q$ and $\sim(p \vee q)$.

Solution

p	q	$\sim p$	$\sim q$	$\sim p \wedge \sim q$	$p \vee q$	$\sim(p \vee q)$
T	T	F	F	F	T	F
T	F	F	T	F	T	F
F	T	T	F	F	T	F
F	F	T	T	T	F	T

Identical

Because the fifth and seventh columns in the table are identical, the two given statements are logically equivalent.

The equivalence established in Example 4 is one of two well-known rules in logic called **DeMorgan's Laws.** Verification of the second of DeMorgan's Laws is left as an exercise.

> **DeMorgan's Laws**
>
> **1.** $\sim(p \vee q) \equiv \sim p \wedge \sim q$
>
> **2.** $\sim(p \wedge q) \equiv \sim p \vee \sim q$

Compound statements that are true, no matter what the truth values of the component statements, are called **tautologies.** One simple example is the statement "p or not p," as shown in the table.

$p \vee \sim p$ is a tautology

p	$\sim p$	$p \vee \sim p$
T	F	T
F	T	T

E.1 Exercises

In Exercises 1–12, classify the sentence as a statement, a nonstatement, or an open statement.

1. All dogs are brown.

2. Can I help you?

3. That figure is a circle.

4. Substitute 4 for x.

5. x is larger than 4.

6. 8 is larger than 4.

7. $x + y = 10$

8. $12 + 3 = 14$

9. Hockey is fun to watch.

10. One mile is greater than 1 kilometer.

11. It is more than 1 mile to the school.

12. Come to the party.

In Exercises 13–20, determine whether the open statement is true for the given values of x.

13. $x^2 - 5x + 6 = 0$
(a) $x = 2$
(b) $x = -2$

14. $x^2 - x - 6 = 0$
(a) $x = 2$
(b) $x = -2$

15. $x^2 \leq 4$
(a) $x = -2$
(b) $x = 0$

16. $|x - 3| = 4$
(a) $x = -1$
(b) $x = 7$

17. $4 - |x| = 2$
(a) $x = 0$
(b) $x = 1$

18. $\sqrt{x^2} = x$
(a) $x = 3$
(b) $x = -3$

19. $\dfrac{x}{x} = 1$
(a) $x = -4$
(b) $x = 0$

20. $\sqrt[3]{x} = -2$
(a) $x = 8$
(b) $x = -8$

In Exercises 21–24, write the verbal form for each of the following.

(a) $\sim p$ (b) $\sim q$ (c) $p \wedge q$ (d) $p \vee q$

21. p: The sun is shining.
q: It is hot.

22. p: The car has a radio.
q: The car is red.

23. p: Lions are mammals.
q: Lions are carnivorous.

24. p: Twelve is less than 15.
q: Seven is a prime number.

In Exercises 25–28, write the verbal form for each of the following.

(a) $\sim p \wedge q$ (b) $\sim p \vee q$

(c) $p \wedge \sim q$ (d) $p \vee \sim q$

25. p: The sun is shining.
q: It is hot.

26. p: The car has a radio.
q: The car is red.

27. p: Lions are mammals.
q: Lions are carnivorous.

28. p: Twelve is less than 15.
q: Seven is a prime number.

In Exercises 29–32, write the symbolic form of the given compound statement. In each case let p represent the statement "It is four o'clock," and let q represent the statement "It is time to go home."

29. It is four o'clock and it is not time to go home.

30. It is not four o'clock or it is not time to go home.

31. It is not four o'clock or it is time to go home.

32. It is four o'clock and it is time to go home.

In Exercises 33–36, write the symbolic form of the given compound statement. In each case let p represent the statement "The dog has fleas," and let q represent the statement "The dog is scratching."

33. The dog does not have fleas or the dog is not scratching.

34. The dog has fleas and the dog is scratching.

35. The dog does not have fleas and the dog is scratching.

36. The dog has fleas or the dog is not scratching.

In Exercises 37–42, write the negation of the given statement.

37. The bus is not blue.

38. Frank is not 6 feet tall.

39. x is equal to 4.

40. x is not equal to 4.

41. The earth is not flat.

42. The earth is flat.

In Exercises 43–48, construct a truth table for the given compound statement.

43. $\sim p \wedge q$

44. $\sim p \vee q$

45. $\sim p \vee \sim q$

46. $\sim p \wedge \sim q$

47. $p \vee \sim q$

48. $p \wedge \sim q$

In Exercises 49–54, use a truth table to determine whether the given statements are logically equivalent.

49. $\sim p \wedge q, \quad p \vee \sim q$

50. $\sim(p \wedge \sim q), \quad \sim p \vee q$

51. $\sim(p \vee \sim q), \quad \sim p \wedge q$

52. $\sim(p \vee q), \quad \sim p \vee \sim q$

53. $p \wedge \sim q, \quad \sim(\sim p \vee q)$

54. $p \wedge \sim q, \quad \sim(\sim p \wedge q)$

In Exercises 55–58, determine whether the statements are logically equivalent.

55. (a) The house is red and it is not made of wood.
(b) The house is red or it is not made of wood.

56. (a) It is not true that the tree is not green.
(b) The tree is green.

57. (a) The statement that the house is white or blue is not true.
(b) The house is not white and it is not blue.

58. (a) I am not 25 years old and I am not applying for this job.
(b) The statement that I am 25 years old and I am applying for this job is not true.

In Exercises 59–62, use a truth table to determine whether the given statement is a tautology.

59. $\sim p \wedge p$

60. $\sim p \vee p$

61. $\sim(\sim p) \vee \sim p$

62. $\sim(\sim p) \wedge \sim p$

63. Use a truth table to verify the second of DeMorgan's Laws:

$$\sim(p \wedge q) \equiv \sim p \vee \sim q$$

E.2 Implications, Quantifiers, and Venn Diagrams

Implications ■ Logical Quantifiers ■ Venn Diagrams

Implications

A statement of the form "If p, then q," is called an **implication** (or a conditional statement) and is denoted by

$$p \rightarrow q.$$

The statement p is called the **hypothesis** and the statement q is called the **conclusion.** There are many different ways to express the implication $p \rightarrow q$, as shown in the following list.

▶ Different Ways of Stating Implications

The implication $p \rightarrow q$ has the following equivalent verbal forms.

1. If p, then q.

2. p implies q.

3. p, only if q.

4. q follows from p.

5. q is necessary for p.

6. p is sufficient for q.

Normally, we think of the implication $p \rightarrow q$ as having a cause-and-effect relationship between the hypothesis p and the conclusion q. However, you should be careful not to confuse the truth value of the component statements with the truth value of the implication. The following truth table should help you keep this distinction in mind.

Implication

p	q	$p \rightarrow q$
T	T	T
T	F	F
F	T	T
F	F	T

Note in the table that the implication $p \rightarrow q$ is false only when p is true and q is false. This is like a promise. Suppose you promise a friend that "If the sun shines, I will take you fishing." The only way you can break your promise is if the sun shines (p is true) and you do not take your friend fishing (q is false). If the sun doesn't shine (p is false), you have no obligation to go fishing, and so the promise cannot be broken.

Example 1 Finding Truth Values of Implications

Give the truth value of each implication.

a. If 3 is odd, then 9 is odd.

b. If 3 is odd, then 9 is even.

c. If 3 is even, then 9 is odd.

d. If 3 is even, then 9 is even.

Solution

	Hypothesis	*Conclusion*	*Implication*
a.	**T**	**T**	**T**
b.	**T**	**F**	**F**
c.	**F**	**T**	**T**
d.	**F**	**F**	**T**

The next example shows how to write an implication as a disjunction.

Example 2 Identifying Equivalent Statements

Use a truth table to show the logical equivalence of the following statements.

a. If I get a raise, I will take my family on a vacation.

b. I will not get a raise *or* I will take my family on a vacation.

Solution

Let p represent the statement "I will get a raise," and let q represent the statement "I will take my family on a vacation." Then, you can represent the statement in part (a) as $p \to q$ and the statement in part (b) as $\sim p \vee q$. The logical equivalence of these two statements is shown in the following truth table.

$p \to q \equiv \sim p \vee q$

p	q	$\sim p$	$\sim p \vee q$	$p \to q$
T	**T**	**F**	**T**	**T**
T	**F**	**F**	**F**	**F**
F	**T**	**T**	**T**	**T**
F	**F**	**T**	**T**	**T**

Identical (columns $\sim p \vee q$ and $p \to q$)

Because the fourth and fifth columns of the truth table are identical, you can conclude that the two statements $p \to q$ and $\sim p \vee q$ are equivalent.

From the table in Example 2 and the fact that $\sim(\sim p) \equiv p$, we can write the **negation of an implication.** That is, because $p \rightarrow q$ is equivalent to $\sim p \vee q$, it follows that the negation of $p \rightarrow q$ must be $\sim(\sim p \vee q)$, which by DeMorgan's Laws can be written as follows.

$$\sim(p \rightarrow q) \equiv p \wedge \sim q$$

For the implication $p \rightarrow q$, there are three important associated implications.

1. The **converse** of $p \rightarrow q$: $q \rightarrow p$
2. The **inverse** of $p \rightarrow q$: $\sim p \rightarrow \sim q$
3. The **contrapositive** of $p \rightarrow q$: $\sim q \rightarrow \sim p$

From the table below, you can see that these four statements yield two pairs of logically equivalent implications. The connective "$\rightarrow$" is used to determine the truth values in the last three columns of the table.

p	q	$\sim p$	$\sim q$	$p \rightarrow q$	$\sim q \rightarrow \sim p$	$q \rightarrow p$	$\sim p \rightarrow \sim q$
T	T	F	F	T	T	T	T
T	F	F	T	F	F	T	T
F	T	T	F	T	T	F	F
F	F	T	T	T	T	T	T

Identical ($p \rightarrow q$ and $\sim q \rightarrow \sim p$) Identical ($q \rightarrow p$ and $\sim p \rightarrow \sim q$)

Example 3 Writing the Converse, Inverse, and Contrapositive

Write the converse, inverse, and contrapositive for the implication "If I get a B on my test, then I will pass the course."

Solution

a. *Converse:* If I pass the course, then I got a B on my test.

b. *Inverse:* If I do not get a B on my test, then I will not pass the course.

c. *Contrapositive:* If I do not pass the course, then I did not get a B on my test.

In Example 3, be sure you see that neither the converse nor the inverse is logically equivalent to the original implication. To see this, consider that the original implication simply states that if you get a B on your test, then you will pass the course. The converse is not true because knowing that you passed the course does not imply that you got a B on the test. (After all, you might have gotten an A on the test.)

A **biconditional statement,** denoted by $p \leftrightarrow q$, is the conjunction of the implications $p \rightarrow q$ and $q \rightarrow p$. Often a biconditional statement is written as "p if and only if q," or in shorter form as "p iff q." A biconditional statement is true when both components are true and when both components are false, as shown in the following truth table.

Biconditional Statement: p if and only if q

p	q	$p \to q$	$q \to p$	$p \leftrightarrow q$	$(p \to q) \wedge (q \to p)$
T	**T**	**T**	**T**	**T**	**T**
T	**F**	**F**	**T**	**F**	**F**
F	**T**	**T**	**F**	**F**	**F**
F	**F**	**T**	**T**	**T**	**T**

The following list summarizes some of the laws of logic that we have discussed up to this point.

Laws of Logic

1. For every statement p, either p is true or p is false. Law of Excluded Middle
2. $\sim(\sim p) \equiv p$ Law of Double Negation
3. $\sim(p \vee q) \equiv \sim p \wedge \sim q$ DeMorgan's Law
4. $\sim(p \wedge q) \equiv \sim p \vee \sim q$ DeMorgan's Law
5. $p \to q \equiv \sim p \vee q$ Law of Implication
6. $p \to q \equiv \sim q \to \sim p$ Law of Contraposition

Logical Quantifiers

Logical quantifiers are words such as *some*, *all*, *every*, *each*, *one*, and *none*. Here are some examples of statements with quantifiers.

Some isosceles triangles are right triangles.

Every painting on display is for sale.

Not all corporations have male chief executive officers.

All squares are parallelograms.

Being able to recognize the negation of a statement involving a quantifier is one of the most important skills in logic. For instance, consider the statement "All dogs are brown." In order for this statement to be false, you do not have to show that *all* dogs are not brown, you must simply find at least one dog that is not brown. So, the negation of the statement is "Some dogs are brown."

Next we list some of the more common negations involving quantifiers.

Negating Statements with Quantifiers

	Statement	*Negation*
1.	All p are q.	Some p are not q.
2.	Some p are q.	No p is q.
3.	Some p are not q.	All p are q.
4.	No p is q.	Some p are q.

When using logical quantifiers, the word *all* can be replaced by the words *each* or *every*. For instance, the following are equivalent.

All p are q. Each p is q. Every p is q.

Similarly, the word *some* can be replaced by the words *at least one*. For instance, the following are equivalent.

Some p are q. At least one p is q.

Example 4 Negating Quantifying Statements

Write the negation of each statement.

a. All students study.

b. Not all prime numbers are odd.

c. At least one mammal can fly.

d. Some bananas are not yellow.

Solution

a. Some students do not study.

b. All prime numbers are odd.

c. No mammals can fly.

d. All bananas are yellow.

Venn Diagrams

Venn diagrams are figures that are used to show relationships between two or more sets of objects. They can help us to interpret quantifying statements. Study the following Venn diagrams in which the circle marked A represents people more than 6 feet tall and the circle marked B represents the basketball players.

1. All basketball players are more than 6 feet tall.

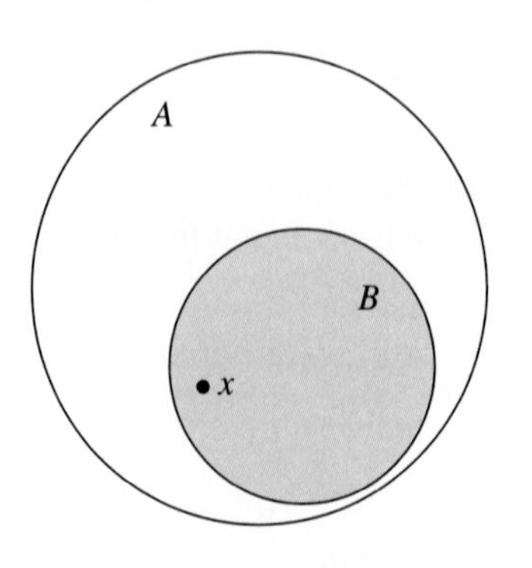

2. Some basketball players are more than 6 feet tall.

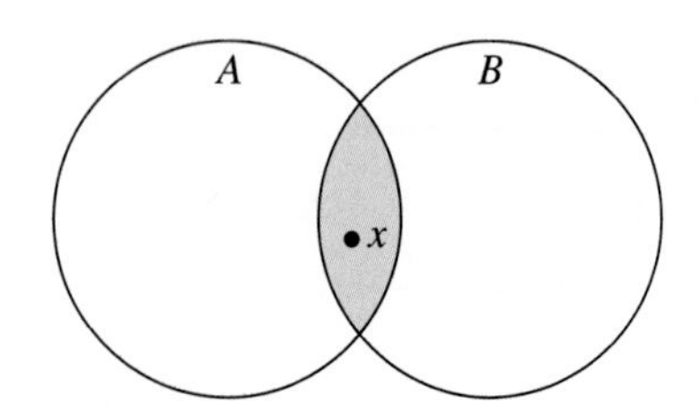

3. Some basketball players are not more than 6 feet tall.

4. No basketball player is more than 6 feet tall.

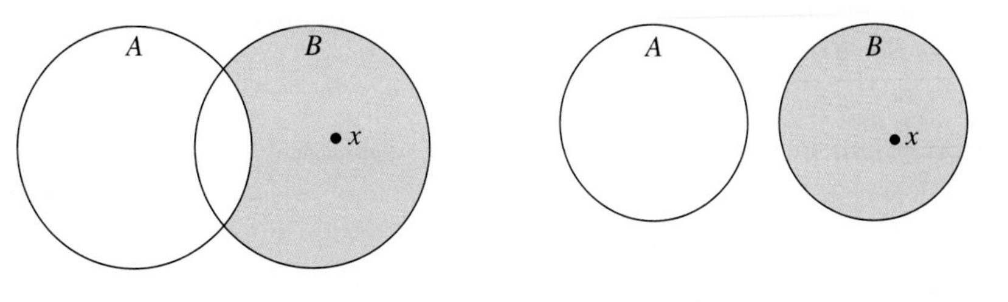

E.2 Exercises

In Exercises 1–4, write the verbal form for each of the following.

(a) $p \rightarrow q$ (b) $q \rightarrow p$ (c) $\sim q \rightarrow \sim p$ (d) $p \rightarrow \sim q$

1. p: The engine is running.
q: The engine is wasting gasoline.

2. p: The student is at school.
q: It is nine o'clock.

3. p: The integer is even.
q: It is divisible by 2.

4. p: The person is generous.
q: The person is rich.

In Exercises 5–10, write the symbolic form of the compound statement. Let p represent the statement "The economy is expanding," and let q represent the statement "Interest rates are low."

5. If interest rates are low, then the economy is expanding.

6. If interest rates are not low, then the economy is not expanding.

7. An expanding economy implies low interest rates.

8. Low interest rates are sufficient for an expanding economy.

9. Low interest rates are necessary for an expanding economy.

10. The economy will expand only if interest rates are low.

In Exercises 11–20, give the truth value of the implication.

11. If 4 is even, then 12 is even.

12. If 4 is even, then 2 is odd.

13. If 4 is odd, then 3 is odd.

14. If 4 is odd, then 2 is odd.

15. If $2n$ is even, then $2n + 2$ is odd.

16. If $2n + 1$ is even, then $2n + 2$ is odd.

17. $3 + 11 > 16$ only if $2 + 3 = 5$.

18. $\frac{1}{6} < \frac{2}{3}$ is necessary for $\frac{1}{2} > 0$.

19. $x = -2$ follows from $2x + 3 = x + 1$.

20. If $2x = 224$, then $x = 10$.

In Exercises 21–26, write the converse, inverse, and contrapositive of the statement.

21. If the sky is clear, then you can see the eclipse.

22. If the person is nearsighted, then he is ineligible for the job.

23. If taxes are raised, then the deficit will increase.

24. If wages are raised, then the company's profits will decrease.

25. It is necessary to have a birth certificate to apply for the visa.

26. The number is divisible by 3 only if the sum of its digits is divisible by 3.

In Exercises 27–40, write the negation of the statement.

27. Paul is a junior or a senior.

28. Jack is a senior and he plays varsity basketball.

29. If the temperature increases, then the metal rod will expand.

30. If the test fails, then the project will be halted.

31. We will go to the ocean only if the weather forecast is good.

32. Completing the pass on this play is necessary if we are going to win the game.

33. Some students are in extracurricular activities.

34. Some odd integers are not prime numbers.

35. All contact sports are dangerous.

36. All members must pay their dues prior to June 1.

37. No child is allowed at the concert.

38. No contestant is over the age of 12.

39. At least one of the $20 bills is counterfeit.

40. At least one unit is defective.

In Exercises 41–48, construct a truth table for the compound statement.

41. $\sim(p \rightarrow \sim q)$

42. $\sim q \rightarrow (p \rightarrow q)$

43. $\sim(q \rightarrow p) \wedge q$

44. $p \rightarrow (\sim p \vee q)$

45. $[(p \vee q) \wedge (\sim p)] \rightarrow q$

46. $[(p \rightarrow q) \wedge (\sim q)] \rightarrow p$

47. $(p \leftrightarrow \sim q) \rightarrow \sim p$

48. $(p \vee \sim q) \leftrightarrow (q \rightarrow \sim p)$

In Exercises 49–56, use a truth table to show the logical equivalence of the two statements.

49. $q \rightarrow p$ $\qquad \sim p \rightarrow \sim q$

50. $\sim p \rightarrow q$ $\qquad p \vee q$

51. $\sim(p \rightarrow q)$ $\qquad p \wedge \sim q$

52. $(p \vee q) \rightarrow q$ $\qquad p \rightarrow q$

53. $(p \rightarrow q) \vee \sim q$ $\qquad p \vee \sim p$

54. $q \rightarrow (\sim p \vee q)$ $\qquad q \vee \sim q$

55. $p \rightarrow (\sim p \wedge q)$ $\qquad \sim p$

56. $\sim(p \wedge q) \rightarrow \sim q$ $\qquad p \vee \sim q$

57. Select the statement that is logically equivalent to the statement "If a number is divisible by 6, then it is divisible by 2."

(a) If a number is divisible by 2, then it is divisible by 6.

(b) If a number is not divisible by 6, then it is not divisible by 2.

(c) If a number is not divisible by 2, then it is not divisible by 6.

(d) Some numbers are divisible by 6 and not divisible by 2.

58. Select the statement that is logically equivalent to the statement "It is not true that Pam is a conservative and a Democrat."

(a) Pam is a conservative and a Democrat.

(b) Pam is not a conservative and not a Democrat.

(c) Pam is not a conservative or she is not a Democrat.

(d) If Pam is not a conservative, then she is a Democrat.

59. Select the statement that is *not* logically equivalent to the statement "Every citizen over the age of 18 has the right to vote."

(a) Some citizens over the age of 18 have the right to vote.

(b) Each citizen over the age of 18 has the right to vote.

(c) All citizens over the age of 18 have the right to vote.

(d) No citizen over the age of 18 can be restricted from voting.

60. Select the statement that is *not* logically equivalent to the statement "It is necessary to pay the registration fee to take the course."

(a) If you take the course, then you must pay the registration fee.

(b) If you do not pay the registration fee, then you cannot take the course.

(c) If you pay the registration fee, then you may take the course.

(d) You may take the course only if you pay the registration fee.

In Exercises 61–70, sketch a Venn diagram and shade the region that illustrates the given statement. Let *A* be a circle that represents people who are happy, and let *B* be a circle that represents college students.

61. All college students are happy.

62. All happy people are college students.

63. No college students are happy.

64. No happy people are college students.

65. Some college students are not happy.

66. Some happy people are not college students.

67. At least one college student is happy.

68. At least one happy person is not a college student.

69. Each college student is sad.

70. Each sad person is not a college student.

In Exercises 71–74, state whether the statement follows from the given Venn diagram. Assume that each area shown in the Venn diagram is non-empty. (*Note:* Use only the information given in the diagram. Do not be concerned with whether the statement is actually true or false.)

71. (a) All toads are green.
(b) Some toads are green.

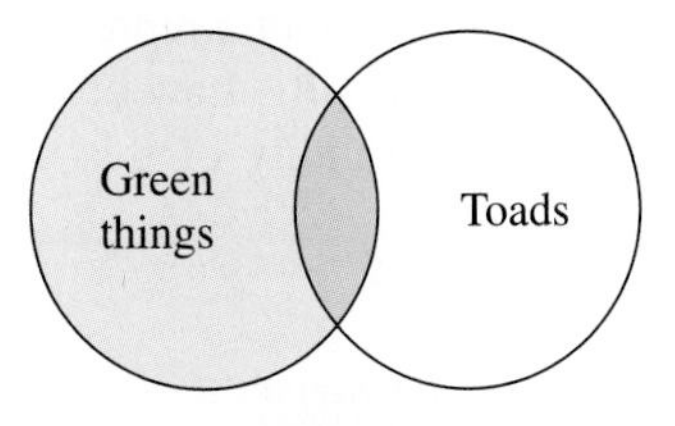

72. (a) All men are company presidents.
(b) Some company presidents are women.

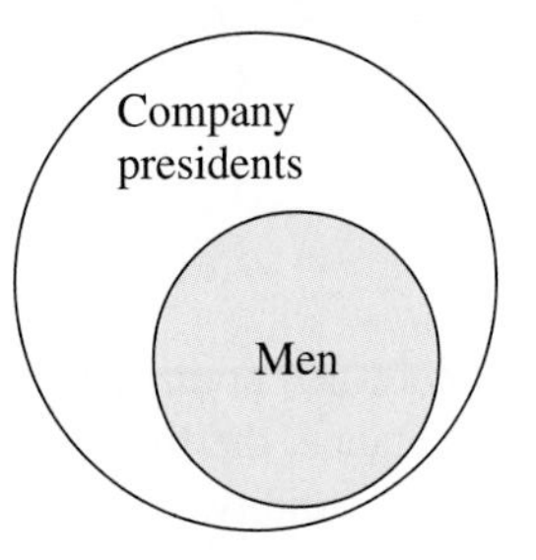

73. (a) All blue cars are old.
(b) Some blue cars are not old.

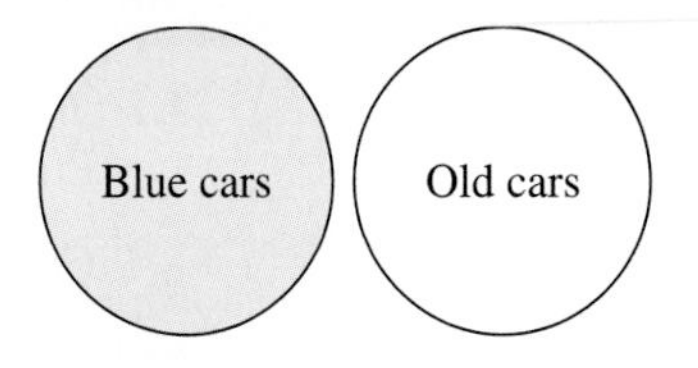

74. (a) No football players are more than 6 feet tall.
(b) Every football player is more than 6 feet tall.

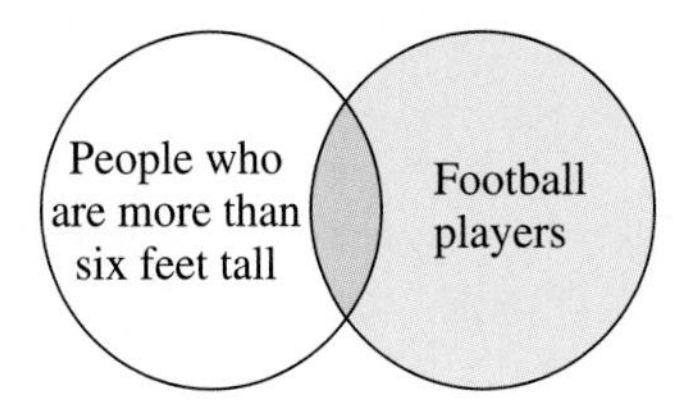

E.3 Logical Arguments

Arguments ■ Venn Diagrams and Arguments ■ Proofs

Arguments

An **argument** is a collection of statements, listed in order. The last statement is called the **conclusion** and the other statements are called the **premises.** An argument is **valid** if the conjunction of all the premises implies the conclusion. The most common type of argument takes the following form.

Premise #1: $p \rightarrow q$

Premise #2: p

Conclusion: q

This form of argument is called the **Law of Detachment** or ***Modus Ponens.*** It is illustrated in the following example.

Example 1 A Valid Argument

Show that the following argument is valid.

Premise #1: If Sean is a freshman, then he is taking algebra.

Premise #2: Sean is a freshman.

Conclusion: So, Sean is taking algebra.

Solution

Let p represent the statement "Sean is a freshman," and let q represent the statement "Sean is taking algebra." Then the argument fits the Law of Detachment, which can be written as $[(p \to q) \wedge p] \to q$. The validity of this argument is shown in the following truth table.

Law of Detachment

p	q	$p \to q$	$(p \to q) \wedge p$	$[(p \to q) \wedge p] \to q$
T	T	T	T	T
T	F	F	F	T
F	T	T	F	T
F	F	T	F	T

Keep in mind that the validity of an argument has nothing to do with the truthfulness of the premises or conclusion. For instance, the following argument is valid—the fact that it is fanciful does not alter its validity.

Premise #1: If I snap my fingers, elephants will stay out of my house.

Premise #2: I am snapping my fingers.

Conclusion: So, elephants will stay out of my house.

Four Types of Valid Arguments

Name		*Pattern*
1. Law of Detachment or *Modus Ponens*	Premise #1:	$p \to q$
	Premise #2:	p
	Conclusion:	q
2. Law of Contraposition or *Modus Tollens*	Premise #1:	$p \to q$
	Premise #2:	$\sim q$
	Conclusion:	$\sim p$
3. Law of Transitivity or Syllogism	Premise #1:	$p \to q$
	Premise #2:	$q \to r$
	Conclusion:	$p \to r$
4. Law of Disjunctive Syllogism	Premise #1:	$p \vee q$
	Premise #2:	$\sim p$
	Conclusion:	q

Example 2 An Invalid Argument

Determine whether the following argument is valid.

Premise #1: If John is elected, the income tax will be increased.

Premise #2: The income tax was increased.

Conclusion: So, John was elected.

Solution

This argument has the following form.

	Pattern	*Implication*
Premise #1:	$p \to q$	$[(p \to q) \land q] \to p$
Premise #2:	q	
Conclusion:	p	

This is not one of the four valid forms of arguments that were listed. You can construct a truth table to verify that the argument is invalid, as follows.

An Invalid Argument

p	q	$p \to q$	$(p \to q) \land q$	$[(p \to q) \land q] \to p$
T	**T**	**T**	**T**	**T**
T	**F**	**F**	**F**	**T**
F	**T**	**T**	**T**	**F**
F	**F**	**T**	**F**	**T**

An invalid argument, such as the one in Example 2, is called a **fallacy.** Other common fallacies are given in the following example.

Example 3 Common Fallacies

Each of the following arguments is invalid.

a. *Arguing from the Converse:* If the football team wins the championship, then students will skip classes. The students skipped classes. So, the football team won the championship.

b. *Arguing from the Inverse:* If the football team wins the championship, then students will skip classes. The football team did not win the championship. So, the students did not skip classes.

c. *Arguing from False Authority:* Wheaties are best for you because Joe Montana eats them.

d. *Arguing from an Example:* Beta brand products are not reliable because my Beta brand snowblower does not start in cold weather.

e. *Arguing from Ambiguity:* If an automobile carburetor is modified, the automobile will pollute. Brand X automobiles have modified carburetors. So, Brand X automobiles pollute.

f. *Arguing by False Association:* Joe was running through the alley when the fire alarm went off. So, Joe started the fire.

Example 4 A Valid Argument

Determine whether the following argument is valid.

Premise #1: You like strawberry pie or you like chocolate pie.

Premise #2: You do not like strawberry pie.

Conclusion: So, you like chocolate pie.

Solution

This argument has the following form.

Premise #1: $p \vee q$

Premise #2: $\sim p$

Conclusion: q

This argument is a disjunctive syllogism, which is one of the four common types of valid arguments.

In a valid argument, the conclusion drawn from the premise is called a **valid conclusion.**

Example 5 Making Valid Conclusions

Given the following two premises, which of the conclusions are valid?

Premise #1: If you like boating, then you like swimming.

Premise #2: If you like swimming, then you are a scholar.

a. Conclusion: If you like boating, then you are a scholar.

b. Conclusion: If you do not like boating, then you are not a scholar.

c. Conclusion: If you are not a scholar, then you do not like boating.

Solution

a. This conclusion is valid. It follows from the Law of Transitivity or Syllogism.

b. This conclusion is invalid. The fallacy stems from arguing from the inverse.

c. This conclusion is valid. It follows from the Law of Contraposition.

Venn Diagrams and Arguments

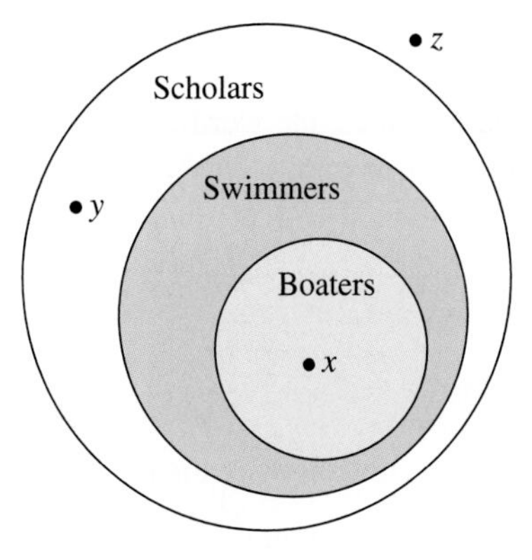

Figure E.1

Venn diagrams can be used to test informally the validity of an argument. For instance, a Venn diagram for the premises in Example 5 is shown in Figure E.1. In this figure, the validity of Conclusion (a) is seen by choosing a boater x in all three sets. Conclusion (b) is seen to be invalid by choosing a person y who is a scholar but does not like boating. Finally, person z indicates the validity of Conclusion (c).

Venn diagrams work well for testing arguments that involve quantifiers, as shown in the next two examples.

Example 6 Using a Venn Diagram to Show That an Argument Is Not Valid

Use a Venn diagram to test the validity of the following argument.

Premise #1:	Some plants are green.
Premise #2:	All lettuce is green.
Conclusion:	So, lettuce is a plant.

Solution

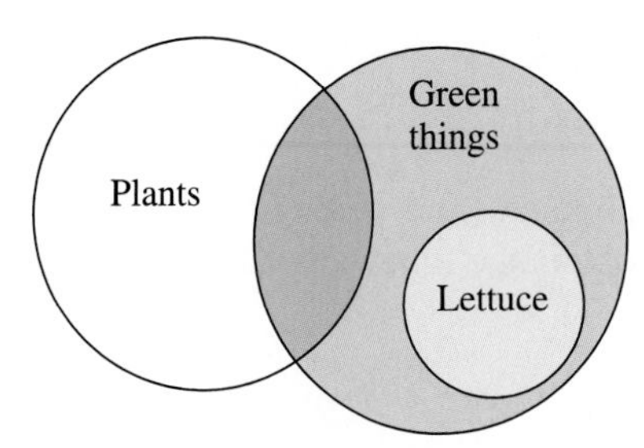

Figure E.2

From the Venn diagram shown in Figure E.2, you can see that this is not a valid argument. Remember that even though the conclusion is true (lettuce is a plant), this does not imply that the argument is true.

When you are using Venn diagrams, you must remember to draw the most general case. For example, in Figure E.2 the circle representing plants is not drawn entirely within the circle representing green things because you are told that only *some* plants are green.

Example 7 Using a Venn Diagram to Show That an Argument Is Valid

Use a Venn diagram to test the validity of the following argument.

Premise #1:	All good tennis players are physically fit.
Premise #2:	Some golfers are good tennis players.
Conclusion:	So, some golfers are physically fit.

Solution

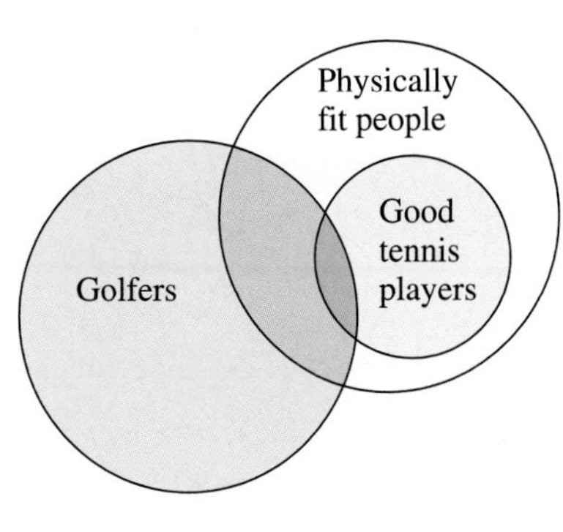

Figure E.3

Because the set of golfers intersects the set of good tennis players, as shown in Figure E.3, the set of golfers must also intersect the set of physically fit people. So, the argument is valid.

Proofs

What does the word *proof* mean to you? In mathematics, the word *proof* is used to mean simply a valid argument. Many proofs involve more than two premises and a conclusion. For instance, the proof in Example 8 involves three premises and a conclusion.

Example 8 A Proof by Contraposition

Use the following three premises to prove that "It is not snowing today."

Premise #1: If it is snowing today, Greg will go skiing.

Premise #2: If Greg is skiing today, then he is not studying.

Premise #3: Greg is studying today.

Solution

Let p represent the statement "It is snowing today," let q represent "Greg is skiing," and let r represent "Greg is studying today." So, the given premises have the following form.

Premise #1: $p \rightarrow q$

Premise #2: $q \rightarrow \sim r$

Premise #3: r

By noting that $r \equiv \sim(\sim r)$, reordering the premises, and writing the contrapositives of the first and second premises, you obtain the following valid argument.

Premise #3: r

Contrapositive of Premise #2: $r \rightarrow \sim q$

Contrapositive of Premise #1: $\sim q \rightarrow \sim p$

Conclusion: $\sim p$

So, you can conclude $\sim p$. That is, "It is not snowing today."

E.3 Exercises

In Exercises 1–4, use a truth table to show that the given argument is valid.

1. Premise #1: $p \rightarrow \sim q$
Premise #2: q
Conclusion: $\sim p$

2. Premise #1: $p \leftrightarrow q$
Premise #2: p
Conclusion: q

3. Premise #1: $p \vee q$
Premise #2: $\sim p$
Conclusion: q

4. Premise #1: $p \wedge q$
Premise #2: $\sim p$
Conclusion: q

In Exercises 5–8, use a truth table to show that the given argument is invalid.

5. Premise #1: $\sim p \rightarrow q$
Premise #2: p
Conclusion: $\sim q$

6. Premise #1: $p \rightarrow q$
Premise #2: $\sim p$
Conclusion: $\sim q$

7. Premise #1: $p \vee q$
Premise #2: q
Conclusion: p

8. Premise #1: $\sim(p \wedge q)$
Premise #2: q
Conclusion: p

In Exercises 9–22, determine whether the argument is valid or invalid.

9. Premise #1: If taxes are increased, then businesses will leave the state.
Premise #2: Taxes are increased.
Conclusion: So, businesses will leave the state.

10. Premise #1: If a student does the homework, then a good grade is certain.
Premise #2: Liza does the homework.
Conclusion: So, Liza will receive a good grade for the course.

11. Premise #1: If taxes are increased, then businesses will leave the state.
Premise #2: Businesses are leaving the state.
Conclusion: So, taxes were increased.

12. Premise #1: If a student does the homework, then a good grade is certain.
Premise #2: Liza received a good grade for the course.
Conclusion: So, Liza did her homework.

13. Premise #1: If the doors are kept locked, then the car will not be stolen.
Premise #2: The car was stolen.
Conclusion: So, the car doors were unlocked.

14. Premise #1: If Jan passes the exam, she is eligible for the position.
Premise #2: Jan is not eligible for the position.
Conclusion: So, Jan did not pass the exam.

15. Premise #1: All cars manufactured by the Ford Motor Company are reliable.
Premise #2: Lincolns are manufactured by Ford.
Conclusion: So, Lincolns are reliable cars.

16. Premise #1: Some cars manufactured by the Ford Motor Company are reliable.
Premise #2: Lincolns are manufactured by Ford.
Conclusion: So, Lincolns are reliable .

17. Premise #1: All federal income tax forms are subject to the Paperwork Reduction Act of 1980.
Premise #2: The 1040 Schedule A form is subject to the Paperwork Reduction Act of 1980.
Conclusion: So, the 1040 Schedule A form is a federal income tax form.

18. Premise #1: All integers divisible by 6 are divisible by 3.
Premise #2: Eighteen is divisible by 6.
Conclusion: So, 18 is divisible by 3.

19. Premise #1: Eric is at the store or the handball court.
Premise #2: He is not at the store.
Conclusion: So, he must be at the handball court.

20. Premise #1: The book must be returned within 2 weeks or you pay a fine.
Premise #2: The book was not returned within 2 weeks.
Conclusion: So, you must pay a fine.

21. Premise #1: It is not true that it is a diamond and it sparkles in the sunlight.
Premise #2: It does sparkle in the sunlight.
Conclusion: So, it is a diamond.

22. Premise #1: Either I work tonight or I pass the mathematics test.
Premise #2: I'm going to work tonight.
Conclusion: So, I will fail the mathematics test.

In Exercises 23–30, determine which conclusion is valid from the given premises.

23. Premise #1: If 7 is a prime number, then 7 does not divide evenly into 21.
Premise #2: Seven divides evenly into 21.
(a) Conclusion: So, 7 is a prime number.
(b) Conclusion: So, 7 is not a prime number.
(c) Conclusion: So, 21 divided by 7 is 3.

24. Premise #1: If the fuel is shut off, then the fire will be extinguished.
Premise #2: The fire continues to burn.
(a) Conclusion: So, the fuel was not shut off.
(b) Conclusion: So, the fuel was shut off.
(c) Conclusion: So, the fire becomes hotter.

25. Premise #1: It is necessary that interest rates be lowered for the economy to improve.
Premise #2: Interest rates were not lowered.
(a) Conclusion: So, the economy will improve.
(b) Conclusion: So, interest rates are irrelevant to the performance of the economy.
(c) Conclusion: So, the economy will not improve.

26. Premise #1: It will snow only if the temperature is below 32° at some level of the atmosphere.
Premise #2: It is snowing.
(a) Conclusion: So, the temperature is below 32° at ground level.
(b) Conclusion: So, the temperature is above 32° at some level of the atmosphere.
(c) Conclusion: So, the temperature is below 32° at some level of the atmosphere.

27. Premise #1: Smokestack emissions must be reduced or acid rain will continue as an environmental problem.
Premise #2: Smokestack emissions have not decreased.
(a) Conclusion: So, the ozone layer will continue to be depleted.
(b) Conclusion: So, acid rain will continue as an environmental problem.
(c) Conclusion: So, stricter automobile emission standards must be enacted.

28. Premise #1: The library must upgrade its computer system or service will not improve.
Premise #2: Service at the library has improved.
(a) Conclusion: So, the computer system was upgraded.
(b) Conclusion: So, more personnel were hired for the library.
(c) Conclusion: So, the computer system was not upgraded.

29. Premise #1: If Rodney studies, then he will make good grades.
Premise #2: If he makes good grades, then he will get a good job.
(a) Conclusion: So, Rodney will get a good job.
(b) Conclusion: So, if Rodney doesn't study, then he won't get a good job.
(c) Conclusion: So, if Rodney doesn't get a good job, then he didn't study.

30. Premise #1: It is necessary to have a ticket and an ID card to get into the arena.
Premise #2: Janice entered the arena.
(a) Conclusion: So, Janice does not have a ticket.
(b) Conclusion: So, Janice has a ticket and an ID card.
(c) Conclusion: So, Janice has an ID card.

In Exercises 31–34, use a Venn diagram to test the validity of the argument.

31. Premise #1: All numbers divisible by 10 are divisible by 5.
Premise #2: Fifty is divisible by 10.
Conclusion: So, 50 is divisible by 5.

32. Premise #1: All human beings require adequate rest.
Premise #2: All infants are human beings.
Conclusion: So, all infants require adequate rest.

33. Premise #1: No person under the age of 18 is eligible to vote.
Premise #2: Some college students are eligible to vote.
Conclusion: So, some college students are under the age of 18.

34. Premise #1: Every amateur radio operator has a radio license.
Premise #2: Jackie has a radio license.
Conclusion: So, Jackie is an amateur radio operator.

In Exercises 35–38, use the premises to prove the given conclusion.

35. Premise #1: If Sue drives to work, then she will stop at the grocery store.
Premise #2: If she stops at the grocery store, then she will buy milk.
Premise #3: Sue drove to work today.
Conclusion: So, Sue will buy milk.

36. Premise #1: If Bill is patient, then he will succeed.
Premise #2: Bill will get bonus pay if he succeeds.
Premise #3: Bill did not get bonus pay.
Conclusion: So, Bill is not patient.

37. Premise #1: If this is a good product, then we should buy it.
Premise #2: Either it was made by XYZ Corporation, or we will not buy it.
Premise #3: It is not made by XYZ Corporation.
Conclusion: So, it is not a good product.

38. Premise #1: If it is raining today, Pam will clean her apartment.
Premise #2: If Pam is cleaning her apartment today, then she is not riding her bike.
Premise #3: Pam is riding her bike today.
Conclusion: It is not raining today.

Answers to Integrated Reviews, Odd-Numbered Exercises, Quizzes, and Tests

Chapter 1

Section 1.1 *(page 9)*

1. (a) $2, \frac{9}{3}$ (b) $-3, 2, \frac{9}{3}$ (c) $-3, 2, -\frac{3}{2}, \frac{9}{3}, 4.5$

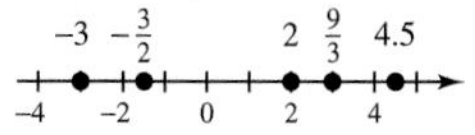

3. (a) $\frac{8}{4}$ (b) $\frac{8}{4}$ (c) $-\frac{5}{2}, 6.5, -4.5, \frac{8}{4}, \frac{3}{4}$ **5.** $2 < 5$

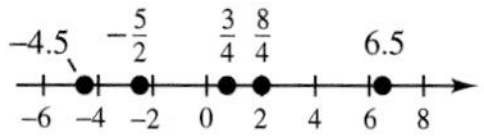

7. $-4 < -1$ **9.** $-2 < \frac{3}{2}$ **11.** $-\frac{9}{2} < -3$

13. $3 > -4$ **15.** $4 > -\frac{7}{2}$

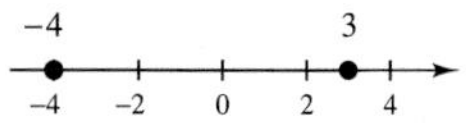

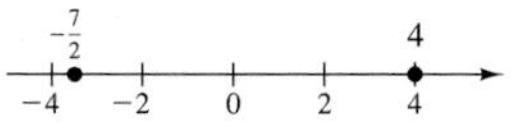

17. $0 > -\frac{7}{16}$ 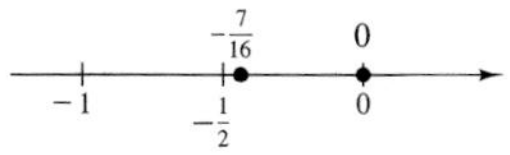 **19.** $-4.6 < 1.5$

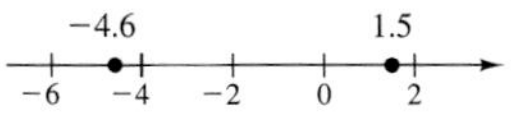

21. $\frac{7}{16} < \frac{5}{8}$ **23.** $-2\pi > -10$

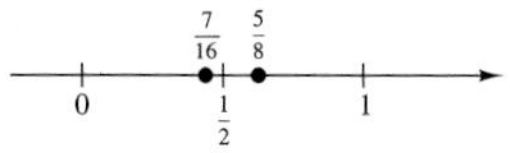

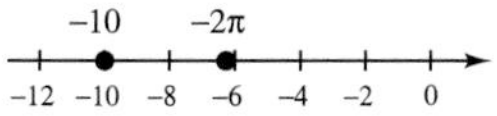

25. 2 **27.** 4

29. -5

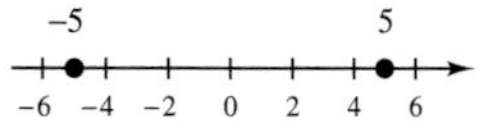

Distance: 5

31. 3.8 **33.** $\frac{5}{2}$

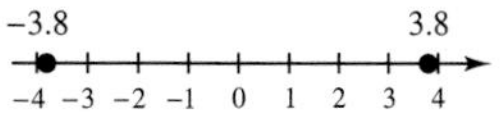

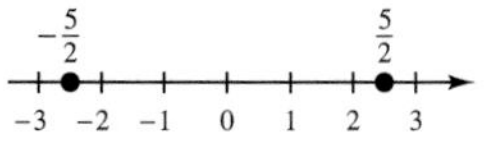

Distance: 3.8 Distance: $\frac{5}{2}$

35. $\frac{5}{2}, \frac{5}{2}$

37. $\frac{4}{3}, \frac{4}{3}$ **39.** 7 **41.** 3.4 **43.** $\frac{7}{2}$ **45.** -4.09

47. -23.6 **49.** 3.2 **51.** $|-15| = |15|$

53. $|-4| > |3|$ **55.** $|32| < |-50|$ **57.** $\left|\frac{3}{16}\right| < \left|\frac{3}{2}\right|$

59. $-|-48.5| < |-48.5|$ **61.** $|-\pi| > -|-2\pi|$

63. **65.**

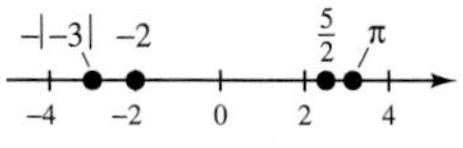

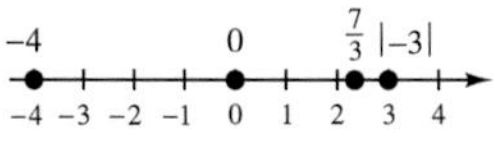

67. $-4.5, 20.5$ **69.** $-5.5, 1.5$

71. (a) $A = \{1°, -4°, -15°, 0°, 5°, 8°, 2°, 3°\}$

(b) $B = \{-3°, -5°, -12°, -20°, -6°, -\frac{4}{3}°, 0°, 2°, -1°, -2°, -9°, -10°, -8°\}$

(c) $C = \{0°, 2°\}$

(d) $-15°, -4\frac{1}{2}°, -4°, 0°, 1°, 2°, 2.5°, 3°, 5°, 7.2°, 8°$

(e) $2°, 0°, -1°, -\frac{4}{3}°, -2°, -3°, -5°, -6°, -8°, -9°, -10°, -12°, -20°$

(f) December 11

73. Two. They are -3 and 3.

75. 3. -10 is 3 units from -7 and 3 is 10 units from -7.

77. $\frac{3}{8} = 0.375$, so 0.35 is the smaller number.

79. True **81.** True

83. False. $\frac{1}{2}$ is not an integer.

Section 1.2 *(page 24)*

1. 9 **3.** -2

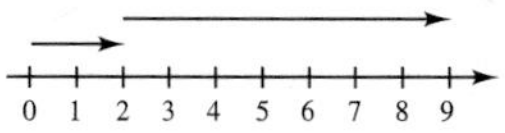

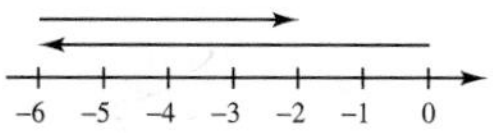

5. -1 **7.** 0 **9.** -1 **11.** -19 **13.** -30

15. -16 **17.** 8 **19.** 4 **21.** 30 **23.** -36

25. -16 **27.** 2225 **29.** 4558 **31.** 898 **33.** 3

35. 0 **37.** 35 **39.** 10 **41.** -103 **43.** -6

45. -610 **47.** 0 **49.** 8 **51.** 8 **53.** -14

55. -50 **57.** 500 **59.** -15 **61.** $2 + 2 + 2 = 6$

63. $(-3) + (-3) + (-3) + (-3) + (-3) = -15$

65. 21 **67.** -32 **69.** 72 **71.** -930 **73.** 90

75. -30 **77.** 90 **79.** 12 **81.** 338

83. -4725 **85.** $-62{,}352$ **87.** 3 **89.** -6

91. Division by zero is undefined. **93.** 27 **95.** -6

97. 0 **99.** 4 **101.** 32 **103.** -32 **105.** 110

107. 4540 **109.** 86 **111.** 1045 **113.** $-532{,}000$

115. Composite **117.** Prime **119.** Prime

121. Composite **123.** Prime **125.** $2 \cdot 2 \cdot 3$

127. $2 \cdot 3 \cdot 5 \cdot 7$ **129.** $2 \cdot 2 \cdot 2 \cdot 2 \cdot 2 \cdot 2 \cdot 3$

131. $3 \cdot 5 \cdot 5 \cdot 7$ **133.** $3 \cdot 5 \cdot 13 \cdot 13$

135. 12°F **137.** \$1,012,000 **139.** (a) \$180 million (b) \$30 million

141. $-24°$ **143.** \$6000 **145.** 57,600 square feet

147. 65 miles per hour **149.** 594 cubic inches

151. (a) $3 + 2 = 5$

(b) Adding two integers with like signs

(c) On the last two plays, the team gained 3 yards and 2 yards for a total of 5 yards.

153. 2; it is divisible only by 1 and itself. Any other even number is divisible by 1, itself, and 2.

155. $\sqrt{1997} < 45$

157. To add two negative integers, add their absolute values and attach the negative sign.

159. Negative

161. The sum of three terms each equal to -5

163. $(2m)n = 2(mn)$. The product of two odd integers is odd.

165. n is a multiple of 4. $\frac{12}{2} = 6$

167. Perfect (< 25): 6
Abundant (< 25): 12, 18, 20, 24
First perfect greater than 25 is 28.

Mid-Chapter Quiz *(page 28)*

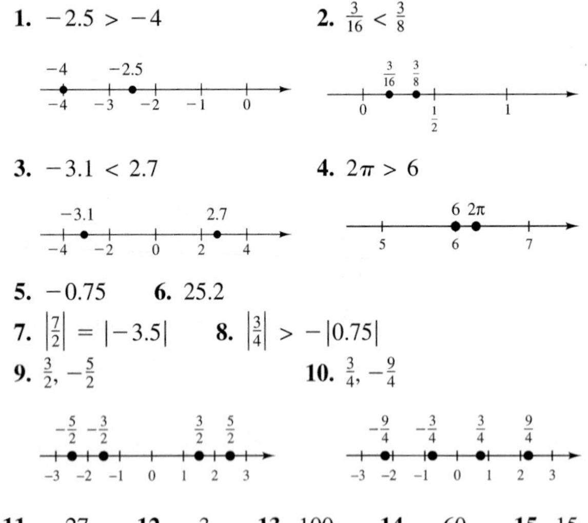

1. $-2.5 > -4$ **2.** $\frac{3}{16} < \frac{3}{8}$

3. $-3.1 < 2.7$ **4.** $2\pi > 6$

5. -0.75 **6.** 25.2

7. $\left|\frac{7}{2}\right| = |-3.5|$ **8.** $\left|\frac{3}{4}\right| > -|0.75|$

9. $\frac{3}{2}, -\frac{5}{2}$ **10.** $\frac{3}{4}, -\frac{9}{4}$

11. -27 **12.** -3 **13.** 100 **14.** -60 **15.** 15

16. -4 **17.** \$450,450 **18.** 128 cubic feet **19.** 15 feet

20. False. The sum of two positive integers is positive.

Section 1.3 *(page 40)*

1. 5 **3.** 4 **5.** 6 **7.** 60 **9.** 1 **11.** $\frac{1}{4}$

13. $\frac{2}{3}$ **15.** $\frac{5}{16}$ **17.** $\frac{2}{25}$ **19.** $\frac{3}{5}$ **21.** $\frac{3}{5}$ **23.** $\frac{3}{5}$

25. $\frac{14}{11}$ **27.** $\frac{3}{8}$ **29.** -1 **31.** $-\frac{1}{2}$ **33.** $\frac{4}{3}$

35. $\frac{6}{16}$ **37.** $\frac{10}{25}$ **39.** $\frac{5}{6}$ **41.** $-\frac{1}{12}$ **43.** $\frac{9}{16}$

45. $-\frac{7}{24}$ **47.** $\frac{4}{3}$ **49.** $-\frac{41}{24}$ **51.** $-\frac{1}{12}$ **53.** $\frac{17}{48}$

55. $-\frac{17}{12}$ **57.** $\frac{5}{6}$ **59.** $\frac{23}{5}$ **61.** $\frac{37}{10}$ **63.** $\frac{26}{3}$

65. $-\frac{115}{11}$ **67.** $\frac{55}{6}$ **69.** $-\frac{17}{16}$ **71.** $-\frac{53}{12}$ **73.** $-\frac{121}{12}$

75. $\frac{3}{10}$ **77.** $\frac{3}{8}$ **79.** $-\frac{3}{8}$ **81.** $\frac{21}{20}$ **83.** $\frac{27}{40}$

85. $-\frac{3}{16}$ **87.** $\frac{12}{5}$ **89.** 1 **91.** $\frac{121}{12}$ **93.** $\frac{56}{3}$

95. $\frac{1}{7}$ **97.** $\frac{7}{4}$ **99.** $\frac{1}{2}$ **101.** $-\frac{8}{27}$

103. Division by zero is undefined. **105.** -90 **107.** $\frac{5}{6}$

109. $-\frac{16}{3}$ **111.** $\frac{5}{2}$ **113.** $\frac{10}{7}$ **115.** 0.75

117. 0.5625 **119.** $0.\overline{6}$ **121.** $0.58\overline{3}$ **123.** $0.\overline{45}$

125. 2.27 **127.** -1.90 **129.** -57.02 **131.** 0.04

133. 39.08 **135.** ≈ 1 **137.** $\$1\frac{5}{8}$

139. $\frac{1013}{40} = 25.325$ tons **141.** $\frac{5}{8}$ **143.** \$11.85

145. 48 breadsticks **147.** \$677.49

149. (a) \$30,600 (b) \$30,600

151. (a) Answers will vary. (b) 4.7 hours

153. (g) December 4 (h) December 4 and 5
(i) December 4 and 5 (j) $-\frac{2}{35} \approx -0.06°$
(k) $-\frac{113}{21} \approx -5.38°$ (l) g, h, i

155. No. Rewrite both fractions with like denominators. Then add their numerators and write the sum over the common denominator.

157. (a) $\frac{1}{3}$ (b) $\frac{1}{6}$ (c)

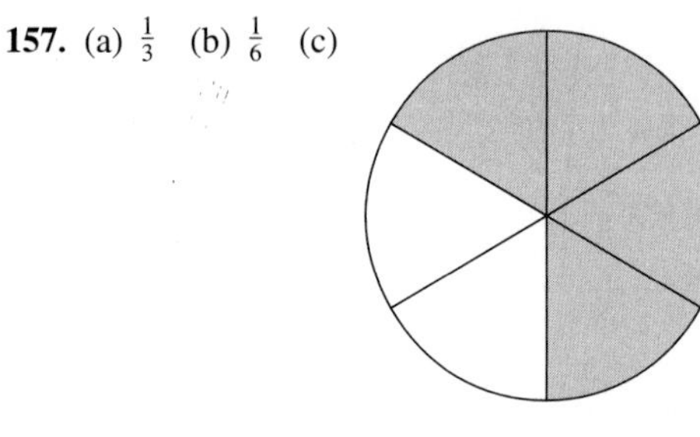

159. There are 12 one-fourths in 3.

161. False **163.** True **165.** True **167.** $\frac{3}{6} + \frac{4}{5} = \frac{13}{10}$

Section 1.4 *(page 51)*

1. 2^5 **3.** $\left(-\frac{1}{4}\right)^3$ **5.** $(-3)(-3)(-3)(-3)(-3)(-3)$

7. $(9.8)(9.8)(9.8)$ **9.** $\left(-\frac{1}{2}\right)\left(-\frac{1}{2}\right)\left(-\frac{1}{2}\right)\left(-\frac{1}{2}\right)\left(-\frac{1}{2}\right)$

11. Negative **13.** Negative **15.** 9 **17.** 64

19. -125 **21.** $\frac{1}{64}$ **23.** -1.728 **25.** 8 **27.** -9

29. 27 **31.** $-\frac{11}{2}$ **33.** 9 **35.** 17 **37.** -64

39. $\frac{7}{3}$ **41.** $\frac{8}{3}$ **43.** $\frac{5}{4}$ **45.** -1 **47.** $\frac{5}{6}$ **49.** 4

51. Division by zero is undefined. **53.** 13 **55.** 210

57. 5840 **59.** 0.0084 **61.** 7.32 **63.** 0

65. 1.19 **67.** 836.94 **69.** $24^2 = (4 \cdot 6)^2 = 4^2 \cdot 6^2$

71. $-3^2 = -(3)(3) = -9$

73. Commutative Property of Multiplication

75. Commutative Property of Addition

77. Additive Identity Property

79. Additive Inverse Property

81. Associative Property of Addition

83. Associative Property of Multiplication

85. Multiplicative Inverse Property

87. Distributive Property **89.** Distributive Property

91. Associative Property of Addition

93. $(u + v)5$ or $5(v + u)$ **95.** $x + 3$ **97.** $6x + 12$

99. $100 + 25y$ **101.** $(3x + 2y) + 5$ **103.** $6(xy)$

105. (a) -50 (b) $\frac{1}{50}$ **107.** (a) 1 (b) -1

109. (a) $-2x$ (b) $\dfrac{1}{2x}$ **111.** (a) $-ab$ (b) $\dfrac{1}{ab}$

113. 48 **115.** 22 **117.** $5(x + 3) = 5x + 15$

119. Division by zero is undefined.

121. $4(2 + x) = 4(x + 2)$ Commutative Property of Addition
$= 4x + 8$ Distributive Property

123. $7x + 9 + 2x$
$= 7x + 2x + 9$ Commutative Property of Addition
$= (7x + 2x) + 9$ Associative Property of Addition
$= (7 + 2)x + 9$ Distributive Property
$= 9x + 9$ Addition of Real Numbers
$= 9(x + 1)$ Distributive Property

125. 36 square units **127.** 0.07, $31,500

129. $11,070 **131.** (a) $x(1 + 0.06) = 1.06x$ (b) $27.51

133. $a + b + 2c + 12$ **135.** No

137. (a) Base (b) Exponent

139. No. $2 \cdot 5^2 = 2 \cdot 25 = 50,\ 10^2 = 100$

141. $12 + (48 \div 6) - 5$

143. Commutative Property of Addition:
$a + b = b + a,\ 3 + x = x + 3$
Commutative Property of Multiplication:
$ab = ba,\ 3x = x(3)$

145. (a) Additive Identity Property: $a + 0 = a$. The addition of any quantity and 0 yields the same quantity.
$7 + 0 = 7$

(b) Additive Inverse Property: $a + (-a) = 0$. Adding any quantity and its opposite yields the additive identity 0.
$7 + (-7) = 0$

Review Exercises *(page 56)*

1. $-\frac{1}{10} < 4$ **3.** $-3 > -7$

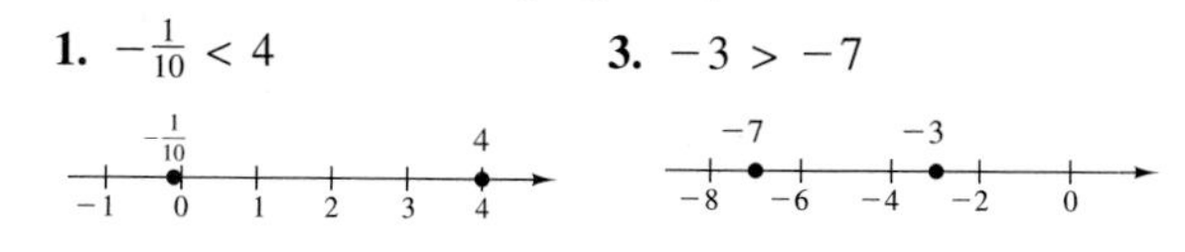

5. $-152, 152$ **7.** $\frac{7}{3}, \frac{7}{3}$ **9.** 8.5 **11.** -8.5

13. $|-84| = |84|$ **15.** $\left|\frac{3}{10}\right| > -\left|\frac{4}{5}\right|$ **17.** 100

19. 11 **21.** 240 **23.** -268 **25.** -38 **27.** 45

29. 1500 **31.** -558 **33.** -18

35. Division by zero is undefined. **37.** 1162

39. -102 **41.** -469 **43.** $-15{,}869$ **45.** 789

47. Prime **49.** Composite **51.** $2 \cdot 3 \cdot 3 \cdot 3 \cdot 7$

53. $2 \cdot 2 \cdot 13 \cdot 31$ **55.** 18 **57.** 21 **59.** $\frac{2}{3} = \frac{10}{15}$

61. $\frac{6}{10} = \frac{15}{25}$ **63.** $\frac{2}{5}$ **65.** $\frac{3}{4}$ **67.** $\frac{1}{9}$ **69.** $\frac{103}{96}$

71. $\frac{5}{4}$ **73.** $\frac{17}{8}$ **75.** $-\frac{1}{12}$ **77.** 1 **79.** $\frac{2}{3}$

81. $\frac{6}{7}$ **83.** Division by zero is undefined. **85.** 21

87. 796.11 **89.** 1841.74 **91.** 343 **93.** -343

95. $2^2 < 2^4$ **97.** $\frac{3}{4} > \left(\frac{3}{4}\right)^2$ **99.** $\frac{81}{625}$ **101.** 160

103. 54 **105.** $\frac{37}{8}$ **107.** 140 **109.** -3 **111.** 7

113. 0 **115.** Additive Inverse Property

117. Commutative Property of Multiplication

119. Multiplicative Identity Property

121. Distributive Property **123.** 0.6

125. False. If the absolute value of the negative integer is less than the positive integer, the sum will be positive.

127. 32,000 miles **129.** (a) $5030, $3090, $4510, $5700
(b) $14,850, $3480
(c) $18,330

131. $36 **133.** $3.52 **135.** (a) $6750 (b) $9250

Chapter Test *(page 59)*

1. (a) $8, \frac{12}{4}$ (b) $-10, 8, \frac{12}{4}$ (c) $-10, 8, \frac{3}{4}, \frac{12}{4}, 6.5$

2. $-\frac{3}{5} > -|-2|$ **3.** -4 **4.** 10 **5.** 10 **6.** 47

7. -160 **8.** 8 **9.** 3 **10.** 1 **11.** $\frac{17}{24}$

12. $-\frac{45}{2}$ **13.** $\frac{7}{12}$ **14.** -27 **15.** $-\frac{4}{9}$ **16.** 33

17. Distributive Property

18. Multiplicative Inverse Property

19. Associative Property of Addition

20. Commutative Property of Multiplication

21. $\frac{5}{12}$ **22.** $-3^4 = -1(3^4)$

23. Exponentiation, multiplication, subtraction

24.

$\frac{2}{3}, \frac{6}{9}$

Chapter 2

Section 2.1 *(page 68)*

Integrated Review *(page 68)*

1. Commutative Property of Multiplication
2. Additive Inverse Property
3. Distributive Property
4. Associative Property of Addition
5. 3 **6.** 8 **7.** $\frac{9}{2}$ **8.** $\frac{2}{7}$ **9.** $-\frac{7}{11}$ **10.** $\frac{10}{3}$
11. $6000 **12.** 15 feet

1. $60t$ **3.** $2.19m$ **5.** Variable: x
Constant: 3
7. Variables: x, z **9.** $4x, 3$ **11.** $3x^2, 5$
13. $\frac{5}{3}, -3y^3$ **15.** $2x, -3y, 1$ **17.** $3(x + 5), 10$
19. $\frac{x}{4}, \frac{5}{x}$ **21.** $\frac{3}{x + 2}, -3x, 4$ **23.** -6 **25.** $-\frac{1}{3}$
27. $-\frac{3}{2}$ **29.** 2π **31.** 4.7 **33.** $y \cdot y \cdot y \cdot y \cdot y$
35. $2 \cdot 2 \cdot x \cdot x \cdot x \cdot x$ **37.** $4 \cdot y \cdot y \cdot z \cdot z \cdot z$
39. $a^2 \cdot a^2 \cdot a^2 = a \cdot a \cdot a \cdot a \cdot a \cdot a$
41. $4 \cdot x \cdot x \cdot x \cdot x \cdot x \cdot x$ **43.** $a \cdot a \cdot a \cdot b \cdot b \cdot b$
45. $(x + y)(x + y)$ **47.** $\left(\frac{a}{3s}\right)\left(\frac{a}{3s}\right)\left(\frac{a}{3s}\right)\left(\frac{a}{3s}\right)$
49. $3 \cdot 3 \cdot 3 \cdot (r + s)(r + s)(r + s)(r + s)$ **51.** $2u^4$
53. $(2u)^4$ **55.** a^3b^2 **57.** $3^3(x - y)^2$ **59.** $\left(\frac{x^2}{2}\right)^3$
61. (a) 0 (b) 7 **63.** (a) 3 (b) 13 **65.** (a) 6 (b) 0 **67.** (a) 3 (b) -20
69. (a) 17 (b) 4 **71.** (a) 0 (b) Division by zero is undefined.
73. (a) 10
(b) Division by zero is undefined.
75. (a) $\frac{15}{2}$ (b) 10 **77.** (a) 175 (b) 140
79. (a)

x	-1	0	1	2	3	4
$3x - 2$	-5	-2	1	4	7	10

(b) 3
(c) $\frac{2}{3}$

81. n^2, 64 square units **83.** $a(a + b)$, 45 square units
85. (a) Square: 2 diagonals
(b) Pentagon: 5 diagonals
(c) Hexagon: 9 diagonals
87. (a) 4, 5, 5.5, 5.75, 5.875, 5.938, 5.9698
Approaches 6.
(b) 9, 7.5, 6.75, 6.375, 6.188, 6.094, 6.047
Approaches 6.
89. (a) $(15 \cdot 12)c = 180c$
Plastic/aluminum chairs: $351
Wood/padded seat chairs: $531
(b) Canopy 1: $215
Canopy 2: $265
Canopy 3: $415
Canopy 4: $565
Canopy 5: $715
91. No. The term includes the sign and is $-3x$.
93. No. When $y = 3$, the expression is undefined.

Section 2.2 *(page 80)*

Integrated Review *(page 80)*

1. $a^m \cdot a^n = a^{m+n}$ **2.** Distributive Property
3. 12 **4.** 120 **5.** -11 **6.** -5760
7. 35 **8.** -350 **9.** $\frac{1}{80}$ **10.** $\frac{45}{16}$
11. 2,362,000 **12.** 52 miles per hour

1. u^6 **3.** $3x^7$ **5.** $5x^7$ **7.** $-15z^5$ **9.** $2xy^2z^2$
11. $-6ab^7$ **13.** t^8 **15.** $5u^5v^5$ **17.** $-8s^3$
19. $a^{10}b^{11}$ **21.** $16u^6v^{11}$ **23.** $(x - 3)^8$
25. $(x - 2y)^6$ **27.** $x^5 \cdot x^3 = x^{5+3} = x^8 \neq x^{15}$
29. $-3x^3 \neq -27x^3 = (-3x)^3$
31. Commutative Property of Addition
33. Associative Property of Multiplication
35. Multiplicative Identity Property
37. Commutative Property of Multiplication
39. Additive Inverse Property
41. Multiplicative Inverse Property
43. Additive Inverse Property, Additive Identity Property
45. $(x + 10) - (x + 10) = 0$
Additive Inverse Property

47. $v(2) = 2v$

Commutative Property of Multiplication

49. $5(t - 2) = 5t + 5(-2)$

Distributive Property

51. $5x \cdot \frac{1}{5x} = 1$

Multiplicative Inverse Property

53. $12 + (8 - x) = (12 + 8) - x$

Associative Property of Addition

55. $-10x + 5y$ **57.** $3x + 6$ **59.** $4x + 4xy + 4y^2$

61. $3x^2 + 3x$ **63.** $-12y^2 + 16y$ **65.** $-u + v$

67. $3x^2 - 4xy$ **69.** ab; ac; $a(b + c) = ab + ac$

71. $2a$; $2(b - a)$; $2a + 2(b - a) = 2b$

73. $6x^2, -3xy, y^2$; $6, -3, 1$ **75.** $16t^3, 3t^3$; $4, -5$

77. $6x^2y, -4x^2y$; $2xy$

79. Variable factors are not alike. $x^2y \neq xy^2$ **81.** $-2y$

83. $-2x + 5$ **85.** $11x + 4$ **87.** $3r + 7$

89. $x^2 - xy + 4$ **91.** $17z + 11$

93. $z^3 + 3z^2 + 3z + 1$ **95.** $-x^2y + 4xy + 12xy^2$

97. $2\left(\frac{1}{x}\right) + 8$ **99.** $11\left(\frac{1}{t}\right) - 2t$

101. False. $3(x - 4) = 3x - 12$ **103.** True **105.** 416

107. 39.9 **109.** $12x$ **111.** $4x$ **113.** $6x^2$

115. $-10z^3$ **117.** $9a$ **119.** $-6x^5$ **121.** $-24x^4y^4$

123. $2x$ **125.** $13s - 2$ **127.** $-2m + 15$

129. $44 + 2x$ **131.** $8x + 26$ **133.** $2x - 17$

135. $10x - 7x^2$ **137.** $3x^2 + 5x$ **139.** $4t^2 - 11t$

141. $26t - 2t^2$ **143.** $\frac{x}{3}$ **145.** $\frac{7z}{5}$ **147.** $-\frac{11x}{12}$

149. x **151.** \$21,589.25 **153.** Area of the square: x^2
Volume of the cube: x^3

155. $4x + 12$ **157.** 9375 square feet

159. $(6x)^4 = (6x)(6x)(6x)(6x)$

$6x^4 = 6x \cdot x \cdot x \cdot x$

161. (a), (c), and (d)

163. To combine like terms, add the respective coefficients and attach the common variable factor(s). $3x^2 - 5x^2 = -2x^2$

165. Remove the innermost symbols first and combine like terms. A symbol of grouping preceded by a *minus* sign can be removed by changing the sign of each term within the symbols.

167. It does not change if the parentheses are removed because multiplication is a higher-order operation than subtraction. Removing the brackets does change the expression because the division would be performed prior to the subtraction.

Mid-Chapter Quiz *(page 84)*

1. (a) 0 (b) 10 (c) 0

2. (a) 2 (b) 0 (c) Division by zero is undefined.

3. (a) -5 (b) $\frac{5}{16}$

4. (a) $(3y)^4$ (b) $2^3(x - 3)^2$

5. x^7 **6.** v^{10}

7. $9y^5$ **8.** $8(x - 4)^6$ **9.** $\frac{10z^3}{21y^4}$ **10.** $\left(\frac{x}{y}\right)^7$

11. Associative Property of Multiplication

12. Distributive Property

13. Multiplicative Inverse Property

14. Commutative Property of Addition **15.** $6x - 2$

16. $-8y + 12$ **17.** $y^2 + 4xy + y$ **18.** $3\left(\frac{1}{u}\right) + 3u$

19. $8a - 7b$ **20.** $-8x - 66$ **21.** $\frac{1}{10}\pi r^4h$

22. 45,700

Section 2.3 *(page 95)*

Integrated Review *(page 95)*

1. Negative **2.** 15, 3

3. False. $-4^2 = -1 \cdot 4 \cdot 4 = -16$

4. True. $(-4)^2 = (-4)(-4) = 16$ **5.** 78

6. 120 **7.** $\frac{3}{4}$ **8.** $\frac{14}{3}$ **9.** $\frac{23}{9}$ **10.** $-\frac{111}{10}$

11. 5 weeks, \$16.50 **12.** 40 meters

1. (d) **3.** (e) **5.** (b) **7.** $x + 5$ **9.** $x - 25$

11. $x - 6$ **13.** $2x$ **15.** $\frac{x}{3}$ **17.** $\frac{x}{50}$ **19.** $\frac{3}{10}x$

21. $3x + 5$ **23.** $8 + 5x$ **25.** $10(x + 4)$

27. $|x + 4|$ **29.** $x^2 + 1$ **31.** A number decreased by 10

33. The product of 3 and a number, increased by 2

35. Seven times a number increased by 4

37. Three times the difference of 2 and a number

39. The sum of a number and 1, divided by 2

41. The square of a number, increased by 5

43. $(x + 3)x = x^2 + 3x$ **45.** $(25 + x) + x = 25 + 2x$

47. $(x - 9)(3) = 3x - 27$ **49.** $\dfrac{8(x + 24)}{2} = 4x + 96$

51. $0.10d$ **53.** $0.06L$ **55.** $\dfrac{100}{r}$ **57.** $15m + 2n$

59. $t = 10.2$ years **61.** $t = 11.9$ years

63.

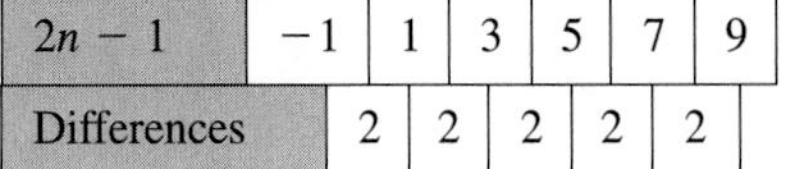

n	0	1	2	3	4	5
$2n - 1$	-1	1	3	5	7	9
Differences	2	2	2	2	2	

65. 3 **67.** $a = 5, b = 4$ **69.** $3x(6x - 1) = 18x^2 - 3x$

71. $\frac{1}{2}(12)(5x^2 + 2) = 30x^2 + 12$

73.

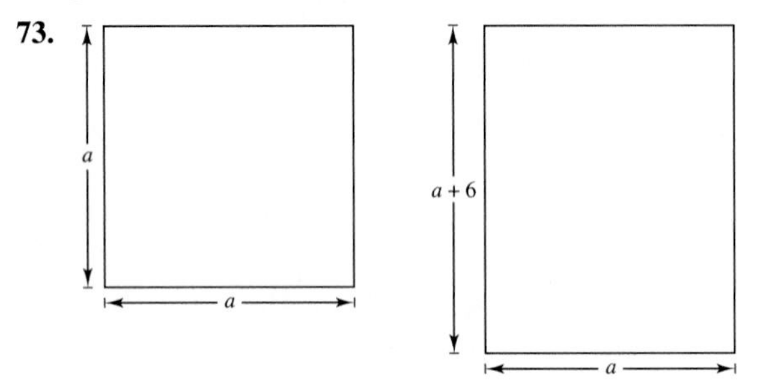

Perimeter of the square: $4a$ centimeters
Area of the square: a^2 square centimeters
Perimeter of the rectangle: $4a + 12$ centimeters
Area of the rectangle: $a(a + 6)$ square centimeters

75. s^2 square inches **77.** $5w$ **79.** (a) $m + n$
(b) 3, 5, 8, 13, 21

81. Subtraction **83.** (a), (b), (e) **85.** $\left(\dfrac{5}{n}\right)3, \dfrac{5}{3n}$

Section 2.4 *(page 105)*

Integrated Review *(page 105)*

1. Negative
2. Positive. The product of an even number of negative factors is positive.

3. $10 + 6$ **4.** Multiplicative Inverse Property

5. t^7 **6.** $-3y^5$ **7.** u^6 **8.** $2a^5b^5$

9. $12a^3b$ **10.** $2(x + 3)^5$

11. Perimeter: $6x$
Area: $\dfrac{9x^2}{4}$

12. Perimeter: $9x - 2$
Area: $5x^2 - 4x$

1. (a) Solution (b) Not a solution **3.** (a) Not a solution (b) Solution

5. (a) Not a solution (b) Solution **7.** (a) Solution (b) Not a solution

9. (a) Solution (b) Not a solution **11.** (a) Solution (b) Solution

13. (a) Not a solution (b) Not a solution **15.** (a) Not a solution (b) Not a solution

17. (a) Not a solution (b) Not a solution **19.** (a) Solution (b) Not a solution

21. (a) Solution (b) Not a solution **23.** (a) Not a solution (b) Solution

25. (a) Solution (b) Not a solution

27.

$5x + 12 = 22$	Given equation
$5x + 12 - 12 = 22 - 12$	Subtract 12 from both sides.
$5x = 10$	Combine like terms.
$\dfrac{5x}{5} = \dfrac{10}{5}$	Divide both sides by 5.
$x = 2$	Solution

29.

$\dfrac{2}{3}x = 12$	Given equation
$\dfrac{3}{2}\left(\dfrac{2}{3}x\right) = \dfrac{3}{2}(12)$	Multiply both sides by $\frac{3}{2}$.
$x = 18$	Solution

31.

$2(x - 1) = x + 3$	Given equation
$2x - 2 = x + 3$	Distributive Property
$-x + 2x - 2 = -x + x + 3$	Subtract x from both sides.
$x - 2 = 3$	Combine like terms.
$x - 2 + 2 = 3 + 2$	Add 2 to both sides.
$x = 5$	Solution

33.

$x = -2(x + 3)$	Given equation
$x = -2x - 6$	Distributive Property
$2x + x = 2x - 2x - 6$	Add $2x$ to both sides.
$3x = 0 - 6$	Additive Inverse Property
$3x = -6$	Combine like terms.
$\dfrac{3x}{3} = \dfrac{-6}{3}$	Divide both sides by 3.
$x = -2$	Solution

35. 2 **37.** 10 **39.** The sum of a number and 8 is 25.

41. Ten times a number decreased by 3 is 8 times the number.

43. The sum of a number and 1 divided by 3 is 8.

45. $x + 6 = 94$ **47.** $3650 + x = 4532$

49. $x + 12 = 45$ **51.** $4(x + 6) = 100$

53. $2x - 14 = \frac{x}{3}$ **55.** $0.32m = 135.36$

57. $2l + 2(\frac{1}{3}l) = 96$ **59.** $3r + 25 = 160$

61. $135 = 2.5x$ **63.** $p + 45 = 375$

65. $750{,}000 - 3D = 75{,}000$ **67.** $1.75n = 2000$

69. 15 dollars **71.** 15 dollars **73.** 6000 feet

75. An equation whose solution set is not the entire set of real numbers is called a conditional equation. The solution set of an identity is all real numbers.

77. Simplifying an expression means removing all symbols of grouping and combining like terms. Solving an equation means finding all values of the variable for which the equation is true.

Simplify: $3(x - 2) - 4(x + 1) = 3x - 6 - 4x - 4$
$= -x - 10$

Solve: $3(x - 2) = 6$
$3x - 6 = 6$
$3x = 12 \rightarrow x = 4$

79. (a) Simplify each side by removing symbols of grouping, combining like terms, and reducing fractions on one or both sides.

(b) Add (or subtract) the same quantity to (from) both sides of the equation.

(c) Multiply (or divide) both sides of the equation by the same nonzero real number.

(d) Interchange the two sides of the equation.

Review Exercises *(page 110)*

1. 4; $-\frac{1}{2}x^3$, $-\frac{1}{2}$ **3.** y^2, 1; $-10yz$, -10; $\frac{2}{3}z^2$, $\frac{2}{3}$

5. $(5z)^3$ **7.** $a^2(b - c)^2$ **9.** (a) 5 (b) 5

11. (a) 4 (b) -2 **13.** x^7 **15.** x^6 **17.** $-2t^6$

19. $-5x^3y^4$ **21.** $-64y^7$

23. Multiplicative Inverse Property

25. Commutative Property of Multiplication

27. Associative Property of Addition

29. $4x + 12y$ **31.** $-10u + 15v$ **33.** $8x^2 + 5xy$

35. $a - 3b$ **37.** $-2a$ **39.** $11p - 3q$ **41.** $\frac{15}{4}s - 5t$

43. $x^2 + 2xy + 4$ **45.** $3x - 3y + 3xy$

47. $3\left(1 + \frac{r}{n}\right)^2$ **49.** $5u - 10$ **51.** $5s - r$

53. $10z - 1$ **55.** $2z - 2$ **57.** $8x - 32$

59. $-2x + 4y$ **61.** $\frac{2}{3}x + 5$ **63.** $2x - 10$

65. $50 + 7x$ **67.** $\frac{x + 10}{8}$ **69.** $x^2 + 64$

71. A number plus 3

73. A number decreased by 2, divided by 3

75. (a) Not a solution (b) Solution **77.** (a) Not a solution (b) Solution

79. (a) Solution (b) Not a solution **81.** (a) Not a solution (b) Solution

83. (a) Solution (b) Solution **85.** $P(\frac{9}{10})^5$ **87.** $0.28I$ **89.** $9x^2$

91. $10s$ **93.** $625n$

95. (a)

n	0	1	2	3	4	5
$n^2 + 3n + 2$	2	6	12	20	30	42
Differences	4	6	8	10	12	
Differences	2	2	2	2		

(b) Third row: entries increase by 2
Fourth row: constant 2

97. $x + \frac{1}{x} = \frac{37}{6}$ **99.** $6x - \frac{1}{2}(6x) = \frac{1}{2}(6x) = 24$

Chapter Test *(page 113)*

1. $2x^2$, 2; $-7xy$, -7; $3y^3$, 3 **2.** $x^3(x + y)^2$

3. Associative Property of Multiplication

4. Commutative Property of Addition

5. Additive Inverse Property

6. Multiplicative Identity Property

7. $3x + 24$ **8.** $-3y + 2y^2$ **9.** c^8 **10.** $-10u^4v$

11. $-a - 7b$ **12.** $8u - 8v$ **13.** $4z - 4$

14. $18 - 2t$ **15.** (a) 25 (b) -31

16. Division by zero is undefined. **17.** $\frac{1}{5}n + 2$

18. (a) Perimeter: $2w + 2(2w - 4)$; Area: $w(2w - 4)$

(b) Perimeter: $6w - 8$; Area: $2w^2 - 4w$

(c) Perimeter: unit of length; Area: square units

(d) Perimeter: 64 feet; Area: 240 square feet

19. $3n + 2m$ **20.** (a) Not a solution (b) Solution

Chapter 3

Section 3.1 *(page 124)*

Integrated Review *(page 124)*

1. (a) $(ab)^n = a^n b^n$ (b) $(a^m)^n = a^{mn}$

2. Associative Property of Addition **3.** u^8

4. $9a^6$ **5.** $-3(x-5)^5$ **6.** $-40r^3s^4$

7. $\frac{2m^3}{5n^4}$ **8.** $\frac{(x+3)^2}{2(x+8)}$ **9.** $-9x + 11y$

10. $8v - 4$ **11.** $\frac{1}{12}$ mile **12.** $30\frac{23}{30}$ tons

1. 8 **3.** 13 **5.** 4 **7.** 3

9. Original equation
Subtract 15 from both sides.
Combine like terms.
Divide both sides by 5.
Simplify.

11. Original equation
Subtract 5 from both sides.
Combine like terms.
Divide both sides by -2.
Simplify.

13. 6 **15.** $-\frac{7}{3}$

17. 3 **19.** 2 **21.** 4 **23.** $\frac{2}{3}$ **25.** 2 **27.** $\frac{1}{3}$

29. -2 **31.** No solution **33.** 1 **35.** $\frac{5}{2}$ **37.** $-\frac{1}{4}$

39. Identity **41.** $\frac{2}{5}$ **43.** 0 **45.** $\frac{2}{3}$ **47.** 30

49. $\frac{5}{3}$ **51.** $\frac{5}{6}$ **53.** Identity **55.** 2 **57.** 36 feet

59. 3 feet, 3 feet, 6 feet **61.** $2\frac{1}{4}$ hours **63.** 150

65. 7 hours 20 minutes **67.** 30 **69.** 35, 37

71. (a)

t	1	1.5	2
Width	300	240	200
Length	300	360	400
Area	90,000	86,400	80,000

t	3	4	5
Width	150	120	100
Length	450	480	500
Area	67,500	57,600	50,000

(b) The area decreases.

73. The red box weighs 6 ounces. If you removed three blue boxes from each side, the scale would still balance. The Addition (or Subtraction) Property of Equality

75. Subtract 5 from each side of the equation. Addition Property of Equality

77. The answer is probably incorrect, because it implies an average speed of approximately 83.5 miles per hour. It is unlikely that a moving van would have an average speed that great.

79. True

Section 3.2 *(page 134)*

Integrated Review *(page 134)*

1. (a) $\frac{1}{5} + \frac{7}{5} = \frac{1+7}{5} = \frac{8}{5}$

(b) $\frac{1}{5} + \frac{7}{3} = \frac{3}{15} + \frac{35}{15} = \frac{38}{15}$

2. Answers vary. Examples are given.
$3x^2 + 2\sqrt{x}$
$\frac{4}{x^2+1}$

3. $4x^6$ **4.** $8y^5$ **5.** $5z^7$ **6.** $(a+3)^7$

7. $3x - 4$ **8.** $-x^2 + 1$ **9.** $-y^4 + 2y^2$

10. $10t - 4t^2$ **11.** 7.5 gallons **12.** (a) \$24,300 (b) \$4301

1. -3 **3.** 10 **5.** 5 **7.** 2 **9.** -10 **11.** 2

13. -5 **15.** Identity **17.** -4 **19.** No solution

21. $\frac{8}{5}$ **23.** 1 **25.** $\frac{1}{2}$ **27.** $\frac{2}{9}$ **29.** 1 **31.** 3

33. 3 **35.** $-\frac{3}{2}$ **37.** $\frac{5}{2}$ **39.** $-\frac{2}{5}$ **41.** $-\frac{10}{3}$

43. No solution **45.** $\frac{1}{6}$ **47.** 50 **49.** $\frac{32}{5}$ **51.** 10

53. 0 **55.** $\frac{16}{3}$ **57.** $\frac{4}{3}$ **59.** $\frac{4}{11}$ **61.** 6 **63.** 5.00

65. 7.71 **67.** 123.00 **69.** 3.51 **71.** 8.99

73. 4.8 hours

75. (a) 97
(b) No. A 100 on the final will yield only 86.25% for the course.

77. 25 quarts **79.** $1\frac{1}{3}$ quarts **81.** $x = 4$ feet

83. (a) Each of the n bricks is 8 inches long. Each of the $(n-1)$ mortar joints is $\frac{1}{2}$ inch wide. The total length is 93 inches.
(b) 11

85. Use the Distributive Property to remove symbols of grouping. Remove the innermost symbols first and combine like terms. Symbols of grouping preceded by a *minus* sign can be removed by changing the sign of each term within the symbols.

$$\begin{aligned} 2x - [3 + (x - 1)] &= 2x - [3 + x - 1] \\ &= 2x - [2 + x] \\ &= 2x - 2 - x = x - 2 \end{aligned}$$

87. Divide both sides by 3.

89. It clears the equation of fractions.

Section 3.3 *(page 145)*

Integrated Review *(page 145)*

1. $-28 < 63$ **2.** 0 **3.** 0

−28 63
−60 −40 −20 0 20 40 60 80

4. -38 **5.** -530 **6.** 29 **7.** $8x - 20$

8. $-xz^2 + 2y^2z$ **9.** (a) 7 (b) 16

10. (a) Division by zero is undefined. (b) 2

11. $14.67 **12.** $5r$

	Percent	*Parts out of 100*	*Decimal*	*Fraction*
1.	40%	40	0.40	$\frac{2}{5}$
3.	7.5%	7.5	0.075	$\frac{3}{40}$
5.	63%	63	0.63	$\frac{63}{100}$
7.	15.5%	15.5	0.155	$\frac{31}{200}$
9.	60%	60	0.60	$\frac{3}{5}$
11.	150%	150	1.50	$\frac{3}{2}$

13. 62% **15.** 20% **17.** 7.5% **19.** 250%

21. 0.125 **23.** 1.25 **25.** 2.50 **27.** 0.0075

29. 80% **31.** 125% **33.** $83\frac{1}{3}\%$ **35.** 35%

37. $37\frac{1}{2}\%$ **39.** $41\frac{2}{3}\%$ **41.** 45 **43.** 77.52

45. 0.42 **47.** 176 **49.** 2100 **51.** 2200

53. 132 **55.** 360 **57.** 72% **59.** 12.5%

61. 2.75% **63.** 500%

	Cost	*Selling Price*	*Markup*	*Markup Rate*
65.	$26.97	$49.95	$22.98	85.2%
67.	$40.98	$74.38	$33.40	81.5%
69.	$69.29	$125.98	$56.69	81.8%
71.	$13,250.00	$15,900.00	$2650.00	20%
73.	$107.97	$199.96	$91.99	85.2%

	List Price	*Sale Price*	*Discount*	*Discount Rate*
75.	$39.95	$29.95	$10.00	25%
77.	$23.69	$18.95	$4.74	20%
79.	$189.99	$159.99	$30.00	15.8%
81.	$119.96	$59.98	$59.98	50%
83.	$995.00	$695.00	$300.00	30.2%

85. $544 **87.** $3435 **89.** $71\frac{2}{3}\%$ **91.** 7.2%

93. $312.50 **95.** $24,409 **97.** 10,210

99. 500

101.

< 15	131.8 million
15–24	56.5 million
25–44	181.9 million
45–64	159.6 million
65–74	90.6 million
> 75	77.4 million

103. (a) 1,345,098

(b) 97,854

(c) The number of men in biology increased at a faster rate.

105. (a) $3(19.50) + x = 99$, $40.50

(b) $24 + x = 80$, $56.00, Second package

(c) $40.50 = p(99)$, 40.9%

(d) $56 = p(80)$, 70%

(e) $x = 99 + 0.05(99)$, $103.95

(f) $19.50 + 60(3.2)x = 92.46$, $0.38

107. A rate is a fixed ratio.

109. Decimal to percent: Multiply by 100.
$0.37 = 37\%$

111. Yes. Multiply by 100 and affix the percent sign.

Section 3.4 *(page 156)*

Integrated Review *(page 156)*

1. $\frac{15}{12} = \frac{5 \cdot 3}{4 \cdot 3} = \frac{5}{4}$ **2.** $\frac{3}{5} \div \frac{x}{2} = \frac{3}{5} \cdot \frac{2}{x} = \frac{6}{5x}$

3. $(3x)y = 3(xy)$ **4.** Additive Identity Property

5. 13 **6.** -122 **7.** 9,300,000 **8.** -4

9. $\frac{77}{5}$ **10.** 8 **11.** $2(n - 10)$ **12.** $\frac{1}{4}b(b + 6)$

1. $\frac{4}{1}$ **3.** $\frac{1}{2}$ **5.** $\frac{2}{3}$ **7.** $\frac{9}{1}$ **9.** $\frac{3}{2}$ **11.** $\frac{2}{3}$ **13.** $\frac{1}{4}$

15. $\frac{7}{15}$ **17.** $\frac{2}{1}$ **19.** $\frac{3}{8}$ **21.** $\frac{3}{50}$ **23.** $\frac{3}{4}$ **25.** $\frac{3}{10}$

27. $0.0395 **29.** $0.0645 **31.** $27\frac{3}{4}$-ounce can

33. 16-ounce package **35.** 2-liter bottle **37.** 12

39. 50 **41.** $\frac{10}{3}$ **43.** 16 **45.** $\frac{3}{16}$

47. $\frac{1}{2}$ **49.** 27 **51.** $\frac{14}{5}$ **53.** $\frac{2}{1}$

55. $\frac{12}{1}$ **57.** $\frac{20}{1}$ **59.** $\frac{3}{2}$ **61.** $\frac{100}{49}$

63. 16 gallons **65.** 250 blocks **67.** $1142

69. 22,691 **71.** $46\frac{2}{3}$ minutes **73.** 20 pints

75. 245 miles **77.** $\frac{5}{2}$ **79.** $6\frac{2}{3}$ feet **81.** 80%

83. $5216 **85.** $0.52 **87.** $57.00

89. No. It is necessary to know one of the following: the total number of students in the class, the number of men in the class, or the number of women in the class.

91. A proportion is a statement that equates two ratios.

Mid-Chapter Quiz *(page 160)*

1. 40 **2.** 8 **3.** $\frac{19}{2}$ **4.** 0 **5.** $-\frac{1}{3}$ **6.** $\frac{40}{13}$

7. 36 **8.** 11 **9.** 5 **10.** -2 **11.** 2.06

12. 51.23 **13.** 15.5 **14.** 42 **15.** 200%

16. 455 **17.** 12 meters × 18 meters **18.** 10 hours

19. 6 square meters, 12 square meters, 24 square meters

20. 93 **21.** $495.37 **22.** 44.1% **23.** $\frac{225}{64}$

Section 3.5 *(page 169)*

Integrated Review *(page 169)*

1. $2n$ is an even integer and $2n + 1$ is an odd integer.

2.
$$2x - 3 = 10$$
$$2x - 3 + 3 = 10 + 3$$
$$2x = 13$$

3. $-28y^3$ **4.** $81x^8$ **5.** $\frac{20u^3}{3}$ **6.** $2y$

7. $13x - 5x^2$ **8.** $-5t + 32$ **9.** $10v - 40$

10. $60 - 10x$ **11.** 6% **12.** 20% off

1. $\frac{2A}{b}$ **3.** $\frac{E}{I}$ **5.** $\frac{V}{wh}$ **7.** $\frac{A - P}{Pt}$ **9.** $\frac{S}{1 + R}$

11. $\frac{2A - ah}{h}$ **13.** $\frac{3V + \pi h^3}{3\pi h^2}$ **15.** $\frac{2(h - v_0 t)}{t^2}$

17. 100π meters^3 **19.** 165 miles **21.** 5.6 hours

23. 2112 feet per second **25.** 784 square feet

27. 8 meters **29.** 30 inches

31. Radius: 3.98 inches
Area: 49.74 square inches

33. 24 square inches

35. 96 cubic inches **37.** $540 **39.** $15,975

41. 11% **43.** $1200 **45.** $4000

47. 0.176 hour ≈ 10.6 minutes **49.** 28 miles

51. 1154 miles per hour **53.** $\frac{1}{3}$ hour

55. (a) Answers will vary.
(b) 48 miles per hour

57. Solution 1: 25 gallons
Solution 2: 75 gallons

59. Solution 1: 5 quarts
Solution 2: 5 quarts

61. 20-cent stamps: 40
33-cent stamps: 60

63. 8 nickels
12 dimes

65. 30 pounds at $2.49 per pound
70 pounds at $3.89 per pound

67. $\frac{8}{7}$ gallon

69.

Corn x	Soybeans $100 - x$	Price per ton of the mixture
0	100	$200
20	80	$185
40	60	$170
60	40	$155
80	20	$140
100	0	$125

(a) Decreases

(b) Decreases

(c) Average of the two prices

71. $1\frac{1}{5}$ hours **73.** Answers will vary. **75.** 15

77. Candidate A: 250 votes
Candidate B: 250 votes
Candidate C: 500 votes

79. Perimeter: units—inches, feet, meters
Area: units squared—square inches, square meters
Volume: units cubed—cubic inches, cubic centimeters

81. The circumference would double; the area would quadruple.

Circumference: $C = 2\pi r$. If r is doubled, you have $C = 2\pi(2r) = 2(2\pi r)$.

Area: $A = \pi r^2$. If r is doubled, you have $A = \pi(2r)^2 = 4\pi r^2$.

83. $\frac{1}{5}$

Section 3.6 *(page 183)*

Integrated Review *(page 183)*

1. Distributive Property

2. $(x + 2) - 4 = x + (2 - 4)$

3. $|a| = -a$ if $a < 0$ **4.** $a \cdot b < 0$

5. $-\frac{1}{2} > -7$ **6.** $-\frac{1}{3} < -\frac{1}{6}$ **7.** $-\pi < -3$

8. $-6 > -\frac{13}{2}$ **9.** Perimeter: $2x^2 + 7x - 3$
Area: $\frac{5}{2}x^3 - \frac{3}{2}x^2$

10. Perimeter: $2y^2$
Area: $y^3 - 2y^2 + y$

11. \$332,050 **12.** 53 miles per hour

1. x is greater than or equal to 3.

3. x is less than or equal to 10.

5. y is greater than $-\frac{3}{2}$ *and* less than or equal to 5.

7. (a) Yes (b) No (c) Yes (d) No

9. (a) No (b) No (c) Yes (d) Yes

11. (a) Yes (b) No (c) No (d) Yes

13. (a) Yes (b) No (c) Yes (d) No

15. b **16.** f **17.** c **18.** a **19.** d **20.** e

21. $t \geq 5$ **23.** $x \leq 2$

25. $x < 3$ **27.** $x > -4$

29. $x \leq 18$ **31.** $x > 6$

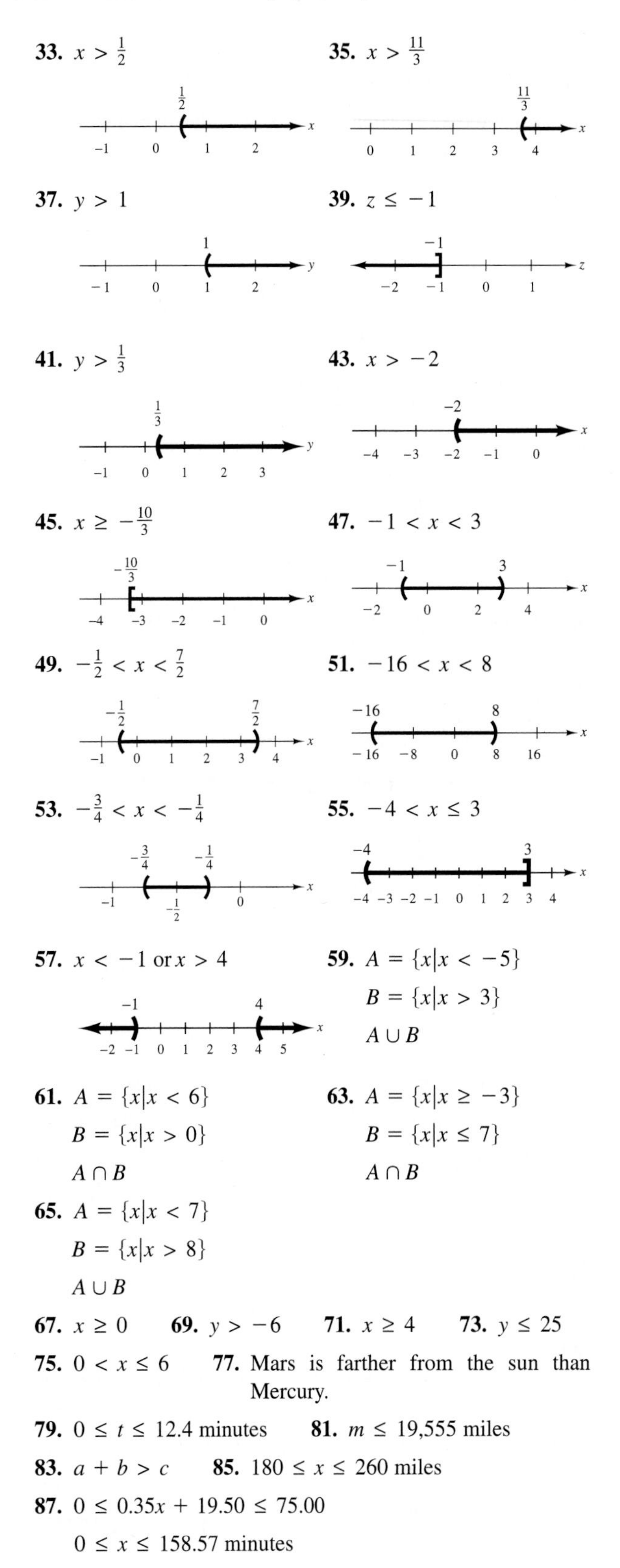

33. $x > \frac{1}{2}$ **35.** $x > \frac{11}{3}$

37. $y > 1$ **39.** $z \leq -1$

41. $y > \frac{1}{3}$ **43.** $x > -2$

45. $x \geq -\frac{10}{3}$ **47.** $-1 < x < 3$

49. $-\frac{1}{2} < x < \frac{7}{2}$ **51.** $-16 < x < 8$

53. $-\frac{3}{4} < x < -\frac{1}{4}$ **55.** $-4 < x \leq 3$

57. $x < -1$ or $x > 4$

59. $A = \{x | x < -5\}$
$B = \{x | x > 3\}$
$A \cup B$

61. $A = \{x | x < 6\}$
$B = \{x | x > 0\}$
$A \cap B$

63. $A = \{x | x \geq -3\}$
$B = \{x | x \leq 7\}$
$A \cap B$

65. $A = \{x | x < 7\}$
$B = \{x | x > 8\}$
$A \cup B$

67. $x \geq 0$ **69.** $y > -6$ **71.** $x \geq 4$ **73.** $y \leq 25$

75. $0 < x \leq 6$ **77.** Mars is farther from the sun than Mercury.

79. $0 \leq t \leq 12.4$ minutes **81.** $m \leq 19{,}555$ miles

83. $a + b > c$ **85.** $180 \leq x \leq 260$ miles

87. $0 \leq 0.35x + 19.50 \leq 75.00$
$0 \leq x \leq 158.57$ minutes

89. Yes. Addition and Subtraction Properties of Inequalities

91. Infinite number. $x - 3 > 0 \rightarrow x > 3$

93. The process is similar except that the direction of the inequality must be reversed if both sides of the inequality are multiplied or divided by a negative number.

95. $<$ **97.** False. **99.** True

$$-\tfrac{1}{2}x + 6 > 0$$
$$-\tfrac{1}{2}x > -6$$
$$x < 12$$

Section 3.7 *(page 190)*

Integrated Review *(page 190)*

1. $2n$ is an even integer and $2n - 1$ is an odd integer.

2. No. $-3x^2 \neq (-3x)^2 = 9x^2$

3. $\dfrac{27}{12} = \dfrac{3 \cdot 9}{3 \cdot 4} = \dfrac{9}{4}$ **4.** $\dfrac{2}{3} \div \dfrac{5}{3} = \dfrac{2}{3} \cdot \dfrac{3}{5} = \dfrac{2}{5}$

5. $3 > -2$ **6.** $-3 < -2$ **7.** $-\frac{1}{2} > -3$

8. $-\frac{1}{3} > -\frac{2}{3}$ **9.** $\frac{1}{2} > \frac{5}{16}$ **10.** $4 < \frac{45}{11}$

11. $\$76,300 - \$75,926 = 374 < 500$

12. $\$39,632 - \$37,800 = 1832 > 500$

1. Yes **3.** No

5. $u - 3 = 7$
$u - 3 = -7$

7. $\frac{1}{2}x + 7 = \frac{3}{2}$
$\frac{1}{2}x + 7 = -\frac{3}{2}$

9. ± 7 **11.** ± 15 **13.** No solution **15.** ± 6

17. 8, 16 **19.** -14, 8 **21.** 13, 19 **23.** No solution

25. -9, 12 **27.** $-\frac{7}{3}$, 1 **29.** 1, $\frac{13}{5}$ **31.** 3, 12

33. -8, 24 **35.** $|x + 3| = 8$ **37.** $|x - 4| = 2$

39. (a) Yes (b) No (c) No (d) Yes

41. (a) No (b) Yes (c) Yes (d) No

43. (a) No (b) No (c) No (d) Yes

45. (a) Yes (b) No (c) Yes (d) No

47. $-1 < z + 2 < 1$ **49.** $5 - h \geq 2$
$5 - h \leq -2$

51.

53.
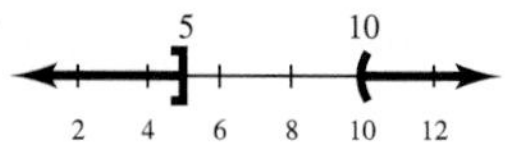

55. $-3 < y < 3$

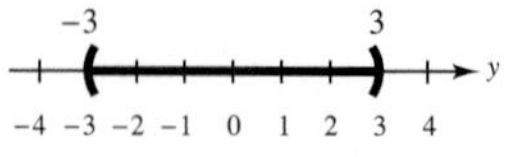

57. $y \leq -2$ or $y \geq 2$

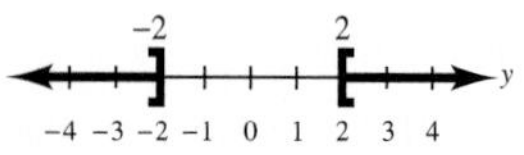

59. $t < -5$ or $t > 5$

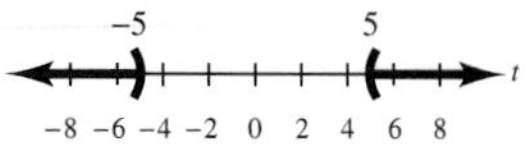

61. $0 \leq y \leq 4$

63. $-5 < x < 3$

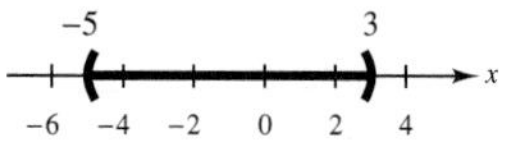

65. $-6 < x < 6$

67. $y < 3$ or $y > 7$

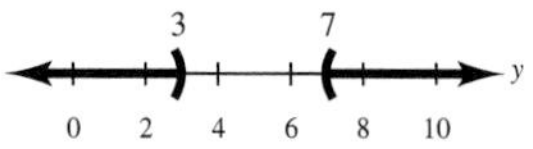

69. $m \leq -5$ or $m \geq 1$

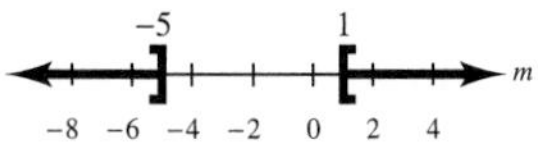

71. c **73.** a **75.** $|x - 1| < 2$ **77.** $|x + 5| \geq 10$

79. $|x| < 2$ **81.** $|x| > 5$ **83.** $|x - 4| > 2$

85.
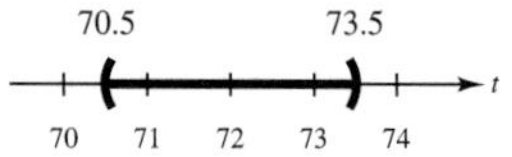

87. (a) $|s - x| \leq 0.005$
(b) $3.495 \leq x \leq 3.5005$

89. $|a|$ represents the two real numbers that are a units from 0 on the real number line.

91. To solve an absolute value equation, rewrite the equation in equivalent forms that can be solved by previously learned methods. If the form is an absolute value of an expression that equals a constant, form two equations by writing the expression without the absolute value signs equal to plus or minus the constant.

Review Exercises *(page 194)*

1. 35 **3.** 28 **5.** Original equation
Add 12 to both sides.
Combine like terms.
Divide both sides by 10.
Simplify.

7. 3 **9.** 3 **11.** 5 **13.** 4 **15.** 3 **17.** $\frac{4}{3}$

19. 20 **21.** 20 **23.** $\frac{5}{3}$ **25.** 7 **27.** $\frac{19}{3}$ **29.** 20

31. 7.99 **33.** 224.31

35.

Percent	Parts out of 100	Decimal	Fraction
35%	35	0.35	$\frac{7}{20}$

37. 20 **39.** 400 **41.** 60% **43.** $\frac{1}{8}$ **45.** $\frac{4}{3}$

47. $\dfrac{7}{2}$ **49.** $-\dfrac{10}{3}$ **51.** 9 **53.** $\dfrac{2A}{r^2}$ **55.** 520 miles

57. 8 hours **59.** 60 miles per hour **61.** $1 \leq x < 4$

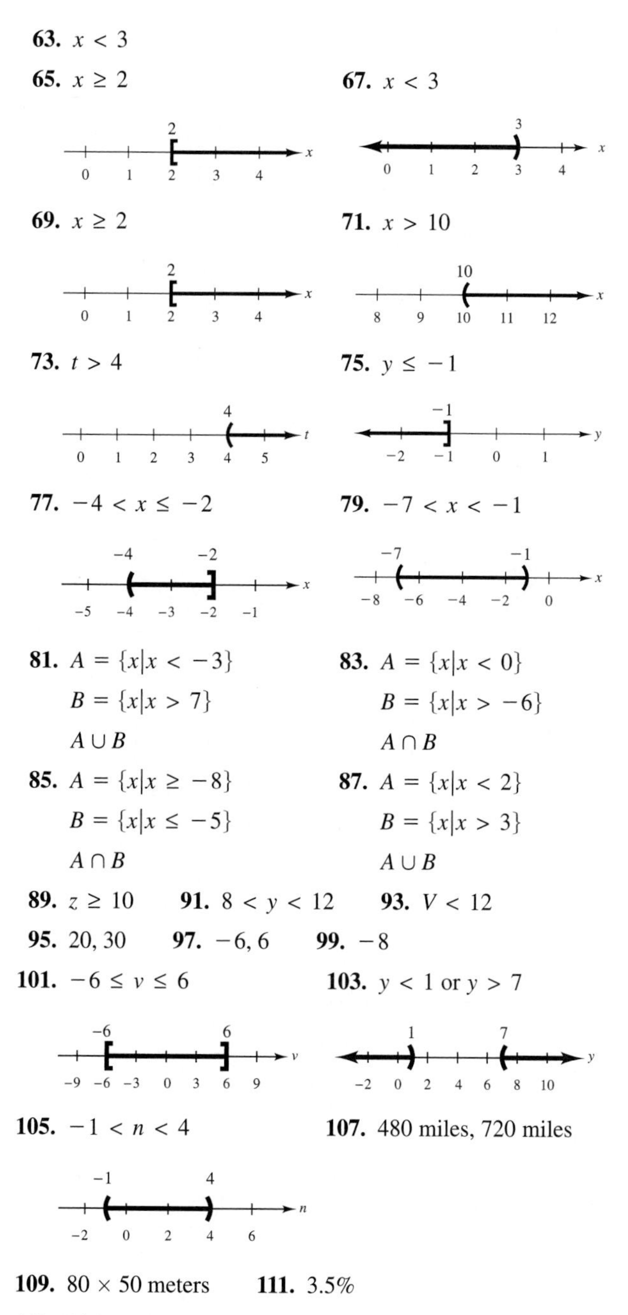

63. $x < 3$

65. $x \geq 2$

67. $x < 3$

69. $x \geq 2$

71. $x > 10$

73. $t > 4$

75. $y \leq -1$

77. $-4 < x \leq -2$

79. $-7 < x < -1$

81. $A = \{x|x < -3\}$
$B = \{x|x > 7\}$
$A \cup B$

83. $A = \{x|x < 0\}$
$B = \{x|x > -6\}$
$A \cap B$

85. $A = \{x|x \geq -8\}$
$B = \{x|x \leq -5\}$
$A \cap B$

87. $A = \{x|x < 2\}$
$B = \{x|x > 3\}$
$A \cup B$

89. $z \geq 10$ **91.** $8 < y < 12$ **93.** $V < 12$

95. 20, 30 **97.** $-6, 6$ **99.** -8

101. $-6 \leq v \leq 6$

103. $y < 1$ or $y > 7$

105. $-1 < n < 4$

107. 480 miles, 720 miles

109. 80×50 meters **111.** 3.5%

113. Living with spouse: 9,525,120
Living with other relatives: 1,587,520
Living alone or with nonrelatives: 8,731,360

115. \$1687 **117.** 214 miles **119.** 9.4 hours

121. 5.3 kilometers per hour **123.** 13 dimes, 17 quarters

125. 30×26 feet **127.** \$475 **129.** \$285,714.29

131. $\frac{30}{11} \approx 2.7$ hours

Chapter Test *(page 198)*

1. $\frac{21}{4}$ **2.** 7 **3.** -10 **4.** 10 **5.** 3, 7

6. $-\frac{1}{3}, 3$ **7.** 11.03 **8.** $2\frac{1}{2}$ hours **9.** $37\frac{1}{2}\%$, 0.375

10. 1200 **11.** 36% **12.** $\frac{5}{9}$ **13.** $\frac{12}{7}$

14. 110 miles **15.** 48 miles per hour

16. $\frac{36}{7} \approx 5.1$ hours **17.** $\frac{S - C}{C}$ **18.** \$6250

19. $x \leq 4$ **20.** $x < -6$

21. $-1 < x \leq 2$ **22.** $-1 \leq x < 5$

23. $-7 \leq x \leq -1$ **24.** $x < -1$ or $x > 2$

Cumulative Test: Chapters 1–3 *(page 199)*

1. $-\frac{3}{4} < \left|-\frac{7}{8}\right|$ **2.** 1200 **3.** $-\frac{11}{24}$ **4.** $-\frac{25}{12}$ **5.** 8

6. 14 **7.** 28 **8.** 5 **9.** 20 **10.** $3^3(x + y)^2$

11. $-2x^2 + 6x$ **12.** Associative Property of Addition

13. $15x^7$ **14.** a^8b^7 **15.** $7x^2 - 6x - 2$ **16.** 6

17. $\frac{52}{3}$ **18.** 5 **19.** $-5 \leq x < 1$

20. $\frac{15{,}000 \text{ miles}}{1 \text{ year}} \cdot \frac{1 \text{ gallon}}{28.3 \text{ miles}} \cdot \frac{\$1.179}{1 \text{ gallon}} \approx \$624.91$ per year

21. $\frac{3}{4}$ **22.** 246, 248

23. \$920 **24.** \$57,000

Chapter 4

Section 4.1 *(page 209)*

Integrated Review *(page 209)*

1. $3x = 7$ is a linear equation since it has the form $ax + b = c$. $x^2 + 3x = 2$ is not of that form, and therefore is not linear.
2. Substitute 3 for x in the equation to verify that it satisfies the equation.

3. -10 **4.** 4 **5.** 14 **6.** 4 **7.** 6

8. $\frac{1}{9}$ **9.** 144 **10.** 200 **11.** \$19,250

12. 8 hours 45 minutes

1. **3.**

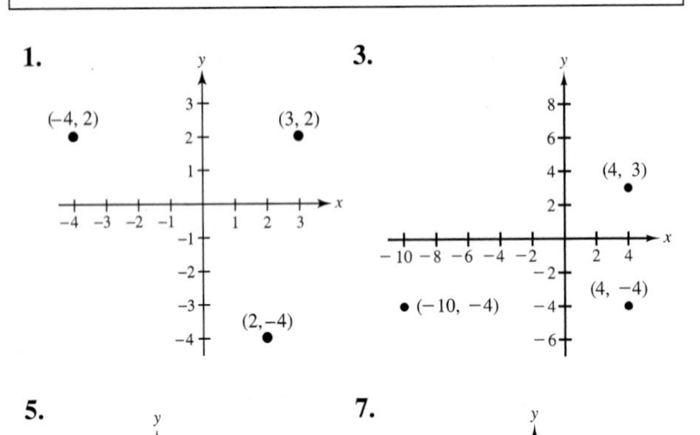

5. **7.**

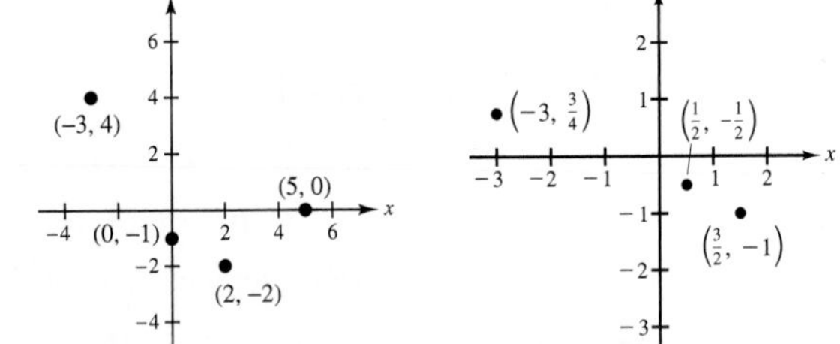

9.

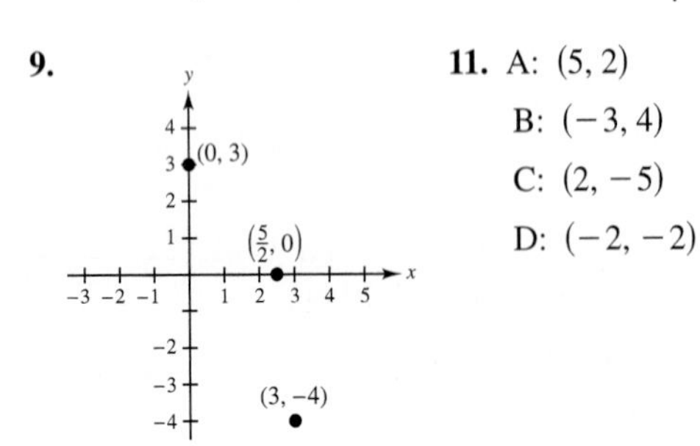

11. A: $(5, 2)$
B: $(-3, 4)$
C: $(2, -5)$
D: $(-2, -2)$

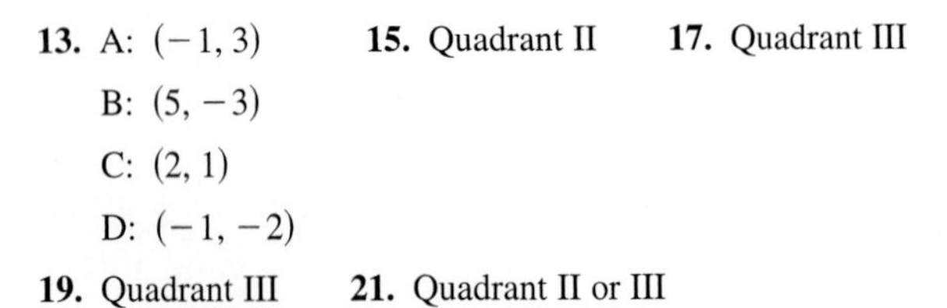

13. A: $(-1, 3)$
B: $(5, -3)$
C: $(2, 1)$
D: $(-1, -2)$

15. Quadrant II **17.** Quadrant III

19. Quadrant III **21.** Quadrant II or III

23. Quadrant III or IV **25.** Quadrant II or IV

27. $(0, 3)$ **29.** $(-5, -10)$

31. **33.** **35.** **37.**

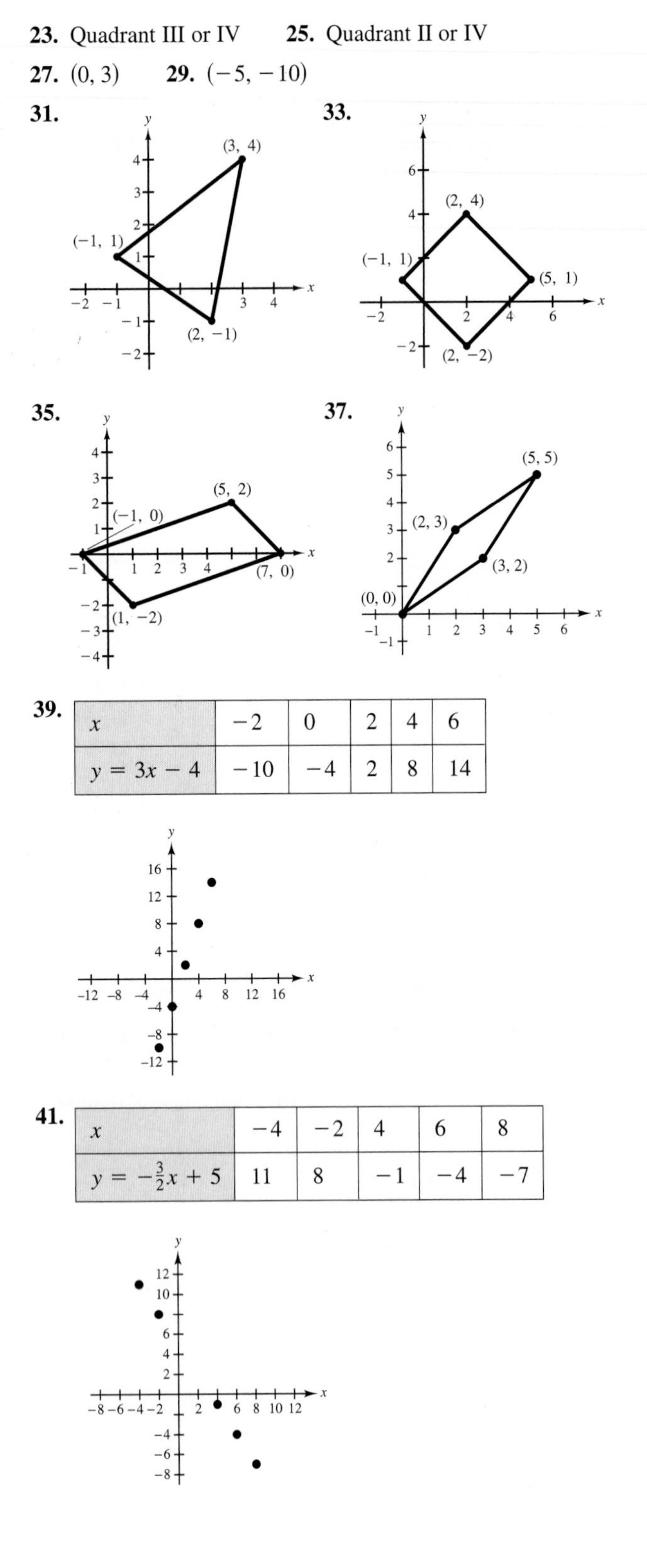

39.

x	-2	0	2	4	6
$y = 3x - 4$	-10	-4	2	8	14

41.

x	-4	-2	4	6	8
$y = -\frac{3}{2}x + 5$	11	8	-1	-4	-7

43.

x	-2	-1	0	1	2
$y = 2x - 1$	-5	-3	-1	1	3

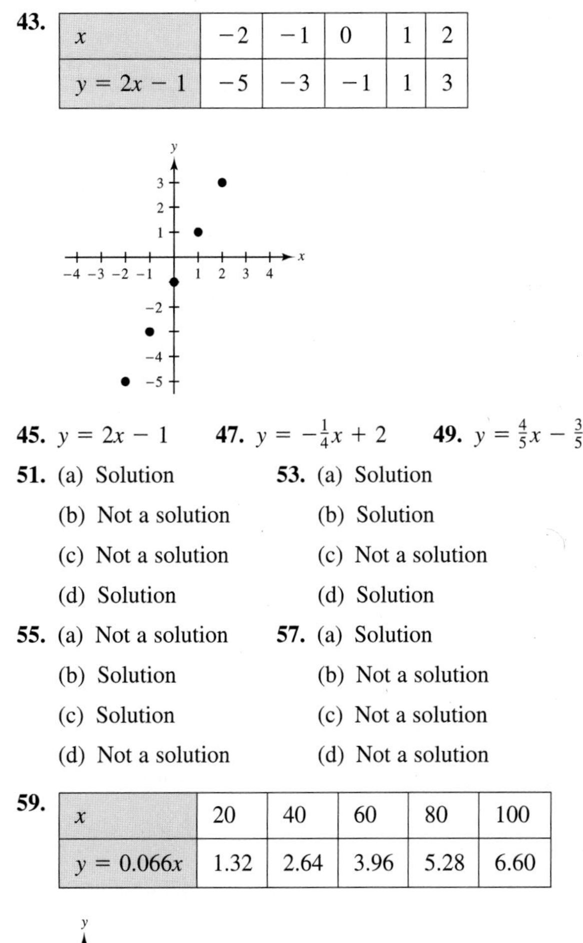

45. $y = 2x - 1$ **47.** $y = -\frac{1}{4}x + 2$ **49.** $y = \frac{4}{5}x - \frac{3}{5}$

51. (a) Solution
(b) Not a solution
(c) Not a solution
(d) Solution

53. (a) Solution
(b) Solution
(c) Not a solution
(d) Solution

55. (a) Not a solution
(b) Solution
(c) Solution
(d) Not a solution

57. (a) Solution
(b) Not a solution
(c) Not a solution
(d) Not a solution

59.

x	20	40	60	80	100
$y = 0.066x$	1.32	2.64	3.96	5.28	6.60

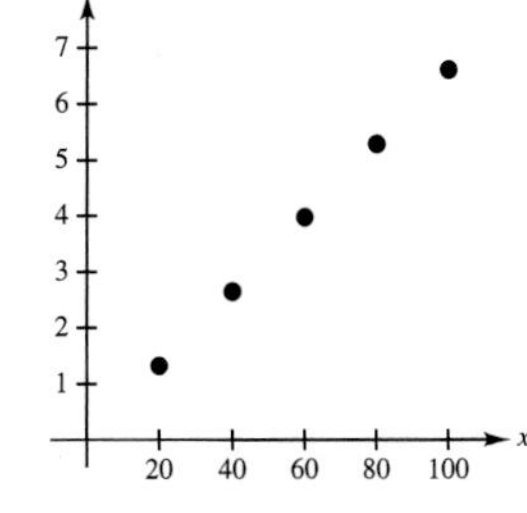

61. $y = 35x + 5000$

x	100	150	200	250	300
$y = 35x + 5000$	8500	10,250	12,000	13,750	15,500

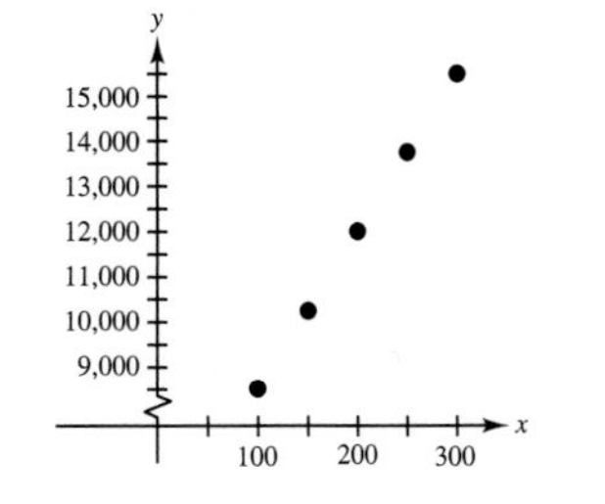

63. (a)

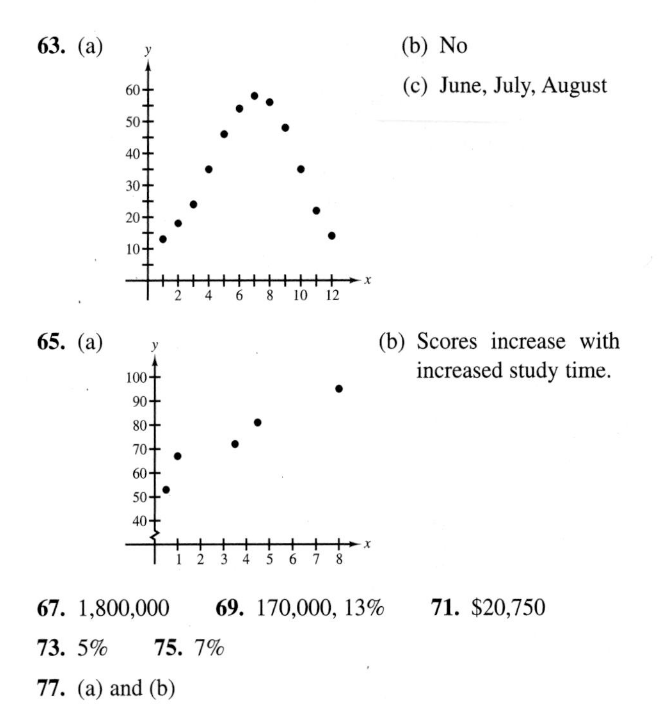

(b) No
(c) June, July, August

65. (a)
(b) Scores increase with increased study time.

67. 1,800,000 **69.** 170,000, 13% **71.** $20,750

73. 5% **75.** 7%

77. (a) and (b)

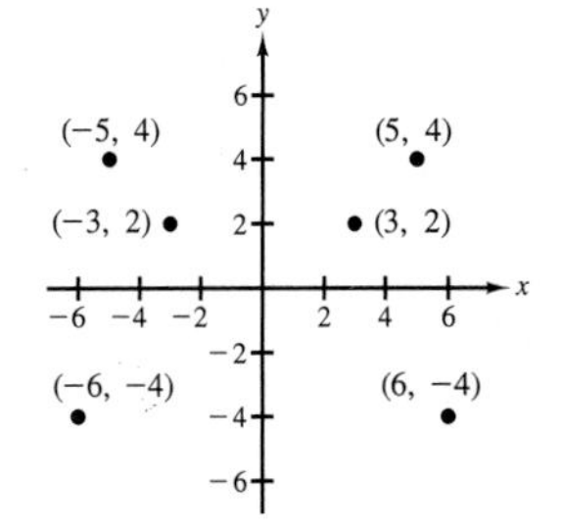

(c) Reflection in the y-axis

79. Order is significant because each number in the pair has a particular interpretation. The first measures horizontal distance and the second measures vertical distance.

81. The x-coordinate of any point on the y-axis is 0. The y-coordinate of any point on the x-axis is 0.

83. Third quadrant: $(-, -)$
Fourth quadrant: $(+, -)$

85. The y-coordinates increase if the coefficient of x is positive and decrease if the coefficient is negative.

Section 4.2 *(page 219)*

Integrated Review *(page 219)*

1. $x - 2 + c > 5 + c$ **2.** $(x - 2)c > 5c$

3. $x\left(\frac{1}{x}\right) = 1$

4. Commutative Property of Addition

5. $-9x + 11y$ **6.** $8z - 4$ **7.** $-y^4 + 2y^2$

8. $10t - 4t^2$ **9.** $3x + 30$ **10.** 0

11. (a) 65 miles

(b) 1 day, since $2(30) > 50.80$

(c) No.
1 day, 208.75 miles
2 days, 115 miles
3 days, 21.25 miles

12. 25×15 inches

1. g **3.** a **5.** h **7.** d

9.

x	-2	-1	0	1	2
y	11	10	9	8	7

11.

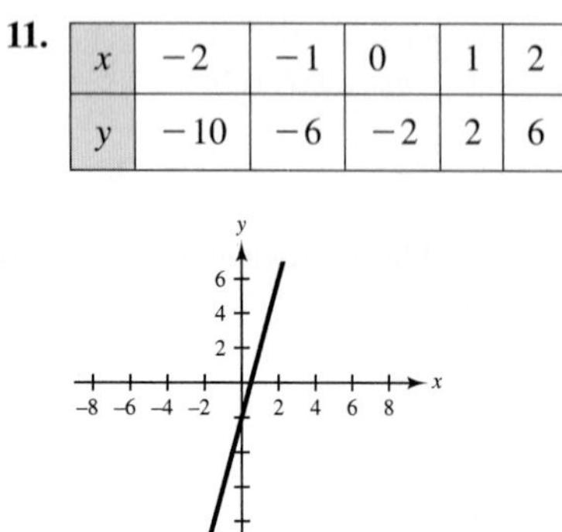

x	-2	-1	0	1	2
y	-10	-6	-2	2	6

13.

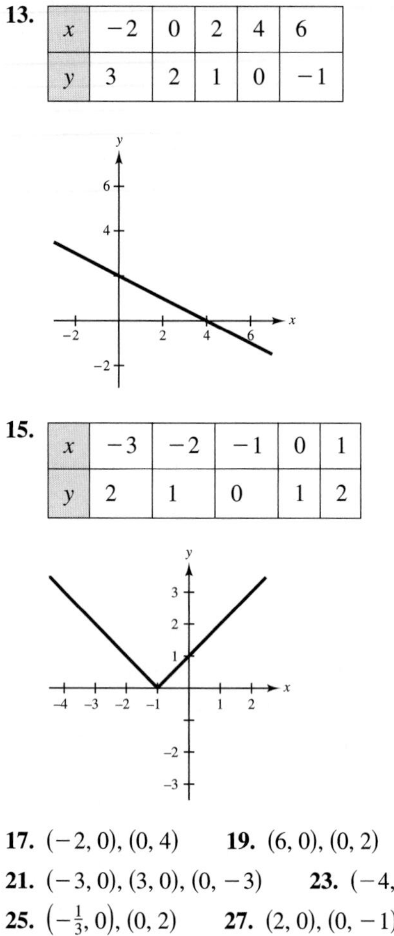

x	-2	0	2	4	6
y	3	2	1	0	-1

15.

x	-3	-2	-1	0	1
y	2	1	0	1	2

17. $(-2, 0), (0, 4)$ **19.** $(6, 0), (0, 2)$

21. $(-3, 0), (3, 0), (0, -3)$ **23.** $(-4, 0), (4, 0), (0, 16)$

25. $\left(-\frac{1}{3}, 0\right), (0, 2)$ **27.** $(2, 0), (0, -1)$

29. $(1, 0), (0, -1)$ **31.** $(2, 0), (0, 4)$

33. $\left(\frac{9}{2}, 0\right), \left(0, \frac{3}{2}\right)$ **35.** $(4, 0), (0, -6)$

37. **39.** **41.** **43.**

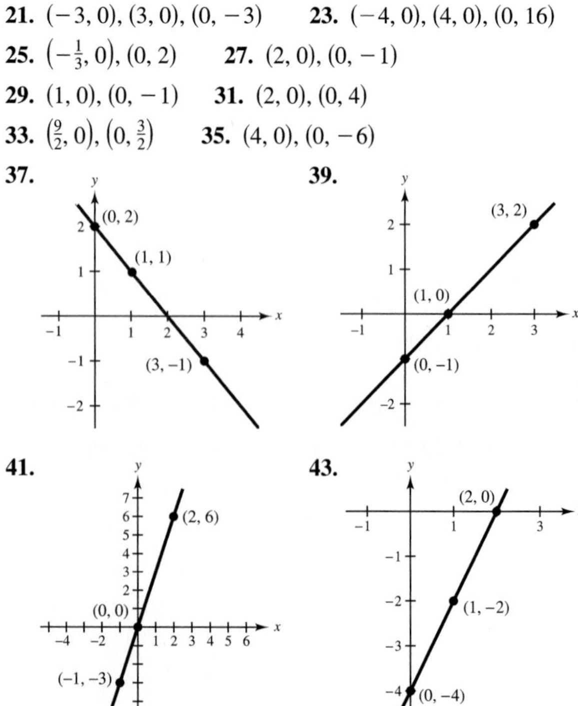

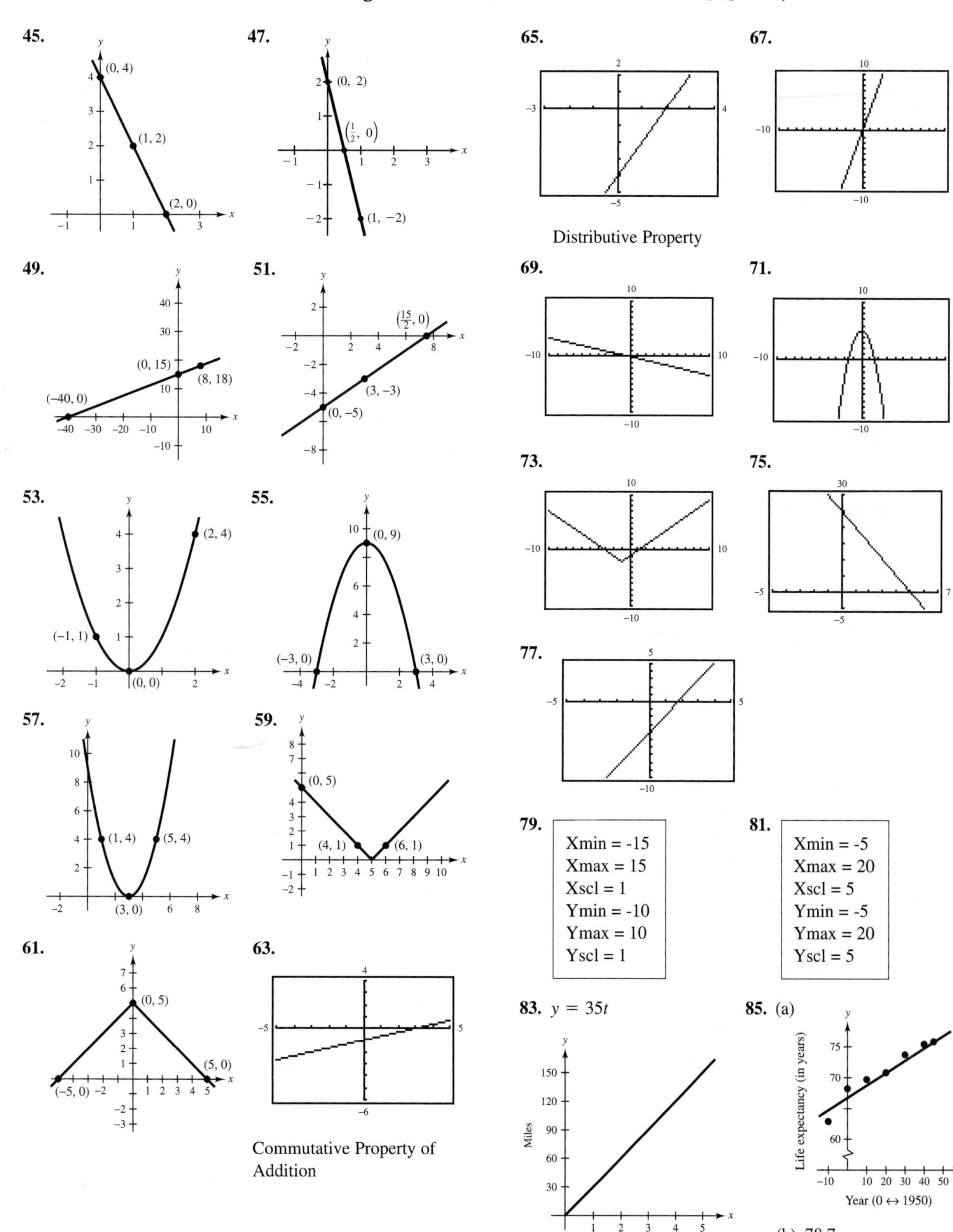
45.
(0, 4)
(1, 2)
(2, 0)
47.
(0, 2)
$\left(\frac{1}{2}, 0\right)$
(1, −2)
49.
(0, 15)
(8, 18)
(−40, 0)
51.
$\left(\frac{15}{2}, 0\right)$
(3, −3)
(0, −5)
53.
(2, 4)
(−1, 1)
(0, 0)
55.
(0, 9)
(−3, 0)
(3, 0)
57.
(1, 4)
(5, 4)
(3, 0)
59.
(0, 5)
(4, 1)
(6, 1)
61.
(0, 5)
(−5, 0)
(5, 0)
63.
Commutative Property of Addition
65.
Distributive Property
67.
69.
71.
73.
75.
77.
79.
Xmin = -15
Xmax = 15
Xscl = 1
Ymin = -10
Ymax = 10
Yscl = 1
81.
Xmin = -5
Xmax = 20
Xscl = 5
Ymin = -5
Ymax = 20
Yscl = 5
83. y = 35t
Miles
Hours
85. (a)
Life expectancy (in years)
Year (0 ↔ 1950)
(b) 78.7

87. Left: Declining balances
Right: Straight-line depreciation

89. Straight-line depreciation is easier to compute. The declining balances method yields a more realistic approximation of the faster rate of depreciation early in the useful lifetime of the equipment.

91. The set of all solutions of an equation plotted on a rectangular coordinate system is called its graph.

93. Make up a table of values showing several solution points. Plot these points on a rectangular coordinate system and connect them with a smooth curve or line.

95. To find the x-intercept(s), let $y = 0$ and solve the equation for x. To find the y-intercept(s), let $x = 0$ and solve the equation for y.

97. Answers will vary.

Section 4.3 *(page 229)*

Integrated Review *(page 229)*

1. $a < c$
Transitive Property

2. $\frac{7x}{7} = \frac{21}{7}$
$x = 3$

3. $11s - 5t$ **4.** $-x^2 + 1$ **5.** $x - 4$

6. $3x^2y - xy^2 - 5xy$ **7.** -3 **8.** $\frac{4}{7}$ **9.** $\frac{9}{2}$

10. 2 **11.** 9.2% **12.** $833\frac{1}{3}$ miles per hour

1. Domain: $\{-4, 1, 2, 4\}$
Range: $\{-3, 2, 3, 5\}$

3. Domain: $\{-9, \frac{1}{2}, 2\}$
Range: $\{-10, 0, 16\}$

5. Domain: $\{-1, 1, 5, 8\}$
Range: $\{-7, -2, 3, 4\}$

7. Function

9. Not a function **11.** Function **13.** Not a function

15. Not a function **17.** Function **19.** Function

21. Not a function **23.** Function **25.** Function

27. Function **29.** Not a function **31.** Function

33. Not a function **35.** Function

37. (a) 1 (b) $\frac{5}{2}$ (c) -2 (d) $-\frac{1}{3}$

39. (a) -1 (b) 5 (c) -7 (d) -2

41. (a) 5 (b) -3 (c) -15 (d) $-\frac{13}{3}$

43. (a) 49 (b) -4 (c) 1 (d) $-\frac{13}{8}$

45. (a) 8 (b) 8 (c) 0 (d) 2

47. (a) 1 (b) 15 (c) 0 (d) 0

49. (a) 4 (b) 0 (c) 12 (d) $\frac{1}{2}$

51. (a) -1 (b) 0 (c) 26 (d) $-\frac{7}{8}$

53. $R = \{4, 3, 2, 1, 0\}$ **55.** $R = \{100\}$

57. $R = \{-8, -1, 0, 1, 8\}$ **59.** $R = \{0, 1, 2\}$

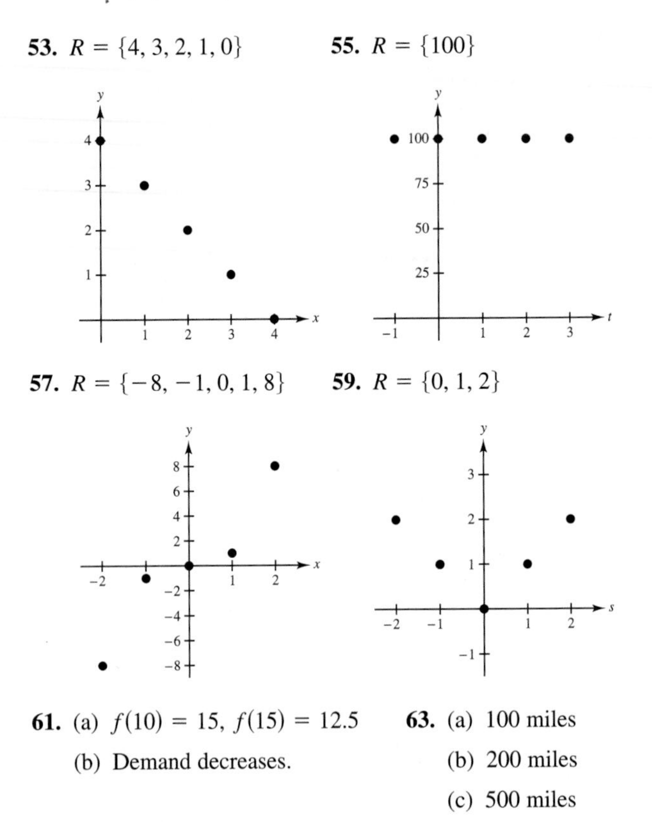

61. (a) $f(10) = 15$, $f(15) = 12.5$
(b) Demand decreases.

63. (a) 100 miles
(b) 200 miles
(c) 500 miles

65. High school enrollment is a function of the year.

67. 12,900,000 **69.** $P = 4s$
P is a function of s.

71. (a) L is a function of t.
(b) $9.5 \le L \le 15$

73. A relation is any set of ordered pairs. A function is a relation in which no two ordered pairs have the same first component and different second components.

Example:

Domain		Range
1	→	4
2	→	5
3	→	6

75. The domain is the set of inputs of a function, and the range is the set of outputs of the function.

77. If the graph of an equation has the property that no vertical line intersects the graph at two (or more) points, then the equation represents y as a function of x.

79. Yes. Example: $f(x) = 10$

Mid-Chapter Quiz *(page 233)*

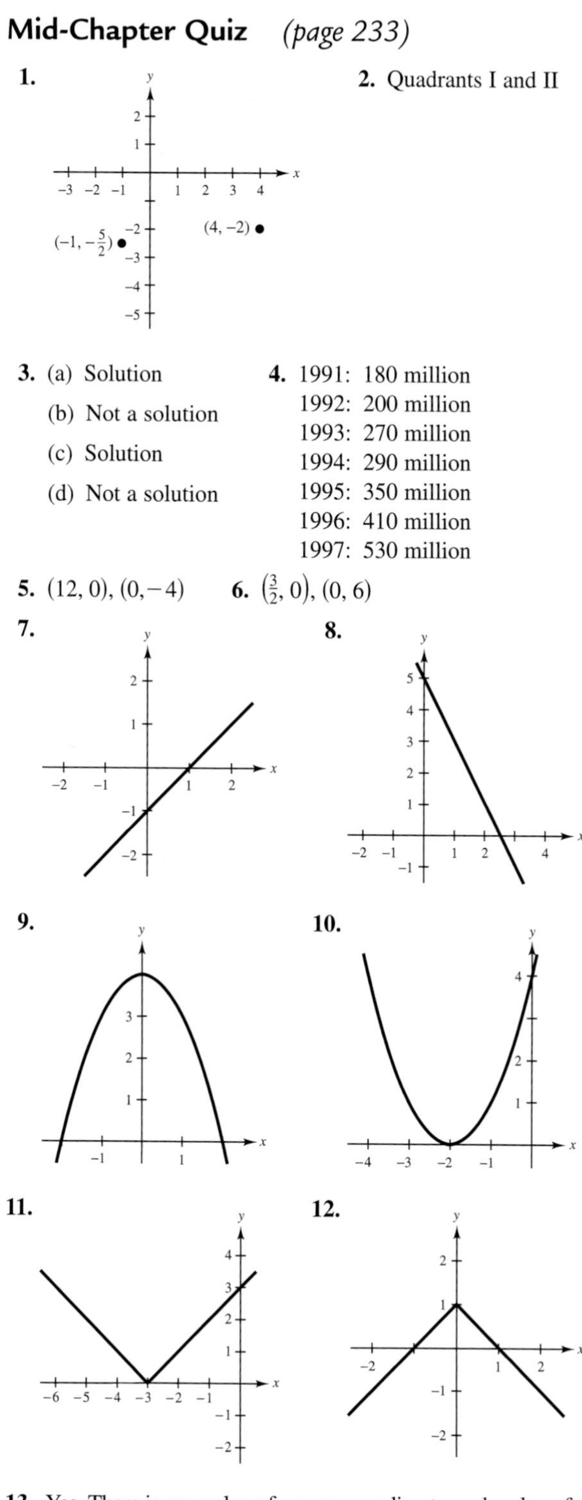

1. (graph)

2. Quadrants I and II

3. (a) Solution (b) Not a solution (c) Solution (d) Not a solution

4. 1991: 180 million
1992: 200 million
1993: 270 million
1994: 290 million
1995: 350 million
1996: 410 million
1997: 530 million

5. $(12, 0), (0, -4)$ 6. $\left(\frac{3}{2}, 0\right), (0, 6)$

7. (graph) 8. (graph)

9. (graph) 10. (graph)

11. (graph) 12. (graph)

13. Yes. There is one value of y corresponding to each value of x.

14. No. There are two values of y corresponding to $x > 0$.

15. (a) -8 (b) -2 (c) 13 (d) -3

16. (a) 6 (b) 6 (c) 0 (d) 0

17. $R = \{0, 2, 6\}$ 18. $s > 0$

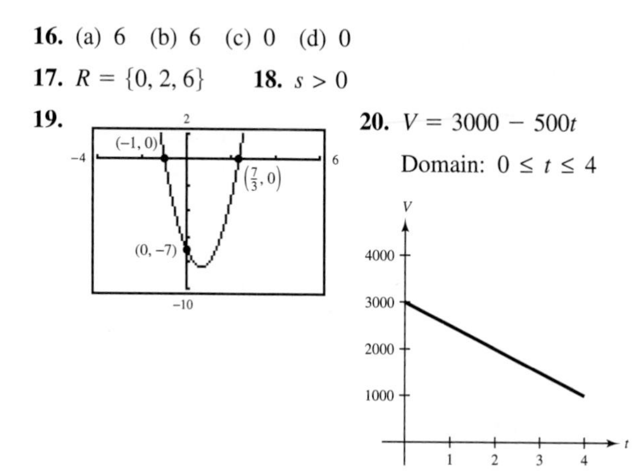

19. (graph)

20. $V = 3000 - 500t$
Domain: $0 \le t \le 4$

Section 4.4 *(page 242)*

Integrated Review *(page 242)*

1. Equivalent 2. $5x = 6 + 2$ 3. x^9
4. y^2z^5 5. u^8v^4 6. a^4b^4 7. $50x^5$
8. $54y^3z^5$ 9. $x + 2$ 10. $x^2 - 4x - 2$
11. 2 feet, 2 feet, 6 feet 12. 1.5 hours

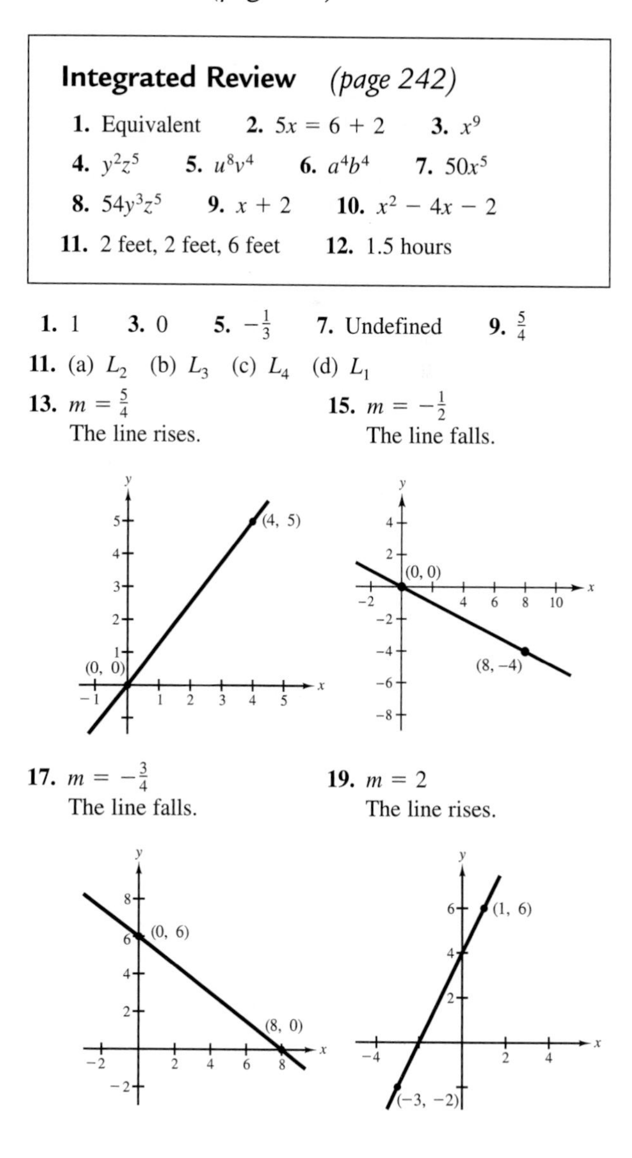

1. 1 3. 0 5. $-\frac{1}{3}$ 7. Undefined 9. $\frac{5}{4}$

11. (a) L_2 (b) L_3 (c) L_4 (d) L_1

13. $m = \frac{5}{4}$
The line rises.

15. $m = -\frac{1}{2}$
The line falls.

17. $m = -\frac{3}{4}$
The line falls.

19. $m = 2$
The line rises.

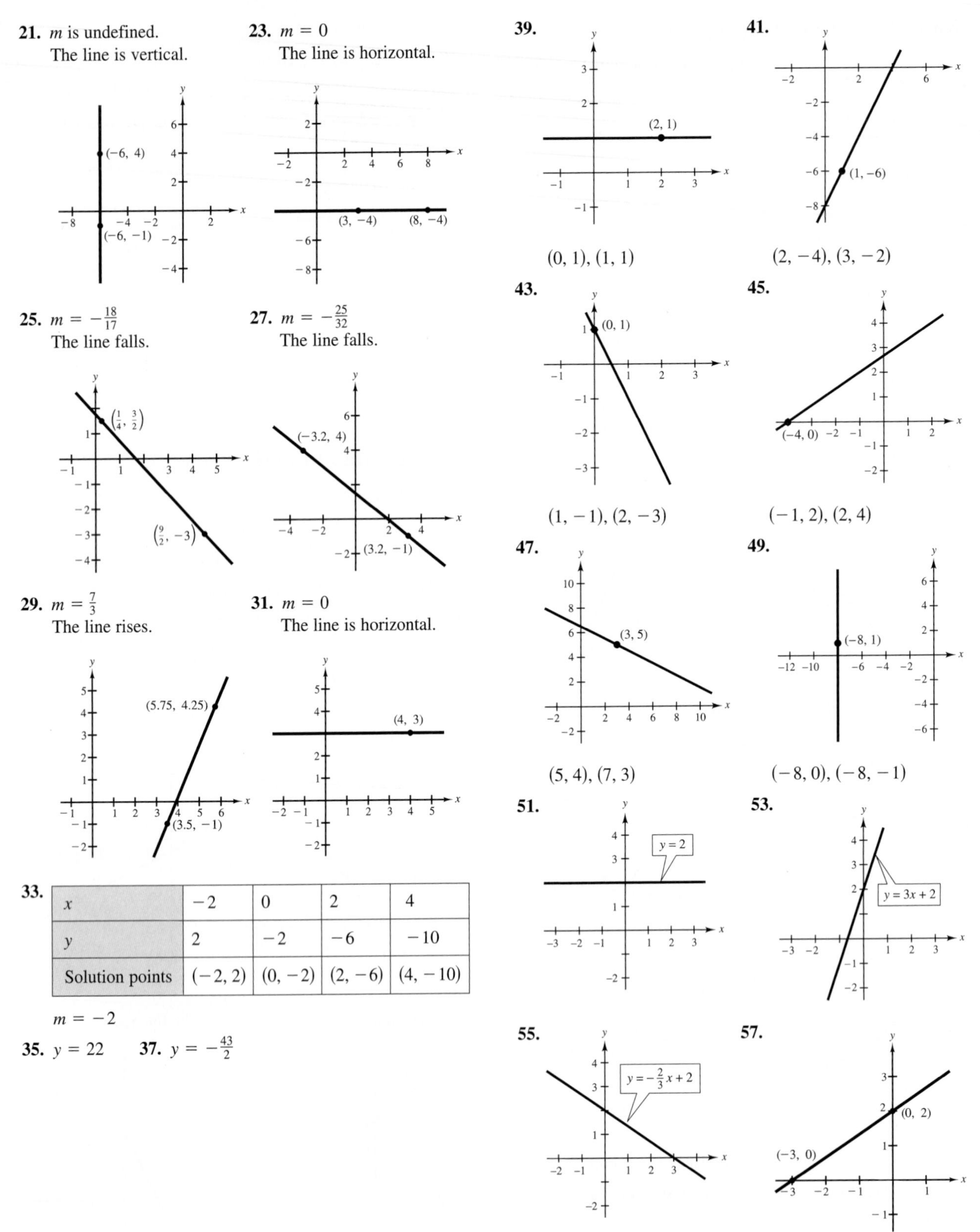

21. m is undefined.
The line is vertical.

23. $m = 0$
The line is horizontal.

25. $m = -\frac{18}{17}$
The line falls.

27. $m = -\frac{25}{32}$
The line falls.

29. $m = \frac{7}{3}$
The line rises.

31. $m = 0$
The line is horizontal.

33.

x	-2	0	2	4
y	2	-2	-6	-10
Solution points	$(-2, 2)$	$(0, -2)$	$(2, -6)$	$(4, -10)$

$m = -2$

35. $y = 22$ **37.** $y = -\frac{43}{2}$

39. $(0, 1), (1, 1)$

41. $(2, -4), (3, -2)$

43. $(1, -1), (2, -3)$

45. $(-1, 2), (2, 4)$

47. $(5, 4), (7, 3)$

49. $(-8, 0), (-8, -1)$

51.

53.

55.

57.

59. **61.**

63. $y = -x$ **65.** $y = -\frac{1}{2}x$

67. $y = 2x - 3$ **69.** $y = \frac{1}{3}x + 2$

71. $y = -\frac{1}{2}x + 1$ **73.** $y = \frac{3}{4}x + \frac{1}{2}$

75. $y = -5$

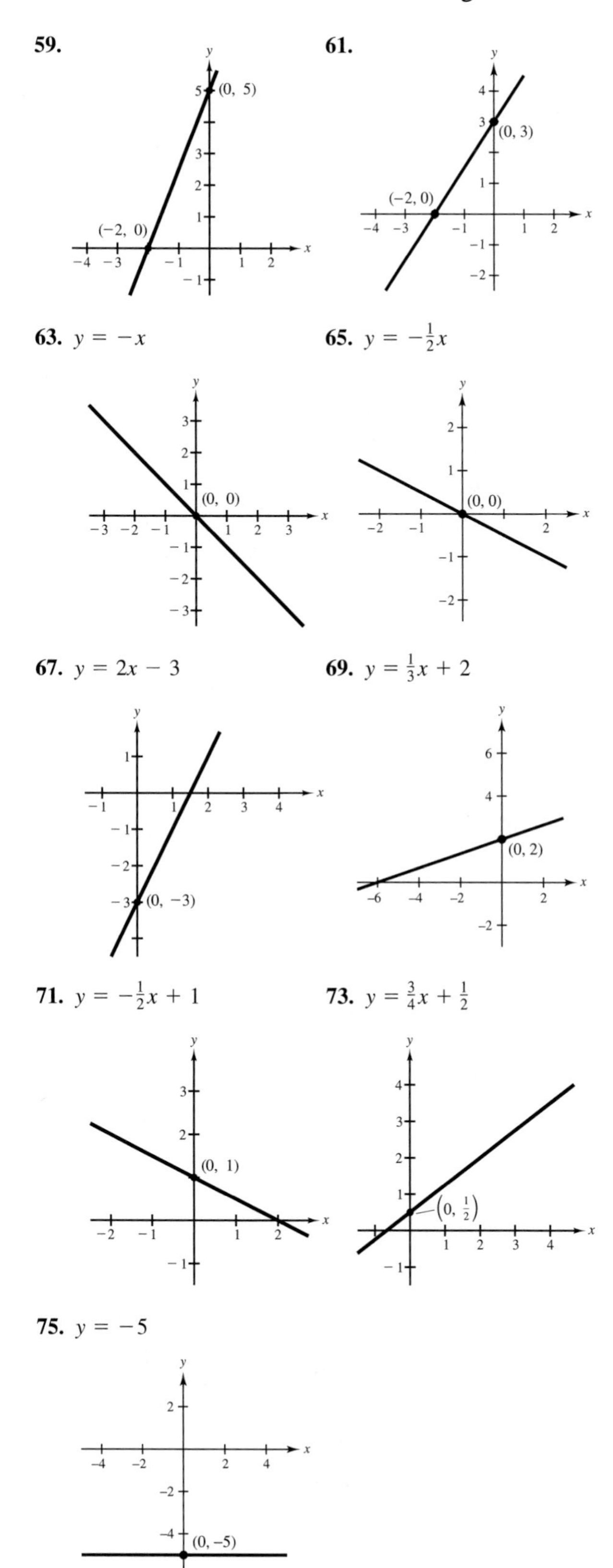

77. Perpendicular **79.** Parallel

81. Parallel **83.** Perpendicular

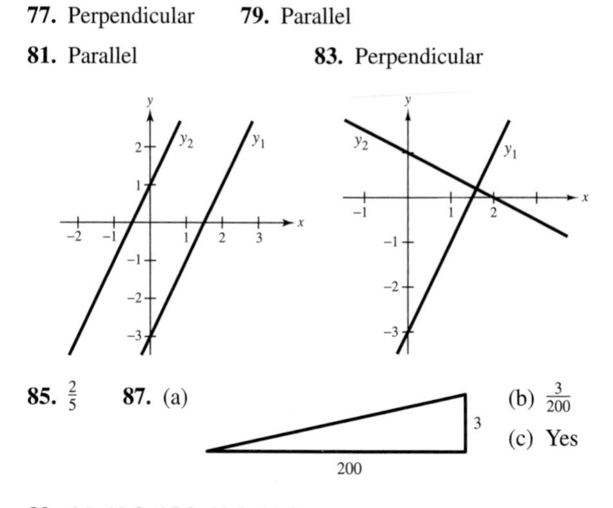

85. $\frac{2}{5}$ **87.** (a)

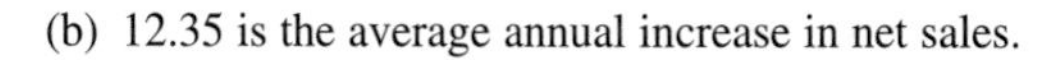

(b) $\frac{3}{200}$

(c) Yes

89. (a) 11.8, 15.2, 11.1, 11.3

(b) 12.35 is the average annual increase in net sales.

91.

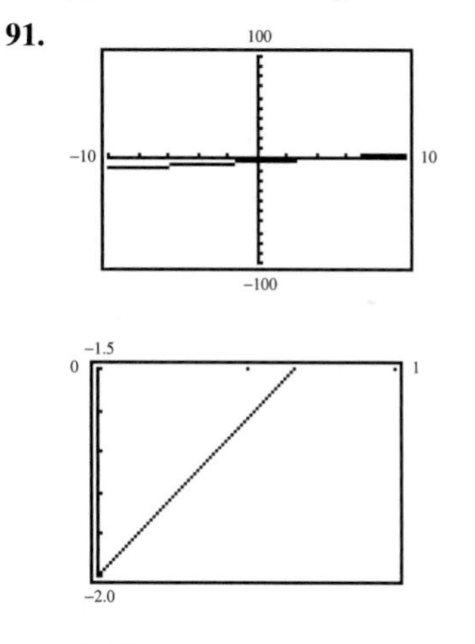

(a) No

(b) No. Use the square feature.

(c) Answers will vary.

93. Yes. The slope is the ratio of the change in y to the change in x.

95. False

97. No. The slopes of nonvertical perpendicular lines have opposite signs. The slopes are the negative reciprocals of each other.

99. The slope

101. If the points lie on the same line, the slopes of the lines between any two pairs of points will be the same.

Section 4.5 *(page 253)*

Integrated Review *(page 253)*

1. 60. The greatest common factor is the product of the common prime factors.

2. 900. The least common multiple is the product of the highest powers of the prime factors of the numbers.

3. $12 - 8x$ **4.** x^3y^3 **5.** $x + 10$ **6.** 1

7. $y = -3x + 4$ **8.** $y = x + 4$

9. $y = \frac{4}{5}x + \frac{2}{5}$ **10.** $y = -\frac{3}{4}x + \frac{5}{4}$

1. $2x + y = 0$

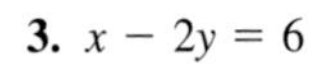

3. $x - 2y = 6$

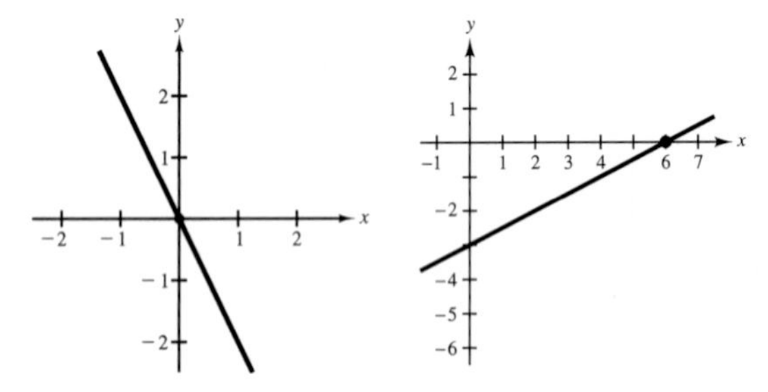

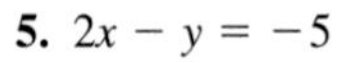

5. $2x - y = -5$

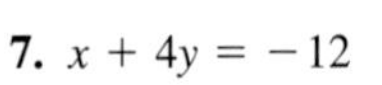

7. $x + 4y = -12$

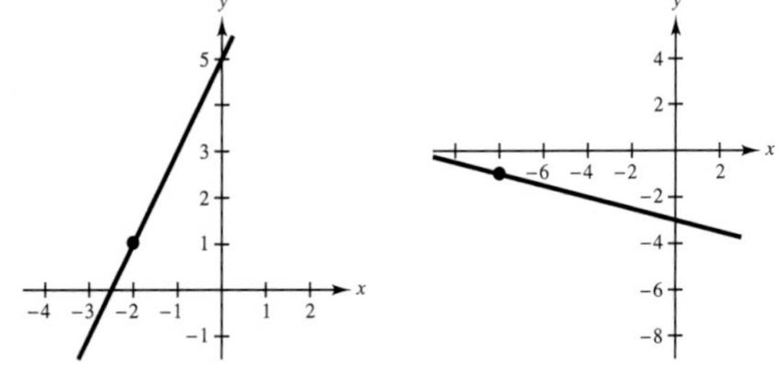

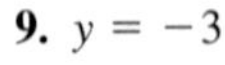

9. $y = -3$

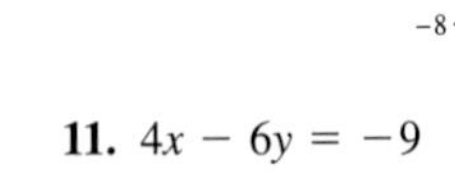

11. $4x - 6y = -9$

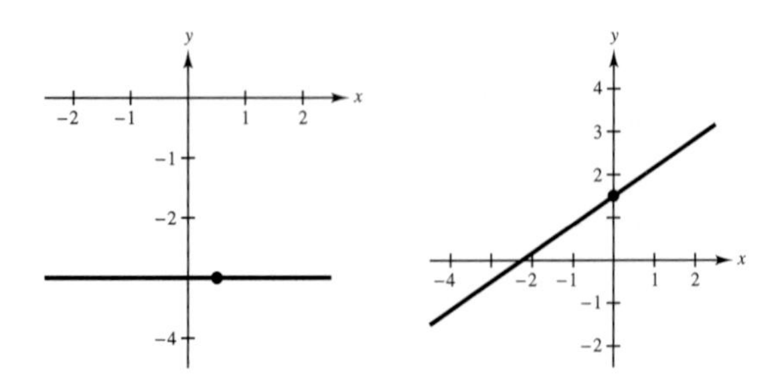

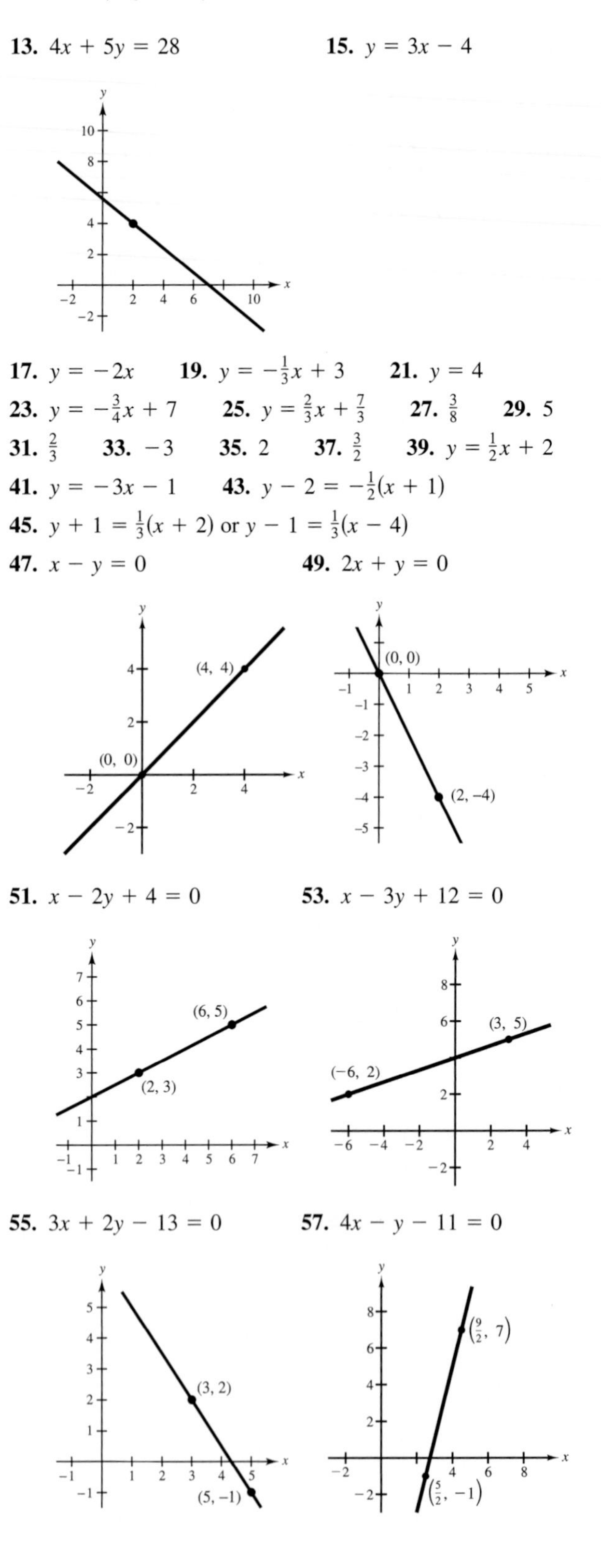

13. $4x + 5y = 28$ **15.** $y = 3x - 4$

17. $y = -2x$ **19.** $y = -\frac{1}{3}x + 3$ **21.** $y = 4$

23. $y = -\frac{3}{4}x + 7$ **25.** $y = \frac{2}{3}x + \frac{7}{3}$ **27.** $\frac{3}{8}$ **29.** 5

31. $\frac{2}{3}$ **33.** -3 **35.** 2 **37.** $\frac{3}{2}$ **39.** $y = \frac{1}{2}x + 2$

41. $y = -3x - 1$ **43.** $y - 2 = -\frac{1}{2}(x + 1)$

45. $y + 1 = \frac{1}{3}(x + 2)$ or $y - 1 = \frac{1}{3}(x - 4)$

47. $x - y = 0$ **49.** $2x + y = 0$

51. $x - 2y + 4 = 0$ **53.** $x - 3y + 12 = 0$

55. $3x + 2y - 13 = 0$ **57.** $4x - y - 11 = 0$

59. $x + y - 3 = 0$ **61.** $3x + 5y - 10 = 0$

63. $2x - y - 6 = 0$ **65.** $3x + 2y - 13 = 0$

67. $3x + 5y - 31 = 0$ **69.** $8x + 6y - 19 = 0$

71. $6x + 5y - 9 = 0$ **73.** (a) $x - y - 1 = 0$
(b) $x + y - 3 = 0$

75. (a) $3x + 4y + 20 = 0$ **77.** (a) $2x + y - 5 = 0$
(b) $4x - 3y + 60 = 0$ (b) $x - 2y + 5 = 0$

79. (a) $y = 0$ **81.** (a) $2x - 3y - 11 = 0$
(b) $x + 1 = 0$ (b) $3x + 2y - 10 = 0$

83. $x = -2$ **85.** $y = \frac{2}{3}$ **87.** $x = 4$

89. $y = -8$

91. **93.**

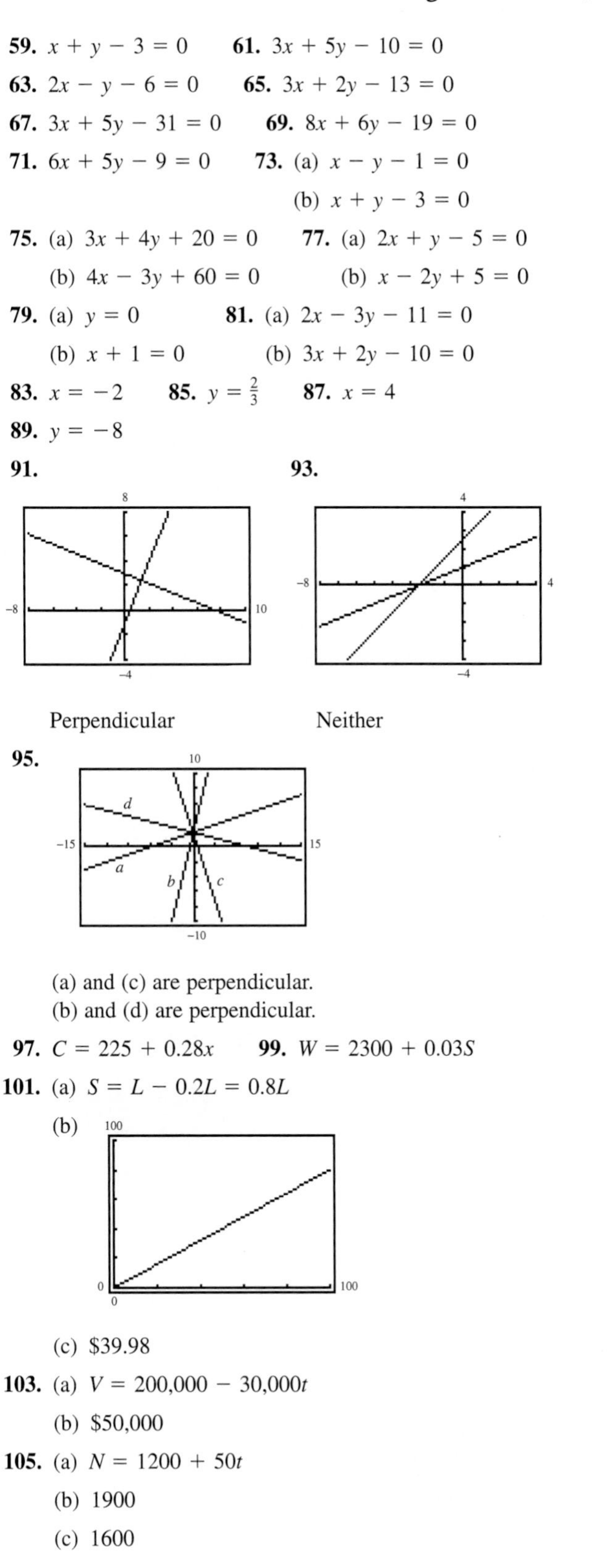

Perpendicular Neither

95.

(a) and (c) are perpendicular.
(b) and (d) are perpendicular.

97. $C = 225 + 0.28x$ **99.** $W = 2300 + 0.03S$

101. (a) $S = L - 0.2L = 0.8L$
(b)
(c) \$39.98

103. (a) $V = 200{,}000 - 30{,}000t$
(b) \$50,000

105. (a) $N = 1200 + 50t$
(b) 1900
(c) 1600

107. (a) (f): $m = -10$; Loan decreases by \$10 per week.
(b) (e): $m = 1.50$; Pay increases \$1.50 per unit.
(c) (g): $m = 0.32$; Amount increases \$0.32 per mile.
(d) (h): $m = -100$; Annual depreciation is \$100.

109. Yes. When different pairs of points are selected, the change in y and the change in x are the lengths of the sides of similar triangles. Corresponding sides of similar triangles are proportional.

111. m is the slope of the line and b is its y-intercept.

113. Set $y = 0$ and solve the resulting equation.

115. Answers will vary.

Section 4.6 *(page 262)*

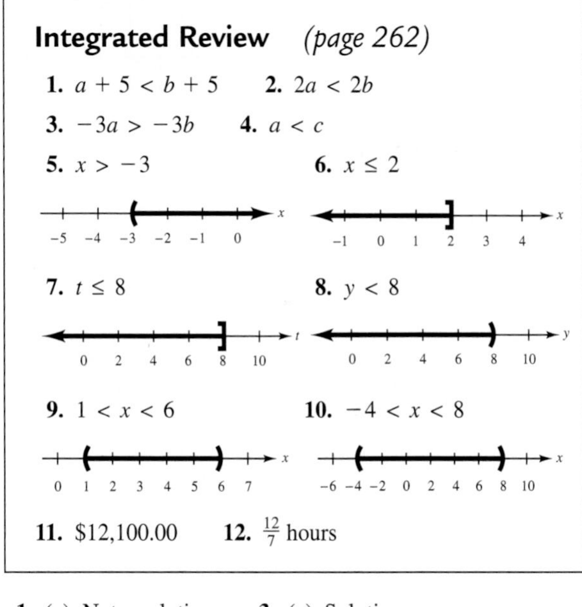

Integrated Review *(page 262)*

1. $a + 5 < b + 5$ **2.** $2a < 2b$

3. $-3a > -3b$ **4.** $a < c$

5. $x > -3$ **6.** $x \leq 2$

7. $t \leq 8$ **8.** $y < 8$

9. $1 < x < 6$ **10.** $-4 < x < 8$

11. \$12,100.00 **12.** $\frac{12}{7}$ hours

1. (a) Not a solution
(b) Solution
(c) Solution
(d) Not a solution

3. (a) Solution
(b) Solution
(c) Solution
(d) Not a solution

5. (a) Solution
(b) Not a solution
(c) Solution
(d) Not a solution

7. (a) Solution
(b) Solution
(c) Solution
(d) Solution

9. Dashed **11.** Solid **13.** b **15.** d

17. c **19.** b

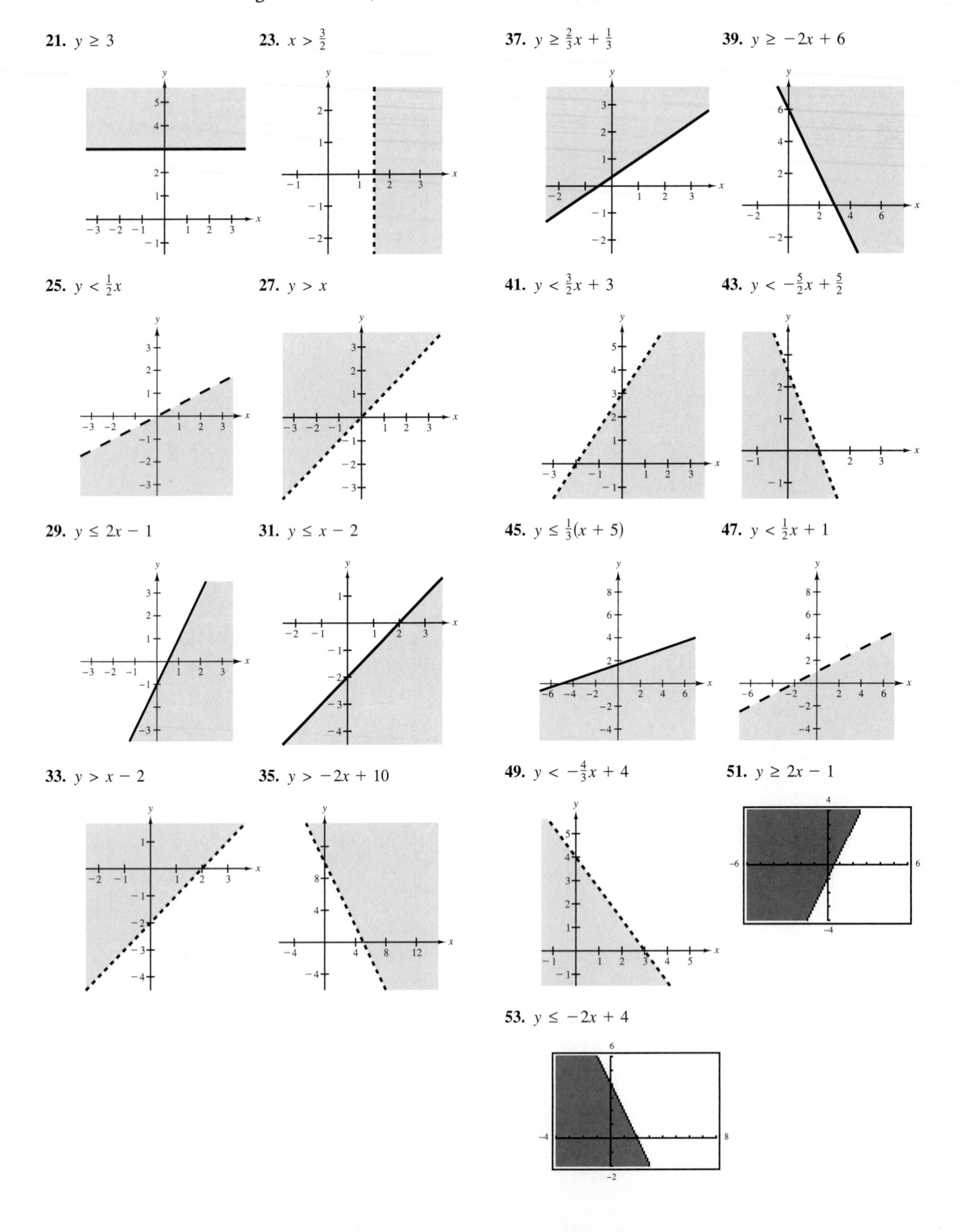

21. $y \geq 3$
23. $x > \frac{3}{2}$
37. $y \geq \frac{2}{3}x + \frac{1}{3}$
39. $y \geq -2x + 6$
25. $y < \frac{1}{2}x$
27. $y > x$
41. $y < \frac{3}{2}x + 3$
43. $y < -\frac{5}{2}x + \frac{5}{2}$
29. $y \leq 2x - 1$
31. $y \leq x - 2$
45. $y \leq \frac{1}{3}(x + 5)$
47. $y < \frac{1}{2}x + 1$
33. $y > x - 2$
35. $y > -2x + 10$
49. $y < -\frac{4}{3}x + 4$
51. $y \geq 2x - 1$
53. $y \leq -2x + 4$

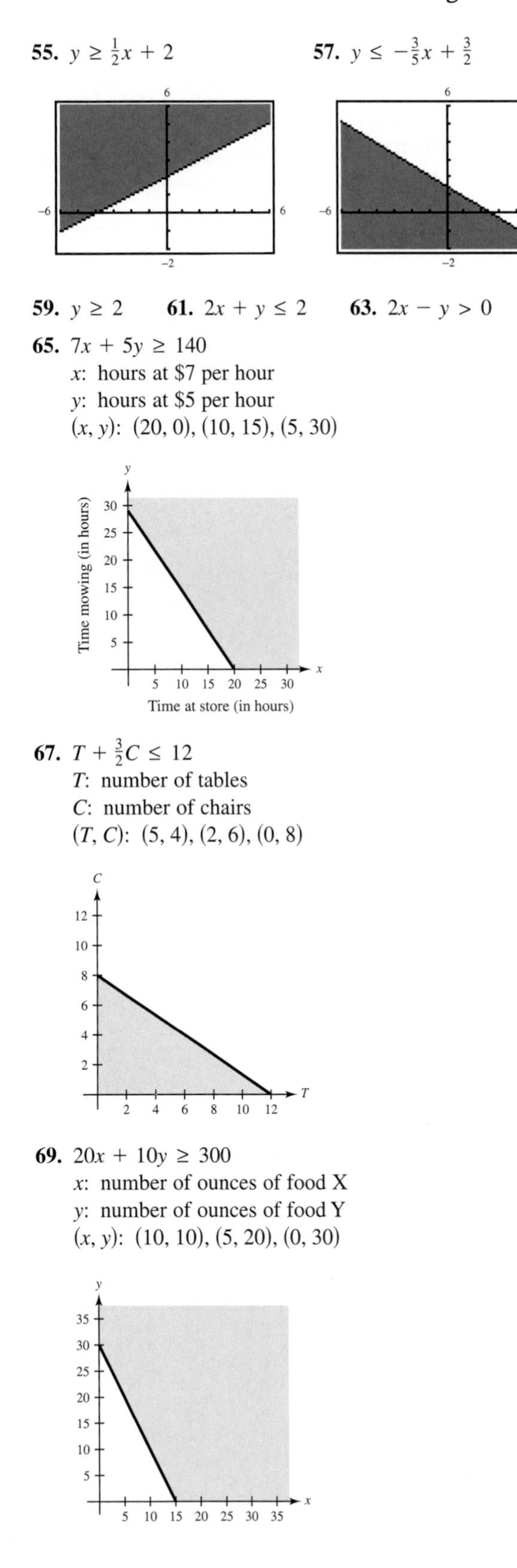

55. $y \geq \frac{1}{2}x + 2$

57. $y \leq -\frac{3}{5}x + \frac{3}{2}$

59. $y \geq 2$ **61.** $2x + y \leq 2$ **63.** $2x - y > 0$

65. $7x + 5y \geq 140$
x: hours at \$7 per hour
y: hours at \$5 per hour
(x, y): $(20, 0), (10, 15), (5, 30)$

67. $T + \frac{3}{2}C \leq 12$
T: number of tables
C: number of chairs
(T, C): $(5, 4), (2, 6), (0, 8)$

69. $20x + 10y \geq 300$
x: number of ounces of food X
y: number of ounces of food Y
(x, y): $(10, 10), (5, 20), (0, 30)$

71. $ax + by < c,\ ax + by \leq c$
$ax + by > c,\ ax + by \geq c$

73. Use dashed lines for the inequalities < and > and solid lines for the inequalities ≤ and ≥.

75. (a) The solution is an unbounded interval on the x-axis.
(b) The solution is a half-plane.

77. Yes

Review Exercises *(page 266)*

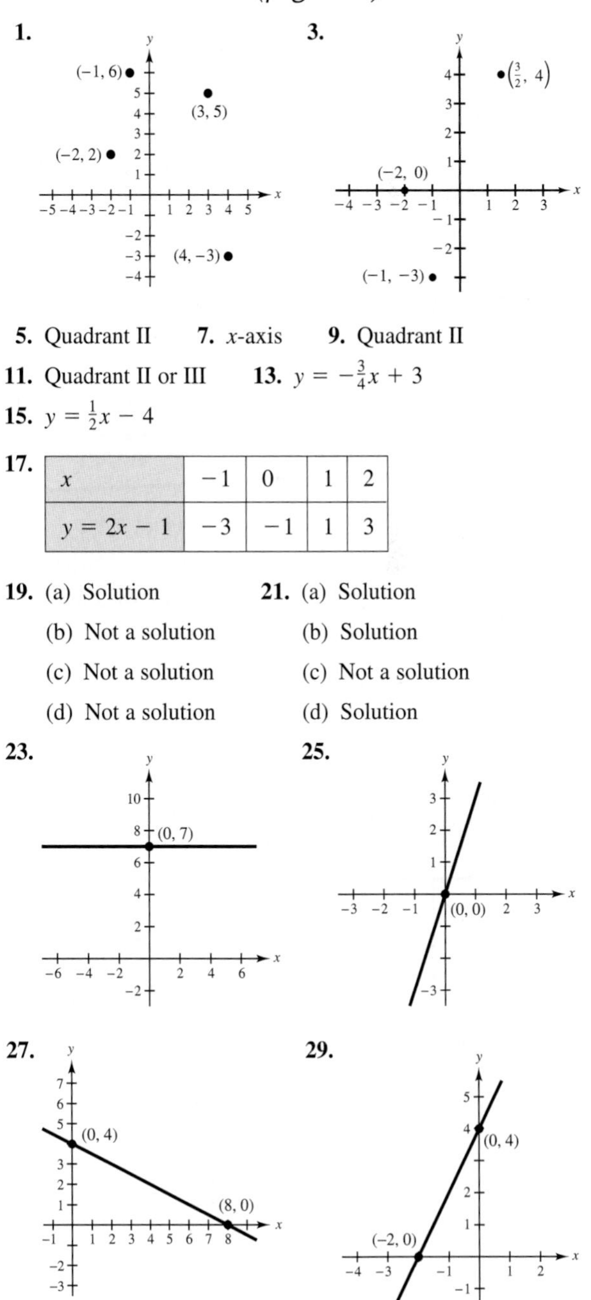

1.

3.

5. Quadrant II **7.** x-axis **9.** Quadrant II

11. Quadrant II or III **13.** $y = -\frac{3}{4}x + 3$

15. $y = \frac{1}{2}x - 4$

17.

x	-1	0	1	2
$y = 2x - 1$	-3	-1	1	3

19. (a) Solution
(b) Not a solution
(c) Not a solution
(d) Not a solution

21. (a) Solution
(b) Solution
(c) Not a solution
(d) Solution

23.

25.

27.

29.

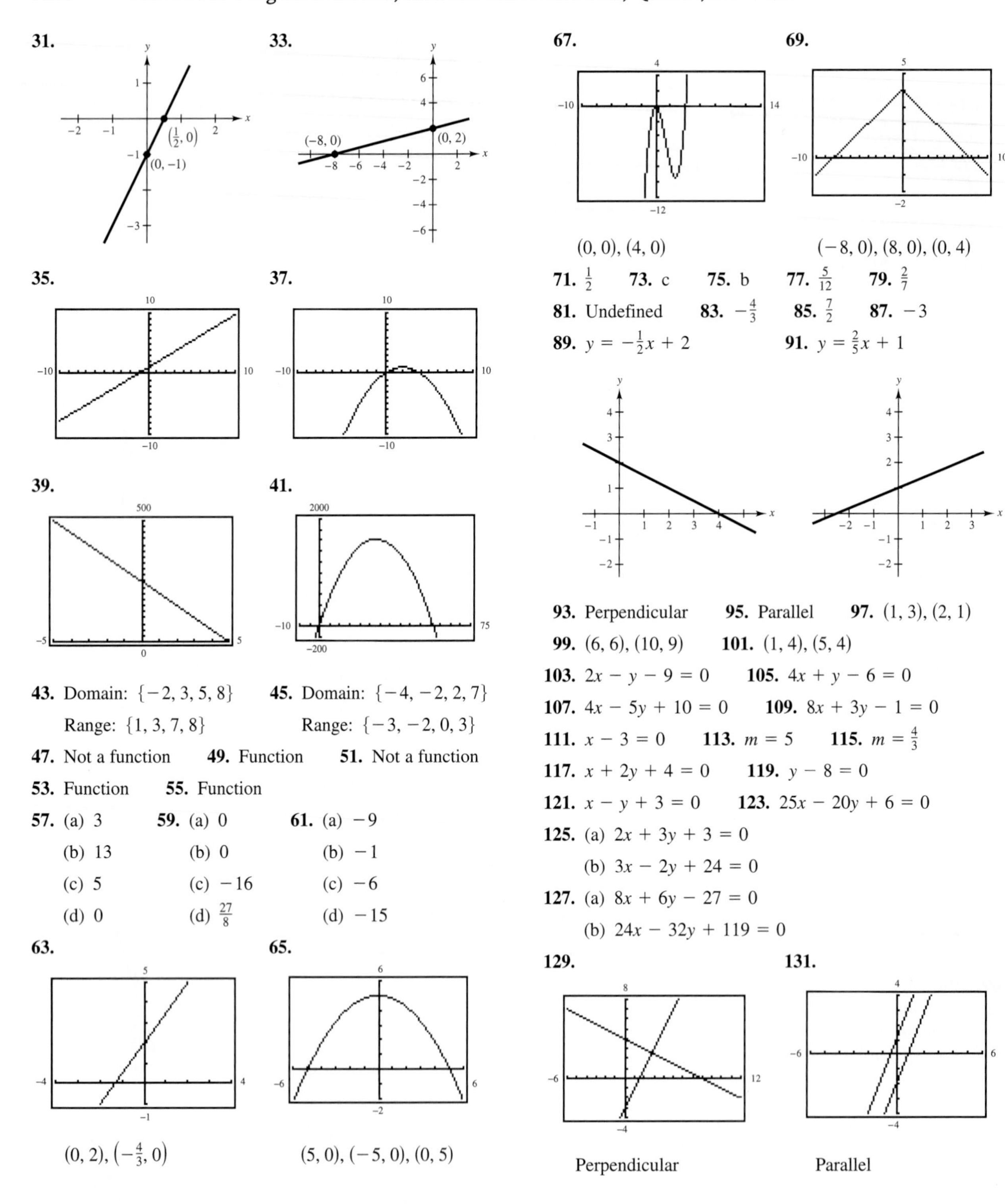

31. $\left(\frac{1}{2}, 0\right)$, $(0, -1)$

33. $(-8, 0)$, $(0, 2)$

35.

37.

39.

41.

43. Domain: $\{-2, 3, 5, 8\}$
Range: $\{1, 3, 7, 8\}$

45. Domain: $\{-4, -2, 2, 7\}$
Range: $\{-3, -2, 0, 3\}$

47. Not a function **49.** Function **51.** Not a function

53. Function **55.** Function

57. (a) 3 (b) 13 (c) 5 (d) 0

59. (a) 0 (b) 0 (c) -16 (d) $\frac{27}{8}$

61. (a) -9 (b) -1 (c) -6 (d) -15

63. $(0, 2)$, $\left(-\frac{4}{3}, 0\right)$

65. $(5, 0)$, $(-5, 0)$, $(0, 5)$

67. $(0, 0)$, $(4, 0)$

69. $(-8, 0)$, $(8, 0)$, $(0, 4)$

71. $\frac{1}{2}$ **73.** c **75.** b **77.** $\frac{5}{12}$ **79.** $\frac{2}{7}$

81. Undefined **83.** $-\frac{4}{3}$ **85.** $\frac{7}{2}$ **87.** -3

89. $y = -\frac{1}{2}x + 2$ **91.** $y = \frac{2}{5}x + 1$

93. Perpendicular **95.** Parallel **97.** $(1, 3)$, $(2, 1)$

99. $(6, 6)$, $(10, 9)$ **101.** $(1, 4)$, $(5, 4)$

103. $2x - y - 9 = 0$ **105.** $4x + y - 6 = 0$

107. $4x - 5y + 10 = 0$ **109.** $8x + 3y - 1 = 0$

111. $x - 3 = 0$ **113.** $m = 5$ **115.** $m = \frac{4}{3}$

117. $x + 2y + 4 = 0$ **119.** $y - 8 = 0$

121. $x - y + 3 = 0$ **123.** $25x - 20y + 6 = 0$

125. (a) $2x + 3y + 3 = 0$
(b) $3x - 2y + 24 = 0$

127. (a) $8x + 6y - 27 = 0$
(b) $24x - 32y + 119 = 0$

129. Perpendicular

131. Parallel

133. (a) Not a solution
(b) Not a solution
(c) Solution
(d) Solution

135. $x \geq 2$

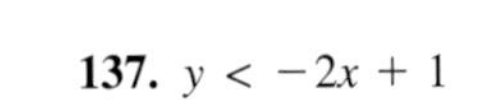

137. $y < -2x + 1$

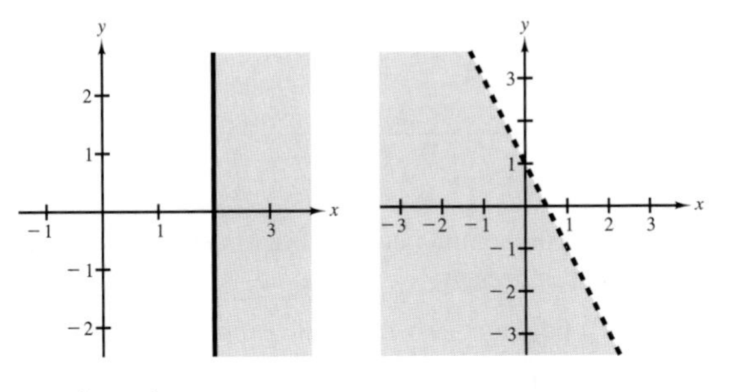

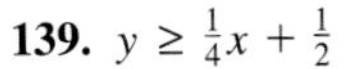

139. $y \geq \frac{1}{4}x + \frac{1}{2}$

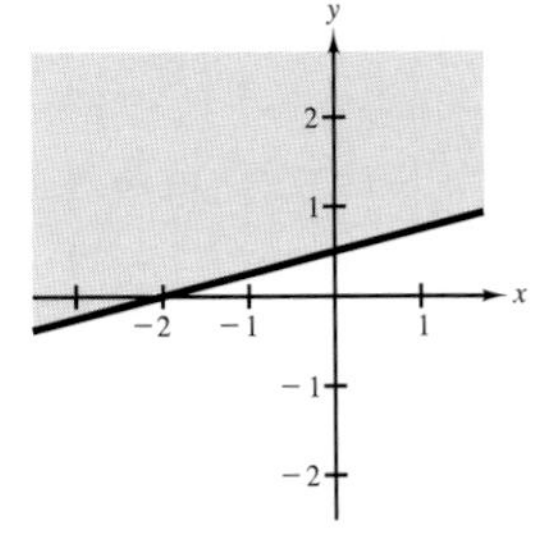

141. $y < 2$ **143.** $y \leq x + 1$

145. (a)

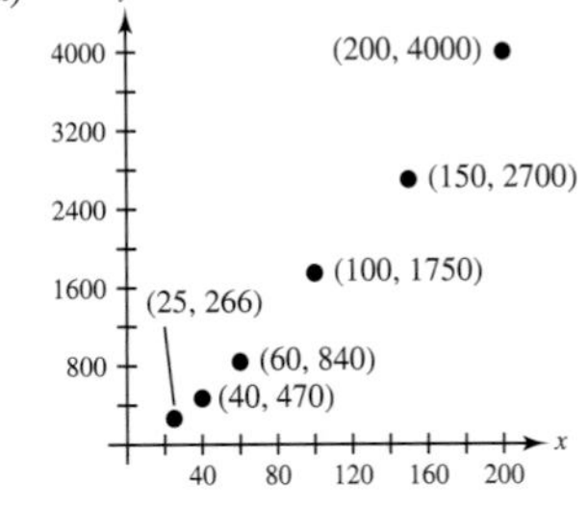

(b) Approximately linear
(c) 2225 lumens, graph

147. $C = 2.25 + 0.75x$

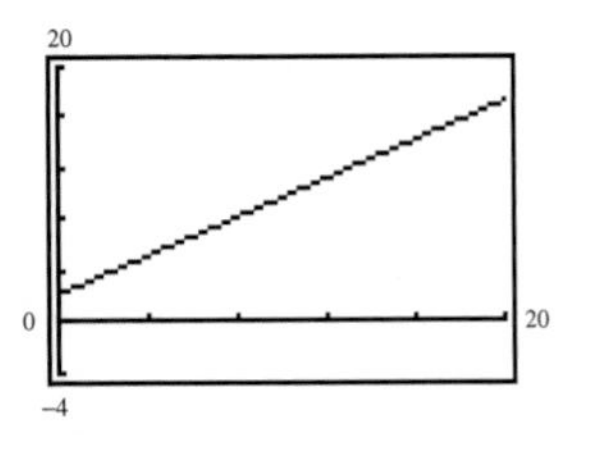

149. $A = x(12 - x),\ 0 < x < 12$

151. (a)

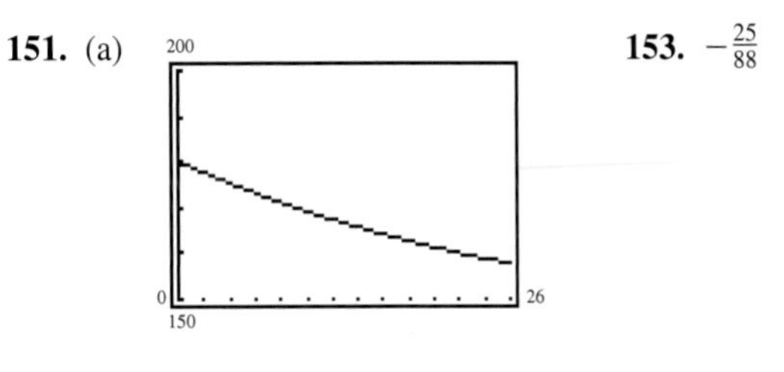

(b) About 157.8 pounds
(c) About 4.3 weeks

153. $-\frac{25}{88}$

155. (a) $C = 5.35x + 16{,}000$
(b) $P = 2.85x - 16{,}000$

Chapter Test *(page 271)*

1.

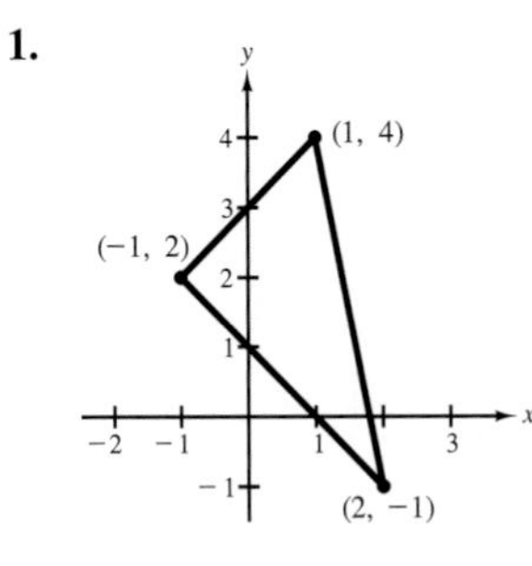

2. (a) Not a solution
(b) Solution
(c) Solution
(d) Not a solution

3. 0 **4.** $(-4, 0), (0, 3)$

5. No. There are two values of y corresponding to $x = 0$.

6. Yes. For each value of x there corresponds one and only one value of y.

7. (a) 0 (b) 0 (c) -16 (d) $-\frac{3}{8}$

8. $\frac{3}{14}$ **9.** $(-2, 2), (-1, 0)$ **10.** $-\frac{5}{3}$

11.

12.

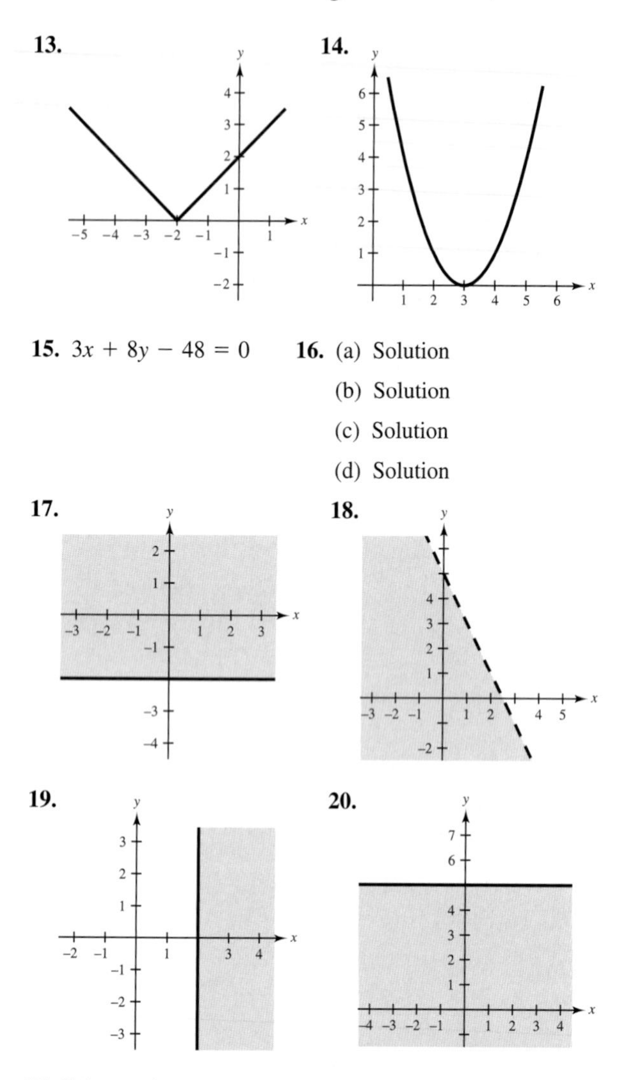

15. $3x + 8y - 48 = 0$

16. (a) Solution
(b) Solution
(c) Solution
(d) Solution

21. Sales are increasing at a rate of 230 units per year.

Chapter 5

Section 5.1 *(page 280)*

Integrated Review *(page 280)*

1. An algebraic expression is a collection of letters (called variables) and real numbers (called constants) combined by using addition, subtraction, multiplication, or division.

2. The terms of an algebraic expression are those parts separated by addition or subtraction.

3. $10x - 10$ **4.** $12 - 8z$ **5.** $-2 + 3x$

6. $-50x + 75$ **7.** $5x - 2y$ **8.** $\frac{1}{6}x + 8$

9. $7x - 16$ **10.** $-12x - 6$

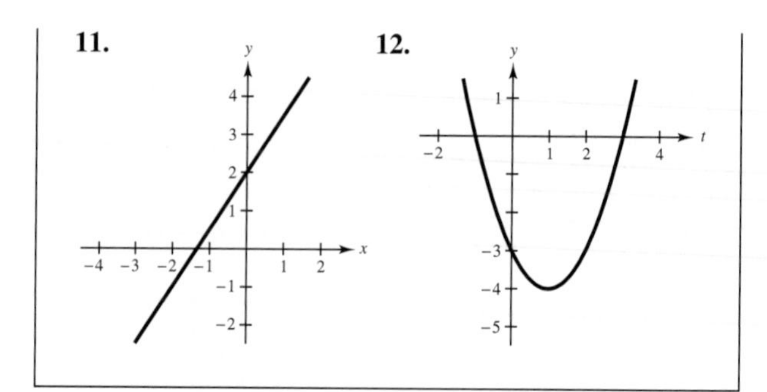

1. Polynomial

3. Not a polynomial because the exponent in the first term is not an integer.

5. Not a polynomial because the exponent is negative.

7. Not a polynomial because the exponent in the first term is negative.

9. Polynomial: $5 - 32x$
Standard form: $-32x + 5$
Degree: 1
Leading coefficient: -32

11. Polynomial: $x^3 - 4x^2 + 9$
Standard form: $x^3 - 4x^2 + 9$
Degree: 3
Leading coefficient: 1

13. Polynomial: $8x + 2x^5 - x^2 - 1$
Standard form: $2x^5 - x^2 + 8x - 1$
Degree: 5
Leading coefficient: 2

15. Polynomial: 10
Standard form: 10
Degree: 0
Leading coefficient: 10

17. Polynomial: $v_0t - 16t^2$
Standard form: $-16t^2 + v_0t$
Degree: 2
Leading coefficient: -16

19. Trinomial **21.** Binomial **23.** Monomial

25. $5x^3 - 10$ **27.** $3y^2$ **29.** $x^6 - 4x^3 - 2$

31. $14x + 6$ **33.** $4z^2 - z - 2$ **35.** $2b^3 - b^2$

37. $13 - 8t^2$ **39.** $4b^2 - 3$ **41.** $\frac{3}{2}y^2 + \frac{5}{4}$

43. $1.6t^3 - 3.4t^2 - 7.3$ **45.** $5x + 13$ **47.** $-x - 28$

49. $2x^3 + 2x^2 + 8$ **51.** $3x^4 - 2x^3 - 3x^2 - 5x$

53. $3x^2 + 2$ **55.** $y^4 + 4$ **57.** $4x^2 + 2x + 2$

59. $5y^3 + 12$ **61.** $4x^2 + 8$ **63.** $9x - 11$

65. $x^2 - 2x + 2$ **67.** $-3x^3 + 1$ **69.** $-u^2 + 5$

71. $-3x^5 - 3x^4 + 2x^3 - 6x + 6$ **73.** $x - 1$

75. $-x^2 - 2x + 3$ **77.** $-2x^4 - 5x^3 - 4x^2 + 6x - 10$

79. $-2x^3$ **81.** $4t^3 - 3t^2 + 15$ **83.** $5x^3 - 6x^2$

85. $3x^3 + 4x + 10$ **87.** $-2x - 20$

89. $3x^3 - 2x + 2$ **91.** $2x^4 + 9x + 2$

93. $8x^3 + 29x^2 + 11$ **95.** $12z + 8$ **97.** $4t^2 + 20$

99. $6v^2 + 90v + 30$ **101.** $10z + 4$ **103.** $2x^2 - 2x$

105. $21x^2 - 8x$ **107.** $6x$

109. (a) $T = -0.03t^2 + 1.55t + 106.28$

(b)

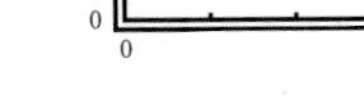

(c) Increasing

111. (a) Length: $2x^2$; Width: $3x + 5$

(b) $4x^2 + 6x + 10$

(c) Girth: $8x + 10$
Length and girth: $2x^2 + 8x + 10$
Yes.

113. (a) Sometimes true. $x^3 - 2x^2 + x + 1$ is a polynomial that is not a trinomial.

(b) True

115. Add (or subtract) their respective coefficients and attach the common variable factor.

117. To subtract one polynomial from another, add the opposite. You can do this by changing the sign of each of the terms of the polynomial that is being subtracted and then adding the resulting like terms.

Section 5.2 *(page 292)*

Integrated Review *(page 292)*

1. The point represented by $(3, -2)$ is located 3 units to the right of the y-axis and 2 units below the x-axis.

2. $(3, 4), (-3, 4), (-3, -4), (3, -4)$ **3.** $\frac{9}{4}x - \frac{5}{2}$

4. $2x - 2$ **5.** $7x - 8$ **6.** $-2y + 14$

7. $-4z + 12$ **8.** $-5u - 5$ **9.** $29,090.91

10. 1 hour, 5 miles

11. **12.**

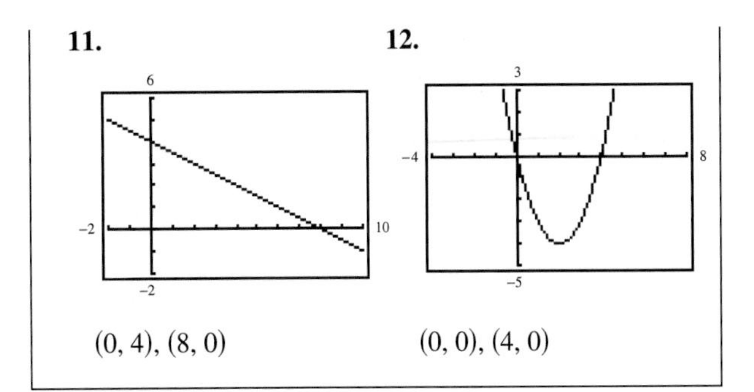

$(0, 4), (8, 0)$ $(0, 0), (4, 0)$

1. $-2x^2$ **3.** $4t^3$ **5.** $\frac{5}{2}x^2$ **7.** $6b^3$ **9.** $3y - y^2$

11. $-x^3 + 4x$ **13.** $6t^2 - 15t$

15. $-12x - 12x^3 + 24x^4$ **17.** $3x^3 - 6x^2 + 3x$

19. $2x^3 - 4x^2 + 16x$ **21.** $4t^4 - 12t^3$

23. $4x^4 - 3x^3 + x^2$ **25.** $-12x^5 + 18x^4 - 6x^3$

27. $30x^3 + 12x^2$ **29.** $12x^5 - 6x^4$ **31.** $x^2 + 7x + 12$

33. $6x^2 - 7x - 5$ **35.** $2x^2 - 5xy + 2y^2$

37. $2x^2 + 6x + 4$ **39.** $-8x^2 + 18x + 18$

41. $3x^2 - 5xy + 2y^2$ **43.** $3x^3 + 6x^2 - 4x - 8$

45. $2x^5 + 12x^3 + 4x^2 + 24$ **47.** $15s + 4$

49. $7x^3 + 32x^2 - 2x - 8$ **51.** $x^2 + 12x + 20$

53. $2x^2 - x - 10$ **55.** $x^3 + 3x^2 + x - 1$

57. $x^4 - 5x^3 - 2x^2 + 11x - 5$ **59.** $x^3 - 8$

61. $x^4 - 6x^3 + 5x^2 - 18x + 6$

63. $3x^4 - 12x^3 - 5x^2 - 4x - 2$ **65.** $x^2 + x - 6$

67. $x^3 + 27$ **69.** $x^4 - x^2 + 4x - 4$

71. $x^5 + 5x^4 - 3x^3 + 8x^2 + 11x - 12$

73. $x^3 - 6x^2 + 12x - 8$ **75.** $x^4 - 4x^3 + 6x^2 - 4x + 1$

77. $x^3 - 12x - 16$ **79.** $4u^3 + 4u^2 - 5u - 3$

81. $x^2 - 9$ **83.** $x^2 - 16$ **85.** $4u^2 - 9$

87. $16t^2 - 36$ **89.** $4x^2 - 9y^2$ **91.** $16u^2 - 9v^2$

93. $4x^4 - 25$ **95.** $x^2 + 12x + 36$ **97.** $t^2 - 6t + 9$

99. $9x^2 + 12x + 4$ **101.** $64 - 48z + 9z^2$

103. $4x^2 - 20xy + 25y^2$ **105.** $36t^2 + 60st + 25s^2$

107. $x^2 + y^2 + 2xy + 2x + 2y + 1$

109. $u^2 + v^2 - 2uv + 6u - 6v + 9$ **111.** $8x$

113. Yes **115.** $x^3 + 6x^2 + 12x + 8$

117. (a) $x^2 - 1$ **119.** (a) $6w$

(b) $x^3 - 1$ (b) $2w^2$

(c) $x^4 - 1$

(d) $x^5 - 1$

121. $x^2 + 7x + 12 = (x + 4)(x + 3)$

123. $2x^2 + 4x = 2x(x + 2)$ **125.** $x^2 + bx + ax + ab$

127. $2x[2(x + 1)] = 4x^2 + 4x$

129. $z(z + 4) = (z + 5)(z + 4) - 5(z + 4)$

131. (a)

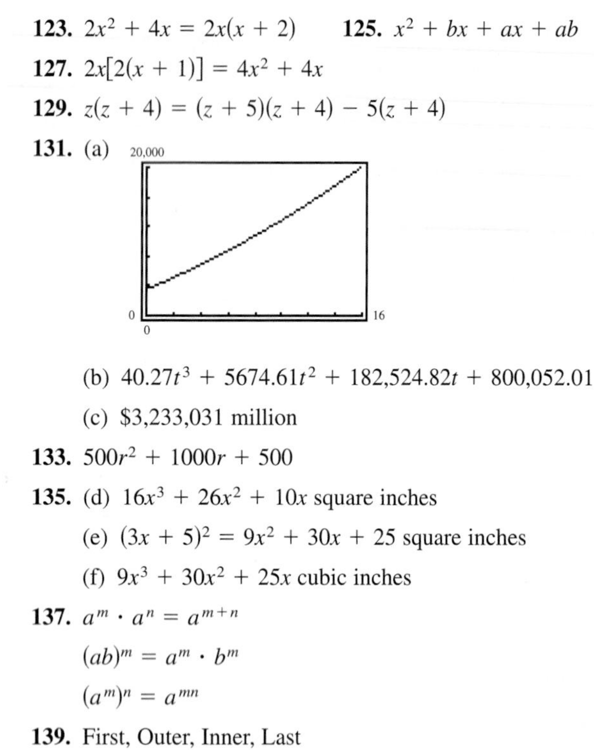

(b) $40.27t^3 + 5674.61t^2 + 182{,}524.82t + 800{,}052.01$

(c) \$3,233,031 million

133. $500r^2 + 1000r + 500$

135. (d) $16x^3 + 26x^2 + 10x$ square inches

(e) $(3x + 5)^2 = 9x^2 + 30x + 25$ square inches

(f) $9x^3 + 30x^2 + 25x$ cubic inches

137. $a^m \cdot a^n = a^{m+n}$

$(ab)^m = a^m \cdot b^m$

$(a^m)^n = a^{mn}$

139. First, Outer, Inner, Last

141. mn. Each term of the first factor must be multiplied by each term of the second factor.

143. False. $(x + 2)^2 = x^2 + 4x + 4$

Mid-Chapter Quiz *(page 296)*

1. Because the exponent of the third term is negative.

2. Degree: 4
Leading coefficient: -3

3. $3x^5 - 3x + 1$

4. False.
$(x - 1)(x + 5) = x^2 + 4x - 5$

5. $y^2 + 6y + 3$ **6.** $-v^3 + v^2 + 6v - 5$

7. $3s - 11$ **8.** $3x^2 + 5x - 4$ **9.** $10r^3$

10. $-2m^4$ **11.** $2y^2 + 7y - 15$

12. $2x^3 + 5x^2 - 14x - 8$ **13.** $16 - 24x + 9x^2$

14. $4u^2 - 9$ **15.** $5x^4 + 3x^3 - 2x + 2$

16. $2x^3 - 4x^2 + 3x + 1$ **17.** $6x^3 - x^2 - 33x - 5$

18. $5x^5 - 21x^4 + 18x^3 + 3x^2 - 9x$

19. $10x + 36$ **20.** $x^2 + 4x$

Section 5.3 *(page 302)*

Integrated Review *(page 302)*

1. The graph of a function is the set of all solution points of the function.

2. Construct a table of solution points, plot these solution points on a rectangular coordinate system, and use the pattern to connect the points with a smooth curve or line.

3. (0, 0), (9, 3) (The answer is not unique.)

4. To find the x-intercept(s), solve the equation $f(x) = 0$ for x. To find the y-intercept, find $f(0)$.

5. x^5 **6.** y^2z^{11} **7.** $\frac{x^6}{y^3}$ **8.** $\frac{2a^3b^3}{3c}$

9. **10.**

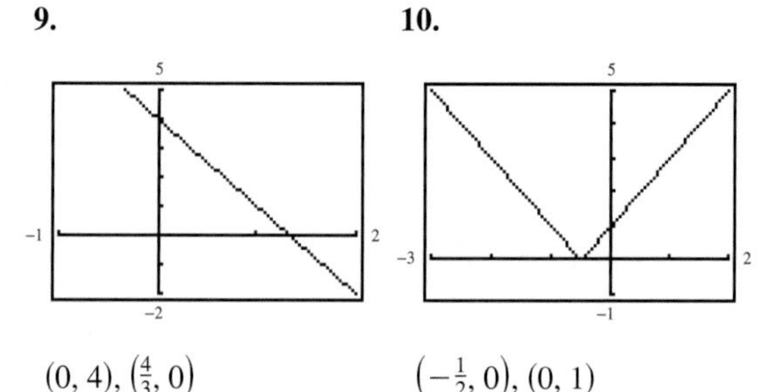

9. $(0, 4), \left(\frac{4}{3}, 0\right)$ 10. $\left(-\frac{1}{2}, 0\right), (0, 1)$

11. **12.**

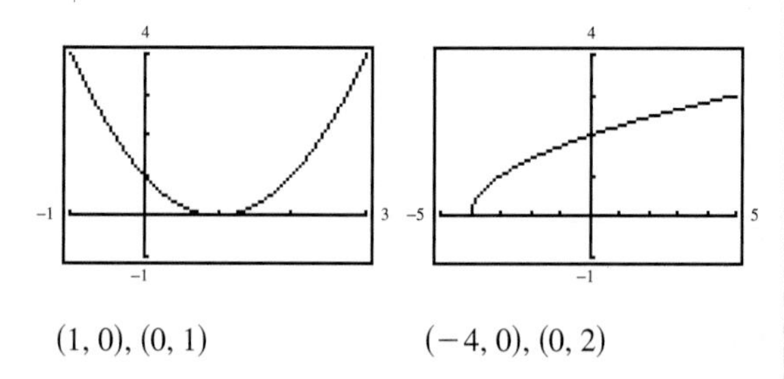

11. $(1, 0), (0, 1)$ 12. $(-4, 0), (0, 2)$

1. $\frac{1}{3^3}$ **3.** $\frac{1}{y^5}$ **5.** $\frac{8}{x^7}$ **7.** $\frac{7}{x^4y}$ **9.** $\frac{z^4}{2}$

11. $\frac{2xy^2}{3}$ **13.** 4^{-1} **15.** x^{-2} **17.** $10t^{-5}$

19. $5x^{-n}$ **21.** $2x^2y^{-4}$ **23.** $\frac{1}{3^2} = \frac{1}{9}$

25. $\frac{1}{(-4)^3} = -\frac{1}{64}$ **27.** $4^2 = 16$ **29.** $2(3^4) = 162$

31. $\frac{3^2}{2^4} = \frac{9}{16}$ **33.** $\frac{3^4}{4^2} = \frac{81}{16}$ **35.** $\left(\frac{3}{2}\right)^2 = \frac{9}{4}$

37. 0.0048 **39.** 41.7265 **41.** 4 **43.** x^2

45. $\frac{1}{u^3}$ **47.** $\frac{x}{y}$ **49.** x^5 **51.** $\frac{1}{y^6}$ **53.** $\frac{1}{x^2}$

55. $\frac{1}{y^6}$ **57.** $\frac{1}{s^2}$ **59.** 1 **61.** $\frac{1}{b^5}$ **63.** $\frac{1}{9x^4y^2}$

65. $\frac{a^6}{64b^9}$ **67.** $-\frac{8}{x}$ **69.** $\frac{10}{x}$ **71.** $\frac{x^2}{9z^4}$ **73.** 1

75. $-\frac{81}{16}$ **77.** $\frac{1}{32x}$ **79.** 1 **81.** 1

83. 9.3×10^7 **85.** 1.637×10^9 **87.** 4.35×10^{-4}

89. 4.392×10^{-3} **91.** 1.6×10^7 **93.** 1,090,000

95. 0.0867 **97.** 0.00852 **99.** 6.21 **101.** 8003.05

103. 4.984×10^{12} **105.** 3.0981×10^6

107. 3.35544×10^{32} **109.** 1.15743×10^{-22}

111. 5.2345679×10^2 **113.** 9.894×10^{13}

115. 8.45 minutes

117.

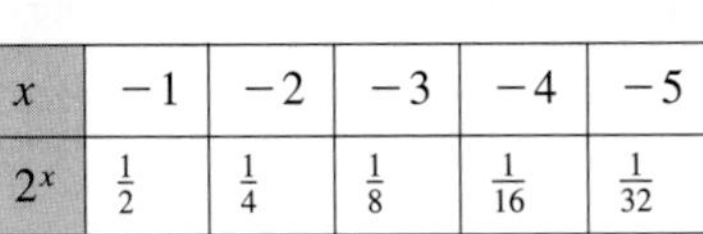

Planet	Mercury	Saturn	Neptune	Pluto
Kilometers	5.83×10^7	1.43×10^9	4.50×10^9	5.90×10^9

119. (a)

x	-1	-2	-3	-4	-5
2^x	$\frac{1}{2}$	$\frac{1}{4}$	$\frac{1}{8}$	$\frac{1}{16}$	$\frac{1}{32}$

(b)

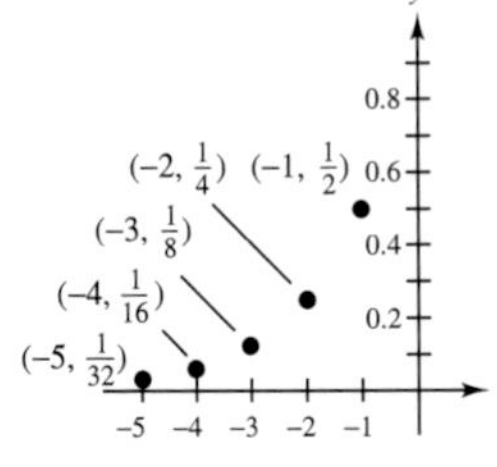

(c) Approaches 0. No.

121. 1.38×10^{-16} **123.** True **125.** False. Let $x = 2$.

127. True **129.** $3.4 \times 10^7 = 34{,}000{,}000$

$3.4 \times 10^{-6} = 0.0000034$

131. $(3 \times 10^5)(4 \times 10^6)$

$= (3 \times 10^5)(10^6 \times 4)$ Commutative Property of Multiplication

$= 3(10^5 \times 10^6)(4)$ Associative Property of Multiplication

$= 3(10^{5+6})(4)$ Property of exponents

$= (3 \cdot 4)10^{11}$ Commutative Property of Multiplication

$= 12 \times 10^{11}$ Multiplication

$= 1.2 \times 10^{12}$ Scientific notation

Section 5.4 *(page 312)*

Integrated Review *(page 312)*

1. $\frac{24x}{18} = \frac{6 \cdot 4x}{6 \cdot 3} = \frac{4x}{3}$

2. Quadrant II. Since the x-coordinate is negative, the point lies to the left of the y-axis. Since the y-coordinate is positive, the point lies above the x-axis.

3. $\frac{2}{3}$ **4.** $\frac{1}{8}$ **5.** $\frac{2}{5}$ **6.** $\frac{25}{6}$ **7.** $-10x^5$

8. $4z^2 - 1$ **9.** $x^2 + 14x + 49$

10. $2x^2 + 3x - 20$

11. $(2n + 1)(2n + 3) = 4n^2 + 8n + 3$

12. 51.4 miles per hour

1. x^3 **3.** $\frac{1}{x^3}$ **5.** $\frac{1}{z^3}$ **7.** $3u$ **9.** $2y^2$ **11.** $\frac{4^4}{x^2}$

13. $\frac{27}{ab}$ **15.** $-3x$ **17.** $\frac{4}{x^3}$ **19.** $4z^2$ **21.** $\frac{8b}{3}$

23. $-\frac{11y}{2}$ **25.** $\frac{3s^3}{2r^2}$ **27.** $\frac{1}{2z}$ **29.** $\frac{x^2}{2y}$ **31.** $\frac{4}{3v^2}$

33. $z + 1$ **35.** $z - 3$ **37.** $3x - \frac{5}{3}$ **39.** $b - 2$

41. $5x - 2$ **43.** $-5z^2 - 2z$ **45.** $4z^2 + \frac{3}{2}z - 1$

47. $m^2 + 3 - \frac{4}{m}$ **49.** $1 - \frac{3}{x}$

51. $3x - 1 + \frac{3}{2x} - \frac{1}{2x^2} + \frac{2}{x^3}$ **53.** $x - 2$ **55.** $x + 5$

57. $y + 2$ **59.** $6t + 1$ **61.** $x^2 - 2x + 5 + \frac{3}{x - 2}$

63. $7 - \frac{11}{x + 2}$ **65.** $x^2 + 2x + 4$ **67.** $x - 3 + \frac{18}{x + 3}$

69. $3x - 1$ **71.** $x^3 + x^2 + x + 1$

73. $4x - 1 + \frac{2}{x + 1}$ **75.** $x^2 + 2x - 3$

77. $3t^2 + t + 1 - \frac{4}{t + 2}$ **79.** $2x$ **81.** $5uv$

83. $-2x + 5$

85. Error; You can only cancel common factors of the numerator and denominator.

87. Valid

89. (a) Yes

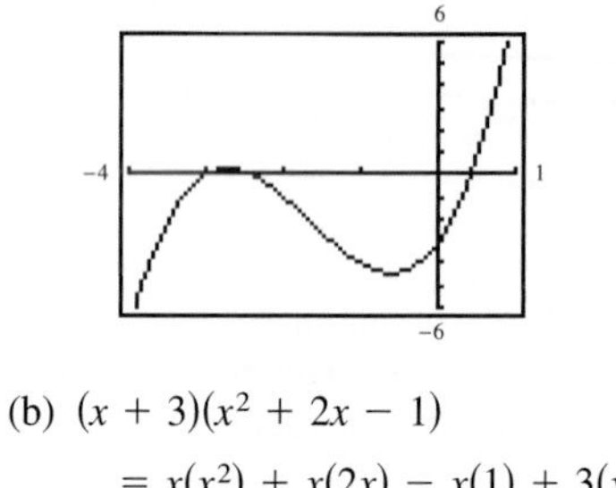

(b) $(x + 3)(x^2 + 2x - 1)$
$= x(x^2) + x(2x) - x(1) + 3(x^2) + 3(2x) - 3(1)$
$= x^3 + 5x^2 + 5x - 3$

(c) $(x^3 + 5x^2 + 5x - 3) \div (x + 3) = x^2 + 2x - 1$

91. (a) $1 + \dfrac{10}{t + 8}$

(b)

t	0	10	20	30	40	50	60
$\dfrac{t + 18}{t + 8}$	2.25	1.56	1.36	1.26	1.21	1.17	1.15

(c) The values approach 1.

93. $x - 3$

95. (a) $x^2 + 2$
(b) $x - 3$
(c) $x + 3$
(d) 11

97. $\dfrac{3x^8}{2x^3} = \left(\dfrac{3}{2}\right)\left(\dfrac{x^8}{x^3}\right) = \dfrac{3}{2}x^{8-3} = \dfrac{3}{2}x^5$

99. The remainder is 0 and the divisor is a factor of the dividend.

Review Exercises *(page 316)*

1. Polynomial: $10x - 4 - 5x^3$
Standard form: $-5x^3 + 10x - 4$
Degree: 3
Leading coefficient: -5

3. Polynomial: $4x^3 - 2x + 5x^4 - 7x^2$
Standard form: $5x^4 + 4x^3 - 7x^2 - 2x$
Degree: 4
Leading coefficient: 5

5. Polynomial: $7x^4 - 1$
Standard form: $7x^4 - 1$
Degree: 4
Leading coefficient: 7

7. Polynomial: -2
Standard form: -2
Degree: 0
Leading coefficient: -2

9. $x^4 + x^2 + 2$ **11.** $3 - 2x$ **13.** $3x - 1$
15. $-2t - 4$ **17.** $3x^3 - 2x + 3$ **19.** $4x^2 - 7$
21. $3x^4 - 3x^3 - 2x^2 + 5x + 4$ **23.** $-x^2 + 2x$
25. $-5x^3 - 5x - 2$ **27.** $7y^2 - y + 6$
29. $2x^4 - 7x^2 + 3$ **31.** $3x^2 + 4x - 14$
33. $2x^2 + 8x$ **35.** $x^2 + 2x - 24$ **37.** $2x^2 + 2x - 12$
39. $12x^2 + 7x - 12$ **41.** $2x^3 + 13x^2 + 19x + 6$
43. $2t^3 - 7t^2 + 9t - 3$ **45.** $u^2 - 6u + 5$
47. $x^2 + 6x + 9$ **49.** $16x^2 - 56x + 49$
51. $\frac{1}{4}x^2 - 4x + 16$ **53.** $u^2 - 36$ **55.** $9t^2 - 1$
57. $4x^2 - 4xy + y^2$ **59.** $4x^2 - 16y^2$ **61.** $\frac{1}{16}$
63. $\frac{1}{36}$ **65.** 9 **67.** 64 **69.** $\frac{125}{27}$ **71.** $-\frac{2}{5}$
73. 9,000,000 **75.** 37,000 **77.** $\dfrac{1}{y^4}$ **79.** $\dfrac{6}{t^2}$
81. $\dfrac{x^6}{7}$ **83.** $\dfrac{2}{xy^3}$ **85.** $\dfrac{1}{t^2}$ **87.** $4y^2$ **89.** $\dfrac{1}{9a^4}$
91. $\dfrac{x^4}{y^6}$ **93.** $\dfrac{1}{t^3}$ **95.** 1 **97.** $\dfrac{25}{y^2}$ **99.** $\dfrac{2}{uv}$
101. $2x - \dfrac{3}{x}$ **103.** $x + 3$ **105.** $x + 2$
107. $8x + 5 + \dfrac{2}{3x - 2}$ **109.** $2x^2 + 4x + 3 + \dfrac{5}{x - 1}$
111. $x^2 - 2$ **113.** $80 - 2x^2$ **115.** $2x^2 + 8x + 8$
117. (a) $4x - 6$ (b) $x^2 - 3x$ **119.** $2x + 3$
121. 0.15 foot **123.** $x^2 - y^2 = (x + y)(x - y)$

Chapter Test *(page 319)*

1. Degree: 4
Leading coefficient: -3
2. $z^4 + 2z^2 - 3$
3. $2z^2 - 3z + 15$ **4.** $7u^3 - 1$ **5.** $-y^2 + 8y + 3$
6. $-6x^2 + 12x$ **7.** $10b^2 + b - 3$ **8.** $9x^3$
9. $2z^3 + z^2 - z + 10$ **10.** $x^2 - 10x + 25$
11. $4x^2 - 9$ **12.** $3x + 5$ **13.** $x^2 + 2x + 3$
14. $2x^2 + 4x - 3 - \dfrac{2}{2x + 1}$ **15.** $\dfrac{2a}{3}$ **16.** $\dfrac{x^4}{9y^6}$
17. (a) $\frac{1}{64}$ (b) $\frac{3}{8}$ (c) 22,500,000,000
18. $4x^2 - x$ **19.** $2x^2 + 11x - 6$ **20.** 384,000,000
21. 1.013×10^5 **22.** $x - 3$

Chapter 6

Section 6.1 *(page 327)*

Integrated Review *(page 327)*

1. A function is a set of ordered pairs in which no two ordered pairs have the same first component and a different second component.

2. The set of first components is the domain of a function. The set of second components is the range.

3.

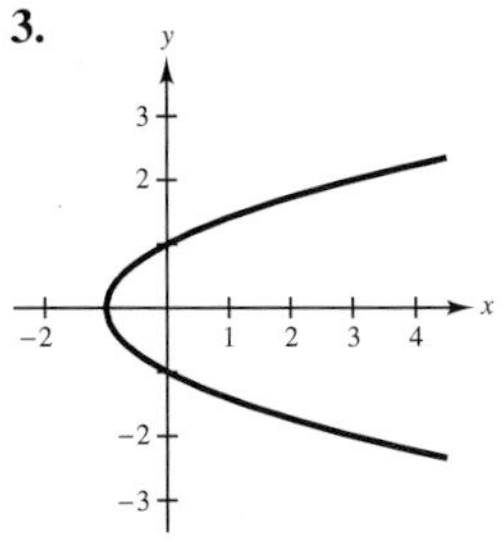

4.

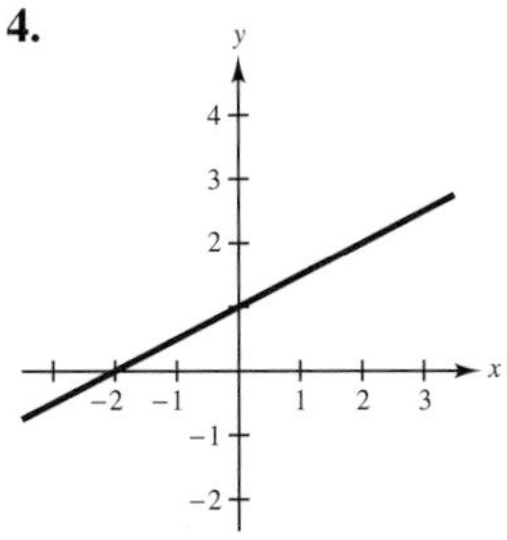

5. (a) 1 (b) 3 (c) $-\frac{1}{2}$ (d) $\frac{1}{4}$

6. (a) 0 (b) 0 (c) 12 (d) $\frac{65}{4}$

7. (a) 1 (b) 3 (c) 0 (d) $\sqrt{21}$

8. (a) 3 (b) 1 (c) 1 (d) 6

9. 3% **10.** 3 hours 45 minutes

11.

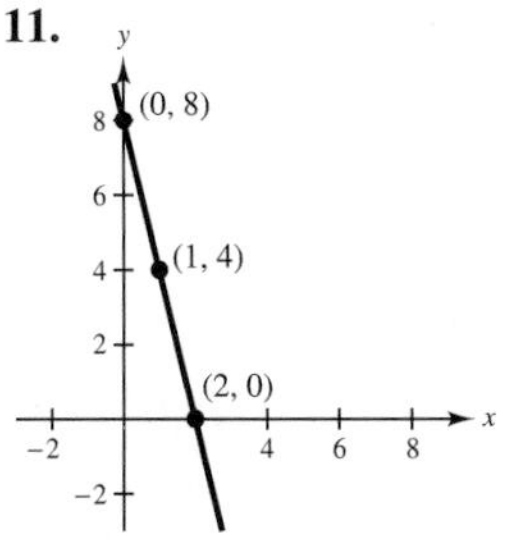

12.

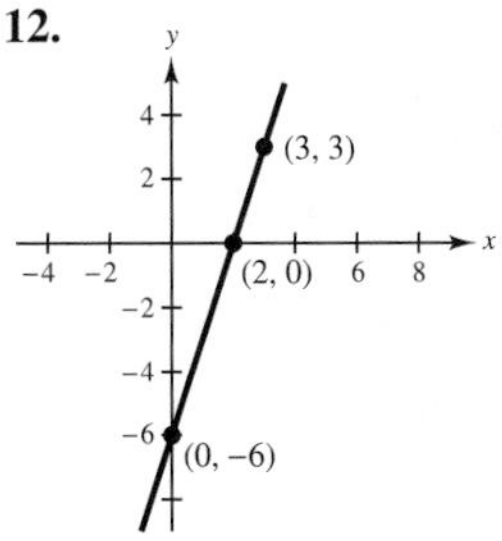

13.

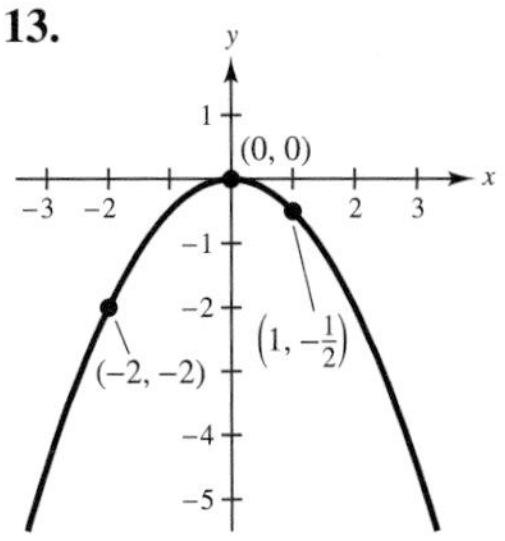

14.

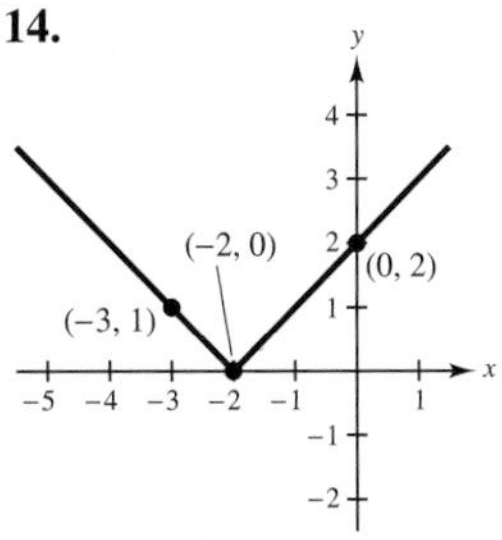

1. 6 **3.** 2 **5.** z^2 **7.** $2x$ **9.** u^2v **11.** $3yz^2$
13. 1 **15.** $14a^2b^2$ **17.** $3(x + 1)$ **19.** $6(z - 1)$

21. $8(t - 2)$ **23.** $-5(5x + 2)$ **25.** $6(4y^2 - 3)$
27. $x(x + 1)$ **29.** $u(25u - 14)$ **31.** $2x^3(x + 3)$
33. No common factor **35.** $2x(6x - 1)$
37. $-5r(2r^2 + 7)$ **39.** $8a^3b^3(2 + 3a)$
41. $10ab(1 + a)$ **43.** $4(3x^2 + 4x - 2)$
45. $25(4 + 3z - 2z^2)$ **47.** $3x^2(3x^2 + 2x + 6)$
49. $5u(2u + 1)$ **51.** $(x - 3)(x + 5)$
53. $(s + 10)(t - 8)$ **55.** $(b + 2)(a^2 - b)$
57. $z^2(z + 5)(z + 1)$ **59.** $(a + b)(2a - b)$
61. $-5(2x - 1)$ **63.** $-3(x - 1000)$
65. $-(x^2 - 2x - 4)$ **67.** $-2(x^2 - 6x - 2)$
69. $(x + 10)(x + 1)$ **71.** $(a - 4)(a + 1)$
73. $(y - 4)(ky + 2)$ **75.** $(t - 3)(t^2 + 2)$
77. $(x + 2)(x^2 + 1)$ **79.** $(2z + 1)(3z^2 - 1)$
81. $(x - 1)(x^2 - 3)$ **83.** $(4 - x)(x^2 - 2)$
85. $x + 3$ **87.** $10y - 1$ **89.** $14x + 5y$

91. $y_1 = y_2$

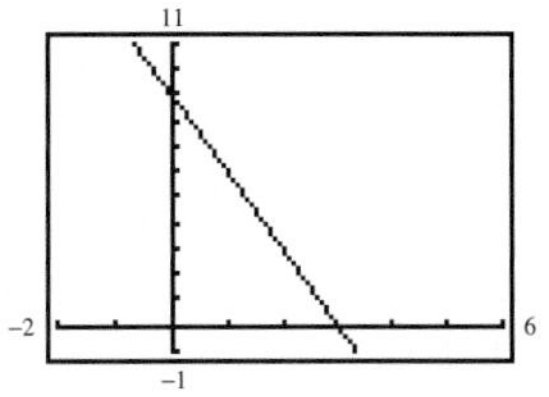

93. $y_1 = y_2$

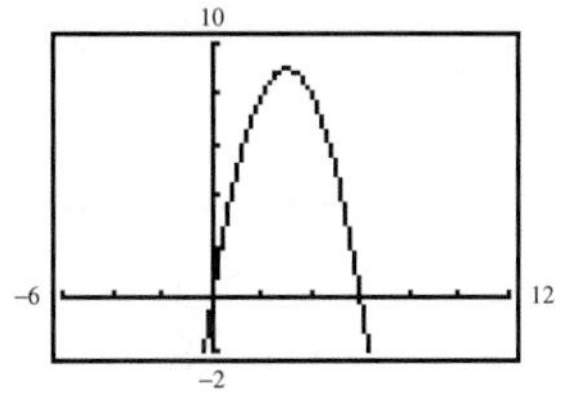

95. $x + 1$ **97.** $6x^2$ **99.** $9x^2\left(6 - \frac{\pi}{2}\right)$

101. $2\pi r(r + h)$ **103.** $kx(Q - x)$

105. $x^2 + x - 6 = (x - 2)(x + 3)$

107. Determine the prime factorization of each term. The greatest common factor contains each common prime factor, repeated the minimum number of times it occurs in any one of the factorizations.

109. Noun: Any one of the expressions that, when multiplied together, yield the product

Verb: To find the expressions that, when multiplied together, yield the given product

111.
$$\begin{aligned} x^3 - 3x^2 - 5x + 15 &= (x^3 - 3x^2) + (-5x + 15) \\ &= x^2(x - 3) - 5(x - 3) \\ &= (x - 3)(x^2 - 5) \end{aligned}$$

Section 6.2 *(page 335)*

Integrated Review *(page 335)*

1. If there are two y-intercepts, then there are two values of y that correspond to $x = 0$.

2. 4 **3.** $y^2 + 2y$ **4.** $-a^3 + a^2$

5. $x^2 - 7x + 10$ **6.** $v^2 + 3v - 28$

7. $4x^2 - 25$ **8.** $x^3 - 4x^2 + 10$ **9.** \$3,975,000

10. \$717 **11.** $x \geq 46$ **12.** $140 \leq x \leq 227.5$

1. $x + 1$ **3.** $a - 2$ **5.** $y - 5$ **7.** $z - 2$

9. $(x + 1)(x + 11)$, $(x - 1)(x - 11)$

11. $(x + 12)(x + 1)$, $(x - 12)(x - 1)$, $(x + 6)(x + 2)$, $(x - 6)(x - 2)$, $(x + 4)(x + 3)$, $(x - 4)(x - 3)$

13. $(x + 2)(x + 4)$ **15.** $(x - 5)(x - 8)$

17. $(z - 3)(z - 4)$ **19.** Prime **21.** $(x + 2)(x - 3)$

23. $(x - 3)(x + 5)$ **25.** Prime **27.** $(u + 2)(u - 24)$

29. $(x + 15)(x + 4)$ **31.** $(x - 8)(x - 9)$

33. $(x + 12)(x - 20)$ **35.** $(x + 2y)(x - y)$

37. $(x + 5y)(x + 3y)$ **39.** $(x - 9z)(x + 2z)$

41. $(a + 5b)(a - 3b)$ **43.** $3(x + 5)(x + 2)$

45. $4(y - 3)(y + 1)$ **47.** Prime **49.** $9(x^2 + 2x - 2)$

51. $x(x - 10)(x - 3)$ **53.** $x^2(x - 2)(x - 3)$

55. $-3x(y - 3)(y + 6)$ **57.** $x(x + 2y)(x + 3y)$

59. $2xy(x + 3y)(x - y)$ **61.** $\pm 8, \pm 16$ **63.** $\pm 4, \pm 20$

65. $\pm 12, \pm 13, \pm 15, \pm 20, \pm 37$ **67.** $2, -10$ **69.** $5, -7$

71. $8, -10$

73. $y_1 = y_2$ **75.** $y_1 = y_2$

77. $(x + 3)(x + 1)$ **79.** $(x + 3)(x + 2)$

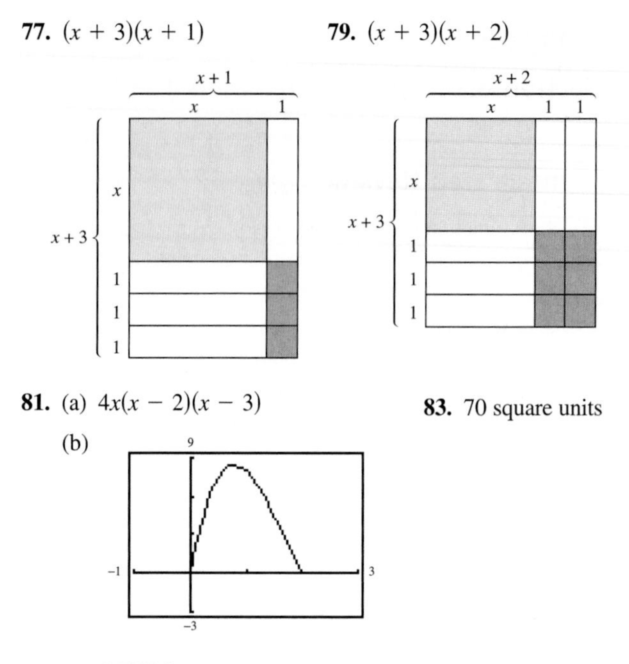

81. (a) $4x(x - 2)(x - 3)$

(b)

0.785 foot

83. 70 square units

85. Because the constant term is positive in the polynomial, the signs in the binomial factors must be the same.

87. The polynomial is not factorable using factors with integer coefficients.

89. The prime number, because there are not as many possible factorizations to examine.

Section 6.3 *(page 343)*

Integrated Review *(page 343)*

1. Prime **2.** The sum of the digits is divisible by 3.

3. $2^2 \cdot 5^3$ **4.** $3^2 \cdot 5 \cdot 7$ **5.** $2^3 \cdot 3^2 \cdot 11$

6. $5^2 \cdot 7 \cdot 13$ **7.** $2x^2 + 9x - 35$

8. $9x^2 - 12x + 4$

9. **10.**

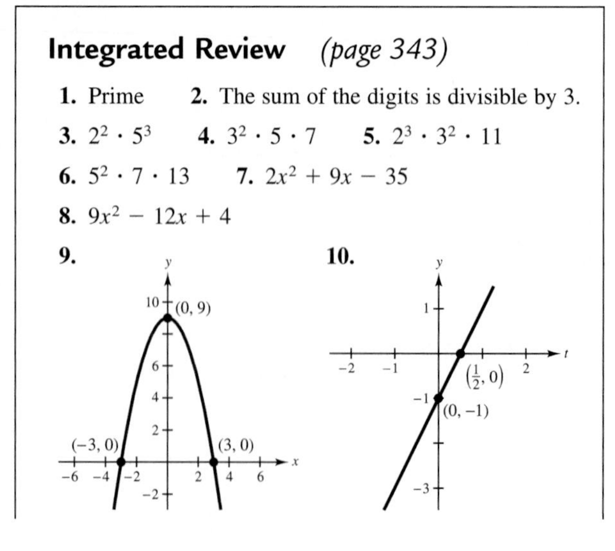

11. (a)

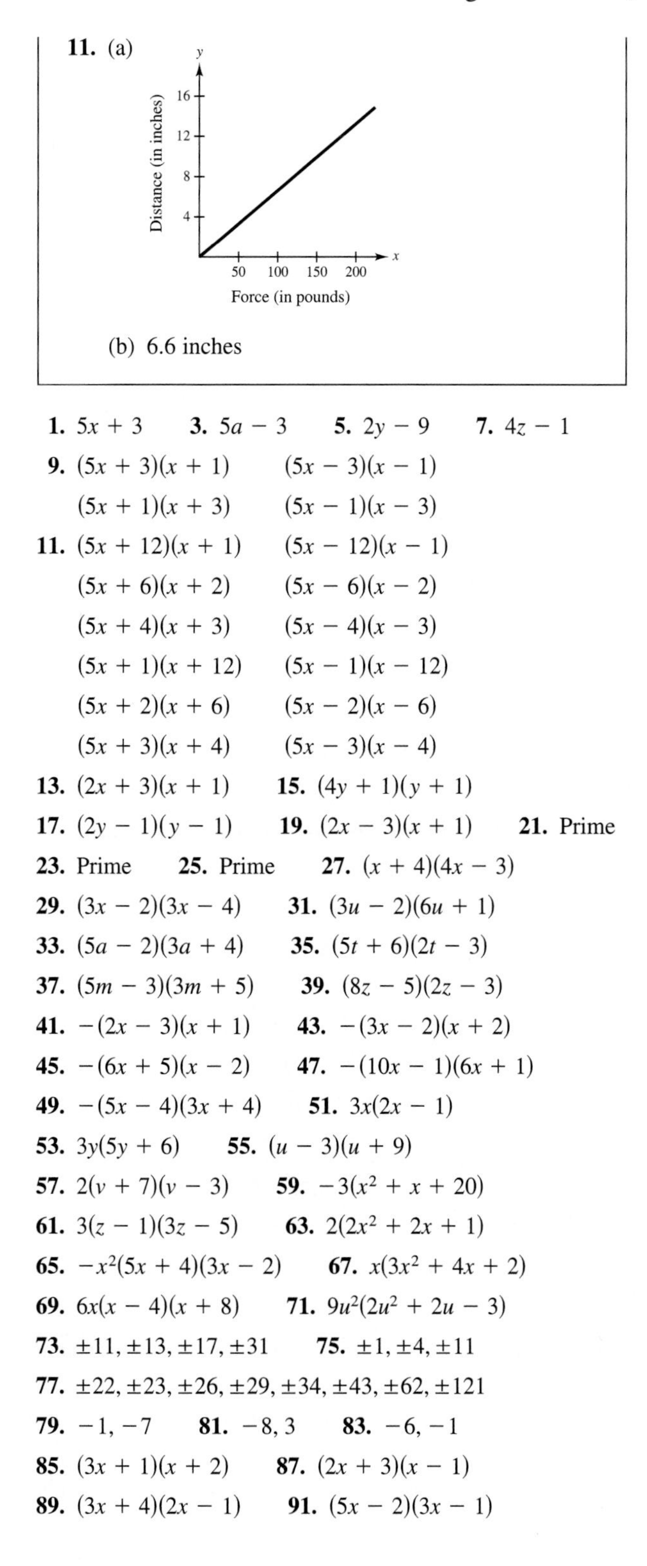

(b) 6.6 inches

1. $5x + 3$ **3.** $5a - 3$ **5.** $2y - 9$ **7.** $4z - 1$

9. $(5x + 3)(x + 1)$ $(5x - 3)(x - 1)$
$(5x + 1)(x + 3)$ $(5x - 1)(x - 3)$

11. $(5x + 12)(x + 1)$ $(5x - 12)(x - 1)$
$(5x + 6)(x + 2)$ $(5x - 6)(x - 2)$
$(5x + 4)(x + 3)$ $(5x - 4)(x - 3)$
$(5x + 1)(x + 12)$ $(5x - 1)(x - 12)$
$(5x + 2)(x + 6)$ $(5x - 2)(x - 6)$
$(5x + 3)(x + 4)$ $(5x - 3)(x - 4)$

13. $(2x + 3)(x + 1)$ **15.** $(4y + 1)(y + 1)$

17. $(2y - 1)(y - 1)$ **19.** $(2x - 3)(x + 1)$ **21.** Prime

23. Prime **25.** Prime **27.** $(x + 4)(4x - 3)$

29. $(3x - 2)(3x - 4)$ **31.** $(3u - 2)(6u + 1)$

33. $(5a - 2)(3a + 4)$ **35.** $(5t + 6)(2t - 3)$

37. $(5m - 3)(3m + 5)$ **39.** $(8z - 5)(2z - 3)$

41. $-(2x - 3)(x + 1)$ **43.** $-(3x - 2)(x + 2)$

45. $-(6x + 5)(x - 2)$ **47.** $-(10x - 1)(6x + 1)$

49. $-(5x - 4)(3x + 4)$ **51.** $3x(2x - 1)$

53. $3y(5y + 6)$ **55.** $(u - 3)(u + 9)$

57. $2(v + 7)(v - 3)$ **59.** $-3(x^2 + x + 20)$

61. $3(z - 1)(3z - 5)$ **63.** $2(2x^2 + 2x + 1)$

65. $-x^2(5x + 4)(3x - 2)$ **67.** $x(3x^2 + 4x + 2)$

69. $6x(x - 4)(x + 8)$ **71.** $9u^2(2u^2 + 2u - 3)$

73. $\pm 11, \pm 13, \pm 17, \pm 31$ **75.** $\pm 1, \pm 4, \pm 11$

77. $\pm 22, \pm 23, \pm 26, \pm 29, \pm 34, \pm 43, \pm 62, \pm 121$

79. $-1, -7$ **81.** $-8, 3$ **83.** $-6, -1$

85. $(3x + 1)(x + 2)$ **87.** $(2x + 3)(x - 1)$

89. $(3x + 4)(2x - 1)$ **91.** $(5x - 2)(3x - 1)$

93. $(3a + 5)(a + 2)$ **95.** $(8x - 3)(2x + 1)$

97. $(3x - 2)(4x - 3)$ **99.** $(u - 2)(6u + 7)$

101. $(2x + 1)(x + 2)$ **103.** $l = 2x + 3$

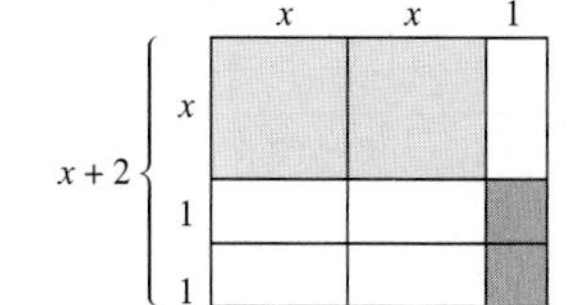

105. $2x + 10$

107. (a) $y_1 = y_2$

(b)

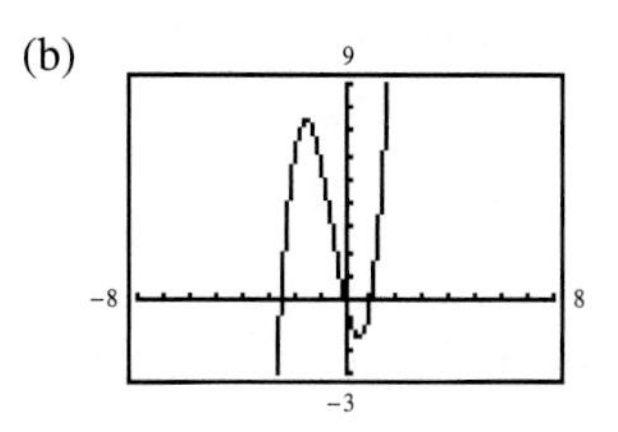

(c) $\left(-\frac{5}{2}, 0\right), (0, 0), (1, 0)$

109. (a) $3x^2 + 16x - 12$

(b) $3x - 2$ and $x + 6$

111. The product of the last terms of the binomials is 15, not -15.

113. Four. $(ax + 1)(x + c), (ax + c)(x + 1),$
$(ax - 1)(x - c), (ax - c)(x - 1)$

115. $2x^3 + 2x^2 + 2x$

Mid-Chapter Quiz *(page 346)*

1. $2x - 3$ **2.** $x - y$ **3.** $y - 6$ **4.** $2x + 1$

5. $10(x^2 + 7)$ **6.** $2a^2b(a - 2b)$ **7.** $(x + 2)(x - 3)$

8. $(t - 3)(t^2 + 1)$ **9.** $(y + 6)(y + 5)$

10. $(u + 6)(u - 5)$ **11.** $x(x - 6)(x + 5)$

12. $2y(x + 8)(x - 4)$ **13.** Prime

14. $(3 + z)(2 - 5z)$ **15.** $(3x - 2)(2x + 1)$

16. $2s^2(5s^2 - 7s + 1)$ **17.** $\pm 7, \pm 8, \pm 13$ **18.** 16, 21

19. $(3x + 1)(x + 6)$ $(3x - 1)(x - 6)$
$(3x + 6)(x + 1)$ $(3x - 6)(x - 1)$
$(3x + 2)(x + 3)$ $(3x - 2)(x - 3)$
$(3x + 3)(x + 2)$ $(3x - 3)(x - 2)$

20. $10(2x + 8)$ **21.** $y_1 = y_2$

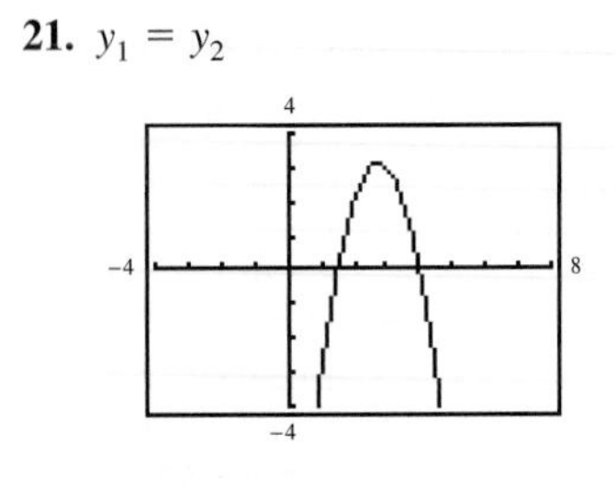

Section 6.4 *(page 353)*

Integrated Review *(page 353)*

1. Quadrant II **2.** Quadrant I or II **3.** $(-4, 0)$
4. $(9, -6)$ **5.** 4 **6.** 1 **7.** -1 **8.** $\frac{16}{9}$
9. $\frac{5}{2}$ **10.** 10 **11.** 6954 **12.** \$12,155

1. $(x + 6)(x - 6)$ **3.** $(u + 8)(u - 8)$
5. $(7 + x)(7 - x)$ **7.** $\left(u + \frac{1}{2}\right)\left(u - \frac{1}{2}\right)$
9. $\left(t + \frac{1}{4}\right)\left(t - \frac{1}{4}\right)$ **11.** $(4y + 3)(4y - 3)$
13. $(10 + 7x)(10 - 7x)$ **15.** $(x + 1)(x - 3)$
17. $-z(10 + z)$ **19.** $2(x + 6)(x - 6)$
21. $2(2 + 5x)(2 - 5x)$ **23.** $(y^2 + 9)(y + 3)(y - 3)$
25. $(1 + x^2)(1 + x)(1 - x)$ **27.** $3(x + 2)(x - 2)(x^2 + 4)$
29. $(3x + 2)(3x - 2)(9x^2 + 4)$ **31.** $(x - 2)^2$
33. $(z + 3)^2$ **35.** $(2t + 1)^2$ **37.** $(5y - 1)^2$
39. $\left(b + \frac{1}{2}\right)^2$ **41.** $\left(2x - \frac{1}{4}\right)^2$ **43.** $(x - 3y)^2$
45. $(2y + 5z)^2$ **47.** $(3a - 2b)^2$ **49.** ± 2 **51.** $\pm\frac{8}{5}$
53. ± 36 **55.** 9 **57.** 4 **59.** $(x - 2)(x^2 + 2x + 4)$
61. $(y + 4)(y^2 - 4y + 16)$ **63.** $(1 + 2t)(1 - 2t + 4t^2)$
65. $(3u + 2)(9u^2 - 6u + 4)$ **67.** $6(x - 6)$
69. $u(u + 3)$ **71.** $5y(y - 5)$ **73.** $5(y + 5)(y - 5)$
75. $y^2(y + 5)(y - 5)$ **77.** $(1 - 2x)^2$ **79.** $(x - 1)^2$
81. $(9x + 1)(x + 1)$ **83.** $2x(2 - x)(1 + x)$
85. $(3t + 4)(3t - 4)$ **87.** $-z(z + 12)$
89. $(t + 10)(t - 12)$ **91.** $u(u^2 + 2u + 3)$ **93.** Prime
95. $2(t - 2)(t^2 + 2t + 4)$ **97.** $2(1 - 2x)(1 + 2x + 4x^2)$
99. $(x^2 + 9)(x + 3)(x - 3)$
101. $(1 + x^2)(1 + x)(1 - x)$ **103.** $(x + 1)(x - 1)(x - 4)$
105. $x(x + 3)(x + 4)(x - 4)$
107. $(2 + y)(2 - y)(y^2 + 2y + 4)(y^2 - 2y + 4)$

109. $y_1 = y_2$ **111.** $y_1 = y_2$

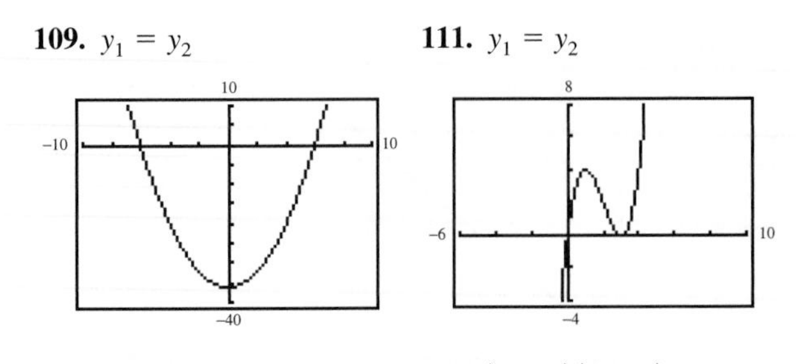

113. 441 **115.** 3599 **117.** $\pi(R - r)(R + r)$
119. $(x + 4)^2 - 2^2 = (x + 6)(x + 2)$
121. $2x^2 + 4x + 2 = (2x + 2)(x + 1)$ **123.** 15×15 feet
125. $a^2 - b^2 = (a + b)(a - b)$ **127.** No. $(x + 2)(x - 2)$
129. False. $a^3 + b^3 = (a + b)(a^2 - ab + b^2)$

Section 6.5 *(page 363)*

Integrated Review *(page 363)*

1. Additive Inverse Property
2. Multiplicative Identity Property
3. Distributive Property
4. Associative Property of Addition
5. (a) 3 (b) 28 **6.** (a) $\frac{3}{2}$ (b) $\frac{7}{6}$ **7.** $-\frac{1}{16}$
8. $\frac{8}{5}$ **9.** $2t^2 - 2t + 1$ **10.** $-8u + 15$
11. \$750 **12.** 70 miles per hour

1. 0, 5 **3.** 2, 3 **5.** $-1, 2$ **7.** $-\frac{1}{3}, \frac{5}{2}$ **9.** $-2, 6$
11. $\frac{100}{7}, 60$ **13.** $-8, 0, \frac{5}{4}$ **15.** $-12, -\frac{3}{2}, 1$
17. $-4, 4$ **19.** $-10, 10$ **21.** $-3, 3$ **23.** $-2, 8$
25. $-13, 5$ **27.** $-2, 0$ **29.** $0, \frac{1}{4}$ **31.** $-3, 4$
33. $-2, 8$ **35.** 1 **37.** -7 **39.** $\frac{3}{2}$ **41.** $-2, 4$
43. $-\frac{1}{2}, 3$ **45.** $-\frac{5}{3}, 1$ **47.** $-5, 3$ **49.** $-2, 7$
51. $-\frac{3}{2}, 1$ **53.** 2, 7 **55.** $-5, 0$ **57.** $-3, -2, 0$
59. $-4, 0, \frac{3}{2}$ **61.** $-3, 2, 3$ **63.** $-4, 1, 4$
65. $-2, -1, 0, 1$
67. $(-3, 0), (1, 0)$; The number of solutions equals the number of x-intercepts.
69. $(-3, 0), (4, 0)$; The number of solutions equals the number of x-intercepts.
71. $(0, 0), (3, 0)$; The number of solutions equals the number of x-intercepts.
73. $\left(-\frac{5}{2}, 0\right), (0, 0), (1, 0)$; The number of solutions equals the number of x-intercepts.

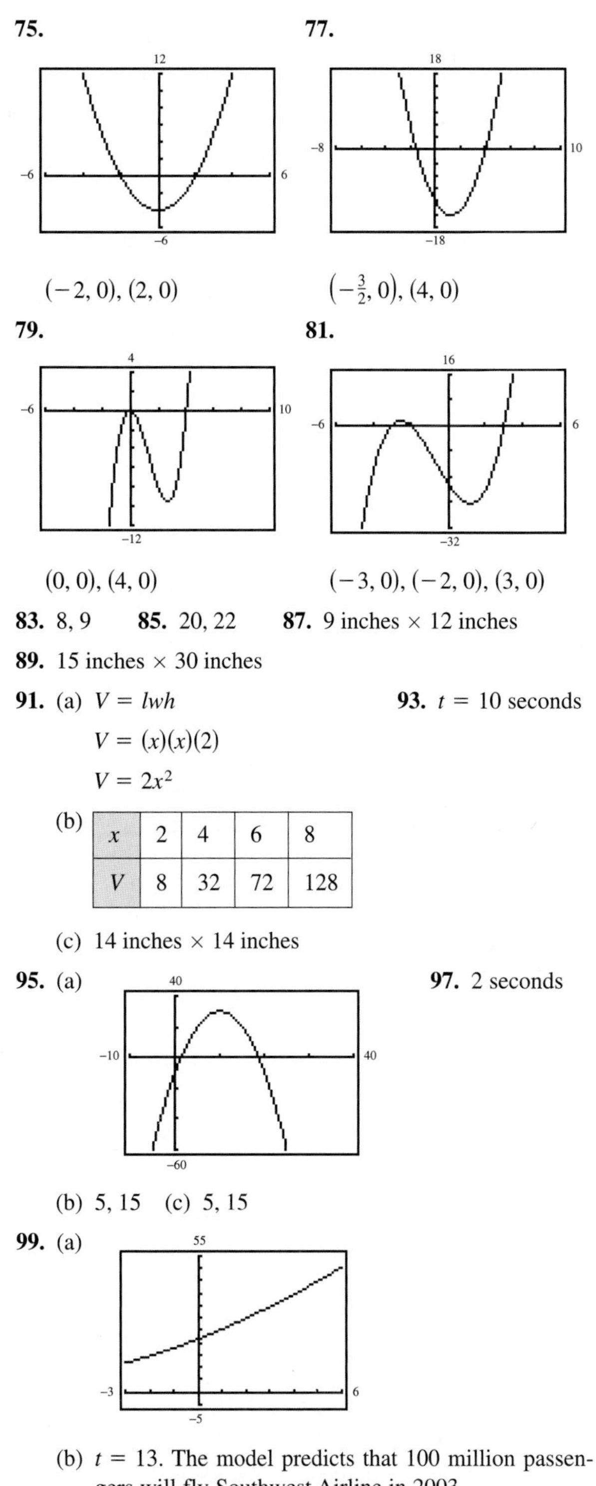

75. $(-2, 0), (2, 0)$ **77.** $\left(-\frac{3}{2}, 0\right), (4, 0)$

79. $(0, 0), (4, 0)$ **81.** $(-3, 0), (-2, 0), (3, 0)$

83. 8, 9 **85.** 20, 22 **87.** 9 inches × 12 inches

89. 15 inches × 30 inches

91. (a) $V = lwh$
$V = (x)(x)(2)$
$V = 2x^2$

93. $t = 10$ seconds

(b)

x	2	4	6	8
V	8	32	72	128

(c) 14 inches × 14 inches

95. (a) (graph)

(b) 5, 15 (c) 5, 15

97. 2 seconds

99. (a) (graph)

(b) $t = 13$. The model predicts that 100 million passengers will fly Southwest Airline in 2003.

101. 0, 1 **103.** If $ab = 0$, then $a = 0$ or $b = 0$.

105. A linear equation has degree 1 and a quadratic equation has degree 2.

107. n

109. False. This is not an application of the Zero-Factor Property because the number of factors whose product is 21 is unlimited.

Review Exercises *(page 368)*

1. 10 **3.** $9ab$ **5.** $3(x - 2)$ **7.** $t(3 - t)$

9. $5x^2(1 + 2x)$ **11.** $4a(2 - 3a^2)$

13. $(x + 1)(x - 3)$ **15.** $(y + 3)(y^2 + 2)$

17. $(x^2 + 1)(x + 2)$ **19.** $(x + 3)(x - 4)$

21. $(x - 7)(x + 4)$ **23.** $(u - 4)(u + 9)$

25. $(x - y)(x + 10y)$ **27.** $(y + 3x)(y - 9x)$

29. $4(x - 2)(x - 4)$ **31.** $\pm 6, \pm 10$ **33.** ± 12

35. $(1 - x)(5 + 3x)$ **37.** $(10 + x)(5 - x)$

39. $(3x + 2)(2x + 1)$ **41.** $3u(2u + 5)(u - 2)$

43. $(2x - 1)(x - 1)$ **45.** $\pm 2, \pm 5, \pm 10, \pm 23$

47. $\pm 4, \pm 7, \pm 11, \pm 17, \pm 28, \pm 59$ **49.** $2, -6$

51. $(x + 1)(x - 1)$ **53.** $(a + 10)(a - 10)$

55. $(5 + 2y)(5 - 2y)$ **57.** $(u + 3)(u - 1)$

59. $(x - 4)^2$ **61.** $(x + 3)^2$ **63.** $(3s + 2)^2$

65. $st(s + t)(s - t)$ **67.** $(a + 1)(a^2 - a + 1)$

69. $(3 - 2t)(9 + 6t + 4t^2)$ **71.** $-4a(2a + 1)^2$

73. $-9, 9$ **75.** 6 **77.** $-1, \frac{3}{4}$ **79.** $0, \frac{3}{2}$ **81.** 0, 4

83. 3, 4 **85.** $-5, -1, 1$ **87.** $3x + 1$

89. $2x^2(10 - \pi)$

91. (a) (graph)

(b) 20 (c) 20

93. 40 inches × 60 inches **95.** 12, 14

Chapter Test *(page 370)*

1. $7x^2(1 - 2x)$ **2.** $(z + 7)(z - 3)$

3. $(t - 5)(t + 1)$ **4.** $(3x - 4)(2x - 1)$

5. $3y(2y + 5)(y + 5)$ **6.** $(2 + 5v)(2 - 5v)$

7. $(2x - 5)^2$ **8.** $(-z - 5)(z + 13)$

9. $(x + 2)(x + 3)(x - 3)$ **10.** $(4 + z^2)(2 + z)(2 - z)$

11. $\frac{1}{5}(2x - 3)$ **12.** ± 6 **13.** 36

14. $3x^2 - 3x - 6 = 3(x + 1)(x - 2)$ **15.** $-4, \frac{3}{2}$

16. 0, 2 **17.** $-3, \frac{2}{3}$ **18.** $-\frac{3}{2}, 2$

19. 7 inches × 12 inches **20.** 2 seconds; $\frac{3}{2}$ seconds

21. 24, 26

Cumulative Test: Chapters 4–6 *(page 371)*

1. Because $x = -2$, the point must lie in Quadrant II or Quadrant III.

2. (a) Not a solution (b) Solution (c) Solution (d) Not a solution

3. **4.**

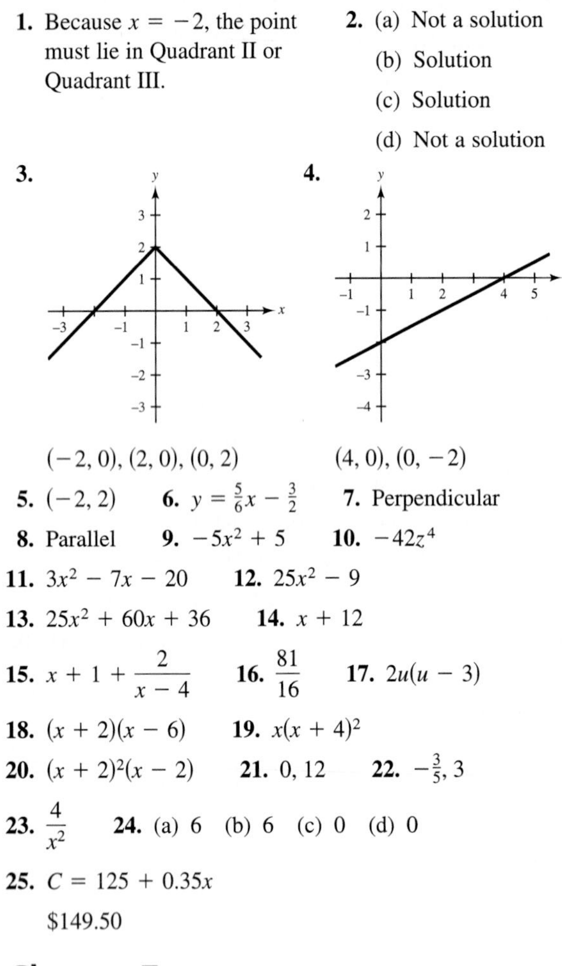

$(-2, 0), (2, 0), (0, 2)$ $(4, 0), (0, -2)$

5. $(-2, 2)$ **6.** $y = \frac{5}{6}x - \frac{3}{2}$ **7.** Perpendicular

8. Parallel **9.** $-5x^2 + 5$ **10.** $-42z^4$

11. $3x^2 - 7x - 20$ **12.** $25x^2 - 9$

13. $25x^2 + 60x + 36$ **14.** $x + 12$

15. $x + 1 + \dfrac{2}{x - 4}$ **16.** $\dfrac{81}{16}$ **17.** $2u(u - 3)$

18. $(x + 2)(x - 6)$ **19.** $x(x + 4)^2$

20. $(x + 2)^2(x - 2)$ **21.** 0, 12 **22.** $-\frac{3}{5}, 3$

23. $\dfrac{4}{x^2}$ **24.** (a) 6 (b) 6 (c) 0 (d) 0

25. $C = 125 + 0.35x$

$149.50

Chapter 7

Section 7.1 *(page 385)*

Integrated Review *(page 385)*

1. **2.**

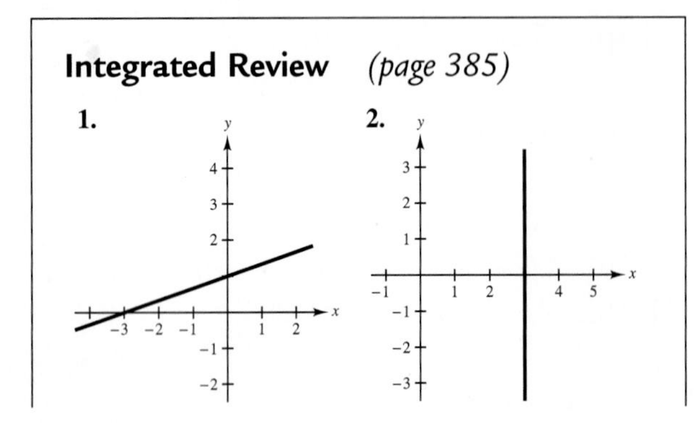

3. $\frac{3}{2}$ **4.** $m = -3$ has the greater absolute value.

5. $\frac{5}{11}$ **6.** $\frac{14}{11}$ **7.** 50 **8.** 64

9. $y = \frac{1}{4}(5 - 3x)$ **10.** $y = \frac{2}{3}(3 - x)$

11. **12.** **13.** **14.**

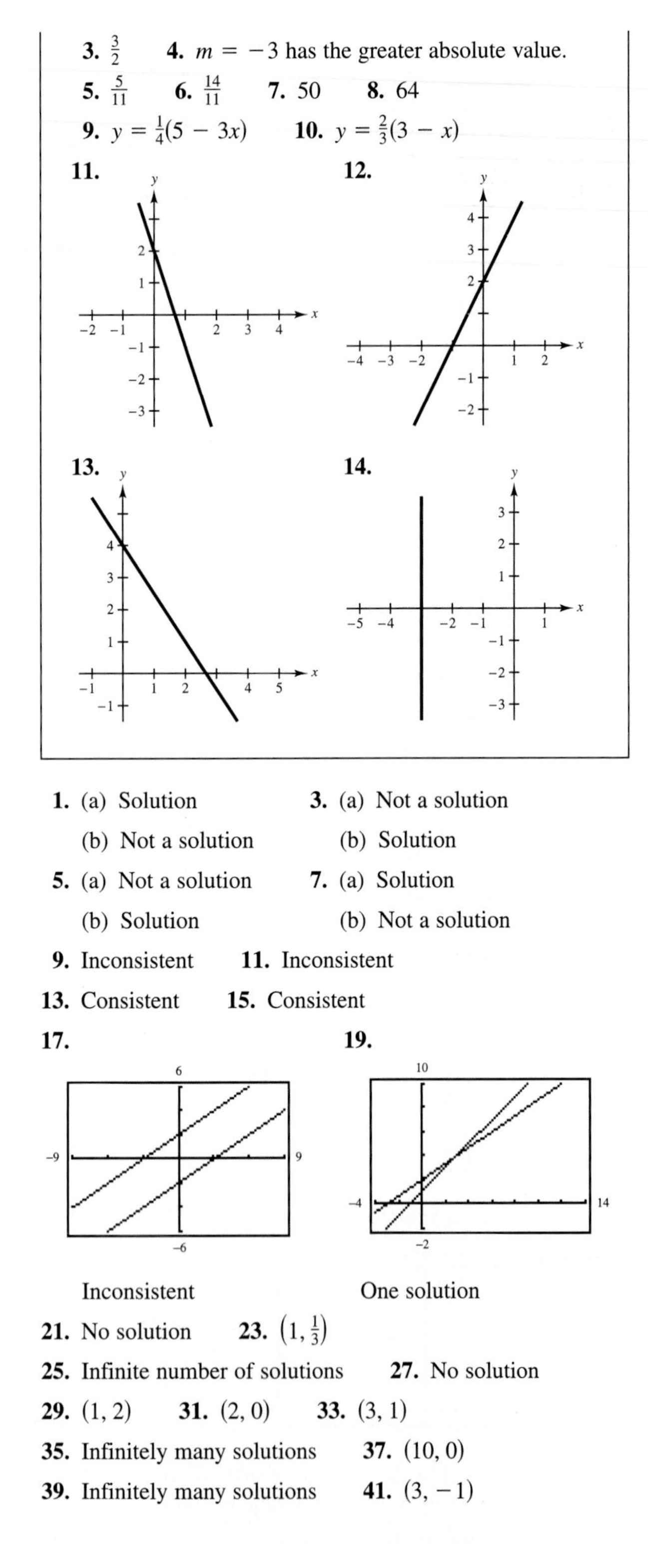

1. (a) Solution (b) Not a solution

3. (a) Not a solution (b) Solution

5. (a) Not a solution (b) Solution

7. (a) Solution (b) Not a solution

9. Inconsistent **11.** Inconsistent

13. Consistent **15.** Consistent

17. Inconsistent **19.** One solution

21. No solution **23.** $\left(1, \frac{1}{3}\right)$

25. Infinite number of solutions **27.** No solution

29. (1, 2) **31.** (2, 0) **33.** (3, 1)

35. Infinitely many solutions **37.** (10, 0)

39. Infinitely many solutions **41.** $(3, -1)$

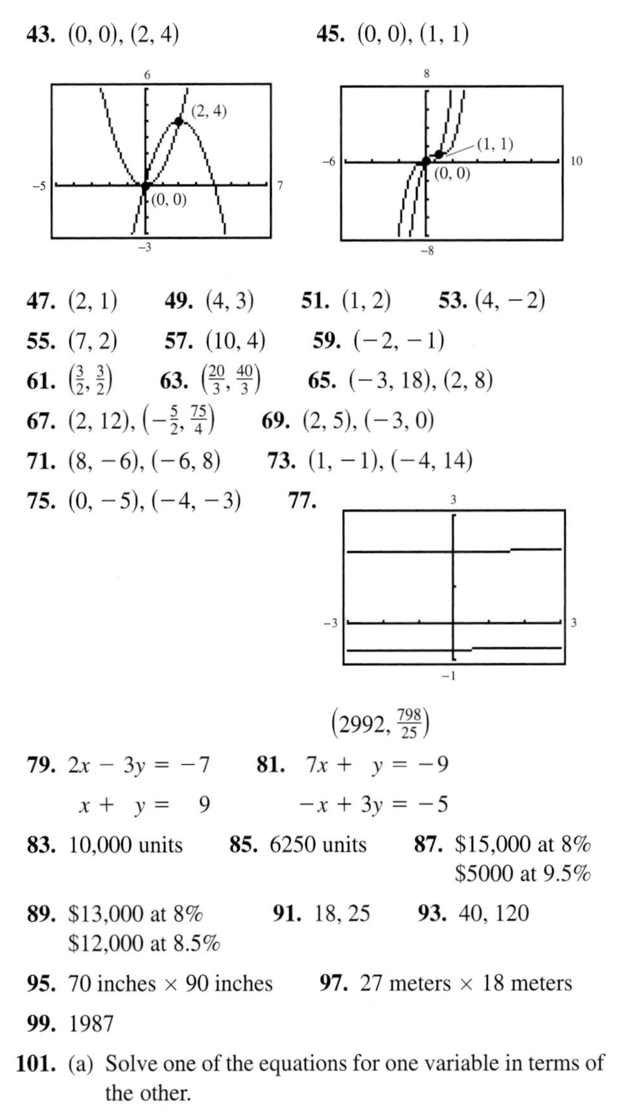

43. (0, 0), (2, 4) **45.** (0, 0), (1, 1)

47. (2, 1) **49.** (4, 3) **51.** (1, 2) **53.** (4, −2)

55. (7, 2) **57.** (10, 4) **59.** (−2, −1)

61. $\left(\frac{3}{2}, \frac{3}{2}\right)$ **63.** $\left(\frac{20}{3}, \frac{40}{3}\right)$ **65.** (−3, 18), (2, 8)

67. (2, 12), $\left(-\frac{5}{2}, \frac{75}{4}\right)$ **69.** (2, 5), (−3, 0)

71. (8, −6), (−6, 8) **73.** (1, −1), (−4, 14)

75. (0, −5), (−4, −3) **77.**

$\left(2992, \frac{798}{25}\right)$

79. $2x - 3y = -7$
$x + y = 9$

81. $7x + y = -9$
$-x + 3y = -5$

83. 10,000 units **85.** 6250 units **87.** \$15,000 at 8%, \$5000 at 9.5%

89. \$13,000 at 8%, \$12,000 at 8.5% **91.** 18, 25 **93.** 40, 120

95. 70 inches × 90 inches **97.** 27 meters × 18 meters

99. 1987

101. (a) Solve one of the equations for one variable in terms of the other.
(b) Substitute the expression found in Step (a) into the other equation to obtain an equation in one variable.
(c) Solve the equation obtained in Step (b).
(d) Back-substitute the solution from Step (c) into the expression obtained in Step (a) to find the value of the other variable.
(e) Check the solution in each of the original equations of the system.

103. After finding a value for one of the variables, substitute this value back into one of the original equations. This is called back-substitution.

105. The graphical method usually yields approximate solutions.

Section 7.2 *(page 395)*

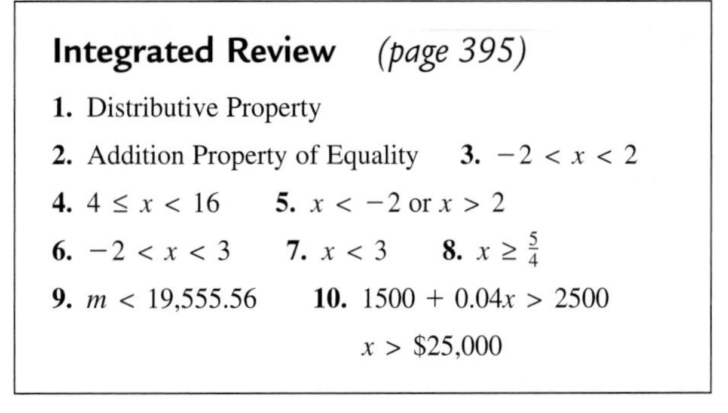

Integrated Review *(page 395)*

1. Distributive Property

2. Addition Property of Equality **3.** $-2 < x < 2$

4. $4 \le x < 16$ **5.** $x < -2$ or $x > 2$

6. $-2 < x < 3$ **7.** $x < 3$ **8.** $x \ge \frac{5}{4}$

9. $m < 19{,}555.56$ **10.** $1500 + 0.04x > 2500$
$x > \$25{,}000$

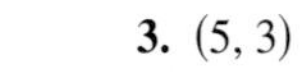

1. (2, 0) **3.** (5, 3)

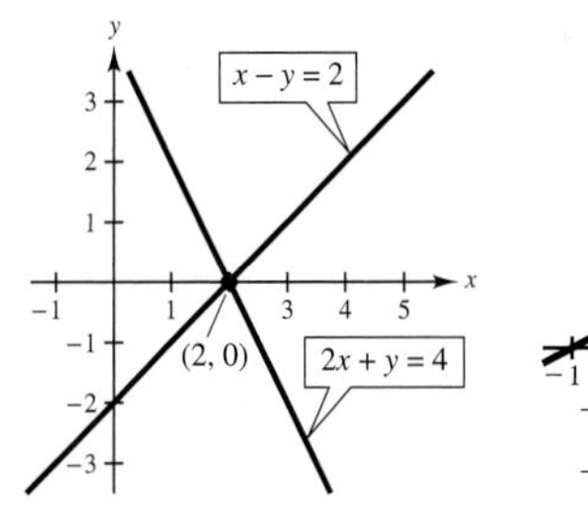

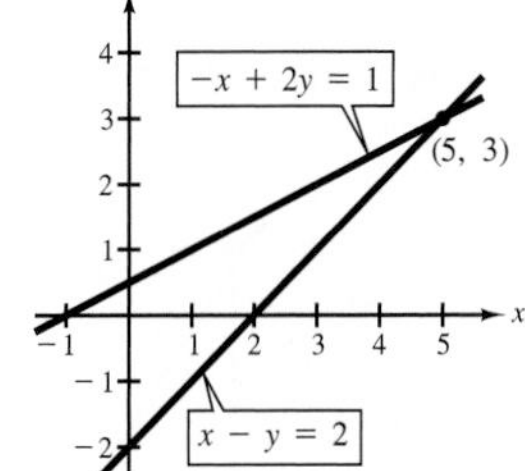

5. (2, −3) **7.** Inconsistent

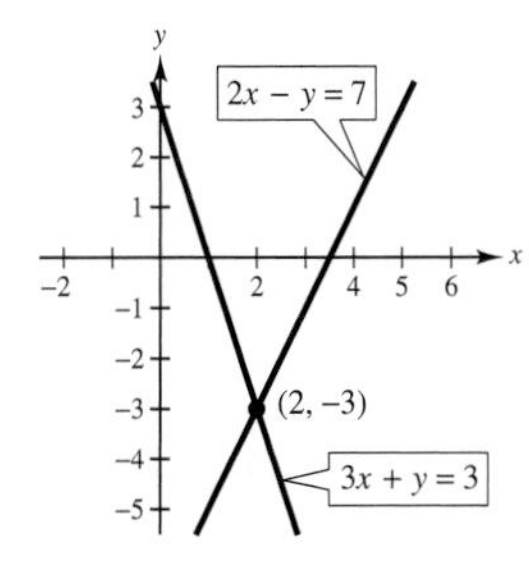

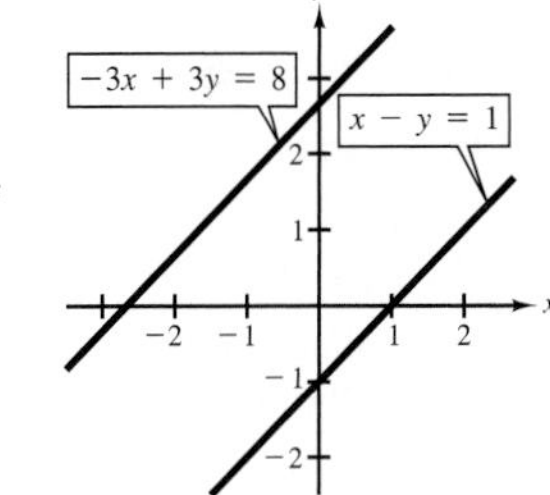

9. Infinitely many solutions **11.** $\left(\frac{1}{2}, \frac{3}{2}\right)$

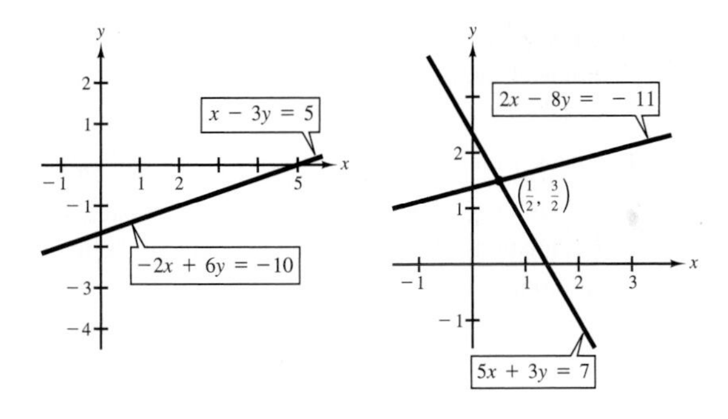

13. (3, 2) **15.** (−2, 5) **17.** (2, 1) **19.** (3, −4)

21. (−1, −1) **23.** (5, −1) **25.** (7, −2)

27. Inconsistent **29.** $\left(\frac{3}{2}, 1\right)$ **31.** (6, 3)

33. $(-2, -1)$ **35.** Infinitely many solutions

37. Inconsistent **39.** $(12.5, 4.948)$ **41.** $(-3, 7)$

43. $(2, 7)$ **45.** $(15, 10)$ **47.** $(4, 3)$ **49.** Consistent

51. Consistent **53.** Inconsistent **55.** $k = 4$

57. $x + 2y = 0$
$4x + 2y = 9$

59. 122 weeks

61. 8% bond: \$15,000
9.5% bond: \$5000

63. 4 hours

65. Speed of plane in still air: 550 miles per hour
Wind speed: 50 miles per hour

67. Adult: 375
Children: 125

69. Regular unleaded: \$1.11
Premium unleaded: \$1.22

71. 40% solution: 12 liters
65% solution: 8 liters

73. \$5.65 variety: 6.1 pounds
\$8.95 variety: 3.9 pounds

75. (a) $y = -1.5x + 3\frac{5}{6}$

(b)

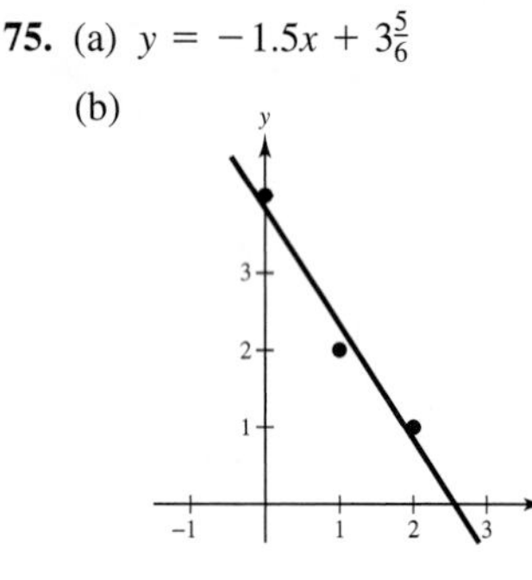

77. Depth: 10 feet; Lengths of sections: 122 feet, 125 feet

79. When adding a nonzero multiple of one equation to another equation to eliminate a variable, you get $0 = 0$ for the second equation.

81. (a) Obtain coefficients for x or y that differ only in sign by multiplying all terms of one or both equations by suitably chosen constants.

(b) Add the equations to eliminate one variable, and solve the resulting equation.

(c) Back-substitute the value obtained in Step (b) into either of the original equations and solve for the other variable.

(d) Check your solution in both of the original equations.

83. When it is easy to solve for one of the variables in one of the equations of the system, it might be better to use substitution.

Section 7.3 *(page 407)*

Integrated Review *(page 407)*

1. One solution

2. Multiply both sides of the equation by the lowest common denominator, 24.

3. $4x^8$ **4.** $8x^{10}y^{15}$ **5.** $\frac{4}{x^{11}}$ **6.** $\frac{3}{t^4}$

7. $-1, 5$ **8.** $\frac{1}{2}$ **9.** $d = 15t$

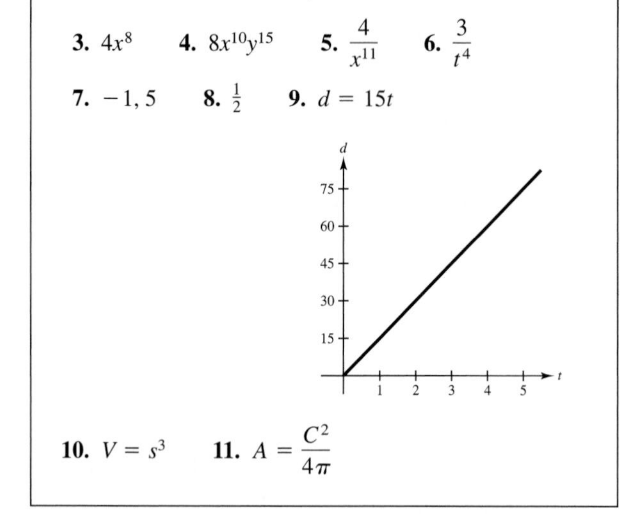

10. $V = s^3$ **11.** $A = \frac{C^2}{4\pi}$

1. (a) Not a solution
(b) Solution
(c) Solution
(d) Not a solution

3. $(22, -1, -5)$

5. $(14, 3, -1)$

7. No. When the first equation was multiplied by -2 and added to the second equation, the constant term should have been -11.

9. $x - 2y = 8$
$y = 14$
Eliminated the x-term from the second equation

11. $x - 2y + 3z = 5$
$-y + 8z = 9$
$2x - 3z = 0$
Eliminated the x-term in Equation 2

13. $(1, 2, 3)$ **15.** $(1, 2, 3)$ **17.** $(2, -3, -2)$

19. No solution **21.** $(-4, 8, 5)$ **23.** No solution

25. $\left(\frac{3}{10}, \frac{2}{5}, 0\right)$ **27.** $(-4, 2, 3)$ **29.** $(-1, 5, 5)$

31. $\left(-\frac{1}{2}a + \frac{1}{2}, \frac{3}{5}a + \frac{2}{5}, a\right)$ **33.** $\left(-\frac{1}{2}a + \frac{1}{4}, \frac{1}{2}a + \frac{5}{4}, a\right)$

35. $(1, -1, 2)$ **37.** $\left(\frac{6}{13}a + \frac{10}{13}, \frac{5}{13}a + \frac{4}{13}, a\right)$

39. $\left(-\frac{1}{2}, 2, 10\right)$ **41.**
$$\begin{aligned} x + 2y - z &= -4 \\ y + 2z &= 1 \\ 3x + y + 3z &= 15 \end{aligned}$$

43. $s = -16t^2 + 144$ **45.** $s = -16t^2 + 48t$

47. $y = 2x^2 + 3x - 4$ **49.** $y = x^2 - 4x + 3$

51. $y = -x^2 + 2x$ **53.** $y = \frac{1}{2}x^2 - \frac{1}{2}x$; Yes

55. $x^2 + y^2 - 4x = 0$ **57.** $x^2 + y^2 - 6x - 8y = 0$

59. $x^2 + y^2 - 2x - 4y - 20 = 0$

61. 20 gallons of spray X
18 gallons of spray Y
16 gallons of spray Z

63. Strings: 50
Winds: 20
Percussion: 8

65. (d)
$$\begin{aligned} x + y + z &= 200 \\ 8x + 15y + 100z &= 4995 \\ x - 4z &= 0 \end{aligned}$$

(e) Students: 140
Nonstudents: 25
Major contributors: 35

(f) No

67. Substitute $y = 3$ into the first equation to obtain $x + 2(3) = 2$ or $x = 2 - 6 = -4$.

69. Answers will vary.

Mid-Chapter Quiz *(page 411)*

1. $(10, 4)$

2.

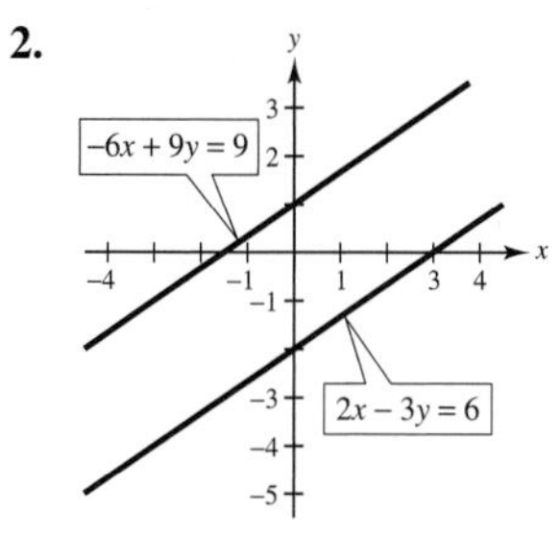

No solution

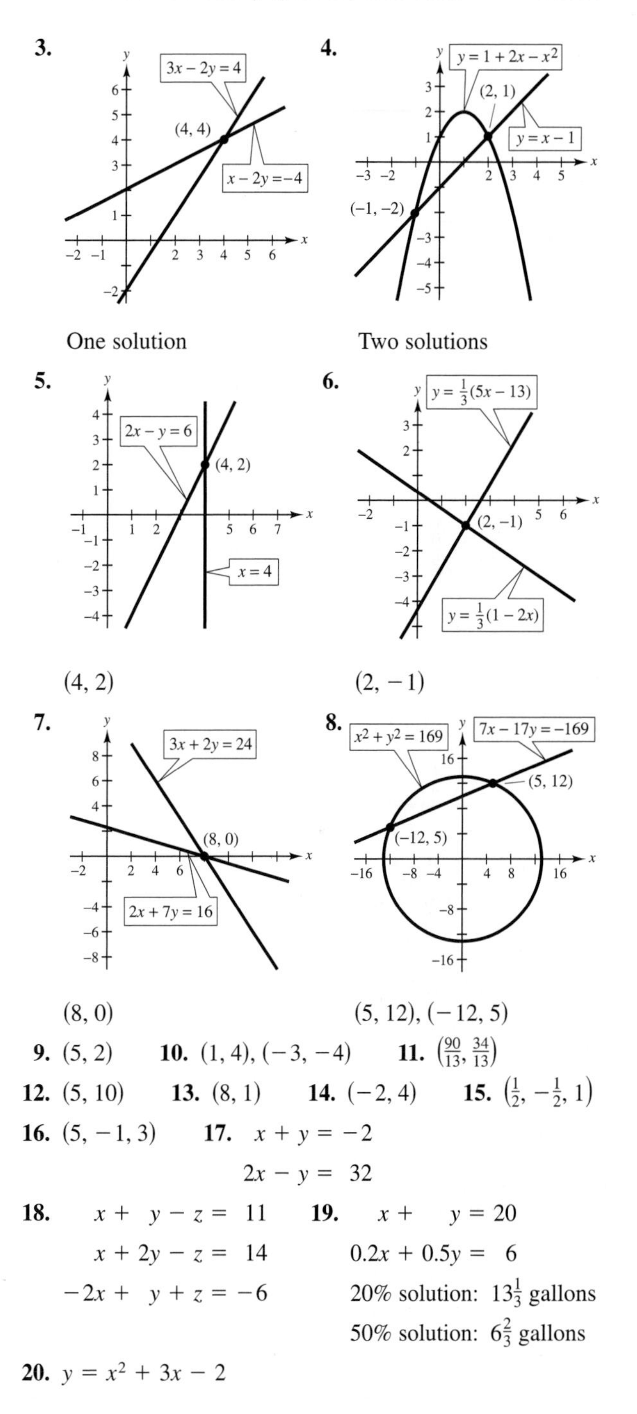

9. $(5, 2)$ **10.** $(1, 4), (-3, -4)$ **11.** $\left(\frac{90}{13}, \frac{34}{13}\right)$

12. $(5, 10)$ **13.** $(8, 1)$ **14.** $(-2, 4)$ **15.** $\left(\frac{1}{2}, -\frac{1}{2}, 1\right)$

16. $(5, -1, 3)$ **17.**
$$\begin{aligned} x + y &= -2 \\ 2x - y &= 32 \end{aligned}$$

18.
$$\begin{aligned} x + y - z &= 11 \\ x + 2y - z &= 14 \\ -2x + y + z &= -6 \end{aligned}$$

19.
$$\begin{aligned} x + y &= 20 \\ 0.2x + 0.5y &= 6 \end{aligned}$$
20% solution: $13\frac{1}{3}$ gallons
50% solution: $6\frac{2}{3}$ gallons

20. $y = x^2 + 3x - 2$

Section 7.4 *(page 420)*

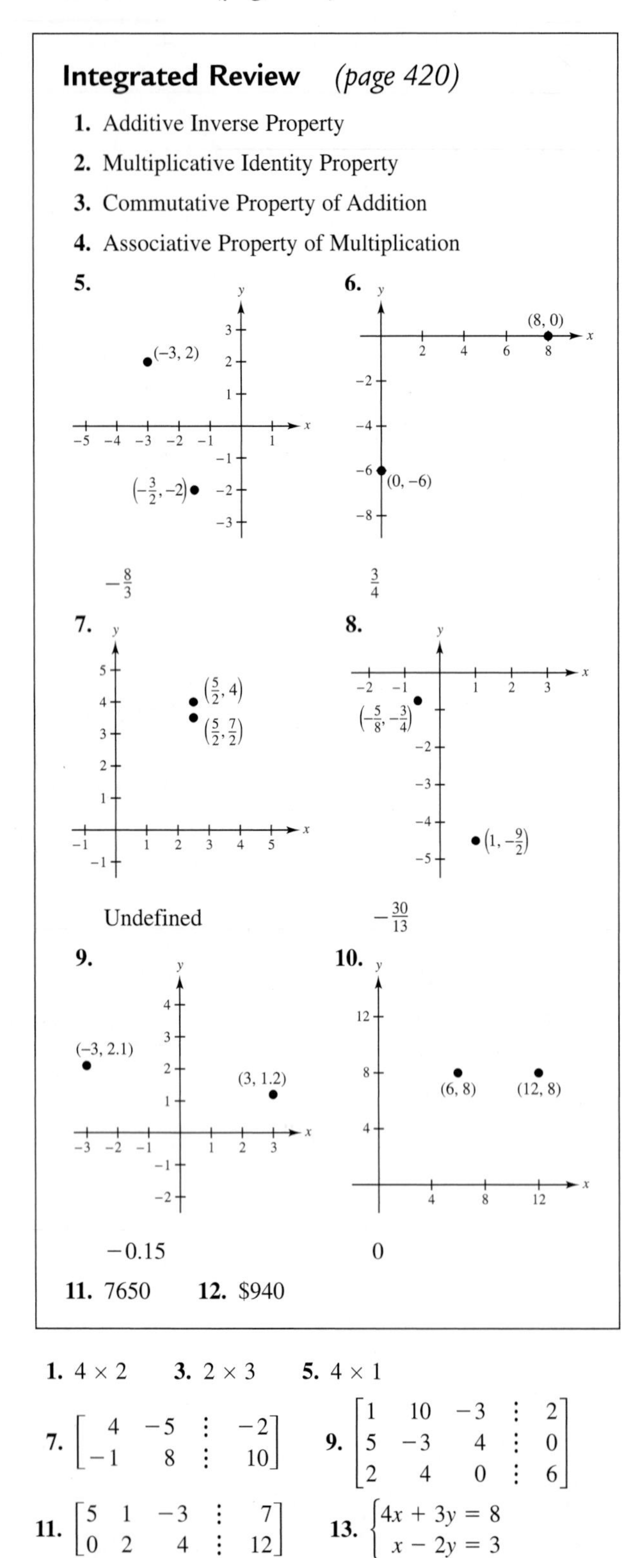

Integrated Review *(page 420)*

1. Additive Inverse Property
2. Multiplicative Identity Property
3. Commutative Property of Addition
4. Associative Property of Multiplication

5. $-\frac{8}{3}$ 6. $\frac{3}{4}$

7. Undefined 8. $-\frac{30}{13}$

9. -0.15 10. 0

11. 7650 12. $940

1. 4×2 **3.** 2×3 **5.** 4×1

7. $\begin{bmatrix} 4 & -5 & \vdots & -2 \\ -1 & 8 & \vdots & 10 \end{bmatrix}$ **9.** $\begin{bmatrix} 1 & 10 & -3 & \vdots & 2 \\ 5 & -3 & 4 & \vdots & 0 \\ 2 & 4 & 0 & \vdots & 6 \end{bmatrix}$

11. $\begin{bmatrix} 5 & 1 & -3 & \vdots & 7 \\ 0 & 2 & 4 & \vdots & 12 \end{bmatrix}$ **13.** $\begin{cases} 4x + 3y = 8 \\ x - 2y = 3 \end{cases}$

15. $\begin{cases} x \quad\quad + 2z = -10 \\ \quad 3y - z = 5 \\ 4x + 2y \quad = 3 \end{cases}$

17. $\begin{cases} 5x + 8y + 2z \quad = -1 \\ -2x + 15y + 5z + w = 9 \\ x + 6y - 7z \quad = -3 \end{cases}$

19. $\begin{bmatrix} 1 & 4 & 3 \\ 0 & 2 & -1 \end{bmatrix}$ **21.** $\begin{bmatrix} 1 & -2 & \frac{2}{3} \\ 2 & 8 & 15 \end{bmatrix}$

23. $\begin{bmatrix} 1 & 1 & 4 & -1 \\ 0 & 5 & -2 & 6 \\ 0 & 3 & 20 & 4 \end{bmatrix}$ **25.** $\begin{bmatrix} 1 & 2 & 3 \\ 0 & 1 & 2 \end{bmatrix}$

$\begin{bmatrix} 1 & 1 & 4 & -1 \\ 0 & 1 & -\frac{2}{5} & \frac{6}{5} \\ 0 & 3 & 20 & 4 \end{bmatrix}$

27. $\begin{bmatrix} 1 & 0 & -\frac{7}{5} \\ 0 & 1 & \frac{11}{10} \end{bmatrix}$ **29.** $\begin{bmatrix} 1 & 1 & 0 & 5 \\ 0 & 1 & 2 & 0 \\ 0 & 0 & 1 & -1 \end{bmatrix}$

31. $\begin{bmatrix} 1 & -1 & -1 & 1 \\ 0 & 1 & 6 & 3 \\ 0 & 0 & 1 & \frac{4}{5} \end{bmatrix}$ **33.** $\begin{bmatrix} 1 & 1 & -1 & 3 \\ 0 & 1 & -4 & 1 \\ 0 & 0 & 0 & 0 \end{bmatrix}$

35. $\begin{cases} x - 2y = 4 \\ \quad y = -3 \end{cases}$ $(-2, -3)$ **37.** $\begin{cases} x + 5y = 3 \\ \quad y = -2 \end{cases}$ $(13, -2)$

39. $\begin{cases} x - y + 2z = 4 \\ \quad y - z = 2 \\ \quad\quad z = -2 \end{cases}$ $(8, 0, -2)$ **41.** $\left(\frac{9}{5}, \frac{13}{5}\right)$ **43.** $(1, 1)$

45. No solution **47.** $(2, -3, 2)$

49. $(2a + 1, 3a + 2, a)$ **51.** $(1, 2, -1)$

53. $(1, -1, 2)$ **55.** $(34, -4, -4)$ **57.** No solution

59. $(-12a - 1, 4a + 1, a)$ **61.** $\left(2, 5, \frac{5}{2}\right)$

63. 8%: $800,000
9%: $500,000
12%: $200,000

65. Certificates of deposit: $250{,}000 - \frac{1}{2}s$
Municipal bonds: $125{,}000 + \frac{1}{2}s$
Blue-chip stocks: $125{,}000 - s$
Growth stocks: s
If $s = \$100{,}000$, then
Certificates of deposit: $200,000
Municipal bonds: $175,000
Blue-chip stocks: $25,000
Growth stocks: $100,000

67. $3.50: 15 pounds
$4.50: 10 pounds
$6.00: 25 pounds

69. 5, 8, 20

71. $y = x^2 + 2x + 4$ **73.** $y = -10.5x^2 + 25.5x - 7$

75. $x^2 + y^2 - 5x - 3y + 6 = 0$

77. (a) $y = -\frac{1}{250}x^2 + \frac{3}{5}x + 6$

(b)

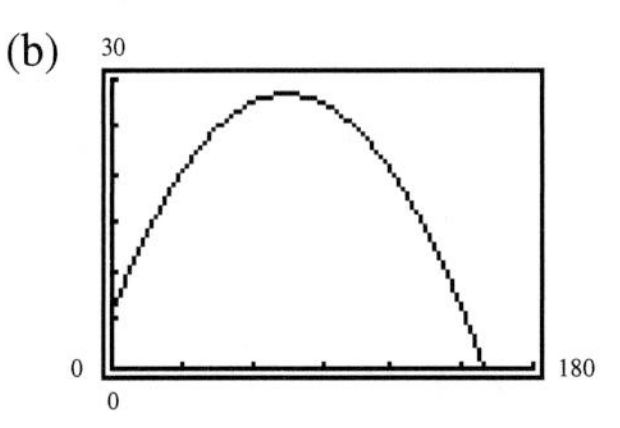

Maximum height: 28.5 ft
The ball struck the ground at approximately (159.4, 0).

79. $\frac{2x^2 - 9x}{(x-2)^3} = \frac{2}{x-2} - \frac{1}{(x-2)^2} - \frac{10}{(x-2)^3}$

81. (a) Interchange two rows.

(b) Multiply a row by a nonzero constant.

(c) Add a multiple of a row to another row.

83. The one matrix can be obtained from the other by using the elementary row operations.

85. There will be a row in the matrix with all zero entries except in the last column.

Section 7.5 *(page 433)*

Integrated Review *(page 433)*

1. pq **2.** $mq + np$ **3.** mn

4. No. $p = -1, q = -1$ is another solution.

5. $-4, 1$ **6.** $-2, 3$ **7.** $\frac{5}{2}$ **8.** ± 4

9. $-4, 2 \pm 2\sqrt{3}i$ **10.** $2, \pm\frac{2\sqrt{3}}{3}i$

11. $\frac{320}{r}$ **12.** $\frac{9}{2}x + 7$

1. 5 **3.** 27 **5.** 0 **7.** 6 **9.** -24

11. -0.16 **13.** -24 **15.** -2 **17.** -30 **19.** 3

21. 0 **23.** -75 **25.** -58 **27.** 102 **29.** -0.22

31. $x - 5y + 2$ **33.** 248 **35.** -32 **37.** -6.37

39. $(1, 2)$ **41.** $(2, -2)$ **43.** $\left(\frac{3}{4}, -\frac{1}{2}\right)$

45. Not possible, $D = 0$ **47.** $\left(\frac{2}{3}, \frac{1}{2}\right)$ **49.** $(-1, 3, 2)$

51. $\left(1, \frac{1}{2}, \frac{3}{2}\right)$ **53.** $(1, -2, 1)$ **55.** Not possible, $D = 0$

57. $\left(\frac{22}{27}, \frac{22}{9}\right)$ **59.** $\left(\frac{51}{16}, -\frac{7}{16}, -\frac{13}{16}\right)$ **61.** 1, 6 **63.** 16

65. 7 **67.** $\frac{31}{2}$ **69.** $\frac{33}{8}$ **71.** 16 **73.** $\frac{53}{2}$

75. 250 square miles **77.** Collinear **79.** Collinear

81. Not collinear **83.** $3x - 5y = 0$

85. $7x - 6y - 28 = 0$ **87.** $9x + 10y + 3 = 0$

89. $32x - 30y + 44 = 0$

91. $y = 2x^2 - 6x + 1$ **93.** $y = \frac{1}{2}x^2 - 2x$

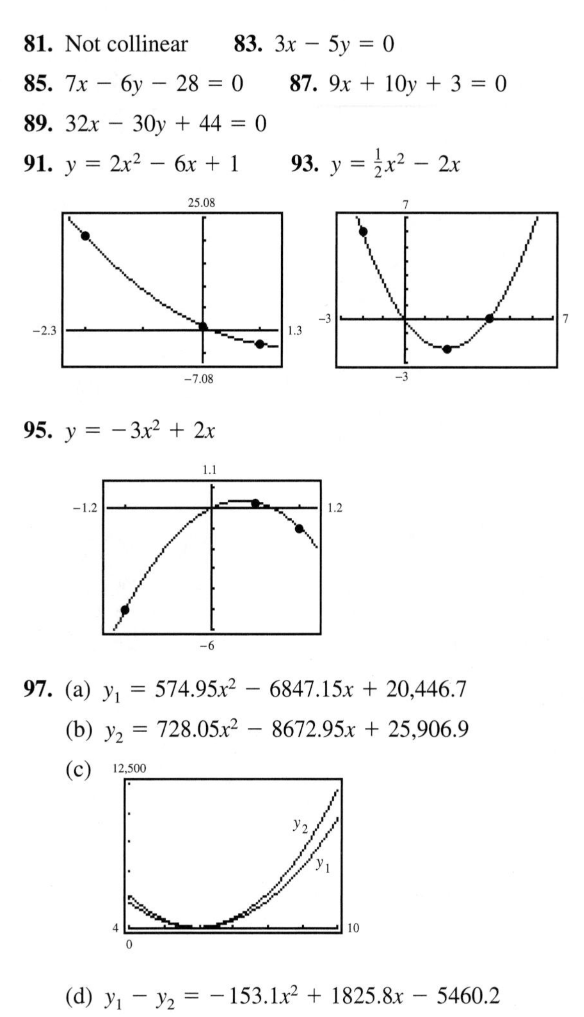

95. $y = -3x^2 + 2x$

97. (a) $y_1 = 574.95x^2 - 6847.15x + 20{,}446.7$

(b) $y_2 = 728.05x^2 - 8672.95x + 25{,}906.9$

(c)

(d) $y_1 - y_2 = -153.1x^2 + 1825.8x - 5460.2$

(e)

The trade deficit is increasing.

99. (a) $\left(\frac{4k-3}{2k-1}, \frac{4k-1}{2k-1}\right)$ (b) $\frac{1}{2}$

101. A determinant is a real number associated with a square matrix.

103. The minor of an entry of a square matrix is the determinant of the matrix that remains after deleting the row and column in which the entry occurs.

Review Exercises *(page 438)*

1. (a) Not a solution **3.** (a) Solution

(b) Solution (b) Not a solution

5. $(1, 1)$ **7.** No solution

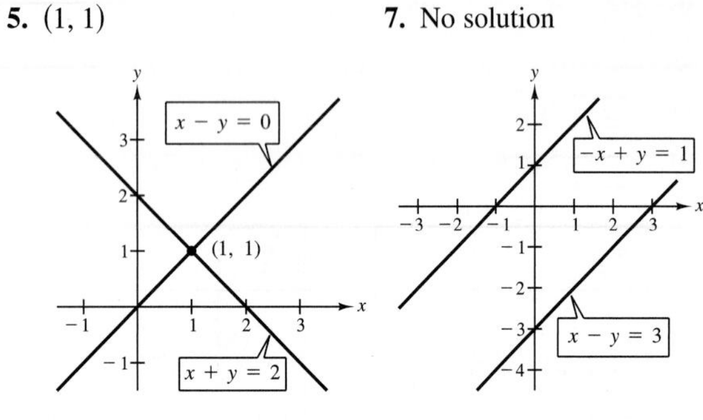

9. $(4, 8)$ **11.** Infinite number of solutions

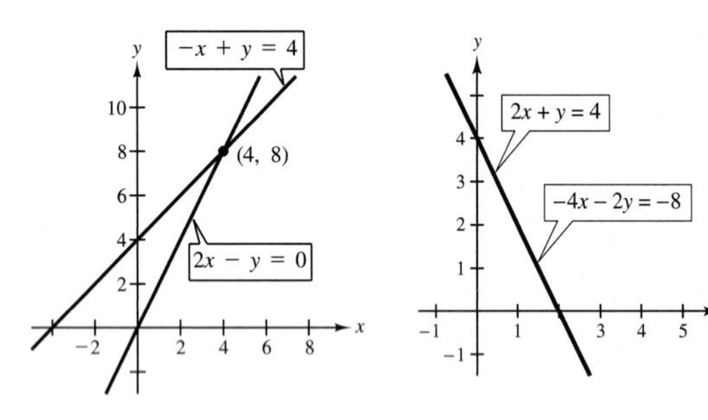

13. $(0, 1)$ **15.** $(3, 4)$

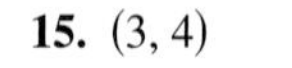

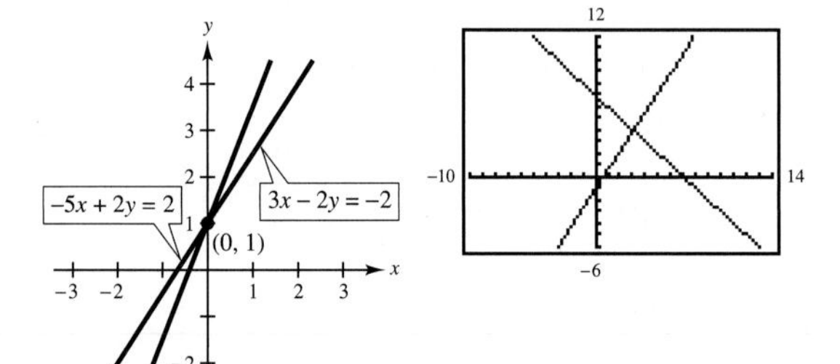

17. $(1, -3), \left(-\frac{1}{3}, -\frac{35}{9}\right)$ **19.** $(2, -1)$

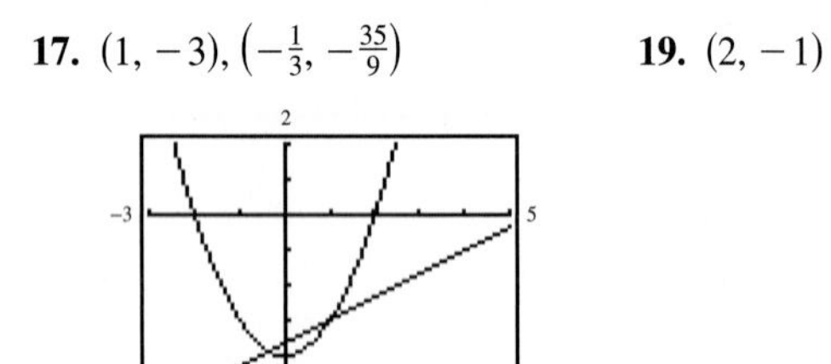

21. No solution **23.** $(-10, -5)$

25. $(-2, 20), (-1, 5)$ **27.** $(-1, 0), (0, -1)$ **29.** $(0, 0)$

31. $\left(\frac{5}{2}, 3\right)$ **33.** $(-0.5, 0.8)$ **35.** $(2, -3, 3)$

37. $(0, 1, -2)$ **39.** $(1, 0, -4)$ **41.** $(10, -12)$

43. $(0.6, 0.5)$ **45.** $\left(\frac{24}{5}, \frac{22}{5}, -\frac{8}{5}\right)$ **47.** $\left(\frac{1}{2}, -\frac{1}{3}, 1\right)$

49. 5 **51.** -51 **53.** 1 **55.** $(-3, 7)$

57. Not possible, $D = 0$ **59.** $(2, -3, 3)$

61. $3x + y = -2$
$6x + y = 0$ **63.** 16,667 units

65. 75% solution: 40 gallons
50% solution: 60 gallons

67. 96 meters × 144 meters **69.** \$9.95 tapes: 400
\$14.95 tapes: 250

71. 250 miles per hour
290 miles per hour **73.** 16, 20, 32

75. $y = 3x^2 + 11x - 20$ **77.** 16 **79.** 7

81. $x - 2y + 4 = 0$ **83.** $2x + 6y - 13 = 0$

Chapter Test *(page 441)*

1. $\left(1, \frac{1}{2}\right)$ **2.** $(2, 4)$ **3.** $(2, 6), (-1, 9)$ **4.** $(3, 2)$

5. $(-2, 2)$ **6.** $\left(\frac{1}{4}, \frac{1}{3}\right)$ **7.** $(2, 2z - 1, z)$ **8.** $(-1, 3, 3)$

9. $(2, 1, -2)$ **10.** $\left(4, \frac{1}{7}\right)$ **11.** $(5, 4)$ **12.** $(5, 1, -1)$

13. $\left(-\frac{11}{5}, \frac{56}{25}, \frac{32}{25}\right)$ **14.** Inconsistent
One solution
Infinitely many solutions

15. -62 **16.** $-\frac{24}{5}$

17. $x + 2y = -1$
$x + y = 2$

18. $x + y = 200$
$4x - y = 0$
40 miles, 160 miles

19. $y = 2x^2 - 3x + 4$ **20.** \$13,000 at 4.5%
\$9000 at 5%
\$3000 at 8% **21.** 12

Chapter 8

Section 8.1 *(page 450)*

Integrated Review *(page 450)*

1. $m = \dfrac{y_2 - y_1}{x_2 - x_1}$ **2.** (a) $m > 0$
(b) $m < 0$
(c) $m = 0$
(d) m is undefined. **3.** 10

4. 12 **5.** $-8x - 10$ **6.** $-2x^2 + 14x$ **7.** $\dfrac{25}{x^4}$

8. $\dfrac{4u^3}{3}$ **9.** 30% solution: $13\frac{1}{3}$ gallons
60% solution: $6\frac{2}{3}$ gallons **10.** \$500

1. $(-\infty, 8) \cup (8, \infty)$ **3.** $(-\infty, -4) \cup (-4, \infty)$

5. $(-\infty, \infty)$ **7.** $(-\infty, \infty)$ **9.** $(-\infty, \infty)$

11. $(-\infty, -3) \cup (-3, 0) \cup (0, \infty)$

13. $(-\infty, -4) \cup (-4, 4) \cup (4, \infty)$

15. $(-\infty, 0) \cup (0, 3) \cup (3, \infty)$

17. $(-\infty, 2) \cup (2, 3) \cup (3, \infty)$

19. $(-\infty, -1) \cup \left(-1, \frac{5}{3}\right) \cup \left(\frac{5}{3}, \infty\right)$

21. (a) 1

(b) -8

(c) Undefined (division by 0)

(d) 0

23. (a) 0

(b) 0

(c) Undefined (division by 0)

(d) Undefined (division by 0)

25. (a) $\frac{25}{22}$

(b) 0

(c) Undefined (division by 0)

(d) Undefined (division by 0)

27. $(0, \infty)$

29. $\{1, 2, 3, 4, \ldots\}$ **31.** $[0, 100)$ **33.** $x + 3$

35. $(3)(x + 16)^2$ **37.** $(x)(x - 2)$ **39.** $x + 2$

41. $\frac{x}{5}$ **43.** $6y,\ y \neq 0$ **45.** $\frac{6x}{5y^3},\ x \neq 0$ **47.** $\frac{x - 3}{4x}$

49. $x,\ x \neq 8, x \neq 0$ **51.** $\frac{1}{2},\ x \neq \frac{3}{2}$ **53.** $-\frac{1}{3},\ x \neq 5$

55. $\frac{1}{a + 3}$ **57.** $\frac{x}{x - 7}$ **59.** $\frac{y(y + 2)}{y + 6},\ y \neq 2$

61. $\frac{x(x + 2)}{x - 3},\ x \neq 2$ **63.** $-\frac{3x + 5}{x + 3},\ x \neq 4$

65. $\frac{x + 8}{x - 3},\ x \neq -\frac{3}{2}$ **67.** $\frac{3x - 1}{5x - 4},\ x \neq -\frac{4}{5}$

69. $\frac{3y^2}{y^2 + 1},\ x \neq 0$ **71.** $\frac{y - 8x}{15},\ y \neq -8x$

73. $\frac{5 + 3xy}{y^2},\ x \neq 0$ **75.** $\frac{u - 2v}{u - v},\ u \neq -2v$

77. $\frac{3(m - 2n)}{m + 2n}$

79. Evaluating both sides when $x = 10$ yields $\frac{3}{2} \neq 9$.

81. Evaluating both sides when $x = 0$ yields $1 \neq \frac{3}{4}$.

83.

x	-2	-1	0	1	2	3	4
$\frac{x^2 - x - 2}{x - 2}$	-1	0	1	2	Undefined	4	5
$x + 1$	-1	0	1	2	3	4	5

$$\frac{x^2 - x - 2}{x - 2} = \frac{(x - 2)(x + 1)}{x - 2} = x + 1,\ x \neq 2$$

85. $\frac{x}{x + 3},\ x > 0$

87. (a) $C = 2500 + 9.25x$

(b) $\overline{C} = \frac{2500 + 9.25x}{x}$

(c) $\{1, 2, 3, 4, \ldots\}$

(d) \$34.25

89. (a) Van: $45(t + 3)$; Car: $60t$

(b) $d = |15(9 - t)|$

(c) $\frac{4t}{3(t + 3)}$

91. π **93.** $\frac{1000(10{,}730 + 1509t)}{3426 + 65t}$

95. Let u and v be polynomials. The algebraic expression u/v is a rational expression.

97. The rational expression is in simplified form if the numerator and denominator have no factors in common (other than ± 1).

99. You can cancel only common factors.

Section 8.2 *(page 459)*

Integrated Review *(page 459)*

1. $u^2 - v^2 = (u + v)(u - v)$

$9t^2 - 4 = (3t + 2)(3t - 2)$

2. $u^2 - 2uv + v^2 = (u - v)^2$

$4x^2 - 12x + 9 = (2x - 3)^2$

3. $u^3 + v^3 = (u + v)(u^2 - uv + v^2)$

$8x^3 + 64 = (2x + 4)(4x^2 - 8x + 16)$

4. $(3x - 2)(x + 5)$. Multiply **5.** $5x(1 - 4x)$

6. $(2 + x)(14 - x)$ **7.** $(3x - 5)(5x + 3)$

8. $(4t + 1)^2$ **9.** $(y - 4)(y^2 + 4y + 16)$

10. $(2x + 1)(4x^2 - 2x + 1)$

11. **12.**

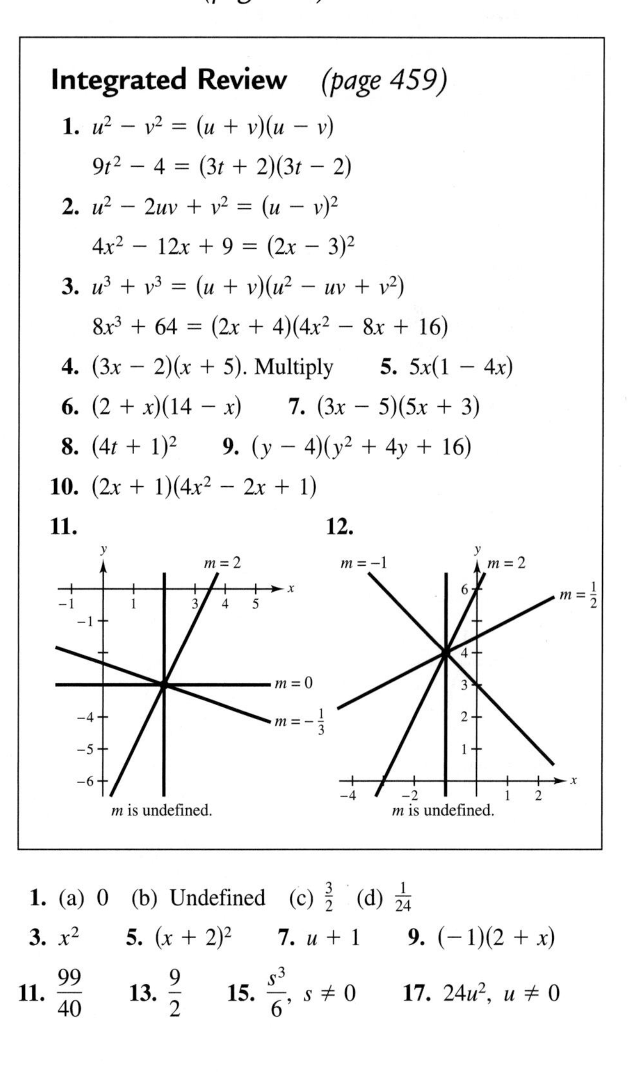

1. (a) 0 (b) Undefined (c) $\frac{3}{2}$ (d) $\frac{1}{24}$

3. x^2 **5.** $(x + 2)^2$ **7.** $u + 1$ **9.** $(-1)(2 + x)$

11. $\frac{99}{40}$ **13.** $\frac{9}{2}$ **15.** $\frac{s^3}{6},\ s \neq 0$ **17.** $24u^2,\ u \neq 0$

19. $24,\ x \neq -\frac{3}{4}$ **21.** $\frac{2uv(u+v)}{3(3u+v)},\ u \neq 0$

23. $-1,\ r \neq 12$ **25.** $-\frac{x+8}{x^2},\ x \neq \frac{3}{2}$

27. $4(r+2),\ r \neq 3, r \neq 2$ **29.** $2t+5,\ t \neq 3, t \neq -2$

31. $\frac{xy(x+2y)}{x-2y}$ **33.** $\frac{(x-y)^2}{x+y},\ x \neq -3y$

35. $\frac{(x-1)(2x+1)}{(3x-2)(x+2)},\ x \neq \pm 5, x \neq -1$

37. $\frac{x^2(x^2-9)(2x+5)(3x-1)}{2(2x+1)(2x+3)(3-2x)},\ x \neq 0, x \neq \frac{1}{2}$

39. $\frac{(x+3)^2}{x},\ x \neq 3, x \neq 4$ **41.** $-\frac{8}{27}$

43. $\frac{4x}{3},\ x \neq 0$ **45.** $\frac{3y^2}{2ux^2},\ v \neq 0$ **47.** $\frac{3}{2(a+b)}$

49. $x^4y(x+2y),\ x \neq 0, y \neq 0, x \neq -2y$

51. $\frac{3x}{10},\ x \neq 0$ **53.** $-\frac{5x(x+1)}{2},\ x \neq 0, x \neq 5, x \neq -1$

55. $\frac{(x+3)(4x+1)}{(3x-1)(x-1)},\ x \neq -3, x \neq -\frac{1}{4}$

57. $x+2,\ x \neq \pm 2, x \neq -3$

59. $-\frac{(x+2)(x^2-3x-10)}{(x+3)(x^2-4x+4)},\ x \neq \pm 2, x \neq 7$

61. $\frac{x+4}{3},\ x \neq -2, x \neq 0$ **63.** $\frac{1}{4},\ x \neq -1, x \neq 0, y \neq 0$

65. $\frac{(x+1)(2x-5)}{x},\ x \neq -1, x \neq -5, x \neq -\frac{2}{3}$

67. $\frac{x^4}{(x^n+1)^2},\ x^n \neq -3, x^n \neq 3, x \neq 0$

69.

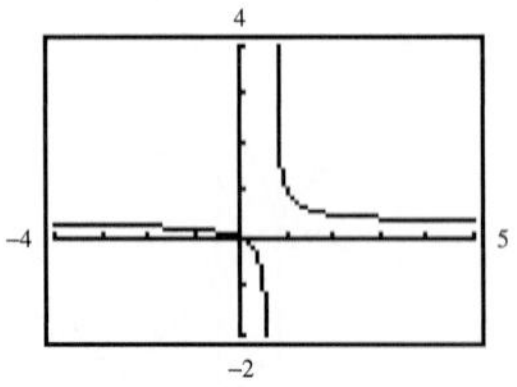

71.

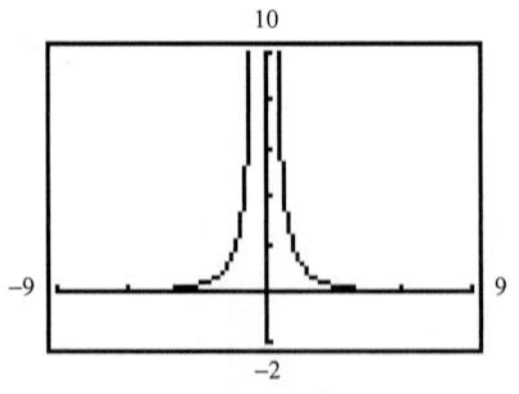

73. $\frac{2w^2+3w}{6}$ **75.** $\frac{x}{4(2x+1)}$ **77.** $\frac{x}{2(2x+1)}$

79. (a) $\frac{1}{20}$ minute (b) $\frac{x}{20}$ minutes (c) $\frac{7}{4}$ minutes

81. (a)

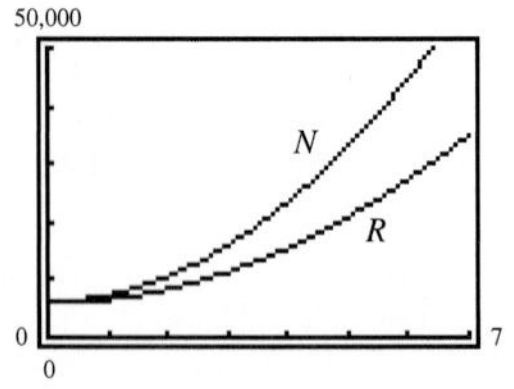

(b) $\frac{6{,}115{,}200 + 590{,}700t^2}{12(6357 + 1070t^2)}$

(c)

Year, t	0	2	4	6
Monthly bill	\$80.16	\$66.42	\$55.25	\$50.84

(d) The number of subscribers was increasing at a faster rate than the revenue.

83. Multiply the rational expression by the reciprocal of the polynomial.

85. Invert the divisor, not the dividend.

Section 8.3 *(page 469)*

Integrated Review *(page 469)*

1. (a) $y = \frac{3}{5}x + \frac{4}{5}$
(b) $y - 2 = \frac{3}{5}(x - 2)$

2. If $m > 0$, the line rises from left to right.
If $m < 0$, the line falls from left to right.

3. $42x^2 - 60x$ **4.** $6 + y - 2y^2$ **5.** $121 - x^2$

6. $16 - 25z^2$ **7.** $x^2 + 2x + 1$ **8.** $2t$

9. $x^3 - 8$ **10.** $2t^3 - 5t^2 - 12t$

11. $P = 12x + 6$, $A = 5x^2 + 9x$ **12.** $P = 12x$, $A = 6x^2$

1. $\frac{3}{2}$ **3.** $-\frac{x}{4}$ **5.** $-\frac{3}{a}$ **7.** $-\frac{2}{9}$ **9.** $\frac{2z^2-2}{3}$

11. $\frac{x+6}{3}$ **13.** $-\frac{4}{3}$ **15.** $1,\ y \neq 6$

17. $\frac{1}{x-3},\ x \neq 0$ **19.** $20x^3$ **21.** $36y^3$

23. $15x^2(x+5)$ **25.** $126z^2(z+1)^4$

27. $56t(t+2)(t-2)$ **29.** $6x(x+2)(x-2)$ **31.** x^2

33. $(u+1)$ **35.** $-(x+2)$

37. $\frac{2n^2(n+8)}{6n^2(n-4)}, \frac{10(n-4)}{6n^2(n-4)}$

39. $\frac{2(x+3)}{x^2(x+3)(x-3)}, \frac{5x(x-3)}{x^2(x+3)(x-3)}$

41. $\frac{3v^2}{6v^2(v+1)}, \frac{8(v+1)}{6v^2(v+1)}$

43. $\frac{(x-8)(x-5)}{(x+5)(x-5)^2}, \frac{9x(x+5)}{(x+5)(x-5)^2}$ **45.** $\frac{25-12x}{20x}$

47. $\frac{7(a+2)}{a^2}$ **49.** $0,\ x \neq 4$ **51.** $\frac{3(x+2)}{x-8}$

53. $\dfrac{5(5x + 22)}{x + 4}$ **55.** $1,\ x \neq \dfrac{2}{3}$ **57.** $\dfrac{1}{2x(x - 3)}$

59. $\dfrac{x^2 - 7x - 15}{(x + 3)(x - 2)}$ **61.** $\dfrac{x - 2}{x(x + 1)}$

63. $\dfrac{5(x + 1)}{(x + 5)(x - 5)}$ **65.** $\dfrac{4}{x^2(x^2 + 1)}$

67. $\dfrac{x^2 + x + 9}{(x - 2)(x - 3)(x + 3)}$ **69.** $\dfrac{4x}{(x - 4)^2}$

71. $\dfrac{y - x}{xy},\ x \neq -y$ **73.** $\dfrac{2(4x^2 + 5x - 3)}{x^2(x + 3)}$

75. $-\dfrac{u^2 - uv - 5u + 2v}{(u - v)^2}$ **77.** $\dfrac{x}{x - 1},\ x \neq -6$

79.

81. $\dfrac{x}{2(3x + 1)},\ x \neq 0$

83. $\dfrac{4 + 3x}{4 - 3x},\ x \neq 0$ **85.** $-4x - 1,\ x \neq 0, x \neq \dfrac{1}{4}$

87. $\dfrac{3}{4},\ x \neq 0, x \neq 3$ **89.** $\dfrac{5(x + 3)}{2x(5x - 2)}$

91. $y - x,\ x \neq 0, y \neq 0, x \neq -y$ **93.** $\dfrac{-(y - 1)(y - 3)}{y(4y - 1)}$

95. $\dfrac{x(x + 6)}{3x^3 + 10x - 30},\ x \neq 0, x \neq 3$ **97.** $-\dfrac{1}{2(h + 2)}$

99.

x	-3	-2	-1	0	1	2	3
$\dfrac{\left(1 - \frac{1}{x}\right)}{\left(1 - \frac{1}{x^2}\right)}$	$\frac{3}{2}$	2	Undef.	Undef.	Undef.	$\frac{2}{3}$	$\frac{3}{4}$
$\dfrac{x}{x + 1}$	$\frac{3}{2}$	2	Undef.	0	$\frac{1}{2}$	$\frac{2}{3}$	$\frac{3}{4}$

Domain of the complex fraction: $(-\infty, -1) \cup (-1, 0) \cup (0, 1) \cup (1, \infty)$. Domain of the simplified fraction: $(-\infty, -1) \cup (-1, \infty)$. The two expressions are equivalent except at $x = 0$ and $x = 1$.

101. $\dfrac{5t}{12}$ **103.** $\dfrac{5x}{24}$ **105.** $\dfrac{11x}{45}, \dfrac{13x}{45}$ **107.** $\dfrac{R_1R_2}{R_1 + R_2}$

109. (a) Upstream: $\dfrac{10}{5 - x}$; Downstream: $\dfrac{10}{5 + x}$

(b) $f(x) = \dfrac{10}{5 - x} + \dfrac{10}{5 + x}$ (c) $f(x) = \dfrac{100}{(5 + x)(5 - x)}$

111. Rewrite each fraction in terms of the lowest common denominator, combine the numerators, and place the result over the lowest common denominator.

113. When the numerators are subtracted, the result should be $(x - 1) - (4x - 11) = x - 1 - 4x + 11$.

Mid-Chapter Quiz *(page 473)*

1. $(-\infty, 0) \cup (0, 4) \cup (4, \infty)$ **2.** (a) 0 (b) $\frac{9}{2}$ (c) Undefined (d) $\frac{8}{9}$

3. $\dfrac{3}{2}y$ **4.** $\dfrac{2u^2}{9v}$ **5.** $-\dfrac{2x + 1}{x}$ **6.** $\dfrac{z + 3}{2z - 1}$

7. $\dfrac{7 + 3ab}{a}$ **8.** $\dfrac{n^2}{m + n}$ **9.** $\dfrac{t}{2}$ **10.** $\dfrac{5x}{x - 2}$

11. $\dfrac{8x}{3(x - 1)(x^2 + 2x - 3)}$ **12.** $\dfrac{4(u - v)^2}{5uv}$

13. $-\dfrac{3t}{2}$ **14.** $\dfrac{2(x + 1)}{3x}$ **15.** $\dfrac{x(1 - 3x)}{4(x + 5)}$

16. $\dfrac{4x^4 + x^3 - 18x^2 + 8}{x^2(x^2 - 4)}$ **17.** $\dfrac{5(2 - x)}{4x - 15}$

18. $\dfrac{2(x^2 + 9)}{x + 3}$ **19.** (a) $\dfrac{6000 + 10.50x}{x}$ (b) \$22.50 **20.** $\dfrac{8(x + 2)}{(x + 4)^2}$

Section 8.4 *(page 481)*

Integrated Review *(page 481)*

1. Quadrants II or III **2.** Quadrants I or II

3. x-axis **4.** $(9, -6)$ **5.** $x < \frac{3}{2}$ **6.** $x < 5$

7. $1 < x < 5$ **8.** $x < 2$ or $x > 8$

9. $x \leq -8$ or $x \geq 16$ **10.** $-24 \leq x \leq 36$

11. 15 minutes, 2 miles **12.** 7.5%: \$15,000; 9%: \$9000

1. (a) Not a solution (b) Not a solution (c) Not a solution (d) Solution

3. (a) Not a solution (b) Solution (c) Solution (d) Not a solution

5. 10 **7.** 8 **9.** $-\frac{9}{32}$ **11.** 10 **13.** $-\frac{2}{9}$

15. $\frac{7}{4}$ **17.** $\frac{43}{8}$ **19.** 61 **21.** $\frac{18}{5}$ **23.** $-\frac{26}{5}$

25. 3 **27.** 3 **29.** $-\frac{11}{5}$ **31.** $\frac{4}{3}$ **33.** ± 6

35. ± 4 **37.** $-9, 8$ **39.** 3, 13 **41.** No solution

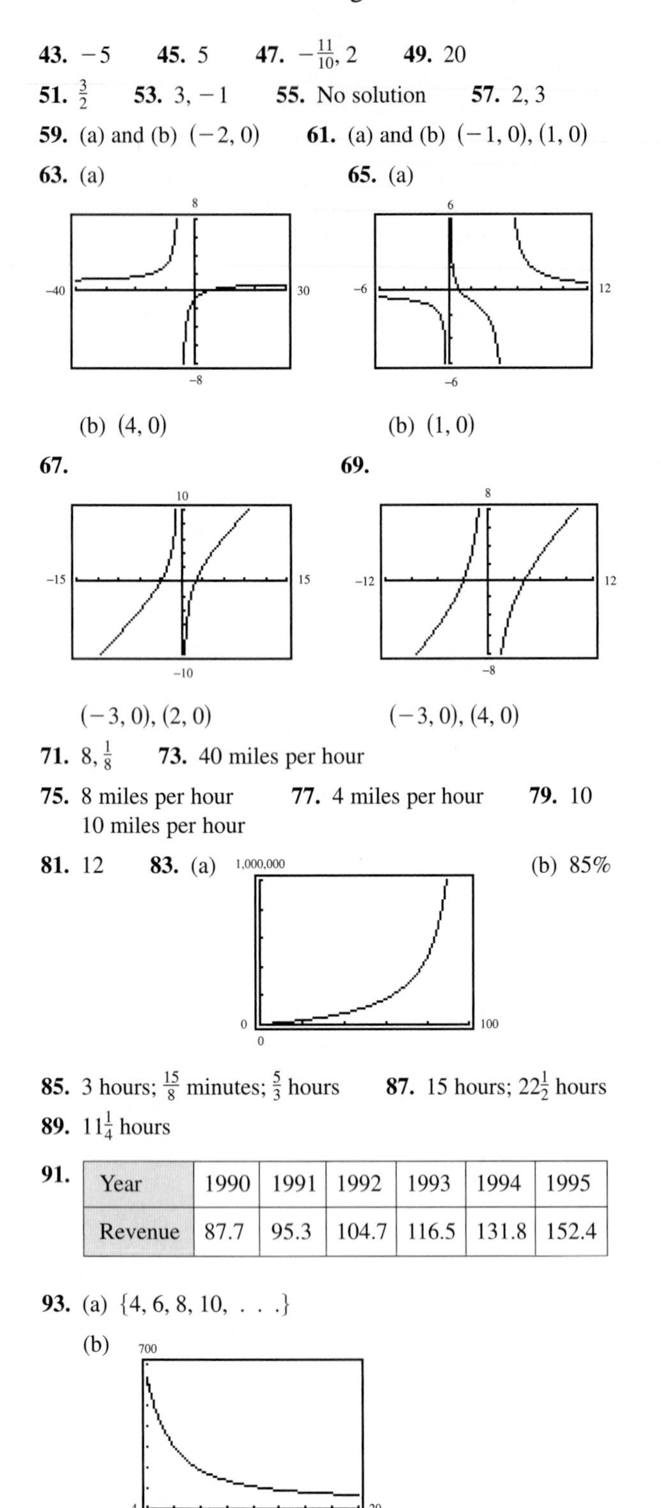

43. -5 **45.** 5 **47.** $-\frac{11}{10}, 2$ **49.** 20

51. $\frac{3}{2}$ **53.** $3, -1$ **55.** No solution **57.** 2, 3

59. (a) and (b) $(-2, 0)$ **61.** (a) and (b) $(-1, 0), (1, 0)$

63. (a)

(b) $(4, 0)$

65. (a)

(b) $(1, 0)$

67.

$(-3, 0), (2, 0)$

69.

$(-3, 0), (4, 0)$

71. $8, \frac{1}{8}$ **73.** 40 miles per hour

75. 8 miles per hour
10 miles per hour **77.** 4 miles per hour **79.** 10

81. 12 **83.** (a)

(b) 85%

85. 3 hours; $\frac{15}{8}$ minutes; $\frac{5}{3}$ hours **87.** 15 hours; $22\frac{1}{2}$ hours

89. $11\frac{1}{4}$ hours

91.

Year	1990	1991	1992	1993	1994	1995
Revenue	87.7	95.3	104.7	116.5	131.8	152.4

93. (a) $\{4, 6, 8, 10, \ldots\}$

(b)

(c) 10d

95. (d) 3 miles per hour, obtained by solving $\dfrac{10}{5-x} + \dfrac{10}{5+x} = 6.25$

(e) Yes. 11.1 hours

97. Multiply both sides of the equation by the lowest common denominator, solve the resulting equation, and check the result. It is important to check the result for any errors or extraneous solutions.

99. (a) Simplify each side by removing symbols of grouping, combining like terms, and reducing fractions on one or both sides.

(b) Add (or subtract) the same quantity to (from) both sides of the equation.

(c) Multiply (or divide) both sides of the equation by the same nonzero real number.

(d) Interchange the two sides of the equation.

101. When the equation involves only two fractions, one on each side of the equation, the equation can be solved by cross-multiplication.

Section 8.5 *(page 491)*

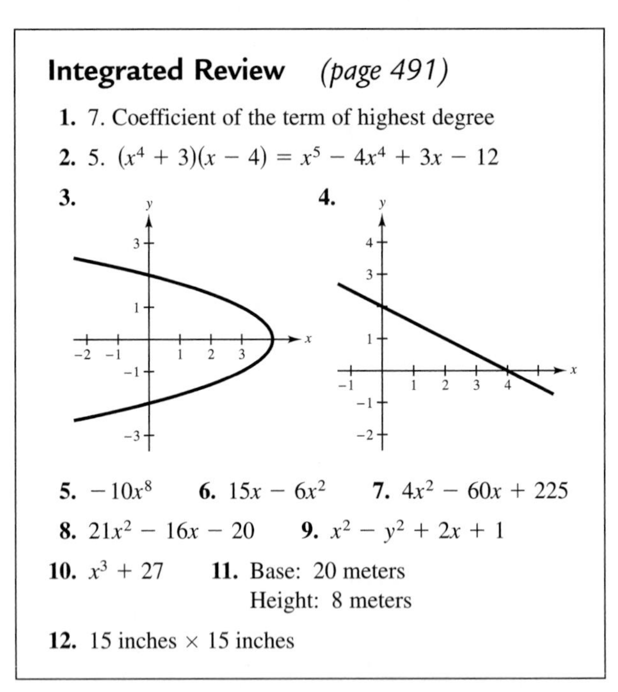

Integrated Review *(page 491)*

1. 7. Coefficient of the term of highest degree

2. 5. $(x^4 + 3)(x - 4) = x^5 - 4x^4 + 3x - 12$

3.

4.

5. $-10x^8$ **6.** $15x - 6x^2$ **7.** $4x^2 - 60x + 225$

8. $21x^2 - 16x - 20$ **9.** $x^2 - y^2 + 2x + 1$

10. $x^3 + 27$ **11.** Base: 20 meters
Height: 8 meters

12. 15 inches × 15 inches

1. (a)

x	0	0.5	0.9	0.99	0.999
y	-4	-8	-40	-400	-4000

x	2	1.5	1.1	1.01	1.001
y	4	8	40	400	4000

x	2	5	10	100	1000
y	4	1	0.4444	0.0404	0.0040

(b)

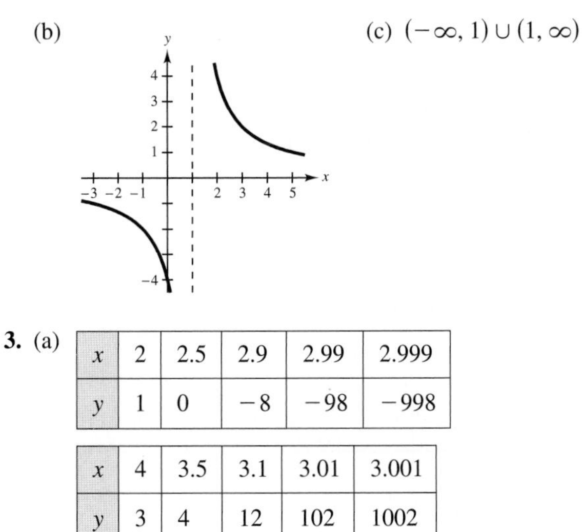

(c) $(-\infty, 1) \cup (1, \infty)$

3. (a)

x	2	2.5	2.9	2.99	2.999
y	1	0	-8	-98	-998

x	4	3.5	3.1	3.01	3.001
y	3	4	12	102	1002

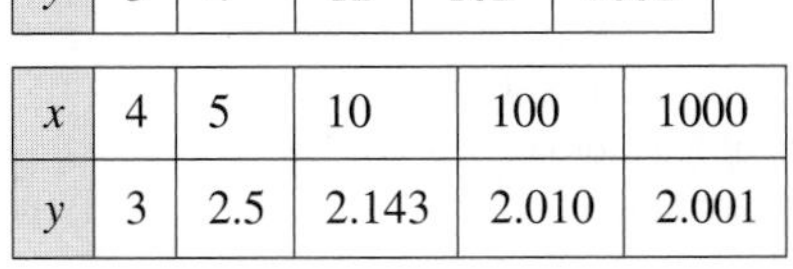

x	4	5	10	100	1000
y	3	2.5	2.143	2.010	2.001

(b)

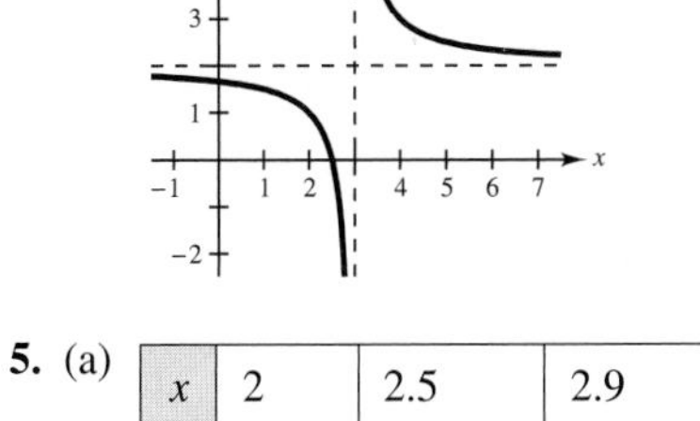

(c) $(-\infty, 3) \cup (3, \infty)$

5. (a)

x	2	2.5	2.9	2.99	2.999
y	-1.2	-2.727	-14.75	-149.7	-1500

x	4	3.5	3.1	3.01	3.001
y	1.714	3.231	15.246	150.25	1500.2

x	4	5	10	100	1000
y	1.714	0.938	0.330	0.030	0.003

(b)

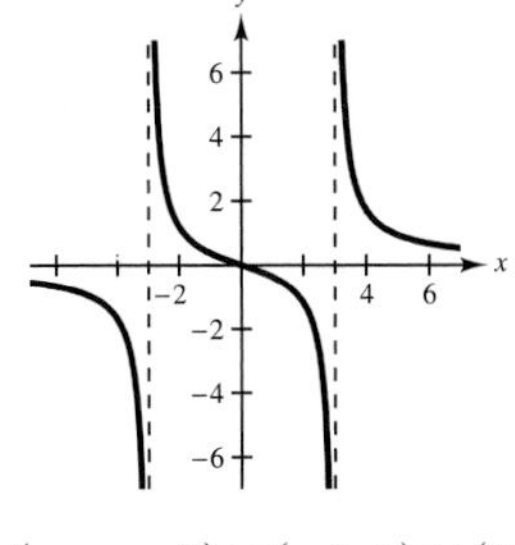

(c) $(-\infty, -3) \cup (-3, 3) \cup (3, \infty)$

7. Domain: $(-\infty, 0) \cup (0, \infty)$
Horizontal asymptote: $y = 0$
Vertical asymptote: $x = 0$

9. Domain: $(-\infty, -8) \cup (-8, \infty)$
Horizontal asymptote: $y = 1$
Vertical asymptote: $x = -8$

11. Domain: $(-\infty, 3) \cup (3, \infty)$
Horizontal asymptote: $y = \frac{2}{3}$
Vertical asymptote: $t = 3$

13. Domain: $\left(-\infty, \frac{1}{3}\right) \cup \left(\frac{1}{3}, \infty\right)$
Horizontal asymptote: $y = \frac{5}{3}$
Vertical asymptote: $x = \frac{1}{3}$

15. Domain: $(-\infty, 0) \cup (0, 1) \cup (1, \infty)$
Horizontal asymptote: $y = 0$
Vertical asymptotes: $t = 0, t = 1$

17. Domain: $(-\infty, \infty)$
Horizontal asymptote: $y = 2$
Vertical asymptote: None

19. Domain: $(-\infty, -1) \cup (-1, 1) \cup (1, \infty)$
Horizontal asymptote: $y = 1$
Vertical asymptotes: $x = -1, x = 1$

21. Domain: $(-\infty, 0) \cup (0, \infty)$
Horizontal asymptote: $y = 1$
Vertical asymptote: $z = 0$

23. Domain: $(-\infty, 0) \cup (0, \infty)$
Horizontal asymptote: None
Vertical asymptote: $x = 0$

25. d **27.** b **29.** d **31.** a

33. **35.**

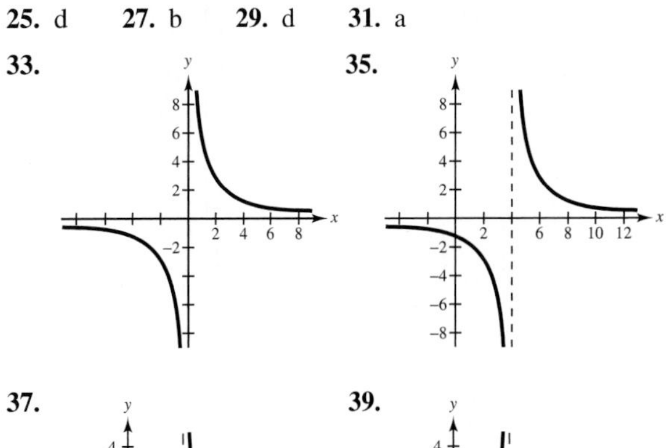

37. **39.**

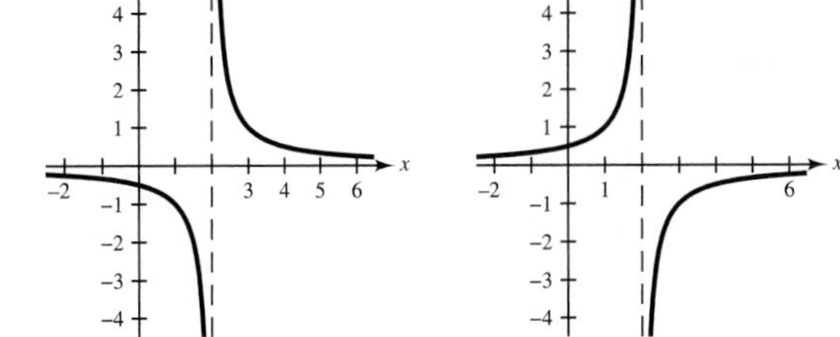

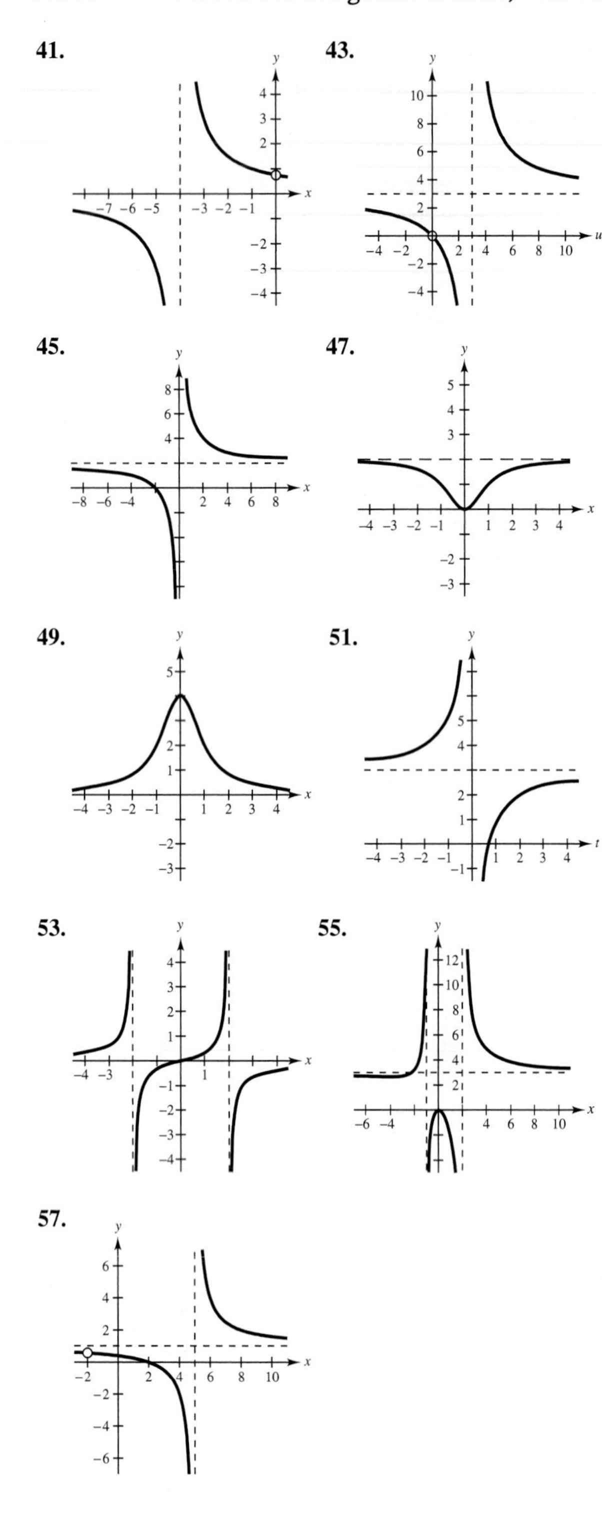

59.

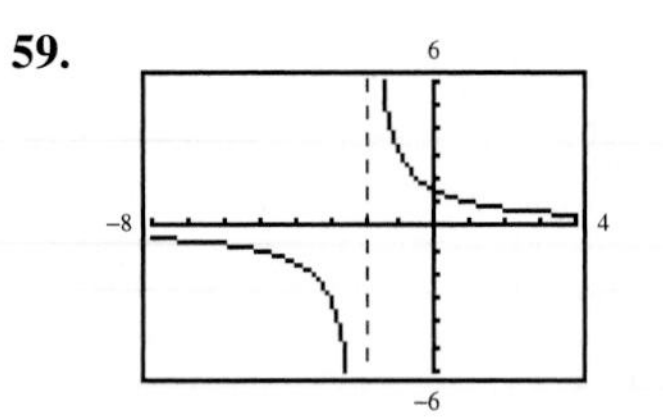

Domain: $(-\infty, -2) \cup (-2, \infty)$
Horizontal asymptote: $y = 0$
Vertical asymptote: $x = -2$

61.

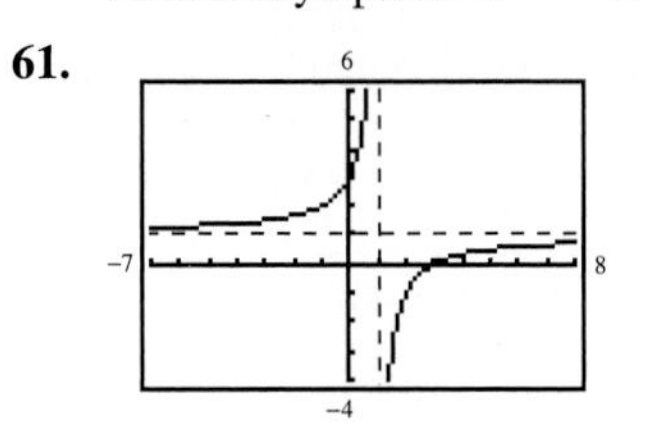

Domain: $(-\infty, 1) \cup (1, \infty)$
Horizontal asymptote: $y = 1$
Vertical asymptote: $x = 1$

63.

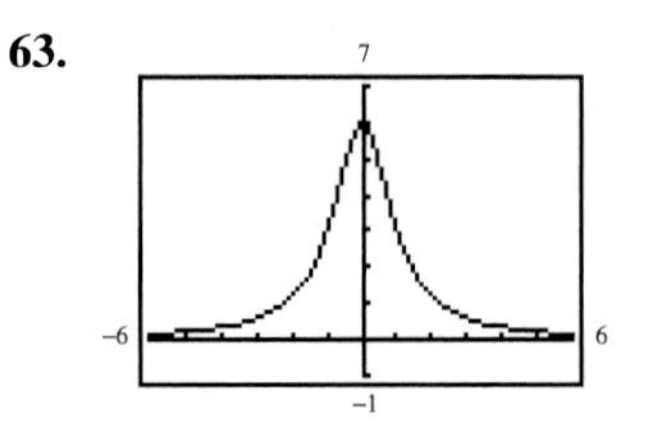

Domain: $(-\infty, \infty)$
Horizontal asymptote: $y = 0$
Vertical asymptote: None

65.

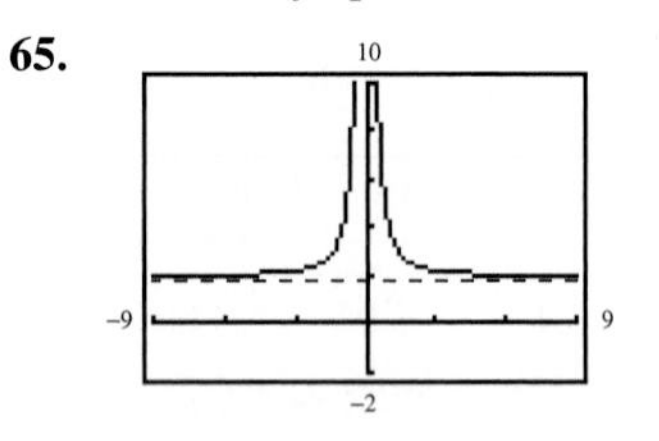

Domain: $(-\infty, 0) \cup (0, \infty)$
Horizontal asymptote: $y = 2$
Vertical asymptote: $x = 0$

67.

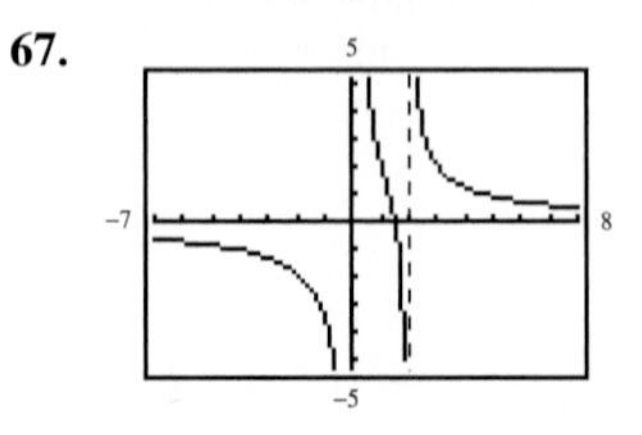

Domain: $(-\infty, 0) \cup (0, 2) \cup (2, \infty)$
Horizontal asymptote: $y = 0$
Vertical asymptotes: $x = 0, x = 2$

69.

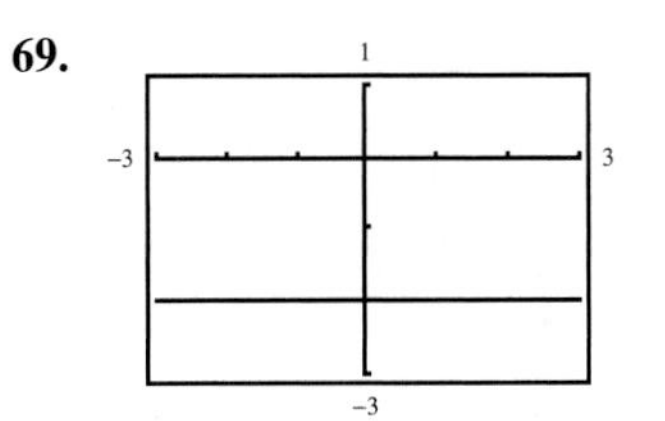

The fraction is not reduced to lowest terms.

71. (a) $\overline{C} = \dfrac{2500 + 0.50x}{x}$ (b) \$3, \$0.75

(c)

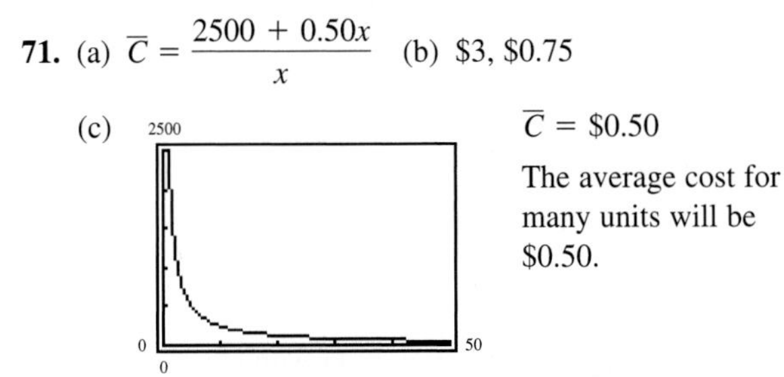

$\overline{C} = \$0.50$

The average cost for many units will be \$0.50.

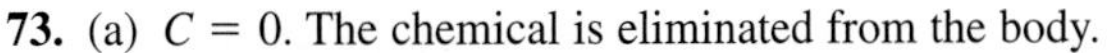

73. (a) $C = 0$. The chemical is eliminated from the body.

(b)

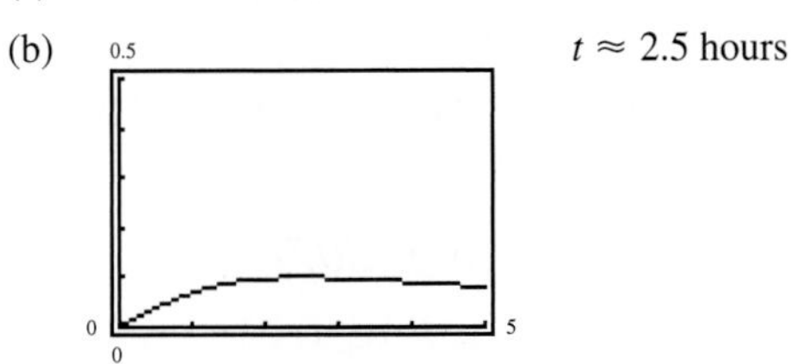

$t \approx 2.5$ hours

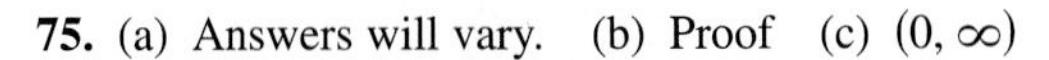

75. (a) Answers will vary. (b) Proof (c) $(0, \infty)$

(d)

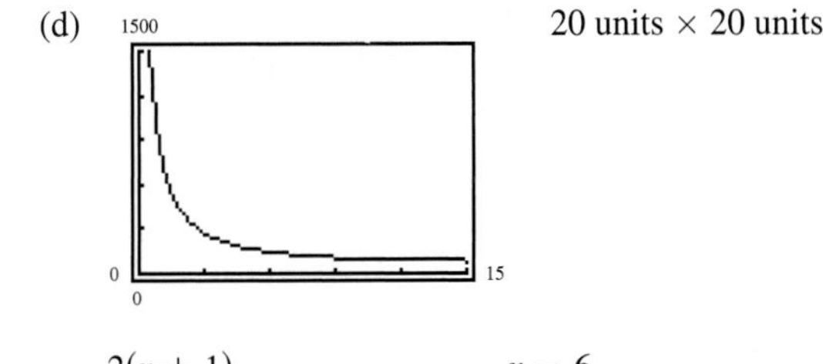

20 units × 20 units

77. $y = \dfrac{2(x + 1)}{x - 3}$ **79.** $y = \dfrac{x - 6}{(x - 4)(x + 2)}$

81. (f)

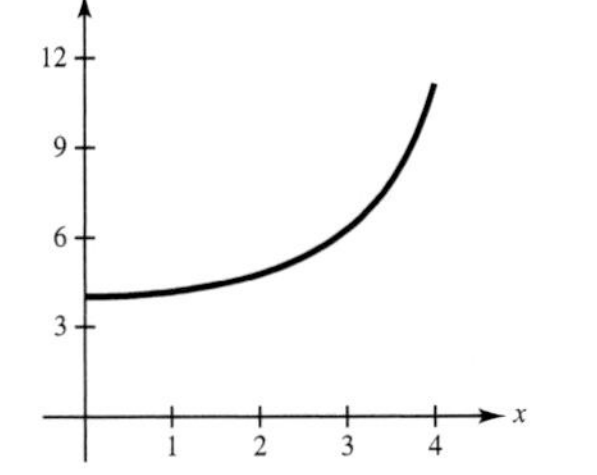

Since $f(x) \le 12$, the domain is $[0, 4]$.

(g) No. An x-intercept would mean that the time necessary for the trip would be zero for some value of x.

(h) $x = 5$. As x approaches 5, the speed for paddling upstream approaches 0.

83. An asymptote of a graph is a line to which the graph becomes arbitrarily close as $|x|$ or $|y|$ increases without bound.

85. No. $f(x) = \dfrac{1}{x^2 + 1}$ has no vertical asymptote.

Section 8.6 *(page 501)*

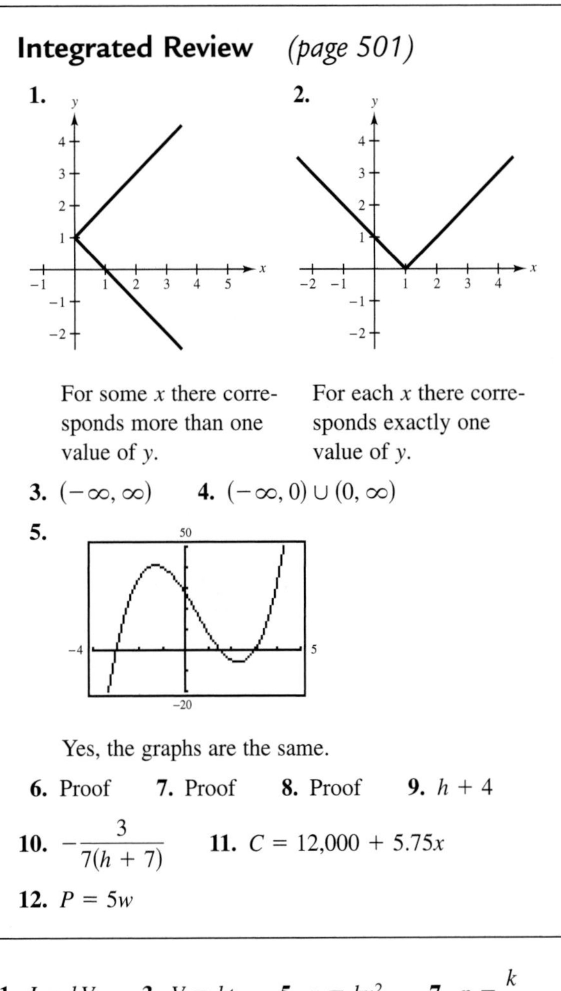

Integrated Review *(page 501)*

1. For some x there corresponds more than one value of y.

2. For each x there corresponds exactly one value of y.

3. $(-\infty, \infty)$ **4.** $(-\infty, 0) \cup (0, \infty)$

5. Yes, the graphs are the same.

6. Proof **7.** Proof **8.** Proof **9.** $h + 4$

10. $-\dfrac{3}{7(h + 7)}$ **11.** $C = 12{,}000 + 5.75x$

12. $P = 5w$

1. $I = kV$ **3.** $V = kt$ **5.** $u = kv^2$ **7.** $p = \dfrac{k}{d}$

9. $P = \dfrac{k}{\sqrt{1 + r}}$ **11.** $A = klw$ **13.** $P = \dfrac{k}{V}$

15. A varies jointly as the base and height.

17. A varies jointly as the length and the width.

19. V varies jointly as the square of the radius and the height.

21. r varies directly as the distance and inversely as the time.

23. $s = 5t$ **25.** $F = \frac{5}{16}x^2$ **27.** $H = \frac{5}{2}u$

29. $n = \dfrac{48}{m}$ **31.** $g = \dfrac{4}{\sqrt{z}}$ **33.** $F = \dfrac{25}{6}xy$

35. $d = \dfrac{120x^2}{r}$ **37.** (a) \$4921.25 (b) Price per unit

39. (a) 2 inches (b) 15 pounds **41.** 18 pounds

43. 32 feet per second per second **45.** $208\frac{1}{3}$ feet **47.** 4

49. No, k is different for each pizza. The 15-inch pizza is the best buy.

51. 667 units **53.** 324 pounds **55.** $\frac{1}{4}$

57. $p = \dfrac{114}{t}$, 17.5% **59.** \$270; the amount invested

61.

x	2	4	6	8	10
$y = kx^2$	4	16	36	64	100

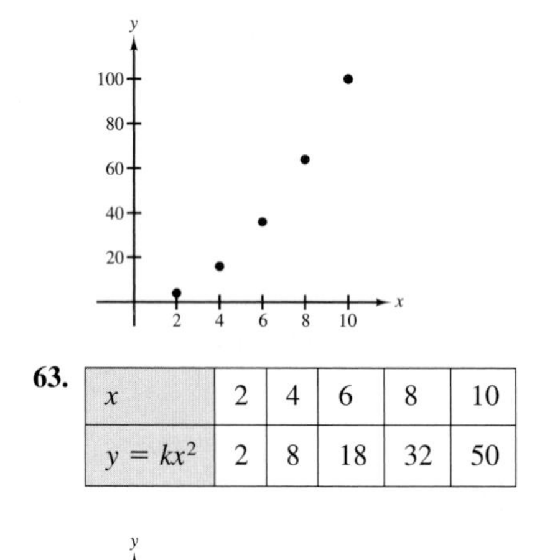

63.

x	2	4	6	8	10
$y = kx^2$	2	8	18	32	50

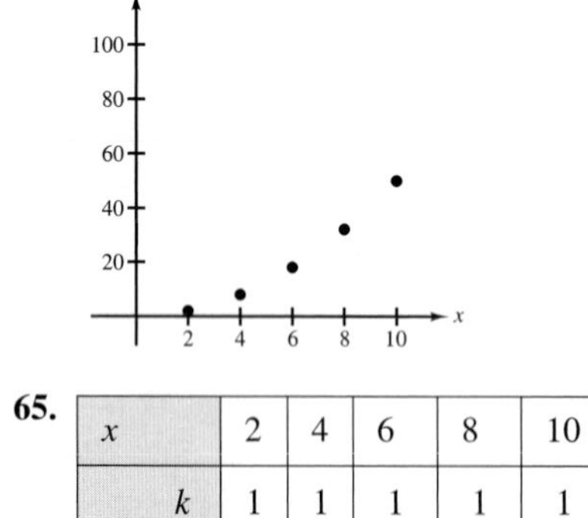

65.

x	2	4	6	8	10
$y = \dfrac{k}{x^2}$	$\frac{1}{2}$	$\frac{1}{8}$	$\frac{1}{18}$	$\frac{1}{32}$	$\frac{1}{50}$

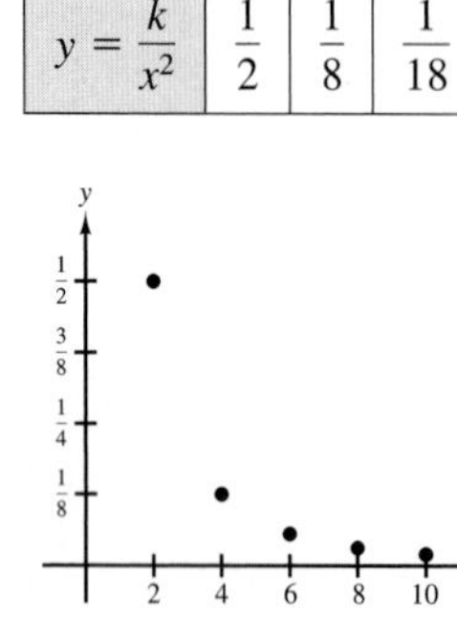

67.

x	2	4	6	8	10
$y = \dfrac{k}{x^2}$	$\frac{5}{2}$	$\frac{5}{8}$	$\frac{5}{18}$	$\frac{5}{32}$	$\frac{1}{10}$

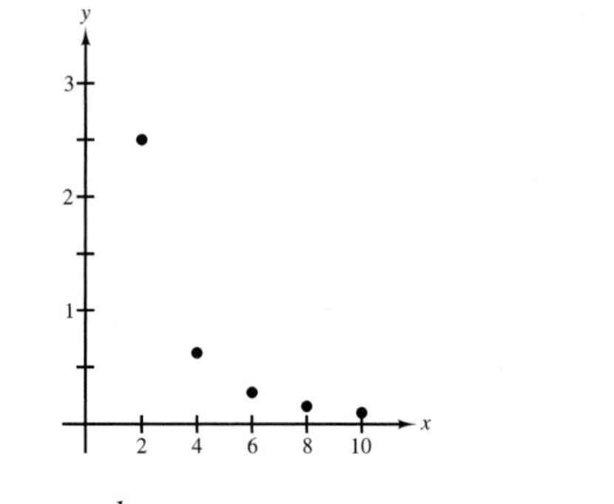

69. $y = \dfrac{k}{x}$ with $k = 4$ **71.** $y = kx$ with $k = -\dfrac{3}{10}$

73. Increase. Because $y = kx$ and $k > 0$, the variables increase or decrease together.

75. y will quadruple. If $y = kx^2$ and x is replaced with $2x$, you have $y = k(2x)^2 = 4kx^2$.

Review Exercises *(page 507)*

1. $(-\infty, 8) \cup (8, \infty)$ **3.** $(-\infty, 1) \cup (1, 6) \cup (6, \infty)$

5. $\dfrac{2x^3}{5}$, $x \neq 0, y \neq 0$ **7.** $\dfrac{b-3}{6(b-4)}$ **9.** -9, $x \neq y$

11. $\dfrac{x}{2(x+5)}$, $x \neq 5$ **13.** $3x^5y^2$ **15.** $\dfrac{8x^3}{5}$, $x \neq 0$

17. $\dfrac{y}{8x}$, $y \neq 0$ **19.** $12z(z-6)$, $z \neq -6$

21. $-\frac{1}{4}$, $u \neq 0, u \neq 3$ **23.** $3x^2$, $x \neq 0$

25. $\dfrac{125y}{x}$, $y \neq 0$ **27.** $\dfrac{x(x-1)}{x-7}$, $x \neq -1, x \neq 1$

29. $\dfrac{6(x+5)}{x(x+7)}$, $x \neq 5, x \neq -5$ **31.** $-\dfrac{7}{9}$ **33.** $-\dfrac{13}{48}$

35. $\dfrac{4x+3}{(x+5)(x-12)}$ **37.** $\dfrac{5x^3-5x^2-31x+13}{(x+2)(x-3)}$

39. $\dfrac{x+24}{x(x^2+4)}$ **41.** $\dfrac{6(x-9)}{(x+3)^2(x-3)}$

43. $\dfrac{3t^2}{5t-2}$, $t \neq 0$

45. $\dfrac{-a^2+a+16}{(4a^2+16a+1)(a-4)}$, $a \neq 0, a \neq -4$

47. Proof **49.** Proof **51.** -120 **53.** $\frac{36}{23}$ **55.** 5

57. $-4, 6$ **59.** $-\frac{16}{3}, 3$ **61.** $-\frac{5}{2}, 1$ **63.** $-2, 2$

65. $-\frac{9}{5}, 3$

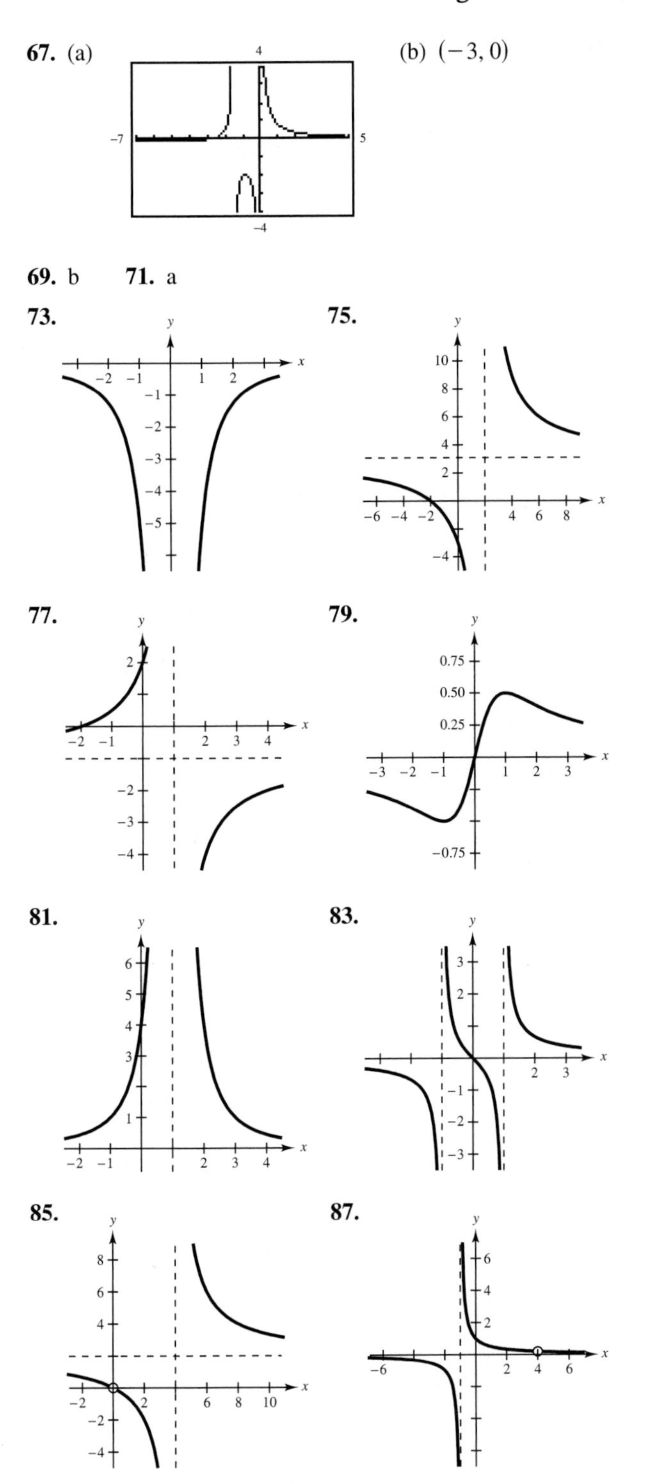

67. (a) (b) $(-3, 0)$

69. b **71.** a

73. **75.** **77.** **79.** **81.** **83.** **85.** **87.**

89. $y = \dfrac{3x}{x-4}$ **91.** $k = 6, y = 6\sqrt[3]{x}$

93. $k = \frac{1}{18}, T = \frac{1}{18}rs^2$ **95.** $(0, 6]$

97. 56 miles per hour **99.** 25 **101.** 4

103. 6 minutes, 40 seconds **105.** (a) 304, 453, 702

(b) 30 years

107. d will increase by a factor of 4. **109.** 158 pounds

Chapter Test *(page 511)*

1. $(-\infty, -5) \cup (-5, 5) \cup (5, \infty)$ **2.** $x^3(x-3)(x+3)^2$

3. (a) $-\dfrac{1}{3}, x \neq 2$ (b) $\dfrac{2a+3}{5}$ **4.** $\dfrac{5z}{3}, z \neq 0$

5. $\dfrac{4}{y+4}, y \neq 2$ **6.** $\dfrac{(2x+3)^2}{x+1}, x \neq \dfrac{3}{2}$

7. $\dfrac{14y^6}{15}, x \neq 0$ **8.** $\dfrac{x^3}{4}, x \neq 0, x \neq -2$

9. $-(3x+1), x \neq 0, x \neq \dfrac{1}{3}$ **10.** $\dfrac{-2x^2+2x+1}{x+1}$

11. $\dfrac{5x^2-15x-2}{(x-3)(x+2)}$ **12.** $\dfrac{5x^3+x^2-7x-5}{x^2(x+1)^2}$

13. (a) (b)

14. 22 **15.** $-\frac{15}{2}, -1$

16.

17. $6\frac{2}{3}$ hours, 10 hours

18. $S = \dfrac{kx^2}{y}$ **19.** $k = \dfrac{1}{4}$ **20.** 240 cubic meters

Chapter 9

Section 9.1 *(page 521)*

Integrated Review *(page 521)*

1. a^{m+n} **2.** $a^m b^m$ **3.** a^{mn} **4.** a^{m-n}

5. $y = 4 - 3x$ **6.** $y = \frac{2}{3}(1 - x)$

7. $y = \frac{1}{3}(4 - x^2)$ **8.** $y = 4 - x^2$

9. $y = \frac{1}{3}(2\sqrt{x} - 15)$ **10.** $y = \frac{1}{5}(6|x| + 10)$

11. $\frac{12}{5}$ hours **12.** $\frac{189}{4} = 47.25$ miles per hour

1. 8 **3.** -7 **5.** -2 **7.** Not a real number

9. 7 **11.** 4.2 **13.** Square root **15.** 8 **17.** 10

19. Not a real number **21.** $-\frac{2}{3}$ **23.** Not a real number

25. 5 **27.** -23 **29.** 5 **31.** 10 **33.** 6

35. $-\frac{1}{4}$ **37.** 11 **39.** -24 **41.** 3

43. Not a real number **45.** Irrational **47.** Rational

49. $16^{1/2} = 4$ **51.** $27^{2/3} = 9$ **53.** $\sqrt[4]{256^3} = 64$

55. 5 **57.** -6 **59.** -8 **61.** $\frac{1}{4}$ **63.** $\frac{1}{9}$

65. $\frac{4}{9}$ **67.** $\frac{3}{11}$ **69.** 9 **71.** -64 **73.** 25

75. $t^{1/2}$ **77.** $x^{7/4}$ **79.** $u^{7/3}$ **81.** $s^{13/2}$

83. $x^{-1} = \dfrac{1}{x}$ **85.** $t^{-9/4} = \dfrac{1}{t^{9/4}}$ **87.** x^3 **89.** $y^{13/12}$

91. $x^{3/4}y^{1/4}$ **93.** $y^{5/2}z^4$ **95.** 3 **97.** $\sqrt[3]{2}$ **99.** $\frac{1}{2}$

101. $\sqrt{c}$ **103.** $\dfrac{3y^2}{4z^{4/3}}$ **105.** $\dfrac{9y^{3/2}}{x^{2/3}}$ **107.** $x^{1/4}$

109. $\sqrt[8]{y}$ **111.** $x^{3/8}$ **113.** $\sqrt{x + y}$

115. $\dfrac{1}{(3u - 2v)^{5/6}}$ **117.** 8.5440 **119.** 9.9845

121. 0.0038 **123.** 4.3004 **125.** 66.7213

127. 1.0420 **129.** 0.7915 **131.** $[0, \infty)$

133. $(0, \infty)$ **135.** $(-\infty, 0]$

137.

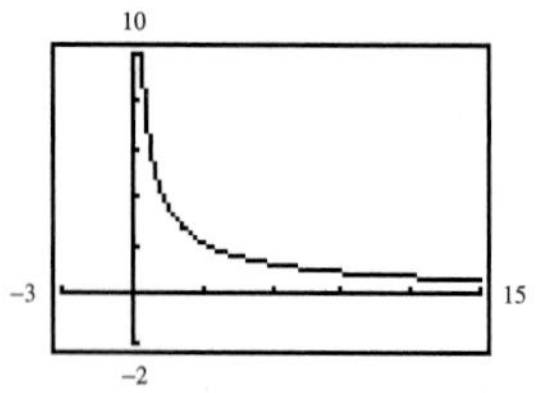

Domain: $(0, \infty)$

139.

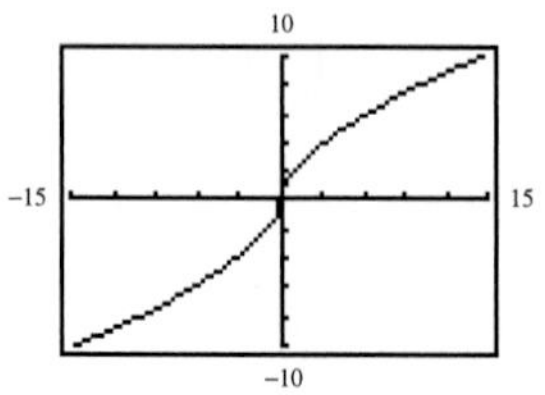

Domain: $(-\infty, \infty)$

141. $2x^{3/2} - 3x^{1/2}$ **143.** $1 + 5y$ **145.** 0.128

147. 23 feet × 23 feet **149.** 10.49 centimeters

151. (a) 15.0 feet

(b) $h = \sqrt{15.0^2 - \left(\dfrac{a}{2}\right)^2}$

(c) 8.29 feet

(d) 25.38 feet

153. Given $\sqrt[n]{a}$, a is the radicand and n is the index.

155. No. $\sqrt{2}$ is an irrational number. Its decimal representation is a nonterminating, nonrepeating decimal.

157. 1, 4, 5, 6, 9; Yes

Section 9.2 *(page 529)*

Integrated Review *(page 529)*

1. Replace the inequality sign with an equal sign and sketch the graph of the resulting equation. (Use a dashed line for < or >, and a solid line for ≤ and ≥.) Test one point in each of the half-planes formed by the graph. If the point satisfies the inequality, shade the entire half-plane to denote that every point in the region satisfies the inequality.

2. The first includes the points on the line $3x + 4y = 4$ and the second does not.

3. $-(x - 3)(x^2 + 1)$ **4.** $(2t + 13)(2t - 13)$

5. $(x - 1)(x - 2)$ **6.** $(x - 1)(2x + 7)$

7. $(x + 1)(11x - 5)$ **8.** $(2x - 7)^2$

9. 816 adults, 384 students **10.** 267 units

1. $2\sqrt{5}$ **3.** $5\sqrt{2}$ **5.** $4\sqrt{6}$ **7.** $6\sqrt{6}$

9. $13\sqrt{7}$ **11.** 0.2 **13.** $0.06\sqrt{2}$ **15.** $1.1\sqrt{2}$

17. $\dfrac{\sqrt{15}}{2}$ **19.** $\dfrac{\sqrt{13}}{5}$ **21.** $3x^2\sqrt{x}$ **23.** $4y^2\sqrt{3}$

25. $3\sqrt{13}|y^3|$ **27.** $2|x|y\sqrt{30y}$ **29.** $8a^2b^3\sqrt{3ab}$

31. $2\sqrt[3]{6}$ **33.** $2\sqrt[3]{14}$ **35.** $2x\sqrt[3]{5x^2}$ **37.** $3|y|\sqrt{2y}$

39. $xy\sqrt[3]{x}$ **41.** $|x|\sqrt[4]{3y^2}$ **43.** $2xy\sqrt[5]{y}$ **45.** $\dfrac{\sqrt[3]{35}}{4}$

47. $\dfrac{\sqrt[5]{15}}{3}$ **49.** $\dfrac{2\sqrt[5]{x^2}}{y}$ **51.** $\dfrac{3a\sqrt[3]{2a}}{b^3}$ **53.** $\dfrac{4a^2\sqrt{2}}{|b|}$

55. $3x^2$ **57.** $\dfrac{\sqrt{3}}{3}$ **59.** $\dfrac{\sqrt{7}}{7}$ **61.** $4\sqrt{3}$

63. $\dfrac{\sqrt[4]{20}}{2}$ **65.** $\dfrac{3\sqrt[3]{2}}{2}$ **67.** $\dfrac{\sqrt{y}}{y}$ **69.** $\dfrac{2\sqrt{x}}{x}$

71. $\dfrac{\sqrt{2x}}{2x}$ **73.** $\dfrac{2\sqrt{3b}}{b^2}$ **75.** $\dfrac{\sqrt[3]{18xy^2}}{3y}$

77. $\dfrac{a^2\sqrt[3]{a^2b}}{b}$ **79.** $2\sqrt{2}$ **81.** $24\sqrt{2} - 6$

83. $-11\sqrt[4]{3} - 5\sqrt[4]{7}$ **85.** $30\sqrt[3]{2}$ **87.** $12\sqrt{x}$

89. $13\sqrt{y}$ **91.** $(10 - z)\sqrt[3]{z}$ **93.** $\dfrac{2\sqrt{5}}{5}$ **95.** $\dfrac{9\sqrt{5}}{5}$

97. $\dfrac{\sqrt{2x}(2x - 3)}{2x}$ **99.** $\sqrt{7} + \sqrt{18} > \sqrt{7 + 18}$

101. $5 > \sqrt{3^2 + 2^2}$ **103.** $3\sqrt{5}$ **105.** $3\sqrt{13}$

107. (a) $5\sqrt{10}$ (b) $400\sqrt{10} \approx 1264.9$ square feet

109. 89.44 cycles per second

111. (a)

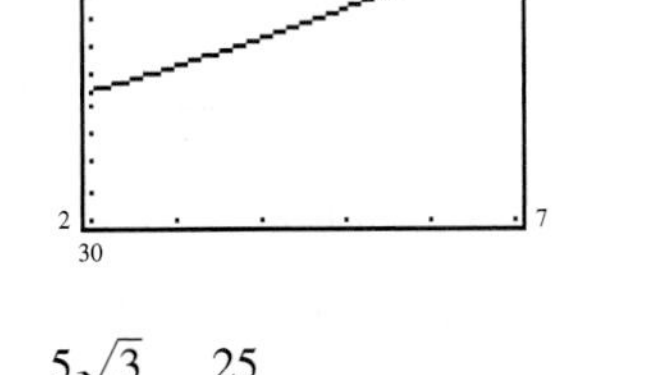

(b) 2004

113. No. $\dfrac{5\sqrt{3}}{3} \neq \dfrac{25}{3}$

115. $\sqrt{6} \cdot \sqrt{15} = \sqrt{9 \cdot 10} = 3\sqrt{10}$

117. No. $\sqrt{2} + \sqrt{18} = \sqrt{2} + 3\sqrt{2} = 4\sqrt{2}$

119. $x < 0$. $\sqrt{(-8)^2} = \sqrt{64} = 8$

Section 9.3 *(page 536)*

Integrated Review *(page 536)*

1. $mn = c$ **2.** The signs are the same.

3. The signs are different. **4.** $m + n = b$

5. $2x - y = 0$ **6.** $x + y - 6 = 0$

7. $y - 3 = 0$ **8.** $x - 4 = 0$

9. $6x + 11y - 96 = 0$ **10.** $x + y - 11 = 0$

11. $\dfrac{360}{r}$ **12.** $2L + 2\left(\dfrac{L}{3}\right)$

1. 4 **3.** $3\sqrt{2}$ **5.** $2\sqrt[3]{9}$ **7.** $2\sqrt[4]{3}$

9. $2\sqrt{5} - \sqrt{15}$ **11.** $2\sqrt{10} + 8\sqrt{2}$ **13.** $3\sqrt{2}$

15. $6 - 2\sqrt{5}$ **17.** $y + 4\sqrt{y}$ **19.** $4\sqrt{a} - a$

21. $2 - 7\sqrt[3]{4}$ **23.** -1

25. $\sqrt{15} + 3\sqrt{3} - 5\sqrt{5} - 15$ **27.** $8\sqrt{5} + 24$

29. $2\sqrt[3]{3} - 3\sqrt[3]{4} + 3\sqrt[3]{6} - 9$ **31.** $2x + 20\sqrt{2x} + 100$

33. $45x - 17\sqrt{x} - 6$ **35.** $9x - 25$

37. $\sqrt[3]{4x^2} + 10\sqrt[3]{2x} + 25$ **39.** $y - 5\sqrt[3]{y} + 2\sqrt[3]{y^2} - 10$

41. $t + 5\sqrt[3]{t^2} + \sqrt[3]{t} - 3$ **43.** $(x + 3)$

45. $(4 - 3x)$ **47.** $\left(2u + \sqrt{2u}\right)$ **49.** $2 - \sqrt{5},\ -1$

51. $\sqrt{11} + \sqrt{3},\ 8$ **53.** $\sqrt{15} - 3,\ 6$

55. $\sqrt{x} + 3,\ x - 9$ **57.** $\sqrt{2u} + \sqrt{3},\ 2u - 3$

59. $2\sqrt{2} - 2,\ 4$ **61.** $\sqrt{x} - \sqrt{y},\ x - y$

63. $\dfrac{1 - 2\sqrt{x}}{3}$ **65.** $\dfrac{-1 + \sqrt{3y}}{4}$ **67.** (a) $2\sqrt{3} - 4$ (b) 0

69. (a) 0 (b) -1 **71.** $-3\left(\sqrt{2} + 2\right)$ **73.** $\dfrac{7\left(5 - \sqrt{3}\right)}{22}$

75. $\dfrac{5 + 2\sqrt{10}}{5}$ **77.** $\dfrac{\sqrt{6} - \sqrt{2}}{2}$ **79.** $\dfrac{-9\left(\sqrt{3} + \sqrt{7}\right)}{4}$

81. $\dfrac{4\sqrt{7} + 11}{3}$ **83.** $\dfrac{2x - 9\sqrt{x} - 5}{4x - 1}$

85. $\dfrac{\left(\sqrt{15} + \sqrt{3}\right)x}{4}$ **87.** $\dfrac{2t^2\left(\sqrt{5} + \sqrt{t}\right)}{5 - t}$

89. $4\left(\sqrt{3a} - \sqrt{a}\right)$ **91.** $\dfrac{3(x - 4)\left(x^2 + \sqrt{x}\right)}{x(x - 1)(x^2 + x + 1)}$

93. $-\dfrac{\sqrt{u + v}\left(\sqrt{u - v} + \sqrt{u}\right)}{v}$

95. **97.**

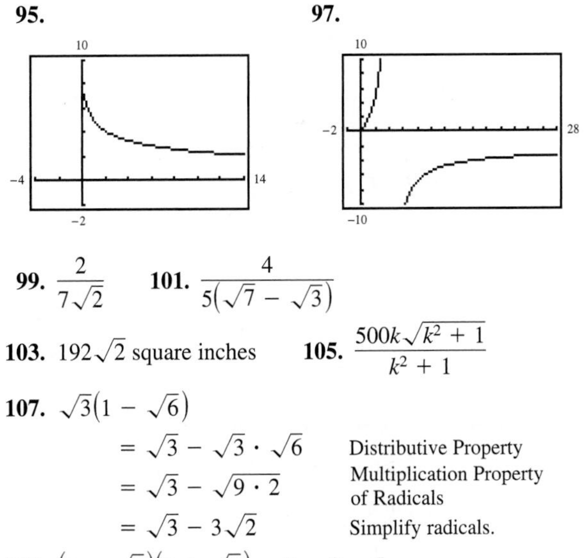

99. $\dfrac{2}{7\sqrt{2}}$ **101.** $\dfrac{4}{5\left(\sqrt{7} - \sqrt{3}\right)}$

103. $192\sqrt{2}$ square inches **105.** $\dfrac{500k\sqrt{k^2 + 1}}{k^2 + 1}$

107. $\sqrt{3}\left(1 - \sqrt{6}\right)$

$= \sqrt{3} - \sqrt{3} \cdot \sqrt{6}$ Distributive Property

$= \sqrt{3} - \sqrt{9 \cdot 2}$ Multiplication Property of Radicals

$= \sqrt{3} - 3\sqrt{2}$ Simplify radicals.

109. $\left(3 - \sqrt{2}\right)\left(3 + \sqrt{2}\right) = 9 - 2 = 7$

Multiplying the number by its conjugate yields the difference of two squares. Squaring a square root eliminates the radical.

Mid-Chapter Quiz *(page 539)*

1. 15 **2.** $\frac{3}{2}$ **3.** 8 **4.** 9 **5.** $3|x|\sqrt{3}$

6. $3|x|\sqrt{x}$ **7.** $\dfrac{2|u|\sqrt{u}}{3}$ **8.** $\dfrac{2\sqrt[3]{2}}{u^2}$ **9.** $4\sqrt{2y}$

10. $6x\sqrt[3]{5x^2} + 4x\sqrt[3]{5x}$ **11.** $16 + 6\sqrt{2}$ **12.** $10 - 4\sqrt{2}$

13. $3 + 5\sqrt{6}$ **14.** $60 + 67\sqrt{3}$ **15.** $\dfrac{\sqrt{21} - \sqrt{7}}{2}$

16. $-3 - 3\sqrt{2}$ **17.** $\frac{4}{3}(3 - \sqrt{6})$

18. $\frac{1}{2}(4\sqrt{3} - 3\sqrt{2} + \sqrt{6} - 4)$ **19.** $1 - \sqrt{4}, -3$

20. $\sqrt{10} + 5, -15$ **21.** $\sqrt{5^2 + 12^2} = \sqrt{169} = 13$

22. $23 + 8\sqrt{2}$ inches

Section 9.4 *(page 546)*

Integrated Review *(page 546)*

1. The function is undefined when the denominator is zero. The domain is all real numbers x such that $x \neq -2$ and $x \neq 3$.

2. $\dfrac{2x^2 + 5x - 3}{x^2 - 9}$ is undefined if $x = -3$.

3. $36x^5y^8$ **4.** 1 **5.** $4rs^2$ **6.** $\dfrac{9x^2}{16y^6}$

7. $-\dfrac{x + 13}{5x^2}$ **8.** $\dfrac{x^2 - 4}{25(x^2 - 9)}$ **9.** $\dfrac{2x + 5}{x - 5}$

10. $-\dfrac{5x - 8}{x - 1}$

11. $\left(\frac{3}{2}, 0\right)$, $(0, -3)$ **12.** $(0, 2)$, $\left(\frac{8}{3}, 0\right)$

1. (a) Not a solution (b) Not a solution (c) Not a solution (d) Solution

3. (a) Not a solution (b) Solution (c) Not a solution (d) Not a solution

5. 400 **7.** 9 **9.** 27 **11.** 49 **13.** No solution

15. 64 **17.** 90 **19.** -27 **21.** $\frac{4}{5}$ **23.** 5

25. No solution **27.** $\frac{44}{3}$ **29.** $\frac{14}{25}$ **31.** No solution

33. 4 **35.** No solution **37.** 7 **39.** -15

41. $-\frac{2}{3}$ **43.** 8 **45.** 1, 3 **47.** $\frac{1}{4}$ **49.** $\frac{1}{2}$ **51.** 4

53. 4 **55.** 216 **57.** 4, -12 **59.** -16

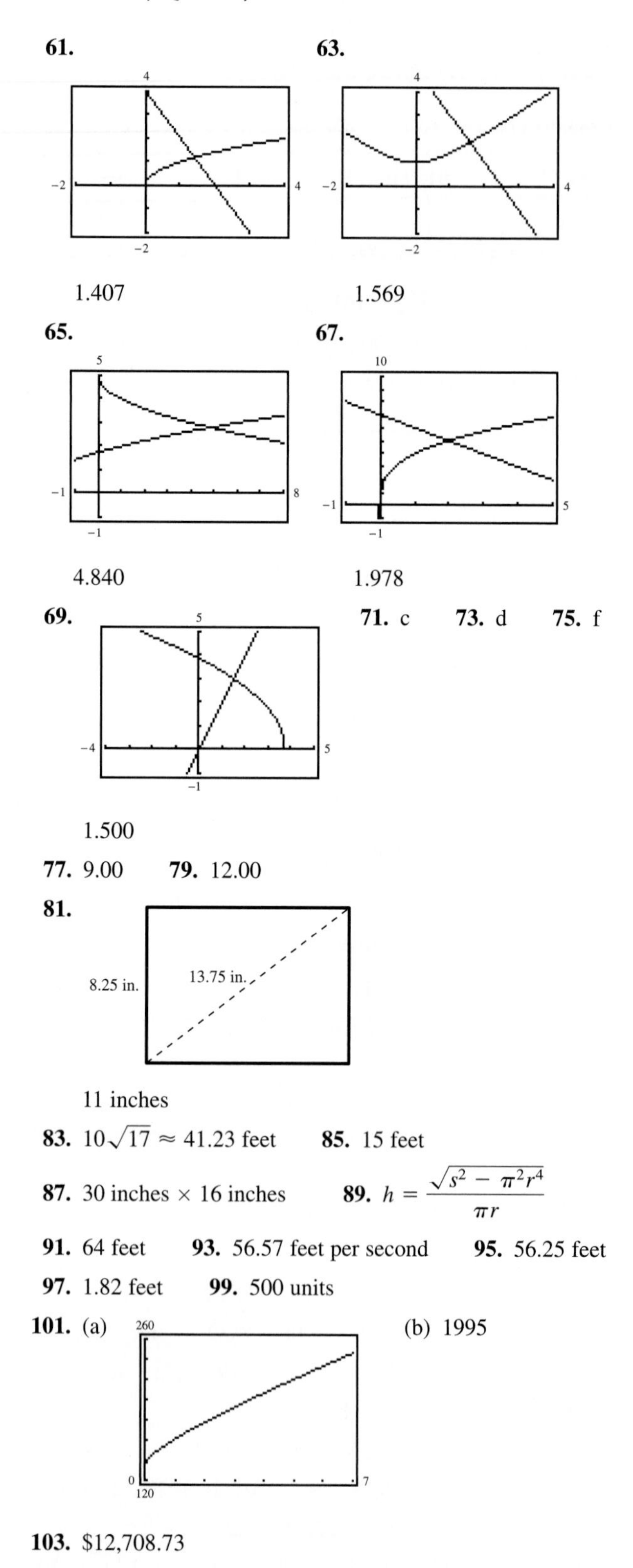

61. 1.407 **63.** 1.569

65. 4.840 **67.** 1.978

69. 1.500 **71.** c **73.** d **75.** f

77. 9.00 **79.** 12.00

81. 11 inches

83. $10\sqrt{17} \approx 41.23$ feet **85.** 15 feet

87. 30 inches × 16 inches **89.** $h = \dfrac{\sqrt{s^2 - \pi^2r^4}}{\pi r}$

91. 64 feet **93.** 56.57 feet per second **95.** 56.25 feet

97. 1.82 feet **99.** 500 units

101. (a) (b) 1995

103. $12,708.73

105. No. It is not an operation that necessarily yields an equivalent equation. There may be extraneous solutions.

107. $(\sqrt{x} + \sqrt{6})^2 \neq (\sqrt{x})^2 + (\sqrt{6})^2$

Section 9.5 *(page 556)*

Integrated Review *(page 556)*

1. $\frac{u}{v} \cdot \frac{w}{z} = \frac{uw}{vz}$

$\frac{3t}{5} \cdot \frac{8t^2}{15} = \frac{3t(8t^2)}{5(15)} = \frac{8t^3}{25}$

2. $\frac{u}{v} \div \frac{w}{z} = \frac{u}{v} \cdot \frac{z}{w}$

$\frac{3t}{5} \div \frac{8t^2}{15} = \frac{3t}{5} \cdot \frac{15}{8t^2} = \frac{9}{8t}$

3. Rewrite the fractions so they have like denominators and then use the rule

$\frac{u}{w} + \frac{v}{w} = \frac{u+v}{w}.$

$\frac{3t}{5} + \frac{8t^2}{15} = \frac{9t}{15} + \frac{8t^2}{15} = \frac{9t + 8t^2}{15}$

4. $\frac{t-5}{5-t} = \frac{-1(5-t)}{5-t} = -1$ **5.** $\frac{x}{5}$

6. $\frac{x}{5(x+y)}$ **7.** $\frac{9}{2(x+3)}$ **8.** $\frac{1}{x-2}$

9. $\frac{x^2 + 2x - 13}{x(x-2)}$ **10.** $\frac{(x+1)(x+3)}{3}$

11. $\frac{7x}{9}, \frac{19x}{18}$ **12.** $\frac{C_1C_2}{C_1 + C_2}$

1. $2i$ **3.** $-12i$ **5.** $\frac{2}{5}i$ **7.** $0.3i$ **9.** $2\sqrt{2}i$

11. $3\sqrt{3}i$ **13.** $\sqrt{7}i$ **15.** 2 **17.** $\sqrt{5}i$

19. $\frac{3\sqrt{2}}{8}i$ **21.** $10i$ **23.** $3\sqrt{2}i$ **25.** $3\sqrt{3}i$

27. -4 **29.** $-3\sqrt{6}$ **31.** -0.44 **33.** $-2\sqrt{3} - 3$

35. $5\sqrt{2} - 4\sqrt{5}$ **37.** $4 + 3\sqrt{2}i$ **39.** -16

41. $-8i$ **43.** $a = 3, b = -4$ **45.** $a = 2, b = -3$

47. $a = -4, b = -2\sqrt{2}$ **49.** $a = 2, b = -2$

51. $10 + 4i$ **53.** $-14 - 40i$ **55.** $-14 + 20i$

57. $9 - 7i$ **59.** $3 + 6i$ **61.** $\frac{13}{6} + \frac{3}{2}i$

63. $-3 + 49i$ **65.** $7 + (3\sqrt{7} - 5)i$ **67.** -36

69. 24 **71.** $-36i$ **73.** $27i$ **75.** -9

77. $-65 - 10i$ **79.** $20 - 12i$ **81.** $4 + 18i$

83. $-20 + 12\sqrt{5}i$ **85.** $-40 - 5i$ **87.** $-14 + 42i$

89. 9 **91.** $-7 - 24i$ **93.** $-21 + 20i$

95. $2 + 11i$ **97.** 5 **99.** 68 **101.** 31

103. 100 **105.** 4 **107.** 2.5 **109.** $-10i$

111. $2 + 2i$ **113.** $-\frac{24}{53} + \frac{84}{53}i$ **115.** $-\frac{6}{5} + \frac{2}{5}i$

117. $\frac{8}{5} - \frac{1}{5}i$ **119.** $1 - \frac{6}{5}i$ **121.** $-\frac{53}{25} + \frac{29}{25}i$

123. (a) Solution (b) Solution **125.** (a) Solution (b) Solution

127. (a) $\left(\frac{-5 + 5\sqrt{3}i}{2}\right)^3 = 125$

(b) $\left(\frac{-5 - 5\sqrt{3}i}{2}\right)^3 = 125$

129. (a) $1, \frac{-1 + \sqrt{3}i}{2}, \frac{-1 - \sqrt{3}i}{2}$

(b) $2, \frac{-2 + 2\sqrt{3}i}{2} = -1 + \sqrt{3}i, \frac{-2 - 2\sqrt{3}i}{2} = -1 - \sqrt{3}i$

(c) $4, \frac{-4 + 4\sqrt{3}i}{2} = -2 + 2\sqrt{3}i, \frac{-4 - 4\sqrt{3}i}{2} = -2 - 2\sqrt{3}i$

131. $2a$ **133.** $2bi$ **135.** $i = \sqrt{-1}$

137. $\sqrt{-3}\sqrt{-3} = (\sqrt{3}i)(\sqrt{3}i) = 3i^2 = -3$ **139.** 13

Review Exercises *(page 560)*

1. 7 **3.** -9 **5.** -2 **7.** -4 **9.** 1.2 **11.** $\frac{5}{6}$

13. $-\frac{1}{5}$ **15.** $2i$ **17.** $49^{1/2} = 7$ **19.** $\sqrt[3]{216} = 6$

21. 81 **23.** -125 **25.** $\frac{1}{16}$ **27.** $-\frac{9}{16}$ **29.** $x^{7/12}$

31. $z^{5/3}$ **33.** $\frac{1}{x^{5/4}}$ **35.** $a\sqrt[3]{b^2}$ **37.** $\sqrt[8]{x}$

39. $\sqrt[3]{3x + 2}$ **41.** 0.04 **43.** 10.63

45. $(-\infty, \infty)$

47. $[0, \infty)$

49. $6\sqrt{10}$ **51.** $5u^2v^2\sqrt{3u}$ **53.** $0.5x^2\sqrt{y}$

55. $2b\sqrt[4]{4a^2b}$ **57.** $2ab\sqrt[3]{6b}$ **59.** $\frac{\sqrt{30}}{6}$ **61.** $\frac{\sqrt{3x}}{2x}$

63. $\frac{\sqrt[3]{4x^2}}{x}$ **65.** $\sqrt{7}$ **67.** $-24\sqrt{10}$

69. $14\sqrt{x} - 9\sqrt[3]{x}$ **71.** $7\sqrt[4]{y+3}$ **73.** $12\sqrt{x} - 2\sqrt[3]{x}$

75. $\frac{2\sqrt{5}}{5}$ **77.** $10\sqrt{3}$ **79.** $5\sqrt{2} + 3\sqrt{5}$

81. $5\sqrt{2} + 2\sqrt{5}$ **83.** $7\sqrt{6} + 6\sqrt{2} - 4\sqrt{3} - 14$

85. $12\sqrt{5} + 41$ **87.** $3 - x$ **89.** $-3(1 + \sqrt{2})$

91. $\frac{6(4 - \sqrt{6})}{5}$ **93.** $-\frac{(\sqrt{2} - 1)(\sqrt{3} + 4)}{13}$

95. $\frac{(\sqrt{x} + 10)^2}{x - 100}$

97. **99.**

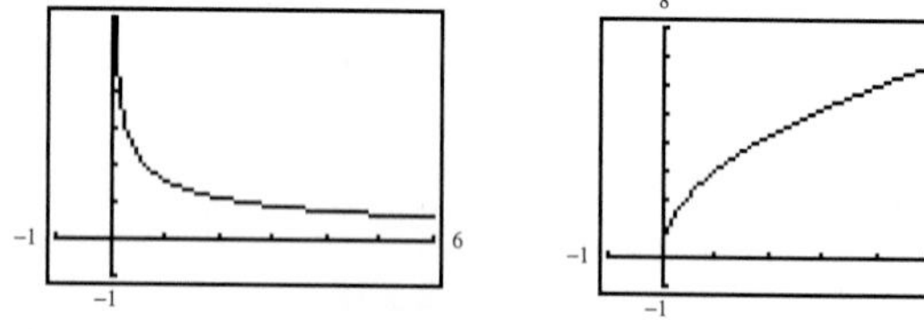

101. 225 **103.** No real solution **105.** 105 **107.** 3

109. 5 **111.** $-5, -3$ **113.** 6 **115.** $\frac{3}{32}$

117. $4\sqrt{3}i$ **119.** $10 - 9\sqrt{3}i$ **121.** $\frac{3}{4} - \sqrt{3}i$

123. $15i$ **125.** $(11 - 2\sqrt{21})i$ **127.** -5

129. $\sqrt{70} - 2\sqrt{10}$ **131.** $x = 2, y = 3$

133. $x = 4, y = 125$ **135.** $8 - 3i$ **137.** $8 + 4i$

139. 25 **141.** $11 - 60i$ **143.** $-\frac{7}{3}i$

145. $-\frac{8}{17} + \frac{2}{17}i$ **147.** $\frac{13}{37} - \frac{33}{37}i$

149. $21 + 12\sqrt{2}$ inches **151.** 1.37 feet

153. 500 watts **155.** 9000 watts **157.** 9.77 feet

Chapter Test *(page 563)*

1. (a) 64 (b) 10 **2.** (a) $\frac{1}{9}$ (b) 6 **3.** (a) $x^{1/3}$ (b) 25

4. (a) $\frac{4}{3}\sqrt{2}$ (b) $2\sqrt[3]{3}$ **5.** (a) $2x\sqrt{6x}$ (b) $2xy^2\sqrt[4]{x}$

6. Multiply the numerator and denominator of a fraction by a factor such that no radical contains a fraction and no denominator of a fraction contains a radical.

$\frac{\sqrt{6}}{2}$

7. $-10\sqrt{3x}$ **8.** $5\sqrt{3x} + 3\sqrt{5}$ **9.** $16 - 8\sqrt{2x} + 2x$

10. $7\sqrt{3}(3 + 4y)$ **11.** 27 **12.** No solution **13.** 9

14. $x = 4, y = 400$ **15.** $x = 3, y = 1$ **16.** $2 - 2i$

17. $-5 - 12i$ **18.** $-8 + 4i$ **19.** $13 + 13i$

20. $-2 - 5i$ **21.** 100 feet

Cumulative Test: Chapters 7–9 *(page 564)*

1. c **2.** e **3.** d **4.** f **5.** a **6.** b

7. $(2, 1)$ **8.** $(3, -2)$ **9.** $(5, 4)$ **10.** $\left(-\frac{1}{5}, -\frac{22}{5}\right)$

11. $\frac{x(x+2)(x+4)}{9(x-4)}$ **12.** $\frac{3x+5}{x(x+3)}$ **13.** $x + y$

14. $-4 + 3\sqrt{2}i$ **15.** $-7 - 24i$ **16.** $t^{1/2}$

17. $35\sqrt{5x}$ **18.** $2x - 6\sqrt{2x} + 9$ **19.** $\sqrt{10} + 2$

20. $\frac{2}{17} - \frac{9}{17}i$ **21.** 2, 5 **22.** 2, 9 **23.** 16 **24.** 4

25. b **26.** d **27.** a **28.** c **29.** 128 feet

30. 50 **31.** $r_2 = \frac{\sqrt{15r_1}}{5}$

32. $16(1 + \sqrt{2}) \approx 38.6$ inches

Chapter 10

Section 10.1 *(page 573)*

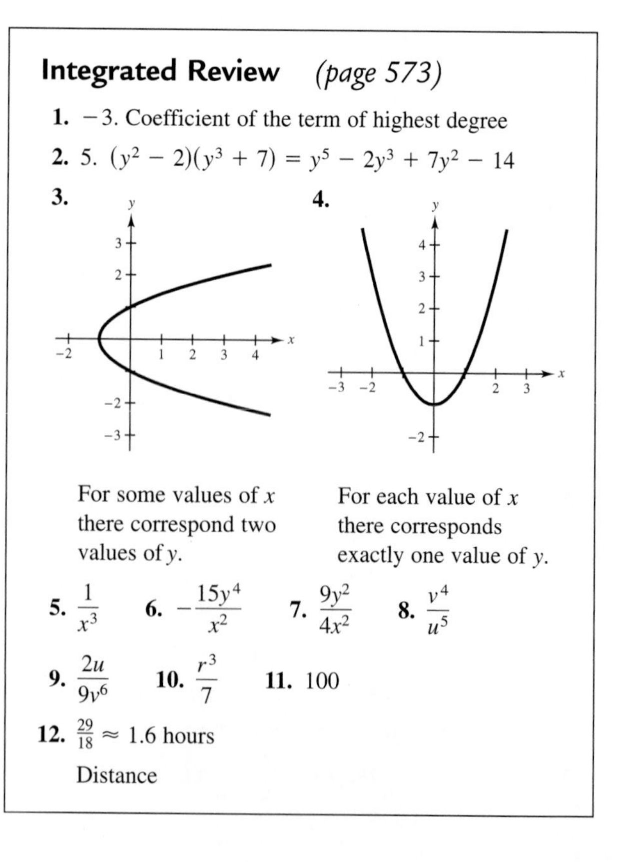

Integrated Review *(page 573)*

1. -3. Coefficient of the term of highest degree

2. 5. $(y^2 - 2)(y^3 + 7) = y^5 - 2y^3 + 7y^2 - 14$

3. **4.**

For some values of x there correspond two values of y.

For each value of x there corresponds exactly one value of y.

5. $\frac{1}{x^3}$ **6.** $-\frac{15y^4}{x^2}$ **7.** $\frac{9y^2}{4x^2}$ **8.** $\frac{v^4}{u^5}$

9. $\frac{2u}{9v^6}$ **10.** $\frac{r^3}{7}$ **11.** 100

12. $\frac{29}{18} \approx 1.6$ hours

Distance

1. $5, 7$ **3.** $-9, 8$ **5.** $-9, 5$ **7.** 6 **9.** $-\frac{4}{3}$

11. $0, 3$ **13.** $9, 12$ **15.** $\frac{5}{3}, 6$ **17.** $1, 6$ **19.** $-\frac{5}{6}, \frac{1}{2}$

21. ± 8 **23.** ± 3 **25.** $\pm\frac{4}{5}$ **27.** ± 8 **29.** $\pm\frac{5}{2}$

31. $\pm\frac{15}{2}$ **33.** $-17, 9$ **35.** $2.5, 3.5$ **37.** $2 \pm \sqrt{7}$

39. $\dfrac{-1 \pm 5\sqrt{2}}{2}$ **41.** $\dfrac{3 \pm 7\sqrt{2}}{4}$ **43.** $\pm 6i$

45. $\pm 2i$ **47.** $\pm\dfrac{\sqrt{17}}{3}i$ **49.** $3 \pm 5i$ **51.** $-\dfrac{4}{3} \pm 4i$

53. $-\dfrac{3}{2} \pm \dfrac{3\sqrt{6}}{2}i$ **55.** $-6 \pm \dfrac{11}{3}i$ **57.** $1 \pm 3\sqrt{3}i$

59. $-1 \pm 0.2i$ **61.** $\dfrac{2}{3} \pm \dfrac{1}{3}i$ **63.** $-\dfrac{7}{3} \pm \dfrac{\sqrt{38}}{3}i$

65. $0, \frac{5}{2}$ **67.** $-4, \frac{3}{2}$ **69.** ± 30 **71.** $\pm 30i$ **73.** ± 3

75. $-5, 15$ **77.** $5 \pm 10i$ **79.** $-2 \pm 3\sqrt{2}i$

81.

$(-3, 0), (3, 0)$

83.

$(-3, 0), (5, 0)$

85.

$(1, 0), (5, 0)$

87.

$(2, 0), \left(-\frac{3}{2}, 0\right)$

89.

$\left(-\frac{4}{3}, 0\right), (4, 0)$

91.

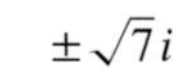

$\pm\sqrt{7}i$

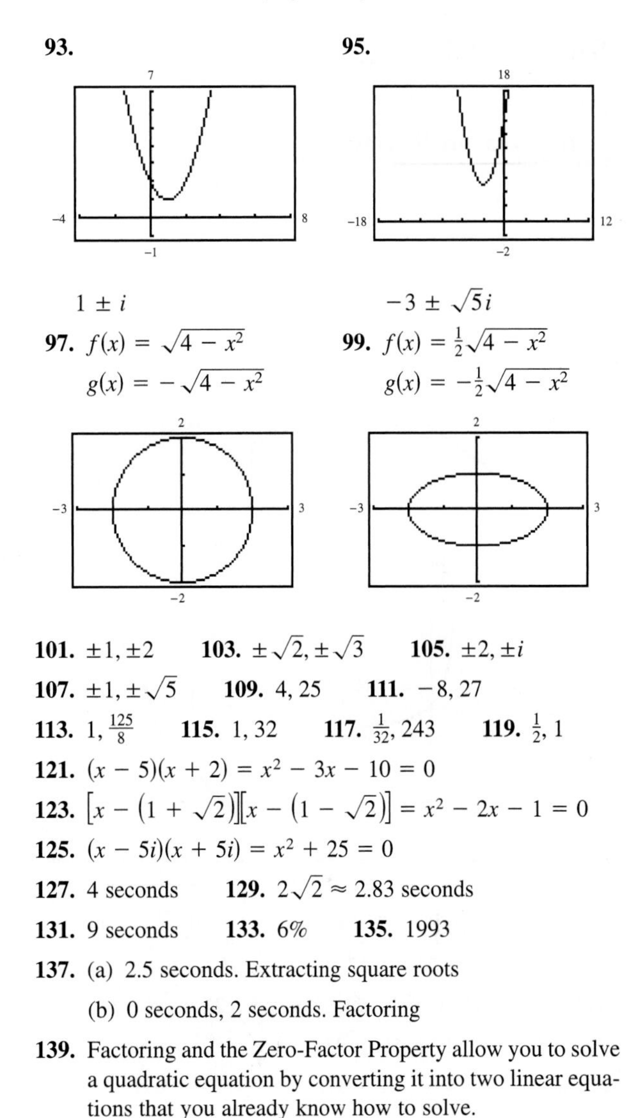

93.

$1 \pm i$

95.

$-3 \pm \sqrt{5}i$

97. $f(x) = \sqrt{4 - x^2}$
$g(x) = -\sqrt{4 - x^2}$

99. $f(x) = \frac{1}{2}\sqrt{4 - x^2}$
$g(x) = -\frac{1}{2}\sqrt{4 - x^2}$

101. $\pm 1, \pm 2$ **103.** $\pm\sqrt{2}, \pm\sqrt{3}$ **105.** $\pm 2, \pm i$

107. $\pm 1, \pm\sqrt{5}$ **109.** $4, 25$ **111.** $-8, 27$

113. $1, \frac{125}{8}$ **115.** $1, 32$ **117.** $\frac{1}{32}, 243$ **119.** $\frac{1}{2}, 1$

121. $(x - 5)(x + 2) = x^2 - 3x - 10 = 0$

123. $\left[x - \left(1 + \sqrt{2}\right)\right]\left[x - \left(1 - \sqrt{2}\right)\right] = x^2 - 2x - 1 = 0$

125. $(x - 5i)(x + 5i) = x^2 + 25 = 0$

127. 4 seconds **129.** $2\sqrt{2} \approx 2.83$ seconds

131. 9 seconds **133.** 6% **135.** 1993

137. (a) 2.5 seconds. Extracting square roots
(b) 0 seconds, 2 seconds. Factoring

139. Factoring and the Zero-Factor Property allow you to solve a quadratic equation by converting it into two linear equations that you already know how to solve.

141. False. The solutions are $x = 5$ and $x = -5$.

143. To solve an equation of quadratic form, determine an algebraic expression u such that substitution yields the quadratic equation $au^2 + bu + c = 0$. Solve this quadratic equation for u and then, through back-substitution, find the solution of the original equation.

Section 10.2 *(page 580)*

Integrated Review *(page 580)*

1. a^4b^4 **2.** a^{rs} **3.** $\frac{b^r}{a^r}$ **4.** $\frac{1}{a^r}$ **5.** 6

6. 3 **7.** $-\frac{2}{3}, 5$ **8.** $-5, 8$

9. **10.**

11. **12.**

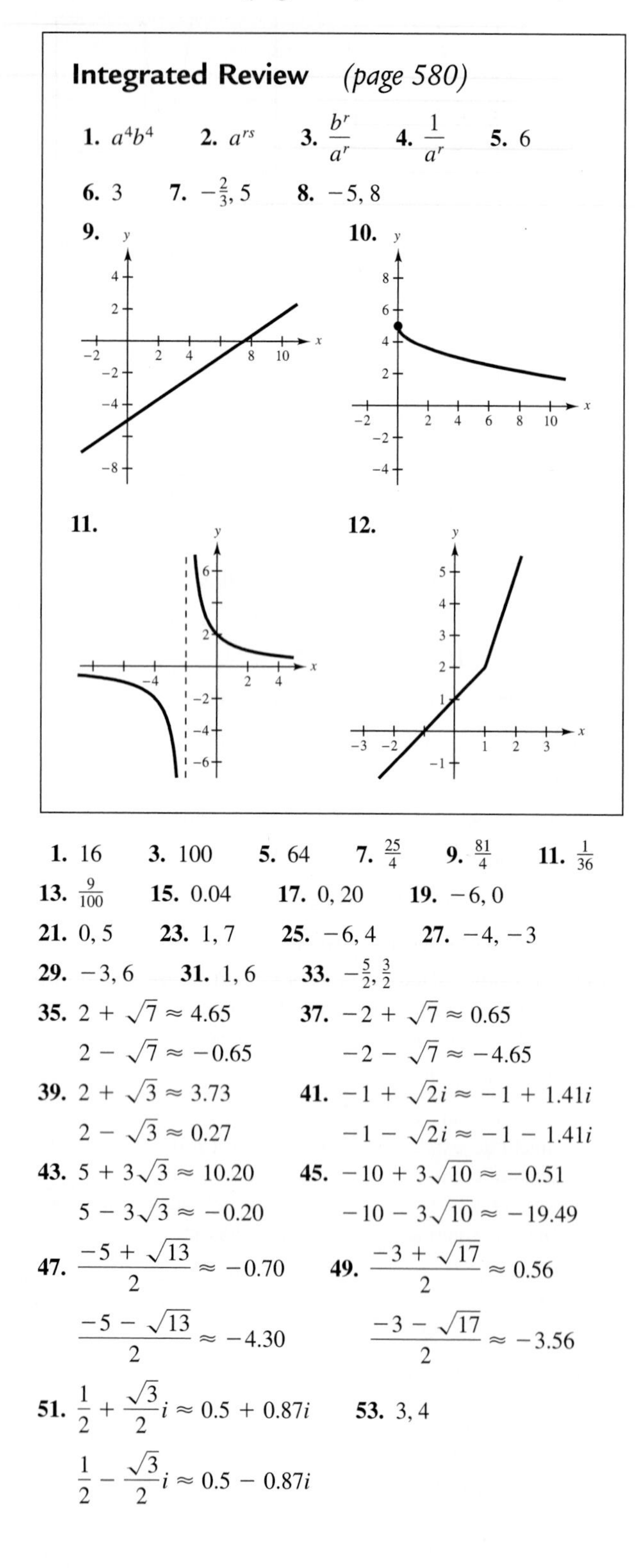

1. 16 **3.** 100 **5.** 64 **7.** $\frac{25}{4}$ **9.** $\frac{81}{4}$ **11.** $\frac{1}{36}$

13. $\frac{9}{100}$ **15.** 0.04 **17.** 0, 20 **19.** $-6, 0$

21. 0, 5 **23.** 1, 7 **25.** $-6, 4$ **27.** $-4, -3$

29. $-3, 6$ **31.** 1, 6 **33.** $-\frac{5}{2}, \frac{3}{2}$

35. $2 + \sqrt{7} \approx 4.65$, $2 - \sqrt{7} \approx -0.65$

37. $-2 + \sqrt{7} \approx 0.65$, $-2 - \sqrt{7} \approx -4.65$

39. $2 + \sqrt{3} \approx 3.73$, $2 - \sqrt{3} \approx 0.27$

41. $-1 + \sqrt{2}i \approx -1 + 1.41i$, $-1 - \sqrt{2}i \approx -1 - 1.41i$

43. $5 + 3\sqrt{3} \approx 10.20$, $5 - 3\sqrt{3} \approx -0.20$

45. $-10 + 3\sqrt{10} \approx -0.51$, $-10 - 3\sqrt{10} \approx -19.49$

47. $\frac{-5 + \sqrt{13}}{2} \approx -0.70$, $\frac{-5 - \sqrt{13}}{2} \approx -4.30$

49. $\frac{-3 + \sqrt{17}}{2} \approx 0.56$, $\frac{-3 - \sqrt{17}}{2} \approx -3.56$

51. $\frac{1}{2} + \frac{\sqrt{3}}{2}i \approx 0.5 + 0.87i$, $\frac{1}{2} - \frac{\sqrt{3}}{2}i \approx 0.5 - 0.87i$

53. 3, 4

55. $\frac{1 + 2\sqrt{7}}{3} \approx 2.10$, $\frac{1 - 2\sqrt{7}}{3} \approx -1.43$

57. $\frac{-3 + \sqrt{137}}{8} \approx 1.09$, $\frac{-3 - \sqrt{137}}{8} \approx -1.84$

59. $\frac{-4 + \sqrt{10}}{2} \approx -0.42$, $\frac{-4 - \sqrt{10}}{2} \approx -3.58$

61. $\frac{-9 + \sqrt{21}}{6} \approx -0.74$, $\frac{-9 - \sqrt{21}}{6} \approx -2.26$

63. $\frac{-1 + \sqrt{10}}{2} \approx 1.08$, $\frac{-1 - \sqrt{10}}{2} \approx -2.08$

65. $\frac{3}{10} + \frac{\sqrt{191}}{10}i \approx 0.30 + 1.38i$, $\frac{3}{10} - \frac{\sqrt{191}}{10}i \approx 0.30 - 1.38i$

67. $\frac{7 + \sqrt{57}}{2} \approx 7.27$, $\frac{7 - \sqrt{57}}{2} \approx -0.27$

69. $-1 + \sqrt{3}i \approx -1 + 1.73i$, $-1 - \sqrt{3}i \approx -1 - 1.73i$

71. $-1 \pm 2i$

73. $1 \pm \sqrt{3}$ **75.** $1 \pm \sqrt{3}$ **77.** $4 \pm 2\sqrt{2}$

79. $\left(-2 \pm \sqrt{5}, 0\right)$

81. $\left(1 \pm \sqrt{6}, 0\right)$

83. $\left(-3 \pm 3\sqrt{3}, 0\right)$

85. $\left(\frac{-1 \pm \sqrt{13}}{2}, 0\right)$

87. (a) $x^2 + 8x$ (b) $x^2 + 8x + 16$ (c) $(x + 4)^2$

89. 4 centimeters, 6 centimeters

91. 15 feet × $46\frac{2}{3}$ feet or 20 feet × 35 feet

93. 271 meters, 129 meters

95. 139 units, 861 units

97. $\frac{25}{4}$. Divide the coefficient of the first-degree term by 2, and square the result to obtain $\left(\frac{5}{2}\right)^2 = \frac{25}{4}$.

99. Yes. $x^2 + 1 = 0$

101. True. Given the solutions $x = r_1$ and $x = r_2$, the quadratic equation can be written as $(x - r_1)(x - r_2) = 0$.

Section 10.3 *(page 588)*

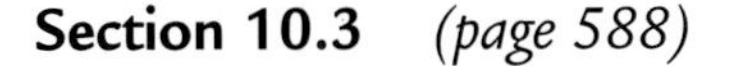

Integrated Review *(page 588)*

1. $\sqrt{a}\sqrt{b}$ **2.** $\dfrac{\sqrt{a}}{\sqrt{b}}$

3. No. $\sqrt{72} = \sqrt{36 \cdot 2} = 6\sqrt{2}$

4. No. $\dfrac{10}{\sqrt{5}} = \dfrac{10\sqrt{5}}{(\sqrt{5})^2} = 2\sqrt{5}$ **5.** $23\sqrt{2}$

6. 150 **7.** 7 **8.** $11 + 6\sqrt{2}$ **9.** $\dfrac{4\sqrt{10}}{5}$

10. $\dfrac{5(1 + \sqrt{3})}{4}$ **11.** 10 inches $\times$ 15 inches

12. 200 units

1. $2x^2 + 2x - 7 = 0$ **3.** $-x^2 + 10x - 5 = 0$

5. 4, 7 **7.** $-2, -4$ **9.** $-\frac{1}{2}$ **11.** $-\frac{3}{2}$ **13.** $-\frac{1}{2}, \frac{2}{3}$

15. $-15, 20$ **17.** $1 \pm \sqrt{5}$ **19.** $-2 \pm \sqrt{3}$

21. $-3 \pm 2\sqrt{3}$ **23.** $5 \pm \sqrt{2}$ **25.** $-\dfrac{3}{4} \pm \dfrac{\sqrt{15}}{4}i$

27. $-\dfrac{1}{3}, 1$ **29.** $\dfrac{-2 \pm \sqrt{10}}{2}$ **31.** $\dfrac{-1 \pm \sqrt{5}}{3}$

33. $\dfrac{-3 \pm \sqrt{21}}{4}$ **35.** $\dfrac{3}{8} \pm \dfrac{\sqrt{7}}{8}i$ **37.** $\dfrac{5 \pm \sqrt{73}}{4}$

39. $\dfrac{3 \pm \sqrt{13}}{6}$ **41.** $\dfrac{-3 \pm \sqrt{57}}{6}$ **43.** $\dfrac{1 \pm \sqrt{5}}{5}$

45. $\dfrac{-1 \pm \sqrt{10}}{5}$

47. Two distinct imaginary solutions

49. Two distinct irrational solutions

51. Two distinct imaginary solutions

53. One (repeated) rational solution

55. Two distinct imaginary solutions **57.** ± 13

59. $-3, 0$ **61.** $\frac{9}{5}, \frac{21}{5}$ **63.** $-\frac{3}{2}, 18$ **65.** $-4 \pm 3i$

67. 8, 16 **69.** $\dfrac{13}{6} \pm \dfrac{13\sqrt{11}}{6}i$ **71.** $\dfrac{-5 \pm 5\sqrt{17}}{12}$

73. $-\frac{11}{6}, \frac{5}{2}$

75. $(0.18, 0), (1.82, 0)$

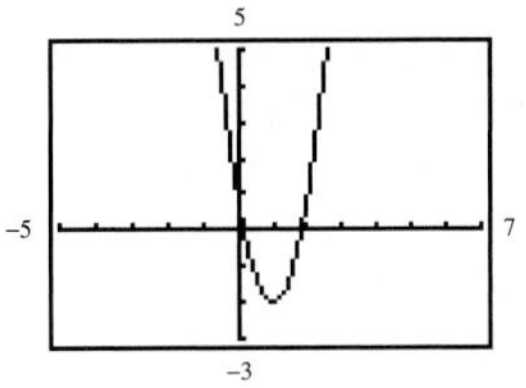

77. $(2.50, 0)$

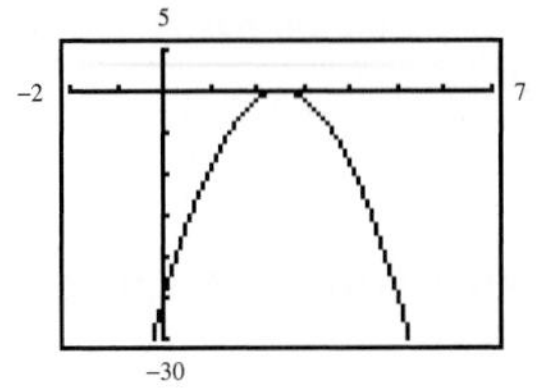

79. $(3.23, 0), (0.37, 0)$

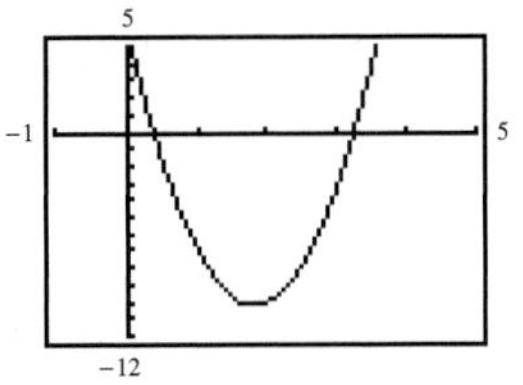

81. $(99.80, 0), (0.20, 0)$

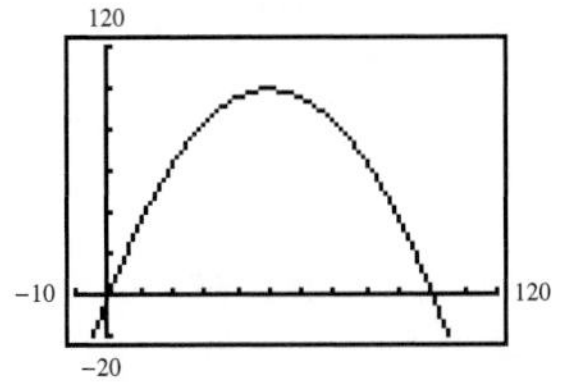

83. No real solutions **85.** Two real solutions

87. $\dfrac{5 \pm \sqrt{185}}{8}$ **89.** $\dfrac{3 + \sqrt{17}}{2}$

91. (a) $c < 9$ (b) $c = 9$ (c) $c > 9$

93. (a) $c < 16$ (b) $c = 16$ (c) $c > 16$

95. 5.1 inches $\times$ 11.4 inches

97. (a) 2.5 seconds (b) $\dfrac{5 + 5\sqrt{3}}{4} \approx 3.4$ seconds

99. (a) (b) 1991 (c) 400,500

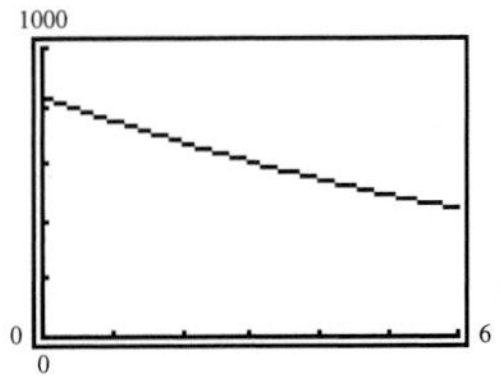

101.

	x_1	x_2	$x_1 + x_2$	x_1x_2
(a)	-2	3	1	-6
(b)	-3	$\frac{1}{2}$	$-\frac{5}{2}$	$-\frac{3}{2}$
(c)	$-\frac{3}{2}$	$\frac{3}{2}$	0	$-\frac{9}{4}$
(d)	$5 + 3i$	$5 - 3i$	10	34

103. (c) $\dfrac{4 + \sqrt{66}}{4} \approx 3.0$ seconds; Quadratic Formula

(d) 84-foot level: 3.5 seconds
100-foot level: 3.7 seconds

105. $b^2 - 4ac$. If the discriminant is positive, the quadratic equation has two real solutions; if it is zero, the equation has one (repeated) real solution; and if it is negative, the equation has no real solutions.

107. The four methods are factoring, extracting square roots, completing the square, and the Quadratic Formula.

Mid-Chapter Quiz *(page 591)*

1. ± 6 **2.** $-4, \frac{5}{2}$ **3.** $\pm 2\sqrt{3}$ **4.** $-1, 7$

5. $-5 \pm 2\sqrt{6}$ **6.** $\dfrac{-3 \pm \sqrt{19}}{2}$ **7.** $-2 \pm \sqrt{10}$

8. $\dfrac{3 \pm \sqrt{105}}{12}$ **9.** $-\dfrac{5}{2} \pm \dfrac{\sqrt{3}}{2}i$ **10.** $-2, 10$

11. $-3, 10$ **12.** $-2, 5$ **13.** $\dfrac{3}{2}$ **14.** $\dfrac{-5 \pm \sqrt{10}}{3}$

15. 36 **16.** $\pm 2i, \pm\sqrt{3}i$

17. **18.**

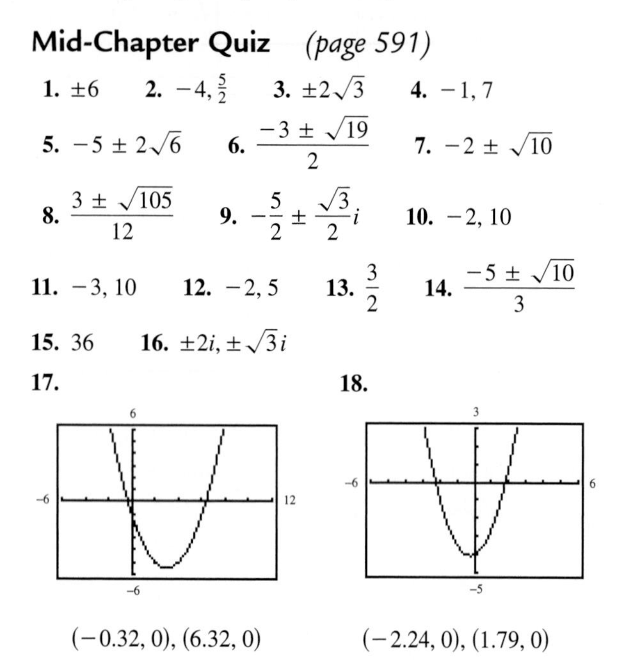

$(-0.32, 0), (6.32, 0)$ $(-2.24, 0), (1.79, 0)$

19. 50 units **20.** 35 meters $\times$ 65 meters

Section 10.4 *(page 599)*

Integrated Review *(page 599)*

1. $(x + b)^2 = x^2 + 2bx + b^2$ **2.** $x^2 + 5x + \frac{25}{4}$

3. $-11x$ **4.** $-41v$ **5.** $6x^2 + 9$ **6.** -4

7. $2|x|y\sqrt{6y}$ **8.** $3\sqrt[3]{5}$ **9.** $\dfrac{2\sqrt{3}b^3}{a^2}$ **10.** 2

11. $\sqrt{5} \approx 2.24$ seconds **12.** $\frac{5}{4}\sqrt{6} \approx 3.06$ seconds

1. e **3.** b **5.** d

7. $y = (x - 0)^2 + 2,\ (0, 2)$ **9.** $y = (x - 2)^2 + 3,\ (2, 3)$

11. $y = (x + 3)^2 - 4,\ (-3, -4)$

13. $y = -(x - 3)^2 - 1,\ (3, -1)$

15. $y = -(x - 1)^2 - 6,\ (1, -6)$

17. $y = 2\left(x + \frac{3}{2}\right)^2 - \frac{5}{2},\ \left(-\frac{3}{2}, -\frac{5}{2}\right)$ **19.** $(4, -1)$

21. $(-1, 2)$ **23.** $\left(-\frac{1}{2}, 3\right)$ **25.** Upward, $(0, 2)$

27. Downward, $(10, 4)$ **29.** Upward, $(0, -6)$

31. Downward, $(3, 0)$ **33.** $(\pm 5, 0), (0, 25)$

35. $(0, 0), (9, 0)$ **37.** $\left(\frac{3}{2}, 0\right), (0, 9)$ **39.** $(0, 3)$

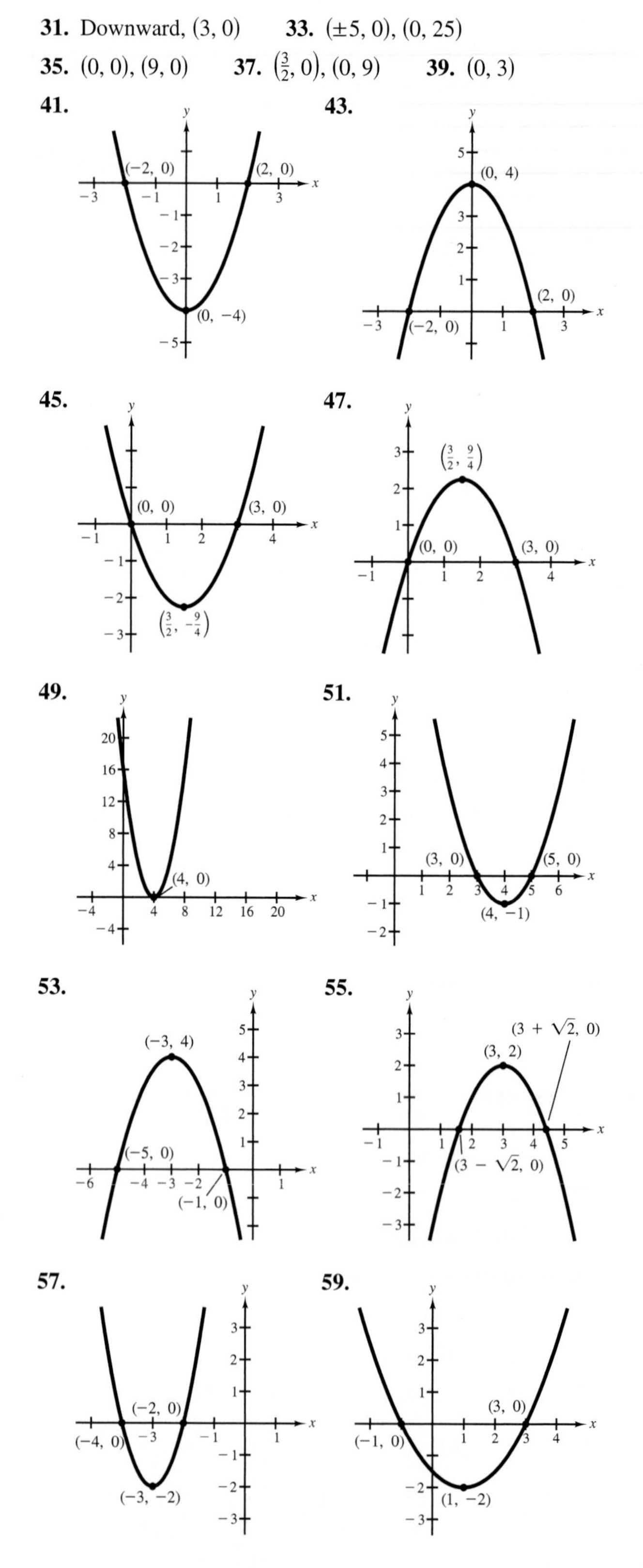

61. **63.**

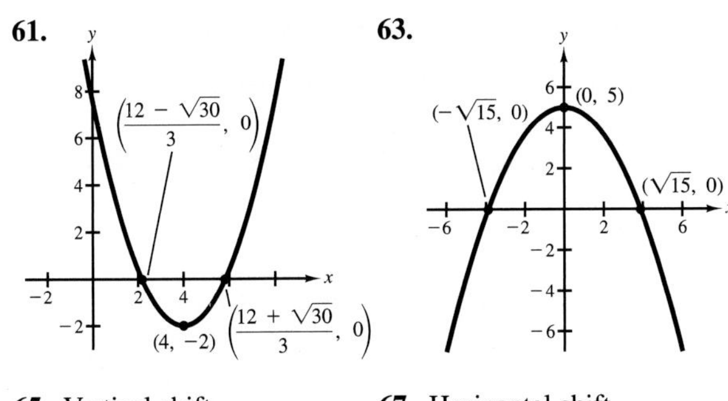

65. Vertical shift 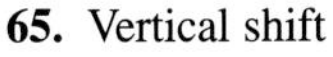 **67.** Horizontal shift

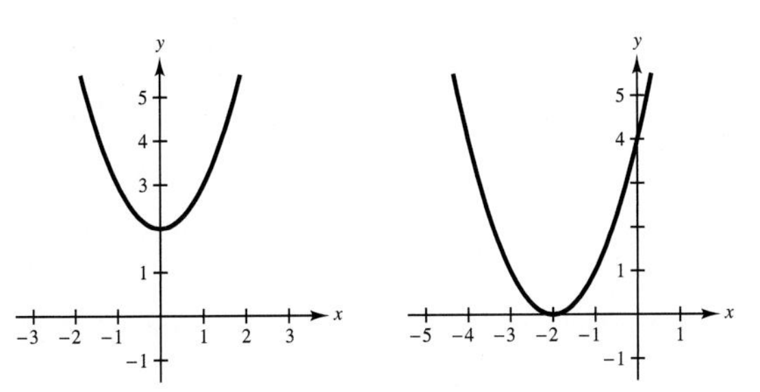

69. Horizontal and vertical shifts

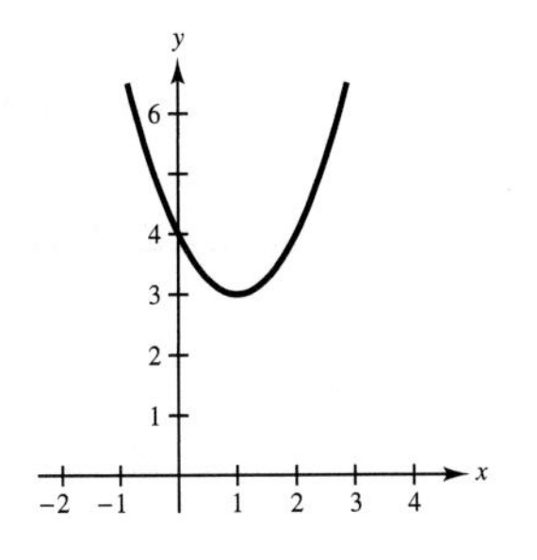

71. Horizontal and vertical shifts

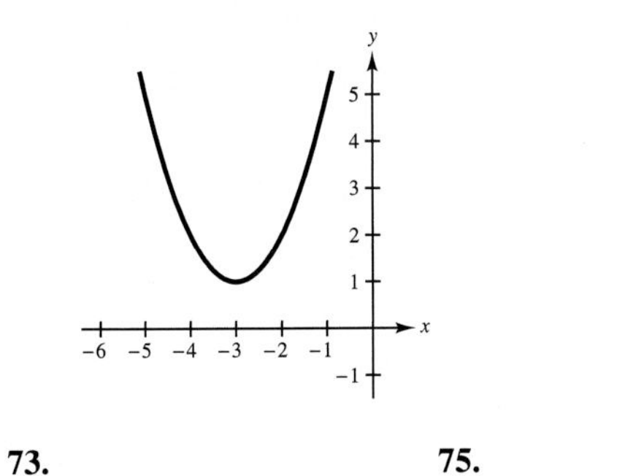

73.

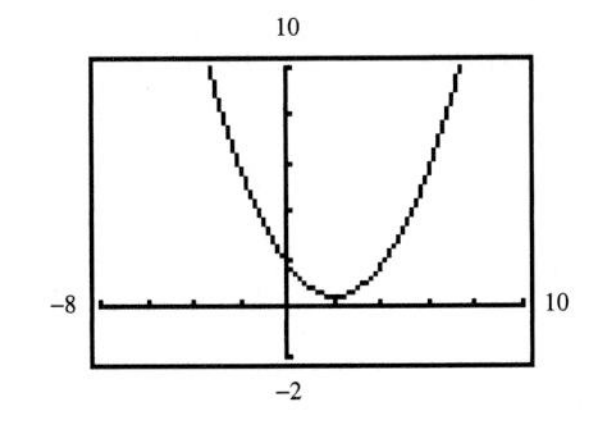

Vertex: (2, 0.5)

75.

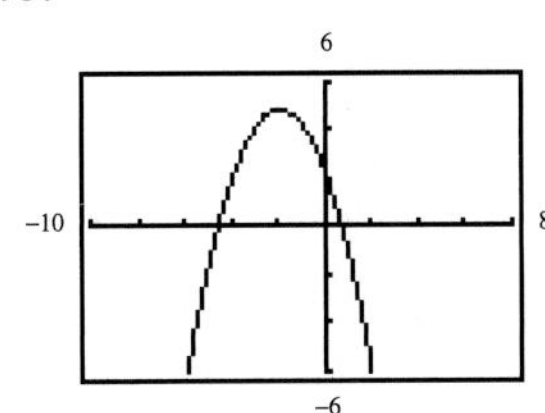

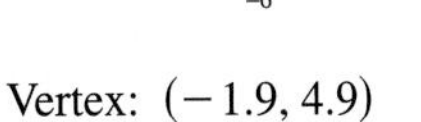
Vertex: (−1.9, 4.9)

77. $y = -x^2 + 4$ **79.** $y = x^2 + 4x + 2$

81. $y = -\frac{1}{2}x^2 + 2x + 4$ **83.** $y = x^2 - 4x + 5$

85. $y = x^2 - 4x$ **87.** $y = \frac{1}{2}x^2 - 3x + \frac{13}{2}$

89. $y = -4x^2 - 8x + 1$

91. Horizontal shift 3 units to the right

93. Horizontal shift 2 units to the right and vertical shift 3 units downward

95. (a) 4 feet

(b) 16 feet

(c) $12 + 8\sqrt{3} \approx 25.9$ feet

97. 14 feet

99. (a)

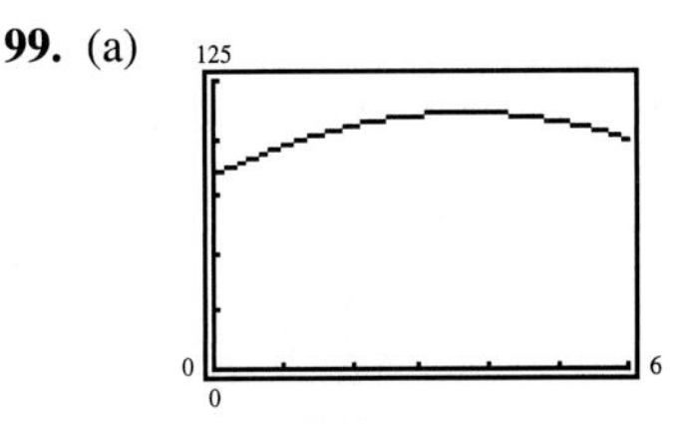

(b) 1993, 110,800

101. (a) Answers will vary.

(b) (50, 3375), 150 radios

(c) Recommend for orders of 150 radios or less

103.

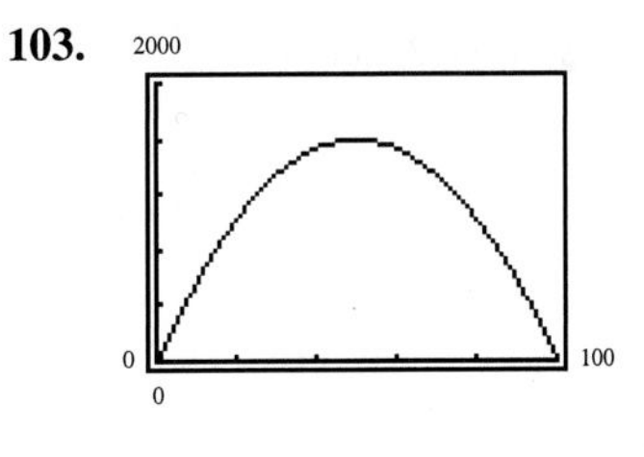

50

105. $y = \frac{1}{2500}x^2$ **107.** Parabola

109. To find any x-intercepts, set $y = 0$ and solve the resulting equation for x. To find any y-intercepts, set $x = 0$ and solve the resulting equation for y.

111. If the discriminant is positive, the parabola has two x-intercepts; if it is zero, the parabola has one x-intercept; and if it is negative, the parabola has no x-intercepts.

113. Find the y-coordinate of the vertex of the graph of the function.

Section 10.5 *(page 609)*

Integrated Review *(page 609)*

1. $m = \dfrac{y_2 - y_1}{x_2 - x_1}$ **2.** (a) $y = mx + b$

(b) $y - y_1 = m(x - x_1)$

(c) $Ax + By + C = 0$

(d) $y - b = 0$

3. $2x + 4y = 0$ **4.** $3x - 4y = 0$

5. $2x - y = 0$ **6.** $x + y - 6 = 0$

7. $22x + 16y - 161 = 0$

8. $134x - 73y + 146 = 0$ **9.** $y - 8 = 0$

10. $x + 3 = 0$ **11.** 8 people

12. 3 miles per hour

1. 18 dozen, \$1.20 per dozen **3.** 16, \$30

5. 108 square inches **7.** 70 feet **9.** 64 inches

11. 180 square kilometers **13.** 440 meters

15. 12 inches × 16 inches **17.** Base: 24 inches
Height: 16 inches

19. 50 feet × 250 feet or 100 feet × 125 feet

21. No.

Area $= \frac{1}{2}(b_1 + b_2)h = \frac{1}{2}x[x + (550 - 2x)] = 43{,}560$

This equation has no real solution.

23. Height: 12 inches
Width: 24 inches **25.** 8% **27.** 6%

29. 2.59% **31.** 48 **33.** 5

35. 15.86 miles or 2.14 miles

37. (a) $d = \sqrt{(3 + x)^2 + (4 + x)^2}$

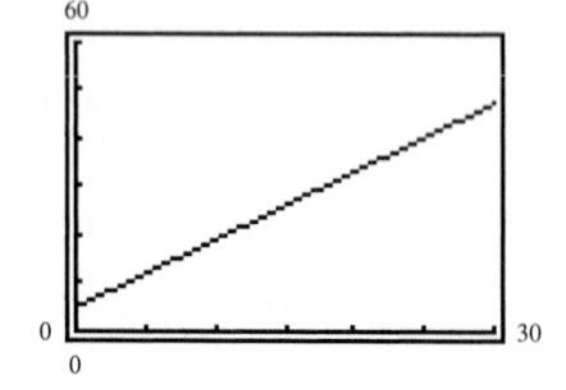

$x \approx 3.5$ when $d = 10$

(b) $\dfrac{-7 + \sqrt{199}}{2} \approx 3.55$ meters

39. 9.1 hours, 11.1 hours **41.** 6.8 days, 9.8 days

43. 3 seconds **45.** 9.5 seconds **47.** 4.7 seconds

49. (a) 3 seconds, 7 seconds **51.** 15, 16

(b) 10 seconds

53. 14, 16 **55.** 21, 23 **57.** 400 miles per hour

59. 46 miles per hour or 65 miles per hour

61. (a) $b = 20 - a$

$A = \pi ab$

$A = \pi a(20 - a)$

(b)

a	4	7	10	13	16
A	201.1	285.9	314.2	285.9	201.1

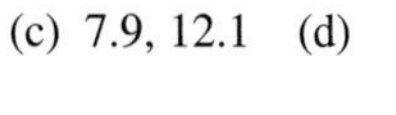
(c) 7.9, 12.1 (d)

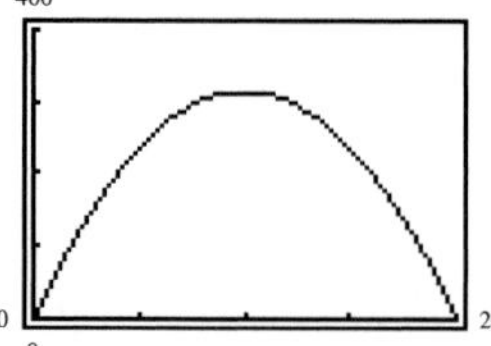

63. (a) Write a verbal model that will describe what you need to know.

(b) Assign labels to each part of the verbal model—numbers to the known quantities and letters to the variable quantities.

(c) Use the labels to write an algebraic model based on the verbal model.

(d) Solve the resulting algebraic equation and check your solution.

65. Dollars **67.** $(x + 4)^2 = 0$

Section 10.6 *(page 620)*

Integrated Review *(page 620)*

1. No. 3.682×10^9 **2.** $[10^6, 10^8)$

3. $6v(u^2 - 32v)$ **4.** $5x^{1/3}(x^{1/3} - 2)$

5. $(x - 10)(x - 4)$ **6.** $(x + 2)(x - 2)(x + 3)$

7. $(4x + 11)(4x - 11)$ **8.** $4x(x^2 - 3x + 4)$

9. $\frac{3}{2}h^2$ **10.** $\frac{1}{3}b^2$ **11.** $5x^2$ **12.** $x^2 + 8x$

1. $0, \frac{5}{2}$ **3.** $\pm\frac{9}{2}$ **5.** $-3, 5$ **7.** $1, 3$ **9.** $\frac{5}{2}$

11. Negative: $(-\infty, 4)$
Positive: $(4, \infty)$ **13.** Negative: $(6, \infty)$
Positive: $(-\infty, 6)$

15. Negative: $(0, 4)$
Positive: $(-\infty, 0) \cup (4, \infty)$

17. Negative: $(-\infty, -2) \cup (2, \infty)$
Positive: $(-2, 2)$

19. Negative: $(-1, 5)$
Positive: $(-\infty, -1) \cup (5, \infty)$

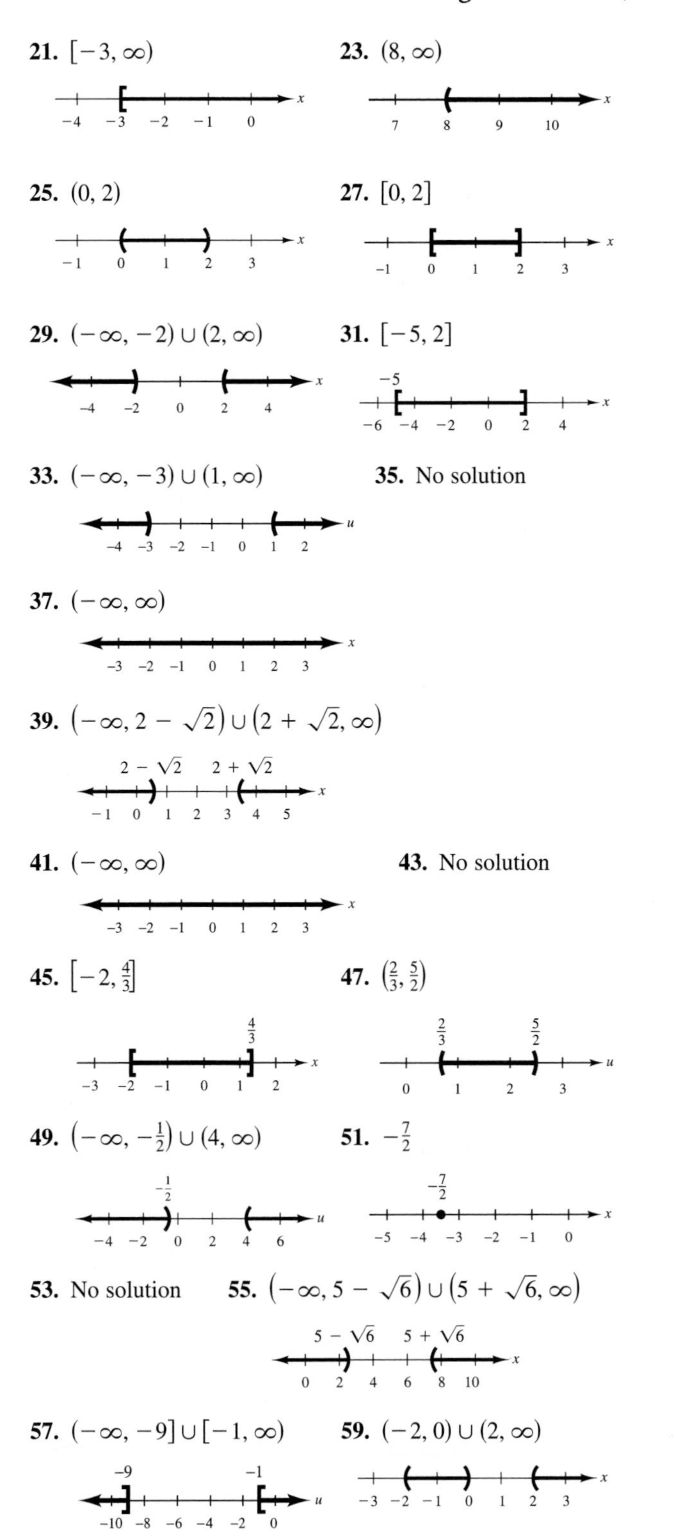

21. $[-3, \infty)$

23. $(8, \infty)$

25. $(0, 2)$

27. $[0, 2]$

29. $(-\infty, -2) \cup (2, \infty)$

31. $[-5, 2]$

33. $(-\infty, -3) \cup (1, \infty)$

35. No solution

37. $(-\infty, \infty)$

39. $(-\infty, 2 - \sqrt{2}) \cup (2 + \sqrt{2}, \infty)$

41. $(-\infty, \infty)$

43. No solution

45. $\left[-2, \frac{4}{3}\right]$

47. $\left(\frac{2}{3}, \frac{5}{2}\right)$

49. $\left(-\infty, -\frac{1}{2}\right) \cup (4, \infty)$

51. $-\frac{7}{2}$

53. No solution

55. $(-\infty, 5 - \sqrt{6}) \cup (5 + \sqrt{6}, \infty)$

57. $(-\infty, -9] \cup [-1, \infty)$

59. $(-2, 0) \cup (2, \infty)$

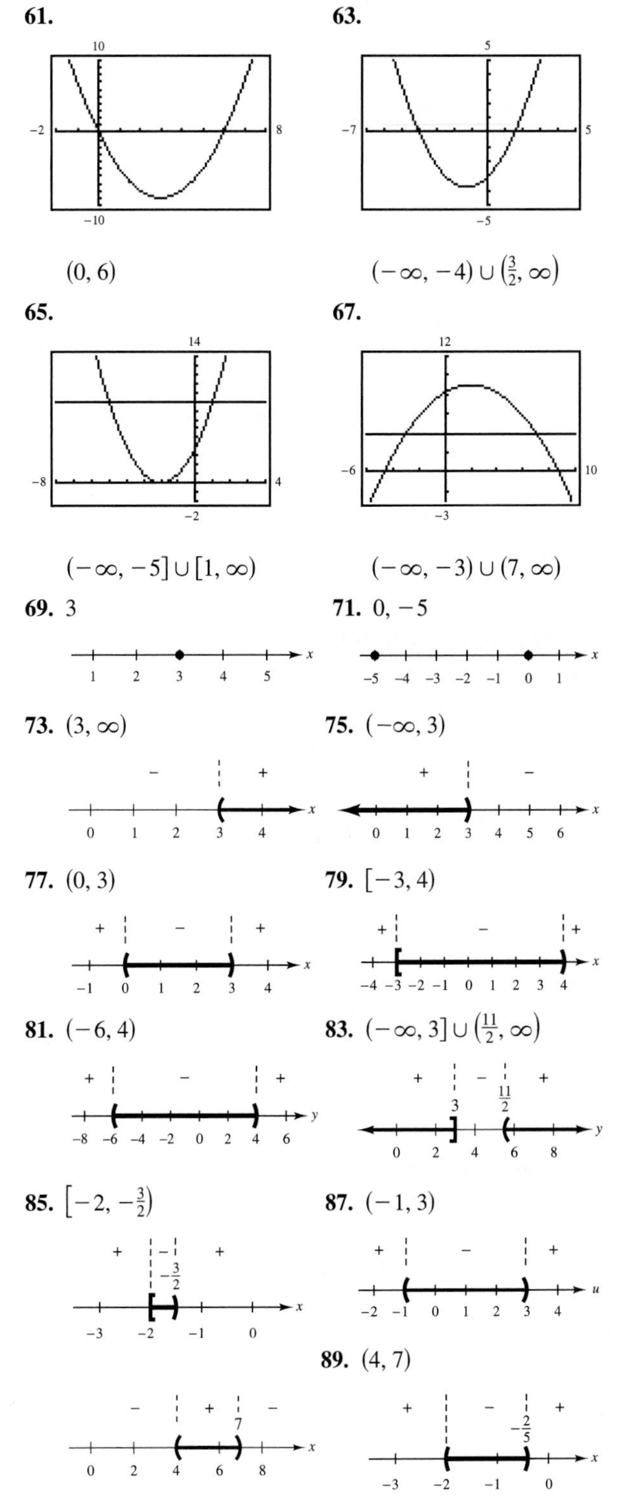

61. $(0, 6)$

63. $(-\infty, -4) \cup \left(\frac{3}{2}, \infty\right)$

65. $(-\infty, -5] \cup [1, \infty)$

67. $(-\infty, -3) \cup (7, \infty)$

69. 3

71. 0, -5

73. $(3, \infty)$

75. $(-\infty, 3)$

77. $(0, 3)$

79. $[-3, 4)$

81. $(-6, 4)$

83. $(-\infty, 3] \cup \left(\frac{11}{2}, \infty\right)$

85. $\left[-2, -\frac{3}{2}\right)$

87. $(-1, 3)$

89. $(4, 7)$

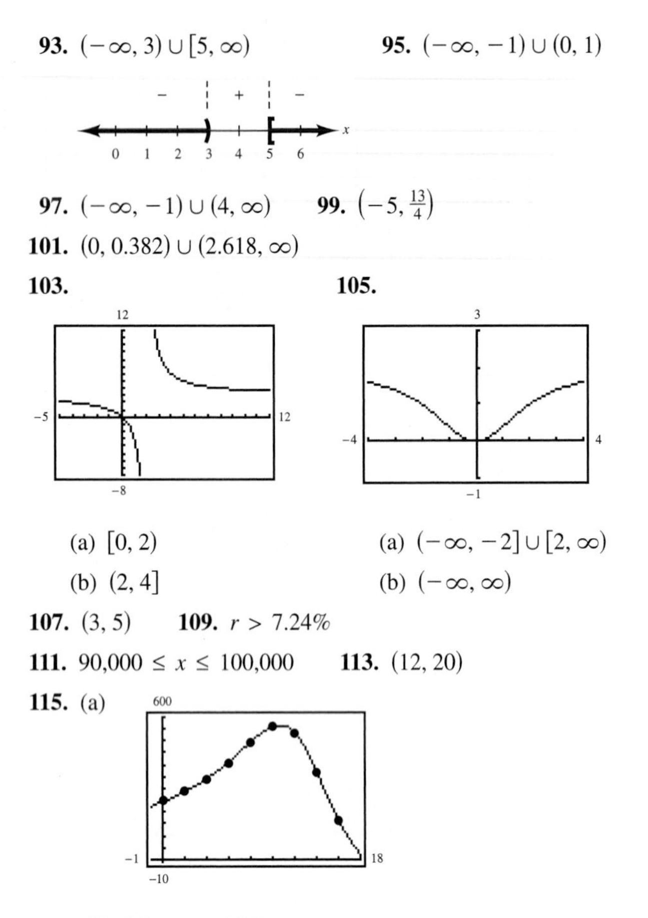

93. $(-\infty, 3) \cup [5, \infty)$ **95.** $(-\infty, -1) \cup (0, 1)$

97. $(-\infty, -1) \cup (4, \infty)$ **99.** $\left(-5, \frac{13}{4}\right)$

101. $(0, 0.382) \cup (2.618, \infty)$

103. **105.**

(a) $[0, 2)$ (a) $(-\infty, -2] \cup [2, \infty)$

(b) $(2, 4]$ (b) $(-\infty, \infty)$

107. $(3, 5)$ **109.** $r > 7.24\%$

111. $90{,}000 \le x \le 100{,}000$ **113.** $(12, 20)$

115. (a)

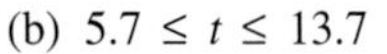

(b) $5.7 \le t \le 13.7$

117. The direction of the inequality is reversed.

119. A polynomial can change signs only at the x-values that make the polynomial zero. The zeros of the polynomial are called the critical numbers, and they are used to determine the test intervals in solving polynomial inequalities.

121. $x^2 + 1 < 0$

Review Exercises *(page 624)*

1. $-12, 0$ **3.** $\pm\frac{1}{2}$ **5.** $-\frac{5}{2}$ **7.** $-9, 10$ **9.** $-\frac{3}{2}, 6$

11. ± 50 **13.** $\pm 2\sqrt{3}$ **15.** $-4, 36$ **17.** $\pm 11i$

19. $\pm 5\sqrt{2}i$ **21.** $-4 \pm 3\sqrt{2}i$ **23.** $\pm\sqrt{5}, \pm i$

25. $1, 9$ **27.** $1, 1 \pm \sqrt{6}$ **29.** 64 **31.** $3 \pm 2\sqrt{3}$

33. $\frac{3}{2} \pm \frac{\sqrt{3}}{2}i$ **35.** $\frac{1}{3} \pm \frac{\sqrt{17}}{3}i$ **37.** $\frac{-5 \pm \sqrt{19}}{2}$

39. $-6, 5$ **41.** $-\frac{7}{2}, 3$ **43.** $\frac{8 \pm 3\sqrt{6}}{5}$

45. $\frac{10}{3} \pm \frac{5\sqrt{2}}{3}i$ **47.** One repeated rational solution

49. Two distinct rational solutions

51. Two distinct rational solutions

53. Two distinct imaginary solutions

55. $(4, -13)$ **57.** $\left(\frac{1}{4}, \frac{23}{8}\right)$

59. **61.**

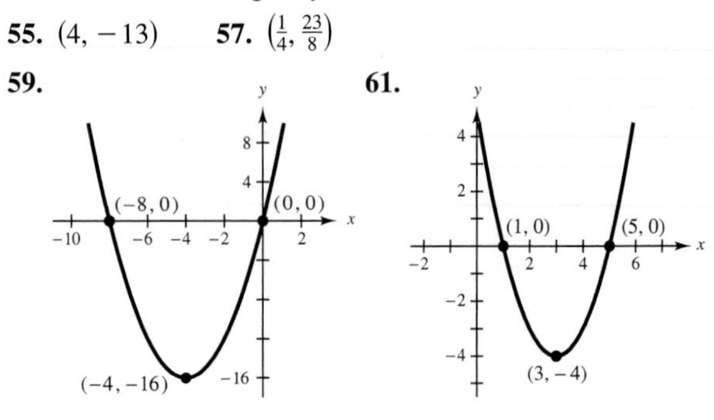

63. Vertical shift

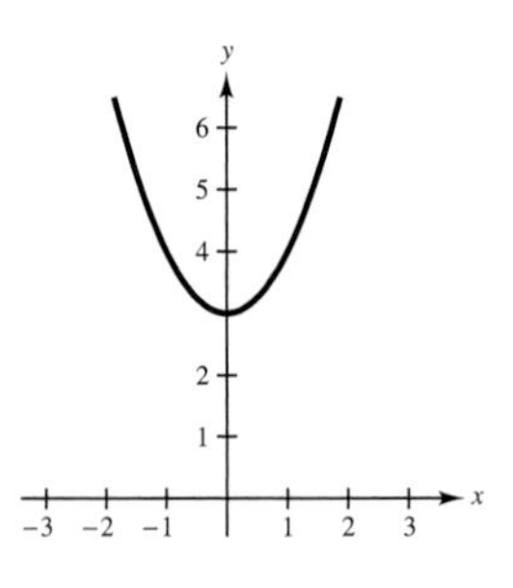

65. Horizontal shift and vertical shift

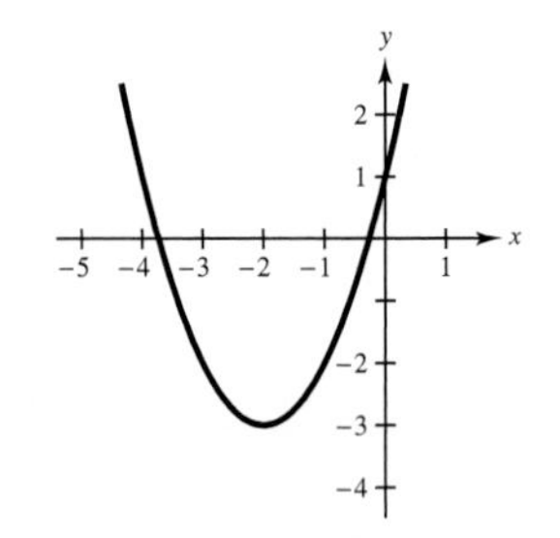

67. $y = -2(x - 3)^2 + 5$ **69.** $y = 2(x - 2)^2 - 5$

71. $y = \frac{1}{16}(x - 5)^2$

73. $(0, 7)$ **75.** $(-\infty, -2] \cup [6, \infty)$

77. $\left(-4, \frac{5}{2}\right)$ **79.** $(-\infty, -3] \cup \left(\frac{7}{2}, \infty\right)$

81. $\left(-6, -\frac{5}{2}\right)$

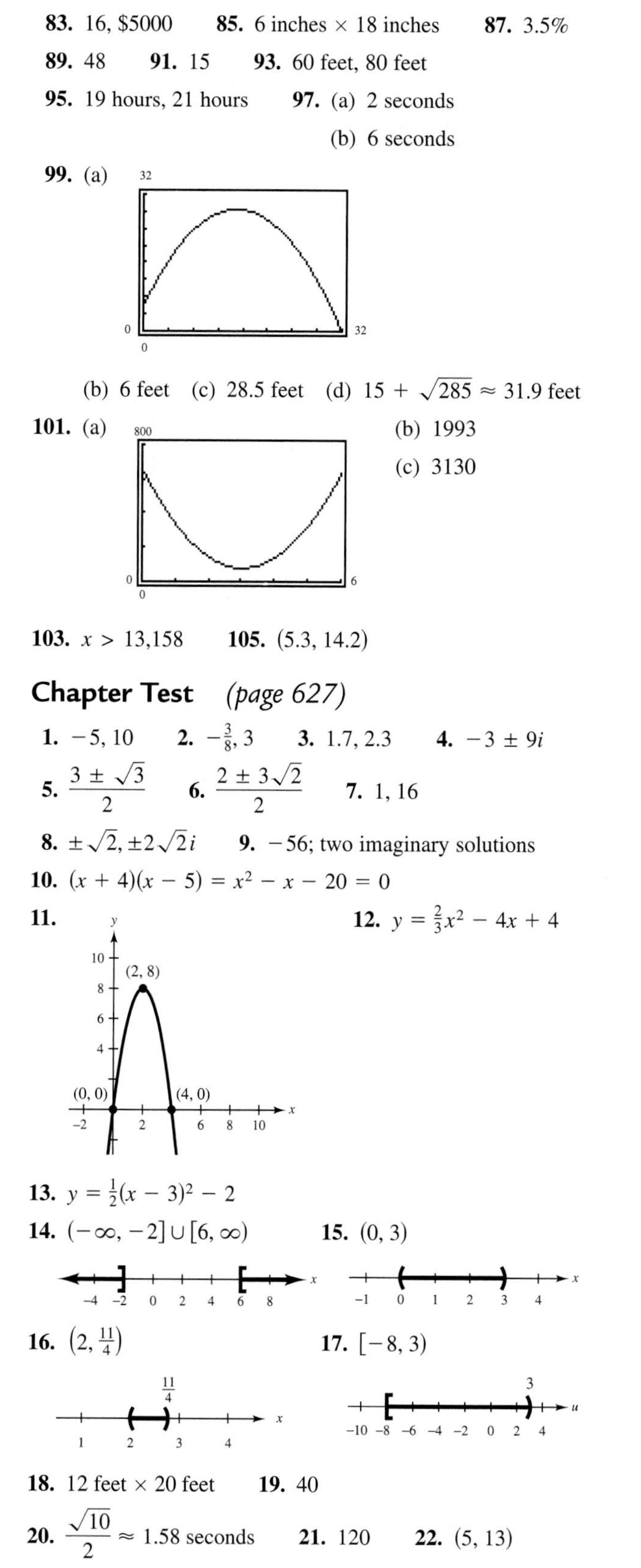

83. 16, \$5000 **85.** 6 inches × 18 inches **87.** 3.5%

89. 48 **91.** 15 **93.** 60 feet, 80 feet

95. 19 hours, 21 hours **97.** (a) 2 seconds (b) 6 seconds

99. (a)

(b) 6 feet (c) 28.5 feet (d) $15 + \sqrt{285} \approx 31.9$ feet

101. (a) (b) 1993 (c) 3130

103. $x > 13{,}158$ **105.** (5.3, 14.2)

Chapter Test *(page 627)*

1. $-5, 10$ **2.** $-\frac{3}{8}, 3$ **3.** 1.7, 2.3 **4.** $-3 \pm 9i$

5. $\dfrac{3 \pm \sqrt{3}}{2}$ **6.** $\dfrac{2 \pm 3\sqrt{2}}{2}$ **7.** 1, 16

8. $\pm\sqrt{2}, \pm 2\sqrt{2}i$ **9.** -56; two imaginary solutions

10. $(x + 4)(x - 5) = x^2 - x - 20 = 0$

11. **12.** $y = \frac{2}{3}x^2 - 4x + 4$

13. $y = \frac{1}{2}(x - 3)^2 - 2$

14. $(-\infty, -2] \cup [6, \infty)$ **15.** (0, 3)

16. $\left(2, \frac{11}{4}\right)$ **17.** $[-8, 3)$

18. 12 feet × 20 feet **19.** 40

20. $\dfrac{\sqrt{10}}{2} \approx 1.58$ seconds **21.** 120 **22.** (5, 13)

Chapter 11

Section 11.1 *(page 638)*

Integrated Review *(page 638)*

1. Test one point in each of the half-planes formed by the graph of $x + y = 5$. If the point satisfies the inequality, shade the entire half-plane to denote that every point in the region satisfies the inequality.

2. The first contains the boundary and the second does not.

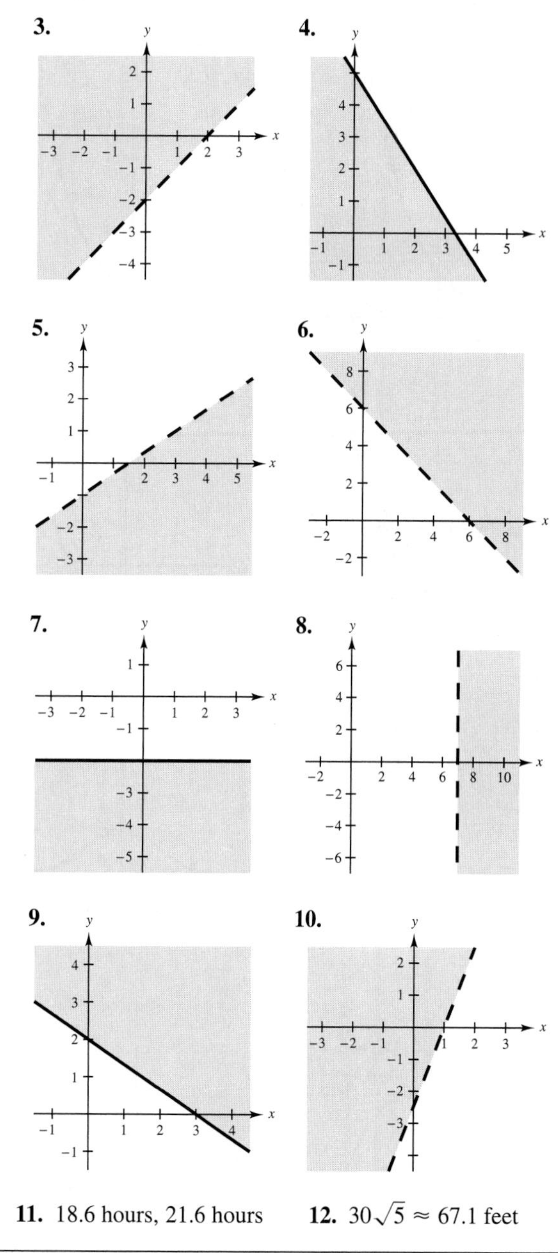

11. 18.6 hours, 21.6 hours **12.** $30\sqrt{5} \approx 67.1$ feet

1. 2^{2x-1} **3.** e^2 **5.** $8e^{3x}$ **7.** $-2e^x$ **9.** 11.036

11. 1.396 **13.** 51.193 **15.** 0.906

17. (a) $\frac{1}{9}$ (b) 1 (c) 3

19. (a) 0.935 (b) 1.225 (c) 1.163

21. (a) 500 (b) 250 (c) 56.657

23. (a) 1000 (b) 1628.895 (c) 2653.298

25. (a) 486.111 (b) 47.261 (c) 0.447

27. (a) 73.891 (b) 1.353 (c) 0.183

29. (a) 333.333 (b) 434.557 (c) 499.381

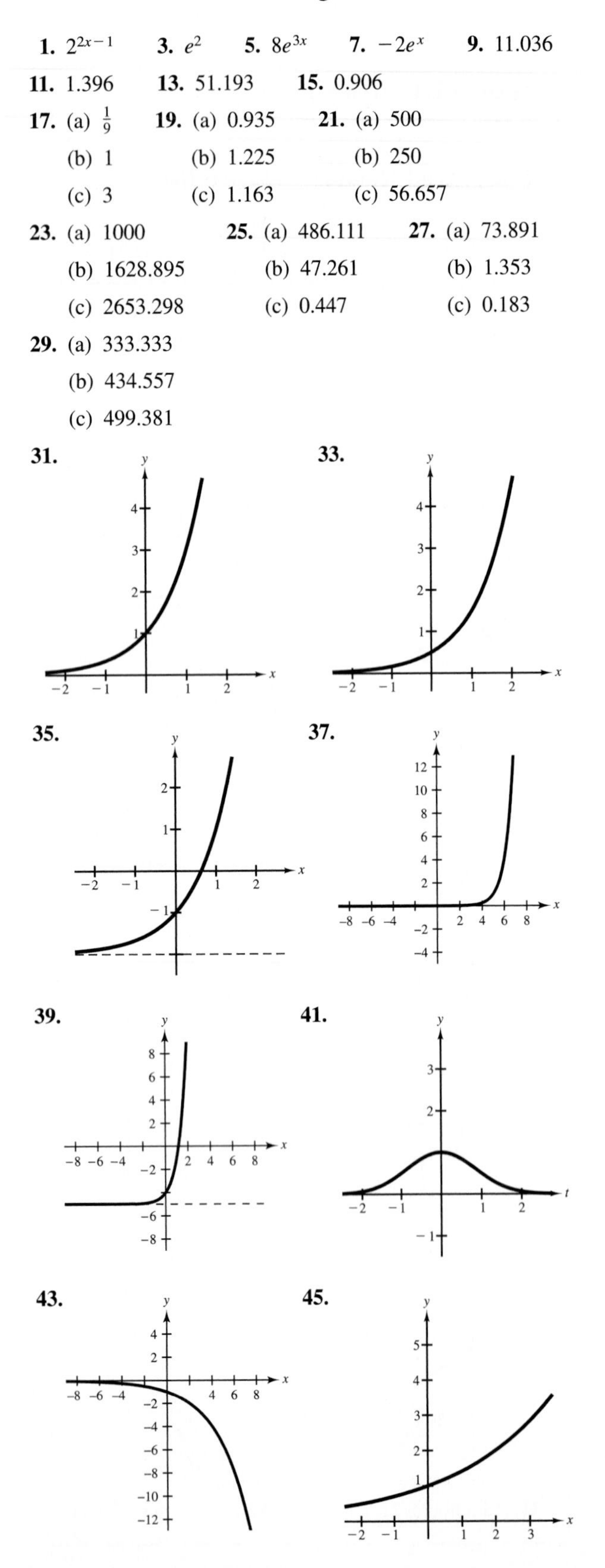

31. **33.** **35.** **37.** **39.** **41.** **43.** **45.**

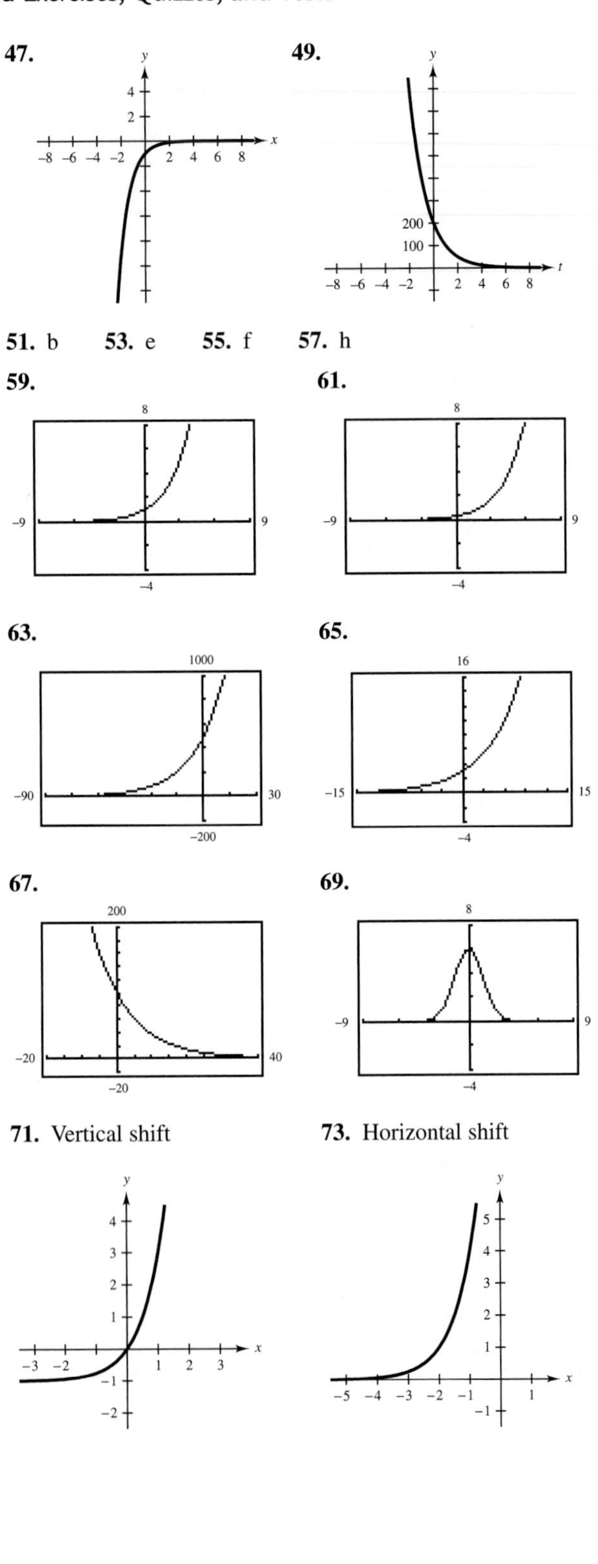

47. **49.**

51. b **53.** e **55.** f **57.** h

59. **61.** **63.** **65.** **67.** **69.**

71. Vertical shift **73.** Horizontal shift

75. Reflection in the x-axis

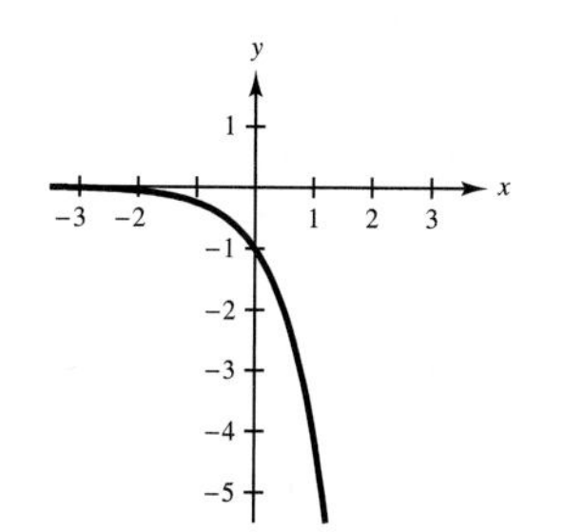

77. (a) Algebraic
(b) Algebraic
(c) Exponential
(d) Algebraic

79. 2.520 grams

81.

n	1	4	12	365	Continuous
A	\$466.10	\$487.54	\$492.68	\$495.22	\$495.30

83.

n	1	4	12
A	\$4734.73	\$4870.38	\$4902.71

n	365	Continuous
A	\$4918.66	\$4919.21

85.

n	1	4	12
A	\$226,296.28	\$259,889.34	\$268,503.32

n	365	Continuous
A	\$272,841.23	\$272,990.75

87.

n	1	4	12
P	\$2541.75	\$2498.00	\$2487.98

n	365	Continuous
P	\$2483.09	\$2482.93

89.

n	1	4	12
P	\$18,429.30	\$15,830.43	\$15,272.04

n	365	Continuous
P	\$15,004.64	\$14,995.58

91. (a) \$22.04
(b) \$20.13

93. (a) \$80,634.95
(b) \$161,269.89

95. $V(t) = 16{,}000\left(\frac{3}{4}\right)^t$

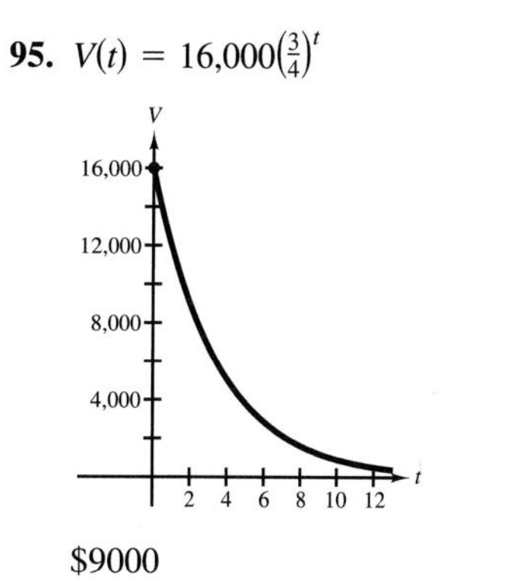

\$9000

97. (a) $A_1 = 500e^{0.06t}$, $A_2 = 500e^{0.08t}$
(b) (c)

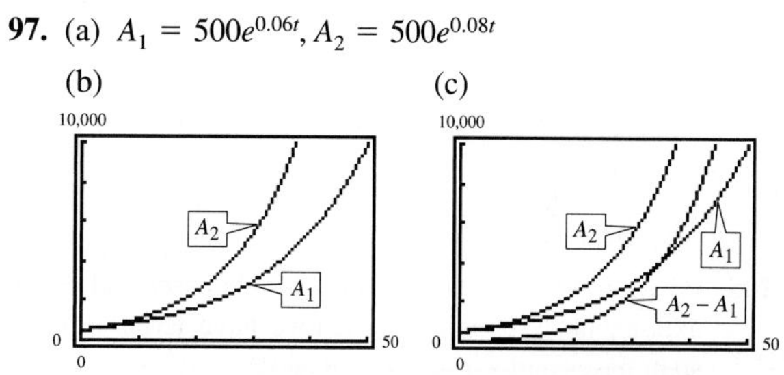

(d) The difference between the functions increases at an increasing rate.

99. (a)

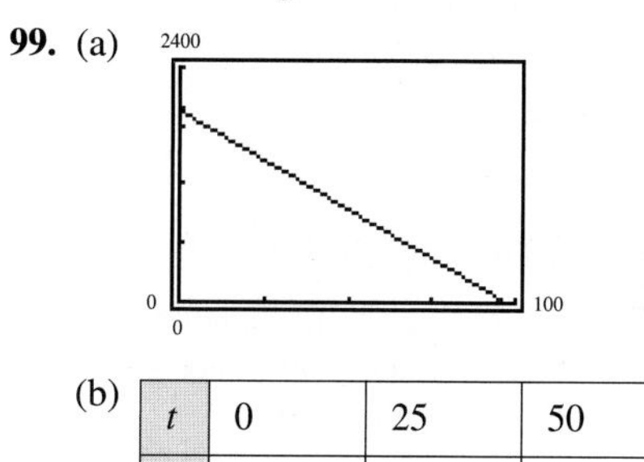

(b)

t	0	25	50	75
h	2000 ft	1450 ft	950 ft	450 ft

(c) Ground level: 97.5 seconds

101. (a) and (b)

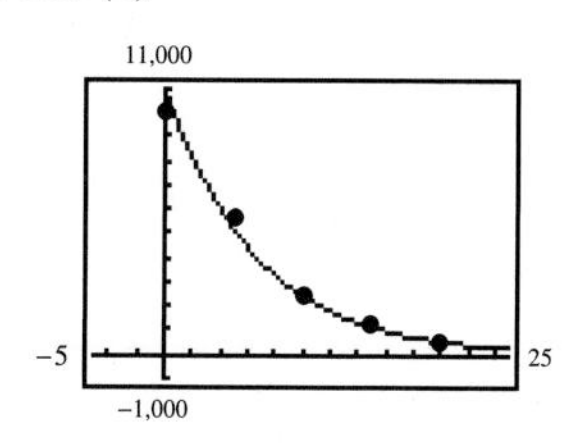

(c)

h	0	5	10	15	20
P	10,332	5583	2376	1240	517
Approx.	10,958	5176	2445	1155	546

(d) 3300 kilograms per square meter
(e) 11.3 kilometers

103. (a)

x	1	10	100	1000	10,000
$\left(1+\frac{1}{x}\right)^x$	2	2.5937	2.7048	2.7169	2.7181

(b)

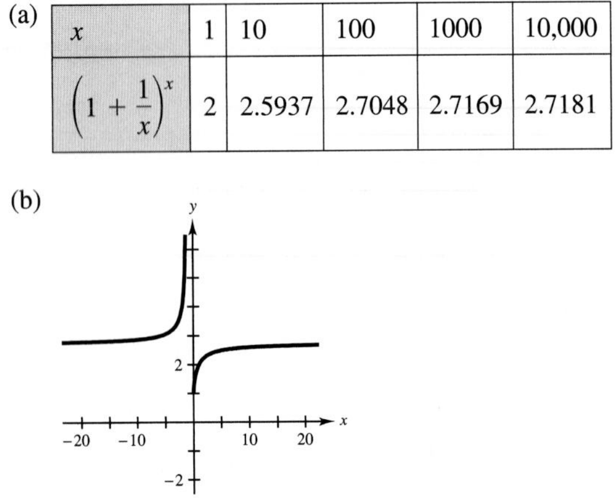

The graph appears to be approaching a horizontal asymptote.

(c) The value approaches e.

105. Polynomials have terms with variable bases and constant exponents. Exponential functions have terms with constant bases and variable exponents.

107. f is an increasing function and g is a decreasing function.

109. False. e is an irrational number.

Section 11.2 *(page 650)*

Integrated Review *(page 650)*

1. y is not a function of x because for some values of x there correspond two values of y. For example, $(4, 2)$ and $(4, -2)$ are solution points.
2. y is a function of x because for each value of x there corresponds exactly one value of y.
3. The domain of f is $-2 \le x \le 2$ and the domain of g is $-2 < x < 2$. g is undefined at $x = \pm 2$.

4. $\{4, 5, 6, 8\}$ **5.** $-2x^2 - 4$ **6.** $30x^3 + 40x^2$

7. $u^2 - 16v^2$ **8.** $9a^2 - 12ab + 4b^2$

9. $t^3 - 6t^2 + 12t - 8$ **10.** $\frac{1}{2}x^2 - \frac{1}{4}x$

11. 100 feet **12.** 13 minutes

1. (a) $2x - 7$ (b) $2x - 10$ (c) 1 (d) 4

3. (a) $2x^2 - 1$ (b) $2x^2 + 20x + 44$ (c) 7 (d) 2

5. (a) $|3x - 3|$ (b) $3|x - 3|$ (c) 0 (d) 3

7. (a) $\sqrt{x+1}$ (b) $\sqrt{x-4} + 5$ (c) 2 (d) 7

9. (a) $\dfrac{x^2}{2 - 3x^2}$ (b) $2(x - 3)^2$ (c) -1 (d) 2

11. (a) -1 (b) -2 (c) -2

13. (a) -1 (b) 1

15. (a) 10 (b) 1 (c) 1

17. (a) 0 (b) 10

19. (a) $(-\infty, \infty)$ (b) $(-\infty, \infty)$

21. (a) $[2, \infty)$ (b) $[0, \infty)$

23. (a) $[-3, \infty)$ (b) $(-\infty, \infty)$

25. (a) $[1, \infty)$ (b) $(-\infty, -5)$

27. No **29.** Yes **31.** No

33. Yes

35. Yes

37. No

39. Yes

41. No

43. $f(g(x)) = f\left(\frac{1}{10}x\right) = 10\left(\frac{1}{10}x\right) = x$

$g(f(x)) = g(10x) = \frac{1}{10}(10x) = x$

45. $f(g(x)) = f(x - 15) = (x - 15) + 15 = x$

$g(f(x)) = g(x + 15) = (x + 15) - 15 = x$

47. $f(g(x)) = f\left[\frac{1}{2}(1 - x)\right] = 1 - 2\left[\frac{1}{2}(1 - x)\right]$
$= 1 - (1 - x) = x$
$g(f(x)) = g(1 - 2x) = \frac{1}{2}[1 - (1 - 2x)] = \frac{1}{2}(2x) = x$

49. $f(g(x)) = f\left[\frac{1}{3}(2 - x)\right] = 2 - 3\left[\frac{1}{3}(2 - x)\right]$
$= 2 - (2 - x) = x$
$g(f(x)) = g(2 - 3x) = \frac{1}{3}[2 - (2 - 3x)] = \frac{1}{3}(3x) = x$

51. $f(g(x)) = f(x^3 - 1) = \sqrt[3]{(x^3 - 1) + 1} = \sqrt[3]{x^3} = x$
$g(f(x)) = g\left(\sqrt[3]{x + 1}\right) = \left(\sqrt[3]{x + 1}\right)^3 - 1$
$= x + 1 - 1 = x$

53. $f(g(x)) = f\left(\frac{1}{x}\right) = \frac{1}{(1/x)} = x$
$g(f(x)) = g\left(\frac{1}{x}\right) = \frac{1}{(1/x)} = x$

55. $f^{-1}(x) = \frac{1}{5}x$ **57.** $f^{-1}(x) = 2x$

59. $f^{-1}(x) = x - 10$ **61.** $f^{-1}(x) = 3 - x$

63. $f^{-1}(x) = \sqrt[7]{x}$ **65.** $f^{-1}(x) = x^3$ **67.** $f^{-1}(x) = \frac{x}{8}$

69. $g^{-1}(x) = x - 25$ **71.** $g^{-1}(x) = \frac{3 - x}{4}$

73. $g^{-1}(t) = 4t - 8$ **75.** $h^{-1}(x) = x^2,\ x \ge 0$

77. $f^{-1}(t) = \sqrt[3]{t + 1}$ **79.** $g^{-1}(s) = \frac{5}{s} - 4$

81. $f^{-1}(x) = x^2 - 3,\ x \ge 0$

83. **85.**

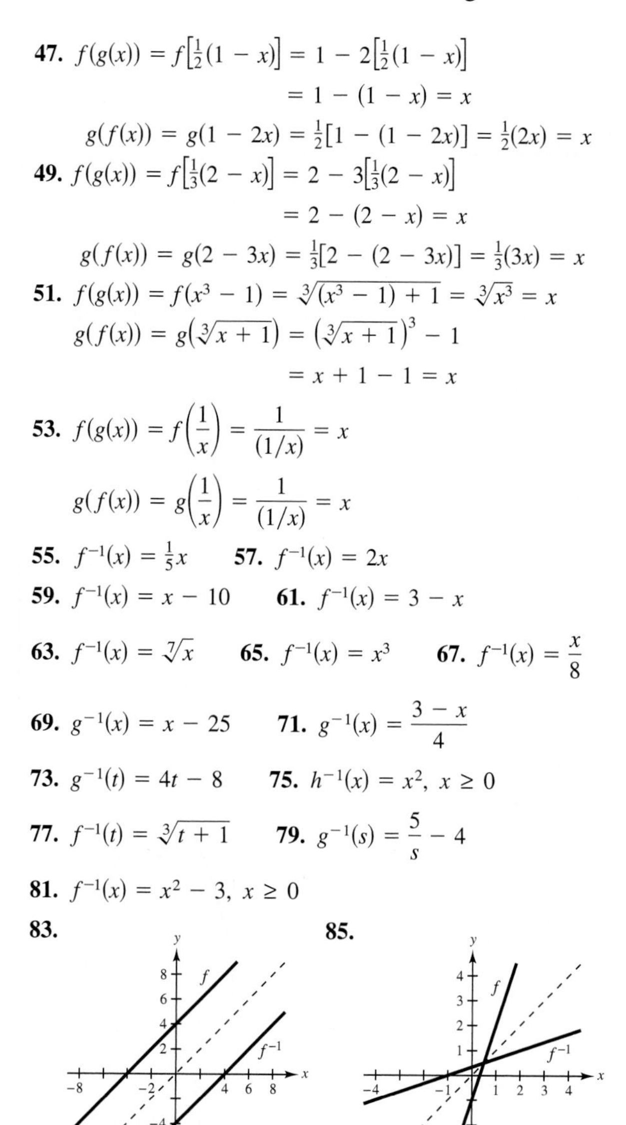

87.

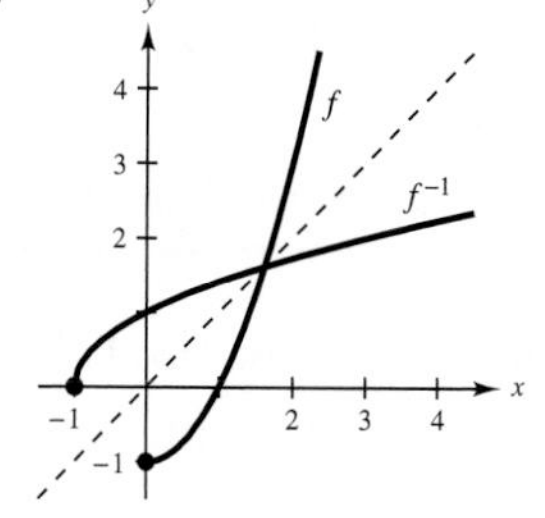

89. b **91.** d

93. **95.** **97.** **99.**

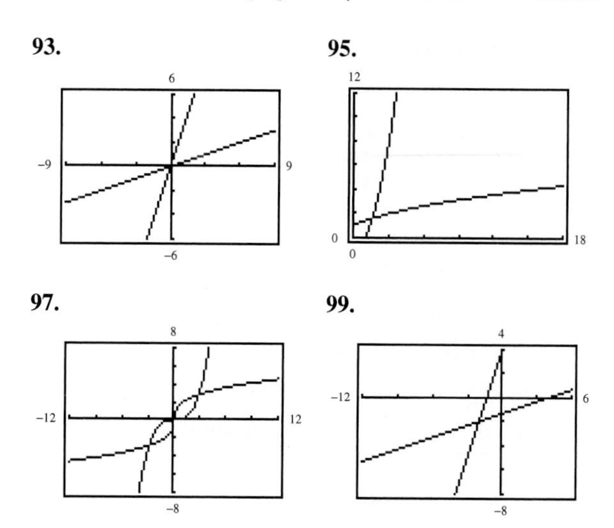

101. $x \ge 2$; $f^{-1}(x) = \sqrt{x} + 2$, domain of f^{-1}: $x \ge 0$

103. $x \ge 0$; $f^{-1}(x) = x - 1$, domain of f^{-1}: $x \ge 1$

105.

x	0	1	3	4
f^{-1}	6	4	2	0

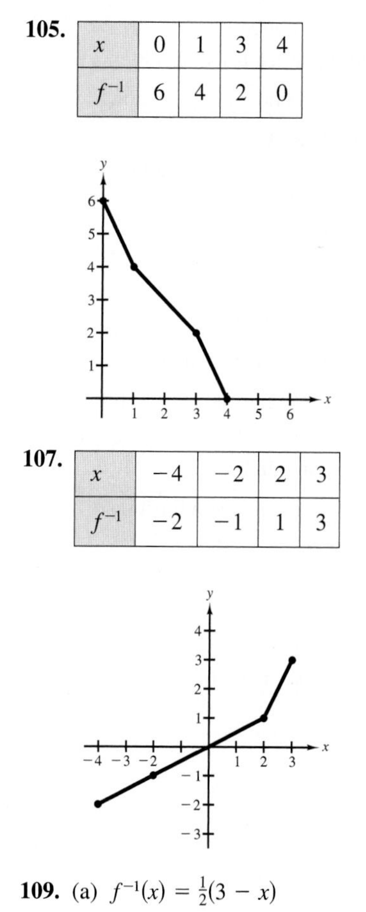

107.

x	-4	-2	2	3
f^{-1}	-2	-1	1	3

109. (a) $f^{-1}(x) = \frac{1}{2}(3 - x)$
(b) $(f^{-1})^{-1}(x) = 3 - 2x$

111. (a) Total cost = cost of \$0.50 compound + cost of \$0.75 compound

$y = 0.50x + 0.75(100 - x)$

(b) $y = 4(75 - x)$
x: total cost
y: number of pounds at \$0.50 per pound

(c) $50 \leq x \leq 75$

(d) 60 pounds

113. (a) $f(g(x)) = 0.02x - 200{,}000$

(b) $g(f(x)) = 0.02(x - 200{,}000)$

$g(f(x))$ represents the bonus, because it gives 2% of sales over \$200,000.

115. (a) $R = p - 2000$

(b) $S = 0.95p$

(c) $(R \circ S)(p) = 0.95p - 2000$;
5% discount followed by the \$2000 rebate

$(S \circ R)(p) = 0.95(p - 2000)$;
5% discount after the price is reduced by the rebate

(d) $(R \circ S)(26{,}000) = 22{,}700$

$(S \circ R)(26{,}000) = 22{,}800$

$R \circ S$ yields the smaller cost because the dealer discount is calculated on a larger base.

117. True

119. False

$f(x) = \sqrt{x - 1}$; Domain: $[1, \infty)$

$f^{-1}(x) = x^2 + 1,\ x \geq 0$; Domain: $[0, \infty)$

121. If $f(x) = 2x$ and $g(x) = x^2$, then $(f \circ g)(x) = 2x^2$ and $(g \circ f)(x) = 4x^2$.

123.
- Interchange the roles of x and y.
- If the new equation represents y as a function of x, solve the new equation for y.
- Replace y by $f^{-1}(x)$.

125. Graphically, a function f has an inverse function if and only if no horizontal line intersects the graph of f at more than one point. This is equivalent to saying that the function f is one-to-one.

Section 11.3 *(page 663)*

Integrated Review *(page 663)*

1. Horizontal shift 4 units to the right

2. Reflection in the x-axis

3. Vertical shift 1 unit upward

4. Horizontal shift 3 units to the left and a vertical shift 5 units downward

5. $2x(x^2 - 3)$ **6.** $(2 - y)(6 + y)$

7. $(t + 5)^2$ **8.** $(5 - u)(1 + u^2)$

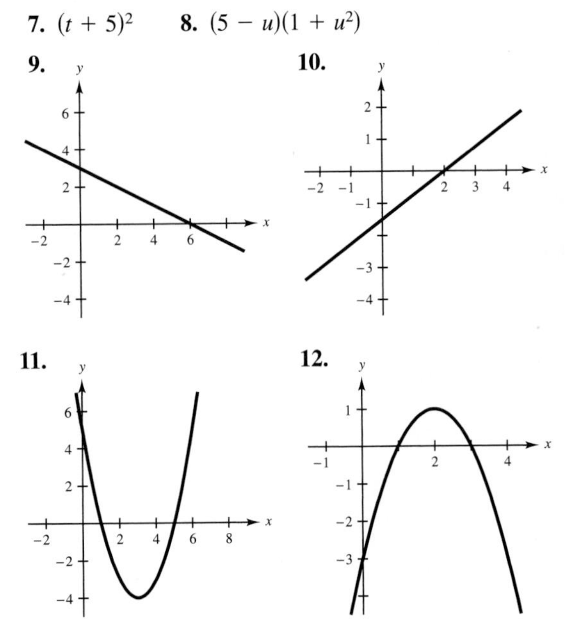

1. $5^2 = 25$ **3.** $4^{-2} = \frac{1}{16}$ **5.** $3^{-5} = \frac{1}{243}$

7. $36^{1/2} = 6$ **9.** $8^{2/3} = 4$ **11.** $2^{1.3} \approx 2.462$

13. $\log_7 49 = 2$ **15.** $\log_3 \frac{1}{9} = -2$ **17.** $\log_8 4 = \frac{2}{3}$

19. $\log_{25} \frac{1}{5} = -\frac{1}{2}$ **21.** $\log_4 1 = 0$

23. $\log_5 9.518 \approx 1.4$ **25.** 3 **27.** 1 **29.** 3

31. -2 **33.** -3 **35.** -4

37. There is no power to which 2 can be raised to obtain -3.

39. 0

41. There is no power to which 5 can be raised to obtain -6.

43. $\frac{1}{2}$ **45.** $\frac{3}{4}$ **47.** 4 **49.** 1.4914 **51.** -0.0706

53. 0.7335

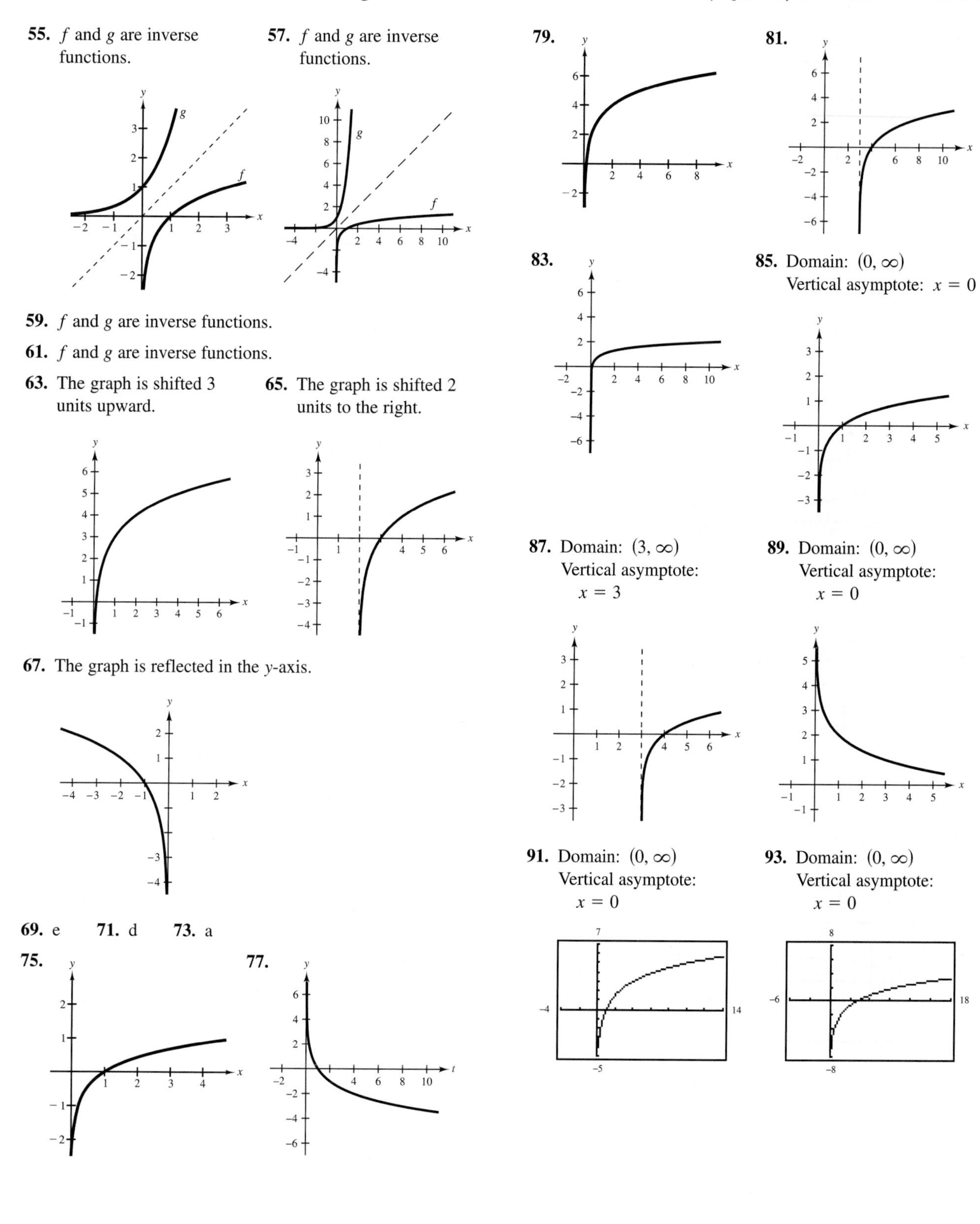
55. f and g are inverse functions.
57. f and g are inverse functions.
59. f and g are inverse functions.
61. f and g are inverse functions.
63. The graph is shifted 3 units upward.
65. The graph is shifted 2 units to the right.
67. The graph is reflected in the y-axis.
69. e
71. d
73. a
75.
77.
79.
81.
83.
85. Domain: (0, ∞)
Vertical asymptote: x = 0
87. Domain: (3, ∞)
Vertical asymptote: x = 3
89. Domain: (0, ∞)
Vertical asymptote: x = 0
91. Domain: (0, ∞)
Vertical asymptote: x = 0
93. Domain: (0, ∞)
Vertical asymptote: x = 0

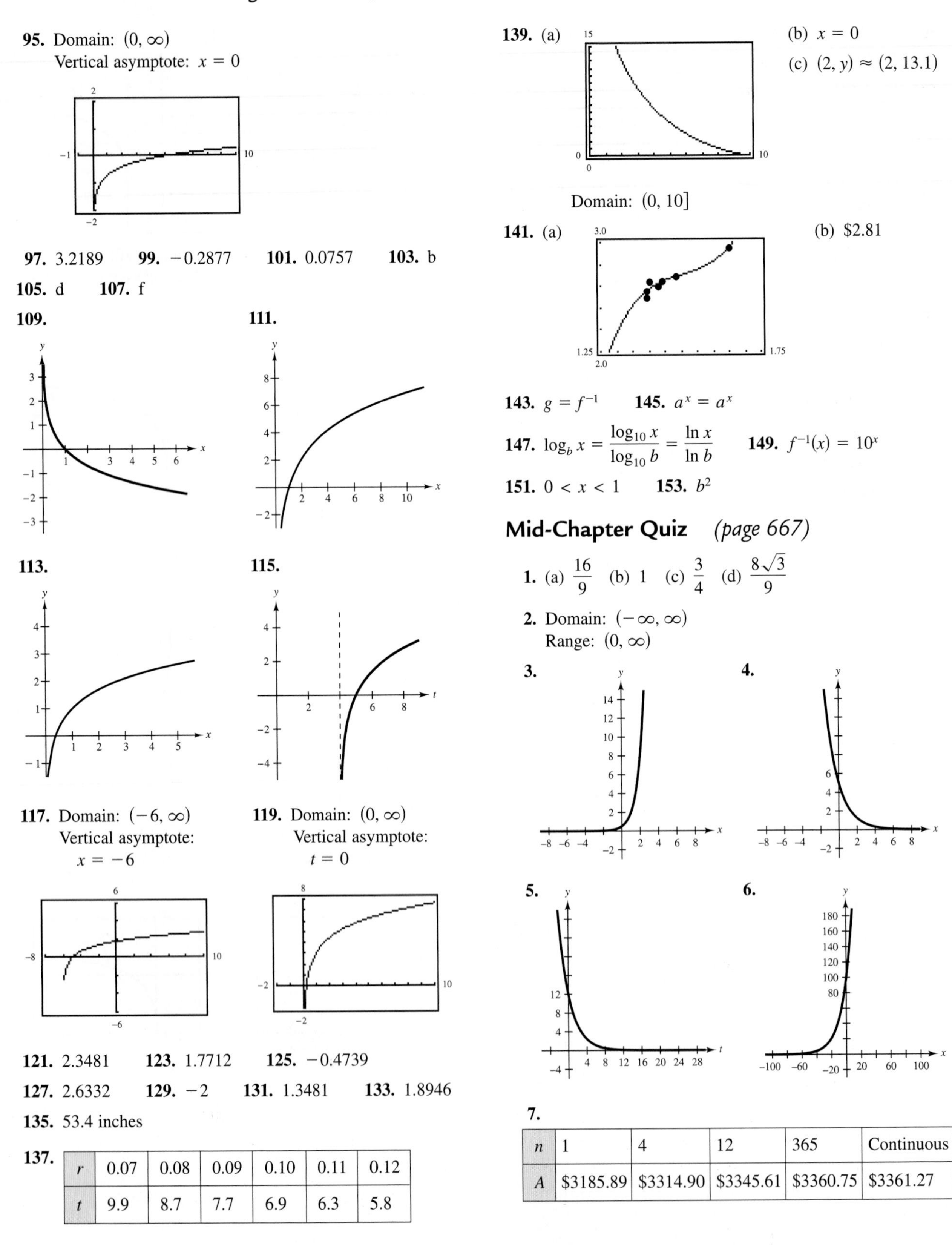

95. Domain: $(0, \infty)$
Vertical asymptote: $x = 0$

97. 3.2189 **99.** -0.2877 **101.** 0.0757 **103.** b

105. d **107.** f

109.

111.

113.

115.

117. Domain: $(-6, \infty)$
Vertical asymptote:
$x = -6$

119. Domain: $(0, \infty)$
Vertical asymptote:
$t = 0$

121. 2.3481 **123.** 1.7712 **125.** -0.4739

127. 2.6332 **129.** -2 **131.** 1.3481 **133.** 1.8946

135. 53.4 inches

137.

r	0.07	0.08	0.09	0.10	0.11	0.12
t	9.9	8.7	7.7	6.9	6.3	5.8

139. (a)
(b) $x = 0$
(c) $(2, y) \approx (2, 13.1)$

Domain: $(0, 10]$

141. (a)
(b) \$2.81

143. $g = f^{-1}$ **145.** $a^x = a^x$

147. $\log_b x = \dfrac{\log_{10} x}{\log_{10} b} = \dfrac{\ln x}{\ln b}$ **149.** $f^{-1}(x) = 10^x$

151. $0 < x < 1$ **153.** b^2

Mid-Chapter Quiz *(page 667)*

1. (a) $\dfrac{16}{9}$ (b) 1 (c) $\dfrac{3}{4}$ (d) $\dfrac{8\sqrt{3}}{9}$

2. Domain: $(-\infty, \infty)$
Range: $(0, \infty)$

3.

4.

5.

6.

7.

n	1	4	12	365	Continuous
A	\$3185.89	\$3314.90	\$3345.61	\$3360.75	\$3361.27

8. \$2.71

9. (a) $2x^3 - 3$ (b) $(2x - 3)^3$ (c) -19 (d) 125

10. $f(g(x)) = 3 - 5\left[\frac{1}{5}(3 - x)\right] = 3 - 3 + x = x$

$g(f(x)) = \frac{1}{5}[3 - (3 - 5x)] = \frac{1}{5}(5x) = x$

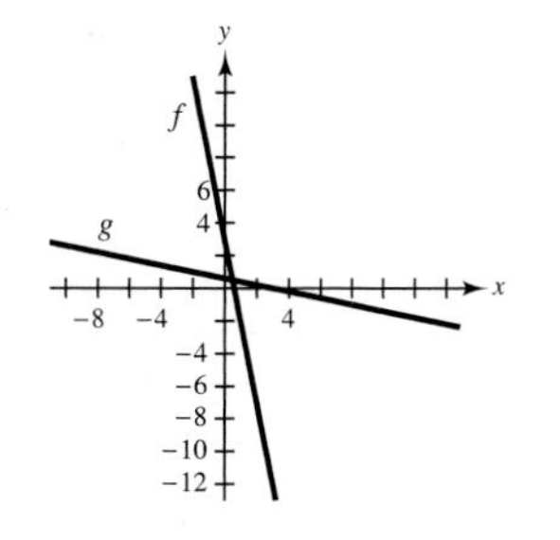

11. $h^{-1}(x) = \frac{1}{10}(x - 3)$ **12.** $g^{-1}(t) = \sqrt[3]{2(t - 2)}$

13. $4^{-2} = \frac{1}{16}$ **14.** $\log_3 81 = 4$ **15.** 3

16. $f^{-1}(x) = g(x)$

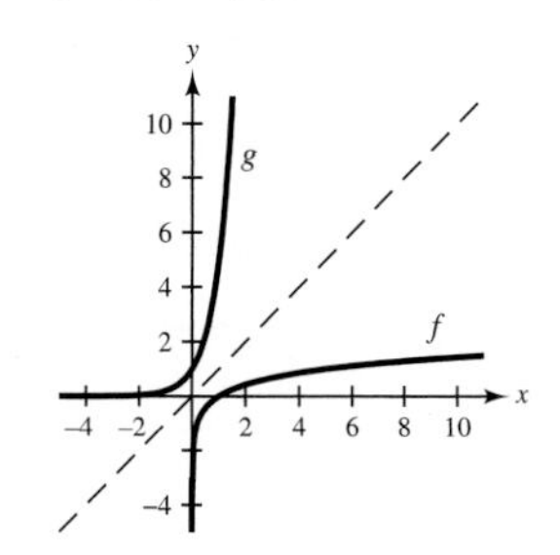

17. **18.**

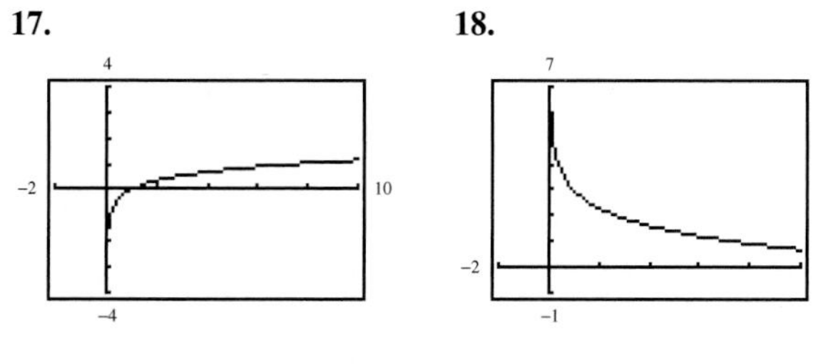

19. $h = 2, k = 1$ **20.** 3.4096

Section 11.4 *(page 673)*

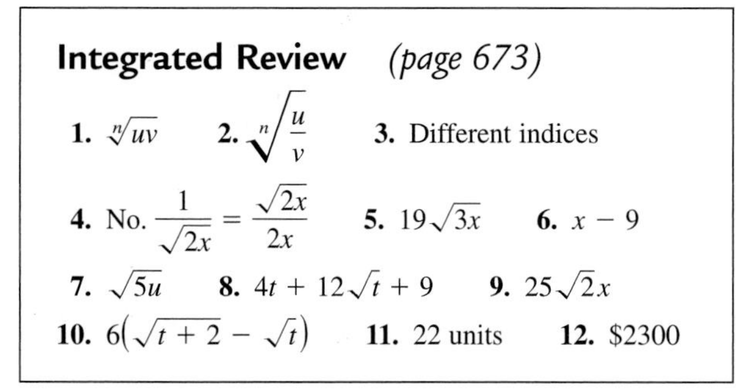

Integrated Review *(page 673)*

1. $\sqrt[n]{uv}$ **2.** $\sqrt[n]{\dfrac{u}{v}}$ **3.** Different indices

4. No. $\dfrac{1}{\sqrt{2x}} = \dfrac{\sqrt{2x}}{2x}$ **5.** $19\sqrt{3x}$ **6.** $x - 9$

7. $\sqrt{5u}$ **8.** $4t + 12\sqrt{t} + 9$ **9.** $25\sqrt{2x}$

10. $6\left(\sqrt{t + 2} - \sqrt{t}\right)$ **11.** 22 units **12.** \$2300

1. 2 **3.** -9 **5.** $\frac{1}{2}$ **7.** 0 **9.** 4 **11.** 2

13. 2 **15.** 1 **17.** 2 **19.** -3 **21.** 12 **23.** 1

25. 1 **27.** 1.2925 **29.** 0.2925 **31.** 0.2500

33. 2.7925 **35.** 0 **37.** 0.954 **39.** 1.556

41. 0.778 **43.** $\log_3 11 + \log_3 x$ **45.** $2 \log_7 x$

47. $-2 \log_5 x$ **49.** $\frac{1}{2}(\log_4 3 + \log_4 x)$ **51.** $\ln 3 + \ln y$

53. $\log_2 z - \log_2 17$ **55.** $\ln 5 - \ln(x - 2)$

57. $2 \ln x + \ln(y - 2)$ **59.** $6 \log_4 x + 2 \log_4(x - 7)$

61. $\frac{1}{3} \log_3(x + 1)$ **63.** $\frac{1}{2}[\ln x + \ln(x + 2)]$

65. $2[\ln(x + 1) - \ln(x - 1)]$ **67.** $\frac{1}{3}[2 \ln x - \ln(x + 1)]$

69. $3 \ln a + \ln(b - 4) - 2 \ln c$

71. $\ln x + \frac{1}{3} \ln y - 4(\ln w + \ln z)$

73. $\log_6 a + \frac{1}{2} \log_6 b + 3 \log_6(c - d)$

75. $\ln(x + y) + \frac{1}{5} \ln(w + 2) - (\ln 3 + \ln t)$ **77.** $\log_{12} \dfrac{x}{3}$

79. $\log_2 3x$ **81.** $\log_{10} \dfrac{4}{x}$ **83.** $\ln b^4,\ b > 0$

85. $\log_5(2x)^{-2},\ x > 0$ **87.** $\ln \sqrt[3]{2x + 1}$ **89.** $\log_3 2\sqrt{y}$

91. $\ln \dfrac{x^2y^3}{z},\ x > 0, y > 0, z > 0$

93. $\ln \dfrac{2^5y^3}{x},\ x > 0, y > 0$ **95.** $\ln(xy)^4,\ x > 0, y > 0$

97. $\ln\left(\dfrac{x}{x + 1}\right)^2,\ x > 0$ **99.** $\log_4 \dfrac{x + 8}{x^3},\ x > 0$

101. $\log_5 \dfrac{\sqrt{x + 2}}{x - 3}$ **103.** $\log_6 \dfrac{(c + d)^5}{\sqrt{m - n}}$

105. $\log_2 \sqrt[5]{\dfrac{x^3}{y^4}},\ y > 0$ **107.** $\log_6 \dfrac{\sqrt[5]{x - 3}}{x^2(x + 1)^3},\ x > 3$

109. $2 + \ln 3$ **111.** $1 + \frac{1}{2} \log_5 2$ **113.** $1 - 2 \log_4 x$

115. **117.**

119. Evaluate when $x = e$ and $y = e$

121. $B = 10(\log_{10} I + 16)$; 60 decibels

123. $E = 1.4 \log_{10} \dfrac{C_2}{C_1}$ **125.** True **127.** True

129. False. $\log_3(u + v)$ does not simplify. **131.** True

133. False. 0 is not in the domain of f.

135. False. $f(x - 3) = \ln(x - 3)$

137. False. If $v = u^2$, then $f(v) = \ln u^2 = 2 \ln u = 2f(u)$.

Section 11.5 *(page 682)*

Integrated Review *(page 682)*

1. No. A system of linear equations has no solutions, one solution, or an infinite number of solutions.

2. The equations represent parallel lines and therefore have no point of intersection.

3. 2 **4.** $5 \pm 2\sqrt{2}$ **5.** $-\frac{1}{2}$ **6.** $\frac{5}{3}$

7. 1, 7 **8.** 47

9. $d = 73t$

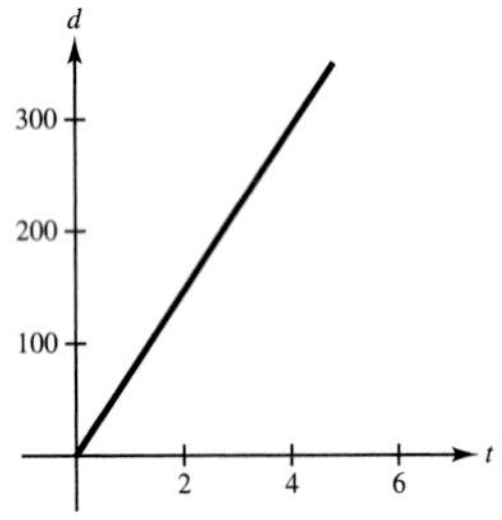

10. $V = 25\pi h$

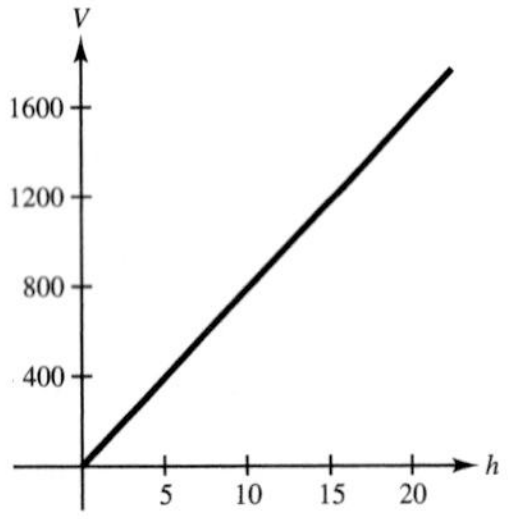

11. $V = 10\pi r^2$

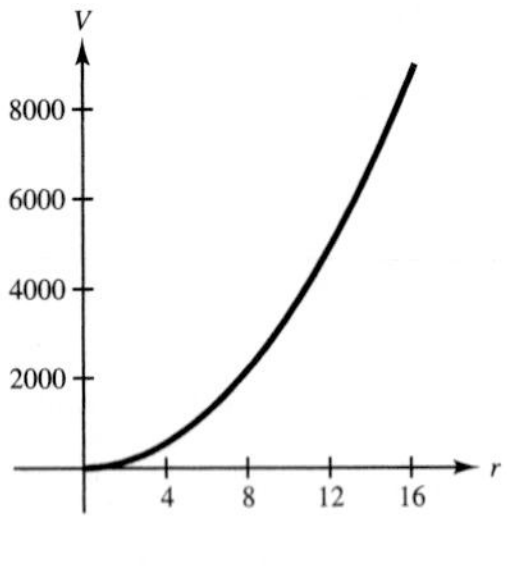

12. $F = 25x$

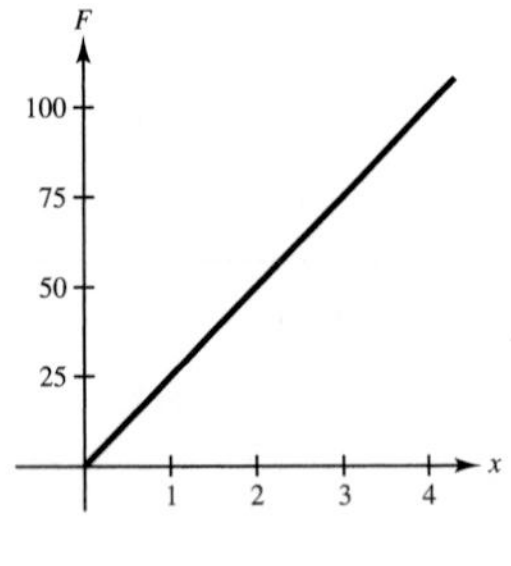

1. (a) Not a solution (b) Solution **3.** (a) Solution (b) Not a solution

5. (a) Not a solution (b) Solution **7.** 5 **9.** 8 **11.** 8

13. $\frac{2}{3}$ **15.** 2 **17.** -3 **19.** -6 **21.** 2

23. $\frac{22}{5}$ **25.** 6 **27.** 9 **29.** 4 **31.** No solution

33. -7 **35.** $2x - 1$ **37.** $2x$, $x > 0$

39. 5.49 **41.** 1.17 **43.** 0.86 **45.** 0.83

47. -2.37 **49.** -3.60 **51.** 2.64

53. 3.00 **55.** 1.23 **57.** 35.35 **59.** 6.80

61. 12.22 **63.** 3.28 **65.** No solution

67. -1.04 **69.** 2.48 **71.** 0.90 **73.** 0.38

75. 0.39 **77.** 8.99 **79.** 9.73 **81.** 4.62

83. 1000.00 **85.** 22.63 **87.** 2187.00 **89.** 6.52

91. 25.00 **93.** 10.04 **95.** ± 20.09 **97.** 3.00

99. 19.63 **101.** 12.18 **103.** 2000.00 **105.** 3.20

107. 4.00 **109.** 0.75 **111.** 5.00 **113.** 2.46

115. 2.29 **117.** 6.00

119.

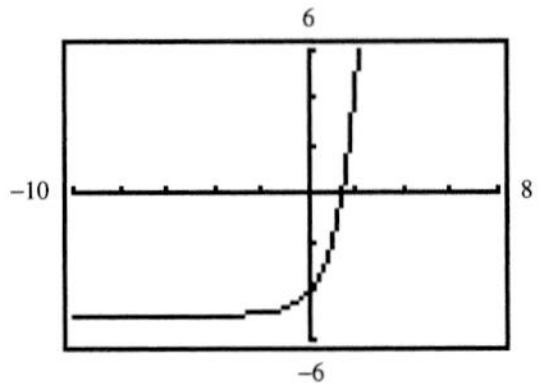

(1.40, 0)

121.

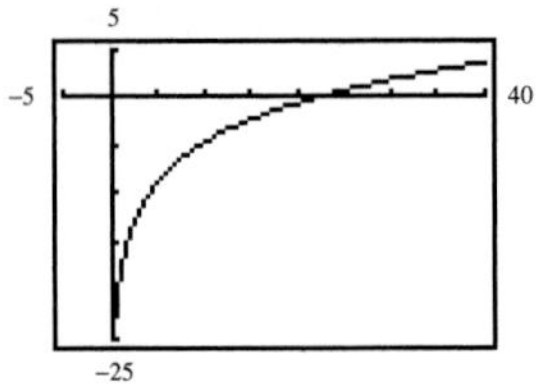

(21.82, 0)

123.

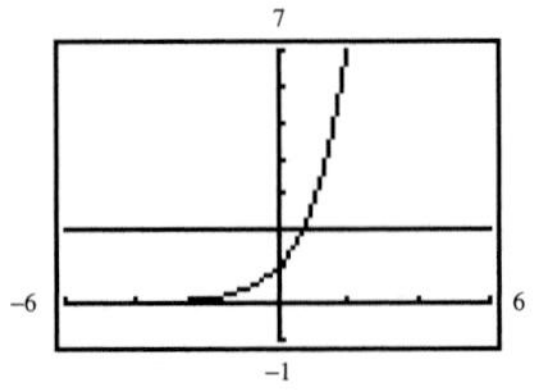

(0.69, 2)

125.

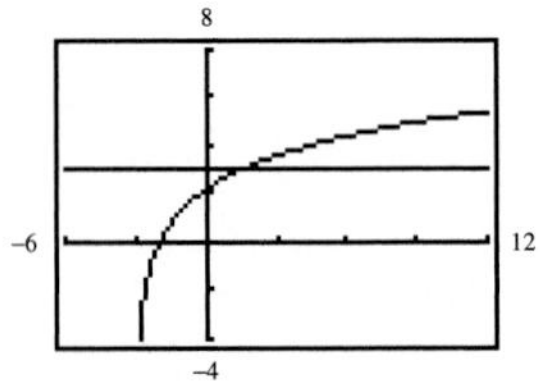

(1.48, 3)

127. 9% **129.** 7.70 years

131. $10^{-8.5}$ watts per square centimeter **133.** 205

135. (a) 3.64 months

(b)

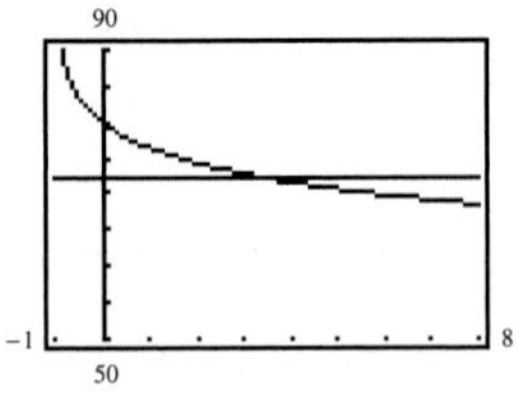

(c) Answers will vary.

137. (a) $k = \frac{1}{4} \ln \frac{8}{15} \approx -0.1572$

(b) $\approx$3.25 hours

(c) $\approx$2.84 hours

139. (c) 7.2%

(d) $6\frac{3}{4}$ years

(e) 8.24%

(f) Double: 11.6 years; Quadruple: 23.1 years

141. $2^{x-1} = 30$

143. To solve an exponential equation, first isolate the exponential expression, then take the logarithms of both sides of the equation, and solve for the variable.

To solve a logarithmic equation, first isolate the logarithmic expression, then exponentiate both sides of the equation, and solve for the variable.

Section 11.6 *(page 692)*

Integrated Review *(page 692)*

1. Direct variation as nth power
2. Inverse variation **3.** Joint variation
4. Joint variation **5.** $(3, 3)$ **6.** $\left(\frac{10}{3}, \frac{5}{3}\right)$
7. $(2, 4), \left(-\frac{1}{2}, \frac{1}{4}\right)$ **8.** $(0, 0), (8, 2)$ **9.** $(1, 2, 1)$
10. $(4, -1, 3)$ **11.** (a) Down
(b) $(0, 0), (4, 0)$
(c) $(2, 4)$

12.

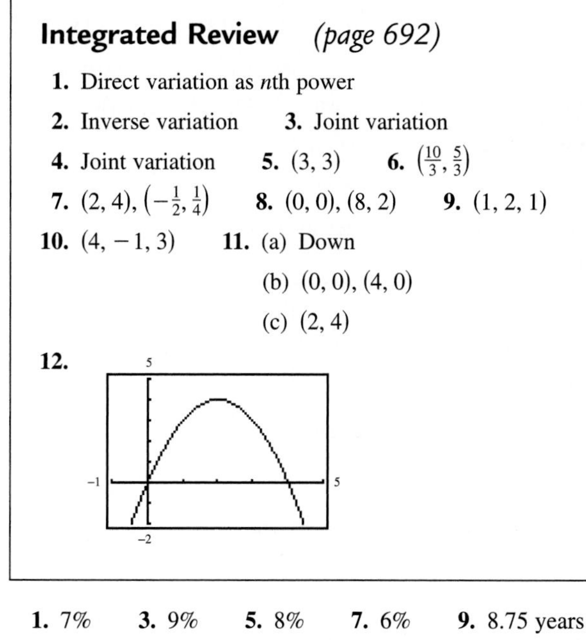

1. 7% **3.** 9% **5.** 8% **7.** 6% **9.** 8.75 years
11. 6.60 years **13.** 9.24 years **15.** 14.21 years
17. Continuous **19.** Quarterly **21.** 8.33%
23. 7.23% **25.** 6.136% **27.** 8.30% **29.** No
31. \$1652.99 **33.** \$626.46 **35.** \$3080.15
37. \$951.23 **39.** \$5496.57 **41.** \$320,250.81
43. Total deposits: \$7200.00; Total interest: \$10,529.42
45. $k = \frac{1}{2}\ln\frac{8}{3} \approx 0.4904$ **47.** $k = \frac{1}{3}\ln\frac{1}{2} \approx -0.2310$
49. $y = 12.2e^{0.0076t}$; 14.9 **51.** $y = 14.7e^{0.0221t}$; 26.1
53. $y = 10.5e^{0.0005t}$; 10.6 **55.** $y = 15.5e^{0.0092t}$; 19.7
57. (a) k is larger in Exercise 51, because the population of Shanghai is increasing faster than the population of Osaka.
(b) k corresponds to r; k gives the annual percentage rate of growth.
59. 3.91 grams **61.** 4.51 grams **63.** 4.08 grams
65. 3.3 grams **67.** 4.43 grams **69.** \$9281
71. The one in Alaska is 63 times as great.
73. The one in Mexico is 40 times as great.
75. 7.04 **77.** 10^7 times
79. (a)

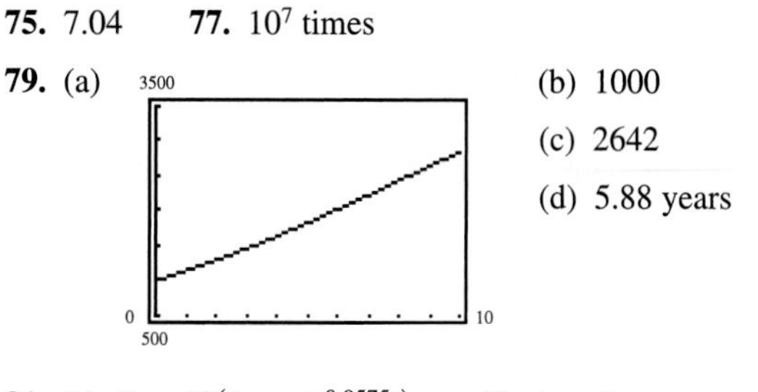

(b) 1000
(c) 2642
(d) 5.88 years

81. (a) $S = 10(1 - e^{-0.0575x})$ **83.** $k < 0$
(b) 3300 units
85. The effective yield of an investment collecting compound interest is the simple interest rate that would yield the same balance at the end of 1 year. To compute the effective yield, divide the interest earned in 1 year by the amount invested.
87. 10

Review Exercises *(page 697)*

1. (a) $\frac{1}{8}$ (b) 2 (c) 4 **3.** (a) 2.718 (b) 0.351 (c) 0.135 **5.** c **7.** a

9. **11.** **13.** **15.** **17.** **19.**

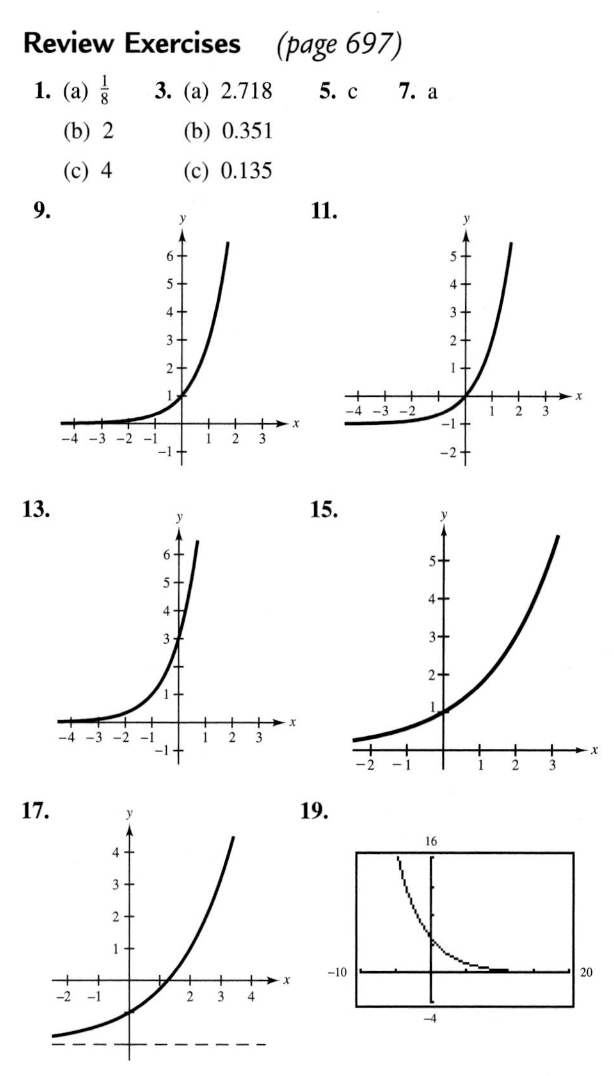

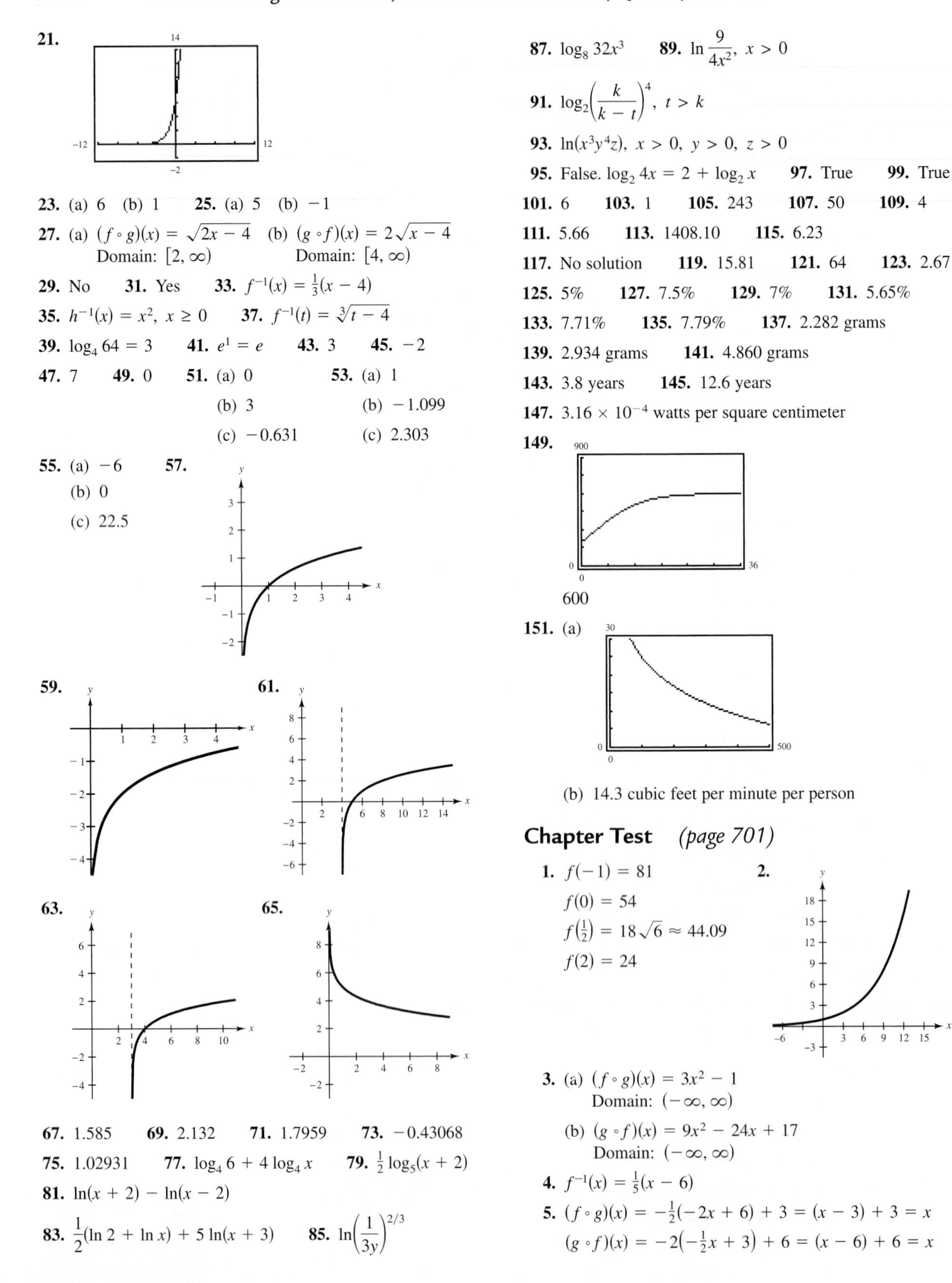

21.

23. (a) 6 (b) 1 **25.** (a) 5 (b) −1

27. (a) $(f \circ g)(x) = \sqrt{2x - 4}$ (b) $(g \circ f)(x) = 2\sqrt{x - 4}$
Domain: $[2, \infty)$ Domain: $[4, \infty)$

29. No **31.** Yes **33.** $f^{-1}(x) = \frac{1}{3}(x - 4)$

35. $h^{-1}(x) = x^2,\ x \geq 0$ **37.** $f^{-1}(t) = \sqrt[3]{t - 4}$

39. $\log_4 64 = 3$ **41.** $e^1 = e$ **43.** 3 **45.** −2

47. 7 **49.** 0 **51.** (a) 0 (b) 3 (c) −0.631 **53.** (a) 1 (b) −1.099 (c) 2.303

55. (a) −6 (b) 0 (c) 22.5 **57.**

59.

61.

63.

65.

67. 1.585 **69.** 2.132 **71.** 1.7959 **73.** −0.43068

75. 1.02931 **77.** $\log_4 6 + 4\log_4 x$ **79.** $\frac{1}{2}\log_5(x + 2)$

81. $\ln(x + 2) - \ln(x - 2)$

83. $\frac{1}{2}(\ln 2 + \ln x) + 5\ln(x + 3)$ **85.** $\ln\left(\frac{1}{3y}\right)^{2/3}$

87. $\log_8 32x^3$ **89.** $\ln \frac{9}{4x^2},\ x > 0$

91. $\log_2\left(\frac{k}{k - t}\right)^4,\ t > k$

93. $\ln(x^3y^4z),\ x > 0,\ y > 0,\ z > 0$

95. False. $\log_2 4x = 2 + \log_2 x$ **97.** True **99.** True

101. 6 **103.** 1 **105.** 243 **107.** 50 **109.** 4

111. 5.66 **113.** 1408.10 **115.** 6.23

117. No solution **119.** 15.81 **121.** 64 **123.** 2.67

125. 5% **127.** 7.5% **129.** 7% **131.** 5.65%

133. 7.71% **135.** 7.79% **137.** 2.282 grams

139. 2.934 grams **141.** 4.860 grams

143. 3.8 years **145.** 12.6 years

147. 3.16×10^{-4} watts per square centimeter

149.

600

151. (a)

(b) 14.3 cubic feet per minute per person

Chapter Test *(page 701)*

1. $f(-1) = 81$
$f(0) = 54$
$f\left(\frac{1}{2}\right) = 18\sqrt{6} \approx 44.09$
$f(2) = 24$

2.

3. (a) $(f \circ g)(x) = 3x^2 - 1$
Domain: $(-\infty, \infty)$

(b) $(g \circ f)(x) = 9x^2 - 24x + 17$
Domain: $(-\infty, \infty)$

4. $f^{-1}(x) = \frac{1}{5}(x - 6)$

5. $(f \circ g)(x) = -\frac{1}{2}(-2x + 6) + 3 = (x - 3) + 3 = x$
$(g \circ f)(x) = -2\left(-\frac{1}{2}x + 3\right) + 6 = (x - 6) + 6 = x$

6. $g = f^{-1}$

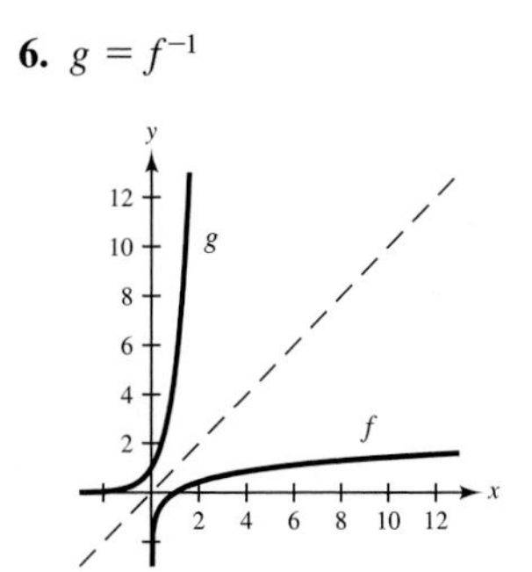

7. $\log_4 5 + 2\log_4 x - \frac{1}{2}\log_4 y$ **8.** $\ln \frac{x}{y^4},\ y > 0$

9. $3 + \log_5 6$ **10.** 64 **11.** 0.973 **12.** 13.733

13. 15.516 **14.** 2 **15.** 8 **16.** 109.196 **17.** 0

18. (a) \$8012.78 **19.** \$10,806.08 **20.** 7%

(b) \$8110.40

21. \$8469.14 **22.** 600 **23.** 1141 **24.** 4.4 years

Chapter 12

Section 12.1 *(page 710)*

Integrated Review *(page 710)*

1. $-7x = 35$

$$\frac{-7x}{-7} = \frac{35}{-7}$$

$$x = -5$$

2. $7x + 63 = 35$

$$7x + 63 - 63 = 35 - 63$$

$$7x = -28$$

$$x = -4$$

3. It is a solution if the equation is true when -3 is substituted for t.

4. Multiply both sides of the equation by the lowest common denominator $x(x + 1)$.

5. $\frac{1}{(x + 10)^2}$ **6.** $18(x - 3)^3$ **7.** $\frac{1}{a^8}$ **8.** $2x$

9. $8x\sqrt{2x}$ **10.** $\frac{5(\sqrt{x} + 2)}{x - 4}$

11. (a) $A = x(2x - 3)$

(b) 250, 0, 12, 0

(c) $\frac{3 + \sqrt{1609}}{4} \approx 10.8$

12. (a) $A = \frac{1}{2}x(x - 4)$

(b)

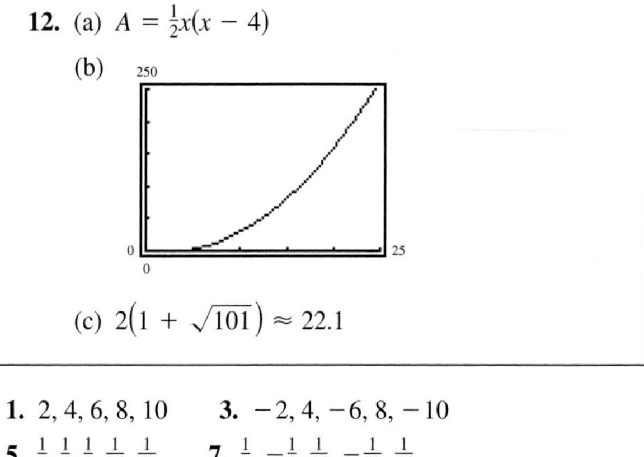

(c) $2(1 + \sqrt{101}) \approx 22.1$

1. 2, 4, 6, 8, 10 **3.** $-2, 4, -6, 8, -10$

5. $\frac{1}{2}, \frac{1}{4}, \frac{1}{8}, \frac{1}{16}, \frac{1}{32}$ **7.** $\frac{1}{4}, -\frac{1}{8}, \frac{1}{16}, -\frac{1}{32}, \frac{1}{64}$

9. $1, -0.2, 0.04, -0.008, 0.0016$ **11.** $\frac{1}{2}, \frac{1}{3}, \frac{1}{4}, \frac{1}{5}, \frac{1}{6}$

13. $\frac{2}{5}, \frac{1}{2}, \frac{6}{11}, \frac{4}{7}, \frac{10}{17}$ **15.** $-1, \frac{1}{4}, -\frac{1}{9}, \frac{1}{16}, -\frac{1}{25}$

17. $\frac{9}{2}, \frac{19}{4}, \frac{39}{8}, \frac{79}{16}, \frac{159}{32}$ **19.** 2, 3, 4, 5, 6

21. $0, 3, -1, \frac{3}{4}, -\frac{1}{4}$ **23.** -72 **25.** $\frac{31}{2520}$ **27.** 5

29. $\frac{1}{132}$ **31.** 53,130 **33.** $\frac{1}{n + 1}$ **35.** $n(n + 1)$

37. $2n$ **39.** c **41.** b

43. 4, 0, 10, −4

45. 10, 0, 10, 0

47. 5, 0, 10, −5

49. $a_n = 3n$

51. $a_n = 3n - 2$ **53.** $a_n = n^2 - 1$

55. $a_n = (-1)^{n+1}2n$ **57.** $a_n = \frac{n + 1}{n + 2}$

59. $a_n = \frac{(-1)^{n+1}}{2^n}$ **61.** $a_n = \frac{1}{2^{n-1}}$ **63.** $a_n = 1 + \frac{1}{n}$

65. $a_n = \frac{1}{n!}$ **67.** 63 **69.** 77 **71.** 100

73. $\frac{3019}{3600}$ **75.** $\frac{437}{60}$ **77.** -48 **79.** $\frac{8}{9}$ **81.** $\frac{182}{243}$

83. 273 **85.** 852 **87.** $\frac{65}{4}$ **89.** 6.5793 **91.** $\sum_{k=1}^{5} k$

93. $\sum_{k=1}^{5} 2k$ **95.** $\sum_{k=1}^{10} \frac{1}{2k}$ **97.** $\sum_{k=1}^{20} \frac{1}{k^2}$ **99.** $\sum_{k=0}^{9} \frac{1}{(-3)^k}$

101. $\sum_{k=1}^{20} \frac{4}{k+3}$ **103.** $\sum_{k=1}^{11} \frac{k}{k+1}$ **105.** $\sum_{k=1}^{20} \frac{2k}{k+3}$

107. $\sum_{k=0}^{6} k!$ **109.** 3.6 **111.** 0.8

113. (a) \$535, \$572.45, \$612.52, \$655.40, \$701.28, \$750.37, \$802.89, \$859.09

(b) \$7487.23 (c)

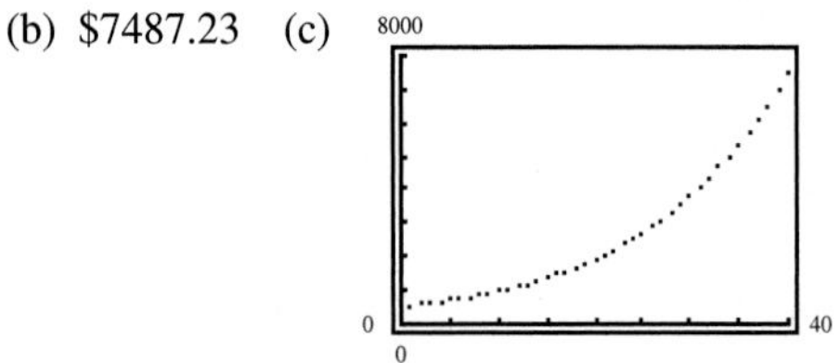

(d) Yes. Investment earning compound interest increases at an increasing rate.

115. $a_5 = 108°$, $a_6 = 120°$

At the point where any two hexagons and a pentagon meet, the sum of the three angles is $a_5 + 2a_6 = 348° < 360°$. Therefore, there is a gap of 12°.

117. 25.7°, 45°, 60°, 72°, 81.8°

119. $a_n = 3n$: 3, 6, 9, 12, . . .

121. Terms in which n is odd **123.** True

Section 12.2 *(page 718)*

Integrated Review *(page 718)*

1. A collection of letters (called variables) and real numbers (called constants) combined with the operations of addition, subtraction, multiplication, and division is called an algebraic expression.
2. The terms of an algebraic expression are those parts separated by addition or subtraction.
3. $2x^3 - 3x^2 + 2$ **4.** $7x^4$ **5.** $(-\infty, \infty)$
6. $(-\infty, \infty)$ **7.** $[-4, 4)$
8. $(-\infty, -6) \cup (-6, 6) \cup (6, \infty)$ **9.** $(2, \infty)$
10. $(-\infty, \infty)$ **11.** \$30,798.61 **12.** \$5395.40

1. 3 **3.** -6 **5.** -12 **7.** $\frac{2}{3}$ **9.** $-\frac{5}{4}$

11. Arithmetic, 2 **13.** Arithmetic, -2

15. Arithmetic, -16 **17.** Arithmetic, 0.8

19. Arithmetic, $\frac{3}{2}$ **21.** Not arithmetic

23. Not arithmetic **25.** Not arithmetic

27. 7, 10, 13, 16, 19 **29.** 6, 4, 2, 0, -2

31. $\frac{3}{2}, 4, \frac{13}{2}, 9, \frac{23}{2}$ **33.** $\frac{8}{5}, \frac{11}{5}, \frac{14}{5}, \frac{17}{5}, 4$ **35.** $4, \frac{15}{4}, \frac{7}{2}, \frac{13}{4}, 3$

37. $a_n = \frac{1}{2}n + \frac{5}{2}$ **39.** $a_n = -25n + 1025$

41. $a_n = -4n + 32$ **43.** $a_n = \frac{3}{2}n + \frac{3}{2}$

45. $a_n = \frac{5}{2}n + \frac{5}{2}$ **47.** $a_n = 4n + 4$

49. $a_n = -10n + 60$ **51.** $a_n = -\frac{1}{2}n + 11$

53. $a_n = -0.05n + 0.40$ **55.** 25, 28, 31, 34, 37

57. 9, 6, 3, 0, -3 **59.** $-10, -4, 2, 8, 14$

61. 100, 80, 60, 40, 20 **63.** 210 **65.** 1425

67. 255 **69.** 62,625 **71.** 35 **73.** 522

75. 1850 **77.** 900 **79.** 12,200 **81.** 243

83. 23 **85.** b **87.** e **89.** c

91. **93.**

95.

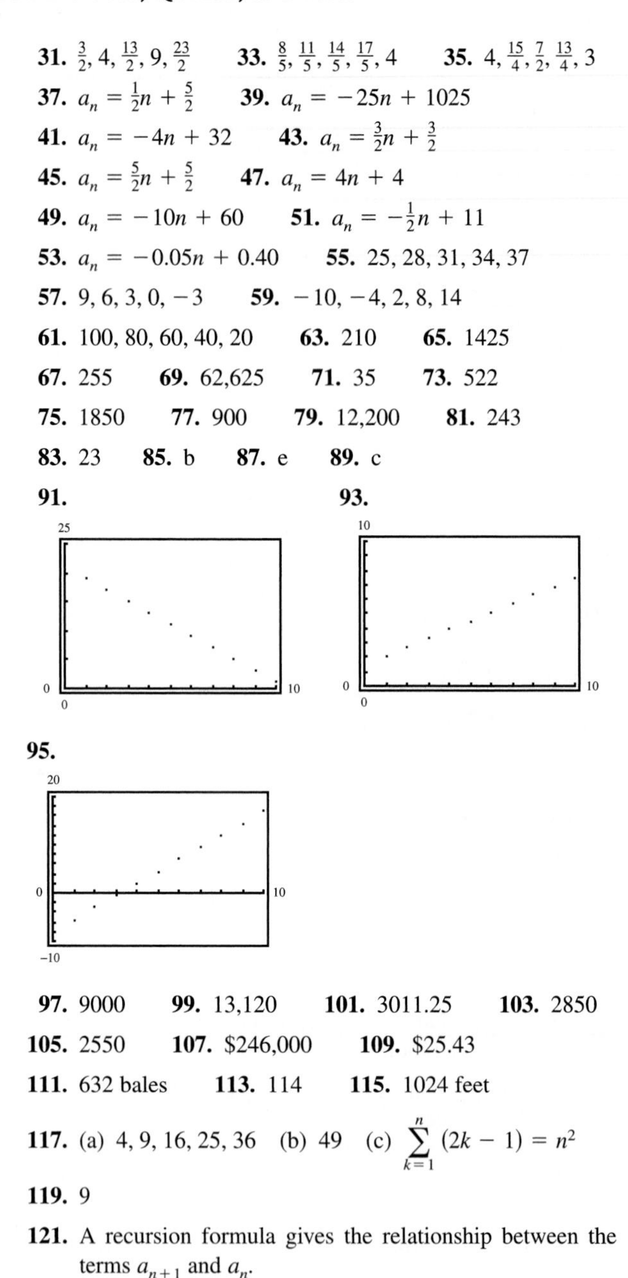

97. 9000 **99.** 13,120 **101.** 3011.25 **103.** 2850

105. 2550 **107.** \$246,000 **109.** \$25.43

111. 632 bales **113.** 114 **115.** 1024 feet

117. (a) 4, 9, 16, 25, 36 (b) 49 (c) $\sum_{k=1}^{n} (2k - 1) = n^2$

119. 9

121. A recursion formula gives the relationship between the terms a_{n+1} and a_n.

123. $\frac{101}{2}(100 + 200)$

Section 12.3 *(page 728)*

Integrated Review *(page 728)*

1. The point is 6 units to the left of the y-axis and 4 units above the x-axis.
2. $(10, 5), (-10, 5), (-10, -5), (10, -5)$
3. The graph of f is the set of ordered pairs $(x, f(x))$, where x is in the domain of f.

4. To find the x-intercept, set $y = 0$ and solve the equation for x. To find the y-intercept, set $x = 0$ and solve the equation for y.

5. $x > \frac{5}{3}$ **6.** $y < 6$ **7.** $35 < x < 60$

8. $-12 < x < 30$ **9.** $x < 1$ or $x > \frac{5}{2}$

10. $-1 < x < 0$ or $x > \frac{5}{2}$

11. $\frac{19\sqrt{2}}{2} \approx 13.4$ inches **12.** $5\sqrt{89} \approx 47.2$ feet

1. 3 **3.** -3 **5.** $-\frac{1}{2}$ **7.** $-\frac{3}{2}$ **9.** π

11. 1.06 **13.** Geometric, $\frac{1}{2}$ **15.** Not geometric

17. Geometric, 2 **19.** Not geometric

21. Geometric, $-\frac{2}{3}$ **23.** Geometric, 1.02

25. 4, 8, 16, 32, 64 **27.** $6, 2, \frac{2}{3}, \frac{2}{9}, \frac{2}{27}$

29. $1, -\frac{1}{2}, \frac{1}{4}, -\frac{1}{8}, \frac{1}{16}$ **31.** $4, -2, 1, -\frac{1}{2}, \frac{1}{4}$

33. 1000, 1010, 1020.1, 1030.30, 1040.60

35. 4000, 3960.40, 3921.18, 3882.36, 3843.92

37. $10, 6, \frac{18}{5}, \frac{54}{25}, \frac{162}{125}$ **39.** $\frac{3}{256}$ **41.** $48\sqrt{2}$

43. 1486.02 **45.** -0.00610 **47.** $\frac{81}{64}$ **49.** $\pm\frac{243}{32}$

51. $\frac{64}{3}$ **53.** $a_n = 2(3)^{n-1}$ **55.** $a_n = 2^{n-1}$

57. $a_n = \left(-\frac{1}{5}\right)^{n-1}$ **59.** $a_n = 4\left(-\frac{1}{2}\right)^{n-1}$

61. $a_n = 8\left(\frac{1}{4}\right)^{n-1}$ **63.** $a_n = 14\left(\frac{3}{4}\right)^{n-1}$

65. $a_n = 4\left(-\frac{3}{2}\right)^{n-1}$ **67.** b **69.** a

71. 1023 **73.** 772.48 **75.** 2.25 **77.** -5460

79. 6.06 **81.** $-14{,}762$ **83.** 16 **85.** 13,120

87. 48 **89.** 1103.57 **91.** 12,822.71 **93.** 2

95. $\frac{2}{3}$ **97.** $\frac{6}{5}$ **99.** 32

101. **103.**

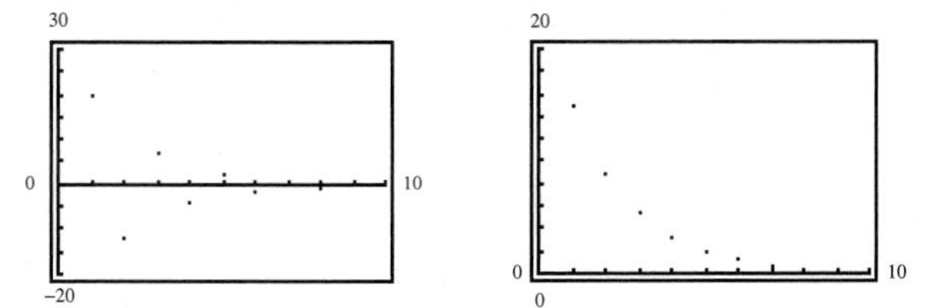

105. (a) $250{,}000(0.75)^n$ (b) \$59,326.17 (c) The first year

107. \$3,623,993 **109.** \$19,496.56 **111.** \$105,428.44

113. \$75,715.32

115. (a) \$5,368,709.11 (b) \$10,737,418.23

117. (a) $P = (0.999)^n$ (b) 69.4%

(c)

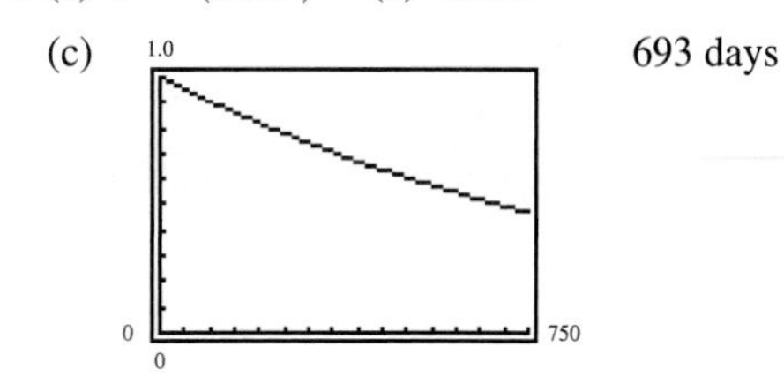

693 days

119. 70.875 square inches **121.** 666.21 feet

123. (a) $a_n = 2^n$ (b) $2 + 2^2 + 2^3 + 2^4 + \cdots + 2^{66}$

(c) It is likely that you have had common ancestors in the last 2000 years.

125. $a_n = a_1 r^{n-1}$ **127.** $a_n = \left(-\frac{2}{3}\right)^{n-1}$

129. An increasing annuity is an investment plan where equal deposits are made in an account at equal time intervals.

Mid-Chapter Quiz *(page 732)*

1. $32, 8, 2, \frac{1}{2}, \frac{1}{8}$ **2.** $-\frac{3}{5}, 3, -\frac{81}{7}, \frac{81}{2}, -135$ **3.** 100

4. 40 **5.** 87 **6.** -32 **7.** $\sum_{k=1}^{20} \frac{2}{3k}$

8. $\sum_{k=1}^{25} \frac{(-1)^{k-1}}{k^3}$ **9.** $\frac{1}{2}$ **10.** -6 **11.** 3 **12.** $\frac{1}{2}$

13. $20 - 3(n - 1)$ **14.** $32\left(-\frac{1}{4}\right)^{n-1}$ **15.** 4075

16. 9030 **17.** 25.947 **18.** 18,392.796 **19.** 9

20. $\frac{16}{15}$ **21.** -0.026 **22.** a_n: upper graph
b_n: lower graph

23. 5.5° **24.** Arithmetic

Section 12.4 *(page 738)*

Integrated Review *(page 738)*

1. No. The matrix must be square.

2. Interchange two rows.
Multiply a row by a nonzero constant.
Add a multiple of one row to another row.

3. Yes **4.** -200 **5.** 32 **6.** -60

7. -126 **8.** 58 **9.** $y = -0.07x^2 + 1.3x + 2$

10. $y = 4x - 9$

1. 15 **3.** 252 **5.** 1 **7.** 1 **9.** 1225

11. 12,650 **13.** 593,775 **15.** 792 **17.** 2,598,960

19. 2,535,650,040 **21.** 5,200,300 **23.** 15 **25.** 35

27. 70 **29.** $a^3 + 6a^2 + 12a + 8$

31. $x^8 + 8x^7y + 28x^6y^2 + 56x^5y^3 + 70x^4y^4 + 56x^3y^5 + 28x^2y^6 + 8xy^7 + y^8$

33. $32x^5 - 80x^4 + 80x^3 - 40x^2 + 10x - 1$

35. $64y^6 + 192y^5z + 240y^4z^2 + 160y^3z^3 + 60y^2z^4 + 12yz^5 + z^6$

37. $x^8 + 8x^6 + 24x^4 + 32x^2 + 16$

39. $x^6 + 18x^5 + 135x^4 + 540x^3 + 1215x^2 + 1458x + 729$

41. $x^6 - 24x^5 + 240x^4 - 1280x^3 + 3840x^2 - 6144x + 4096$

43. $x^4 + 4x^3y + 6x^2y^2 + 4xy^3 + y^4$

45. $u^3 - 6u^2v + 12uv^2 - 8v^3$

47. $81a^4 + 216a^3b + 216a^2b^2 + 96ab^3 + 16b^4$

49. $32x^{10} - 80x^8y + 80x^6y^2 - 40x^4y^3 + 10x^2y^4 - y^5$

51. 120 **53.** -1365 **55.** 1760 **57.** 54 **59.** 70

61. 1.172 **63.** 510,568.785

65. $\frac{1}{32} + \frac{5}{32} + \frac{10}{32} + \frac{10}{32} + \frac{5}{32} + \frac{1}{32}$

67. $\frac{1}{256} + \frac{12}{256} + \frac{54}{256} + \frac{108}{256} + \frac{81}{256}$

69. The difference between consecutive entries increases by 1.

2, 3, 4, 5

71. $n + 1$

73. The signs of the terms alternate in the expansion of $(x - y)^n$.

75. They are the same.

Section 12.5 *(page 747)*

Integrated Review *(page 747)*

1. $g(x) = 2(5^x)$ is exponential since it has a constant base and variable exponent.

2. Using the law of exponents $a^m \cdot a^n = a^{m+n}$, you have $e^2 \cdot e^{-x^2} = e^{2+(-x^2)} = e^{2-x^2}$.

3. $4^3 = 64$ **4.** $3^{-4} = \frac{1}{81}$ **5.** $e^0 = 1$

6. $e^{1.6094\ldots} \approx 5$ **7.** $\dfrac{\ln 50}{\ln 3} \approx 3.56$

8. $6 \ln 2 \approx 4.16$ **9.** 69 **10.** $e^{10} - 3 \approx 22{,}023.47$

11.

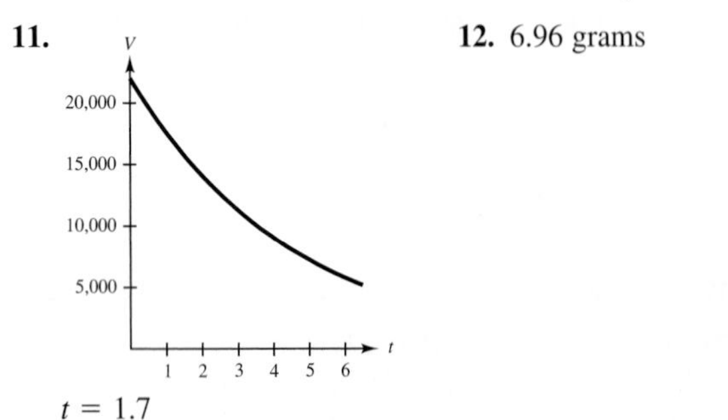

$t = 1.7$

12. 6.96 grams

1. 5 **3.** 9 **5.** 8 **7.** 10 **9.** 8 **11.** 6

13. 7 **15.** 6 **17.** 6 **19.** 260 **21.** 6,760,000

23. (a) 900 (b) 720 (c) 400 **25.** 18 **27.** 720

29. ABCD, ABDC, ACBD, ACDB, ADBC, ADCB, BACD, BADC, BCAD, BCDA, BDAC, BDCA, CABD, CADB, CBAD, CBDA, CDAB, CDBA, DABC, DACB, DBAC, DBCA, DCAB, DCBA

31. AB, BA, AC, CA, AD, DA, BC, CB, BD, DB, CD, DC

33. 720 **35.** 64,000 **37.** 40,320 **39.** 5040

41. {A, B}, {A, C}, {A, D}, {A, E}, {A, F}, {B, C}, {B, D}, {B, E}, {B, F}, {C, D}, {C, E}, {C, F}, {D, E}, {D, F}, {E, F}

43. 1140 **45.** 126 **47.** 220 **49.** 3003

51. (a) 15 (b) 6 **53.** (a) 70 (b) 16

55. 21 **57.** 9 **59.** 35

61. Let E_1 and E_2 be two events that can occur in m_1 ways and m_2 ways, respectively. The number of ways the two events can occur is $m_1 \cdot m_2$.

63. Permutation: The ordering of five students for a picture

Combination: The selection of three students from a group of five students for a class project

Section 12.6 *(page 756)*

Integrated Review *(page 756)*

1. 0 **2.** 1 **3.** x **4.** $\log_a u + \log_a v$

5. $\log_a u - \log_a v$ **6.** $n \log_a u$

7. $2 \log_2 x + \log_2 y$ **8.** $\frac{1}{2}\log_2(x^2 + 1)$

9. $\ln 7 - \ln(x - 3)$ **10.** $2[\ln(u + 2) - \ln(u - 2)]$

11. (a)

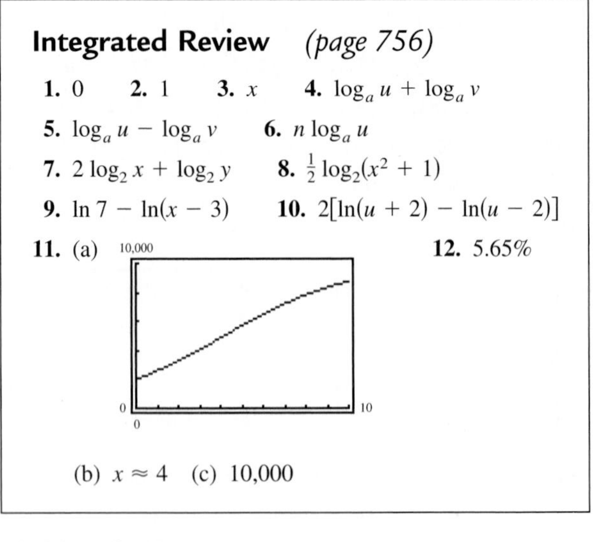

(b) $x \approx 4$ (c) 10,000

12. 5.65%

1. 26 **3.** 10

5. {ABC, ACB, BAC, BCA, CAB, CBA}

7. {WWW, WWL, WLW, WLL, LWW, LWL, LLW, LLL}

9. 0.65 **11.** 0.18 **13.** $\frac{3}{8}$ **15.** $\frac{7}{8}$ **17.** $\frac{1}{2}$

19. $\frac{3}{13}$ **21.** $\frac{1}{6}$ **23.** $\frac{5}{6}$ **25.** 0.9 **27.** 0.243

29. 0.609 **31.** (a) $\frac{1}{5}$ (b) $\frac{1}{3}$ (c) 1

33. (a) 0.8 (b) 0.2 **35.** $\frac{14}{65}$ **37.** $\frac{1917}{6565}, \frac{4648}{6565}$

39. 0.4375

41.

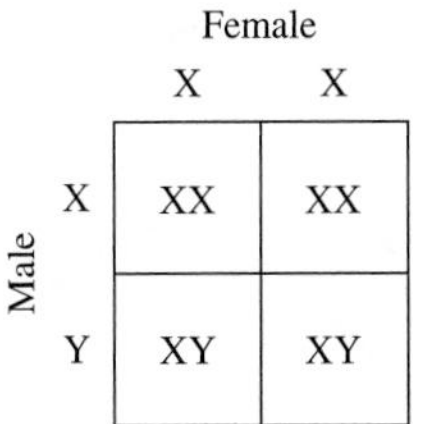

		Female X	Female X
Male	X	XX	XX
Male	Y	XY	XY

(a) Probability of a girl $= \frac{2}{4} = \frac{1}{2}$

Probability of a boy $= \frac{2}{4} = \frac{1}{2}$

(b) Because the probabilities are the same, it is equally likely that a newborn will be a boy or a girl.

43. $\frac{1}{24}$ **45.** $\frac{1}{100{,}000}$ **47.** $\frac{1}{45}$

49. $\frac{1}{45}$ **51.** $\frac{1}{210}$ **53.** $\frac{33}{66{,}640}$

55. (d) 120

(e) Without replacement, since each person receives 1 gift

(f) $\frac{1}{150}, \frac{1}{105}$

57. The probability that the event does not occur is $1 - \frac{3}{4} = \frac{1}{4}$.

59. Over an extended period, it will rain 40% of the time under the given weather conditions.

Review Exercises *(page 761)*

1. 8, 11, 14, 17, 20 **3.** $1, \frac{3}{4}, \frac{5}{8}, \frac{9}{16}, \frac{17}{32}$ **5.** $a_n = 2n - 1$

7. $a_n = \dfrac{n}{(n+1)^2}$ **9.** a **11.** b **13.** d **15.** 28

17. $\frac{4}{5}$ **19.** $\displaystyle\sum_{k=1}^{4} (5k - 3)$ **21.** $\displaystyle\sum_{k=1}^{6} \frac{1}{3k}$ **23.** -2.5

25. 127, 122, 117, 112, 107 **27.** $\frac{5}{4}, 2, \frac{11}{4}, \frac{7}{2}, \frac{17}{4}$

29. 5, 8, 11, 14, 17 **31.** $80, \frac{155}{2}, 75, \frac{145}{2}, 70$ **33.** $4n + 6$

35. $-50n + 1050$ **37.** 486 **39.** $\frac{2525}{2}$ **41.** 2527.5

43. $\frac{3}{2}$ **45.** 10, 30, 90, 270, 810

47. 100, -50, 25, -12.5, 6.25 **49.** 3, 6, 12, 24, 48

51. $a_n = \left(-\frac{2}{3}\right)^{n-1}$ **53.** $a_n = 24(2)^{n-1}$

55. $a_n = 12\left(-\frac{1}{2}\right)^{n-1}$ **57.** 8190 **59.** -1.928

61. 19.842 **63.** 116,169.54 **65.** 8 **67.** 12

69. 2.275×10^6 **71.** 56 **73.** 1

75. 91,390 **77.** 177,100

79. $x^{10} + 10x^9 + 45x^8 + 120x^7 + 210x^6 + 252x^5 + 210x^4 + 120x^3 + 45x^2 + 10x + 1$

81. $81x^4 - 216x^3y + 216x^2y^2 - 96xy^3 + 16y^4$

83. $u^{18} + 9u^{16}v^3 + 36u^{14}v^6 + 84u^{12}v^9 + 126u^{10}v^{12} + 126u^8v^{15} + 84u^6v^{18} + 36u^4v^{21} + 9u^2v^{24} + v^{27}$

85. $-61{,}236$ **87.** 280 **89.** 5100 **91.** 462

93. (a) $a_n = 85{,}000(1.012)^n$ (b) 154,328

95. 8 **97.** 3003 **99.** $\frac{1}{3}$ **101.** $\frac{1}{24}$ **103.** 0.346

Chapter Test *(page 764)*

1. $1, -\frac{2}{3}, \frac{4}{9}, -\frac{8}{27}, \frac{16}{81}$ **2.** 35 **3.** -45

4. $\displaystyle\sum_{k=1}^{12} \frac{2}{3k+1}$ **5.** 12, 16, 20, 24, 28

6. $a_n = -100n + 5100$ **7.** 3825 **8.** $-\frac{3}{2}$

9. $a_n = 4\left(\frac{1}{2}\right)^{n-1}$ **10.** 1020 **11.** $\frac{3069}{1024}$ **12.** 1

13. 12 **14.** \$47,868.33 **15.** 1140

16. $x^5 - 10x^4 + 40x^3 - 80x^2 + 80x - 32$ **17.** 56

18. 26,000 **19.** 12,650 **20.** 0.25 **21.** $\frac{3}{26}$ **22.** $\frac{1}{6}$

Cumulative Test: Chapters 10–12 *(page 765)*

1. $5 \pm 5\sqrt{2}i$ **2.** $\dfrac{-3 \pm \sqrt{3}}{3}$ **3.** 2 **4.** 6

5. $y = -2x^2 + 8x - 5$ **6.** $(-\infty, 0] \cup \left[\frac{7}{2}, \infty\right)$

7. $(-3, -1) \cup (0, 1)$

8. **9.**

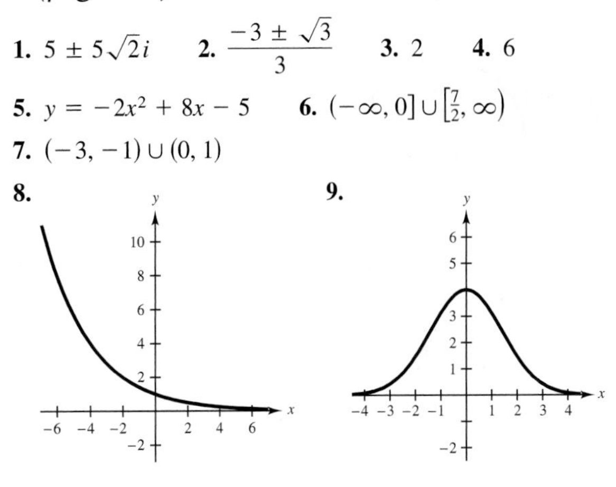

10. 8 **11.** 5 **12.** 0 **13.** -4

14. These are inverse functions so their graphs are symmetric about the line $y = x$.

15. $\log_2 \dfrac{(xy)^3}{z}$ **16.** $4\sqrt{2}$ **17.** 0.8673 **18.** 0.4660

19. 5 **20.** $a_n = 10 - 2n$ **21.** $a_n = (-1)^n\left(\frac{2}{3}\right)^{n-1}$

22. 920 **23.** 6 **24.** $z^4 - 12z^3 + 54z^2 - 108z + 81$

25. V: 25,000, 20,000, 15,000, 10,000, 5,000; t: 1, 2, 3, 4, 5, 6

$t = 2.3$ years

26. 7.5%

27. \$402,493 **28.** 120 **29.** $\frac{1}{4}$

Appendix A *(page A9)*

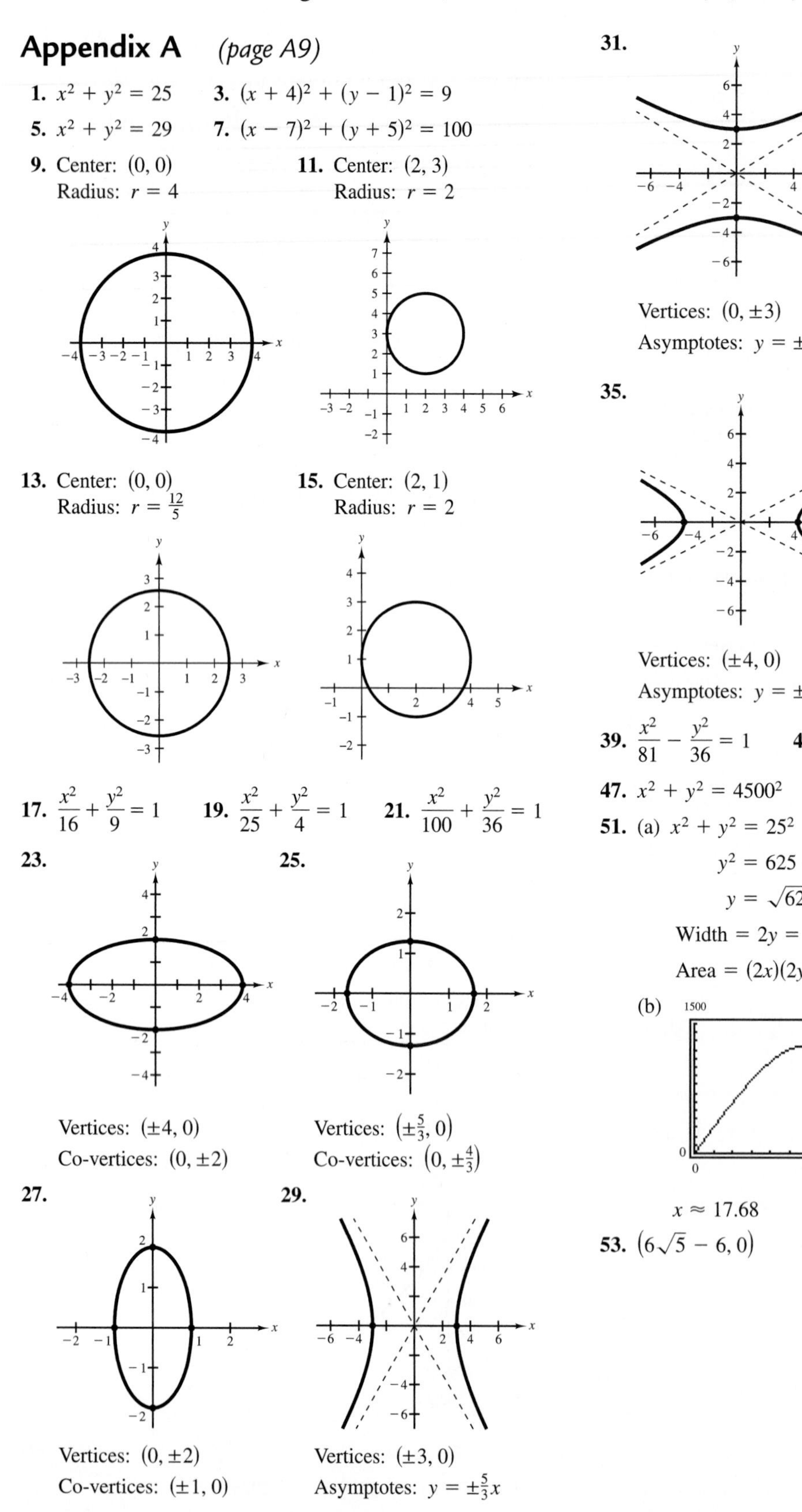

1. $x^2 + y^2 = 25$ **3.** $(x + 4)^2 + (y - 1)^2 = 9$

5. $x^2 + y^2 = 29$ **7.** $(x - 7)^2 + (y + 5)^2 = 100$

9. Center: $(0, 0)$
Radius: $r = 4$

11. Center: $(2, 3)$
Radius: $r = 2$

13. Center: $(0, 0)$
Radius: $r = \frac{12}{5}$

15. Center: $(2, 1)$
Radius: $r = 2$

17. $\dfrac{x^2}{16} + \dfrac{y^2}{9} = 1$ **19.** $\dfrac{x^2}{25} + \dfrac{y^2}{4} = 1$ **21.** $\dfrac{x^2}{100} + \dfrac{y^2}{36} = 1$

23. Vertices: $(\pm 4, 0)$
Co-vertices: $(0, \pm 2)$

25. Vertices: $\left(\pm\frac{5}{3}, 0\right)$
Co-vertices: $\left(0, \pm\frac{4}{3}\right)$

27. Vertices: $(0, \pm 2)$
Co-vertices: $(\pm 1, 0)$

29. Vertices: $(\pm 3, 0)$
Asymptotes: $y = \pm\frac{5}{3}x$

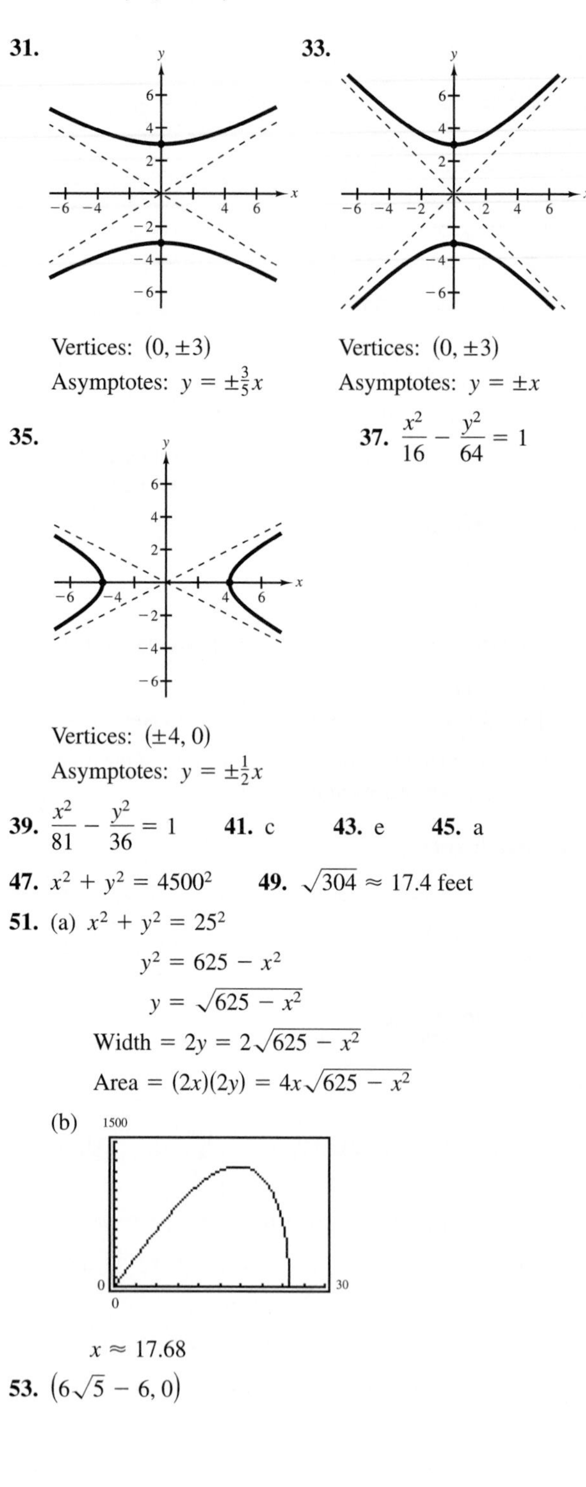

31. Vertices: $(0, \pm 3)$
Asymptotes: $y = \pm\frac{3}{5}x$

33. Vertices: $(0, \pm 3)$
Asymptotes: $y = \pm x$

35. Vertices: $(\pm 4, 0)$
Asymptotes: $y = \pm\frac{1}{2}x$

37. $\dfrac{x^2}{16} - \dfrac{y^2}{64} = 1$

39. $\dfrac{x^2}{81} - \dfrac{y^2}{36} = 1$ **41.** c **43.** e **45.** a

47. $x^2 + y^2 = 4500^2$ **49.** $\sqrt{304} \approx 17.4$ feet

51. (a) $x^2 + y^2 = 25^2$

$$y^2 = 625 - x^2$$

$$y = \sqrt{625 - x^2}$$

Width $= 2y = 2\sqrt{625 - x^2}$

Area $= (2x)(2y) = 4x\sqrt{625 - x^2}$

(b)

$x \approx 17.68$

53. $\left(6\sqrt{5} - 6, 0\right)$

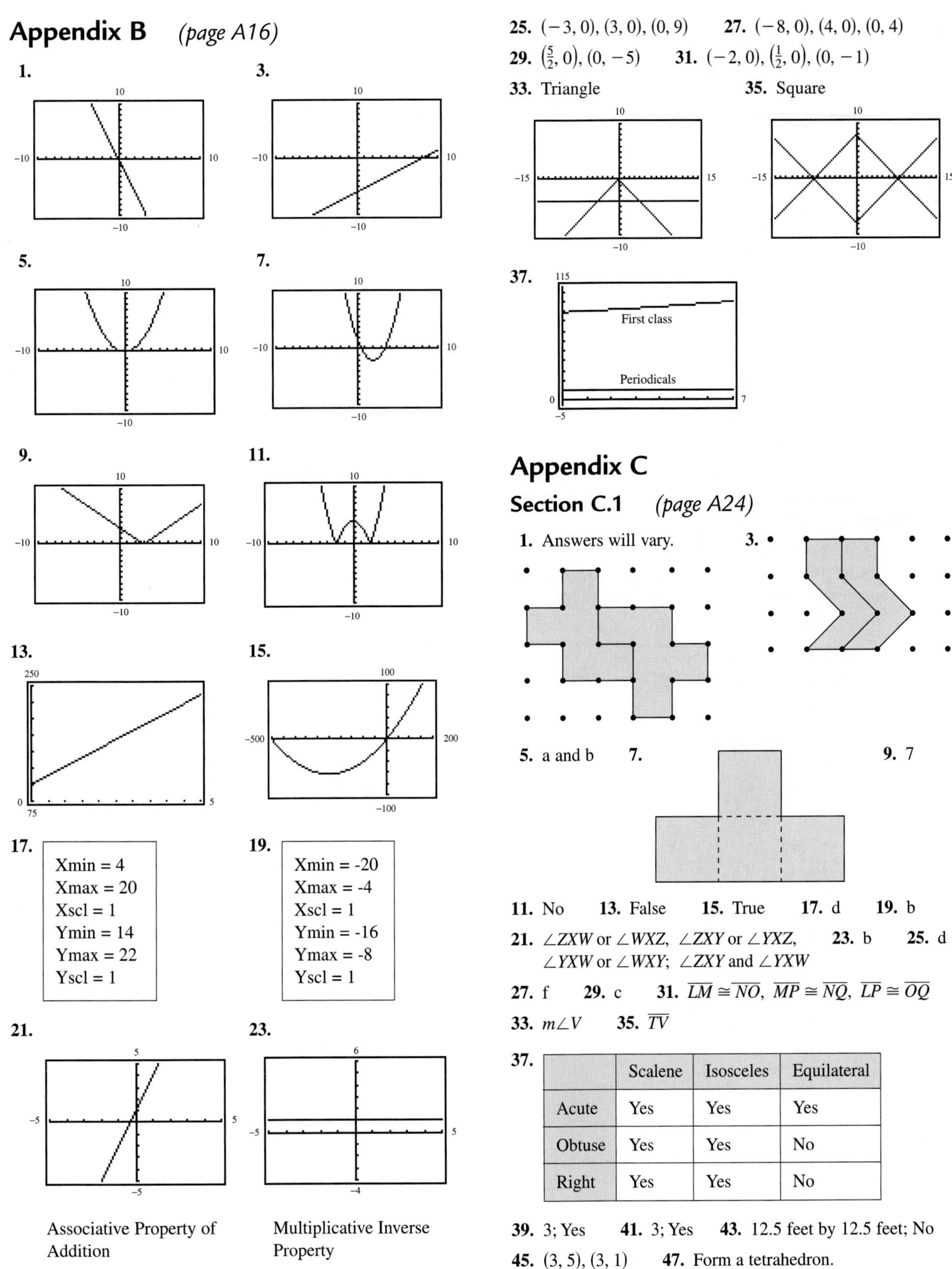

Appendix B *(page A16)*

1. **3.** **5.** **7.** **9.** **11.** **13.** **15.**

17.
Xmin = 4
Xmax = 20
Xscl = 1
Ymin = 14
Ymax = 22
Yscl = 1

19.
Xmin = -20
Xmax = -4
Xscl = 1
Ymin = -16
Ymax = -8
Yscl = 1

21. Associative Property of Addition

23. Multiplicative Inverse Property

25. $(-3, 0), (3, 0), (0, 9)$ **27.** $(-8, 0), (4, 0), (0, 4)$

29. $\left(\frac{5}{2}, 0\right), (0, -5)$ **31.** $(-2, 0), \left(\frac{1}{2}, 0\right), (0, -1)$

33. Triangle **35.** Square

37.

Appendix C

Section C.1 *(page A24)*

1. Answers will vary. **3.**

5. a and b **7.** **9.** 7

11. No **13.** False **15.** True **17.** d **19.** b

21. $\angle ZXW$ or $\angle WXZ$, $\angle ZXY$ or $\angle YXZ$, $\angle YXW$ or $\angle WXY$; $\angle ZXY$ and $\angle YXW$ **23.** b **25.** d

27. f **29.** c **31.** $\overline{LM} \cong \overline{NO}$, $\overline{MP} \cong \overline{NQ}$, $\overline{LP} \cong \overline{OQ}$

33. $m\angle V$ **35.** $\overline{TV}$

37.

	Scalene	Isosceles	Equilateral
Acute	Yes	Yes	Yes
Obtuse	Yes	Yes	No
Right	Yes	Yes	No

39. 3; Yes **41.** 3; Yes **43.** 12.5 feet by 12.5 feet; No

45. $(3, 5), (3, 1)$ **47.** Form a tetrahedron.

Section C.2 *(page A31)*

1. **3.** **5.**

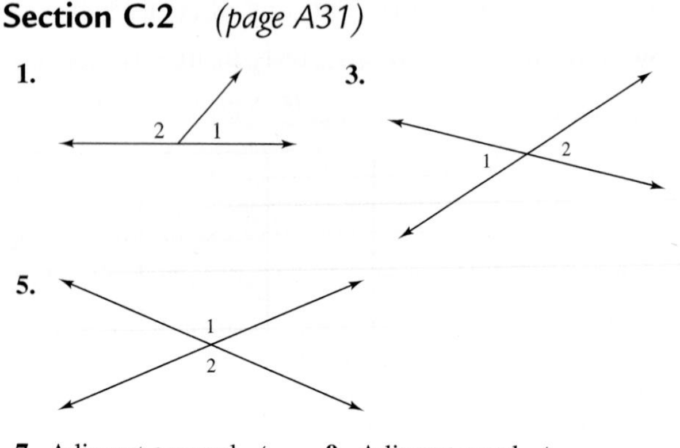

7. Adjacent ≅ suppl. ∠ **9.** Adjacent suppl. ∠

11. Adjacent compl. ∠ **13.** False **15.** False

17. True **19.** 110° **21.** 55° **23.** 35° **25.** c

27. ∠3 and ∠5 *or* ∠4 and ∠6

29. ∠4 and ∠5 *or* ∠3 and ∠6

31. $m\angle 1 = 110°$ because it forms a linear pair with the given angle; $m\angle 2 = 110°$ by the Alternate Exterior Angles Theorem

33. $m\angle 1 = 70°$ by the Consecutive Interior Angles Theorem; $m\angle 2 = 70°$ because it forms a linear pair with the given angle, or by the Alternate Interior Angles Theorem

35. $a = 30°, b = 20°$ **37.** ∠2, ∠5, ∠7

39. $m\angle 1 = m\angle 3 = 70°, m\angle 4 = m\angle 6 = 135°,$
$m\angle 2 = 110°, m\angle 5 = 45°, m\angle 7 = 25°, m\angle 8 = 155°$

41. 35° **43.** 40°

45. True. The third angle must be $180° - 2(60°) = 60°$.

47. $m\angle 1 = 30°, m\angle 2 = 60°, m\angle 3 = 50°, m\angle 4 = 35°,$
$m\angle 5 = 90°, m\angle 6 = 55°, m\angle 7 = 55°, m\angle 8 = 125°,$
$m\angle 9 = 35°$

49. **51.**

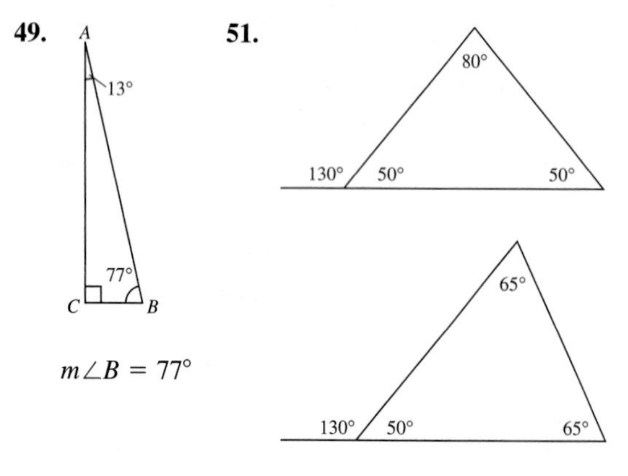

$m\angle B = 77°$

53. 30°, 60°, 90° **55.** 38°, 59°, 83°

Appendix D *(page A43)*

1.

Stems	*Leaves*
7	0 5 5 5 7 7 8 8 8
8	1 1 1 1 2 3 4 5 5 5 5 7 8 9 9 9
9	0 2 8
10	0 0

3.

Stems	*Leaves*
5	2 5 9
6	2 3 6 6 7
7	0 1 2 3 4 7 8 8 9
8	0 1 3 4 5 7 9
9	0 0 2 3 3 3 5 6 8 9
10	0 0

5. Frequency Distribution

Interval	*Tally*
[15, 22)	卌 |||
[22, 29)	卌 |
[29, 36)	卌
[36, 43)	||||
[43, 50)	卌 ||

Histogram

7.

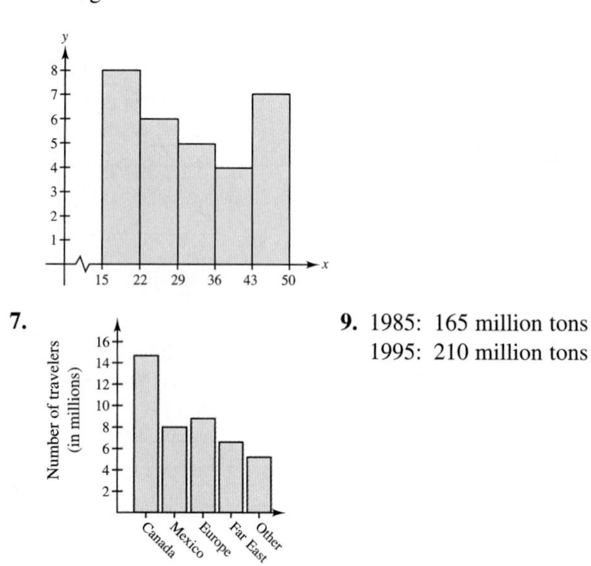

9. 1985: 165 million tons
1995: 210 million tons

11. Recycled waste

13. Total waste equals the sum of the other three quantities.

15.

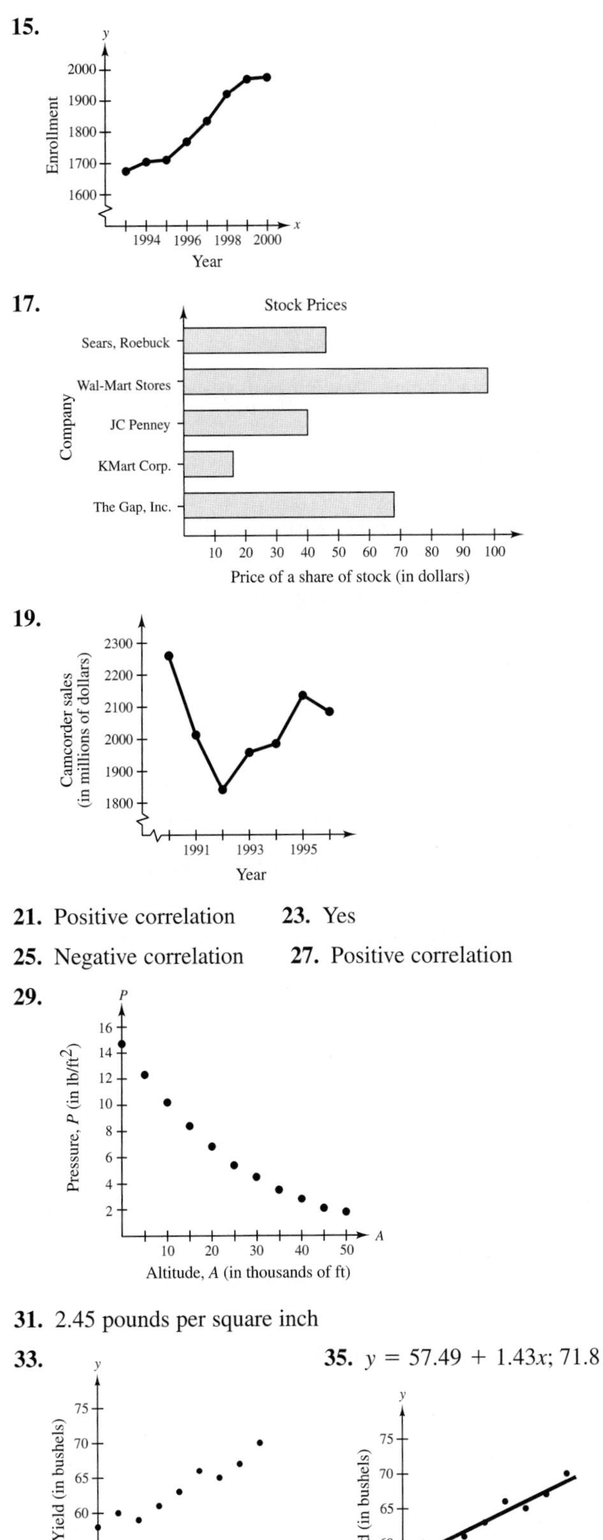

17.

19.

21. Positive correlation **23.** Yes

25. Negative correlation **27.** Positive correlation

29.

31. 2.45 pounds per square inch

33.

35. $y = 57.49 + 1.43x$; 71.8

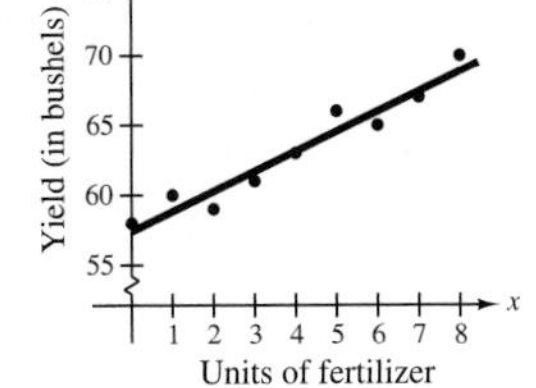

37.

39. $v = 1117.3 - 4.1h$; 1006.6

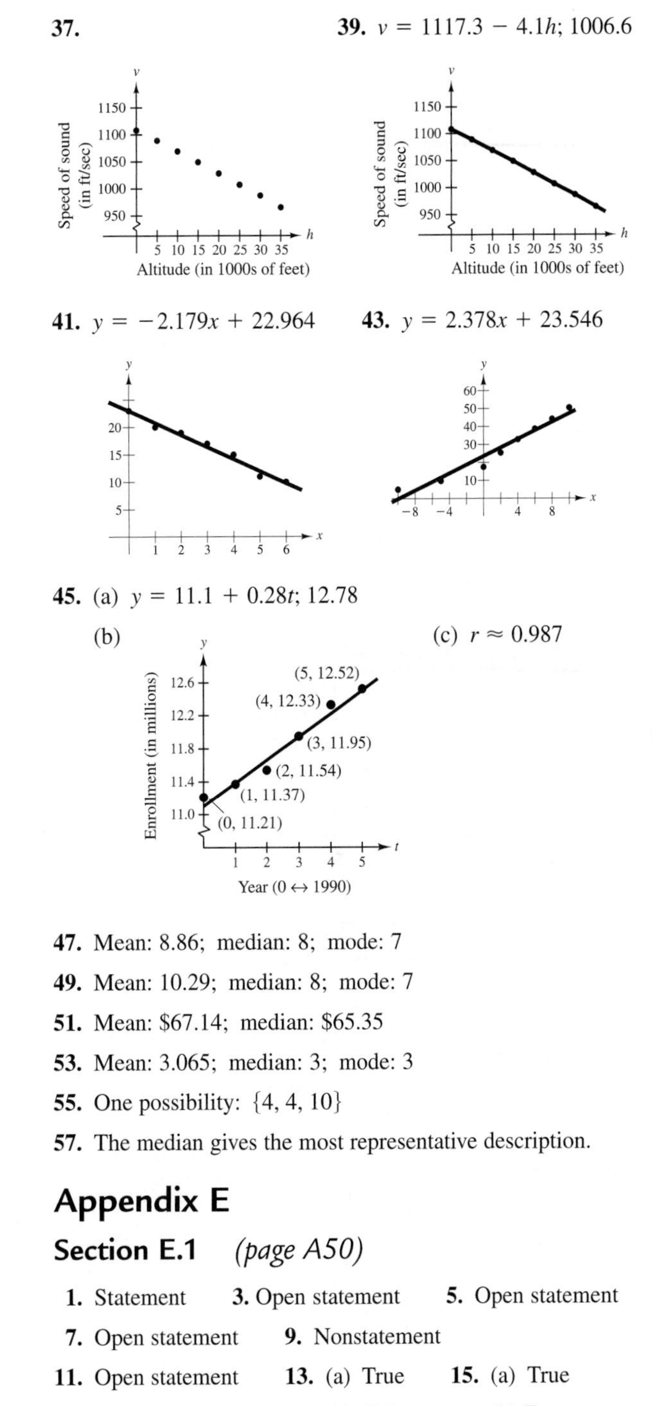

41. $y = -2.179x + 22.964$ **43.** $y = 2.378x + 23.546$

45. (a) $y = 11.1 + 0.28t$; 12.78

(b)

(c) $r \approx 0.987$

47. Mean: 8.86; median: 8; mode: 7

49. Mean: 10.29; median: 8; mode: 7

51. Mean: \$67.14; median: \$65.35

53. Mean: 3.065; median: 3; mode: 3

55. One possibility: {4, 4, 10}

57. The median gives the most representative description.

Appendix E

Section E.1 *(page A50)*

1. Statement **3.** Open statement **5.** Open statement

7. Open statement **9.** Nonstatement

11. Open statement **13.** (a) True (b) False **15.** (a) True (b) True

17. (a) False (b) False **19.** (a) True (b) False

21. (a) The sun is not shining.

(b) It is not hot.

(c) The sun is shining and it is hot.

(d) The sun is shining or it is hot.

23. (a) Lions are not mammals.
(b) Lions are not carnivorous.
(c) Lions are mammals and lions are carnivorous.
(d) Lions are mammals or lions are carnivorous.

25. (a) The sun is not shining and it is hot.
(b) The sun is not shining or it is hot.
(c) The sun is shining and it is not hot.
(d) The sun is shining or it is not hot.

27. (a) Lions are not mammals and lions are carnivorous.
(b) Lions are not mammals or lions are carnivorous.
(c) Lions are mammals and lions are not carnivorous.
(d) Lions are mammals or lions are not carnivorous.

29. $p \wedge \sim q$ **31.** $\sim p \vee q$ **33.** $\sim p \vee \sim q$

35. $\sim p \wedge q$ **37.** The bus is blue.

39. x is not equal to 4. **41.** The earth is flat.

43.

p	q	$\sim p$	$\sim p \wedge q$
T	T	F	F
T	F	F	F
F	T	T	T
F	F	T	F

45.

p	q	$\sim p$	$\sim q$	$\sim p \vee \sim q$
T	T	F	F	F
T	F	F	T	T
F	T	T	F	T
F	F	T	T	T

47.

p	q	$\sim q$	$p \vee \sim q$
T	T	F	T
T	F	T	T
F	T	F	F
F	F	T	T

49. Not logically equivalent **51.** Logically equivalent

53. Logically equivalent **55.** Not logically equivalent

57. Logically equivalent **59.** Not a tautology

61. A tautology

63. Identical: $\sim(p \wedge q)$ and $\sim p \vee \sim q$

p	q	$\sim p$	$\sim q$	$p \wedge q$	$\sim(p \wedge q)$	$\sim p \vee \sim q$
T	T	F	F	T	F	F
T	F	F	T	F	T	T
F	T	T	F	F	T	T
F	F	T	T	F	T	T

Section E.2 *(page A57)*

1. (a) If the engine is running, then the engine is wasting gasoline.
(b) If the engine is wasting gasoline, then the engine is running.
(c) If the engine is not wasting gasoline, then the engine is not running.
(d) If the engine is running, then the engine is not wasting gasoline.

3. (a) If the integer is even, then it is divisible by 2.
(b) If it is divisible by 2, then the integer is even.
(c) If it is not divisible by 2, then the integer is not even.
(d) If the integer is even, then it is not divisible by 2.

5. $q \rightarrow p$ **7.** $p \rightarrow q$ **9.** $p \rightarrow q$ **11.** True

13. True **15.** False **17.** True **19.** True

21. Converse:
If you can see the eclipse, then the sky is clear.

Inverse:
If the sky is not clear, then you cannot see the eclipse.

Contrapositive:
If you cannot see the eclipse, then the sky is not clear.

23. Converse:
If the deficit increases, then taxes were raised.

Inverse:
If taxes are not raised, then the deficit will not increase.

Contrapositive:
If the deficit does not increase, then taxes were not raised.

25. Converse:
It is necessary to apply for the visa to have a birth certificate.

Inverse:
It is not necessary to have a birth certificate to not apply for the visa.

Contrapositive:
It is not necessary to apply for the visa to not have a birth certificate.

27. Paul is not a junior and not a senior.

29. The temperature will increase and the metal rod will not expand.

31. We will go to the ocean and the weather forecast is not good.

33. No students are in extracurricular activities.

35. Some contact sports are not dangerous.

37. Some children are allowed at the concert.

39. None of the $20 bills are counterfeit.

41.

p	q	$\sim q$	$p \to \sim q$	$\sim(p \to \sim q)$
T	T	F	F	T
T	F	T	T	F
F	T	F	T	F
F	F	T	T	F

43.

p	q	$q \to p$	$\sim(q \to p)$	$\sim(q \to p) \wedge q$
T	T	T	F	F
T	F	T	F	F
F	T	F	T	T
F	F	T	F	F

45.

p	q	$\sim p$	$p \vee q$	$(p \vee q) \wedge (\sim p)$	$[(p \vee q) \wedge (\sim p)] \to q$
T	T	F	T	F	T
T	F	F	T	F	T
F	T	T	T	T	T
F	F	T	F	F	T

47.

p	q	$\sim p$	$\sim q$	$p \leftrightarrow (\sim q)$	$(p \leftrightarrow \sim q) \to \sim p$
T	T	F	F	F	T
T	F	F	T	T	F
F	T	T	F	T	T
F	F	T	T	F	T

49.

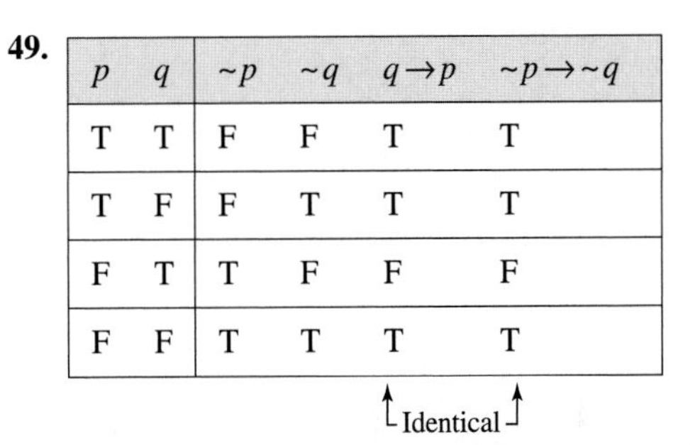

p	q	$\sim p$	$\sim q$	$q \to p$	$\sim p \to \sim q$
T	T	F	F	T	T
T	F	F	T	T	T
F	T	T	F	F	F
F	F	T	T	T	T

Identical ($q \to p$ and $\sim p \to \sim q$)

51.

p	q	$\sim q$	$p \to q$	$\sim(p \to q)$	$p \wedge \sim q$
T	T	F	T	F	F
T	F	T	F	T	T
F	T	F	T	F	F
F	F	T	T	F	F

Identical ($\sim(p \to q)$ and $p \wedge \sim q$)

53.

p	q	$\sim p$	$\sim q$	$p \to q$	$(p \to q) \vee \sim q$
T	T	F	F	T	T
T	F	F	T	F	T
F	T	T	F	T	T
F	F	T	T	T	T

$p \vee \sim p$
T
T
T
T

Identical ($p \vee \sim p$ and $(p \to q) \vee \sim q$)

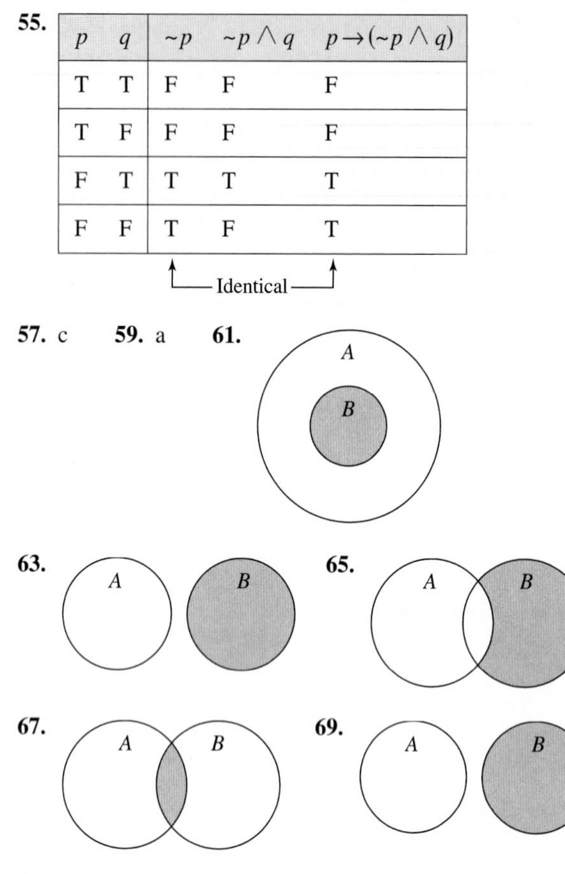

55.

p	q	$\sim p$	$\sim p \wedge q$	$p \rightarrow (\sim p \wedge q)$
T	T	F	F	F
T	F	F	F	F
F	T	T	T	T
F	F	T	F	T

Identical

57. c **59.** a **61.**

63. **65.** **67.** **69.**

71. (a) Statement does not follow.
(b) Statement follows.

73. (a) Statement does not follow.
(b) Statement does not follow.

Section E.3 *(page A64)*

1.

p	q	$\sim p$	$\sim q$	$p \rightarrow \sim q$	$(p \rightarrow \sim q) \wedge q$
T	T	F	F	F	F
T	F	F	T	T	F
F	T	T	F	T	T
F	F	T	T	T	F

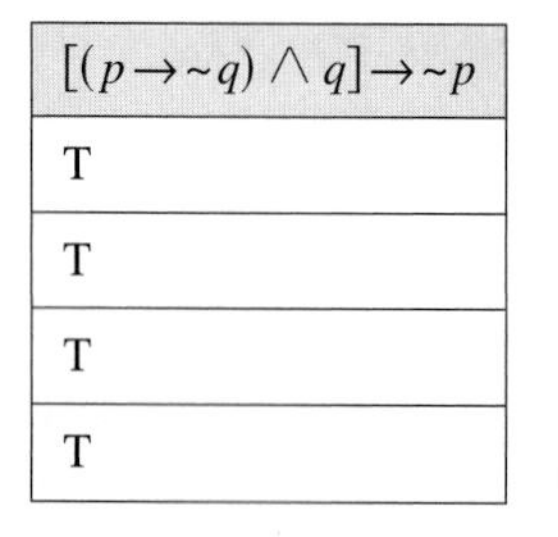

$[(p \rightarrow \sim q) \wedge q] \rightarrow \sim p$
T
T
T
T

3.

p	q	$\sim p$	$p \vee q$	$(p \vee q) \wedge \sim p$
T	T	F	T	F
T	F	F	T	F
F	T	T	T	T
F	F	T	F	F

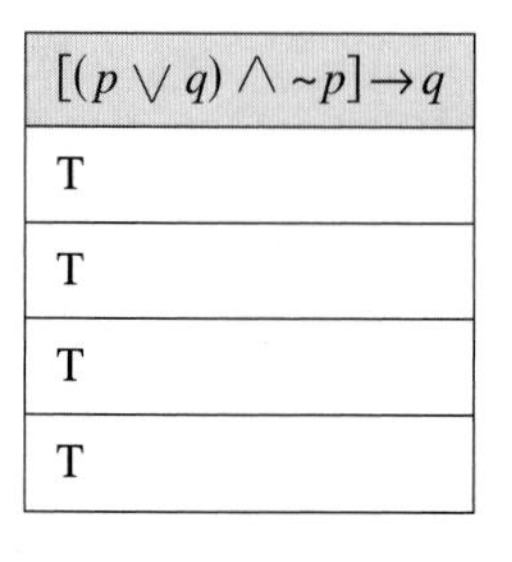

$[(p \vee q) \wedge \sim p] \rightarrow q$
T
T
T
T

5.

p	q	$\sim p$	$\sim q$	$\sim p \rightarrow q$	$(\sim p \rightarrow q) \wedge p$
T	T	F	F	T	T
T	F	F	T	T	T
F	T	T	F	T	F
F	F	T	T	F	F

$[(\sim p \rightarrow q) \wedge p] \rightarrow \sim q$
F
T
T
T

7.

p	q	$p \vee q$	$(p \vee q) \wedge q$	$[(p \vee q) \wedge q] \rightarrow p$
T	T	T	T	T
T	F	T	F	T
F	T	T	T	F
F	F	F	F	T

9. Valid **11.** Invalid **13.** Valid **15.** Valid
17. Invalid **19.** Valid **21.** Invalid
23. b **25.** c **27.** b **29.** c

31. Valid

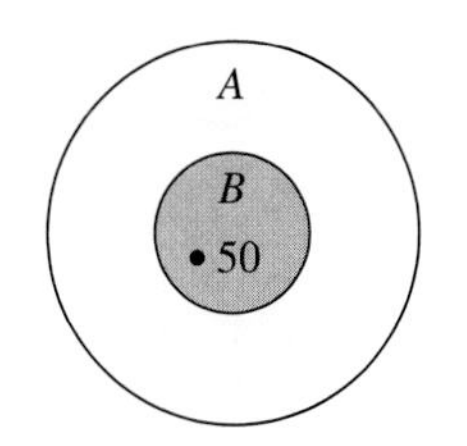

A: All numbers divisible by 5
B: All numbers divisible by 10

33. Invalid

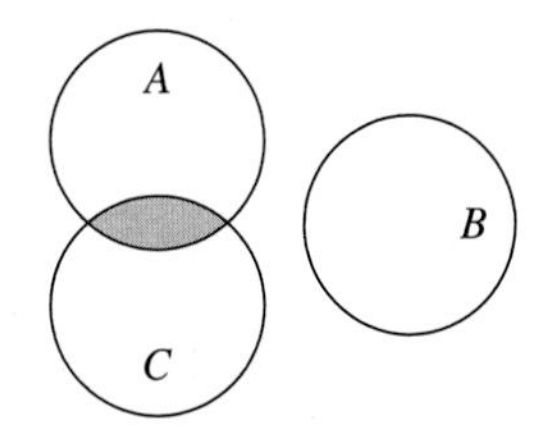

A: People eligible to vote
B: People under the age of 18
C: College students

35. Let p represent the statement "Sue drives to work," let q represent "Sue will stop at the grocery store," and let r represent "Sue will buy milk."

First write:

Premise #1: $p \rightarrow q$
Premise #2: $q \rightarrow r$
Premise #3: p

Reorder the premises:

Premise #3: p
Premise #1: $p \rightarrow q$
Premise #2: $q \rightarrow r$
Conclusion: r

Then we can conclude r. That is, "Sue will buy milk."

37. Let p represent "This is a good product," let q represent "We will buy it," and let r represent "The product was made by XYZ Corporation."

First write:

Premise #1: $p \rightarrow q$
Premise #2: $r \vee \sim q$
Premise #3: $\sim r$

Note that $p \rightarrow q \equiv \sim q \rightarrow \sim p$, and reorder the premises:

Premise #2: $r \vee \sim q$
Premise #3: $\sim r$
(Conclusion from Premise #2, Premise #3: $\sim q$)
Premise #1: $\sim q \rightarrow \sim p$
Conclusion: $\sim p$

Then we can conclude $\sim p$. That is, "It is not a good product."

Index of Applications

BIOLOGY AND LIFE SCIENCE APPLICATIONS

BUSINESS APPLICATIONS

CHEMISTRY AND PHYSICS APPLICATIONS

CONSTRUCTION APPLICATIONS

CONSUMER APPLICATIONS

GEOMETRY APPLICATIONS

INTEREST RATE APPLICATIONS

MISCELLANEOUS APPLICATIONS

TIME AND DISTANCE APPLICATIONS

U.S. DEMOGRAPHICS APPLICATIONS

Index

T

U

V

W

X

Y

Z